Introductory Statistics: Exploring the World Through Data

Third Edition

Robert Gould

University of California, Los Angeles

Rebecca Wong

West Valley College

Colleen Ryan

Moorpark Community College

Director, Portfolio Management: Deirdre Lynch
Senior Portfolio Manager: Suzanna Bainbridge
Portfolio Management Assistant: Morgan Danna
Content Producer: Tamela Ambush
Managing Producer: Karen Wernholm
Producer: Shana Siegmund
Associate Content Developer: Sneh Singh
Manager, Courseware QA: Mary Durnwald
Manager, Content Development: Robert Carroll

Product Marketing Manager: Emily Ockay
Field Marketing Manager: Andrew Noble
Marketing Assistant: Shannon McCormack
Senior Author Support/Technology Specialist: Joe Vetere
Manager, Rights and Permissions: Gina Cheselka
Manufacturing Buyer: Carol Melville, LSC Communications
Cover Design, Full Service Vendor, Composition: Pearson CSC
Full Service Project Management: Chere Bemelmans - Pearson CSC
Cover Image: Paul Scott/EyeEm/Getty Images

Library of Congress Cataloging-in-Publication Data
Names: Gould, Robert, 1965- author. | Ryan, Colleen N. (Colleen Nooter),
    1939- author.
Title: Introductory statistics : exploring the world through data / Robert Gould (University of California, Los Angeles),
    Colleen Ryan (California Lutheran University).
Description: Third edition. | Boston : Pearson, [2020] | Includes index.
Identifiers: LCCN 2018045613 | ISBN 9780135188927 (third edition)
Subjects: LCSH: Statistics–Textbooks.
Classification: LCC QA276.12 .G687 2020 | DDC 519.5–dc23 LC record available at https://lccn.loc.gov/2018045613

17 2022

ISBN 13: 978-0-13-516300-9 (Instructor Edition)
ISBN 10: 0-13-516300-5
ISBN 13: 978-0-13-518892-7 (Student Edition)
ISBN 10: 0-13-518892-X

Dedication

To my parents and family, my friends, and my colleagues who are also friends. Without their patience and support, this would not have been possible.

—Rob

To Nathaniel and Allison, to my students, colleagues, and friends. Thank you for helping me be a better teacher and a better person.

—Rebecca

To my teachers and students, and to my family who have helped me in many different ways.

—Colleen

About the Authors

Robert Gould

Robert L. Gould (Ph.D., University of California, Los Angeles) is a leader in the statistics education community. He has served as chair of the American Statistical Association's (ASA) Statistics Education Section, chair of the American Mathematical Association of Two-Year Colleges/ASA Joint Committee, and has served on the National Council of Teacher of Mathematics/ASA Joint Committee. He served on a panel of co-authors for the *2005 Guidelines for Assessment and Instruction in Statistics Education (GAISE) College Report* and is co-author on the revision for the *GAISE K-12 Report*. As lead principal investigator of the NSF-funded Mobilize Project, he led the development of the first high school level data science course, which is taught in the Los Angeles Unified School District and several other districts. Rob teaches in the Department of Statistics at UCLA, where he directs the undergraduate statistics program and is director of the UCLA Center for Teaching Statistics. In recognition for his activities in statistics education, in 2012 Rob was elected Fellow of the American Statistical Association.

In his free time, Rob plays the cello and enjoys attending concerts of all types and styles.

Rebecca Wong

Rebecca K. Wong has taught mathematics and statistics at West Valley College for more than twenty years. She enjoys designing activities to help students explore statistical concepts and encouraging students to apply those concepts to areas of personal interest.

Rebecca earned at B.A. in mathematics and psychology from the University of California, Santa Barbara, an M.S.T. in mathematics from Santa Clara University, and an Ed.D. in Educational Leadership from San Francisco State University. She has been recognized for outstanding teaching by the National Institute of Staff and Organizational Development and the California Mathematics Council of Community Colleges.

When not teaching, Rebecca is an avid reader and enjoys hiking trails with friends.

Colleen Ryan

Colleen N. Ryan has taught statistics, chemistry, and physics to diverse community college students for decades. She taught at Oxnard College from 1975 to 2006, where she earned the Teacher of the Year Award. Colleen currently teaches statistics part-time at Moorpark Community College. She often designs her own lab activities. Her passion is to discover new ways to make statistical theory practical, easy to understand, and sometimes even fun.

Colleen earned a B.A. in physics from Wellesley College, an M.A.T. in physics from Harvard University, and an M.A. in chemistry from Wellesley College. Her first exposure to statistics was with Frederick Mosteller at Harvard.

In her spare time, Colleen sings, has been an avid skier, and enjoys time with her family.

Contents

CHAPTER **9**

Inferring Population Means 435

CHAPTER **10**

Associations between Categorical Variables 507

CHAPTER **11**

Multiple Comparisons and Analysis of Variance 558

CHAPTER **12**

Experimental Design: Controlling Variation 604

CHAPTER **13**

Inference without Normality 639

CHAPTER 14

Inference for Regression 690

Preface

About This Book

We believe firmly that analyzing data to uncover insight and meaning is one of the most important skills to prepare students for both the workplace and civic life. This is not a book about "statistics," but is a book about understanding our world and, in particular, understanding how statistical inference and data analysis can improve the world by helping us see more clearly.

Since the first edition, we've seen the rise of a new science of data and been amazed by the power of data to improve our health, predict our weather, connect long-lost friends, run our households, and organize our lives. But we've also been concerned by data breaches, by a loss of privacy that can threaten our social structures, and by attempts to manipulate opinion.

This is not a book meant merely to teach students to interpret the statistical findings of others. We *do* teach that; we all need to learn to critically evaluate arguments, particularly arguments based on data. But more importantly, we wish to inspire students to examine data and make their own discoveries. This is a book about *doing*. We are not interested in a course to teach students to memorize formulas or to ask them to mindlessly carry out procedures. Students must learn to think critically with and about data, to communicate their findings to others, and to carefully evaluate others' arguments.

What's New in the Third Edition

As educators and authors, we were strongly inspired by the spirit that created the Guidelines for Assessment and Instruction in Statistics Education (GAISE) (http://amstat.org/asa/education/Guidelines-for-Assessment-and-Instruction-in-Statistics-Education-Reports.aspx), which recommends that we

- teach statistical thinking, which includes teaching statistics as an investigative process and providing opportunities for students to engage in multivariate thinking;

- focus on conceptual understandings;

- integrate real data with a context and purpose;

- foster active learning;

- use technology to explore concepts and to analyze data;

- use assessments to improve and evaluate student learning.

These have guided the first two editions of the book. But the rise of data science has led us to rethink how we engage students with data, and so, in the third edition, we offer some new features that we hope will prepare students for working with the complex data that surrounds us.

More precisely, you'll find

- An emphasis on what we call the Data Cycle, a device to guide students through the statistical investigation process. The Data Cycle includes four phases: Ask Questions, Consider Data, Analyze Data, and Interpret Data. A new marginal icon indicates when the Data Cycle is particularly relevant.

- An increased emphasis on formulating "statistical investigative questions" as an important first step in the Data Cycle. Previous editions have emphasized the other three steps, but we feel students need practice in formulating questions that will help them interpret data. To formulate questions is to engage in mathematical and statistical modeling, and this edition spends more time teaching this important skill.

- The end-of-chapter activities have been replaced by a series of "Data Projects." These are self-guided activities that teach students important "data moves" that will help them navigate through the large and complex data sets that are so often found in the real world.

- The addition of a "Data Moves" icon. Some examples are based on extracts of data from much larger data sets. The Data Moves icon points students to these data sets and also indicate the "data moves" used to extract the data. We are indebted to Tim Erickson for the phrase "data moves" and the ideas that motivate it.

- A smoother and more refined approach to simulations in Chapter 5.

- Updated technology guides to match current hardware and software.

- Hundreds of new exercises.

- New and updated examples in each chapter.

- New and updated data sets, with the inclusion of more large data.

Approach

Our text is concept-based, as opposed to method-based. We teach useful statistical methods, but we emphasize that applying the method is secondary to understanding the concept.

In the real world, computers do most of the heavy lifting for statisticians. We therefore adopt an approach that frees the instructor from having to teach tedious procedures and leaves more time for teaching deeper understanding of concepts. Accordingly, we present formulas as an aid to understanding the concepts, rather than as the focus of study.

We believe students need to learn how to

- Determine which statistical procedures are appropriate.

- Instruct the software to carry out the procedures.

- Interpret the output.

We understand that students will probably see only one type of statistical software in class. But we believe it is useful for students to compare output from several different sources, so in some examples we ask them to read output from two or more software packages.

Coverage

The first two-thirds of this book are concept-driven and cover exploratory data analysis and inferential statistics—fundamental concepts that every introductory statistics student should learn. The final third of the book builds on that strong conceptual foundation and is more methods-based. It presents several popular statistical methods and more fully explores methods presented earlier, such as regression and data collection.

Our ordering of topics is guided by the process through which students should analyze data. First, they explore and describe data, possibly deciding that graphics and numerical summaries provide sufficient insight. Then they make generalizations (inferences) about the larger world.

Chapters 1–4: Exploratory Data Analysis. The first four chapters cover data collection and summary. Chapter 1 introduces the important topic of data collection and compares and contrasts observational studies with controlled experiments. This chapter also teaches students how to handle raw data so that the data can be uploaded to their statistical software. Chapters 2 and 3 discuss graphical and numerical summaries of single variables based on samples. We emphasize that the purpose is not just to produce a graph or a number but, instead, to explain what those graphs and numbers say about the world. Chapter 4 introduces simple linear regression and presents it as a technique for providing graphical and numerical summaries of relationships between two numerical variables.

We feel strongly that introducing regression early in the text is beneficial in building student understanding of the applicability of statistics to real-world scenarios. After completing the chapters covering data collection and summary, students have acquired the skills and sophistication they need to describe two-variable associations and to generate informal hypotheses. Two-variable associations provide a rich context for class discussion and allow the course to move from fabricated problems (because one-variable analyses are relatively rare in the real world) to real problems that appear frequently in everyday life. We return to regression in Chapter 14, when we discuss statistical inference in the context of regression, which requires quite a bit of machinery. We feel that it would be a shame to delay until the end of the course all the insights that regression without inference can provide.

Chapters 5–8: Inference. These chapters teach the fundamental concepts of statistical inference. The main idea is that our data mirror the real world, but imperfectly; although our estimates are uncertain, under the right conditions we can quantify our uncertainty. Verifying that these conditions exist and understanding what happens if they are not satisfied are important themes of these chapters.

Chapters 9–11: Methods. Here we return to the themes covered earlier in the text and present them in a new context by introducing additional statistical methods, such as estimating population means, analyzing categorical variables, and analyzing relations between a numerical and a categorical variable. We also introduce multiple comparisons and use them to motivate the need for the statistical method of ANOVA.

Chapters 12–14: Special Topics. Students who have covered all topics up to this point will have a solid foundation in statistics. These final chapters build on that foundation and offer more details, as we explore the topics of designing controlled experiments, survey sampling, additional contexts for hypothesis testing, and using regression to make inferences about a population.

In Chapter 12 we provide guidance for reading scientific literature. Even if your schedule does not allow you to cover Chapter 12, we recommend using Section 12.3 to offer students the experience of critically examining real scientific papers.

Organization

Our preferred order of progressing through the text is reflected in the Contents, but there are some alternative pathways as well.

10-week Quarter. The first eight chapters provide a full, one-quarter course in introductory statistics. If time remains, cover Sections 9.1 and 9.2 as well, so that students can solidify their understanding of confidence intervals and hypothesis tests by revisiting the topic with a new parameter.

Proportions First. Ask two statisticians, and you will get three opinions on whether it is best to teach means or proportions first. We have come down on the side of proportions for a variety of reasons. Proportions are much easier to find in popular news media (particularly around election time), so they can more readily be tied to students' everyday lives. Also, the mathematics and statistical theory are simpler; because there's no need to provide a separate estimate for the population standard deviation, inference is based on the Normal distribution, and no further approximations (that is, the *t*-distribution) are required. Hence, we can quickly get to the heart of the matter with fewer technical diversions.

The basic problem here is how to quantify the uncertainty involved in estimating a parameter and how to quantify the probability of making incorrect decisions when posing hypotheses. We cover these ideas in detail in the context of proportions. Students can then more easily learn how these same concepts are applied in the new context of means (and any other parameter they may need to estimate).

Means First. Conversely, many people feel that there is time for only one parameter and that this parameter should be the mean. For this alternative presentation, cover Chapters 6, 7, and 9, in that order. On this path, students learn about survey sampling and the terminology of inference (population vs. sample, parameter vs. statistic) and then tackle inference for the mean, including hypothesis testing.

To minimize the coverage of proportions, you might choose to cover Chapter 6, Section 7.1 (which treats the language and framework of statistical inference in detail), and then Chapter 9. Chapters 7 and 8 develop the concepts of statistical inference more slowly than Chapter 9, but essentially, Chapter 9 develops the same ideas in the context of the mean.

If you present Chapter 9 before Chapters 7 and 8, we recommend that you devote roughly twice as much time to Chapter 9 as you have devoted to previous chapters, because many challenging ideas are explored in this chapter. If you have already covered Chapters 7 and 8 thoroughly, Chapter 9 can be covered more quickly.

Features

We've incorporated into this text a variety of features to aid student learning and to facilitate its use in any classroom.

Integrating Technology

Modern statistics is inseparable from computers. We have worked to make this textbook accessible for any classroom, regardless of the level of in-class exposure to technology, while still remaining true to the demands of the analysis. We know that students sometimes do not have access to technology when doing homework, so many exercises provide output from software and ask students to interpret and critically evaluate that given output.

Using technology is important because it enables students to handle real data, and real data sets are often large and messy. The following features are designed to guide students.

- **TechTips** outline steps for performing calculations using TI-84® (including TI-84 + C®) graphing calculators, Excel®, Minitab®, and StatCrunch®. We do not want students to get stuck because they don't know how to reproduce the results we show in the book, so whenever a new method or procedure is introduced, an icon, Tech , refers students to the TechTips section at the end of the chapter. Each set of TechTips contains at least one mini-example, so that students are not only learning to use the technology but also practicing data analysis and reinforcing ideas discussed in the text. Most of the provided TI-84 steps apply to all TI-84 calculators, but some are unique to the TI-84 + C calculator. Throughout the text, screenshots of TI calculators are labeled "TI-84" but are, in fact, from a TI-84 Plus C Silver Edition.

- All **data sets** used in the exposition and exercises are available at http://www.pearsonhighered.com/mathstatsresources/.

Guiding Students

- Each chapter opens with a **Theme**. Beginners have difficulty seeing the forest for the trees, so we use a theme to give an overview of the chapter content.

- Each chapter begins by posing a real-world **Case Study**. At the end of the chapter, we show how techniques covered in the chapter helped solve the problem presented in the Case Study.

- **Margin Notes** draw attention to details that enhance student learning and reading comprehension.

> **Caution** notes provide warnings about common mistakes or misconceptions.

> **Looking Back** reminders refer students to earlier coverage of a topic.

> **Details** clarify or expand on a concept.

- **KEY POINT** **Key Points** highlight essential concepts to draw special attention to them. Understanding these concepts is essential for progress.

- **Snapshots** break down key statistical concepts introduced in the chapter, quickly summarizing each concept or procedure and indicating when and how it should be used.

- **Data Moves** point students toward more complete source data.

- An abundance of worked-out **examples** model solutions to real-world problems relevant to students' lives. Each example is tied to an end-of-chapter exercise so that students can practice solving a similar problem and test their understanding. Within the exercise sets, the icon TRY indicates which problems are tied to worked-out examples in that chapter, and the numbers of those examples are indicated.

- The **Chapter Review** that concludes each chapter provides a list of important new terms, student learning objectives, a summary of the concepts and methods discussed, and sources for data, articles, and graphics referred to in the chapter.

Active Learning

- Each chapter ends in a **Data Project**. These are activities designed for students to work alone or in pairs. Data analysis requires practice, and these sections, which grow increasingly more complex, are intended to guide students through basic "data moves" to help them find insight in complex data.

- All exercises are located at the end of the chapter. **Section Exercises** are designed to begin with a few basic problems that strengthen recall and assess basic knowledge, followed by mid-level exercises that ask more complex, open-ended questions. **Chapter Review Exercises** provide a comprehensive review of material covered throughout the chapter.

The exercises emphasize good statistical practice by requiring students to verify conditions, make suitable use of graphics, find numerical values, and interpret their findings in writing. All exercises are paired so that students can check their work on the odd-numbered exercise and then tackle the corresponding even-numbered exercise. The answers to all odd-numbered exercises appear in the back of the student edition of the text.

Challenging exercises, identified with an asterisk (*), ask open-ended questions and sometimes require students to perform a complete statistical analysis.

- Most chapters include select exercises, marked with a **g** within the exercise set, to indicate that problem-solving help is available in the **Guided Exercises** section. If students need support while doing homework, they can turn to the Guided Exercises to see a step-by-step approach to solving the problem.

Acknowledgments

We are grateful for the attention and energy that a large number of people devoted to making this a better book. We extend our gratitude to Chere Bemelmans, who handled production, and to Tamela Ambush, content producer. Many thanks to John Norbutas for his technical advice and help with the TechTips. We thank Deirdre Lynch, editor-in-chief, for signing us up and sticking with us, and we are grateful to Emily Ockay for her market development efforts.

We extend our sincere thanks for the suggestions and contributions made by the following reviewers of this edition:

Beth Burns, *Bowling Green State University*
Rod Elmore, *Mid Michigan Community College*
Carl Fetteroll, *Western New England University*
Elizabeth Flynn, *College of the Canyons*
David French, *Tidewater Community College*
Terry Fuller, *California State University, Northridge*
Kimberly Gardner, *Kennesaw State University*
Ryan Girard, *Kauai Community College*
Carrie Grant, *Flagler College*

Deborah Hanus, *Brookhaven College*
Kristin Harvey, *The University of Texas at Austin*
Abbas Jaffary, *Moraine Valley Community College*
Tony Jenkins, *Northwestern Michigan College*
Jonathan Kalk, *Kauai Community College*
Joseph Kudrle, *University of Vermont*
Matt Lathrop, *Heartland Community College*
Raymond E. Lee, *The University of North Carolina at Pembroke*
Karen McNeal, *Moraine Valley Community College*

Tejal Naik, *West Valley College*
Hadley Pridgen, *Gulf Coast State College*
John M. Russell, *Old Dominion University*
Amy Salvati, *Adirondack Community College*
Marcia Siderow, *California State University, Northridge*
Kenneth Strazzeri, *George Mason University*
Amy Vu, *West Valley College*
Rebecca Walker, *Guttman Community College*

We would also like to extend our sincere thanks for the suggestions and contributions made by the following reviewers, class testers, and focus group attendees of the previous edition.

Arun Agarwal, *Grambling State University*
Anne Albert, *University of Findlay*
Michael Allen, *Glendale Community College*
Eugene Allevato, *Woodbury University*
Dr. Jerry Allison, *Trident Technical College*
Polly Amstutz, *University of Nebraska*
Patricia Anderson, *Southern Adventist University*
MaryAnne Anthony-Smith, *Santa Ana College*
David C. Ashley, *Florida State College at Jacksonville*
Diana Asmus, *Greenville Technical College*
Kathy Autrey, *Northwestern State University of Louisiana*
Wayne Barber, *Chemeketa Community College*
Roxane Barrows, *Hocking College*
Jennifer Beineke, *Western New England College*
Diane Benner, *Harrisburg Area Community College*
Norma Biscula, *University of Maine, Augusta*
K.B. Boomer, *Bucknell University*

Mario Borha, *Loyola University of Chicago*
David Bosworth, *Hutchinson Community College*
Diana Boyette, *Seminole Community College*
Elizabeth Paulus Brown, *Waukesha County Technical College*
Leslie Buck, *Suffolk Community College*
R.B. Campbell, *University of Northern Iowa*
Stephanie Campbell, *Mineral Area College*
Ann Cannon, *Cornell College*
Rao Chaganty, *Old Dominion University*
Carolyn Chapel, *Western Technical College*
Christine Cole, *Moorpark College*
Linda Brant Collins, *University of Chicago*
James A. Condor, *Manatee Community College*
Carolyn Cuff, *Westminster College*
Phyllis Curtiss, *Grand Valley State University*
Monica Dabos, *University of California, Santa Barbara*
Greg Davis, *University of Wisconsin, Green Bay*
Bob Denton, *Orange Coast College*
Julie DePree, *University of New Mexico–Valencia*
Jill DeWitt, *Baker Community College of Muskegon*

Paul Drelles, *West Shore Community College*
Keith Driscoll, *Clayton State University*
Rob Eby, *Blinn College*
Nancy Eschen, *Florida Community College at Jacksonville*
Karen Estes, *St. Petersburg College*
Mariah Evans, *University of Nevada, Reno*
Harshini Fernando, *Purdue University North Central*
Stephanie Fitchett, *University of Northern Colorado*
Elaine B. Fitt, *Bucks County Community College*
Michael Flesch, *Metropolitan Community College*
Melinda Fox, *Ivy Tech Community College, Fairbanks*
Joshua Francis, *Defiance College*
Michael Frankel, *Kennesaw State University*
Heather Gamber, *Lone Star College*
Debbie Garrison, *Valencia Community College, East Campus*
Kim Gilbert, *University of Georgia*
Stephen Gold, *Cypress College*
Nick Gomersall, *Luther College*
Mary Elizabeth Gore, *Community College of Baltimore County–Essex*

Ken Grace, *Anoka Ramsey Community College*
Larry Green, *Lake Tahoe Community College*
Jeffrey Grell, *Baltimore City Community College*
Albert Groccia, *Valencia Community College, Osceola Campus*
David Gurney, *Southeastern Louisiana University*
Chris Hakenkamp, *University of Maryland, College Park*
Melodie Hallet, *San Diego State University*
Donnie Hallstone, *Green River Community College*
Cecil Hallum, *Sam Houston State University*
Josephine Hamer, *Western Connecticut State University*
Mark Harbison, *Sacramento City College*
Beverly J. Hartter, *Oklahoma Wesleyan University*
Laura Heath, *Palm Beach State College*
Greg Henderson, *Hillsborough Community College*
Susan Herring, *Sonoma State University*
Carla Hill, *Marist College*
Michael Huber, *Muhlenberg College*
Kelly Jackson, *Camden County College*
Bridgette Jacob, *Onondaga Community College*
Robert Jernigan, *American University*
Chun Jin, *Central Connecticut State University*
Jim Johnston, *Concord University*
Maryann Justinger, Ed.D., *Erie Community College*
Joseph Karnowski, *Norwalk Community College*
Susitha Karunaratne, *Purdue University North Central*
Mohammed Kazemi, *University of North Carolina–Charlotte*
Robert Keller, *Loras College*
Omar Keshk, *Ohio State University*
Raja Khoury, *Collin County Community College*
Brianna Killian, *Daytona State College*
Yoon G. Kim, *Humboldt State University*
Greg Knofczynski, *Armstrong Atlantic University*
Jeffrey Kollath, *Oregon State University*
Erica Kwiatkowski-Egizio, *Joliet Junior College*
Sister Jean A. *Lanahan, OP, Molloy College*
Katie Larkin, *Lake Tahoe Community College*
Michael LaValle, *Rochester Community College*
Deann Leoni, *Edmonds Community College*
Lenore Lerer, *Bergen Community College*
Quan Li, *Texas A&M University*
Doug Mace, *Kirtland Community College*

Walter H. Mackey, *Owens Community College*
Keith McCoy, *Wilbur Wright College*
Elaine McDonald-Newman, *Sonoma State University*
William McGregor, *Rockland Community College*
Bill Meisel, *Florida State College at Jacksonville*
Bruno Mendes, *University of California, Santa Cruz*
Wendy Miao, *El Camino College*
Robert Mignone, *College of Charleston*
Ashod Minasian, *El Camino College*
Megan Mocko, *University of Florida*
Sumona Mondal, *Clarkson University*
Kathy Mowers, *Owensboro Community and Technical College*
Mary Moyinhan, *Cape Cod Community College*
Junalyn Navarra-Madsen, *Texas Woman's University*
Azarnia Nazanin, *Santa Fe College*
Stacey O. Nicholls, *Anne Arundel Community College*
Helen Noble, *San Diego State University*
Lyn Noble, *Florida State College at Jacksonville*
Keith Oberlander, *Pasadena City College*
Pamela Omer, *Western New England College*
Ralph Padgett Jr., *University of California – Riverside*
Nabendu Pal, *University of Louisiana at Lafayette*
Irene Palacios, *Grossmont College*
Ron Palcic, *Johnson County Community College*
Adam Pennell, *Greensboro College*
Patrick Perry, *Hawaii Pacific University*
Joseph Pick, *Palm Beach State College*
Philip Pickering, *Genesee Community College*
Victor I. Piercey, *Ferris State University*
Robin Powell, *Greenville Technical College*
Nicholas Pritchard, *Coastal Carolina University*
Linda Quinn, *Cleveland State University*
William Radulovich, *Florida State College at Jacksonville*
Mumunur Rashid, *Indiana University of Pennsylvania*
Fred J. Rispoli, *Dowling College*
Danielle Rivard, *Post University*
Nancy Rivers, *Wake Technical Community College*
Corlis Robe, *East Tennesee State University*
Thomas Roe, *South Dakota State University*
Alex Rolon, *North Hampton Community College*
Dan Rowe, *Heartland Community College*

Ali Saadat, *University of California – Riverside*
Kelly Sakkinen, *Lake Land College*
Carol Saltsgaver, *University of Illinois–Springfield*
Radha Sankaran, *Passaic County Community College*
Delray Schultz, *Millersville University*
Jenny Shook, *Pennsylvania State University*
Danya Smithers, *Northeast State Technical Community College*
Larry Southard, *Florida Gulf Coast University*
Dianna J. Spence, *North Georgia College & State University*
René Sporer, *Diablo Valley College*
Jeganathan Sriskandarajah, *Madison Area Technical College–Traux*
David Stewart, *Community College of Baltimore County–Cantonsville*
Linda Strauss, *Penn State University*
John Stroyls, *Georgia Southwestern State University*
Joseph Sukta, *Moraine Valley Community College*
Sharon 1. Sullivan, *Catawba College*
Lori Thomas, *Midland College*
Malissa Trent, *Northeast State Technical Community College*
Ruth Trygstad, *Salt Lake Community College*
Gail Tudor, *Husson University*
Manuel T. Uy, *College of Alameda*
Lewis Van Brackle, *Kennesaw State University*
Mahbobeh Vezvaei, *Kent State University*
Joseph Villalobos, *El Camino College*
Barbara Wainwright, *Sailsbury University*
Henry Wakhungu, *Indiana University*
Jerimi Ann Walker, *Moraine Valley Community College*
Dottie Walton, *Cuyahoga Community College*
Jen-ting Wang, *SUNY, Oneonta*
Jane West, *Trident Technical College*
Michelle White, *Terra Community College*
Bonnie-Lou Wicklund, *Mount Wachusett Community College*
Sandra Williams, *Front Range Community College*
Rebecca Wong, *West Valley College*
Alan Worley, *South Plains College*
Jane-Marie Wright, *Suffolk Community College*
Haishen Yao, *CUNY, Queensborough Community College*
Lynda Zenati, *Robert Morris Community College*
Yan Zheng-Araujo, *Springfield Community Technical College*
Cathleen Zucco-Teveloff, *Rider University*
Mark A. Zuiker, *Minnesota State University, Mankato*

MyLab Statistics Online Course for *Introductory Statistics: Exploring the World Through Data, 3e*

(Access Code Required)

MyLab™ Statistics is available to accompany Pearson's market-leading text offerings. To give students a consistent tone, voice, and teaching method, each text's flavor and approach is tightly integrated throughout the accompanying MyLab Statistics course, making learning the material as seamless as possible.

NEW! Integrated Review

This MyLab includes a full suite of supporting Integrated Review resources for the Gould, *Introductory Statistics* course, including pre-made, assignable (and editable) quizzes to assess the prerequisite skills needed for each chapter, and personalized remediation for any gaps in skills that are identified. Each student, therefore, receives just the help that he or she needs—no more, no less.

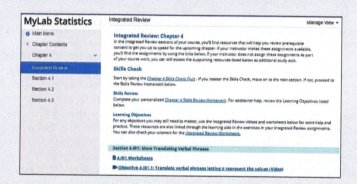

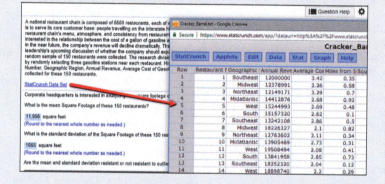

NEW! Data Projects

Data Projects from the text are assignable in MyLab Statistics and provide opportunities for students to practice statistical thinking beyond the classroom. **StatCrunch Projects** that either span the entire curriculum or focus on certain key concepts are also assignable in MyLab Statistics and encourage students to apply concepts to real situations and make data-informed decisions.

UPDATED! Conceptual Questions

The Conceptual Question Library in MyLab Statistics includes 1,000 assignable questions that assess conceptual understanding. These questions are now correlated by chapter to make it easier than ever to navigate and assign these types of questions.

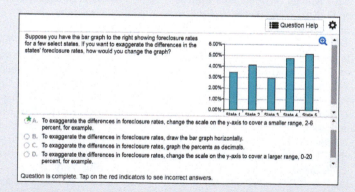

Resources for Success

Student Resources

StatCrunch

StatCrunch® is powerful web-based statistical software that allows users to collect, crunch, and communicate with data. The vibrant online community offers tens of thousands of shared data sets for students and instructors to analyze, in addition to all of the data sets in the text or online homework. StatCrunch is integrated directly into MyLab Statistics or it can be purchased separately. Learn more at www.statcrunch.com.

Video Resources

Chapter Review videos walk students through solving some of the more complex problems and review key concepts from each chapter. Data Cycle of Everyday Things videos demonstrate for students that data collection and data analysis can be applied to answer questions about everyday life. StatTalk Videos, hosted by fun-loving statistician Andrew Vickers, demonstrate important statistical concepts through interesting stories and real-life events. Assessment questions for each video are also available.

Data Sets

All data sets from the textbook are available in MyLab Statistics. They can be analyzed in StatCrunch or downloaded for use in other statistical software programs.

Statistical Software Support

Instructors and students can copy data sets from the text and MyLab Statistics exercises directly into software such as StatCrunch or Excel®. Students can also access instructional support tools including tutorial videos, Study Cards, and manuals for a variety of statistical software programs including, StatCrunch, Excel, Minitab®, JMP®, R, SPSS, and TI 83/84 calculators.

Student Solutions Manual

Written by James Lapp, this manual provides detailed, worked-out solutions to all odd-numbered text exercises. It is available in print and can be downloaded from MyLab Statistics. (ISBN-13: 978-0-13-518923-8; ISBN-10: 0-13-518923-3)

Instructor Resources

Instructor's Edition

Includes answers to all text exercises, as well as a set of Instructor Notes at the front of the text that offer chapter-by-chapter teaching suggestions and commentary. (ISBN-13: 978-0-13-516300-9; ISBN-10: 0-13-516300-5)

Instructor Solutions Manual

Written by James Lapp, the Instructor Solutions Manual contains worked-out solutions to all text exercises. It can be downloaded from MyLab Statistics or from www.pearson.com.

PowerPoint Slides

PowerPoint slides provide an overview of each chapter, stressing important definitions and offering additional examples. They can be downloaded from MyLab Statistics or from www.pearson.com.

TestGen

TestGen® (www.pearson.com/testgen) enables instructors to build, edit, print, and administer tests using a computerized bank of questions developed to cover all the objectives of the text. TestGen is algorithmically based, allowing instructors to create multiple but equivalent versions of the same question or test, and modify test bank questions or add new questions. It is available for download from Pearson's online catalog, www.pearson.com. The questions are also assignable in MyLab Statistics.

Learning Catalytics

Now included in all MyLab Statistics courses, this student response tool uses students' smartphones, tablets, or laptops to engage them in more interactive tasks and thinking during lecture. Learning Catalytics™ fosters student engagement and peer-to-peer learning with real-time analytics. Access pre-built exercises created specifically for statistics.

Question Libraries

In addition to StatCrunch Projects and the Conceptual Question Library, MyLab Statistics also includes a Getting Ready for Statistics library that contains more than 450 exercises on prerequisite topics.

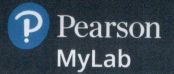

Statistical Software Bundle Options

Minitab and Minitab Express™

Bundling Minitab software with educational materials ensures students have access to the software they need in the classroom, around campus, and at home. And having 12-month access to Minitab and Minitab Express ensures students can use the software for the duration of their course. ISBN 13: 978-0-13-445640-9 ISBN 10: 0-13-445640-8

JMP Student Edition

An easy-to-use, streamlined version of JMP desktop statistical discovery software from SAS Institute, Inc. is available for bundling with the text. (ISBN-13: 978-0-13-467979-2 ISBN-10: 0-13-467979-2)

XLSTAT™

An Excel add-in that enhances the analytical capabilities of Excel. XLSTAT is used by leading businesses and universities around the world. It is available to bundle with this text. For more information go to www.pearsonhighered.com/xlstatupdate. (ISBN-13: 978-0-321-75940-5; ISBN-10: 0-321-75940-0)

Index of Applications

1 Introduction to Data

Statistics is the science of data, so we must learn the types of data we will encounter and the methods for collecting data. The method used to collect data is very important because it determines what types of conclusions we can reach and, as you'll learn in later chapters, what types of analyses we can do. By organizing the data we've collected, we can often spot patterns that are not otherwise obvious.

This text will teach you to examine data to better understand the world around you. If you know how to sift data to find patterns, can communicate the results clearly, and understand whether you can generalize your results to other groups and contexts, you will be able to make better decisions, offer more convincing arguments, and learn things you did not know before. Data are everywhere, and making effective use of them is such a crucial task that one prominent economist has proclaimed statistics one of the most important professions of the decade (*McKinsley Quarterly* 2009).

The use of statistics to make decisions and convince others to take action is not new. Some statisticians date the current practice of statistics back to the mid-nineteenth century. One famous example occurred in 1854, when the British were fighting the Russians in the brutal Crimean War. A British newspaper had criticized the military medical facilities, and a young but well-connected nurse, Florence Nightingale, was appointed to study the situation and, if possible, to improve it.

Nightingale carefully recorded the numbers of deaths, the causes of the deaths, and the times and dates of the deaths. She organized these data graphically, and these graphs enabled her to see a very important pattern: A large percentage of deaths were due to contagious disease, and many deaths could be prevented by improving sanitary conditions. Within six months, Nightingale had reduced the death rate by half. Eventually she convinced Parliament and military authorities to completely reorganize the medical care they provided. Accordingly, she is credited with inventing modern hospital management.

In modern times, we have equally important questions to answer. Do cell phones cause brain tumors? Are alcoholic drinks healthful in moderation? Which diet works best for losing weight? What percentage of the public is concerned about job security? **Statistics**—the science (and art!) of collecting and analyzing observations to learn about ourselves, our surroundings, and our universe—helps answer questions such as these.

Data are the building blocks of statistics. This chapter introduces some of the basic types of data and explains how we collect them, store them, and organize them. You'll be introduced to the Data Cycle, a guide for how to interact with data in a productive way. These ideas and skills will provide a basic foundation for your study of the rest of the text.

CASE STUDY

Dangerous Habit?

Will your coffee habit give you cancer? A court in California considered whether Californians' morning cup of coffee should include a health warning. In 1986, Californians voted for the Safe Drinking Water and Toxic Enforcement Act, which requires that products that contain harmful chemicals be labeled as hazardous. Coffee contains a chemical, acrylamide, that, in the official jargon "is known to the State of California to cause cancer." In 2010, a lawyer sued the coffee industry to force companies to either label their product as hazardous or to remove the chemical from their product. As of the date of publication of this book, the lawsuit continues. Complicating this lawyer's efforts is the fact that recent research suggests that drinking coffee is possibly beneficial to our health and maybe even reduces the risk of cancer.

Does coffee cause cancer? Does it prevent cancer? Why are there conflicting opinions? In this chapter we explore questions such as these, and consider the different types of evidence needed to make *causal* claims, such as the claim that drinking coffee will give you cancer.

SECTION 1.1

What Are Data?

The study of statistics rests on two major concepts: variation and data. **Variation** is the more fundamental of these concepts. To illustrate this idea, draw a circle on a piece of paper. Now draw another one, and try to make it look just the same. Now another. Are all three exactly the same? We bet they're not. They might be slightly different sizes, for instance, or slightly different versions of round. This is an example of variation. How can you reduce this variation? Maybe you can get a penny and outline the penny. Try this three times. Does variation still appear? Probably it does, even if you need a magnifying glass to see, say, slight variations in the thickness of the penciled line.

    Data are observations that you or someone else (or some*thing* else) records. The drawings in Figure 1.1 are data that record our attempts to draw three circles that look the same. Analyzing pictorial data such as these is not easy, so we often try to quantify such observations—that is, to turn them into numbers. How would you measure whether these three circles are the same? Perhaps you would compare diameters or circumferences, or somehow try to measure how and where these circles depart from being perfect circles. Whatever technique you chose, these measurements could also be considered data.

Details

Data Are What Data Is
If you want to be "old school" grammatically correct, then the word *data* is plural. So we say "data *are*" and not "data *is*." The singular form is *datum*. However, this usage is changing over time, and some dictionaries now say that *data* can be used as both a singular and a plural noun.

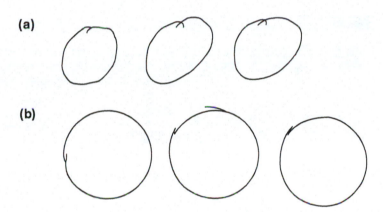

(a)

(b)

◀ **FIGURE 1.1 (a)** Three circles drawn by hand. **(b)** Three circles drawn using a coin. It is clear that the circles drawn by hand show more variability than the circles drawn with the aid of a coin.

    Data are more than just numbers, though. David Moore, a well-known statistician, defined data as "numbers in context." By this he meant that data consist not only of the numbers we record but also of the story behind the numbers. For example,

$$10.00, 9.88, 9.81, 9.81, 9.75, 9.69, 9.5, 9.44, 9.31$$

are just numbers. But in fact these numbers represent "Weight in pounds of the ten heaviest babies in a sample of babies born in North Carolina in 2004." Now these numbers have a context and have been elevated into data. See how much more interesting data are than numbers?

    These data were collected by the state of North Carolina in part to help researchers understand the factors that contribute to low-weight and premature births. If doctors understand the causes of premature birth, they can work to prevent it—perhaps by helping expectant mothers change their behavior, perhaps by medical intervention, and perhaps by a combination of both.

KEY POINT	Data are "numbers in context."

Data play a pivotal role in our economy, culture, and everyday lives. Much of this textbook is concerned with data collected for what we might call "professional" purposes, such as answering scientific questions or making business decisions. But in fact, data are everywhere. Google, for example, saves every search you make and combines this with data on which links you click in order to improve the way it presents information (and, of course, to determine which advertisements will appear on your search results).

In this book you'll see data from a variety of sources. For example, thanks to small, portable sensors, you can now join the "Personal Data Movement." Members of this movement record data about their daily lives and analyze it in order to improve their health, to run faster, or just to make keepsakes—a modern-day scrapbook of personal data visualizations. Maybe you or a friend uses a smart watch to keep track of your runs. One of the authors of this text carries a FitBit to record his daily activity. From this he learned that he typically takes 2500 more steps on days that he lectures than on days that he does not.

Speaking of Twitter, did you know every tweet in the "twitterverse" is saved and can be accessed? Twitter, like many other websites, provides what's called an API for accessing data. API stands for Application Program Interface, and it's basically a language that allows programmers to communicate with websites in order to access data that the website wishes to make public. For example, the statistical analysis software package StatCrunch makes use of an API provided by Twitter to create a "Word Wall" on whatever keywords you choose to type or show you currently trending tags. See Figures 1.2 and 1.3.

▼ FIGURE 1.2 Trending Tags on a day in August 2017. Can you guess what day of the week this was?

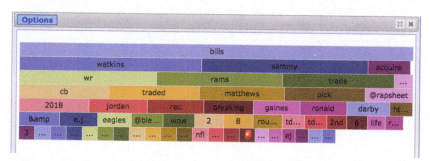

▲ FIGURE 1.3 A StatCrunch-generated "word wall" showing most common words appear in tweets that include the word *Bills*.

Another common way to store data on the Internet is using HTML tables. HTML (Hyper Text Markup Language) is another example of a software language that tells your browser how to display a web page. HTML tells the browser which words are "headers," which are paragraphs, and which should be displayed in a table. For example, while reading a Wikipedia article about coffee, the authors came across a table showing coffee production by country. These data are stored in an HTML file. This table is relatively small, and so it is simple to enter it into statistical analysis software. But other tables are quite large, and software packages must be employed to "scrape" the data (Figure 1.4).

Sometimes you can find data by turning to your government. The website data.gov has over 197,000 data sets available. The city of Miami, Florida, is one of many cities around the United States that provides data on a variety of topics. Figure 1.5 shows the first few rows of a data set that provides salaries for roughly 28,000 employees of the city of Miami.

Data provided through an open data portal, such as these data, can be stored in a variety of formats. These data can be downloaded as "CSV," "CSV for Excel," "JSON," "XML," and some other formats as well. You don't need to worry about these, but

Production [edit]

Top ten green coffee producers in 2014

Rank	Country	Teragrams[81]
1	Brazil	2.8
2	Vietnam	1.4
3	Colombia	0.7
4	Indonesia	0.6
5	Ethiopia	0.4
6	India	0.3
7	Honduras	0.3
8	Guatemala	0.2
9	Peru	0.2
10	Uganda	0.2
	World	8.8

StatCrunchThis Output

URL:
https://en.wikipedia.org/wiki/Coffee

Table 2

StatCrunch | Applets | Edit | Data | Stat | Graph | Help

Row	Rank	Country	Teragrams[8	var4	var5	var6	var7	var8
1	1	Brazil	2.8					
2	2	Vietnam	1.4					
3	3	Colombia	0.7					
4	4	Indonesia	0.6					
5	5	Ethiopia	0.4					
6	6	India	0.3					
7	7	Honduras	0.3					
8	8	Guatemala	0.2					
9	9	Peru	0.2					
10	10	Uganda	0.2					
11		World	8.8					
12								
13								
14								

▲ **FIGURE 1.4** Coffee production HTML table from en.wikipedia.org/wiki/Coffee (viewed on August 11, 2017; left) and the same table "scraped" by StatCrunch (right).

Title	Department	Annual Salaı	Gross Pay Lı	Gross Year 1	Gross Year 1
W&S SEWEF	WATER AND	42044.6	3411.87	45630.98	44282.92
AIRPORT OP	AVIATION	50068.72	2086.67	36306	35244.41
ASST W&S S	WATER AND	96744.96	3764.42	60230.72	58444.64
BUS OPERA1	TRANSPORT	46332	2183.58	38911.15	37764.66
CORRECTIO	CORRECTIO	102386.44	4079.85	65808.99	63897.41
LANDSCAPE	PARKS, REC	43264	1707.46	29852.11	28977.41
TREATMENT	WATER AND	66755.52	4175.56	60779.63	58977.12
CORRECTIO	CORRECTIO	77477.92	3023.38	49120.06	47671.78
ENVIRONME	REGULATOR	63657.36	2486.82	39789.12	38613.92
BUS OPERA1	TRANSPORT	50616.8	2701.54	43895.25	42599.25
SR SYSTEMS	INFORMATIC	113020.96	4410.42	70266.72	68180.16
ADMINISTRA	PARKS, REC	83850.26	3563.23	50379.58	48889.04
BUS MAINT	TRANSPORT	49576.8	2501.1	46157.56	44803.71
POLICE OFF:	POLICE	85223.06	5221.32	82099.06	79708.32
SEAPORT EN	SEAPORT	63283.74	2979.46	60668.57	58903.53
POLICE REC	POLICE	49352.42	1941.63	33237.68	32284.86

◀ **FIGURE 1.5** The first few rows of public employee salary data sets provided by the city of Miami, Florida.

for most applications, "CSV," which stands for "comma-separated values" will be understandable by most data analysis packages. (For example, Excel, Minitab, and StatCrunch can all import CSV files.)

You won't have to scrape and download data on your own in order to do the examples and exercises in this textbook (unless you want to, of course). The data you need are provided for you, ready to upload into one of several common statistical analysis packages (Excel, Minitab, StatCrunch, and the TI-84 graphing calculator). However, this book includes projects that might lead you to uncharted waters, and so you should be aware that different data storage types exist.

In the next section, you'll see that you can store data in different structures, and some structures are particularly helpful in some circumstances.

What Is Data Analysis?

In this text you will study the science of data. Most important, you will learn to analyze data. What does this mean? You are analyzing data when you examine data of some sort and explain what they tell us about the real world. In order to do this, you must first learn about the different types of data, how data are stored and structured, and how they are summarized. The process of summarizing data takes up a big part of this text;

indeed, we could argue that the entire text is about summarizing data, either through creating a visualization of the data or distilling them down to a few numbers that we hope capture their essence.

> **KEY POINT**   Data analysis involves creating summaries of data and explaining what these summaries tell us about the real world.

Classifying and Storing Data

▲ **FIGURE 1.6** A photo of Carhenge, Nebraska.

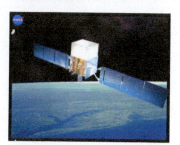

▲ **FIGURE 1.7** Satellites in NASA's Earth Observing Mission record ultraviolet reflections and transmit these data back to Earth. Such data are used to construct images of our planet. Earth Observer (http://eos .gsfc.nasa.gov/).

> 🔖 **Details**
>
> **More Grammar**
> We're using the word *sample* as a noun—it is an object, a collection of data that we study. Later we'll also use the word *sample* as a verb—that is, to describe an action. For example, we'll sample ice cream cones to measure their weight.

The first step in understanding data is to understand the different types of data you will encounter. As you've seen, data are numbers in context. But that's only part of the story; data are also recorded observations. Your photo from your vacation to Carhenge in Nebraska is data (Figure 1.6). The ultraviolet images streaming from the Earth Observer Satellite system are data (Figure 1.7). These are just two examples of data that are not numbers. Statisticians work hard to help us analyze complex data, such as images and sound files, just as easily as we study numbers. Most of the methods involve recoding the data into numbers. For example, your photos can be digitized in a scanner, converted into a very large set of numbers, and then analyzed. You might have a digital camera that gives you feedback about the quality of a photo you've taken. If so, your camera is not only collecting data but also analyzing it!

Almost always, our data sets will consist of characteristics of people or things (such as gender and weight). These characteristics are called **variables**. Variables are not "unknowns" like those you studied in algebra. We call these characteristics variables because they have variability: The values of the variable can be different from person to person.

> **KEY POINT**   Variables in statistics are different from variables in algebra. In statistics, variables record characteristics of people or things.

When we work with data, they are grouped into a collection, which we call either a **data set** or a **sample**. The word *sample* is important, because it implies that the data we see are just one part of a bigger picture. This "bigger picture" is called a **population**. Think of a population as the Data Set of Everything—it is the data set that contains all of the information about everyone or everything with respect to whatever variable we are studying. Quite often, the population is really what we want to learn about, and we learn about it by studying the data in our sample. However, many times it is enough just to understand and describe the sample. For example, you might collect data from students in your class simply because you want to know about the students in your class, not because you wish to use this information to learn about all students at your school. Sometimes, data sets are so large that they effectively *are* the population, as you'll soon see in the data reflecting births in North Carolina.

Two Types of Variables

While variables can be of many different types, there are two basic types that are very important to this book. These basic types can be broken into small subcategories, which we'll discuss later.

Numerical variables describe quantities of the objects of interest. The values will be numbers. The weight of an infant is an example of a numerical variable.

Categorical variables describe qualities of the objects of interest. These values will be categories. The sex of an infant is an example of a categorical variable. The possible values are the categories "male" and "female." Eye color of an infant is another example; the categories might be brown, blue, black, and so on. You can often identify categorical variables because their values are *usually* words, phrases, or letters. (We say "usually" because we sometimes use numbers to represent a word or phrase. Stay tuned.)

EXAMPLE 1 Crash-Test Results

The data in Table 1.1 are an excerpt from crash-test dummy studies in which cars are crashed into a wall at 35 miles per hour. Each row of the data set represents the observed characteristics of a single car. This is a small sample of the database, which is available from the National Transportation Safety Administration. The *head injury* variable reflects the risk to the passengers' heads. The higher the number, the greater the risk.

Make	Model	Doors	Weight	Head Injury
Acura	Integra	2	2350	599
Chevrolet	Camaro	2	3070	733
Chevrolet	S-10 Blazer 4X4	2	3518	834
Ford	Escort	2	2280	551
Ford	Taurus	4	2390	480
Hyundai	Excel	4	2200	757
Mazda	626	4	2590	846
Volkswagen	Passat	4	2990	1182
Toyota	Tercel	4	2120	1138

◄ **TABLE 1.1** Crash-test results for cars.

QUESTION For each variable, state whether it is numerical or categorical.

SOLUTION The variables *make* and *model* are categorical. Their values are descriptive names. The units of *doors* are, quite simply, the number of doors. The units of *weight* are pounds. The variables *doors* and *weight* are numerical because their values are measured quantities. The units for *head injury* are unclear; head injury is measured using some scale that the researchers developed.

TRY THIS! Exercise 1.3

Coding Categorical Data with Numbers

Sometimes categorical variables are "disguised" as numerical. The *smoke* variable in the North Carolina data set (Table 1.2) has numbers for its values (0 and 1), but in fact those numbers simply indicate whether or not the mother smoked. Mothers were asked, "Did you smoke?" and if they answered "Yes," the researchers coded this categorical response with a 1. If they answered "No," the response was coded with a 0. These particular numbers represent categories, not quantities. *Smoke* is a categorical variable.

Coding is used to help both humans and computers understand what the values of a variable represent. For example, a human would understand that a "yes" under the

Weight	Female	Smoke
7.69	1	0
0.88	0	1
6.00	1	0
7.19	1	0
8.06	1	0
7.94	1	0

▲ **TABLE 1.2** Data for newborns with coded categorical variables.

"Smoke" column would mean that the person was a smoker, but to the computer, "yes" is just a string of symbols. If instead we follow a convention where a 1 means "yes" and a 0 means "no," then a human understands that the 1s represent smokers, and a computer can easily add the values together to determine, for example, how many smokers are in the sample.

This approach for coding categorical variables is quite common and useful. If a categorical variable has only two categories, as do *gender* and *smoke*, then it is almost always helpful to code the values with 0 and 1. To help readers know what a "1" means, rename the variable with either one of its category names. A "1" then means the person belongs to that category, and a 0 means the person belongs to the other category. For example, instead of calling a variable *gender*, we rename it *female*. And then if the baby is a boy we enter the code 0, and if it's a girl we enter the code 1.

Sometimes your computer does the coding for you without your needing to know anything about it. So even if you see the words *female* and *male* on your computer, the computer has probably coded these with values of 0 and 1 (or vice versa).

Storing Your Data

The format in which you record and store your data is very important. Computer programs will require particular formats, and by following a consistent convention, you can be confident that you'll remember the qualities of your own data set if you need to revisit it months or even years later. Data are often stored in a spreadsheet-like format in which each row represents the object (or person) of interest. Each column represents a variable. In Table 1.3, each row represents a movie. The column heads are variables: *Title, Rating, Runtime, Critics Rating*. (The "rating" is the Motion Pictures Association of America rating to indicate the movie's intended audience. The Critics Rating is a score from 0 to 100 from the website Rotten Tomatoes. High scores are good.) This format is sometimes referred to as the **stacked data** format.

Title	Rating	Runtime	Critics Rating
Cars 2	G	106	39
Alvin and the Chipmunks: Chipwrecked	G	87	12
Monsters University	G	104	78
Alice Through the Looking Glass	PG	113	30
Chasing Mavericks	PG	116	31
Despicable Me 2	PG	98	73
Cloudy with a Chance of Meatballs 2	PG	95	70
Hotel Transylvania	PG	91	45

▲ **TABLE 1.3** Data for a few movies rated G or PG.

When you collect your own data, the stacked format is almost always the best way to record and store your data. One reason is that it allows you to easily record several different variables for each subject. Another reason is that it is the format that most software packages will assume you are using for most analyses. (The exceptions are the TI-84 and Excel.)

Some technologies, such as the TI calculators, require, or at least accommodate, data stored in a different format, called **unstacked data**. Unstacked data tables are also common in some books and media publications. In this format, each column represents a variable from a different group. For example, one column could represent the length in minutes of movies rated G and another column could represent the length of movies rated PG. The data set, then, is a single variable (*Runtime*) broken into distinct groups. The groups are determined by a categorical variable (in this case, *Rating*.) Table 1.4

Rated G	Rated PG
112	113
90	116
95	95
	98
	91

▲ **TABLE 1.4** Movie runtime (in minutes) by rating group (unstacked).

shows an example of the *Runtime* variable in Table 1.3. Figure 1.8 shows the same data in TI-84 input format.

The great disadvantage of the unstacked format is that it can store only two variables at a time: the variable of interest (for example, Runtime) and a categorical variable that tells us which group the observation belongs in (for example, Rating). However, most of the time we record many variables for each observation. For example, recording a movie's title, rating, running time and critics' rating in the stacked format enables us to display as many variables as we wish.

NORMAL FLOAT AUTO REAL RADIAN MP					
L1	L2	L3	L4	L5	2
112	113	------	------	------	
90	116				
95	95				
------	98				
	91				

L2(6)=

▲ **FIGURE 1.8** TI-84 data input screen (unstacked data).

EXAMPLE **2** Personal Data Collection

Using a sensor worn around her wrist, Safaa recorded the amount of sleep she got on several nights. She also recorded whether it was a weekend or a weeknight. For the weekends, she recorded (in hours): 8.1, 8.3. For the weeknights she recorded 7.9, 6.5, 8.2, 7.0, 7.3.

QUESTION Write these data in both the stacked format and the unstacked format.

SOLUTION In the stacked format, each row represents a unit of observation, and each column measures a characteristic of that observation. For Safaa, the unit of observation was a night of sleep, and she measured two characteristics: time and whether it was a weekend. In stacked format, her data would look like this:

Time	Weekend
8.1	Yes
8.3	Yes
7.9	No
6.5	No
8.2	No
7.0	No
7.3	No

(Note that you might have coded the "Weekend" variable differently. For example, instead of entering "Yes" or "No," you might have written either "Weekend" or "Weeknight" in each row.)

In the unstacked format, the numerical observations appear in separate columns, depending on the value of the categorical variable:

Weekend	Weeknight
8.1	7.9
8.3	6.5
	8.2
	7.0
	7.3

See the Tech Tips to review how to enter data like these using your technology.

> ❗ **Caution**
>
> **Look at the Data Set!**
> The fact that different people use different formats to store data means that the first step in any data investigation is to look at the data set. In most real-life situations, stacked data are the more useful format because this format permits you to work with several variables at the same time.

TRY THIS! Exercise 1.11

Investigating Data

▶ **FIGURE 1.9** The Data Cycle reminds us of the four stages of a statistical investigation. (Designed by Alyssa Brode for the Mobilize Project at UCLA.)

Now that you've seen some examples of data, it's time to learn what to do when you interact with data, as this book will ask you to do. To help guide you, consider the Data Cycle (see Figure 1.9).

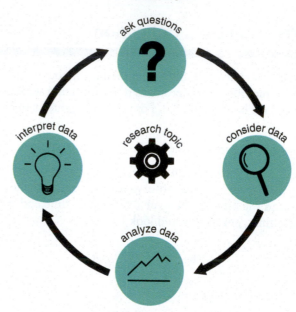

The Data Cycle

The Data Cycle is a representation of a statistical investigation cycle: the stages we go through when analyzing data. In real life, you might not necessarily go in this order of the stages listed here. In fact, you are likely to alternate between various stages. But it is useful to plan your analysis in this order.

As we move through the book, we'll deepen our understanding of what each stage entails. For now, the goal is to show you how the Data Cycle is used. Don't worry if you don't understand all of the details in what follows. We'll go into detail in subsequent chapters.

The cycle revolves around a research topic. Research topics can be very broad and serious; for example, "What makes a runner faster or slower?" or "What effect do taxes have on the economy?" Research topics might be geared toward answering some pointed, important questions: "Does human activity contribute to global warming?" or "Do cell phones cause cancer?" The first step is to break this big topic into smaller steps. These smaller steps should be phrased as questions.

Ask Questions. The trick here is to ask *good* questions, and you'll get better at that as you read on. Good questions are questions that can be answered with data. Better questions are questions that address the research topic and increase your understanding of the issues. Often in this book, we will ask you to consider a data set and to pose questions that you want answered.

For example, Table 1.5 shows a few randomly selected rows from a data set containing all 19,212 runners in the 2017 Los Angeles Marathon. Most of the variables are

Bib.Number	Age	Place	Gender.Place	5k.Split	15k.Split	Clock.Time	Net.Time	Hometown	Gender
8752	36	3,718	2,874	1,607	4,825	15,695	15,539	Pasadena, CA	M
11478	31	14,785	5,585	1,881	6,814	23,487	22,816	Victorville, CA	F
3372	47	2,246	1,839	1,530	4,763	14,368	14,330	Danville, CA	M

▲ **TABLE 1.5** Three randomly selected runners from the 2017 Los Angeles Marathon.

concerned with time given in seconds, for example, the time from when the race begins to when the runner crosses the finish line (*Net.Time*). The variable *15k.Split* gives the time it took this runner to run the first 15 kilometers. You might have some ideas about what qualities lead to faster and slower runners. This will be our research topic: Can we better understand what qualities are associated with running a fast Los Angeles Marathon?

Consider the variables provided and write down two questions you would like to know about this marathon; focus your attention on questions that could be answered with these variables (assuming you can see the full data set).

Perhaps your questions might have concerned the variables themselves. What does *Bib.Number* mean? What's the difference between *Clock.Time* and *Net.Time*? A natural question to ask is, "Who won?" but, since we don't have the runners' names, this can't be answered. So instead you might ask, "What was the fastest time?" (8938 minutes, or about 2 hours 29 minutes). You might also have wondered how different the speeds were for men and women. Or perhaps you wondered if older people ran this race slower than younger people?

These are all examples of important questions, although only the last two address the research topic. In Chapter 2 you'll learn about Statistical Questions, which are a particular type of question that can be very productive in a statistical investigation.

Consider Data. In this stage, we consider which data are available to answer the question. In fact, many statistical investigations begin at this stage: you are given data and need to generate useful questions to help you understand what the data are about. This was the case with the L.A. Marathon data.

When considering if your data are helpful for answering the questions you've posed, it's very important to understand the context of the data. Who are what was observed? What variables were measured, and how were they measured? What were the units of measurement? Who collected the data? How did they collected the data? Why did they collected the data? When did they collect the data? Where did they collect the data.

Sometimes these questions cannot be answered—we simply don't have the information at hand. If so, this could be a reason to distrust the data and proceed with extreme caution. If you collected the data yourself, be sure to record this information so that future analysts can understand and make use of your efforts.

In this case, the data came from the website http://www.lamarathon.com/race-week-end/results. (The data were modified somewhat for pedagogical purposes.) The data were collected on every runner in the race, as is typical for these large events. We aren't provided with information about how the data were collected, but it seems reasonable to assume that the data came from entry forms and from the race officials themselves.

- Who or what was observed? All participants in the 2017 Los Angeles Marathon.

- What variables were measured, and how? We may have to do a little research into the context in order to understand what the variables here mean. For example, with a little bit of googling, we can learn that in a large running event such as a marathon, it often takes a runner awhile after the start of the race before she or he gets to the actual start line. For this reason, the "Clock Time," the time from when the race began and the runner crossed the finish line, is often longer than the "Net Time" the time it took the runner to go from the start line to the finish line. "Bib Number" is the runner's assigned ID number. The units for the times are in seconds.

- Who collected the data? The data are official results, and we can assume they were assembled and collected by the race officials.

- When and where did they collect the data? The running times were collected on the day of the race, March 19, 2017. These times were collected at the start and finish line of the race.

Data Moves ▶ These data were scraped from a web page that provides results of marathons from across the country. A "script" was written in the statistical programming language R to convert the many pages of HTML files on this website into a file ready to be analyzed.

File ▶ lamarathon.csv

Analyze Data. This is the primary topic of this book. You'll learn in Chapter 2 that the first step in an analysis is to visualize the data and that, sometimes, this visualization is enough to answer the question of interest. Let's consider the question, "How different were the speeds for men and women?" This is a tricky question, because the running times varied considerably for men and women. Some women were faster than some men; some men were faster than some women. To help us answer, we refine this question to "What was the typical difference in speed between men and women in this race?" You'll see in Chapter 2 that a visualization such as the one in Figure 1.10 is a helpful first step toward answering questions like ours. Figure 1.10 displays the *distributions* of the *Net.Time* variable for both men and women. (You'll learn about how to read such visualizations in Chapter 2.) We've added a vertical bar to indicate the location of the *mean* time for women (top) and men (bottom). You'll learn about how, when, and why to use the mean in Chapter 3. From the graph, it looks like the mean running time for women is a bit over 20,000 seconds (a bit more than 5.5 hours) and for men is less than 20,000 seconds.

▶ **FIGURE 1.10** Visualization of running times for women (left) and men (right). The vertical line indicates the mean running time for each group.

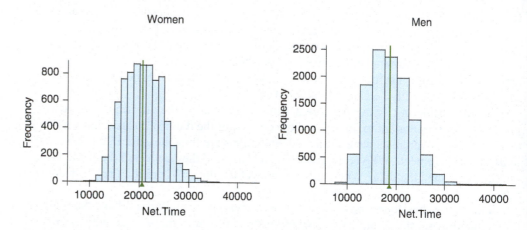

Interpret Data. The final step is to interpret your analysis. *Interpret* is a fancy word for "answer your question!" Our question was, "What was the typical difference in speed between men and women in this race?" In Chapter 3 you'll see that the mean value gives us one way of measuring this notion of "typical," and from Figure 1.10 we can roughly judge that the means differ by about 2000 seconds. (This is actually a tricky judgment call from this graphic, but you'll soon learn some tools that help you judge these distances.) Our answer to the question is: Typically, the men were 2000 seconds faster than the women in this race.

Examples 3 and 4 will give you practice with the "Ask Questions" part of the Data Cycle.

EXAMPLE **3** At the Movies

Data Moves ▶ James Molyneux merged data from several sources to compile this data set. Merging related datasets can often lead to greater insights.
File ▶ movieratings.csv

The data file movieratings.csv, compiled by statistician James Molyneux, contains data on almost 5000 movies. The following variables are provided:

Title, year, runtime, mpaa_rating, studio, color, director, language, country, aspect_ratio, n_post_face, n_critics, n_audience, reviews_num, audience_rating, critics_rating, budget, gross, imdbi_id

The Data Cycle begins with questioning. You might not know precisely what these variables mean or how the data were measured, but use your general knowledge about movies.

QUESTION   Which variables would you consider in order to answer this question: Do the critics tend to rate movies lower than the audience rates them?

SOLUTION We see that there are two variables, *audience_rating* and *critics_rating*, that would likely help us determine if critics tend to rate movies differently than "regular" people.

TRY THIS! Exercise 1.15

EXAMPLE 4 Movie Ratings

The variables provided in the datafile movieratings.csv include the following:

Title, year, runtime, mpaa_rating, studio, color, director, language, country, aspect_ratio, n_post_face, n_critics, n_audience, reviews_num, audience_rating, critics_rating, budget, gross, imdb_id

Often we are provided with a "data dictionary" that gives us some guidelines as to what the variables mean. Here's a simplified data dictionary: *runtime* is the length of the movie; *color* is whether the movie is in color or black-and-white; *mpaa_rating* is the Motion Pictures Association of America guide to the age level of the movie—G, PG, PG-13, and so on; *critics_rating* and *audience_rating* are average scores of critics from the Rotten Tomatoes website.

QUESTION Which of these questions cannot be answered with the data in movieratings.csv?

a. Do critics rate R-rated movies more highly than G-rated movies?

b. Are comedies shorter than dramas?

c. Do audiences prefer shorter movies over longer movies?

d. Do movies that have a large budget receive higher audience ratings?

SOLUTION Question (b) cannot be answered with these data because it requires knowing the *genre* (comedy, drama, documentary, etc.) of the movie, and we do not have a variable that provides this information.

TRY THIS! Exercise 1.19

The Data Cycle is not meant to be a hard-and-fast rule that, if followed rigorously, is promised to bring you success and happiness. Think of it, instead, as a guiding principle, a device to help you if you get stuck. For example, if you've just done an analysis and are thinking, "Am I done?" take a look at the Data Cycle. After the analysis comes the interpretation, and so, unless you've done an interpretation, you're not yet done!

SECTION 1.4

Organizing Categorical Data

Once we have a data set, we next need to organize and display the data in a way that helps us see patterns. This task of organization and display is not easy, and we discuss it throughout the entire text. In this section we introduce the topic for the first time, in the context of categorical variables.

With categorical variables, we are usually concerned with knowing how often a particular category occurs in our sample. We then (usually) want to compare how often a category occurs for one group with how often it occurs for another (liberal/conservative, man/woman). To do these comparisons, you need to understand how to calculate percentages and other rates.

	Male	Female
Not Always	2	3
Always	3	7

▲ **TABLE 1.6** This two-way table shows counts for 15 youths who responded to a survey about wearing seat belts.

Male	Not Always
1	1
1	1
1	0
1	0
1	0
0	1
0	1
0	1
0	0
0	0
0	0
0	0
0	0
0	0
0	0

▲ **TABLE 1.7** This data set is equivalent to the two-way summary shown in Table 1.6. We highlighted in red those who did not always wear a seat belt (the risk takers).

> ! **Caution**
>
> **Two-Way Tables Summarize Categorical Variables**
>
> It is tempting to look at a two-way table like Table 1.6 and think that you are looking at numerical variables, because you see numbers. But the values of the variables are actually categories (gender and whether or not the subject always wears a seat belt). The numbers you see are summaries of the data.

A common method for summarizing two potentially related categorical variables is to use a two-way table. **Two-way tables** show how many times each combination of categories occurs. For example, Table 1.6 is a two-way table from the Youth Behavior Risk Survey that shows gender and whether or not the respondent always (or almost always) wears a seat belt when riding in or driving a car. The actual Youth Behavior Risk Survey has over 10000 respondents, but we are practicing on a small sample from this much larger data set.

The table tells us that two people were male and did not always wear a seat belt. Three people were female and did not always wear a seat belt. These counts are also called frequencies. A **frequency** is simply the number of times a value is observed in a data set.

Some books and publications discuss two-way tables as if they displayed the original data collected by the investigators. However, two-way tables do not consist of "raw" data but, rather, are summaries of data sets. For example, the data set that produced Table 1.6 is shown in Table 1.7.

To summarize this table, we simply count how many of the males (a 1 in the Male column) also do not always wear seat belts (a 1 in the Not Always column). We then count how many both are male and always wear seat belts (a 1 in the Male column, a 0 in the Not Always column), how many both are female and don't always wear seat belts (a 0 in the Male column, a 1 in the Not Always column), and finally, how many both are female and always wear a seat belt (a 0 in the Male column, a 0 in the Not Always column).

Example 5 illustrates that summarizing the data in a two-way table can make it easy to compare groups.

EXAMPLE 5 Percentages of Seat Belt Wearers

The 2011 Youth Behavior Risk Survey is a national study that asks American youths about potentially risky behaviors. We show the two-way summary again. All of the people in the table were between 14 and 17 years old. The participants were asked whether they wear a seat belt while driving or riding in a car. The people who said "always" or "almost always" were put in the Always group. The people who said "sometimes" or "rarely" were put in the Not Always group.

	Male	Female
Not Always	2	3
Always	3	7

QUESTIONS

a. How many men are in this sample? How many women? How many people do not always wear seat belts? How many always wear seat belts?

b. What percentage of the sample are men? What percentage are women? What percentage don't always wear seat belts? What percentage always wear seat belts?

c. Are the men in the sample more likely than the women in the sample to take the risk of not wearing a seat belt?

SOLUTIONS

a. We can count the men by adding the first column: $2 + 3 = 5$ men. Adding the second column gives us the number of women: $3 + 7 = 10$.

We get the number who do not always wear seat belts by adding the first row: $2 + 3 = 5$ people don't always wear seat belts. Adding the second row gives us the number who always wear seat belts: $3 + 7 = 10$.

b. This question asks us to convert the numbers we found in part (a) to percentages. To do this, we divide the numbers by 15, because there were 15 people in the sample. To convert to percentages, we multiply this proportion by 100%.

The proportion of men is $5/15 = 0.333$. The percentage is $0.333 \times 100\% = 33.3\%$. The proportion of women must be $100\% - 33.3\% = 66.7\%$ $(10/15 \times 100\% = 66.7\%)$.

The proportion who do not always wear seat belts is $5/15 = 0.333$, or 33.3%. The proportion who always wear seat belts is $100\% - 33.3\% = 66.7\%$.

c. You might be tempted to answer this question by counting the number of males who don't always wear seat belts (2 people) and comparing that to the number of females who don't always wear seat belts (3 people). However, this is not a fair comparison because there are more females than males in the sample. Instead, we should look at the percentage of those who don't always wear seat belts in each group. This question should be reworded as follows:

Is the percentage of males who don't always wear seat belts greater than the percentage of females who don't always wear seat belts?

Because 2 out of 5 males don't always wear seat belts, the percentage of males who don't always wear seat belts is $(2/5) \times 100\% = 40\%$.

Because 3 out of 10 females don't always wear seat belts, the percentage of females who don't always wear seat belts is $(3/10) \times 100\% = 30\%$.

In fact, females in this sample engage in this risky behavior less often than males. Among all U.S. youth, it is estimated that about 28% of males do not always wear their seat belt, compared to 23% of females.

TRY THIS!   Exercise 1.21

The calculations in Example 5 took us from frequencies to percentages. Sometimes, we want to go in the other direction. If you know the total number of people in a group and are given the percentage that meets some qualification, you can figure out *how many* people in the group meet that qualification.

EXAMPLE 6 Numbers of Seat Belt Wearers

A statistics class has 300 students, and they are asked whether they always ride or drive with a seat belt.

QUESTIONS

a. Suppose that 30% of the students do not always wear a seat belt. How many students is this?

b. Suppose we know that in another class, 20% of the students do not always wear seat belts, and this is equal to 43 students. How many students are in the class?

SOLUTIONS

a. We need to find 30% of 300. When working with percentages, first convert the percentage to its decimal equivalent:

$$30\% \text{ of } 300 = 0.30 \times 300 = 90$$

Therefore, 90 students don't always wear seat belts.

b. The question tells us that 20% of some unknown larger number (call it y) must be equal to 43.

$$0.20y = 43$$

Divide both sides by 0.20 and you get

$$y = 215$$

There are 215 total students in the class, and 43 of them don't always wear seat belts.

TRY THIS!   Exercise 1.29

Sometimes, you may come across data summaries that are missing crucial information. Pay close attention when tables give counts of things. Suppose we wanted to know which city was most at risk for a crime such as burglary? Table 1.8 gives the number of burglaries in 2009 reported in several major as reported by the Federal Bureau of Investigation. We show only the ten cities with the greatest number of burglaries.

► **TABLE 1.8** The ten U.S. cities with the greatest numbers of burglaries reported.

State	City	Number of Burglaries
Texas	Houston	19,858
California	Los Angeles	16,160
Nevada	Las Vegas	14,876
New York	New York City	14,100
Illinois	Chicago	13,152
Arizona	Phoenix	12,798
Texas	San Antonio	11,633
Texas	Dallas	11,121
Indiana	Indianapolis	11,085
Tennessee	Memphis	10,272

It looks like Houston, Texas, is the most dangerous city in terms of burglaries and Memphis, Tennessee, one of the least dangerous (among these top ten). But this table is missing a crucial piece of information: the number of people who live in the city. A city with 1 million people is probably going to have more burglaries than a town of 40,000.

How do we control for the difference in populations? Table 1.9 includes the information missing from Table 1.8: the population counts.

State	City	Population	Number of Burglaries
Texas	Houston	2,275,221	19,858
California	Los Angeles	3,962,726	16,160
Nevada	Las Vegas	1,562,134	14,876
New York	New York City	8,550,861	14,100
Illinois	Chicago	2,728,695	13,152
Arizona	Phoenix	1,559,744	12,798
Texas	San Antonio	1,463,586	11,633
Texas	Dallas	1,301,977	11,121
Indiana	Indianapolis	863,675	11,085
Tennessee	Memphis	657,936	10,272

▲ **TABLE 1.9** The same cities with population sizes included.

With this extra information we can figure out which city is most risky based on its size. For example, what percentage of residents reported burglaries in Houston? There were 2,275,221 residents and 19,858 burglaries. And so the percentage burgled is $(19858/2275221) \times 100\% = 0.87\%$, after rounding.

Sometimes, with percentages as small as this, we understand the numbers more easily if instead of reporting the "per cent" we report "per 1000" or even "per 10,000. We call such numbers **rates**. To get the burglary rate per 1000 residents, instead of multiplying $(19858/2275221)$ by 100 we multiply by 1000: $(19858/2275221) \times 1000 = 8.7$ burglaries per 1000 people. These results are shown in Table 1.10 (rounded to the hundredths place for easier reading).

State	City	Population	Number of Burglaries	Burglaries per 1000 residents
Texas	Houston	2,275,221	19,858	8.73
California	Los Angeles	3,962,726	16,160	4.08
Nevada	Las Vegas	1,562,134	14,876	9.52
New York	New York City	8,550,861	14,100	1.65
Illinois	Chicago	2,728,695	13,152	4.82
Arizona	Phoenix	1,559,744	12,798	8.21
Texas	San Antonio	1,463,586	11,633	7.95
Texas	Dallas	1,301,977	11,121	8.54
Indiana	Indianapolis	863,675	11,085	12.83
Tennessee	Memphis	657,936	10,272	15.61

▲ **TABLE 1.10** The ten cities with the most burglaries with burglaries per 1000 residents.

We now see that Memphis has the highest burglary rate among these cities and New York City the lowest!

EXAMPLE 7 Comparing Rates of Stolen Cars

Which model of car has the greatest risk of being stolen? The Highway Loss Data Institute reports that the Ford F-250 pickup truck is the most stolen car; 7 F-250s are reported stolen out of every 1000 that are insured. By way of contrast, the Jeep Compass is the least stolen; only 0.5 Jeep Compass is reported stolen for every 1000 insured (Insurance Institute for Highway Safety 2013).

QUESTION   Why does the Highway Loss Data Institute report theft rates rather than the number of each type of car stolen?

SOLUTION   We need to take into account the fact that some cars are more popular than others. Suppose there were many more Jeep Compasses than Ford F-250s. In that case, we might see a greater number of stolen Jeeps, simply because there are more of them to steal. By looking at the *theft rate*, we adjust for the total number of cars of that particular kind on the road.

TRY THIS!   Exercise 1.31

KEY POINT   In order for us to compare groups, the groups need to be similar. When the data consist of counts, then percentages or rates are often better for comparisons because they take into account possible differences among the sizes of the groups.

Collecting Data to Understand Causality

Often, the most important questions in science, business, and everyday life are questions about **causality**. These are usually phrased in the form of "what if" questions. What if I take this medicine; will I get better? What if I change my Facebook profile; will my profile get more hits?

Questions about causality are frequently in the news. The *Los Angeles Times* reported that many people believe a drink called peanut milk can cure gum disease and slow the onslaught of baldness. The BBC News (2010) reported that "[h]appiness wards off heart disease." Statements such as these are everywhere we turn these days. How do we know whether to believe these claims?

The methods we use to collect data determine what types of conclusions we can make. Only one method of data collection is suitable for making conclusions about causal relationships, but as you'll see, that doesn't stop people from making such conclusions anyway. In this section we talk about three methods commonly used to collect data in an effort to answer questions about causality: anecdotes, observational studies, and controlled experiments.

Most questions about causality can be understood in terms of two variables: the **treatment variable** and the **outcome variable**. (The outcome variable is also sometimes called the **response variable** because it responds to changes in the treatment.) We are essentially asking whether the treatment variable causes changes in the outcome variable. For example, the treatment variable might record whether or not a person drinks peanut milk, and the outcome variable might record whether or not that person's gum disease improved. Or the treatment variable might record whether or not a person is generally happy, and the outcome variable might record whether or not that person suffered from heart disease in a ten-year period.

People who receive the treatment of interest (or have the characteristic of interest) are said to be in the **treatment group**. Those who do not receive that treatment (or do not have that characteristic) are in the **comparison group**, which is also called the **control group.**

Anecdotes

Peanut milk is a drink invented by Jack Chang, an entrepreneur in San Francisco, California. He noticed that after he drank peanut milk for a few months, he stopped losing hair and his gum disease went away. According to the *Los Angeles Times* (Glionna 2006), another regular drinker of peanut milk says that the beverage caused his cancer to go into remission. Others have reported that drinking the beverage has reduced the severity of their colds, has helped them sleep, and has helped them wake up.

This is exciting stuff! Peanut milk could very well be something we should all be drinking. But can peanut milk really solve such a wide variety of problems? On the face of it, it seems that there's evidence that peanut milk has cured people of illness. The *Los Angeles Times* reports the names of people who claim that it has. However, the truth is that this is simply not enough evidence to justify any conclusion about whether the beverage is helpful, harmful, or without any effect at all.

These testimonials are examples of anecdotes. An **anecdote** is essentially a story that someone tells about her or his own (or a friend's or relative's) experience. Anecdotes are an important type of evidence in criminal justice because eyewitness testimony can carry a great deal of weight in a criminal investigation. However, for answering questions about groups of people with great variability or diversity, anecdotes are essentially worthless.

The primary reason why anecdotes are not useful for reaching conclusions about cause-and-effect relationships is that the most interesting things that we study have so much variety that a single report can't capture the variety of experience. For example, have you ever bought something because a friend recommended it, only to find that after a few weeks it fell apart? If the object was expensive, such as a car, you might have been angry at your friend for recommending such a bad product. But how do you know whose experience was more typical, yours or your friend's? Perhaps the car is in fact a very reliable model, and you just got a lemon.

A very important question to ask when someone claims that a product brings about some kind of change is to ask, "Compared to what?" Here the claim is that drinking peanut milk will make you healthier. The question to ask is, "Healthier compared to what?" Compared to people who don't drink peanut milk? Compared to people who take medicine for their particular ailment? To answer these questions, we need to examine the health of these other groups of people who do not drink peanut milk.

Anecdotes do not give us a comparison group. We might know that a group of people believe that peanut milk made them feel better, but we don't know how the milk drinkers' experiences compare to those of people who did not drink peanut milk.

> **KEY POINT**   When someone makes a claim about causality, ask, "Compared to what?"

Another reason for not trusting anecdotal evidence is a psychological phenomenon called the placebo effect. People often react to the idea of a treatment, rather than to the treatment itself. A **placebo** is a harmless pill (or sham procedure) that a patient believes is actually an effective treatment. Often, the patient taking the pill feels better, even though the pill actually has no effect whatsoever. In fact, a survey of U.S. physicians published in the *British Medical Journal* (Britt 2008) found that up to half of physicians prescribe sugar pills—placebos—to manage chronic pain. This psychological wish fulfillment—we feel better because we think we *should* be feeling better—is called the **placebo effect**.

Observational Studies

The identifying mark of an **observational study** is that the subjects in the study are put into the treatment group or the control group either by their own actions or by the decision of someone else who is not involved in the research study. For example, if we wished to study the effects on health of smoking cigarettes (as many researchers have), then our treatment group would consist of people who had chosen to smoke, and the control group would consist of those who had chosen not to smoke.

Observational studies compare the outcome variable in the treatment group with the outcome variable in the control group. Thus, if many more people are cured of gum disease in the group that drinks peanut milk (treatment) than in the group that does not (control), then we would say that drinking peanut milk is associated with improvement in gum disease; that is, there is an **association** between the two variables. If fewer people in the happy group tend to have heart disease than those in the not-happy group, we would say that happiness is associated with improved heart health.

Note that we do not conclude that peanut milk *caused* the improvement in gum disease. In order for us to draw this conclusion, the treatment group and the control group must be very similar in every way except that one group gets the treatment and the other doesn't. For example, if we knew that the group of people who started drinking peanut milk and the group that did not drink peanut milk were alike in every way—both groups had the same overall health; were roughly the same ages; included the same mix of genders and races, education levels; and so on—then if the peanut milk group members are healthier after a year, we would be fairly confident in concluding that peanut milk is the reason for their better health.

Unfortunately, in observational studies this goal of having very similar groups is *extremely* difficult to achieve. *Some* characteristic is nearly always different in one group than in the other. This means that the groups may experience different outcomes because of this different characteristic, not because of the treatment. A difference between the two groups that could explain why the outcomes were different is called a **confounding variable** or **confounding factor**.

For example, early observational studies on the effects of smoking found that a greater percentage of smokers than of nonsmokers had lung cancer. However, some scientists argued that genetics was a confounding variable (Fisher 1959). They maintained that the smokers differed genetically from the nonsmokers. This genetic difference made some people more likely to smoke and more susceptible to lung cancer.

This was a convincing argument for many years. It not only proposed a specific difference between the groups (genetics) but also explained how that difference might come about (genetics makes some people smoke more, perhaps because it tastes better to them or because they have addictive personalities). And the argument also explained why this difference might affect the outcome (the same genetics cause lung cancer). Therefore, the skeptics said, genetics—and not smoking—might be the cause of lung cancer.

Later studies established that the skeptics were wrong about genetics. Some studies compared pairs of identical twins in which one twin smoked and the other did not. These pairs had the same genetic makeup, and still a higher percentage of the smoking twins had cancer than of the nonsmoking twins. Because the treatment and control groups had the same genetics, genetics could not explain why the groups had different cancer rates. When we compare groups in which we force one of the variables to be the same, we say that we are *controlling for* that variable. In these twin studies, the researchers controlled for genetics by comparing people with the same genetic makeup (Kaprio and Koskenvuo 1989).

A drawback of observational studies is that we can never know whether there exists a confounding variable. We can search very hard for it, but the mere fact that we don't find a confounding variable does not mean it isn't there. For this reason, we can never make cause-and-effect conclusions from observational studies.

> **KEY POINT**   We can never draw cause-and-effect conclusions from observational studies because of potential confounding variables. A single observational study can show only that there is an *association* between the treatment variable and the outcome variable.

EXAMPLE 8 Does Poverty Lower IQ?

"Chronic Poverty Can Lower Your IQ, Study Shows" is a headline from the online magazine *Daily Finance*. The article (Nisen 2013) reported on a study published in the journal *Science* (Mani et al. 2013) that examined the effects of poverty on problem-solving skills from several different angles. In one part of the study, researchers observed sugar cane farmers in rural India both before and after harvest. Before the harvest, the farmers typically have very little money and are often quite poor. Researchers gave the farmers IQ exams before the harvest and then after the harvest, when they had more money, and found that the farmers scored much higher after the harvest.

QUESTION   Based on this evidence alone, can we conclude that poverty lowers people's IQ scores? If yes, explain why. If no, suggest a possible confounding factor.

SOLUTION   No, we cannot. This is an observational study. The participants are in or out of the treatment group ("poverty") because of a situation beyond the researchers' control. A possible confounding variable is nutrition; before harvest, without much money, the farmers are perhaps not eating well, and this could lower their IQ scores.

(In fact, the researchers considered this confounding variable. They determined that nutrition was relatively constant both before and after the harvest, so it was ruled out as a confounding variable. But other confounding variables may still exist.)

TRY THIS!   Exercise 1.47

Controlled Experiments

In order to answer cause-and-effect questions, we need to create a treatment group and a control group that are alike in every way possible, except that one group gets a treatment and the other does not. As you've seen, this cannot be done with observational studies because of confounding variables. In a **controlled experiment**, researchers take control by assigning subjects to the control or treatment group. If this assignment is done correctly, it ensures that the two groups can be nearly alike in every relevant way except whether or not they receive the treatment under investigation.

Well-designed and well-executed controlled experiments are the only means we have for definitively answering questions about causality. However, controlled experiments are difficult to carry out (this is one reason why observational studies are often done instead). Let's look at some of the attributes of a well-designed controlled experiment.

A well-designed controlled experiment has four key features:

- The sample size must be large so that we have opportunities to observe the full range of variability in the humans (or animals or objects) we are studying.

- The subjects of the study must be assigned to the treatment and control groups at random.

- Ideally, the study should be "double-blind," as explained later.

- The study should use a placebo if possible.

These features are all essential in order to ensure that the treatment group and the control group are as similar as possible.

To understand these key design features, imagine that a friend has taken up a new exercise routine in order to lose weight. By exercising more, he hopes to burn more calories and so lose weight. But he notices a strange thing: the more he exercises, the hungrier he gets, and so the more he eats. If he is eating more, can he lose weight? This is a complex issue, because different people respond differently both to exercise and to food. How can we know whether exercise actually leads to weight loss? (See Rosenkilde et al. (2012) for a study related to the hypothetical one presented here.) To think about how you might answer this question, suppose you select a group of slightly overweight young men to participate in your study. For a comparison group, you might ask some of them not to exercise at all during the study. The men in the treatment group, however, you will ask to exercise for about 30 minutes each day at a moderate level.

Sample Size

A good controlled experiment designed to determine whether exercise leads to weight loss should have a large number of people participate in the study. People react to changes in their activity level in a variety of ways, so the effects of exercise can vary from person to person. To observe the full range of variability, you therefore need a large number of people. How many people? This is a hard question to answer, but in general, the more the better. You should be critical of studies with very few participants.

Random Assignment

The next step is to assign people to the treatment group and the comparison group such that the two groups are similar in every way possible. As we saw when we discussed observational studies, letting the participants choose their own group doesn't work, because people who like to exercise might differ in important ways (such as level of motivation) that would affect the outcome.

Instead, a good controlled experiment uses **random assignment** to assign participants to groups. One way of doing this is to flip a coin. Heads means the participant goes into the treatment group, and tails means she or he goes into the comparison group (or the other way around—as long as you're consistent). In practice, the randomizing might instead be done with a computer or even with the random-number generator on a calculator, but the idea is always the same: No human determines group assignment. Rather, assignment is left to chance.

If both groups have enough members, random assignment will "balance" the groups. The variation in weights, the mix of metabolisms and daily calorie intake, and the mix of most variables will be similar in both groups. Note that by "similar" we don't mean exactly the same. We don't expect both groups to have exactly the same percentage of people who like to exercise, for example. Except in rare cases, random variation results in slight differences in the mixes of the groups. But these differences should be small.

Whenever you read about a controlled experiment that does not use random assignment, remember that there is a very real possibility that the results of the study are invalid. The technical term for what happens with nonrandomized assignment is **bias**. We say that a study exhibits bias when the results are influenced in one particular direction. A researcher who puts the heaviest people in the exercise group, for example, is biasing the outcome. It's not always easy, or even possible, to predict what the effects of the bias will be, but the important point is that the bias creates a confounding variable and makes it difficult, or impossible, to determine whether the treatment we are investigating really affects the outcome we're observing.

 KEY POINT   Random assignment (assignment to treatment groups by a randomization procedure) helps balance the groups to minimize bias. This helps make the groups comparable.

Blinding

So far, we've recruited a large number of men and randomly assigned half to exercise and half to remain sedentary. In principle, these two groups will be very similar. However, there are still two potential differences.

First, we might know who is in which group. This means that when we interact with a participant, we might consciously or unconsciously treat that person differently, depending on which group he or she belongs to. For example, if we believe strongly that exercise helps with weight loss, we might give special encouragement or nutrition advice to people who are in the exercise group that we don't give to those in the comparison group. If we do so, then we have biased the study.

To prevent this from happening, researchers should be **blind** to assignment. This means that an independent party—someone who does not regularly see the participants and who does not participate in determining the results of the study—handles the assignment to groups. The researchers who measure the participants' weight loss do not know who is in which group until the study has ended; this ensures that their measurements will not be influenced by their prejudices about the treatment.

Second, we must consider the participants themselves. If they know they are in the treatment group, they may behave differently than they would if they knew nothing about their group assignment. Perhaps they will work harder at losing weight.

Or perhaps they will eat much more because they believe that the extra exercise allows them to eat anything they want.

To prevent this from happening, the participants should also not know whether they are in the treatment group or the comparison group. In some cases, this can be accomplished by not even telling the participants the intent of the study; for example, the participants might not know whether the goal is to examine the effect of a sedentary lifestyle on weight or whether it is about the effects of exercise. (However, ethical considerations often forbid the researchers from engaging in deception.)

When neither the researchers nor the participants know whether the participants are in the treatment or the comparison group, we say that the study is **double-blind**. The double-blind format helps prevent the bias that can result if one group acts differently from the other because they know they are being treated differently or because the researchers treat the groups differently or evaluate them differently because of what the researchers hope or expect.

Placebos

The treatment and comparison groups might differ in still another way. People often react not just to a particular medical treatment, but also to the very idea that they are getting medical treatment. This means that patients who receive a pill, a vaccine, or some other form of treatment often feel better even when the treatment actually does absolutely nothing. Interestingly, this placebo effect also works in the other direction: if they are told that a certain pill might cause side effects (for example, a rash), some patients experience the side effects even though the pill they were given is just a sugar pill.

To neutralize the placebo effect, it is important that the comparison group receive attention similar to what the treatment group receives, so that both groups feel they are being treated the same by the researchers. In our exercise study, the groups behave very differently. However, the sedentary group might receive weekly counseling about lifestyle change or might be weighed and measured just as frequently as the treatment group. If, for instance, we were studying whether peanut milk improves baldness, we would require the comparison group to take a placebo drink so that we could rule out any placebo effect and thus perform a valid comparison between the treatment and control groups.

> **Details**
>
> **The Real Deal**
> A study similar to the one described here was carried out in Denmark in 2012. The researchers found that young, overweight men who exercised 30 minutes per day lost more body fat than those who exercised 60 minutes per day. Both groups of exercisers lost more body fat than a group that did not exercise at all. One conclusion is that more intense exercise for overweight people leads to an increase in appetite, and so moderate levels of exercise are best for weight loss. (Rosenkilde, et al. 2012).

> **KEY POINT**
>
> The following qualities are the "gold standard" for experiments:
>
> *Large sample size*. This ensures that the study captures the full range of variation among the population and allows small differences to be noticed.
>
> *Controlled and randomized*. Random assignment of subjects to treatment or comparison groups helps to minimize bias.
>
> *Double-blind*. Neither subjects nor researchers know who is in which group.
>
> *Placebo* (if appropriate). This format controls for possible differences between groups that occur simply because some subjects are more likely than others to expect their treatment to be effective.

EXAMPLE 9 Brain Games

Brain-training video games, such as Nintendo's Brain Age, claim to improve basic intelligence skills, such as memory. A study published in the journal *Nature* investigated whether playing such games can actually boost intelligence (Owen et al. 2010). The researchers explain that 11,430 people logged onto a web page and were randomly assigned to one of three groups. Group 1 completed six training tasks that emphasized "reasoning, planning and problem-solving." Group 2 completed games that emphasized a broader range of cognitive skills. Group 3 was a control group and didn't play any of these games; instead, members were prompted to answer "obscure" questions. At the end of six weeks, the participants were compared on several different measures of thinking skills. The results? The control group did just as well as the treatment groups.

QUESTION Which features of a well-designed controlled experiment does this study have? Which features are missing?

SOLUTION Sample size: The sample size of 11,430 is quite large. Each of the three groups will have about 3800 people.

Randomization: The authors state that patients were randomly assigned to one of the three groups.

Double-blind format: Judging on the basis of this description, there was no double-blind format. It's possible (indeed, it is likely) that the researchers did not know, while analyzing the outcome, to which treatment group individuals had been assigned. But we do not know whether *participants* were aware of the existence of the three different groups and how they differed.

Placebo: The control group participated in a "null" game, in which they simply answered questions. This activity is a type of placebo because the participants could have thought that this null game was a brain game.

TRY THIS! Exercise 1.49

> **❗ Caution**
>
> **At Random**
> The concept of randomness is used in two different ways in this section. *Random assignment* is used in a controlled experiment. Subjects are randomly assigned to treatment and control groups in order to achieve a balance between groups. This ensures that the groups are comparable to each other and that the only difference between the groups is whether or not they receive the treatment under investigation. *Random selection* occurs when researchers select subjects from some larger group via a random method. We must employ random selection if we wish to extend our results to a larger group.

Extending the Results

In both observational studies and controlled experiments, researchers are often interested in knowing whether their findings, which are based on a single collection of people or objects, will extend to the world at large.

The researchers in Example 9 concluded that brain games are not effective, but might it just be that the games weren't effective for those people who decided to participate? Maybe if the researchers tested people in another country, for example, the findings would be different.

It is usually not possible to make generalizations to a larger group of people unless the subjects for the study are representative of the larger group. The only way to collect a sample that is representative is to collect the objects we study at random. We will discuss how to collect a random sample, and why we can then make generalizations about people or objects who were not in the sample, in Chapter 7.

Selecting subjects using a random method is quite common in polls and surveys (which you'll also study in Chapter 7), but it is much less common in other types of studies. Most medical studies, for example, are not conducted on people selected randomly, so even when a cause-and-effect relationship emerges between the treatment and the response, it is impossible to say whether this relationship will hold for a larger (or different) group of people. For this reason, medical researchers often work hard to replicate their findings in diverse groups of people.

Statistics in the News

When reading in a newspaper or blog about a research study that relies on statistical analysis, you should ask yourself several questions to evaluate how much faith you can put in the conclusions reached in the study:

1. *Is this an observational study or a controlled experiment?*

   If it's an observational study, then you can't conclude that the treatment caused the observed outcome.

2. *If the study is a controlled experiment, was there a large sample size? Was randomization used to assign participants to treatment groups? Was the study double-blind? Was there a placebo?*

   See the relevant section of this chapter for a review of the importance of these attributes.

3. *Was the paper published in a peer-reviewed journal? What is the journal's reputation?*

   "Peer-reviewed" means that each paper published in the journal is rigorously evaluated by at least two anonymous researchers familiar with the field. The best journals are very careful about the quality of the research they report on. They have many checkpoints to make sure that the science is as good as it can be. (But remember, this doesn't mean the science is perfect. If you read a medical journal regularly, you'll see much debate from issue to issue about certain results.) Other journals, by contrast, sometimes allow sloppy research results, and you should be very wary of these journals.

4. *Did the study follow people for a long enough time?*

   Some treatments take a long time to work, and some illnesses take a long time to show themselves. For example, many cost-conscious people like to refill water bottles again and again with tap water. Some fear that drinking from the same plastic bottle again and again might lead to cancer. If this is true, it might take a very long time for a person to get cancer from drinking out of the same bottle day after day. So researchers who wish to determine whether drinking water from the same bottle causes cancer should watch people for a very long time.

Often it is hard to get answers to all these questions from a newspaper article. Fortunately, the Internet has made it much easier to find the original papers, and your college library or local public library will probably have access to many of the most popular journals.

Even when a controlled experiment is well designed, things can still go wrong. One common way in which medical studies go astray is that people don't always do what their doctor tells them to do. Thus, people randomized to the treatment group might not actually take their treatments. Or people randomized to the Atkins diet might switch to Weight Watchers because they don't like the food on the Atkins diet. A good research paper will report on these difficulties and will be honest about the effect on the researchers' conclusions.

EXAMPLE 10 Does Skipping Breakfast Make You Gain Weight?

The New York Times reported on a seven-year study about the eating habits of a sample of 50,000 adults in the United States (Rabin 2017). The participants were all members of the Seventh Day Adventist religion. According to the report, "breakfast eaters" were more likely to keep their weight down after seven years than were "breakfast skippers."

QUESTION Is this most likely an observational study or a controlled experiment? Why? Does this mean that if you skip breakfast you will gain weight?

SOLUTION This is most likely an observational study. The treatment variable is whether or not someone eats breakfast. Although you could possibly assign people to eat or not eat breakfast for a short time, it is very unlikely this could be sustained for seven years and with such a large number of participants. Therefore, most likely researchers simply observed the habits that the participants voluntarily adopted. (This is, in fact, how the study was conducted.)

Because this is an observational study, we cannot conclude that the *treatment variable* (skipping breakfast) affects the *outcome variable* (weight gain). Possibly, people who regularly skip breakfast have other lifestyle characteristics that might cause them to gain weight. A potential confounding variable is that people who skip breakfast regularly might be under considerable time pressure, and this stress results in overeating at other times of day.

TRY THIS!   Exercise 1.53

EXAMPLE **11** Crohn's Disease

Crohn's disease is a bowel disease that causes cramping, abdominal pain, fever, and fatigue. A study reported in the *New England Journal of Medicine* (Columbel et al. 2010) tested two medicines for the disease: injections of infliximab (Inflix) and oral azathioprine (Azath). The participants were randomized into three groups. All groups received an injection and a pill (some were placebos but still a pill and an injection). One group received Inflix injections alone (with placebo pills), one received Azath pills alone (with placebo injections), and one group received both injections and pills. A good outcome was defined as the disease being in remission after 26 weeks. The accompanying table shows the summary of the data that this study yielded.

	Combination	Inflix Alone	Azath Alone
Remission	96	75	51
Not in Remission	73	94	119

QUESTIONS

a. Compare the percentages in remission for the three treatments. Which treatment was the most effective and which was the least effective for this sample?

b. Can we conclude that the combination treatment causes a better outcome? Why or why not?

SOLUTIONS

a. For the combination: 96/169, or 56.8%, success
   For the Inflix alone: 75/169, or 44.4%, success
   For the Azath alone: 51/170, or 30%, success

The combination treatment was the most effective for this sample, and Azath alone was the least effective.

b. Yes, we can conclude that the combination of drugs causes a better outcome than the single drugs. The study was placebo-controlled and randomized. The sample size was reasonably large. Blinding was not mentioned, but at least, thanks to the placebos, the patients did not know what treatment they were getting.

TRY THIS!   Exercise 1.55

Dangerous Habit?

Does drinking coffee cause cancer? This question turns out to be difficult to answer. It is true that coffee contains acrylamide, but it is difficult to know if it contains it in large enough quantities to cause harm in humans. Acrylamide is produced by some plant-based foods when they are heated up or cooked. (Not just coffee, but also french fries and potato chips.) Controlled studies carried out on rats and mice *did* show that ingesting acrylamide through water increased their cancer risk. But controlled studies are not possible or ethical with humans, and according to the American Cancer Society, attempts to find significant differences in cancer rates between those who eat acrylamide-rich diets and those who do not have either failed to find an increased risk or have found mixed results. In 2016, the International Agency for Research on Cancer, an agency run by the World Health Organization (WHO), downgraded the risks of coffee from "probably carcinogenic for humans" to "not classifiable as to its carcinogeneticity to humans." In slightly plainer language, this means that the current evidence does not allow one to conclude that coffee causes cancer in humans.

This does not mean that scientists have proved that drinking coffee is safe. As time goes on, we will no doubt learn more about the risks, or lack of risks, of drinking coffee as potential confounding variables are ruled out.

Sources:

https://www.cancer.org/cancer/cancer-causes/acrylamide.html

https://www.iarc.fr/en/media-centre/iarcnews/2016/DebunkMyth.php

DATAPROJECT ▸ How Are Data Stored?

1 OVERVIEW

To begin your data exploration, you'll learn about how data are stored and how some files can be downloaded onto your computer and then uploaded into statistical software.

2 GOAL

Download an interesting data file and understand how it is structured.

3 YOUR CITY, YOUR DATA

Many governments throughout the world are participating in "open data" initiatives. Open data are data collected, usually at taxpayer expense, to help inform policy and decision making. Because taxpayers pay for the collection, advocates of open data argue that the public should be able to view and use the data.

To see if your local government has open data, perform an Internet search using the words "open data" and the name of your city or county, or a city near you. Most medium-to-large cities participate, as do many smaller towns and cities, although some have much more data than others.

Most cities that have open data will have many, many different data sets. Not all of these will be interesting to you, and not all of them contain useful data. Frankly, some won't contain any data at all.

Approach this project with the mind-set of an explorer. The structure and quality of the websites offering these data sets varies drastically. Some (such as Santa Monica, CA) make it really easy; a few make it really hard. Some will claim they have data, but what they really provide are tools to analyze data that they won't allow you to download or view. Others will provide photos or screenshots of data but not anything you can analyze. So be brave and open-minded and expect to encounter some dead-ends!

Project: Once you've identified a government with open data, find a data set or theme that interests you. For example, you might be interested in response times of the fire department, crimes, how much government employees are paid, or the health ratings of restaurants. Some cities organize data by theme, such as Finance and Public Safety. You might have to do some poking around to find a data set that is interesting.

Your goal is to download a data file. But not just any data file; you need one we can that your statistical software will understand. To do this, you'll want to look for some sort of option to "Export" or "Download" a file. In some cases, these words might not be used and you're expected to simply click on a link.

There are three things you need to pay attention to: the format of the data, the structure of the data, and the existence of *metadata*.

Format refers to the types of files provided and the software that can open them. For example, files that end in .txt—text files—can usually be opened by many types of software, whereas files that end in .docx can be opened only by Microsoft Word. StatCrunch wants files that end in .xlsx and .xls (Excel spreadsheets); .ods (Open Office spreadsheet); or .csv, .txt and .tsc (text files).

Data files often use a *delimiter* to determine where one entry ends and another begins. For example, if a data file contains this:

1345

you don't know whether it is the number one thousand three hundred forty-five or the four digits 1, 3, 4, 5 or the two numbers 13 and 45 or anything else. A delimiter tells you where one value ends and the next begins. A comma-delimited file, which usually ends in the extension .csv, uses commas to do this:

1, 34, 5

for the values one, thirty-four, and five. A tab-delimited file (.tsv or .txt) uses tabs to show the same thing:

1   34   5

Metadata is documentation that helps you understand the data. Sometimes it is called a "codebook" or "data dictionary." It should answer the who, what, when, where, why, and how questions of the data.

But most important, it should tell you what the variable names mean and what the values mean. With any luck, it will also tell you the data type (for example, is it a number or a character?).

Warning: you might have to look at a great many data files before you find one that is useful.

Assignment: Download a data set and write a report answering these questions:

1. What is the URL of the page from which you downloaded your data?

2. What file types are available for your data? What file type did you download?

3. How many variables are there? How many observations?

4. What does a row in the data set represent? In other words, what is the basic unit of observation?

5. Why were these data collected? What is their purpose? How were they collected?

6. In terms of megabytes (or gigabytes), how large is the file? (This information might not be available on the website; you might need to determine using your computer's operating system.)

7. Why were *you* interested in this data set? Give some examples of what you want to learn from these data.

8. Try to upload the data to StatCrunch. Describe what happens. Did you get what you expected? Did you get an error? Describe the outcome.

CHAPTER REVIEW

KEY TERMS

statistics, 2
variation, 3
data, 3
variables, 6
data set, 6
sample, 6
population, 6
numerical variable, 7
categorical variable, 7

stacked data, 8
unstacked data, 8
two-way table, 14
frequency, 14
rate, 17
causality, 18
treatment variable, 18
outcome variable (or response
     variable), 18

treatment group, 18
comparison group (or control
     group), 18
anecdotes, 18
placebo, 19
placebo effect, 19
observational
     study, 19
association, 19

confounding variable
     (or confounding
     factor), 20
controlled experiment, 21
random assignment, 22
bias, 22
blind, 22
double-blind, 23
random selection, 24

LEARNING OBJECTIVES

After reading this chapter and doing the assigned homework problems, you should

- Be able to distinguish between numerical and categorical variables and understand methods for coding categorical variables.

- Know how to find and use rates (including percentages) and understand when and why they are more useful than counts for describing and comparing groups.

- Understand when it is possible to infer a cause-and-effect relationship from a research study and when it is not.

- Be able to explain how confounding variables prevent us from inferring causation and suggest confounding variables that are likely to occur in some situations.

- Be able to distinguish between observational studies and controlled experiments.

SUMMARY

Statistics is the science (and art) of collecting and analyzing observations (called data) and communicating your discoveries to others. Often, we are interested in learning about a population on the basis of a sample taken from that population.

Statistical investigations often progress through four stages (which we call the Data Cycle): Asking questions, considering the data available to answer the questions, analyzing the data, and interpreting the analysis to answer the questions.

With categorical variables, we are often concerned with comparing rates or frequencies between groups. A two-way table is sometimes a useful summary. Always be sure that you are making valid comparisons by comparing proportions or percentages of groups or that you are comparing the appropriate rates.

Many studies are focused on questions of causality: If we make a change to one variable, will we see a change in the other?

Anecdotes are not useful for answering such questions. Observational studies can be used to determine whether associations exist between treatment and outcome variables, but because of the possibility of confounding variables, observational studies cannot support conclusions about causality. Controlled experiments, if they are well designed, do allow us to draw conclusions about causality.

A well-designed controlled experiment should have the following attributes:

A large sample size
Random assignment of subjects to a treatment group and to a control group
A double-blind format
A placebo

SOURCES

BBC News. 2010. Happiness wards off heart disease, study suggests. February 18, 2010. http://news.bbc.co.uk/2/hi/health/8520549.stm.

Britt, C. 2008. U.S. doctors prescribe drugs for placebo effects, survey shows. *British Medical Journal*. October 23, 2008. Bloomberg.com.

Colombel, S., et al. 2010. Infliximab, azathioprine, or combination therapy for Crohn's disease. *New England Journal of Medicine*, vol. 362 (April 15): 1383–1395.

Fisher, R. 1959. *Smoking: The cancer controversy*. Edinburgh, UK: Oliver and Boyd.

Glionna, J. 2006. Word of mouth spreading about peanut milk. *Los Angeles Times*, May 17.

Insurance Institute for Highway Safety. 2013. http://www.iihs.org/news/

Kahleova, H., Lloren, J. I., Maschak, A., Hill, M., Fraser, G. (2017) Meal frequency and timing are associated with changes in body index in Adventist Health Study 2. *Journal of Nutrition* doi: 10.3945/jn.116.244749.

Kaprio, J., and M. Koskenvuo. 1989. Twins, smoking and mortality: A 12-year prospective study of smoking-discordant twin pairs. *Social Science and Medicine*, vol. 29, no. 9: 1083–1089.

Mani, A., et al. 2013. Poverty impedes cognitive function. *Science*, vol. 341 (August 30): 976–980.

McKinsley Quarterly. 2009. Hal Varian on how the Web challenges managers. Business Technology Office, January.

Nisen, M. 2013. Chronic poverty can lower your IQ, study shows. *Daily Finance*, August 31. http://www.dailyfinance.com/

Owen, A., et al. 2010. Letter: Putting brain training to the test. *Nature*, April 20, 2010. doi:10.1038/nature09042.

Rabin, R. C. 2017. The case for a breakfast feast. *The New York Times*, August 21.

Rosenkilde, M., Auerbach, P., Holm Reichkendler, M., Ploug, T., Merete STallknecht, B., Sjodin, A., "Body fat loss and compensatory

mechanisms in response to different doses of aerobic exercise—a randomized controlled trial in overweight sedentary males", Am J Physiol REgul Integr Comp Physiol 303: R571-R579, 2012. doi: 10.1152/ajpregu.00141.2012.

Rosenkilde, M., et al. 2012. Body fat loss and compensatory mechanisms in response to different doses of aerobic exercise in a randomized controlled trial in overweight sedentary males. *American Journal of Physiology: Regulatory, Integrative and Comparative Physiology,* vol. 303 (September): R571–R579.

SECTION EXERCISES

SECTION 1.2

The data in Table 1A were collected from one of the authors' statistics classes. The first row gives the variable, and each of the other rows represents a student in the class.

Female	Commute Distance (Miles)	Hair Color	Ring Size	Height (inches)	Number of Aunts	College Units Acquired	Living Situation
0	0	Brown	9.5	71	5	35	Dorm
0	0	Black	8	66	0	20	Dorm
1	0	Brown	7.5	63	3	0	Dorm
0	14	Brown	10	65	2	30	Commuter
1	17	Brown	6	70	1	15	Commuter
1	0	Blonde	5.5	60	0	12	Dorm
0	0	Black	12	76	4	42	Dorm
1	0	Brown	5	70	7	18	Dorm
1	21	Brown	8	64	2	16	Commuter
0	13	Brown	7.5	63	4	40	Commuter
1	0	Brown	8.5	61.5	3	44	Dorm

▲ TABLE 1A

1.1 Variables In Table 1A, how many variables are there?

1.2 People In Table 1A, there are observations on how many people?

1.3 (Example 1) Are the following variables, from Table 1A, numerical or categorical? Explain.

  a. Living situation

  b. Commute distance

  c. Number of aunts

1.4 Are the following variables, from Table 1A, numerical or categorical? Explain.

  a. Ring size

  b. Hair color

  c. Height

1.5 Give an example of another numerical variable we might have recorded for the students whose data are in Table 1A?

1.6 Give an example of another categorical variable we might have recorded for the students whose data are in Table 1A?

1.7 Coding What do the numbers 1 and 0 mean for the variable *Female*, in Table 1A (which is coded categorical data)? Often, it does not make sense, or is not even possible, to add a categorical variable. Does it make sense for *Female*? If so, what does the sum represent?

1.8 Coding Suppose you decided to code living situation using Dorm as the label for the column. How many ones and how many zeroes would there be?

1.9 Coding Explain why the variable *Female*, in Table 1A, is categorical, even though its values are numbers. Often, it does not make sense, or is not even possible, to add the values of a categorical variable. Does it make sense for *Male?* If so, what does the sum represent?

1.10 Coding Students who have accumulated fewer than 30 units are called Freshmen.

  a. Create a new categorical variable, named *Freshman*, that classifies each student in Table 1A as a freshman (less than 30 units) or not a freshman. Call this variable *Freshman*. Report the coded values in a column in the same order as those in the table.

b. Was the original variable (*college units*) numerical or categorical?

c. Is your new coded column (*Freshman*) from part a numerical or categorical?

TRY 1.11 Facebook Wall Posts (Example 2) A student shared data from the StatCrunch Friend Data Application. Data on gender and number of wall posts for a sample of friends are shown below. (Source: StatCrunch, Facebook Friend Data, posted 2/13/14)

Male	Wall Posts
1	1916
1	183
1	836
0	9802
1	95
1	512
0	153
0	1221

a. Is the format of this data set stacked or unstacked?

b. Explain the coding. What do 1 and 0 represent?

c. If you answered "stacked" in part a, then unstack the data into two columns labeled Male and Female. If you answered "unstacked," then stack the data into one column and choose a appropriate name for the stacked variable.

1.12 Age of Marriage A student did a survey on the age of marriage for married male and female students.

Men	Women
29	24
23	24
30	32
32	35
25	23

a. Is the format of the data set stacked or unstacked?

b. If you answered "stacked" then unstack the data into two columns. If you answered "unstacked" then stack the data into one column; choose an appropriate name for the stacked variable and use coding.

c. There are two variables here: *Gender* and *Age*. Which of them is numerical, and which is categorical?

1.13 Snacks Emmanuel, a student at a Los Angeles high school, kept track of the calorie content of all the snacks he ate for one week. He also took note of whether the snack was mostly "sweet" or "salty."

The sweet snacks: 90, 310, 500, 500, 600, 90

The salty snacks: 150, 600, 500, 550

Write these data as they might appear in (a) stacked format with codes and (b) unstacked format.

1.14 Movies A sample of students were questioned to determine how much they would be willing to pay to see a movie in a theater that served dinner at the seats. The male students responded (in dollars): 10, 15, 15, 25, and 12. The female students responded: 8, 30, 15, and 15. Write these data as they might appear in (a) stacked format with codes and (b) unstacked format.

SECTION 1.3

Use the data in Table 1A to answer questions 1.15 through 1.18.

TRY 1.15 Investigating Data (Example 3) Suppose you wanted to know whether living situation was associated with number of units the student had acquired. Could you do that with this data table? If so, which variables would you use?

1.16 Investigating Data Suppose you wanted to know whether the men or the women tended to be taller. Could you do that with this data table? If so, which variables would you use?

1.17 Investigating Data Suppose you wanted to know whether living situation was associated with number of hours of study per week. Could you do that with this data table? If so, which variables would you use?

1.18 Investigating Data Suppose you wanted to know whether ring size and height were associated. Could you do that with this data table? If so, which variables would you use?

TRY 1.19 Investigating Data (Example 4) A data set on Shark Attacks Worldwide posted on StatCrunch records data on all shark attacks in recorded history including attacks before 1800. Variables contained in the data include time of attack, date, location, activity the victim was engaged in when attacked, type of injuries sustained by the victim, whether or not the injury was fatal, and species of shark. Which of the following questions could not be answered using this data set? (Source: www.sharkattackfile.net)

a. In what month do most shark attacks occur?

b. Are shark attacks more likely to occur in warm temperature or cooler temperatures?

c. Attacks by which species of shark are more likely to result in a fatality?

d. What country has the most shark attacks per year?

1.20 Investigating Data Suppose a surfer wanted to learn if surfing during a certain time of day made one less likely to be attacked by a shark. Using the Shark Attacks Worldwide data set, which variables could the surfer use in order to answer this question?

SECTION 1.4

TRY 1.21 Hands (Example 5) A survey was done of men's and women's hands to see if the ring finger appeared longer than the index finger or not. Yes means the ring finger is longer, and No means the ring finger appears shorter or the same length as the index finger. The students in this survey were not told the theory that men are more likely to have a longer ring finger than women due to more testosterone.

	Men	Women
Yes	33	32
No	7	13

a. What percentage of the men said Yes?

b. What percentage of the women said Yes?

c. What percentage of the people who said Yes were men?

d. If a large group of 250 men had the same rate of responses as the men in this sample, how many men of the 250 would say yes?

1.22 Hands-Biased A survey was done of men's and women's hands to see if the ring finger appeared longer than the index finger or not. Yes means the ring finger is longer, and No means the ring finger appears shorter or the same length as the index finger. The students in this survey were told the theory that men are more likely to have a longer ring finger than women because of additional testosterone.

	Men	Women
Yes	23	13
No	4	14

a. What percentage of the men said No?

b. What percentage of the women said No?

c. What percentage of the people who said No were men?

d. If a large group of 600 men had the same rate of responses as the men in this sample, how many men of the 600 would say No?

1.23 Finding and Using Percentages

a. A statistics class is made up of 15 men and 23 women. What percentage of the class is male?

b. A different class has 234 students, and 64.1% of them are men. How many men are in the class?

c. A different class is made up of 40% women and has 20 women in it. What is the total number of students in the class?

1.24 Finding and Using Percentages

a. A hospital employs 346 nurses, and 35% of them are male. How many male nurses are there?

b. An engineering firm employs 178 engineers, and 112 of them are male. What percentage of these engineers are *female*?

c. A large law firm is made up of 65% male lawyers, or 169 male lawyers. What is the total number of lawyers at the firm?

1.25 Women Find the frequency, proportion, and percentage of women in Table 1A on page 31.

1.26 Brown-Haired People Find the frequency, proportion, and percentage of brown-haired people in Table 1A on page 31.

g 1.27 Two-Way Table from Data Make a two-way table from Table 1A for gender and living situation. Put the labels Male and Female across the top and Dorm and Commuter on the side and then tally the data. *See page 38 for guidance.*

a. Report how many are in each cell.

b. Find the sums for each column and row and the grand total and put them into your table. The grand total is the total number of people and is put in the lower right corner.

c. What percentage of the females live in a dorm?

d. What percentage of the people living in a dorm are female?

e. What percentage of people live in a dorm?

f. If the distribution of females remained roughly the same and you had 70 females, how many of them would you expect to be living in the dorm?

1.28 Two-Way Table from Data Make a two-way table from Table 1 for gender and hair color. Put the labels Male and Female across the top and Brown, Black, Blonde, and Red and then tally the data.

a. Report how many are in each cell.

b. Find the sums of each row and column and the grand total and put them into your table.

c. What percentage of the females have brown hair?

d. What percentage of the people who have brown hair are female?

e. What percentage of the people have brown hair?

f. If the distribution of hair color of females remained roughly the same and you had 60 females, how many of them would have brown hair?

TRY 1.29 Occupation Growth (Example 6) The *2017 World Almanac and Book of Facts* reported that the U.S. occupation projected to grow the most is personal care aide. By 2024 there will be a need for 160,328 personal care aides, a growth of about 26% over 2014 levels. How many personal care aides were there in 2014?

1.30 Chocolate Sales The *2017 World Almanac and Book of Facts* reported that in 2016, M&Ms had sales of approximately $3.48 million and that this accounted for 12.95% of the total chocolate candy sales. What was the total amount of chocolate candy sales?

TRY 1.31 Incarceration Rates (Example 7) The table gives the prison population and total population for a sample of states in 2014–15. (Source: *The 2017 World Almanac and Book of Facts*)

State	Prison Population	Total Population
California	136,088	39,144,818
New York	52,518	19,795,791
Illinois	48,278	12,859,995
Louisiana	38,030	4,670,724
Mississippi	18,793	2,992,333

Find the number of people in prison per thousand residents in each state and rank each state from the highest rate (rank 1) to the lowest rank (rank 6). Compare these rankings of rates with the ranks of total numbers of people in prison. Of the states in this table, which state has the highest prison population? Which state has the highest rate of imprisonment? Explain why these two answers are different.

g 1.32 Population Density The accompanying table gives the 2018 population and area (in square kilometers) of five U.S. cities. *See page 39 for guidance.* (Source: www.citymayors.com).

City	Population	Area (square km)
Miami	4,919,000	2891
Detroit	3,903,000	3267
Atlanta	3,500,000	5083
Seattle	2,712,000	1768
Baltimore	2,076,000	1768

a. Determine and report the ranking of the population density (people per square kilometer) by dividing the population of each city by its area. Use rank 1 for the highest density.

b. If you wanted to live in the city (of these six) with the lowest population density, which would you choose?

c. If you wanted to live in the city (of these six) with the highest population density, which would you choose?

1.33 Health Insurance The accompanying table gives the population (in hundred thousands) and number of people not covered by health insurance (in hundred thousands) for the United States. Find the percentage of people not covered by health insurance for each of the given years and describe the trend. (Source: *2017 World Almanac and Book of Facts*)

Year	Uninsured	Total Population
1990	34,719	249,778
2000	36,586	279,282
2015	29,758	316,574

1.34 Cable TV Subscriptions The accompanying table gives the number of cable television subscribers (in millions) and the number of households with televisions (in millions) in the United States. Find the percentage of TV owners with cable subscriptions for each year and comment on the trend over time. (Source: *2017 World Almanac and Book of Facts*)

Year	# Subscribers	# Households with TVs
2012	103.6	114.7
2013	103.3	114.1
2014	103.7	115.7
2015	100.2	116.5
2016	97.8	116.4

1.35 Percentage of Elderly The projected U.S. population is given for different decades. The projected number of people 65 years of age or older is also given. Find the percentage of people 65 or over and comment on the trend over time. Numbers are in millions of people (Source: *2017 World Almanac and Book of Facts*)

Year	Population	Older Population
2020	334	54.8
2030	358	70.0
2040	380	81.2
2050	400	88.5

1.36 Marriage and Divorce The marriage and divorce rates are given per 1000 people in various years. Find the divorce rate as a percentage of the marriage rate and comment on the trend over time. (Source: https://www.cdc.gov/nchs)

Year	Marriage	Divorce
2000	8.2	4.0
2005	7.6	3.6
2010	6.8	3.6
2014	6.9	3.2

1.37 Course Enrollment Rates Two sections of statistics are offered, the first at 8 a.m. and the second at 10 a.m. The 8 a.m. section has 25 women, and the 10 a.m. section has 15 women. A student claims this is evidence that women prefer earlier statistics classes than men do. What information is missing that might contradict this claim?

1.38 Pedestrian Fatalities In 2015, the National Highway Traffic Safety Administration reported the number of pedestrian fatalities in San Francisco County was 24 and that the number in Los Angeles County was 209. Can we conclude that pedestrians are safer in San Francisco than in Los Angeles? Why or why not? If you answered

no, what additional data would allow us to make a conclusion about which county is safer for pedestrians? (Source: https://cdan.nhtsa.gov)

SECTION 1.5

For Exercises 1.39 through 1.44, indicate whether the study is an observational study or a controlled experiment.

1.39 Patients with high blood pressure are asked to keep food diaries recording all items they eat for a one-week period. Researchers analyze the food diary data for trends.

1.40 Patients with multiple sclerosis are randomly assigned a new drug or a placebo and are then given a test of coordination after six months.

1.41 A researcher is interested in the effect of music on memory. She randomly divides a group of students into three groups: those who will listen to quiet music, those who will listen to loud music, and those who will not listen to music. After the appropriate music is played (or not played), she gives all the students a memory test.

1.42 Patients with Alzheimer's disease are randomly divided into two groups. One group is given a new drug, and the other is given a placebo. After six months they are given a memory test to see whether the new drug fights Alzheimer's better than a placebo.

1.43 A group of boys is randomly divided into two groups. One group watches violent cartoons for one hour, and the other group watches cartoons without violence for one hour. The boys are then observed to see how many violent actions they take in the next two hours, and the two groups are compared.

1.44 A local public school encourages, but does not require, students to wear uniforms. The principal of the school compares the grade point averages (GPAs) of students at this school who wear uniforms with the GPAs of those who do not wear uniforms to determine whether those wearing uniforms tend to have higher GPAs.

1.45 Vitamin C and Cancer The blog *NHS Choices* (February 10, 2014) noted that "there has been increasing anecdotal evidence that vitamin C may still be useful as an anticancer medicine if used in high concentrations and given directly into the vein (intravenously)." Explain what it means that there is "increasing anecdotal evidence" that Vitamin C may be a useful anticancer medicine. How does anecdotal evidence contrast with scientific evidence? What kind of conclusions, if any, can be made from anecdotal evidence?

1.46 Aloe Vera You can find many testimonials on the Internet that drinking aloe vera juice helps with digestive ailments. From these testimonials can we conclude that aloe vera juice causes digestive problems to go away? Why or why not?

TRY **1.47 Effects of Tutoring on Math Grades (Example 8)**
A group of educators want to determine how effective tutoring is in raising students' grades in a math class, so they arrange free tutoring for those who want it. Then they compare final exam grades for the group that took advantage of the tutoring and the group that did not. Suppose the group participating in the tutoring tended to receive higher grades on the exam. Does that show that the tutoring worked? If not, explain why not and suggest a confounding variable.

1.48 Treating Depression A doctor who believes strongly that antidepressants work better than "talk therapy" tests depressed patients by treating half of them with antidepressants and the other half with talk therapy. After six months the patients are evaluated on a scale of 1 to 5, with 5 indicating the greatest improvement.

a. The doctor is concerned that if his most severely depressed patients do not receive the antidepressants, they will get much worse. He therefore decides that the most severe patients will be assigned to receive the antidepressants. Explain why this will affect his ability to determine which approach works best.

b. What advice would you give the doctor to improve his study?

c. The doctor asks you whether it is acceptable for him to know which treatment each patient receives and to evaluate them himself at the end of the study to rate their improvement. Explain why this practice will affect his ability to determine which approach works best.

d. What improvements to the plan in part c would you recommend?

1.49 Try Exercise and Language Learning (Example 9) In a 2017 study designed to investigate the effects of exercise on second-language learning, 40 subjects were randomly assigned to one of two conditions: an experimental group that engaged in simultaneous physical activity while learning vocabulary in a second language and a control group that learned the vocabulary in a static learning environment. Researchers found that learning second-language vocabulary while engaged in physical activity led to higher performance than learning in a static environment. (Source: Liu et al., "It takes biking to learn: Physical activity improves learning a second language," *PLoS One*, May 18, 2017, https://doi.org/10.1371/journal.pone.0177624)

a. What features of a well-designed controlled experiment does this study have? Which features are missing?

b. Assuming that the study was properly conducted, can we conclude that the physical activity while learning caused the higher performance in learning second-language vocabulary? Explain.

1.50 Pneumonia Vaccine for Young Children A study reported by Griffin et al. compared the rate of pneumonia between 1997 and 1999 before pneumonia vaccine (PCV7) was introduced and between 2007 and 2009 after pneumonia vaccine was introduced. Read the excerpts from the abstract, and answer the question that follows it. (Source: Griffin et al., "U.S. hospitalizations for pneumonia after a decade of pneumococcal vaccination," *New England Journal of Medicine*, vol. 369 [July 11, 201]: 155–163)

> We estimated annual rates of hospitalization for pneumonia from any cause using the Nationwide Inpatient Sample database. . . . Average annual rates of pneumonia-related hospitalizations from 1997 through 1999 (before the introduction of PCV7) and from 2007 through 2009 (well after its introduction) were used to estimate annual declines in hospitalizations due to pneumonia.
>
> The annual rate of hospitalization for pneumonia among children younger than 2 years of age declined by 551.1 per 100,000 children . . . which translates to 47,000 fewer hospitalizations annually than expected on the basis of the rates before PCV7 was introduced.

Results for other age groups were similar. Does this show that pneumonia vaccine caused the decrease in pneumonia that occurred? Explain.

1.51 Does Fish Oil Lower Asthma Risk? The *New England Journal of Medicine* reported on a study of fish oil consumption in pregnant mothers and the subsequent development of asthma in their children. Read the excerpts from the abstract and answer the questions that follow: (Source: Bisgaard et al., "Fish oil–derived fatty acids in pregnancy and wheeze and asthma in offspring,"

New England Journal of Medicine, vol. 375 [December 2016]: 2530–2539, doi:10.1056/NEJMoa1503734

> Methods: We randomly assigned 736 pregnant women at 24 weeks of gestation to receive fish oil or a placebo (olive oil) daily. Neither the investigators nor the participants were aware of group assignments during follow-up for the first 3 years of the children's lives, after which there was a 2-year follow-up period during which only the investigators were unaware of group assignments.
>
> Results: A total of 695 children were included in the trial, and 95.5% completed the 3-year, double-blind follow-up period. The risk of persistent wheeze or asthma in the treatment group was 16.9%, versus 23.7% in the control group, corresponding to a relative reduction of 30.7%.

a. Was this a controlled experiment or an observational study? Explain how you know.

b. Assuming the study was properly conducted, can we conclude that the lower rate of asthma was caused by the mother's consumption of fish oil?

1.52 Association between Glycemic Load and Acne? An article in the *Journal of the Academy of Nutrition and Dietetics* reported on a study of diet in subjects with moderate to severe acne. Read the excerpts from the abstract and answer the questions that follow. (Source: Burris et al., "Differences in dietary glycemic load and hormones in New York City adults with no or moderate/severe acne," *Journal of the Academy of Nutrition and Dietetics*, vol. 117 [September 2017]: 1375–1383)

> Methods: Sixty-four participants (no acne, n = 32; moderate/severe acne, n = 32) were included in this study. Participants completed a 5-day food record, had blood drawn and completed a questionnaire to evaluate food-aggravated acne beliefs and acne-specific quality of life.
>
> Results: Participants with moderate/severe acne consumed greater total carbohydrate compared to participants without acne. Participants with moderate/severe acne had greater insulin compared to participants without acne. Although there were no differences between groups, 61% of participants reported food-influenced acne.

a. Was this a controlled experiment or an observational study? Explain how you know.

b. Assuming the study was properly conducted, can we conclude that higher consumption of carbohydrates causes more severe acne? Explain.

TRY **1.53 Milk and Cartilage (Example 10)** Cartilage is a smooth, rubber-like padding that protects the long bones in the body at the joints. A study by Lu et. al. in *Arthritis Care & Research* found that women who drank one glass of milk daily had 32% thicker, healthier cartilage than women who did not. Researchers obtained information on milk consumption through questionnaires and measured cartilage through x-rays. In the article, researched conclude, "Our study suggested that frequent milk intake may be associated with reduced OA progression in women." (Source: Lu et al., "Milk consumption and progression of medial tibiofemoral knee osteoarthritis: Data from the osteoarthritis initiative," *Arthritis Care & Research*, vol. 66 [June 2014]: 802–809, https://doi.org/10.10002/acr.22297)

Does this study show drinking milk causes increased cartilage production? Why or why not?

1.54 Autism and MMR Vaccine An article by Wakefield et al. in the British medical journal *Lancet* claimed that autism was caused by the measles, mumps, and rubella (MMR) vaccine. This vaccine is typically given to children twice, at about the age of 1 and again at about 4 years of age. In the article 12 children with autism who had all received the vaccines shortly before developing autism were studied. The article was later retracted by *Lancet* because the conclusions were not justified by the design of the study.

Can you conclude that the MMR vaccine causes Autism from this study? Explain why *Lancet* might have felt that the conclusions (MMR causes autism) were not justified by listing potential flaws in the study, as described above. (Source: A. J. Wakefield et al., "Ileal lymphoid-nodular hyperplasia, non-specific colitis, and pervasive developmental disorder in children." Lancet, vol. 351 (February 1998): 637–641)

TRY **1.55 Diet and Depression (Example 11)** An article in the journal *BMC Medicine* reported on a study designed to study the effect of diet on depression. Subjects suffering from moderate to severe depression were randomly assigned to one of two groups: a diet intervention group and a social support control group. The 33 subjects in the diet intervention group received counseling and support to adhere to a "ModiMedDiet," based primarily on a Mediterranean diet. The 34 subjects in the social support group participated in a "befriending" protocol, where trained personnel engaged in conversation and activities with participants. At the end of a 12-week period, 11 of the diet intervention group achieved remission from depression compared to 3 of the control group.

	Diet (Intervention)	Support (Control)
Remission	11	3
No Remission	22	31

a. Find and compare the sample percentage of remission for each group.

b. Was this a controlled experiment or an observational study? Explain.

c. Can we conclude that the diet caused a remission in depression? Why or why not?

1.56 Effect of Confederates on Compliance A study was conducted to see whether participants would ignore a sign that said, "Elevator may stick between floors. Use the stairs." The study was done at a university dorm on the ground floor of a three-level building. Those who used the stairs were said to be compliant, and those who used the elevator were said to be noncompliant. There were three possible situations, two of which involved confederates. A confederate is a person who is secretly working with the experimenter. In the first situation, there was no confederate. In the second situation, there was a compliant confederate (one who used the stairs), and in the third situation, there was a noncompliant confederate (one who used the elevator). The subjects tended to imitate the confederates. What more do you need to know about the study to determine whether the presence or absence of a confederate causes a change in the compliance of subjects? (Source: Wogalter et al. [1987], reported in Shaffer and Merrens, *Research Stories in Introductory Psychology* [Boston: Allyn and Bacon, 2001])

1.57 A Salad a Day Keeps Stroke Away? The *Harvard Heart Letter* reported on a study that examined the diets of 1226 older women over 15 years. They discovered that the more vegetables the women consumed, the lower their risk of dying of cardiovascular disease. From this study can we conclude that eating a diet high in vegetables prevents cardiovascular disease? Why or why not?

1.58 Does Drinking Sugary Beverages Lead to Dementia? The September 2017 issue of *Alzheimer's and Dementia* reported on a study that found an association between drinking sugary drinks and lower brain volume. Is this likely to be a conclusion from observational studies or randomized experiments? Can we conclude that drinking sugary beverages causes lower brain volume? Why or why not?

CHAPTER REVIEW EXERCISES

1.59 Secondhand Smoke Exposure and Young Children Researchers wanted to assess whether a theory-based, community health worker–delivered intervention for household smokers will lead to reduced secondhand smoke exposure to children in Chinese families. Smoking parents or caregivers who had a child aged 5 years or younger at home were randomly assigned to the intervention group that received information on smoking hygiene and the effects of secondhand smoke exposure delivered by community health workers or to the comparison group who received no additional information regarding secondhand smoke. At a 6-month follow-up, researchers assessed whether or not families had adopted any smoking restrictions at home. The results are shown in the following table. (Source: Abdullah et al., "Secondhand smoke exposure reduction intervention in Chinese households of young children: A randomized controlled trial," Academic Pediatrics, vol. 15 (November–December 2015): 588–598, https://doi.org/10.1016/j.acap.2015.06.008)

	Intervention	Control
Smoking Restrictions	61	37
No restrictions	37	45
Total	98	82

a. What percentage of those receiving the intervention adopted smoking restrictions at home?

b. What percentage of those in the control group adopted smoking restrictions at home?

c. Based on this data, do you think the intervention may have been effective in promoting the adoption of smoking restrictions in the home?

1.60 Coffee Consumption The August 27, 2017, issue of *Science Daily* reported that higher coffee consumption is associated with a lower risk of death. This was based on an observational study of nearly 20,000 participants. Researchers found that participants who consumed at least 4 cups of coffee per day had a 64% lower risk of mortality than those who never or almost never consumed coffee. Does this mean that a person can reduce his or her chance of death by increasing the amount of coffee consumed?

1.61 Speeding Tickets College students who were drivers were asked if they had ever received a speeding ticket (yes or no). The results are shown in the table, along with gender.

a. There are two variables in the table, state what they are and whether each is categorical or numerical.

b. Make a two-way table of the results with Male and Female across the top and Yes and No at the left edge.

c. Compare the percentages of men and women who have received speeding tickets.

Gender	Ticket	Gender	Ticket
m	y	f	y
m	y	f	y
m	y	f	n
m	y	f	n
m	y	f	n
m	y	f	n
m	n	f	n
m	n	f	n
m	n	f	n
m	n	f	n
f	y	f	n
f	y	f	n
f	y		

1.62 100 MPH College students who were drivers were asked if they had ever driven a car 100 mph or more (yes or no). The results are shown in the table, along with gender.

a. There are two variables in the table, state what they are and whether each is categorical or numerical.

b. Make a two-way table of the results with Male and Female across the top and Yes and No at the left edge.

c. Compare the percentages of men and women who have driven 100 mph or more.

Gender	100 + mph	Gender	100 + mph
m	y	f	y
m	y	f	y
m	y	f	n
m	y	f	n
m	y	f	n
m	y	f	n
m	n	f	n
m	n	f	n
m	n	f	n
f	y	f	n
f	y	f	n
f	y	f	n

* **1.63 Writing: Vitamin D** Describe the design of a controlled experiment to determine whether the use of vitamin D supplements reduces the chance of broken bones in women with osteoporosis (weak bones). Assume you have 200 women with osteoporosis to work with. Your description should include all the features of a controlled experiment. Also decide how the results would be determined.

* **1.64 Writing: Strokes** People who have had strokes are often put on "blood thinners" such as aspirin or Coumadin to help prevent a second stroke. Describe the design of a controlled experiment to determine whether aspirin or Coumadin works better in preventing second strokes. Assume you have 300 people who have had a first stroke to work with. Include all the features of a good experiment. Also decide how the results would be determined.

1.65 Yoga and High-Risk Adolescents Can mindful yoga have a beneficial impact on alcohol use in high-risk adolescents? Read excerpts from the research published in *The Journal of Child and Family Studies* and answer the questions that follow. (Source: Fishbein et al., "Behavioral and psychophysiological effects of a yoga intervention on high-risk adolescents: A randomized control trial," *Journal of Child and Family Studies*, vol. 25 [February 2016]: 518–529, https://doi.org/10.1007/s10826-015-0231-6)

> Abstract: We designed a 20-session mindful yoga intervention for adolescents attending a school for students at high risk of dropping out. The 69 participants were randomly assigned to control and intervention groups. Survey data were collected before and after the yoga curriculum. At the post test, students in the yoga condition exhibited trends toward decreased alcohol use as compared to control students.

a. Identify the treatment variable and the response variable.

b. Was this a controlled experiment or an observational study?

c. Based on this study, can you conclude that yoga caused a decrease in alcohol use? Why or why not?

1.66 Neurofeedback and ADHD Some studies have indicated that neurofeedback may be an effective treatment for ADHD. Read excerpts from the research published in *The Lancet Psychiatry* and answer the questions that follow. (Source: Schönenberg et al., "Neurofeedback, sham neurofeedback, and cognitive-behavioral group therapy in adults with attention-deficit hyperactivity disorder: A triple-blind, randomised, controlled trial, "*The Lancet Psychiatry*, vol. 4 [September 2017]: 673–684)

> Methods: We did a concurrent, triple-blind, randomised, controlled trial using adults with ADHD, aged 18 to 60 years. Participants were randomly assigned to three groups: a neurofeedback group which received 30 true neurofeedback sessions over 15 weeks, a sham neurofeedback group which received 15 sham (fake) followed by 15 true neurofeedback sessions over 15 weeks, or a meta-cognitive group therapy group which received 12 sessions over 12 weeks. The primary outcome was symptom score on the Conners' adult ADHD rating scale, assessed before treatment, at midtreatment (after 8 weeks), after treatment (after 16 weeks), and 6 months later.

> Results: Self-reported ADHD symptoms decreased substantially for all treatment groups between pretreatment and the end of 6 month follow-up, independent of treatment condition. There were no significant differences in outcomes between any of the groups.

a. Identify the treatment variable and the response variable.

b. Was this a controlled experiment or an observational study?

c. Based on this study, would you agree that neurofeedback may be an effective treatment for ADHD? Why or why not?

1.67 Virtual Reality and Fall Risk A study was conducted to assess whether 5 weeks of training with virtual reality (VR) can reduce the risk of falls in adults. Thirty-four older adults underwent 15 VR training sessions consisting of walking on a treadmill with a VR simulation. At the end of the VR training program, participants showed improved mobility and gait speed. In the abstract the authors conclude that "[t]readmill training with VR appears to be an effective and practical clinical tool to improve mobility and reduce falls in older adults." Do these results indicate that VR training can cause improvement in mobility and gait speed among older adults? What essential component of both controlled experiments and observational studies is missing from this study? (Source: Shema et al., "Improved mobility and reduced fall risk in older adults after five weeks of virtual reality training," Journal of Alternative Medical Research, 9(2), 171–175.)

1.68 Ear Infections Babies 6 to 23 months of age with inner ear infections were given antibiotics. The children were randomly assigned to receive antibiotics for a full 10 days or to receive antibiotics for 5 days and then a placebo for 5 days. There were 229 children assigned the shorter course, and 77 of them had "clinical failure" whereas of the 238 assigned to the longer course of antibiotics, 39 had clinical failure. (Source: Hoberman et al., "Shortened antimicrobial treatment for acute otitis media in young children," New England Journal of Medicine, vol. 375 [December 2016]: 2446–2456)

a. Compare the percentage of clinical failure in each group and state which group did better.

b. Create a two-way table with 10 days and 5 days across the top and failure and success down the left side. Fill in all four numbers.

c. Was this an observational study or a controlled experiment? How do you know?

d. Can you conclude that the treatment caused the difference? Why or why not?

1.69 Effects of Light Exposure A study carried out by Baturin and colleagues looked at the effects of light on female mice. Fifty mice were randomly assigned to a regimen of 12 hours of light and 12 hours of dark (LD), while another 50 mice were assigned to 24 hours of light (LL). Researchers observed the mice for two years, beginning when the mice were 2 months old. Four of the LD mice

and 14 of the LL mice developed tumors. The accompanying table summarizes the data. (Source: Baturin et al., "The effect of light regimen and melatonin on the development of spontaneous mammary tumors in mice," Neuroendocrinology Letters, vol. 22 [December 2001]: 441–447)

	LD	LL
Tumors	4	14
No tumors	46	36

a. Determine the percentage of mice that developed tumors from each group (LL and LD). Compare them and comment.

b. Was this a controlled experiment or an observational study? How do you know?

c. Can we conclude that light for 24 hours a day causes an increase in tumors in mice? Why or why not?

TRY 1.70 Scared Straight The idea of sending delinquents to "Scared Straight" programs has appeared recently in several media programs (such as *Dr. Phil*) and on a program called *Beyond Scared Straight*. So it seems appropriate to look at a randomized experiment from the past. In 1983, Roy Lewis reported on a study in California. Each male delinquent in the study (all were aged 14–18) was randomly assigned to either Scared Straight or no treatment. The males who were assigned to Scared Straight went to a prison, where they heard prisoners talk about their bad experiences there. Then the males in both the experimental and the control group were observed for 12 months to see whether they were rearrested. The table shows the results. (Source: Lewis, "Scared straight—California style: Evaluation of the San Quentin Squires program," *Criminal Justice and Behavior*, vol. 10 [June 1983]: 209–226)

	Scared Straight	No Treatment
Rearrested	43	37
Not rearrested	10	18

a. Report the rearrest rate for the Scared Straight group and for the No Treatment group, and state which is higher.

b. This experiment was done in the hope of showing that Scared Straight would cause a lower arrest rate. Did the study show that? Explain.

UIDED EXERCISES

g 1.27 Two-Way Table from Data Make a two-way table from Table 1A for gender and living situation. Put the labels Male and Female across the top and Dorm and Commuter on the side and then tally the data. Guidance is given for parts a and b.

a. Report how many are in each cell by using STEP 1 and 2 of the guidance given below.

b. Report the totals by using STEP 3 of the guidance.

Female	Gender	Living Situation	Checked
0	Male	Dorm	✓
0	Male	Dorm	✓
1	Female	Dorm	✓

Female	Gender	Living Situation	Checked
0	Male	Commuter	✓
1	Female	Commuter	
1	Female	Dorm	
0	Male	Dorm	
1	Female	Dorm	
1	Female	Commuter	
0	Male	Commuter	
1	Female	Dorm	

Guidance

Step 1 ▶ Refer to the part of the spreadsheet given. First we must "decode" the column labeled "Female" into Gender. 0 = Male and 1 = Female. For each cell in the two-way table, make a tally mark for each person who has both characteristics that belong to that cell. After making a tally mark, cross off that row of the table or put check mark next to that row in the table so you will know which rows you have already tallied. The first four tally marks are given. (So far we have counted two males who live in a dorm, one female who lives in a dorm, and one male who commutes.)

	Male	Female
Dorm	‖	∣
Commuter	∣	

Step 2 ▶ When you have finished tallying, check to see that you have a total of 11 tally marks, one for each row of the data table. Then change your tally marks into numbers. Note that one number is given for you so you can check to ensure that you get the same value.

	Male	Female
Dorm		4
Commuter		

Step 3 ▶ Put in the totals: Put the total number of males in the bottom cell of the "Male" column and the total number of females in the bottom cell of the "Female" column. Put the total number of dorm students in the cell far-right cell of the dorm row and the total number of commuter students in the far-right cell of the commuter row. Note that the total of males and females should be 11 and the total of dorm and commuter should be 11. Note that two totals are given for you so you can check to ensure that you get the same value.

	Male	Female	Total
Dorm		4	7
Commuter			
Total	5		

Go back to the original question for the other parts.

g 1.32 Population Density The accompanying table gives the 2018 population and area (in square kilometers) of six U.S cities. (Source: www.citymayors.com).

City	Population	Area (square km)
Miami	4,919,000	2891
Detroit	3,903,000	3267
Atlanta	3,500,000	5083
Seattle	2,712,000	1768
Baltimore	2,076,000	1768

a. Determine and report the ranking of the population density (people per square kilometer) by dividing the population of each city by its area. Use rank 1 for the highest density.

b. If you wanted to live in the city (of these six) with the lowest population density, which would you choose?

c. If you wanted to live in the city (of these six) with the highest population density, which would you choose?

Step 1 ▶ Find the population density for each city by dividing the population (people) by area (square kilometers). The unit for the resulting number will be people per square kilometer. For example, to find the population density of Atlanta, calculate and round your answer off to the nearest whole number. The population density of Atlanta is people per square kilometer. Two cities are done for you.

City	Population	Area (square km)	Pop. Density
Miami	4,919,000	2891	
Detroit	3,903,000	3267	
Atlanta	3,500,000	5083	689
Seattle	2,712,000	1768	
Baltimore	2,076,000	1768	1174

Step 2 ▶ When you have found the population density for all 5 cities, rank them with 1 representing the city with the highest density and 5 representing the city with the lowest density.

Step 3 ▶ Notice that the ranks will not be the same as the ranks for the populations. For example, even though Detroit has a higher population than Seattle, it will rank lower in population density. Why are the population density ranks different than the population ranks?

2 Picturing Variation with Graphs

Any collection of data exhibits variation. The most important tool for organizing this variation is called the distribution of the sample, and visualizing this distribution is the first step in every statistical investigation. We can learn much about a numerical variable by focusing on three components of the distribution: the shape, the center, and the variability, or horizontal spread. Examining a graph of a distribution can lead us to deeper understanding of the situation that produced the data.

One of the major concepts of statistics is that although individual events are hard to predict, large numbers of events usually exhibit predictable patterns. The search for patterns is a key theme in science and business. An important first step in this search is to identify and visualize the key features of your data.

Using graphics to see patterns and identify important trends or features is not new. One of the earliest statistical graphs dates back to 1786, when a Scottish engineer named William Playfair published a paper examining whether there was a relationship between the price of wheat and wages. To help answer this question, Playfair produced a graph (shown in Figure 2.1) that is believed to be the first of its kind. This graph became the prototype of two of the most commonly used tools in statistics: the bar chart and the histogram.

Graphics such as these can be extraordinarily powerful ways of organizing data, detecting patterns and trends, and communicating findings. The graphs that we use have changed somewhat since Playfair's day, but graphics are of fundamental importance when analyzing data. The first step in any statistical analysis is to make

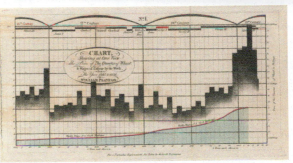

▲ **FIGURE 2.1** Playfair's chart explores a possible relationship between wages and the price of wheat. (Source: Playfair 1786)

a picture of some kind in order to check our intuition against the data. If our intuitions are wrong, it could very well be because the world works differently than we thought. Thus, by making and examining a display of data (as illustrated in this chapter's Case Study), we gain some insight into how the world works.

In Chapter 1 we discussed some of the methods used to collect data. In this chapter we cover some of the basic graphics used in analyzing the data we collect. Then, in Chapter 3, we'll comment more precisely on measuring and comparing key features of our data.

CASE STUDY

Student-to-Teacher Ratio at Colleges

Are private four-year colleges better than public four-year colleges? That depends on what you mean by "better." One measure of quality that many people find useful (and there are many other ways to measure quality) is the student-to-teacher ratio: the number of students enrolled divided by the number of teachers. For schools with small student-to-teacher ratios, we expect class sizes to be small; students can get extra attention in a small class.

The data in Table 2.1 on the next page were collected from some schools that award four-year degrees. The data are for the 2013–14 academic year; 83 private colleges and 44 state-supported (public) colleges were sampled. Each ratio was rounded to the nearest whole number for simplicity. For example, the first private college listed has a student-to-teacher ratio of 10, which means that there are about ten students for every teacher. What differences do you expect between public and private schools? What similarities do you anticipate?

It is nearly impossible to compare the two groups without imposing some kind of organization on the data. In this chapter you will see several ways in which we can graphically organize groups of data like these so that we can compare the two types of colleges. At the end of this chapter, you'll see what the right graphical summaries can tell us about how these types of colleges compare.

Private Colleges								
10	9	16	17	10	12	12	20	11
10	10	14	19	12	9	11	22	15
5	23	11	20	16	14	6	10	10
17	14	12	10	11	13	17	22	10
12	12	11	12	11	12	15	9	
12	14	32	11	13	17	20	17	
13	9	11	12	8	18	8	12	
12	13	11	13	12	12	14	13	
11	17	15	14	14	10	15	18	
11	15	12	6	12	7	15	12	

Public Colleges				
16	23	19	18	20
11	14	21	20	15
19	19	19	20	15
20	18	26	14	19
18	25	15	23	
22	19	26	16	
23	17	19	12	
22	18	15	14	
16	19	19	17	
25	13	25	18	

▲ TABLE 2.1 Ratio of students to teachers at private and public colleges.
(Source: http://nces.ed.gov/ipeds/)

Visualizing Variation in Numerical Data

Some common statistical questions we wish to ask about a variable are "What's a typical value? What's common? What's unusual? What's extreme?" More often we have questions that compare groups based on a variable: "Which group is bigger? Which group has more variability?" An essential tool for answering questions such as these is the distribution. The **distribution of a sample** of data is simply a way of organizing the data. Think of the distribution of the sample as a list that records (1) the values that were observed and (2) the frequencies of these values. **Frequency** is another word for the *count* of how many times the value occurred in the collection of data.

> **KEY POINT** The distribution of a sample is one of the central organizational concepts of data analysis. The distribution organizes data by recording all the values observed in a sample, as well as how many times each value was observed.

Distributions are important because they capture much of the information we need in order to make comparisons between groups, examine data for errors, and learn about real-world processes. Distributions enable us to examine the variation of the data in our sample, and from this variation we can often learn more about the world.

The answer to every statistical question begins with a visualization. By creating an appropriate graphic, we can see patterns that might otherwise escape our notice.

For example, here are some raw data from the National Collegiate Athletic Association (NCAA), available online. This set of data shows the number of goals scored by NCAA female soccer players in Division III in the 2016–17 season. (Division III schools are colleges or universities that are not allowed to offer scholarships to athletes.) To get us started, we show only first-year students.

11, 14, 16, 13, 13, 10, 13, 11, 16, 21, 13, 19, 10, 10, 14, 13, 10, 13,
15, 10, 15, 13, 11, 19, 11, 11, 16, 10, 12, 11, 14, 11, 10, 14, 10, 19, 12

This list includes only the values. A distribution lists the values and also the frequencies. The distribution of this sample is shown in Table 2.2.

Some statistical questions we might ask are "How many goals are typically scored by a player?" "Is 19 an unusually large number of goals?" It's hard, though not impossible, to answer these questions with a table because it is hard to see patterns. A picture would make it easier to compare the numbers of goals for different groups. For example, in a season, do men typically score more goals or fewer goals than women do?

When examining distributions, we use a two-step process:

1. See it.

2. Summarize it.

In this section we explain how to visualize the distribution. In the next section, we discuss the characteristics you should look for to help you summarize it.

All the methods we use for visualizing distributions are based on the same idea: Make some sort of mark that indicates how many times each value occurred in our data set. In this way, we get a picture of the sample distribution so that we can see at a glance which values occurred and how often.

Two very useful methods for visualizing distributions of numerical variables are dotplots and histograms. Dotplots are simpler; histograms are more commonly used and perhaps more useful.

Dotplots

In constructing a **dotplot**, we simply put a dot above a number line where each value occurs. We can get a sense of frequency by seeing how high the dots stack up. Figure 2.2 shows a dotplot for the number of goals for first-year female soccer players.

With this simple picture, we can see more than we could from Table 2.2. We can see from this dotplot that most women scored 20 or fewer goals in the 2016–17 season, but one scored more. We also get a sense that, for at least this group of soccer players, scoring 19 or more goals per season is unusual.

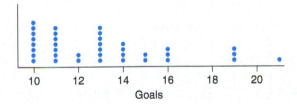

▲ **FIGURE 2.2** Dotplot of the number of goals scored by first-year women soccer players in NCAA Division III, 2016–17. Each dot represents a soccer player. Note that the horizontal axis begins at 10.

Value	Frequency
10	8
11	7
12	2
13	7
14	4
15	2
16	3
17	0
18	0
19	3
20	0
21	1

▲ **TABLE 2.2** Distribution of the number of goals scored by first-year women soccer players in NCAA Division III in 2016–17.

 Data Moves ▶ These data are extracted from a more complete set, which contains data on both men and women and all three NCAA divisions.

Data ▶ ncaasoccer.csv

📌 **Details**

Making Dotplots
Dotplots are easy to make with pen and paper, but don't worry too much about recording values to great accuracy. The purpose of a plot like this is to help us see the overall shape of the distribution, not to record details about individual observations.

SNAPSHOT ▶ The Dotplot

WHAT IS IT? ▶	A graphical summary.
WHAT DOES IT DO? ▶	Shows a picture of the distribution of a numerical variable.
HOW DOES IT DO IT? ▶	Each observation is represented by a dot on a number line.
HOW IS IT USED? ▶	To see patterns in samples of data with variation.

Histograms

While dotplots have one dot for each observation in the data set, **histograms** produce a smoother graphic by grouping observations into intervals, called bins. These groups are formed by dividing the number line into bins of equal width and then counting how many observations fall into each bin. To represent the bins, histograms display vertical bars, where the height of each bar is proportional to the number of observations inside that bin.

For example, with the goals scored in the dotplot in Figure 2.2, we could create a series of bins that go from 10 to 12, 12 to 14, 14 to 16, and so on. Fifteen women scored between 10 and 12 goals during the season, so the first bar has a height of 15. (Count the dots in Figure 2.2 for 10 and 11 goals.) The second bin contains nine observations and consequently has a height of 9. The finished graph is shown in Figure 2.3. (Note that some statisticians use the word *interval* in place of *bin*. You might even see another word that means the same thing: *class*.) Figure 2.3 shows, among other things, that six women scored between 14 and 16 goals, three between 16 and 18, and so on.

◀ **FIGURE 2.3** Histogram of goals for female first-year soccer players in NCAA Division III, 2016–17. The first bar, for example, tells us that 15 players scored between 10 and 12 goals during the season.

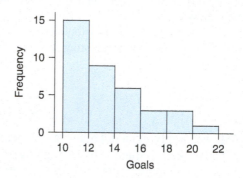

Making a histogram requires paying attention to quite a few details. For example, we need to decide on a rule for what to do if an observation lands exactly on the boundary of two bins. In which bin would we place an observation of 12 goals? A common rule is to decide always to put "boundary" observations in the bin on the right, but we could just as well decide always to put them in the bin on the left. The important point is to be consistent. The graphs here use the right-hand rule and put boundary values in the bin to the right.

Figure 2.4 shows two more histograms of the same data. Even though they display the same data, they look very different—both from each other and from Figure 2.3. Why?

Changing the width of the bins in a histogram changes its shape. Figure 2.3 has bins with a width of 2 goals. In contrast, Figure 2.4a has much smaller bins, and Figure 2.4b has wider bins. Note that when we use small bins, we get a spiky histogram. When we use wider bins, the histogram gets less spiky. Using wide bins hides more detail. If you chose very wide bins, you would have no details at all. You would see just one big rectangle!

How large should the bins be? Too small and you see too much detail (as in Figure 2.4a). Too large and you don't see enough (as in Figure 2.4b). Most computer software will automatically make a good choice. The software package that produced

◀ **FIGURE 2.4** Two more histograms of goals scored in one season the same data as in Figure 2.3. **(a)** This histogram has narrow bins and is spiky. **(b)** This histogram has wide bins and offers less detail.

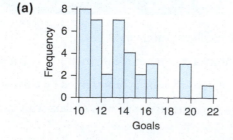

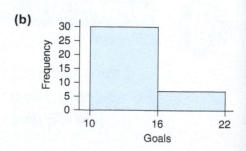

Figure 2.3, StatCrunch, automatically chose a binwidth of 2. Still, if you can, you should try different sizes to see how different choices change your impression of the distribution of the sample. Fortunately, most statistical software packages make it quite easy to change the bin width.

Relative Frequency Histograms   A variation on the histogram (and statisticians, of course, love variation) is to change the units of the vertical axis from frequencies to relative frequencies. A **relative frequency** is simply a proportion. So instead of reporting that the first bin had 15 observations in it, we would report that the proportion of observations in the first bin was $15/37 = 0.405$. We divide by 37 because there were a total of 37 observations in the data set. Figure 2.5 is the same as the first histogram shown for the distribution of goals (Figure 2.3); however, Figure 2.5 reports relative frequencies, and Figure 2.3 reports frequencies.

Using relative frequencies does not change the shape of the graph; it just communicates different information to the viewer. Rather than answering the question, "How many players scored between 10 and 23 goals?" (15 players), it now answers the question, "What *proportion* of players scored between 10 and 12 goals?" (0.405).

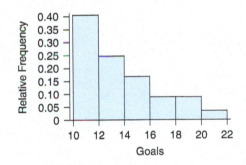

◄ **FIGURE 2.5** Relative frequency histogram of goals scored by first-year women soccer players in NCAA Division III, 2016–17.

SNAPSHOT ► The Histogram

WHAT IS IT? ►	A graphical summary for numerical data.
WHAT DOES IT DO? ►	Shows a picture of the distribution of a numerical variable.
HOW DOES IT DO IT? ►	Observations are grouped into bins, and bars are drawn to show how many observations (or what proportion of observations) lie in each bin.
HOW IS IT USED? ►	By smoothing over details, histograms help our eyes pick up more important, large-scale patterns. Be aware that making the bins wider hides detail, and that making the bins smaller can show too much detail. The vertical axis can display the frequency, the relative frequency, or percentages.

EXAMPLE 1 Visualizing Bar Exam Pass Rates at Law Schools

In order to become a lawyer, you must pass your state's bar exam. When you are choosing a law school, it therefore makes great sense to choose one that has a high percentage of graduates passing the bar exam. The Internet Legal Research Group website provides the pass rate for 196 law schools in the United States in 2013 (Internet Legal Research Group).

QUESTION   Is 80% a good pass rate for a law school?

SOLUTION   This is a subjective question. An 80% pass rate might be good enough for a prospective student, but another way of looking at this is to determine whether there are many schools that do better than 80% or whether 80% is a typical pass rate or whether there are very few schools that do so well. Questions such as these can often be answered by considering the distribution of our sample of data, so our first step is to choose an appropriate graphical presentation of the distribution.

Either a dotplot or a histogram would show us the distribution. Figure 2.6a shows a dotplot. It's somewhat difficult (but not impossible) to read a dotplot with 196 observations; there's too much detail to allow us to answer broad questions such as these. It is easier to use a histogram, as in Figure 2.6b. In this histogram, each bar has a width of 10 percentage points, and the *y*-axis tells us how many law schools had a pass rate within those limits.

▶ **FIGURE 2.6 (a)** A dotplot for the bar-passing rate for 196 law schools. Each dot represents a law school, and the dot's location indicates the bar-passing rate for that school. **(b)** A histogram shows the same data as in part (a), except that the details have been smoothed.

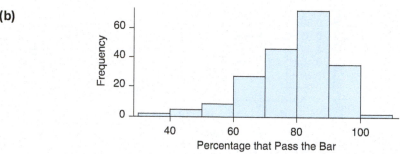

(a)

(b)

We see that about 75 law schools had a pass rate between 80% and 90%. Another 35 or so had a pass rate over 90%. Adding these together, 110 (75 + 35) had a pass rate higher than 80%, and this is 110/196 = 0.561, or about 56% of the schools. This tells us that a pass rate of 80% is not all that unusual; a majority of law schools do this well or better, and a large percentage (47%) do worse.

TRY THIS!   Exercise 2.5

Stemplots

Stemplots, which are also called stem-and-leaf plots, are useful for visualizing numerical variables when you don't have access to technology and the data set is not large. Stemplots are also useful if you want to be able to easily see the actual values of the data.

To make a **stemplot**, divide each observation into a "stem" and a "leaf." The **leaf** is the last digit in the observation. The **stem** contains all the digits that precede the leaf. For the number 60, the *6* is the stem and the *0* is the leaf. For the number 632, the *63* is the stem and the *2* is the leaf. For the number 65.4, the *65* is the stem and the *4* is the leaf.

A stem-and-leaf plot can help us understand data such as drinking behaviors. Alcohol is a big problem at many colleges and universities. For this reason, a collection of college students who said that they drink alcohol were asked how many alcoholic drinks they had consumed in the last seven days. Their answers were

1, 1, 1, 1, 1, 2, 2, 2, 3, 3, 3, 3, 3, 3, 4, 5, 5, 5, 6, 6, 6, 6, 8, 10, 10, 15, 17, 20, 25, 30, 30, 40

For one-digit numbers, imagine a 0 at the front. The observation of 1 drink becomes 01, the observation of 2 drinks becomes 02, and so on. Then each observation is just two digits; the first digit is the stem, and the last digit is the leaf. Figure 2.7 shows a stemplot of these data.

If you rotate a stemplot 90 degrees counterclockwise, it does not look too different from a histogram. Unlike histograms, stemplots display the actual values of the data. With a histogram, you know only that the values fall somewhere within an interval.

Stemplots are often organized with the leaves in order from lowest to highest, which makes it easier to locate particular values, as in Figure 2.7. This is not necessary, but it makes the plot easier to use.

From the stemplot, we see that most students drink a moderate amount in a week but that a few drink quite a bit. Forty drinks per week is almost six per day, which qualifies as problem drinking by some physicians' definitions.

Figure 2.8 shows a stemplot of some exam scores. Note the empty stems at 4 and 5, which show that there were no exam grades between 40 and 59. Most of the scores are between 60 and 100, but one student scored very low relative to the rest of the class.

Stem	Leaves
0	111112223333345556668
1	0057
2	05
3	00
4	0

▲ FIGURE 2.7 A stemplot for alcoholic drinks consumed by college students. Each digit on the right (the leaves) represents a student. Together, the stem and the leaf indicate the number of drinks for an individual student.

Stem	Leaves
3	8
4	
5	
6	0257
7	00145559
8	0023
9	0025568
10	00

▲ FIGURE 2.8 A stemplot for exam grades. Two students had scores of 100, and no students scored in the 40s or 50s.

SNAPSHOT ▸ The Stemplot

WHAT IS IT? ▸	A graphical summary for numerical data.
WHAT DOES IT DO? ▸	Shows a picture of the distribution of a numerical variable.
HOW DOES IT DO IT? ▸	Numbers are divided into leaves (the last digit) and stems (the preceding digits). Stems are written in a vertical column, and the associated leaves are "attached."
HOW IS IT USED? ▸	In very much the same way as a histogram. It is often useful when technology is not available and the data set is not large.

SECTION 2.2

Summarizing Important Features of a Numerical Distribution

When examining a distribution, pay attention to describing the shape of the distribution, the **typical value (center)** of the distribution, and the **variability (spread)** in the distribution. The notion of "typical" is subjective, but the value at the center of a distribution often comes close to most people's notion of the "typical value." The variability is reflected in the amount of horizontal spread the distribution has.

KEY POINT When examining distributions of numerical data, pay attention to the shape, the center, and the horizontal spread.

Figure 2.9 compares distributions for two groups. You've already seen histogram (a)—it's the histogram for the goals scored in 2017 by first-year women soccer players in Division III. Histogram (b) shows goals scored for first-year male soccer players in Division III in the same year. How do these two distributions compare?

▶ **FIGURE 2.9** Distributions of the goals scored for **(a)** first-year women and **(b)** first-year men in Division III soccer in 2017.

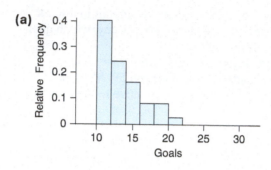
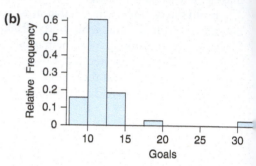

1. *Shape.* Are there any interesting or unusual features about the distributions? Are the shapes very different? (If so, this might be evidence that men play the game differently than women do.)

2. *Center.* What is the typical value of each distribution? Is the typical number of goals scored in a season different for men than for women?

3. *Spread.* The horizontal spread presents the variation in goals per game for each group. How do the amounts of variation compare? If one group has low variation, it suggests that the soccer skills of the members are pretty much the same. Lots of variability might mean that there is a wider variety of skill levels.

Let's consider these three aspects of a distribution one at a time.

Shape

You should look for three basic characteristics of a distribution's shape:

1. Is the distribution symmetric or skewed?

2. How many mounds appear? One? Two? None? Many?

3. Are unusually large or small values present?

Symmetric or Skewed?

A symmetric distribution is one in which the left-hand side of the graph is roughly a mirror image of the right-hand side. The idealized distributions in Figure 2.10 show two possibilities. Figure 2.10a is a **symmetric distribution** with one mound. (Statisticians often describe a distribution with this particular shape as a **bell-shaped distribution**. Bell-shaped distributions play a major role in statistics, as you will see throughout this text.)

▶ **FIGURE 2.10** Sketches of **(a)** a symmetric distribution and **(b)** a right-skewed distribution.

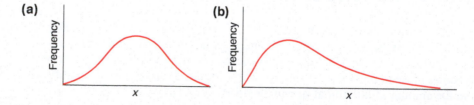

Figure 2.10b represents a nonsymmetric distribution with a skewed shape that has one mound. Note that it has a "tail" that extends out to the right (toward larger values). Because the tail goes to the right, we call it a **right-skewed distribution**. This is a typical shape for the distribution of a variable in which most values are relatively small but there are also a few very large values. If the tail goes to the left, it is a **left-skewed distribution**, where most values are relatively large but there are also a few very small values.

Figure 2.11 shows a histogram of 123 college women's heights. How would you describe the shape of this distribution? This is a good real-life example of a symmetric distribution. Note that it is not perfectly symmetric, but you will never see "perfect" in real-life data.

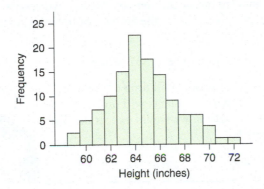

◄ **FIGURE 2.11** Histogram of heights of women. (Source: Brian Joiner in Tufte 1983, 141)

Suppose we asked a sample of people how many hours of TV they watched in a typical week. Would you expect a histogram of these data to be bell-shaped? Probably not. The smallest possible value for this data set would be 0, and most people would probably cluster near a common value. However, a few people probably watch quite a bit more TV than most other people. Figure 2.12 shows a histogram based on actual data. We've added an arrow to emphasize that the tail of this distribution points to the right; this is a right-skewed distribution.

Figure 2.13 shows a left-skewed distribution of test scores. This is the sort of distribution that you should hope your next exam will have. Most people scored pretty high, so the few people who scored low create a tail on the left-hand side. A very difficult test, one in which most people scored very low and only a few did well, would be right-skewed.

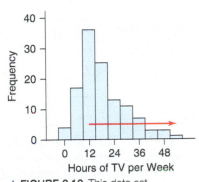

▲ **FIGURE 2.12** This data set on TV hours viewed per week is skewed to the right. (Source: Minitab Program)

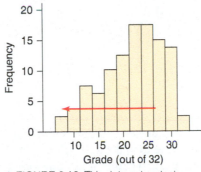

▲ **FIGURE 2.13** This data set on test scores is skewed to the left.

Another circumstance in which we often see skewed distributions is when we collect data on people's income. When we graph the distribution of incomes of a large group of people, we quite often see a right-skewed distribution. You can't make less than 0 dollars per year. Most people make a moderate amount of money, but there is no upper limit on how much a person can make, and a few people in any large sample will make a very large amount of money.

Example 2 shows that you can often make an educated guess about the shape of a distribution even without collecting data.

EXAMPLE **2** Roller Coaster Endurance

A morning radio show is sponsoring a contest in which contestants compete to win a car. About 40 contestants are put on a roller coaster, and whoever stays on it the longest wins. Suppose we make a histogram of the amount of time the contestants stay on (measured in hours or even days).

QUESTION   What shape do we expect the histogram to have, and why?

SOLUTION   Probably most people will drop out relatively soon, but a few will last for a very long time. The last two contestants will probably stay for a very long time indeed. Therefore, we would expect the distribution to be right-skewed.

TRY THIS!   Exercise 2.9

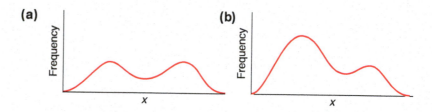

How Many Mounds?   What do you think would be the shape of the distribution of heights if we included men in our sample as well as women? The distributions of women's heights by themselves and men's heights by themselves are usually symmetric and have one mound. But because we know that men tend to be taller than women, we might expect a histogram that combines men's and women's heights to have two mounds.

The statistical term for a one-mound distribution is **unimodal distribution**, and a two-mound distribution is called a **bimodal distribution**. Figure 2.14a shows a bimodal distribution. A **multimodal distribution** has more than two modes. The modes do not have to be the same height (in fact, they rarely are). Figure 2.14b is perhaps the more common bimodal distribution.

▶ **FIGURE 2.14** Idealized bimodal distributions. **(a)** Modes of roughly equal height. **(b)** Modes that differ in height.

(a)

Frequency

x

(b)

Frequency

x

These sketches are idealizations. In real life you won't see distributions this neat. You will have to make a decision about whether a histogram is close enough to be called symmetric and whether it has one mound, two mounds, no mounds, or many mounds. The existence of multiple mounds is sometimes a sign that very different groups have been combined into a single collection (such as combining men's heights with women's heights). When you see multimodal distributions, you may want to look back at the original data and see whether you can examine the groups separately, if separate groups exist. At the very least, whenever you see more than one mound, you should ask yourself, "Could these data be from different groups?"

EXAMPLE 3 Two Marathons, Merged

Data were collected on the finishing times for two different marathons. One marathon consisted of a small number of elite runners: the 2012 Olympic Games. The other marathon included a large number of amateur runners: a marathon in Portland, Oregon.

QUESTION What shape would you expect the distribution of this sample to have?

SOLUTION We expect the shape to be bimodal. The elite runners would tend to have faster finishing times, so we expect one mound on the left for the Olympic runners and another mound on the right. Figure 2.15 is a histogram of the data.

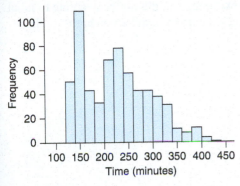

▲ **FIGURE 2.15** Histogram of finishing times for two marathons.

There appears to be one mound centered at about 150 minutes (2.5 hours) and another centered at about 225 minutes (about 3.8 hours).

TRY THIS! Exercise 2.11

When we view a histogram, our understanding of the shape of a distribution is affected by the width of the bins. Figure 2.15 reveals the bimodality of the distribution partly because the width of the bins is such that we see the right level of detail. If we had made the bins too big, we would have got less detail and might not have seen the bimodal structure. Figure 2.16 shows what would happen. Experienced data analysts usually start with the bin width the computer chooses and then also examine histograms made with slightly wider and slightly narrower bins. This lets them see whether any interesting structure emerges.

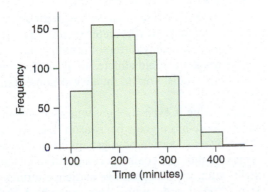

◀ **FIGURE 2.16** Another histogram of the same running times as in Figure 2.15. Here, the bins are wider and "wash out" detail, so the distribution no longer looks bimodal, even though it should.

Do Extreme Values Occur? Sometimes when you make a histogram or dot-plot, you see extremely large or extremely small observations. When this happens, you should report these values and, if necessary, take action (although *reporting* is often action enough). Extreme values can appear when an error is made in entering data. For example, we asked students in one of our classes to record their weights in pounds. Figure 2.17 shows the distribution. The student who wrote 1200 clearly made a mistake. He or she probably meant to write 120.

Extreme values such as these are called **outliers**. The term *outlier* has no precise definition. Sometimes you may think an observation is an outlier, but another person might disagree, and this is fine. However, if there is no gap between the observation in question and the bulk of the histogram, then the observation probably should not be considered an outlier. Outliers are points that don't fit the pattern of the rest of the data, so if a large percentage of the observations are extreme, it might not be accurate to label them as outliers. After all, if lots of points don't fit a pattern, maybe you aren't seeing the right pattern!

▶ **FIGURE 2.17** Histogram of weights with an extreme value.

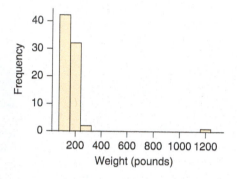

KEY POINT Outliers are values so large or small that they do not fit into the pattern of the distribution. There is no precise definition of the term *outlier*. Outliers can be caused by mistakes in data entry, but genuine outliers are sometimes unusually interesting observations.

Sometimes outliers result from mistakes, and sometimes they do not. The important thing is to make note of them when they appear and to investigate further, if you can. You'll see in Chapter 3 that the presence of outliers can affect how we measure center and spread.

Center

An article in the *New York Times* claimed that the "typical amount of an item pawned in the United States is around $150." Clearly, the amounts of pawned items must differ from item to item, so how can we describe them with a single number? Or consider this question: "How long does it take to run a 100-meter sprint?" You know that it depends on the runner and the race; there are many different possible times. So we might answer, "It depends. I've seen some track-and-field athletes do it in 11 seconds and some much faster." But if pressed for more details, you might answer something like "The typical time is about 10.5 seconds."

The need for a "typical" value is even stronger when we're comparing groups. Do college male soccer players score more goals per season than do female college soccer players? The amount that players scored in both groups varies. To answer this question we need to know the typical number of goals in each of these groups.

For example, a movie-rating website provides audience ratings for a large number of movies, rated on a 0 to 10 point scale, with 10 being the best. Figure 2.18 shows a histogram depicting the distribution of scores for films that were in English. What's the typical rating? The center of this distribution is about 6.5 points, so we might say that the typical rating is (roughly) 6.5.

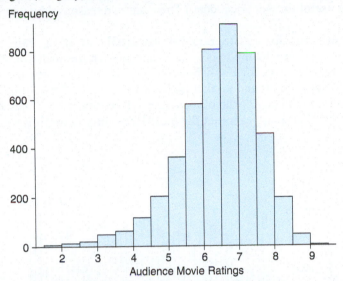

◀ **FIGURE 2.18** The distribution of audience ratings for films in English. A value of 0 is the worst possible movie, and 10 is the best.

Now you might not agree the center is exactly at 6.5, but the concept we want you to understand is that if someone asked you, "What rating do people assign to English-language movies?" the value you would answer, after consulting this histogram, is going to be one of the values towards the center of this distribution. For example, you're probably not going to answer 2 or 9, which are values towards the extremes.

If the distribution is bimodal or multimodal, it may not make sense to seek a "typical" value for a data set. If the data set combines two very different groups, then it might be more useful to find separate typical values for each group. What is the typical finishing time of the runners in Figure 2.15? There is no single typical time because there are two distinct groups of runners. The elite group have their typical time, and the amateurs have a different typical time. However, it *does* make sense to ask about the typical test score for the student scores in Figure 2.13, because there is only one mound and only one group of students.

> **! Caution**
>
> **Multimodal Centers**
> Be careful about giving a single typical value for a multimodal or bimodal distribution. These distributions sometimes indicate the existence of different and diverse groups combined into the same data set. It may be better to report a different typical value for each group, if possible.

EXAMPLE 4   Typical Bar-Passing Rate for Law Schools

Examine the distribution of bar-passing rates for law schools in the United States (see Figure 2.6b, which is repeated here for convenience).

QUESTION   What is a typical bar-passing rate for a law school?

SOLUTION   We show Figure 2.6b again. The center is somewhere in the range of 80% to 90% passing, so we would say that the typical bar-passing rate is between 80% and 90%.

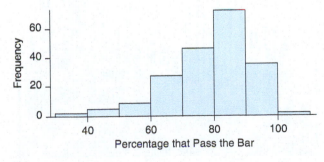

◀ **FIGURE 2.6b** (repeated)
Percentage passing the bar exam from different law schools.

TRY THIS!   Exercise 2.13

Why Not the Mode? The most frequently occurring value is called the *mode*. In general, it is not wise to use the mode for numerical data, particularly if trying to read it from a histogram. (You'll see shortly that the mode is useful when analyzing categorical data, but is not useful for numerical data.) There are two reasons why you shouldn't use the mode.

The first reason is that a histogram can obscure the location of the mode, so our impression of where the mode is located depends on the width of the bins. The histogram of critics ratings for English-language films gives us the impression that there is a value between 80 and 90 that is the most "popular," meaning it is a value that occurred more than any of the other values. However, if I use narrower bins, as in Figure 2.19, then it appears that the most frequently occurring value is now around 92.

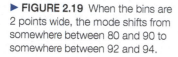

► **FIGURE 2.19** When the bins are 2 points wide, the mode shifts from somewhere between 80 and 90 to somewhere between 92 and 94.

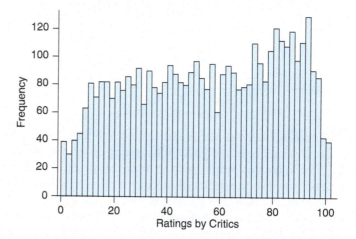

The second reason to avoid using the mode is that there can be many, many different numerical values reported in a data set. This leads to at least two problems. One is that there can be many different modes within the same data set. Another is that the mode can change radically if there is just a minor change in the data.

To see this, consider the dotplot of critics' ratings in Figure 2.20. This shows us that the most frequent value, the mode, is at 93. But it is the most frequent by just one "vote." If a single critic had changed her mind, say, changing the 93 to a 94, there would then be a tie for mode between the values of 93 and 86. If two critics had changed their minds, the mode would have jumped from 93 to 86. The mode is a very unstable measure of the "typical" when analyzing numerical data.

► **FIGURE 2.20** A dotplot for the critics' ratings of English-language movies.

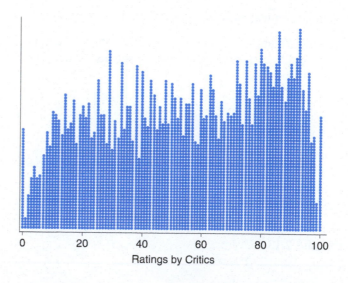

Variability

The third important feature to consider when examining a distribution is the amount of variation in the data. If all the values of a numerical variable are the same, then the histogram (or dotplot) will be skinny. On the other hand, if there is great variety, the histogram will be spread out, thus displaying greater variability.

Here's a very simple example. Figure 2.21a shows a family of four people who are all very similar in height. Note that the histogram of these heights (Figure 2.21b) is quite skinny; in fact, it is just a single bar!

(a)   **(b)**

▲ **FIGURE 2.21** Family One with a Small Variation in Height

Figure 2.22a, on the other hand, shows a family that exhibits large variation in height. The histogram for this family (Figure 2.22b) is more spread out.

(a)   **(b)**

▲ **FIGURE 2.22** Family Two with a Large Variation in Height

EXAMPLE 5 The Price of College

It's no secret that not-for-profit private four-year colleges typically cost more than public colleges. But the distributions of costs at these two types of institutions differ in other ways as well. Figure 2.23 shows the total tuition and fees in 2014 for public and private schools.

QUESTION How does the variability in costs compare between these types of institutions?

► **FIGURE 2.23** The total tuition and fees in 2014 for public (top) and private schools (bottom).

Data Moves ► The full data set can be found in the Integrated Post-secondary Education Data System (IPEDS) at nces .ed.gov/ipeds. We've extracted a few variables of interest and given them more descriptive names.

Data ► fouryearsample.csv

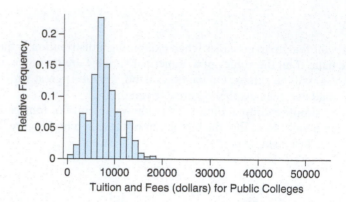

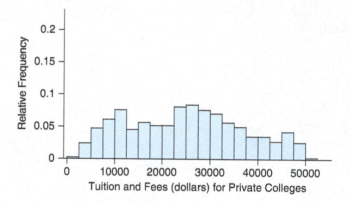

SOLUTION   The variability among private schools is much greater than among public schools. Private school costs span about $50,000, while all the four-year public schools' costs are contained within a range of $20,000.

TRY THIS!   Exercise 2.15

Describing Distributions

When you are asked to describe a distribution or to compare two distributions, your description should include the center of the distribution (What is the typical value?), the spread (How much variability is there?), and the shape. If the shape is bimodal, it might not be appropriate to mention the center, and you should instead identify the approximate location of the mounds. You should also mention any unusual features, such as extreme values.

EXAMPLE 6 Body Piercings

How common are body piercings among college students? How many piercings does a student typically have? One statistics professor asked a large class of students to report (anonymously) the number of piercings they possessed.

QUESTION   Describe the distribution of body piercings for students in a statistics class.

SOLUTION   The first step is to "see it" by creating an appropriate graphic of the distribution of the sample. Figure 2.24 shows a histogram for these data. To summarize this distribution, we examine the shape, the center, and the spread and comment on any unusual features.

The distribution of piercings is right-skewed, as we might expect, given that many people will have 0, 1, or 2 piercings but a few people are likely to have more. The typical number of piercings (the center of the distribution) seems to be about 1, although a majority of students have none. The number of piercings ranges from 0 to about 10. An interesting feature of this distribution is that it appears to be multimodal. There are three peaks: at 0, 2, and 6, which are even numbers. This makes sense because piercings often come in pairs. But why is there no peak at 4? (The authors do not know.) What do you think the shape of the distribution would look like if it included only the men? Only the women?

TRY THIS!   Exercise 2.17

SECTION 2.3

Visualizing Variation in Categorical Variables

When visualizing data, we treat categorical variables in much the same way as numerical variables. We visualize the distribution of the categorical variable by displaying the values (categories) of the variable and the number of times each value occurs.

To illustrate, consider the Statistics Department at the University of California, Los Angeles (UCLA). UCLA offers an introductory statistics course every summer, and it needs to understand what sorts of students are interested in this class. In particular, understanding whether the summer students are mostly first-year students (eager to complete their general education requirements) or seniors (who put off the class as long as they could) can help the department better plan its course offerings.

Table 2.3 shows data from a sample of students in an introductory course offered during the 2013 summer term at UCLA. The "unknown" students are probably not enrolled in any university (adult students taking the course for business reasons or high school students taking the class to get a head start).

Class is a categorical variable. Table 2.4 on the next page summarizes the distribution of this variable by showing us all the values in our sample and the frequency with which each value appears. Note that we added a row for first-year students.

Two types of graphs that are commonly used to display the distribution of a sample of categorical data are bar charts and pie charts. Bar charts look, at first glance, very similar to histograms, but they have several important differences, as you will see.

Bar Charts

A **bar chart** (also called **bar graph** or **bar plot**) shows a bar for each observed category. The height of the bar is proportional to the frequency of that category. Figure 2.25a on the next page shows a bar graph for the UCLA statistics class data. The vertical axis

Student ID	Class
1	Senior
2	Junior
3	Unknown
4	Unknown
5	Senior
6	Graduate
7	Senior
8	Senior
9	Unknown
10	Unknown
11	Sophomore
12	Junior
13	Junior
14	Sophomore
15	Unknown
16	Senior
17	Unknown
18	Unknown
19	Sophomore
20	Junior

▲ TABLE 2.3 Identification of classes for students in statistics.

Class	Frequency
Unknown	7
First-year student	0
Sophomore	3
Junior	4
Senior	5
Graduate	1
Total	**20**

▲ **TABLE 2.4** Summary of classes for students in statistics.

▶ **FIGURE 2.25** **(a)** Bar chart showing numbers of students in each class enrolled in an introductory statistics section. The largest "class" is the group made up of seven unknowns. First-year students are not shown because there were none in the data set. **(b)** The same information as shown in part (a) but now with relative frequencies. The unknowns are about 0.35 (35%) of the sample.

displays frequency. We see that the sample has one graduate student and four juniors. We could also display relative frequency if we wished (Figure 2.25b). The shape does not change; only the numbers on the vertical axis change.

Note that there are no first-year students in the sample. We might expect this of a summer course, because entering students are unlikely to take courses in the summer before they begin college, and students who have completed a year of college are generally no longer first-year students (assuming they have completed enough units).

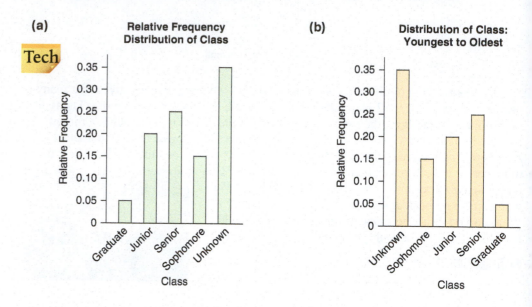

Bar Charts versus Histograms Bar charts and histograms look a lot alike, but they have some very important differences.

- In a bar chart, it sometimes doesn't matter in which order you place the bars. Quite often, the categories of a categorical variable have no natural order. If they do have a natural order, you might want to sort them in that order. For example, in Figure 2.26a we've sorted the categories into a fairly natural order, from "Unknown" to "Graduate." In Figure 2.26b we've sorted them from most frequent to least frequent. Either choice is acceptable.

▶ **FIGURE 2.26** **(a)** Bar chart of classes using natural order. **(b)** Pareto chart of the same data. Categories are ordered with the largest frequency on the left and arranged so the frequencies decrease to the right.

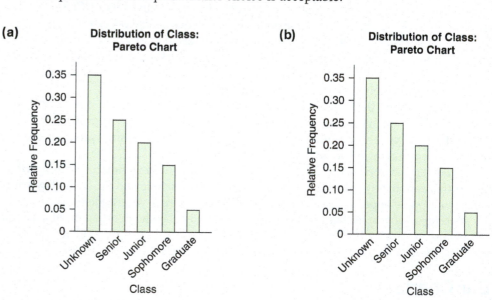

Figure 2.26b is useful because it shows more clearly that the most populated category consists of "Unknown." This result suggests that there might be a large demand for this summer course outside of the university. Bar charts that are sorted from most frequent to least frequent are called **Pareto charts**. (These charts were invented by the Italian economist and sociologist Vilfredo Pareto, 1848–1943.) They are often an extremely informative way of organizing a display of categorical data.

- Another difference between histograms and bar charts is that in a bar chart it doesn't matter how wide or narrow the bars are. The widths of the bars have no meaning.

- A final important difference is that a bar chart has gaps between the bars. This indicates that it is impossible to have observations between the categories. In a histogram, a gap indicates that no values were observed in the interval represented by the gap.

SNAPSHOT ▸ The Bar Chart

WHAT IS IT? ▸	A graphical summary for categorical data.
WHAT DOES IT DO? ▸	Shows a picture of the distribution of a categorical variable.
HOW DOES IT DO IT? ▸	Each category is represented by a bar. The height of the bar is proportional to the number of times that category occurs in the data set.
HOW IS IT USED? ▸	To see patterns of variation in categorical data. The categories can be presented in order of most frequent to least frequent, or they can be arranged in another meaningful order.

Pie Charts

Pie charts are another popular format for displaying relative frequencies of data. A **pie chart** looks, as you would expect, like a pie. The pie is sliced into several pieces, and each piece represents a category of the variable. The area of the piece is proportional to the relative frequency of that category. The largest piece in the pie in Figure 2.27 belongs to the category "Unknown" and takes up about 35% of the total pie.

Some software will label each slice of the pie with the percentage occupied. This isn't always necessary, however, because a primary purpose of the pie chart is to help us judge how frequently categories occur relative to one another. For example, the pie chart in Figure 2.27 shows us that "Unknown" occupies a fairly substantial portion of the whole data set. Also, labeling each slice gets cumbersome and produces cluttered graphs if there are many categories.

Although pie charts are very common (we bet that you've seen them before), they are not commonly used by statisticians or in scientific settings. One reason for this is that the human eye has a difficult time judging how much area is taken up by the wedge-shaped slices of the pie chart. Thus, in Figure 2.27, the "Sophomore" slice looks only slightly smaller than the "Junior" slice. But you can see from the bar chart (Figure 2.26) that they're actually noticeably different. Pie charts are extremely difficult to use to compare the distribution of a variable across two different groups (such as comparing males and females). Also, if there are many different categories, it can be difficult to provide easy-to-read labels for the pie charts.

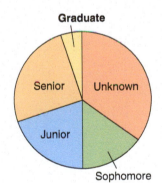

▲ **FIGURE 2.27** Pie chart showing the distribution of the categorical variable *Class* in a statistics course.

SNAPSHOT ▸ The Pie Chart

WHAT IS IT?	▶	A graphical summary for categorical data.
WHAT DOES IT DO?	▶	Shows the proportion of observations that belong to each category.
HOW DOES IT DO IT?	▶	Each category is represented by a wedge in the pie. The area of the wedge is proportional to the relative frequency of that category.
HOW IS IT USED?	▶	To understand which categories are most frequent and which are least frequent. Sometimes it is useful to label each wedge with the proportion of occurrence.

SECTION 2.4

Summarizing Categorical Distributions

The concepts of *shape*, *center*, and *spread* that we used to summarize numerical distributions sometimes don't make sense for categorical distributions because we can often order the categories any way we please. The center and the shape would be different for every ordering of categories. However, we can still talk about typical outcomes and the variability in the sample.

The Mode

When describing a distribution of a categorical variable, pay attention to which category occurs most often. This value, the one with the tallest bar in the bar chart, can sometimes be considered the "typical" outcome. There might be a tie for first place, and that's okay. It just means there's not as much variability in the sample. (Read on to see what we mean by that.)

The category that occurs most often is called the **mode**. This meaning of the word *mode* is similar to its meaning when we use it with numerical variables. However, one big difference between categorical and numerical variables is that we call a categorical variable bimodal only if two categories are nearly tied for most frequent outcomes. (The two bars don't need to be exactly the same height, but they should be very close.) Similarly, a categorical variable's distribution is multimodal if more than two categories all have roughly the tallest bars. For a numerical variable, the heights of the mounds do not need to be the same height for the distribution to be multimodal.

For an example of a mode, let's examine this study from the Pew Research Center (www.pewsocialtrends.org, April 9, 2008). The researchers interviewed 2413 Americans in 2008 and asked them what economic class they felt they belonged in: the lower class, the middle class, or the upper class. Figure 2.28 shows the results.

EXAMPLE **7** A Matter of Class

In 2012, the Pew survey (www.pewsocialtrends.org, August 22, 2012) asked a new group of 2508 Americans which economic class they identified with. The bar chart in Figure 2.29 shows the distribution of the responses in 2012.

QUESTION What is the typical response? How would you compare the responses in 2012 with the responses in 2008?

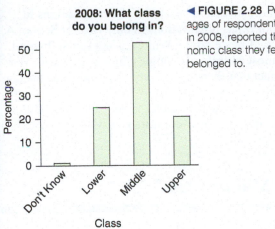

2008: What class do you belong in?

◄ **FIGURE 2.28** Percentages of respondents who, in 2008, reported the economic class they felt they belonged to.

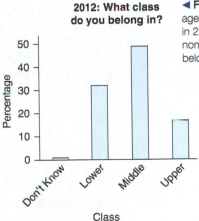

2012: What class do you belong in?

◄ **FIGURE 2.29** Percentages of respondents who, in 2012, reported the economic class they felt they belonged to.

SOLUTION   The mode, the typical response, is still the middle class. However, in 2012 fewer than 50% identified themselves as members of the middle class. This is a lower percentage than in 2008. It also appears that the percentage who identified themselves as members of the lower class increased from 2008, while the percentage of those identifying themselves as members of the upper class decreased.

TRY THIS!   Exercise 2.37

We see that the mode is the middle class; over half of the people (in fact, 53%) identified themselves as members of the middle class. The remaining people were almost equally divided between the lower class (25%) and the upper class (21%).

Variability

When thinking about the variability of a categorical distribution, it is sometimes useful to think of the word *diversity*. If the distribution has a lot of diversity (many observations spread across many different categories), then its variability is high. Bar charts of distributions with high variability sometimes have several different modes or several categories that are close contenders for being the mode. On the other hand, if all the observations are in one single category, then diversity is low. If you are examining a bar chart and there is a single category that is clearly the only mode—a category with far more observations than the other categories—then variability is low.

For example, Figure 2.30 shows bar charts of the ethnic composition of two schools in the Los Angeles City School System. Which school has the greater variability in ethnicity?

(a)

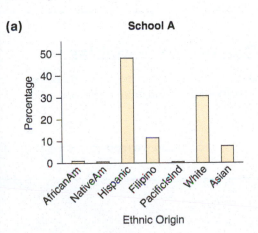

School A

(b)

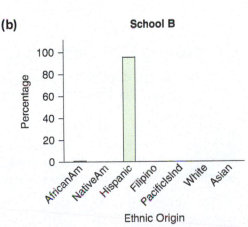

School B

◄ **FIGURE 2.30** Percentages of students at two Los Angeles schools who are identified with several ethnic groups. Which school has more ethnic variability?

School A (Figure 2.30a) has much more diversity, because it has observations in more than four categories, whereas School B has observations in only two. At School A, the mode is clearly Hispanic, but the second-place group, Whites, is not too far behind.

School B (Figure 2.30b) consists almost entirely of a single ethnic group, with very small numbers in the other groups. The fact that there is one very clear mode means that School B has lower variability than School A.

Comparing variability graphically for categorical variables is not easy to do. But sometimes, as in the case of Figure 2.30, there are clear-cut instances where you can generally make some sort of useful comparison.

EXAMPLE 8 The Shrinking Middle Class?

Consider the question that the Pew Foundation asked: "Which economic class, Lower, Middle, or Upper, do you identify with?" Imagine asking this question of a new group of people. Provide a sketch of a bar chart that shows (a) what it would look like if there was the least amount of variability and (b) what it would look like if there was maximum variability. One hint: there are multiple answers possible for part (a), but only one for part (b). There are four responses possible: Don't Know, Lower, Middle and Upper.

SOLUTION There would be the least amount of variability if everyone provided the same answer. It doesn't matter what that answer is, but let's say it is "Middle." In this case, there would be only one bar for the Middle class, as in Figure 2.31a, and that bar would have 100% of the respondents.

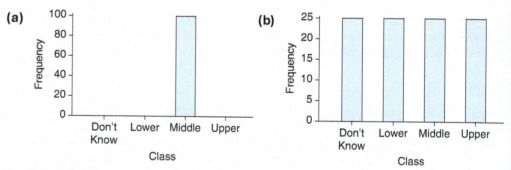

▲ **FIGURE 2.31** **(a)** A possible bar graph that shows the least amount of variability; everyone is in the same class. **(b)** A possible bar graph that shows the maximum amount of variability.

The maximum amount of variability occurs when everyone is as different as possible. In this case, each category would have the same proportion of respondents. And so all four bars would have the same height of 25% as in Figure 2.31b.

TRY THIS!   Exercise 2.39

Describing Distributions of Categorical Variables

When describing a distribution of categorical data, you should mention the mode (or modes) and say something about the variability. Example 9 illustrates what we mean.

KEY POINT   When summarizing graphs of categorical data, report the mode or modes and describe the variability (diversity).

EXAMPLE 9 Causes of Death

According to some experts, about 51.5% of babies currently born in the United States are male. But among people between 100 and 104 years old, there are four times as many women as men (U.S. Census Bureau 2000). How does this happen? One possibility is that the percent of boys born has changed over time. Another possibility is presented in the two bar charts in Figure 2.32. These bar charts show the numbers of deaths per 100,000 people in one year, for people aged 15 to 24 years, for both males and females.

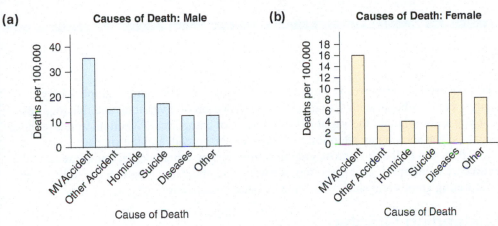

▲ **FIGURE 2.32** The number of deaths per 100,000 males **(a)** and females **(b)** for people 15 to 24 years old in a one-year period.

QUESTION   Compare the distributions depicted in Figure 2.32. Note that the categories are put into the same order on both graphs.

SOLUTION   First, note that the bar graphs are not on the same scale, as you can see by comparing the values on the *y*-axes. This presentation is typical of most software and is sometimes desirable, because otherwise one of the bar charts might have such small bars that we couldn't easily discern differences.

Although motor vehicle accidents are the mode for both groups, males show a consistently high death rate for all other causes of death, whereas females have relatively low death rates in the categories for other accident, homicide, and suicide. In other words, the cause of death for females is less variable than that for males. It is also worth noting that the death rates are higher for males in every category. For example, roughly 16 out of every 100,000 females died in a motor vehicle accident in one year, while roughly 35 out of every 100,000 males died in car accidents in the same year.

TRY THIS!   Exercise 2.43

We can also make graphics that help us compare two distributions of categorical variables. When comparing two groups across a categorical variable, it is often useful to put the bars side by side, as in Figure 2.33. This graph makes it easier to compare rates of death for each cause. The much higher death rate for males is made clear.

▶ **FIGURE 2.33** Death rates of males and females, graphed side by side.

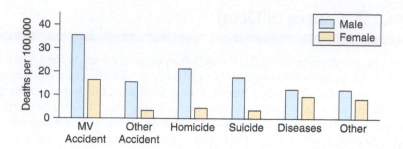

Interpreting Graphs

The first step in every investigation of data is to make an appropriate graph. Many analyses of data begin with visualizing the distribution of a variable, and this requires that you know whether the variable is numerical or categorical. When interpreting these graphics, you should pay attention to the center, the spread, and the shape.

Often, you will come across graphics produced by other people, and these can take extra care to interpret. In this section, we warn you about some potential pitfalls when interpreting graphs and show you some unusual visualizations of data.

Misleading Graphs

A well-designed statistical graphic can help us discover patterns and trends and can communicate these patterns clearly to others. However, our eyes can play tricks on us, and manipulative people can take advantage of this to use graphs to give false impressions.

The most common trick—one that is particularly effective with bar charts—is to change the scale of the vertical axis so that it does not start at the origin (0). Figure 2.34a shows the number of violent crimes per year in the United States as reported by the FBI (www.fbi.gov). The graphic seems to indicate a dramatic drop in crime since 2007.

However, note that the vertical axis starts at 1,200,000 (that is, 1.20 million) crimes. Because the origin begins at 1.20 million and not at 0, the bars are all shorter than they would be if the origin were at 0. The drop in 2010 seems particularly dramatic, in part because the height of the bar for 2010 is less than half the height of the bar for 2009. What does this chart look like if we make the bars the correct height? Figure 2.34b shows the same data, but to the correct scale. It's still clear that there has been a decline, but the decline doesn't look nearly so dramatic now, does it? Why not?

The reason is that when the origin is correctly set to 0, as in Figure 2.34b, it is clear that the percentage decline has not been so great. For instance, the number of crimes is clearly lower in 2010 than in 2009, but it is not half as low, as Figure 2.34a might suggest.

▶ **FIGURE 2.34** **(a)** This bar chart shows a dramatic decline in the number of violent crimes since 2007. The origin for the vertical axis begins at 1.20 million, not at 0. **(b)** This bar chart reports the same data as part (a), but here the vertical axis begins at the origin (0).

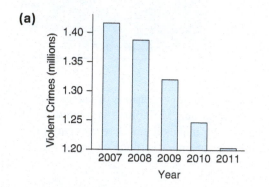

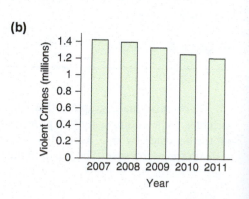

(a)   **(b)**   **(c)**   **(d)**

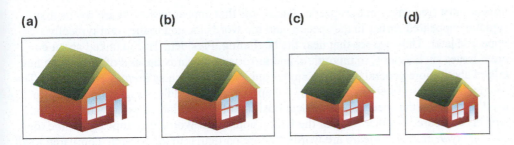

◄ **FIGURE 2.35** Deceptive graphs: Image **(a)** represents 7.1 million homes sold in 2005, image **(b)** represents 6.5 million homes sold in 2006, image **(c)** represents 5.8 million homes sold in 2007, and image **(d)** represents 4.9 million homes sold in 2008. (Source: *L.A. Times*, April 30, 2008)

Most of the misleading graphics you will run across exploit a similar theme: Our eye tends to compare the relative sizes of objects. Many newspapers and magazines like to use pictures to represent data. For example, the plot in Figure 2.35 attempts to illustrate the number of homes sold for some past years, with the size of each house representing the number of homes sold that year. Such graphics can be very misleading, because the pictures are often not to scale with the actual numbers.

In Figure 2.35 the *heights* of the homes are indeed proportional to the sales numbers, but our eye reacts to the *areas* instead. The smallest house is 69% as tall as the largest house (because 4.9 is 69% of 7.1), but the area of the smallest house is only about 48% of that of the largest house, so our tendency to react to area rather than to height exaggerates the difference.

The Future of Statistical Graphics

The Internet allows for a great variety of graphical displays of data that take us beyond simple visualizations of distributions. Many statisticians, computer scientists, and designers are experimenting with new ways to visualize data. Most exciting is the rise of interactive displays. The *State of the Union Visualization*, for example (http://stateoftheunion .onetwothree.net), makes it possible to compare the content of State of the Union speeches. Every U.S. president delivers a State of the Union address to Congress near the beginning of each year. This interactive graphic enables users to compare words from different speeches and "drill down" to learn details about particular words or speeches.

For example, Figure 2.36 is based on the State of the Union address that President Donald Trump delivered on February 28, 2017. The largest words are the words that

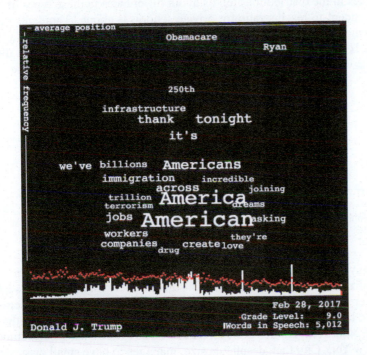

◄ **FIGURE 2.36** President Trump's 2017 State of the Union speech, visualized as a "word cloud." This array shows us the most commonly used words, approximately where these words occurred in the speech, how often they occurred, and how unusual they are compared to the content of other State of the Union speeches. (Courtesy of Brad Borevitz, http://onetwothree.net. Used with permission.)

appear most frequently in the speech. The words that appear to the left are words that typically appeared earlier in the speech, and the words located to the right typically appeared later. Thus, we see that near the beginning, President Trump talked about concrete issues such as jobs, immigration, and terrorism. Towards the end of the speech he talked about more general topics, as suggested by words such as *dreams* and *joining*. The words *Americans*, *America,* and *American* appear in the middle, suggesting they were used throughout the speech. The vertical position of a word indicates how unusual that word is compared to other State of the Union addresses. Words that appear near the top, such as *Ryan* and *Obamacare* are words that are particular to this speech, compared to all other State of the Union addresses. Words near the bottom, such as *drug*, *create*, and *love* are words that occur relatively frequently throughout all State of the Union addresses.

Figure 2.36 is packed with information. On the bottom, in red, you see dots that represent the grade level required to read and understand the speech. Each dot represents a different speech, and the speeches are sorted from first (George Washington) to most recent (Donald Trump). Near the bottom right we see that the grade level of the speech is 9.0, and that this is considerably lower than it was 100 years ago but consistent with more recent speeches. The white "bars" represent the number of words in the State of the Union address. This speech contained 5012 words, which is fairly typical of the last few years. The tallest spike represents President Jimmy Carter's speech in 1981, which had 33,613 words and a grade level of 15.3!

CASE STUDY REVISITED

Student-to-Teacher Ratio at Colleges

How do public colleges compare to private colleges in student-to-teacher ratios? The data were presented at the beginning of the chapter. That list of raw data makes it hard to see patterns, so it is very difficult to compare groups. But because the student-to-teacher ratio is a numerical variable, we can display these distributions as two histograms to enable us to make comparisons between public and private schools.

In Figure 2.37, the distribution of student-to-teacher ratios for the private schools is shown in the left histogram, and that for public schools is shown in the right histogram. The differences in the distributions tell us about differences in these types of colleges. The typical student-to-teacher ratio for private schools is about 13 students per teacher, while for public schools it's closer to 20. Both groups have similar spread, although the private-schools group has an outlier. This outlier makes the private-schools distribution more right-skewed, while the public-schools distribution is more symmetric.

▶ **FIGURE 2.37** Histograms of the student-to-teacher ratio at samples of private colleges and public colleges.

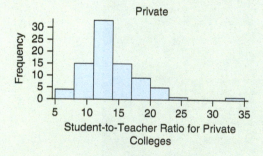

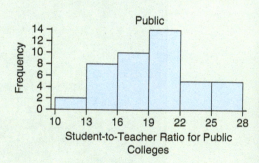

By the way, many of the schools with the smallest student-to-teacher ratios are highly specialized. The private school with a student-to-teacher ratio of 5 is a Catholic seminar that enrolls only 130 or so would-be priests. In case you are wondering which public college has the smallest student-to-teacher ratio in Figure 2.37, it is the City University of New York, Hunter College, a quite large (more than 16,000 students) college in New York City that, despite its size, manages to have small class sizes.

DATAPROJECT ▸ Asking Questions

1 OVERVIEW

In Chapter 1, you found a data set that interests you. Now it's time to explore. The first step is to ask the right questions.
Data: Data chosen in Chapter 1 Project.

2 GOALS

Know the components of a good statistical question; write three good statistical questions for your chosen data set.

3 CURIOUSER AND CURIOUSER

When examining data sets, many questions arise, and these range from the ordinary to the profound. For example, What does that abbreviation mean? What does variable X34 mean? How can this data set tell me what I want to know?

This project will help you phrase a particular type of question, called a statistical investigative question, or SIQ. An SIQ is a question that can be answered by examining one or more variables in a data set. While we're just getting started, we'll focus on one-variable questions.

Project: Begin by reminding yourself why you chose the data set that you did. (This was one of the questions you answered in Chapter 1.) If you found the data interesting, it was likely because you had some questions that you hoped could be answered with the data. What were those questions? (A good way of phrasing your questions is to begin with "I wonder . . . .") Posing a good statistical investigative question is not easy, but it might help to consider these three types of questions: summary questions, comparison questions, and relationship questions. To illustrate, imagine that we're considering a data set of fire truck response times in Los Angeles.

Summary questions are about a variable or group of variables. For example, "I wonder what was the slowest response time?" "I wonder what was a typical response time?"

Comparison questions compare different groups. For example, "I wonder if response times were slower on weekends than on weekdays?" "I wonder if response times were faster in wealthier neighborhoods than in poorer neighborhoods?"

Relationship questions ask about associations between two or more variables. "I wonder if response times vary with time of day?" asks about how the response time variable is related to the time of day.

Try writing one example of each type of question for your data.

Not all SIQs are created equal. Some are more useful than others. Some are impossible to answer. Here's one way of rating your question:

1. *Do the variables exist in the data set to answer your question?*
   If my question is, "I wonder if response times vary with the hour of the day?" then it is clear that my variables of interest are response time and hour of the day. But do these variables exist? Here's a gray area: Perhaps I have a variable that is Time Of Day, and it has values like 03:02:22. My question is about the hour of the day, and I don't have a variable that just has the hour ("03"). In this case, the information is there, even if the variable isn't, so you get partial credit.
   Score yourself 1 point if all variables exist.
   Score yourself 1/2 point if at least one variable has the information you need but not in the form you need it.

2. *Is the population of interest clear?*
   Consider these two questions: "I wonder what the typical response time was for fire trucks in Los Angeles between 2001 and 2017?" and "I wonder what the typical response was for fire trucks?" The first question indicates the population that my data came from: Los Angeles in the years 2001 to 2017. The last question doesn't tell us anything about the population. Note that we're not requiring that the population be correct; you need only be clear about what you think the population should be. For example, you might say, "Do all fire trucks in the United States respond faster on weekdays than on weekends?" You might not have the data to answer this, but if this is the population you care about, then

it's important to state your question that way so that you can decide later which data sets are required to answer the question.

Score 1 point if the population of interest is clear.

3. *Does the question require considering the entire group of data or just a single observation?*

Can the answer be given by selecting a single observation from the data set? An example would be "I wonder what the fastest response time was." A better question would be "I wonder how different the fastest response time was from the slowest response time." This requires looking at two values. Or, "I wonder what the typical response time was," which requires considering all the observations.

If your question can be answered with a single observation, score 0 points. Otherwise, score 1 point.

4. *Can your question be answered with the data at hand?*

For example, I might ask, "I wonder if response times are faster in Los Angeles than in San Diego?" On one hand, I have a response time variable, but I don't have any data from San Diego.

If you have all of the data you need, score 1 point. If you have only some of the data, score 1/2 point. If you have none of the data, score 0 points.

5. *Finally, are you interested in the answer to the question?*

This is very subjective, but give yourself 1 point if you think it's interesting, 0 points if not, and 1/2 point if somewhere in between. For example, I could ask the question, "I wonder what twice the average response time is?" But why? Is there any reason for this? If not, then it gets 0 points.

Assignment

1. For each of the three questions you wrote, score the quality and discuss the reasons for your score. Rewrite the questions to improve the score.

2. Now look at your actual variables. Create one of each type of question (summary, comparison, relationship) using the actual variables in your data set. Is this possible? If not, explain.

CHAPTER REVIEW

KEY TERMS

distribution of a sample, *42*
frequency, *42*
dotplot, *43*
histogram, *44*
relative frequency, *45*
stemplot (or stem-and-leaf plot), *46*

leaf, *46*
stem, *46*
typical value (center), *47*
variability, *47*
symmetric distribution, *48*
bell-shaped distribution, *48*

right-skewed distribution, *49*
left-skewed distribution, *49*
unimodal distribution, *50*
bimodal distribution, *50*
multimodal distribution, *50*
outlier, *52*

bar graph (bar chart), *57*
Pareto chart, *59*
pie chart, *59*
mode (in categorical
    variables), *60*

LEARNING OBJECTIVES

After reading this chapter and doing the assigned homework problems, you should

- Understand that a distribution of a sample of data displays a variable's values and the frequencies (or relative frequencies) of those values.

- Know how to make graphs of distributions of numerical and categorical variables and how to interpret the graphs in context.

- Be able to compare centers and spreads of distributions of samples informally.

- Pose statistical questions for numerical and categorical variables.

SUMMARY

The first step toward answering a statistical investigative question should almost always be to make plots of the distributions of the relevant variables in your data set. In the "Consider Data" step you should identify whether the variables are numerical or categorical so that you can choose an appropriate graphical representation. Sometimes, this graph is the only analysis you will need, as you may be able to answer your question from the graph. Questions about the "typical" value can often be answered, at least approximately, by locating the center of the distribution, and questions about the variability can also be answered by examining the shape and horizontal spread.

If the variable is numerical, you can make a dotplot, histogram, or stemplot. Pay attention to the shape (Is it skewed or symmetric? Is it unimodal or multimodal?), to the center (What is a typical outcome?), and to the spread (How much variability is present?). You should also look for unusual features, such as outliers.

Be aware that many of these terms are deliberately vague. You might think a particular observation is an outlier, but another person might not agree. That's okay, because the purpose isn't to determine whether such points are "outliers" but to indicate whether further investigation is needed. An outlier might, for example, be caused by a typing error someone made when entering the data.

If you see a bimodal or multimodal distribution, ask yourself whether the data might contain two or more groups combined into the single graph.

If the variable is categorical, you can make a bar chart, a Pareto chart (a bar chart with categories ordered from most frequent to least frequent), or a pie chart. Pay attention to the mode (or modes) and to the variability.

SOURCES

Minitab 15 Statistical Software. 2007. [Computer software]. State College, PA: Minitab, Inc. http://www.minitab.com
National Center for Educational Statistics. http://www.nces.ed.gov/ipeds/
Playfair, W. 1786. *Commercial and political atlas: Representing, by copperplate charts, the progress of the commerce, revenues, expenditure, and debts of England, during the whole of the eighteenth century*. London: Corry. Reprinted in H. Wainer and I. Spence (eds.), *The commercial

and political atlas and statistical breviary*, New York: Cambridge University Press, 2005. http://www.math.yorku.ca/SCS/Gallery/milestone/refs.html#Playfair:1786
Tufte, E. 1983. *The visual display of quantitative information* (1st ed.), Cheshire, CT: Graphics Press. Pg. 141. Photo by Brian Joiner.
U.S. Census Bureau. 2000. http://ceic.mt.gov/C2000/SF12000/Pyramid/pptab00.htm

SECTION EXERCISES

SECTIONS 2.1 AND 2.2

2.1 Pulse Rates The dotplot shown is for resting pulse rates of 125 people (according to NHANES, National Health and Nutrition Examination Survey).

Pulse (beats per minute)

Many sources say the resting pulse should not be more than 100.

a. How many of the 125 people had resting pulse rates of more than 100?

b. What percentage of people had resting pulse rates of more than 100?

2.2 Glucose A dotplot of the glucose readings from 132 people (from NHANES) is shown. Some doctors recommend that glucose readings not be above 120 mg/dl.

a. How many of these people have glucose readings above 120 mg/dl?

b. What percentage of these people have glucose readings above 120 mg/dl?

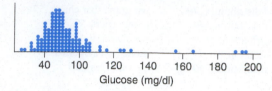

2.3 Pulse Rates The histogram shown is for pulse rates for 125 people. Convert the vertical axis to relative frequency and show the value that would replace each of the values on the vertical axis. (Source: NHANES)

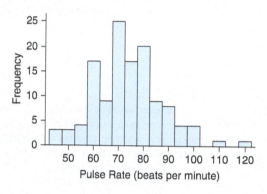

2.4 Fast Food Calories The histogram shown is for number of calories in a sample of fast-food items. (Source: StatCrunch survey).

a. What is the bin width of the histogram?

b. Would this graph be best described as unimodal or bimodal?

c. Approximately what percentage of the fast-food items contained fewer than 300 calories?

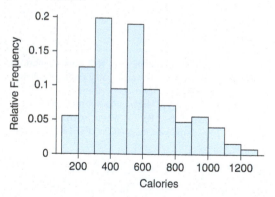

TRY **2.5 Pulse Rates (Example 1)** The resting pulse rates (in beats per minute) of a sample of 91 undergraduate students were recorded. The data are displayed in the following histogram and dotplots. Based on these graphs, would you consider a resting pulse rate of 90 beats per minutes to be unusually high? Why or why not? (Source: Minitab Data Library)

(A)

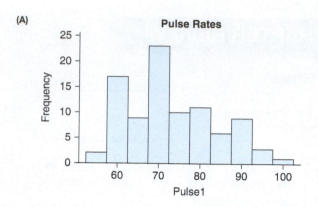

(B)

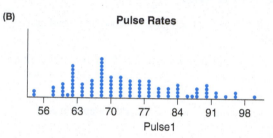

2.6 Post Office Customers A post office branch recorded the number of customers served per day for a period of time. The data are displayed in the following histogram and dotplot. Based on these graphs, would it be unusual for this branch to serve 250 customers in a day? Why or why not? (Source: Minitab Data Library)

(A)

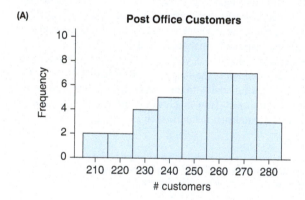

(B)

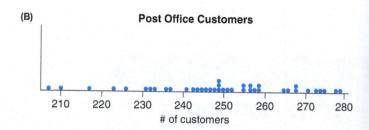

2.7 Cereals The following dotplots show the number of calories in a sample of cereals from two manufacturers: "G" = General Mills and "K" = Kelloggs. (Source: StatCrunch)

a. Write a few short sentences that compare the center and the spread for each dotplot.

b. Based on this sample, cereals from which manufacturer tend to have more variation in calories?

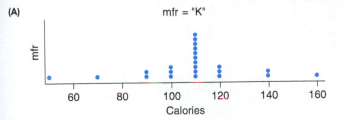

(A)   mfr = "K"

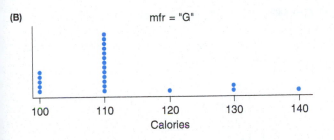

(B)   mfr = "G"

2.8 Coins' Weights The weights of coins (in grams) were obtained for coins in the United States and coins for other countries (including Mexico, Brazil, Canada, and several in Europe).

a. Write a few sentences to compare the distributions of weights of coins from the United States with those from other countries. Be sure to mention the shape, the center, and the spread.

b. Which group typically has heavier coins? (Source: http://www.eeps .com/zoo/cages/Coins.txt)

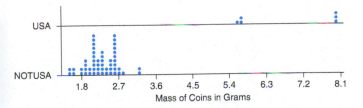

2.9 Sleep (Example 2) An instructor asks a class of 120 students how many hours they slept during the previous night. What shape do you think this distribution will have? Explain.

2.10 Parking Tickets A group of 80 drivers in a large city is asked for the number of parking tickets they received in the previous month. What shape do you think this distribution will have? Explain.

2.11 Arm spans (Example 3) According to the ancient Roman architect Vitruvius, a person's arm span (the distance from fingertip to fingertip with the arms stretched wide) is approximately equal to his or her height. For example, people 5 feet tall tend to have an arm span of 5 feet. Explain, then, why the distribution of arm span for a class containing roughly equal numbers of men and women might be bimodal.

★ 2.12 Tuition The distribution of in-state annual tuition for all colleges and universities in the United States is bimodal. What is one possible reason for this bimodality?

TRY **2.13 Pulse Rates (Example 4)** From the histogram in Exercise 2.3, approximately what is the typical pulse rate in this sample?

2.14 Fast Food Calories From the histogram shown in Exercise 2.4, what is the typical number of calories for a fast food item in this sample?

TRY **2.15 BMI (Example 5)** The histograms show the Body Mass Index for 90 females and 89 males according to NHANES. Compare the distributions of BMIs for women and men. Be sure to compare the shapes, the centers, and the amount of variation for the two groups.

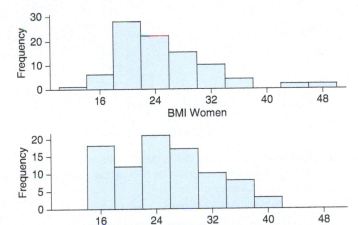

2.16 Triglycerides The histograms show triglyceride levels for 87 men and 99 women. Triglycerides are a form of fat found in blood.

a. Compare the distributions of triglyceride levels for men and women. (Be sure to compare the shapes, the centers, and the spreads.)

b. Triglyceride levels under 150 are good; levels above 500 are very high and may be a health concern. Which group seems to have better triglyceride levels? Explain. (Source: NHANES)

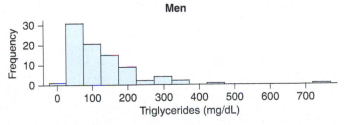

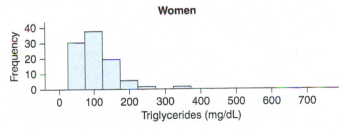

TRY **2.17 Education (Example 6)** In 2012, the General Social Survey (GSS), a national survey conducted nearly every year, reported the number of years of formal education for 2018 people. The following histogram shows the distribution of data.

a. Describe and interpret the distribution of years of formal education. Mention any unusual features.

b. Assuming that those with 16 years of education completed a bachelor's degree, estimate how many of the people in this sample have a bachelor's degree or higher.

c. The sample includes 2018 people. What percentage of people in this sample have a bachelor's degree or higher? How does this compare with Wikipedia's estimate that 27% have a bachelor's degree?

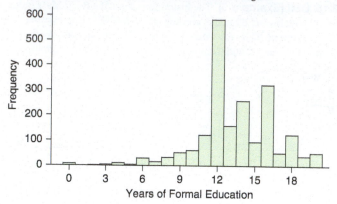

2.18 Siblings The histogram shows the distribution of the numbers of siblings (brothers and sisters) for 2000 adults surveyed in the 2012 General Social Survey.

a. Describe the shape of the distribution.

b. What is the typical number of siblings, approximately?

c. About how many people in this survey have no siblings?

d. What percentage of the 2000 adults surveyed have no siblings?

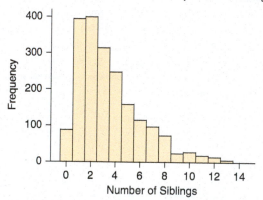

2.19 Monthly Car Costs The histograms show the monthly costs for operating two makes of cars: Ford and BMW. Which make typically has higher monthly costs? Which make has more variation in costs? (Source: Edmonds.com)

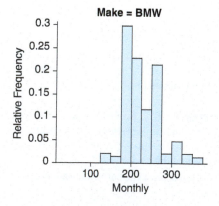

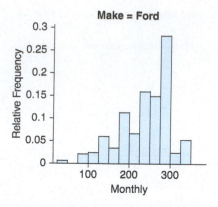

2.20 Car MPG The histograms show the miles per gallon (mpg) for two makes of cars: Acura and BMW. Compare the distributions. Which make has a higher mpg? Which make has more variation in mpg? (Source: Edmonds.com)

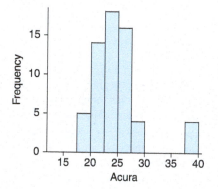

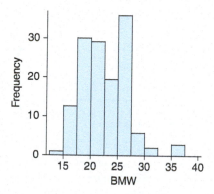

2.21 Matching Histograms Match each of the three histograms to the correct situation.

1. Assessed value of a representative collection of houses in a city

2. Number of bedrooms in the houses

3. Height of house (in stories) for houses in a region that allows up to three stories but not more.

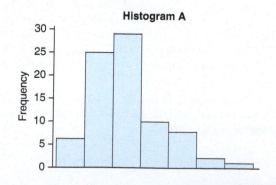

Histogram B

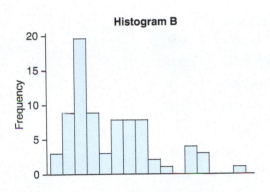

Histogram C

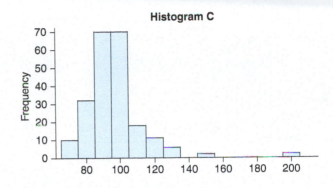

Histogram C

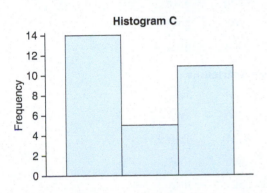

2.22 Matching Match each histogram given with the following descriptions of data.

1. Number of cups of coffee consumed by one person during a day.
2. Maximum speed driven in a car by college students who drive.
3. Number of times last week that a college student had breakfast.

Histogram A

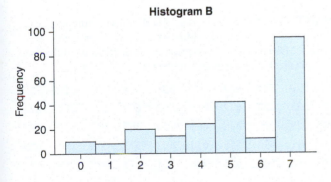

Histogram B

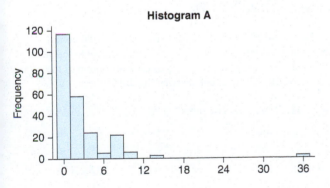

2.23 Matching Match each description with the correct histogram.

1. Heights of students in a large UCLA statistics class that contains about equal numbers of men and women.
2. Numbers of hours of sleep the previous night in the same large statistics class.
3. Numbers of driving accidents for students at a large university in the United States.

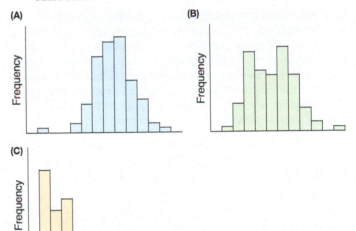

2.24 Matching Match each description with the correct histogram.

1. Quantitative SAT scores for 1000 students who have been accepted for admission to a private university.
2. Weights of over 500 adults, about half of whom are men and half of whom are women.
3. Ages of all 39 students in a community college statistics class that is made up of full-time students and meets during the morning.

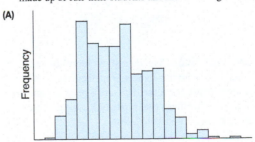

(B)

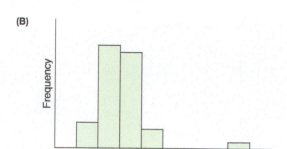

(C)

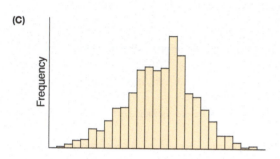

g 2.25 Comparing Weights of Olympic Hockey and Soccer Players The weights of Olympic hockey players and soccer players from recent Olympic games (in pounds) are shown in the following two tables. Make appropriate graphs and use your graphs to write a brief comparison of the distributions of the players' weights. Pose a statistical question that can be answered with your graphs. Include appropriate graphics. *See page 84 for guidance.* (Source: teamusa.usahockey.com. ussoccer.com)

Olympic Hockey Team Weights	
212	180
215	181
196	202
200	183
213	189
219	219
207	190
198	190
221	205
212	196
	200
	205

Olympic Soccer Team Weights	
150	185
175	190
145	190
200	160
155	170
140	180
185	145
155	175
165	150
185	175
165	185
195	

2.26 Rents in San Francisco The data show monthly rents for studio and one-bedroom apartments in San Francisco. Write a statistical question appropriate for these data and a brief comparison of the distribution of rents for these two types of housing units. Include appropriate graphics. (Source: Zillow)

Studio Apartments

3280	2960	3039
2435	3716	3200
3327	2984	3395
2910	2500	2750
2800	2995	2635
3005	3030	3635
2674	3395	2795
3250	2650	3315
3610	2050	2095
2935	3114	2500

One-Bedroom Apartments

3280	3654	3680
3312	3290	3575
3405	3154	3210
2905	3490	4095
3085	3503	3035
3945	4174	2850
3435	3226	1850
3265	3830	3400
3227	3774	3399
3350	3560	3295

2.27 Textbook Prices The table shows prices of 50 college textbooks in a community college bookstore, rounded to the nearest dollar. Make an appropriate graph of the distribution of the data, pose a statistical question, and describe the distribution.

76	19	83	45	88	70	62	84	85	87
86	37	88	45	75	83	126	56	30	33
26	88	30	30	25	89	32	48	66	47
115	36	30	60	36	140	47	82	138	50
126	66	45	107	112	12	97	96	78	60

2.28 SAT scores The following table shows a random sample of 50 quantitative SAT scores of first-year students admitted at a university. Make an appropriate graph of the distribution of the data, pose a statistical question, and describe the distribution.

649	570	734	653	652	538	674	705	729	737
672	583	729	677	618	662	692	692	672	624
669	529	609	526	665	724	557	647	719	593
624	611	490	556	630	602	573	575	665	620
629	593	665	635	700	665	677	653	796	601

2.29 Animal Longevity The following table in Exercise 2.30 shows the average life span for some mammals in years, according to infoplease.com. Graph these average life spans and describe the distribution. What is a typical life span? Identify the three outliers and report their life spans. If you were to include humans in this graph, where would the data point be? Humans live an average of about 75 years.

2.30 Animal Gestation Periods The accompanying table also shows the gestation period (in days) for some animals. The gestation period is the length of pregnancy. Graph the gestation period and describe the distribution. If there are any outliers, identify the animal(s) and give the gestation periods. If you were to include humans in this graph, where would the data point be? The human gestation period is about 266 days.

Animal	Gestation (days)	Life Span (years)
Baboon	187	12
Bear, grizzly	225	25
Beaver	105	5
Bison	285	15
Camel	406	12
Cat, domestic	63	12
Chimp	230	20
Cow	284	15
Deer	201	8
Dog, domestic	61	12
Elephant, African	660	35
Elephant, Asian	645	40
Elk	250	16
Fox, red	52	7
Giraffe	457	10
Goat	151	8
Gorilla	258	20
Hippo	238	41
Horse	330	20
Leopard	98	12
Lion	100	15
Monkey, rhesus	166	15
Moose	240	12
Pig, domestic	112	10
Puma	90	12
Rhino, black	450	15
Sea Lion	350	12
Sheep	154	12
Squirrel, gray	44	10
Tiger	105	16
Wolf, maned	63	5
Zebra, Grant's	365	15

2.31 Tax Rate A StatCrunch survey asked people what they thought the maximum income tax rate should be in the United States. Make separate dotplots of the responses from Republicans and Democrats. If possible, put one above the other, using the same horizontal axis. Then compare the groups by commenting on the shape, the center, and the spread of each distribution. The data are at this text's website. (Source: StatCrunch Survey: Responses to Taxes in the U.S. Owner: scsurvey)

2.32 Pets A StatCrunch survey asked people whether they preferred cats or dogs and how many pets they had. (Those who preferred both cats and dogs and those who preferred another type of pet are not included in these data.) Make two dotplots showing the distribution of the numbers of pets: one for those preferring dogs and one for those preferring cats. If possible, put one dotplot above the other, using the same horizontal axis. Then compare the distributions. (Source: Stat-Crunch: Pet Ownership Survey Responses. Owner: chitt71)

Cat	Dog	
15	6	2
5	1	1
1	1	1
1	2	1
5	2	1
3	1	1
1	2	7
2	1	1
3	1	1
6	2	1
1	1	1
1	3	1
1	4	1
4	1	2
0	4	1
1	3	2
3	4	0
0	2	2
2	1	4
	2	2
	1	3
	1	1
	3	3
	1	2
	1	2
	2	1
	7	
	4	

2.33 Law School Tuition Data are shown for the cost of one year of law school at 30 of the top law schools in the United States in 2013. The numbers are in thousands of dollars. Make a histogram of the costs, pose a statistical question, and describe the distribution. If there are any outliers, identify the school(s). (Source: http://grad-schools.usnews.rankingsandreviews.com. Accessed via StatCrunch.)

Yale University	53.6
Harvard University	50.9
Stanford University	50.8
University of Chicago	50.7
Columbia University	55.5
University of Pennsylvania	53.1
New York University	51.1
Duke University	50.8
Cornell University	55.2
Georgetown University	48.8
Vanderbilt University	46.8
Washington University	47.5
University of Boston	44.2
University of Southern California	50.6
George Washington University	47.5
University of Notre Dame	46.0
Boston College	43.2
Washington and Lee University	43.5
Emory University	46.4
Fordham University	49.5
Brigham Young University	21.9
Tulane University	45.2
American University	45.1
Wake Forest University	39.9
College of William and Mary	37.8
Loyola Marymount University	44.2
Baylor University	46.4
University of Miami	42.9
Syracuse University	45.7
Northeastern University	43.0

2.34 Text Messages Recently, 115 users of StatCrunch were asked how many text messages they sent in one day. Make a histogram to display the distribution of the numbers of text messages sent, pose a statistical question, and describe the distribution. Some sample data are shown. The full data set is at this text's website. (Source: StatCrunch Survey: Responses to How Often Do You Text? Owner: Webster West)

Sent Texts	
1	50
1	6
0	5
5	300
5	30

2.35 Beer, Calories Data are available on the number of calories in 12 ounces of beer for 101 different beers. Make a histogram to show the distribution of the numbers of calories, pose a statistical question, and describe the distribution. The first few entries are shown in the accompanying table. (Source: beer100.com, accessed via StatCrunch. Owner: Webster West)

Brand	Brewery	% Alcohol	Calories/12 oz
Anchor Porter	Anchor	5.60	209
Anchor Steam	Anchor	4.90	153
Anheuser Busch Natural Light	Anheuser Busch	4.20	95
Anheuser Busch Natural Ice	Anheuser Busch	5.90	157
Aspen Edge	Adolph Coors	4.10	94
Blatz Beer	Pabst	4.80	153

2.36 Beer, Alcohol Data are available on the percent alcohol in 101 different beers. Make a histogram of the data, pose a statistical question, and describe that distribution. (Source: beer100.com, accessed via StatCrunch. Owner: Webster West)

SECTIONS 2.3 AND 2.4

TRY **2.37 Changing Multiple-Choice Answers When Told *Not to Do So* (Example 7)** One of the authors wanted to determine the effect of changing answers on multiple-choice tests. She studied the tests given by another professor, who had told his students before their exams that if they had doubts about an answer they had written, they would be better off *not changing* their initial answer. The author went through the exams to look for erasures, which indicate that the first choice was changed. In these tests, there is only one correct answer for each question. Do the data support the view that students should not change their initial choice of an answer?

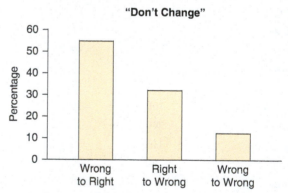

2.38 Preventable Deaths According to the World Health Organization (WHO), there are five leading causes of preventable deaths in the world. They are shown in the following graph.

a. Estimate how many preventable deaths result from high blood pressure.

b. Estimate how many preventable deaths result from tobacco use.

c. Does this graph support the theory that the greatest rate of preventable death comes from overweight and obesity, as some people have claimed?

d. This is a bar chart with a special name because of the decreasing order of the bars. What is that name?

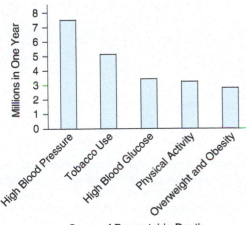

Cause of Preventable Death

2.39 Ice Cream Preference (Example 8) Suppose a group of school-aged children is asked, "Which of these three ice cream flavors do you like the most: vanilla, chocolate, or strawberry?" Describe a bar graph of their responses that would have the least amount of variability. Also describe a bar graph that would have the most variability.

2.40 Community College Applicants Applicants to California community colleges are asked to indicate one of these education goals at the time of application: transfer to a four-year institution, an AA degree, a CTE certificate, job retraining, or personal enrichment. In a group of 500 applications, describe a bar graph of these data that would have the least amount of variability. Also describe a bar graph that would have the most variability.

2.41 Gun Availability Pew Research conducted a survey in 2017 asking gun owners what percentage of time they had a loaded gun easily within reach when at home. The results for male gun owners are displayed in a bar chart and a pie chart.

a. Which period was the most frequent response?

b. Use the graphs to estimate the difference in the percentage responding "All of the Time" and "Never." Is this easier to estimate using the bar chart or the pie chart?

Gun Availability for Men

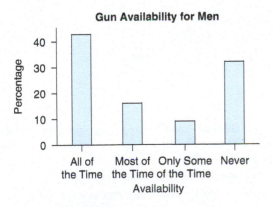

Gun Availability for Men

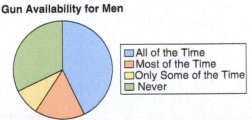

2.42 Entry-Level Education The Bureau of Labor Statistics tracks data on the percentage of jobs that require various levels of education. The 2016 Entry Level Education data are displayed in the following bar chart and pie chart.

a. Approximately what percentage of jobs require some type of college degree (Associate's, Bachelor's, or Graduate)?

b. Use the graphs to estimate the difference between the percentage of jobs that require a high school diploma and those that require no formal education. Is this easier to estimate using the bar chart or the pie chart?

Entry Level Education Requirement

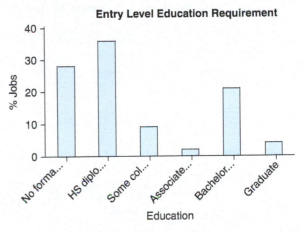

Entry level Education

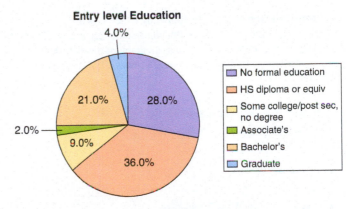

TRY 2.43 Obesity Among Adults (Example 9) Data on obesity rates for adults in the United States are displayed in the bar plot for three age groups. (Source: 2017 World Almanac and Book of Facts)

a. Which age group typically has the highest rate of obesity?

b. Comment on any similarities and differences in the obesity rates of men and women.

Obesity by Age

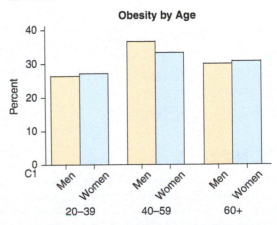

2.44 Fitness Among Adults Data on the percentage of adults living in the United States who meet the federal aerobic fitness (AF) and muscle-strengthening (MS) standards are displayed in the bar plot. (Source: 2017 World Almanac and Book of Facts)

a. Comment on the fitness similarities and differences in these four regions.

b. Comment on the similarities and difference in aerobic and muscle-strengthening fitness among U.S. adults.

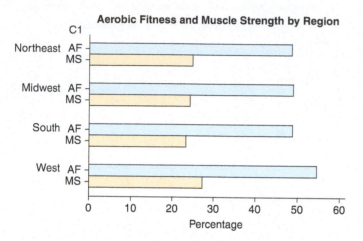

Aerobic Fitness and Muscle Strength by Region

2.45 Internet Browsers The following table gives the percent of the market controlled by the most popular Internet browsers in 2016. Sketch or otherwise create an appropriate graph of the distribution and comment on its important features. (Source: 2017 World Almanac and Book of Facts)

Browser	Percentage of Market
Chrome	58%
Firefox	14%
Explorer	10%
Safari	10%
Others	8%

2.46 Commercial Radio The following table gives the number of commercial U.S. radio stations for the four most popular formats. Sketch or otherwise create an appropriate graph of the distribution and comment on its important features. (Source: 2017 World Almanac and Book of Facts)

Format	Number of Stations
Country	2126
News Talk	1353
Spanish	877
Classic Hits	871

SECTION 2.5

2.47 Garage The accompanying graph shows the distribution of data on whether houses in a large neighborhood have a garage. (A 1 indicates the house has a garage, and a 0 indicates it does not have a garage.) Is this a bar graph or a histogram? How could the graph be improved?

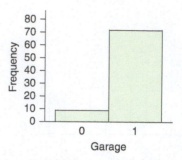

2.48 Body Image A student has gathered data on self-perceived body image, where 1 represents "underweight," 2 represents "about right," and 3 represents "overweight." A graph of these data is shown. What type of graph would be a better choice to display these data, and why?

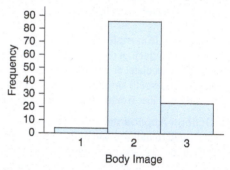

2.49 Pie Chart of Sleep Hours The pie chart reports the number of hours of sleep "last night" for 118 college students. What would be a better type of graph for displaying these data? Explain why this pie chart is hard to interpret.

Number of Hours of Sleep

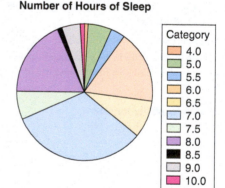

Category
4.0
5.0
5.5
6.0
6.5
7.0
7.5
8.0
8.5
9.0
10.0

2.50 Age and Gender The following graph shows the ages of females (labeled 1) and males (labeled 0) who are majoring in psychology in a four-year college.

a. Is this a histogram or a bar graph? How do you know?

b. What type(s) of graph(s) would be more appropriate?

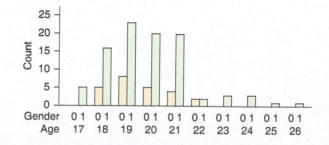

2.51 Musicians Survey: StatCrunch Graph The accompanying graph is a special histogram with additional information; it was made using StatCrunch. People who studied music as children were asked how many hours a day they practiced when they were teenagers, and whether they still play now that they are adults. To understand the graph, look at the third bar (spanning 1.0 to 1.5); it shows that there were seven people (the light red part of the bar) who practiced between 1.0 and 1.5 hours and did not still play as adults, and there were two people (the light blue part of the bar) who practiced 1.0 to 1.5 hours and still play as adults. Comment on what the graph shows. What other types of graphs could be used for this data set?

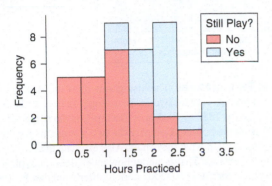

2.52 Cell Phone Use Refer to the accompanying bar chart, which shows the time spent on a typical day talking on the cell phone for some men and women. Each person was asked to choose the one of four intervals that best fit the amount of time he or she spent on the phone (for example, "0 to 4 hours" or "12 or more hours").

* a. Identify the two variables. Then state whether they are categorical or numerical and explain.

 b. Is the graph a bar chart or a histogram? Which would be the better choice for these data?

 c. If you had the actual number of hours for each person, rather than just an interval, what type of graph should you use to display the distribution of the actual numbers of hours?

 d. Compare the modes of the two distributions and interpret what you discover: What does this say about the difference between men's and women's cell phone use?

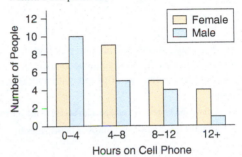

CHAPTER REVIEW EXERCISES

2.53 Sleep The following table shows the first few entries for the number of hours of sleep in one night for a sample of college students. The data are stacked and coded, where 0 represents a male student and 1 represents a female student. What would be appropriate graphs to compare the distributions of hours of sleep in one night for male and female college students if you had the complete data set? Explain. Write a statistical question that can be answered with the complete data.

Female	Sleep
1	6
0	5
0	6
1	5.5
1	7

2.54 Majors The following table shows the first few entries for the majors of a sample of college students. The data are stacked and coded, where 0 represents a female student and 1 represents a male student. What would be appropriate graphs to compare the distribution of majors for male and female students if you had the complete data set? Explain. Write a statistical question that can be answered with the complete data.

Male	Major
0	Psychology
1	Business
0	Business
0	Economics
1	Psychology

2.55 Hormone Replacement Therapy The use of the drug Prempro, a combination of two female hormones that many women take after menopause, is called hormone replacement therapy (HRT). In July 2002, a medical article reported the results of a study that was done to determine the effects of Prempro on many diseases. (Source: Writing Group for the Women's Health Initiative Investigators, "Risks and Benefits of Estrogen Plus Progestin in Healthy Postmenopausal Women," *JAMA* 388 [2002]: 321–33)

The study was placebo-controlled, randomized, and double-blind. From studies like these, it is possible to make statements about cause and effect. The figure shows comparisons of disease rates in the study.

 a. For which diseases was the disease rate higher for those who took HRT? And for which diseases was the rate lower for those who took HRT?

 b. Why do you suppose we compare the rate per 10,000 women (per year), rather than just reporting the numbers of women observed who got the disease?

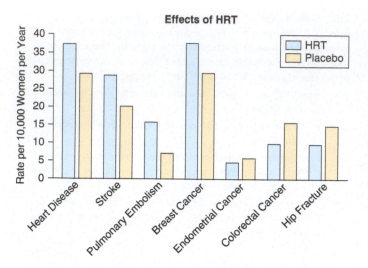

Effects of HRT

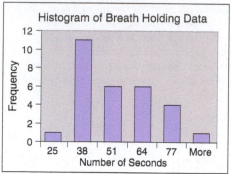

went from a low of 25 seconds to a high of 90 seconds, as you can see in the stemplot. Suggest improvements to the following histogram generated by Excel, assuming that what is wanted is a histogram of the data (not a bar chart).

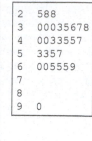

Histogram of Breath Holding Data

2	588
3	00035678
4	0033557
5	3357
6	005559
7	
8	
9	0

2.56 E-Music The bar graph shows information reported in *Time* magazine (May 28, 2012). For each country, the percentage of music sold over the Internet and the percentage of people with access to the Internet are displayed.

a. Which two countries have the largest percentage with access to the Internet?

b. Which two countries have the largest percentage of music sales by Internet?

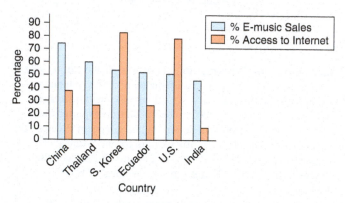

2.57 Hormone Replacement Therapy Again The following bar chart shows a comparison of breast cancer rates for those who took HRT and those who took a placebo. Explain why the graph is deceptive, and indicate what could be done to make it less so.

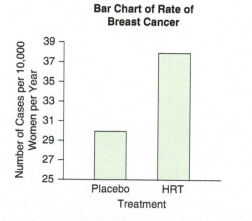

2.58 Holding Your Breath A group of students held their breath as long as possible and recorded the times in seconds. The times

* **2.59 Global Temperatures** The histograms show the average global temperature per year for two 26-year ranges in degrees Fahrenheit. The range for 1880 to 1905 is on the top, and the range for 1980 to 2005 is on the bottom. Compare the two histograms for the two periods, and explain what they show. Also estimate the difference between the centers. That is, about how much does the typical global temperature for the 1980–2005 period differ from that for the 1880–1905 period? (Source: Goddard Institute for Space Studies, NASA, 2009)

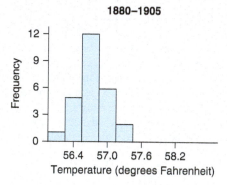

1880–1905

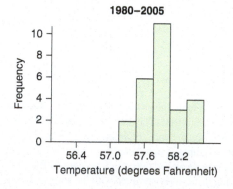

1980–2005

* **2.60 Employment after Law School** Accredited law schools were ranked from 1 for the best (Harvard) down to number 181 by the Internet Legal Research Group. When you decide on a law school to attend, one of the things you might be interested in is whether, after graduation, you will be able to get a job for which your law degree is required. We split the group of 181 law schools in half, with the top-ranked schools in one group and the lower-ranked schools in the other. The histograms show the distribution of the percentages of graduates who, 9 months after graduating,

have obtained jobs that require a law degree. Compare the two histograms.

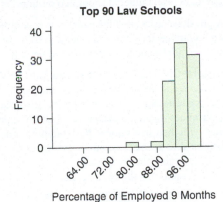

Top 90 Law Schools

Percentage of Employed 9 Months after Graduation

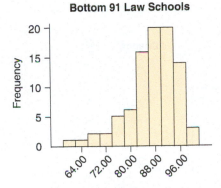

Bottom 91 Law Schools

Percentage of Employed 9 Months after Graduation

2.61 Opinions on Nuclear Energy People were asked whether they favored nuclear energy in 2010 and in 2016. The data were sorted by political affiliation and are shown in the following bar graph.

a. How have opinions on nuclear energy changed from 2010 to 2016? (Source: Gallup.com)

b. Which political party showed the most change in opinion on nuclear energy in this period?

c. How could the bar graph have been organized differently to make it easier to compare opinion changes within political parties?

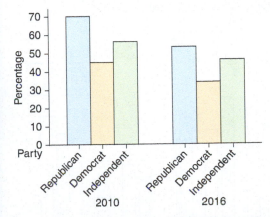

2.62 Stem Cell Research People were asked whether they thought stem cell research was morally acceptable. The graph shows the results of two surveys, one done in 2002 and one done on 2017.

What does the graph tell us about changes in opinions about stem cell research? Explain. (Source: Gallup.com)

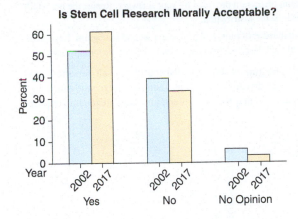

Is Stem Cell Research Morally Acceptable?

2.63 Create a dotplot that has at least 10 observations and is right-skewed.

2.64 Create a dotplot that has at least 10 observations and does not have skew.

2.65 Traffic Cameras College students Jeannette Mujica, Ricardo Ceja Zarate, and Jessica Cerda conducted a survey in Oxnard, California, of the number of cars going through a yellow light at intersections with and without traffic cameras that are used to automatically fine drivers who run red lights. The cameras were very noticeable to drivers. The amount of traffic was constant throughout the study period (the afternoon commute.) The data record the number of cars that crossed the intersection during a yellow light for each light cycle. A small excerpt of the data is shown in the following table; see this text's website for all the data. What differences, if any, do you see between intersections with cameras and those without? Use an appropriate graphical summary, and write a comparison of the distributions.

# Cars	Cam
1	Cam
2	Cam
1	No Cam
3	No Cam

2.66 Ideal Weight Thirty-nine students (26 women and 13 men) reported their ideal weight (in most cases, not their current weight). The table shows the data.

Women					
110	115	123	130	105	110
130	125	120	115	120	120
120	110	120	150	110	130
120	118	120	135	130	135
90	110				
Men					
160	130	220	175	190	190
135	170	165	170	185	155
160					

a. Explain why the distribution of ideal weights is likely to be bimodal if men and women are both included in the sample.

b. Make a histogram combining the ideal weights of men and women. Use the default histogram provided by your software. Report the bin width and describe the distribution.

c. Vary the number of bins, and print out a second histogram. Report the bin width and describe this histogram. Compare the two histograms.

2.67 MPH The graphs show the distribution of self-reported maximum speeds ever driven by male and female college students who drive. Compare the shapes, the centers, and the spreads and mention any outliers.

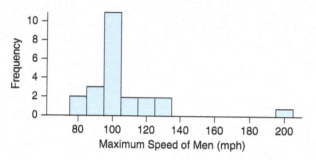

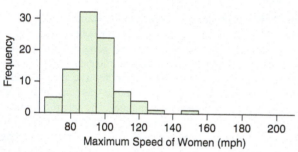

2.68 Shoe Sizes The graph shows shoe sizes for men and women. Compare shapes, centers, and spreads, and mention if there are outliers.

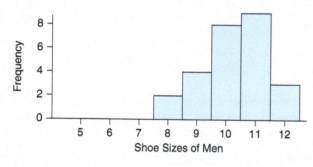

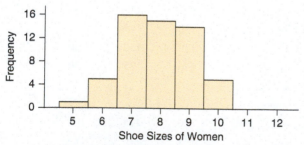

2.69 CEO Salaries Predict the shape of the distribution of the salaries of 25 chief executive officers (CEOs). A typical value is about 50 million per year, but there is an outlier at about 200 million.

2.70 Cigarettes A physician asks all of his patients to report the approximate number of cigarettes they smoke in a day. Predict the shape of the distribution of number of cigarettes smoked per day.

2.71 Changing Multiple-Choice Answers When Told to Do So One of the authors wanted to determine the effect of changing answers on multiple-choice tests. She had advised her students that if they had changed their minds about a previous answer, they should replace their first choice with their new choice. By looking for erasures on the exam, she was able to count the number of changed answers that went from wrong to right, from right to wrong, and from wrong to wrong. The results are shown in the bar chart.

a. Do the data support her view that it is better to replace your initial choice with the revised choice?

b. Compare this bar chart with the one in Exercise 2.37. Does changing answers generally tend to lead to higher or to lower grades?

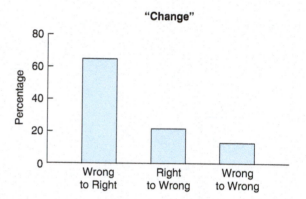

2.72 ER Visits for Injuries The graph shows the rates of visits to the emergency room (ER) for injuries by gender and by age. Note that we are concerned with the rate per 100 people of that age and gender in the population. (Source: National Safety Council 2004)

a. Why does the National Safety Council give us rates instead of numbers of visits?

b. For which ages are the males more likely than the females to have an ER visit for an injury? For which ages are the males and females similar? For which ages do the females have more visits?

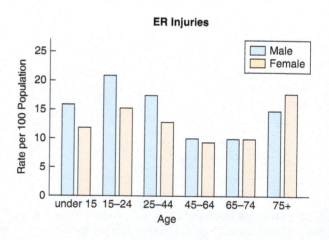

2.73 Social Media Use The Pew Research Center documents the variety of ways in which Americans use social media. The bar chart shows the frequency of use for three social media sites among adult social media users.

a. Which site is used most regularly by social media users? Explain.

b. Which site is used least regularly by social media users? Explain.

c. Which site shows the least amount of variation in use? Explain.

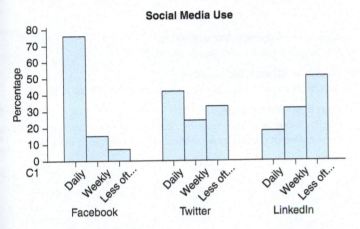

2.74 Social Media Use The Pew Research Center gathered data in 2018 on American social media use by gender. The bar graph shows the percentage of Internet users who use these social media platforms.

a. Of these three social media sites, which is most frequently used by men? By women?

b. Comment on the similarities and differences in frequency of use of these three social media sites between men and women.

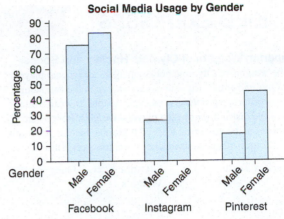

2.75 Choosing a Graph For each situation, describe the type of statistical graph that could be made to answer the statistical question posted.

a. We have the systolic blood pressures by gender of a large number of elderly people.

Our statistical question is, "What percentage of these people have blood pressure over 130 mmHG?"

b. We have demographic data on everyone who bought their ticket to a local football game online.

Our statistical question is, "Which zip code had the lowest participation from women?"

2.76 Choosing a Graph For each situation, describe the type of statistical graph that could be made to answer the statistical question posted:

a. We have the systolic blood pressures by gender of a large number of elderly people.

Our statistical question is, "Do these elderly men tend to have higher systolic blood pressure than these elderly women?"

b. We have demographic data on everyone who bought their ticket to a local football game online.

Our statistical question is, "Which zip code bought the greatest number of tickets?"

gUIDED EXERCISES

g 2.25 Comparing Weights of Olympic Hockey and Soccer Players The weights of Olympic hockey players and soccer players from recent Olympic games (in pounds) are shown in the following tables. Make appropriate graphs and use your graphs to write a brief comparison of the distributions of the players' weights. Pose a statistical question that can be answered with your graphs. Include appropriate graphics. (Source: teamusa.usahockey.com, ussoccer.com)

Olympic Hockey Team Weights	
212	180
215	181
196	202
200	183
213	189
219	219
207	190
198	190
221	205
212	196
	200
	205

Olympic Soccer Team Weights	
150	185
175	190
145	190
200	160
155	170
140	180
185	145
155	175
165	150
185	175
165	185
195	

Step 1 ▶ Create graphs.
Since the data are numerical, dotplots or histograms using the same axis could be made to compare the data. Often it is helpful to "stack" the graphs using the same horizontal axis with one graph above the other as shown in the figure.

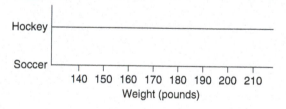

Step 2 ▶ Examine the shape.
What is the shape of the data for the Olympic hockey team? What is the shape of the data for the Olympic soccer team?

Step 3 ▶ Examine the center.
Which group seems to weigh more: the hockey players or the soccer players?

Step 4 ▶ Examine the variation.
Which group has a wider spread of data?

Step 5 ▶ Check for outliers.
Were there any usually heavy or unusually light athletes in the data?

Step 6 ▶ Summarize.
Finally, in one or more sentences, compare the shape, the center, and the variation in the two groups. Mention any outliers, if there were any.

TechTips

General Instructions for All Technology

EXAMPLE ▶ Use the following ages to make a histogram:

7, 11, 10, 10, 16, 13, 19, 22, 42

Columns

Data sets are generally put into columns, not rows. The columns may be called variables, lists, or something similar. Figure 2a shows a column of data.

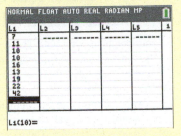

▲ **FIGURE 2a** TI-84 Data Screen

TI-84

Resetting the Calculator (Clearing the Memory)

If you turn off the TI-84, it does not reset the calculator. All the previous data and choices remain. If you are having trouble and want to start over, you can reset the calculator.

1. **2nd Mem** (with the + sign)
2. **7** for **Reset**
3. **1** for **All RAM**
4. **2** for **Reset**

It will say **RAM cleared** if it has been done successfully.

Entering Data into the Lists

1. Press **STAT** and select **EDIT** (by pressing **ENTER** when **EDIT** is highlighted).
2. If you find that there are data already in **L1** (or any list you want to use), you have three options:
   a. Clear the entire list by using the arrow keys to highlight the **L1** label and then pressing **CLEAR** and then **ENTER**. Do not press **DELETE**, because then you will no longer have an **L1**. If you delete **L1**, to get it back you can **Reset** the calculator.
   b. Delete the individual entries by highlighting the top data entry, then pressing **DELETE** several times until all the data are erased. (The numbers will scroll up.)
   c. Overwrite the existing data. CAUTION: Be sure to **DELETE** data in any cells not overwritten.
3. Type the numbers from the example into **L1** (List1). After typing each number, you may press **ENTER** or use ▼ (the arrow down on the keypad). Double-check your entries before proceeding.

Histogram

1. Press **2nd**, **STATPLOT** (which is in the upper-left corner of the keypad).
2. If more than the first plot is **On**, press **4** (for **PlotsOff**) and **ENTER** to turn them all off.
3. Press **2nd**, **STATPLOT**, and **1** (for **Plot 1**).
4. Turn on **Plot1** by pressing **ENTER** when **On** is flashing; see Figure 2b.

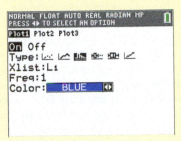

▲ **FIGURE 2b** TI-84

5. Use the arrows on the keypad to get to the histogram icon (highlighted in Figure 2b) and press **ENTER** to choose it. Use the down arrow to get to Xlist and press **2nd** and **1** for **L1**. The settings shown in Figure 2b will lead to a histogram of the data in List 1 (L1).
6. Press **ZOOM** and **9** (**ZoomStat**) to create the graph.
7. To see the numbers shown in Figure 2c, press **TRACE** (in the top row on the keypad) and move around using the group of four arrows on the keypad to see other numbers.

Figure 2c shows a histogram of the numbers in the example.

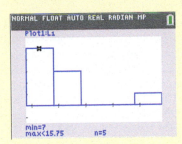

▲ **FIGURE 2c** TI-84 Histogram

The TI-84 cannot make stemplots, dotplots, or bar charts.

Downloading Numerical Data from a Computer into a TI-84

Before you can use your computer with your TI-84, you must install (on your computer) the software needed for data transfer and a driver for the calculator. You need to do this only once. If you have done so and are ready to download data, go to step 3.

Texas Instruments software for computer-to-calculator communication are available for download at the web site https://education. ti.com. There are two versions to consider. The older, TI Connect™ software works with all the TI-83 and TI-84 series calculators, except for the new TI-84 plus CE. The newer, TI Connect™ CE software works with all the TI-83 and TI-84 series calculators that have USB ports.

Both versions are free, and both can be installed on the same computer.

1. Downloading and running the setup program(s) for TI Connect™ and/or TI Connect™ CE. At the https://education. ti.com website, select **Downloads**, then **All Software, OS and Apps.** For **Technology,** select your calculator model. For **View,** select **Connectivity Software**. Click Find. Select and download the setup file for the software appropriate for your calculator and computer. Also download and save the PDF Guidebook(s) for future reference. On your computer, navigate to your Downloads folder and run the setup program(s). Follow the on-screen instructions.

2. Installing the calculator driver on the computer.

    When you connect your calculator to your computer (using the USB cable) for the first time, the necessary driver will be automatically installed.

This completes the computer installation.

3. Connect the calculator to the computer and start either TI Connect™ or TI Connect™ CE. On a Windows 10 computer, the new, start shortcut will show up under **Recently Added Apps** in the **Start Menu**.

You are now ready to input the data that is to be downloaded to the TI-84 calculator.

The procedure is different for the two apps. For the TI Connect™ CE, do step 4. For the TI Connect™, go to steps 5 through 9.

4. Downloading data using the newer, TI Connect™ CE. Refer to Figure 2d. **Actions > Import Data (.csv) to**. Navigate to the folder containing the data file, select (single click) the file, and click **Open**. Click **CONTINUE**. On the next screen, choose L1 for **NAME ON CALCULATOR**, select your calculator, and click SEND. On your calculator, **STAT > EDIT** to verify the download.

    Caution: Since the TI-84 calculators can only handle numerical data, the .csv (comma separated value) file must have only numerical data, and no column labels. Otherwise you may get an error statement, or worse, data entries of zero.

▲ **FIGURE 2d** TI Data Editor TI Connect™ CE Home Screen

5. Inputting data using the older, TI Connect™.

    Click on **TI Data Editor**. You have two options here: step 6a. Import file, or step 6b. Copy and Paste

6a. Importing a file.
    The TI Connect™ app will import data from .csv (comma separated value) and .txt files. Refer to Figure 2e. Choose **File > Import**. Navigate to the folder containing the data file, single click on the file and click **Open**. The data will now appear in the TI Connect™ screen as a list. See the Caution note in step 4.

6b. Copy and Paste.

    On your computer, open the file containing the data (using Excel, Notepad, etc.). **Copy** the column of *numerical data* to your clipboard. See the Caution note in step 4.

    Refer to Figure 2e. In the Data Editor screen, click on the icon of a blank sheet of paper, which is located at the far left of the Tool Bar. This will insert a new blank column for the data in the Data Editor screen. (For future reference, clicking on the icon again will insert another blank column.)

    Click on the first cell in the newly inserted column. It will turn blue indicating that it is active. Now **Paste** (the column of numbers) from the clipboard.

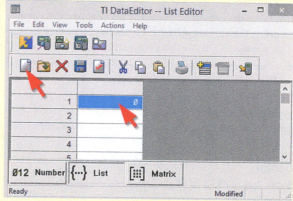

▲ **FIGURE 2e** TI Data Editor

7. Choose a name for the inputted data.

   In the Data Editor screen, while the data column is active, choose **File** and **Properties**. On the next screen, Variable Properties, choose a **Variable Name**, such as L_1.

8. Connect the TI-84 to the computer and turn it on.

9. Downloading the data to the TI-84 using TI Connect™.

   In the Data Editor screen, Figure 2e, click **Actions**, then **Send All Lists**. When you get the Warning, click **Replace** to overwrite the old data. On your calculator **STAT > EDIT** to verify the download.

MINITAB

Entering the Data

When you open Minitab, you will see a blank spreadsheet for entering data. Type the data from the example into **C1,** column 1. Be sure your first number is put into row 1, not above row 1 in the label region. Be sure to enter only numbers. (If you want a label for the column, type it in the label region *above* the numbers.) Alternatively you can paste in data from the computer clipboard. Or you can choose **Open > File** to open a .xls, .xlsx, .txt, .dat, or .csv formatted data file. Double-check your entries before proceeding.

All Minitab Graphs

After making the graph, double click on what you want to change, such as labels.

Histogram

1. Click **Graph > Histogram**
2. Leave the default option **Simple** and click **OK**.
3. Double click **C1** (or the name for the column) and click **OK**. (Another way to get **C1** in the big box is to click **C1** and click **Select**.)
4. After obtaining the histogram, if you want different bins (intervals), double click on the *x*-axis and look for **Binning**.

Figure 2f shows a Minitab histogram of the ages.

Stemplot
Click **Graph > Stem-and-Leaf**

Dotplot
Click **Graph > Dotplot**

Bar Chart
Click **Graph > Bar Chart**

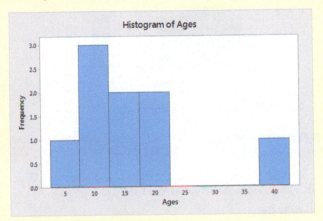

▲ **FIGURE 2f** Minitab Histogram

EXCEL

When you open Excel and choose (click on) **Blank workbook,** you will see a blank spreadsheet for entering data. Before entering data for the first time, click the **Data** tab (top of screen, middle); you should see **Data Analysis** just below and on the far right. If you do not see it, you will need to load the Data Analysis Toolpak (instructions follow). Now click the **Add-Ins** tab; you should see the XLSTAT icon ≥ just below and on the far left. If you do not see it, you should install XLSTAT (instructions follow). You will need both of these add-ins in order to perform all of the statistical operations described in this text.

Data Analysis Toolpak

1. Click **File > Options > Add-Ins**.
2. In **Manage Box,** select **Excel Add-Ins** and click **Go**.
3. In **Add-ins available** box, check **Analysis Toolpak** and click **OK**.

XLSTAT

1. Close Excel.
2. Download XLSTAT from www.myPearsonstore.com.
3. Install XLSTAT.
4. Open Excel, click **Add-Ins tab**, and click the XLSTAT icon ≥.

You only need to install the Data Analysis Toolpak once. The Data Analysis tab should be available now every time Excel is opened. For XLSTAT, however, step 4 may need to run each time Excel is opened (if you expect to run XLSTAT routines).

Entering Data

See Figure 2g. Enter the data from the example into column A with **Ages** in cell A1. Double-check your entries before proceeding.

▲ **FIGURE 2g** Excel Data Screen

Histogram

1. Click **XLSTAT, > Visualizing data > Histograms**.
2. When the box under **Data** is activated, drag your cursor over the column containing the data including the label **Ages**.
3. Select **Continuous, Sheet,** and Check **Sample labels.** Click **OK, Continue,** and **OK**.

Figure 2h shows the histogram.

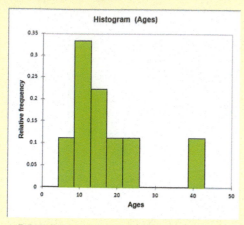

▲ **FIGURE 2h** XLSTAT Histogram

Dotplot and Stemplot

1. Click **XLSTAT, > Describing data > Descriptive statistics**.
2. When the box under **Quantitative Data** is activated, drag your cursor over the column containing the data including the label **Ages**. Select **Sheet** and Check **Sample labels.**

Dotplot

3. Click **Options, Charts, Charts(1)** and select **Scatter-grams, Options,** and **Horizontal**.
4. Click **OK** and **Continue**.

Stemplot

3. Click **Options, Charts, Charts(1)** and select **Stem-and-leaf plots**.
4. Click **OK** and **Continue**.

Bar Chart

After typing a summary of your data in table form (including labels), drag your cursor over the table to select it, click **Insert, Column** (in the **Charts** group), and select either the first option under **2-D Column** or the first option under **2-D Bar**.

STATCRUNCH

For Help: After logging in, click on **Help** or **Resources** and **Watch StatCrunch Video Tutorials on YouTube**.

Entering Data

1. Click **Open StatCrunch** and you will see a spreadsheet as shown in Figure 2i.
2. Enter the data from the example into the column labeled var1.
3. If you want labels on the columns, click on the variable label, such as **var1**, and backspace to remove the old label and type the new label. Double-check your entries before proceeding.

Pasting Data

1. If you want to paste data from your computer clipboard, click in an empty cell in row 1 and press **Ctrl+V** on your keyboard.
2. If your clipboard data include labels, click on the **var1** cell instead, and then press **Ctrl+V** on your keyboard.

Loading Data

1. To load from an existing data file, **Data > Load > From file > on my computer > Browse**. Then navigate to the folder that contains the file and select the file.

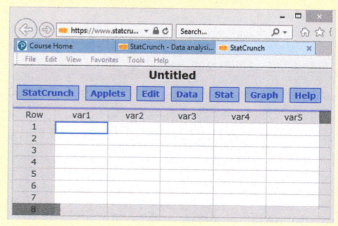

▲ **FIGURE 2i** StatCrunch Data Table

Histogram

1. Click **Graph > Histogram**
2. Under **Select columns**, click the variable you want a histogram for.
3. Click **Compute!**
4. To edit the graph, click on the hamburger icon in the lower-left corner.
5. To copy the graph for pasting into a document for submission, click **Options** and **Copy** and then paste it into a document.

Figure 2j shows the StatCrunch histogram of the ages.

Stemplot
Click **Graph > Stem and Leaf**

Dotplot
Click **Graph > Dotplot**

Bar Chart
Click **Graph > Bar Plot > with data**

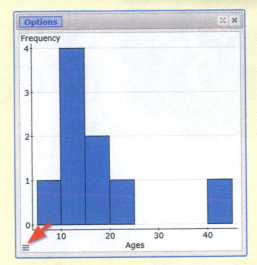

▲ **FIGURE 2j** StatCrunch Histogram

3 Numerical Summaries of Center and Variation

THEME

The complexity of numerical distributions can often be neatly summarized. In many cases, two numbers—one to measure the typical value and one to measure the variability—are all we need to summarize a set of data and make comparisons between groups. We can use many different ways of measuring what is typical and of measuring variability; which method is best to use depends on the shape of the distribution.

oogle the phrase *average American* and you'll find lots of entertaining facts. For example, the average American lives within 3 miles of a McDonald's, showers 10.4 minutes a day, and prefers smooth peanut butter to chunky. The average American woman is 5'4" tall and weighs 140 pounds, while the average American female fashion model is considerably taller and weighs much less: 5'11" and 117 pounds.

Whether or not these descriptions are correct, they are attempting to describe a typical American. The reason for doing this is to try to understand a little better what Americans are like, or perhaps to compare one group (American women) to another (female fashion models).

These summaries can seem odd because we all know that people are too complex to be summarized with a single number. Characteristics such as weight, distance from a McDonald's, and length of a shower vary quite a bit from person to person. If we're describing the "typical" American, shouldn't we also describe how much variation exists among Americans? If we're making comparisons between groups, as we do in this chapter's Case Study, how do we do it in a way that takes into account the fact that individuals may vary considerably?

In Chapter 2, we talked about looking at graphs of distributions of data to get an intuitive, informal sense of the typical value (the center) and the amount of variation (the spread). In this chapter, we explore ways of making these intuitive concepts more precise by assigning numbers to them. We will see how this step makes it much easier to compare and interpret sets of data, for both symmetric and skewed distributions. These measures are important tools that we will use throughout the text.

CASE STUDY

Living in a Risky World

Our perception of how risky an activity is plays a role in our decision making. If you think that flying is very risky, for example, you will be more willing to put up with a long drive to get where you want to go. A team of University of California, Los Angeles psychologists were interested in understanding how people perceive risk and whether a simple reporting technique was enough to detect differences in perceptions between groups. The researchers asked over 500 subjects to consider various activities and rate them in terms of how risky they thought the activities were. The ratings were on a scale of 0 (no risk) to 100 (greatest possible risk). For example, subjects were asked to assign a value to the following activities: "Use a household appliance" and "Receive a diagnostic X-ray every 6 months." One question of interest to the researchers was whether men and women would assign different risk levels to these activities (Carlstrom et al. 2000).

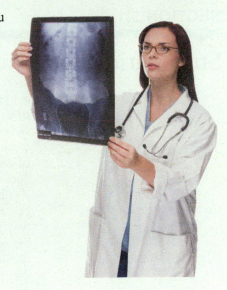

Figure 3.1a on the next page shows a histogram for the perceived risk of using household appliances, and Figure 3.1b shows a histogram for the risk of twice-annual X-rays. (Women are represented in the left panels, men in the right.) What differences do you see between the genders? How would you quantify these differences? In this chapter, you'll learn several techniques for answering these questions. And at the end of the chapter, we'll use these techniques to compare perceived risk between men and women.

▶ **FIGURE 3.1** Histograms showing the distributions of perceived risk by gender, for two activities. **(a)** Perceived risk of using everyday household appliances. **(b)** Perceived risk of receiving a diagnostic X-ray twice a year. The left panel in each part is for women, and the right panel is for men.

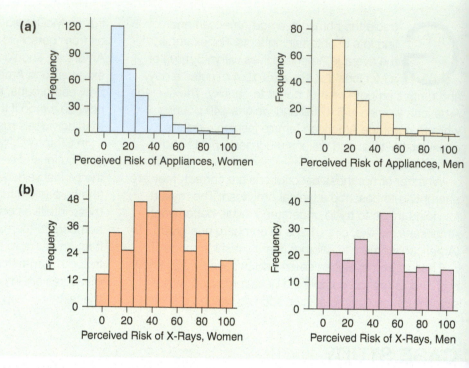

SECTION 3.1

Summaries for Symmetric Distributions

As you learned in Chapter 2, statistical questions are often about the center and spread of a distribution. The somewhat informal and subjective notions of center and spread developed in the last chapter are here made more precise. In this chapter, we'll see how to measure the center and spread with numbers. As it turns out, there are different ways of measuring these characteristics, and which way you choose requires that you first consider visualizations of the data.

In this chapter, the two different ways in which we will think about the concept of center are (1) center as the balancing point (or center of mass) and (2) center as the halfway point. In this section we introduce the idea of the center as a balancing point, useful for symmetric distributions. Then in Section 3.3, we introduce the idea of the center as the halfway point, useful for skewed distributions. Each of these approaches results in a different measure, and our choice also affects the method we use to measure variability.

The Center as Balancing Point: The Mean

The most commonly used measure of center is the mean. The **mean** of a collection of data is the arithmetic average. The mean can be thought of as the balancing point of a distribution of data, and when the distribution is symmetric, the mean closely matches our concept of the "typical value."

Visualizing the Mean   Different groups of statistics students often have different backgrounds and levels of experience. Some instructors collect student data to help them understand whether one class of students might be very different from another. For example, if one class is offered earlier in the morning, or later in the evening, it might attract a different composition of students. An instructor at Peoria Junior College in Illinois collected data from two classes, including the students' ACT scores. (ACT is a standardized national college entrance exam.) Figure 3.2 shows the distribution of self-reported ACT scores for one statistics class.

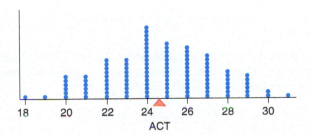

◀ FIGURE 3.2 The distribution of ACT scores for one class of statistics students. The mean is indicated here with a fulcrum (triangle). The mean is at the point on the dotplot that would balance if the points were placed on a seesaw. (Source: StatCrunch, Statistical_Data_499)

The balancing point for this distribution is roughly in the middle, because the distribution is fairly symmetric. The mean ACT for these students is calculated to be 24.6. You'll see how to do this calculation shortly. At the moment, rather than worrying about how to get the precise number, note that 24.6 is about the point where the distribution would balance if it were on a seesaw.

When the distribution of the data is more or less symmetric, the balancing point is roughly in the center, as in Figure 3.2. However, when the distribution is not symmetric, as in Figure 3.3, the balancing point is off-center, and the average may not match what our intuition tells us is the center.

> **↻ Looking Back**
>
> **Symmetric Distributions**
> Recall that symmetric distributions are those in which the left-hand side of the graph of the distribution is roughly a mirror image of the right-hand side.

◀ FIGURE 3.3 Winnings in the 2018 season of the top 100 ranked tennis players. (Source: https://www.usatoday.com/sports/tennis/players/)

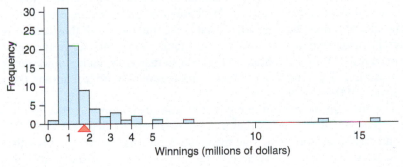

Figure 3.3 shows a very skewed distribution: the winnings of the top-ranked professional tennis players in the 2018 season. (The top winner was Rafael Nadal at just under 16 million dollars. The lowest was Guillermo Garcia-Lopez, at about 0.5 million dollars. We should note that only 78 out of the 100 players had reported winnings.) Where would you place the center of this distribution? Because the distribution is skewed right, the balancing point is fairly high; the mean is at 1.8 million dollars. However, when you consider that about 75% of the players made less than this amount, you might not think that 1.8 million dollars is what the "typical" player made. In other words, in this case the mean might not represent our idea of atypical winning.

EXAMPLE 1 Math Scores

Figure 3.4 shows the distribution of math achievement scores for 46 countries as determined by the National Assessment of Education Progress. The scores are meant to measure the accomplishment of each country's eighth-grade students with respect to mathematical achievement.

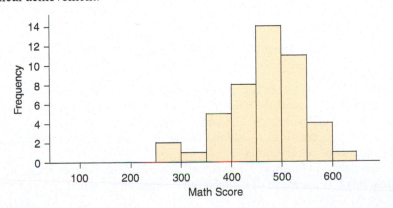

◀ FIGURE 3.4 Distribution of international math scores for 46 countries. The scores measure the mean level of mathematical achievement for a country's eighth-grade students. (Source: "International Math Scores" 2007)

QUESTION   Based on the histogram, approximately what value do you think is the mean math achievement score?

SOLUTION   The mean score is at the point where the histogram would balance if it were placed on a seesaw. The distribution looks roughly symmetric, and the balancing point looks to be somewhere between 400 and 500 points.

CONCLUSION   The mean international math achievement score looks to be about 450 points. By the way, the U.S. score was 504, and the highest three scores were from Singapore (highest at 605), South Korea, and Hong Kong. This tells us that the typical score, among the 46 countries, on the international math achievement test was between 400 and 500 points.

TRY THIS!   Exercise 3.3

The Mean in Context

As mentioned, the mean tells us one answer to the question of what is a typical value for a variable. We know that different countries had different math achievement scores, but typically, what is the math score of these countries? When you report the mean of a sample of data, you should not simply report the number but, rather, should report it in the context of the data so that your reader understands what you are measuring: The typical score on the international math test among these countries is between 400 and 500 points.

Knowing the typical value of one group enables us to compare this group to another group. For example, that same instructor at Peoria Junior College recorded ACT scores for a second classroom and found that the mean there was also 24.6. This tells us that these two classrooms were comparable, at least in terms of their mean ACT score. However, one classroom had a slightly younger mean age: 26.9 years compared to 27.2 years.

> **KEY POINT**   The mean of a collection of data is located at the "balancing point" of a distribution of data. The mean is one representation of the "typical" value of a variable.

Calculating the Mean for Small Data Sets

The mean of a sample is used so often in statistics that it has its own symbol: $\bar{x}$, which is pronounced "x-bar." To calculate the mean, find the (arithmetic) **average** of the numbers; that is, simply add up all the numbers and divide that sum by the number of observations. Formula 3.1 shows you how to calculate the mean, or average.

$$\textbf{Formula 3.1:}\quad \text{Mean} = \bar{x} = \frac{\sum x}{n}$$

The symbol $\sum$ is the Greek capital sigma, which stands for *summation*. The x that comes after $\sum$ represents the value of a single observation. Therefore, $\sum x$ means that you should add all the values. The letter n represents the number of observations. Therefore, this equation tells us to add the values of all the observations and divide that sum by the number of observations.

The mean shown in Formula 3.1 is sometimes called the **sample mean** in order to make it clear that it is the mean of a collection (or sample) of data.

EXAMPLE **2** Gas Buddy

According to GasBuddy.com (a website that invites people to submit prices at local gas stations), the prices of 1 gallon of regular gas at 12 service stations near the downtown area of Austin, TX, were as follows one winter day in 2018:

$2.19, $2.19, $2.39, $2.19, $2.24, $2.39, $2.27, $2.29, $2.17, $2.29, $2.30, $2.29

A dotplot (not shown) indicates that the distribution is fairly symmetric.

QUESTION   Find the mean price of a gallon of regular gas at these service stations. Explain what the value of the mean signifies in this context (in other words, interpret the mean).

SOLUTION   Add the 12 numbers together to get $27.20. We have 12 observations, so we divide $27.20 by 12 to get $2.266, which we round to $2.27.

$$\bar{x} = \frac{2.19 + 2.19 + 2.39 + 2.19 + 2.24 + 2.39 + 2.27 + 2.29 + 2.17 + 2.29 + 2.30 + 2.29}{12}$$

$$= \frac{27.20}{12} = 2.27$$

CONCLUSION   The typical price of a gallon of gas at these gas stations in Austin, Texas, was $2.27 on this particular day.

TRY THIS!   Exercise 3.7a

Calculating the Mean for Larger Data Sets   Formula 3.1 tells you how to compute the mean if you have a small set of numbers that you can easily type into a calculator. But for large data sets, you are better off using a computer or a statistical calculator. That is true for most of the calculations in this text, in fact. For this reason, we will often just display what you would see on your computer or calculator and describe in the TechTips section the exact steps used to get the solution.

For example, the histogram in Figure 3.5 shows the distribution of the amount of particulate matter, or smog, in the air in 371 cities in the United States in 2017, as reported by the Environmental Protection Agency (EPA). (The units are micrograms of particles per cubic meter.) When inhaled, these particles can affect the heart and lungs, so you would prefer your city to have a fairly low amount of particulate matter. (The EPA says that levels over 15 micrograms per cubic meter are unsafe.) Looking at the histogram, you can estimate the mean value using the fact that, since the distribution is somewhat symmetric, the average will be about in the middle (about 8 micrograms per cubic meter). If you were given the list of 371 values, you could find the mean using Formula 3.1. But you'll find it easier to use the preprogrammed routines of your calculator or software. Example 3 demonstrates how to do this.

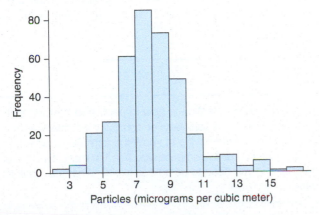

◄ **FIGURE 3.5** Levels of particulate matter for 371 U.S. cities in 2017. Because the distribution is fairly symmetric, the balancing point is roughly in the middle at about 8 micrograms per cubic meter.

EXAMPLE 3 Mean Smog Levels

We used four different statistical software programs to find the mean particulate level for the 371 cities. The data were uploaded into StatCrunch, Minitab, Excel, and the TI-84 calculator.

QUESTION   For each of the computer outputs shown in Figure 3.6, find the mean particulate matter. Interpret the mean.

► **FIGURE 3.6(a)** StatCrunch output.

Column	n	Mean	Variance	Std. dev.	Std. err.	Median	Range	Min	Max	Q1	Q3
pm25_daily	371	7.9471698	4.7615798	2.1821044	0.11328922	7.8	13.7	2.3	16	6.6	9.1

Summary statistics:

► **FIGURE 3.6(b)** Minitab output.

Descriptive Statistics: PM2pt5DailyMean

Statistics

Variable	N	N*	Mean	SE Mean	StDev	Minimum	Q1	Median	Q3	Maximum	IQR
PM2pt5DailyMean	371	141	7.9472	0.1133	2.1821	2.3000	6.6000	7.8000	9.1000	16.0000	2.5000

► **FIGURE 3.6(c)** TI-84 calculator output.

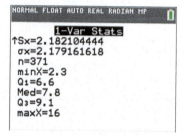

► **FIGURE 3.6(d)** Excel output.

PM2pt5DailyMean	
Mean	7.94717
Standard Error	0.11329
Median	7.8
Mode	7.2
Standard Deviation	2.1821
Sample Variance	4.76158
Kurtosis	1.65638
Skewness	0.78283
Range	13.7
Minimum	2.3
Maximum	16
Sum	2948.4
Count	371

SOLUTION   We have already seen, from Figure 3.5, that the distribution is close to symmetric, so the mean is a useful measure of center. With a large data set like this, it makes sense to use a computer to find the mean. Minitab and many other statistical software packages produce a whole slew of statistics with a single command, and your job is to choose the correct value.

CONCLUSION   The software outputs give us a mean of 7.9 micrograms per cubic meter. We interpret this to mean that the typical level of particulate matter for these cities is 7.9 micrograms per cubic meter. For StatCrunch, Minitab, and Excel the mean is not hard to find; it is labeled clearly. For the TI output, the mean is labeled as $\bar{x}$, "*x*-bar."

TRY THIS!   Exercise 3.11

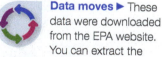

Data moves ► These data were downloaded from the EPA website. You can extract the variable analyzed here from the file airquality.csv, which contains many different measures of air quality.

Data ► More data are included in the file airquality.csv

KEY POINT   In this text, pay more attention to how to apply and interpret statistics than to individual formulas. Your calculator or computer will nearly always find the correct values without your having to know the formula. However, you need to tell the calculator *what* to compute, you need to make sure the computation is meaningful, and you need to be able to explain what the result tells you about the data.

SNAPSHOT ▸ The Mean of a Sample

WHAT IS IT? ▸	A numerical summary.
WHAT DOES IT DO? ▸	Measures the center of the distribution of a sample of data.
HOW DOES IT DO IT? ▸	The mean identifies the "balancing point" of the distribution, which is the arithmetic average of the values.
HOW IS IT USED? ▸	The mean represents the typical value in a set of data when the distribution is roughly symmetric.

Measuring Variation with the Standard Deviation

The mean amount of particulate matter in the air, 7.9 micrograms per cubic meter, does not tell the whole story. Even though the mean of the 371 cities is at a safe level (less than 15 micrograms per cubic meter), this does not imply that any particular city—yours for example—is at a healthful level. Are most cities close to the mean level of 7.9? Or do cities tend to have levels of particulate matter far from 7.9? Values in a data set vary, and this variation is measured informally by the horizontal spread of the distribution of data. A measure of variability, coupled with a measure of center, helps us understand whether most observations are close to the typical value or far from it.

Visualizing the Standard Deviation The histograms in Figure 3.7 show daily high temperatures in degrees Fahrenheit recorded over one recent year at two locations in the United States. The histogram shown in Figure 3.7a records data collected in Provo, Utah, a city far from the ocean and at an elevation of 4500 feet.

The histogram shown in Figure 3.7b records data collected in San Francisco, California, which sits on the Pacific coast and is famously chilly in the summer. (Mark Twain is alleged to have said, "The coldest winter I ever spent was a summer in San Francisco.")

The distributions of temperatures in these cities are similar in several ways. Both temperature distributions are fairly symmetric. You can see from the histograms that both cities have about the same mean temperature. For San Francisco the mean daily high temperature was about 65 degrees; for Provo it was about 67 degrees. But note the difference in the spread of temperatures!

One effect of the ocean on coastal climate is to moderate the temperature: the highs are not that high, and the lows not too low. This suggests that a coastal city should have less spread in the distribution of temperatures. How can we measure this spread, which we can see informally from the histograms?

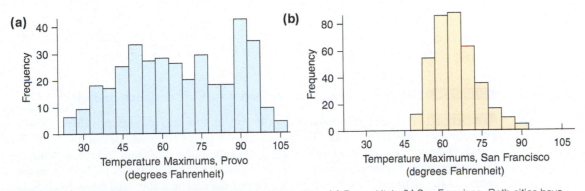

▲ **FIGURE 3.7** Distributions of daily high temperatures in two cities: **(a)** Provo, Utah; **(b)** San Francisco. Both cities have about the same mean temperature, although the variation in temperatures is much greater in Provo than in San Francisco. (Sources: "Provo Temperatures" 2007; "San Francisco Temperatures" 2007)

Note that in San Francisco, most days were fairly close to the mean temperature of 65 degrees; rarely was it more than 10 degrees warmer or cooler than 65. (In other words, rarely was it colder than 55 or warmer than 75 degrees.) Provo, on the other hand, had quite a few days that were more than 10 degrees warmer or cooler than average.

The **standard deviation** is a number that measures how far away the typical observation is from the mean. Distributions such as San Francisco's temperatures have smaller standard deviations because more observations are fairly close to the mean. Distributions such as Provo's have larger standard deviations because more observations are farther from the mean.

As you'll soon see, for most distributions, a majority of observations are within one standard deviation of the mean value.

> **KEY POINT**   The standard deviation should be thought of as the typical distance of the observations from their mean.

EXAMPLE **4**   Comparing Standard Deviations from Histograms

Each of the three graphs in Figure 3.8 shows a histogram for a distribution of the same number of observations, and all the distributions have a mean value of about 3.5.

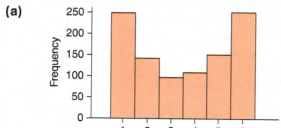
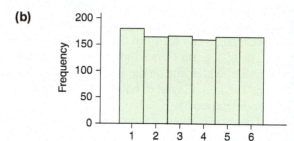
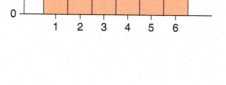
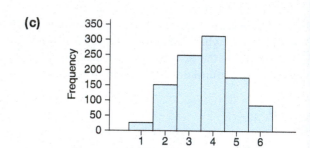

▶ **FIGURE 3.8** All three of these histograms have the same mean, but each has a different standard deviation.

QUESTION   Which distribution has the largest standard deviation, and why?

SOLUTION   All three groups have the same minimum and maximum values. However, the distribution shown in Figure 3.8a has the largest standard deviation. Why? The standard deviation measures how widely spread the points are from the mean. Note that the histogram in Figure 3.8a has the greatest number of observations farthest from the mean (at the values of 1 and 6). Figure 3.8c has the smallest standard deviation because so many of the data are near the center, close to the mean, which we can see because the taller bars in the center show us that there are more observations there.

CONCLUSION   Figure 3.8a has the largest standard deviation, and Figure 3.8c has the smallest standard deviation.

TRY THIS!   Exercise 3.15

The Standard Deviation in Context   The standard deviation is somewhat more abstract, and harder to understand, than the mean. Fortunately, in a symmetric, unimodal distribution, a handy rule of thumb helps make this measure of spread more comprehensible. In these distributions, the majority of the observations (in fact, about two-thirds of them) are less than one standard deviation from the mean.

For temperatures in San Francisco (see Figure 3.7), the standard deviation is about 8 degrees, and the mean is 65 degrees. This tells us that in San Francisco, on a majority of days, the high temperature is within 8 degrees of the mean temperature of 65 degrees—that is, usually it is no colder than $65 - 8 = 57$ degrees and no warmer than $65 + 8 = 73$ degrees. In Provo, the standard deviation is substantially greater: 21 degrees. On a typical day in Provo, the high temperature is within 21 degrees of the mean. Provo has quite a bit more variability in temperature.

EXAMPLE 5   Standard Deviation of Smog Levels

The mean particulate matter in the 371 cities graphed in Figure 3.5 is 7.9 micrograms per cubic meter, and the standard deviation is 2.2 micrograms per cubic meter.

QUESTION   Find the level of particulate matter one standard deviation above the mean and one standard deviation below the mean. Keeping in mind that the EPA says that levels over 15 micrograms per cubic meter are unsafe, what can we conclude about the air quality of most of the cities in this sample?

SOLUTION   The typical city has a level of 7.9 micrograms per cubic meter, and because the distribution is unimodal and (roughly) symmetric, most cities have levels within 2.2 micrograms per cubic meter of this value. In other words, most cities have levels of particulate matter between

$$7.9 - 2.2 = 5.7 \text{ micrograms per cubic meter and}$$
$$7.9 + 2.2 = 10.1 \text{ micrograms per cubic meter}$$

CONCLUSION   Because the value of 10.1 (one standard deviation above the mean) is lower than 15, most cities are below the safety limit. (The three cities reporting the highest levels of particulate matter were Hanford-Corcoran, California; Visalia-Porterville, California; and Hilo, Hawaii. The three cities reporting the lowest levels of particulate matter were Clearlake, California; Dickinson, North Dakota; and Williston, North Dakota.)

As you'll soon see, we can say even more. In a few pages you'll learn about the Empirical Rule, which tells us that about 95% of all cities should be within two standard deviations of the mean particulate level.

TRY THIS!   Exercise 3.17

Calculating the Standard Deviation The formula for the standard deviation is somewhat more complicated than that for the mean, and a bit more work is necessary to calculate it. A calculator or computer is pretty much required for all but the smallest data sets. Just as the mean of a sample has its own symbol, the standard deviation of a sample of data is represented by the letter *s*.

$$\text{Formula 3.2:}\quad \text{Standard deviation} = s = \sqrt{\frac{\sum(x - \bar{x})^2}{n - 1}}$$

Think of this formula as a set of instructions. Essentially, the instructions say that we need first to calculate how far away each observation is from the mean. This distance, including the positive or negative sign, $(x - \bar{x})$, is called a **deviation**. We square these deviations so that they are all positive numbers, and then we essentially find the average. (If we had divided by *n*, and not *n* − 1, it would have been the average. We'll demonstrate in the Chapter 9 exercises why we divide by *n* − 1 and not *n*.) Finally, we take the square root, which means that we're working with the same units as the original data, not with squared units.

EXAMPLE 6 A Gallon of Gas

From the website GasBuddy.com, we collected the prices of a gallon of regular gas at 12 gas stations in a neighborhood in Austin, Texas, for one day in January 2018.

$2.19, $2.19, $2.39, $2.19, $2.24, $2.39, $2.27, $2.29, $2.17, $2.29, $2.30, $2.29

QUESTION Find the standard deviation for the prices. Explain what this value means in the context of the data.

SOLUTION We show this result two ways. The first way is by hand, which illustrates how to apply Formula 3.2. The second way uses a statistical calculator. The first step is to find the mean. We did this earlier in Example 2, using Formula 3.1, which gave us a mean value of $2.27. We substitute this value for $\bar{x}$ in Formula 3.2.

Table 3.1 shows the first two steps. First we find the deviations (in column 2). Next we square each deviation (in column 3). The numbers are sorted so we can more easily compare the differences.

The sum of the squared deviations—the sum of the unrounded values that were used in column 3—is approximately 0.061. Dividing this by 11 (because *n* − 1 = 12 − 1 = 11), we get 0.005545455. The last step is to take the square root of this. The result is our standard deviation:

$$s = \sqrt{0.005545455} = 0.07446781$$

To recap:

$$s = \sqrt{\frac{\sum(x - \bar{x})^2}{n - 1}} = \sqrt{\frac{0.061}{12 - 1}} = \sqrt{\frac{0.061}{11}} = \sqrt{0.005545455}$$
$$= 0.07446781, \text{ or about 7 cents}$$

x	x − x̄	(x − x̄)²
2.19	−0.08	0.0064
2.19	−0.08	0.0064
2.39	0.12	0.0144
2.19	−0.08	0.0064
2.24	−0.03	0.0009
2.39	0.12	0.0144
2.27	0.00	0.0000
2.29	0.02	0.0004
2.17	−0.10	0.0100
2.29	0.02	0.0004
2.30	0.03	0.0009
2.29	0.02	0.0004

▲ TABLE 3.1

Tech

When you are doing these calculations, your final result will be more accurate if you do not round any of the intermediate results as we did in this example, to better illustrate the process. For this reason, it is far easier, and more accurate, to use a statistical calculator or statistical software to find the standard deviation. Figure 3.9 shows the standard deviation as $Sx = 0.0743863787$, which we round to about 0.07 dollar (or, if you prefer, 7 cents). Note that the value reported by the calculator in Figure 3.9 is not exactly the same as the figure we obtained by hand. In part, this is because we used an approximate value for the mean in our hand calculations.

CONCLUSION The standard deviation is about 7 cents, or $0.07. Therefore, at most of these gas stations, the price of a gallon of gas is within 7 cents of $2.27.

TRY THIS! Exercise 3.19b

```
NORMAL FLOAT AUTO REAL RADIAN MP
        1-Var Stats
 x̄=2.266666667
 Σx=27.2
 Σx²=61.7142
 Sx=.0743863787
 σx=.0712195354
 n=12
 minX=2.17
↓Q₁=2.19
```

▲ **FIGURE 3.9** The standard deviation is denoted "Sx" in the TI calculator output.

One reason why we suggest using statistical software rather than the formulas we present is that we nearly always look at data using several different statistics and approaches. We nearly always begin by making a graph of the distribution. Usually the next step is to calculate a measure of the center and then a measure of the spread. It does not make sense to have to enter the data again every time you want to examine them; it is much better to enter them once and use the functions on your calculator (or software).

SNAPSHOT ▶ The Standard Deviation of a Sample

WHAT IS IT? ▶	A numerical summary.
WHAT DOES IT DO? ▶	Measures the spread of a distribution of a sample of data.
HOW DOES IT DO IT? ▶	It measures the typical distance of the observations from the mean.
HOW IS IT USED? ▶	To measure the amount of variability in a sample when the distribution is fairly symmetric.

Variance, a Close Relative of the Standard Deviation Another way of measuring spread—a way that is closely related to the standard deviation—is the variance. The **variance** is simply the standard deviation squared, and it is represented symbolically by s^2.

Formula 3.3: $\text{Variance} = s^2 = \dfrac{\sum (x - \bar{x})^2}{n - 1}$

In Example 5, the standard deviation of the concentration of particulate matter in the cities in our sample was 2.6 micrograms per cubic meter. The variance is therefore $2.6 \times 2.6 = 6.76$ micrograms squared per cubic meter squared. The standard deviation in daily high temperatures in Provo is 21 degrees, so the variance is $21 \times 21 = 441$ degrees squared.

For most applications, the standard deviation is preferred over the variance. One reason is that the units for the variance are always squared (degrees squared in the last paragraph), which implies that the units used to measure spread are different from the units used to measure center. The standard deviation, on the other hand, has the same units as the mean.

SECTION 3.2

What's Unusual? The Empirical Rule and z-Scores

Finding the standard deviation and the mean is a useful way to compare different samples and to compare observations from one sample with those in another sample.

The Empirical Rule

The **Empirical Rule** is a rough guideline, a rule of thumb, that helps us understand how the standard deviation measures variability. This rule says that if the distribution is unimodal and symmetric, then

- Approximately 68% of the observations (roughly two-thirds) will be within one standard deviation of the mean.

- Approximately 95% of the observations will be within two standard deviations of the mean.

- Nearly all the observations will be within three standard deviations of the mean.

When we say that about 68% of the observations are within one standard deviation of the mean, we mean that if we count the observations that are between the mean minus one standard deviation and the mean plus one standard deviation, we will have counted about 68% of the total observations.

The Empirical Rule is illustrated in Figure 3.10 in the context of the data on particulate matter in 371 U.S. cities, introduced in Example 3. Suppose we did not have access to the actual data and knew only that the distribution is unimodal and symmetric, that the mean particulate matter is 7.9 micrograms per cubic meter, and that the standard deviation is 2.2 micrograms per cubic meter. The Empirical Rule predicts that about 68% of the cities will fall between 5.7 micrograms per cubic meter ($7.9 - 2.2 = 5.7$) and 10.1 micrograms per cubic meter ($7.9 + 2.2 = 10.1$).

The Empirical Rule predicts that about 95% of the cities will fall within two standard deviations of the mean, which means that about 95% of the cities will be between 3.5 and 12.3 micrograms per cubic meter ($7.9 - (2 \times 2.2) = 3.5$ and $7.9 + (2 \times 2.2) = 12.3$). Finally, nearly all cities, according to the Empirical Rule, will be between 1.3 and 14.5 micrograms per cubic meter. This is illustrated in Figure 3.10.

▶ **FIGURE 3.10** The Empirical Rule predicts how many observations we will see within one standard deviation of the mean (68%), within two standard deviations of the mean (95%), and within three standard deviations of the mean (almost all).

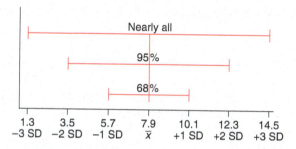

KEY POINT	In a relatively large collection of observations, if the distribution is unimodal and roughly symmetric, then about 68% of the observations are within one standard deviation of the mean; about 95% are within two standard deviations of the mean, and almost all observations are within three standard deviations of the mean. Not all unimodal, symmetric distributions are the same, so your actual outcomes might differ from these values, but the Empirical Rule works well enough in a surprisingly large number of situations.

EXAMPLE 7   Comparing the Empirical Rule to Actual Smog Levels

Because the distribution of levels of particulate matter (PM) in 371 U.S. cities is roughly unimodal and symmetric, the Empirical Rule predicts that about 68% of the cities will have particulate matter levels between 5.7 and 10.1 micrograms per cubic meter, about 95% of the cities will have PM levels between 3.5 and 12.3 micrograms per cubic meter, and nearly all will have PM levels between 1.3 and 14.5 micrograms per cubic meter.

QUESTION Figure 3.11 shows the actual histograms for the distribution of PM levels for these 371 cities. The location of the mean is indicated, as well as the boundaries for points within one standard deviation of the mean (a), within two standard deviations of the mean (b), and within three standard deviations of the mean (c). Using these figures, compare the actual number of cities that fall within each boundary to the number predicted by the Empirical Rule.

SOLUTION From Figure 3.11a, by counting the heights of the bars between the two boundaries, we find that there are about $61 + 85 + 73 + 49 = 268$ cities that actually lie between these two boundaries. (No need to count very precisely; we're after approximate numbers here.) The Empirical Rule predicts that about 68% of the cities, or $0.68 \times 371 = 252$ cities, will fall between these two boundaries. The Empirical Rule is pretty accurate in this case, although it underpredicts a little.

From Figure 3.11b, we count that there are about $21 + 27 + 61 + 85 + 73 + 49 + 20 + 8 = 344$ cities within two standard deviations of the mean. The Empirical Rule predicts that about 95% of the cities, or $0.95 \times 371 = 352$ cities, will fall between these two boundaries. So again, the Empirical Rule is not too far off.

Finally, from Figure 3.11c, we clearly see that all but a few cities (3 or so) are within three standard deviations of the mean. We summarize these findings in a table.

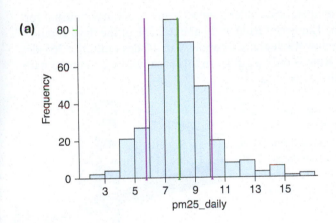

(a)

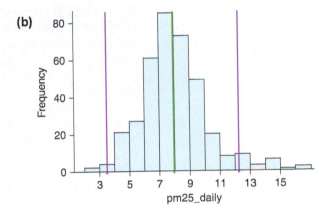

(b)

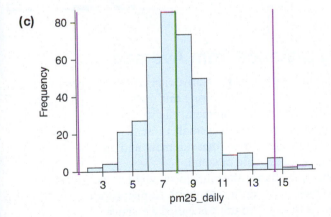

(c)

▲ **FIGURE 3.11** **(a)** Boundaries are placed one standard deviation below the mean (5.7) and one standard deviation above the mean (10.1). The middle line in all three figures indicates the mean (7.9). **(b)** The Empirical Rule predicts that about 95% of observations will fall between these two boundaries (3.5 and 12.3 micrograms per cubic meter). **(c)** The Empirical Rule predicts that nearly all observations will fall between these two boundaries (1.3 and 14.5 micrograms per cubic meter).

CONCLUSION

Interval	Empirical Rule Prediction	Actual Number
within 1 SD of mean	252	268
within 2 SD of mean	352	344
within 3 SD of mean	nearly all	all but 3

TRY THIS! Exercise 3.29

EXAMPLE 8 Temperatures in San Francisco

We have seen that the mean daily high temperature in San Francisco is 65 degrees Fahrenheit and that the standard deviation is 8 degrees.

QUESTION Using the Empirical Rule, decide whether it is unusual in San Francisco to have a day when the maximum temperature is colder than 49 degrees.

SOLUTION One way of answering this question is to find out about how many days in the year we would expect the high temperature to be colder than 49 degrees. The Empirical Rule says that about 95% of the days will have temperatures within two standard deviations of the average—that is, within $2 \times 8 = 16$ degrees of 65 degrees. Therefore, on most days the temperature will be between $65 - 16 = 49$ degrees and $65 + 16 = 81$ degrees. Because only (approximately) 5% of the days are outside this range, we know that days either warmer or cooler than this are rare. According to the Empirical Rule, which assumes a roughly symmetric distribution, about half the days outside the range—2.5%—will be colder, and half will be warmer. This shows that about 2.5% of 365 days (roughly 9 or 10 days) should have a maximum temperature colder than 49 degrees.

CONCLUSION A maximum temperature colder than 49 degrees is fairly unusual for San Francisco. According to the Empirical Rule, only about 2.5% of the days should have maximum temperatures below 49 degrees, or about 9 or 10 days out of a 365-day year. Of course, the Empirical Rule is only a guide. In this case, we can compare with the data to see just how often these cooler days occurred.

Your concept of "unusual" might be different from ours. The main idea is that "unusual" is rare, and in selecting two standard deviations, we chose to define temperatures that occur 2.5% of the time or less as rare and therefore unusual. You might very reasonably set a different standard for what you wish to consider unusual.

TRY THIS! Exercise 3.31

z-Scores: Measuring Distance from Average

The question, "How unusual is this?" is perhaps the statistician's favorite question. (It is just as popular as "Compared to what?") Answering this question is complicated because the answer depends on the units of measurement. Eighty-four is a big value if we are measuring a man's height in inches, but it is a small value if we are measuring his weight in pounds. Unless we know the units of measurement and the objects being measured, we can't judge whether a value is big or small.

One way around this problem is to change the units to standard units. **Standard units** measure a value relative to the sample rather than with respect to some absolute measure. A measurement converted to standard units is called a *z-score*.

Visualizing z-Scores Specifically, a standard unit measures how many standard deviations away an observation is from the mean. In other words, it measures a distance, but instead of measuring in feet or miles, it counts the number of standard deviations. A measurement with a z-score of 1.0 is one standard deviation *above* the mean. A measurement with a z-score of −1.4 is 1.4 standard deviations *below* the mean.

Figure 3.12 shows a dotplot of the heights (in inches) of 247 adult men. The average height is 70 inches, and the standard deviation is 3 inches (after rounding). Below

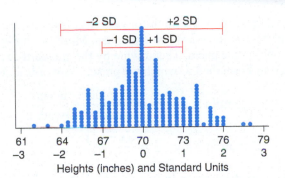

the dotplot is a ruler that marks off how far from average each observation is, measured in terms of standard deviations. The average height of 70 inches is marked as 0 because 70 is zero standard deviations away from the mean. The height of 76 is marked as 2 because it is two standard deviations above the mean. The height of 67 is marked as −1 because it is one standard deviation *below* the mean.

We would say that a man from this sample who is 73 inches tall has a *z*-score of 1.0 standard unit. A man who is 67 inches tall has a *z*-score of −1.0 standard unit.

Using z-Scores in Context   *z*-Scores enable us to compare observations in one group with those in another, even if the two groups are measured in different units or under different conditions. For instance, some students might choose their math class on the basis of which professor they think is an easier grader. So if one student gets a 65 on an exam in a hard class, how do we compare her score to that of a student who gets a 75 in an easy class? If we converted to standard units, we would know how far above (or below) the average each test score was, so we could compare these students' performances.

EXAMPLE 9 Exam Scores

Maria scored 80 out of 100 on her first stats exam in a course and 85 out of 100 on her second stats exam. On the first exam, the mean was 70, and the standard deviation was 10. On the second exam, the mean was 80, and the standard deviation was 5.

QUESTION   On which exam did Maria perform better when compared to the whole class?

SOLUTION   On the first exam, Maria is 10 points above the mean because 80-70 is 10. Because the standard deviation is 10 points, she is one standard deviation above the mean. In other words, her *z*-score for the first exam is 1.0.

On the second exam, she is 5 points above average because 85-80 is 5. Because the standard deviation is 5 points, she is one standard deviation above the mean. In other words, her *z*-score is again 1.0.

CONCLUSION   The second exam was a little easier; on average, students scored higher, and there was less variability in the scores. But Maria scored one standard deviation above average on both exams, so she did equally well on both when compared to the whole class.

TRY THIS!   Exercise 3.35

Calculating the z-Score It's straightforward to convert to z-scores when the result is a whole number, as in the last few examples. More generally, to convert a value to its z-score, first subtract the mean. Then divide by the standard deviation. This simple recipe is summarized in Formula 3.4, which we present in words and symbols:

$$\text{Formula 3.4a:} \quad z = \frac{(\text{value} - \text{mean})}{\text{standard deviation}}$$

$$\text{Formula 3.4b:} \quad z = \frac{(x - \bar{x})}{s}$$

Let's apply this to the data shown in Figure 3.12. What is the z-score of a man who is 75 inches tall? Remember that the mean height is 70 inches and the standard deviation is 3 inches. Formula 3.4 says first to subtract the mean height.

$$75 - 70 = 5 \text{ inches}$$

Next divide by the standard deviation:

$$5/3 = 1.67 \text{ (rounding off to two decimal digits)}$$

$$z = \frac{x - \bar{x}}{s} = \frac{75 - 70}{3} = \frac{5}{3} = 1.67$$

This person has a z-score of 1.67. In other words, we would say that he is 1.67 standard deviations taller than average.

EXAMPLE 10 Daily Temperatures

The mean daily high temperature in San Francisco is 65 degrees F, and the standard deviation is 8 degrees. On one day, the high temperature is 49 degrees.

QUESTION What is this temperature in standard units? Assuming that the Empirical Rule applies, does this seem unusual?

SOLUTION
$$z = \frac{x - \bar{x}}{s} = \frac{49 - 65}{8} = \frac{-16}{8} = -2.00$$

CONCLUSION This is an unusually cold day. From the Empirical Rule, we know that 95% of z-scores are between −2 and 2 standard units, so it is fairly unusual to have a day as cold as or colder than this one.

TRY THIS! Exercise 3.37

SNAPSHOT ▶ The z-Score

WHAT IS IT?	▶	A standardized observation.
WHAT DOES IT DO?	▶	Converts a measurement into standard units.
HOW DOES IT DO IT?	▶	By measuring how many standard deviations away a value is from the sample mean.
HOW IS IT USED?	▶	To compare values from different groups, such as two exam scores from different exams, or to compare values measured in different units, such as inches and pounds.

SECTION 3.3

Summaries for Skewed Distributions

As we noted earlier, for a skewed distribution, the center of balance is not a good way of measuring a "typical" value. Another concept of center, which is to think of the center as being the location of the *middle* of a distribution, works better in these situations. You saw one example of this in Figure 3.3, which showed that the mean tennis winnings was quite a bit higher than what a majority of the players actually won. Figure 3.13 shows another example of a strongly right-skewed distribution. This is the distribution of incomes of over 42,000 New York State residents, drawn from a survey done by the U.S. government in 2016. The mean income of $50,334 is marked with a triangle. However, note that the mean doesn't seem to match up very closely with what we think of as typical. The mean seems to be too high to be typical. In fact, a majority (about 69%) of residents earn less than this mean amount.

Data moves ▶ These data are extracted from factfinder.census.gov, which provides data on 284 variables.

Data ▶ The file censusincome.csv provides the data for six of the 284 variables for six states.

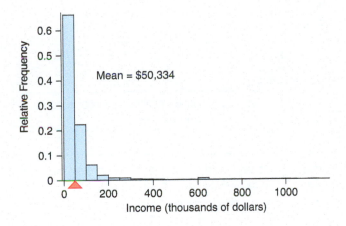

◀ FIGURE 3.13 The distribution of annual incomes for a collection of New York State residents is right-skewed. Thus the mean is somewhat greater than what many people would consider a typical income. The location of the mean is shown with a triangle.

In skewed distributions, the mean can often be a poor measure of "typical." Instead of using the mean and the standard deviation in these cases, we measure the center and spread differently.

The Center as the Middle: The Median

The median provides an alternative way of determining a typical observation. The **median** of a sample of data is the value that would be right in the middle if you were to sort the data from smallest to largest. The median cuts a distribution down the middle, so about 50% of the observations are below it, and about 50% are above it.

Visualizing the Median One of the authors found herself at the grocery store, trying to decide whether to buy ham or turkey for sandwiches. Which is more healthful? Figure 3.14 shows a dotplot of the percentage of fat for each of ten types of sliced ham. The vertical line marks the location of the median at 23.5%. Note that five observations

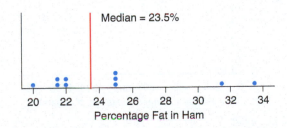

◀ FIGURE 3.14 A dotplot of percentage fat from ham has a median of 23.5%. This means that half the observations are below 23.5% and half are above it.

lie below the median and five lie above it. The median cuts the distribution exactly in half. In Example 12, you'll see how the median percentage of fat in sliced turkey compares to that in sliced ham.

Finding the median for the distribution of incomes of the 42,000 New York State residents, shown again in Figure 3.15, is slightly more complicated because we have many more observations. The median (shown with the red vertical line) cuts the total area of the histogram in half. The median is at $30,000, and very close to 50% of the observations are below this value and just about 50% are above it.

▶ **FIGURE 3.15** The distribution of incomes of New York State residents, with the median indicated by a vertical line. The median has about 50% of the observations above it and about 50% below it. Residents who reported incomes of 0 dollars were excluded.

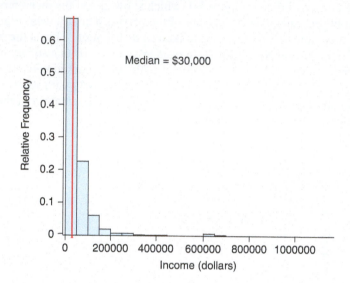

Median = $30,000

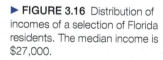

KEY POINT   The median is the value that cuts a distribution in half. The median value represents a "typical" value in a data set.

The Median in Context The median is used for the same purpose as the mean: to give us a typical value of a set of data. Knowing the typical value of one group helps us to compare it to another. For example, as we've seen, the typical median income of this sample of New York State residents is $30,000. How does the typical income in New York compare to the typical income in Florida? A representative sample of 44,000 Florida residents (where many New Yorkers go to retire) has a median income of $27,000, which is only slightly less than the median income in New York (Figure 3.16).

▶ **FIGURE 3.16** Distribution of incomes of a selection of Florida residents. The median income is $27,000.

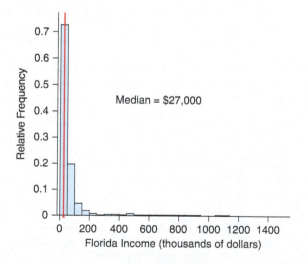

Median = $27,000

The typical person in our data set of New York incomes makes more than the typical person in our Florida data set, as measured by the median. Because the median for New York is $30,000, we know that more than half of New York residents in the sample make more than the median Florida salary of $27,000.

The median is often reported in news stories when the discussion involves variables with distributions that are skewed. For example, you may hear reports of "median housing costs" and "median salaries."

Calculating the Median   To calculate the value of the median, follow these steps:

1. Sort the data from smallest to largest.

2. If the set contains an odd number of observed values, the median is the middle observed value.

3. If the set contains an even number of observed values, the median is the average of the two middle observed values. This places the median precisely halfway between the two middle values.

EXAMPLE **11** Twelve Gas Stations

The prices of a gallon of regular gas at 12 Austin, Texas, gas stations in January 2018 (see Example 6) were the following:

$2.19, $2.19, $2.39, $2.19, $2.24, $2.39, $2.27, $2.29, $2.17, $2.29, $2.30, $2.29

QUESTION   Find the median price for a gallon of gas and interpret the value.

SOLUTION   First, we sort the data from smallest to largest.

2.17, 2.19, 2.19, 2.19, 2.24, 2.27, 2.29, 2.29, 2.29, 2.30, 2.39, 2.39

Because the data set contains an even number of observations (12), the median is the average of the two middle observations, the sixth and seventh: 2.27 and 2.29.

2.17, 2.19, 2.19, 2.19, 2.24, 2.27, 2.29, 2.29, 2.29, 2.30, 2.39, 2.39
                                         Med

CONCLUSION   The median is $2.28, which is the typical price of a gallon of gas at these 12 gas stations.

TRY THIS!   Exercise 3.43a

Example 12 demonstrates how to find the median in a sample with an odd number of observations.

EXAMPLE **12** Sliced Turkey

Figure 3.14 showed that the median percentage of fat from the various brands of sliced ham for sale at a grocery store was 23.5%. How does this compare to the median percentage of fat for the turkey? Here are the percentages of fat for the available brands of sliced turkey:

14, 10, 20, 20, 40, 20, 10, 10, 20, 50, 10

QUESTION   Find the median percentage of fat and interpret the value.

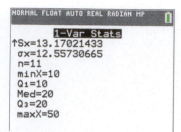

```
NORMAL FLOAT AUTO REAL RADIAN MP
       1-Var Stats
↑Sx=13.17021433
 σx=12.55730665
 n=11
 minX=10
 Q₁=10
 Med=20
 Q₃=20
 maxX=50
```

▲ **FIGURE 3.17** Some TI-84 output for the percentage of fat in the turkey.

SOLUTION The data are sorted and displayed below. Because we have 11 observations, the median is the middle observation, 20.

$$10 \quad 10 \quad 10 \quad 10 \quad 14 \quad 20 \quad 20 \quad 20 \quad 20 \quad 40 \quad 50$$
$$\text{Med}$$

CONCLUSION The median for the turkey is 20% fat. Thus, the typical percentage of fat for these types of sliced turkey is 20%. This is (slightly) less than that for the typical sliced ham, which has 23.5% fat. Figure 3.17 provides TI-84 output that confirms our calculation.

TRY THIS! Exercise 3.45

SNAPSHOT ▸ The Median of a Sample

WHAT IS IT? ▸	A numerical summary.
WHAT DOES IT DO? ▸	Measures the center of a distribution.
HOW DOES IT DO IT? ▸	It is the value that has roughly the same number of observations above it and below it.
HOW IS IT USED? ▸	To measure the typical value in a data set, particularly when the distribution is skewed.

Measuring Variability with the Interquartile Range

The standard deviation measures how spread out observations are with respect to the mean. But if we don't use the mean, then it doesn't make sense to use the standard deviation. When a distribution is skewed and you are using the median to measure the center, an appropriate measure of variation is called the interquartile range. The **interquartile range (IQR)** tells us, roughly, how much space the middle 50% of the data occupy.

Visualizing the IQR To find the IQR, we cut the distribution into four parts with roughly equal numbers of observations. The distance taken up by the middle two parts is the interquartile range.

The dotplot in Figure 3.18 shows the distribution of weights for a class of introductory statistics students. The vertical lines slice the distribution into four parts so that each part has about 25% of the observations. The IQR is the distance between the first "slice" (at about 121 pounds) and the third slice (at about 160 pounds). This is an interval of 39 pounds ($160 - 121 = 39$).

▶ **FIGURE 3.18** The distribution of weights (in pounds) of students in a class is divided into four sections so that each section has roughly 25% of the observations. The IQR is the distance between the outer vertical lines (at 121 pounds and 160 pounds).

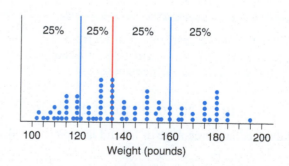

Figure 3.19 shows distributions for the same students, but this time the weights are separated by gender. The vertical lines are located so that about 25% of the data are below the leftmost line, and about 25% are above the rightmost line. This means that about half the data lie between these two boundaries. The distance between these boundaries is the IQR. You can see that the IQR for the males, about 38 pounds, is much larger than the IQR for the females, which is about 20 pounds. The females have less variability in their weights.

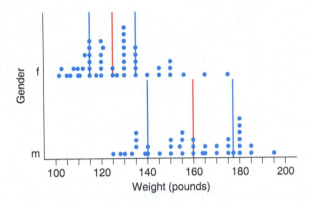

◄ **FIGURE 3.19** The dotplot of Figure 3.18 with weights separated by gender (the women on top). The men have a larger interquartile range than the women.

The IQR focuses only on the middle 50% of the data. We can change values outside of this range without affecting the IQR. Figure 3.20 shows the men's weights, but this time we've changed one of the men's weights to be very small. The IQR is still the same as in Figure 3.19.

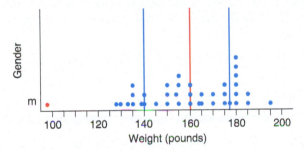

◄ **FIGURE 3.20** The men's weights are given with a fictitious point (in red) below 100. Moving an extremely large (or small) value does not change the interquartile range.

The Interquartile Range in Context The IQR for the incomes of the New Yorkers in the data set previously shown in Figure 3.15 is $47,000, as shown in Figure 3.21 at the top of the next page. This tells us that the middle 50% of people in our data set had incomes that varied by as much as $47,000. Compare this to the incomes from Florida, which have an IQR of $37,000. There is less variability among the Floridians; Floridians are more similar (at least in terms of their incomes).

An IQR of $47,000 for New Yorkers seems like a pretty large spread. However, considered in the context of the entire distribution (see Figure 3.21), which includes incomes near $0 and as large as one million dollars, the IQR looks fairly small. The reason is that lots of people (half of our data set) have incomes in this fairly narrow interval.

Calculating the Interquartile Range Calculating the interquartile range involves two steps. First, you must determine where to "cut" the distribution. These points are called the **quartiles**, because they cut the distribution into quarters (fourths). The **first quartile (Q1)** has roughly one-fourth, or 25%, of the observations at or below it. The **second quartile (Q2)** has about 50% at or below it; actually, Q2 is just another name for the median. The **third quartile (Q3)** has about 75% of the observations at or below it. The second step is the easiest: To find the interquartile range you simply find the interval between Q3 and Q1—that is, Q3 − Q1.

To find the quartiles,

⚓ Details

Software and Quartiles
Different software packages don't always agree with each other on the values for Q1 and Q3, and therefore they might report different values for the IQR. The reason is that several different accepted methods exist for calculating Q1 and Q3. So don't be surprised if the software on your computer gives different values from your calculator.

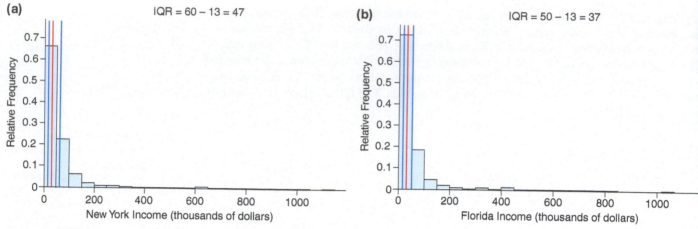

▲ **FIGURE 3.21 (a)** The distribution of incomes for New Yorkers. **(b)** The distribution of income for Floridians. In both figures, vertical bars are drawn to divide the distribution into four areas, each with about 25% of the observations. The IQR is the distance between the outer vertical lines, and it is wider for the New York incomes. The middle vertical line indicates the median income.

- First find the median, which is also called the second quartile, Q2. The median cuts the data into two regions.
- The first quartile (Q1) is the median of the lower half of the sorted data. (Do not include the median observation in the lower half if you started with an odd number of observations.)
- The third quartile (Q3) is the median of the upper half of the sorted data. (Again, do not include the median itself if your full set of data has an odd number of observations.)

Formula 3.5: Interquartile Range $=$ Q3 $-$ Q1

EXAMPLE **13** Heights of Children

A group of eight children have the following heights (in inches):

$$48.0, 48.0, 53.0, 53.5, 54.0, 60.0, 62.0, \text{ and } 71.0$$

They are shown in Figure 3.22.

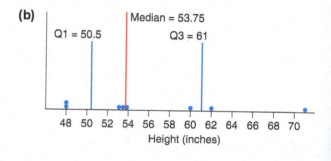

▲ **FIGURE 3.22 (a)** Eight children sorted by height. **(b)** A dotplot of the heights of the children. The median and the quartiles are marked with vertical lines. Note that two dots, or 25% of the data, appear in each of the four regions.

QUESTION   Find the interquartile range for the distribution of the children's heights.

SOLUTION   As before, we first explain how to do the calculations by hand and then show the output of technology.

First we find Q2 (the median). Note that the data are sorted and that there are four observed values below the median and four observed values above the median.

Duncan	Charlie	Grant	Aidan		Sophia	Seamus	Cathy	Drew
48	48	53	53.5		54	60	62	71

Q2

53.75

To find Q1, examine the numbers below the median and find the median of them, as shown.

	Duncan	Charlie		Grant	Aidan
	48	48		53	53.5

Q1

50.50

To find Q3, examine the numbers above the median and find the median of them.

Sophia	Seamus		Cathy	Drew
54	60		62	71

Q3

61.00

Together, these values are as follows:

Duncan	Charlie		Grant	Aidan		Sophia	Seamus		Cathy	Drew
48	48		53	53.5		54	60		62	71
	Q1					Q2			Q3	
	50.50					53.75			61.00	

Here's how we calculated the values:

$$Q1 = \frac{48 + 53}{2} = \frac{101}{2} = 50.50 \qquad \text{(Halfway between Charlie and Grant)}$$

$$Q2 = \frac{53.5 + 54}{2} = \frac{107.5}{2} = 53.75 \qquad \text{(Halfway between Aidan and Sophia)}$$

$$Q3 = \frac{60 + 62}{2} = \frac{122}{2} = 61.00 \qquad \text{(Halfway between Seamus and Cathy)}$$

The last step is to subtract:

$$IQR = Q3 - Q1 = 61.00 - 50.50 = 10.50$$

Figure 3.22b shows the location of Q1, Q2, and Q3. Note that 25% of the data (two observations) lie in each of the four regions created by the vertical lines.

Figure 3.23 shows the TI-84 output. The TI-84 does not calculate the IQR directly; you must subtract Q3 − Q1 yourself. The IQR is 61 − 50.5 = 10.5, which is the same as the IQR done by hand above.

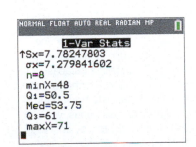

▲ FIGURE 3.23 Some output of a TI-84 for eight children's heights.

CONCLUSION   The interquartile range of the heights of the eight children is 10.5 inches.

TRY THIS!   Exercise 3.43b

Finding the Range, Another Measure of Variability

Another measure of variability is similar to the IQR but much simpler. The **range** is the distance spanned by the entire data set. It is very simple to calculate: It is the largest value minus the smallest value.

Formula 3.6:   Range = maximum − minimum

For the heights of the eight children (Example 13), the range is 71.0 − 48.0 = 23.0 inches.

The range is useful for a quick measurement of variability because it's very easy to calculate. However, because it depends on only two observations—the largest and the smallest—it is very sensitive to any peculiarities in the data. For example, if someone makes a mistake when entering the data and enters 710 inches instead of 71 inches, then the range will be very wrong. The IQR, on the other hand, depends on many observations and is therefore more reliable.

SNAPSHOT ▶ The Interquartile Range

WHAT IS IT? ▶	A numerical summary.
WHAT DOES IT DO? ▶	It measures the spread of the distribution of a data set.
HOW DOES IT DO IT? ▶	It computes the distance taken up by the middle half of the sorted data.
HOW IS IT USED? ▶	To measure the variability in a sample, particularly when the distribution is skewed.

SECTION 3.4

Comparing Measures of Center

Which should you choose, the mean (accompanied by the standard deviation) or the median (with the IQR)? In the Data Cycle, in between the "asking questions" and "analyzing data" phases, is a "consider data" phase. During this phase, we examine the data, often through data visualizations, to determine which analyses would be most appropriate. The choice between the mean and the median can often be made by examining the shape of the distribution.

Look at the Shape First

This decision begins with a picture. The shape of the distribution will determine which measures are best for communicating the typical value and the variability in the distribution.

Data moves ▶ The app iTunes stores all data in a database using a format called "xml," or Extensible Markup Language. If you use iTunes, you can see your own library by choosing "File > Library > Export Library" from the iTunes menu. Some statistical software packages allow you to read in data in this format, although this often requires some programming.

File ▶ itunessample

EXAMPLE 14   MP3 Song Lengths

One of the authors created a data set of the songs on his mp3 player. He wants to describe the distribution of song lengths.

QUESTION   What measures should he use for the center and spread: the mean (250.2 seconds) with the standard deviation (152.0 seconds) or the median (226 seconds) with the interquartile range (117 seconds)? Interpret the appropriate measures. Refer to the histogram in Figure 3.24.

SOLUTION Before looking at the histogram, you should think about what shape you expect the graph to be. No song can be shorter than 0 seconds. Most songs on the radio are about 4 minutes (240 seconds), with some a little longer and some a little shorter. However, a few songs, particularly classical tracks, are quite long, so we might expect the distribution to be right-skewed. This suggests that the median and the IQR are the best measures to use.

Figure 3.24 confirms that, as we predicted, the distribution is right-skewed, so the median and interquartile range would be the best measures to use.

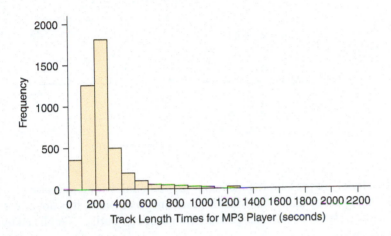

► **FIGURE 3.24** The distribution of lengths of songs (in seconds) on the mp3 player of one of the authors.

CONCLUSION The median length is 226 seconds (roughly 3 minutes and 46 seconds), and the interquartile range is 117 seconds (close to 2 minutes). In other words, the typical track on the author's mp3 player is about 4 minutes, but there's quite a bit of variability, with the middle 50% of the tracks differing by about 2 minutes.

TRY THIS! Exercise 3.55

Sometimes, you don't have the data themselves, so you can't make a picture. If so, then you must deduce a shape for the distribution that is reasonable and choose the measure of center on the basis of this deduction.

When a distribution is right-skewed, as it is with the mp3 song lengths, the mean is generally larger than the median. You can see this in Figure 3.24; the right tail means the balancing point must be to the right of the median. With the same reasoning, we can see that in a left-skewed distribution, the mean is generally less than the median. In a symmetric distribution, the mean and the median are approximately the same.

> **KEY POINT**
> In a symmetric distribution, the mean and the median are approximately the same. In a right-skewed distribution, the mean tends to be greater than the median, and in a left-skewed distribution, the mean tends to be less than the median.

The Effect of Outliers

Even when a distribution is mostly symmetric, the presence of one or more outliers can have a large effect on the mean. Usually, the median is a more representative measure of center when an outlier is present.

Looking Back

Outliers

Recall from Chapter 2 that an outlier is an extremely large or small observation relative to the bulk of the data.

The average height of the eight children from Example 13 was 56.2 inches. Imagine that we replace the tallest child (who is 71 inches tall) with basketball player Shaquille O'Neal, whose height is 85 inches. Our altered data set is now

48, 48, 53, 53.5, 54, 60, 62, and 85 inches

In order to keep the balance of the data, the mean has to shift higher. The mean of this new data set is 57.9 inches—over 1 inch higher. The median of the new data set, however, is the same: 53.75 inches, as shown in Figure 3.25.

▶ **FIGURE 3.25** The effect of changing the tallest child's height into Shaquille O'Neal's height. Note that the mean (shown with triangles) changes, but the median (the vertical line) stays the same.

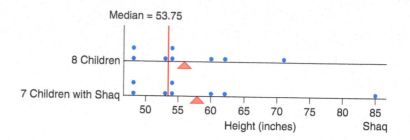

When outliers are present, the median is a good choice for a measure of center. In technical terms, we say that the median is **resistant to outliers**; it is not affected by the size of an outlier and does not change even if a particular outlier is replaced by an even more extreme value.

> **KEY POINT**
> The median is resistant to outliers; it is not affected by the size of the outliers. It is therefore a good choice for a measure of the center if the data contain outliers and you want to reduce their effect.

EXAMPLE 15 Fast Food

A (very small) fast-food restaurant has five employees. Each employee's annual income is about $16,000 per year. The owner, on the other hand, makes $100,000 per year.

QUESTION Find both the mean and the median. Which would you use to represent the typical income at this business—the mean or the median?

SOLUTION Figure 3.26 shows a dotplot of the data. The mean income is $30,000, and the median is $16,000. Nearly all the employees earned less than the mean amount!

▶ **FIGURE 3.26** Dotplot of salaries for five employees and their boss at a fast-food company. The triangle indicates the mean salary.

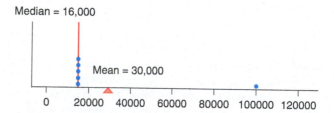

CONCLUSION Given the choice between mean and median, the median is better at showing the typical income.

Why are the mean and the median so different? Because the owner's salary of $100,000 is an outlier.

TRY THIS! Exercise 3.57

Many Modes: Summarizing Center and Spread

What should you do if the distribution is bimodal or has several modes? Unfortunately, the answer is "It's complicated."

You learned in Chapter 2 that multiple modes in a graphical display of a distribution sometimes indicate that the data set combines different groups. For example, perhaps a data set containing heights includes both men and women. The distribution could very well be bimodal because we're combining two groups of people who differ markedly in terms of their heights. In this case, and in many other contexts, it is more useful to separate the groups and report summary measures separately for each group. If we know which observations belong to the men and which to the women, then we can separate the data and compute the mean height for men separately from the mean height for women.

For example, Figure 3.27 shows a histogram of the finishing times of female marathon runners. The most noticeable feature of this distribution is that there appear to be two modes. When confronted with this situation, a natural question to ask is, "Are two different groups of runners represented in this data set?"

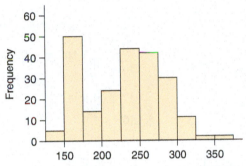

Time for Two Marathons, Women Only (minutes)

As it turns out, the answer is "yes." Table 3.2 shows the first few lines of the data set. From this, we see that the times belong to runners from two different events. One event was the 2012 Olympics, which includes the best marathoners in the world. The other event was an amateur marathon in Portland, Oregon, in 2013. (The data are available in the file marathontimes.csv.)

Figure 3.28 shows the data separately for each of these events. We could now compute measures for center and spread separately for the Olympic and amateur events. However, you will sometimes find yourself in situations where a bimodal distribution occurs but it *does* make sense to compute a single measure of center. We can't give you advice for what to do in all situations. Our best advice is always to ask, "Does my summary make sense?"

◄ **FIGURE 3.27** Marathon times reported for two marathons amateur and Olympic athletes. Note the two modes. Only women runners were included.

↻ **Looking Back**

Bimodality
Recall from Chapter 2 that a mode is represented by a major mound in a graph (such as a histogram) of a single numerical variable. A bimodal distribution has two major mounds.

Time (minutes)	Event
185.1	Olympics
202.2	Olympics
214.5	Amateur
215.7	Amateur

▲ **TABLE 3.2** Four lines of marathon times for women.

(a)

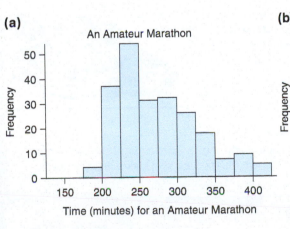

An Amateur Marathon

Time (minutes) for an Amateur Marathon

(b)

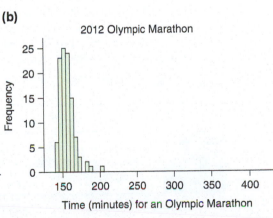

2012 Olympic Marathon

Time (minutes) for an Olympic Marathon

◄ **FIGURE 3.28** Women's times for a marathon, separated into two groups: **(a)** amateur athletes and **(b)** Olympic athletes.

Comparing Two Groups with Different-Shaped Distributions

Sometimes you'll have situations in which you wish to compare two groups, but one group has a symmetric distribution and the other a skewed distribution. Which measure, mean or median, should you use for the comparison?

When comparing two distributions, you should always use the same measures of center and spread for both distributions. Otherwise, the comparison is not valid.

EXAMPLE 16 Marathon Times, Revisited

In Figure 3.27 we lumped all of the marathon runners' finishing times into one group. But in fact, our data set had a variable that told us which time belonged to an Olympic runner and which to an amateur runner, so we could separate the data into groups. Figure 3.28 shows the same data, separated by groups.

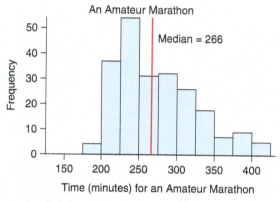
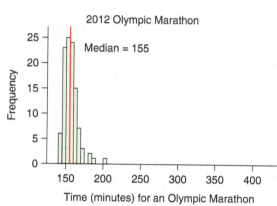

▲ **FIGURE 3.29** Women's times for a marathon, separated into two groups: **(a)** amateur athletes and **(b)** Olympic athletes. The median is shown for each group.

QUESTION   Typically, which group has the fastest finishing times?

SOLUTION   The distribution of Olympic runners is right-skewed, so the median would be the best measure. Although the distribution of amateur runners is more symmetric, we report the median because we want to compare the typical running time of the amateurs to the typical running time of the Olympic runners. The median of the women Olympic runners is 154.8 minutes (about 2.6 hours), and the median for the amateur women runners is 266.3 minutes (about 4 hours and 24 minutes). Figure 3.29 shows the location of each group's median running time.

CONCLUSION   The typical woman Olympic runner finishes the marathon considerably faster: a median time of 154.8 minutes (about 2.6 hours) compared to 266.3 minutes (about 4 hours and 24 minutes) for the amateur athlete.

TRY THIS!   Exercise 3.59

KEY POINT   When you are comparing groups, if any one group is strongly skewed or has outliers, it is usually best to compare the medians and interquartile ranges for all groups.

Using Boxplots for Displaying Summaries

Boxplots are a useful graphical tool for visualizing a distribution, especially when comparing two or more groups of data. A boxplot shows us the distribution divided into fourths. The left edge of the box is at the first quartile (Q1) and the right edge is at the third quartile (Q3). Thus, the middle 50% of the sorted observations lie inside the box. Therefore, the length of the box is proportional to the IQR.

A vertical line inside the box marks the location of the median. Horizontal lines, called whiskers, extend from the ends of the box to the smallest and largest values, or nearly so. (We'll explain soon.) Thus, the entire length of the boxplot spans most, or all, of the range of the data.

Figure 3.30 compares a dotplot (with the quartiles marked with vertical lines) and a boxplot for the price of gas at stations in Austin, Texas, as discussed in Examples 6 and 11.

Unlike many of the graphics used to visualize data, boxplots are relatively easy to draw by hand, assuming that you've already found the quartiles. Still, most of the time you will use software or a graphing calculator to draw the boxplot. Most software packages produce a variation of the boxplot that helps identify observations that are extremely large or small compared to the bulk of the data.

These extreme observations are called potential outliers. Potential outliers are different from the outliers we discussed in Chapter 2 because sometimes points that look extreme in a boxplot are not that extreme when shown in a histogram or dotplot. They are called *potential* outliers because you should consult a histogram or dotplot of the distribution before deciding whether the observation is too extreme to fit the pattern of the distribution. (Remember, whether or not an observation is an outlier is a subjective decision.)

Potential outliers are identified by this rule: They are observations that are a distance of more than 1.5 interquartile ranges below the first quartile (the left edge of a horizontal box) or above the third quartile (the right edge).

To allow us to see these potential outliers, the whiskers are drawn from the edge of each box to the most extreme observation that is not a potential outlier. This implies that before we can draw the whiskers, we must identify any potential outliers.

(a)

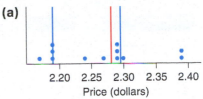

(b)

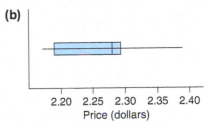

▲ **FIGURE 3.30** Two views of the same data: **(a)** A dotplot with Q1, Q2, and Q3 indicated and **(b)** a boxplot for the price of regular, unleaded gas at stations in Austin, Texas.

KEY POINT   Whiskers in a boxplot extend to the most extreme values that are not potential outliers. Potential outliers are points that are more than 1.5 IQRs from the edges of the box.

Figure 3.31 is a boxplot of temperatures in San Francisco (see Examples 8 and 10). From the boxplot, we can see that

$$IQR = 70 - 59 = 11$$
$$1.5 \times IQR = 1.5 \times 11 = 16.5$$
$$\text{Right limit} = 70 + 16.5 = 86.5$$
$$\text{Left limit} = 59 - 16.5 = 42.5$$

Any points below 42.5 or above 86.5 would be potential outliers.

The whiskers go to the most extreme values that are not potential outliers. On the left side of the box, observations smaller than 42.5 would be potential outliers. However, there are no observations that small. The smallest is 49, so the whisker extends to 49.

▶ **FIGURE 3.31** Boxplot of maximum daily temperatures in San Francisco.

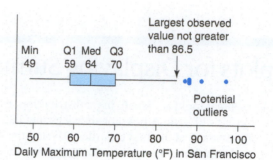

On the right, several values in the data set are larger than 86.5. The whisker extends to the largest temperature that is less than (or equal to) 86.5, and the larger values are shown in Figure 3.31 with dots. These represent days that were unusually warm in San Francisco. Figure 3.32 shows a boxplot made with a TI-84.

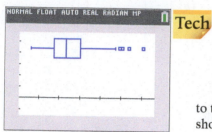

▲ **FIGURE 3.32** TI-84 output for a boxplot of San Francisco temperatures.

EXAMPLE 17 Skyscraping

Figure 3.33 shows the distribution of the 960 tallest buildings in the world as measured by the number of floors in the building. Some summary statistics: Q1 is 50 floors, the median is 55 floors, and Q3 is 64 floors. The tallest building is the Burj Khalifa in Dubai, with 162 floors.

QUESTION Sketch the boxplot. Describe how you determined where to draw the whiskers. Are there outliers? Are the outliers mostly very short buildings or very tall buildings?

▶ **FIGURE 3.33** Number of floors for the 960 tallest buildings in the world. The vertical lines indicate (from left to right) the first quartile, the median and the third quartile.

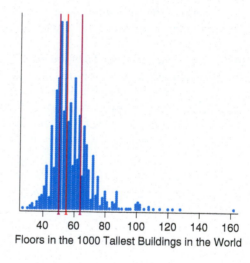

Floors in the 1000 Tallest Buildings in the World

Data Moves ▶ These data were accessed from phorio.com using *StatCrunchThis*.

Data ▶ tall_buildings.csv

SOLUTION The left edge of the box is at the first quartile, 50 floors, and the right edge is at the third quartile, 64 floors. We draw a line inside the box at the median of 55 floors.

Potential outliers on the left side of the box are those more than 1.5 × IQR below Q1. The IQR is Q3 − Q1 = 64 − 50 = 14. Thus, 1.5 × IQR = 1.5 × 14 = 21. This means that potential outliers on the left must be 21 floors less than 50, or 50 − 21 = 29, floor or below. There is no 29-floor building in this data set, so we draw the whisker to the next-shortest at 30 floors. This leaves the shortest building with 27 floors as a potential outlier.

On the right, potential outliers are more than 21 floors above the third quartile, so any building with more than 64 + 21 = 85 floors is a potential outlier. So we draw the right-hand-side whisker to the tallest building that has 85 or fewer floors. Because there are buildings with 85 floors, this whisker extends to 85. All the buildings with more floors we indicate with dots or stars to show that they are potential outliers.

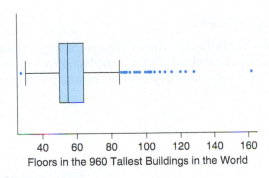

◀ **FIGURE 3.34** Boxplot summarizing the distribution of the number of floors of the world's tallest buildings.

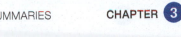

CONCLUSION   The boxplot is shown in Figure 3.34.
We see there are quite a few potential outliers on the tall side, and just one is a shorter-than-usual building.

TRY THIS!   Exercise 3.71

Investigating Potential Outliers

What do you do with potential outliers? The first step is always to investigate. A potential outlier might not be an outlier at all. Or a potential outlier might tell an interesting story, or it might be the result of an error in entering data.

Figure 3.35a is a boxplot of the National Assessment of Educational Progress international math scores for 42 countries (International Math Scores 2007). One country (South Africa, as it turns out) is flagged as a potential outlier. However, if we examine a histogram, shown in Figure 3.35b, we see that this outlier is really not that extreme. Most people would not consider South Africa to be an outlier in this distribution because it is not separated from the bulk of the distribution in the histogram.

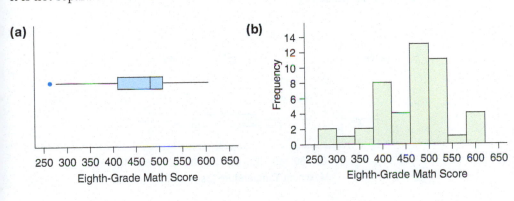

◀ **FIGURE 3.35** **(a)** Distribution of international math scores for eighth-grade achievement. The boxplot indicates a potential outlier. **(b)** The histogram of the distribution of math scores shows that although South Africa's score of 264 might be the lowest, it is not that much lower than the bulk of the data.

Figure 3.36 shows a boxplot and histogram for the fuel economy (in city driving) of the 2010 model sedans from Ford, Toyota, and GM, in miles per gallon, as listed on their websites. Two potential outliers appear, which are far enough from the bulk of the distribution as shown in the histogram that many people would consider them real outliers. These outliers turn out to be hybrid cars: the Ford Fusion and the Toyota Prius. These hybrids run on both electricity and gasoline, so they deliver much better fuel economy.

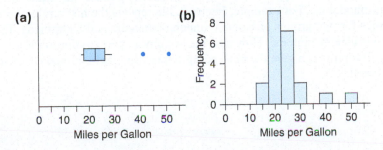

◀ **FIGURE 3.36** Distribution of fuel economies for cars from three manufacturers. **(a)** The boxplot identifies two potential outliers. **(b)** The histogram confirms that these points are indeed more extreme than the bulk of the data. (Sources: www.gm.com and www.fordvehicles.com)

Horizontal or Vertical?

Boxplots do not have to be horizontal. Many software packages (such as Minitab) provide you with the option of making vertical boxplots. (See Figure 3.37a). Which direction you choose is not important. Try both to see which is more readable.

▶ **FIGURE 3.37** **(a)** Default output of Minitab for a boxplot of the San Francisco temperatures. **(b)** Boxplot with a horizontal orientation.

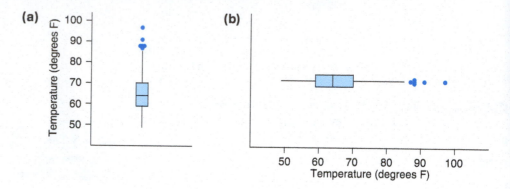

Using Boxplots to Compare Distributions

Boxplots are often a very effective way of comparing two or more distributions. How do temperatures in Provo compare with those in San Francisco? Figure 3.38 shows boxplots of daily maximum temperatures for San Francisco and Provo. At a glance, we can see how these two distributions differ and how they are similar. Both cities have similar typical temperatures (the median temperatures are about the same). Both distributions are fairly symmetric (because the median is in the center of the box, and the boxplots are themselves fairly symmetric). However, the amount of variation in daily temperatures is much greater in Provo than in San Francisco. We can see this easily because the box is wider for Provo's temperatures.

▶ **FIGURE 3.38** Boxplots of daily maximum temperatures in Provo and San Francisco emphasize the difference in variability of temperature in the two cities.

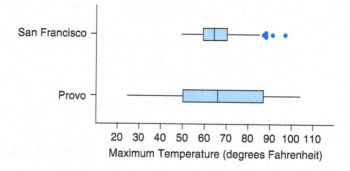

Also note that although both cities do have days that reach about 100 degrees, these days are unusual in San Francisco—they're flagged as potential outliers—but merely fall in the upper 25% for Provo.

Things to Watch for with Boxplots

Boxplots are best used only for unimodal distributions because they hide bimodality (or any multimodality). For example, Figure 3.39a repeats the histogram of marathon running times for two groups of women runners: amateurs and Olympians. The distribution is clearly bimodal. However, the boxplot in part (b) doesn't show us the bimodality. Boxplots can give the misleading impression that a bimodal distribution is really unimodal.

(a)
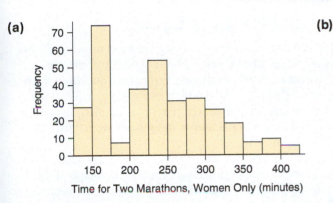

(b)
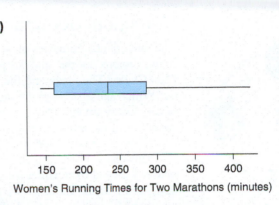

▲ **FIGURE 3.39 (a)** Histogram of finishing times (in seconds) for two groups of women marathon runners: Olympic athletes and amateurs. The graph is bimodal because the elite athletes tend to run faster, and therefore, there is a mode around 160 minutes and another mode around 240 minutes. **(b)** The boxplot hides this interesting feature.

Boxplots should *not* be used for very small data sets. It takes at least five numbers to make a boxplot, so if your data set has fewer than five observations, you can't make a boxplot.

Finding the Five-Number Summary

Boxplots are not really pictures of a distribution in the way that histograms are. Instead, boxplots help us visualize the location of various summary statistics. The boxplot is a visualization of a numerical summary called the **five-number summary**. These five numbers are

the minimum, Q1, the median, Q3, and the maximum

For example, for daily maximum temperatures in San Francisco, the five-number summary is

49, 59, 64, 70, 97

as you can see in Figure 3.31.

Note that a boxplot is not just a picture of the five-number summary. Boxplots always show the maximum and minimum values, but sometimes they also show us potential outliers.

SNAPSHOT ▶ The Boxplot

WHAT IS IT? ▶ A graphical summary.

WHAT DOES IT DO? ▶ Provides a visual display of numerical summaries of a distribution of numerical data.

HOW DOES IT DO IT? ▶ The box stretches from the first quartile to the third quartile, and a vertical line indicates the median. Whiskers extend to the largest and smallest values that are not potential outliers, and potential outliers are indicated with special marks.

HOW IS IT USED? ▶ Boxplots are useful for comparing distributions of different groups of data.

Numerical Summaries and the Data Cycle

The numerical summaries presented in this chapter are among the most important tools in our toolbox for the analysis phase of the Data Cycle. (Figure 3.40 illustrates the Data Cycle.) The mean and the median each helps us answer statistical questions such as "Which age group ran the marathon fastest?" and "Which treatment group had the best recovery time?" The standard deviation and the interquartile range provide useful measures of the amount of variability in our sample and so help us answer statistical questions about consistency and precision.

This chapter also helped motivate the need for the important "consider data" phase. We need to consider the available data carefully and examine the shape of the distribution of the variables we plan to use in our analysis, because the shape determines whether the median and the IQR are more useful than the mean and the standard deviation. In the Chapter 3 Data Project, you'll get a chance to use both visual and numerical summaries to pose and answer questions about the Los Angeles Marathon.

The Data Cycle

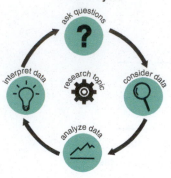

▲ **FIGURE 3.40** The phases of a statistical investigation process.

CASE STUDY REVISITED

Living in a Risky World

How do the men and women of this study compare when it comes to assigning risk to using a household appliance and to getting an annual X-ray at the doctor's? In Chapter 2 we compared groups graphically, and this is still the first step. But in this chapter, we learned about methods for comparing groups numerically, and this will enable us to be more precise in our comparison.

Our first step is to examine the pictures of the distributions to decide which measures would be most appropriate. (We repeat Figure 3.1.)

▶ **FIGURE 3.1a (repeated)** Histograms showing the distributions of perceived risk of using appliances. The women's data are shown on the left, and the men's data are shown on the right.

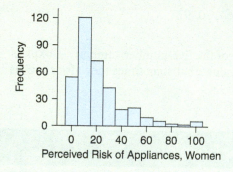
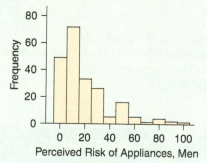

Risk of Appliances

The histograms for the perceived risk of using appliances (Figure 3.1a) do not show large differences between men and women. We can see that both distributions are right-skewed, and both appear to have roughly the same typical value, although the women's typical value might be slightly higher than the men's. Because the distribution is right-skewed, we compute the median and the IQR to compare the two groups. (Table 3.3 summarizes these comparisons.) The men assigned a median risk of 10 to using household appliances, and the women assigned a median risk of 15. We see that, first impressions aside, these women tend to feel that using appliances is a riskier activity than do these men. Also, more differences in opinion occurred among these women than among these men. The IQR was 25 for women and 20 for men. Thus, the middle 50% of the women varied by as much as 25 points in how risky they saw this activity; there was less variability for the men.

	Risk of Appliances	
	Median	IQR
Men	10	20
Women	15	25

	Risk of X-Rays	
	Mean	SD
Men	46.8	20.0
Women	47.8	20.8

▲ **TABLE 3.3** Comparison of perceived risks for men and women.

Risk of X-Rays

Both distributions for the perceived risk level of X-rays (Figure 3.1b) were fairly symmetric, so it makes sense to compare the two groups using the mean and standard deviation. The mean risk level for men was 46.8 and for women was 47.8. Typically, men and women feel roughly the same about the risk associated with X-rays. The standard deviations are about the same, too: men have a standard deviation of 20.0 and women of 20.8. From the Empirical Rule, we know that a majority (about two-thirds) of men in this sample rated the risk level between 26.8 and 66.8. The majority of women rated it between 27.0 and 68.6.

► FIGURE 3.1b (repeated)
Histograms showing the distributions of perceived risk of X-rays. The women's data are shown on the left, and the men's data are shown on the right.

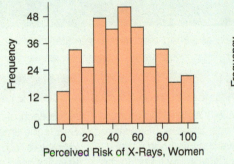 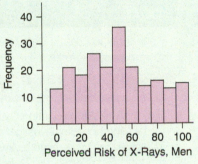

The comparisons are summarized in Table 3.3.

DATAPROJECT ▸ The Statistical Investigation Cycle

The Data Cycle

1 OVERVIEW

We engage in an entire statistical investigation cycle.

2 GOAL

Know how to use the Data Cycle as a road map to navigate statistical investigations.

3 THERE AND BACK AGAIN

The Statistical Investigation Cycle is a four-step process for tackling complex investigations. Because "Statistical Investigation Cycle" is an awkward phrase, we prefer the term *Data Cycle*. The Data Cycle is illustrated in the figure above.

The Data Cycle is not meant to be a rule or a recipe that you follow. Think of it instead as a guide. If you are stuck, ask yourself which phase you are in, and then ask yourself what you need to do to get to the next phase.

We usually start the cycle by asking questions, and usually these are statistical investigation questions (SIQs). The next step is to consider which data you can find that would help you answer these questions. But often we begin with the "consider data" step, as you will in this project, and go backwards to pose SIQs that can be addressed by these data.

You've learned about analyzing data in this chapter and Chapter 2, and so here you'll practice making graphical and numerical summaries that address your questions. Finally, you'll use your analyses to answer the questions—that's the "interpret data" phase. Often, we hope, this leads to more questions, and the cycle begins again.

Project: This time we're going to work with data that we collected. You can find the lamarathondata.csv file on StatCrunch. These data come from the 2017 Los Angeles Marathon at http://www.lamarathon.com/race-weekend/results. We've taken the liberty of making some changes and adding some variables. Take a moment to examine the data set and understand its basic structure: how many variables, how many observations, what type of variables, and what do the variable names mean?

We're guessing the more mysterious variable names (unless you are a runner) are the ones that end in *Split* as well as the *Bib.Number* variable. The Split times are the number of seconds it took the runner to

reach each of these milestones. For example, the *5k. Split* is the time it took the runner to run the first 5 kilometers. *Bib.Number* is simply the number the runner was assigned and is a unique runner ID. *Div.Place* is the place in the runner's age group. One interesting feature about large races such as this is that the time it takes a runner to get to the starting line from the moment the race begins can be considerable. For this reason, there are two race times provided. *Clock.Time* is the time (in seconds) to run the race from the moment the race began. *Net.Time* is the time to run the race from the moment the runner crossed the start line.

Assignment: Write a report based on the following activities:

1. Create at least three SIQs, with one question in each of the categories (summary, comparison, relationship). Score yourself using the rubric from Chapter 2, and rewrite the questions to improve the score, if possible.

2. For each question, create a statistical graphic that might address the question. (If your questions score high, it should be clear which variables you need to use to make your graphics.) Using only this graphic, answer your question as best as you can.

3. For the summary and comparison question, create numerical summaries to address these questions. Use these to write sentences that answer your summary and comparison statistical questions.

4. Comment on how well you were able to answer your questions; were your answers incomplete? Did you have to make assumptions? Did these raise even more questions in your mind?

5. For each of the questions above, indicate which phase or phases of the Data Cycle you were engaged with.

CHAPTER REVIEW

KEY TERMS

mean, 92
sample mean, 94
average, 94
standard deviation, 98
deviation, 100

variance, 101
Empirical Rule, 102
standard units, 104
z-score, 104
median, 107

interquartile range (IQR), 110
quartiles, 111
first quartile (Q1), 111
second quartile (Q2), 111
third quartile (Q3), 111

range, 114
resistant to outliers, 116
boxplot, 119
potential outlier, 119
five-number summary, 123

LEARNING OBJECTIVES

After reading this chapter and doing the assigned homework problems, you should

- Understand how measures of center and spread are used to describe characteristics of real-life samples of data.

- Understand when it is appropriate to use the mean and the standard deviation and when it is better to use the median and the interquartile range.

- Understand the mean as the balancing point of the distribution of a sample of data and the median as the point that has roughly 50% of the distribution below it.

- Be able to write comparisons between samples of data in context.

SUMMARY

The first step in any statistical investigation is to make a picture of the distribution of a numerical variable using a dotplot or histogram. Before computing any summary statistics, you must examine a graph of the distribution to determine the shape and whether or not there are outliers. As noted in Chapter 2, you should report the shape, the center, and the variability of every distribution.

If the shape of the distribution is symmetric and there are no outliers, you can describe the center and the spread by using either the median with the interquartile range or the mean with the standard deviation, although it is customary to use the mean and the standard deviation.

If the shape is skewed or there are outliers, you should use the median with the interquartile range.

If the distribution is multimodal, try to determine whether the data consist of separate groups; if so, you should analyze these groups separately. Otherwise, see whether you can justify using a single measure of center and spread.

If you are comparing two distributions and one of the distributions is skewed or has outliers, then it is usually best to compare the median and interquartile ranges for both groups.

The choices are summarized in Table 3.4.

Converting observations to z-scores changes the units to standard units, and this enables us to compare individual observations from different groups.

Formulas

Formula 3.1:   Mean $= \bar{x} = \dfrac{\sum x}{n}$

The mean is the measure of center best used if the distribution is symmetric.

Formula 3.2:   Standard deviation $= s = \sqrt{\dfrac{\sum(x - \bar{x})^2}{n - 1}}$

The standard deviation is the measure of variability best used if the distribution is symmetric.

Shape	Summaries of Center and Variability
If distribution is bimodal or multimodal	Try to separate groups, but if you cannot, decide whether you can justify using a single measure of center. If a single measure will not work, then report the approximate locations of the modes.
If any group's distribution is strongly skewed or has outliers	Use medians and interquartile ranges for all groups.
If all groups' distributions are roughly symmetric	Use means and standard deviations for all groups.

▲ **TABLE 3.4**   Preferred measures to report when summarizing data or comparing two or more groups.

Formula 3.3:   Variance $= s^2 = \dfrac{\sum(x - \bar{x})^2}{n - 1}$

The variance is another measure of variability used if the distribution is symmetric.

Formula 3.4b: $z = \dfrac{x - \bar{x}}{s}$

This formula converts an observation to standard units.

Formula 3.5:   Interquartile range $= Q3 - Q1$

The interquartile range is the measure of variability best used if the distribution is skewed.

Formula 3.6:   Range $=$ maximum $-$ minimum

The range is a crude measure of variability.

SOURCES

Carlstrom, L. K., J. A. Woodward, and C. G. Palmer. 2000. Evaluating the simplified conjoint expected risk model: Comparing the use of objective and subjective information. *Risk Analysis*, vol. 20: 385–392.

Environmental Protection Agency. Particulate matter. http://www.epa.gov/airtrends/factbook.html

International math scores. 2007. http://www.edweek.org/ew/articles/2007/05/02/35air.h26.html.

Provo temperatures. 2007. http://www.pgjr.alpine.k12.ut.us/science/james/provo2006.html.

San Francisco temperatures. 2007. http://169.237.140.1/calludt.cgi/WXDATAREPORT.

SECTION EXERCISES

SECTION 3.1

3.1 Earnings A sociologist says, "Typically, men in the United States still earn more than women." What does this statement mean? (Pick the best choice.)

a. All men make more than all women in the United States.

b. All U.S. women's salaries are less varied than all men's salaries.

c. The center of the distribution of salaries for U.S. men is greater than the center for women.

d. The highest-paid people in the United States are men.

3.2 Houses A real estate agent claims that all things being equal, houses with swimming pools tend to sell for less than those without swimming pools. What does this statement mean? (Pick the best choice.)

a. There are fewer homes with swimming pools than without.

b. The typical price for homes with pools is smaller than the typical price for homes without pools.

c. There's more variability in the price of homes with pools than in the price of those without.

d. The most expensive houses sold do not have pools.

TRY 3.3 Exercise Hours (Example 1) The histogram shows the self-reported number of exercise hours in a week for 50 students. Without doing calculations, what is the approximate mean number of exercise hours for these 50 students? Explain.

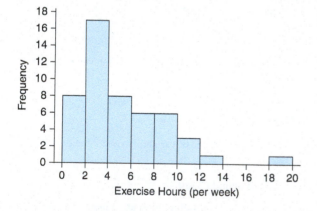

3.4 Sodium The following histogram shows the sodium level for 132 people (from the National Health and Nutrition Examination Survey [NHANES]). The units are milliequivalents per liter. Without calculating, give the approximate mean of this distribution.

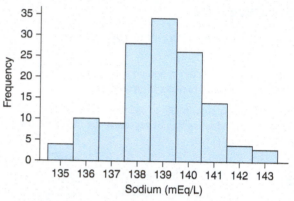

3.5 Tall Buildings The table shows the location and number of floors in some of the tallest buildings in the world. (Source: Infoplease.com)

City	# Floors
Dubai	163
Mecca	120
Hong Kong	108
Taipei	101
Shanghai	101

a. Find and interpret (report in context) the mean number of floors in this data set.

b. Find and interpret the standard deviation of the number of floors in this data set.

c. Which of the given observations is farthest from the mean and therefore contributes most to the standard deviation?

3.6 Roller Coasters The table shows the names and heights of some of the tallest roller coasters in the United States. (Source: Today.com)

Name	Height (in feet)
Kingda Ka	456
Top Thrill Dragster	420
Superman	415
Fury 325	325
Millennium Force	310

a. Find and interpret (report in context) the mean height of these roller coasters.

b. Find and interpret the standard deviation of the height of these roller coasters.

c. If the Kingda Ka coaster was only 420 feet high, how would this affect the mean and standard deviation you calculated in (a) and (b). Now recalculate the mean and standard deviation using 420 as the height of Kingda Ka. Was your prediction correct?

▼3.7 River Lengths (Example 2) The table shows the lengths (in miles) of major rivers in North America. (Source: *World Almanac and Book of Facts 2017*)

River	Length (in miles)
Arkansas	1459
Colorado	1450
Mackenzie	2635
Mississippi-Missouri-Red Rock	3710
Rio Grande	1900

a. Find and interpret (report in context) the mean, rounding to the nearest tenth mile. Be sure to include units for your answer.

b. Find the standard deviation, rounding to the nearest tenth mile. Be sure to include units for your answer. Which river contributes most to the size of the standard deviation? Explain.

c. If the St. Lawrence River (length 800 miles) were included in the data set, explain how the mean and standard deviation from parts (a) and (b) would be affected? Then recalculate these values including the St. Lawrence River to see if your prediction was correct.

3.8 Children of First Ladies This list represents the number of children for the first six "first ladies" of the United States. (Source: *2009 World Almanac and Book of Facts*)

Martha Washington	0
Abigail Adams	5
Martha Jefferson	6
Dolley Madison	0
Elizabeth Monroe	2
Louisa Adams	4

a. Find the mean number of children, rounding to the nearest tenth. Interpret the mean in this context.

b. According to eh.net/encyclopedia, women living around 1800 tended to have between 7 and 8 children. How does the mean of these first ladies compare to that?

c. Which of the first ladies listed here had the number of children that is farthest from the mean and therefore contributes most to the standard deviation?

d. Find the standard deviation, rounding to the nearest tenth.

3.9 200-Meter Run The table show the gold medal Olympic times (in seconds) for the 200-meter run. Data are shown for the first five Olympics of the 1900s and five more recent Olympics in the 2000s. (Source: *World Almanac and Book of Facts 2017*)

Olympic Year	Time	Olympic Year	Time
1900	22.2	2000	20.1
1904	21.6	2004	19.8
1908	22.6	2008	19.3
1912	21.7	2012	19.3
1920	22.0	2016	19.8

a. Find and interpret (report in context) the mean and standard deviation of the winning times for the first five Olympics of the 1900s. Round to the nearest hundredth of a second.

b. Find the mean and standard deviation of the winning times for the more recent Olympics.

c. Compare the winning times of the early 1900s and the 2000s Olympics. Are recent winners faster or slower than those of the early 1900s? Which group has less variation in its winning times?

3.10 Olympic Swimming Times The table shows the 100-meter backstroke and the 100-meter butterfly gold medal Olympic times (in seconds) for five recent Olympics.

100-Meter Backstroke	100-Meter Butterfly
53.7	52.0
54.1	51.3
52.6	50.1
52.2	51.2
52.0	50.4

a. Find and interpret (report in context) the mean and standard deviation of the gold medal times for each stroke. Round to the nearest hundredth of a second.

b. Compare the mean and the standard deviation for the two strokes. Which stoke tends to have a faster gold medal time? Which has more variation in winning times?

TRY 3.11 Wedding Costs by Gender (Example 3) StatCrunch did a survey asking respondents their gender and how much they thought should be spent on a wedding. The following table shows Minitab descriptive statistics for wedding costs, split by gender.

a. How many people were surveyed?

b. Compare the results for men and women. Which group thought more should be spent on a wedding? Which group had more variation in their responses?

Descriptive Statistics: Amount

Statistics

Variable	Gender	N	Mean	StDev	Minimum	Q1	Median	Q3	Maximum
Amount	Female	117	35,378	132,479	0	5,000	10,000	20,000	1,000,000
	Male	68	54,072	139,105	2	5,000	10,000	30,000	809,957

3.12 Wedding Costs by Experience StatCrunch did a survey asking respondents how much they thought should be spent on a wedding and whether or not they had already had a wedding. The following table shows Minitab descriptive statistics for wedding costs. Compare the means and the standard deviations for those who had a wedding and those who had not. Which group thought more should be spent on a wedding? Which group had more variation in its responses?

Descriptive Statistics: Amount

Statistics

Variable	Had Wedding	N	N*	Mean	SE Mean	StDev	Minimum	Q1	Median	Q3	Maximum
Amount	No	130	0	44,060	12,122	138,207	0	5,000	10,000	25,000	1,000,000
	Yes	55	0	37,970	17,232	127,794	0	5,000	8,000	20,000	809,957

3.13 Surfing College students and surfers Rex Robinson and Sandy Hudson collected data on the self-reported numbers of days surfed in a month for 30 longboard surfers and 30 shortboard surfers.

Longboard: 4, 9, 8, 4, 8, 8, 7, 9, 6, 7, 10, 12, 12, 10, 14, 12, 15, 13, 10, 11, 19, 19, 14, 11, 16, 19, 20, 22, 20, 22
Shortboard: 6, 4, 4, 6, 8, 8, 7, 9, 4, 7, 8, 5, 9, 8, 4, 15, 12, 10, 11, 12, 12, 11, 14, 10, 11, 13, 15, 10, 20, 20

a. Compare the means in a sentence or two.

b. Compare the standard deviations in a sentence or two.

3.14 State College Tuition The tuition costs (in dollars) for a sample of four-year state colleges in California and Texas are shown below. Compare the means and the standard deviations of the data and compare the state tuition costs of the two states in a sentence or two. (Source: calstate.edu, texastribune.com)

CA: 7040, 6423, 6313, 6802, 7048, 7460

TX: 7155, 7504, 7328, 8230, 7344, 5760

TRY 3.15 Winter Temperatures San Jose and Denver (Example 4) The histograms below were created from data on the daily high temperature in San Jose and Denver during a winter month. Compare the two distributions. Which city do you think has the higher typical temperature? Which city has more variation in temperature? Explain. (Source: Accuweather.com)

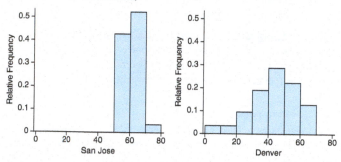

3.16 Summer Temperatures L.A. and NYC The histograms below were created from data on the daily high temperature in Los Angeles and New York City during a summer month. Compare the two distributions. Which city do you think has the higher typical temperature? Which city has more variation in temperature? Explain. (Source: Accuweather.com)

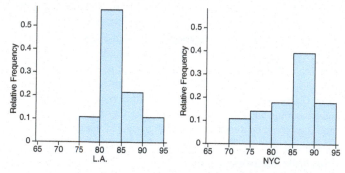

TRY 3.17 Weight Gain during Pregnancy (Example 5) The mean weight gain for women during a full-term pregnancy is 30.2 pounds. The standard deviation of weight gain for this group is 9.9 pounds, and the shape of the distribution of weight gains is symmetric and unimodal. (Source: BMJ 2016; 352 doi: https://doi.org/10.1136/bmj.i555)

a. State the weight gain for women one standard deviation below the mean and for one standard deviation above the mean.

b. Is a weight gain of 35 pounds more or less than one standard deviation from the mean?

3.18 Birth Length The mean birth length for U.S. children born at full term (after 40 weeks) is 52.2 centimeters (about 20.6 inches). Suppose the standard deviation is 2.5 centimeters and the distributions are unimodal and symmetric. (Source: www.babycenter.com)

a. What is the range of birth lengths (in centimeters) of U.S.-born children from one standard deviation below the mean to one standard deviation above the mean?

b. Is a birth length of 54 centimeters more than one standard deviation above the mean?

TRY 3.19 Orange Juice Prices (Example 6) From Amazon.com, the prices of 10 varieties of orange juice (59- to 64-ounce containers) sold were recorded: $3.88, $2.99, $3.99, $2.99, $3.69, $2.99, $4.49, $3.69, $3.89, $3.99.

a. Find and interpret the mean price of orange juice sold on this site. Round to the nearest cent.

b. Find the standard deviation for the prices. Round to the nearest cent. Explain what this value means in the context of the data.

3.20 Sibling Ages Four siblings are 2, 6, 9, and 10 years old.

a. Calculate the mean of their current ages. Round to the nearest tenth.

b. Without doing any calculation, predict the mean of their ages 10 years from now. Check your prediction by calculating their mean age in 10 years (when they are 12, 16, 19, and 20 years old).

c. Calculate the standard deviation of their current ages. Round to the nearest tenth.

d. Without doing any calculation, predict the standard deviation of their ages 10 years from now. Check your prediction by calculating the standard deviation of their ages in 10 years.

e. Adding 10 years to each of the siblings ages had different effects on the mean and the standard deviation. Why did one of these values change while the other remained unchanged? How does adding the same value to each number in a data set affect the mean and standard deviation?

3.21 Olympics In the most recent summer Olympics, do you think the standard deviation of the running times for all men who ran the 100-meter race would be larger or smaller than the standard deviation of the running times for the men's marathon? Explain.

3.22 Weights Suppose you have a data set with the weights of all members of a high school soccer team and all members of a high school academic decathlon team (a team of students selected because they often answer quiz questions correctly). Which team do you think would have a larger standard deviation of weights? Explain.

3.23 Home Prices (SC and TN) The prices (in $ thousand) of a sample of three-bedroom homes for sale in South Carolina and Tennessee are shown in the following table. Write a report that compares the prices of these homes. In your report, answer the questions of which state had the most expensive homes and which had the most variability in home prices. (Source: Zillow.com)

South Carolina	Tennessee
292	200
323	205
130	400
190	138
110	190
183	292
185	127

Data continued on next page.

South Carolina	Tennessee
160	183
205	215
165	280
334	200
160	220
180	125
134	160
221	302

3.24 Home Prices (FL and GA) The prices (in $ thousand) of a sample of three-bedroom homes for sale in Florida and Georgia are shown in the table. (Source: Zillow.com)

a. In which state are homes typically more expensive? Support your answer using an appropriate statistic.

b. Which state has more variation in home prices? Support your answer using an appropriate statistic.

c. Notice that one home price in the Florida data ($432.4 thousand) seems a unusually high when compared with the other home prices. What would happen to the standard deviation if this home price was removed from the data? Calculate the standard deviation, omitting this value from the data. Was your prediction correct?

Florida Home Prices ($ thousands)	Georgia Home Prices ($ thousands)
159.9	265
139.9	170
149.9	148
187.9	182
199	220.5
345	282.9
209.9	230
229.9	123.9
319.9	149.9
234.9	359.9
312.3	385
432.4	199.2
256	254.9

3.25 Drinkers The number of alcoholic drinks per week is given for adult men and women who drink. The data are at this text's website. (Source: Alcohol data from adults survey results, accessed via StatCrunch. Owner: rosesege)

a. Compare the mean number of drinks for men and women.

b. Compare the standard deviation of the number of drinks of men and of women.

c. Remove the outliers of 70 and 48 drinks for the men, and compare the means again. What effect did removing the outliers have on the mean?

d. What effect do you think removing the two outliers would have on the standard deviation, and why?

3.26 Smoking Mothers The birth weights (in grams) are given for babies born to 22 mothers who smoked during their pregnancy

and to 35 mothers who did not smoke. Seven pounds is about 3200 grams. (Source: Smoking Mothers, Holcomb 2006, accessed via StatCrunch. Owner: kupresanin99)

a. Compare the means and standard deviations in context.

b. Remove the outlier of 896 grams for the smoking mothers, make the comparison again, and comment on the effect of removing the outlier.

Smoke?		Smoke?	
No	Yes	No	Yes
3612	3276	4312	3108
3640	1974	4760	2030
3444	2996	2940	3304
3388	2968	4060	2912
3612	2968	4172	
3080	5264	2968	
3612	3668	2688	
3080	3696	4200	
3388	3556	3920	
4368	2912	2576	
3612	2296	2744	
3024	1008	3864	
2436	896	2912	
4788	2800	3668	
3500	2688	3640	
4256	3976	3864	
3640	2688	3556	
4256	2002		

3.27 Standard Deviation Is it possible for a standard deviation to be equal to zero? Explain.

3.28 Standard Deviation Is it possible for a standard deviation to be negative? Explain.

SECTION 3.2

TRY **3.29 Major League Baseball Runs (Example 7)** The histogram shows the number of runs scored by major league baseball teams for three seasons. The distribution is roughly unimodal and symmetric, with a mean of 687 and a standard deviation of 66 runs. An interval one standard deviation above and below the mean is marked on the histogram.

a. According to the Empirical Rule, approximately what percent of the data should fall in the interval from 621 to 753 (that is, one standard deviation above and below the mean)?

b. Use the histogram to estimate the actual percent of teams that fall in this interval. How did your estimate compare to the value predicted by the Empirical Rule?

c. Between what two values would you expect to find about 95% of the teams?

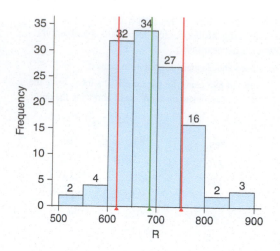

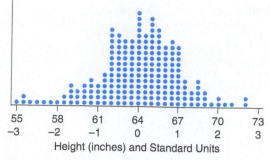

a. What is the z-score for a height of 58 inches (4 feet 10 inches)?

b. What is the height of a woman with a z-score of 1?

3.34 Heights Refer to the dotplot in the previous question.

a. What is the height of a woman with a z-score of −1?

b. What is the z-score for a woman who is 70 inches tall (5 feet 10 inches)?

TRY **3.35 Unusual IQs (Example 9)** Wechsler IQ tests have a mean of 100 and a standard deviation of 15. Which is more unusual: an IQ above 110 or an IQ below 80?

3.36 Lengths of Pregnancy Distributions of gestation periods (lengths of pregnancy) for humans are roughly bell-shaped. The mean gestation period for humans is 272 days, and the standard deviation is 9 days for women who go into spontaneous labor. Which is more unusual, a baby being born 9 days early or a baby being born 9 days late? Explain.

TRY **3.37 Low-Birth-Weight Babies (Example 10)** Babies born weighing 2500 grams (about 5.5 pounds) or less are called low-birth-weight babies, and this condition sometimes indicates health problems for the infant. The mean birth weight for U.S.-born children is about 3462 grams (about 7.6 pounds). The mean birth weight for babies born one month early is 2622 grams. Suppose both standard deviations are 500 grams. Also assume that the distribution of birth weights is roughly unimodal and symmetric. (Source: www.babycenter.com)

a. Find the standardized score (z-score), relative to all U.S. births, for a baby with a birth weight of 2500 grams.

b. Find the standardized score for a birth weight of 2500 grams for a child born one month early, using 2622 as the mean.

c. For which group is a birth weight of 2500 grams more common? Explain what that implies. Unusual z-scores are far from 0.

3.38 Birth Lengths Babies born after 40 weeks' gestation have a mean length of 52.2 centimeters (about 20.6 inches). Babies born one month early have a mean length of 47.4 centimeters. Assume both standard deviations are 2.5 centimeters and the distributions are unimodal and symmetric. (Source: www.babycenter.com)

a. Find the standardized score (z-score), relative to all U.S. births, for a baby with a birth length of 45 centimeters.

b. Find the standardized score of a birth length of 45 centimeters for babies born one month early, using 47.4 as the mean.

c. For which group is a birth length of 45 centimeters more common? Explain what that means.

3.39 Men's Heights Assume that men's heights have a distribution that is symmetric and unimodal, with a mean of 69 inches and a standard deviation of 3 inches.

3.30 Residential Energy Consumption (East) Data on residential energy consumption per capita (measured in million BTU) had a mean of 70.8 and a standard deviation of 7.3 for the states east of the Mississippi River. Assume that the distribution of residential energy use if approximately unimodal and symmetric.

a. Between which two values would you expect to find about 68% of the per capita energy consumption rates?

b. Between which two values would you expect to find about 95% of the per capita energy consumption rates?

c. If an eastern state had a per capita residential energy consumption rate of 54 million BTU, would you consider this unusual? Explain.

d. Indiana had a per capita residential energy consumption rate of 80.5 million BTU. Would you consider this unusually high? Explain.

g **3.31 Pollution Index (Example 8)** In 2017 a pollution index was
TRY calculated for a sample of cities in the eastern states using data on air and water pollution. Assume the distribution of pollution indices is unimodal and symmetric. The mean of the distribution was 35.9 points with a standard deviation of 11.6 points. (Source: numbeo. com) see Guidance page 142.

a. What percentage of eastern cities would you expect to have a pollution index between 12.7 and 59.1 points?

b. What percentage of eastern cities would you expect to have a pollution index between 24.3 and 47.5 points?

c. The pollution index for New York, in 2017 was 58.7 points. Based on this distribution, was this unusually high? Explain.

3.32 Pollution Index In 2017 a pollution index was calculated for a sample of cities in the western states using data on air and water pollution. Assume the distribution of pollution indices is unimodal and symmetric. The mean of the distribution was 43.0 points with a standard deviation of 11.3 points. (Source: numbeo.com)

a. What percentage of western cities would you expect to have a pollution index between 31.7 and 54.3 points?

b. What percentage of western cities would you expect to have a pollution index between 20.4 and 65.6?

c. The pollution index for San Jose in 2017 was 51.9 points. Based on this distribution, was this unusually high? Explain.

3.33 Heights and z-Scores The dotplot shows heights of college women; the mean is 64 inches (5 feet 4 inches), and the standard deviation is 3 inches.

a. What men's height corresponds to a z-score of 2.00?

b. What men's height corresponds to a z-score of −1.50?

3.40 Women's Heights Assume that women's heights have a distribution that is symmetric and unimodal, with a mean of 64 inches and a standard deviation of 2.5 inches.

a. What women's height corresponds with a z-score of −1.00?

b. Professional basketball player Evelyn Akhator is 75 inches tall and plays in the WNBA (women's league). Professional basketball player Draymond Green is 79 inches tall and plays in the NBA (men's league). Compared to his or her peers, who is taller? (See problem 3.39 for data on men's heights.)

SECTION 3.3

Note: Reported interquartile ranges will vary depending on technology.

3.41 Name two measures of the center of a distribution, and state the conditions under which each is preferred for describing the typical value of a single data set.

3.42 Name two measures of the variation of a distribution, and state the conditions under which each measure is preferred for measuring the variability of a single data set.

3.43 Marvel Movies (Example 11 and 13) The top ten movies based on Marvel comic book characters for the U.S. box office as of fall 2017 are shown in the following table, with domestic gross rounded to the nearest hundred million. (Source: ultimatemovieranking.com)

a. Sort the domestic gross income from smallest to largest. Find the median by averaging the two middle numbers. Interpret the median in context.

b. Using the sorted data, find Q1 and Q3. Then find the interquartile range and interpret it in context.

c. Find the range of the data. Explain why the IQR is preferred over the range as a measure of variability.

Movie	Domestic Gross ($ millions)
The Avengers (2012)	677
Spiderman (2002)	602
Spiderman 2 (2004)	520
Avengers: Age of Ultron (2015)	471
Iron Man 3 (2013)	434
Spiderman 3 (2007)	423
Captain America: Civil War (2016)	408
Guardians of the Galaxy Vol. 2 (2017)	389
Iron Man (2008)	384
Deadpool (2016)	363

3.44 DC Movies The top seven movies based on DC comic book characters for the U.S. box office as of fall 2017 are shown in the following table, rounded to the nearest hundred million. (Source: ultimatemovieranking.com)

a. Find and interpret the median in context.

b. Find and interpret the IQR in context.

c. Find the range of the data. Explain why the IQR is preferred over the range as a measure of variability.

Movie	Adjusted Domestic Gross ($ millions)
The Dark Knight (2008)	$643
Batman (1989)	$547
Superman (1978)	$543
The Dark Knight Rises (2012)	$487
Wonder Woman (2017)	$407
Batman Forever (1995)	$366
Superman II (1981)	$346

TRY **3.45 Top Seven Marvel Movies (Example 12)** Use the data in Exercise 3.43, find and interpret the median domestic gross of the top seven Marvel movies.

3.46 Top Five DC Movies Use the data in Exercise 3.44, find and interpret the median domestic gross of the top five DC movies.

3.47 Total Energy Consumption Data was collected on the total energy consumption per capita (in million BTUs) for all the states. A summary of the data is shown in the following table. (Source: eia.gov)

Summary statistics:

Column	Min	Q1	Median	Q3	Max
Total BTU	188.6	237.7	305.1	390	912.2

a. What percentage of the states consumed more than 390 million BTUs per capita?

b. What percentage of the states consumed more than 237.7 million BTUs per capita?

c. What percentage of the states consumed less than 305.1 million BTUs per capita?

d. Find and interpret in context the IQR for this data set.

3.48 Industrial Energy Consumption Data was collected on the industrial energy consumption per capita (in million BTUs) for all the states. A summary of the data is shown in the following table. (Source: eia.gov)

Summary statistics:

Column	Min	Q1	Median	Q3	Max
Industrial BTU	10.4	47.1	91.2	143.1	634.8

a. What percentage of the states consumed fewer than 47.1 million BTUs per capita?

b. What percentage of the states consumed fewer than 143.1 million BTUs per capita?

c. Complete this sentence: 50% of the states consumed more than ___ million BTUs per capita.

d. Is there more variability in Total Energy Consumption or in Industrial Energy Consumption for the states? (See problem 3.47 for data on total energy consumption.)

SECTION 3.4

3.49 Outliers

a. In your own words, describe to someone who knows only a little statistics how to recognize when an observation is an outlier. What action(s) should be taken with an outlier?

b. Which measure of the center (mean or median) is more resistant to outliers, and what does "resistant to outliers" mean?

3.50 Center and Variation When you are comparing two sets of data and one set is strongly skewed and the other is symmetric, which measures of the center and variation should you choose for the comparison?

3.51 An Error A dieter recorded the number of calories he consumed at lunch for one week. As you can see, a mistake was made on one entry. The calories are listed in increasing order:

331, 374, 387, 392, 405, 4200

When the error is corrected by removing the extra 0, will the median calories change? Will the mean? Explain without doing any calculations.

3.52 Baseball Strike In 1994, Major League Baseball (MLB) players went on strike. At the time, the average salary was $1,049,589, and the median salary was $337,500. If you were representing the owners, which summary would you use to convince the public that a strike was not needed? If you were a player, which would you use? Why was there such a large discrepancy between the mean and median salaries? Explain. (Source: www.usatoday.com)

3.53 Home Prices Home prices in San Luis Obispo County for a recent month are shown in the histogram. (Source: StatCrunch)

a. Describe the shape of the distribution.

b. Because of the shape, what measures of center and spread should be used to describe the distribution?

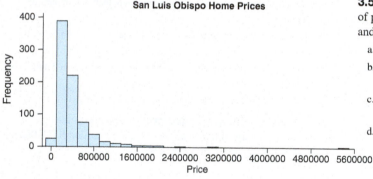
San Luis Obispo Home Prices

3.54 Youth Heights The National Longitudinal Survey records the heights of a representative sample of youths aged 14 to 20. The histograms show data for the heights of males and females. If you were comparing the heights of males and females, which measures of center and spread would you use? Why?

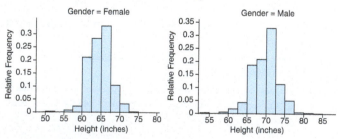

TRY 3.55 Senator Years in Office (Example 14) The following histograms show the number of years in office for Democratic and Republican U.S. senators. (Source: Infogalactic.com)

a. Describe the shape of each histogram.

b. Because of the shapes, what measures of center should be compared: the means or the medians?

c. Because of the shapes, what measures of spread should be compared: the standard deviations or the interquartile ranges?

d. Use the appropriate measures to compare the distributions of years in office for the two political parties.

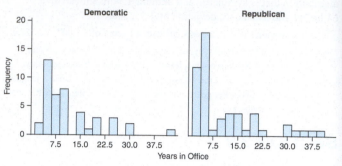

Descriptive Statistics: Years in Office

Statistics

Variable	Party	N	N*	Mean	StDev	Minimum	Q1	Median	Q3	Maximum
Years in Office	Democratic	44	0	11.750	8.835	2.000	4.000	9.000	16.000	42.000
	Republican	54	0	10.944	10.117	2.000	4.000	6.000	14.000	40.000

3.56 MLB Player Ages The following histograms show the ages of professional baseball players for two teams: the Chicago Cubs and the Oakland Athletics.

a. Describe the shape of each histogram.

b. Because of the shapes, which measures of center should be compared: the means or the medians?

c. Because of the shapes, which measures of spread should be compared: the standard deviations or the interquartile ranges?

d. Use the appropriate measures to compare the distributions of player ages.

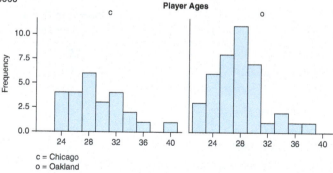
Player Ages

c = Chicago
o = Oakland

Statistics

Variable	Team	Mean	SE Mean	Minimum	Q1	Median	Q3	Maximum
age	c	28.8000	0.8347	23.0000	25.0000	28.0000	32.0000	39.0000
	o	27.2250	0.5757	21.0000	25.0000	27.0000	29.0000	37.0000

TRY 3.57 Death Row: South (Example 15) The following table shows the numbers of capital prisoners (prisoners on death row) in 2017 in the southern U.S. states. (Source: http://www.deathpenaltyinfo.org)

a. Find the median number of prisoners and interpret (using a sentence in context).

b. Find the interquartile range (showing Q1 and Q3 in the process) to measure the variability in the number of prisoners.

c. What is the mean number of capital prisoners?

d. Why is the mean so much larger than the median?

e. What is better to report the median, instead of the mean, as a typical measure?

State	Capital Prisoners
Alabama	191
Arkansas	32
Florida	374
Georgia	61
Kentucky	33
Louisiana	73
Maryland	0
Mississippi	48
North Carolina	152
Oklahoma	47
South Carolina	41
Tennessee	62
Texas	243
Virginia	5
West Virginia	0

3.58 Death Row: West The following table shows the numbers of capital prisoners (prisoners on death row) in 2017 in the western U.S. states. (Source: http://www.deathpenaltyinfo.org)

a. Find the median number of prisoners and interpret (using a sentence in context).

b. Find the interquartile range (showing Q1 and Q3 in the process) to measure the variability in the number of prisoners.

c. What is the mean number of capital prisoners?

d. Why is the mean so much larger than the median?

e. What is better to report the median, instead of the mean, as a typical measure?

State	Capital Prisoners
Alaska	0
Arizona	125
California	746
Colorado	3
Hawaii	0
Idaho	8
Montana	2
Nevada	82
New Mexico	2
Oregon	33
Utah	9
Washington	8
Wyoming	1

g 3.59 The Consumer Price Index (CPI) (Example 16) indicates cost of living for a typical consumer and is used by government economists as an economic indicator. The following data shows the CPI for large urban areas in midwestern and western states in the United States. see Guidance page 143

Midwest:

| 227.8 | 223.3 | 220.5 | 218.7 | 222.3 | 226.6 | 230.6 | 219.3 |

West:

| 216.9 | 240.0 | 260.2 | 244.6 | 128 | 244.2 | 269.4 | 258.6 | 249.4 |

Compare the CPI of the two regions. Start with a graph to determine shape; then compare appropriate measures of center and spread and mention any potential outliers.

3.60 Heights of Sons and Dads The data at this text's website give the heights of 18 male college students and their fathers, in inches.

a. Make histograms and describe the shapes of the two data sets from the histograms.

b. Fill in the following table to compare descriptive statistics.

	Mean	Median	Standard Deviation	Interquartile Range
Sons	____	____	____	____
Dads	____	____	____	____

c. Compare the heights of the sons and their dads, using the means and the standard deviations.

d. Compare the heights of the sons and their dads, using the medians and the interquartile ranges.

e. Which pair of statistics is more appropriate for comparing these samples: the mean and the standard deviation or the median and the interquartile range? Explain.

3.61 Mean from a Histogram The histogram shows the lengths of ring fingers (in millimeters) for a sample of eighth-graders. (Source: AMSTAT Census at School)

a. Without doing any calculations, approximate the mean finger length for the sample.

b. Approximate the mean by completing the work that is started below. Note the left-hand side of the bin is being used in this approximation:

$$\bar{x} = \frac{1(60) + 8(70) + 8(80) + \ldots}{27}$$

c. Explain why the method used in part (b) is an approximation of the mean rather than the actual mean.

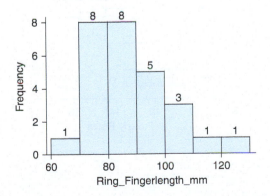

3.62 Mean from a Histogram The histogram shows the lengths of index fingers (in millimeters) for a sample of eighth-graders. (Source: AMSTAT Census at School)

a. Use the histogram to approximate the mean ring finger length for the sample.

b. Approximate the mean by completing the work that is started below. Note that the left-hand side of each bin is being used in this approximation:

$$\bar{x} = \frac{3(60) + 9(70) + 5(80) + \cdots}{27}$$

c. Explain why the method used in part (b) is an approximation of the mean rather than the actual mean.

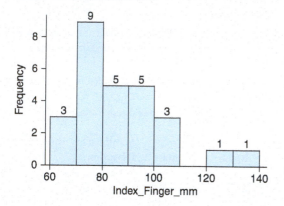

SECTION 3.5

3.63 Poverty Rates The following boxplot shows the poverty rates (the proportion of the population below the government's official poverty level) for the 50 states and the District of Columbia. The regions are the West (W), South (S), NE (Northeast), and MW (Midwest). (Source: *2017 World Almanac*)

a. List the regions from highest to lowest median poverty rate.

b. List the regions from lowest to highest interquartile range.

c. Do any of the regions have a state with an unusually low or an unusually high poverty rate? Explain.

d. Which region has the least amount of variability in poverty rate? Explain.

e. Why is the interquartile range a better measure of the variability for these data than the range is?

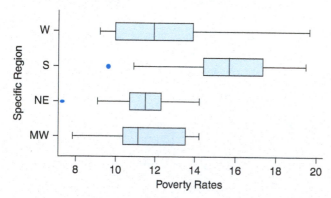

3.64 Regional Population Density The figure shows the population density (people per square mile) for the 50 states in the United States, based on an estimate from the U.S. Census Bureau. The regions are the Midwest (MW), Northeast (NE), South (S), and West (W). In the West, the potential outlier is California, and in the South, the potential outlier is Maryland.

Why is it best to compare medians and interquartile ranges for these data rather than comparing means and standard deviations? List the approximate median number of people per square mile for each location; for example, the median for the MW is between 50 and 100. Also arrange the regions from lowest interquartile range (on the left) to highest.

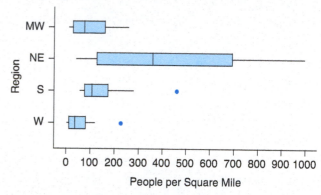

3.65 Professional Sport Ticket Prices The following boxplot shows the average ticket price for professional hockey (NHL), football (NFL), basketball (NBA), and baseball (MLB) for the 2017 seasons.

a. Which sport has the most expensive ticket prices? Which sport has the least expensive ticket prices?

b. Compare the ticket prices for professional hockey and basketball. In your comparison, compare the price of a typical ticket, the amount of variation in ticket prices, and the presence of any outliers in the data. (Source: vividtickets.com)

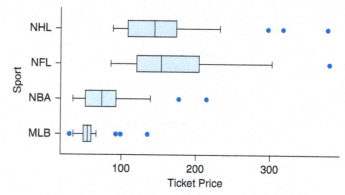

3.66 BA Attainment The following boxplot shows the percentage of the population that has earned a bachelor's (BA) degree in the western (W) and eastern (E) United States. Estimate and interpret the median for each group. (Source: *2017 World Almanac and Book of Facts*)

a. In which region does a greater percentage of the population have a BA degree?

b. Which region has more variation in BA attainment?

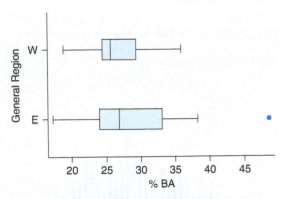

3.67 Matching Boxplots and Histograms

a. Report the shape of each of the following histograms

b. Match each histogram with the corresponding boxplot (A, B, or C).

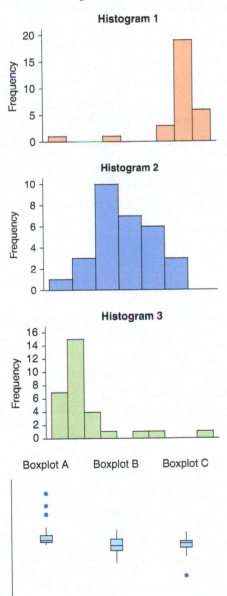

Boxplot A   Boxplot B   Boxplot C

3.68 Matching Boxplots and Histograms Match each of the histograms (X, Y, and Z) with the corresponding boxplot (C, M, or P). Explain your reasoning.

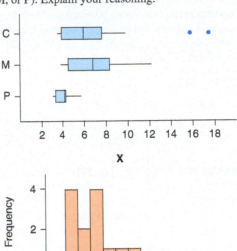

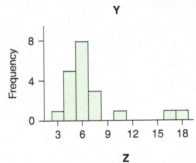

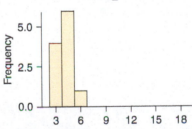

3.69 Public Libraries Data at this text's website show the number of central public libraries in each of the 50 states and the District of Columbia. A summary of the data is shown in the following table. Should the maximum and minimum values of this data set be considered potential outliers? Why or why not? You can check your answer by using technology to make a boxplot using fences to identify potential outliers. (Source: Institute of Museum and Library Services)

Summary statistics

Column	n	Mean	Std. dev.	Median	Min	Max	Q1	Q3
Central Public Libraries	51	175.76471	170.37319	112	1	756	63	237

3.70 Gas Taxes Data at this text's website show the gas taxes for each of the 50 states and the District of Columbia. A summary of the data is shown in the following table. Should the maximum and minimum values of this data set be considered potential outliers? Why or why not? You can check your answer by using technology to make a boxplot using fences to identify potential outliers. (Source: *2017 World Almanac and Book of Facts*)

Summary statistics

Column	n	Std. dev.	Median	Min	Max	Q1	Q3
Gas Taxes (ct/gal)	51	8.1011009	46.4	30.7	68.7	40.2	51

TRY **3.71 Roller Coaster Heights (Example 17)** The dotplot shows the distribution of the heights (in feet) of a sample of roller coasters. The five-number summary of the data is given in the following table. Sketch a boxplot of the data. Explain how you determined the length of the whiskers.

Minimum	Q1	Median	Q3	Maximum
2.438	8.526	18.288	33.223	128.016

Roller Coaster Height

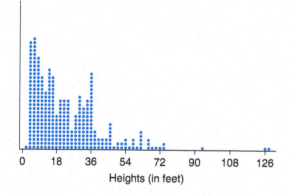

Heights (in feet)

3.72 Passing the Bar Exam The following dotplot shows the distribution of passing rates for the bar exam law schools in the United States in.

The five number summary is

0.60, 0.84, 0.90, 0.94, 1.00

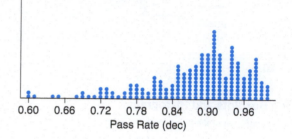

Pass Rate (dec)

Draw the boxplot and explain how you determined where the whiskers go.

* **3.73 Exam Scores** The five-number summary for a distribution of final exam scores is

40, 78, 80, 90, 100

Explain why it is not possible to draw a boxplot based on this information. (*Hint:* What more do you need to know?)

* **3.74 Exam Scores** The five-number summary for a distribution of final exam scores is

60, 78, 80, 90, 100

Is it possible to draw a boxplot based on this information? Why or why not?

CHAPTER REVIEW EXERCISES

3.75 Gas Taxes (South) The following table shows the gas tax (in cents per gallon) in each of the southern U.S. states. (Source: *2017 World Almanac and Book of Facts*)

a. Find and interpret the median gas tax using a sentence in context.

b. Find and interpret the interquartile range.

c. What is the mean gas tax?

d. Note that the mean for this data set is greater than the median. What does this indicate about the shape of the data? Make a graph of the data and discuss the shape of the data.

State	Gas Tax (cents/gallon)	State	Gas Tax (cents/gallon)
Alabama	39.3	Maryland	51
Arkansas	40.2	Mississippi	37.2
Delaware	41.4	N. Carolina	53.7
District of Columbia	41.9	S. Carolina	35.2
Florida	55	Tennessee	39.8
Georgia	49.4	Texas	38.4
Kentucky	44.4	Virginia	40.7
Louisiana	38.4	W. Virginia	51.6

3.76 Gas Taxes (West) The following table shows the gas tax (in cents per gallon) in each of the western U.S. states. (Source: *2017 World Almanac and Book of Facts*)

a. Find and interpret the median gas tax using a sentence in context.

b. Find and interpret the interquartile range.

c. What is the mean gas tax?

d. Note that the mean and the median for this data set are very similar. What does this indicate about the shape of the data? Make a graph of the data and discuss the shape of the data.

State	Gas Tax (cents/gallon)
Alaska	30.7
Arizona	37.4
California	58.8
Colorado	40.4
Hawaii	60.4
Idaho	50.4
Montana	46.2
Nevada	52.3
New Mexico	37.3
Oklahoma	35.4
Oregon	49.5
Utah	47.8
Washington	62.9
Wyoming	42.4

3.77 Final Exam Grades The data that follow are final exam grades for two sections of statistics students at a community college. One class met twice a week relatively late in the day; the other class met four times a week at 11 a.m. Both classes had the same instructor and covered the same content. Is there evidence that the performances of the classes differed? Answer by making appropriate plots (including side-by-side boxplots) and reporting and comparing appropriate summary statistics. Explain why you chose the summary statistics that you used. Be sure to comment on the shape of the distributions, the center, and the spread, and be sure to mention any unusual features you observe.

11 a.m. grades: 100, 100, 93, 76, 86, 72.5, 82, 63, 59.5, 53, 79.5, 67, 48, 42.5, 39

5 p.m. grades: 100, 98, 95, 91.5, 104.5, 94, 86, 84.5, 73, 92.5, 86.5, 73.5, 87, 72.5, 82, 68.5, 64.5, 90.75, 66.5

3.78 Speeding Tickets College students Diane Glover and Esmeralda Olguin asked 25 men and 25 women how many speeding tickets they received in the last three years.

Men: 14 men said they had 0 tickets, 9 said they had 1 ticket, 1 had 2 tickets, and 1 had 5 tickets.

Women: 18 said they had 0 tickets, 6 said they had 1 ticket, and 1 said she had 2 tickets.

Is there evidence that the men and women differed? Answer by making appropriate plots and comparing appropriate summary statistics. Be sure to comment on the shape of the distributions and to mention any unusual features you observe.

3.79 Heights The following graph shows the heights for a large group of adults. Describe the distribution, and explain what might cause this shape. (Source: www.amstat.org)

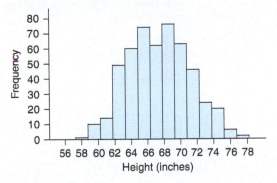

3.80 Marathon Times The following histogram of marathon times includes data for men and women and also for both an Olympic marathon and an amateur marathon. Greater values indicate slower runners. (Sources: www.forestcityroadraces.com and www.runnersworld.com)

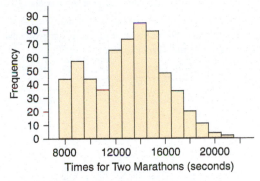

a. Describe the shape of the distribution.

b. What are two different possible reasons for the two modes?

c. Knowing that there are usually fewer women who run marathons than men and that more people ran in the amateur marathon than in the Olympic marathon, look at the size of the mounds and try to decide which of the reasons stated in part b is likely to cause this. Explain.

3.81 Streaming Video A StatCrunch survey asked people how many hours of video they watched daily. The data can be found at this text's website. Who watches more hours of video, males or females? Support your answer with appropriate graphs and summary statistics. (Source: StatCrunch, Responses to Video Streaming survey. Owner: scsurvey)

3.82 Retirement Age A StatCrunch survey asked at what age should a person consider retirement. The data can be found at this text's website. Who thinks people should retire at a younger age, males or females? Support your answer with appropriate graphs and summary statistics. (Source: StatCrunch, Responses to Retirement Age survey. Owner: scsurvey)

3.83 Chain Restaurant Calories *The New York Times* collected data on the number of calories in meals at Chipotle restaurant. The distribution of calories was symmetric and unimodal and a graph of the distribution is shown below.

a. Use the graph to estimate the mean number of calories in a Chipotle meal.

b. One method for estimating the standard deviation of a symmetric unimodal distribution is to approximate the range and then divide the range by 6. This is because nearly all the data should be within 3 standard deviations of the mean in each direction. Use this method to find an approximation of the standard deviation.

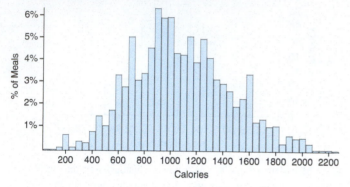

3.84 12th-Grade Sleep The histogram shows hours of sleep on a school night for a sample of 12th-grade students.

a. Use the histogram and left-hand side of each bin to estimate the mean hours (to the nearest tenth) of sleep for 12th-graders on a school night.

b. When comparing the mean and the median for this data set, would you expect these two values to be fairly similar or would you expect that one is much greater than the other? Explain.

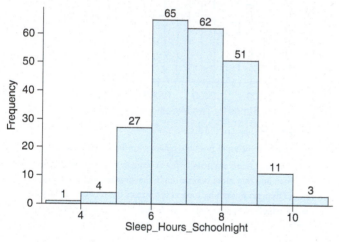

For exercises 3.85 through 3.88, construct two sets of numbers with at least five numbers in each set with the following characteristics:

3.85 The means are the same, but the standard deviation of one of the sets is smaller than that of the other. Report the mean and both standard deviations.

3.86 The means are the same, but the standard deviation of one of the sets is larger than that of the other. Report the mean and both standard deviations.

3.87 The means are different, but the standard deviations are the same. Report the standard deviation and both means.

3.88 The mean of set A is smaller than that of set B, but the median of set B is smaller than that of set A. Report the mean and the median of both sets of data.

3.89 Educational Attainment The tables below show the percentage of the population in western and southern states of the U.S. with a bachelor's degree. Write a short report comparing the education attainment of these two regions. In your report, answer the question which region has the highest educational attainment (as measured by percentage of the population with a bachelor's degree) and which region has the most variability.

Western States

26.6	25.6	29.9	35.9	29.6	23.9	27.4	21.8	25.3	22.7	29.2	28.5	31	23.8

Southern States

18.9	28.7	48.5	25.3	27.5	21	21.4	37.3	19.6	26.5	24.3	23	25.5	34	17.3

3.90 Unemployment Rates The tables below show the 2015 unemployment rates for states in the northeastern and midwestern regions of the United States. Compare the unemployment rates for the two regions, commenting on the typical unemployment rate of each region and then comparing the amount of variation in the unemployment rate for each region. (Source: *2017 World Almanac and Book of Facts*)

Northeast States

5.6	5	3.4	5.6	5.3	5.1	5	3.7

Midwest States

5.9	4.8	3.7	4.2	5.4	3.7	5	3	2.7	4.9
3.1	4.6								

3.91 Professional Sports Ticket Prices The following histograms show ticket prices for professional hockey (NHL) and professional football (NFL) tickets.

a. Based on the shape of the distributions, which measure of center should be used to compare the prices: the mean or the median?

b. Write a short report comparing the ticket prices for professional hockey and football. In your report comment on the typical ticket prices and the variability in ticket prices.

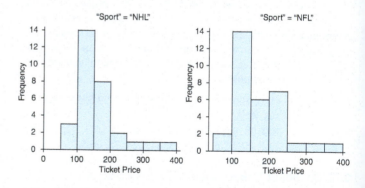

Summary statistics for Ticket Price:
Group by: Sport

Sport	n	Mean	Std. dev.	Median	Min	Max	Q1	Q3
NHL	30	158.63333	68.812831	144.5	89	378	107	174
NFL	32	168.6875	64.759823	152.5	86	380	120.5	204.5

3.92 Professional Sports Ticket Prices The following histograms show ticket prices for professional baseball (MLB) and professional basketball (NBA) tickets.

a. Based on the shape of the distributions, which measure of center should be used to compare the prices: the mean or the median?

b. Write a short report comparing the ticket prices for professional baseball and basketball. In your report comment on the typical ticket prices and the variability in ticket prices.

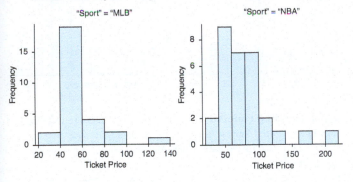

Summary statistics for Ticket Price:
Group by: Sport

Sport ⬍	n ⬍	Mean ⬍	Std. dev. ⬍	Median ⬍	Min ⬍	Max ⬍	Q1 ⬍	Q3 ⬍
MLB	28	59.142857	20.260734	55.5	32	135	50	59.5
NBA	30	80.366667	40.075346	73.5	35	215	52	93

3.93 Exam Scores An exam score has a mean of 80 and a standard deviation of 4.

a. Find and interpret in context an exam score that corresponds with a z-score of 2.

b. What exam score corresponds with a z-score of −1.5?

3.94 Boys' Heights Three-year-old boys in the United States have a mean height of 38 inches and a standard deviation of 2 inches. How tall is a three-year-old boy with a z-score of −1.0? (Source: www.kidsgrowth.com)

3.95 SAT and ACT Scores Quantitative SAT scores have a mean of 500 and a standard deviation of 100, while ACT scores have a mean of 21 and a standard deviation of 5. Assuming both types of scores have distributions that are unimodal and symmetric, which is more unusual: a quantitative SAT score of 750 or an ACT score of 28? Show your work.

3.96 Children's Heights Mrs. Diaz has two children: a 3-year-old boy 43 inches tall and a 10-year-old girl 57 inches tall. Three-year-old boys have a mean height of 38 inches and a standard deviation of 2 inches, and 10-year-old girls have a mean height of 54.5 inches and a standard deviation of 2.5 inches. Assume the distributions of boys' and girls' heights are unimodal and symmetric. Which of Mrs. Diaz's children is more unusually tall for his or her age and gender? Explain, showing any calculations you perform. (Source: www.kidsgrowth.com)

3.97 Broadway Ticket Prices The following histogram shows the average ticket prices of 28 Broadway shows in the 2017 season. The median ticket price was $97.33. (Source: BroadwayWorld.com)

a. Describe the shape of the distribution.

b. Would the mean of the data be greater than, less than, or about the same as the median of the data? Explain.

c. Would a majority of prices be greater than or less than the mean price?

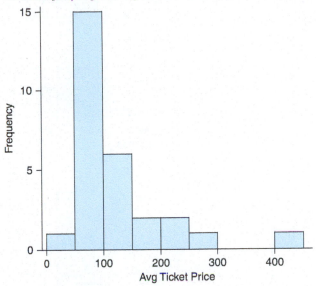

3.98 Students' Ages Here are the ages of some students in a statistics class: 17, 19, 35, 18, 18, 20, 27, 25, 41, 21, 19, 19, 45, and 19. The teacher's age is 66 and should be included as one of the ages when you do the calculations. The figure shows a histogram of the data.

a. Describe the distribution of ages by giving the shape, the numerical value for an appropriate measure of the center, and the numerical value for an appropriate measure of spread, as well as mentioning any outliers.

b. Make a rough sketch (or copy) of the histogram, and mark the approximate locations of the mean and of the median. Why are they not at the same location?

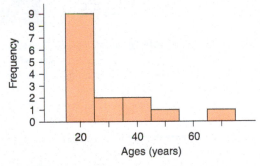

3.99 Race Times The following graph shows the time it took grade 11 and 12 student athletes to run the 100-meter race (in seconds) in a recent year. Write one or two investigative questions that could be answered by analyzing these graphs.

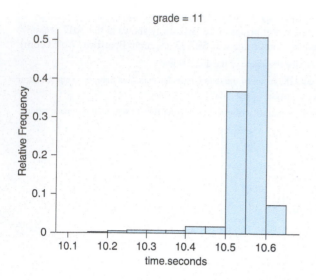

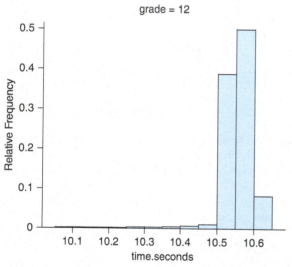

3.100 Professional Basketball Player Weights The graphs below show the weights of professional basketball players. One graph displays the weights of Centers (C) and the other graph displays the weights of Shooting Guards (SG). Write one or two investigative questions that could be answered by analyzing these graphs.

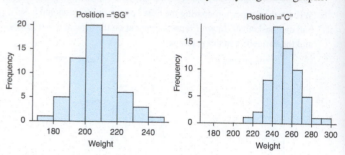

3.101 Building Heights Data was collected on the tallest buildings in the United States. A portion of the data table is shown below. Write one or two investigative questions that could be answered using these data.

City	Height (m)	Height (Ft)	# Stories	Year Built	Materials	Type
New York Ci	472.4	1550	95	2020	concrete	residential / hotel / retail
Chicago	442.1	1451	108	1974	steel	office
New York Ci	437.7	1436	113	-		residential / hotel
New York Ci	435.3	1428	82	2019	concrete	residential

3.102 Cereal Data Data were collected on cereals stocked by a supermarket. A portion of the data table is shown below. Data includes shelf location (top, middle, bottom), name, manufacturer, type (hot or cold), and amount of calories, sodium and fiber per serving. Write one or two investigative questions that could be answered using these data.

shelf	name	mfr	type	calories	sodium	fiber
Top	100%_Bran	N	C	70	130	10
Top	100%_Natural_Bran	Q	C	120	15	2
Top	All-Bran	K	C	70	260	9
Top	All-Bran_with_Extra_	K	C	50	140	14

gUIDED EXERCISES

g 3.31 Pollution Index (Example 8) In 2017 a pollution index was calculated for a sample of cities in the eastern states using data on air and water pollution. Assume the distribution of pollution indices is unimodal and symmetric. The mean of the distribution was 35.9 points with a standard deviation of 11.6 points. (Source: numbeo.com)

QUESTIONS Answer these questions by following the numbered steps.

a. What percentage of eastern cities would you expect to have a pollution index between 12.7 and 59.1 points?
b. What percentage of eastern cities would you expect to have a pollution index between 24.3 and 47.5 points?
c. The pollution index for New York, in 2017 was 58.7 points. Based on this distribution, was this unusually high? Explain.

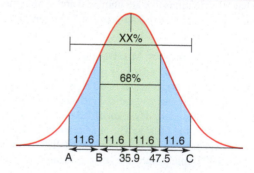

The green area is 68% of the area under the curve. The green and blue areas together shade XX% of the area. The numbers without percentage signs are crime rates, with a mean of 35.9 and a standard deviation of 11.6.

Step 1 ▶ Draw a diagram, using the Empirical Rule.

Reproduce Figure A, which is a sketch of the distribution of the pollution indices. What percentage of the data should occur within two standard deviations of the mean? Include this number in the figure, where it now says XX.

Step 2 ▶ Why 35.9?

How was the number 35.9, shown in the sketch, obtained?

Step 3 ▶ A, B, and C

Fill in the numbers for the pollution indices for areas A, B, and C.

Step 4 ▶ Read the answers for questions a and b from your graph.

What percentage of eastern cities would you expect to have a pollution index between 12.7 and 59.1 points?
b. What percentage of eastern cities would you expect to have a pollution index between 24.3 and 47.5 points?

Step 5 ▶ Unusual?

Many people consider any numbers outside the boundaries of A and C (the interval containing approximately 95% of the data) to be unusual. Use your graph to locate 58.7 on the horizontal axis. Use the location to answer question C.

c. The pollution index for New York, in 2017 was 58.7 points. Based on this distribution, was this unusually high? Explain.

3.59 The Consumer Price Index (CPI) (Example 16) indicates cost of living for a typical consumer and is used by government economists as an economic indicator. The following data show the CPI for large urban areas in midwestern and western states in the U.S.

Midwest

227.8	223.3	220.5	218.7	222.3	226.6	230.6	219.3

West

216.9	240.0	260.2	244.6	128	244.2	269.4	258.6	249.4

QUESTION Compare the CPI of the two regions. Start with a graph to determine shape; then compare appropriate measures of center and spread and mention any potential outliers.

Step 1 ▶ Histograms

Make histograms of the two sets of data separately. (You may want to use the same horizontal axes so that you can see the comparison easily.)

Step 2 ▶ Shapes

Report the shapes of the two data sets.

Step 3 ▶ Measures to Compare

If either data set is skewed or has an outlier (or more than one), you should compare medians and interquartile ranges for *both* groups. If both data sets are roughly symmetric, you should compare means and standard deviations. Which measures should be compared with these two data sets?

Step 4 ▶ Compare Centers

Compare the centers (means or medians) in the following sentence: The _____ (mean or median) Consumer Price Index for the midwestern states was ___ and the _____ (mean or median) Consumer Price Index for the western states was _____. This shows that the CPI was larger for the _____ states.

Step 5 ▶ Comparing Variations

Compare the variations in the following sentence: The _____ (standard deviation or interquartile range) for the CPI for midwestern states was ____ and for western states the ____ (standard deviation or interquartile range) was _____. This shows that the _____ states tended to have more variation as measured by the _____ (standard deviation or interquartile range).

Step 6 ▶ Outliers

Report any outliers and state to which group(s) they belong.

Step 7 ▶ Final Comparison

Finally, in a sentence or two, make a complete comparison of CPI for the two regions.

TechTips

EXAMPLE ► Analyze the data given by finding descriptive statistics and making boxplots. The tables give calories per ounce for sliced ham and turkey. Table 3A shows unstacked data, and Table 3B shows stacked data. We coded the meat types with numerical values (1 for ham and 2 for turkey), but you could also use descriptive terms, such as *ham* and *turkey*.

Ham	Turkey
21	35
25	25
35	25
35	25
25	25
30	25
30	29
35	29
40	23
30	50
	25

▲ TABLE 3A

Cal	Meat
21	1
25	1
35	1
35	1
25	1
30	1
30	1
35	1
40	1
30	1
35	2
25	2
25	2
25	2
25	2
25	2
29	2
29	2
23	2
50	2
25	2

▲ TABLE 3B

TI-84

Enter the unstacked data (Table 3A) into **L1** and **L2**.

For Descriptive Comparisons of Two Groups
Follow the steps twice, first for **L1** and then for **L2**.

Finding One-Variable Statistics
1. Press **STAT**, choose **CALC** (by using the right arrow on the keypad), and choose **1** (for **1-Var Stats**).
2. Specify **L1** (or the list containing the data) by pressing **2ND**, **1**, and **ENTER**. Then press **ENTER**, **ENTER**.
3. Output: On your calculator, you will need to scroll down using the down arrow on the keypad to see all of the output.

Making Boxplots
1. Press **2ND**, **STATPLOT**, **4** (**PlotsOff**), and **ENTER** to turn the plots off. This will prevent you from seeing old plots as well as the new ones.
2. Press **2ND**, **STATPLOT**, and **1**.

3. Refer to Figure 3a. Turn on **Plot1** by pressing **ENTER** when **On** is flashing. (**Off** will no longer be highlighted.)

▲ **FIGURE 3a** TI-84 Plot Selection Screen

4. Use the arrows on the keypad to locate the boxplot with outliers (as shown highlighted in Figure 3a), and press **ENTER**. (If you accidentally choose the other boxplot, there will never be any separate marks for potential outliers.)

5. Use the down arrow on the keypad to get to the **XList**. Choose **L₁** by pressing **2ND** and **1**.

6. Press **GRAPH**, **ZOOM**, and **9** (**Zoomstat**) to make the graph.

7. Press **TRACE** and move around with the arrows on the keypad to see the numerical labels.

Making Side-by-Side Boxplots

For side-by-side boxplots, turn on **Plot2**, choose the boxplot with outliers, and enter **L₂** (**2ND** and **2**) for **XLIST**. Then, when you choose **GRAPH**, **ZOOM**, and **9**, you should see both boxplots. Press **TRACE** to see numbers.

Figure 3b shows side-by-side boxplots with the boxplot from the turkey data on the bottom and the boxplot for the ham data on the top.

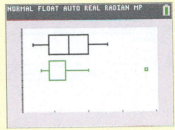

▲ **FIGURE 3b** TI-84 Boxplots

MINITAB

For Comparisons of Two Groups

Input the data from the text's website or manually enter the data given. You may use unstacked data entered into two different columns (Table 3A), or you may use stacked data and use descriptive labels such as Ham and Turkey or codes such as 1 and 2 (Table 3B).

Finding Descriptive Statistics: One-Column Data or *Unstacked* Data in Two or More Columns

1. **Stat > Basic Statistics > Display Descriptive Statistics**.

2. Double click on the column(s) containing the data, such as **Ham** and **Turkey**, to put it (them) in the **Variables** box.

3. Ignore the **By variables (optional)** box.

4. Click on **Statistics**; you can choose what you want to add, such as the interquartile range, then click **OK**.

5. Click **OK**.

Finding Descriptive Statistics: *Stacked and Coded*

1. **Stat > Basic Statistics > Display Descriptive Statistics**.

2. See Figure 3c: Double-click on the column(s) containing the stack of data, **Calories**, to put it in the **Variables** box.

3. When the **By variables (optional)** box is activated (by clicking in it), double click the column containing the categorical labels, here **Meat**.

4. Click on **Statistics**; you can choose what you want to add, such as the interquartile range. You can also make boxpots by clicking **Graphs**. Click **OK**.

5. Click **OK**.

Making Boxplots

1. **Graph > Boxplot**.

2. For a single boxplot, choose **One Y**, **Simple**, and click **OK**. See Figure 3d.

3. Double click the label for the column containing the data, **Ham** or **Turkey**, and click **OK**.

4. For side-by-side boxplots

   a. If the data are unstacked, choose **Multiple Y's, Simple**, shown in Figure 3d. Then double click both labels for the columns and click **OK**.

   b. If the data are stacked, choose **One Y, With Groups** (the top right in Figure 3d) and click **OK**. Then see Figure 3e. Double click on the label for the data stack (such as **Calories**), then click in the **Categorical variables ...** box and double click the label for codes or words defining the groups, such as **Meat**.

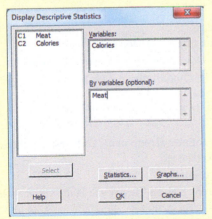

▲ **FIGURE 3c** Minitab Descriptive Statistics Input Screen

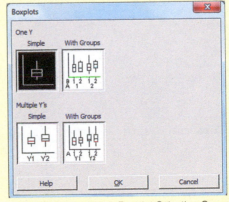

▲ **FIGURE 3d** Minitab Boxplot Selection Screen

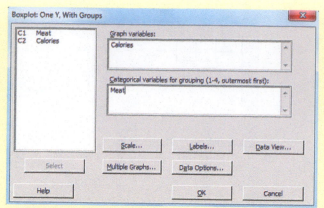

▲ FIGURE 3e Minitab Input Screen for Boxplots: One Y, with Groups

5. Labeling and transposing the boxplots. If you want to change the labeling, double click on what you want to change after the

boxplot(s) are made. To change the orientation of the boxplot to horizontal, double click on the *x*-axis and select **Transpose value and category scales**.

Figure 3f Shows Minitab boxplots of the ham and turkey data, without transposition.

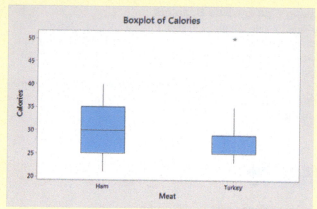

▲ FIGURE 3f Minitab Boxplots

EXCEL

Entering Data

Input the unstacked data from the text's website or enter them manually. You may have labels in the first row such as **Ham** and **Turkey**.

Finding Descriptive Statistics

1. Click **Data, Data Analysis, Descriptive Statistics**, and **OK**.
2. See Figure 3g: In the dialogue screen, for the **Input Range** highlight the cells containing the data (one column only) and then Check **Summary Statistics**. If you include the label in the top cell such as A1 in the **Input Range**, you need to check **Labels in First Row**. Click **OK**.

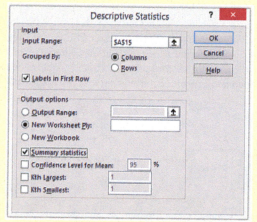

▲ FIGURE 3g Excel Input Screen for Descriptive Statistics

Comparing Two Groups

To compare two groups, use unstacked data and do the preceding analysis twice. (Choosing **Output Range** and selecting appropriate cells on the same sheet makes comparisons easier.)

Boxplots (Requires XLSTAT Add-in)

1. Click **XLSTAT, > Describing data > Descriptive statistics. Descriptive statistics**.
2. When the box under **Quantitative data** is activated, drag your cursor over the column containing the data, including the label at the top such as **Ham**.
3. Click **Charts(1)**.
4. Click **Box plots, Options, Outliers** and choose **Horizontal (or Vertical)**.
5. Click **OK** and **Continue**. See step 6 in the side-by-side instructions.

Side-by-Side Boxplots

Use unstacked data with labels in the top row.

1. Click **XLSTAT > Describing data > Descriptive statistics**.
2. When the box under **Quantitative data** is activated, drag your cursor over all the columns containing the unstacked data, including labels at the top such as **Ham** and **Turkey**.
3. Click **Charts(1)**.
4. Click **Box plots, Options, Group plots, Outliers,** and choose **Horizontal** (or **Vertical**).
5. Click **OK** and **Continue**.
6. When you see the small labels **Turkey** and **Ham**, you may drag them to where you want them, and you can increase the font size.

Figure 3h shows boxplots for the ham and turkey data. The red crosses give the locations of the two means.

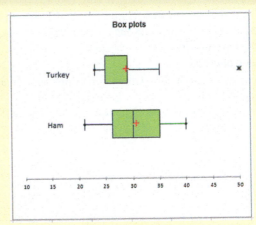

▲ **FIGURE 3h** XLSTAT Boxplots

For Comparisons of Two Groups

Input the data from the text's website or enter the data manually. I loaded both unstacked (Table 3A) and stacked (Table 3B). You need to input only one, either stacked or unstacked. Label the columns.

Finding Summary Statistics (*stacked* or *unstacked* data)

1. **Stat > Summary Stats > Columns** will open an input dialog window similar to Figure 3i.

2. If the data are *stacked*, click on **Calories** in the box titled **Select column(s):** and ignore any other choices that may be listed in the box. **Calories** (and only **Calories**) should appear in the box to the right. Skip the **Where:** box. In the **Group by:** box, click on the **v** symbol, then click on **Meat** from the drop down list.

   If the data are *unstacked*, in the **Select column(s):** box choose **Ham** by single clicking it, then while pressing the keyboard **shift** key click **Turkey** (ignore any other choices in the **Select column(s):** box). Both **Ham** and **Turkey** should be displayed in the box to the right. Skip the **Where:** and **Group by:** boxes.

3. To include IQR in the output, go to the **Statistics:** box, click on **n** and drag down to **IQR**.

4. Click **Compute!** to get the summary statistics.

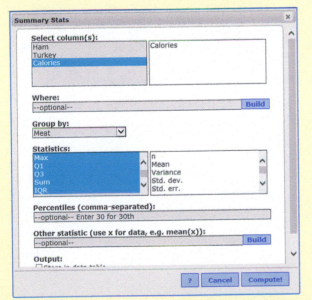

▲ **FIGURE 3i** StatCrunch Input Screen for Summary Statistics

Making Boxplots (*stacked* or *unstacked data*)

1. **Graph > Boxplot** will open an input dialog window similar to Figure 3j.

2. *Unstacked* data: In the **Select column(s):** box, click **Ham**, then while pressing the keyboard **shift**, click **Turkey** (ignore any other choices in the **Select column(s):** box). Both **Ham** and **Turkey** should be displayed in the box to the right. Skip **Where:, Group by:** and **Grouping options:**.

   Stacked and coded data: Click on **Calories** in the **Select column(s):** box. Ignore any other choices in the **Select column(s):** box. **Calories** should be displayed in the box to the right. Skip the **Where:** box. In the **Group by:** box, click on the **v** symbol, then click **Meat** in the drop down list. Skip **Grouping options**.

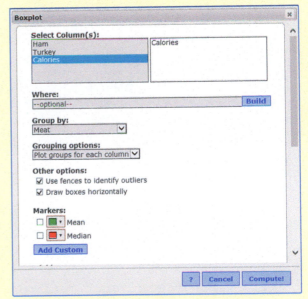

▲ **FIGURE 3j** StatCrunch Input Screen for Boxplots

147

3. Check **Use fences to identify outliers** to make sure that the outliers show up as separate marks. You may also check **Draw boxes horizontally** if that is what you want.

4. Click **Compute!**

5. To copy your graph, click **Options** and **Copy**, and paste it into a document.

Figure 3k shows boxplots of the ham and turkey data; the top box comes from the turkey data (calories per ounce).

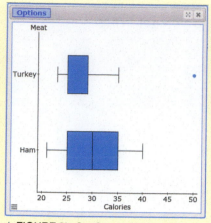

▲ **FIGURE 3k** StatCrunch Boxplots

4

Regression Analysis: Exploring Associations between Variables

THEME

Relationships between two numerical variables can be described in several ways, but, as always, the first step is to understand what is happening visually. When it fits the data, the linear model can help us understand how average values of one variable vary with respect to another variable, and we can use this knowledge to make predictions about one of the variables on the basis of the other.

One of the foremost applications of statistics is to predict outcomes. For example, when you visit a shopping website, can the owners of the website predict what you are interested in buying and whether or not you will buy it? If an item is up for auction on eBay, can we predict how much it will eventually sell for? How much money will a soon-to-be-released movie make? If a patient of a known weight is given a drug, how long before the drug takes affect? Questions such as these rely on the fact that the outcome we wish to predict is related in someway to other variables that we can measure. What you buy when you visit Amazon might be predicted from websites you have recently visited or Google searches you have recently performed. How much an item is sold for on eBay might be related to the number of bids that were placed in the first day of bidding.

Questions such as these are often answered through fitting a regression model. This is an old technique. For example, Francis Galton (1822–1911) was fascinated by the question of whether and to what extent genetics plays a role in determining basic human physical characteristics. He examined the heights of thousands of father–son pairs to determine the nature of the relationship between their heights. If a father is 6 inches taller than average, how much taller than average will his son be? How certain can we be of the answer? Will there be much variability? If there's a lot of variability, then perhaps factors other than the father's genetic material play a role in determining height.

In this chapter we consider associations between models and show how regression can be applied to make predictions about one variable based on knowing the value of another. As in previous chapters, graphs play a major role in revealing patterns in data, and graphs become even more important when we have two variables, not just one. For this reason, we'll start by using graphs to visualize associations between two numerical variables, and we'll see how to quantify these relationships.

CASE STUDY

Forecasting Home Prices

After a spectacular plunge beginning in 2007, home prices in many cities began a steady climb in 2012, creating new challenges for affordable housing. To those wishing to break into the market, for those who want to sell their home, or for anyone wondering about the price of a home, a variety of websites predict the cost that you will pay for a particular home or the amount that you will receive if you sell your home. For example, Zillow.com lets you enter a zip code and see the predicted cost of any home in that zipcode.

It is one thing to list the prices of homes that are already on the market, but how can a website predict the price of a home that is not for sale? The solution is a technique called regression. Regression is a powerful and flexible method that allows analysts to predict an output for a given set of inputs. In this chapter we study the most basic type of regression, so-called simple linear regression. You'll see how it is used to make predictions and how regression gives us insight into relationships between numerical variables.

Visualizing Variability with a Scatterplot

At what age do men and women first marry? How does this vary among the 50 states in the United States? These are questions about the relationship between two variables: age at marriage for women, and age at marriage for men. The primary tool for examining two-variable relationships, when both variables are numerical, is the **scatterplot**. In a scatterplot, each point represents one observation. The location of the point depends on the values of the two variables. For example, we might expect that states where men marry later would also have women marrying at a later age. Figure 4.1 shows a scatterplot of these data, culled from U.S. Census data. Each point represents a state (and one point represents Washington, D.C.) and shows us the typical age at which men and women marry in that state. The two points in the lower-left corner represent Idaho and Utah, where the typical woman first marries at about 23 years of age, the typical man at about age 25. The point in the upper-right corner represents Washington, D.C.

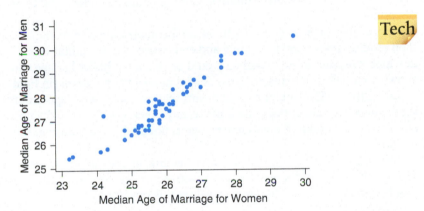

◄ **FIGURE 4.1** A scatterplot of typical marrying ages for men and women in the United States. The points represent the 50 states and the District of Columbia.

When examining histograms (or other pictures of distributions for a single variable), we look for center, spread, and shape. When studying scatterplots, we look for **trend** (which is like center), **strength** (which is like spread), and **shape** (which is like, well, shape). Let's take a closer look at these characteristics.

Recognizing Trend

The trend of an association is the general tendency of the scatterplot as you scan from left to right. Usually trends are either increasing (uphill, /) or decreasing (downhill, \), but other possibilities exist. Increasing trends are called **positive associations** (or **positive trends**), and decreasing trends are **negative associations** (or **negative trends**).

Figure 4.2 shows examples of positive and negative trends. Figure 4.2a reveals a positive trend between the age of a used car and the miles it was driven (mileage). The positive trend matches our common sense: We expect older cars to have been driven farther because generally, the longer a car is owned, the more miles it travels. Figure 4.2b

> **! Caution**
>
> **About the Lower Left Corner**
> In scatterplots we do not require that the lower left corner be (0, 0). The reason is that we want to zoom in on the data and not show a substantial amount of empty space.

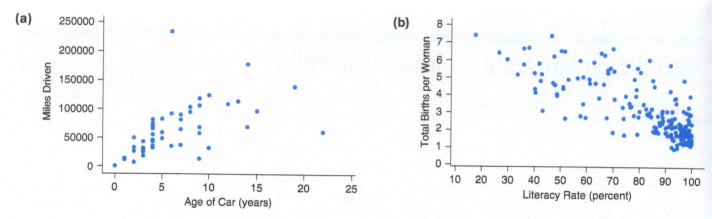

▲ **FIGURE 4.2** Scatterplots with **(a)** positive trend and **(b)** negative trend. (Sources: (a) the authors; (b) United Nations, Statistics Division, http://unstats.un.org)

shows a negative trend—the birthrate of a country against that country's literacy rate. The negative trend suggests that countries with higher literacy rates tend to have a lower rate of childbirth.

Sometimes, the absence of a trend can be interesting. For example, running a marathon requires considerable training and endurance, and we might expect that the speed at which a person runs a marathon would be related to his or her age. But Figure 4.3a shows no trend at all between the ages of runners in a marathon (in Forest City, Canada) and their times. The lack of a trend means that no matter what age group we examine, the runners have about the same times as any other age group, and so we conclude that at least for this group of elite runners, age is not associated with running speed in the marathon.

Figure 4.3b shows simulated data reflecting an association between two variables that cannot be easily characterized as positive or negative—for smaller *x*-values the trend is negative (\\), and for larger *x*-values it is positive (/).

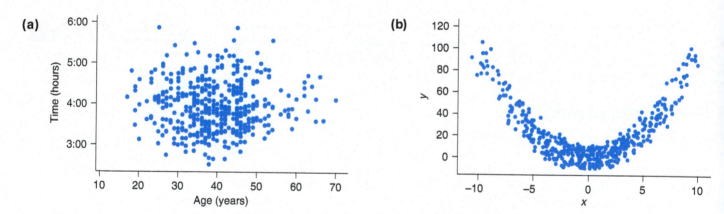

▲ **FIGURE 4.3** Scatterplots with **(a)** no trend and **(b)** a changing trend. (Sources: (a) Forest City Marathon, http://www.forestcityroadraces.com/; (b) simulated data)

Seeing Strength of Association

Weak associations result in a large amount of scatter in the scatterplot. A large amount of scatter means that points have a great deal of spread in the vertical direction. This vertical spread makes it somewhat harder to detect a trend. Strong associations have little vertical variation.

Figure 4.4 on the next page enables us to compare the strengths of two associations. Figure 4.4a shows the association between height and weight for a sample of active adults. Figure 4.4b involves the same group of adults, but this time we examine the association between waist size and weight. Which association is stronger?

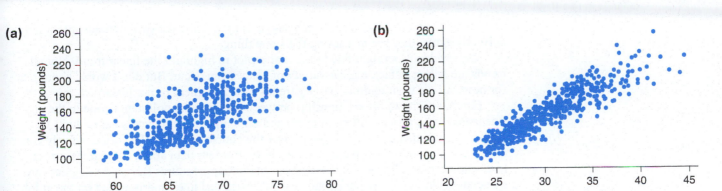

▲ **FIGURE 4.4** **(a)** A graph showing a relatively weaker association. **(b)** A graph showing a stronger association, because the points have less vertical spread. (Source: Heinz et al. 2003)

The association between waist size and weight is the stronger one (Figure 4.4b). To see this, in Figure 4.4a, consider the data for people who are 65 inches tall. Their weights vary anywhere from about 120 pounds to 230 pounds, a range of 110 pounds. If you were using height to predict weight, you could be off by quite a bit. Compare this with the data in Figure 4.4b for people with a waist size of 30 inches. Their weights vary from about 120 pounds to 160 pounds, only a 40-pound range. The association between waist size and weight is stronger than that between height and weight because there is less vertical spread, so tighter predictions can be made. If you had to guess someone's weight, and could ask only one question before guessing, you'd do a better job if you asked about the person's waist size than if you asked about his or her height.

Labeling a trend as strong, very strong, or weak is a subjective judgment. Different statisticians might have different opinions. Later in this section, we'll see how we can measure strength with a number.

Identifying Shape

The simplest shape for a trend is **linear**. Fortunately, linear trends are relatively common in nature. Linear trends always increase (or decrease) at the same rate. They are called linear because the trend can be summarized with a straight line. In real life, the observations don't fall perfectly on a straight line but tend to vary above and below a line. For this reason, scatterplots of linear trends often look roughly football-shaped, as shown in Figure 4.4a.

Figure 4.5 shows a linear trend from data gathered by one of the authors, who used a FitBit pedometer to count his daily steps. He learned that his cell phone could also count his steps, though it used a different mechanism for measuring his movements and a different algorithm to turn these measurements into counts. Did the two methods tend to produce the same count? Figure 4.5 shows that there's a linear trend between these

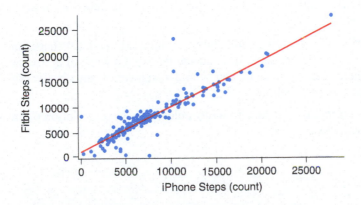

◄ **FIGURE 4.5** A line has been inserted to emphasize the linear trend.

two methods of counting steps, which suggests that, while the two methods might not perfectly agree, they are measuring the same thing.

We've added a straight line to the scatterplot to highlight the linear trend. Not all trends are linear; in fact, a great variety of shapes can occur. But don't worry about that for now: All we want to do is classify trends as either linear or not linear.

Figure 4.6 shows the relationship between levels of the pollutant ozone in the air (measured in parts per million, or ppm) and air temperature (degrees Fahrenheit) in Upland, California, near Los Angeles, over the course of a year. The trend is fairly flat at first and then becomes steeper. For temperatures less than about 55 degrees, ozone levels do not vary all that much. However, higher temperatures (above 55 degrees) are associated with much greater ozone levels. The curved line superimposed on the graph shows the nonlinear trend.

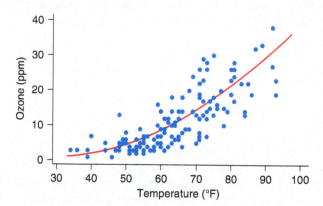

▲ **FIGURE 4.6** Ozone (ppm) is associated with temperature (degrees Fahrenheit) in a nonlinear way. (Source: Breiman, from the R earth package. See Faraway 2005.)

Nonlinear trends are more difficult to summarize than linear trends. This chapter does not cover nonlinear trends—you'll have to read Chapter 14 for that. In this chapter, our focus is on linear trends, and so it is very important that you first examine a scatterplot to be sure that the trend is linear. If you apply the techniques in this chapter to a nonlinear trend, you might reach disastrously incorrect conclusions!

> **KEY POINT**   When examining associations, look for the trend, the strength of the trend, and the shape of the trend.

Writing Clear Descriptions of Associations

Good communication skills are vital for success in general, and preparing you to clearly describe patterns in data is an important goal of this text. Here are some tips to help you describe two-variable associations.

- A written description should always include (1) trend, (2) shape, and (3) strength (not necessarily in that order) and should explain what all of these mean in (4) the context of the data. You should also mention any observations that do not fit the general trend.

Example 1 demonstrates how to write a clear, precise description of an association between numerical variables.

EXAMPLE 1 Age and Mileage of Used Cars

Figure 4.2a on page 152 displays an association between the age and mileage of a sample of used cars.

QUESTION Describe the association.

SOLUTION The association between the age and mileage of used cars is positive and linear. This means that older cars tend to have greater mileage. The association is moderately strong; some scatter is present but not enough to hide the shape of the relationship. There is one exceptional point: One car is only about 6 years old but has been driven many miles.

TRY THIS! Exercise 4.5

The description in Example 1 is good because it mentions trend (a "positive" association), shape ("linear"), and strength ("moderately strong") and does so in context ("older cars tend to have greater mileage").

- It is very important that your descriptions be precise. For example, it would be wrong to say that older cars have greater mileage. This statement implies that all older cars in the data set have greater mileage, which isn't true. The one exceptional car (upper-left corner of the plot) is relatively new (about 6 years old) but has a very high mileage (about 250,000 miles). Some older cars have relatively few miles on them. To be precise, you could say older cars *tend* to have higher mileage. The word *tend* indicates that you are describing a trend that has variability, so the trend you describe is not true of all individuals but instead is a characteristic of the entire group.

- When writing a description of a relationship, you should also mention unusual features, such as outliers, small clusters of points, or anything else that does not seem to be part of the general pattern. Figure 4.7 includes an outlier. These data are from a statistics class in which students reported their weights and heights. One student wrote the wrong height.

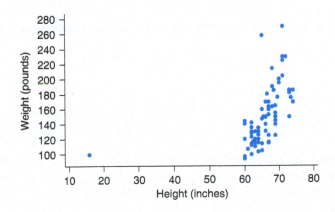

◄ **FIGURE 4.7** A fairly strong, positive association between height and weight for a statistics class. One student reported the wrong height. (Source: R. Gould, UCLA, Department of Statistics)

Data Moves ▶ The original data are provided in web-browser tables, and were downloaded using the programming language R. The data provide a rich set of variables, but do not provide the variable *speed*. We created this variable using a "transform" data move by dividing the length of a marathon (26.2188 miles) by the *Net.Time* variable. (Source: https://www.lamarathon.com/race-weekend/results)

Data file ▶ lamarathons.csv

Asking Statistical Questions About Regression

Once we have two or more variables in play, the number and variety of questions we can pose increases excitingly. For example, is the speed of a marathon runner related to his or her age? If so, what's the nature of that relationship? Does speed gradually decrease as one gets older, or is the decline sudden? Or is there a "prime" age at which people tend to run fastest, and those younger and older than that age tend to run more slowly? Figure 4.8, consisting of data from the Los Angeles Marathon, tells us that the answer for at least this group of runners is, well, complicated. There's great variety of running speeds at all ages and no clear trend!

The preceding questions concern the existence of a relationship and its shape. In other contexts, we often find that questions concerning the strength of relationships are productive and useful. For example, is height or the circumference of the waist a better predictor of a person's weight? For relationships that we've determined are linear, we're sometimes interested in the rate of change. For example, if we compare two people whose waist sizes differ by 1 inch, on average, how does their weight differ?

Often the questions you ask depend on the context, and there's really a great many of questions we could pose of data such as those in this chapter.

▶ FIGURE 4.8 The relationship between speed and age in the Los Angeles Marathon for one recent year. While the fastest runners do seem to be fairly young, between the age of 20 and 40, there's no clear overall trend.

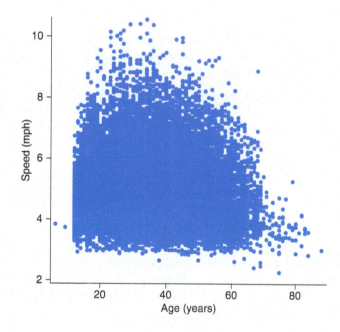

Measuring Strength of Association with Correlation

The **correlation coefficient** is a number that measures the strength of the linear association between two numerical variables—for example, the relationship between people's heights and weights. We can't emphasize enough that the correlation coefficient *makes sense only if the trend is linear and if both variables are numerical.*

The correlation coefficient, represented by the letter *r*, is always a number between −1 and +1. Both the value and the sign (positive or negative) of *r* have information we can use. If the value of *r* is close to −1 or +1, then the association is very strong; if close to 0, the association is weak. If the value of the correlation coefficient is positive, then the trend is positive; if the value is negative, the trend is negative.

Visualizing the Correlation Coefficient

Figure 4.9 presents a series of scatterplots that show associations of gradually decreasing strength. The strongest linear association appears in Figure 4.9a; the points fall exactly along a line. Because the trend is positive and perfectly linear, the correlation coefficient is equal to 1.

The next scatterplot, Figure 4.9b, shows a slightly weaker association. The points are more spread out vertically. We can see a linear trend, but the points do not fall exactly along a line. The trend is still positive, so the correlation coefficient is also positive. However, the value of the correlation coefficient is less than 1 (it is 0.98).

The remaining scatterplots show weaker and weaker associations, and their correlation coefficients gradually decrease. The last scatterplot, Figure 4.9f, shows no

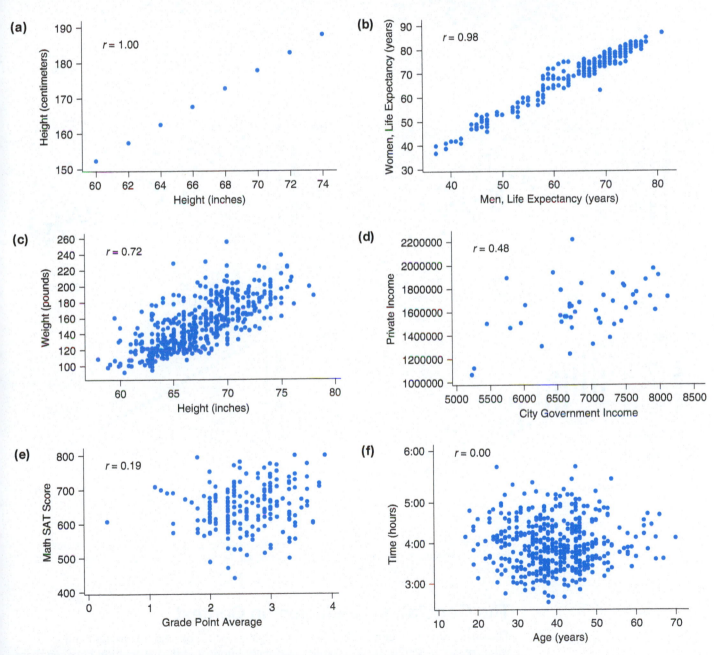

▲ FIGURE 4.9 Scatterplots with gradually decreasing positive correlation coefficients. (Sources: (a) simulated data; (b) www.overpopulation.com; (c) Heinz 2003; (d) Bentow and Afshartous; (e) Fathom™ Sample Document, Educational Testing Service [ETS] validation study; (f) Forest City Marathon)

association at all between the two variables, and the correlation coefficient has a value close to 0.

The next set of scatterplots (Figure 4.10) starts with data from a marathon (having a correlation coefficient close to 0), and the negative correlations gradually get stronger. The last figure has a correlation coefficient of exactly −1.

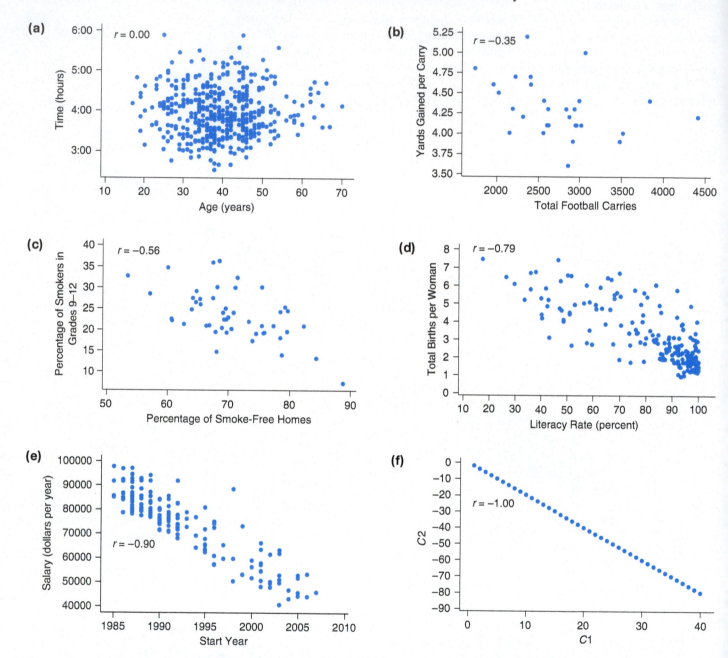

▲ **FIGURE 4.10** Scatterplots with increasingly negative correlations. (Sources: **(a)** Forest City Marathon; **(b)** http://wikipedia.org; **(c)** Centers for Disease Control; **(d)** UN statistics; **(e)** Minitab Student 12 file "Salary," adjusted for inflation; **(f)** simulated data)

The Correlation Coefficient in Context

The correlation between the number of Google searches per day for *zombie* and *vampire* is $r = 0.924$. If we are told, or already know, that the association between these variables is linear, then we know that the trend is positive and strong. The fact that the correlation is close to 1 means that there is not much scatter in the scatterplot.

College admission offices sometimes report correlations between students' Scholastic Aptitude Test (SAT) scores and their first-year grade-point averages (GPAs). If the association is linear and the correlation is high, this justifies using the SAT to make admissions decisions because a high correlation would indicate a strong association between SAT scores and academic performance. A positive correlation means that students who score above average on the SAT tend to get above-average grades. Conversely, those who score below average on the SAT tend to get below-average grades. Note that we're careful to say "tend to." Certainly, some students with low SAT scores do very well, and some with high SAT scores struggle to pass their classes. The correlation coefficient does not tell us about individual students; it tells us about the overall trend.

More Context: Correlation Does Not Mean Causation!

Some of the most important research questions in science, business, and government involve cause and effect. Quite often, we want to know if an intervention changes things. Does giving a patient a new medicine improve their condition? Does changing a tax law improve the economy? Does changing the placement of an advertisement on a web page cause more people to click on that ad?

When two numerical variables are involved, the first step in deciding issues of causality is to make a scatterplot. Almost always, if there is a direct causal relationship, a scatterplot will show an association (although that association may be weak). And for this reason, a common statistical investigation question in these contexts is "Is there an association?"

For example, participants in a study designed to determine if the type of diet affected the amount of weight lost were asked a series of questions every month. These questions were designed to measure how well the participants followed their assigned diet. Researchers wondered if adhering to the diet caused weight loss. Figure 4.11 suggests that there is weak but negative correlation between the amount of weight lost and the average adherence to the diet over the year; subjects with greater adherence tended to lose more weight.

However, one important step of the Data Cycle is that we consider the data, and in this context that means considering how the data were collected. As we learned in Chapter 1, in an observational study such as this, the presence of potential confounding factors makes it impossible to conclude there is a cause-and-effect relationship based solely on an association between the variables. For the diet data, subjects determined their own level of adherence, and this might have been affected by other events in their lives, such as stress, that could also have affected their weight loss.

For another, more fanciful example, consider this: a positive correlation exists between the number of blankets sold in Canada per week and the number of brush fires in Australia per week. Are brush fires in Australia caused by cold Canadians? Probably not. The correlation is likely to be the result of weather. When it is winter in Canada, people buy blankets. When it is winter in Canada, it is summer in Australia (which is located in the Southern Hemisphere), and summer is brush fire season.

> **Caution**

Linearity Is Important
The correlation coefficient is interpretable only for linear associations.

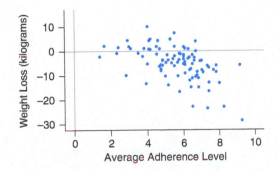

◄ FIGURE 4.11 The correlation between weight loss and average adherence does not necessarily imply that adherence to a diet causes weight loss. We can conclude that adherence *may* cause weight loss.

If you learn nothing else from this book, remember this: no matter how tempting, do *not* conclude that a cause-and-effect relationship between two variables exists just because there is a correlation, no matter how close to $+1$ or -1 that correlation might be. Always consider the method of data collection, and ensure that the method supports your conclusion.

> **KEY POINT**   Correlation alone does not imply causation.

Finding the Correlation Coefficient

↻ Looking Back

z-Scores
Recall that *z*-scores show how many standard deviations a measurement is from the mean. To find a *z*-score from a measurement, first subtract the mean and then divide the difference by the standard deviation.

The correlation coefficient is best determined through the use of technology. We calculate a correlation coefficient by first converting each observation to a *z*-score, using the appropriate variable's mean and standard deviation. For example, to find the correlation coefficient that measures the strength of the linear relation between weight and height, we first convert each person's weight and height to *z*-scores. The next step is to multiply the observations' *z*-scores together. If both are positive or both negative—meaning that both *z*-scores are above average or both are below average—then the product is a positive number. In a strong positive association, most of these products are positive values. In a strong negative association, however, observations above average on one variable tend to be below average on the other variable. In this case, one *z*-score is negative and one positive, so the product is negative. Thus, in a strong negative association, most *z*-score products are negative.

To find the correlation coefficient, add the products of *z*-scores together and divide by $n - 1$ (where n is the number of observed pairs in the sample). In mathematical terms, we get

$$\text{Formula 4.1:} \quad r = \frac{\sum z_x z_y}{n - 1}$$

The following example illustrates how to use Formula 4.1 in a calculation.

EXAMPLE 2 Heights and Weights of Six Women

Figure 4.12a shows the scatterplot for heights and weights of six women.

QUESTION   Using the data provided, find the correlation coefficient of the linear association between heights (inches) and weights (pounds) for these six women.

Heights	61	62	63	64	66	68
Weights	104	110	141	125	170	160

◀ **FIGURE 4.12a** Scatterplot showing heights and weights of six women.

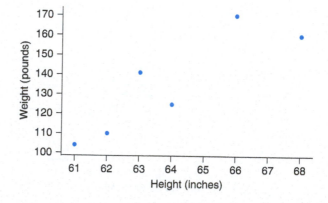

SOLUTION   Before proceeding, we verify that the conditions hold. Figure 4.12a suggests that a straight line is an acceptable model; a straight line through the data might summarize the trend, although this is hard to see with so few points.

Next, we calculate the correlation coefficient. Ordinarily, we use technology to do this, and Figure 4.12b shows the output from StatCrunch, which gives us the value $r = 0.88093364$.

Data set from:	StatCrunch	Data	Stat	Graphics	Help
	Row	Height	Weight	var3	
● ● ● Correlation ⚠		61	104		
Options		62	110		
		63	141		
Correlation between Height and Weight is:		64	125		
0.88093364		66	170		
		68	160		
	6				
	7				
	8				

◀ **FIGURE 4.12b** StatCrunch, like all statistical software, lets you calculate the correlation between any two columns you choose.

Because the sample size is small, we can confirm this output using Formula 4.1. It is helpful to go through the steps of this calculation to better understand how the correlation coefficient measures linear relationships between variables.

The first step is to calculate average values of height and weight and then to determine the standard deviation for each.

For the height: $\bar{x} = 64$ and $s_x = 2.608$

For the weight: $\bar{y} = 135$ and $s_y = 26.73$

Next we convert all the points to pairs of z-scores and multiply them together. For example, for the woman who is 68 inches tall and weighs 160 pounds,

$$z_x = \frac{x - \bar{x}}{s_x} = \frac{68 - 64}{2.608} = \frac{4}{2.608} = 1.53$$

$$z_y = \frac{y - \bar{y}}{s_y} = \frac{160 - 135}{26.73} = \frac{25}{26.73} = 0.94$$

The product is

$$z_x \times z_y = 1.53 \times 0.94 = 1.44$$

Note that this product is positive and shows up in the upper-right quadrant in Figure 4.12c.

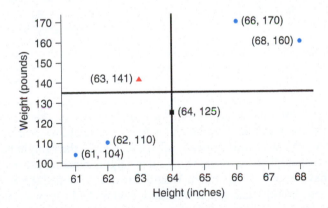

◀ **FIGURE 4.12c** The same scatterplot as in 4.12a but with the plot divided into quadrants based on average height and weight. Points represented with blue circles contribute a positive value to the correlation coefficient (positive times positive is positive, or negative times negative equals a positive). The red triangle represents an observation that contributes negatively (a negative z-score times a positive z-score is negative), and the black square contributes nothing because one of the z-scores is zero.

Figure 4.12c can help you visualize the rest of the process. The two blue circles in the upper-right portion represent observations that are above average in both variables, so both z-scores are positive. The two blue circles in the lower-left region represent observations that are below average in both variables; the products of the two negative *z*-scores are positive, so they add to the correlation. The red triangle has a positive *z*-score for weight (it is above average) but a negative *z*-score for height, so the product is negative. The black square is a point that makes no contribution to the correlation coefficient. This person is of average height, so her *z*-score for height is 0.

The correlation between height and weight for these six women comes out to be about 0.881.

CONCLUSION The correlation coefficient for the linear association of weights and heights of these six women is *r* = 0.881. Thus, there is a strong positive correlation between height and weight for these women. Taller women tend to weigh more.

TRY THIS! Exercise 4.21a

Understanding the Correlation Coefficient

The correlation coefficient has a few features you should know about when interpreting a value of *r* or deciding whether you should compute the value.

- *Changing the order of the variables does not change* r. This means that if the correlation between life expectancy for men and women is 0.977, then the correlation between life expectancy for women and men is also 0.977. This makes sense because the correlation measures the strength of the linear relationship between *x* and *y*, and that strength will be the same no matter which variable gets plotted on the horizontal axis and which on the vertical.

  Parts (a) and (b) of Figure 4.13 have the same correlation; we've just swapped axes.

(a) 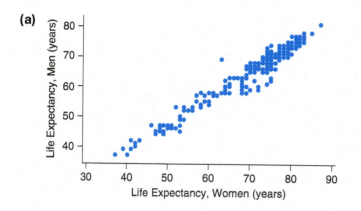 **(b)**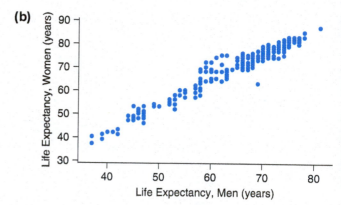

▲ **FIGURE 4.13** Scatterplots showing the relationship between men's and women's life expectancies for various countries. **(a)** Women's life expectancy is plotted on the *x*-axis. **(b)** Men's life expectancy is plotted on the *x*-axis. (Sources: http://www.overpopulation.com and Fathom™ Sample Documents)

- *Adding a constant or multiplying by a positive constant does not affect* r. The correlation between the heights and weights of the six women in Example 2 was 0.881. What would happen if all six women in the sample had been asked to wear 3-inch platform heels when their heights were measured? Everyone would have been 3 inches taller. Would this have changed the value of *r*? Intuitively, you should sense that it wouldn't. Figure 4.14a shows a scatterplot of the original data, and Figure 4.14b shows the data with the women in 3-inch heels.

  We haven't changed the strength of the relationship. All we've done is shift the points on the scatterplot 3 inches to the right. But shifting the points doesn't change the relationship between height and weight. We can verify that the

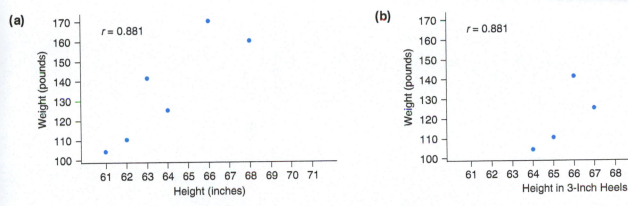

▲ **FIGURE 4.14** **(a)** A repeat of the scatterplot of height and weight for six women. **(b)** The same women in 3-inch heels. The correlation remains the same.

correlation is unchanged by looking at the formula. The heights will have the same z-scores both before and after the women put on the shoes; since everyone "grows" by the same amount, everyone is still the same number of standard deviations away from the average, which also "grows" by 3 inches. As another example, if science found a way to add 5 years to the life expectancy of men in all countries in the world, the correlation between life expectancies for men and women would still be the same.

More generally, we can add a constant (a fixed value) to all the values of one variable, or of both variables, and not affect the correlation coefficient.

For the very same reason, we can multiply either or both variables by positive constants without changing r. For example, to convert the women's heights from inches to feet, we multiply their heights by 1/12. Doing this does not change how strong the association is; it merely changes the units we're using to measure height. Because the strength of the association does not change, the correlation coefficient does not change.

- *The correlation coefficient is unitless.* Height is measured in inches and weight in pounds, but r has no units because the z-scores have no units. This means that we will get the same value for correlation whether we measure height in inches, meters, or fathoms.

- *Linear, linear, linear.* We've said it before, but we'll say it again: we're talking only about linear relationships here. The correlation can be misleading if you do not have a linear relationship. Parts (a) through (d) of Figure 4.15 illustrate the fact that different nonlinear patterns can have the same correlation. All these graphs have $r = 0.817$, but the graphs have very different shapes. The take-home message is that the correlation alone does not tell us much about the shape of a graph. We must also know that the relationship is linear to make sense of the correlation.

Remember: *Always* make a graph of your data. If the trend is nonlinear, the correlation (and, as you'll see in the next section, other statistics) can be very misleading.

> **⚠ Caution**
>
> **Correlation Coefficient and Linearity**
> A value of r close to 1 or −1 does *not* tell you that the relationship is linear. You must check visually; otherwise, your interpretation of the correlation coefficient might be wrong.

> **KEY POINT**   The correlation coefficient does not tell you whether an association is linear. However, if you already know that the association is linear, then the correlation coefficient tells you how strong the association is.

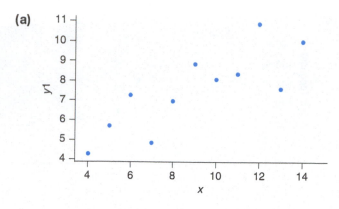

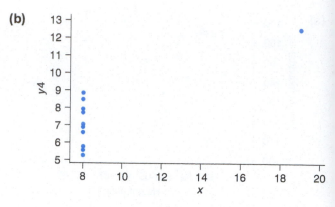

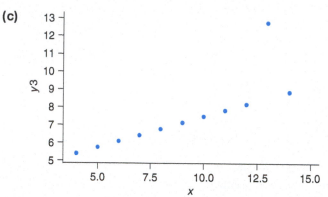

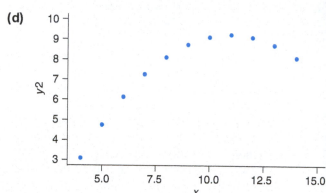

▲ **FIGURE 4.15 (a–d)** Four scatterplots with the same correlation coefficient of 0.817 have very different shapes. The correlation coefficient is meaningful only if the trend is linear. (Source: Anscombe 1973)

SNAPSHOT ► **The Correlation Coefficient**

WHAT IS IT? ►	Correlation coefficient.
WHAT DOES IT DO? ►	Measures the strength of a linear association.
HOW DOES IT DO IT? ►	By comparing z-scores of the two variables. The products of the two z-scores for each point are averaged.
HOW IS IT USED? ►	The sign tells us whether the trend is positive ($+$) or negative ($-$). The value tells us the strength. If the value is close to 1 or -1, then the points are tightly clustered about a line; if the value is close to 0, then there is no linear association.
	Note: The correlation coefficient can be interpreted only with linear associations.

SECTION 4.3

Modeling Linear Trends

How much more do people tend to weigh for each additional inch in height? How much value do cars lose each year as they age? Are home run hitters good for their teams? Can we predict how much space a book will take on a bookshelf just by knowing how many pages are in the book? It's not enough to remark that a trend exists. To make a prediction based on data, we need to measure the trend and the strength of the trend.

To measure the trend, we're going to perform a bit of statistical sleight of hand. Rather than interpret the data themselves, we will substitute a model of the data and interpret the model. The model consists of an equation and a set of conditions that describe when the model will be appropriate. Ideally, this equation is a very concise and accurate description of the data; if so, the model is a good fit. When the model is a good fit to the data, any understanding we gain about the model accurately applies to our understanding of the real world. If the model is a bad fit, however, then our understanding of real situations might be seriously flawed.

The Regression Line

The **regression line** is a tool for making predictions about future observed values. It also provides us with a useful way of summarizing a linear relationship. Recall from Chapter 3 that we could summarize a sample distribution with a mean and a standard deviation. The regression line works the same way: It reduces a linear relationship to its bare essentials and enables us to analyze a relationship without being distracted by small details.

Review: Equation of a Line The regression line is given by an equation for a straight line. Recall from algebra that equations for straight lines contain a **y-intercept** and a **slope**. The equation for a straight line is

$$y = mx + b$$

The letter m represents the slope, which tells how steep the line is, and the letter b represents the y-intercept, which is the value of y when $x = 0$.

Statisticians write the equation of a line slightly differently and put the intercept first; they use the letter a for the intercept and b for the slope and write

$$y = a + bx$$

We often use the names of variables in place of x and y to emphasize that the regression line is a model about two real-world variables. We will sometimes write the word *predicted* in front of the y-variable to emphasize that the line consists of predictions for the y-variable, not actual values. A few examples should make this clear.

Visualizing the Regression Line Can we know how wide a book is on the basis of the number of pages in the book? A student took a random sample of books from his shelf, measured the width of the spine (in millimeters, mm), and recorded the number of pages. Figure 4.16 illustrates how the regression line captures the basic trend of a linear association between width of the book and the number of pages for this sample. The equation for this line is

$$\text{Predicted Width} = 6.22 + 0.0366\ \text{Pages}$$

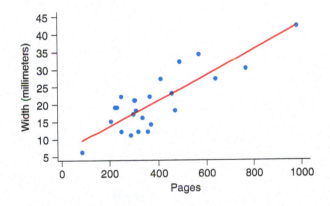

◀ FIGURE 4.16 The regression line summarizes the relationship between the width of the book and the number of pages for a small sample of books. (Source: E. Onaga, 2005, UCLA, Department of Statistics)

Data Move ▶ The Lahman Baseball Database contains many separate files linked by a common identification code, such as the name of a baseball player or a team. We merged several files related to batting and kept only the rows pertaining to the 2016 season and only rows for players whose home run count was in the top 25%.

Data file ▶ Batting2016.csv, from www.baseball1.com

In baseball, two numbers used to measure how good a batter is are the number of runs batted in (RBIs) and the number of home runs. (A home run occurs when the batter rounds all three bases with one hit and scores a run. An RBI occurs when a player is already on base, and a batter hits the ball far enough for the on-base runner to score a run.) Some baseball fans believe that players who hit a lot of home runs might be exciting to watch but are not that good for the team. (Some believe that home-run hitters tend to strike out more often; this takes away scoring opportunities for the team.) Figure 4.17 shows the relationship between the number of home runs and RBIs for the top 25% of home-run hitters in the major leagues in the 2016 season. The association seems fairly linear, and the regression line can be used to predict how many RBIs a player will earn, given the number of home runs hit. The data suggest that players who hit a large number of home runs tend to score a large number of points through RBIs. The equation for this regression line is

$$\text{Predicted RBIs} = 23.83 + 2.13 \text{ Home Runs}$$

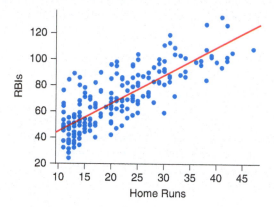

▶ FIGURE 4.17 The regression line summarizes the relationship between RBIs and home runs for the top home run hitters in the 2016 Major League Baseball season.

Regression in Context Suppose you have a 10-year-old car and want to estimate how much it is worth. One of the more important uses of the regression line is to make predictions about what *y*-values can be obtained for particular *x*-values. Figure 4.18 suggests that the relationship between age and value is linear, and the regression line that summarizes this relationship is

$$\text{Predicted Value} = 21375 - 1215 \text{ Age}$$

We can use this equation to predict approximately how much a 10-year-old car is worth:

$$\text{Predicted Value} = 21375 - 1215 \times 10$$
$$= 21375 - 12150$$
$$= 9225$$

▶ FIGURE 4.18 The regression line summarizes the relationship between the value of a car, according to the Kelley Blue Book, and the car's age for a small sample of students' cars. (Source: C. Ryan 2006, personal communication.)

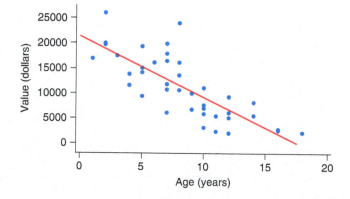

The regression line predicts that a 10-year-old car will be valued at about $9225. As we know, many factors other than age affect the value of a car, and perhaps with more information we might make a better prediction. However, if the only thing we know about a car is its age, this may be the best prediction we can get. It is also important to keep in mind that this sample comes from one particular group of students in one particular statistics class and is not representative of all used cars on the market in the United States.

Using the regression line to make predictions requires certain assumptions. We'll go into more detail about these assumptions later, but for now, just use common sense. This predicted value of $9225 is useful only if our data are valid. For instance, if all the cars in our data set are Toyotas, and our 10-year-old car is a Chevrolet, then the prediction is probably not useful.

EXAMPLE **3** Book Width

A college instructor who has far too many books on his shelf is wondering whether he has room for one more. He has about 20 mm of space left on his shelf, and he can tell from the online bookstore that the book he wants has 598 pages. The regression line is

$$\text{Predicted Width} = 6.22 + 0.0366 \text{ Pages}$$

QUESTION   Will the book fit on his shelf?

SOLUTION   Assuming that the data used to fit this regression line are representative of all books, we would predict the width of the book corresponding to 598 pages to be

$$\text{Predicted Width} = 6.22 + 0.0366 \times 598$$
$$= 6.22 + 21.8868$$
$$= 28.1068 \text{ mm}$$

CONCLUSION   The book is predicted to be 28 mm wide. Even though the actual book width is likely to differ somewhat from 28 mm, it seems that the book will probably not fit on the shelf.

TRY THIS!   Exercise 4.29

Common sense tells us that not all books with 598 pages are exactly 28 mm wide. There is a lot of variation in the width of a book for a given number of pages.

Finding the Regression Line   In almost every case, we'll use technology to find the regression line. However, it is important to know how the technology works and to be able to calculate the equation when we don't have access to the full data set.

To understand how technology finds the regression line, imagine trying to draw lines through the scatterplots in Figures 4.16 through 4.18 to best capture the linear trend. We could have drawn almost any line, and some of them would have looked like pretty good summaries of the trend. What makes the regression line special? How do we find the intercept and slope of the regression line?

The regression line is chosen because it is the line that comes closest to most of the points. More precisely, the square of the vertical distances between the points and the line, on average, is bigger for any other line we might draw than for the regression line. Figure 4.19 shows these vertical distances with black lines. The regression line is sometimes called the "best fit" line because, in this sense, it provides the best fit to the data.

📌 Details

Least Squares
The regression line is also called the least squares line because it is chosen so that the sum of the squares of the differences between the observed y-value, y, and the value predicted by the line, $\hat{y}$, is as small as possible. Mathematically, this means that the slope and intercept are chosen so that $\Sigma(y - \hat{y})^2$ is as small as possible.

▶ **FIGURE 4.19** The "best fit" regression line, showing the vertical distance between each observation and the line. For any other line, the average of the squared distances is larger.

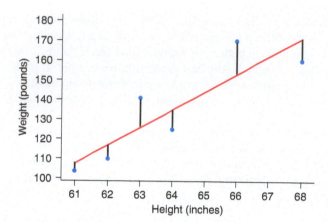

To find this best fit line, we need to find the slope and the intercept. The slope, b, of the regression line is the ratio of the standard deviations of the two variables, multiplied by the correlation coefficient:

Formula 4.2a: The slope, $b = r\dfrac{s_y}{s_x}$

Once we have found the slope, we can find the intercept. Finding the intercept, a, requires that we first find the means of the variables, $\bar{x}$ and $\bar{y}$:

Formula 4.2b: The intercept, $a = \bar{y} - b\bar{x}$

Now put these quantities into the equation of a line, and the regression line is given by

Formula 4.2c: The regression line, Predicted $y = a + bx$

EXAMPLE 4   SAT Scores and GPAs

A college in the northeastern United States used SAT scores to help decide which applicants to admit. To determine whether the SAT was useful in predicting success, the college examined the relationship between the SAT scores and first-year GPAs of admitted students. Figure 4.20a shows a scatterplot of the math SAT scores and first-year GPAs for a random sample of 200 students. The scatterplot suggests a weak, positive linear association: students with higher math SAT scores tend to get higher first-year GPAs. The average math SAT score of this sample was 649.5 with a standard deviation of 66.3. The average GPA was 2.63 with a standard deviation of 0.58. The correlation between GPA and math SAT score was 0.194.

QUESTIONS

a. Find the equation of the regression line that best summarizes this association. Note that the x-variable is the math SAT score and the y-variable is the GPA.

b. Using the equation, find the predicted GPA for a person in this group with a math SAT score of 650.

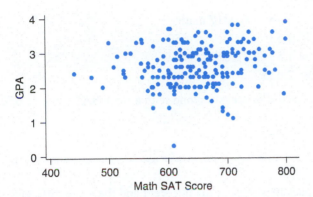

◀ **FIGURE 4.20a** Students with higher math SAT scores tend to have higher first-year GPAs, but this linear association is fairly weak. (Source: ETS validation study of an unnamed college and Fathom™ Sample Documents, 2007, Key Curriculum Press)

SOLUTIONS

a. With access to the full data set at this text's website, we can use technology to find (and plot) the regression line. Still, when summary statistics are provided (means and standard deviations of the two variables as well as their correlation), it is not time-consuming to use Formula 4.2 to find the regression line.

Figure 4.20b shows StatCrunch output for the regression line. (StatCrunch provides quite a bit more information than we will use in this chapter.)

According to technology, the equation of the regression line is

$$\text{Predicted GPA} = 1.53 + 0.0017 \text{ Math}$$

Tech

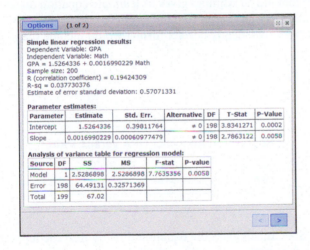

◀ **FIGURE 4.20b** StatCrunch output for the regression line to predict GPA from math SAT score. The structure of the output is typical for many statistical software packages.

We now check this calculation by hand, using Formula 4.2.

We are given that

For math SAT scores: $\bar{x} = 649.5$, $s_x = 66.3$

For GPAs: $\bar{y} = 2.63$, $s_y = 0.580$

and $r = 0.194$

First we must find the slope:

$$b = r\frac{s_y}{s_x} = 0.194 \times \frac{0.580}{66.3} = 0.001697$$

Now we can use the slope to find the intercept:

$$a = \bar{y} - b\bar{x} = 2.63 - 0.001697 \times 649.5 = 2.63 - 1.102202 = 1.53$$

Rounding off yields

$$\text{Predicted GPA} = 1.53 + 0.0017 \text{ math}$$

b. Predicted GPA $= 1.53 + 0.0017$ math
$= 1.53 + 0.0017 \times 650$
$= 1.53 + 1.105$
$= 2.635$

CONCLUSION We would expect someone with a math SAT score of 650 to have a GPA of about 2.64.

TRY THIS! Exercise 4.33

Different software packages present the intercept and the slope differently. Therefore, you need to learn how to read the output of the software you are using. Example 5 shows output from several packages.

EXAMPLE 5 Technology Output for Regression

Figure 4.21 shows outputs from Minitab, StatCrunch, the TI-84, and Excel for finding the regression equation in Example 4 for GPA and math SAT scores.

QUESTION For each software package, explain how to find the equation of the regression line from the given output.

CONCLUSION Figure 4.21a: Minitab gives us a simple equation directly: GPA $= 1.53 + 0.00170$ Math (after rounding).

However, the more statistically correct format would include the adjective "Predicted":

$$\text{Predicted GPA} = 1.53 + 0.00170 \text{ Math}$$

▶ **FIGURE 4.21a** Minitab output.

Regression Analysis: GPA versus Math

The regression equation is
GPA = 1.526 + 0.001699 Math

Figure 4.21b: StatCrunch gives the equation directly near the top, but it also lists the intercept and slope separately in the table near the bottom.

▶ **FIGURE 4.21b** StatCrunch output.

Options (1 of 2)

Simple linear regression results:
Dependent Variable: GPA
Independent Variable: Math
GPA = 1.5264336 + 0.0016990229 Math
Sample size: 200
R (correlation coefficient) = 0.19424309
R-sq = 0.037730376
Estimate of error standard deviation: 0.57071331

Parameter estimates:

Parameter	Estimate	Std. Err.	Alternative	DF	T-Stat	P-Value
Intercept	1.5264336	0.39811764	≠ 0	198	3.8341271	0.0002
Slope	0.0016990229	0.00060977479	≠ 0	198	2.7863122	0.0058

Analysis of variance table for regression model:

Source	DF	SS	MS	F-stat	P-value
Model	1	2.5286898	2.5286898	7.7635356	0.0058
Error	198	64.49131	0.32571369		
Total	199	67.02			

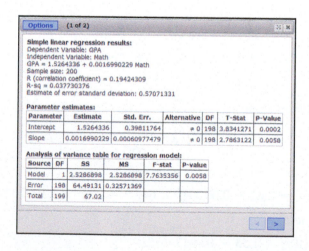

	Coefficients	Standard Error	t Stat	P-valve	Lower 95%	Upper 95%	Lower 95.0%	Upper 95.0%
Intercept	1.526434	0.398118	3.834127	0.000169	0.741339	2.311528	0.741339	2.311528
Variable 1	0.001699	0.00061	2.786312	0.00585	0.000497	0.002902	0.000497	0.002902

▲ FIGURE 4.21c Excel output.

Figure 4.21c: Excel shows the coefficients in the column labeled "Coefficients." The intercept is in the first row, labeled "Intercept," and the slope is in the row labeled "Variable 1."

TRY THIS! Exercise 4.35

Interpreting the Regression Line

An important use of the regression line is to make predictions about the value of y that we will see for a given value of x. However, the regression line provides more information than just a predicted y-value. The regression line can also tell us about the rate of change of the mean value of y with respect to x, and it can help us understand the underlying theory behind cause-and-effect relationships.

Choosing x and y: Order Matters In Section 4.2, you saw that the correlation coefficient is the same no matter which variable you choose for x and which you choose for y. With regression, however, order matters.

Consider the collection of data about book widths. We used it earlier to predict the width of a book, given the number of pages. The equation of the regression line for this prediction problem (shown in Figure 4.22a) is

$$\text{Predicted Width} = 6.22 + 0.0366 \text{ Pages}$$

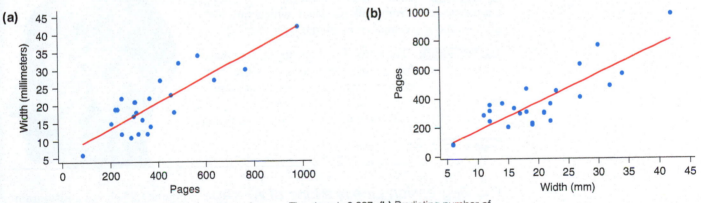

▲ FIGURE 4.22 (a) Predicting width from number of pages. The slope is 0.037. (b) Predicting number of pages from width. The slope is 19.6.

But what if we instead wanted to predict how many pages there are in a book on the basis of the width of the book?

To do this, we would switch the order of the variables and use *Pages* as our y-variable and *Width* as our x-variable. Then the slope is calculated to be 19.6 (Figure 4.22b).

It is tempting to think that because we are flipping the graph over when we switch x and y, we can just flip the slope over to get the new slope. If this were true, then we could find the new slope simply by calculating 1/(old slope). However, that approach doesn't work. That would give us a slope of

$$\frac{1}{0.0366} = 27.3, \text{ which is not the same as the correct value of } 19.6.$$

How, then, do we know which variable goes where?

We use the variable plotted on the horizontal axis to make predictions about the variable plotted on the vertical axis. For this reason, the x-variable is called the **explanatory variable**, the **predictor variable**, or the **independent variable**. The y-variable is called the **response variable**, the **predicted variable**, or the **dependent variable**. These names reflect the different roles played by the x- and y-variables in regression. Which variable is which depends on what the regression line will be used to predict.

You'll see many pairs of terms used for the x- and y-variables in regression. Some are shown in Table 4.1.

x-Variable	*y*-Variable
Predictor variable	Predicted variable
Explanatory variable	Response variable
Independent variable	Dependent variable

▲ **TABLE 4.1** Terms used for the *x*- and *y*-variables.

EXAMPLE 6 Moving Along

Neurologists have a variety of exams they give to patients to determine if they are suffering from cognitive impairment. Illnesses that cause cognitive decline (such as Alzheimer's) have other effects as well, and perhaps these more easily observed symptoms might be useful for predicting the cognitive level of a patient. Researchers (Janbourian et al. 2014) recruited 400 people between the ages of 18 and 65 and recorded how long it took them to walk 50 meters in a well-lit indoor hallway. Each person was also given a series of tests designed to test memory and basic problem-solving abilities. These tests were combined into a signal "psychometric score."

QUESTION   When making a scatterplot to predict psychometric score from walking speed, which variable goes on the x-axis and which on the y-axis?

CONCLUSION   The researchers hope to measure walking speed to predict a patient's psychometric score. Therefore walking speed is the predictor (independent) variable and is plotted on the x-axis, and psychometric score is the response (dependent) variable and is plotted on the y-axis.

TRY THIS   Exercise 4.41

The Regression Line Is a Line of Averages
Figure 4.23 shows a histogram of the weights of a sample of 507 active people. A useful statistical question is, "What is the typical weight?"

One way of answering this question is to calculate the mean of the sample. The distribution of weights is a little right-skewed but not terribly so, and so the mean will probably give us a good idea of what is typical. The mean weight of this group is 152.5 pounds.

Now, we know that an association exists between height and weight and that shorter people tend to weigh less. If we know someone's height, then, what weight would we

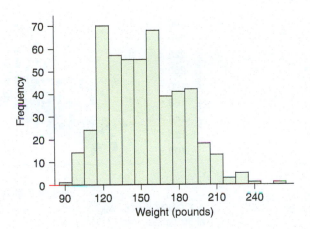

guess for that person? Surely not the average weight of the whole group! We can do better than that. For instance, what's the typical weight of someone 66 inches tall? To answer this, it makes sense to look only at the weights of those people in our sample who are about 66 inches tall. To make sure we have enough 66-inch-tall people in our sample, let's include everyone who is *approximately* 66 inches tall. So let's look at a slice of people whose height is between 65.5 inches and 66.5 inches. We come up with 47 people. Some of their weights are

$$130, 150, 142, 149, 118, 189 \ldots$$

Figure 4.24a shows this slice, which is centered at 66 inches.

The mean of these numbers is 141.5 pounds. We put a special symbol on the plot to record this point—a triangle at the point (66, 141.5), shown in Figure 4.24b.

The reason for marking this point is that if we wanted to predict the weights of those who were 66 inches tall, one good answer would be 141.5 pounds.

What if we wanted to predict the weight of someone who is 70 inches tall? We could take a slice of the sample and look at those people who are between 69.5 inches and 70.5 inches tall. Typically, they're heavier than the people who are about 66 inches tall. Here are some of their weights:

$$189, 197, 151, 207, 157 \ldots$$

Their mean weight is 173.0 pounds. Let's put another special triangle symbol at (70, 173.0) to record this. We can continue in this fashion, and Figure 4.24c shows where the mean weights are for a few more heights.

Note that the means fall (nearly) on a straight line. What could be the equation of this line? Figure 4.24d shows the regression line superimposed on the scatterplot with the means. They're nearly identical.

In theory, these means should lie exactly on the regression line. However, when working with real data, we often find that the theory doesn't always fit perfectly.

The series of graphs in Figure 4.24 illustrates a fundamental feature of the regression line: It is a line of means. You plug in a value for x, and the regression line "predicts" the mean y-value for everyone in the population with that x-value.

KEY POINT	When the trend is linear, the regression line connects the points that represent the mean value of y for each value of x.

(a)

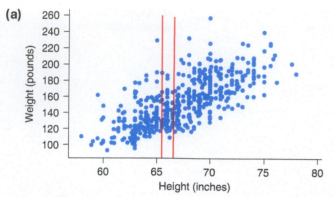

▲ **FIGURE 4.24a** Heights and weights with a slice at 66 inches.

(b)

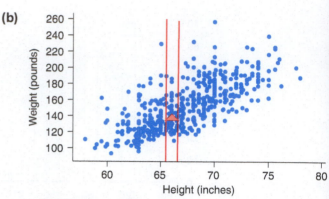

▲ **FIGURE 4.24b** Heights and weights with the average weight at 66 inches, which is 141.5 pounds.

(c)

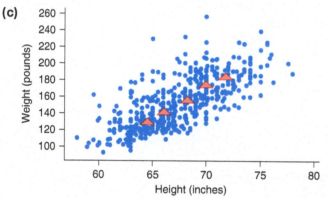

▲ **FIGURE 4.24c** Heights and weights with more means marked.

(d)

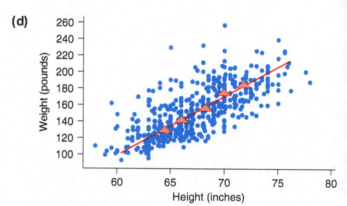

▲ **FIGURE 4.24d** Heights and weights with means, and a straight line superimposed on the scatterplot.

EXAMPLE 7 Funny-Car Race Finishing Times

A statistics student collected data on National Hot Rod Association driver John Force. The data consist of a sample of funny-car races. The drivers race in several different trials and change speed frequently. One question of interest is whether the fastest speed (in miles per hour, mph) driven during a trial is associated with the time to finish that trial (in seconds). If a driver's fastest speed cannot be maintained and the driver is forced to slow down, then it might be best to avoid going that fast. An examination of Figure 4.25, the scatterplot of finishing time (response variable) versus fastest speed (predictor) for a sample of trials for John Force, shows a reasonably linear association. The regression line is

$$\text{Predicted Time} = 7.84 - 0.0094 \,(\text{Highest mph})$$

▶ **FIGURE 4.25** A scatterplot for the funny-car race. (Source: J. Hettinga, 2005, UCLA, Department of Statistics)

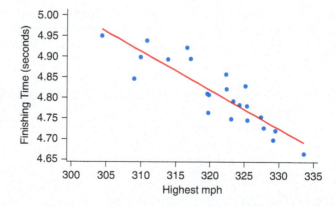

QUESTION   What is the predicted mean finishing time when the driver's fastest speed is 318 mph?

SOLUTION   We predict by estimating the mean finishing time when the speed is 318 mph. To find this, plug 318 into the regression line:

$$\text{Predicted Time} = 7.84 - 0.0094 \times 318 = 4.85 \text{ seconds}$$

CONCLUSION   We predict that the finishing time will be 4.85 seconds.

TRY THIS!   Exercise 4.43

Interpreting the Slope   Once we've established that the mean value of y depends on the value of x, a good statistical question to ask is, "How do typical y-values compare for different x-values?" When the association is linear, the slope provides an answer to that question. The slope tells us how the mean y-values compare when the x-value are one unit apart. For example, how different is the mean finishing time when the driver's fastest speed is 1 mph faster? The slope in Example 7 tells us that the means differ by a slim 0.0094 second. (Of course, races can be won or lost by such small differences.) What if the fastest time is 10 mph faster? Then the mean finishing time is $10 \times 0.0094 = 0.094$ second faster. We can see that in a typical race, if the driver wants to lose 1 full second from racing time, he or she must go over 100 mph faster, which is clearly not possible.

We should pay attention to whether the slope is 0 or very close to 0. When the slope is 0, we say that no (linear) relationship exists between x and y. The reason is that a 0 slope means that no matter what value of x you consider, the predicted value of y is always the same. The slope is 0 whenever the correlation coefficient is 0.

For example, the slope for runners' ages and their marathon times (see Figure 4.3a) is very close to 0 (0.000014 second per year of age).

KEY POINT   For linear trends, a slope of 0 means there is no association between x and y.

SNAPSHOT ► The Slope

WHAT IS IT? ►	The slope of a regression line.
WHAT DOES IT DO? ►	Tells us how different the mean y-value is for observations that are 1 unit apart on the x-variable.
HOW DOES IT DO IT? ►	The regression line tells us the average y-values for any value of x. The slope tells us how these average y-values differ for different values of x.
HOW IS IT USED? ►	To measure the linear association between two variables.

EXAMPLE **8** Math and Critical Reading SAT Scores

A regression analysis to study the relationship between SAT math and SAT critical reading scores resulted in this regression model:

$$\text{Predicted Critical Reading} = 398.81 + 0.3030 \text{ Math}$$

This model was based on a sample of 200 students.

The scatterplot in Figure 4.26 shows a linear relationship with a weak positive trend.

▶ **FIGURE 4.26** The regression line shows a weak positive trend between critical reading SAT score and math SAT score.

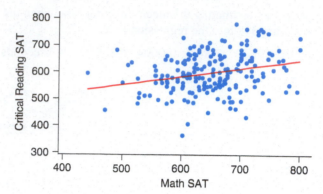

QUESTION   Interpret the slope of this regression line.

SOLUTION   The slope tells us that students who score 10 points higher on the math SAT had an average critical reading SAT score that was

$$10 \times 0.3030 = 3.03 \text{ points higher}$$

CONCLUSION   Students who score 10 points higher on the math SAT score, on average, 3.03 points higher on critical reading.

TRY THIS!   Exercise 4.61a

Interpreting the Intercept   The intercept tells us the predicted mean *y*-value when the *x*-value is 0. Quite often, this is not terribly helpful. Sometimes it is ridiculous. For example, the regression line to predict weight, given someone's height, tells us that if a person is 0 inches tall, then his or her predicted weight is negative 231.5 pounds!

Before interpreting the intercept, ask yourself whether it makes sense to talk about the *x*-variable taking on a 0 value. For the SAT data, you might think it makes sense to talk about getting a 0 on the math SAT. However, the lowest possible score on the SAT math portion is 200, so it is not possible to get a score of 0. (One lesson statisticians learn early is that you must know something about the data you analyze—knowing only the numbers is not enough!)

EXAMPLE **9**   The Integrated Postsecondary Education Data System

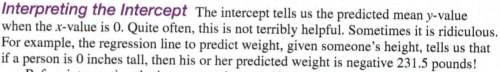

Do men and women graduate at a similar rate from two-year colleges? Figure 4.27 shows the association between male and female two-year graduation rates at public, nonprofit community colleges for a recent year. (Source, IPEDS, https://nces.ed.gov/ ipeds/use-the-data. Accessed March 5 2108.) The units are in percentage points. The regression line is

$$\text{Predicted Female Graduation Rate} = 3.8 + 0.9 \text{ (Male Graduation Rate)}$$

QUESTION   Interpret the intercept and slope.

CONCLUSION   The intercept is the predicted mean graduation rate for women at two-year colleges at which the graduation rate for men was 0 percent. The regression line predicts that the graduation rate will be 3.8 percent. There are in fact a small number of colleges that report graduation rates of 0 percent in this database, and it is possible that this is not because no one graduates but because of how the rates are reported.

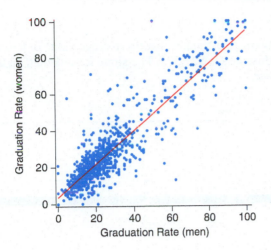

◀ FIGURE 4.27 Graduation rates for men and women at public two-year colleges.

The slope tells us that schools that graduate an additional 1 percent of men tend to graduate, on average, an additional 0.9 percentage point of women.

TRY THIS!   Exercise 4.65

SNAPSHOT ▶ The Intercept

WHAT IS IT? ▶	The intercept of a regression line.
WHAT DOES IT DO? ▶	Tells us the average y-value for all observations that have a zero x-value.
HOW DOES IT DO IT? ▶	The regression line is the best fit to a linear association, and the intercept is the best prediction of the y-value when the x-value is 0.
HOW IS IT USED? ▶	It is not always useful. Often, it doesn't make sense for the x-variable to assume the value 0.

EXAMPLE 10 Age and Value of Cars

Figure 4.28 shows the relationship between age and Kelley Blue Book value for a sample of cars. These cars were owned by students in one of the author's statistics classes.

The regression line is

$$\text{Predicted Value} = 21375 - 1215 \text{ Age}$$

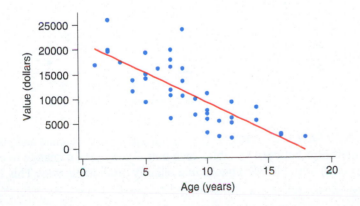

◀ FIGURE 4.28 The age of a car and its value for a chosen sample.

QUESTION Interpret the slope and intercept.

CONCLUSION The intercept estimates that the average value of a new car (0 years old) owned by students in this class is $21,375. The slope tells us that, on average, cars lost $1215 in value each year.

TRY THIS! Exercise 4.67

Evaluating the Linear Model

Regression is a very powerful tool for understanding linear relationships between numerical variables. However, we need to be aware of several potential interpretation pitfalls so that we can avoid them. We will also discuss methods for determining just how well the regression model fits the data.

Pitfalls to Avoid

You can avoid most pitfalls by simply making a graph of your data and examining it closely before you begin interpreting your linear model. This section offers some advice for sidestepping a few subtle complications that might arise.

Don't Fit Linear Models to Nonlinear Associations Regression models are useful only for linear associations. (Have you heard that somewhere before?) If the association is not linear, a regression model can be misleading and deceiving. For this reason, before you fit a regression model, you should always make a scatterplot to verify that the association seems linear.

Figure 4.29 shows an example of a bad regression model. The association between mortality rates for several industrialized countries (deaths per 1000 people) and wine consumption (liters per person per year) is nonlinear. The regression model is

$$\text{Predicted Mortality} = 7.69 - 0.0761 \text{ (Wine Consumption)}$$

but it provides a poor fit. The regression model suggests that countries with middle values of wine consumption should have higher mortality rates than they actually do.

► **FIGURE 4.29** The straight-line regression model is a poor fit to this nonlinear relationship. (Source: Leger et al. 1979)

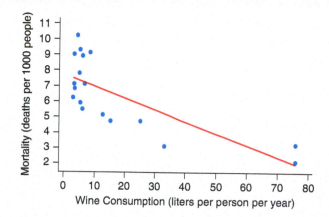

Correlation Is Not Causation One important goal of science and business is to discover relationships between variables such that if you make a change to one variable, you know the other variable will change in a reliably predictable way. This is what is

meant by "x causes y": make a change in x, and a change in y will usually follow. For example, the distance it takes to stop your car depends on how fast you were traveling when you first applied the brakes (among other things), the amount of memory an mp3 file takes on your hard drive depends on the length of the song, and the size of your phone bill depends on how many minutes you talked. In these cases, a strong causal relationship exists between two variables, and if you were to collect data and make a scatterplot, you would see an association between those variables.

In statistics, however, we are often faced with the reverse situation, in which we see an association between two variables and wonder whether there is a cause-and-effect relationship. The correlation coefficient for the association could be quite strong, but as we saw earlier, correlation does not mean cause and effect. A strong correlation or a good-fitting regression line is not evidence of a cause-and-effect relationship.

> **KEY POINT** An association between two variables is not sufficient evidence to conclude that a cause-and-effect relationship exists between the variables, no matter how strong the correlation or how well the regression line fits the data.

Be particularly careful about drawing cause-and-effect conclusions when interpreting the slope of a regression line. For example, for the SAT data,

$$\text{Predicted Critical Reading} = 398.81 + 0.3030 \text{ Math}$$

Even if this regression line fits the association very well, it does not give us sufficient evidence to conclude that if you improve your math score by 10 points, your critical reading score will go up by 3.03 points. As you learned in Chapter 1, because these data were not collected from a controlled experiment, the presence of confounding factors could prevent you from making a causal interpretation.

Beware of the algebra trap. In algebra, you were taught to interpret the slope to mean that "as x increases by 1 unit, y goes up by b units." However, quite often with data, the phrase "as x increases" doesn't make sense. When looking at the height and weight data, where x is height and y is weight, to say "x increases" means that people are growing taller! This is not accurate. It is much more accurate to interpret the slope as making comparisons between groups. For example, when comparing people of a certain height with those who are 1 inch taller, you can see that the taller individuals tend to weigh, on average, b pounds more.

When can we conclude that an association between two variables means a cause-and-effect relationship is present? Strictly speaking, never from an observational study and only when the data were collected from a controlled experiment. (Even in a controlled experiment, care must be taken that the experiment was designed correctly.) However, for many important questions, controlled experiments are not possible. In these cases, we can sometimes make conclusions about causality after a number of observational studies have been collected and examined and if there is a strong theoretical explanation for why the variables should be related in a cause-and-effect fashion. For instance, it took many years of observational studies to conclude that smoking causes lung cancer, including studies that compared twins—one twin who smoked and one who did not—and numerous controlled experiments on lab animals.

Beware of Outliers Recall that when calculating sample means, we must remember that outliers can have a big effect. Because the regression line is a line of means, you might think that outliers would have a big effect on the regression line. And you'd be right. You should always check a scatterplot before performing a regression analysis to be sure there are no outliers.

The graphs in Figure 4.30 illustrate this effect. Both graphs in Figure 4.30 show crime rates (number of reported crimes per 100,000 people) versus population density (people per square mile) in 2000. Figure 4.30a includes all 50 states and the District of

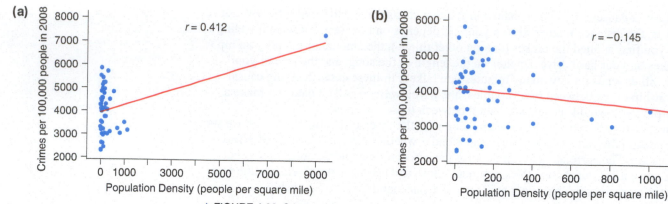

▲ **FIGURE 4.30** Crime rates and population densities in the states. Part **(a)** includes Washington, D.C., and part **(b)** does not. (Source: Joyce 2009)

> **Looking Back**
>
> **Outliers**
> You learned about outliers for one variable in Chapter 3. Outliers are values that lie a great distance from the bulk of the data. In two-variable associations, outliers are points that do not fit the trend or are far from the bulk of the data.

Columbia. The District is an outlier and has a strong influence on the regression line. Figure 4.30b excludes the District of Columbia.

Without the District of Columbia, the slope of the regression line actually changes sign! Thus, whether you conclude that a positive or a negative association occurs between crime and population density depends on whether this one city is included. These types of observations are called **influential points** because their presence or absence has a big effect on conclusions. When you have influential points in your data, it is good practice to try the regression and correlation with and without these points (as we did) and to comment on the difference.

Regressions of Aggregate Data Researchers sometimes do regression analysis based on what we call **aggregate data**. Aggregate data are those for which each plotted point is from a summary of a group of individuals. For example, in a study to examine the relationship between SAT math and critical reading scores, we might use the *mean* of each of the 50 states rather than the scores of individual students. The regression line provides a summary of linear associations between two variables. The mean provides a summary of the center of a distribution of a variable. What happens if we have a collection of means of two variables and we do a regression with these?

This is a legitimate activity, but you need to proceed with caution. For example, Figure 4.31 shows scatterplots of critical reading and math SAT scores. However, in Figure 4.31a, each point represents an individual: an SAT score for each first-year student enrolled in one northeastern university for a fall semester. The scatterplot in Figure 4.31b seems to show a much stronger relationship, because in this case each point is an aggregate. Specifically, each point represents a state in the United States: the mean SAT score for all students in the state taking SAT tests.

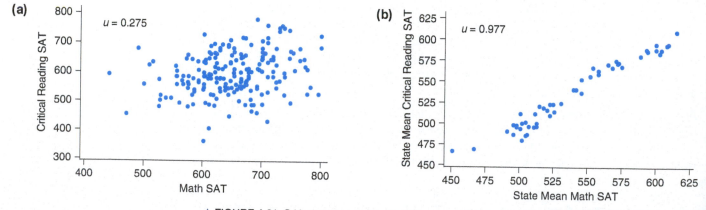

▲ **FIGURE 4.31** Critical reading and math SAT scores. **(a)** Scores for individuals. **(b)** Means for states.

We can still interpret Figure 4.31b, as long as we're careful to remember that we are talking about states, not people. We *can* say a strong correlation exists between a state's mean math SAT score and a state's mean critical reading SAT score. We *cannot* say that there is a strong correlation between individual students' math and critical reading scores.

Don't Extrapolate! **Extrapolation** means that we use the regression line to make predictions beyond the range of our data. This practice can be dangerous because although the association may have a linear shape for the range we're observing, that might not be true over a larger range. This means that our predictions might be wrong outside the range of observed *x*-values.

Figure 4.32a shows a graph of height versus age for children between 2 and 9 years old from a large national study. We've superimposed the regression line on the graph (Figure 4.32a), and it looks like a pretty good fit. However, although the regression model provides a good fit for children ages 2 through 9, it fails when we use the same model to predict heights for older children.

The regression line for the data shown in Figure 4.32a is

$$\text{Predicted Height} = 80.7 + 6.22 \text{ Age}$$

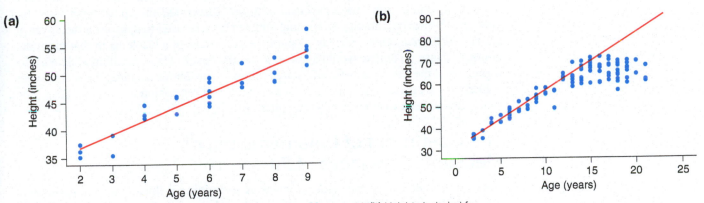

▲ FIGURE 4.32 (a) Ages and heights for children between 2 and 9 years old. (b) Heights included for people up to age 21 years. The straight line is not valid in the upper range of ages. (Source: National Health and Nutrition Examination Survey, Centers for Disease Control)

However, we observed only children between the ages of 2 and 9. Can we use this line to predict the height of a 20-year-old?

The regression model predicts that the mean height of 20-year-old is 80.78 inches:

$$\text{Predicted Height} = 31.78 + 2.45 \text{ Age} = 31.78 + 2.45 \times 20 = 80.78: \text{ nearly 7 feet!}$$

We can see from Figure 4.32b that the regression model provides a poor fit if we include people over the age of 9. Beyond that age, the trend is no longer linear, so we get bad predictions from the model.

It is often tempting to use the regression model beyond the range of data used to build the model. Don't. Unless you are very confident that the linear trend continues without change beyond the range of observed data, you must collect new data to cover the new range of values.

KEY POINT   Don't extrapolate!

The Origin of the Word Regression (Regression toward the Mean)

The first definition in the *Oxford Dictionary* says that *regression* is a "backward movement or return to a previous state." So why are we are using it to describe predictions of one numerical variable (such as value of a car) from another numerical variable (such as age of the car)?

The term *regression* was coined by Francis Galton, who used the regression model to study genetic relationships. He noticed that even though taller-than-average fathers tended to have taller-than-average sons, the sons were somewhat closer to average than the fathers were. Also, shorter-than-average fathers tended to have sons who were closer to the average than their fathers. He called this phenomenon regression toward mediocrity, but later it came to be known as **regression toward the mean**.

You can see how regression toward the mean works by examining the formula for the slope of the regression line:

$$b = r \frac{s_y}{s_x}$$

This formula tells us that fathers who are one standard deviation taller than average (s_x inches above average) have sons who are not one standard deviation taller than average (s_y) but are instead r times s_y inches taller than average. Because r is a number between -1 and 1, r times s_y is usually smaller than s_y. Thus, the "rise" will be less than the "run" in terms of standard deviations.

The *Sports Illustrated* jinx is an example of regression toward the mean. According to the jinx, athletes who make the cover of *Sports Illustrated* end up having a really bad year after appearing. Some professional athletes have refused to appear on the cover of *Sports Illustrated*. (Once, the editors published a picture of a black cat in that place of honor, because no athlete would agree to grace the cover.) However, if an athlete's performance in the first year is several standard deviations above average, the second year is likely to be closer to average. This is an example of regression toward the mean. For a star athlete, closer to average can seem disastrous.

The Coefficient of Determination, r^2, Measures Goodness of Fit

If we are convinced that the association we are examining is linear, then the regression line provides the best numerical summary of the relationship. But how good is "best"? The correlation coefficient, which measures the strength of linear relationships, can also be used to measure how well the regression line summarizes the data.

The **coefficient of determination** is simply the correlation coefficient squared: r^2. In fact, this statistic is often called *r-squared*. Usually, when reporting *r*-squared, we multiply by 100% to convert it to a percentage. Because r is always between -1 and 1, *r*-squared is always between 0% and 100%. A value of 100% means the relationship is perfectly linear and the regression line perfectly predicts the observations. A value of 0% means there is no linear relationship and the regression line does a very poor job.

For example, when we predicted the width of a book from the number of pages in the book, we found the correlation between these variables to be $r = 0.9202$. So the coefficient of determination is $0.9202^2 = 0.8468$, which we report as 84.7%.

What does this value of 84.7% mean? A useful interpretation of *r*-squared is that it measures how much of the variation in the response variable is explained by the explanatory variable. For example, 84.7% of the variation in book widths was explained by the number of pages. What does this mean?

Figure 4.33 shows a scatterplot (simulated data) with a constant value for y ($y = 6240$) no matter what the x-value is. You can see that there is no variation in y, so there is also nothing to explain.

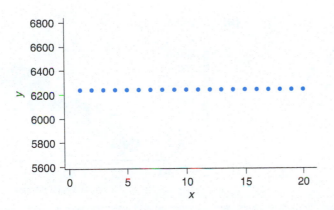

Figure 4.34 shows the height in both inches and in centimeters for several people. Here, we see variation in the *y*-variable because people naturally vary in height. Note that the points fall perfectly on a line because we're simply recording the same variable—height—for each person. The only difference is that the *x*-variable shows height in inches and the *y*-variable in centimeters. In other words, if you are given an *x*-value (a person's height in inches), then you know the *y*-value (the person's height in centimeters) precisely. Thus, all the variation in *y* is explained by the regression model. In this case, the coefficient of determination is 100%; all variation in *y* is perfectly explained by the best-fit line.

Real data are messier. Figure 4.35 shows a plot of the age and value of some cars. The regression line has been superimposed to remind us that there is, in fact, a linear trend and that the regression line does capture it. The regression model explains some of the variation in *y*, but as we can see, it's not perfect; plugging the value of *x* into the regression line gives us an imperfect prediction of what *y* will be. In fact, for these data, $r = -0.778$, so with this regression line we've explained $(-0.778)^2 = 0.605$, or about 60.5%, of the variation in *y*.

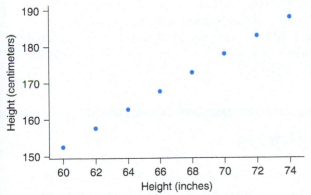

▲ FIGURE 4.34 Heights of people in inches and in centimeters (cm) with a correlation of 1. The coefficient of determination is 100%.

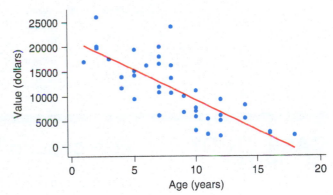

▲ FIGURE 4.35 Age and value of cars; the correlation is −0.778, and the coefficient of determination is 60.5%.

The practical implication of *r*-squared is that it helps determine which explanatory variable would be best for making predictions about the response variable. For example, is waist size or height a better predictor of weight? We can see the answer to this question from the scatterplots in Figure 4.36, which show that the linear relationship is stronger (has less scatter) for waist size.

The *r*-squared for predicting weight from height is 51.4% (Figure 4.36a), and the *r*-squared for predicting weight from waist size is 81.7% (Figure 4.36b). We can explain more of the variation in these people's weights by using their waist sizes than by using their heights, and therefore, we can make better (more precise) predictions using waist size as the predictor.

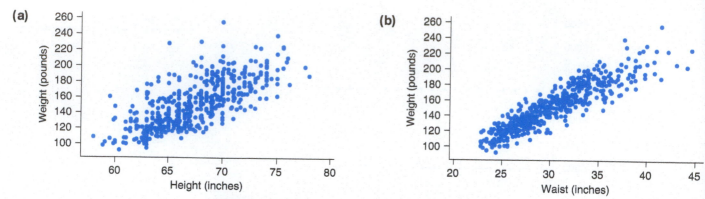

(a)

(b)

▲ **FIGURE 4.36** Scatterplots of height versus weight **(a)** and waist size versus weight **(b)**. Waist size has the larger coefficient of determination.

KEY POINT   If the association is linear, the larger the coefficient of determination (*r*-squared), the smaller the variation or scatter about the regression line, and the more accurate the predictions tend to be.

SNAPSHOT ▸ R-Squared

WHAT IS IT? ▸   *r*-squared or coefficient of determination.

WHAT DOES IT DO? ▸   Measures how well the data fit the linear model.

HOW DOES IT DO IT? ▸   If the points are tightly clustered around the regression line, r^2 has a value near 1, or 100%. If the points are more spread out, it has a value closer to 0. If there is no linear trend, and the points are a formless blob, it has a value near 0.

HOW IS IT USED? ▸   Works only with linear associations. Large values of *r*-squared tell us that predicted *y*-values are likely to be close to actual values. The coefficient of determination shows us the percentage of variation that can be explained by the regression line.

CASE STUDY REVISITED

Forecasting Home Prices

How does a company such as Zillow predict the value of a home? While we don't have access to the proprietary algorithm that Zillow uses, we can show how regression can help. To predict the price of a home, you need access to a great amount of historical data on home sales. The following scatterplots were all taken from data provided by the state of Colorado for a large number of sales dating back decades. In the interest of providing easily viewed graphs, we focus on only two townships. Figure 4.37 shows the fairly linear association between the value (the amount the house sold for in U.S. dollars) and the size of the home (in square feet) with the linear regression line superimposed. (You'll notice the fit doesn't seem quit perfect. That's because of a feature common in price data: the bigger homes have more variability in prices. If you take a future statistics course, you'll learn about approaches for adjusting for this feature, which statisticians call *heteroscedasticity*.)

The regression line is *predicted price* = −61506 + 288 × *Size*, which tells us that, according to this model, each additional square foot is valued at an average of $288 in this market. As you saw in this chapter, we can also use this to predict the price of a home for any given size. So a 2500-square-foot home would be predicted to cost −61506 + 288 × 2500 = $658,494.

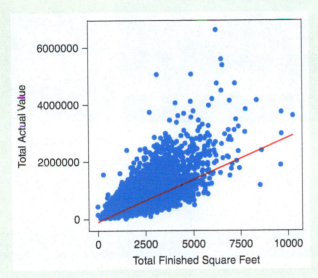

◀ **FIGURE 4.37** Regression line relating the value of homes to the total size.

A quick glance at the scatterplot shows us that our prediction has plenty of room for improvement. Homes that are 2500 square feet have a tremendous variability in prices, and so judging by the wide vertical spread of this plot, our prediction is likely to be fairly far off the true price of the home.

We can improve our prediction by providing the model with more information. For example, an old real estate joke says that the three most important components driving the price of a home are "location, location, and location." So we might expect homes in the two different townships represented in this data set to have different costs. Figure 4.38 shows which townships the homes belong to, and shows that the regression lines are different in these townships. A 2500-square-foot home in township "1N" is now predicted to cost about $740,000, while the same sized home in "3N" is predicted to cost $505,000. The fact that there is less variability about the two regression lines than there was when we ignored the location of the home tells us that a model that includes both the township and the size of the home will likely provide a better prediction than size alone.

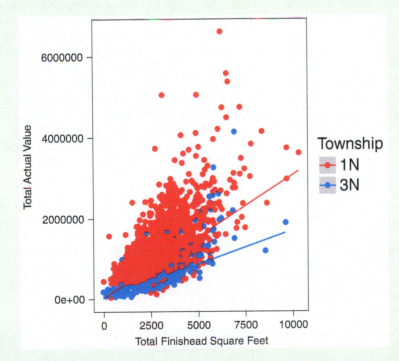

◀ **FIGURE 4.38** Regression lines for two separate townships.

This technique of using more than one predictor in a linear regression model is called multiple regression. Multiple regression allows us to add many predictor variables, and doing so often results in more precise predictions.

DATAPROJECT ▸ Data Moves

1 OVERVIEW

Chapter 4 introduced two important tools for analyzing two-variable associations when both variables are numerical: the scatterplot and the regression line. In this project we're going to introduce *data moves*, which are actions you take on a data set to help prepare or carry out a data analysis. They can enhance the statistical investigation questions (SIQs) that you can ask of a data set.

2 GOAL

Pose comparison and relation statistical investigation questions with two numerical variables. Create new data sets with fewer variables; create new variables from old ones.

3 DATA MOVES TO SHRINK AND TO GROW

Data sets with many variables can be overwhelming. And often, you don't need all those variables to answer the questions you are interested in. Suppose our primary question is whether runners' paces change during the race. We don't need variables like *Home Town* or *Gender Place* to answer our questions. But the marathon data set has quite a few "split time" variables that show the time it takes for runners to reach certain distances as measured from the start of the race, and those are the variables we'll focus on.

Do runners maintain a constant speed, or do they speed up or slow down in places?" We could rephrase this into a more precise SIQ as "Did the runners in the 2017 Los Angeles Marathon run faster, on average, in the first five kilometers than in the last five?"

Project: We're going to walk through some steps you can take to help you answer questions such as these. We're going to show you first how to make your data set more manageable by selecting the variables you are interested in and also how to make your data set larger by creating new variables that help you answer your questions.

A quick inspection of the data shows that we don't have a variable that tells us the runners' speed. We know their time and their distance, though, and so we can find the speed.

But before we do that, let's simplify the data set. One common data move is to select a few variables that you plan to work with. Or, if you want to look at it in a half-empty sort of way, we often want to discard variables that we know we will not use.

Our question involves the time for the first 5 kilometers, and this will require the variable *5k.Split*. We also want to know about the last 5 kilometers. You might not know this, but a little internet search will reveal that a marathon has 42.2 kilometers. The last 5 kilometers are therefore kilometers 37.2 through 42.2. Now our data don't provide us with that. But they do give us both the 35-kilometer split and the 40-kilometer split, and so perhaps we can refine our question to look at a 5-kilometer stretch near the end. Maybe this is not as fulfilling as knowing the speed at the very last 5 kilometer stretch, but it is the best we can do. (Of course, there are other approaches. What do you think about comparing the first 5 kilometers with the last 2.2?) This is a common spin around the Data Cycle: We start with a question, we consider data, and before we do the analysis, we realize we have to cycle back and refine our question.

It's best to err on the side of caution and leave more variables than you think you'll need. In StatCrunch, we can create a new data set using these commands:

Data > Arrange > Subset

We then select the variables we want and check "Open in a new data set" under Options. And, like that, we have our first data move!

We kept *Bib.Number* because it is always wise to keep an ID variable. We also kept *Age* and *Gender*, in case we wanted to investigate questions about age.

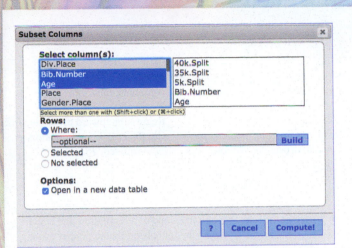

Now that we've shrunk the data, it's time to make it bigger! This data move is sometimes called *feature creation*. In data science parlance, a "feature" is similar to a "variable." We want to calculate the speed of the runners in the first 5 kilometers. Because speed = (distance/time), we need to create a new variable that does that. The distance is 5 kilometers, and the time is given in the variable *5K.Split*.

Select **Data > Compute > Expression.**

Then type in the formula in the "Expression" both and give the new variable a name:

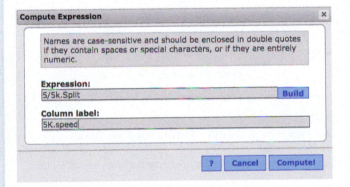

Hitting **Compute**! creates a new column with units of kilometers/second.

We can create the speed for kilometers 35 to 40 as follows:

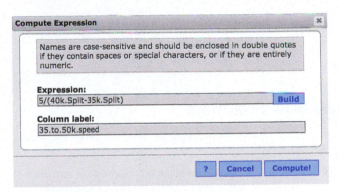

We are now prepared to answer the question: Is the typical speed faster in the first 5 kilometers than in a 5-kilometer stretch near the end of the race? We'll leave it to you to explore how to get StatCrunch to make graphical and numerical summaries that can answer this question.

Assignment

Pose two SIQs that we can ask about the LA Marathon data. One question should be a comparison of two numerical variables; the other about the relationship between two numerical variables.

Create a new data set that has only the variables you need to answer your questions plus at least one additional variable that you created from other variables in the data set.

Write a short report that answers your statistical questions, using appropriate graphs and numerical investigative summaries. (Remember, associations between numerical variables can often be summarized with a regression line. If a regression line won't summarize the relationship you're asking about, explain why.) Include a discussion of why and how you created any additional features.

CHAPTER REVIEW

KEY TERMS

Scatterplot, *151*
Trend, *151*
Strength, *151*
Shape, *151*
Positive association (positive trend), *151*

Negative association (negative trend), *151*
Linear, *153*
Correlation coefficient, *156*
Regression line, *165*
y-intercept, *165*

Slope, *165*
Explanatory variable, predictor variable, independent variable, *172*
Response variable, predicted variable, dependent variable, *172*

Influential point, *180*
Aggregate data, *180*
Extrapolation, *181*
Regression toward the mean, *182*
Coefficient of determination, r^2, *r*-squared, *182*

LEARNING OBJECTIVES

After reading this chapter and doing the assigned homework problems, you should

- Be able to write a concise and accurate description of an association between two numerical variables based on a scatterplot.

- Understand how to use a regression line to summarize a linear association between two numerical variables.

- Interpret the intercept and the slope of a regression line in context, and know how to use the regression line to predict mean values of the response variable.

- Critically evaluate a regression model.

SUMMARY

The first step in looking at the relationship between two numerical variables is to make a scatterplot and learn as much about the association as you can. Examine trend, strength, and shape. Look for outliers.

Regression lines and correlation coefficients can be interpreted meaningfully only if the association is linear. A correlation coefficient close to 1 or –1 does *not* mean the association is linear. When the association is linear, the correlation coefficient can be calculated by Formula 4.1:

$$\textbf{Formula 4.1:} \quad r = \frac{\sum z_x z_y}{n - 1}$$

A regression line can summarize the relationship in much the same way as a mean and a standard deviation can summarize a distribution. Interpret the slope and intercept, and use the regression line to make predictions about the mean *y*-value for any given value of *x*. The coefficient of determination, r^2, indicates the strength of the association by measuring the proportion of variation in *y* that is explained by the regression line.

The regression line is given by

$$\textbf{Formula 4.2c:} \quad \text{Predicted } y = a + bx$$

where the slope, *b*, is

$$\textbf{Formula 4.2a:} \quad b = r \frac{s_y}{s_x}$$

and the intercept, *a*, is

$$\textbf{Formula 4.2b:} \quad a = \bar{y} - b\bar{x}$$

When interpreting a regression analysis, be careful:

Don't extrapolate.
Don't make cause-and-effect conclusions if the data are observational.
Beware of outliers, which may (or may not) strongly affect the regression line.
Proceed with caution when dealing with aggregated data.

SOURCES

Anscombe, F. J. 1973. Graphs in statistical analysis. *American Statistician,* vol. 21, no. 1: 17–21.

Bentow, S., and D. Afshartous. Statistical analysis of Brink's data. www.stat.ucla.edu/cases.

Faraway, J. J. 2005. *Extending the linear model with R.* Boca Raton, FL: CRC Press.

Heinz, G., L. Peterson, R. Johnson, and C. Kerk. 2003. Exploring relationships in body dimensions. *Journal of Statistics Education,* vol. 11, no. 2. http://www.amstat.org/publications/jse/v11n2/datasets.heinz.html.

Jabourian, A., et al. 2014. Gait velocity is an indicator of cognitive performance in healthy middle-aged adults. *PLoS ONE,* vol. 9 (August): e103211. doi:10.1371/journal.pone.0103211.

Joyce, C. A., ed. *The 2009 world almanac and book of facts.* New York: World Almanac.

Leger, A., A. Cochrane, and F. Moore. 1979. Factors associated with cardiac mortality in developed countries with particular reference to the consumption of wine. *The Lancet,* vol. 313 (May): 1017–1020.

SECTION EXERCISES

SECTION 4.1

4.1 GPA Predictors The scatterplots show SAT scores and GPA in college for a sample of students. The top graph uses the critical reading SAT score to predict GPA in college and the bottom graph shows math SAT to predict GPA. Which is the better predictor of GPA for these students, critical reading SAT or math SAT? Explain your answer.

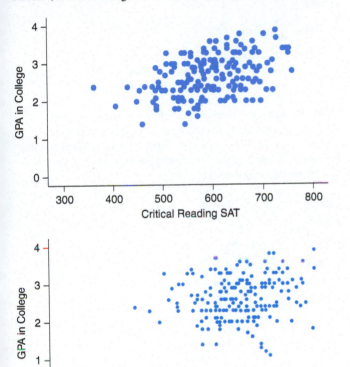

4.2 Salary and Employment The first graph shows the years a person was employed before working at the company and the salary at the company. The second graph shows the years employed at the company and the salary. Which graph shows a stronger relationship and could do a better job predicting salary at the company? (Source: Minitab 14)

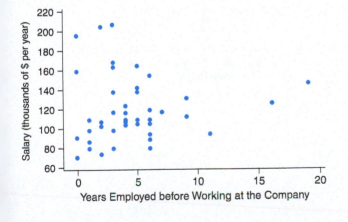

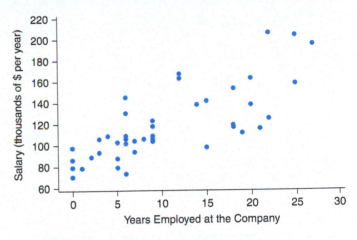

4.3 Age and Credits The scatterplot below shows data on age of a sample students and the number of college credits attained. Comment on the strength, direction, and shape of the trend.

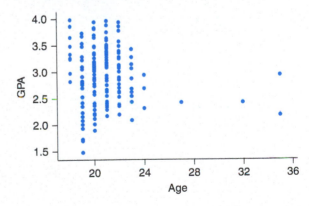

4.4 Age and GPA The scatterplot shows data on age and GPA for a sample of college students. Comment on the trend of the scatterplot. Is the trend positive, negative, or near zero?

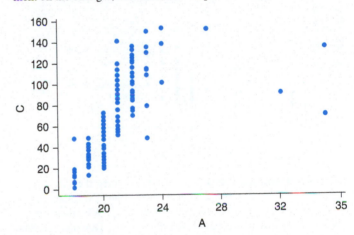

TRY 4.5 Credits and GPA (Example 1) The scatterplot shows data on credits attained and GPA for a sample of college students. Comment on the trend of the scatterplot. Is the trend positive, negative, or near zero?

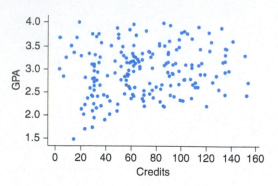

4.6 Salary and Education The scatterplot shows data on salary and years of education for a sample of workers. Comment on the trend of the scatterplot. Is the trend positive, negative, or near zero?

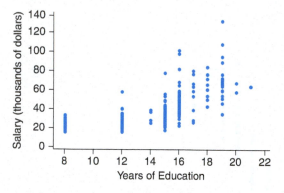

4.7 Sisters and Brothers The scatterplot shows the numbers of brothers and sisters for a large number of students. Do you think the trend is somewhat positive or somewhat negative? What does the direction (positive or negative) of the trend mean? Does the direction make sense in this context?

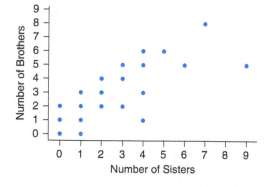

4.8 House Price and Area Describe the trend in the scatterplot of house price and area for some houses. State which point appears to be an outlier that does not fit the rest of the data.

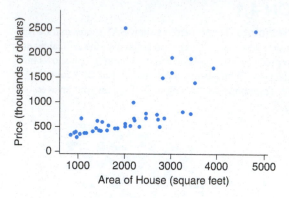

4.9 Work and TV The scatterplot shows the number of work hours and the number of TV hours per week for some college students who work. There is a very slight trend. Is the trend positive or negative? What does the direction of the trend mean in this context? Identify any unusual points.

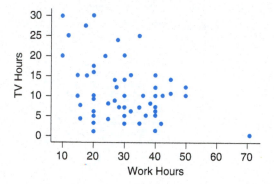

4.10 Work and Sleep The scatterplot shows the number of hours of work per week and the number of hours of sleep per night for some college students. Does the graph show a strong increasing trend, a strong decreasing trend, or very little trend? Explain.

4.11 Age and Sleep The scatterplot shows the age and number of hours of sleep "last night" for some students. Do you think the trend is slightly positive or slightly negative? What does that mean?

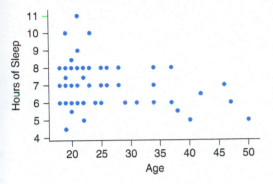

4.12 Height and Weight for Women The figure shows a scatterplot of the heights and weights of some women taking statistics. Describe what you see. Is the trend positive, negative, or near zero? Explain.

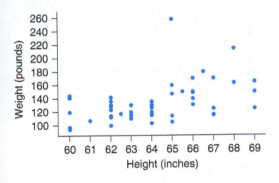

SECTION 4.2

4.13 College Tuition and ACT

a. The first scatterplot shows the college tuition and percentage acceptance at some colleges in Massachusetts. Would it make sense to find the correlation using this data set? Why or why not?

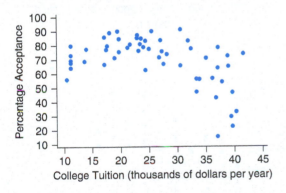

b. The second scatterplot shows the composite grade on the ACT (American College Testing) exam and the English grade on the same

exam. Would it make sense to find the correlation using this data set? Why or why not?

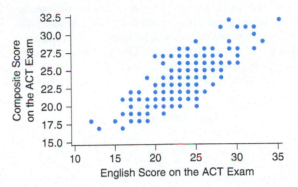

4.14 Ages of Women Who Give Birth The figure shows a scatterplot of birthrate (live births per 1000 women) and the age of the mother in the United States. Would it make sense to find the correlation for this data set? Explain. According to this graph, at approximately what age does the highest fertility rate occur?
(Source: Wendel and Wendel (eds.), *Vital statistics of the United States: Births, life expectancy, deaths, and selected health data*, 2nd ed. [Lanham, MD: Bernan Press, 2006])

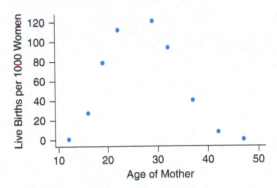

4.15 Law School The scatterplot shows the LSAT (Law School Aptitude Test) scores for a sample of law schools and the percent of students who were employed immediately after law school graduation. Do you think the correlation coefficient among these variables is positive, negative, or near zero? Give a reason for your choice.
(Source: Internet Legal Research Group)

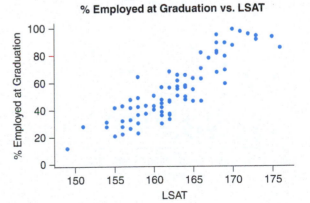

% Employed at Graduation vs. LSAT

4.16 Medical School The scatterplot shows the acceptance rate and selectivity index for a sample of medical schools. The acceptance rate is the percentage of applicants who were accepted into the medical school. The selectivity index is a measure based on GPA, test scores, and acceptance rates. A higher index indicates a more selective school. Do you think the correlation coefficient among these variables is positive, negative, or near zero? Give a reason for your choice. (Source: Accepted.com)

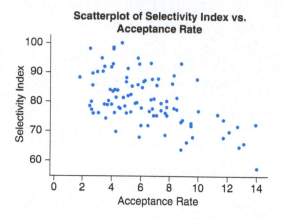

4.17 Matching Pick the letter of the graph that goes with each numerical value listed below for the correlation. Correlations:

0.767 _____

0.299 _____

−0.980 _____

(A)

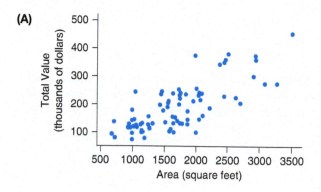

(B)

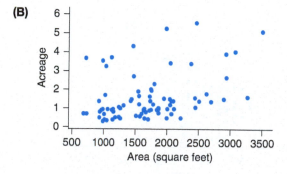

(C)

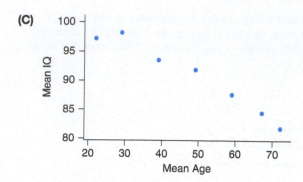

4.18 Matching Pick the letter of the graph that goes with each numerical value listed below for the correlation. Correlations:

−0.903 _____

0.374 _____

0.777 _____

(A)

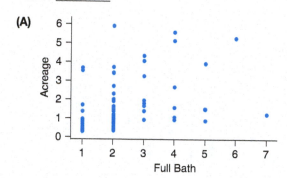

(B)

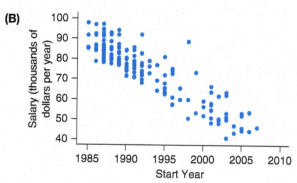

(C)

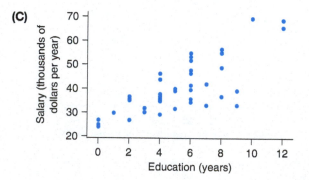

4.19 Matching Match each of the following correlations with the corresponding graph.

0.87 _____

−0.47 _____

0.67 _____

(A)

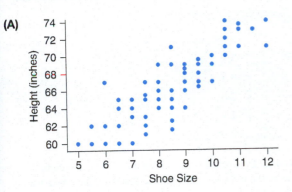

(B)

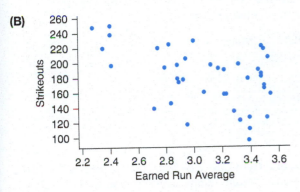

(Source: StatCrunch: 2011 MLB Pitching Stats according to owner: IrishBlazeFighter)

(C)

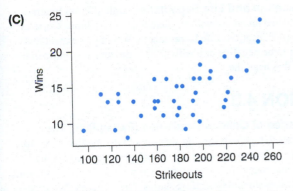

(Source: StatCrunch: 2011 MLB Pitching Stats according to owner: IrishBlazeFighter.)

4.20 Matching Match each of the following correlations with the corresponding graph.

−0.51 _____

0.98 _____

0.18 _____

(A)

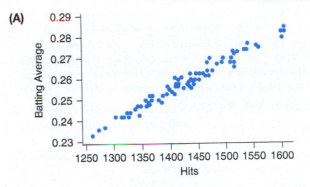

(B)

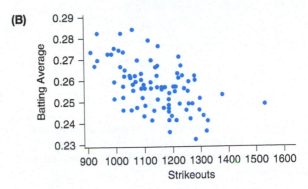

(C)

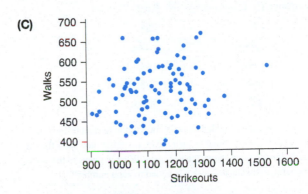

TRY **4.21 Airline Tickets (Example 2)** The distance (in kilometers) and price (in dollars) for one-way airline tickets from San Francisco to several cities are shown in the table.

Destination	Distance (km)	Price ($)
Chicago	2960	229
New York City	4139	299
Seattle	1094	146
Austin	2420	127
Atlanta	3440	152

a. Find the correlation coefficient for this data using a computer or statistical calculator. Use distance as the *x*-variable and price as the *y*-variable.

b. Recalculate the correlation coefficient for this data using price as the *x*-variable and distance as the *y*-variable. What effect does this have on the correlation coefficient?

c. Suppose a $50 security fee was added to the price of each ticket. What effect would this have on the correlation coefficient?

d. Suppose the airline held an incredible sale, where travelers got a round-trip ticket for the price of a one-way ticket. This means that the distances would be doubled while the ticket price remained the same. What effect would this have on the correlation coefficient?

4.22 Train Tickets The table for part (a) shows distances between selected cities and the cost of a business class train ticket for travel between these cities.

a. Calculate the correlation coefficient for the data shown in the table by using a computer or statistical calculator.

Distance (in miles)	Cost (in $)
439	281
102	152
215	144
310	293
406	281

b. The table for part (b) shows the same information, except that the distance was converted to kilometers by multiplying the number of miles by 1.609. What happens to the correlation when the numbers are multiplied by a constant?

Distance (in kilometers)	Cost
706	281
164	152
346	144
499	293
653	281

c. Suppose a surcharge is added to every train ticket to fund track maintenance. A fee of $20 is added to each ticket, no matter how long the trip is. The following table shows the new data. What happens to the correlation coefficient when a constant is added to each number?

Distance (in miles)	Cost (in $)
439	301
102	172
215	164
310	313
406	301

4.23 SAT and GPA In Exercise 4.1 there is a graph of the relationship between SAT score and college GPA. SAT score was the predictor and college GPA was the response variable. If you reverse the variables so that college GPA was the predictor and SAT score was the response variable, what effect would this have on the numerical value of the correlation coefficient?

4.24 House Price The correlation between house price (in dollars) and area of the house (in square feet) for some houses is 0.91. If you found the correlation between house price in *thousands* of dollars and area in square feet for the same houses, what would the correlation be?

4.25 Rate My Professor Seth Wagerman, a former professor at California Lutheran University, went to the website RateMyProfessors.com and looked up the quality rating and also the "easiness" of the six full-time professors in one department. The ratings are 1 (lowest quality) to 5 (highest quality) and 1 (hardest) to 5 (easiest). The numbers given are averages for each professor. Assume the trend is linear, find the correlation, and comment on what it means.

Quality	Easiness
4.8	3.8
4.6	3.1
4.3	3.4
4.2	2.6
3.9	1.9
3.6	2.0

4.26 Cousins Five people were asked how many female first cousins they had and how many male first cousins. The data are shown in the table. Assume the trend is linear, find the correlation, and comment on what it means.

Female	Male
2	4
1	0
3	2
5	8
2	2

4.27 GPA and Gym Use *USA Today College* published an article with the headline "Positive Correlation Found between Gym Usage and GPA." Explain what a positive correlation means in the context of this headline.

4.28 Education and Life Expectancy United Press International published an article with the headline "Study Finds Correlation between Education, Life Expectancy." Would you expect this correlation to be negative or positive? Explain your reasoning in the context of this headline.

SECTION 4.3

TRY **4.29 Salaries of College Graduates (Example 3)** The scatterplot shows the median starting salaries and the median mid-career salaries for graduates at a selection of colleges. (Source: *The Wall Street Journal*, Salary increase by salary type, http://online.wsj.com/public/resources/documents/info-Salaries_for_Colleges_by_Type-sort.html. Accessed via StatCrunch. Owner: Webster West)

a. As the data are graphed, which is the independent and which the dependent variable?

b. Why do you suppose median salary at a school is used instead of the mean?

c. Using the graph, estimate the median mid-career salary for a median starting salary of $60,000.

d. Use the equation to predict the median mid-career salary for a median starting salary of $60,000.

e. What other factors besides starting salary might influence mid-career salary?

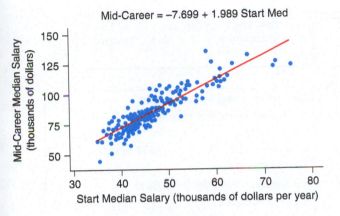

Mid-Career = −7.699 + 1.989 Start Med

4.30 Mother and Daughter Heights The graph shows the heights of mothers and daughters. (Source: StatCrunch: Mother and Daughter Heights.xls. Owner: craig_slinkman)

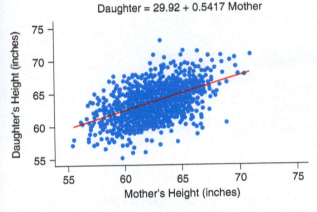

Daughter = 29.92 + 0.5417 Mother

a. As the data are graphed, which is the independent variable and which the dependent variable?

b. From the graph, approximate the predicted height of the daughter of a mother who is 60 inches (5 feet) tall.

c. From the equation, determine the predicted height of the daughter of a mother who is 60 inches tall.

d. Interpret the slope.

e. What other factors besides mother's height might influence the daughter's height?

4.31 Median Weekly Earning by Gender The scatterplot shows the median weekly earning (by quarter) for men and women in the United States for the years from 2005 through 2017. The correlation is 0.974. (Source: Bureau of Labor Statistics)

a. Use the scatterplot to estimate the median weekly income for women in a quarter in which the median pay for men is about $850.

b. Use the regression equation shown above the graph to get a more precise estimate of the median pay for women in a quarter in which the median pay for men is $850.

c. What is the slope of the regression equation? Interpret the slope of the regression equation.

d. What is the y-intercept of the regression equation? Interpret the y-intercept of the regression equation or explain why it would be inappropriate to do so.

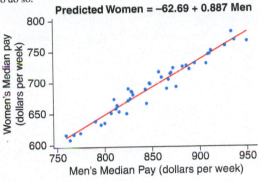

Predicted Women = −62.69 + 0.887 Men

4.32 Home Prices and Size The scatterplot shows the size (in square feet) and selling prices for homes in a certain zip code in California. (Source: realtor.com)

a. Use the graph to estimate the selling price of a home with 2000 square feet.

b. Use the equation to predict the selling price for a home with 2000 square feet.

c. What is the slope of the regression equation? Interpret the slope of the regression equation.

d. What is the y-intercept of the regression equation? Interpret the y-intercept of the regression equation or explain why it would be inappropriate to do so.

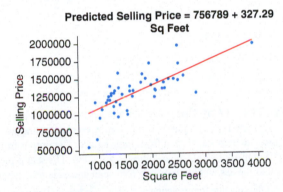

Predicted Selling Price = 756789 + 327.29 Sq Feet

TRY *4.33 Height and Arm Span for Women (Example 4) TI-84 output from a linear model for predicting arm span (in centimeters) from height (in inches) is given in the figure. Summary statistics are also provided.

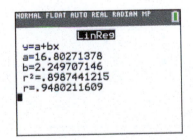

	Mean	Standard Deviation
Height, x	63.59	3.41
Arm span, y	159.86	8.10

To do parts a through c, assume that the association between arm span and height is linear.

a. Report the regression equation, using the words *height* and *arm span*, not x and y, employing the output given.

b. Verify the slope by using the formula $b = r\dfrac{s_y}{s_x}$.

c. Verify the y-intercept using $a = \bar{y} - b\bar{x}$.

d. Using the regression equation, predict the arm span (in centimeters) for someone 64 inches tall.

4.34 Hand and Foot Length for Women The computer output shown below is for predicting foot length from hand length (in centimeters) for a group of women. Assume the trend is linear. Summary statistics for the data are shown in the table below.

	Mean	Standard Deviation
Hand, x	17.682	1.168
Foot, y	23.318	1.230

```
The regression equation is
Y = 5.67 + 0.998 X
Pearson correlation of HandL and FootL = 0.948
```

a. Report the regression equation, using the words *hand* and *foot*, not x and y.

b. Verify the slope by using the formula $b = r\dfrac{s_y}{s_x}$.

c. Verify the y-intercept by using the formula $a = \bar{y} - b\bar{x}$.

d. Using the regression equation, predict the foot length (in centimeters) for someone who has a hand length of 18 centimeters.

TRY 4.35 Height and Arm Span for Men (Example 5) Measurements were made for a sample of adult men. A regression line was fit to predict the men's arm span from their height. The output from several different statistical technologies is provided. The scatterplot confirms that the association between arm span and height is linear.

a. Report the equation for predicting arm span from height. Use words such as *arm span*, not just x and y.

b. Report the slope and the intercept from each technology, using all the digits given.

(A)

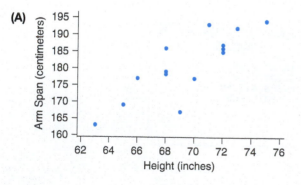

(B)
```
The regression equation is
Armspan = 6.2 + 2.51 Height
```

(C)

Simple linear regression results:
Dependent Variable: Armspan
Independent Variable: Height
Armspan = 6.2408333 + 2.514674 Height
Sample size: 15
R (correlation coefficient) = 0.8681
R-sq = 0.7535989
Estimate of error standard deviation: 5.409662

(D)

	Coefficients
Intercept	6.240833
X Variable	2.514674

(E)

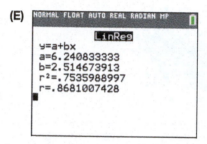

4.36 Hand Length and Foot Length for Men Measurements were made for a sample of adult men. Assume that the association between their hand length and foot length is linear. Output for predicting foot length from hand length is provided from several different statistical technologies.

a. Report the equation for predicting foot length from hand length. Use the variable names *FootL* and *HandL* in the equation, rather than x and y.

b. Report the slope and the intercept from each technology, using all the digits given.

(A)
```
The regression equation is
FootL = 15.8 + 0.563 HandL
```

(B)

Simple linear regression results:
Dependent Variable: FootL
Independent Variable: HandL
FootL = 15.807631 + 0.5626551 HandL
Sample size: 17
R (correlation coefficient) = 0.404
R-sq = 0.1632489
Estimate of error standard deviation: 1.6642156

(C)

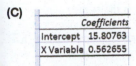

	Coefficients
Intercept	15.80763
X Variable	0.562655

(D)

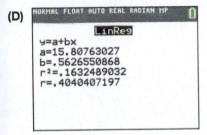

4.37 Comparing Correlation for Arm Span and Height The correlation between height and armspan in a sample of adult women was found to be $r = 0.948$. The correlation between arm span and height in a sample of adult men was found to be $r = 0.868$. Assuming both associations are linear, which association—the association between height and arm span for women, or the association between height and arm span for men—is stronger? Explain.

*4.38 Age and Weight for Men and Women The scatterplot shows a solid blue line for predicting weight from age of men; the dotted red line is for predicting weight from age of women. The data were collected from a large statistics class.

a. Which line is higher and what does that mean?

b. Which line has a steeper slope and what does that mean?

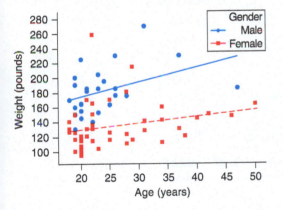

4.39 Singles and Doubles Winning Percentage The following graph shows the winning percentages in singles matches and doubles matches for a sample of male professional tennis players. (Source: tennis.com)

a. Based on this scatterplot, would you say there is a strong linear association between these two variables?

b. Would the numerical value of the correlation between these two variables be close to negative one, positive one, or zero? Give a reason for your answer.

c. Based on this graph, do you think one can accurately predict a professional tennis player's doubles winning percentage based on his singles winning percentage?

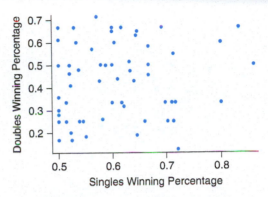

4.40 Seesaw The figure shows a scatterplot of the height of the left seat of a seesaw and the height of the right seat of the same seesaw. Estimate the numerical value of the correlation, and explain the reason for your estimate.

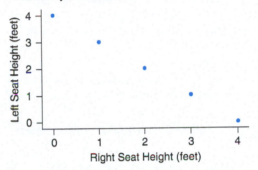

TRY 4.41 Choosing the Predictor and Response Variables (Example 6) Indicate which variable you think should be the predictor (x) and which variable should be the response (y). Explain your choices.

a. You have collected data on used cars for sale. The variables are price and odometer readings of the cars.

b. Research is conducted on monthly household expenses. Variables are monthly water bill and household size.

c. A personal trainer gathers data on the weights and time spent in the gym for each of her clients.

4.42 Choosing the Predictor and the Response Variable Indicate which variable you think should be the predictor (x) and which variable should be the response (y). Explain your choices.

a. A researcher measures subjects' stress levels and blood pressures.

b. Workers who commute by car record the length of their commutes (in miles) and the amount spent monthly on gasoline purchases.

c. Amusement parks record the heights and maximum speeds of roller coasters.

TRY 4.43 Percentage of Smoke-Free Homes and Percentage of High School Students Who Smoke (Example 7) The following figure shows a scatterplot with the regression line. The data are for the 50 states. The predictor is the percentage of smoke-free homes. The response is the percentage of high school students who smoke. The data came from the Centers for Disease Control and Prevention.

a. Explain what the trend shows.

b. Use the regression equation to predict the percentage of students in high school who smoke, assuming that there are 70% smoke-free homes in the state. Use 70 not 0.70.

Predicted Pct. Smokers = 56.32 − 0.464 (Pct. Smoke-free)

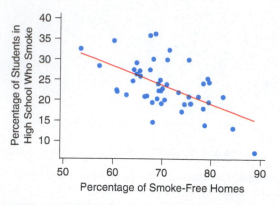

4.44 Effect of Adult Smoking on High School Student Smoking

The following figure shows a scatterplot with a regression line. The data are for the 50 states. The predictor is the percentage of adults who smoke. The response is the percentage of high school students who smoke. (The point in the lower left is Utah.)

a. Explain what the trend shows.

b. Use the regression equation to predict the percentage of high school students who smoke, assuming that 25% of adults in the state smoke. Use 25, not 0.25.

Predicted Pct. Smokers = −0.838 + 1.124 (Pct. Adult Smoke)

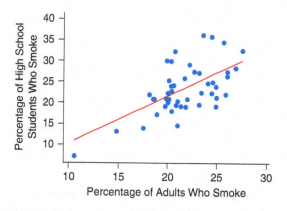

4.45 Car Insurance and Age

The following graph shows the average car insurance premium for a sample of ages. (Source: valuepenguin.com)

a. Explain what the graph tells us about insurance rates for drivers at different ages. Explain why insurance rates might follow this trend.

b. Would it be appropriate to do a linear regression analysis on these data? Why or why not?

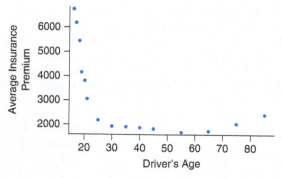

4.46 Life Insurance and Age

The graph shows the monthly premiums for a 10-year $250,000 male life insurance policy by age of purchase. For example, a 20-year-old male could purchase such a policy for about $10 per month, while a 50-year-old male would pay about $24 per month for the same policy.

a. Explain what the graph tells us about life insurance rates for males at different ages. Explain why life insurance rates might follow this trend.

b. Would it be appropriate to do a linear regression analysis on these data? Why or why not?

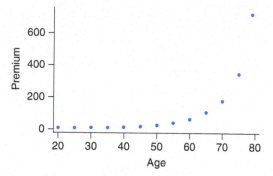

4.47 How Is the Time of a Flight Related to the Distance of the Flight?

g The following table gives the distance from Boston to each city (in thousands of miles) and gives the time for one randomly chosen, commercial airplane to make that flight. Do a complete regression analysis that includes a scatterplot with the line, interprets the slope and intercept, and predicts how much time a nonstop flight from Boston to Seattle would take. The distance from Boston to Seattle is 3000 miles. *See page 209 for guidance.*

City	Distance (1000s of miles)	Time (hours)
St. Louis	1.141	2.83
Los Angeles	2.979	6.00
Paris	3.346	7.25
Denver	1.748	4.25
Salt Lake City	2.343	5.00
Houston	1.804	4.25
New York	0.218	1.25

4.48 Distance and Train Ticket Price

The following table gives the distance from Boston to each city and the cost of a train ticket from Boston to that city for a certain date.

City	Distance (in miles)	Ticket Price (in $)
Washington, D.C.	439	181
Hartford	102	73
New York	215	79
Philadelphia	310	293
Baltimore	406	175
Charlotte	847	288
Miami	1499	340
Roanoke	680	219
Atlanta	1086	310

Data continued on next page.

City	Distance (in miles)	Ticket Price (in $)
Tampa	1349	370
Montgomery	1247	373
Columbus	776	164
Indianapolis	950	245
Detroit	707	189
Nashville	1105	245

a. Use technology to produce a scatterplot. Based on your scatterplot do you think there is a strong linear relationship between these two variables? Explain.

b. Compute r and write the equation of the regression line. Use the words "Ticket Price" and "Distance" in your equation. Round off to two decimal places.

c. Provide an interpretation of the slope of the regression line.

d. Provide an interpretation of the y-intercept of the regression line or explain why it would not be appropriate to do so.

e. Use the regression equation to predict the cost of a train ticket from Boston to Pittsburgh, a distance of 572 miles.

4.49 Do States with Higher Populations Have More Millionaires? The following table gives the number of millionaires (in thousands) and the population (in hundreds of thousands) for the states in the northeastern region of the United States in 2008. The numbers of millionaires come from *Forbes Magazine* in March 2007.

a. Without doing any calculations, predict whether the correlation and slope will be positive or negative. Explain your prediction.

b. Make a scatterplot with the population (in hundreds of thousands) on the x-axis and the number of millionaires (in thousands) on the y-axis. Was your prediction correct?

c. Find the numerical value for the correlation.

d. Find the value of the slope and explain what it means in context. Be careful with the units.

e. Explain why interpreting the value for the intercept does not make sense in this situation.

State	Millionaires	Population
Connecticut	86	35
Delaware	18	8
Maine	22	13
Massachusetts	141	64
New Hampshire	26	13
New Jersey	207	87
New York	368	193
Pennsylvania	228	124
Rhode Island	20	11
Vermont	11	6

4.50 Movie Ratings The following table give the Rotten Tomatoes and Metacritic scores for the several movies produced in 2017. Both of these ratings systems give movies a score using a scale from 0 to 100. (Source: vox.com)

a. Use technology to make a scatterplot using Rotten Tomatoes as the independent variable and Metacritic as the dependent variable. Based

on your scatterplot, do you think there is a strong linear association between these variables?

b. Compute the correlation coefficient, r, and write the equation of the regression line. Use the words "Rotten Tomatoes" and "Metacritic" in your equation. Round to two decimal places.

c. Provide an interpretation of the slope of the regression line.

d. Provide an interpretation of the y-intercept of the regression line or explain why it would not be appropriate to do so.

Movie	Rotten Tomatoes	Metacritic
Beauty and the Beast	70	65
Wonder Woman	92	76
Guardians of the Galaxy Vol. 2	82	67
Spider-Man: Homecoming	92	73
Despicable Me 3	61	49
Logan	93	77
The Fate of the Furious	66	56
Dunkirk	93	94
The LEGO Batman Movie	90	75
Get Out	99	84
The Boss Baby	51	50
Pirates of the Caribbean: Dead Men Tell No Tales	29	39
Kong: Skull Island	76	62
Cars 3	68	59
War for the Planet of the Apes	93	82
Split	74	62
Transformers: The Last Knight	15	28
Fifty Shades Darker	10	33
Girls Trip	88	71
Baby Driver	94	86

4.51 Pitchers The table shows the Earned Run Average (ERA) and WHIP rating (walks plus hits per inning) for the top 40 Major League Baseball pitchers in the 2017 season. Top pitchers will tend to have low ERA and WHIP ratings. (Source: ESPN.com)

a. Make a scatterplot of the data and state the sign of the slope from the scatterplot. Use WHIP to predict ERA.

b. Use linear regression to find the equation of the best-fit line. Show the line on the scatterplot using technology or by hand.

c. Interpret the slope.

d. Interpret the y-intercept or explain why it would be inappropriate to do so.

WHIP	ERA	WHIP	ERA
0.87	2.25	1.21	3.55
0.95	2.31	1.22	3.64
0.9	2.51	1.22	3.66
1.02	2.52	1.27	3.82
1.15	2.89	1.15	3.83
0.97	2.9	1.16	3.86

Data continued on next page.

WHIP	ERA	WHIP	ERA
1.18	2.96	1.27	3.89
1.04	2.98	1.35	3.9
1.31	3.09	1.28	3.92
1.07	3.2	1.42	4.03
1.13	3.28	1.26	4.07
1.1	3.29	1.36	4.13
1.35	3.32	1.28	4.14
1.17	3.36	1.22	4.15
1.32	3.4	1.33	4.16
1.23	3.43	1.37	4.19
1.25	3.49	1.2	4.24
1.22	3.53	1.25	4.26
1.19	3.53	1.3	4.26
1.21	3.54	1.37	4.26

4.52 Text Messages The following table shows the number of text messages sent and received by some people in one day. (Source: StatCrunch: Responses to survey How often do you text? Owner: Webster West. A subset was used.)

a. Make a scatterplot of the data, and state the sign of the slope from the scatterplot. Use the number sent as the independent variable.

b. Use linear regression to find the equation of the best-fit line. Graph the line with technology or by hand.

c. Interpret the slope.

d. Interpret the intercept.

Sent	Received	Sent	Received
1	2	10	10
1	1	3	5
0	0	2	2
5	5	5	5
5	1	0	0
50	75	2	2
6	8	200	200
5	7	1	1
300	300	100	100
30	40	50	50

SECTION 4.4

4.53 Answer the questions using complete sentences.

a. What is an influential point? How should influential points be treated when doing a regression analysis?

b. What is the coefficient of determination and what does it measure?

c. What is extrapolation? Should extrapolation ever be used?

4.54 Answer the questions using complete sentences.

a. An economist noted the correlation between consumer confidence and monthly personal savings was negative. As consumer confidence increases, would we expect monthly personal savings to increase, decrease, or remain constant?

b. A study found a correlation between higher education and lower death rates. Does this mean that one can live longer by going to college? Why or why not?

4.55 If there is a positive correlation between number of years studying math and shoe size (for children), does that prove that larger shoes cause more studying of math or vice versa? Can you think of a confounding variable that might be influencing both of the other variables?

4.56 Suppose that the growth rate of children looks like a straight line if the height of a child is observed at the ages of 24 months, 28 months, 32 months, and 36 months. If you use the regression obtained from these ages and predict the height of the child at 21 years, you might find that the predicted height is 20 feet. What is wrong with the prediction and the process used?

4.57 Coefficient of Determination If the correlation between height and weight of a large group of people is 0.67, find the coefficient of determination (as a percentage) and explain what it means. Assume that height is the predictor and weight is the response, and assume that the association between height and weight is linear.

4.58 Coefficient of Determination Does a correlation of -0.70 or $+0.50$ give a larger coefficient of determination? We say that the linear relationship that has the larger coefficient of determination is more strongly correlated. Which of the values shows a stronger correlation?

4.59 Investing Some investors use a technique called the "Dogs of the Dow" to invest. They pick several stocks that are performing poorly from the Dow Jones group (which is a composite of 30 well-known stocks) and invest in these. Explain why these stocks will probably do better than they have done before.

4.60 Blood Pressure Suppose a doctor telephones those patients who are in the highest 10% with regard to their recently recorded blood pressure and asks them to return for a clinical review. When she retakes their blood pressures, will those new blood pressures, as a group (that is, on average), tend to be higher than, lower than, or the same as the earlier blood pressures, and why?

TRY **4.61 Salary and Year of Employment (Example 8)** The equation for the regression line relating the salary and the year first employed is given above the figure.

a. Report the slope and explain what it means.

b. Either interpret the intercept (4,255,000) or explain why it is not appropriate to interpret the intercept.

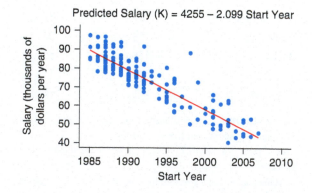

Predicted Salary (K) = 4255 − 2.099 Start Year

4.62 MPG: Highway and City The following figure shows the relationship between the number of miles per gallon on the highway and that in the city for some cars.

a. Report the slope and explain what it means.

b. Either interpret the intercept (7.792) or explain why it is not appropriate to interpret the intercept.

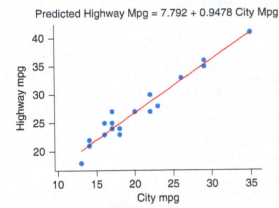

Predicted Highway Mpg = 7.792 + 0.9478 City Mpg

4.63 Cost of Turkeys The following table shows the weights and prices of some turkeys at different supermarkets.

a. Make a scatterplot with weight on the *x*-axis and cost on the *y*-axis. Include the regression line on your scatterplot.

b. Find the numerical value for the correlation between weight and price. Explain what the sign of the correlation shows.

c. Report the equation of the best-fit straight line, using weight as the predictor (*x*) and cost as the response (*y*).

d. Report the slope and intercept of the regression line, and explain what they show. If the intercept is not appropriate to report, explain why.

e. Add a new point to your data: a 30-pound turkey that is free. Give the new value for *r* and the new regression equation. Explain what the negative correlation implies. What happened?

f. Find and interpret the coefficient of determination using the original data.

Weight (pounds)	Price
12.3	$17.10
18.5	$23.87
20.1	$26.73
16.7	$19.87
15.6	$23.24
10.2	$ 9.08

4.64 Wine Calories The table shows the calories in a five-ounce serving and the % alcohol content for a sample of wines. (Source: healthalicious.com)

Calories	% alcohol
122	10.6
119	10.1
121	10.1
123	8.8
129	11.1
236	15.5

a. Make a scatterplot using % alcohol as the independent variable and calories as the dependent variable. Include the regression line on your scatterplot. Based on your scatterplot do you think there is a strong linear relationship between these variables?

b. Find the numerical value of the correlation between % alcohol and calories. Explain what the sign of the correlation means in the context of this problem.

c. Report the equation of the regression line and interpret the slope of the regression line in the context of this problem. Use the words *calories* and *% alcohol* in your equation. Round to two decimal places.

d. Find and interpret the value of the coefficient of determination.

e. Add a new point to your data: a wine that is 20% alcohol that contains 0 calories. Find *r* and the regression equation after including this new data point. What was the effect of this one data point on the value of *r* and the slope of the regression equation?

TRY **4.65 Teacher Pay and Expenditure Per Student (Example 9)** The scatterplot shows the average teacher pay and the per pupil expenditure for each of the 50 states and the District of Columbia. The regression equation is also shown. (Source: *The 2017 World Almanac and Book of Facts*).

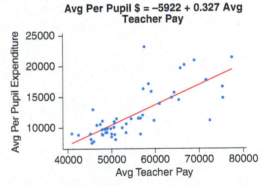

Avg Per Pupil $ = −5922 + 0.327 Avg Teacher Pay

a. From the scatterplot is the correlation between average teacher pay and per pupil expenditure positive, negative, or near zero?

b. What is the slope of the regression equation? Interpret the slope in the context of the problem.

c. What is the *y*-intercept of the regression equation? Interpret the *y*-intercept or explain why it would be inappropriate to do so for this problem.

d. Use the regression equation to estimate the per pupil expenditure for a state with an average teacher pay of $60,000.

4.66 Teacher Pay and High School Graduation Rates The scatterplot shows the average teacher pay and high school graduation percentage rate for each of the 50 states and the District of Columbia. The regression equation is also shown. (Source: *2017 World Almanac Book of Facts* and higheredinfo.org)

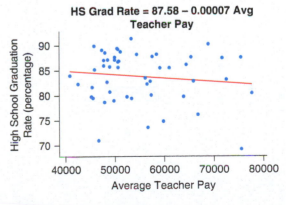

HS Grad Rate = 87.58 − 0.00007 Avg Teacher Pay

a. Based on the scatterplot is the correlation between average teacher pay and high school graduation rate positive, negative, or near zero?

b. Should the regression equation be used to predict the high school graduation rate for a state with an average teacher salary of $60,000? If so, predict the graduation rate. If not, explain why the regression equation should not be used to make this prediction.

TRY 4.67 Does Having a Job Affect Students' Grades? (Example 10) Grades on a political science test and the number of hours of paid work in the week before the test were recorded. The instructor was trying to predict the grade on a test from the hours of work. The following figure shows a scatterplot and the regression line for these data.

a. Referring to the figure, state whether you think the correlation is positive or negative, and explain your prediction.

b. Interpret the slope.

c. Interpret the intercept.

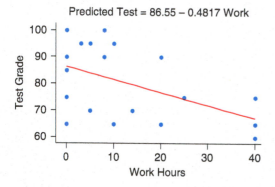

Predicted Test = 86.55 − 0.4817 Work

4.68 Weight of Trash and Household Size Data were collected that included information on the weight of the trash (in pounds) on the street for one week and the number of people who live in the house. The following figure shows a scatterplot with the regression line.

a. Is the trend positive or negative? What does that mean?

b. Now calculate the correlation between the weight of trash and the number of people. (Use *r*-squared from the figure and take the square root of it.)

c. Report the slope. For each additional person in the house, there are, on average, how many additional pounds of trash?

d. Either report the intercept or explain why it is not appropriate to interpret it.

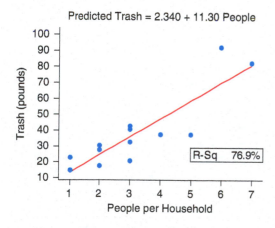

Predicted Trash = 2.340 + 11.30 People

4.69 Predicting Home Runs Data on the number of home runs, strikeouts, and batting averages for a sample of 50 Major League Baseball players were obtained. Regression analyses were conducted on the relationships between home runs and strikeouts and between home runs and batting averages. The StatCrunch results are shown below. (Source: mlb.com)

Simple linear regression results:
Dependent Variable: Home Runs
Independent Variable: Strikeouts
Home Runs = 0.092770565 + 0.22866236 Strikeouts
Sample size: 50
R (correlation coefficient) = 0.63591835
R-sq = 0.40439215
Estimate of error standard deviation: 8.7661607

Simple linear regression results:
Dependent Variable: Home Runs
Independent Variable: Batting Average
Home Runs = 45.463921 − 71.232795 Batting Average
Sample size: 50
R (correlation coefficient) = −0.093683651
R-sq = 0.0087766264
Estimate of error standard deviation: 11.30876

Based on this sample, is there a stronger association between home runs and strikeouts or home runs and batting average? Provide a reason for your choice based on the StatCrunch results provided.

4.70 Predicting 3-Point Baskets Data on the 3-point percentage, field-goal percentage, and free-throw percentage for a sample of 50 professional basketball players were obtained. Regression analyses were conducted on the relationships between 3-point percentage and field-goal percentage and between 3-point percentage and free-throw percentage. The StatCrunch results are shown below. (Source: nba.com)

Simple linear regression results:
Dependent Variable: 3 Point %
Independent Variable: Field Goal %
3 Point % = 40.090108 − 0.091032596 Field Goal %
Sample size: 50
R (correlation coefficient) = −0.048875984
R-sq = 0.0023888618
Estimate of error standard deviation: 7.7329785

Simple linear regression results:
Dependent Variable: 3 Point %
Independent Variable: Free Throw %
3 Point % = −8.2347225 + 0.54224127 Free Throw %
Sample size: 50
R (correlation coefficient) = 0.57040364
R-sq = 0.32536031
Estimate of error standard deviation: 6.3591944

Based on this sample, is there a stronger association between 3-point percentage and field-goal percentage or 3-point percentage and free-throw percentage? Provide a reason for your choice based on the StatCrunch results provided.

4.71 4th-Grade Reading and Math Scores Data from the National Data shown in the table are the 4th-grade reading and math scores for a sample of states from the National Assessment of Educational Progress. The scores represent the percentage of 4th-graders in each state who scored at or above basic level in reading and math. A scatterplot of the data suggests a linear trend. (Source: nationsreportcard.gov)

4th-Grade Reading Scores	4th-Grade Math Scores	4th-Grade Reading Scores	4th-Grade Math Scores
65	75	68	78
61	78	61	79
62	79	69	80
65	79	68	77
59	72	75	89
72	82	71	84
74	81	68	83
70	82	75	84
56	69	63	78
75	85	71	85

a. Find and report the value for the correlation coefficient and the regression equation for predicting the math score from the reading score. Use the words *Reading* and *Math* in your regression equation and round off to two decimal places. Then find the predicted math score for a state with a reading score of 70.

b. Find and report the value of the correlation coefficient regression equation for predicting the reading score from the math score. Then find the predicted reading score for a state with a math score of 70.

c. Discuss the effect of changing the choice of dependent and independent variable on the value of r and on the regression equation.

4.72 SAT Scores The following table shows the average SAT Math and Critical Reading scores for students in a sample of states. A scatterplot for these two variables suggests a linear trend. (Source: qsleap.com)

SAT Math Score	SAT Critical Reading Score	SAT Math Score	SAT Critical Reading Score
580	563	463	450
596	599	494	494
561	556	488	487
589	590	592	597
494	494	581	574
525	530	470	486
500	521	579	575
551	544	523	524
489	502	518	516
498	504	414	388
597	608	502	510
557	563	509	497
576	569	591	605
523	521	589	586
499	504		

a. Find and report the value for the correlation coefficient and the regression equation for predicting the math score from the critical reading score, rounding off to two decimal places. Then find the predicted math score for a state with a critical reading score of 600.

b. Find and report the value of the correlation coefficient and the regression equation for predicting the critical reading score from the math score. Then find the predicted reading score for a state with a math score of 600.

c. Discuss the effect of changing the choice of dependent and independent variable on the value of r and on the regression equation.

g *4.73 Test Scores Assume that in a political science class, the teacher gives a midterm exam and a final exam. Assume that the association between midterm and final scores is linear. The summary statistics have been simplified for clarity see Guidance on page 209.

Midterm: Mean = 75,   Standard deviation = 10
Final: Mean = 75,   Standard deviation = 10
Also, $r = 0.7$ and $n = 20$.

According to the regression equation, for a student who gets a 95 on the midterm, what is the predicted final exam grade? What phenomenon from the chapter does this demonstrate? Explain. *See page 209 for guidance.*

***4.74 Test Scores** Assume that in a sociology class, the teacher gives a midterm exam and a final exam. Assume that the association between midterm and final scores is linear. Here are the summary statistics:

Midterm: Mean = 72,   Standard deviation = 8
Final: Mean = 72,   Standard deviation = 8
Also, $r = 0.75$ and $n = 28$.

a. Find and report the equation of the regression line to predict the final exam score from the midterm score.

b. For a student who gets 55 on the midterm, predict the final exam score.

c. Your answer to part (b) should be higher than 55. Why?

d. Consider a student who gets a 100 on the midterm. Without doing any calculations, state whether the predicted score on the final exam would be higher, lower, or the same as 100.

CHAPTER REVIEW EXERCISES

* **4.75 Heights and Weights of People** The following table shows the heights and weights of some people. The scatterplot shows that the association is linear enough to proceed.

Height (inches)	Weight (pounds)
60	105
66	140
72	185
70	145
63	120

a. Calculate the correlation, and find and report the equation of the regression line, using height as the predictor and weight as the response.

b. Change the height to centimeters by multiplying each height in inches by 2.54. Find the weight in kilograms by dividing the weight in pounds by 2.205. Retain at least six digits in each number so there will be no errors due to rounding.

c. Report the correlation between height in centimeters and weight in kilograms, and compare it with the correlation between the height in inches and weight in pounds.

d. Find the equation of the regression line for predicting weight from height, using height in centimeters and weight in kilograms. Is the equation for weight in pounds and height in inches the same as or different from the equation for weight in kilograms and height in centimeters?

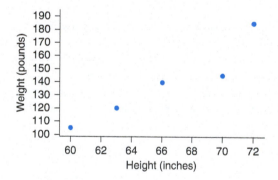

* **4.76 Heights and Weights of Men** The table shows the heights (in inches) and weights (in pounds) of 14 college men. The scatterplot shows that the association is linear enough to proceed.

Height (inches)	Weight (pounds)	Height (inches)	Weight (pounds)
68	205	70	200
68	168	69	175
74	230	72	210
68	190	72	205
67	185	72	185
69	190	71	200
68	165	73	195

a. Find the equation for the regression line with weight (in pounds) as the response and height (in inches) as the predictor. Report the slope and

the intercept of the regression line, and explain what they show. If the intercept is not appropriate to report, explain why.

b. Find the correlation between weight (in pounds) and height (in inches).

c. Find the coefficient of determination and interpret it.

d. If you changed each height to centimeters by multiplying heights in inches by 2.54, what would the new correlation be? Explain.

e. Find the equation with weight (in pounds) as the response and height (in inches) as the predictor, and interpret the slope.

f. Summarize what you found: Does changing units change the correlation? Does changing units change the regression equation?

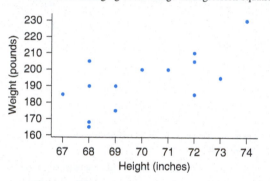

4.77 Fast-Food Calories, Carbs, and Sugar The data shows the number of calories, carbohydrates (in grams) and sugar (in grams) found in a selection of menu items at McDonald's. Scatterplots suggest the relationship between calories and both carbs and sugars is linear. The data are also available on this text's website. (Source: shapefit.com)

Calories	Carbs (in grams)	Sugars (in grams)
530	47	9
520	42	10
720	52	14
610	47	10
600	48	12
540	45	9
740	43	10
240	32	6
290	33	7
340	37	7
300	32	6
430	35	7
380	34	7
430	35	6
440	35	7
430	34	7
750	65	16
590	51	14
510	55	10
350	42	8

Data continued on next page.

Calories	Carbs (in grams)	Sugars (in grams)
670	58	11
510	44	9
610	57	11
450	43	9
360	40	5
360	40	5
430	41	6
480	43	6
430	43	7
390	39	5
500	44	11
670	68	12
510	54	10
630	56	7
480	42	6
610	56	8
450	42	6
540	61	14
380	47	12
340	37	8
260	30	7
340	34	5
260	27	4
360	32	3
280	25	2
330	26	3
190	12	0
750	65	16

a. Calculate the correlation coefficient and report the equation of the regression line using carbs as the predictor and calories as the response variable. Report the slope and interpret it in the context of this problem. Then use your regression equation to predict the number of calories in a menu item containing 55 grams of carbohydrates.

b. Calculate the correlation coefficient and report the equation of the regression line using sugar as the predictor and calories as the response variable. Report the slope and interpret it in the context of this problem. Then use your regression equation to predict the number of calories in a menu item containing 10 grams of sugars.

c. Based on your answers to parts (a) and (b), which is a better predictor of calories for these data: carbs or sugars? Explain your choice using appropriate statistics.

4.78 Granola Bars The following table shows the fat content (in grams) and calories for a sample of granola bars. (Source: calorielab.com)

Fat (in grams)	Calories
7.6	370
3.3	106.1
18.7	312.4

Fat (in grams)	Calories
3.8	113.1
5	117.8
5.5	131.9
7.2	140.6
6.1	118.8
4.6	124.4
3.9	105.1
6.1	136
4.8	124
4.4	119.3
7.7	192.6

a. Use technology to make a scatterplot of the data. Use fat as the independent variable (x) and calories as the dependent variable (y). Does there seem to be a linear trend to the data?

b. Compute the correlation coefficient and the regression equation, using fat as the independent variable and calories as the dependent variable.

c. What is the slope of the regression equation? Interpret the slope in the context of this problem.

d. What is the y-intercept of the regression equation? Interpret the y-intercept in the context of this problem or explain why it would be inappropriate to do so.

e. Find and interpret the coefficient of determination.

f. Use the regression equation to predict the calories in a granola bar containing 7 grams of fat.

g. Would it be appropriate to use the regression equation to predict the calories in a granola bar containing 25 grams of fat? If so, predict the number of calories in such a bar. If not, explain why it would be inappropriate to do so.

h. Looking at the scatterplot there is a granola bar in the sample that has an extremely high number of calories given the moderate amount of fat it contains. Remove its data from the sample and recalculate the correlation coefficient and regression equation. How did removing this unusual point change the value of r and the regression equation?

4.79 Shoe Size and Height The scatterplot shows the shoe size and height for some men (M) and women (F).

a. Why did we not extend the red line (for the women) all the way to 74 inches, instead stopping at 69 inches?

b. How do we interpret the fact that the blue line is above the red line?

c. How do we interpret the fact that the two lines are (nearly) parallel?

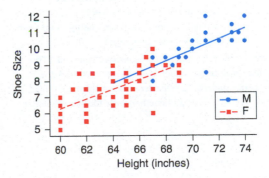

4.80 Age and Sleep The following scatterplot shows the age in years and the number of hours of sleep for some men (M) and women (F).

a. How do we interpret the fact that both lines have a negative slope?

b. How do we interpret the fact that the slopes are the same for both lines?

c. How do we interpret the fact that the lines are nearly the same?

d. Why is the line for the men shorter than the line for the women?

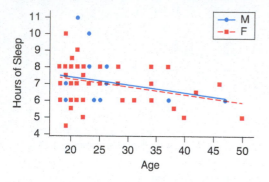

4.81 Age and Weight The following scatterplot shows the age and weight for some women. Some of them exercised regularly, and some did not. Explain what it means that the blue line (for those who did not exercise) is a bit steeper than the red line (for those who did exercise). (Source: StatCrunch: 2012 Women's final. Owner: molly7son@yahoo.com)

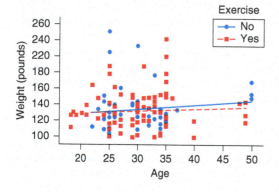

4.82 Heights and Test Scores

a. The following figure shows hypothetical data for a group of children. By looking at the figure, state whether the correlation between height and test score is positive, negative, or near zero.

b. The shape and color of the each marker show what grade these children were in at the time they took the test. Look at the six different groupings (for grades 1, 2, 3, 4, 5, and 6) and decide whether the correlation (the answer to part [a]) would stay the same if you controlled for grade (that is, if you looked only within specific grades).

c. Suppose a school principal looked at this scatterplot and said, "This means that taller students get better test scores, so we should give more assistance to shorter students." Do the data support this conclusion? Explain. If yes, say why. If no, give another cause for the association.

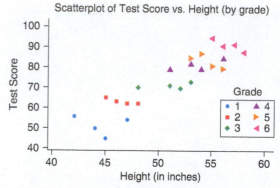

4.83 Law School Selectivity and Employment The acceptance rate for a sample of law schools and the percentage of students employed at graduation are on this text's website. A low acceptance rate means the law school is highly selective in admitting students. (Source: Internet Legal Research Group)

a. Do more selective law schools have better job-placement success? Make a graph that could addresses this question, report whether the trend is linear, and write a sentence answering the research question. If the trend is not linear, comment on what it shows and do not go on to part (b).

b. If the trend is linear, do the following:

I. Write the regression equation.

II. Interpret the slope of the regression equation.

III. Interpret the y-intercept of the regression equation or explain why it would be inappropriate to do so.

IV. Find and interpret the value of the coefficient of determination.

V. Use the regression equation to predict the percentage of students employed at graduation for a law school with a 50% acceptance rate.

4.84 Law School LSAT Score and Bar Exam The LSAT is a standardized test required for entrance to most law schools. The high LSAT score for admitted students and the percentage of students passing the bar exam immediately after law school graduation for a sample of law schools is found on this text's website. (Source: Internet Legal Research Group)

a. Do law schools that have high LSAT-scoring students also have a higher rate of students who pass the bar exam? Make a graph and report whether the trend is linear. Interpret the association if any. If the trend is not linear, comment on what it shows and do not go on to part (b).

b. If the trend is linear, do the following:

I. Write the regression equation.

II. Interpret the slope of the regression equation.

III. Interpret the y-intercept of the regression equation or explain why it would be inappropriate to do so.

IV. Find and interpret the value of the coefficient of determination.

V. Use the regression equation to predict the percent of students employed at graduation for a law school with a high LSAT score for admitted students of 150.

4.85 Snack Food Calories Data on the fat, carbohydrate, and calorie content for a sample of popular snack foods are found on this text's website. Use the data to determine which is a better predictor of the number of calories in these snack foods: fat or carbohydrates?

4.86 Salary and Education Does education pay? The salary per year in dollars, the number of years employed (YrsEm), and the number of years of education after high school (Educ) for the employees of a company were recorded. Determine whether number of years employed or number of years of education after high school is a better predictor of salary. Explain your thinking. Data are at this text's website. (Source: Minitab File)

4.87 Film Budgets and Box Office Move studios try to predict how much money their movies will make. One possible predictor is the amount of money spent on the production of the movie. The table shows the budget and amount of money made for a sample of movies made in 2017. The budget (amount spent making the movie) and gross (amount earned by ticket sales) are shown in the table. Make a scatterplot of the data and comment on what you see. If appropriate, do a complete linear regression analysis. If it is not appropriate to do so, explain why not. (Source: IMDB)

Film	Gross(in $ millions)	Budget (in $ millions)
Wonder Woman	412.6	149
Beauty and the Beast	504	160
Guardians of the Galaxy Vol. 2	389.8	200
Spider-Man: Homecoming	334.2	175
It	327.5	35
Despicable Me 3	264.6	80
Logan	226.3	97
The Fate of the Furious	225.8	250
Dunkirk	188	100
The LEGO Batman Movie	175.8	80
Thor Ragnarok	310.7	180
Get Out	175.5	5
Dead Men Tell No Tales	172.6	230
Cars 3	152	175

4.88 Fuel-Efficient Cars The following table gives the number of miles per gallon in the city and on the highway for some of the most fuel efficient cars according to Consumer Reports. Make a scatterplot of the data using city mileage as the predictor variable. Find the regression equation and use it to predict the highway mileage for a fuel-efficient car that gets 40 miles per gallon in city driving. Would it be appropriate to use the regression equation to predict the highway mileage for a fuel-efficient car that got 60 miles per gallon in city driving? If so, make the prediction. If not, explain why it would be inappropriate to do so.

Model	City Mileage	Highway Mileage
Toyota Prius 3	43	59
Hyundai Ioniq	42	60
Toyota Prius Prime	38	62
Kia Niro	33	52
Toyota Prius C	37	48
Chevrolet Malibu	33	49

Model	City Mileage	Highway Mileage
Ford Fusion	35	41
Hyundai Sonata	31	46
Toyota Camry	32	43
Ford C-Max	35	38

4.89 Tall Buildings The following scatterplot shows information about the world's tallest 169 buildings. *Stories* means *floors*.

a. What does the trend tell us about the relationship between stories and height (feet)?

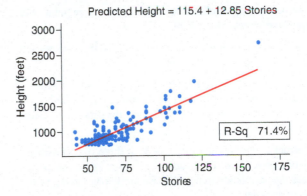

Predicted Height = 115.4 + 12.85 Stories

R-Sq 71.4%

b. The regression line for predicting the height (in feet) from the number of stories is shown above the graph. What height would you predict for a building with 100 stories?

c. Interpret the slope.

d. What, if anything, do we learn from the intercept?

e. Interpret the coefficient of determination.

(This data set is available at this text's website, and it contains several other variables. You might want to check to see whether the year the building was constructed is related to its height, for example.)

4.90 Poverty and High School Graduation Rates Poverty rates and high school graduation rates for the 50 states and the District of Columbia are graphed below. (Source: *2017 World Almanac and Book of Facts*)

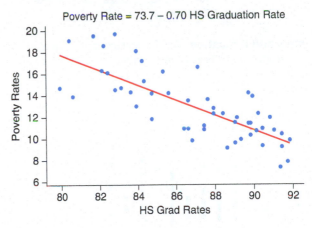

Poverty Rate = 73.7 − 0.70 HS Graduation Rate

a. What does the trend tell us about the relationship between poverty and high school graduation rates?

b. Interpret the slope of the regression equation.

c. The value of the coefficient of determination for this data set is 61.1%. Explain what this means in the context of the problem.

d. Data on bachelor's degree attainment and advanced degree attainment are available on this text's website. Which level of education (high school, bachelor's degree, or advanced degree) is most closely associated with the state poverty level?

For 4.91 through 4.94 show your points in a rough scatterplot and give the coordinates of the points.

* **4.91** Construct a small set of numbers with at least three points with a perfect positive correlation of 1.00.

* **4.92** Construct a small set of numbers with at least three points with a perfect negative correlation of −1.00.

* **4.93** Construct a set of numbers (with at least three points) with a strong negative correlation. Then add one point (an influential point) that changes the correlation to positive. Report the data and give the correlation of each set.

* **4.94** Construct a set of numbers (with at least three points) with a strong positive correlation. Then add one point (an influential point) that changes the correlation to negative. Report the data and give the correlation of each set.

4.95 The following figure shows a scatterplot of the educational level of twins. Describe the scatterplot. Explain the trend and mention any unusual points. (Source: www.stat.ucla.edu)

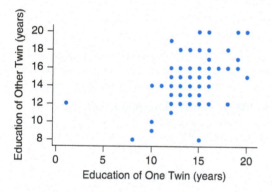

4.96 Wages and Education The figure shows a scatterplot of the wages and educational level of some people. Describe what you see. Explain the trend and mention any unusual points. (Source: www.stat.ucla.edu)

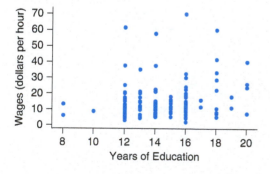

4.97 Do Students Taking More Units Study More Hours? The following figure shows the number of units that students were enrolled in and the number of hours (per week) that they reported studying. Do you think there is a positive trend, a negative trend, or no noticeable trend? Explain what this means about the students.

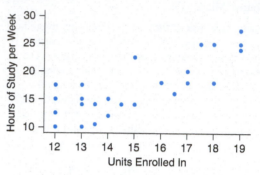

4.98 Hours of Exercise and Hours of Homework The following scatterplot shows the number of hours of exercise per week and the number of hours of homework per week for some students. Explain what it shows.

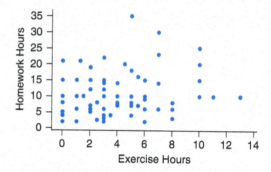

4.99 Children's Ages and Heights The following figure shows information about the ages and heights of several children. Why would it not make sense to find the correlation or to perform linear regression with this data set? Explain.

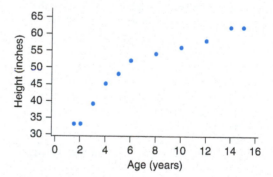

4.100 Blackjack Tips The following figure shows the amount of money won by people playing blackjack and the amount of tips they gave to the dealer (who was a statistics student), in dollars.

Would it make sense to find a correlation for this data set? Explain.

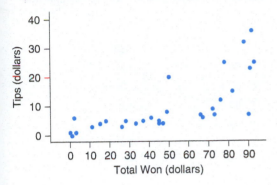

* **4.101 Decrease in Cholesterol** A doctor is studying cholesterol readings in his patients. After reviewing the cholesterol readings, he calls the patients with the highest cholesterol readings (the top 5% of readings in his office) and asks them to come back to discuss cholesterol-lowering methods. When he tests these patients a second time, the average cholesterol readings tend to have gone down somewhat. Explain what statistical phenomenon might have been partly responsible for this lowering of the readings.

* **4.102 Test Scores** Suppose that students who scored much lower than the mean on their first statistics test were given special tutoring in the subject. Suppose that they tended to show some improvement on the next test. Explain what might cause the rise in grades other than the tutoring program itself.

gUIDED EXERCISES

g 4.47 How Is the Time of a Flight Related to the Distance of the Flight? The table gives the distance from Boston to each city (in thousands of miles) and gives the time for one randomly chosen commercial airplane to make that flight. Do a complete regression analysis that includes a scatterplot with the line, interprets the slope and intercept, and predicts how much time a nonstop flight from Boston to Seattle would take. The distance from Boston to Seattle is 3000 miles.

City	Distance (1000s of miles)	Time (hours)
St. Louis	1.141	2.83
Los Angeles	2.979	6.00
Paris	3.346	7.25
Denver	1.748	4.25
Salt Lake City	2.343	5.00
Houston	1.804	4.25
New York	0.218	1.25

Step 1 ▶ Make a scatterplot.
Be sure that *distance* is the *x*-variable and *time* is the *y*-variable, because we are trying to predict time from distance. Graph the line using technology. See the TechTips starting on page 210.

Step 2 ▶ Is the linear model appropriate?
Does it seem that the trend is linear, or is there a noticeable curve?

Step 3 ▶ Find the equation.
Find the equation for predicting time (in hours) from miles (in thousands).

Step 4 ▶ Slope
Interpret the slope in context.

Step 5 ▶ Intercept
Interpret the intercept in context. Although there are no flights with a distance of zero, try to explain what might cause the added time that the intercept represents.

Step 6 ▶ Time to Seattle
Answer the question. About how long should it take to fly nonstop from Boston to Seattle?

4.73 Test Scores Assume that in a political science class, the teacher gives a midterm exam and a final exam. Assume that the association between midterm and final scores is linear. The summary statistics have been simplified for clarity.

Midterm:	Mean = 75,	Standard deviation = 10
Final:	Mean = 75,	Standard deviation = 10

Also, $r = 0.7$ and $n = 20$.

For a student who gets 95 on the midterm, what is the predicted final exam grade? Assume the graph is linear.

Step 1 ▶ Find the equation of the line to predict the final exam score from the midterm score.
Standard form: $y = a + bx$

a. First find the slope: $b = r\left(\dfrac{s_{final}}{s_{midterm}}\right)$

b. Then find the *y*-intercept, *a*, from the equation

$$a = \bar{y} - b\bar{x}$$

c. Write out the following equation:

$$\text{Predicted } y = a + bx$$

However, use "Predicted Final" instead of "Predicted *y*" and "Midterm" in place of *x*.

Step 2 ▶ Use the equation to predict the final exam score for a student who gets 95 on the midterm.

Step 3 ▶ Your predicted final exam grade should be less than 95. Why?

TechTips

General Instructions for All Technology

Download data from the text's website, or enter data manually using two columns of equal length. Refer to TechTips in Chapter 2 for a review of entering data. Each row represents a single observation, and each column represents a variable. All technologies will use the example that follows.

EXAMPLE ▶ Analyze the six points in the data table with a scatterplot, correlation, and regression. Use heights (in inches) as the *x*-variable and weight (in pounds) as the *y*-variable.

Height	Weight
61	104
62	110
63	141
64	125
66	170
68	160

TI-84

These steps assume you have entered the heights into **L1** and the weights into **L2**.

Making a Scatterplot

1. Press **2ND**, **STATPLOT** (which is the button above **2ND**), **4**, and **ENTER**; to turn off plots made previously.

2. Press **2ND**, **STATPLOT**, and **1** (for **Plot1**).

3. Refer to Figure 4a: Turn on **Plot1** by pressing **ENTER** when **On** is highlighted.

▲ **FIGURE 4a** TI-84 Plot1 Dialogue Screen

4. Use the arrows on the keypad to get to the scatterplot (first of the six plots) and press **ENTER** when the scatterplot is high-lighted. Be careful with the **Xlist** and **Ylist**. To get **L1**, press **2ND** and **1**. To get **L2**, press **2ND** and **2**.

5. Press **GRAPH**, **ZOOM** and **9** (**ZoomStat**) to create the graph.

6. Press **TRACE** to see the coordinates of the points, and use the arrows on the keypad to go to other points. Your output will look like Figure 4b, but without the line.

7. To get the output with the line in it, shown in Figure 4b: **STAT**, **CALC**, **8:LinReg(a + bx)**, **L1**, **ENTER**, **L2**, **ENTER**, **ENTER**, **Y1** (You get the **Y1** by pressing **VARS**, **Y-VARS**, **1: Function**, **1:Y1**), **ENTER**, **ENTER**.

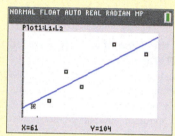

▲ **FIGURE 4b** TI-84 Plot with Line

8. Press **GRAPH**, **ZOOM** and **9**.

9. Press **TRACE** to see the numbers, and use the arrows on the keypad to get to other numbers.

Finding the Correlation and Regression Equation Coefficients

Before finding the correlation, you must turn the diagnostics on, as shown here.

Press **2ND**, **CATALOG**, and scroll down to **DiagnosticOn** and press **ENTER** twice. The diagnostics will stay on unless you **Reset** your calculator or change the batteries.

1. Press **STAT**, choose **CALC**, and **8** (for **LinReg (a + bx)**).

2. Press **2ND L1** (or whichever list is X, the predictor), press **ENTER**, press **2ND L2** (or whichever list is Y, the response), and press **ENTER, ENTER, ENTER**.

Figure 4c shows the output.

▲ **FIGURE 4c** TI-84 Output

Making a Scatterplot

1. **Graph > Scatterplot**
2. Leave the default **Simple** and click **OK.**
3. Double-click the column containing the weights so that it goes under the **Y Variables**. Then double-click the column containing the heights so that it goes under the **X Variables**.
4. Click **OK.** After the graph is made, you can edit the labels by clicking on them.

Finding the Correlation

1. **Stat > Basic Statistics > Correlation**
2. Double-click both the predictor column and the response column (in either order).
3. Click **OK.** You will get 0.881.

Finding the Regression Equation Coefficients

1. **Stat > Regression > Regression > Fit Regression Model**
2. Put in the **Responses:** (y) and **Continuous Predictors:** (x) columns.
3. Click **OK.** You may need to scroll up to see the regression equation. It will be easier to understand if you have put in labels for the columns, such as "Height" and "Weight." You will get: Weight = −443 + 9.03 Height.

To Display the Regression Line on a Scatterplot

1. **Stat > Regression > Fitted Line Plot**
2. Double-click the **Response (Y)** column and then double-click the **Predictor (X)** column.
3. Click **OK.** Figure 4d shows the fitted line plot.

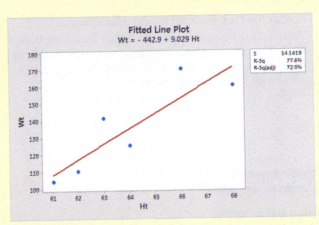

▲ **FIGURE 4d** Minitab Fitted Line Plot

Making a Scatterplot

1. Select (highlight) the two columns containing the data, with the predictor column to the *left* of the response column. You may include the labels at the top or not include them.
2. Click **Insert**, in **Charts** click the picture of a scatterplot, and click the upper-left option shown here:

Scatter

3. Note that the lower-left corner of the chart is not at the origin, (0, 0). If you want to zoom in or out on the data by changing the minimum value for an axis, right-click on the axis numbers, click **Format axis**, in **Axis Options** change the **Minimum** to the desired value. You may want to do this twice: once for the x-axis and once for the y-axis. Then close the **Format Axis** menu.
4. When the chart is active (click on it), **Chart Tools** are shown at the top of screen, right of center. Click **Design**, then **Add Chart Elements**, then **Axis Titles**, and **Chart Title** to add appropriate labels. After the labels are added, you can click on them to change the spelling or add words. Delete the labels on the right-hand side, such as **Series 1**, if you see any.

Finding the Correlation

1. Click on **Data**, click on **Data Analysis**, select **Correlation**, and click **OK.**
2. For the **Input Range**, select (highlight) both columns of data (if you have highlighted the labels as well as the numbers, you must also click on the **Labels in first row**).
3. Click **OK.** You will get 0.880934.

   (Alternatively, just click the f_x button, for **category** choose **statistical**, select **CORREL**, click **OK**, and highlight the two columns containing the numbers, one at a time. The correlation will show up on the dialogue screen, and you do *not* have to click **OK.**)

Finding the Coefficients of the Regression Equation

1. Click on **Data, Data Analysis, Regression**, and **OK.**
2. For the **Input Y Range**, select the column of numbers (not words) that represents the response or dependent variable. For the **Input X Range**, select the column of numbers that represents the predictor or independent variable.
3. Click **OK.**

   A large summary of the model will be displayed. Look under **Coefficients** at the bottom. For the **Intercept** and the slope (next to **XVariable1**), see Figure 4e, which means the regression line is

$$y = -442.9 + 9.03x$$

	Coefficients
Intercept	-442.8824
X Variable 1	9.0294

▲ **FIGURE 4e** Excel Regression Output

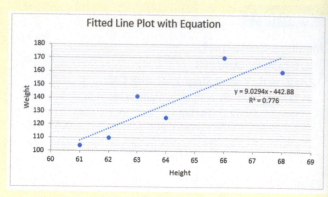

▲ **FIGURE 4f** Excel Fitted Line Plot with Equation

To Display the Regression Line on a Scatterplot

4. After making the scatterplot, under **Chart Tools** click **Design**. In the **Chart Layouts** group, click the triangle to the right of **Quick Layout**. Choose Layout 9 (the option in the lower-right portion, which shows a line in it and also *fx*).

Refer to Figure 4f.

STATCRUNCH

Making a Scatterplot

1. **Graph > Scatter plot**
2. Select an **X variable** and a **Y variable** for the plot.
3. Click **Compute!** to construct the plot.
4. To copy the graph, click **Options** and **Copy**.

Finding the Correlation and Coefficients for the Equation

1. **Stat > Regression > Simple Linear**
2. Select the **X variable** and **Y variable** for the regression.
3. Click **Compute!** to view the equation and numbers, which are shown in Figure 4g.

Plotting the Regression Line on a Scatterplot

1. **Stat > Regression > Simple Linear**
2. Select your columns for X and Y.
3. Click **Compute!**
4. Click the **>** in the lower-right corner (see Figure 4h).
5. To copy the graph, click **Options** and **Copy**.
6. To edit the graph, click on the hamburger in the lower-left corner.

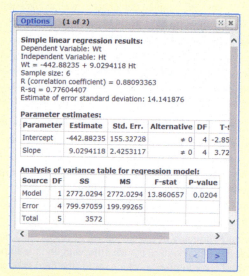

▲ **FIGURE 4g** StatCrunch Regression Output

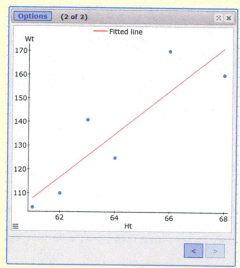

▲ **FIGURE 4h** StatCrunch Fitted Line Plot

Modeling Variation with Probability

Probabilities are long-run relative frequencies used to describe processes where the outcome depends on chance, such as flipping a coin or rolling a die. Theoretical probabilities are based on specific assumptions (usually based on a theory) about the chance process. Empirical probabilities are based on observation of the actual process. The two types of probabilities are closely related, and we need both of them to analyze samples of data in the real world.

n 1969, the United States was fighting the Vietnam War and drafting men to serve in the military. To determine who was chosen, government officials wrote the days of the year (January 1, January 2, and so on) on capsules. The capsules were placed in a large container and mixed up. They were then drawn out one at a time. The first date chosen was assigned the rank 1, the second date was assigned the rank 2, and so on. Men were drafted on the basis of their birthday. Those whose birthday had rank 1 were drafted first, then those whose birthday had rank 2, and so on until the officials had enough men.

Although the officials thought that this method was random, some fairly convincing evidence indicates that it was not (Starr 1997). Figure 5.1a shows boxplots with the actual ranks for each month. Figure 5.1b shows what the boxplots might have looked like if the lottery had been truly random. In Figure 5.1b, each month has roughly the same rank. However, in Figure 5.1a, a few months had notably lower ranks than the other months. Bad news if

you were born in December—you were more likely to be called up first.

What went wrong? The capsules, after having dates written on them, were clustered together by month as they were put into the tumbler. But the capsules weren't mixed up enough to break up these clusters. The mixing wasn't adequate to create a truly random mix.

It's not easy to generate true randomness, and humans have a hard time recognizing random events when they see them. Probability gives us a tool for understanding randomness. Probability helps us answer the question, "How often does something like this happen by chance?" By answering that question, we create an important link between our data and the real world. In previous chapters, you learned how to organize, display, and summarize data to see patterns and trends. Probability is a vital tool because it gives us the ability to generalize our understanding of data to the larger world. In this chapter, we'll explore issues of randomness and probability: What is randomness? How do we measure it? And how do we use it?

(a)

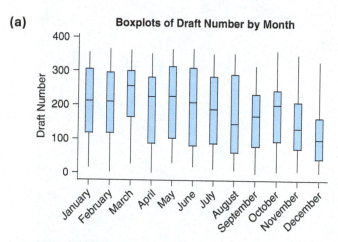

(b)

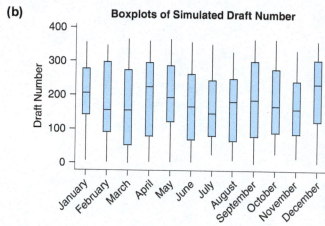

▲ FIGURE 5.1 Boxplots of **(a)** actual Vietnam draft numbers by month and **(b)** what might have happened if the draft had really been random.

CASE STUDY

SIDS or Murder?

In November 2000, Sally Clark was convicted in England of killing her two children. Her children had died several years apart, and the initial diagnosis was sudden infant death syndrome (SIDS). SIDS is the abrupt, unexplained death of an infant under 1 year of age. It is estimated that in the United States, about 5500 infants die of SIDS every year. Although some risk factors have been identified (most are related to the mother's health during pregnancy), the cause or causes are as yet unknown.

Clark was convicted of murder on the basis of the expert testimony of Sir Roy Meadow, a prominent British physician who was an expert on SIDS. Dr. Meadow quoted a published statistic: the probability of an infant dying of SIDS is 1/8543.

That is, in the United Kingdom, one child in every 8543 dies of SIDS. Dr. Meadow concluded that the probability of two children in the same family dying of SIDS was therefore $(1/8543) \times (1/8543)$, or about 1 in 73 million. The jury agreed that this event—two children in the same family dying of SIDS—was so unlikely that they convicted Ms. Clark of murder, and she was sent to prison.

But then in 2003, Sally Clark's conviction was overturned, and she was released from prison. Dr. Meadow was accused of professional misconduct. Why did Dr. Meadow multiply the two probabilities together? Why was the verdict overturned? We will answer these questions at the end of the chapter.

SECTION 5.1

What Is Randomness?

What exactly is randomness? You might see a definition similar to "Haphazard; for no apparent reason or purpose." You probably use the word to describe events that seem to happen with no purpose at all. Sometimes the word is used to describe things that have no apparent pattern. However, in statistics the word *random* has a more precise meaning, as you will see.

We can do a small experiment to compare our natural understanding of randomness with real-life randomness. We asked a student to imagine flipping a coin and to write down the resulting heads (H) and tails (T). To compare this to a real random sequence, we flipped an actual coin 20 times and recorded the results. Which of the sequences below do you think is real, and which is invented by the student?

H	H	T	H	T	H	H	T	T	T	H	T	H	H	H	T	H	T	T	H
T	T	T	H	H	H	T	T	T	T	T	H	H	H	H	T	T	H	T	H

The first row is the one made up by the student, and the second row records the results of actually tossing a coin 20 times. Is it possible to tell by comparing the two sequences? Not always, but this time the student did something that most people do in this situation. Most of the time, he wrote very short "streaks." (A streak is any sequence of consecutive heads or tails. TT is a streak of two tails.) Only once did he write as many as three heads or three tails in a row. It's as if, after writing three heads in a row, he thought, "Wait! If I put a fourth, no one will believe it is random."

However, the real, truly random sequence had one streak of five tails (beginning at the seventh flip) and another streak with four heads. These long streaks are examples of the way chance creates things that look like a pattern or structure but really are not.

> **KEY POINT**
> People are not good at creating truly random samples or random experiments, so we need to rely on outside mechanisms such as coin flips or random number tables.

Real randomness is hard to achieve without help from a computer or some other randomizing device. If a computer is not available to generate random numbers, another useful approach is to use a random number table. (An example is provided in Appendix A of this text.) A random number table provides a sequence of digits from 0 to 9 in a random order. Here, **random** means that no predictable pattern occurs and that no digit is more likely to appear than any other. (Of course, if you use the same table often enough, it might seem predictable to you, but to an outsider, it will not seem predictable.)

For example, if we are doing a controlled experiment, we might assign each subject in our study a random number from this table as the subjects come into our office. The odd-numbered subjects might then go into the control group, and the even-numbered subjects might go into the treatment group.

To use a random number table to simulate coin flipping, we assign a number to each outcome. For example, suppose we let even numbers (0, 2, 4, 6, 8) represent tails and let odd numbers (1, 3, 5, 7, 9) represent heads. Now choose any row of the table you wish and any column. For example, we chose column 11 and row 30 because that's the date of the day we wrote this paragraph (November 30, or 11/30). Read off the next 20 digits, but translate them into "heads" and "tails." What's the longest streak you get?

► **TABLE 5.1** Lines 28 through 32 (indicated at the left side of the table) from the random number table in Appendix A. The red 7 in the 11th column and the 30th line, or row, is our starting point.

Line						
28	31498	85304	22393	21634	34560	77404
29	93074	27086	62559	86590	18420	33290
30	90549	53094	76282	53105	45531	90061
31	11373	96871	38157	98368	39536	08079
32	52022	59093	30647	33241	16027	70336

EXAMPLE 1 Simulating Randomness

Let's play a game. You roll a six-sided die until the first 6 appears. You win one point for every roll. You pass the die to your opponent, who also rolls until the first 6 appears. Whoever has the most points, wins.

QUESTION Simulate rolling the die until a 6 appears. Use Table 5.1, and start at the very first entry of line 28 in Table 5.1. How many rolls did it take?

SOLUTION We'll let the digits 1 through 6 represent the outcome shown on the die. We'll ignore the digits 7, 8, 9, and 0. Starting at line 28, we get these random digits: 3, 1, 4, (ignore, ignore, ignore), 5, 3, (ignore), 4, 2, 2, 3, (ignore), 3, 2, 1, 6

CONCLUSION We rolled the die 12 times before the first 6 appeared.

TRY THIS! Exercise 5.1

Computers and calculators have random number generators that come close to true randomness. Computer-generated random numbers are sometimes called pseudo-random numbers because they are generated on the basis of a seed value—a number that starts the random sequence. If you input the same seed number, you will always see the same sequence of pseudo-random numbers. However, it is not possible to predict what number will come next in the sequence, and in this sense the generated numbers are considered random. For most practical work (and certainly for everything we cover in this text), these pseudo-random numbers are as random as we need. (You should be aware, though, that not all statistics packages produce equally convincing sequences of random numbers.)

You Try It Now is a good time to take a moment to examine the statistical software you're using in this course to determine how (and if) you can generate random numbers.

For example, in StatCrunch, you can generate random digits by selecting *Random numbers* from the *Applets* menu. In the box that says "Minimum value:" enter 0, and in the box that says "Maximum value:" enter 9. Enter any number you wish for the sample size, and check the box that says "Allow repeats (with replacement)". Finally, click on "Compute!" Now that we have generated these digits, we can use them the same way we used the random number table in Example 1. Figure 5.2 shows an example generating 25 digits.

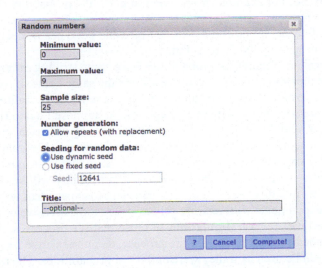

◀ **FIGURE 5.2** When we hit *Compute!*, StatCrunch will generate 25 random integers between 0 and 9, inclusive. For example: 4, 8, 9, 1, 2, 6, 9 . . . .

Less technological ways exist for generating random outcomes, such as actually flipping a coin. When you play a card game, you carefully shuffle the cards to mix them up. In the games of Mah Jong and tile rummy, players create randomness by scrambling the tiles on the tabletop. In many board games, players either roll dice or spin a spinner to determine how far they will move their game pieces. In raffles, tickets are put into a basket, the basket is spun, and then a ticket is drawn by someone who does so without looking into the basket.

Such physical randomizations must be done with care. A good magician can flip a coin so that it always comes up heads. A child learns quickly that a gentle tap of a spinner can move the spinner to the desired number. A deck of cards needs to be shuffled at least seven times in order for the result to be considered random (as mathematician and magician Persi Diaconnis proved). Many things that we think are random might not be. This is the lesson the U.S. government learned from its flawed Vietnam draft lottery, which was described in the chapter introduction. Quite often, statisticians are employed to check whether random processes, such as the way winners are selected for state lotteries, are truly random.

Theoretical, Empirical, and Simulation-Based Probabilities

Probability is used to measure how often random events occur. When we say, "The probability of getting heads when you flip a coin is 50%," we mean that flipping a coin results in heads about half the time. This is sometimes called the "frequentist" approach to probability because in it, probabilities are defined as relative frequencies.

We will use three different approaches to studying probability. The first is called **theoretical probability**. The theoretical probability of an outcome is the relative frequency if we repeated the action *infinitely* many times. Of course, it is impossible to do anything infinitely many times. That's why this approach is called *theoretical*. But we can often apply mathematical reasoning to determine a theoretical probability for a random event. When we say that the theoretical probability of getting heads when flipping

Looking Back

Relative Frequencies
Relative frequencies, introduced in Chapter 2, are the same as proportions.

a coin is 50%, we mean that if we flip that coin infinitely many times, exactly half of the flips will be heads. To determine this "50%" value, we made an assumption that each side of the coin is equally likely to land faceup after a flip.

For many interesting problems it is difficult, and sometimes impossible, to find the theoretical probability solely through mathematical reasoning. The other two approaches to probability—empirical probability and simulations—are helpful for these cases. **Empirical probabilities** are based on outcomes of an experiment or by observing the real-life process. For example, if you want to know the probability of getting heads when you flip a coin, just flip the coin a large number of times and record the relative frequency of heads. If you flip a coin 10 times and get 6 heads, declare the empirical probability of heads to be 6/10 or 0.6.

Empirical and theoretical probabilities have some striking differences. Chief among them is that theoretical probabilities are always the same value; if we all agree on the theory, then we all agree on the values for the theoretical probabilities. Empirical probabilities, however, change with every experiment. Suppose I again flip the exact same coin 10 times, and this time I get 3 heads. This time my empirical probability of heads is 0.3. Last time it was 0.6. Empirical probabilities are themselves random and vary from experiment to experiment. In Section 5.4 you'll learn that if the number of repetitions of the experiment is very large, then the empirical probability will be close to the theoretical probability and this variation from experiment to experiment will be small.

When experiments with real-life processes are too expensive or time-consuming, we can sometimes rely on **simulations** of those real-life processes. If the simulation is a close enough match to the real-life process, then the relative frequencies from the simulation can be used as if they were empirical probabilities. The random number table introduced earlier is one potential tool for carrying out simulations. Most commonly, computers are used since they can quickly generate random numbers.

Probabilities found through simulations are similar to empirical probabilities. They too vary from experiment to experiment. And, like empirical probabilities, if the number of trials is large, the probability found through simulation will be close to the theoretical probability. We will often use the same term, "empirical," to refer to both simulation-based and empirical probabilities.

> **KEY POINT**
>
> Theoretical probabilities are based on theory and tell us how often an event would occur if an experiment were repeated infinitely many times. Empirical probabilities tell us how often an event occurred in an actual set of experiments or observations. Simulations can sometimes substitute for actual experiments if the simulations are designed to closely match the real-life processes for which we wish to calculate probabilities.

SECTION 5.2

Finding Theoretical Probabilities

We can use empirical and simulation-based probabilities as an estimate of theoretical probabilities, and if we do our experiment enough times, the estimate can be pretty good. But how good is pretty good? To understand how to use empirical probabilities to estimate and to verify theoretical probabilities, we will first learn how to calculate theoretical probabilities.

Facts about Theoretical Probabilities

Probabilities are always numbers between 0 and 1 (including 0 and 1). Probabilities can be expressed as fractions, decimals, or percentages: 1/2, 0.50, and 50% are all used to represent the probability that a coin comes up heads.

Some values have special meanings. If an event never happens, then its probability is 0. If you purchase a lottery ticket after all the prizes have been given out, the probability is 0 that you will win one of those prizes. If the probability of an event happening is 1, then that event always happens. If you flip a coin, the probability of a coin landing heads or tails is 1.

Another useful property to remember is that the probability that an event will *not* happen is 1 minus the probability that it will happen. If there is a 0.90 chance that it will rain, then there is a $1 - 0.9 = 0.10$ chance that it will not. If there is a 1/6 chance of rolling a "1" on a die, then there is a $1 - (1/6) = 5/6$ probability that you will not get a 1.

We call such a "not event" a **complement**. The complement of the event "it rains today" is the event "it does not rain today." The complement of the event "coin lands heads" is "coin lands tails." The complement of the event "a die lands with a 1 on top" is "a die lands with a 2, 3, 4, 5, or 6 on top"; in other words, the die lands with a number that is *not* 1 on top.

Events are usually represented by uppercase letters: A, B, C, and so on. For example, we might let A represent the event "it rains tomorrow." Then the notation P(A) means "the probability that it will rain tomorrow." In sentence form, the notation P(A) = 0.50 translates into English as "The probability that it will rain tomorrow is 0.50, or 50%."

A common misinterpretation of probability is to think that large probabilities mean that the event will certainly happen. For example, suppose your local weather reporter predicts a 90% chance of rain tomorrow. Tomorrow, however, it doesn't rain. Was the weather reporter wrong? Not necessarily. When the weather reporter says there is a 90% chance of rain, it means that on 10% of the days like tomorrow it does not rain. Thus, a 90% chance of rain means that on 90% of all days just like tomorrow, it rains, but on 10% of those days it does not.

Summary of Probability Rules

Rule 1: A probability is always a number from 0 to 1 (or 0% to 100%) inclusive (which means 0 and 1 are allowed). It may be expressed as a *fraction*, a *decimal*, or a *percentage*.

In symbols: For any event A,

$$0 \leq P(A) \leq 1$$

Rule 2: The probability that an event will not occur is 1 minus the probability that the event will occur. In symbols, for any event A,

$$P(A \text{ does } not \text{ occur}) = 1 - P(A \text{ does occur})$$

The symbol A^c is used to represent the complement of A. With this notation, we can write Rule 2 as

$$P(A^c) = 1 - P(A)$$

Finding Theoretical Probabilities with Equally Likely Outcomes

In some situations, all the possible outcomes of a random experiment occur with the same frequency. We call these situations "equally likely outcomes." For example, when you flip a coin, heads and tails are equally likely. When you roll a die, 1, 2, 3, 4, 5, and 6 are all equally likely, assuming, of course, that the die is balanced correctly.

When we are dealing with equally likely outcomes, it is sometimes helpful to list all of the possible outcomes. A list that contains *all* possible (and equally likely) outcomes is called the **sample space**. We often represent the sample space with the letter *S*.

Details

Precision Dice
The dice that casinos use are very different from the dice you use to play board games at home. Casino dice are precisely made. One casino claims, for instance, that each hole is exactly 17/1000 inch deep and filled with a material that is exactly the same density as the cube itself.

An **event** is any collection of outcomes in the sample space. For example, the sample space S for rolling a die is the numbers 1, 2, 3, 4, 5, 6. The event "get an even number on the die" consists of the even outcomes in the sample space: 2, 4, and 6.

When the outcomes are equally likely, the probability that a particular event occurs is just the number of outcomes that make up that event divided by the total number of equally likely outcomes in the sample space. In other words, it is the number of outcomes resulting in the event divided by the number of outcomes in the sample space.

Summary of Probability Rules

Rule 3:

$$\text{Probability of A} = P(A) = \frac{\text{Number of outcomes in A}}{\text{Number of all possible outcomes}}$$

This is true *only* for equally likely outcomes.

For example, suppose 30 people are in your class, and one person will be selected at random by a raffle to win a prize. What is the probability that you will win? The sample space is the list of the names of the 30 people. The event A is the event that contains only one outcome: your name. The probability that you win is 1/30, because there is only one way for you to win and there are 30 different ways that this raffle can turn out. We write this using mathematical notation as follows:

$$P(\text{you win prize}) = 1/30$$

We can be even more compact:

Let A represent the event that you win the raffle. Then

$$P(A) = 1/30$$

One consequence of Rule 3 is that the probability that *something* in the sample space will occur is 1. In symbols, $P(S) = 1$. This is because

$$P(S) = \frac{\text{Number of outcomes in S}}{\text{Number of outcomes in S}} = 1$$

EXAMPLE 2 Ten Dice in a Bowl

Reach into a bowl that contains 5 red dice, 3 green dice, and 2 white dice (Figure 5.3). But assume that, unlike what you see in Figure 5.3, the dice have been well mixed.

QUESTION What is the probability of picking (a) A red die? (b) A green die? (c) A white die?

SOLUTIONS The bowl contains 10 dice, so we have 10 possible outcomes. All are equally likely (assuming all the dice are equal in size, they are mixed up within the bowl, and we do not peek when choosing).

a. Five dice are red, so the probability of picking a red die is 5/10, 1/2, 0.50, or 50%. That is,

$$P(\text{red die}) = 1/2, \text{ or } 50\%$$

b. Three dice are green, so the probability of picking a green die is 3/10, or 30%. That is,

$$P(\text{green die}) = 3/10, \text{ or } 30\%$$

c. Two dice are white, so the probability of picking a white die is 2/10, 1/5, or 20%. That is,

$$P(\text{white die}) = 1/5, \text{ or } 20\%$$

◄ **FIGURE 5.3** Ten dice in a bowl.

Note that the probabilities add up to 1, or 100%, as they must.

TRY THIS!   Exercise 5.11

Example 3 shows that it is important to make sure the outcomes in your sample space are equally likely.

EXAMPLE 3 Adding Two Dice

Roll two dice and add the outcomes. Assume each side of each die is equally likely to appear face up when rolled. Event A is the event that the sum of the two dice is 7.

QUESTION   What is the probability of event A? In other words, find P(A).

SOLUTION   This problem is harder because it takes some work to list all of the equally likely outcomes, which are shown in Table 5.2.

Table 5.2 lists 36 possible equally likely outcomes. Here are the outcomes in event A:

$$(1, 6), (2, 5), (3, 4), (4, 3), (5, 2), (6, 1)$$

There are six outcomes for which the dice add to 7.

| **Die 1** | 1 | 1 | 1 | 1 | 1 | 1 | 2 | 2 | 2 | 2 | 2 | 2 |
| **Die 2** | 1 | 2 | 3 | 4 | 5 | 6 | 1 | 2 | 3 | 4 | 5 | 6 |

| **Die 1** | 3 | 3 | 3 | 3 | 3 | 3 | 4 | 4 | 4 | 4 | 4 | 4 |
| **Die 2** | 1 | 2 | 3 | 4 | 5 | 6 | 1 | 2 | 3 | 4 | 5 | 6 |

| **Die 1** | 5 | 5 | 5 | 5 | 5 | 5 | 6 | 6 | 6 | 6 | 6 | 6 |
| **Die 2** | 1 | 2 | 3 | 4 | 5 | 6 | 1 | 2 | 3 | 4 | 5 | 6 |

▲ **TABLE 5.2** All possible equally-likely outcomes for two six-sided dice.

CONCLUSION   The probability of rolling a sum of 7 is 6/36, or 1/6.

TRY THIS!   Exercise 5.15

It is important to make sure that the outcomes in the sample space are equally likely. A common mistake when solving Example 3 is listing all the possible *sums* instead of listing all the equally likely outcomes of the two dice. If we made that mistake here, our list of sums would look like this:

$$2, 3, 4, 5, 6, 7, 8, 9, 10, 11, 12$$

This list has 11 sums, and only 1 of them is a 7, so we would incorrectly conclude that the probability of getting a sum of 7 is 1/11.

Why didn't we get the correct answer of 1/6? The reason is that the outcomes we listed—2, 3, 4, 5, 6, 7, 8, 9, 10, 11, 12—are not equally likely. For instance, we can get a sum of 2 in only one way: roll two "aces," for $1 + 1$. Similarly, we have only one way to get a 12: roll two 6s, for $6 + 6$.

However, there are six ways of getting a 7: (1, 6), (2, 5), (3, 4), (4, 3), (5, 2), and (6, 1). In other words, a sum of 7 happens more often than a sum of 2 or a sum of 12. The outcomes 2, 3, 4, 5, 6, 7, 8, 9, 10, 11, 12 are not equally likely.

Usually it is not practical to list all the outcomes in a sample space—or even just those in the event you're interested in. For example, if you are dealing out 5 playing cards from a 52-card deck, the sample space has 2,598,960 possibilities. You really do not want to list all of those outcomes. Mathematicians have developed rules for counting the number of outcomes in complex situations such as these. These rules do not play an important role in introductory statistics, and we do not include them in this text.

Combining Events with "AND" and "OR"

As you saw in Chapter 4, we are often interested in studying more than one variable at a time. Real people, after all, often have several attributes we want to study, and we frequently want to know the relationship among these variables. The words AND and OR can be used to combine events into new, more complex events. The real people in Figure 5.4a, for example, have two attributes we decided to examine. They are either wearing a hat or not. Also, they are either wearing glasses or not. In the photo, the people who are wearing hats AND glasses are raising their hands.

Another way to visualize this situation is with a **Venn diagram**, as shown in Figure 5.4b. The rectangle represents the sample space, which consists of all possible outcomes if we were to select a person at random. The ovals represent events—for example, the event that someone is wearing glasses. The people who "belong" to *both* events are in the intersection of the two ovals.

The word **AND** creates a new event out of two other events. The probability that a randomly selected person in this photo is wearing a hat is 3/6, because three of the six people are wearing a hat. The probability that a randomly selected person wears glasses

> **! Caution**
>
> **"Equally Likely Outcomes" Assumption**
> Just wishing it were true doesn't make it true. The assumption of equally likely outcomes is not always true. And if this assumption is wrong, your theoretical probabilities will not be correct.

> **! Caution**
>
> **Venn Diagrams**
> The areas of the regions in Venn diagrams have no numerical meaning. A large area does not contain more outcomes than a small area.

▶ **FIGURE 5.4** **(a)** Raise your hand if you are wearing glasses AND a hat. **(b)** The people wearing both glasses AND a hat (Maria and David) appear in the intersection of the two circles in this Venn diagram.

is 4/6. The probability that a randomly selected person is wearing a hat AND glasses is 2/6, because only two people are in both groups. We could write this, mathematically, as

$$P(\text{wears glasses AND wears a hat}) = 2/6$$

KEY POINT The word AND creates a new event out of two events A and B. The new event consists of *only* those outcomes that are in *both* event A and event B.

In most situations, you will not have a photo to rely on. A more typical situation is given in Table 5.3, which records frequencies of two attributes for the people in a random sample of a recent U.S. census (www.census.gov). The two attributes are highest educational level and current marital status. ("Single" means never married. All other categories refer to the respondent's current status. Thus, a person who was divorced but then remarried is categorized as "Married.")

Education Level	Single	Married	Divorced	Widow/Widower	Total
Less HS	17	70	10	28	125
High school	68	240	59	30	397
College or higher	27	98	15	3	143
Total	112	408	84	61	665

▲ **TABLE 5.3** Education and marital status for 665 randomly selected U.S. residents. "Less HS" means did not graduate from high school.

EXAMPLE 4 Education AND Marital Status

Suppose we will select a person at random from the collection of 665 people categorized in Table 5.3.

QUESTIONS

a. What is the probability that the person is married?

b. What is the probability that the person has a college education or higher?

c. What is the probability that the person is married AND has a college education or higher?

SOLUTIONS The sample space has a total of 665 equally likely outcomes.

a. In 408 of those outcomes, the person is married. So the probability that a randomly selected respondent is married is 408/665, or 61.4% (approximately).

b. In 143 of those outcomes, the person has a college education or higher. So the probability that the selected person has a college education or higher is 143/665, or 21.5%.

c. There are 665 possible outcomes. In 98 of them, the respondents both are married AND have a college degree or higher. So the probability that the selected person is married AND has a college degree is 98/665, or 14.7%.

TRY THIS! Exercise 5.19

Using "OR" to Combine Events

The people in Figure 5.5a were asked to raise their hands if they were wearing glasses OR wearing a hat. Note that people who are wearing both also have their hands raised. If we selected one of these people at random, the probability that this person is wearing

(a)

Mike Rena Maria Alan   John   David

(b)

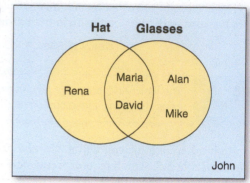

▲ FIGURE 5.5 **(a)** Raise your hand if you are wearing a hat OR glasses. This photograph illustrates the inclusive OR.
(b) In this Venn diagram, note the orange region for "raise your hand if you are wearing a hat OR glasses."

glasses OR wearing a hat would be 5/6, because we would count people who wear glasses, people who wear a hat, and people who wear both glasses AND a hat.

In a Venn diagram, OR events are represented by shading all relevant events. Here Mike, Rena, Maria, Alan, and David appear in the yellow area because each is wearing glasses OR wearing a hat.

The last example illustrates a special meaning of the word OR. This word is used slightly differently in mathematics and probability than you may use it in English. In statistics and probability, we use the **inclusive OR**. For example, the people in the photo shown in Figure 5.5a were asked to raise their hands if they had a hat OR glasses. This means that the people who raise their hands have a hat only, have glasses only, or have both a hat AND glasses.

> **KEY POINT**   The word OR creates a new event out of the events A and B. The new event consists of all outcomes that are only in A, that are only in B, or that are in both.

EXAMPLE 5 OR with Marital Status

Again, select someone at random from Table 5.3.

QUESTION   What is the probability the person is single OR married?

SOLUTION   The event of interest occurs if we select a person who is married, a person who is single, or a person who is both married AND single. There are 665 possible equally likely outcomes. Of these, 112 are single and 408 are married (and none are both!) Thus, there are $112 + 408 = 520$ people who are single OR married.

CONCLUSION   The probability that the selected person is single OR married is 520/665, or 78.2%.

TRY THIS!   Exercise 5.21

> **! Caution**
>
> **AND**
> P(A AND B) will always be less than (or equal to) P(A) and less than (or equal to) P(B). If this isn't the case, you've made a mistake!

EXAMPLE 6 Education OR Marital Status

Select someone at random from Table 5.3 on page 223 (which is shown again on the next page).

QUESTION   What is the probability that the person is married OR has a college degree or higher? (Assume that everyone in the "College or higher" category has a college degree.)

SOLUTION   Table 5.3 gives us 665 possible outcomes. The event of interest occurs if we select someone who is married, or someone who has a college degree or higher, or someone who both is married AND has a college degree or higher. There are 408 married people, and 143 people with a college degree or higher.

But wait a minute: there are not 408 + 143 different people who are married OR have a college degree or higher—some of these people got counted twice! The people who are both married AND have a college degree or higher were counted once when we looked at the married people, and they were counted a second time when we looked at the college and graduate school graduates. We can see from Table 5.4 that 98 people *both* are married AND have a college degree or higher. So we counted 98 people too many. Thus, we have 408 + 143 − 98 = 453 different outcomes in which the person is married or has a college degree or higher. Table 5.4 is the same as Table 5.3 except for the added ovals, which are meant to be interpreted as in a Venn diagram.

Education Level	Single	Married	Divorced	Widow/Widower	Total
Less HS	17	70	10	28	125
High school	68	240	59	30	397
College or higher	27	98	15	3	143
Total	112	408	84	61	665

▲ **TABLE 5.4** Here we reprint Table 5.3, with ovals for Married and for College added.

The numbers in bold type represent the people who are married OR have a college degree or higher. This Venn-like treatment emphasizes that one group (of 98 people) is in both categories and reminds us not to count them twice.

Another way to say this is that we count the 453 distinct outcomes by adding the numbers in the ovals in the table but not adding any of them more than once:

$$70 + 240 + 98 + 27 + 15 + 3 = 453$$

CONCLUSION   The probability that the randomly selected person is married OR has a college degree or higher is 453/665, or 68.1%.

TRY THIS!   Exercise 5.23

Mutually Exclusive Events

Did you notice that the first example of an OR (single OR married) was much easier than the second (married OR has a college degree)? In the second example, we had to be careful not to count some of the people twice. In the first example, this was not a problem. Why?

The answer is that in the first example, we were counting people who were married OR single, and no person can be in both categories. It is impossible to be simultaneously married AND single. When two events have no outcomes in common—that is, when it is impossible for both events to happen at once—they are called **mutually exclusive events**. The events "person is married" and "person is single" are mutually exclusive.

The Venn diagram in Figure 5.6 shows two mutually exclusive events. There is no intersection between these events; it is impossible for both event A AND event B to happen at once. This means that the probability that both events occur at the same time is 0.

▶ **FIGURE 5.6** In a Venn diagram, two mutually exclusive events have no overlap.

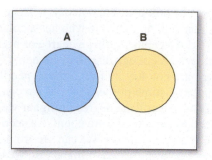

EXAMPLE 7 Mutually Exclusive Events: Education and Marital Status

Imagine selecting a person at random from those represented in Table 5.4.

QUESTION Name two mutually exclusive events, and name two events that are not mutually exclusive. Remember that "marital status" means a person's *current* marital status.

SOLUTION For mutually exclusive events, we can choose any two columns or any two rows. It is impossible for someone to be both divorced AND married (at the same time). It is impossible for someone to have less than a high school education AND to have a high school education. The events "person has a HS education" and "person has less than a HS education" are mutually exclusive because no one can be in both categories at once. The probability that a randomly selected person has a HS education AND has less than a HS education is 0. We could have chosen other pairs of events as well.

To find two events that are not mutually exclusive, find events that have outcomes in common. There are 30 people who have a HS education AND are widows/widowers. Therefore, these events are *not* mutually exclusive. The events "person has HS education" AND "person is a widow/widower" are not mutually exclusive.

TRY THIS! Exercise 5.27

Summary of Probability Rules

Rule 4: The probability that event A happens OR event B happens is

> (the probability that A happens) plus (the probability that B happens) minus (the probability that both A AND B happen)

If A and B are mutually exclusive events (for example, A is the event that the selected person is single, and B is the event that the person is married), then P(A AND B) = 0. In this case, the rule becomes simpler:

Rule 4a: If A and B are mutually exclusive events, the probability that event A happens OR event B happens is the sum of the probability that A happens and the probability that B happens.

Rule 4 in symbols:

> Always: P(A OR B) = P(A) + P(B) − P(A AND B)

Rule 4a in symbols:

> Only if A and B are mutually exclusive: P(A OR B) = P(A) + P(B)

EXAMPLE 8 Rolling a Six-Sided Die

Roll a fair, six-sided die.

QUESTIONS

a. Find the probability that the die shows an even number OR a number greater than 4 on top.

b. Find the probability that the die shows an even number OR the number 5 on top.

SOLUTIONS

a. We can do this in two ways. First, we note that six equally likely outcomes are possible. The even numbers are (2, 4, 6), and the numbers greater than 4 are (5, 6). Thus, the event "even number OR number greater than 4" has four different ways of happening: roll a 2, 4, 5, or 6. We conclude that the probability is 4/6.

The second approach is to use Rule 4. The probability of getting an even number is 3/6. The probability of getting a number greater than 4 is 2/6. The probability of getting both an even number AND a number greater than 4 is 1/6 (because the only way for this to happen is to roll a 6). So

$$P(\text{even OR greater than 4}) = P(\text{even}) + P(\text{greater than 4}) - P(\text{even AND greater than 4})$$
$$= 3/6 + 2/6 - 1/6$$
$$= 4/6$$

b. $P(\text{even OR roll 5}) = P(\text{even}) + P(\text{roll 5}) - P(\text{even AND roll 5})$

It is impossible for the die to be both even AND a 5 because 5 is an odd number. So the events "get a 5" and "get an even number" are mutually exclusive. Therefore, we get

$$P(\text{even number OR a 5}) = 3/6 + 1/6 - 0 = 4/6$$

CONCLUSIONS

a. The probability of rolling an even number OR a number greater than 4 is 4/6 (or 2/3).

b. The probability of rolling an even number OR a 5 is 4/6 (or 2/3).

TRY THIS! Exercise 5.29

EXAMPLE 9 What's New in Science?

The Pew Research Center examined the top science-related Facebook pages to determine the types of posts made. They found that 29% of posts were about new discoveries, 21% were practical applications of science, 16% were advertisements, 12% were explanations of scientific concepts, 7% were reposts, 5% were not related to science, and 10% were "other." Figure 5.7 displays these findings as a bar graph. (Pew Research Center, March, 2018).

QUESTION What's the probability that a randomly selected posting from these pages is an advertisement, is not related to science, OR falls into the "other" category?

► **FIGURE 5.7** The height of the bars indicates the percentage of posts in each category.

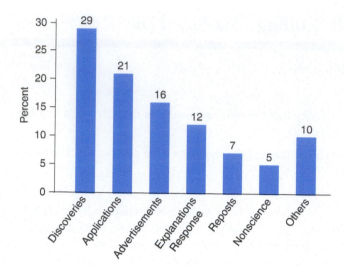

> **Details**
>
> **Probability Distributions**
> Figure 5.7 is an example of a probability distribution. Probability distributions provide us with the probabilities for all possible outcomes. You'll learn about these in Chapter 6.

SOLUTION   Because the categories of responses are mutually exclusive:

P(Advertisements OR Nonscience OR Others) = P(Advertisements) + P(Nonscience) + P(Others) = 16% + 5% + 10% = 31%

CONCLUSION   The probability that a randomly selected post falls into one of these three categories is 31%, greater than the probability that a post will be about a new discovery.

TRY THIS!   Exercise 5.35

Associations in Categorical Variables

Judging on the basis of our sample in Table 5.3, is there an association between marital status and having a college education? If so, we would expect the proportion of married people to be different for those who had a college education and those who did not have a college education. (Perhaps we would find different proportions of marital status for each category of education.)

In other words, if there is an association, we would expect the probability that a randomly selected college-educated person is married to be different from the probability that a person with less than a college education is married.

Conditional Probabilities

Language is important here. The probability that "a college-educated person is married" is different from the probability that "a person is college-educated AND is married." In the AND case, we're looking at everyone in the collection and wondering how many have both a college degree AND are married. But when we ask for the probability that a college-educated person is married, we're taking it as a given that the person is college-educated. We're *not* saying, "Choose someone from the whole collection." We're saying, "Just focus on the people with the college degrees. What proportion of those people are married?"

Probabilities such as these, where we focus on just one group of objects and imagine taking a random sample from that group alone, are called **conditional probabilities**.

For example, in Table 5.5 (which repeats Table 5.3), we've highlighted in red the people with college degrees. In this row, there are 143 people. If we select someone at random from among those 143 people, the probability that the person will be married is 98/143 (or about 0.685). We call this a conditional probability because we're finding the probability of being married *conditioned* on having a college education (that is, we are assuming we're selecting only from college-educated people).

Education Level	Single	Married	Divorced	Widow/Widower	Total
Less HS	17	70	10	28	125
HS	68	240	59	30	397
College or higher	27	98	15	3	143
Total	112	408	84	61	665

TABLE 5.5 What's the probability that a person with a college degree or higher is married? To find this, focus on the row shown in red and imagine selecting a person from this row. We assume everyone in the *College or higher* row has a college degree.

"Given That" versus "AND"   Often, conditional probabilities are worded with the phrase *given that*, as in "Find the probability that a randomly selected person is married given that the person has a college degree." But you might also see it phrased as in the last paragraph: "Find the probability that a randomly selected person with a college degree is married." The latter phrasing is more subtle, because it implies that we're supposed to assume the selected person has a college degree: we must assume we are *given that* the person has a college degree.

Figure 5.8a shows a Venn diagram representing all of the data. The green overlap region represents the event of being married AND having completed college. By way of contrast, Figure 5.8b shows only those with college educations; it emphasizes that if we wish to find the probability of being married, given that the person has a college degree, we need to focus on only those with college degrees.

(a)
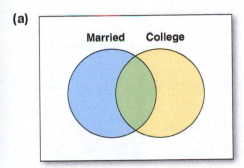

(b)
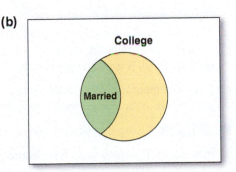

◄ FIGURE 5.8 (a) The probability of being married AND having a college degree; (b) the probability that a person with a college degree is married.

The mathematical notation for a conditional probability might seem a little unusual. We write

$$P(\text{person is married} \mid \text{person has college degree}) = 98/143$$

The vertical bar inside the probability notation is *not* a division sign. You should think of pronouncing it as "given that." This sentence reads, "The probability that the person is married, given that we know this person has a college degree, is 98/143." Some statisticians like to think of the vertical bar as meaning "within" and would translate this as "The probability that we randomly select a married person from within those who have a college degree." Either way you think about it is fine; use whichever makes the most sense to you.

> **KEY POINT**   P(A│B) means to find the probability that event A occurs but to restrict your consideration to those outcomes of A that occur within event B. It means "the probability event A will occur, given that event B has occurred."

EXAMPLE 10 Data Security and Age

The probabilities below are based on a random sample from the U.S. population collected by the Pew Research Center (January, 2017). The "young" people are aged 18 to 29 years, while the "seniors" are those 65 years old or older. Consider these probabilities:

a. The probability that a randomly selected young person will say they feel less secure about their data security than in the past is 41%. Define event A to be "the person is young" and event B to be "The person feels less secure about their data security now than in the past."

b. The probability that a random selected person is young and will say they feel more secure about their data security than in the past is 4%. Define event A to be "the person is young" and event B to be "The person feels more secure about their data security now than in the past."

c. The probability that a random selected senior will say they feel less secure about their data security than in the past is 56%. Define event A to be "the person is a senior" and event B to be "The person feels less secure about their data security now than in the past."

QUESTION   For each of these three statements, determine whether the events in the question are used in a conditional probability or an AND probability. Explain. Write the statement using probability notation.

a. This statement is asking about a conditional probability. It says that, restricting ourselves to those between the ages of 18 and 29, what's the probability we'll select someone with this belief. We are "given that" the group we're sampling from is in this age group. This is a conditional probability.

b. This statement asks us to assume nothing and, once that person is selected, consider whether they are in the age range of 18 to 29 and feel more secure about their data security. This is an AND probability.

c. This statement is also a conditional probability. (Note that older people were more likely to be concerned about their data security.)

SOLUTION   Using probability notation, these statements are

a. P(B│A), or P(person feels less secure │ person is young)

b. P(A and B), or P(person is young AND feels more secure about data security)

c. P(B│A), or P(person feels less secure │ person is senior)

TRY THIS   Exercise 5.41

Finding Conditional Probabilities
If you are given a table like Table 5.5, you can find conditional probabilities as we did earlier: by isolating the group from which you are sampling. However, a formula exists that is useful for times when you do not have such complete information.

The formula for calculating conditional probabilities is

$$P(A│B) = \frac{P(A \text{ AND } B)}{P(B)}$$

Example 11 shows how this formula is used.

EXAMPLE 11 Education and Marital Status

Suppose a person is randomly selected from those represented in Table 5.3 on page 223.

QUESTION Find the probability that a person with less than a high school degree (and no higher degrees) is married. Use the table, but then confirm your calculation with the formula.

SOLUTION We are asked to find P(married | less than high school degree)—in other words, the probability a person with less than a HS degree is married. We are told to imagine randomly selecting a person from only those who have less than a high school degree. There are 125 such people, of whom 70 are married.

$$P(\text{married} | \text{less HS}) = 70/125 = 0.560$$

The formula confirms this:

$$P(\text{married} | \text{less HS}) = \frac{P(\text{married AND less HS})}{P(\text{less HS})} = \frac{\dfrac{70}{665}}{\dfrac{125}{665}} = \frac{70}{125} = 0.560$$

Interestingly, the probability that a college graduate or higher is married (0.685) is greater than the probability that someone with less than a high school education is married (0.560).

TRY THIS! Exercise 5.43

With a little algebra, we can discover that this formula can serve as another way of finding AND probabilities:

$$P(A \text{ AND } B) = P(A)P(B | A)$$

We'll make use of this formula later.

Summary of Probability Rules

Rule 5a: $P(A | B) = \dfrac{P(A \text{ AND } B)}{P(B)}$

Rule 5b: $P(A \text{ AND } B) = P(B) P(A | B)$ and $P(A \text{ AND } B) = P(A) P(B | A)$

Both forms of Rule 5b are true because it doesn't matter which event is called A and which is called B.

Flipping the Condition A common mistake with conditional probabilities is thinking that $P(A | B)$ is the same as $P(B | A)$. These are not the same.

$$P(B | A) \neq P(A | B)$$

A second common mistake is to confuse conditional probabilities with fractions and think that $P(B | A) = 1/P(A | B)$.

$$P(B | A) \neq \frac{1}{P(A | B)}$$

For example, using the data in Table 5.3, we earlier computed that P(married | college) = 98/143 = 0.685. What if we wanted to know the probability that a randomly selected married person is college-educated?

$$P(\text{college} | \text{married}) = ?$$

From Table 5.3, we can see that if we know the person is married, there are 408 possible outcomes. Of these 408 married people, 98 are college-educated, so

$$P(\text{college}\,|\,\text{married}) = 98/408, \text{ or about } 0.240$$

Clearly, P(college | married), which is 0.240, does not equal P(married | college), which is 0.685.

Also, it is *not* true that P(married | college) = 1/P(college | married) = 408/98 = 4.16, a number bigger than 1! It is impossible for a probability to be bigger than 1, so obviously,

$$P(A\,|\,B) \text{ does not equal } \frac{1}{P(B\,|\,A)}$$

> **KEY POINT**
> P(B|A) ≠ P(A|B)

Independent and Dependent Events

We saw that the probability that a person is married differs, depending on whether we know that person is a college grad or that she or he has less than a high school education. If we randomly select from different educational levels, we get a different probability of marriage. Another way of putting this is to say that marital status and education level are **associated**. We know they are associated because the conditional probabilities change depending on which educational level we condition on.

We call variables or events that are *not* associated **independent events**. Independent variables, and independent events, play a very important role in statistics. Let's first talk about independent events.

Two events are independent if knowledge that one event has happened tells you nothing about whether or not the other event has happened. In mathematical notation,

A and B are independent events means P(A | B) = P(A).

In other words, if the event "a person is married" is independent from "a person has a college degree," then the probability that a married person has a college degree, P(college | married), is the same as the probability that any person selected from the sample will have a college degree. We already found that

P(college | married) = 0.240, and P(college) = 143/665 = 0.215

These probabilities are close—so close, in fact, that we might be willing to conclude they are "close enough." (You'll learn in Chapter 10 about how to make decisions like these.) We might conclude that the events "a randomly selected person is married" and "a randomly selected person is college-educated" are independent. However, for now we will assume that *probabilities have to be exactly the same for us to conclude independence.* We conclude that completing college and being married are *not* independent.

It doesn't matter which event you call A and which B, so events are also independent if P(B | A) = P(B).

> **KEY POINT**
> To say that events A and B are independent means that P(A|B) = P(A). In words, knowledge that event B occurred does not change the probability of event A occurring.

EXAMPLE **12** Dealing a Diamond

Figure 5.9 shows a standard deck of playing cards. When playing card games, players nearly always try to avoid showing their cards to the other players. The reason for this is that knowing the other players' cards can sometimes give you an advantage. Suppose you are wondering whether your opponent has a diamond. If you find out that one of the cards he holds is red, does this provide useful information?

QUESTION   Suppose a deck of cards is shuffled and one card is dealt facedown on the table. Are the events "the card is a diamond" and "the card is red" independent?

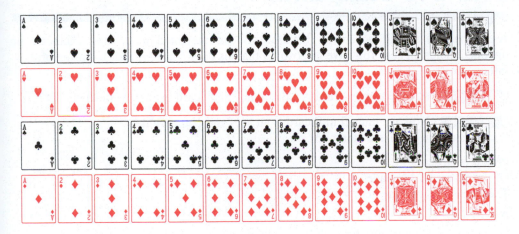

► **FIGURE 5.9** Fifty-two playing cards in a standard deck.

SOLUTION   To answer this, we must apply the definition of independent events and find

$$P(\text{card is a diamond})$$

and compare it to

$$P(\text{card is a diamond} \mid \text{card is red})$$

If these probabilities are different, then the events are not independent; they are associated events.

First we find P(card is a diamond).

Out of a total of 52 cards, 13 are diamonds. Therefore,

$$P(\text{card is a diamond}) = 13/52, \text{ or } 1/4$$

Now suppose we know the card is red. What's the probability that a red card is a diamond? That is, find

$$P(\text{card is a diamond} \mid \text{card is red})$$

The number of equally likely possible outcomes is reduced from 52 to 26 because there are 26 red cards. You are now limited to the 26 cards in the middle two rows of the picture. There are still 13 diamonds. Therefore, the probability that the card is a diamond, given that it is red, is $13/26 = 1/2$.

$$P(\text{card is diamond}) = 1/4$$

$$P(\text{card is diamond} \mid \text{card is red}) = 1/2$$

These probabilities are *not* equal.

CONCLUSION The events "select a diamond" and "select a red card" are associated, because P(select a diamond | color is red) is not the same as P(select a diamond). This means that if you learn, somehow, that an opponent's card is red, then you have gained some information that will be useful in deciding whether he has a diamond.

Note that we could also have compared P(card is red) to P(card is red | card is a diamond), and we would have reached the same conclusion.

TRY THIS! Exercise 5.49

EXAMPLE **13** Dealing an Ace

A playing card is dealt facedown. This time, you are interested in knowing whether your opponent holds an ace. You have discovered that his card is a diamond. Is this helpful information?

QUESTION Are the events "card is a diamond" and "card is an ace" independent?

SOLUTION Now we must find P(card is an ace) and compare it to P(card is an ace | card is a diamond).

P(card is an ace):
Of the 52 cards, 4 are aces.
Therefore, P(card is an ace) = 4/52 = 1/13.

P(card is an ace | card is a diamond):
There are 13 diamonds in the deck, and only one of these 13 is an ace.
Therefore, P(card is an ace | card is a diamond) = 1/13.

We find that P(card is an ace) = 1/13 = P(card is an ace | card is a diamond).

CONCLUSION The events "card is a diamond" and "card is an ace" are independent. This means the information that your opponent's card is a diamond will not help you determine whether it is an ace.

Note that we could also have compared P(card is a diamond) to P(card is adiamond | card is an ace), and we would have reached the same conclusion.

TRY THIS! Exercise 5.51

Intuition about Independence

Sometimes you can use your intuition to make an educated guess about whether two events are independent. For example, flip a coin twice. You should know that P(second flip is heads) = 1/2. But what if you know that the first flip was also a head? Then you need to find

P(second flip is heads | first flip was heads)

Intuitively, we know that the coin always has a 50% chance of coming up heads. The coin doesn't know what happened earlier. Thus,

P(second is heads | first is heads) = 1/2 = P(second is heads)

The two events "second flip comes up heads" and "first flip comes up heads" are independent.

Although you can sometimes feel very confident in your intuition, you should check your intuition with data whenever possible.

EXAMPLE 14   Education and Widows

Suppose we select a person at random from the sample of people asked about education and marital status in Table 5.3. Is the event "person selected has HS education" independent from the event "person selected is widowed"? Intuitively, we would think so. After all, why should a person's educational level affect whether his or her spouse dies first?

QUESTION   Check, using the data in Table 5.3 on page 223, whether these two events are independent.

SOLUTION   To check independence, we need to check whether $P(A|B) = P(A)$. It doesn't matter which event we call A and which we call B, so let's check to see whether
$P(\text{person selected has HS education} | \text{person is widowed}) = P(\text{person has HS education})$
From the table, we see that there are only 61 widows, so

$$P(HS | \text{widowed}) = 30/61 = 0.492$$

$$P(HS) = 397/665 = 0.597$$

The two probabilities are not equal.

CONCLUSION   The events are associated. If you know the person is widowed, then the person is less likely to have a high school education than he or she would be if you knew nothing about his or her marital status. Our intuition was wrong, at least as far as these data are concerned. It is possible, of course, that these data are not representative of the population as a whole or that our conclusion of association is incorrect because of chance variation in the people sampled.

TRY THIS!   Exercise 5.53

Sequences of Independent and Associated Events

A common challenge in probability is to find probabilities for sequences of events. By *sequence,* we mean events that take place in a certain order. For example, a married couple plans to have two children. What is the probability that the first will be a boy and the second a girl? When dealing with sequences, it is helpful to determine first whether the events are independent or associated.

If the two events are associated, then our knowledge of conditional probabilities is useful, and we should use Probability Rule 5b:

$$P(A \text{ AND } B) = P(A) P(B|A)$$

If the events are independent, then we know $P(B|A) = P(B)$, and this rule simplifies to $P(A \text{ AND } B) = P(A) P(B)$. This formula is often called the **multiplication rule**.

Summary of Probability Rules

Rule 5c: Multiplication Rule. If A and B are independent events, then

$$P(A \text{ AND } B) = P(A) P(B)$$

Independent Events   When two events are independent, the multiplication rule speeds up probability calculations for events joined by AND.

For example, suppose that 51% of all babies born in the United States are boys. Then $P(\text{first child is boy}) = 51\%$. What is the probability that a family planning to have two children will have two boys? In other words, how do we find the sequence probability

$P(\text{first child is boy AND second child is boy})$?

Researchers have good reason to suspect that the genders of children in a family are independent (if you do not include identical twins). Because of this, we can apply the multiplication rule:

P(first child is boy AND second child is boy) = P(first is boy) P(second is boy)

$$= 0.51 \times 0.51 = 0.2601$$

The same logic could be applied to finding the probability that the first child is a boy and the second is a girl:

P(first child is boy AND second child is girl) = P(first is boy) P(second is girl)

$$= 0.51 \times 0.49 = 0.2499$$

EXAMPLE 15 Three Coin Flips

Toss a fair coin three times. A fair coin is one in which each side is equally likely to land up when the coin is tossed.

QUESTION   What is the probability that all three tosses are tails? What is the probability that the first toss is heads AND the next two are tails?

SOLUTION   Using mathematical notation, we are trying to find P(first toss is tails AND second is tails AND third is tails). We know these events are independent (this is theoretical knowledge; we "know" this because the coin cannot change itself on the basis of its past). This means the probability is

P(first is tails) $\times$ P(second is tails) $\times$ P(third is tails) $= 1/2 \times 1/2 \times 1/2 = 1/8$

Also, P(first is heads AND second is tails AND third is tails) is

P(heads) $\times$ P(tails) $\times$ P(tails) $= 1/2 \times 1/2 \times 1/2 = 1/8$

CONCLUSION   The probability of getting three tails is the same as that of getting first heads and then two tails: 1/8.

TRY THIS!   Exercise 5.55

EXAMPLE 16 Ten Coin Flips

Suppose I toss a coin 10 times and record whether each toss lands heads or tails. Assume that each side of the coin is equally likely to land up when the coin is tossed.

QUESTION   Which sequence is the more likely outcome?

Sequence A: HTHTHTHTHT

Sequence B: HHTTTHTHHH

SOLUTION   Because these are independent events, the probability that sequence A happens is

$$P(H)P(T)P(H)P(T)P(H)P(T)P(H)P(T)P(H)P(T) = \frac{1}{2} \times \frac{1}{2} \times \frac{1}{2} \times \frac{1}{2} \times \frac{1}{2} \times \frac{1}{2}$$

$$\times \frac{1}{2} \times \frac{1}{2} \times \frac{1}{2} \times \frac{1}{2}$$

$$= \left(\frac{1}{2}\right)^{10} = 0.0009766$$

The probability that sequence B happens is

$$P(H)P(H)P(T)P(T)P(T)P(H)P(T)P(H)P(H)P(H) = \frac{1}{2} \times \frac{1}{2} \times \frac{1}{2} \times \frac{1}{2} \times \frac{1}{2} \times \frac{1}{2}$$
$$\times \frac{1}{2} \times \frac{1}{2} \times \frac{1}{2} \times \frac{1}{2}$$
$$= \left(\frac{1}{2}\right)^{10} = 0.0009766$$

CONCLUSION   Even though sequence A looks more improbable than B because it alternates between heads and tails, both outcomes have the same probability!

TRY THIS!   Exercise 5.57

Another common probability question asks about the likelihood of "at least one" of a sequence happening a certain way.

EXAMPLE **17** Life on Mars

According to a 2018 Marist survey, 68% of American adults think that there is intelligent life on other planets. Suppose we select three people at random, with replacement, from the population of all American adults. (Selecting "with replacement" means that, once a person is selected, they are eligible to be selected again.)

QUESTIONS

a. What is the probability that all three believe in intelligent life on other planets?

b. What is the probability that none of the three believes in intelligent life on other planets?

c. What is the probability that at least one of the selected people believes in intelligent life on other planets?

SOLUTIONS

a. We are asked to find P(first believes AND second believes AND third believes). Because the people were selected at random from the population, these are independent events. (One person's answer won't affect the probability that the next person selected will answer one way or the other.) Because these are independent events, the probability we seek is just:

P(first believes) $\times$ P(second believes) $\times$ P(third believes)
= 0.68 $\times$ 0.68 $\times$ 0.68 = 0.3144

b. The probability that none is a believer is trickier to determine. This event occurs if the first person is not a believer AND the second is not AND the third is not. So we need to find:

P(first does not believe AND second does not believe AND third does not believe)
= P(first not) $\times$ P(second not) $\times$ P(third not)
= (1 − 0.68) $\times$ (1 − 0.68) $\times$ (1 − 0.68) = 0.0328

c. The probability that at least one person believes in life on other planets is the probability that one person believes OR two believe OR all three believe. The calculation is easier if you realize that "at least one person believes" is the complement of "none believes" because it includes all categories except "none." And so

1 − 0.0328 = 0.9672

CONCLUSION The probability that all three randomly selected people believe in intelligent life on other planets is 0.3144. The probability that none of the three believe is 0.0328. And the probability that at least one believes is 0.9672.

TRY THIS! Exercise 5.59

> **! Caution**
>
> **False Assumptions of Independence**
>
> If your assumption that A and B are independent events is wrong, P(A AND B) can be very wrong if you compute with the multiplication rule!

Watch Out for Incorrect Assumptions of Independence
Do not use the multiplication rule if events are not independent. For example, suppose we wanted to find the probability that a randomly selected person is female AND has long hair (say, more than 6 inches long).

About half of the population is female, so P(selected person is female) = 0.50. Suppose that about 35% of everyone in the population has long hair; then P(selected person has long hair) = 0.35. If we used the multiplication rule, we would find that P(selected person has long hair AND is female) = 0.35 × 0.5 = 0.175.

This relatively low probability of 17.5% makes it sound somewhat unusual to find a female with long hair. The reason is that we assumed that having long hair and being female are independent. This is a bad assumption: a woman is more likely than a man to have long hair. Thus, "has long hair" and "is female" are associated, not independent, events. Therefore, once we know that the chosen person is female, there is a greater chance that the person has long hair.

Associated Events with "AND"
If events are not independent, then we rely on Probability Rule 5b: P(A AND B) = P(A) P(B | A). Of course, this assumes that we know the value of P(B | A).

For example, the famous cycling race the Tour de France has been plagued with accusations that cyclists have taken steroids to boost their performance. Testing for these steroids is difficult, partly because these substances occur naturally in the human body and partly because different individuals have different levels than other individuals. Even within an individual, the level of steroid varies during the day. Also, testing for the presence of steroids is expensive and time-consuming. For this reason, racers are chosen randomly for drug tests.

Let's imagine that 2% of the racers have taken an illegal steroid. Also, let's assume that if they took the drug, there is a 99% chance that the test will return a "positive" reading. (A "positive" reading here is a negative for the athlete, who will be disqualified.)

Keep in mind that these tests are not perfect. Even if the cyclist did not take a steroid, there is some probability that the test will return a positive. Let's suppose that, given that an athlete did *not* take a steroid, there is still a 3% chance that the test will return a positive (and the athlete will be unjustly disqualified).

What is the probability that a randomly chosen cyclist will be steroid-free and will still test positive for steroids? In other words, we need to find, for a randomly chosen cyclist,

P(cyclist does not take steroids AND tests positive).

To summarize, we have these pieces of information for a randomly selected athlete at this event:

$$P(\text{took steroid}) = 0.02$$

$$P(\text{did not take steroid}) = 0.98$$

$$P(\text{tests positive} \mid \text{took steroid}) = 0.99$$

$$P(\text{tests positive} \mid \text{did not take steroid}) = 0.03$$

Note that this is a sequence of events. First, the athlete either takes or doesn't take a steroid. Then a test is given, and the results are recorded.

According to Rule 5b,

$$P(A \text{ AND } B) = P(A)\,P(B\,|\,A)$$

$$P(\text{did not take steroid AND tests positive}) = P(\text{did not take steroid}) \times$$

$$P(\text{tests positive}\,|\,\text{did not take steroid})$$

$$= 0.98 \times 0.03$$

$$= 0.0294$$

We see that roughly 2.9% of all cyclists tested will both be steroid-free AND test positive for steroids.

Many people find it useful to represent problems in which events occur in sequence with a tree diagram. We can represent this sequence of all possible outcomes in the tree diagram shown in Figure 5.10.

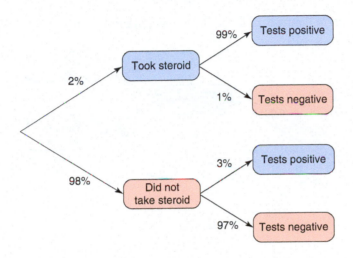

◀ FIGURE 5.10 Tree diagram showing probabilities for the sequence of events in which an athlete either takes or does not take a drug and then is tested for the presence of the drug.

The tree diagram shows all possible outcomes after selecting a cyclist (who either took steroids or did not) and then testing (and the test will be either positive or negative). Because the events "took steroid" and "tests positive" are associated, the probabilities of testing positive are different for the "took steroid" branch and for the "did not take steroid" branch.

We want to find P(did not take steroid AND positive result), so we simply multiply along that branch of the tree that begins with "did not take steroid" and ends with "tests positive":

$$P(\text{did not take steroid AND tests positive}) = P(\text{did not take steroid}) \times$$

$$P(\text{tests positive}\,|\,\text{did not take steroid})$$

$$= 0.98 \times 0.03$$

$$= 0.0294$$

EXAMPLE **18** Airport Screeners

At many airports you are not allowed to take water through the security checkpoint. This means that security screeners must check for people who accidentally pack water in their bags. Suppose that 5% of people accidentally pack a bottle of water in their bags. Also suppose that if there is a bottle of water in a bag, the security screeners will catch it 95% of the time.

QUESTION What is the probability that a randomly chosen person with a backpack has a bottle of water in the backpack and security finds it?

SOLUTION We are asked to find P(packs water AND Security finds the water). This is a sequence of events. First, the person packs the water into the backpack. Later, security finds (or does not find) the water.

We are given

$$P(\text{packs water}) = 0.05$$

$$P(\text{Security finds water} \mid \text{water packed}) = 0.95$$

Therefore,

$$P(\text{packs water AND Security finds it}) = P(\text{packs water}) \times$$

$$P(\text{Security finds water} \mid \text{water packed})$$

$$= 0.05 \times 0.95$$

$$= 0.0475$$

The tree diagram in Figure 5.11 helps show how to find this probability.

▶ **FIGURE 5.11** Tree diagram showing probabilities for the sequence of events in which a traveler either packs or does not pack water and then security either finds or does not find the water.

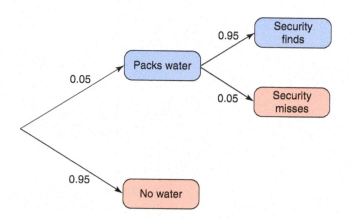

CONCLUSION There is about a 5% (4.75%) chance that a randomly selected traveler will both have packed water and will have the water found by security.

TRY THIS! Exercise 5.61

SECTION 5.4

Finding Empirical and Simulated Probabilities

Empirical probabilities are probabilities based on carrying out real-life random actions and observing the relative frequency that an event of interest happens. For instance, a softball player's batting average—the number of successful hits divided by the number of attempts—can be thought of as the empirical probability that the batter will get a hit next time at the plate. The percentage of times you board the bus and discover there are no empty seats is the empirical probability that there will be no empty seats the next time you catch the bus.

Finding empirical probabilities can be slow and laborious, and for this reason it is often useful to instead perform a simulation that can be carried out rapidly. Simulations can also be helpful when the situation is too complex to find a theoretical probability. In this next example we carry out the real-life random process in order to find an empirical probability. After that, we show how we can instead simulate the process. By the end of this section, we'll see that a simulated probability, if the simulation is repeated enough times, can provide a suitable estimate of the theoretical probability.

EXAMPLE 19 Two Heads Are Better than None

We wish to know the probability of getting exactly 2 heads if we flip a fair coin 3 times. (A "fair coin" is one that is equally likely to land heads as tails.) To find an empirical probability of this event, we flipped a real coin 3 times and noted whether or not there were exactly two heads. We repeated this for five trials. The results are shown below.

Trial	Outcome	Two Heads?
1	HTH	Yes
2	HHT	Yes
3	THH	Yes
4	THT	No
5	HHH	No

QUESTION   What is the empirical probability of getting exactly 2 heads in 3 flips of a fair coin based on these data?

SOLUTION   In the five trials, we got exactly 2 heads on three of the trials, and so for these data, the empirical probability is 3/5, or 0.60. (You'll soon see that with only five trials, our empirical probability can be quite different from the theoretical probability.)

TRY THIS!   Exercise 5.63

Our "experiment" in Example 19 had several components:

- A basic random action (the coin flip)
- A trial, which consisted in this case of three repetitions of that random action
- An event that we were interested in (in this case, the event of getting exactly 2 heads in 3 flips)

When creating a simulation of this experiment, we need to make sure the simulation matches these components of the real-life experiment.

Designing Simulations

How would we design and carry out a simulation to estimate the probability of getting 2 heads in 3 flips of a fair coin? This random process consists of a simple random action—flipping a coin—repeated several times, in this case, three times. The key to a successful simulation is to be able to find an easier-to-carry-out random action with the same probabilities as this simple random action. In this case, because there is a 50% chance that a fair coin will land heads, this means finding something that has a 50% chance of success. The next example shows one way this can be done.

EXAMPLE 20 Simulating Success

Using a random number table, here are several different methods we might use to simulate a single flip of a coin.

A. Each digit will represent a flip of the coin. Let the odd digits (1, 3, 5, 7, 9) represent heads, and the even digits represent tails.

B. Only the digits 0 and 1 will represent a flip of the coin. The other digits will be ignored. Let the digit 1 represent heads, 0 represent tails.

C. Each digit will represent a flip of the coin. Let the digits 0, 1, 2, 3 represent heads and 4, 5, 6, 7, 8, 9 represent tails.

QUESTION　Which of these methods is a valid simulation? Explain.

SOLUTION　Methods (A) and (B) are both valid simulations, because they both will create a 50% probability of getting a "heads" on the simulated coin. Because each digit in the random number table is equally likely to occur, simulation (A) will work because there is a 50% chance we will see one of the digits 1, 3, 5, 7, or 9. Simulation (B) works because we're ignoring all but 0 and 1, and so the digits 1 and 0 will each occur 50% of the time. However, (C) is invalid because there are only four ways of simulating heads and six ways of simulating tails. And so the probability of getting heads using method (C) is 40%, not 50%.

TRY THIS!　Exercise 5.65

Now that we have a method for simulating the simple action of flipping a coin, we can use a random number table to carry out a simulation for the probability of getting exactly 2 heads in 3 flips of a coin. We'll use the random number table shown in Table 5.6.

▶ **TABLE 5.6** The first five lines of the random number table from Appendix A.

Line						
01	21033	32522	19305	90633	80873	19167
02	17516	69328	88389	19770	33197	27336
03	26427	40650	70251	84413	30896	21490
04	45506	44716	02498	15327	79149	28409
05	55185	74834	81172	89281	48134	71185

Our simulation has several components:

A basic random action that we will simulate (a flip of a coin)

A trial that consists of one or more repetitions of that action (three flips of a coin)

An event of interest (getting exactly 2 heads)

We'll use method (A) from Example 20 to represent the basic action of flipping a coin. A trial therefore consists of reading off three digits in a row. At the end of each group of three digits, we record whether our event occurred, that is, whether we got exactly two "heads." For this simulation, we'll carry out 10 trials. (Really we should do many more; several hundred at least.) Table 5.7 shows the results using the first line of the random numbers in Table 5.6.

▶ **TABLE 5.7** Each row shows the results of one trial of our simulation. The red rows indicate trials in which the desired outcome (exactly two heads) occurred.

Trial	Random Numbers	Translation	Number of Heads	Did Event Occur?
1	2 1 0	THT	1	No
2	3 3 3	HHH	3	No
3	2 5 2	THT	1	No
4	2 1 9	THH	2	Yes
5	3 0 5	HTH	2	Yes
6	9 0 6	HTT	1	No
7	3 3 8	HHT	2	Yes
8	0 8 7	TTH	1	No
9	3 1 9	HHH	3	No
10	1 6 7	HTH	2	Yes

The event we're interested in happened four times in the 10 trials (shown in color in Table 5.7), so our simulated probability is 4/10 or 0.40. This is different from our empirical probability earlier (which was 0.60). In fact, this is an event in which one can calculate the theoretical probability, and the theoretical probability, 0.375, is different from both our empirical probability and our simulated probability. We'll see soon that we can get closer agreement between our simulations and the theoretical probability if we use a large number of trials.

Summary of Steps for a Simulation

It is helpful at this point to summarize the steps we went through to carry out our simulation. The same steps will be useful in many other situations.

1. Identify the basic random action and the probability of a successful outcome.

2. Determine how to simulate this random action.

3. Decide how you will simulate one trial.

4. Carry out a trial, and record whether or not the event you are interested in occurred.

5. Repeat a trial many times, at least 1000, and count the number of times your event occurred.

6. The number of times the event occurred, divided by the number of trials, is your empirical probability.

Usually, simulations are carried out using technology since a computer or calculator can execute these steps very rapidly and allow us to do many thousands of trials in a very short time.

EXAMPLE 21 Dice Simulation

Use a simulation to find the approximate probability that a fair, six-sided die will land with the 6 showing on top. Do 10 trials, using these random numbers:

Line				
01	44687	75032	83408	10239
02	80016	58250	91419	56315

QUESTION What is your empirical probability of getting a 6? Compare this to the theoretical probability. Show all steps.

SOLUTION

Step 1: The random action is throwing a die, and the probability of a successful outcome is 1/6, because the probability that a fair, six-sided die lands with the 6 on top is 1/6 (assuming the die is well balanced so that each outcome is equally likely.)

Step 2: We simulate this random action using digits in the random number table. Each number in the table will represent the roll of a die. The number in the table will represent the number that comes up on the die. We'll ignore the digits 0, 7, 8, and 9 because these are impossible outcomes when rolling a six-sided die.

Step 3: The event we're interested in is whether we see a 6 after a single toss.

Step 4: A trial consists of reading a single digit from the table.

Step 5: The first row of Table 5.8 shows the results of one trial. Our simulated die landed on a 4. We record "No" because the event we are studying did not happen. Note that we simply skip the digits 0, 7, 8, and 9.

Step 6: The remaining nine trials are shown in the table.

Step 7: A 6 occurred on one of the 10 trials (the third trial).

Trial	Outcome	Did Event Occur?
1	4	No
2	4	No
3	6	Yes
4	5	No
5	3	No
6	2	No
7	3	No
8	4	No
9	1	No
10	2	No

▲ TABLE 5.8 Simulations for rolling a six-sided die.

CONCLUSION Our empirical probability of getting a 6 on the roll of a balanced die is 1/10, or 0.10. In contrast, the theoretical probability is 1/6, or about 0.167.

TRY THIS! Exercise 5.67

The Law of Large Numbers

The **Law of Large Numbers** is a famous mathematical theorem that tells us that if our simulation is designed correctly, then the more trials we do, the closer we can expect our empirical probability or simulation-based probability to be to the true probability. The Law of Large Numbers shows that as we approach infinitely many trials, the true probability and the simulation-based probability approach the same value.

> **KEY POINT**   The Law of Large Numbers states that if an experiment with a random outcome is repeated a large number of times, the empirical probability of an event is likely to be close to the true probability. The larger the number of repetitions, the closer together these probabilities are likely to be.

The Law of Large Numbers is the reason why simulations are useful: Given enough trials, and assuming that our simulations are a good match to real life, we can get a good approximation of the true probability.

Table 5.9 shows the results of a very simple simulation. We used a computer to simulate flipping a coin, and we were interested in observing the frequency at which the "coin" comes up heads. We show the results at the end of each trial so that you can see how our approximation gets better as we perform more trials. For example, on the first trial we got a head, so our empirical probability is $1/1 = 1.00$. On the second and third trials we got heads, so the empirical probability of heads is $3/3 = 1.00$. On the fourth trial we got tails, and up to that point we've had 3 heads in 4 trials. Therefore, our empirical probability is $3/4 = 0.75$.

We can continue this way, but it is easier to show you the results by making a graph. Figure 5.12a shows a plot of the empirical probabilities against the number of flips of a coin for one particular series of coin flips. Figure 5.12b shows the same plot for another series of coin flips. The two plots differ because of chance, but despite the differences, both "settle" on an empirical probability of 0.50 after a large number of coin flips. This is just as the Law of Large Numbers predicts.

Trial	Outcome	Empirical Probability of Heads
1	H	$1/1 = 1.00$
2	H	$2/2 = 1.00$
3	H	$3/3 = 1.00$
4	T	$3/4 = 0.75$
5	H	$4/5 = 0.80$
6	H	$5/6 = 0.83$
7	T	$5/7 = 0.71$
8	H	$6/8 = 0.75$
9	T	$6/9 = 0.67$
10	H	$7/10 = 0.70$

▲ **TABLE 5.9** Simulations of heads and tails with cumulative simulated probabilities

▶ **FIGURE 5.12** The Law of Large Numbers predicts that after many flips, the proportion of heads we get from flipping a real coin will get close to the true probability of getting heads. Because these simulated probabilities are "settling down" to about 0.50, this supports the theoretical probability of 0.50.

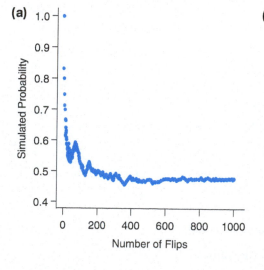

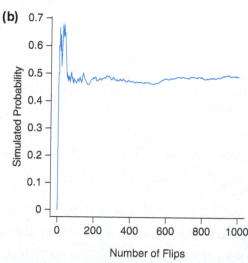

Note that with a small number of trials (say, less than 75 or so), our simulated probability was relatively far away from the theoretical value of 0.50. Also, when the number of trials is small, the simulated probabilities can change a lot with each additional "coin flip." But eventually things settle down, and the simulated probabilities get close to what we know to be the theoretical probability. If you were to simulate a coin flip just 20 times, you might not expect your simulated probability to be too close to the theoretical probability. But after 1000 flips, you'd expect it to be very close.

How Many Trials Should I Do in a Simulation?

In our examples of simulations we did 10 trials, which is not very many. Figure 5.13 shows four different series of coin flips. One of these series is the same as in Figure 5.12b. The percentage of heads at any given number of flips is different for each series because of chance. For example, in one series the first flip was tails, and so the percentage of heads is 0 at the beginning of the series. Whereas for another series, the first flip was heads, and so this series begins at 100%.

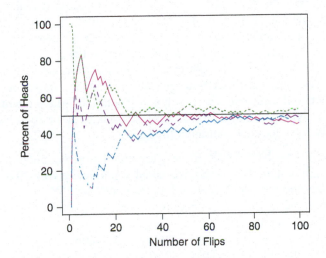

◀ **FIGURE 5.13** The percentage of heads for four different series of coin flips. There is great variability in the percentage of heads when the number of flips is small, but this variability gradually decreases as the number of flips increases.

Notice than after 20 flips the percentage of heads, and thus the simulated probability of heads, ranges from about 30% to nearly 70%. (The horizontal line indicates 50%). However, after 100 flips, the empirical probabilities are very similar. This tells us that if we wish to use an empirical or simulated probability to estimate a theoretical probability, we'll be in much closer agreement with each other, and quite possibly closer to the theoretical probability, if we use a large number of trials. And so the answer to the question, "How many trials should I do in a simulation?" is that it depends on how precise you wish to be. But in general, you want to do as many as you can!

What If My Simulation Doesn't Give the Theoretical Value I Expect?

You have no guarantee that your empirical probability will give you exactly the theoretical value. As you noticed if you did the above activity, even though it's not unusual to get 50% heads in 10 tosses, more often you get some other proportion.

When you don't get the "right" value, three explanations are possible:

1. Your theoretical value is incorrect.

2. Your empirical or simulated probability is just varying—that's what empirical probabilities do. You can make it vary less and get closer to the theoretical value by doing more trials.

3. Your simulation is not a good match to reality.

How do we choose between these alternatives? First, make sure your simulation is correct. From here on, we'll assume so. Now it's a matter of which of the first two alternatives is correct. This is one of the central questions of statistics: Are the data consistent with our expectations, or do they suggest that our expectations are wrong? We will return to this question in almost every chapter in this text.

Some Subtleties with the Law

The Law of Large Numbers (LLN) is one law that cannot be broken. Nevertheless, many people think the law has been broken when it really hasn't, because interpreting the LLN takes some care.

The LLN tells us, for example, that with many flips of a coin, our empirical probability of heads will be close to 0.50. It tells us nothing about the *number* of heads we will get after some number of tosses and nothing about the order in which the heads and tails appear.

Streaks: Tails Are Never "Due."

Streaks: Tails Are Never "Due."   Many people mistakenly believe the LLN means that if they get a large number of heads in a row, then the next flip is more likely to come up tails. For example, if you just flipped five heads in a row, you might think that the sixth is more likely to be a tail than a head so that the empirical probability will work out to be 0.50. Some people might incorrectly say, "Tails are due."

This misinterpretation of the LLN has put many a gambler into debt. The LLN is patient. It says that the empirical probability will equal the true probability after *infinitely* many trials. That's a lot of trials. Thus, a streak of 10 or 20 or even 100 heads, though extremely rare, does not contradict the LLN.

How Common Are Streaks?

How Common Are Streaks?   Streaks are much more common than most people believe. At the beginning of the chapter, we asked you to toss a coin 20 times (or to simulate 20 tosses) and see how long your longest streak was. We now show results of a simulation to find the (empirical) probability that the longest streak is of length 6 or longer in 20 tosses of a coin.

In our 1000 trials, we saw 221 streaks that were a length of 6 or greater. The longest streak we saw was a length of 11. The results are shown in Figure 5.14. Our empirical probability of getting a longest streak of a length of 6 or more was 0.221.

There is another reason why you should not believe that heads are "due" after a long streak of tails. If this were the case, the coin would somehow have to know whether to come up heads or tails. How is the coin supposed to keep track of its past?

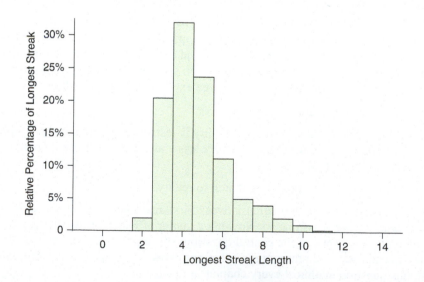

▶ **FIGURE 5.14** Longest streaks in 20 flips of a coin. In 1000 trials, 221 trials had streak lengths of 6 or greater.

CASE STUDY
REVISITED

SIDS or Murder?

Sally Clark was convicted of murdering two children on the basis of the testimony of physician/expert Dr. Meadow. Dr. Meadow testified that the probability of two children in the same family dying of SIDS was extremely low and that, therefore, murder was a more plausible explanation. In Dr. Meadow's testimony, he assumed that the event "one baby in a family dies of SIDS" AND "a second baby in the family dies of SIDS" were independent. For this reason, he applied the multiplication rule to find the probability of two babies dying of SIDS in the same family. The probability of one baby dying was $1/8543$, so the probability of two independent deaths was $(1/8543) \times (1/8543)$, or about 1 in 73 million.

However, as noted in a press release by the Royal Statistical Society, these events may not be independent:

"This approach (multiplying probabilities) . . . would only be valid if SIDS cases arose independently within families, an assumption that would need to be justified empirically. Not only was no such empirical justification provided in the case, but there are very strong *a priori* reasons for supposing that the assumption will be false. There may well be unknown genetic or environmental factors that predispose families to SIDS, so that a second case within the family becomes much more likely."

Dr. Meadow made several other errors in statistical reasoning, which are beyond the scope of this chapter but quite interesting nonetheless. There's a nice discussion that's easy to read at http://www.richardwebster.net/cotdeaths.html. For a summary and a list of supporting references, including a video by a statistician explaining the statistical errors, see http://en.wikipedia.org/wiki/Sally_Clark.

Sally Clark was released from prison but died in March 2007 of alcohol poisoning at the age of 43. Her family believes her early death was caused in part by the stress inflicted by her trial and imprisonment.

DATAPROJECT ▸ Subsetting Data

1 OVERVIEW

Sometimes we wish to focus our investigation on particular rows of a data set. We can do this by "subsetting" the data so that only rows that meet a specified condition are included in the new data set.

2 GOAL

Create new data sets with fewer rows than the original.

3 DATA MOVES TO ZOOM IN

Some statistical investigative questions are about subsets of your data set. For these questions, we want to "zoom in" to focus only on the relevant observations. For example, we might want to examine only Libertarians, or only people taller than 5 feet.

Project: More specifically, what's the typical running time for runners of the 2017 Los Angeles Marathon who were in their 20s? For women, what's the association between the age of a runner and her running time?

The first question concerns only runners between the ages of 20 and 29. One approach to answering a question like this is to create a new data set with only these runners in it. We are selecting only certain rows: those for which the value of the *Age* variable is greater or equal to 20 and less than 30.

The data move in which we select a subset of the rows is called, unsurprisingly, *subsetting*. Think of it as "zooming in" on the data so that we focus on only a certain type of observation. In this case, the type of observation is runners in their 20s.

StatCrunch offers several ways to do this, but the approach we recommend is to select

Data > Arrange > Subset

This opens a dialogue box (see Figure 5.15). Notice that all the columns are automatically selected for us.

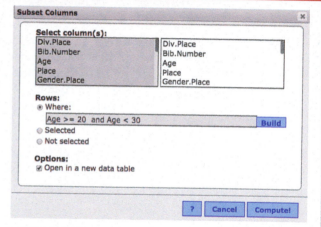

▲ **FIGURE 5.15** This dialogue box will create a new data set in which the runners are in their 20s.

Because we want only those people who are 20 years old or older and younger than 30, we type this expression into the "Where:" box:

Age >= 20 and Age < 30

(Typing **Age >= 20 and Age <= 29** would do the same thing.) Be sure to check the box to "Open in a new data table."

The result is a new data table that has only runners in their 20s. We can now summarize the running time. To select only the women, we use the same approach, but we change the condition for selecting rows, as in Figure 5.16.

Assignment

Report the typical running time for runners in their 20s. Create a graph that can be used to examine the association between age and running time for women runners. Next, pose and answer your own statistical investigative question that can be answered by creating a subset of the data.

▲ FIGURE 5.16 To subset based on a categorical variable, you need to know the values of the variable that you wish to select. Clicking **Build** will give you some guidance.

CHAPTER REVIEW

KEY TERMS

Random, *215*
Probability, *217*
Theoretical probability, *217*
Empirical probabilities, *218*
Simulations, *218*

Complement, *219*
Sample space, *219*
Event, *220*
Venn diagram, *222*
AND, *222*

Inclusive OR, *224*
Mutually exclusive events, *225*
Conditional probabilities, *229*
Associated, *232*
Independent events, *232*

Multiplication rule, *235*
Law of Large Numbers, *244*

LEARNING OBJECTIVES

After reading this chapter and doing the assigned homework problems, you should

- Understand that humans can't reliably create random numbers or sequences.

- Understand that a probability is a long-term relative frequency.

- Know the difference between empirical and theoretical probabilities—and know how to calculate them.

- Be able to determine whether two events are independent or associated and understand the implications of making incorrect assumptions about independent events.

- Understand that the Law of Large Numbers enables us to use empirical probabilities to estimate and test theoretical probabilities.

- Know how to design a simulation to estimate empirical probabilities.

SUMMARY

Random samples or random experiments must be generated with the use of outside mechanisms such as computer algorithms or by relying on random number tables. Human intuition cannot be relied on to produce reliable "randomness."

Probability is based on the concept of long-run relative frequencies: If an action is repeated infinitely many times, how often does a particular event occur? To find theoretical probabilities, we calculate these relative frequencies on the basis of assumptions about the situation and rely on mathematical rules. In finding empirical probabilities, we actually carry out the action many times or, alternatively, rely on a simulation (using a computer or random number table) to quickly carry out the action many times. The empirical probability is the proportion of times a particular event was observed to occur. The Law of Large Numbers tells us that the empirical probability becomes closer to the true probability as the number of repetitions is increased.

Theoretical Probability Rules

Rule 1: A probability is always a number from 0 to 1 (or 0% to 100%) inclusive (which means 0 and 1 are allowed). It may be expressed as a fraction, a decimal, or a percentage.

$$0 \leq P(A) \leq 1$$

Rule 2: For any event A,

$$P(A \text{ does } not \text{ occur}) = 1 - P(A \text{ does occur})$$

A^c is the complement of A:

$$P(A^c) = 1 - P(A)$$

Rule 3: For equally likely outcomes,

$$P(A) = \frac{\text{Number of outcomes in A}}{\text{Number of all possible outcomes}}$$

Rule 4: Always: $P(A \text{ OR } B) = P(A) + P(B) - P(A \text{ AND } B)$
Rule 4a: Only if A and B are mutually exclusive,

$$P(A \text{ OR } B) = P(A) + P(B)$$

Rule 5a: Conditional probabilities

Probability of A given that B occurred: $P(A|B) = \dfrac{P(A \text{ AND } B)}{P(B)}$

Rule 5b: Always: $P(A \text{ AND } B) = P(B)\,P(A|B)$
Rule 5c: Multiplication Rule. If A and B are independent events, then

$$P(A \text{ AND } B) = P(A)\,P(B)$$

This applies for any (finite) number of events. For example, $P(A \text{ AND } B \text{ AND } C \text{ AND } D) = P(A)\,P(B)\,P(C)\,P(D)$ if A, B, C, D are independent of each other.

SOURCES

MetLife Survey of the American teacher. 2013. http://www.harrisinteractive.com

Starr, N. 1997. Nonrandom risk: The 1970 draft lottery. *Journal of Statistics Education*, vol. 5, no. 3. doi:10.1080/10691898.1997.11910534. Vietnam-era draft data can be found, along with supporting references, at www.amstat.org/publications/jse/datasets/draft.txt.

Pew Research Center. 2010. *Social media and young adults*. http://pewinternet.org.

Pew Research Center, 2017. *Americans and Cybersecurity*.

Marist Poll, February, 2018. *Are Americans Poised for an Alien Invasion?*

Pew Research Center, 2018. *The Science People See on Social Media*.

SECTION EXERCISES

SECTION 5.1

5.1 Simulation (Example 1) If we flip a coin 10 times, what percentage of the time will the coin land on heads? A first step to answering this question is to simulate 10 flips. Use the random number table in Appendix A to simulate flipping a coin 10 times. Let the digits 0, 1, 2, 3, 4 represent heads and the digits 5, 6, 7, 8, 9 represent tales. Begin with the first digit in the fifth row.

   a. Write the sequence of 10 random digits.

   b. Change the sequence of 10 random digits to a sequence of heads and tails, writing H for the digits 0, 1, 2, 3, 4 and the T for the digits 5, 6, 7, 8, 9. What was the longest streak of heads in your list?

   c. What percentage of the flips were heads?

5.2 Simulation Suppose you are carrying out a randomized experiment to test if there is a difference in the amount of information remembered between students who take notes using a computer versus those who take notes by hand using pen and paper. You have 20 college student participants and want each to have an equal chance of being assigned to the computer group or the pen and paper group. Let the even digits (0, 2, 4, 6, 8) represent assignment to the computer group and the odd digits (1, 3, 5, 7, 9) represent assignment to the pen and paper group. Begin with the first digit in the third row of the random number table in Appendix A.

   a. Write the sequence of 20 random digits.

   b. Change the sequence of 20 random digits to a sequence of C for computer (digits 0, 2, 4, 6, 8) and P for paper (digits 1, 3, 5, 7, 9).

   c. What percentage of the 20 participants were assigned to the computer group?

   d. Describe another way the random number table could have been used to randomly assign participants to one of the two groups.

5.3 Empirical versus Theoretical A Monopoly player claims that the probability of getting a 4 when rolling a six-sided die is 1/6 because the die is equally likely to land on any of the six sides. Is this an example of an empirical probability or a theoretical probability? Explain.

5.4 Empirical versus Theoretical A person was trying to figure out the probability of getting two heads when flipping two coins. He flipped two coins 10 times, and in 2 of these 10 times, both coins landed heads. On the basis of this outcome, he claims that the probability of two heads is 2/10, or 20%. Is this an example of an empirical probability or a theoretical probability? Explain.

5.5 Empirical versus Theoretical A student flips a coin 10 times and sees that it landed on tails 4 times. Based on this, the student says that the probability of getting a tail is 40%. Is the student referring to an empirical probability or a theoretical probability? Explain.

5.6 Empirical versus Theoretical A bag of candy contains 3 red candies and 7 brown candies. A friend says the probability of reaching the bag without looking and pulling out a red candy is 30% because 3 out of 10 candies are red. Is this an example of an empirical probability or a theoretical probability?

SECTION 5.2

5.7 Medical Group A medical practice group consists of seven doctors, four women and three men. The women are Drs. Town, Wu, Hein, and Lee. The men are Drs. Marland, Penner, and Holmes. Suppose new patients are randomly assigned to one of the doctors in the group.

   a. List the equally likely outcomes that could occur when a patient is assigned to one of the doctors.

   b. What is the probability that the new patient is assigned to a female doctor? Write your answer as a fraction and as a percentage rounded to one decimal place.

   c. What is the probability that the new patient will be assigned to a male doctor? Write your answer as a fraction and as a percentage rounded to one decimal place.

   d. Are the events described in parts (b) and (c) complements? Why or why not?

5.8 Teacher Effectiveness A recent study found that highly experienced teachers may be associated with higher student achievement. Suppose fourth-grade students at an elementary school are randomly assigned to one of eight teachers. Teachers Nagle, Crouse, Warren, Tejada, and Tran are considered highly experienced. Teachers Cochran, Perry, and Rivas are considered less experienced. (Source: Papay and Kraft, "Productivity returns to experience in the teacher labor market," *Journal of Public Economics*, vol. 130 [2015]: 105–119.

   a. List the equally likely outcomes that could occur when a student is assigned to a teacher.

   b. What is the probability that a fourth-grade student at this school is assigned to a highly experienced teacher?

   c. What event is the complement of the event described in part (b)? What it the probability of this event?

5.9 Probability For each of the values, state whether the number could be the probability of an event. Give a reason for your answers.

   a. 0.26

   b. −0.26

   c. 2.6

   d. 2.6%

   e. 26

5.10 Probability For each of the values, state whether the number could be the probability of an event. Give a reason for your answers.

   a. 99%

   b. 0.9

   c. 9.9

   d. 0.0099

   e. −0.90

TRY **5.11 Cards (Example 2)** There are four suits: clubs (♣), diamonds (♦), hearts (♥), and spades (♠), and the following cards appear in each suit: ace, 2, 3, 4, 5, 6, 7, 8, 9, 10, jack, queen, king. The jack, queen, and king are called face cards because they have a drawing of a face on them. Diamonds and hearts are red, and clubs and spades are black.

If you draw 1 card randomly from a standard 52-card playing deck, what is the probability that it will be the following:

a. A heart

b. A red card

c. An ace

d. A face card (jack, queen, or king)

e. A three

5.12 Playing Cards Refer to exercise 5.11 for information about cards. If you draw one card randomly from a standard 52-card playing deck, what is the probability that it will be the following:

a. A black card

b. A diamond

c. A face card (jack, queen, or king)

d. A nine

e. A king or queen

5.13 Guessing on Tests

a. On a true/false quiz in which you are guessing, what is the probability of guessing correctly on one question?

b. What is the probability that a guess on one true/false question will be incorrect?

5.14 Guessing on Tests Consider a multiple-choice test with a total of four possible options for each question.

a. What is the probability of guessing correctly on one question? (Assume that there are three incorrect options and one correct option.)

b. What is the probability that a guess on one question will be incorrect?

TRY **5.15 Four Coin Tosses (Example 3)** The sample space given here shows all possible sequences for tossing a fair coin 4 times. The sequences have been organized by the number of tails in the sequence.

0 Tails	1 Tail	2 Tails	3 Tails	4 Tails
HHHH	THHH	TTHH	HTTT	TTTT
	HTHH	THTH	THTT	
	HHTH	THHT	TTHT	
	HHHT	HTTH	TTTH	
		HTHT		
		HHTT		

a. How many outcomes are in the sample space?

b. Assuming all of the outcomes in the sample space are equally likely, find each of the probabilities:

i. all tails in 4 tosses

ii. only 1 tail in 4 tosses

iii. at most 1 tail in 4 tosses

5.16 Three Children The sample space shows all possible sequences of child gender for a family with 3 children. The table is organized by the number of girls in the family.

0 Girls	1 Girl	2 Girls	3 Girls
BBB	GBB	BGG	GGG
	BGB	GBG	
	BBG	GGB	

a. How many outcomes are in the sample space?

b. If we assume all outcomes in the sample space are equally likely, find the probability of having the following numbers of girls in a family of 3 children:

i. all 3 girls

ii. no girls

iii. exactly 2 girls

5.17 Birthdays What is the probability that a baby will be born on a Friday OR a Saturday OR a Sunday if all the days of the week are equally likely as birthdays?

5.18 Playing Cards If *one* card is selected from a well-shuffled deck of 52 cards, what is the probability that the card will be a club OR a diamond OR a heart? What is the probability of the complement of this event? (Refer to exercise 5.11 for information about cards.)

TRY **5.19 Vacations (Example 4)** The Gallup poll asked respondents if they had taken a vacation in the last year. The respondents were separated into two groups: those who had graduated from college and those who had not. Numbers in the table are based on sample sizes of 250 in each group. (Source: Gallup.com)

	Took a Vacation		
	Yes	No	Total
College Graduate	200	50	250
Not a College Graduate	135	115	250
Total	335	165	500

a. If a person is randomly selected from this group, find the probability for the following:

i. the person is a college graduate

ii. the person took a vacation

iii. the person is a college graduate and took a vacation

b. Which group is more likely to have taken a vacation in the last year: college graduates or non–college graduates? Support your answer with appropriate statistics.

5.20 Reading The Pew Research Center asked a sample of adults if they had read a book in any format in the previous 12 months. The results are shown in the table. (Source: Pewinternet.org)

	Read a Book in the Previous 12 Months		
	Yes	No	Total
Men	534	251	785
Women	566	169	735
Total	1100	420	1520

a. If a person is randomly selected from this group, find the probability of the following:

i. the person is male

ii. the person has read a book in the previous 12 months

iii. the person is a male and has read a book in the previous 12 months

b. Which group is more likely to have read a book in the previous 12 months: men or women? Support your answer with appropriate statistics.

TRY **5.21 Marijuana Legalization (Example 5)** A Gallup poll asked a sample of voters if marijuana should be legalized. Voters' responses and political party affiliation are in the table. (Source: Gallup.com)

Political Affiliation	Favor Legalization of Marijuana		
	Yes	**No**	**Total**
Republican	142	136	278
Democrat	222	86	308
Independent	296	146	442
Total	660	368	1028

a. If a person is randomly selected from this group, find the probability that the person is a Republican.

b. If a person is randomly selected from this group, find the probability that the person is a Democrat.

c. If a person is randomly selected from this group, find the probability that the person is a Republican or a Democrat.

d. Are "Republican" and "Democrat" mutually exclusive events? Give a reason for your answer.

e. Are "Republican and "Democrat" complementary events in this data set? Give a reason for your answer.

5.22 Marijuana Legalization Use the data in exercise 5.21 to answer the following:

a. If a person is chosen randomly from this group, what is the probability that the person is an Independent and said "Yes"?

b. If a person is chosen randomly from this group, what is the probability that the person a Republican who said "No"?

c. Are the events "Yes" and "Democrat" mutually exclusive? Give a reason for your answer.

d. Are the events "Yes" and "No" mutually exclusive? Give a reason for your answer.

e. Are the events "Yes" and "No" complementary events for this data set? Give a reason for your answer.

5.23 Gender Discrimination in Tech (Example 6) A Pew Research poll asked respondents to fill in the blank to this question: Compared to other industries there is _____ discrimination against women in the tech industry. Responses separated by gender are shown in the following table. The results are shown using a sample size of 280 men and 150 women. (Source: Pewresearch.org) *See page 263 for guidance.*

Response	Gender		
	Men	**Women**	**Total**
Less	53	15	68
About the same	174	85	259
More	53	50	103
Total	280	150	430

a. If a person is selected at random from this group, find the probability that the person is a man or said "about the same" (or both)?

b. If a person is selected at random from this group, find the probability the person is a man and said "about the same."

5.24 Gender Discrimination in Tech Use the data in exercise 5.23 to answer the following:

a. What is the probability that a randomly selected person is a woman and said "more."

b. What is the probability that a randomly selected person is a woman or said "more" (or both)?

5.25 Mutually Exclusive Suppose a person is selected at random from a large population.

a. Label each pair of events as mutually exclusive or not mutually exclusive.

  i. The person has traveled to Mexico; the person has traveled to Canada.

  ii. The person is single; the person is married.

b. Give an example of two events that are mutually exclusive when a person is selected at random from a large population.

5.26 Mutually Exclusive Suppose a student is selected at random from a large college population.

a. Label each pair of events as mutually exclusive or not mutually exclusive.

  i. The students is a Chemistry major; the student works on campus.

  ii. The student is a full-time student; the student is only taking one 3-unit class.

b. Give an example of two events that are not mutually exclusive when a student is selected at random from a large college population.

TRY **5.27 Vacations (Mutually Exclusive) (Example 7)** Referring to the table given in exercise 5.19, name a pair of mutually exclusive events that could result when one person is selected at random from the entire group.

5.28 Vacations (Not Mutually Exclusive) Refer to the table in exercise 5.19. Suppose we select one person at random from this group. Name a pair of events that are *not* mutually exclusive.

TRY **5.29 Fair Die (Example 8)** Roll a fair six-sided die.

a. What is the probability that the die shows an even number or a number greater than 4 on top?

b. What is the probability the die shows an odd number or a number less than 3 on top?

5.30 Fair Die Roll a fair six-sided die.

a. What is the probability that the die shows an even number or a number less than 4 on top?

b. What is the probability the die shows an odd number or a number greater than 4 on top?

5.31 Grades Assume that the only grades possible in a history course are A, B, C, and lower than C. The probability that a randomly selected student will get an A in a certain history course is 0.18, the probability that a student will get a B in the course is 0.25, and the probability that a student will get a C in the course is 0.37.

a. What is the probability that a student will get an A OR a B?

b. What is the probability that a student will get an A OR a B OR a C?

c. What is the probability that a student will get a grade lower than a C?

5.32 Changing Multiple-Choice Answers One of the authors did a survey to determine the effect of students changing answers while taking a multiple-choice test on which there is only one correct answer for each question. Some students erase their initial choice and replace it with another. It turned out that 61% of the changes were from incorrect answers to correct and that 26% were from correct to incorrect. What percentage of changes were from incorrect to incorrect?

5.33 Commuting A college conducted a student survey to learn about commute patterns. Students were given a choice of three options: car, bus, or other. When looking at the survey results, 42% of students responded "car," and 23% responded "bus." Assuming all students answered this survey questions, what percentage of the students responded "other"?

5.34 Political Parties Political science researchers often classify voters according to their political party preference, using four categories: Democrat, Republican, Other political parties (including Libertarians and Independents, for example), and Decline to State/No Party Preference. The political party breakdown in California is 45% Democrat, 26% Republican, and 6% Other political parties. What percentage of voters are Decline to state/no party preference? (Source: Public Policy Institute of California)

TRY **5.35 Super Powers (Example 9)** A 2018 Marist poll asked respondents what superpower they most desired. The distribution of responses are shown in the table.

Superpower	Percentage
Travel in time	29%
Read minds	20%
Ability to fly	17%
Teleport	15%
Invisibility	12%
None of these	5%
Unsure	3%

a. What percentage of those surveyed wanted to be able to fly or teleport?

b. If there were 1200 people surveyed, how many wanted to be able to read minds or travel in time?

5.36 Online Presence A 2018 Pew poll asked U.S. adults how often they go online. The responses are shown in the table.

Almost constantly	26%
Several times a day	43%
About once a day	8%
Several times a week	6%
Less often	5%

a. What percentage of respondents go online less than once a day?

b. In a group of 500 U.S. adults, how many would you expect go online almost constantly or several times a day?

* **5.37 Thumbtacks** When a certain type of thumbtack is tossed, the probability that it lands tip up is 60%. All possible outcomes when two thumbtacks are tossed are listed. U means the tip is up, and D means the tip is down.

UU   UD   DU   DD

a. What is the probability of getting two Ups?

b. What is the probability of getting exactly one Up?

c. What is the probability of getting at least one Up (one or more Ups)?

d. What is the probability of getting at most one Up (one or fewer Ups)?

* **5.38 Thumbtacks** When a certain type of thumbtack is tossed, the probability that it lands tip up is 60%, and the probability that it lands tip down is 40%. All possible outcomes when two thumbtacks are tossed are listed. U means the tip is Up, and D means the tip is Down.

UU   UD   DU   DD

a. What is the probability of getting exactly one Down?

b. What is the probability of getting two Downs?

c. What is the probability of getting at least one Down (one or more Downs)?

d. What is the probability of getting at most one Down (one or fewer Downs)?

* **5.39 Multiple-Choice Exam** An exam consists of 12 multiple-choice questions. Each of the 12 answers is either right or wrong. Suppose the probability a student makes fewer than 3 mistakes on the exam is 0.48 and the probability that a student makes from 3 to 8 (inclusive) mistakes is 0.30. Find the probability that a student makes the following:

a. More than 8 mistakes

b. 3 or more mistakes

c. At most 8 mistakes

d. Which two of these three events are complementary, and why?

* **5.40 Driving Exam** A driving exam consists of 30 multiple-choice questions. Each of the answers is either right or wrong. Suppose that the probability of making fewer than 7 mistakes is 0.23 and the probability of making from 7 to 15 mistakes is 0.41. Find the probability of making the following:

a. 16 or more mistakes

b. 7 or more mistakes

c. At most 15 mistakes

d. Which two of these three events are complementary? Explain.

SECTION 5.3

TRY **5.41 Equal Rights for Women (Example 10)** A recent Pew Research poll asked respondents to fill in the blank to this question: "The country ____ when it comes to giving equal rights to women" with one of three choices. The results are shown in the following table using a sample size of 100 men and 100 women. (Source: pewresearch.org)

	Hasn't Gone Far Enough	Has Been about Right	Has Gone Too Far	Total
Men	42	44	14	100
Women	57	34	9	100
Total	99	78	23	200

a. A person is selected randomly from only the women in the group. We want to find the probability that a female responded, "Has been about right." Which of the following statements best describes the question?

  i. P(Has been about right | female)

  ii. P(female | has been about right)

  iii. P(female and responded "has been about right")

b. Find the probability that a person randomly selected from only the women in this group responded "has been about right."

5.42 Equal Rights for Women A person is selected randomly from the entire group whose responses are summarized in the table for exercise 5.41. We want to find the probability that the person selected is a male who said "hasn't gone far enough."

a. Which of the following statements best described the question?

  i. P(hasn't gone far enough | male)

  ii. P(male | hasn't gone far enough)

  iii. P(male and responded "hasn't gone far enough")

b. Find the probability that a person randomly selected from the entire group is a male who responded "hasn't gone far enough."

5.43 Frequent Stress (Example 11) A Gallup poll asked people with and without children under 18 years old if they frequently experienced stress. The results are shown in the table below. (Source: Gallup.com)

	Frequently Experience Stress		
	Yes	**No**	**Total**
With child under 18 years	319	231	550
No child under 18 years	195	305	500
Total	514	536	1050

a. Find the probability that a randomly chosen person from this group said "Yes," given that the person has a child under 18 years old.

b. Find the probability that a randomly chosen person from this group said "Yes," given that the person does not have a child under 18 years old.

c. Find the probability that a randomly chosen person said "Yes" or had a child under 18 years old.

5.44 Frequent Stress Use the data in exercise 5.43 for this problem. Note: Your answers to each of these three questions should not be the same.

a. Find the probability that a randomly selected person did not have a child under 18 years old, given that they said No.

b. Find the probability that a randomly selected person said No, given that the person did not have a child under 18 years old.

c. Find the probability that a randomly selected person from the group did not have a child under 18 years old and said No.

5.45 Independent? Suppose a person is chosen at random. Use your understanding about the world of basketball to decide whether the event that the person is taller than 6 feet and the event that the person plays professional basketball are independent or associated. Explain.

5.46 Independent? About 12% of men and 10% of women are left-handed. If we select a person at random, are the event that the person is male and the event that the person is left-handed independent or associated?

5.47 Independent? Suppose a person is chosen at random. Use your knowledge about the world to decide whether the event that the person has brown eyes and the event that the person is female are independent or associated. Explain.

5.48 Independent? Ring sizes typically range from about 3 to about 14. Based on what you know about gender differences, if we randomly select a person, are the event that the ring size is smaller than 5 and the event that the person is a male independent or associated? Explain.

TRY 5.49 Equal Rights Poll (Example 12) Refer to the table in Exercise 5.41. Suppose a person is randomly selected from this group. Is being female independent of answering "Hasn't Gone Far Enough"?

5.50 Equal Rights Poll Assume a person is selected randomly from the group of people represented in the table in exercise 5.41. The probability the person says "Hasn't Gone Far Enough" given that the person is a woman is 57/100 or 57%. The probability that

person is a woman given that the person says "Hasn't Gone Far Enough" is 57/99 or 57.6%. The probability the person says "Hasn't Gone Far Enough" and is a woman is 57/200 or 28.5%. Why is the last probability the smallest?

TRY 5.51 Hand Folding (Example 13) When people fold their hands together with interlocking fingers, most people are more comfortable with one of two ways. In one way, the right thumb ends up on top, and in the other way, the left thumb is on top. The table shows the data from one group of people. M means man, and W means woman; Right means the right thumb is on top, and Left means the left thumb is on top. Judging on the basis of this data set, are the events "right thumb on top" and male independent or associated? Data were collected in a class taught by one of the authors but were simplified for clarity. The conclusion remains the same as that derived from the original data. *See page 263 for guidance.*

	M	**W**
Right	18	42
Left	12	28

5.52 Dice When two dice are rolled, is the event "the first die shows a 1 on top" independent of the event "the second die shows a 1 on top"?

TRY 5.53 TV News Source and Gender (Example 14) A 2018 Pew Research Center report asked people who got their news from television which television sector they relied on primarily for their news: local TV, network TV, or cable TV. The results were used to generate the data in the table below.

	Local TV	**Network TV**	**Cable TV**	**Total**
Men	66	48	58	
Women	82	54	56	
Total				

a. Include the row totals, column totals, and the grand total in the table. Show the complete table with the totals.

b. Determine whether, for this sample, choice of cable TV is independent of being male. Explain your answer in the context of this problem.

5.54 TV News Source and Gender Using the table in exercise 5.53, determine whether being female is independent of choice of local TV. Explain your answer in the context of this problem.

TRY 5.55 Coin (Example 15) Imagine flipping three fair coins.

a. What is the theoretical probability that all three come up heads?

b. What is the theoretical probability that the first toss is tails AND the next two are heads?

5.56 Die Imagine rolling a fair six-sided die three times.

a. What is the theoretical probability that all three rolls of the die show a 1 on top?

b. What is the theoretical probability that the first roll of the die shows a 6 AND the next two rolls both show a 1 on the top.

TRY 5.57 Die Sequences (Example 16) Roll a fair six-sided die five times, and record the number of spots on top. Which sequence is more likely? Explain.

Sequence A: 66666

Sequence B: 16643

5.58 Babies Assume that babies born are equally likely to be boys (B) or girls (G). Assume a woman has six children, none of whom are twins. Which sequence is more likely? Explain.

Sequence A: GGGGGG

Sequence B: GGGBBB

TRY **5.59 Vacations (Example 17)** According to a recent Gallup poll, 62% of Americans took a vacation away from home in 2017. Suppose two Americans are randomly selected.

a. What is the probability that both took a vacation away from home in 2017?

b. What is the probability that neither took a vacation away from home in 2017?

c. What is the probability that at least one of them took a vacation away from home in 2017?

5.60 Landlines and Cell Phones According to the National Center for Health Statistics, 52% of U.S. households no longer have a landline and instead only have cell phone service. Suppose three U.S. households are selected at random.

a. What is the probability that all three have only cell phone service?

b. What is the probability that at least one has only cell phone service?

TRY * **5.61 Cervical Cancer (Example 18)** According to a study published in *Scientific American*, about 8 women in 100,000 have cervical cancer (which we'll call event C), so P(C) = 0.00008. Suppose the chance that a Pap smear will detect cervical cancer when it is present is 0.84. Therefore,

$$P(\text{test pos} \mid C) = 0.84$$

What is the probability that a randomly chosen woman who has this test will both have cervical cancer AND test positive for it?

* **5.62 Cervical Cancer** About 8 women in 100,000 have cervical cancer (C), so P(C) = 0.00008 and P(no C) = 0.99992. The chance that a Pap smear will incorrectly indicate that a woman without cervical cancer has cervical cancer is 0.03. Therefore,

$$P(\text{test pos} \mid \text{no C}) = 0.03$$

What is the probability that a randomly chosen women who has this test will both be free of cervical cancer and test positive for cervical cancer (a false positive)?

SECTION 5.4

TRY **5.63 Rolling Sixes (Example 19)**

What's the probability of getting at least one six when you roll two dice? The table below shows the outcome of five trials in which two dice were rolled.

a. List the trials that had at least one 6.

b. Based on these data, what's the empirical probability of rolling at least one 6 with two dice?

Trial	Outcome
1	2, 5
2	3, 6
3	6, 1
4	4, 6
5	4, 3

5.64 Rolling a Sum of 7 What's the probability of rolling two numbers whose sum is 7 when you roll two dice? The table below shows the outcome of ten trials in which two dice were rolled.

a. List the trials that had a sum of 7.

b. Based on these data, what's the empirical probability of rolling two numbers whose sum is 7?

Trial	Outcome
1	3, 1
2	1, 2
3	6, 5
4	6, 4
5	5, 2
6	6, 6
7	3, 2
8	2, 1
9	4, 6
10	1, 6

TRY **5.65 Multiple-Choice Test (Example 20)** A multiple-choice test has 30 questions. Each question has three choices, but only one choice is correct. Using a random number table, which of the following methods is a valid simulation of a student who circles his or her choices randomly? Explain. (Note: there might be more than one valid method.)

a. The digits 1, 2, and 3 represent the student's attempt on one question. All other digits are ignored. The 1 represents the correct choice, the 2 and 3 represent incorrect choices.

b. The digits 0, 1, 4 represent the student's attempt on one question. All other digits are ignored. The 0 represents the correct choice, the 1 and 4 represent incorrect choices.

c. Each of the 10 digits represents the student's attempt on one question. The digits 1, 2, 3 represent a correct choice; 4, 5, 6, 7, 8, 9 and 0 represent an incorrect choice.

5.66 True/False Test A true/false test has 20 questions. Each question has two choices (true or false), and only one choice is correct. Which of the following methods is a valid simulation of a student who guesses randomly on each question. Explain. (Note: there might be more than one valid method.)

a. Twenty digits are selected using a row from a random number table. Each digit represents one question on the test. If the number is even the answer is correct. If the number is odd, the answer is incorrect.

b. A die is rolled 20 times. Each roll represents one question on the test. If the die lands on a 6, the answer is correct; otherwise the answer is incorrect.

c. A die is rolled 20 times. Each roll represents one question on the test. If the die lands on an odd number, the answer is correct. If the die lands on an even number, the answer is incorrect.

TRY **5.67 Simulating Coin Flips (Example 21)**

a. Use the line of random numbers below to simulate flipping a coin 20 times. Use the digits 0, 1, 2, 3, 4 to represent heads and the digits 5, 6, 7, 8, 9 to represent tails.

11164   36318   75061   37674

b. Based on these 20 trials, what is the simulated probability of getting heads? How does this compare with the theoretical probability of getting heads?

c. Suppose you repeated your simulation 1000 times and used the simulation to find the simulated probability of getting heads. How would the simulated probability compare with the theoretical probability of getting heads?

5.68 Simulating Rolling a Die

a. Explain how you could use digits from a random number table to simulate rolling a fair six-sided die.

b. Carry out your simulation beginning with line 3 of the random number table in Appendix A. Repeat your simulation 5 times.

c. Use your simulation to find the empirical probability of rolling a 6. How does the empirical probability compare with the theoretical probability?

d. Suppose you repeated your simulation 500 times. How would the empirical probability of rolling a 6 based on your 500 simulations compare with the theoretical probability of rolling a 6?

5.69 Law of Large Numbers Refer to Histograms A, B, and C, which show the relative frequencies from experiments in which a fair six-sided die was rolled. One histogram shows the results for 20 rolls, one the results for 100 rolls, and another the results for 10,000 rolls. Which histogram do you think was for 10,000 rolls, and why?

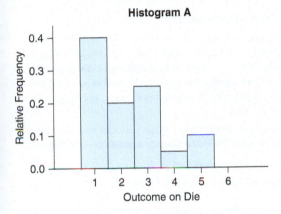

Histogram A

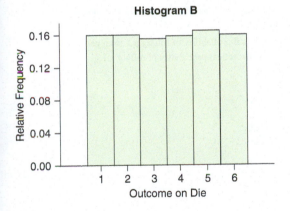

Histogram B

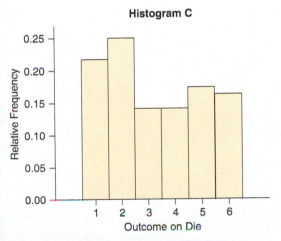

Histogram C

5.70 Law of Large Numbers The table shows the results of rolling a fair six-sided die.

Outcome on Die	20 Trials	100 Trials	1000 Trials
1	8	20	167
2	4	23	167
3	5	13	161
4	1	13	166
5	2	16	172
6	0	15	167

Using the table, find the empirical probability of rolling a 1 for 20, 100, and 1000 trials. Report the theoretical probability of rolling a 1 with a fair six-sided die. Compare the empirical probabilities to the theoretical probability, and explain what they show.

5.71 Coin Flips Imagine flipping a fair coin many times. Explain what should happen to the proportion of heads as the number of coin flips increases.

5.72 Coin Flips, Again Refer to the following figure.

a. After a large number of flips, the overall proportion of heads "settles down" to nearly what value?

b. Approximately how many coin flips does it take before the proportion of heads settles down?

c. What do we call the law that causes this settling down of the proportion?

d. From the graph, determine whether the first flip was heads or tails.

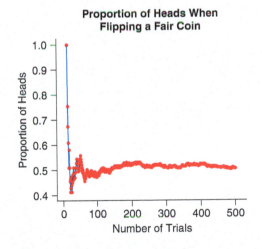

Proportion of Heads When Flipping a Fair Coin

5.73 Law of Large Numbers: Gambling Betty and Jane are gambling. They are cutting cards (picking a random place in the deck to see a card). Whoever has the higher card wins the bet. If the cards have the same value (for example, they are both eights), they try again. Betty and Jane do this a 100 times. Tom and Bill are doing the same thing but are betting only 10 times. Is Bill or Betty more likely to end having very close to 50% wins? Explain. You may refer to the graph to help you decide. It is one simulation based on 100 trials.

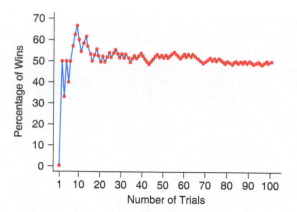

* **5.74 LLN: Grandchildren** Consider two pairs of grandparents. The first pair has 4 grandchildren, and the second pair has 32 grandchildren. Which of the two pairs is more likely to have between 40% and 60% boys as grandchildren, assuming that boys and girls are equally likely as children? Why?

5.75 LLN: Coin If you flip a fair coin repeatedly and the first four results are tails, are you more likely to get heads on the next flip, more likely to get tails again, or equally likely to get heads or tails?

5.76 LLN: Die The graph shows the average when a six-sided die is rolled repeatedly. For example, if the first two rolls resulted in a 6 and a 2, the average would be 4. If the next trial resulted in a 1, the new average would be $(6 + 2 + 1)/3 = 3$. Explain how the graph demonstrates the Law of Large Numbers.

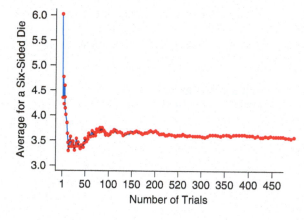

5.77 Jury Duty A jury is supposed to represent the population. We wish to perform a simulation to determine an empirical probability that a jury of 12 people has 5 or fewer women. Assume that about 50% of the population is female, so the probability that a person who is chosen for the jury is a woman is 50%. Using a random number table, we decide that each digit will represent a juror. The digits 0 through 5, we decide, will represent a female chosen, and 6 through 9 will represent a male. Why this is a bad choice for this simulation?

5.78 Left-handed Some estimates say that 10% of the population is left-handed. We wish to design a simulation to find an empirical probability that if five babies are born on a single day, one or more will be left-handed. Suppose we decide that the even digits (0, 2, 4, 6, 8) will represent left-handed babies and the odd digits will represent right-handed babies. Explain what is wrong with the stated simulation method, and provide a correct method.

* **5.79 Simulation: Four-Sided Die**

a. Explain how you could use a random number table (or the random numbers generated by software or a calculator) to simulate rolling a fair *four-sided* die 20 times. Assume you are interested in the probability of rolling a 1. Then report a line or two of the random number table (or numbers generated by a computer or calculator) and the values that were obtained from it.

b. Report the empirical probability of rolling a 1 on the four-sided die from part (a), and compare it with the theoretical probability of rolling a 1.

* **5.80 Simulation: Six-Sided Die**

a. Explain how you could use a random number table to simulate rolling a fair six-sided die 20 times. Assume you wish to find the probability of rolling a 1. Then report a line or two of the random number table (or numbers generated by a computer or calculator) and the values that were obtained from it.

b. Report the empirical probability of rolling a 1 from part (a), and compare it with the theoretical probability of rolling a 1.

CHAPTER REVIEW EXERCISES

5.81 Gun Laws According to a Gallup poll conducted in 2016, 627 out of 1012 Americans surveyed said they were dissatisfied with the country's current gun laws. If an American is selected at random, what is the probability that the person is dissatisfied with the current gun laws?

5.82 Global Warming According to a Gallup poll conducted in 2017, 723 out of 1018 Americans surveyed said they believed global warming is occurring. If an American is selected at random, what is the probability that the person believes that global warming is occurring?

5.83 Independent Variables Use your general knowledge to label the following pairs of variables as independent or associated. Explain.

a. For a sample of adults, gender and shoe size
b. For a sample of football teams, win/loss record for the coin toss at the beginning of the game and number of cheerleaders for the team

5.84 Independent Variables Use your general knowledge to label the following pairs of variables as independent or associated. Explain.

a. The outcome on flips of two separate, fair coins
b. Breed of dog and weight of dog for dogs at a dog show

5.85 Death Penalty According to a Pew Research poll conducted in 2016, 55% of men and 43% of women support the death penalty in cases of murder. Suppose these are accurate percentages. Now suppose a random man and a random woman meet.

a. What is the probability that both support the death penalty?
b. What is the probability that neither support the death penalty?
c. What is the probability that only one of them supports the death penalty?
d. What is the probability that at least one of them supports the death penalty?

5.86 Women's Rights A 2017 Pew Research poll asked people if they agreed with this statement: The United States hasn't gone far enough when it comes to giving women equal rights with men. 42% of men agreed with the statement and 57% of women agreed with the statement. Suppose these are accurate percentages. Now suppose a random man and woman meet.

a. What is the probability that they both agree with the statement?
b. What is the probability that neither believes with the statement?
c. What is the probability that at least one of them agrees with the statement?
d. What is the probability that only one of them agrees with the statement?

5.87 Online Dating A 2016 Pew Research poll reported that 27% of young adults aged 18 to 24 had used an online dating site. Assume the percentage is accurate.

a. If two young adults are randomly selected, what is the probability that both have used an online dating site?
b. If the two young adults chosen were Facebook friends, explain why this would not be considered independent with regard to online dating.

5.88 Online Shopping A 2016 Pew Research poll reported that 80% of Americans shop online. Assume the percentage is accurate.

a. If two Americans are randomly selected, what is the probability that both shop online?
b. If the two Americans selected are a married couple, explain why they would not be considered independent with regard to online shopping.

* **5.89 Birthdays** Suppose all the days of the week are equally likely as birthdays. Alicia and David are two randomly selected, unrelated people.

a. What is the probability that they were both born on Monday?
b. What is the probability that Alicia OR David was born on Monday?

* **5.90 Pass Rate of Written Driver's Exam** In California, about 92% of teens who take the written driver's exam fail the first time they take it (www.teendrivingcourse.com). Suppose that Sam and Maria are randomly selected teenagers taking the test for the first time.

a. What is the probability that they both pass the test?
b. What is the probability that Sam OR Maria passes the test?

5.91 Streaming Television In 2017 the Pew Research Center asked young adults aged 18 to 29 about their media habits. When asked, "What is the primary way you watch television?" 61% said online streaming service, 31% said cable/satellite subscription, and 5% said digital antenna. Suppose the Pew Research Center polled another sample of 2500 young adults from this age group and the percentages were the same as those in 2017.

a. How many would say online streaming services?
b. How many would say cable/satellite subscription?
c. How many would say cable/satellite subscription or digital antenna?
d. Are the responses "online streaming service," "cable/satellite subscription," and "digital antenna" mutually exclusive? Why or why not?

5.92 Reading Habits In 2016 the Pew Research Center asked a sample of American adults which of the following they had done in the previous 12 months: read a book in any format, read a print book, read an e-book, and listened to an audio book. The percentages of who had engaged in each of these activities is shown in the table.

% of U.S. Adults Who Say They Have _____ in the Previous 12 Months	
Read a book in any format	73%
Read a print book	65%
Read an e-book	28%
Listened to an audio book	14%

Suppose Pew Research Center did another survey polling 1200 people and the percentages were the same as those in 2016.

a. How many would have read a book in any format?
b. How many would have read an e-book?
c. Are the events listed in the table mutually exclusive? Why or why not?

5.93 Baseball A 2018 Marist poll found that interest in baseball has been declining recently in the United States. A random sample of U.S. adults were asked how much baseball they intended to watch this season. The results are shown in the following table.

Watch a great deal	7%
Watch a good amount	8%
Watch a little	29%
Not at all	56%

a. What percentage of those surveyed intend to watch at least some baseball this season?

b. In a group of 400 people, how many would you expect would not watch any baseball this season?

5.94 Reading Habits A 2016 Pew Research poll asked respondents about their reading habits during the past year. The results are shown in the table.

Read no books	26%
Read both print and digital books	28%
Read digital only	6%
Read print only	38%

a. What percentage of those surveyed had read at least one digital book in the past year?

b. In a group of 150 people, how many would you expect had not read a book in the past year?

5.95 Opinion about College A 2017 Pew Research poll found that 72% of Democrats and 36% of Republicans felt that colleges and universities have a positive effect on the way things are going in the United States. If 1500 Democrats and 1500 Republicans were surveyed, how many from each group felt that colleges and universities have a positive effect on the country?

5.96 Cell Phone Security A 2017 Pew Research poll found that 28% of cell phone users do not use a screen lock for security on their smartphones. If 500 smartphone users were surveyed, how many do not use a screen lock on their smartphones?

5.97 Coin Flips Let H stand for heads and let T stand for tails in an experiment where a fair coin is flipped twice. Assume that the four outcomes listed are equally likely outcomes:

HH, HT, TH, TT

What are the probabilities of getting the following:

a. 0 heads

b. Exactly 1 head

c. Exactly 2 heads

d. At least one 1 head

e. Not more than 2 heads

5.98 Cubes A bag contains a number of colored cubes: 10 red, 5 white, 20 blue, and 15 black. One cube is chosen at random. What is the probability that the cube is the following:

a. black

b. red or white

c. not blue

d. neither red nor white

e. Are the events described in parts (b) and (d) complements? Why or why not?

5.99 Marijuana Legalization A Gallup poll conducted in 2017 asked people, "Do you think marijuana use should be legal?"

In response, 75% of Democrats, 51% of Republicans, and 67% of Independents said Yes. Assume that anyone who did not answer Yes answered No. Suppose the number of Democrats polled was 400, the number of Republicans polled was 300, and the number of Independents polled was 200.

	Democrats	Republicans	Independents	Total
Yes	300			
No				
Total				

a. Complete the two-way table with counts (not percentages). The first entry is done for you. Suppose a person is randomly selected from this group.

b. What is the probability that the person is a Democrat who said Yes?

c. What is the probability that the person is a Republican who said No?

d. What is the probability that the person said No, given that the person is a Republican?

e. What is the probability that the person is a Republican given that the person said No?

f. What is the probability that the person is a Democrat or a Republican?

5.100 Online Dating The Pew Research Center asked a sample of Americans, "Do you know someone who has entered a long-term relationship via online dating?" The table gives the total number of people by level of educational attainment (rounded) and the percentage who said Yes.

	Total Count	% Yes
High school grad or less	600	18%
Some college	500	30%
College grad	800	46%

a. Make a two-way table of counts (not percentages). The table is started below.

	Yes	No	Total
High school grad or less	108		
Some college			
College grad			
Total			

Suppose a person is randomly selected from this group.

b. What is the probability the person knows someone who has entered a long-term relationship via online dating?

c. If you select from the people who said Yes, what is the probability that the person has some college or is a college grad?

d. What is the probability the person said Yes and is a college grad?

e. As educational attainment increases, what is the trend in the probability that the person knows someone who has entered a long-term relationship via online dating?

5.101 CA Bar Exam In order to practice law, lawyers must pass the bar exam. In California, the passing rate for first-time bar exam test takers who attended an accredited California law school was 70%. Suppose two test-takers from this group are selected at random.

a. What is the probability that they both pass the bar exam?

b. What is the probability that only one passes the bar exam?

c. What is the probability that neither passes the bar exam?

5.102 Driving Tests In addition to behind-the-wheel tests, states require written tests before issuing drivers licenses. The failure rate for the written driving test in Florida is about 60%. (Source: tampabay.com) Suppose three drivers' license test-takers in Florida are randomly selected. Find the probability of the following:

a. all three fail the test

b. none fail the test

c. only one fails the test

5.103 California Recidivism and Gender Women return to prison at a lower rate than men do (58.0% for women, compared to 68.6% for men) in California. For a randomly chosen prisoner, are the event that the person returns to prison and the event that the person is male independent?

5.104 Blue Eyes About 17% of American men have blue eyes and 17% of American women have blue eyes. If we randomly select an American, are the event that the person has blue eyes and the event that the person is male independent?

5.105 Construct a two-way table with 80 men and 100 women in which both groups show an equal percentage of right-handedness.

5.106 Construct a two-way table with 80 men and 100 women in which the percentage of left-handedness is higher for women than for men.

* **5.107 Law of Large Numbers** A famous study by Amos Tversky and Nobel laureate Daniel Kahneman asked people to consider two hospitals. Hospital A is small and has 15 babies born per day. Hospital B has 45 babies born each day. Over one year, each hospital recorded the number of days that it had more than 60% girls born. Assuming that 50% of all babies are girls, which hospital had the most such days? Or do you think both will have about the same number of days with more than 60% girls born? Answer, and explain. (Source: Tversky, *Preference, belief, and similarity: Selected Writings*, ed. [Cambridge, MA: MIT Press], 205)

* **5.108 Law of Large Numbers** A certain professional basketball player typically makes 80% of his basket attempts, which is considered to be good. Suppose you go to several games at which this player plays. Sometimes the player attempts only a few baskets, say, 10. Other times, he attempts about 60. On which of those nights is the player most likely to have a "bad" night, in which he makes much fewer than 80% of his baskets?

* **5.109 Climate Change and Political Party** In 2016 a Pew Research poll asked a sample of Americans if they agreed with this statement: The Earth is warming mostly due to human activity. Responses and the political affiliation of respondents were recorded and are shown in the following table.

	Conservative Republican	Moderate/ Liberal Republican	Moderate/ Conservative Democrat	Liberal Democrat	Total
Yes	65	77	257	332	
No	368	149	151	88	
Total					

a. Fill in the row and column totals for the table.

b. If one person is chosen randomly from the group, what is the probability that the person is a Conservative Republican?

c. If one person is chosen randomly from the group, what is the probability that the person agrees with the statement?

5.110 Climate Change and Political Party Refer to the table in exercise 5.109.

a. If one person is chosen randomly from the group, what is the probability that the person is a Republican (conservative or moderate/ liberal)?

b. If one person is chosen randomly from the group, what is the probability that the person disagrees with the statement?

c. Do a greater percentage of Republicans or Democrats agree with the statement?

5.111 Climate Change: AND Refer to the table in exercise 5.109. Suppose one person is selected at random from this group. Find the probability the person is a Conservative Republican and agrees with the statement.

5.112 Climate Change: AND Refer to the table in exercise 5.109. Suppose one person is selected at random from this group. Find the probability the person agrees with the statement and is a liberal Democrat.

5.113 Climate Change: OR Refer to the table in exercise 5.109. Suppose one person is selected at random from this group. What is the probability the person is a liberal or moderate/conservative Democrat. Are these events mutually exclusive? Why or why not?

5.114 Climate Change: OR Refer to the table in exercise 5.109. Suppose one person is selected at random from this group. What is the probability the person is a conservative or a moderate/ liberal Republican? Are these events mutually exclusive? Why or why not?

5.115 Climate Change: OR Refer to the table in exercise 5.109. Suppose one person is selected at random from this group. Find the probability the person is a Liberal Democrat or said No. Are these events mutually exclusive? Why or why not?

5.116 Climate Change: OR Refer to the table in exercise 5.109. Suppose one person is selected at random from this group. Find the probability the person said yes or is a moderate/conservative Democrat. Are these events mutually exclusive? Why or why not?

5.117 Climate Change: Conditional Probability Refer to the table in exercise 5.109. Suppose one person is selected at random from this group.

a. Find the probability that the person agreed with the statement, given that the person is a liberal Democrat.

b. Find the probability that the person agreed with the statement, given that the person is a conservative Republican.

c. Find the probability that the person is a conservative Republican, given that the person agreed with the statement.

5.118 Climate Change: Conditional Probability Refer to the table in exercise 5.109. Suppose one person is selected at random from this group.

a. Find P(no | conservative Republican)

b. Find P(no | Republican)

c. Find P(liberal Democrat | yes)

5.119 Mutually Exclusive Suppose a person is randomly selected. Label each pair of events as *mutually exclusive* or *not mutually exclusive*.

a. The person is taller than 70 inches; the person is male

b. The person does not own a pet; the person owns a guinea pig

5.120 Mutually Exclusive Suppose a person is randomly selected. Label each pair of events as *mutually exclusive* or *not mutually exclusive*.

a. The person is 40 years old; the person is not old enough to drink alcohol legally

b. The person plays tennis; the person plays the cello.

5.121 Rolling Three of a Kind In the game Yahtzee, five dice are rolled at the same time. Bonus points are awarded if at least three of the dice land on the same number. The table below shows the outcomes for five trials.

a. List the outcomes that had at least three of the dice land on the same number.

b. Based on these data, what's the empirical probability of rolling at least three of the same number?

Trial	Outcome
1	3, 5, 1, 3, 3
2	5, 1, 6, 6, 4
3	6, 6, 4, 4, 6
4	4, 5, 4, 5, 3
5	4, 2, 2, 1, 1

5.122 Multiple-Choice Test A multiple-choice test has 10 questions. Each question has four choices, but only one choice is correct. Which of the following methods is a valid simulation of a student who guesses randomly on each question. Explain. (Note: there might be more than one valid method.)

a. Ten digits are selected using a random number table. Each digit represents one question on the test. If the digit is even, the answer is correct. If the digit is odd, the answer is incorrect.

b. The digits 1, 2, 3, 4 represent the students attempt on one question. All other digits are ignored. The 1 represents a correct choice. The digits 2, 3, and 4 represent an incorrect choice.

c. The digits 1, 2, 3, 4, 5, 6, 7, 8 represent the student's attempt on one question. The digits 0 and 9 are ignored. The digits 1 and 2 represent a correct choice and the digits 3, 4, 5, 6, 7, 8 represent an incorrect choice.

* 5.123 Simulating Guessing on a Multiple-Choice Test Suppose a student takes a 10-question multiple-choice quiz, and for each question on the quiz there are five possible options. Only one option is correct. Now suppose the student, who did not study, guesses at random for each question. A passing grade is 3 (or more) correct. We wish to design a simulation to find the probability that a student who is guessing can pass the exam.

a. In this simulation, the random action consists of a student guessing on a question that has five possible answers. We will simulate this by selecting a single digit from the random number table given in this exercise.

In this table, we will let 0 and 1 represent correct answers, and 2 through 9 will represent incorrect answers. Explain why this is a correct approach for the exam questions with five possible answers.

(This completes the first two steps of the simulation summary given in Section 5.4.)

b. A trial, in this simulation, consists of picking 10 digits in a row. Each digit represents one guess on a question on the exam. Write the sequence of numbers from the first trial. Also translate this to correct and incorrect answers by writing R for right and W for wrong. (This completes step 4.)

c. We are interested in knowing whether there were 3 or more correct answers chosen. Did this occur in the first trial? (This completes step 5.)

d. Perform a second simulation of the student taking this 10-question quiz by guessing randomly. Use the second line of the table given. What score did your student get? Did the event of interest occur this time?

e. Repeat the trial twice more, using lines 3 and 4 of the table. For each trial, write the score and whether or not the event occurred.

f. On the basis of these four trials, what is the empirical probability of passing the exam by guessing?

1 1 3 7 3	9 6 8 7 1
5 2 0 2 2	5 9 0 9 3
1 4 7 0 9	9 3 2 2 0
3 1 8 6 7	8 5 8 7 2

* 5.124 Simulating Guessing on a True/False Test Perform a simulation of a student guessing on a true/false quiz with 10 questions. Use the same four lines of the random number table that are given for the preceding question. Write out each of the seven steps outlined in Section 5.4. Be sure to explain which numbers you will use to represent correct answers and which numbers for incorrect answers. Explain why your choice is logical. Do four repetitions, each trial consisting of 10 questions. Find the empirical probability of getting more than 5 correct out of 10.

5.125 Red Light/Green Light A busy street has three traffic lights in a row. These lights are not synchronized, so they run independently of each other. At any given moment, the probability that a light is green is 60%. Assuming that there is no traffic, follow the steps below to design a simulation to estimate the probability that you will get three green lights.

a. Identify the action with a random outcome, and give the probability that it is a success.

b. Explain how you will simulate this action using the random number table in Appendix A. Which digits will represent green, and which nongreen? If you want to get the same results we did, use all the possible one-digit numbers (0, 1, 2, 3, 4, 5, 6, 7, 8, and 9), and let the first few represent the green lights. How many and what numbers would represent green lights, and what numbers would represent nongreen lights?

c. Describe the event of interest.

d. Explain how you will simulate a single trial.

e. Carry out 20 repetitions of your trial, beginning with the first digit on line 11 of the random number table. For each trial, list the random digits, the outcomes they represent, and whether or not the event of interest happened.

f. What is the empirical probability that you get three green lights?

5.126 Soda A soda-bottling plant has a flaw in that 20% of the bottles it fills do not have enough soda in them. The sodas are sold in six-packs. Follow these steps to carry out a simulation to find the probability that three or more bottles in a six-pack will not have enough soda.

a. Identify the action with a random outcome, and explain how you will simulate this outcome using the random number table in Appendix A. If you want to get the same answers we got, use all the possible one digit numbers (0, 1, 2, 3, 4, 5, 6, 7, 8, and 9), and use some at the beginning of the list of numbers to represent bad and the rest to represent good. What numbers would represent bad and what numbers would represent good, and why?

b. Describe how you will simulate a single trial.

c. Describe the event of interest—that is, the event for which you wish to estimate a probability.

d. Carry out 10 trials, beginning with the first digit on line 15 of the random number table in Appendix A. For each trial, list the digits chosen, the outcomes they represent, and whether or not the event of interest occurred.

e. What is the experimental probability that you get three or more "bad" bottles in a six-pack?

gUIDED EXERCISES

5.23 Gender Discrimination in Tech A Pew Research poll asked respondents to fill in the blank to this question: Compared to other industries there is _____ discrimination against women in the tech industry. Responses separated by gender are shown in the following table. The results are shown using a sample size of 280 men and 150 women. (Source: Pewresearch.org)

Response	Gender Men	Women	Total
Less	53	1 5	68
About the same	174	85	259
More	53	50	103
Total	280	150	430

QUESTION If a person is selected at random from this group, find the probability that the person is a man or said "about the same" (or both)?

Step 1 ▶ What is the probability that the person selected from the table is male?

Step 2 ▶ What is the probability that the person said "about the same"?

Step 3 ▶ If being male and saying "about the same" were mutually exclusive, you could just add the probabilities from step 1 and step 2 to find the probability that the person is male OR said "about the same." Are they mutually exclusive? Why or why not?

Step 4 ▶ What is the probability the person is male AND said "about the same"?

Step 5 ▶ To find the probability that the person is male OR said "about the same," why should you subtract the probability the person is male AND said "about the same" as shown in the calculation below?

P(Male OR said About the Same) = P(Male) + P(Said About the Same) − P(Male AND Said About the Same)

Step 6 ▶ Do the calculation using the formula given in step 5.

Step 7 ▶ Report the answer in a sentence.

5.51 Hand Folding When people fold their hands together with interlocking fingers, most people are more comfortable doing it in one of two ways. In one way, the right thumb ends up on top, and in the other way, the left thumb is on top. The table shows the data from one group of people.

	M	W
Right	18	42
Left	12	28

M means man, W means woman, Right means the right thumb is on top, and Left means the left thumb is on top.

QUESTION Say a person is selected from this group at random. Are the events "right thumb on top" and "male" independent or associated?

To answer, we need to determine whether the probability of having the right thumb on top given that you are a man is equal to the probability of having the right thumb on top (for the entire group). If so, the variables are independent.

Step 1 ▶ Figure out the marginal totals and put them into the table.

Step 2 ▶ Find the overall probability that the person's right thumb is on top.

Step 3 ▶ Find the probability that the right thumb is on top given that the person is a man. (What percentage of men have the right thumb on top?)

Step 4 ▶ Finally, are the variables independent? Why or why not?

TechTips

For All Technology

EXAMPLE: GENERATING RANDOM INTEGERS ▶ Generate four random integers from 1 to 6 for simulating the results of rolling a six-sided die.

TI-84

Seed First before the Random Integers

If you do not seed the calculator, everyone might get the same series of "random" numbers.

1. Enter the last four digits of your Social Security number or cell phone number and press **STO>**.
2. Then press **MATH**, choose **PROB**, and Press **ENTER** (to choose 1:rand). Press **ENTER** again.

You only need to seed the calculator once, unless you **Reset** the calculator. (If you want the same sequence later on, you can seed again with the same number.)

Random Integers

1. Press **MATH**, choose **PROB**, and press **5** (to choose 5:randInt).
2. Press **1**, **ENTER**, **6**, **ENTER**, **4**, **ENTER**, **ENTER**, and **ENTER**.

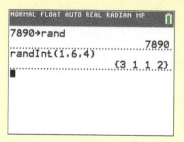

▲ **FIGURE 5a** TI-84

The first two digits (1 and 6 in Figure 5a) determine the smallest and largest integers, and the third digit (4 in Figure 5a) determines the number of random integers generated. The four numbers in the braces in Figure 5a are the generated random integers. Yours will be different. To get four more random integers, press **ENTER** again.

MINITAB

Random Integers

1. **Calc > Random Data > Integer**
2. See Figure 5b. Enter:
   Number of rows of data to generate, **4**
   Store in column(s), **c1**
   Minimum value, **1**
   Maximum value, **6**
3. Click **OK**.

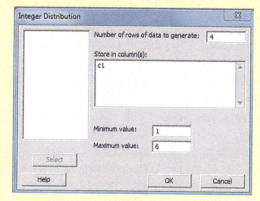

▲ **FIGURE 5b** Minitab

EXCEL

Random Integers

1. Click *fx*, select a category **All**, and **RANDBETWEEN**.
2. See Figure 5c.
   Enter: **Bottom, 1; Top, 6**.
   Click **OK**.
   You will get one random integer in the active cell in the spreadsheet.
3. To get more random integers, put the cursor at the lower-right corner of the cell containing the first integer until you see a black cross (+), and drag downward until you have as many as you need.

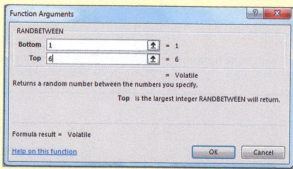

▲ **FIGURE 5c** Excel

Random Integers

1. **Data** > **Simulate** > **Discrete Uniform**

2. See Figure 5d.
   Enter **Rows**, **4**; **Columns**, **1**; **Minimum**, **1**; **Maximum**, **6** and leave **Split across columns** and **Use dynamic seed**.

3. Click **Compute!**

   You will get four random integers (from 1 to 6) in the first empty column.

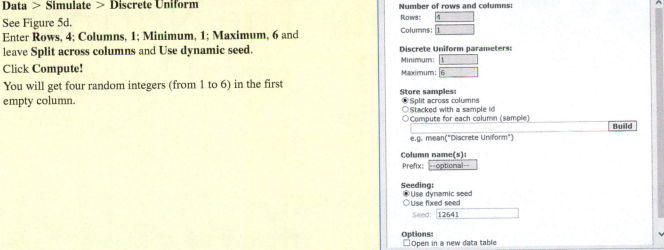

▲ **FIGURE 5d** StatCrunch

6 Modeling Random Events: The Normal and Binomial Models

THEME

Probability distributions describe random outcomes in much the same way as the distributions we discussed in Chapter 2 describe samples of data. A probability distribution tells us the possible outcomes of a random trial and the probability that each of those outcomes will occur.

andom events can seem chaotic and unpredictable. If you flip a coin 10 times, there's no way you can say with absolute certainty how many heads will appear. There's no way your local weather forecaster can tell you with certainty whether it will rain tomorrow. Still, if we watch enough of these random events, patterns begin to emerge. We begin to learn how often different outcomes occur and to gain an understanding of what is common and what is unusual.

Science fiction writer Isaac Asimov is often quoted as having said, "The most exciting phrase to hear in science, the one that heralds new discoveries, is not Eureka! (I found it!) but rather 'hmm . . . that's funny. . . .'" When an outcome strikes us "funny" or unusual, it is because something unlikely occurred. This is exciting because it means a discovery has been made; the world is not the way we thought it was!

However, to know whether something unusual has happened, we first have to know how often it usually occurs. In other words, we need to know the probability. In the case study, you'll see a "homemade" example, in which one of the authors found that five ice cream cones ordered at a local McDonald's weighed more than the advertised amount. Was this a common occurrence, or did McDonald's perhaps deliberately understate the weight of the cones so that no one would ever be disappointed?

In this chapter we'll introduce a new tool to help us characterize probabilities for random events: the probability distribution. We'll examine the Normal probability distribution and the binomial probability distribution, two very useful tools for answering the question, "Is this unusual?"

CASE STUDY

You Sometimes Get More Than You Pay for

A McDonald's restaurant near the home of one of the authors sells ice cream cones that, according to the "fact sheet" provided, weigh 3.18 ounces (converted from grams) and contain 150 calories. Do the ice cream cones really weigh exactly 3.18 ounces? To get 3.18 ounces for every cone would require a very fine-tuned machine or an employee with a *very* good sense of timing. Thus, we expect some natural variation in the weight of these cones. In fact, one of the authors bought five ice cream cones on different days. She found that each of the five cones weighed more than 3.18 ounces. Such an outcome, five out of five over the advertised weight, might occur just by chance. But how often? If such an outcome (five out of five) rarely happens, it's pretty surprising that it happened to us. In that case, perhaps McDonald's actually puts in more than 3.18 ounces. By the end of this chapter, after learning about the binomial and Normal probability distributions, you will be able to measure how surprising this outcome is—or is not.

SECTION 6.1

Probability Distributions Are Models of Random Experiments

A **probability distribution**, sometimes called a **probability distribution function (pdf)**, is a tool that helps us by keeping track of the outcomes of a random experiment and the probabilities associated with those outcomes. For example, suppose the playlist on your mp3 player has 10 songs: 6 are Rock, 2 are Country, 1 is Hip-hop, and 1 is Opera. Put your player on shuffle. What is the probability that the first song played is Rock?

Outcome	Probability
Rock	6/10
Not Rock	4/10

▲ **TABLE 6.1** Probability distribution of songs.

The way the question is worded ("What is the probability . . . is Rock"?) means that we care about only two outcomes: Is the song classified as Rock or is it not? We could write the probabilities as shown in Table 6.1.

Table 6.1 is a very simple probability distribution. It has two important features: It tells us all the possible outcomes of our random experiment (Rock or Not Rock), and it tells us the probability of each of these outcomes. All probability distributions have these two features, though they are not always listed so clearly.

> **KEY POINT** A probability distribution tells us (1) all the possible outcomes of a random experiment and (2) the probability of the outcomes.

Statisticians sometimes refer to probability distributions as **probability models**. The word *model* reminds us that the probabilities might not perfectly match the real-life, long-run frequencies, but we hope that they provide a good-enough match to be useful.

In Chapter 1, we classified variables as either numerical or categorical. It is now useful to break the numerical variables down into two more categories. **Discrete outcomes** (or discrete variables) are numerical values that you can list or count. An example is the number of phone numbers stored on the phones of your classmates. **Continuous outcomes** (or continuous variables) cannot be listed or counted because they occur over a range. The length of time your next phone call will last is a continuous variable. Refer to Figure 6.1 for a visual comparison of these terms.

This distinction is important because if we can list the outcomes, as we can for a discrete variable, then we have a nice way of displaying the probability distribution. However, if we are working with a continuous variable, then we can't list the outcomes, and we have to be a bit more clever in describing the probability distribution function. For this reason, we treat discrete values separately from continuous variables.

🔄 Looking Back

Distributions of a Sample
Probability distributions are similar to distributions of a sample, which were introduced in Chapter 2. A distribution of a sample tells us the values in the sample and their frequency. A probability distribution tells us the possible values of a random experiment and their probability.

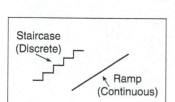

▲ **FIGURE 6.1** Visual representation of discrete and continuous changes in elevation. Note that for the staircase (discrete outcomes), you can count the stairs.

EXAMPLE 1 Discrete or Continuous?

Consider these variables:

a. The weight of a submarine sandwich you're served at a deli.

b. The elapsed time from when you left your house to when you arrived in class this morning.

c. The number of people in the next passing car.

d. The blood-alcohol level of a driver pulled over by the police in a random sobriety check. (Blood-alcohol level is measured as the percentage of the blood that is alcohol.)

e. The number of eggs laid by a randomly selected salmon as observed in a fishery.

QUESTION   Identify each of these numerical variables as continuous or discrete.

SOLUTION   The continuous variables are variables a, b, and d. Continuous variables can take on any value in a spectrum. For example, the sandwich might weigh 6 ounces or 6.1 ounces or 6.0013 ounces. The amount of time it takes you to get to class could be 5000 seconds or 5000.4 seconds or 5000.456 seconds. Blood-alcohol content can be any value between 0 and 1, including 0.0013 (or 0.13%), 0.0013333, 0.001357, and so on.

Variables c and e, on the other hand, are discrete quantities—that is, numbers that can be counted.

TRY THIS!   Exercise 6.1

Discrete Probability Distributions Can Be Tables or Graphs

In a statistics class at University of California, Los Angeles, approximately 40% male and 60% female. Let's arbitrarily code the males as 0 and the females as 1. If we select a person at random, what is the probability that the person is female?

Creating a probability distribution for this situation is as easy as listing both outcomes (0 and 1) and their probabilities (0.40 and 0.60). The easiest way to do this is in a table (see Table 6.2). However, we could also do it in a graph (Figure 6.2).

Female	Probability
0	0.40
1	0.60

▲ **TABLE 6.2** Probability distribution of gender in a class.

◀ **FIGURE 6.2** Probability distribution for selecting a person at random from a particular statistics class and recording whether the person is male (0) or female (1).

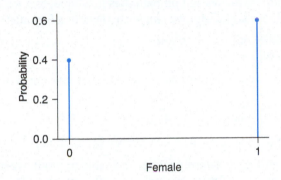

Note that the probabilities sum to 1 (0.40 + 0.60 = 1.0). This is true of all probability distributions since they give the probabilities for all possible outcomes.

Amazon.com invites customers to rate books on a scale of 1 to 5 stars. At a visit to the site in 2018, *Lincoln in the Bardo,* by George Saunders, had received 1477 customer reviews. Suppose we randomly select a reviewer and base our decision whether to buy the book on how many stars that person awarded the book. Table 6.3 and Figure 6.3 illustrate two different ways of representing the probability distribution for this random event. For example, we see that the most likely outcome is that the reviewer gives the book 5 stars. The probability that a randomly selected reviewer will have given the book 5 stars is 44%. (Notice that the second-highest probability is for 1 star. Why do you think these two extremes are so well represented here? We'll consider one possibility in Chapter 7.)

Number of Stars	Probability
5	0.44
4	0.15
3	0.10
2	0.11
1	0.20

▲ **TABLE 6.3** Probability distribution of number of stars in review. Note the probabilities add to 1.

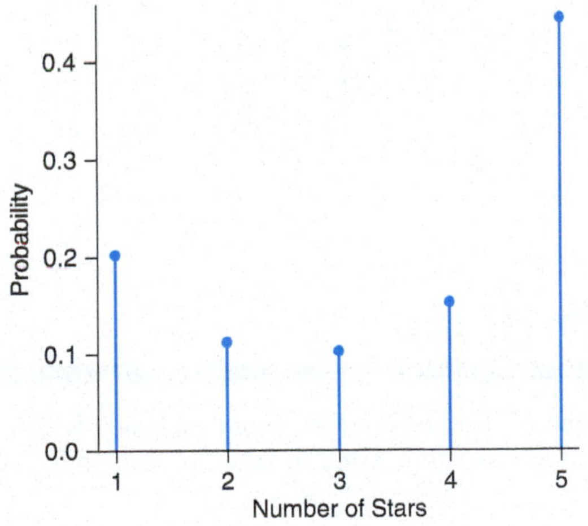

◀ **FIGURE 6.3** Probability distribution for the number of stars given, as a rating, by a randomly selected reviewer for a book on amazon.com.

Note that the probabilities in Table 6.3 add to 1. This is true of all probability distributions: When you add up the probabilities for all possible outcomes, they add to 1. They have to because there are no other possible outcomes.

Discrete Distributions Can Also Be Equations

What if we have too many outcomes to list in a table? For example, suppose a married couple decides to keep having children until they have a girl. How many children will they have, assuming that boys and girls are equally likely and that the gender of one birth doesn't depend on any of the previous births? It could very well turn out that their first child is a girl and they therefore have only one child, or that the first is a boy but the second is a girl. Or, just possibly, they might never have a girl. The value of this experiment could be any number 1, 2, 3, . . ., up to infinity. (Okay, in reality, it's impossible to have that many children. But we can imagine!)

We can't list all these values and probabilities in a table, and we can only hint at what the graph might look like. But we *can* write them in a formula:

The probability of having x children is $(1/2)^x$ (where x is a whole number bigger than 0).

For example, the probability that they have 1 child (that is, the first is a girl) is $(1/2)^1 = 1/2$.

The probability that they have 4 children is $(1/2)^4 = 1/16$.

The probability that they have 10 children is small: $(1/2)^{10} = 0.00098$.

In this text, we will give the probabilities in a table or graph whenever convenient. In fact, even if it's not especially convenient, we will often provide a graph to encourage you to visualize what the probability distribution looks like. Figure 6.4 is part of the graph of the probability distribution for the number of children a couple can have if they continue to have children until the first girl. We see that the probability that the first child is a girl is 0.50 (50%). The probability that the couple has two children is half this: 0.25. The probabilities continue to decrease, and each probability is half the one before it.

▶ **FIGURE 6.4** Partial probability distribution of the number of children born until the first girl. (Only the first 10 outcomes are shown.)

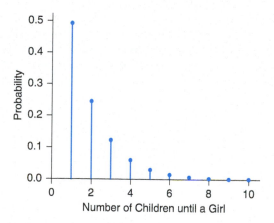

EXAMPLE 2 Playing Dice

Roll a fair six-sided die. A fair die is one in which each side is equally likely to end up on top. You will win $4 if you roll a 5 or a 6. You will lose $5 if you roll a 1. For any other outcome, you will win or lose nothing.

QUESTION   Create a table that shows the probability distribution for the amount of money you will win. Draw a graph of this probability distribution function.

SOLUTION   There are three outcomes: you win $4, you win 0 dollars, you win −$5. (Winning negative five dollars is the same as losing $5.)

You win $4 if you roll a 5 or 6, so the probability is $2/6 = 1/3$.
You win $0 if you roll a 2, 3, or 4, so the probability is $3/6 = 1/2$
You "win" −$5 if you roll a 1, so the probability is 1/6.

We can put the probability distribution function in a table (Table 6.4), or we can represent the pdf as a graph (Figure 6.5).

Winnings	Probability
−5	1/6
0	1/2
4	1/3

▲ **TABLE 6.4** Probability distribution function of the dice game.

◀ **FIGURE 6.5** Probability distribution function of the dice game.

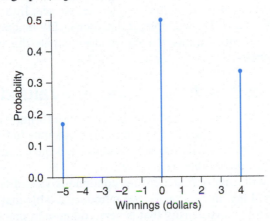

TRY THIS!   Exercise 6.5

Continuous Probabilities Are Represented as Areas under Curves

Finding probabilities for continuous outcomes is more complicated, because we cannot simply list all the possible values we might see. What we can list is the *range of values* we might see.

For example, suppose you want to know the probability that you will wait in line for between 3 and 4 minutes when you go to the coffee shop. You can't list all possible outcomes that could result from your visit: 1.0 minute, 1.00032 minutes, 2.00000321 minutes. It would take (literally) an eternity. But you can specify a range. Suppose this particular coffee shop has done extensive research and knows that everyone gets helped in under 5 minutes. Therefore, all customers get helped within the range of 0 to 5 minutes.

If we want to find probabilities concerning a continuous-valued experiment, we also need to give a range for the outcomes. For example, the manager wants to know the probability that a customer will wait less than 2 minutes; this gives a range of 0 to 2 minutes.

The probabilities for a continuous-valued random experiment are represented as areas under curves. The curve is called a **probability density curve**. The total area under the curve is 1 because this represents the probability that the outcome will be somewhere on the *x*-axis. To find the probability of waiting between 0 and 2 minutes, we find the area under the density curve and between 0 and 2 (Figure 6.6).

◀ **FIGURE 6.6** Probability distribution of times waiting in line at a particular coffee shop.

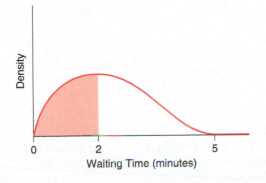

The y-axis in a continuous-valued pdf is labeled "Density." How the density is determined isn't important. What *is* important to know is that the units of density are scaled so that the area under the entire curve is 1.

You might wonder where this curve came from. How did we know the distribution was exactly this shape? In practice, it is very difficult to know which curve is correct for a real-life situation. That's why we call these probability distributions "models." They are meant to mimic real-life probability, but they are not exactly the same as real life. Although we can never really know with certainty if they are correct, we can compare the probabilities in the models to the actual frequencies we see. If they are close, then our probability model is good. For example, if this probability model predicts that 45% of customers get coffee within 2 minutes, then we can compare this prediction to an actual sample of customers.

Finding Probabilities for Continuous-Valued Outcomes

Calculating the area under a curve is not easy. If you have a formula for the probability density, then you can sometimes apply techniques from calculus to find the area. However, for many commonly used probability densities, basic calculus is not helpful, and computer-based approximations are required.

In this book, you will usually find areas for continuous-valued outcomes by using a table or by using technology. In Section 6.2 we introduce a table that can be used to find areas for one type of probability density that is very common in practice: the normal curve.

EXAMPLE **3** Waiting for the Bus

The bus that runs near the home of one of the authors arrives every 12 minutes. If the author arrives at the bus stop at a randomly chosen time, then the probability distribution for the number of minutes he must wait for the bus is shown in Figure 6.7.

▶ **FIGURE 6.7** Probability distribution function showing the number of minutes the author must wait for the bus if he arrives at a randomly determined time.

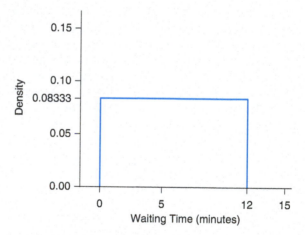

QUESTION (a) Find the probability that the author will have to wait between 4 and 10 minutes for the bus. (*Hint:* Remember that the area of a rectangle is the product of the lengths of the two sides.) (b) Find the probability the author will have to wait between 0 and 12 minutes. Why is this probability not exactly equal to 1? Why should it be?

SOLUTION (a) The distribution shown in Figure 6.7 is called a uniform distribution. Finding areas under this curve is easy because the curve is just a rectangular shape. The area we need to find is shown in Figure 6.8. The area of a rectangle is width times height. The width is $(10 - 4) = 6$, and the height is 0.08333.

◄ **FIGURE 6.8** The shaded area represents the probability that the author will wait between 4 and 10 minutes if he arrives at the bus stop at a randomly determined time.

The probability that the author must wait between 4 and 10 minutes is $6 \times 0.08333 = 0.4998$, or about 0.500. Visually, we see that about half of the area in Figure 6.8 is shaded.

There is approximately a 50% chance that the author must wait between 4 and 10 minutes. **(b)** The question asks for the area of the entire rectangle, which is $12 \times 0.08333 = 0.999996$. The probability is not exactly equal to 1 because of rounding error. It should be exactly equal to 1 because the range of 0 to 12 minutes includes all possible outcomes, and so the total area under this curve must be equal to 1.

TRY THIS!   Exercise 6.11

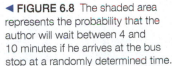

SECTION 6.2

The Normal Model

The **Normal model** is the most widely used probability model for continuous numerical variables. One reason is that many numerical variables in which researchers have historically been interested have distributions for which the Normal model provides a very close fit. Also, an important mathematical theorem called the Central Limit Theorem (to be introduced in Chapter 7) links the Normal model to several key statistical ideas, which provides good motivation for learning this model.

We begin by showing you what the Normal model looks like. Then we discuss how to find probabilities by finding areas underneath the Normal curve. We will illustrate these concepts with examples and discuss why the Normal model is appropriate for these situations.

Visualizing the Normal Distribution

Figure 6.9 on the next page shows several histograms of measurements taken from a sample of about 1400 adult men in the United States. All these graphs have similar shapes: They are unimodal and symmetric. We have superimposed smooth curves over the histograms that capture this shape. You could easily imagine that if we continued to collect more and more data, the histogram would eventually fill in the curve and match the shape almost exactly.

🔁 Looking Back

Unimodal and Symmetric Distributions
Symmetric distributions have histograms whose right and left sides are roughly mirror images of each other. Unimodal distributions have histograms with one mound.

(a)

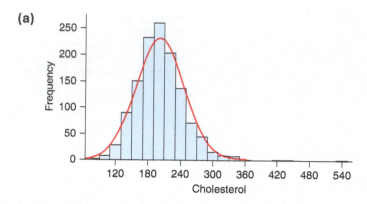

(b)

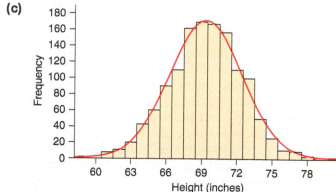

(c)

▶ **FIGURE 6.9** Measurements for a sample of men: **(a)** cholesterol levels, **(b)** diastolic blood pressure readings, and **(c)** height. For all three, a very similar shape appears: nearly symmetric and mound-shaped.

 Details

The Bell Curve
The Normal or Gaussian curve is also called the bell curve.

The curve drawn on these histograms is called the **Normal curve**, or the **Normal distribution**. It is also sometimes called the Gaussian distribution, after Karl Friedrich Gauss (1777–1855), the mathematician who first derived the formula. Statisticians and scientists recognized that this curve provided a model that pretty closely described a good number of continuous-valued data distributions. Today, even though we have many other distributions to model real-life data, the Normal curve is still one of the most frequently used probability distribution functions in science.

Center and Spread   In Chapters 2 and 3 we discussed the center and spread of a distribution of *data*. These concepts are also useful for studying distributions of *probability*. The mean of a probability distribution sits at the balancing point of the probability distribution. The standard deviation of a probability distribution measures the spread of the distribution by telling us how far away, typically, the values are from the mean. The conceptual understanding you developed for the mean and the standard deviation of a sample still apply to probability distributions.

The notation we use is slightly different so that we can distinguish means and standard deviations of probability distributions from means and standard deviations of data. The **mean of a probability distribution** is represented by the Greek character μ (mu, pronounced "mew"), and the **standard deviation of a probability distribution** is represented by the character σ (sigma). These Greek characters are used to avoid confusion of these concepts with their counterparts for samples of data, $\bar{x}$ and s.

The Mean and the Standard Deviation of a Normal Distribution   The exact shape of the Normal distribution is determined by the values of the mean and the standard deviation. Because the Normal distribution is symmetric, the mean is in the exact center of the distribution. The standard deviation determines whether the Normal curve is wide and low (large standard deviation) or narrow and tall (small standard deviation). Figure 6.10 shows two Normal curves that have the same mean but different standard deviations.

A Normal curve with a mean of 69 inches and a standard deviation of 3 inches provides a very good match to the distribution of heights of all adult men in the United

Looking Back

Mean and Standard Deviation
In Chapter 3, you learned that the symbol for the mean of a *sample* of data is $\bar{x}$ and the symbol for the standard deviation of a sample of data is s.

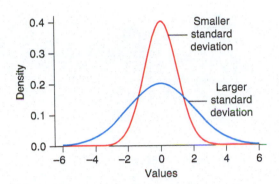

◄ **FIGURE 6.10** Two Normal curves with the same mean but different standard deviations.

States. Surprisingly, a Normal curve with the same standard deviation of 3 inches, but a smaller mean of about 64 inches, describes the distribution of adult women's heights. Figure 6.11 shows what these Normal curves look like.

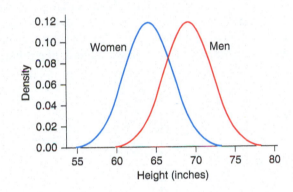

◄ **FIGURE 6.11** Two Normal curves. The blue curve represents the distribution of women's heights; it has a mean of 64 inches and a standard deviation of 3 inches. The red curve represents the distribution of men's heights; it has the same standard deviation, but the mean is 69 inches.

The only way to distinguish among different Normal distributions is by their means and standard deviations. We can take advantage of this fact to write a shorthand notation to represent a particular Normal distribution. The notation $N(\mu, \sigma)$ represents a Normal distribution that is centered at the value of μ (the mean of the distribution) and whose spread is measured by the value of σ (the standard deviation of the distribution). For example, in Figure 6.11, the distribution of women's heights is $N(64, 3)$, and the distribution of men's heights is $N(69, 3)$.

KEY POINT   The Normal distribution is symmetric and unimodal ("bell-shaped"). The notation $N(\mu, \sigma)$ tells us the mean and the standard deviation of the Normal distribution.

Finding Normal Probabilities

The Normal model $N(64, 3)$ gives a good approximation of the distribution of adult women's heights in the United States (where height is measured in inches). Suppose we were to select an adult woman from the United States at random and record her height. What is the probability that she is taller than a specified height?

Because height is a continuous numerical variable, we can answer this question by finding the appropriate area under the Normal curve. For example, Figure 6.12 shows a Normal curve that models the distribution of heights of women in the population—the same curve as in Figure 6.11. (We will often leave the numerical scale off the vertical axis from now on since it is not needed for doing calculations.) The area of the shaded region gives us the probability of selecting a woman taller than 62 inches. The entire area under the curve is 1, as it is for all probability distributions.

▶ **FIGURE 6.12** The area of the shaded region represents the probability of finding a woman taller than 62 inches from a N(64, 3) distribution.

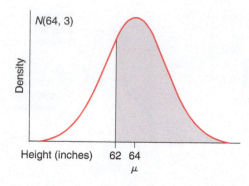

In fact, Figure 6.12 represents both the probability of selecting a woman taller than 62 inches and the probability of selecting a woman *62 inches tall or taller*. Because the areas for both regions (the one that is strictly greater than 62, and the other that includes 62) are the same, the probabilities are also the same. This is a convenient feature of continuous variables: We don't have to be too picky about our language when working with probabilities. This is in marked contrast with discrete variables, as you will soon see.

What if we instead wanted to know the probability that the chosen woman would be between 62 inches and 67 inches tall? That area would look like Figure 6.13.

▶ **FIGURE 6.13** The shaded area represents the probability that a randomly selected woman is between 62 and 67 inches tall. The probability distribution shown is the Normal distribution with mean 64 inches and standard deviation 3 inches.

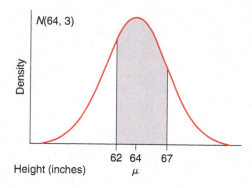

Figure 6.14 shows the area corresponding to the probability that this randomly selected woman is less than 62 inches tall.

▶ **FIGURE 6.14** The shaded area represents the probability that the randomly selected woman is less than 62 inches tall.

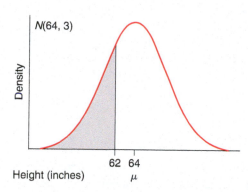

KEY POINT

When you are finding probabilities with Normal models, the first and most helpful step is to sketch the curve, label it appropriately, and shade in the region of interest.

We recommend that you always begin problems that concern Normal models by drawing a picture. One advantage of drawing a picture is that sometimes it is all you need. For example, according to the McDonald's "Fact Sheet," each serving of ice cream in a cone weighs 3.18 ounces. Now, we know that in real life it is not possible to serve exactly 3.18 ounces. Probably the employees operating the machines (or the machines themselves) actually dispense a little more or a little less than 3.18 ounces. Suppose that the amount of ice cream dispensed follows a Normal distribution with a mean of 3.18 ounces. What is the probability that a hungry customer will actually get less than 3.18 ounces?

Figure 6.15a shows the situation. From it we easily see that the area to the left of 3.18 is exactly half of the total area. Thus, we know that the probability of getting less than 3.18 ounces is 0.50. (We also know this because the Normal curve is symmetric, so the mean—the balancing point—must sit right in the middle. Therefore, the probability of getting a value less than the mean is 0.50.)

What if the true mean is actually larger than 3.18 ounces? How will that affect the probability of getting a cone that weighs less than 3.18 ounces? Imagine "sliding" the Normal curve to the right, which corresponds to increasing the mean. Does the area to the left of 3.18 go up or down as the curve slides to the right? Figure 6.15b shows that the area below 3.18 is now smaller than 50%. The larger the mean amount of ice cream dispensed, the less likely it is that a customer will complain about getting too little.

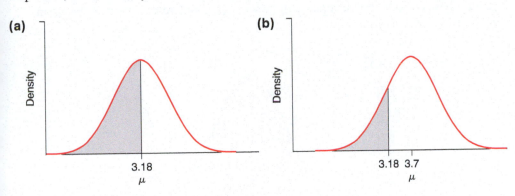

(a)

Density

3.18
μ

(b)

Density

3.18 3.7
μ

◀ **FIGURE 6.15   (a)** A $N(3.18, 0.6)$ curve, showing that the probability of getting a cone that weighs less than 3.18 ounces is 50%. **(b)** A Normal curve with the same standard deviation (0.6) but a larger mean (3.7). The area below 3.18 is now much smaller.

Finding Probability with Technology

Finding the area of a Normal distribution is best done with technology. Most calculators and many software packages will show you how to do this. An Internet search for "Normal Probability Calculator" will also reveal a variety of options. Here we show you how to use StatCrunch to find Normal probabilities.

Figure 6.16 is a screenshot from the StatCrunch calculator for Normal probabilities. It shows the probability that a randomly selected woman is between 62 and 67 inches tall if the $N(64, 3)$ model is a valid description of the distribution of women's heights.

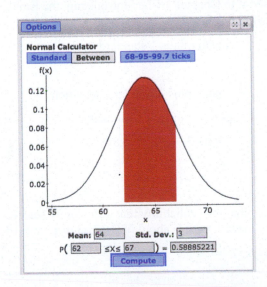

Tech

◀ **FIGURE 6.16** Screenshot showing a calculation of the area under a $N(64, 3)$ curve between 62 and 67. The area, and so the probability, is about 58.9%.

```
NORMAL FLOAT AUTO REAL RADIAN MP
normalcdf(62,67,64,3)
                    .5888522734
```

▲ **FIGURE 6.17** TI-84 output showing that the probability that the woman is between 62 and 67 inches tall is about 59%.

This calculator asks us to enter in the mean and standard deviation of our Normal model. We can then choose "Between" (as we did here) to calculator the probability of seeing an observation that lies between two values, or we can choose "Standard." Standard would let us find "one way" probabilities; for example, the probability that a randomly selected woman's height will be less than 62 inches or that it will be greater than 62 inches (or any other value we enter).

Figure 6.17 shows output from a TI-84 for the same calculation.

EXAMPLE 4 Baby Seals

Some research has shown that the mean length of a newborn Pacific harbor seal is 29.5 inches ($\mu = 29.5$ inches) and that the standard deviation is $\sigma = 1.2$ inches. Suppose that the lengths of these seal pups follow the Normal model.

QUESTION Using the StatCrunch output in Figure 6.18, find the probability that a randomly selected harbor seal pup will be within 1 standard deviation of the mean length of 29.5 inches.

► **FIGURE 6.18** StatCrunch Output: The shaded region represents the area under the curve between 28.3 inches and 30.7 inches. In other words, it represents the probability that the length of a randomly selected seal pup is within 1 standard deviation of the mean.

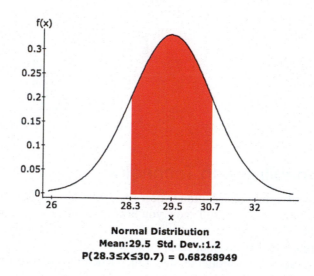

Normal Distribution
Mean:29.5 Std. Dev.:1.2
P(28.3≤X≤30.7) = 0.68268949

SOLUTION The phrase "within one standard deviation of the mean length" is one you will see often. (We used it in Chapter 3 when introducing the Empirical Rule.) It means that the pup's length will be somewhere between

mean minus 1 standard deviation

and

mean plus 1 standard deviation.

Because 1 standard deviation is 1.2 inches, this means the length must be between

$$29.5 - 1.2 = 28.3 \text{ inches}$$

and

$$29.5 + 1.2 = 30.7 \text{ inches}$$

From the results, we see that the probability that a randomly selected seal pup is within 1 standard deviation of mean length is about 68% (from Figure 6.18, it is 68.2689%).

TRY THIS! Exercise 6.17

Without Technology: Apply The Empirical Rule

In Chapter 3 we mentioned the Empirical Rule, which is not so much a rule as a guideline for helping you understand how values in a sample of data are distributed. The Empirical can Rule be applied to any symmetric, unimodal distribution. For any unimodal, symmetric distribution, the Empirical Rule is approximate—and sometimes very approximate. But in the special case that the distribution you are studying is the Normal model, the Empirical Rule is (nearly) exact.

In Example 4, we found that the area between −1 and +1 in a standard Normal distribution was 68%. This is exactly what the Empirical Rule predicts. Using a calculator, we can also find the probability that a randomly selected observation will be between −2 and 2 standard deviations of the mean if that observation comes from the Normal model.

Figure 6.19 shows a sketch of the $N(0, 1)$ model, with the region between −2 and 2 (in standard units) shaded. The result is very close to 95%, as the Empirical Rule predicts.

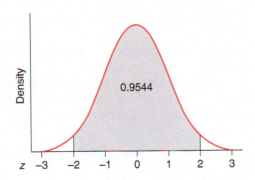

◀ **FIGURE 6.19** The area between z-scores of −2.00 and +2.00 on the standard Normal curve.

These facts from the Empirical Rule help us interpret the standard deviation in the context of the Normal distribution. For example, because the heights of women are Normally distributed and have a standard deviation of about 3 inches, we know that a majority of women (in fact, 68%) have heights within 3 inches of the mean: between 61 inches and 67 inches. Because nearly all women are within 3 standard deviations of the mean, we know we should not expect many women to be taller than three standard deviations above the mean, which corresponds to $64 + 3 \times 3 = 73$ inches tall (73 inches is 6 feet 1 inch). Such women are very rare.

We can apply the Empirical Rule to find probabilities when they involve simple multiples of the standard deviation. Example 5 shows how you might do this.

> **↻ Looking Back**
>
> **The Empirical Rule**
> The Empirical Rule says that if a distribution of a sample of data is unimodal and roughly symmetric, then about 68% of the observations are within 1 standard deviation of the mean, about 95% are within 2 standard deviations of the mean, and nearly all are within 3 standard deviations of the mean.

EXAMPLE 5 Bigger Baby Seals

We again assume that the length of harbor seal pups is Normally distributed with mean of 29.5 inches and a standard deviation of 1.2 inches.

QUESTION   What's the probability that a randomly selected harbor seal pup is longer than 30.7 inches?

SOLUTION   We are asked to find the area above 30.7 inches in a $N(29.5, 1.2)$ distribution. This example illustrates two useful practices for solving problems such as these. First, always draw a sketch. Second, always convert the values to standard units.

Figure 6.20 shows a sketch of the Normal distribution. We've labeled the mean and 30.7 inches on the x-axis. We've also put tickmarks at the whole-number standard units: the mean ± 1 standard deviation, ± 2 standard deviations, and ± 3 standard deviations.

▶ **FIGURE 6.20** Sketch of Normal curve. We wish to find the area to the right of 30.7. The sketch includes tick marks at each additional standard deviation.

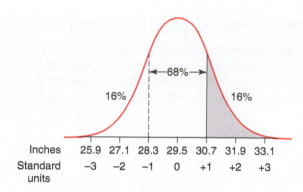

| Inches | 25.9 | 27.1 | 28.3 | 29.5 | 30.7 | 31.9 | 33.1 |
| Standard units | −3 | −2 | −1 | 0 | +1 | +2 | +3 |

For example, to place the tick mark at the mean minus 1, compute $29.5 - 1.2 = 28.3$. To place the tick mark at the mean plus 1 standard deviation, compute $29.5 + 1.2 = 30.7$.

By creating this sketch, we discover the happy fact that the value we are investigating—30.7—is precisely 1 standard deviation above the mean. In other words, 30.7 inches is 1.0 standard unit.

Because the Empirical Rule states that the area between −1 and 1 is 68%, the area *outside* −1 and 1 must be $100\% - 68\% = 32\%$.

The Normal distribution is symmetric, and so we know that the area above +1 must be the same as the area below −1. This means that half of the 32% must be below −1 and half must be above +1. Therefore, 16% is above +1.

CONCLUSION The probability the seal pup will be longer than 30.7 inches is 16%. The technology confirms this, as shown in Figure 6.21.

▶ **FIGURE 6.21** StatCrunch Normal Probability calculator.

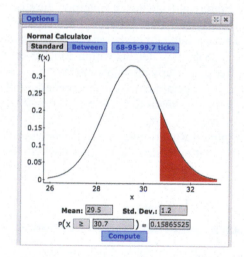

TRY THIS! Exercise 6.25

Without Technology: The Standard Normal

Example 5 illustrates a principle that's very useful for finding probabilities from the Normal distribution without technology. This principle is the recognition that we don't need to refer to values in our distribution in the units in which they were measured. We can instead refer to them in standard units. In other words, we can ask for the probability that the man's height is between 66 inches and 72 inches (measured units), but another way to say the same thing is to ask for the probability that his height is within 1 standard deviation of the mean, or between −1 and +1 standard units.

You can still use the Normal model if you change the units to standard units, but you must also convert the mean and the standard deviation to standard units. This is easy, because the mean is 0 standard deviations away from itself, and any point 1 standard deviation away from the mean is 1 standard unit. Thus, if the Normal model was a good model, then when you convert to standard units, the $N(0, 1)$ model is appropriate.

This model—the Normal model with mean 0 and standard deviation 1—has a special name: the **standard Normal model**.

KEY POINT	$N(0, 1)$ is the standard Normal model: a Normal model with a mean of 0 ($\mu = 0$) and a standard deviation of 1 ($\sigma = 1$).

The standard Normal model is a useful concept, because it allows us to find probabilities for any Normal model. All we need to do is first convert to standard units. We can then look up the areas in a published table that lists useful areas for the $N(0, 1)$ model. One such table is available in Appendix A.

z	.00	.01	.02	.03	.04	.05	.06	.07	.08	.09
0.9	.8159	.8186	.8212	.8238	.8264	.8289	.8315	.8340	.8365	.8389
1.0	**.8413**	.8438	.8461	.8485	.8508	.8531	.8554	.8577	.8599	.8621
1.1	.8643	.8665	.8686	.8708	.8729	.8749	.8770	.8790	.8810	.8830
1.2	.8849	.8869	.8888	.8907	.8925	.8944	.8962	.8980	.8997	.9015

▲ **TABLE 6.5** Excerpt from the Normal Table in Appendix A. This excerpt shows areas to the left of *z* in a standard Normal distribution. For example, the area to the left of *z* = 1.00 is 0.8413, and the area to the left of *z* = 1.01 is 0.8438.

Table 6.5 shows an excerpt from this table. The values within the table represent areas (probabilities). The numbers along the left margin, when joined to the numbers across the top, represent *z*-scores. For instance, the boldface value in this table represents the area under the curve *and to the left* of 1.00 standard unit. This represents the probability that a randomly selected person has a height *less than* 1 standard unit. Figure 6.22 shows what this area looks like.

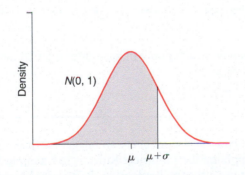

◄ **FIGURE 6.22** The area of the shaded region represents the probability that a randomly selected person (or thing) has a value less than 1.00 standard deviation above the mean, which is about 84%.

> **Details**

z-Scores

In Chapter 3 we gave the formula for a z-score in terms of the mean and standard deviation of a sample:

$$z = \frac{x - \bar{x}}{s}$$

The same idea works for probability distributions, but we must change the notation to indicate that we are using the mean and the standard deviation of a probability distribution:

$$z = \frac{x - \mu}{\sigma}$$

For example, imagine we want to find the probability that a randomly selected woman is shorter than 62 inches. This person is selected from a population of women whose heights follow a $N(64, 3)$ distribution. Our strategy is the following:

1. Convert 62 inches to standard units. Call this number z.

2. Look up the area below z in the table for the $N(0, 1)$ distribution.

EXAMPLE **6** Small Pups

Small newborn seal pups have a lower chance of survival than do larger newborn pups. Suppose that the length of a newborn seal pup follows a Normal distribution with a mean length of 29.5 inches and a standard deviation of 1.2 inches.

QUESTION   What is the probability that a newborn pup selected at random is shorter than 28.0 inches?

SOLUTION   Begin by converting the length 28.0 inches to standard units.

$$z = \frac{28 - 29.5}{1.2} = \frac{-1.5}{1.2} = -1.25$$

Next sketch the area that represents the probability we wish to find (see Figure 6.23). We want to find the area under the Normal curve and to the left of 28 inches, or, in standard units, to the left of -1.25.

▶ **FIGURE 6.23** A standard Normal distribution, showing the shaded area that represents the probability of selecting a seal pup shorter than -1.25 standard deviations below the mean.

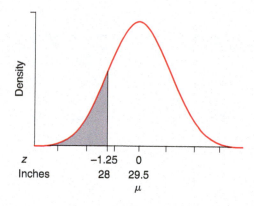

We can look this up in the standard Normal table in Appendix A. Table 6.6 shows the part we are interested in. We see that the area to the left of a z-score of -1.25 is 10.56%.

z	.00	.01	.02	.03	.04	.05	.06	.07	.08	.09
−1.3	.0968	.0951	.0934	.0918	.0901	.0885	.0869	.0853	.0838	.0823
−1.2	.1151	.1131	.1112	.1093	.1075	**.1056**	.1038	.1020	.1003	.0985

▲ **TABLE 6.6** Part of the Standard Normal Table. The value printed in boldface type is the area under the standard Normal density curve to the left of −1.25.

The probability that a newborn seal pup will be shorter than 28 inches is about 11% (rounding up from 10.56%).

TRY THIS!   Exercise 6.27

Note that in Example 6, had the z-score been an integer, such as -1, we could have found the solution by applying the Empirical Rule, as in Example 5. But with a value such as -1.25, we must use either technology or the standard Normal table.

EXAMPLE 7   A Range of Seal Pup Lengths

Tech

Again, suppose that the $N(29.5, 1.2)$ model is a good description of the distribution of seal pups' lengths.

QUESTION   What is the probability that this randomly selected seal pup will be between 27 inches and 31 inches long?

SOLUTION   This question is slightly tricky when using a table. The table gives us only the area *below* a given value. How do we find the area *between* two values?

We proceed in two steps. First we find the area less than 31 inches. Second, we "chop off" the area below 27 inches. The area that remains is the region between 27 and 31 inches. This process is illustrated in Figure 6.24.

To find the area less than 31 inches (shown in Figure 6.24a), we convert 31 inches to standard units:

$$z = \frac{(31 - 29.5)}{1.2} = 1.25$$

(a)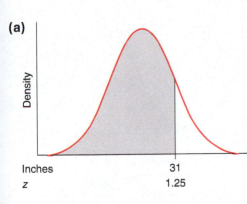

(b)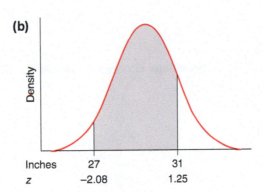

◄ **FIGURE 6.24** Steps for finding the area between 27 and 31 inches under a $N(29.5, 1.2)$ distribution. **(a)** The area below 31. **(b)** We chop off the area below 27, and the remaining area (shaded) is what we are looking for.

Using the standard Normal table in Appendix A, we find that this probability is 0.8944. Next, we find the area below 27 inches:

$$z = \frac{(27 - 29.5)}{1.2} = -2.08$$

Again, using the standard Normal table, we find this area to be 0.0188. Finally, we subtract (or "chop off") the smaller area from the big one:

$$
\begin{array}{r}
0.8944 \\
-0.0188 \\
\hline
0.8756
\end{array}
$$

CONCLUSION   The probability that a newborn seal pup will be between 27 inches and 31 inches long is about 88%. Figure 6.25 confirms that answer.

TRY THIS!   Exercise 6.31

```
NORMAL FLOAT AUTO REAL RADIAN MP
normalcdf(27,31,29.5,1.2)
                   .8757398007
```

▲ **FIGURE 6.25** TI-84 output for finding the probability that a newborn seal pup will be between 27 and 31 inches long.

Finding Measurements from Percentiles for the Normal Distribution

So far we have discussed how to find the probability that you will randomly select an object with a value that is within a certain range. Thus, in Example 6 we found that if newborn seal pups' lengths follow a $N(29.5, 1.2)$ distribution, then the probability that we will randomly select a pup shorter than 28 inches is (roughly) 11%. We found this by finding the area under the Normal curve that is to the left of 28 inches.

📌 **Details**

Inverse Normal
Statisticians sometimes refer to finding measurements from percentiles of Normal distributions as "finding inverse Normal values."

Sometimes, though, we wish to turn this around. We are given a probability, but we want to find the value that corresponds to that probability. For instance, we might want to find the length of a seal pup such that the probability that we'll see any pups shorter than that is 11%. Such a number is called a **percentile**. The 11th percentile for seal pup lengths is 28 inches, the length that has 11% of the area under the Normal curve to its left.

Finding measurements from percentiles is simple with the right technology. The screenshot in Figure 6.26a shows how to use StatCrunch to find the height of a women in the 25th percentile, assuming that women's heights follow a $N(64, 3)$ distribution. Because being in the 25th percentile means that 25% of the women are shorter than this value, we choose the $\leq$ sign and enter 0.25 into the box as shown.

After clicking "Compute," Figure 6.26b shows us that a women in the 25th percentile for height would be about 62.0 inches tall.

▶ **FIGURE 6.26 (a)** StatCrunch set up to calculate the 25th percentile in a Normal distribution. **(b)** The 25th percentile is 61.976531, or about 62 inches.

(a)
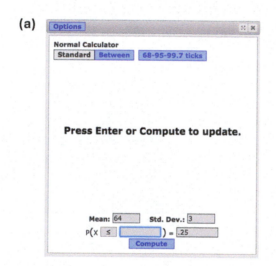

(b)
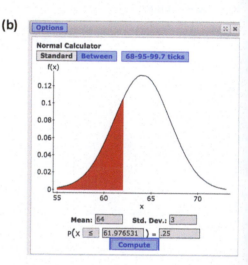

Things are a little trickier if you don't have technology. You first need to use the standard Normal curve, $N(0, 1)$, to find the z-score from the percentile. Then you must convert the z-score to the proper units. So without technology, finding a measurement from a percentile is a two-step process:

Step 1. Find the z-score from the percentile.

Step 2. Convert the z-score to the proper units.

Examples 8 and 9 illustrate this procedure.

EXAMPLE **8** Inverse Normal or Normal?

Suppose that the amount of money people keep in their online PayPal account follows a Normal model. Consider these two situations:

a. A PayPal customer wonders how much money he would have to put into the account to be in the 90th percentile.

b. A PayPal employee wonders what the probability is that a randomly selected customer will have less than $150 in his account.

QUESTION For each situation, identify whether the question asks for a measurement or a Normal probability.

SOLUTIONS

a. This situation gives a percentile (the 90th) and asks for the measurement (in dollars) that has 90% of the other values below it. This is an inverse Normal question.

b. This situation gives a measurement ($150) and asks for a probability.

TRY THIS!   Exercise 6.41

EXAMPLE 9 Finding Measurements from Percentiles by Hand

Assume that women's heights follow a Normal distribution with mean 64 inches and standard deviation 3 inches: $N(64, 3)$. Earlier, we used technology to find that the 25th percentile was approximately 62 inches.

QUESTION   Using the Normal table in Appendix A, confirm that the 25th percentile height is 62 inches.

SOLUTION   The question asks us to find a measurement (the height) that has 25% of all women's heights below it. Use the Normal table in Appendix A. This gives probabilities and percentiles for a Normal distribution with mean 0 and standard deviation 1: a $N(0, 1)$ or standard Normal distribution.

Step 1: Find the z-score from the percentile. To do this, you must first find the probability within the table. For the 25th percentile, use a probability of 0.25. Usually, you will not find exactly the value you are looking for, so settle for the value that is as close as you can get. This value, 0.2514, is bolded in Table 6.7, which is an excerpt from the Normal table in Appendix A.

z	.00	.01	.02	.03	.04	.05	.06	.07	.08	.09
−0.7	.2420	.2389	.2358	.2327	.2296	.2266	.2236	.2206	.2177	.2148
−0.6	.2743	.2709	.2676	.2643	.2611	.2578	.2546	**.2514**	.2483	.2451

▲ **TABLE 6.7** Part of the Standard Normal Table.

You can now see that the z-score corresponding to a probability of 0.2514 is −0.67. This relation between the z-score and the probability is shown in Figure 6.27.

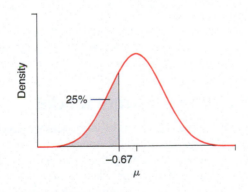

◄ **FIGURE 6.27** The percentile for 0.25 is −0.67 standard unit, because 25% of the area under the standard Normal curve is below −0.67.

Step 2: Convert the z-score to the proper units.
    A z-score of −0.67 tells us that this value is 0.67 standard deviation below the mean. We need to convert this to a height in inches.

One standard deviation is 3 inches, so 0.67 standard deviation is

$$0.67 \times 3 = 2.01 \text{ inches}$$

The height is approximately 2.0 inches below the mean. The mean is 64.0 inches, so the 25th percentile is

$$64.0 - 2.0 = 62.0$$

CONCLUSION   The woman's height at the 25th percentile is 62 inches, assuming that women's heights follow a $N(64, 3)$ distribution.

TRY THIS!   Exercise 6.47

Figure 6.28 shows some percentiles and the corresponding z-scores to help you visualize percentiles.

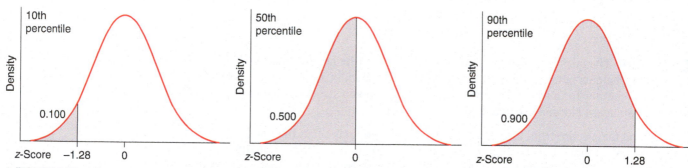

▲ **FIGURE 6.28** z-Scores and percentiles—the 10th, 50th, and 90th percentiles. A percentile corresponds to the percentage in the area to the left under the curve.

SNAPSHOT ▸ The Normal Model

WHAT IS IT?	▸	A model of a distribution for some numerical variables.
WHAT DOES IT DO?	▸	Provides us with a model for finding probabilities for many real-life numerical variables.
HOW DOES IT DO IT?	▸	The probabilities are represented by the area underneath the bell-shaped curve.
HOW IS IT USED?	▸	If the Normal model is appropriate, it can be used for finding probabilities or for finding measurements associated with particular percentiles.

Appropriateness of the Normal Model

The Normal model does not fit all distributions of numerical variables. For example, if we are randomly selecting people who submitted tax returns to the federal government, we cannot use the Normal model to find the probability that someone's income is higher than the mean value. The reason is that incomes are right-skewed, so the Normal model will not fit.

How do we know whether the Normal model is appropriate? Unfortunately, there is no checklist. However, the Normal model is a good first-choice model if you suspect that the distribution is symmetric and has one mode. Once you have collected data, you can check to see whether the Normal model closely matches the data.

In short, the Normal model is appropriate if it produces results that match what we see in real life. If the data we collect match the Normal model fairly closely, then the model is appropriate. Figure 6.29a shows a histogram of the actual heights from a sample of more than 1400 women from the National Health and Nutrition Examination Survey (NHANES, www.cdc.gov/nchs/nhanes), with the Normal model superimposed over the histogram. Note that the model, though not perfect, is a pretty good description of the shape of the distribution. Compare this to Figure 6.29b, which shows the distribution of weights for the same women. The Normal model is not a very good fit for these data. The Normal model has the peak at the wrong place; specifically, the Normal model is symmetric, whereas the actual distribution is right-skewed.

Statisticians have several ways of checking whether the Normal model is a good fit to the population, but the easiest thing for you to do is to make a histogram of your data and see whether it looks unimodal and symmetric. If so, the Normal model is likely to be a good model.

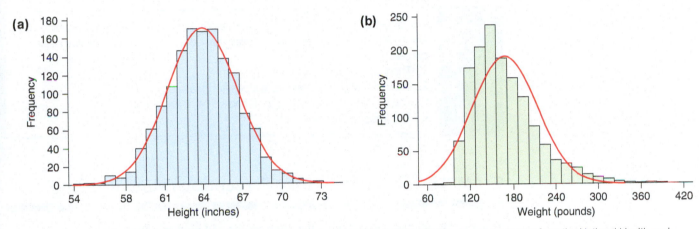

▲ FIGURE 6.29   (a) A histogram of data from a large sample of adult women in the United States drawn at random from the National Health and Nutrition Examination Survey. The red curve is the Normal curve, which fits the shape of the histogram very well, indicating that the Normal model would be appropriate for these data. (b) A histogram of weights for the same women. Here the Normal model is a bad fit to the data.

SECTION 6.3

The Binomial Model

The Normal model applies to many real-life, continuous-valued numerical variables. The **binomial probability model** is useful in many situations with discrete-valued numerical variables (typically counts, whole numbers). As with the Normal model, we will explain what the model looks like and how to calculate probabilities with it. We will also provide examples and discuss why the binomial model is appropriate to the situations.

The classic application of the binomial model is counting heads when flipping a coin. Let's say I flip a coin 10 times. What is the probability I get 1 head? 2 heads? 10 heads? This is a situation that the binomial model fits quite well, and the probabilities of these outcomes are given by the binomial model. If we randomly select 10 people, what's the probability that exactly 5 will be Republicans? If we randomly select 100 students, what's the probability that 10 or fewer will be on the Dean's List? These are examples of situations where the binomial model applies.

How do you recognize a binomial model? The first sign that your random experiment is a candidate for the binomial model is that the outcome you are interested in is a count. If that's the case, then all four of the following characteristics must be present:

1. *A fixed number of trials.* We represent this number with the letter n. For example, if we flip a coin 10 times, then $n = 10$.

2. *Only two outcomes are possible at each trial.* We will call these two outcomes "success" and "failure." For example, we might consider the outcome of heads to be a success. Or we might be selecting people at random and counting the number of males; in this case, of the two outcomes "male" and "female," "male" might be considered a success.

3. *The probability of success is the same at each trial.* We represent this probability with the letter p. For example, the probability of getting heads after a coin flip is $p = 0.50$ and does not change from flip to flip.

4. *The trials are independent.* The outcome of one trial does not affect the outcome of any other trial.

If all four of these characteristics are present, the binomial model applies, and you can easily find the probabilities by looking at the binomial probability distribution.

> **KEY POINT**
>
> The binomial model provides probabilities for random experiments in which you are counting the number of successes that occur. Four characteristics must be present:
> 1. There are a number of trials: n.
> 2. The only two outcomes are success and failure.
> 3. The probability of success, p, is the same at each trial.
> 4. The trials are independent.

EXAMPLE 10 Extrasensory Perception (Mind Reading)

Zener cards are cards used to test whether people can read minds (telepathy). Each card in a Zener deck has one of five designs: a star, a circle, a plus sign, a square, or three wavy lines (Figure 6.30). In an experiment, one person, the "sender," selects a card at random, looks at it, and thinks about the symbol on the card. Another person, the "receiver," cannot see the card (and in some studies cannot even see the sender) and guesses which of the symbols was chosen. A researcher records whether the guess was correct. The card is then placed back in the deck, the deck is shuffled, and another card is drawn. Suppose this happens 10 times (10 guesses are made). The receiver gets 3 guesses correct, and the researcher wants to know the probability of this happening if the receiver is simply guessing.

▶ **FIGURE 6.30** Zener cards (ESP cards) show one of five shapes. A deck has equal numbers of each shape.

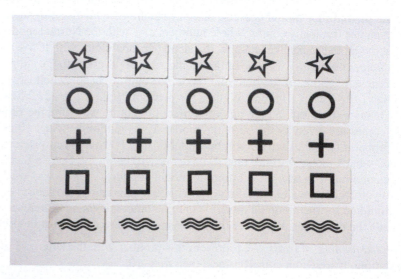

QUESTION   Explain why this is a binomial experiment.

SOLUTION   First, we note that we are counting something: the number of successful guesses. We need to check that the experiment meets the four characteristics of a binomial model. (1) The experiment consists of a fixed number of trials: $n = 10$. (2) The outcome of each trial is success or failure: The receiver either gets the right answer or does not. (3) The probability of a success at a trial is $p = 1/5 = 0.20$, because there are 5 shapes and so the receiver has a 1-in-5 chance of getting it correct, if we assume the receiver is guessing. (4) As long as the cards are put back in the deck and reshuffled (thoroughly), the probability of a success is the same for each trial, and each trial is independent.

All four characteristics are satisfied, so the experiment fits the binomial model.

TRY THIS!   Exercise 6.59

EXAMPLE 11 Why Not Binomial?

The following four experiments are almost, but not quite, binomial experiments:

a. Record the number of different eye colors in a group of 50 randomly selected people.

b. A married couple decide to have children until a girl is born but to stop at five children if they do not have any girls. How many children will the couple have?

c. Suppose the probability that a flight will arrive on time (within 15 minutes of the scheduled arrival time) at O'Hare Airport in Chicago is 85%. How many flights arrive on time out of 300 flights scheduled to land on a day in January?

d. A student guesses on every question of a test that has 10 multiple-choice questions and 10 true/false questions. Record the number of questions the student gets right.

QUESTION   For each situation, explain which of the four characteristics is not met.

SOLUTIONS

a. This is not a binomial experiment because there are more than two eye colors, so more than two outcomes may occur at each trial. However, if we reduced the eye colors to two categories by, say, recording whether the eye color was brown or not brown, then this would be a binomial experiment.

b. This is not a binomial experiment because the number of trials is not fixed before the children are born. The number of "trials" depends on when (or whether) the first girl is born. The number of trials varies depending on what happens—the word *until* tells you that.

c. This experiment is not binomial because the flights are not independent. If the weather is bad, the chance of arriving on time for all flights is lower. Therefore, if one flight arrives late, then another flight is more likely to arrive late.

d. This is not a binomial experiment because the probability of success on each trial is not constant; the probability of success is lower on multiple-choice questions than on true/false questions. Therefore, criterion 3 is not met. However, if the test were subdivided into two sections (multiple-choice and true/false), then each separate section could be called a binomial experiment if we assumed the student was guessing.

TRY THIS!   Exercise 6.61

Visualizing the Binomial Distribution

All binomial models have the four characteristics listed above, but the list gives us flexibility in n and p. For example, if we had flipped the coin 6 times instead of 10, it would still be a binomial experiment. Also, if the probability of a success were 0.6 instead of 0.5, we would still have a binomial experiment. For different values of n and p, we have different binomial experiments, and the binomial distribution looks different in each case.

Figure 6.31 shows that the binomial distribution for $n = 3$ and $p = 0.5$ is symmetric. We can read from the graph that the probability of getting exactly 2 successes (2 heads in 3 flips of a coin) is about 0.38, and the probability of getting no successes is the same as the probability of getting all successes.

▶ **FIGURE 6.31** Binomial distribution with $n = 3$, $p = 0.50$.

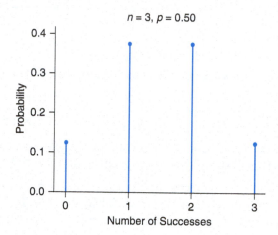

If n is bigger but p remains fixed at 0.50, the distribution is still symmetric as shown in Figure 6.32 for $n = 15$.

If the probability of success is not 50%, the distribution might not be symmetric. Figure 6.33 shows the distribution for $p = 0.3$, which means we're less likely to get a large number of successes than a smaller number, so the probability "spikes" are taller for smaller numbers of successes. The plot is now right-skewed.

However, even if the distribution is not symmetric, if we increase the number of trials, it becomes symmetric. The shape of the distribution depends on both n and p.

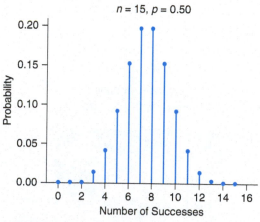

▲ **FIGURE 6.32** Binomial distribution with $n = 15$, $p = 0.50$.

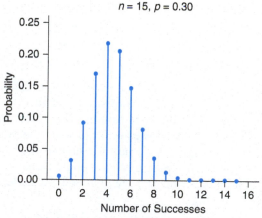

▲ **FIGURE 6.33** Binomial distribution with $n = 15$, $p = 0.30$.

If we keep $p = 0.3$ but increase n to 100, we get a more symmetric shape, as shown in Figure 6.34.

Binomial distributions have the interesting property that if the number of trials is large enough, the distributions are symmetric.

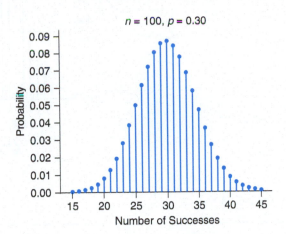

$n = 100, p = 0.30$

◀ **FIGURE 6.34** Binomial distribution with $n = 100$, $p = 0.30$. Note that we show x only for values between 15 and 45. The shape is symmetric, even though p is not 0.50.

 KEY POINT   The shape of a binomial distribution depends on both n and p. Binomial distributions are symmetric when $p = 0.5$, but they are also symmetric when n is large, even if p is close to 0 or 1.

Finding Binomial Probabilities

Because the binomial distribution depends only on the values of n and p, once you have identified an experiment as binomial and have identified the values of n and p, you can find any probability you wish. We use the notation $b(n, p, x)$ to represent the **binomial probability** of getting x successes in a binomial experiment with n trials and probability of success p. For example, imagine tossing a coin 10 times. If each side is equally likely, what is the probability of getting 4 heads? We represent this with $b(10, 0.50, 4)$.

The easiest and most accurate way to find binomial probabilities is to use technology. Statistical calculators and software have the binomial distribution built in and can easily calculate probabilities for you.

EXAMPLE **12** Stolen Bicycles

 Tech

According to the website bicyclelaw.com, only 5% of stolen bicycles are returned to owners.

QUESTION   Accepting for the moment that the four characteristics of a binomial experiment are satisfied, write the notation for the probability that exactly 10 bicycles will be returned if 235 are stolen. What is this probability?

SOLUTION   The number of trials is 235, the probability of success is 0.05, and the number of successes is 10. Therefore, we can write the binomial probability as $b(235, 0.05, 10)$.

Figure 6.35 shows the command and output for finding this probability using a TI-84 (part a) and using StatCrunch (part b). The command *binompdf* stands for "binomial probability distribution function." The results show us that the probability that exactly 10 bicycles will be returned, assuming that this is in fact a binomial experiment, is about 11.1%.

▶ **FIGURE 6.35** Calculations for binomial trial using **(a)** TI-84 and **(b)** StatCrunch.

(a)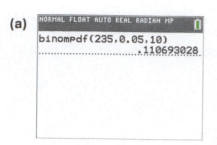

(b)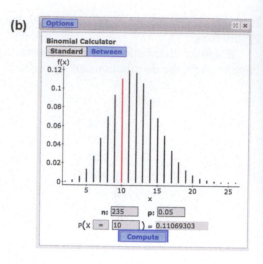

TRY THIS!   Exercise 6.65

Although we applied the binomial model to the stolen bicycles example, we had to make the assumption that the trials are independent. A "trial" in this context consists of a bicycle being stolen, and a "success" occurs if the bike is returned. If one person stole several bikes on a single day from a single location, this assumption of independent trials would be wrong because if police found one of the bikes, they would have a good chance of finding the others that the thief took. Sometimes we have no choice but to make certain assumptions to complete a problem, but we need to be sure to check our assumptions when we can, and we must be prepared to change our conclusion if our assumptions were wrong.

Another approach to finding binomial probabilities is to use a table. Published tables are available that list binomial probabilities for a variety of combinations of values for n and p. One such table is provided in Appendix A. This table lists binomial probabilities for values of n between 2 and 15 and for several different values of p.

EXAMPLE 13   Recidivism in Texas

The three-year recidivism rate of parolees in Texas is about 20% (https://bjafactsheets. iir.com/State/TX). In other words, 20% of released prisoners return to prison within three years of their release. Suppose a prison in Texas released 15 prisoners.

QUESTION   Assuming that whether one prisoner returns to prison is independent of whether any of the others returns, use Table 6.8, which shows binomial probabilities for $n = 15$ and for various values of p, to find the probability that exactly 6 out of 15 will end up back in prison within three years.

SOLUTION   Substituting the numbers, you can see that we are looking for $b(15, 0.20, 6)$. Referring to Table 6.8, you can see—by looking in the table in the row for $x = 6$, and the column for $p = 0.2$—that the probability that exactly 8 parolees will be back in prison within three years is 0.043, or about a 4.3% chance.

x	0.1	0.2	0.25	0.3	0.4	0.5	0.6	0.7	0.75	0.8	0.9
6	.002	.043	.092	.147	.207	.153	.061	.012	.003	.001	.000
7	.000	.014	.039	.081	.177	.196	.118	.035	.013	.003	.000
8	.000	.003	.013	0.35	.118	.196	.177	.081	.039	.014	.000
9	.000	.001	.003	.012	.061	.153	.207	.147	.092	.043	.002
10	.000	.000	.001	.003	.024	.092	.186	.206	.165	.103	.010
11	.000	.000	.000	.001	.007	.042	.127	.219	.225	.188	.043
12	.000	.000	.000	.000	.002	.014	.063	.170	.225	.250	.129
13	.000	.000	.000	.000	.000	.003	.022	.092	.156	.231	.267
14	.000	.000	.000	.000	.000	.000	.005	.031	.067	.132	.343
15	.000	.000	.000	.000	.000	.000	.000	.005	.013	.035	.206

▲ **TABLE 6.8** Binomial probabilities with a sample of 15 and x-values of 6 or higher. The probability for b(15, 0.20, 6) is shown in red.

Using a TI-84 or using StatCrunch, as shown in Figure 6.36 on the next page, we can see another way to get the same answer.

(a)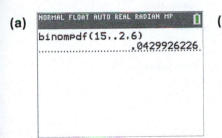

(b)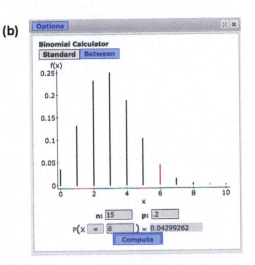

◀ **FIGURE 6.36** Output for calculating b(15, 0.3, 8) using **(a)** TI-84 and **(b)** StatCrunch.

TRY THIS! Exercise 6.67

Finding (Slightly) More Complex Probabilities

EXAMPLE 14 ESP with 10 Trials

For a test of psychic abilities, researchers have asked the sender to draw 10 cards at random from a large deck of Zener cards (see Example 9). Assume that the cards are replaced in the deck after each use and the deck is shuffled. Recall that this deck contains equal numbers of 5 unique shapes. The receiver guesses which card the sender has drawn.

QUESTIONS

a. What is the probability of getting *exactly* 5 correct answers (out of 10 trials) if the receiver is simply guessing (and has no psychic ability)?

b. What is the probability that the receiver will get *5 or more* of the cards correct out of 10 trials?

c. What is the probability of getting *fewer than 5* correct in 10 trials with the ESP cards?

SOLUTIONS

a. In Example 9 we identified this as a binomial experiment. With that done, we must now identify n and p. The number of trials is 10, so $n = 10$. If the receiver is guessing, then the probability of a correct answer is $p = 1/5 = 0.20$. Therefore, we wish to find $b(10, 0.2, 5)$.

Figure 6.37a gives the TI-84 output, where you can see that the probability of getting 5 right out of 10 is only about 0.0264. Figure 6.37b shows Minitab output for the same question.

► **FIGURE 6.37** Technology output for $b(10, 0.2, 5)$. **(a)** Output from a TI-84. **(b)** Output from Minitab.

(a)

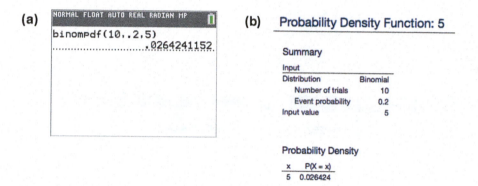

(b)   **Probability Density Function: 5**

Summary

Input

Distribution	Binomial
Number of trials	10
Event probability	0.2
Input value	5

Probability Density

x	P(X = x)
5	0.026424

Figure 6.38a shows a graph of the pdf. The probability $b(10, 0.2, 5)$ is quite small. The graph shows that it is unusual to get exactly 5 correct when the receiver is guessing.

► **FIGURE 6.38(a)** The probability distribution for the numbers of successes for 10 trials with the Zener deck, assuming guessing.

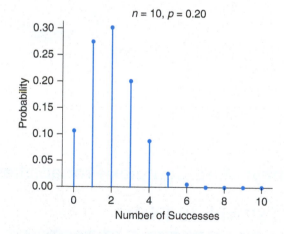

b. The phrase *5 or more* means we need the probability that the receiver gets 5 correct **or 6 or 7 or 8 or 9 or** 10 correct. The outcomes 5 correct, 6 correct, and so on are mutually exclusive because if you get exactly 5 correct, you cannot possibly also get exactly 6 correct. Therefore, we can find the probability of 5 or more correct by adding the individual probabilities together:

$$b(10, 0.2, 5) + b(10, 0.2, 6) + b(10, 0.2, 7) + b(10, 0.2, 8) + b(10, 0.2, 9) + b(10, 0.2, 10)$$
$$= 0.026 \quad + \quad 0.006 \quad + \quad 0.001 \quad + \quad 0.000 \quad + \quad 0.000 \quad + \quad 0.000$$
$$= 0.033$$

These probabilities are circled in Figure 6.38b.

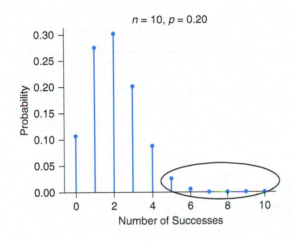

◀ **FIGURE 6.38(b)** Summing the probabilities represented by the circled bars gives us the probability of getting 5 or more correct.

c. The phrase *fewer than 5 correct* means 4, 3, 2, 1, or 0 correct. These probabilities are circled in Figure 6.38c.

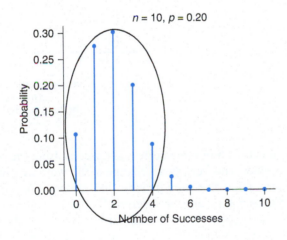

◀ **FIGURE 6.38(c)** Summing the probabilities represented by the circled bars gives the probability of getting fewer than 5 correct.

The event that we get fewer than 5 correct is the complement of the event that we get 5 or more correct, as you can see in Figure 6.39, which shows all possible numbers of successes with 10 trials. *Fewer than 5* is the same event as *4 or fewer* and is shown in the left oval in the figure.

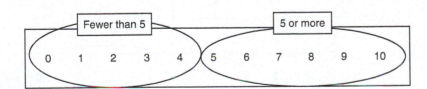

◀ **FIGURE 6.39** The possible numbers of successes out of 10 trials with binomial data. Note that *fewer than 5* is the complement of *5 or more*.

Because we know the probability of 5 or more, we can find its complement by subtracting from 1:

$$1 - 0.033 = 0.967$$

CONCLUSIONS

a. The probability of exactly 5 correct is 0.026.

b. The probability of 5 or more correct is 0.033.

c. The probability of fewer than 5 (that is, of 4 or fewer) is 0.967.

TRY THIS! Exercise 6.75

Most technology also offers you the option of finding binomial probabilities for x *or fewer*. In general, probabilities of *x or fewer* are called **cumulative probabilities**. Figure 6.40a shows the cumulative binomial probabilities provided by the TI-84; notice the "c" in binom**c**df. Figure 6.40b shows the same calculations using StatCrunch.

(a)  **(b)**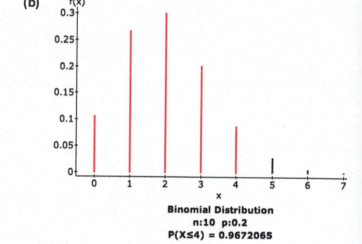

▶ **FIGURE 6.40** Output to calculate binomial trials for $n = 10$, $p = 0.2$, and 4 or fewer successes using **(a)** TI-84 and **(b)** StatCrunch.

The probability of getting 5 or more correct is different from the probability of getting more than 5 correct. This very small change in the wording gives very different results. Figure 6.41 shows that *5 or more* includes the outcomes 5, 6, 7, 8, 9, and 10, whereas *more than 5* does not include the outcome 5.

When finding Normal probabilities, we did not have to worry about such subtleties of language because for a continuous numerical variable, the probability of getting 5 or more is the same as the probability of getting more than 5.

▶ **FIGURE 6.41** Interpretation of words for discrete counts with n of 10 trials. **(a)** Results for *5 or more*; **(b)** results for *more than 5*. Note that they are different.

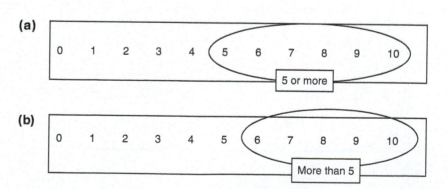

Finding Binomial Probabilities by Hand

When the number of trials is small, we can find probabilities by relying on the probability rules in Chapter 5 and listing all possible outcomes. In fact, there is a general pattern we can exploit to find a formula that will give us the probabilities for any value of n or p. Listing all of the outcomes will help us see why this formula works.

Suppose we are testing someone using the Zener cards that have five different shapes. After each card is guessed, it is returned to the deck, and the deck is shuffled before the next card is drawn. We want to count how many correct guesses a potential psychic makes in four trials. We can have five different outcomes for this binomial experiment: 0 correct answers, 1 correct, 2 correct, 3 correct, and all 4 correct.

The first step is to list all possible outcomes after four attempts. At each trial, the guesser is either right or wrong. Using R for right and W for wrong, we list all possible sequences of right or wrong results in four trials.

4 Right	3 Right	2 Right	1 Right	0 Right
RRRR	RRRW	RRWW	RWWW	WWWW
	RRWR	RWRW	WRWW	
	RWRR	RWWR	WWRW	
	WRRR	WWRR	WWWR	
		WRWR		
		WRRW		

There are 16 possible outcomes, although these are not all equally likely.

Now we find the probabilities of getting 4 right, 3 right, 2 right, 1 right, and 0 right. Because there are five shapes, and all five are equally likely to be chosen, if the receiver is simply guessing and has no psychic ability, then the probability of a right answer at a single trial is $1/5 = 0.2$, and the probability of guessing wrong is 0.8.

Four right means "right AND right AND right AND right." Successive trials are independent (because we replace the card and reshuffle every time), so we can multiply the probabilities using the multiplication rule:

$$P(RRRR) = 0.2 \times 0.2 \times 0.2 \times 0.2 = 0.0016$$

This probability is just $b(4, 0.2, 4) = 1(0.0016) = 1(0.2)^4$. (We multiply by 1 because there is only one way that we can get all 4 right to happen, which is also why there is only one outcome listed in the "4 right" column.)

The probability of getting 3 right and 1 wrong includes all four options in the second group. The probability for each of these options is obtained by calculating the probability of 3 right and 1 wrong, and all of these probabilities will be the same. To get the total probability, therefore, we multiply by 4:

$$P(RRRW) = 0.2 \times 0.2 \times 0.2 \times 0.8 = 0.0064$$
$$P(3 \text{ right and } 1 \text{ wrong, in any order}) = 4(0.0064) = 0.0256$$
$$b(4, 0.2, 3) = 4(0.2)^3(0.8)^1 = 0.0256$$

The probability of getting 2 right and 2 wrong includes all six options in the third group. The probability for each of these options is obtained by calculating the probability of 2 right and 2 wrong; then, to get the total, we multiply by 6:

$$P(RRWW) = 0.2 \times 0.2 \times 0.8 \times 0.8 = 0.0256$$
$$P(2 \text{ right and } 2 \text{ wrong, in any order}) = 6(0.0256) = 0.1536$$
$$b(4, 0.2, 2) = 6(0.2)^2(0.8)^2 = 0.1536$$

Looking Back

And

The multiplication rule was Probability Rule 5c in Chapter 5 and applies only to independent events: $P(A \text{ AND } B) = P(A) \ P(B)$.

The probability of getting 1 right and 3 wrong includes all four options in the fourth group. We obtain the probability for each of the options by calculating the probability of 1 right and 3 wrong; then we multiply by 4 to get the total:

$$P(RWWW) = 0.2 \times 0.8 \times 0.8 \times 0.8 = 0.1024$$
$$P(1 \text{ right and 3 wrong, in any order}) = b(4, 0.2, 1) = 4(0.1024) = 0.4096$$
$$b(4, 0.2, 1) = 4(0.2)^1(0.8)^3 = 0.4096$$

Finally, the probability of getting all four wrong is

$$P(WWWW) = b(4, 0.2, 0) = 0.8 \times 0.8 \times 0.8 \times 0.8 = 0.4096$$

Because there is only one way for this to happen, we multiply 0.4096 by 1. Thus,

$$b(4, 0.2, 0) = 1(0.8)^4 = 0.4096$$

Table 6.9 summarizes the results.

If you add the probabilities, you will see that they add to 1, as they should, because this list includes all possible outcomes that can happen.

Figure 6.42 shows a graph of the probability distribution. Note that the graph is right-skewed because the probability of success is less than 0.50. Also note that the probability of getting 4 out of 4 right with the Zener cards is very small.

Number Right	Probability
4 right	0.0016
3 right	0.0256
2 right	0.1536
1 right	0.4096
0 right	0.4096

▲ TABLE 6.9 Summary of the probabilities of all possible numbers of successes in four trials with the Zener cards.

▶ FIGURE 6.42 Probability distribution using the Zener cards with four trials.

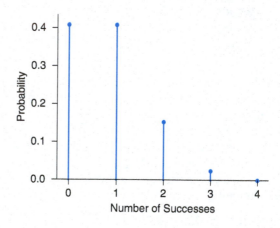

You might compare these probabilities with those in the binomial table. You will find that they agree, if you round off the numbers we found "by hand" to three decimal places.

The Formula This approach of listing all possible outcomes is tedious when $n = 4$, and the tedium increases for larger values of n. (For $n = 5$ we have to list 32 possibilities, and for $n = 6$ there are 64.) However, mathematicians have derived a formula that finds probabilities for the binomial distribution. The binomial table in this text and the results of your calculator are based on this formula:

$P(x \text{ successes in } n \text{ trials of a binomial experiment}) =$

(number of different ways of getting x successes in n trials) $p^x(1 - p)^{(n-x)}$

For example, for our alleged psychic who gets four guesses with the Zener cards, the probability of 3 successes ($x = 3$) in four trials ($n = 4$) is

$$\text{(number of ways of getting 3 successes in 4 trials) } (0.2)^3(0.8)^1$$
$$= 4 (0.2)^3(0.8)^1$$
$$= 0.0256$$

The Shape of the Binomial Distribution: Center and Spread

Unlike the Normal distribution, the mean and the standard deviation of the binomial distribution can be easily calculated. Their interpretation is the same as with all distributions: the mean tells us where the distribution balances, and the standard deviation tells us how far values are, typically, from the mean.

For example, in Figure 6.43, the binomial distribution is symmetric, so the mean sits right in the center at 7.5 successes.

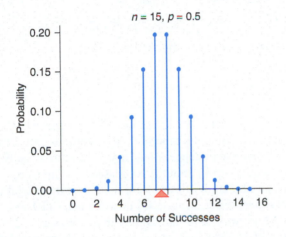

◄ **FIGURE 6.43** A binomial distribution with $n = 15$, $p = 0.50$. The mean is at 7.5 successes.

If the distribution is right-skewed, the mean will be just to the right of the peak, closer to the right tail, as shown in Figure 6.44. This is a binomial distribution with $n = 15$, $p = 0.3$, and the mean sits at the balancing point: 4.5 successes.

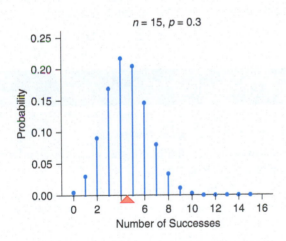

◄ **FIGURE 6.44** A binomial distribution with $n = 15$, $p = 0.3$. The mean is at the balancing point of 4.5 successes.

The mean number of successes, μ, of a binomial probability distribution can be found with a simple formula:

Formula 6.1a: $\mu = np$

In words, the mean number of successes in a binomial experiment is the number of trials times the probability of a success.

This formula should make intuitive sense. If you toss a coin 100 times, you would expect about half of the tosses to be heads. Thus the "typical" outcome is $(1/2) \times 100 = 50$, or $np = 100 \times 0.5$.

The standard deviation for the number of successes, σ, which measures the spread, is less intuitive:

Formula 6.1b: $\sigma = \sqrt{np(1 - p)}$

For example, Figure 6.44 shows a binomial distribution with $n = 15$, $p = 0.30$. The mean is $15 \times 0.3 = 4.5$. The standard deviation is $\sqrt{15 \times 0.3 \times 0.7} = \sqrt{3.15} = 1.775$.

> **KEY POINT**
>
> For a binomial experiment, the mean is
> $$\mu = np$$
> For a binomial experiment, the standard deviation is
> $$\sigma = \sqrt{np(1 - p)}$$

Interpreting the Mean and the Standard Deviation   The mean of any probability distribution, including the binomial, is sometimes called the **expected value**. This name gives some intuitive understanding. If you were to actually carry out a binomial experiment, you would *expect* about μ successes. If I flip a coin 100 times, I *expect* about 50 heads. If, in an ESP study, 10 trials are made and the probability of success at each trial is 0.20, we *expect* about 2 successes due to chance ($10 \times 0.2 = 2$).

Will I get exactly 50 heads? Will the ESP receiver get exactly 2 cards correct? Sometimes, yes. Usually, no. Although we expect μ successes, we usually get μ give or take some amount. That give-or-take amount is what is measured by σ. In 100 tosses of a fair coin, we expect $\mu = 50$ heads, give or take $\sigma = \sqrt{100 \times 0.5 \times 0.5} = 5$ heads. We expect 50, but we will not be surprised if we get between 45 and 55 heads. In the Zener card experiment with 10 trials, we expect the receiver to guess about 2 cards correctly, but in practice we expect him or her to get 2 give or take 1.3 because

$$\sqrt{10 \times 0.2 \times 0.82} = \sqrt{1.6} = 1.26$$

SNAPSHOT ▸ The Binomial Distribution

WHAT IS IT? ▸	A distribution for some discrete variables.
WHAT DOES IT DO? ▸	Gives probabilities for the number of successes observed in a fixed number of trials in a binomial experiment.
HOW DOES IT DO IT? ▸	If the conditions of a binomial experiment are met, once you identify n (the number of trials), p (the probability of success), and x (the number of successes), it gives the probability.
HOW IS IT USED? ▸	The probabilities are generally provided in the form of a table or a formula, but if you need to calculate them, use a calculator or technology.

EXAMPLE 15   Basketball Free-Throw Shots

The NBA leader in free throws for the 2018 season was Stephen Curry of the Golden State Warriors. Curry made approximately 92% of his free throws. Assume that free throw shots are independent; that is, success or failure on one shot does not affect the chance of success on another shot.

QUESTION If Stephen Curry has 600 free throws in an upcoming season, how many would you expect him to sink, give or take how many?

SOLUTION This is a binomial experiment. (Why?) The number of trials is $n = 600$, and the probability of success at each trial is $p = 0.92$. You would expect Curry to sink 92% of 600, or 552, free throws. Using Formula 6.1 for the mean yields

$$\mu = np = 600 \times (0.92) = 552$$

The give-or-take amount is measured by the standard deviation as given in Formula 6.1b:

$$\sigma = \sqrt{600 \times 0.92 \times 0.08} = 6.6453$$

You should expect him to hit about 552 free throws, give or take about 7.0.

$$552 + 7 = 559$$

$$552 - 7 = 545$$

CONCLUSION You would expect Stephen Curry to sink between 454 and 559 free throw shots.

TRY THIS! Exercise 6.79.

Surveys: An Application of the Binomial Model

Perhaps the most common application of the binomial model is in survey sampling. Imagine a large population of people, say the 100 million or so registered voters in the United States. Some percentage of them, call it p, have a certain characteristic. If we choose ten people at random (with replacement), we can ask what the probability is that all of them, or three of them, or six of them have that characteristic.

EXAMPLE **16** News Survey

The Pew Research Center says that 23% of people are "news integrators": people who get their news both from traditional media (television, radio, newspapers, magazines) and from the Internet (www.people-press.org). Suppose we take a random sample of 100 people.

QUESTIONS If, as is claimed, 23% of the population are news integrators, then how many people in our sample should we expect to be integrators? Give or take how many? Would you be surprised if 34 people in the sample turned out to be integrators?

SOLUTION Assuming that all four of the characteristics of a binomial distribution are satisfied, then we would expect 23% of our sample of 100 people to be integrators— that is, 23 people. The standard deviation is $\sqrt{100 \times 0.23 \times 0.77} = 4.2$. Thus, we should expect 23 people, give or take about 4.2 people, to be integrators. This means that we shouldn't be surprised if we got as many as $23 + 4.2 = 27.2$, or about 27, people. However, 34 people is quite a bit more than 1 standard deviation above what we expect. In fact, it is almost 3 standard deviations away, so it would be a surprisingly large number of people.

TRY THIS! Exercise 6.81.

The binomial model works well for surveys when people are selected *with* replacement (which means that, once they are selected, they have a chance of being selected again). If we are counting the people who have a certain characteristic, then the four characteristics of the binomial model are usually satisfied:

1. A fixed number of people are surveyed. In Example 16, $n = 100$.

2. The outcome is success (integrator) or failure (not an integrator).

3. The probability of a success is the same at each trial: $p = 0.23$.

4. The trials are independent. (This means that if one person is found to be an integrator, no one else in the sample will change their response.)

With surveys, we usually don't report the number of people who have an interesting characteristic; instead, we report the percentage of our sample. We would not report that 23 people in our sample of 100 were integrators; we would report that 23% of our sample were integrators. But the binomial distribution still applies because we are simply converting our counts to percentages.

In reality, surveys don't select people with replacement. The most basic surveys sample people without replacement, which means that, strictly speaking, the probability of a success is different after each trial. Imagine that the first person selected is an integrator. Now there are fewer integrators left in the population, so it is no longer true that 23% of the population are integrators; the percentage is slightly less. However, as you can imagine, this is not a problem if the population is very large. In fact, if the population size is very large relative to the sample size (at least 10 times bigger), then this difference is so slight that characteristic 3 is essentially met.

Taking a random sample, either with or without replacement, of a large and diverse population such as all U.S. voters is quite complicated. In practice, the surveys that we read about in the papers or hear about on the news use a modified approach that is slightly different from what we've discussed here. Random selection is still at the heart of these modified methods, though, and the binomial distribution often provides a good approximation for probabilities, even under these more complex schemes.

CASE STUDY REVISITED

You Sometimes Get More Than You Pay For

McDonald's claims that its ice cream cones weigh 3.18 ounces. However, one of the authors bought five cones and found that all five weighed more than that. Is this surprising?

Suppose we assume that the amount of ice cream dispensed follows a Normal distribution centered on 3.18 ounces. In other words, typically, a cone weighs 3.18 ounces, but sometimes a cone weighs a little more, and sometimes a little less. If this is the case, then, because the Normal distribution is symmetric and centered on its mean, the probability that a cone weighs more than 3.18 ounces is 0.50.

If the probability that a cone weighs more than 3.18 ounces is 0.50, what is the probability that five cones all weigh more? We can think of this as a binomial experiment if we assume the trials are independent. (This seems like a reasonable assumption.) We can then calculate $b(5, 0.5, 5)$ to find that the probability is 0.031.

If the typical McDonald's ice cream cone really weighs 3.18 ounces, then our outcome is fairly surprising: it happens only about 3% of the time. This means that it is fairly rare and raises the possibility that this particular McDonald's might deliberately dispense more than 3.18 ounces.

DATAPROJECT ▶ Generating Random Numbers

1 OVERVIEW

In Chapter 6 you learned about an important model, the Normal model, which describes the shape of some variables' distributions. In this project, we'll explore some data moves to help you improve your judgment about whether the Normal model is appropriate for a variable. The same technique can be used whenever you want to understand what a sample of data might look like if it came from a distribution with a particular shape.

2 GOALS

Know how to generate data from a given probability model. Understand how to work with pseudo-random numbers and know the importance of setting a "seed."

3 AM I NORMAL?

The Normal distribution is sometimes called the "bell-shaped curve" because, well, it is shaped somewhat like a bell. And this indicates one of the primary ways of classifying real-life distributions as either Normal or non-Normal: take a look. Does it look like a bell-shaped curve? StatCrunch, like other packages, makes this task a little easier by allowing you to superimpose a Normal model curve on top of your histogram.

Another approach is with a computer simulation. This approach is particularly useful for smaller sample sizes, where histograms can sometimes look skewed even if the population the sample was drawn from is symmetric. Here's the strategy: draw a random sample of the same number of observations as your data from a population that you *know* is Normal. Make a histogram. Repeat this lots of times. You'll get a sense of whether the histogram from your data is similar to, or different from, the simulated histograms. If it's similar, then your population distribution might be Normal. If it's different, then it's probably not.

Project The data set pitcher_quality.csv was compiled by a fantasy sports website and includes information about major league baseball pitchers. One traditional measure of quality of a pitcher is the "earned run average" or ERA. It is the average number of runs per game scored against a particular pitcher.

Make a histogram of the ERAs of the pitchers in this data set. Notice that the histogram is slightly asymmetric. There are only 58 observations in this data set, which is small compared to some data sets we've examined. Sometimes, with small datasets, the shape of the distribution of a sample can vary greatly from sample to sample. We'll be able to explore this variation through a simulation.

More to the point, we want to know if the distribution of this sample is consistent with a sample drawn from a Normal model. In other words, if we collected data on many, many more pitchers, would the complete data set follow the Normal model or have a different shape?

First, compute the summary statistics for the ERA. We need to know the sample size, the mean, and the standard deviation.

Second, select Data > Simulate. A drop-down list shows you about a dozen different models you can use to generate random data. Select the Normal model. Figure 6.46 shows you what this dialogue box looks like.

The very first thing you should do when simulating data is to select what's called a "seed." In the Simulate Normal dialogue box, under "Seeding," check the "Use fixed seed" button. Then, in the box, type in any number you like. This number is the seed.

You see, the numbers that you're about to generate are not really random. They're called "pseudo random" because, if you know the seed used to generate the random numbers, you can get the same "random" numbers every time! But if you don't know the seed, then they behave like random numbers do. Therefore, using a seed, we have more control over our simulation and can repeat it whenever we want, and we can explain to others how to get the same results we got.

We'll want to generate several samples of the same size as our original data. So enter 58 rows and 19 columns. Under "Normal parameters" enter the mean and standard deviation of our original data from the table of summary statistics you computed. Click

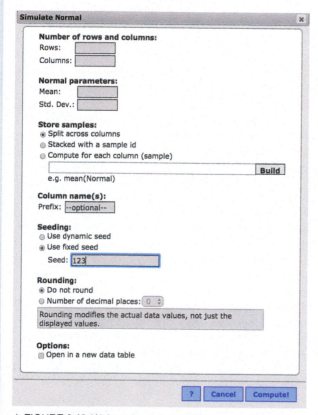

Data	Stat	Graph	Help
Load		>	
Properties			differential	Norma
Save		⌘S	0.45	4.486
Export			-0.25	3.4632
Validate			-0.39	3.7302
Row Selection		>	-0.09	3.579
Compute		>	-0.2	3.855
Simulate		>	0.19	3.8286

Simulate menu items:
- Bernoulli
- Beta
- Binomial
- Cauchy
- Chi-Square
- Discrete Uniform
- Exponential
- F
- Gamma
- Hypergeometric
- Normal
- Poisson
- T
- Uniform
- Weibull

Data values: 3.83, 3.67; 3.09, 3.9; 3.64, 3.91; 2.96, 3.93; 4.13, 3.63; 4.19, 3.88

▲ **FIGURE 6.45** You can simulate data from a variety of models.

Simulate Normal

Number of rows and columns:
Rows:
Columns:

Normal parameters:
Mean:
Std. Dev.:

Store samples:
◉ Split across columns
◎ Stacked with a sample id
◎ Compute for each column (sample)
[   ] [Build]
e.g. mean(Normal)

Column name(s):
Prefix: [--optional--]

Seeding:
◎ Use dynamic seed
◉ Use fixed seed
   Seed: [123]

Rounding:
◉ Do not round
◎ Number of decimal places: [0 ⇕]
Rounding modifies the actual data values, not just the displayed values.

Options:
☐ Open in a new data table

[?] [Cancel] [Compute!]

▲ **FIGURE 6.46** We're using the seed 123.

"Compute!". This will create 19 new columns of "data" in your file.

(If you don't believe us about the seed, open the Simulate Normal dialogue box and generate 19 new Normal samples using the same seed. You'll get the same numbers.)

To compare the histogram of your data, which may or may not be from a Normal distribution, with the 19 generated variables that definitely are from a Normal distribution, make a histogram of all 20 variables on a single graph. To do this, choose

Graph > Histogram

and select the ERA variable and the 19 generated variables. To overlay Normal distributions, under "Type:" select "Density" and under "Display options: Overlay distrib.:" select "Normal." Don't hit "Compute!" just yet!

At the bottom of the dialogue box, under "For multiple graphs," enter 5 rows per page and 4 columns per page. Now you can click "Compute!"

Does the ERA histogram look like it belongs with these other 19? If so, we can say that it is consistent with the Normal model.

Assignment: Looking back (way back) at the data you chose for the Chapter 1 project, which variables do you think will follow the Normal model and which will not? Choose one of each, and use these techniques to compare the histograms of these variables to histograms that come from a Normal model.

Another tool that is slightly more useful for determining Normality is the "QQ plot." Although QQ plots aren't covered in this course, they're easy to use; you can easily use StatCrunch to generate a QQ plot for any variable in your dataset. If the QQ plot follows a straight line, then the data are from a Normal model. Use the techniques for generating random samples to create 20 QQ plots: 19 that truly come from a Normal distribution and 1 from the ERA data. Then compare the Normal QQ plots to your ERA data. Do you think your variable follows the Normal model?

CHAPTER REVIEW

KEY TERMS

Probability distribution, 267
Probability distribution function
    (pdf), 267
Probability model, 268
Discrete outcomes, 268
Continuous
    outcomes, 268

Probability density curve, 271
Normal curve, 274
Normal distribution, 274
Mean of a probability
    distribution, μ, 274
Standard deviation of a probabil-
    ity distribution, σ, 274

Normal model: Notation,
    $N(\mu, \sigma)$, 275
Standard Normal model
    Notation, $N(0, 1)$, 281
Percentile, 284
Binomial probability model
    Notation, $b(n, p, x)$, 291

n is the number of trials
p is the probability of success
    on one trial
x is the number of successes
Cumulative probabilities, 296
Expected value, 300

LEARNING OBJECTIVES

After reading this chapter and doing the assigned homework problems,
you should

- Be able to distinguish between discrete and continuous variables.

- Know when a Normal model is appropriate and be able to apply the model to find probabilities.

- Know when the binomial model is appropriate and be able to apply the model to find probabilities.

SUMMARY

Probability models try to capture the essential features of real-world experiments and phenomena that we want to study. In this chapter, we focused on two very useful models: the Normal model and the binomial model.

The Normal model is an example of a model of probabilities for continuous numerical variables. The Normal probability model is also called the Normal distribution and the Gaussian curve. It can be a useful model when a histogram of data collected for a variable is unimodal and symmetric. Probabilities are found by finding the area under the appropriate region of the Normal curve. These areas are best calculated using technology. If technology is not available, you can also convert measures to standard units and then use the table of areas for the standard Normal distribution, provided in Appendix A.

The binomial model is an example of a model of probabilities for discrete numerical outcomes. The binomial model applies to binomial experiments, which occur when we are interested in counting the number of times some event happens. These four characteristics must be met for the binomial model to be applied:

1. There must be a fixed number of trials, n.
2. Each trial has exactly two possible outcomes.
3. Each of the trials must have the same probability of "success." This probability is represented by the letter p.
4. The trials must be independent of one another.

You can find binomial probabilities with technology or sometimes with a table, such as the one in Appendix A.

It is important to distinguish between continuous and discrete numerical variables, because if the variable has discrete numerical outcomes, then the probability of getting, say, *5 or more* (5 OR 6 OR 7 OR . . .) is different from the probability of getting *more than 5* (6 OR 7 OR . . .). This is not the case for a continuous numerical variable.

Formulas

For converting to standardized scores: $z = \dfrac{x - \mu}{\sigma}$

x is a measurement.
μ is the mean of the probability distribution.
σ is the standard deviation of the probability distribution.

For the mean of a binomial model:

Formula 6.1a: $\mu = np$

μ is the binomial mean.
n is the number of trials.
p is the probability of success of one trial.

For the standard deviation of a binomial model:

Formula 6.1b: $\sigma = \sqrt{np(1 - p)}$

σ is the binomial standard deviation.
n is the number of trials.
p is the probability of success of one trial.

SOURCES

Men's cholesterol levels, blood pressures, and heights throughout this chapter: NHANES (www.cdc.gov/nchs/nhanes).

Women's heights and weights throughout this chapter: NHANES (www.cdc.gov/nchs/nhanes).

SECTION EXERCISES

SECTION 6.1

6.1–6.4 Directions Determine whether each of the following variables would best be modeled as continuous or discrete.

TRY 6.1 (Example 1)
   a. The height of a high-rise apartment building
   b. The number of floors in a high-rise apartment building

6.2   a. The number of cars passing through an intersection in one hour
   b. The weight of a person

6.3   a. The height of a person in inches
   b. The weight of a person in pounds

6.4   a. The weight of a car in pounds
   b. The weight of a car in kilograms

TRY 6.5 Loaded Die (Example 2) A magician has shaved an edge off one side of a six-sided die, and as a result, the die is no longer "fair." The figure shows a graph of the probability density function (pdf). Show the pdf in table format by listing all six possible outcomes and their probabilities.

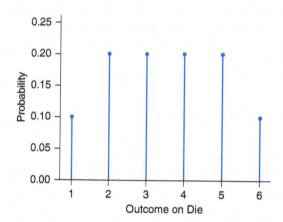

6.6 Fair Die Toss a fair six-sided die. The probability density function (pdf) in table form is given. Make a graph of the pdf for the die.

Number of Spots	1	2	3	4	5	6
Probability	1/6	1/6	1/6	1/6	1/6	1/6

* 6.7 Distribution of Two Thumbtacks When a certain type of thumbtack is flipped, the probability of its landing tip up (U) is 0.60 and the probability of its landing tip down (D) is 0.40.

Now suppose we flip two such thumbtacks: one red, one blue. Make a list of all the possible arrangements using U for up and D for down, listing the red one first; include both UD and DU. Find the probabilities of each possible outcome, and record the result in table form. Be sure the total of all the probabilities is 1.

6.8 Two Children Make a list of all possible outcomes for gender when a family has two children. Assume that the probability of having a boy is 0.50 and the probability of having a girl is also 0.50. Find the probability of each outcome in your list.

6.9 Two Thumbtacks
   a. From your answers in Exercise 6.7, find the probability of getting 0 ups, 1 up, or 2 ups when flipping two thumbtacks, and report the distribution in a table.
   b. Make a probability distribution graph of this.

6.10 Two Children Using your list of outcomes in Problem 6.8:
   a. Find the probability of having 0, 1, or 2 girls in a family of two children and display the probability distribution in a table.
   b. Make a graph of the probability distribution.

TRY 6.11 Snow Depth (Example 3) Eric wants to go skiing tomorrow, but only if there are 3 inches or more of new snow. According to the weather report, any amount of new snow between 1 inch and 6 inches is equally likely. The probability density curve for tomorrow's new snow depth is shown. Find the probability that the new snow depth will be 3 inches or more tomorrow. Copy the graph, shade the appropriate area, and calculate its numerical value to find the probability. The total area is 1.

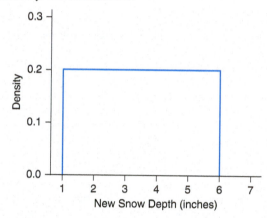

6.12 Snow Depth Refer to Exercise 6.11. What is the probability that the amount of new snow will be between 3 and 6 inches? Copy the graph from Exercise 6.11, shade the appropriate area, and report the numerical value of the probability.

SECTION 6.2

6.13 Applying the Empirical Rule with z-Scores The Empirical Rule applies rough approximations to probabilities for any unimodal, symmetric distribution. But for the Normal distribution we can be more precise. Use the figure and the fact that the Normal curve is symmetric to answer the questions. Do not use a Normal table or technology.

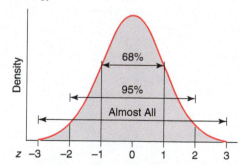

According to the Empirical Rule,

a. Roughly what percentage of *z*-scores are between −2 and 2?

    i. almost all   iii. 68%

    ii. 95%   iv. 50%

b. Roughly what percentage of *z*-scores are between −3 and 3?

    i. almost all   iii. 68%

    ii. 95%   iv. 50%

c. Roughly what percentage of *z*-scores are between −1 and 1.

    i. almost all   iii. 68%

    ii. 95%   iv. 50%

d. Roughly what percentage of *z*-scores are greater than 0?

    i. almost all   iii. 68%

    ii. 95%   iv. 50%

e. Roughly what percentage of *z*-scores are between 1 and 2?

    i. almost all   iii. 50%

    ii. 13.5%   iv. 2%

6.14 Length of Pregnancy Assume that the lengths of pregnancy for humans is approximately Normally distributed, with a mean of 267 days and a standard deviation of 10 days. Use the Empirical Rule to answer the following questions. Do not use the technology or the Normal table. Begin by labeling the horizontal axis of the graph with lengths, using the given mean and standard deviation. Three of the entries are done for you.

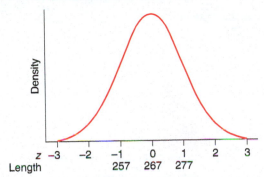

a. Roughly what percentage of pregnancies last more than 267 days?

    i. almost all   iii. 68%

    ii. 95%   iv. 50%

b. Roughly what percentage of pregnancies last between 267 and 277 days?

    i. 34%   iii. 2.5%

    ii. 17%   iv. 50%

c. Roughly what percentage of pregnancies last less than 237 days?

    i. almost all   iii. 34%

    ii. 50%   iv. about 0%

d. Roughly what percentage of pregnancies last between 247 and 287 days?

    i. almost all   iii. 68%

    ii. 95%   iv. 50%

e. Roughly what percentage of pregnancies last longer than 287 days?

    i. 34%   iii. 2.5%

    ii. 17%   iv. 50%

f. Roughly what percentage of pregnancies last longer than 297 days?

    i. almost all   iii. 34%

    ii. 50%   iv. about 0%

6.15 SAT Scores Quantitative SAT scores are approximately Normally distributed with a mean of 500 and a standard deviation of 100. On the horizontal axis of the graph, indicate the SAT scores that correspond with the provided *z*-scores. (See the labeling in Exercise 6.14.) Answer the questions using *only* your knowledge of the Empirical Rule and symmetry.

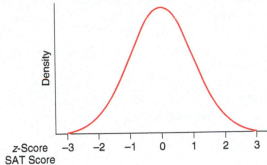

a. Roughly what percentage of students earn quantitative SAT scores greater than 500?

    i. almost all   iii. 50%   v. about 0%

    ii. 75%   iv. 25%

b. Roughly what percentage of students earn quantitative SAT scores between 400 and 600?

    i. almost all   iii. 68%   v. about 0%

    ii. 95%   iv. 34%

c. Roughly what percentage of students earn quantitative SAT scores greater than 800?

    i. almost all   iii. 68%   v. about 0%

    ii. 95%   iv. 34%

d. Roughly what percentage of students earn quantitative SAT scores less than 200?

    i. almost all   iii. 68%   v. about 0%

    ii. 95%   iv. 34%

e. Roughly what percentage of students earn quantitative SAT scores between 300 and 700?

    i. almost all   iii. 68%   v. 2.5%

    ii. 95%   iv. 34%

f. Roughly what percentage of students earn quantitative SAT scores between 700 and 800?

    i. almost all   iii. 68%   v. 2.5%

    ii. 95%   iv. 34%

6.16 Women's Heights Assume that college women's heights are approximately Normally distributed with a mean of 65 inches and a standard deviation of 2.5 inches. On the horizontal axis of the graph, indicate the heights that correspond to the *z*-scores provided. (See the labeling in Exercise 6.14.) Use only the Empirical Rule to choose your answers. Sixty inches is 5 feet, and 72 inches is 6 feet.

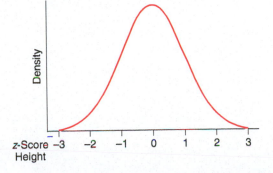

a. Roughly what percentage of women's heights are greater than 72.5 inches?

   i. almost all   iii. 50%   v. about 0%

   ii. 75%   iv. 25%

b. Roughly what percentage of women's heights are between 60 and 70 inches?

   i. almost all   iii. 68%   v. about 0%

   ii. 95%   iv. 34%

c. Roughly what percentage of women's heights are between 65 and 67.5 inches?

   i. almost all   iii. 68%   v. about 0%

   ii. 95%   iv. 34%

d. Roughly what percentage of women's heights are between 62.5 and 67.5 inches?

   i. almost all   iii. 68%   v. about 0%

   ii. 95%   iv. 34%

e. Roughly what percentage of women's heights are less than 57.5 inches?

   i. almost all   iii. 68%   v. about 0%

   ii. 95%   iv. 34%

f. Roughly what percentage of women's heights are between 65 and 70 inches?

   i. almost all   iii. 47.5%   v. 2.5%

   ii. 95%   iv. 34%

6.17 Women's Heights (Example 4) Assume college women's heights are approximately Normally distributed with a mean of 65 inches and a standard deviation of 2.5 inches. Choose the Stat-Crunch output for finding the percentage of college women who are taller than 67 inches and report the correct percentage. Round to one decimal place.

a.

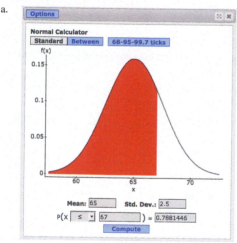

b.

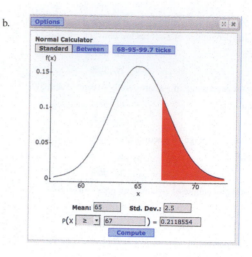

6.18 SAT Scores Quantitative SAT scores are approximately Normally distributed with a mean of 500 and a standard deviation of 100. Choose the correct StatCrunch output for finding the probability that a randomly selected person scores less than 450 on the quantitative SAT and report the probability as a percentage rounded to one decimal place.

a.

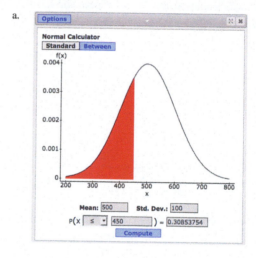

b.

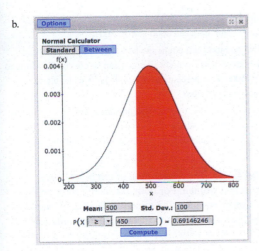

6.19 Standard Normal Use the table or technology to find the answer to each question. Include an appropriately labeled sketch of the Normal curve for each part. Shade the appropriate region. A section of the Normal table is provided.

a. Find the area in a Standard Normal curve to the left of 1.13.

b. Find the area in a Standard Normal curve to the right of 1.13.

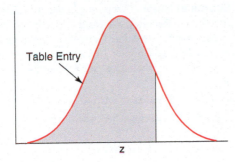

Format of the Normal Table: The area given is the area to the *left* of (less than) the given z-score.

z	.00	.01	.02	.03	.04
0.9	.8159	.8186	.8212	.8238	.8264
1.0	.8413	.8438	.8461	.8485	.8508
1.1	.8643	.8665	.8686	**.8708**	.8729

6.20 Standard Normal Use the table or technology to find the answer to each question. Include an appropriately labeled sketch of the Normal curve for each part. Shade the appropriate region. A section of the Normal table is provided in the previous exercise.

a. Find the area to the left of a z-score of 0.92.

b. Find the area to the right of a z-score of 0.92.

6.21 Standard Normal Use technology or a Normal table to find each of the following. Include an appropriately labeled sketch of the Normal curve for each part with the appropriate area shaded.

a. Find the probability that a z-score will be 2.03 or less.

b. Find the probability that a z-score will be −1.75 or more.

c. Find the probability that a z-score will be between −1.25 and 1.40.

6.22 Standard Normal Use technology or a Normal table to find each of the following. Include an appropriately labeled sketch of the Normal curve for each part with the appropriate area shaded.

a. Find the probability that a z-score will be 2.12 or greater.

b. Find the probability that a z-score will be less than −0.74.

c. Find the probability that a z-score will between 1.25 and 2.37.

6.23 Extreme Positive z-Scores For each question, find the area to the right of the given z-score in a standard Normal distribution. In this question, round your answers to the nearest 0.000. Include an appropriately labeled sketch of the $N(0, 1)$ curve.

a. $z = 4.00$

b. $z = 10.00$ (*Hint:* Should this tail proportion be larger or smaller than the answer to part a? Draw a picture and think about it.)

c. $z = 50.00$

d. If you had the *exact* probability for these tail proportions, which would be the largest and which would be the smallest?

e. Which is equal to the area in part b: the area below (to the left of) $z = -10.00$ or the area above (to the right of) $z = -10.00$?

6.24 Extreme Negative z-Scores For each question, find the area to the right of the given z-score in a standard Normal distribution. In *this* question, round your answers to the nearest 0.000. Include an appropriately labeled sketch of the $N(0, 1)$ curve.

a. $z = -4.00$

b. $z = -8.00$

c. $z = -30.00$

d. If you had the *exact* probability for these right proportions, which would be the largest and which would be the smallest?

e. Which is equal to the area in part b: the area below (to the left of) $z = 8.00$ or the area above (to the right of) $z = 8.00$?

TRY **6.25 St. Bernard Dogs (Example 5)** According to dogtime.com, the mean weight of an adult St. Bernard dog is 150 pounds. Assume the distribution of weights is Normal with a standard deviation of 10 pounds.

a. Find the standard score associated with a weight of 170 pounds.

b. Using the Empirical Rule and your answer to part a, what is the probability that a randomly selected St. Bernard weighs more than 170 pounds?

c. Use technology to confirm your answer to part b is correct.

d. Almost all adult St. Bernard's will have weights between what two values?

6.26 Whales Whales have one of the longest gestation periods of any mammal. According to whalefacts.org, the mean gestation period for a whale is 14 months. Assume the distribution of gestation periods is Normal with a standard deviation of 1.2 months.

a. Find the standard score associated with a gestational period of 12.8 months.

b. Using the Empirical Rule and your answer to part a, what percentage of whale pregnancies will have a gestation period between 12.8 months and 14 months?

c. Would it be unusual for a whale to have a gestation period of 18 months? Why or why not?

g TRY **6.27 Boys' Foot Length (Example 6)** According to the Digital Human Modeling Project, the distribution of foot lengths of 16- to 17-year-old boys is approximately Normal with a mean of 25.2 centimeters and a standard deviation of 1.2 centimeters. In the United States, a man's size 11 shoe fits a foot that is 27.9 centimeters long. What percentage of boys of this age group will wear a size 11 shoe or larger? *See page 316 for guidance.*

6.28 Women's Foot Length According to the Digital Human Modeling Project, the distribution of foot lengths of women is approximately Normal with a mean of 23.1 centimeters and a standard deviation of 1.1 centimeters. In the United States, a women's shoe size of 6 fits feet that are 22.4 centimeters long. What percentage of women in the United States will wear a size 6 or smaller?

6.29 Boys' Foot Length Suppose a shoe store stocks shoes in men's sizes 7 through 12. These shoes will fit men with feet that are 24.6 to 28.8 centimeters long. What percentage of boys aged 16 to 17 will **not** be able to find shoes that fit in this store? Use the statistics for the mean and the standard deviation of foot length found in Exercise 6.27.

6.30 Women's Foot Length Suppose a shoe store stocks shoes in women's sizes 5 though 9. These shoes will fit women with feet that are 21.6 through 25 centimeters long. What percentage of women will be able to find shoes that fit in this store? Use the statistics for the mean and the standard deviation of foot length found in Exercise 6.28.

TRY **6.31 Birth Weights (Example 7)** According to the British Medical Journal, the distribution of weights of newborn babies is approximately Normal, with a mean of 3390 grams and a standard deviation of 550 grams. Use a technology or a table to answer these questions. For each include an appropriately labeled and shaded Normal curve.

a. What is the probability at newborn baby will weigh more than 4000 grams?

b. What percentage of newborn babies weigh between 3000 and 4000 grams?

c. A baby is classified as "low birth weight" if the baby weighs less than 2500 grams at birth. What percentage of newborns would we expect to be "low birth weight"?

6.32 Birth Lengths According to National Vital Statistics, the average length of a newborn baby is 19.5 inches with a standard deviation of 0.9 inches. The distribution of lengths is approximately Normal. Use technology or a table to answer these questions. For each include an appropriately labeled and shaded Normal curve.

a. What is the probability that a newborn baby will have a length of 18 inches or less?

b. What percentage of newborn babies will be longer than 20 inches?

c. Baby clothes are sold in a "newborn" size that fits infants who are between 18 and 21 inches long. What percentage of newborn babies will not fit into the "newborn" size either because they are too long or too short?

6.33 White Blood Cells The distribution of white blood cell count per cubic millimeter of whole blood is approximately Normal with mean 7500 and standard deviation 1750 for healthy patients. Use technology or a table to answer these questions. For each include an appropriately labeled and shaded Normal curve.

a. What is the probability that a randomly selected person will have a white blood cell count between 6000 and 10,000?

b. An elevated white blood cell count can be a sign of infection somewhere in the body. A white blood cell count can be considered elevated if it is over 10,500. What percentage of people have white blood cell counts in this elevated range?

c. A white blood cell count below 4500 is considered low. People in this range may be referred for additional medical testing. What is the probability that a randomly selected person has a white blood cell count below 4500?

6.34 Red Blood Cells The distribution of red blood cell counts is different for men and women. For both, the distribution is approximately Normal. For men, the middle 95% range from 4.5 to 5.7 million cells per microliter and for women, the middle 95% have red blood cells counts between 3.9 and 5.0 million cells per microliter.

a. What is the mean and the standard deviation of red blood cell counts for men? Explain your reasoning.

b. What is the mean and the standard deviation of red blood cell counts for women? Explain your reasoning.

c. Which gender shows more variation in red blood cell counts? Support your answer with appropriate statistics.

6.35 SAT Scores in Illinois According to the 2017 SAT Suite of Assessments Annual Report, the average SAT math score for students in Illinois was 556. Assume the scores are Normally distributed with a standard deviation of 100. Answer the following including an appropriately labeled and shaded Normal curve for each question.

a. What percentage of Illinois Math SAT takers scored 600 or more?

b. What percentage of Illinois Math SAT takers scored between 600 and 650?

c. Suppose students who scored in the top 5% of test takers in the state were eligible for a special scholarship program. What SAT math score would qualify students for this scholarship program?

6.36 SAT Scores in Florida According to the 2017 SAT Suite of Assessments Annual Report, the average ERW (English, Reading, Writing) SAT score in Florida was 520. Assume the scores are Normally distributed with a standard deviation of 100. Answer the following including an appropriately labeled and shaded Normal curve for each question.

a. What is the probability that an ERW SAT taker in Florida scored 500 or less?

b. What percentage of ERW SAT takers in Florida scored between 500 and 650?

c. What ERW SAT score would correspond with the 40th percentile in Florida?

6.37 Arm Span (Men) According to Anthropometric Survey data, the distribution of arm spans for males is approximately Normal with a mean of 71.4 inches and a standard deviation of 3.3 inches.

a. What percentage of men have arm spans between 66 and 76 inches?

b. Professional basketball player, Kevin Durant, has an arm span of almost 89 inches. Find the *z*-score for Durant's arm span. What percentage of males have an arm span at least as long as Durant's?

6.38 Arm Span (Women) According to Anthropometric Survey data, the distribution of arm spans for females is approximately

Normal with a mean of 65.4 inches and a standard deviation of 3.2 inches.

a. What percentage of women have arm spans less than 61 inches?

b. Olympic swimmer, Dana Vollmer, won a bronze medal in the 100-meter butterfly stroke at the 2016 Olympics. An estimate of her arm span is 73 inches. What percentage of females have an arm span at least as long as Vollmer's?

6.39 New York City Weather New York City's mean minimum daily temperature in February is 27°F (http://www.ny.com). Suppose the standard deviation of the minimum temperature is 6°F and the distribution of minimum temperatures in February is approximately Normal. What percentage of days in February has minimum temperatures below freezing (32 °F)?

6.40 Chicago Weather The average winter daily temperature in Chicago has a distribution that is approximately Normal, with a mean of 28 degrees and a standard deviation of 8 degrees. What percentage of winter days in Chicago have a daily temperature of 35 degrees or warmer? (Source: wunderground.com)

6.41 Probability or Measurement (Inverse)? (Example 8) The Normal model $N(69, 3)$ describes the distribution of male heights in the United States. Which of the following questions asks for a probability, and which asks for a measurement? Identify the type of problem and then answer the given question. *See page 316 for guidance.*

a. To be a member of the Tall Club of Silicon Valley a man must be at least 74 inches tall. What percentage of men would qualify for membership in this club?

b. Suppose the Tall Club of Silicon Valley wanted to admit the tallest 2% of men. What minimum height requirement should the club set for its membership criteria?

6.42 Probability or Measurement (Inverse)? The Normal model $N(150, 10)$ describes the distribution of scores on the LSAT, a standardized test required by most law schools. Which of the following questions asks for a probability, and which asks for a measurement? Identify the type of problem and then answer the given question.

a. A law school applicant scored at the 60th percentile on the LSAT. What was the applicant's LSAT score?

b. A law school applicant scored 164 on the LSAT. This applicant scored higher than what percentage of LSAT test takers?

6.43 Inverse Normal, Standard In a standard Normal distribution, if the area to the left of a z-score is about 0.6666, what is the approximate z-score?

First locate, inside the table, the number closest to 0.6666. Then find the z-score by adding 0.4 and 0.03; refer to the table. Draw a sketch of the Normal curve, showing the area and the z-score.

z	.00	.01	.02	.03	.04	.05
0.4	.6554	.6591	.6628	**.6664**	.6700	.6736
0.5	.6915	.6950	.6985	.7019	.7054	.7088
0.6	.7257	.7291	.7324	.7357	.7389	.7422

6.44 Inverse Normal, Standard In a standard Normal distribution, if the area to the left of a z-score is about 0.1000, what is the approximate z-score?

6.45 Inverse Normal, Standard Assume a standard Normal distribution. Draw a separate, well-labeled Normal curve for each part.

a. Find the z-score that gives a left area of 0.7123.

b. Find the z-score that gives a left area of 0.1587.

6.46 Inverse Normal, Standard Assume a standard Normal distribution. Draw a separate, well-labeled Normal curve for each part.

a. Find an approximate z-score that gives a left area of 0.7000.

b. Find an approximate z-score that gives a left area of 0.9500.

TRY **6.47 Gestation of Hippos (Example 9)** The length of gestation for hippopotami is approximately Normal, with a mean of 270 days and a standard deviation of 7 days.

a. What percentage of hippos have a gestation period less than 260 days?

b. Complete this sentence: Only 6% of hippos will have a gestational period longer than ___ days.

c. In 2017, Fiona the Hippo was born at the Cincinnati Zoo, 6 weeks premature. This means her gestational period was only about 228 days. What percentage of hippos have a gestational period of 228 days or less?

6.48 Weights of Newborn Hippos The weight of newborn hippopotami is approximately Normal, with a mean of 88 pounds and a standard deviation of 10 pounds.

a. What is the probability that a newborn hippo weighs between 90 and 110 pounds?

b. Suppose baby hippos that weigh at the 5th percentile or less at birth are unlikely to survive. What weight corresponds with the 5th percentile for newborn hippos?

c. Fiona the Hippo was born at the Cincinnati Zoo in 2017, 6 weeks premature, and weighed only 29 pounds at birth. What percentage of baby hippos are born weighing 29 pounds or less?

6.49 Medical School MCAT Scores on the 2017 MCAT, an exam required for all medical school applicants, were approximately Normal with a mean score of 505 and a standard deviation of 9.4.

a. Suppose an applicant had an MCAT score of 520. What percentile corresponds with this score?

b. Suppose to be considered at a highly selective medical school an applicant should score in the top 10% of all test takers. What score would place an applicant in the top 10%?

6.50 Medical School GPA The distribution of grade point averages GPAs for medical school applicants in 2017 were approximately Normal, with a mean of 3.56 and a standard deviation of 0.34. Suppose a medical school will only consider candidates with GPAs in the top 15% of the applicant pool. An applicant has a GPA of 3.71. Does this GPA fall in the top 15% of the applicant pool?

6.51 Women's Heights Suppose college women's heights are approximately Normally distributed with a mean of 65 inches and a population standard deviation of 2.5 inches. What height is at the 20th percentile? Include an appropriately labeled sketch of the Normal curve to support your answer.

6.52 Men's Heights Suppose college men's heights are approximately Normally distributed with a mean of 70.0 inches and a population standard deviation of 3 inches. What height is at the 20th percentile? Include an appropriately labeled Normal curve to support your answer.

6.53 Inverse SATs Critical reading SAT scores are distributed as $N(500, 100)$.

   a. Find the SAT score at the 75th percentile.

   b. Find the SAT score at the 25th percentile.

   c. Find the interquartile range for SAT scores.

   d. Is the interquartile range larger or smaller than the standard deviation? Explain.

6.54 Inverse Women's Heights College women have heights with the following distribution (inches): $N(65, 2.5)$.

   a. Find the height at the 75th percentile.

   b. Find the height at the 25th percentile.

   c. Find the interquartile range for heights.

   d. Is the interquartile range larger or smaller than the standard deviation? Explain.

6.55 Girls' and Women's Heights According to the National Health Center, the heights of 6-year-old girls are Normally distributed with a mean of 45 inches and a standard deviation of 2 inches.

   a. In which percentile is a 6-year-old girl who is 46.5 inches tall?

   b. If a 6-year-old girl who is 46.5 inches tall grows up to be a woman at the same percentile of height, what height will she be? Assume women are distributed as $N(64, 2.5)$.

6.56 Boys' and Men's Heights According to the National Health Center, the heights of 5-year-old boys are Normally distributed with a mean of 43 inches and a standard deviation of 1.5 inches.

   a. In which percentile is a 5-year-old boy who is 46.5 inches tall?

   b. If a 5-year-old boy who is 46.5 inches tall grows up to be a man at the same percentile of height, what height will he be? Assume adult men's heights (inches) are distributed as $N(69, 3)$.

6.57 Cats' Birth Weights The average birth weight of domestic cats is about 3 ounces. Assume that the distribution of birth weights is Normal with a standard deviation of 0.4 ounce.

   a. Find the birth weight of cats at the 90th percentile.

   b. Find the birth weight of cats at the 10th percentile.

6.58 Elephants' Birth Weights The average birth weight of elephants is 230 pounds. Assume that the distribution of birth weights is Normal with a standard deviation of 50 pounds. Find the birth weight of elephants at the 95th percentile.

SECTION 6.3

TRY **6.59 Gender of Children (Example 10)** A married couple plans to have four children, and they are wondering how many boys they should expect to have. Assume none of the children will be twins or other multiple births. Also assume the probability that a child will be a boy is 0.50. Explain why this is a binomial experiment. Check all four required conditions.

6.60 Coin Flip A coin will be flipped four times, and the number of heads recorded. Explain why this is a binomial experiment. Check all four required conditions.

TRY **6.61 Rolling a Die (Example 11)** A die is rolled 5 times, and the number of spots for each roll is recorded. Explain why this is not a binomial experiment. Name a condition for use of the binomial model that is not met.

6.62 Twins In Exercise 6.59 you are told to assume that none of the children will be twins or other multiple births. Why? Which of the conditions required for a binomial experiment would be violated if there were twins?

6.63 Free Throws Professional basketball player Draymond Green has a free-throw success rate of 70%. Suppose Green takes as many free throws as he can in one minute. Why would it be inappropriate to use the binomial model to find the probability that he makes at least 5 shots in one minute? What condition or conditions for use of the binomial model is or are not met?

6.64 On-Time Arrivals Alaska Airlines has an on-time arrival rate of 88%. Assume that in one day, this airline has 1200 flights. Suppose we pick one day in December and find the number of on-time Alaska Airline arrivals. Why would it be inappropriate to use the binomial model to find the probability that at least 1100 of the 1200 flights arrive on time? What condition or conditions for use of the binomial model is or are not met?

TRY **6.65 Identifying n, p, and x (Example 12)** For each situation, identify the sample size n, the probability of a success p, and the number of success x. When asked for the probability, state the answer in the form $b(n, p, x)$. There is no need to give the numerical value of the probability. Assume the conditions for a binomial experiment are satisfied.

A 2017 Gallup poll found that 53% of college students were very confident that their major will lead to a good job.

   a. If 20 college students are chosen at random, what's the probability that 12 of them were very confident their major would lead to a good job?

   b. If 20 college students are chosen at random, what's the probability that 10 of them are *not* confident that their major would lead to a good job?

6.66 Identifying n, p, and x For each situation, identify the sample size n, the probability of a success p, and the number of success x. When asked for the probability, state the answer in the form $b(n, p, x)$. There is no need to give the numerical value of the probability. Assume the conditions for a binomial experiment are satisfied.

Since the Surgeon General's Report on Smoking and Health in 1964 linked smoking to adverse health effects, the rate of smoking the United States have been falling. According to the Centers for Disease Control and Prevention in 2016, 15% of U.S. adults smoked cigarettes (down from 42% in the 1960s).

   a. If 30 Americans are randomly selected, what is the probability that exactly 10 are smokers?

   b. If 30 Americans are randomly selected, what is the probability that exactly 25 are *not* smokers?

TRY **6.67 Dog Owners (Example 13)** According to the American Veterinary Medical Association, 36% of Americans own a dog.

   a. Find the probability that exactly 4 out of 10 randomly selected Americans own a dog.

   b. In a random sample of 10 Americans, find the probability that 4 or fewer own a dog.

6.68 Cat Owners According to the American Veterinary Medical Association, 30% of Americans own a cat.

   a. Find the probability that exactly 2 out of 8 randomly selected Americans own a cat.

   b. In a random sample of 8 Americans, find the probability that more than 3 own a cat.

6.69 Passports According to data from the U.S. State Department, the percentage of Americans who have a passport has risen dramatically. In 2007, only 27% of Americans had a passport; in 2017 that percentage had risen to 42%. Assume that currently 42% of Americans have a passport.

Suppose 50 Americans are selected at random.

a. Find the probability that fewer than 20 have a passport.

b. Find the probability that at most 24 have a passport.

c. Find the probability that at least 25 have a passport.

6.70 Travel According to a survey conducted by OnePoll, a marketing research company, 10% of Americans have never traveled outside their home state. Assume this percentage is accurate. Suppose a random sample of 80 Americans is taken.

a. Find the probability that more than 12 have never travelled outside their home state.

b. Find the probability that at least 12 have never travelled outside their home state.

c. Find the probability that at most 12 have never travelled outside their home state.

6.71 Wisconsin Graduation Wisconsin has the highest high school graduation rate of all states at 90%.

a. In a random sample of 10 Wisconsin high school students, what is the probability that 9 will graduate?

b. In a random sample of 10 Wisconsin high school students, what is the probability than 8 or fewer will graduate?

c. What is the probability that at least 9 high school students in our sample of 10 will graduate?

6.72 Colorado Graduation Colorado has a high school graduation rate of 75%.

a. In a random sample of 15 Colorado high school students, what is the probability that exactly 9 will graduate?

b. In a random sample of 15 Colorado high school students, what is the probability that 8 or fewer will graduate?

c. What is the probability that at least 9 high school students in our sample of 15 will graduate?

6.73 Cell Phones According to the Centers of Disease Control and Prevention, 52% of U.S. households had no landline and only had cell phone service. Suppose a random sample of 40 U.S. households is taken.

a. Find the probability that exactly 20 the households sampled only have cell phone service.

b. Find the probability that fewer than 20 households only have cell phone service.

c. Find the probability that at most 20 households only have cell phone service.

d. Find the probability that between 20 and 23 households only have cell phone service.

6.74 Landlines According to the Centers of Disease Control and Prevention, 44% of U.S. households still had landline phone service. Suppose a random sample of 60 U.S. households is taken.

a. Find the probability that exactly 25 of the households sampled still have a landline.

b. Find the probability that more than 25 households still have a landline.

c. Find the probability that at least 25 households still have a landline.

d. Find the probability that between 20 and 25 households still have a landline.

TRY 6.75 Drones (Example 14) The use of drones, aircraft without onboard human pilots, is becoming more prevalent in the United States. According to a 2017 Pew Research Center report, 59% of American had seen a drone in action. Suppose 50 Americans are randomly selected.

a. What is the probability that at least 25 had seen a drone?

b. What is the probability that more than 30 had seen a drone?

c. What is the probability that between 30 and 35 had seen a drone?

d. What is the probability that more than 30 had *not* seen a drone?

6.76 Drones A 2017 Pew Research Center report on drones found that only 24% of Americans felt that drones should be allowed at events, like concerts or rallies. Suppose 100 Americans are randomly selected.

a. What is the probability that exactly 25 believe drones should be allowed at these events?

b. Find the probability that more than 30 believe drones should be allowed at these events.

c. What is the probability that between 20 and 30 believe drones should be allowed at these events?

d. Find the probability that at most 70 do *not* believe drones should be allowed at these events.

6.77 Texting While Walking According to a report by the American Academy of Orthopedic Surgeons, 29% of pedestrians admit to texting while walking. Suppose two pedestrians are randomly selected.

a. If the pedestrian texts while walking, record a T. If not, record an N. List all possible sequences of Ts and Ns for the two pedestrians.

b. For each sequence, find the probability that it will occur by assuming independence.

c. What is the probability that neither pedestrian texts while walking?

d. What is the probability that both pedestrians text while walking?

e. What is the probability that exactly one of the pedestrians texts while walking?

6.78 Texting While Driving According to a study by the Colorado Department of Transportation, 25% of Colorado drivers admit to using their cell phones to send texts while driving. Suppose two Colorado drivers are randomly selected.

a. If the driver texts while driving, record a T. If not, record an N. List all possible sequences of Ts and Ns for the two drivers.

b. For each sequence, find the probability that it will occur by assuming independence.

c. What is the probability that both drivers text while driving?

d. What is the probability that neither driver texts while driving?

e. What is the probability that exactly one of the drivers texts while driving?

TRY 6.79 Library Use (Example 15) According to the Pew Research Center, 53% of millennials (those born between 1981 and 1997) reported using a library or bookmobile within the last year. Suppose that a random sample of 200 millennials is taken.

a. Complete this sentence: We would expect ____ of the sample to have used a library or bookmobile within the last year, give or take ____.

b. Would it be surprising to find that 190 of the sample have used a library or bookmobile within the last year? Why or why not?

6.80 Free Throws Professional basketball LeBron James is a 74% free-throw shooter. Assume that free throw shots are independent. Suppose, over the course of a season, James attempts 600 free throws.

a. Find the mean and the standard deviation for the expected number of free throws we expect Curry to make.

b. Would it be surprising if he only made 460 of his free throws? Why or why not?

TRY **6.81 Toronto Driving Test (Example 16)** In Toronto, Canada, 55% of people pass the drivers' road test. Suppose that every day, 100 people independently take the test.

a. What is the number of people who are expected to pass?

b. What is the standard deviation for the number expected to pass?

c. After a great many days, according to the Empirical Rule, on about 95% of these days, the number of people passing will be as low as _____ and as high as _____. (*Hint:* Find two standard deviations below and two standard deviations above the mean.)

d. If you found that on one day, 85 out of 100 passed the test, would you consider this to be a very high number?

6.82 Drivers' Test in Small Towns Toronto drivers have been going to small towns in Ontario (Canada) to take the drivers' road test, rather than taking the test in Toronto, because the pass rate in the small towns is 90%, which is much higher than the pass rate in Toronto. Suppose that every day, 100 people independently take the test in one of these small towns.

a. What is the number of people who are expected to pass?

b. What is the standard deviation for the number expected to pass?

c. After a great many days, according to the Empirical Rule, on about 95% of these days the number of people passing the test will be as low as _____ and as high as _____.

d. If you found that on one day, 89 out of 100 passed the test, would you consider this to be a very high number?

CHAPTER REVIEW EXERCISES

6.83 Discrete or Continuous? Determine whether each of the following variables would best be modeled as continuous or discrete:

a. Number of girls in a family

b. Height of a tree

c. Commute time

d. Concert attendance

6.84 Probability Distribution In a game of chance, players draw one cube out of a bag containing 3 red cubes, 2 white cubes, and 1 blue cube. The player wins $5 if a blue cube is drawn, the player loses $2 if a white cube is drawn. If a red cube is drawn, the player does not win or lose anything.

a. Create a table that shows the probability distribution for the amount of money a player will win or lose when playing this game.

b. Draw a graph of the probability distribution you created in part a.

6.85 Birth Length A study of U.S. births published on the website *Medscape from WebMD* reported that the average birth length of babies was 20.5 inches and the standard deviation was about 0.90 inch. Assume the distribution is approximately Normal. Find the percentage of babies with birth lengths of 22 inches or less.

6.86 Birth Length A study of U.S. births published on the website *Medscape from WebMD* reported that the average birth length of babies was 20.5 inches and the standard deviation was about 0.90 inch. Assume the distribution is approximately Normal. Find the percentage of babies who have lengths of 19 inches or less at birth.

6.87 Males' Body Temperatures A study of human body temperatures using healthy men showed a mean of 98.1 °F and a standard deviation of 0.70 °F. Assume the temperatures are approximately Normally distributed.

a. Find the percentage of healthy men with temperatures below 98.6 °F (that temperature was considered typical for many decades).

b. What temperature does a healthy man have if his temperature is at the 76th percentile?

6.88 Females' Body Temperatures A study of human body temperatures using healthy women showed a mean of 98.4 °F and a standard deviation of about 0.70 °F. Assume the temperatures are approximately Normally distributed.

a. Find the percentage of healthy women with temperatures below 98.6 °F (this temperature was considered typical for many decades).

b. What temperature does a healthy woman have if her temperature is at the 76th percentile?

6.89 Medical Licensing Medical school graduates who want to become doctors must pass the U.S. Medical Licensing Exam (USMLE). Scores on this exam are approximately Normal with a mean of 225 and a standard deviation of 15. Use the Empirical Rule to answer these questions.

a. Roughly what percentage of USMLE scores will be between 210 and 240?

b. Roughly what percentage of USMLE scores will be below 210?

c. Roughly what percentage of USMLE scores will be above 255?

6.90 Medical Licensing See problem 6.89 for information about USMLE scores.

a. What USMLE score would correspond with a z-score of −2?

b. What USMLE score corresponds with a z-score of 1?

c. Find the *z*-score that corresponds with a USMLE score of 250. Would a score of 250 be considered unusually high? Why or why not?

6.91 Systolic Blood Pressures Systolic blood pressures are approximately Normal with a mean of 120 and a standard deviation of 8.

a. What percentage of people have a systolic blood pressure above 130?

b. What is the range of systolic blood pressures for the middle 60% of the population?

c. What percentage of people have a systolic blood pressure between 120 and 130?

d. Suppose people with systolic blood pressures in the top 15% of the population have their blood pressures monitored more closely by health care professionals. What blood pressure would qualify a person for this additional monitoring?

6.92 Temperatures in Los Angeles The distribution of spring high temperatures in Los Angeles is approximately Normal, with a mean of 75 degrees and a standard deviation of 2.5 degrees.

a. What is the probability that the high temperature is less than 70 degrees in Los Angeles on a day in spring?

b. What percentage of Spring day in Los Angeles have high temperatures between 70 and 75 degrees?

c. Suppose the hottest spring day in Los Angeles had a high temperature of 91 degrees. Would this be considered unusually high, given the mean and the standard deviation of the distribution? Why or why not?

6.93 Stress According to a 2017 Gallup poll, 44% of Americans report they frequently feel stressed. Suppose 200 Americans are randomly sampled. Find the probability of the following:

a. Fewer than 80 frequently feel stressed

b. At least 90 frequently feel stressed

c. Between 80 and 100 frequently feel stressed

d. At most 75 frequently feel stressed

6.94 Stress According to a 2017 Gallup poll, 17% of Americans report they rarely feel stressed. Suppose 80 Americans are randomly sampled. Find the probability of the following:

a. Wxactly 15 rarely feel stressed

b. More than 20 rarely feel stressed

c. At most 10 rarely feel stressed

6.95 Voice-Controlled Assistants Voice-controlled video assistants are being incorporated into a wide variety of consumer products, including smartphones, tablets, and stand-alone devices such as the Amazon Echo or Google Home. A Pew Research poll found that 46% of Americans reported using a voice-controlled digital assistant. Suppose a group of 50 Americans is randomly selected.

a. Find the probability that more than half of the sample uses a voice-controlled digital assistant.

b. Find the probability that at most 20 use a voice-controlled digital assistant.

c. In a group of 50 Americans, how many would we expect use one of these devices?

d. Find the standard deviation for this binomial distribution. Using your answers to parts c and d, would it be surprising to find that fewer than 10 used one of these devices? Why or why not?

6.96 Reading According to the Pew Research Center, 73% of Americans have read at least one book during the past year. Suppose 200 Americans are randomly selected.

a. Find the probability that more than 150 have read at least one book during the past year.

b. Find the probability that between 140 and 150 have read at least one book during the past year.

c. Find the mean and the standard deviation for this binomial distribution.

d. Using your answer to part c, complete this sentence: It would be surprising to find that fewer than ___ people in the sample had read at least one book in the last year.

6.97 Jury Duty According to a Pew poll, 67% of Americans believe that jury duty is part of good citizenship. Suppose 500 Americans are randomly selected.

a. Find the probability that more than half believe that jury duty is part of good citizenship.

b. In a group of 500 Americans, how many would we expect hold this belief?

c. Would it be surprising to find that more than 450 out of the 500 American randomly selected held this belief? Why or why not?

6.98 Marijuana Support for the legalization of marijuana has continued to grow among Americans. A 2017 Gallup poll found that 64% of Americans now say that marijuana use should be legal. Suppose a random sample of 150 Americans is selected.

a. Find the probability that at most 110 people support marijuana legalization.

b. Find the probability that between 90 and 110 support marijuana legalization.

c. Complete this sentence: In a group of 150, we would expect ____ to support marijuana legalization, give or take ___.

6.99 Low Birth Weights, Normal *and* Binomial Babies weighing 5.5 pounds or less at birth are said to have low birth weights, which can be dangerous. Full-term birth weights for single babies (not twins or triplets or other multiple births) are Normally distributed with a mean of 7.5 pounds and a standard deviation of 1.1 pounds.

a. For one randomly selected full-term single-birth baby, what is the probability that the birth weight is 5.5 pounds or less?

b. For two randomly selected full-term, single-birth babies, what is the probability that both have birth weights of 5.5 pounds or less?

c. For 200 random full-term single births, what is the approximate probability that 7 or fewer have low birth weights?

d. If 200 independent full-term single-birth babies are born at a hospital, how many would you expect to have birth weights of 5.5 pounds or less? Round to the nearest whole number.

e. What is the standard deviation for the number of babies out of 200 who weigh 5.5 pounds or less? Retain two decimal digits for use in part f.

f. Report the birth weight for full-term single babies (with 200 births) for two standard deviations below the mean and for two standard deviations above the mean. Round both numbers to the nearest whole number.

g. If there were 45 low-birth-weight full-term babies out of 200, would you be surprised?

* 6.100 Quantitative SAT Scores, Normal *and* Binomial

The distribution of the math portion of SAT scores has a mean of 500 and a standard deviation of 100, and the scores are approximately Normally distributed.

a. What is the probability that one randomly selected person will have an SAT score of 550 or more?

b. What is the probability that four randomly selected people will all have SAT scores of 550 or more?

c. For 800 randomly selected people, what is the probability that 250 or more will have scores of 550 or more?

d. For 800 randomly selected people, on average how many should have scores of 550 or more? Round to the nearest whole number.

e. Find the standard deviation for part d. Round to the nearest whole number.

f. Report the range of people out of 800 who should have scores of 550 or more from two standard deviations below the mean to two standard deviations above the mean. Use your rounded answers to part d and e.

g. If 400 out of 800 randomly selected people had scores of 550 or more, would you be surprised? Explain.

6.101 Babies' Birth Length, Inverse Babies in the United States have a mean birth length of 20.5 inches with a standard deviation of 0.90 inch. The shape of the distribution of birth lengths is approximately Normal.

a. How long is a baby born at the 20th percentile?

b. How long is a baby born at the 50th percentile?

c. How does your answer to part b compare to the mean birth length? Why should you have expected this?

6.102 Birth Length and z-Scores, Inverse Babies in the United States have a mean birth length of 20.5 inches with a standard deviation of 0.90 inch. The shape of the distribution of birth lengths is approximately Normal.

a. Find the birth length at the 2.5th percentile.

b. Find the birth length at the 97.5th percentile.

c. Find the z-score for the length at the 2.5th percentile.

d. Find the z-score for the length at the 97.5th percentile.

gUIDED EXERCISES

g 6.27 Boys' Foot Length (Example 6) According to the Digital Human Modeling Project, the distribution of foot lengths of 16- to 17-year-old boys is approximately Normal with a mean of 25.2 centimeters and a standard deviation of 1.2 centimeters. In the United States, a man's size 11 shoe fits a foot that is 27.9 centimeters long.

QUESTION What percentage of boys of this age group will wear a size 11 shoe or larger?

Step 1 ▶ Find the z-score.
To find the z-score for 27.9, subtract the mean and divide by the standard deviation. Report the z-score.

Step 2 ▶ Explain the location of 25.2.
Refer to the Normal curve. Explain why the foot length of 25.2 is right below the z-score of 0. The tick marks on the axis mark the location of z-scores that are integers from −3 to 3.

Step 3 ▶ Label with foot lengths.
Carefully sketch a copy of the curve. Pencil in the foot lengths of 21.6, 22.8, 24, 26.4, 27.6 and 28.8 in the correct places. (Note: These are lengths that are 1, 2, and 3 standard deviations above and below the mean.)

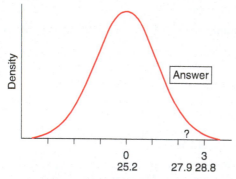

Step 4 ▶ Add the line, z-score and shading.
Draw a vertical line through the curve at the location of 27.9 at a point between 27.5 and 28.8 (indicated on the graph with "?"). Above the 27.9 (?) put in the corresponding z-score. We want to find the percentage of boys in this age group with feet longer than 27.9. Therefore, shade the area to the right of this boundary because numbers to the right are larger.

Step 5 ▶ Use the table for the left area.
Use the following excerpt from the Normal table to find and report the area to the left of the z-score that was obtained from a foot length of 27.9. This is the area of the unshaded region.

Step 6 ▶ Answer.
Because you want the area to the right of the z-score, you will have to subtract the area in step 5 from 1. This is the area of the shaded region. Put it where the box labeled "Answer" is. Check to see that the number makes sense. For example, if the shading is less than half the area, the answer should not be more than 0.5000.

Step 7 ▶ Sentence: Finally write a sentence stating what you found.

z	.00	.01	.02	.03	.04	.05	.06	.07	.08	.09
2.1	.9821	.9826	.9830	.9834	.9838	.9842	.9846	.9850	.9854	.985?
2.2	.9861	.9864	.9868	.9871	.9875	**.9878**	.9881	.9884	.9887	.989?
2.3	.9893	.9896	.9898	.9901	.9904	.9906	.9909	.9911	.9913	.991?

g 6.41 Probability or Measurement (Inverse)? (Example 8) The Normal model $N(69, 3)$ describes the distribution of male heights in the United States. Which of the following questions asks for a probability, and which asks for a measurement? Identify the type of problem and then answer the given question.

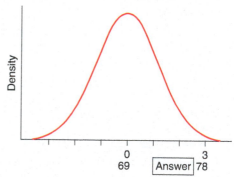

QUESTION   Suppose the Tall Club of Silicon Valley wanted to admit the tallest 2% of men. What minimum height requirement should the club set for its membership criteria? Answer this question by following the numbered steps.

Step 1 ▸ Think about it.
Will the height be above the mean or below it? Explain.

Step 2 ▸ Label z-scores.
Label the curve with integer z-scores. The tick marks represent the position of integer z-scores from -3 to 3.

Step 3 ▸ Draw a vertical line in the graph and shade the sketch.
The tallest 2% of men are in the right tail of the distribution. Draw a vertical line at the approximate location of the height that separates the tallest 2% of men (to the right of the line) from the shortest 98% of men (to the left of the line) and shade the area to the right of the line. The line should be somewhere between a z-score of 2 and 3.

Step 4 ▸ Find the height.
The height can be found using technology (an inverse Normal problem) or using a Normal table. To use a Normal table, locate the area closest to 0.9800. Report the z-score for that area. Then find the height that corresponds to the z-score using the formula

$$x = \mu + z\sigma$$

Step 5 ▸ Add the height to the sketch.
Add the height on the sketch where it says "Answer."

Step 6 ▸ Write a sentence.
Finally, write a sentence stating what you found.

TechTips

For All Technology

All technologies will use the two examples that follow.

EXAMPLE A: NORMAL ▶ Wechsler IQs have a mean of 100 and standard deviation of 15 and are Normally distributed.

   a. Find the probability that a randomly chosen person will have an IQ between 85 and 115.

   b. Find the probability that a randomly chosen person will have an IQ that is 115 or less.

   c. Find the Wechsler IQ at the 75th percentile.

Note: If you want to use technology to find areas from standard units (*z*-scores), use a mean of 0 and a standard deviation of 1.

EXAMPLE B: BINOMIAL ▶ Imagine that you are flipping a fair coin (one that comes up heads 50% of the time in the long run).

   a. Find the probability of getting 28 or fewer heads in 50 flips of a fair coin.

   b. Find the probability of getting exactly 28 heads in 50 flips of a fair coin.

TI-84

NORMAL

a. Between Two Values

1. Press **2ND DISTR** (located below the four arrows on the keypad).
2. Select **2:normalcdf** and press **ENTER**.
3. Enter **lower: 85**, **upper: 115**, μ: **100**, σ: **15**. For **Paste**, press **ENTER**. Then press **ENTER** again.

Your screen should look like Figure 6a, which shows that the probability that a randomly selected person will have a Wechsler IQ between 85 and 115 is equal to 0.6827.

▲ **FIGURE 6a** TI-84 normal**c**df (c stands for "cumulative")

b. Some Value or Less

1. Press **2ND DISTR**.
2. Select **2:normalcdf** and press **ENTER**.
3. Enter: **– 1000000, 115, 100, 15,** press **ENTER** and press **ENTER** again.

Caution: The negative number button (–) is to the left of the **ENTER** button and is not the same as the minus button that is above the plus button.

The probability that a person's IQ is 115 *or less* has an *indeterminate* lower (left) boundary, for which you may use negative 1000000 or any extreme value that is clearly out of the range of data. Figure 6b shows the probability that a randomly selected person will have an IQ of 115 or less.

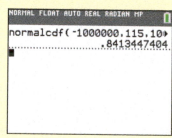

▲ **FIGURE 6b** TI-84 normalcdf with indeterminate left boundary

(If you have an indeterminate upper, or right boundary, then to find the probability that the person's IQ is 85 or more, for example, use an upper, or right boundary (such as 1000000) that is clearly above all the data.)

c. Inverse Normal

If you want a measurement (such as an IQ) from a proportion or percentile:

1. Press **2ND DISTR**.
2. Select **3:invNorm** and press **ENTER**.
3. Enter (left) **area: .75**, μ: **100**, σ: **15**. For **Paste**, press **ENTER**. Then **ENTER** again.

Figure 6c shows the Wechsler IQ at the 75th percentile, which is 110. Note that the 75th percentile is entered as **.75**.

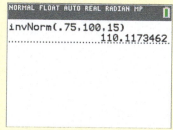

▲ **FIGURE 6c** TI-84 Inverse Normal

BINOMIAL

a. Cumulative (or Fewer)

1. Press **2ND DISTR.**

2. Select **B:binomcdf** (you will have to scroll down to see it) and press **ENTER**. (On a TI-83, it is **A:binomcdf.**)

3. Enter **trials: 50, p: .5, x value: 28**. For **Paste**, press **ENTER**. Then press **ENTER** again.

The answer will be the probability for *x or fewer*. Figure 6d shows the probability of 28 *or fewer* heads out of 50 flips of a fair coin. (You could find the probability of 29 *or more* heads by subtracting your answer from 1.)

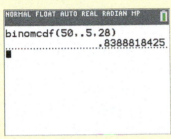

▲ **FIGURE 6d** TI-84 binom**c**df (cumulative)

b. Individual (Exact)

1. Press **2ND DISTR.**

2. Select **A:binompdf** and press **ENTER**. (On a TI-83, it is **0:binompdf.**)

3. Enter **trials: 50, p: .5, x value: 28**. For **Paste**, press **ENTER**. Then press **ENTER** again.

Figure 6e shows the probability of *exactly* 28 heads out of 50 flips of a fair coin.

▲ **FIGURE 6e** TI-84 binom**p**df (individual)

MINITAB

NORMAL

a. Between Two Values

1. Enter the upper boundary, **115**, in the top cell of an empty column; here we use column C1, row 1. Enter the lower boundary, **85**, in the cell below; here column C1, row 2.

2. **Calc > Probability Distributions > Normal.**

3. See Figure 6f: Choose **Cumulative probability**. Enter: **Mean**, 100; **Standard deviation**, 15; **Input column**, C1; **Optional storage**, C2.

4. Click **OK**.

5. Subtract the lower probability from the larger shown in column C2.

$0.8413 - 0.1587 = 0.6826$ is the probability that a Wechsler IQ is between 85 and 115.

b. Some Value or Less

1. The probability of an IQ of 115 *or less*, 0.8413, is shown in column C2, row 1. (In other words, do as in part a earlier, except do *not* enter the lower boundary, 85.)

c. Inverse Normal

If you want a measurement (such as an IQ or height) from a proportion or percentile:

1. Enter the decimal form of the left proportion (.75 for the 75th percentile) into a cell in an empty column in the spreadsheet; here we used column C1, row 1.

2. Calc > Probability Distributions > Normal.

3. See Figure 6g: Choose **Inverse cumulative probability**. Enter: **Mean**, 100; **Standard deviation**, 15; **Input column**, C1; and **Optional storage**, c2 (or an empty column).

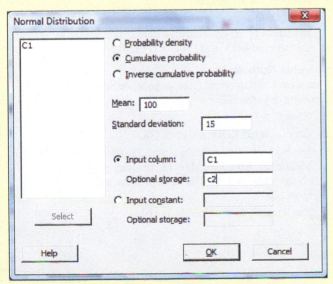

▲ **FIGURE 6f** Minitab Normal

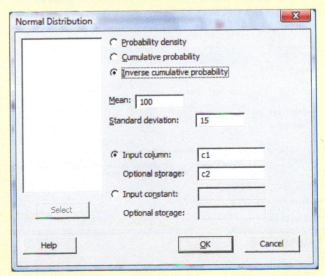

▲ **FIGURE 6g** Minitab Inverse Normal

4. Click **OK**.

You will get **110**, which is the Wechsler IQ at the 75th percentile.

BINOMIAL

a. Cumulative (or Fewer)

1. Enter the upper bound for the number of successes in an empty column; here we used column C1, row 1. Enter **28** to get the probability of 28 or fewer heads.

2. **Calc > Probability Distributions > Binomial**.

3. See Figure 6h. Choose **Cumulative probability**. Enter: **Number of trials**, 50; **Event probability**, .5; **Input column**, c1; **Optional storage**, c2 (or an empty column).

4. Click **OK**.

Your answer will be 0.8389 for the probability of 28 *or fewer* heads.

b. Individual (Exact)

1. Enter the number of successes at the top of column 1, **28** for 28 heads.

2. **Calc > Probability Distributions > Binomial**.

3. Choose **Probability** (at the top of Figure 6h) instead of **Cumulative Probability** and enter: **Number of trials**, **50**; **Event probability**, **.5**; **Input column**, **c1**; **Optional storage**, **c2** (or an empty column).

4. Click **OK**.

Your answer will be 0.0788 for the probability of *exactly* 28 heads.

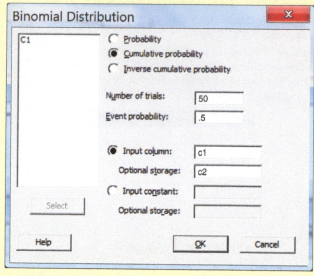

▲ **FIGURE 6h** Minitab Binomial

EXCEL

NORMAL

Unlike the TI-84, Excel makes it easier to find the probability that a random person has an IQ of 115 or less than to find the probability that a random person has an IQ between 85 and 115. This is why, for Excel, part b appears before part a.

b. Some Value or Less

1. Click *fx* (and **select a category All**).

2. Choose **NORM.DIST**.

3. See Figure 6i. Enter: **X**, 115; **Mean**, 100; **Standard_dev**, 15; **Cumulative**, **true** (for 115 *or less*). The answer is shown as 0.8413. Click **OK** to make it show up in the active cell on the spreadsheet.

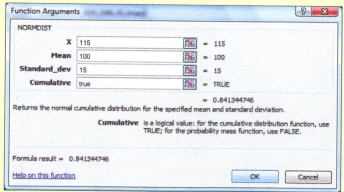

▲ **FIGURE 6i** Excel Normal

a. Between Two Values

If you want the probability of an IQ between 85 and 115:

1. First, follow the instructions given for part b. *Do not change the active cell in the spreadsheet.*

2. You will see **=NORMDIST(115,100,15,true)** in the *fx* box. Click in this box, to the right of **true)**, and put in a minus sign.

3. Now repeat the steps for part b, starting by clicking *fx*, except enter 85 instead of 115 for X. The answer, **0.682689**, will be shown in the active cell.

   (Alternatively, just repeat steps 123 for part b, using 85 instead of 115. Subtract the smaller probability value from the larger (0.8413 − 0.1587 = 0.6826).)

c. Inverse Normal

If you want a measurement (such as an IQ or height) from a proportion or percentile:

1. Click *fx*.

2. Choose **NORM.INV** and click **OK**.

3. See Figure 6j. Enter: **Probability**, .75 (for the 75th percentile); **Mean**, 100: **Standard_dev**, 15. You may read the answer off the screen or click **OK** to see it in the active cell in the spreadsheet.

   The IQ at the 75th percentile is 110.

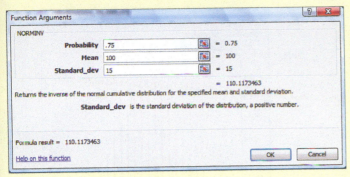

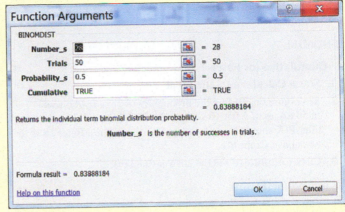

▲ **FIGURE 6j** Excel Inverse Normal

▲ **FIGURE 6k** Excel Binomial

BINOMIAL

a. Cumulative (or Fewer)

1. Click f_x.
2. Choose **BINOM.DIST** and click **OK**.
3. See Figure 6k. Enter: **Number_s, 28**; **Trials, 50**; **Probability_s, .5**; and **Cumulative, TRUE** (for the probability of 28 *or fewer*).

   The answer (**0.8389**) shows up in the dialogue box and in the active cell when you click **OK**.

b. Individual (Exact)

1. Click f_x.
2. Choose **BINOM.DIST** and click **OK**.
3. Use the numbers in Figure 6k, but enter **False** in the Cumulative box. This will give you the probability of getting *exactly* 28 heads in 50 tosses of a fair coin.

   The answer (**0.0788**) shows up in the dialogue box and in the active cell when you click **OK**.

STATCRUNCH

NORMAL

Part b is shown here before part a.

b. Some Value or Less

1. **Stat > Calculators > Normal**
2. See Figure 6l. To find the probability of having a Wechsler IQ of 115 or less, Enter: **Mean, 100**; **Std Dev, 15**. Make sure that the symbol to the right of **P(X** is ≤ (for less than or equal to). Enter the **115** in the box above **Compute**.
3. Click **Compute** to see the answer, **0.8413**.

a. Between Two Values

1. **Stat > Calculators > Normal**
2. See Figure 6l. Click **Between**.
3. Enter: **Mean, 100**; **Std Dev, 15**. For **P**(enter **85**, then in the next box enter **115**.
4. Click **Compute**. See the answer **= 0.6826**.

c. Inverse Normal

If you want a measurement (such as an IQ or height) from a proportion or percentile:

1. **Stat > Calculators > Normal**
2. See Figure 6m. To find the Wechsler IQ at the 75th percentile, enter: **Mean, 100**; **Std. Dev., 15**. Make sure that the arrow to the right of **P(X** points to the left, and enter **0.75** in the box to the right of the = sign.

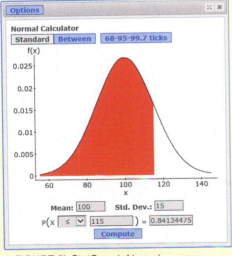

▲ **FIGURE 6l** StatCrunch Normal

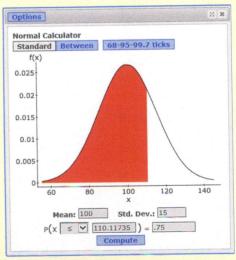

▲ **FIGURE 6m** StatCrunch Inverse Normal

321

3. Click **Compute** and the answer (**110**) is shown above **Compute**.

BINOMIAL

a. Cumulative (or Fewer)

1. **Stat > Calculators > Binomial**
2. See Figure 6n. To find the probability of 28 or fewer heads in 50 tosses of a fair coin, enter: **n, 50;** and **p, 0.5**. The symbol after **P(X** should be ≤ (for *less than or equal to*). Enter **28** in the box above **Compute**.
3. Click **Compute** to see the answer (**0.8389**).

b. Individual (Exact)

1. **Stat > Calculators > Binomial**
2. To find the probability of exactly 28 heads in 50 tosses of a fair coin, use a screen similar to Figure 6n, but to the right of **P(X** choose the equals sign. You will get **0.0788**.

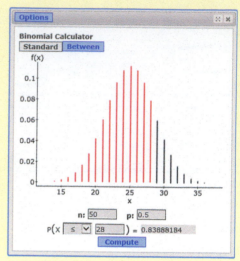

▲ **FIGURE 6n** StatCrunch Binomial

7 Survey Sampling and Inference

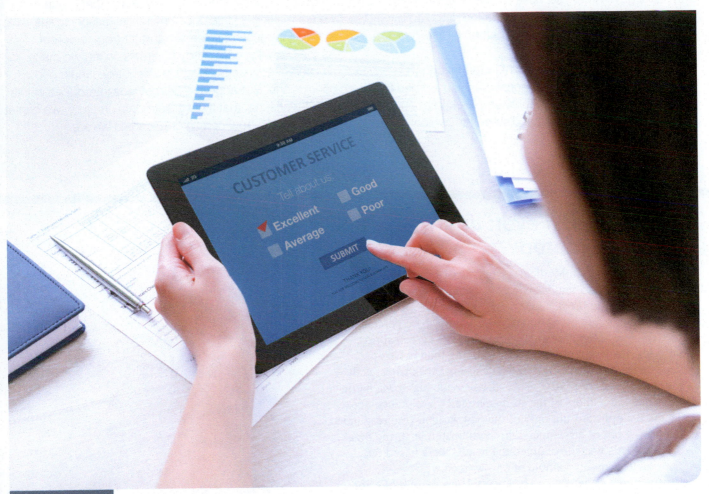

If survey subjects are chosen randomly, then we can use their answers to infer how the entire population would answer. We can also quantify how far off our estimate is likely to be.

Somewhere in your town or city, possibly at this very moment, people are participating in a survey. Perhaps they are filling out a customer satisfaction card at a restaurant. Maybe their television is automatically transmitting information about which show is being watched so that marketers can estimate how many people are viewing their ads. They may even be texting in response to a television survey. Most of you will receive at least one phone call from a survey company that will ask whether you are satisfied with local government services or plan to vote for one candidate over another. The information gathered by these surveys is used to piece together, bit by bit, a picture of the larger world.

You've reached a pivotal point in the text. In this chapter, the data summary techniques you learned in Chapters 2 and 3, the probability you learned about in Chapter 5, and the Normal distribution, which you studied in Chapter 6, are all combined to enable us to generalize what we learn about a small sample to a larger group. Politicians rely on surveys of 1000 voters not because they care how those 1000 individuals will vote. Surveys are important to politicians only if they help them learn about *all* potential voters. In this and later chapters, we study ways to understand and measure just how reliable this projection from sample to the larger world is.

Whenever we draw a conclusion about a large group based on observations of some parts of that group, we are making an inference. Inferential reasoning lies at the foundation of science but is far from foolproof. As the following case study illustrates, when we make an inference, we can never be absolutely certain of our conclusions. But applying the methods introduced in this chapter ensures that if we collect data carefully, we can at least measure how certain or uncertain we are.

CASE STUDY

Spring Break Fever: Just What the Doctors Ordered?

In 2006, the American Medical Association (AMA) issued a press release ("Sex and intoxication among women more common on spring break according to AMA poll") in which it concluded, among other things, that "eighty-three percent of the [female, college-attending] respondents agreed spring break trips involve more or heavier drinking than occurs on college campuses and 74 percent said spring break trips result in increased sexual activity." This survey made big news, particularly since the authors of the study claimed these percentages reflected the opinions not only of the 644 women who responded to the survey but of all women who participated in spring break.

The AMA's website claimed the results were based on "a nationwide random sample of 644 women who . . . currently attend college. . . . The survey has a margin of error of +/−4 percentage points at the 95 percent level of confidence." It all sounds very scientific, doesn't it?

However, some survey specialists were suspicious. After Cliff Zukin, a specialist who was president of the American Association for Public Opinion Research, corresponded with the AMA, it changed its website posting to say the results were based not on a random sample but instead on "a nationwide sample of 644 women . . . who are part of an *online survey panel* . . . [emphasis added]." "Margin of error" is no longer mentioned.

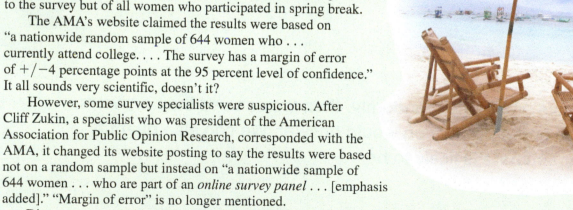

Disagreements over how to interpret these results show just how difficult inference is. In this chapter you'll see why the method used to collect data is so important to inference and how we use probability, under the correct conditions, to calculate a margin of error to quantify our uncertainty. At the end of the chapter, you'll see why the AMA changed its report.

Learning about the World through Surveys

Surveys are probably the most often encountered application of statistics. Most news shows, newspapers, and magazines report on surveys or polls several times a week—and during a major election, several times a day. We can learn quite a bit through a survey if the survey is done correctly. Many useful and interesting statistical investigation questions refer to groups that are too large for us to measure completely. For example, what percentage of schoolchildren in the United States are vaccinated? Asking all parents in the U.S. is not possible in a timely fashion. Surveys give us a method for estimating this percentage without asking each and every parent of a school-aged child.

Survey Terminology

A **population** is a group of objects or people we wish to study. Usually, this group is large—say, the group of all U.S. citizens, or all U.S. citizens between the ages of 13 and 18, or all senior citizens. However, it might be smaller, such as all phone calls made on your cell phone in January. We wish to know the value of a **parameter**, a numerical value that characterizes some aspect of this population. For example, political pollsters want to know what percentage of all registered voters say they will vote in the next election. Drunk-driving opponents want to know what percentage of all teenagers with driver's licenses have drunk alcohol while driving. Designers of passenger airplanes want to know the mean length of passengers' legs so that they can put the rows of seats as close together as possible without causing discomfort.

In this text we focus on two frequently used parameters: the mean of a population and the population proportion. This chapter deals with population proportions.

If the population is relatively small, we can find the exact value of the parameter by conducting a census. A **census** is a survey in which every member of the population is measured. For example, if you wish to know the percentage of people in your classroom who are left-handed, you can perform a census. The classroom is the population, and the parameter is the percentage of left-handers. We sometimes try to take a census with a large population (such as the U.S. Census), but such undertakings are too expensive for nongovernmental organizations and are filled with complications caused by trying to track down and count people who may not want to be found. (For example, the U.S. Census tends to undercount poor, urban-dwelling residents, as well as undocumented immigrants.)

In fact, most populations we find interesting are too large for a census. For this reason, we instead observe a smaller sample. A **sample** is a collection of people or objects taken from the population of interest.

Once a sample is collected, we measure the characteristic we're interested in. A **statistic** is a numerical characteristic of a sample of data. We use statistics to estimate parameters. For instance, we might be interested in knowing what proportion of all registered voters will vote in the next national election. The proportion of all registered voters who will vote in the next election is our *parameter*. Our method to estimate this parameter is to survey a small sample of registered voters. The proportion of the sample who say they will vote in the next election is a *statistic*.

Statistics are sometimes called **estimators**, and the numbers that result are called **estimates**. For example, our *estimator* is the proportion of people in a sample who say they will vote in the next election. When we conduct this survey, we find, perhaps, that 0.75 of the sample say they will vote. This number, 0.75, is our *estimate*.

KEY POINT A statistic is a number that is based on data and used to estimate the value of a characteristic of the population. Thus, it is sometimes called an estimator.

Statistical inference is the science of drawing conclusions about a population on the basis of observing only a small subset of that population. Statistical inference always involves uncertainty, so an important component of this science is measuring our uncertainty.

EXAMPLE 1 Pew Poll: Age and the Internet

In February 2014 (about the time of Valentine's Day), the Pew Research Center surveyed 1428 cell phone users in the United States who were married or in a committed partnership. The survey found that 25% of cell phone owners felt that their spouse or partner was distracted by her or his cell phone when they were together.

QUESTIONS Identify the population and the sample. What is the parameter of interest? What is the statistic?

SOLUTION The population that the Pew Research Center wanted to study consists of all American adults who were married or in a committed partnership and owned a cell phone. The sample, which was taken from the population consists of 1428 such people. The parameter of interest is the percentage of all adults in the United States who were married or in a committed partnership and felt that their spouse or partner was distracted by her or his cell phone when they were together. The statistic, which is the percentage of the sample who felt this way, is 25%.

TRY THIS! Exercise 7.1

An important difference between statistics and parameters is that statistics are knowable. Any time we collect data, we can find the value of a statistic. In Example 1, we *know* that 25% of those surveyed felt that their partner was distracted by the cell phone. In contrast, a parameter is typically unknown. We do not know for certain the percentage of *all* people who felt this way about their partners. The only way to find out would be to ask everyone, and we have neither the time nor the money to do this. Table 7.1 compares the known and the unknown in this situation.

▶ **TABLE 7.1** Some examples of unknown quantities we might wish to estimate, and their knowable counterparts.

Unknown	Known
Population All cell phone owners in a committed relationship	*Sample* A small number of cell phone owners in a committed relationship
Parameter Percentage of all cell phone owners in a committed relationship who felt that their partner was distracted when they were together	*Statistic* Percentage of the sample who felt their partner was distracted when they were together

Statisticians have developed notation for keeping track of parameters and statistics. In general, Greek characters are used to represent population parameters. For example, μ (mu, pronounced "mew," like the beginning of *music*) represents the mean of a population. Also, σ (sigma) represents the standard deviation of a population. Statistics (estimates based on a sample) are represented by English letters: $\bar{x}$ (pronounced "*x*-bar") is the mean of a sample, and s is the standard deviation of a sample, for instance.

One frequently encountered exception is the use of the letter p to represent the proportion of a population and $\hat{p}$ (pronounced "*p*-hat") to indicate the proportion of a sample. Table 7.2 summarizes this notation. You've seen most of these symbols before, but this table organizes them in a new way that is important for statistical inference.

Statistics (based on data)		Parameters (typically unknown)	
Sample mean	$\bar{x}$ (x-bar)	Population mean	μ (mu)
Sample standard deviation	s	Population standard deviation	σ (sigma)
Sample variance	s^2	Population variance	σ^2
Sample proportion	$\hat{p}$ (p-hat)	Population proportion	p

◀ **TABLE 7.2** Notation for some commonly used statistics and parameters.

EXAMPLE 2 Parameter or Statistic?

Consider these two sentences from *The New York Times* (November 5, 2017):
"In Election Day exit polls, 81% of those who described themselves as 'conservative' said they had voted for Donald J. Trump."
"Winston Churchill painted more than 500 pictures in his lifetime."

QUESTION Is the number 81% a statistic or a parameter? Is the number 500 a statistic or a parameter? Explain your choice. For each situation, describe the population.

SOLUTION Election Day exit polls are interviews of a few voters. News agencies carry these out so that they can understand how an election is going before the results are actually tallied. The 81% describes a sample of people, and is therefore a statistic. The population is the set of all people who voted on election day.

The number 500, on the other hand, describes all of Churchill's paintings. The population is the collection of all paintings by Winston Churchill, and this number describes that collection. So the number 500 is a parameter.

TRY THIS! Exercise 7.3

EXAMPLE 3 Getting the Notation Right

The City of Los Angeles provides an open data set of response times for emergency vehicles. Each row of the data set represents an emergency vehicle that has been sent to a particular emergency. A random sample of 1000 of these rows shows that the mean response time was 8.25 minutes. In addition, the proportion of vehicles that were ambulances was 0.328.
Match each of the following statistics to the correct notation.

QUESTIONS Which is correct notation for the statistic 8.25?

a. $\mu = 8.25$

b. $p = 8.25$

c. $\bar{x} = 8.25$

d. $\hat{p} = 8.25$

Which is correct notation for the statistic 0.328?

a. $\mu = 0.328$

b. $p = 0.328$

c. $\bar{x} = 0.328$

d. $\hat{p} = 0.328$

SOLUTION Because 8.25 is describing an average, the correct notation is (c). The notation for (a) would be correct if we had the mean response time for all vehicles. In other words, the notation in (a) is appropriate for a parameter, and (c) is appropriate for a statistic.

Because 0.328 is the proportion of a sample, the correct notation is (d). If the proportion of ambulances among *all* vehicles was 0.328, then (b) would be the appropriate notation.

TRY THIS!   Exercise 7.9

EXAMPLE **4** Name That Population?

The Department of Health in Colorado conducts a "Healthy Kids Colorado Survey." The stated purpose is "to better understand youth health and what factors support youth to make healthy choices." The survey is administered to randomly selected students chosen from all middle and high school students in Colorado.

QUESTION Identify the sample and the population. The Department of Health said that 26.4% of students responding get at least one hour of physical activity per day. Is this a statistic or a parameter?

SOLUTION The sample consists of a group of students who attend middle schools and high schools in Colorado. They are selected from the population which consists of all middle and high school students in Colorado. The figure "26.4%" is describing the sample (because it is based on those who responded to the survey and so were in the sample). For this reason, it is a statistic. As we will soon see, statistics such as these are used to estimate population parameters.

TRY THIS!   Exercise 7.11

What Could Possibly Go Wrong? The Problem of Bias

Unfortunately, it is far easier to conduct a bad survey than to conduct a good survey. One of the many ways in which we can reach a wrong conclusion is to use a survey method that is biased.

Informally speaking, a method is **biased** if it has a tendency to produce, on average, the wrong value. Bias can enter a survey in three ways. The first is through **sampling bias**, which results from taking a sample that is not representative of the population. A second way is **measurement bias**, which comes from asking questions that do not produce a true answer. For example, if we ask people their income, they are likely to inflate the value. In this case, we will get a positive (or "upward") bias: our estimate will tend to be too high. Measurement bias occurs when measurements tend to record values larger (or smaller) than the true value.

The third way occurs because some statistics are naturally biased. For example, if you use the statistic $10\bar{x}$ to estimate the mean, you'll typically get estimates that are ten times too big. Therefore, even when no measuring or sampling bias is present, you must also take care to use an estimator that is not biased.

Measurement Bias In February 2010, the *Albany Times Union* newspaper reported on two surveys to determine the opinions of New York State residents on taxing soda (Crowley 2010). The Quinnipiac University Polling Institute asked, "There is a proposal for an 'obesity tax' or a 'fat tax' on non-diet sugary soft drinks. Do you support or oppose such a measure?" Forty percent of respondents said they supported the tax. Another firm, Kiley and Company, asked, "Please tell me whether you feel the state should take that step in order to help balance the budget, should seriously consider it, should consider it only as a last resort, or should definitely not consider taking that step: 'Imposing a new 18 percent tax on sodas and other soft drinks containing sugar, which would also reduce childhood obesity.'" Fifty-eight percent supported the tax when asked this question. One or both of these surveys have measurement bias.

A famous example occurred in 1993, when, on the basis of the results of a Roper Organization poll, many U.S. newspapers published headlines similar to this one from *The New York Times*: "1 in 5 in New Survey Express Some Doubt About the Holocaust" (April 20, 1993). Almost a year later, *The New York Times* reported that this alarmingly high percentage of alleged Holocaust doubters could be due to measurement error. The actual question respondents were asked contained a double negative: "Does it seem possible, or does it seem impossible to you, that the Nazi extermination of the Jews never happened?" When Gallup repeated the poll but did not use a double negative, only 9% expressed doubts (*The New York Times* 1994).

Sampling Bias   Writing good survey questions to reduce measurement bias is an art and a science. This text, however, is more concerned with sampling bias, which occurs when the estimation method consistently produces a sample that is not representative of the population.

Have you ever heard of Alfred Landon? Unless you're a political science student, you probably haven't. In 1936, Landon was the Republican candidate for U.S. president, running against Franklin Delano Roosevelt. The *Literary Digest*, a popular news magazine, conducted a survey with over 10 million respondents and predicted that Landon would easily win the election with 57% of the vote. The fact that you probably haven't heard of Landon suggests that he didn't win—and, in fact, he lost big, setting a record at the time for the fewest electoral votes ever received by a major-party candidate. What went wrong? The *Literary Digest* used a biased method to produce its sample. The journal relied largely on polling its own readers, and its readers were more well-to-do than the general public and more likely to vote for a Republican. The reputation of the *Literary Digest* was so damaged that two years later it disappeared and was absorbed into *Time* magazine.

The U.S. presidential elections of 2004 and 2008 both had candidates who claimed to have captured the youth vote; both times, the candidates claimed the polls were biased. The reason given was that the surveys used to estimate candidate support relied on landline phones, and many young voters don't own landlines, relying instead on their cell phones. Reminiscent of the 1936 *Literary Digest* poll, these surveys were potentially biased because their sample systematically excluded an important part of the population: those who do not use landlines (Cornish 2007).

In fact, the Pew Foundation conducted a study after the 2010 congressional elections. This study found that polls that excluded cell phones had a sampling bias in favor of Republican candidates.

You can easily find biased sampling methods on many websites. Ratings of businesses, restaurants, or friends' vacation photos are a type survey in which customers are asked to provide their opinion or rating (usually out of 1 to 5 stars). You might rely on websites such as Yelp, Urban Spoon, or Open Table to choose a restaurant for a special occasion. But you should be aware that the provided ratings are not necessarily representative of the feelings of *all* restaurants' patrons. People who leave ratings are often motivated to do so because of strong positive or negative experiences. And so the ratings can be biased and not representative of the restaurant's customers as a whole. This bias is perhaps one reason the graph in Figure 6.3 has peaks at 5 and 1. Those who really loved or really disliked the book were more compelled to enter reviews.

Because of response bias, you should always question what type of people were included in a survey. But the other side of this coin is that you should also question what type of people were left out. Was the survey conducted at a time of day that meant that working people were less likely to participate? Were only landline phones used, thereby excluding people who had only cell phones? Was the question that was asked potentially embarrassing, so that people might have refused to answer? All of these circumstances can bias survey results.

KEY POINT
When reading about a survey, it is important to know:
1. what percentage of people who were asked to participate actually did so; and
2. whether the researchers chose people to participate in the survey or people themselves chose to participate.

If a large percentage of those chosen to participate refused to answer questions, or if people themselves chose whether to participate, the conclusions of a survey are suspect.

Simple Random Sampling Saves the Day

How do we collect a sample with a method that has as little bias as possible? Only one way reliably works: take a random sample.

As we explained in Chapter 5, statisticians have a precise definition of *random*. A random sample does not mean that we stand on a street corner and stop whomever we like to ask them to participate in our survey. (Statisticians call this a **convenience sample**, for obvious reasons.) A random sample must be taken in such a way that every person in our population is equally likely to be chosen.

A truly random sample is difficult to achieve. (And that's a big understatement!) Pollsters have invented many clever ways of pulling this off, often with great success. One basic method that's easy to understand (but somewhat difficult to put into practice) is **simple random sampling (SRS)**.

In SRS, we draw subjects from the population at random and without replacement. **Without replacement** means that once a subject is selected for a sample, that subject cannot be selected again. This is like dealing cards from a deck. Once a card is dealt for a hand, no one else can get the same card. A result of this method is that every sample of the same fixed size is equally likely to be chosen. As a result, we can produce unbiased estimations of the population parameters of interest and can measure the precision of our estimator.

It can't be emphasized enough that if our sample is not random, there's really nothing we can learn about the population. We can't measure a survey's precision, and we can't know how large or small the bias might be. An unscientific survey is a useless survey for the purposes of learning about a population.

In theory, we can take an SRS by assigning a number to each and every member of the population. We then use a random number table or other random number generator to select our sample, ignoring numbers that appear twice.

> **Details**
>
> Simple random sampling is not the only valid method for statistical inference. Statisticians collect representative samples using other methods, as well (for example, sampling **with replacement**). What these methods all have in common is that they take samples randomly.

EXAMPLE 5 Taking a Simple Random Sample

Alberto, Justin, Michael, Audrey, Brandy, and Nicole are in a class.

QUESTION Select an SRS of three names from these six names.

SOLUTION First assign each person a number, as shown:

Alberto	1
Justin	2
Michael	3
Audrey	4
Brandy	5
Nicole	6

Next, select three of these numbers without replacement. Figure 7.1 shows how this is done in StatCrunch, and almost all statistical technologies let you do this quite easily.

Using technology, we got these three numbers: 1, 2, and 6. These correspond to Alberto, Justin, and Nicole.

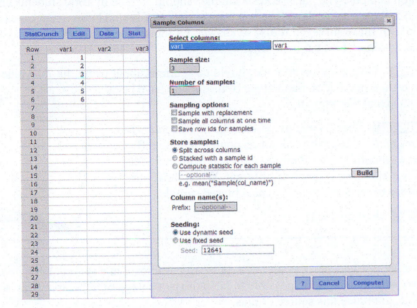

◀ **FIGURE 7.1** StatCrunch will randomly select, without replacement, three numbers from the six shown in the var1 column.

If technology is not available, a random number table, such as the one provided in Appendix A, can be used. Here are two lines from such a table:

77598   29511   98149   63991
31942   04684   69369   50814

You can start at any row or column you please. Here, we choose to start at the upper left (shown in bold face). Next, read off digits from left to right, skipping digits that are not in our population. Because no one has the number 7, skip this number, twice. The first person selected is number 5: Brandy. Then skip 9 and 8 and select number 2: Justin. Skip 9 and 5 (because you already selected Brandy) and select number 1: Alberto.

CONCLUSION   Using technology, we got a sample consisting of Alberto, Justin, and Nicole. Using the random number table, we got a different sample: Brandy, Justin, and Alberto.

TRY THIS!   Exercise 7.17

Sampling in Practice

Sampling in Practice In practice, simple random samples are difficult to collect and often inefficient. In most situations, we can't make a list of all people in the United States and assign each of them a number. To get around this, statisticians use alternative techniques.

In addition, random sampling does not cure all ills. **Nonresponse bias** can still be a problem, and the possibility always exists that methods of taking the random sample are flawed (as they are if only landline telephones are used when many in the population use cell phones).

EXAMPLE 6 Survey on Sexual Harassment

A newspaper at a large college wants to determine whether sexual harassment is a problem on campus. The paper takes a simple random sample of 1000 students and asks each person whether he or she has been a victim of sexual harassment on campus. About 35% of those surveyed refuse to answer. Of those who do answer, 2% say they have been victims of sexual harassment.

QUESTION Give a reason why we should be cautious about using the 2% value as an estimate for the population percentage of those who have been victims of sexual harassment.

CONCLUSION There is a large percentage of students who did not respond. Those who did not respond might be different from those who did, and if their answers had been included, the results could have been quite different. When those surveyed refuse to respond, it can create a biased sample.

TRY THIS! Exercise 7.21

There are always some people who refuse to participate in a survey, but a good researcher will do everything possible to keep the percentage of nonresponders as small as possible, to reduce this source of bias.

SECTION 7.2

Measuring the Quality of a Survey

A frequent complaint about surveys is that a survey based on 1000 people can't possibly tell us what the entire country is thinking. This complaint raises interesting questions: How do we judge whether our estimators are working? What separates a good estimation method from a bad one?

It's difficult, if not impossible, to judge whether any particular survey is good or bad based only on the outcome of the survey. Sometimes we can find obvious sources of bias, but often we don't know whether a survey has failed unless we later learn the true parameter value. (This sometimes occurs in elections when we learn that a survey must have had bias because it severely missed predicting the actual outcome.) Instead, statisticians evaluate the *method* used to estimate a parameter, not the outcome of a particular survey.

KEY POINT Statisticians evaluate the method used for a survey, not the outcome of a single survey.

Before we talk about how to judge surveys, imagine the following scenario: We are not taking just one survey of 1000 randomly selected people. We are sending out an army of pollsters. Each pollster surveys a random sample of 1000 people, and they all use the same method for collecting the sample. Each pollster asks the same question and produces an estimate of the proportion of people in the population who would answer yes to the question. When the pollsters return home, we get to see not just a single estimate (as happens in real life) but a great many estimates. Because each estimate is based on a separate random collection of people, each one will differ slightly. We expect some of these estimates to be closer to the mark than others just because of random variation. What we really want to know is how the group did as a whole. For this reason, we talk about evaluating *estimation methods*, not estimates.

An estimation method is a lot like a golfer. To be a good golfer, we need to get the golf ball in the cup. A good golfer is both *accurate* (tends to hit the ball near the cup) and *precise* (even when she misses, she doesn't miss by very much.)

It is possible to be precise and yet be inaccurate, as shown in Figure 7.2b. Also, it is possible to aim in the right direction (be accurate) but be imprecise, as shown in Figure 7.2c. (Naturally, some of us are bad at both, as shown in Figure 7.2d.) But the best golfers can both aim in the right direction and manage to be very consistent, which Figure 7.2a shows us.

> **! Caution**
>
> **Estimator and Estimates**
> We often use the word **estimator** to mean the same thing as "estimation method." An **estimate**, on the other hand, is a number produced by our estimation method.

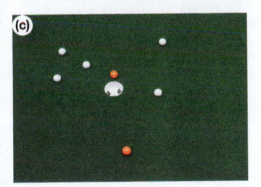

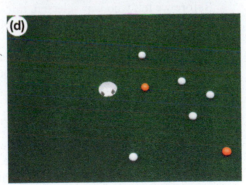

◀ **FIGURE 7.2** **(a)** Shots from a golfer with good aim and precision; the balls are tightly clustered and centered around the cup. **(b)** Shots from a golfer with good precision but poor aim; the balls are close together but centered to the right of the cup. **(c)** Shots from a golfer with good aim — the balls are centered around the cup — but bad precision. **(d)** The worst-case scenario: bad precision *and* bad aim.

Think of the cup as the population parameter, and think of each golf ball as an estimate, a value of $\hat{p}$, that results from a different survey. We want an estimation method that aims in the right direction. Such a method will, on average, get the correct value of the population parameter. We also need a precise method so that if we repeated the survey, we would arrive at nearly the same estimate.

The aim of our method, in other words, the method's *accuracy,* is measured in terms of the *bias*. The *precision* is measured by a number called the *standard error.* Discussion of simulation studies in the next sections will help clarify how accuracy and precision are measured. These simulation studies show how bias and standard error are used to quantify the uncertainty in our inference.

Using Simulations to Understand the Behavior of Estimators

The three simulations that follow will help measure how well the sample proportion works as an estimator of the population proportion.

In the first simulation, imagine doing a survey of 4 people in a very small population of only 8 people. You'll see that the estimator of the population proportion is accurate (no bias) but, because of the small sample size, not terribly precise.

In the second simulation, the first simulation is repeated, using a larger population and sample. The estimator is still unbiased, and you will see a perhaps surprising change in precision. Finally, the third simulation will reveal that using a much larger sample size makes the result even more precise.

To learn how our estimation method behaves, we're going to create a very unusual, unrealistic situation: We're going to create a world in which we know the truth. In this world, there are two types of people: those who like dogs and those who like cats. No one likes both. Exactly 25% of the population are Cat People, and 75% are Dog People. We're going to take a random sample of people from this world and see what proportion of our sample are Cat People. Then we'll do it again. And again. From this repetition, we'll see some interesting patterns emerge.

Simulation 1: Statistics Vary from Sample to Sample To get started, let's create a very small world. This world has 8 people named 1, 2, 3, 4, 5, 6, 7, and 8. People 1 and 2 are Cat People. Figure 7.3 shows this population.

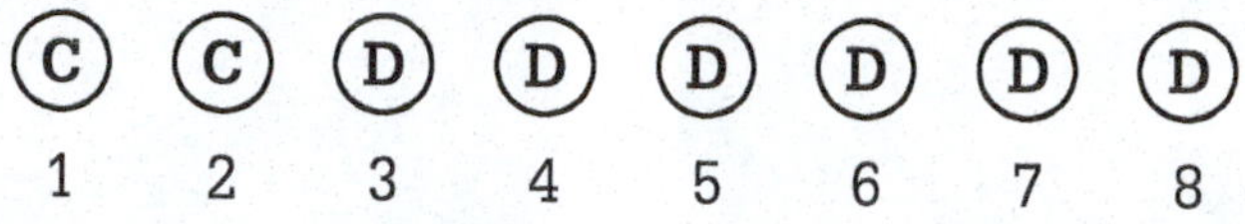

▶ **FIGURE 7.3** The entire population of our simulated world; 25% are Cat People.

$$Ⓒ \quad Ⓒ \quad Ⓓ \quad Ⓓ \quad Ⓓ \quad Ⓓ \quad Ⓓ \quad Ⓓ$$
$$1 \quad\;\; 2 \quad\;\; 3 \quad\;\; 4 \quad\;\; 5 \quad\;\; 6 \quad\;\; 7 \quad\;\; 8$$

From this population, we use the random number table to generate four random numbers between 1 and 8. When a person's number is chosen, he or she steps out of the population and into our sample.

Before we tell who was selected, think for a moment about what you expect to happen. What proportion of our sample will be Cat People? Is it possible for 0% of the sample to be Cat People? For 100%?

Below is our random sample. Note that we sampled without replacement, as in a real survey. We don't want the same person to be in our sample twice.

6	8	4	5
D	D	D	D

None of those selected are Cat People, as Figure 7.4 indicates. The proportion of Cat People in our sample is 0.0. We call this the *sample proportion* because it comes from the sample, not the population.

▶ **FIGURE 7.4** The first sample, which has 0% Cat People. The sample is shown in the box.

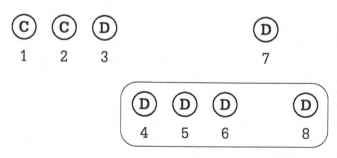

Let's take another random sample. It is possible that we will again get 0, but it is also possible that we will get a different proportion.

7	2	6	3
D	C	D	D

This time, our sample proportion is 0.25.
One more time:

2	8	6	5
C	D	D	D

Again, our sample proportion is 0.25.

The difference between when the true value of the parameter, 0.25, and our estimate based on the sample, is called the *error*. The word *error* has a technical meaning in this context; it doesn't mean a mistake. It's simply a measure of how far away our estimate is from the true value. Table 7.3 shows what has happened so far, along with the errors.

◄ **TABLE 7.3** The results of three repetitions of our simulation.

Repetition	Population Parameter	Sample Statistics	Error
1	$p = 0.25$ Cat People	$\hat{p} = 0.0$ Cat People	$\hat{p} - p = (0.0 - 0.25) = -0.25$
2	$p = 0.25$ Cat People	$\hat{p} = 0.25$ Cat People	$\hat{p} - p = (0.25 - 0.25) = 0.0$
3	$p = 0.25$ Cat People	$\hat{p} = 0.25$ Cat People	$\hat{p} - p = (0.25 - 0.25) = 0.0$

First, notice that the population proportion, p, never changes. It can't, because in our made-up world the population has the same 8 people, and the same 2 are Cat People. However, the sample proportion, $\hat{p}$, can be different in each sample. In fact, $\hat{p}$ is random because it depends on a random sample.

And because $\hat{p}$ is random, the error is also random. It can be positive, if the sample has more than 25% Cat People; negative, if the sample has fewer than 25% Cat People; and 0 if we get the percentage exactly right.

KEY POINT   No matter how many different samples we take, the value of p (the population proportion) is always the same, but the value of $\hat{p}$ changes from sample to sample.

This simulation is, in fact, a random experiment and $\hat{p}$ is our outcome. Because it is random, $\hat{p}$ has a probability distribution. The probability distribution of $\hat{p}$ has a special name: **sampling distribution**. This term reminds us that $\hat{p}$ is not just any random outcome; it is a statistic we use to estimate a population parameter.

Because our world has only 8 people in it and we are taking samples of 4 people, we can write down all of the possible outcomes. There are only 70. By doing this, we can see exactly how often $\hat{p}$ will be 0.0, how often 0.25, and how often 0.50. (Notice that it can never be more than 50%.) These probabilities are listed in Table 7.4, which presents the sampling distribution for $\hat{p}$. Figure 7.5 visually represents this sampling distribution.

Value of $\hat{p}$	Probability of Seeing That Value
0.0	0.21429
0.25	0.57143
0.50	0.21429

▲ **TABLE 7.4** The sampling distribution for $\hat{p}$, based on our random sample.

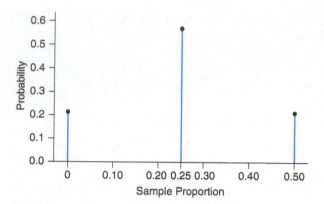

From Table 7.4 and Figure 7.5, we learn several things:

1. Our estimator, $\hat{p}$, is not always the same as our parameter, p. Sometimes $\hat{p}$ turns out to be 0, sometimes it is 0.50, and sometimes it hits the target value of 0.25.

2. The mean of this distribution is 25% — the same value as p.

3. Even though $\hat{p}$ is not always the "true" value, p, we are never more than 25 percentage points away from the true value.

Why are these observations important? Let's consider each one separately.

The first observation reminds us that statistics based on random samples are random. Thus we cannot know ahead of time, with certainty, exactly what estimates our survey will produce.

The second observation tells us that, *on average*, the error in our estimator is 0. Figure 7.5 shows us that for some samples, our error will be +0.25 ($\hat{p}$ is 0.50), but just as often it will be −0.25 ($\hat{p}$ is 0.0). And quite often it will be 0.0 ($\hat{p}$ is 0.25). If we average these errors, the result is 0.0. Visually, you can see that the balancing point of the sampling distribution shown in Figure 7.5 is at 0.25, which tells us that the mean value of the estimator is 0.25 and the mean value of the error is 0.0.

The technical term for the mean error is *bias*. When the bias is 0, we say that the estimator is **unbiased.** A survey that uses an unbiased estimator tends to produce the correct parameter value on average. "On average" means that some surveys will produce estimates that are a bit too large (positive errors), and others will produce estimates that are a bit too small (negative error). But the general tendency will be to produce the correct value.

The third observation is about precision. We know that our estimator is, on average, the same as the parameter, but the sampling distribution tells us how far away, typically, the estimator might stray from average. **Precision** is reflected in the spread of the sampling distribution and is measured by using the standard deviation of the sampling distribution. In this simulation, the standard deviation is 0.16366, or roughly 16%. The standard deviation of a sampling distribution has a special name: the **standard error (SE).**

The standard error measures how much our estimator typically varies from sample to sample. Thus, in this simulated survey, if we survey 4 people, we usually get 25% (or 0.25) Cat People, but this typically varies by plus or minus 16.4% (16.4 percentage points). Looking at the graph in Figure 7.5, we might think that the variability is typically plus or minus 25 percentage points, but we must remember that the standard deviation measures how spread out observations are from the mean value. Many observations are identical to the average value, so the typical, or "standard," deviation from the mean value of 25% is only 16.4 percentage points.

> Bias is the average error across all possible outcomes of a random sample. The error is the difference between the estimate from a particular sample and the true value of the parameter. Bias is a measure of accuracy. An unbiased estimator has perfect accuracy.
>
> Precision is measured using the standard deviation of the sampling distribution, which is called the standard error. When the standard error is small, we say that the estimator is precise.

SNAPSHOT ▶ Sampling Distribution

WHAT IS IT? ▶	A special name for the probability distribution of a statistic.
WHAT DOES IT DO? ▶	Gives us probabilities for a statistic.
WHAT IS IT USED FOR? ▶	It tells us how often we can expect to see particular values of our estimator, and it also gives us important characteristics of the estimator, such as bias and precision.
HOW IS IT USED? ▶	It is used for making inferences about a population.

Simulation 2: The Size of the Population Does Not Affect Precision

The first simulation was very simple, because our made-up world had only 8 people. In our first simulation, the bias was 0, which is good; this means we have an accurate estimator. However, the precision was fairly poor (we had a large standard error). How can we improve precision? To understand, we need a slightly more realistic simulation.

This time, we'll use the same world but make it somewhat bigger. Let's assume we have 1000 people and 25% are Cat People ($p = 0.25$). (In other words, there are 250 Cat People.) We take a random sample of 10 people and find the sample proportion, $\hat{p}$, of Cat People.

Because we've already seen how this is done, we're going to skip a few steps and show the results. This time the potential outcomes are too numerous to list, so instead we just do a simulation:

1. Take a random sample, without replacement, of 10 people.

2. Calculate $\hat{p}$: the proportion of Cat People in our sample.

3. Repeat steps 1 and 2 a total of 10,000 times. Each time, calculate $\hat{p}$ and record its value.

Here are our predictions:

1. We predict that $\hat{p}$ will not be the same value every time because it is based on a random sample, so the value of $\hat{p}$ will vary randomly.

2. We predict that the mean outcome, the typical value for $\hat{p}$, will be 0.25—the same as the population parameter—because our estimator is unbiased.

3. Precision: This one is left to you. Do you think the result will be more precise or less precise than in the last simulation? In the last simulation, only 4 people were sampled, and the variation, as measured by the standard error, was about 0.16. This time more people (10) are being sampled, but the population is much larger (1000). Will the standard error be larger (less precise) or smaller (more precise) than 0.16?

> **Details**
>
> **Simulations and Technology**
> Don't take our word for it. You can carry out this simulation using technology. See the TechTips to learn how to do this using StatCrunch.

After carrying out the 10,000 samples we make a graph of our 10,000 $\hat{p}$s. Figure 7.6 shows a histogram of these. Figure 7.6 is an approximation of the sampling distribution; it is not the actual sampling distribution, because the histogram is based on a simulation. Still, with 10,000 replications, it is a very good approximation of the actual sampling distribution.

▶ **FIGURE 7.6** Simulation results for $\hat{p}$. This histogram is a simulation of the sampling distribution. The true value of p is 0.25. Each sample is based on 10 people, and we repeated the simulation 10,000 times. The mean of the distribution is indicated by the triangle.

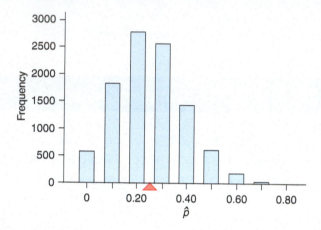

The center of the estimated distribution is at 0.2501, which indicates that essentially no bias exists, because the population parameter is 0.25.

We can estimate the standard error by finding the standard deviation of our simulated $\hat{p}$s. This turns out to be about 0.1356, or 13.56 percentage points.

The value of the standard error tells us that if we were to take another sample of 10 people, we would expect to get about 25% Cat People, give or take 13.6 percentage points.

From Figure 7.6 we learn important information:

1. The bias of $\hat{p}$ is still 0, even though we used a larger population and a larger sample.

2. The variation of $\hat{p}$ is less; this estimator is more precise than with our sample of four people, even though the population is larger. In general, the precision has *nothing* to do with the size of the *population* but only with the size of the *sample*.

Many people are surprised to learn that precision is not affected by population size. How can the level of precision for a survey in a town of 10,000 people be the same as for one in a country of 210 *million* people?

Figure 7.7 provides an analogy. The bowls of soup represent two populations: a big one (a country, perhaps) and a small one (a city). Our goal is to taste each soup (take a sample from the population) to judge whether we like it. If both bowls are well stirred, the size of the bowl doesn't matter—using the same-size spoon, we can get the same amount of taste from either bowl.

▶ **FIGURE 7.7** The bowls of soup represent two populations, and the sample size is represented by the spoons. The precision of an estimate depends only on the size of the sample, not on the size of the population.

The precision of an estimator does not depend on the size of the population; it depends only on the sample size. An estimator based on a sample size of 10 is just as precise in a population of 1000 people as in a population of a million.

Simulation 3: Large Samples Produce More Precise Estimators

How do the simulation and bias change if we increase the sample size? We'll do another simulation with the same population (1000 people and 25% Cat People), but this time, instead of sampling 10 people, we'll sample 100.

Figure 7.8 shows the result. Note that the center of this estimated sampling distribution is still at 25%. Also, our estimation method remains unbiased. However, the shape looks pretty different. First, because many more outcomes are possible for $\hat{p}$, this histogram looks as though it belongs more to a continuous-valued random outcome than to a discrete value. Second, it is much more symmetric than Figure 7.6. You will see in Section 7.3 that the shape of the sampling distribution of $\hat{p}$ depends on the size of the random sample.

An important point to note is that this estimator is much more precise because it uses a larger sample size. By sampling more people, we get more information, so we can end up with a more precise estimate. The estimated standard error, which is simply the standard deviation of the distribution shown in Figure 7.8, is now 0.042, or 4.2 percentage points.

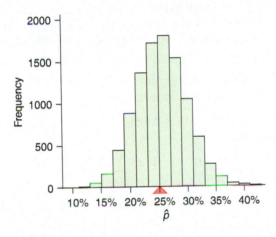

◄ **FIGURE 7.8** Simulated sampling distribution of a sample proportion of Cat People, based on a random sample of 100 people. The simulation was repeated 10,000 times.

Table 7.5 shows a summary of the three simulations.

◄ **TABLE 7.5** Increasing sample size results in increasing precision (measured as decreasing standard error).

Simulation	Population Size	Sample Size		Mean	Standard Error	
1	8	4		25%	16.4%	
2	1000	10	Increasing Sample Size	25%	13.5%	Increasing Precision
3	1000	100		25%	4.2%	

Here is what we learned from Figure 7.8, which is based on sample sizes of 100 "people":

1. The estimator $\hat{p}$ is unbiased for all sample sizes (as long as we take random samples).

2. The precision improves as the sample size gets larger.

3. The shape of the sampling distribution is more symmetric for larger sample sizes.

Surveys based on larger sample sizes have smaller standard errors (SEs) and therefore better precision. Increasing the sample size improves precision.

Finding the Bias and the Standard Error

We've shown that we can estimate bias and precision by running a simulation. But we can also do this mathematically, without running a simulation. Bias and standard error are easy to find for a sample proportion under certain conditions.

The bias of $\hat{p}$ is 0, and the standard error is

$$\text{Formula 7.1a: } SE = \sqrt{\frac{p(1 - p)}{n}}$$

if the following two conditions are met:

Condition 1. The sample must be randomly selected from the population of interest, either with or without replacement. The population parameter to be estimated is the proportion of people (or objects) with some characteristic. This proportion is denoted as p.

Condition 2. If the sampling is without replacement, the population needs to be much larger than the sample size; at least 10 times bigger is a good rule of thumb.

EXAMPLE 7 Pet World

Suppose that in Pet World, the population is 1000 people and 25% of the population are Cat People. Cat People love cats but hate dogs. We are planning a survey in which we take a random sample of 100 people, without replacement. We calculate the proportion of people in our sample who are Cat People.

QUESTION   What value should we expect for this sample proportion? What's the standard error? How do we interpret these values?

SOLUTION   The sample proportion is unbiased, so we expect it to be the same as the population proportion: 0.025.

The standard error is $\sqrt{\dfrac{p(1 - p)}{n}} = \sqrt{\dfrac{0.25 \times 0.75}{100}} = \sqrt{\dfrac{0.1875}{100}}$

$$= \sqrt{0.001875} = 0.04330, \text{ or about 4.3\%}$$

This formula is appropriate because the population size is big with respect to the sample size. The population size is 1000, and the sample size is 100; $100 \times 10 = 1000$, so the population is ten times larger than the sample size.

CONCLUSION   We interpret the values to mean that if we were to take a survey of 100 people from Pet World, we would expect about 25% of them to be Cat People, give or take about 4.3%. The "give or take" means that if you were to draw a sample of 100 and I were to draw a sample of 100, our sample proportions would typically differ from the expected 25% by about 4.3 percentage points.

TRY THIS!   Exercise 7.31

Real Life: We Get Only One Chance

In simulations, we could repeat the survey many times to understand what might happen. In real life, we get just one chance. We take a sample, calculate $\hat{p}$, and then have to live with it.

It is important to realize that bias and precision are both measures of what would happen if we could repeat our survey many times. Bias indicates the typical outcome of surveys repeated again and again. If the bias is 0, we will typically get the right value. If the bias is 0.10, then our estimate will characteristically be 10 percentage points too high. Precision measures how much our estimator will vary from the typical value if we do the survey again. To put it slightly differently, if someone else does the survey, precision helps determine how different her or his estimate could be from ours.

How small must the standard error be for a "good" survey? The answer varies, but the basic rule is that the precision should be small enough to be useful. A typical election poll has a sample of roughly 1000 registered voters and a standard error of about 1.5 percentage points. If the candidates are many percentage points apart, this is good precision. However, if they are neck and neck, this might not be good enough. In Section 7.4, we will discuss how to make decisions about whether the standard error is small enough.

In real life, we don't know the true value of the population proportion, p. This means we can't calculate the standard error. However, we can come pretty close by using the sample proportion. If p is unknown, then a useful approximation to the true standard error is:

$$\textbf{Formula 7.1b:}\quad SE_{est} = \sqrt{\frac{\hat{p}(1-\hat{p})}{n}}$$

The Central Limit Theorem for Sample Proportions

Remember that a probability tells us how often an event happens if we repeat an experiment an infinite number of times. For instance, the sampling distribution of $\hat{p}$ gives the probabilities of where our sample proportions will fall; that is, it tells us how often we would see particular values of $\hat{p}$ if we could repeat our survey infinitely many times. In the simulation, we repeated our fake survey 10,000 times. Ten thousand is a lot, but it's a far cry from infinity.

In the three simulations in Section 7.2, we saw that the shape of the sampling distribution (or our estimated version, based on simulations) changed as the sample size increased (compare Figures 7.5, 7.6, and 7.8). If we used an even larger sample size than 100 (the sample size for the last simulation), what shape would the sampling distribution have? As it turns out, we don't need a simulation to tell us. For this statistic, and for some others, a mathematical theorem called the **Central Limit Theorem (CLT)** gives us a very good approximation of the sampling distribution without our needing to do simulations.

The Central Limit Theorem is helpful because sampling distributions are important. They are important because they, along with the bias and standard error, enable us to measure the quality of our estimation methods. Sampling distributions give us the probability that an estimate falls a specified distance from the population value. For example, we don't want to know simply that 18% of our customers are likely to buy new cell phones in the next year. We also want to know the probability that the true percentage might be higher than some particular value, say, 25%.

Meet the Central Limit Theorem for Sample Proportions

The Central Limit Theorem has several versions. The one that applies to estimating proportions in a population tells us that if some basic conditions are met, then the sampling distribution of the sample proportion is close to the Normal distribution.

More precisely, when estimating a population proportion, p, we must have the same conditions that were used in finding bias and precision, and one new condition as well:

Condition 1. *Random and Independent.* The sample is collected randomly from the population, and observations are independent of each other. The sample can be collected either with or without replacement.

Condition 2. *Large Sample.* The sample size, n, is large enough that the sample expects at least 10 successes (yes's) and 10 failures (no's).

Condition 3. *Big Population.* If the sample is collected without replacement, then the population size must be much (at least ten times) bigger than the sample size.

The sampling distribution for $\hat{p}$ is then approximately Normal, with mean p (the population proportion) and standard deviation the same as the standard error, as given in Formula 7.1a:

$$SE = \sqrt{\frac{p(1-p)}{n}}$$

Looking Back

Normal Notation
Recall that the notation N(mean, standard deviation) designates a particular Normal distribution.

KEY POINT

The Central Limit Theorem for Sample Proportions tells us that if we take a random sample from a population, and if the sample size is large and the population size much larger than the sample size, then the sampling distribution of $\hat{p}$ is approximately

$$N\left(p, \sqrt{\frac{p(1-p)}{n}}\right)$$

If you don't know the value of p, then you can substitute the value of $\hat{p}$ to calculate the estimated standard error.

Figure 7.9 illustrates the CLT for proportions. Figure 7.9a is based on simulations in which the sample size was just 10 people, which is too small for the CLT to apply. In this case, the simulated sampling distribution does not look Normal; it is right-skewed and has large gaps between values. Figure 7.9b is based on simulations of samples of 100 observations. Because the true population proportion is $p = 0.25$, a sample size of 100 is large enough for the CLT to apply, and our simulated sampling distribution looks very close to the Normal model. Figure 7.9b is actually a repeat of Figure 7.8

► **FIGURE 7.9** **(a)** Revision of Figure 7.6, a histogram of 10,000 sample proportions, each based on $n = 10$ with a population percentage p equal to 25%. **(b)** Revision of Figure 7.8, a histogram of 10,000 sample proportions, each based on $n = 100$ with a population percentage p equal to 25%.

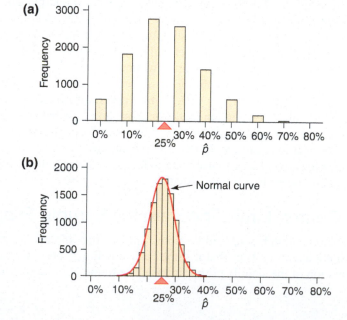

with the Normal curve superimposed. Now that the graphs' horizontal axes are on the same scale, we can see that the sample size of 100 gives better precision than the sample size of 10—the distribution is narrower.

The Normal curve shown in Figure 7.9b has a mean of 0.25 because $p = 0.25$, and it has a standard deviation (also called the standard error) of 0.0433 because

$$\sqrt{\frac{0.25 \times 0.75}{100}} = 0.0433$$

Before illustrating how to use the CLT, we show how to check conditions to see whether the CLT applies.

Checking Conditions for the Central Limit Theorem

The first condition requires that the sample be collected randomly and that observations be independent of each other. There is no way to check this just by looking at the data; you have to trust the researcher's report on how the data were collected, or if you are the researcher, you must take care to use valid random sampling methods.

The second condition dictates that the sample size must be large enough. This we *can* check by looking at the data. The CLT says that the sample size needs to be sufficiently large to get at least 10 successes and 10 failures in our sample. If the probability of a success is p, then we would expect about np successes and $n(1 - p)$ failures. One problem, though, is that we usually don't know the value of p. In this case, we instead check that

$$n\hat{p} \geq 10 \quad \textbf{and} \quad n(1 - \hat{p}) \geq 10$$

For example, if our sample has 100 people and we are estimating the proportion of females in the population, and if our sample has 49% females, then we need to verify that both $100(0.49) \geq 10$ and $100(0.51) \geq 10$.

The third condition applies only to random samples taken without replacement, as is the case when using Simple Random Sampling. In this case, the population must be at least 10 times bigger than the sample. In symbols, if N is the number of people in the population and n is the number in the sample, then

$$N \geq 10n$$

If this condition is not met, and the sample was collected without replacement, then the actual standard error will be a little smaller than what our formula says it should be.

In most real-life applications, the population size is much larger than the sample size. Over 300 million people live in the United States, so the typical survey of 1000 to 3000 people easily meets this condition.

You can see how these conditions are used in the examples that follow.

 KEY POINT The Central Limit Theorem for proportions requires (1) a random sample with independent observations, (2) a large sample, and (3) if SRS is used, a population with at least 10 times as many members as are in the sample.

Using the Central Limit Theorem

The following examples use the CLT to find the probability that the sample proportion will be near (or far from) the population value.

EXAMPLE 8 Pet World Revisited

Let's return to Pet World. The population is 1000 people, and the percent of Cat People in the population is 25%. We'll take a random sample of 100 people.

QUESTION What is the approximate probability that the percentage in our sample will be bigger than 29%? Begin by checking conditions for the CLT.

SOLUTION Although the problem statement gives us values in terms of percentages, we do our calculations using proportions. First we check conditions to see whether the Central Limit Theorem can be applied. The sample size is large enough because $np = 100(0.25) = 25$ is greater than 10, and $n(1 - p) = 100(0.75) = 75$, which is also greater than 10. Also, the population size is 10 times larger than the sample size, because $1000 = 10(100)$. Thus $N \geq 10(n)$; the population is just large enough. We are told that the sample was collected randomly.

According to the CLT, the sampling distribution will be approximately Normal. The mean is the same as the population proportion: $p = 0.25$. The standard deviation is the same as the standard error from Formula 7.1a:

$$SE = \sqrt{\frac{p(1 - p)}{n}} = \sqrt{\frac{0.25 \times 0.75}{100}} = \sqrt{\frac{0.1875}{100}} = \sqrt{0.001875} = 0.0433$$

We can use technology to find the probability of getting a value larger than 0.29 in a $N(0.25, 0.0433)$ distribution. Or we can standardize.

In standard units, 0.29 is

$$z = \frac{0.29 - 0.25}{0.0433} = 0.924 \text{ standard unit}$$

In a $N(0,1)$ distribution, the probability of getting a number bigger than 0.924 is, from Table A in the appendix, about 0.18, or 18%. Figure 7.10 shows the results using technology.

▶ **FIGURE 7.10** Output from StatCrunch. There is about an 18% chance that $\hat{p}$ will be more than 4 percentage points above 25%.

Tech

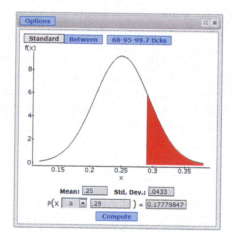

CONCLUSION With a sample size of 100, there is about an 18% chance that $\hat{p}$ will be more than 4 percentage points above 25%.

TRY THIS! Exercise 7.41

> # SNAPSHOT ▸ The Sample Proportion: $\hat{p}$ (p-hat)
>
> **WHAT IS IT?** ▸ The proportion of people or objects in a sample that have a particular characteristic in which we are interested.
>
> **WHAT IT IS USED FOR?** ▸ To estimate the proportion of people or objects in a population that have that characteristic.
>
> **WHY DO WE USE IT?** ▸ If the sample is drawn at random from the population, then the sample proportion is unbiased and has standard error $\sqrt{\dfrac{p(1-p)}{n}}$.
>
> **HOW IS IT USED?** ▸ If, in addition to everything above, the sample size is fairly large, then we can use the Normal distribution to find probabilities concerning the sample proportion.

EXAMPLE 9 Presidential Election Survey

In a hotly contested U.S. election, two candidates for president, a Democrat and a Republican, are running neck and neck; each candidate has 50% of the vote. Suppose a random sample of 1000 voters are asked whether they will vote for the Republican candidate.

QUESTIONS What proportion of the sample should be expected to express support for the Republican? What is the standard error for this sample proportion? Does the Central Limit Theorem apply? If so, what is the approximate probability that the sample proportion will fall within two standard errors of the population value of $p = 0.50$?

SOLUTION Because we have collected a random sample, the sample proportion has no bias (assuming there are no problems collecting the sample). Therefore, we expect that 50% of our sample will support the Republican candidate.

Because the sample size, $n = 1000$, is small relative to the population (which is over 100 million), we can calculate the standard error with

$$SE = \sqrt{\frac{(0.50)(0.50)}{1000}} = 0.0158$$

We can interpret this to mean that we expect our sample proportion to be 0.50, give or take 0.0158, or 50% give or take 1.58 percentage points.

Because the sample size is fairly large (the expected numbers for successes and failures are both equal to $np = 1000 \times 0.50 = 500$, which is larger than 10), the CLT tells us we can use the Normal distribution—in particular, $N(0.50, 0.0158)$.

We are asked to find the probability that the sample proportion will fall within two standard errors of 0.50. In other words, that it will fall somewhere between

$$0.50 - 2SE$$

and

$$0.50 + 2SE$$

> ### ↻ Looking Back
>
> **Empirical Rule**
> Recall that the Empirical Rule says that roughly 68% of observations should be within one standard deviation of the mean, about 95% within two standard deviations of the mean, and nearly all within three standard deviations of the mean. In this context, the standard error is the standard deviation for the sampling distribution.

Because this is a Normal distribution, we know the probability will be very close to 95% (according to the Empirical Rule). But let's calculate the result anyway.

$$0.50 - 2SE = 0.50 - 2(0.0158) = 0.50 - 0.0316 = 0.4684$$

$$0.50 + 2SE = 0.50 + 0.0316 = 0.5316$$

That is, we want to find the area between 0.4684 and 0.5316 in a $N(0.5, 0.0158)$ distribution. Figure 7.11 shows the result using technology, which tells us this probability is 0.9545.

▶ **FIGURE 7.11** The probability that a sample proportion based on a random sample of 1000 people taken from a population in which $p = 0.50$ has about a 95% chance of falling within two standard errors of 0.50.

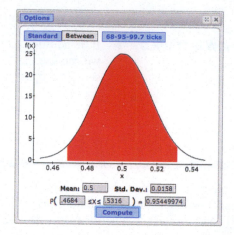

CONCLUSION If each candidate truly has 50% of the vote, then we'd expect our sample proportion to be about 0.50 (or 50%). There is about a 95% chance that the sample proportion falls within two standard errors of 50%.

TRY THIS! Exercise 7.43

The conclusion from Example 6 is useful because it implies that, in general, we can predict where $\hat{p}$ will fall, relative to p. It indicates that $\hat{p}$ is very likely to fall within two standard errors of the true value, as long as the sample size is large enough. If, in addition, we have a small standard error, we know that $\hat{p}$ is quite likely to fall close to p.

KEY POINT If the conditions of a survey sample satisfy those required by the CLT, then the probability that a sample proportion will fall within two standard errors of the population value is 95%.

EXAMPLE 10 Morse and the Proportion of Es

Samuel Morse (1791–1872), the inventor of Morse code, claimed that the letter used most frequently in the English language was E and that the proportion of Es was 0.12. Morse code translates each letter of the alphabet into a combination of "dots" and "dashes," and it was used by telegraph operators, before the days of radio or telephones, to transmit messages around the world. It was important that the most frequently used letters be the easiest for the telegraph operator to type. In Morse code, the letter E is simply "dot."

To check whether Morse was correct about the proportion of Es, we took a simple random sample with replacement from a modern-day book. Our sample consisted of 876 letters, and we found 118 Es, so $\hat{p} = 0.1347$.

QUESTION   Assume that the true proportion of Es in the population is, as Morse claimed, 0.12. Find the probability that, if we were to take another random sample of 876 letters, the sample proportion would be greater than or equal to 0.1347. As a first step, check that the Central Limit Theorem can be applied in this case.

SOLUTION   To check whether we can apply the Central Limit Theorem, we need to make sure the sample size is large enough. Because $p = 0.12$, we check

$$np = 876(0.12) = 105.12, \text{ which is larger than } 10$$

and

$$n(1 - p) = 876(0.88) = 770.88, \text{ which is also larger than } 10$$

The book we sampled from contains far more than 8760 letters, so the population size is much larger than the sample size.

We can therefore use the Normal model for the distribution of sample proportions. The mean of this distribution is

$$p = 0.12$$

The standard error is

$$SE = \sqrt{\frac{p(1 - p)}{n}} = \sqrt{\frac{0.12(0.880)}{876}} = 0.010979$$

$$z = \frac{\hat{p} - p}{SE} = \frac{0.1347 - 0.12}{0.010979} = \frac{0.0147}{0.010979} = 1.339$$

We therefore need to determine the probability of getting a z-score of 1.339 or larger. We can find this with the Normal table; it is the area to the right of a z-score of 1.34. We can also use technology (Figure 7.12) to find the area to the right of 0.1347 in a $N(0.12, 0.012)$ distribution. This probability is represented by the shaded area in Figure 7.13.

▲ **FIGURE 7.12** TI-84 output

◀ **FIGURE 7.13** The shaded area represents the probability of finding a sample proportion of 0.1347 or larger from a population with a proportion of 0.12.

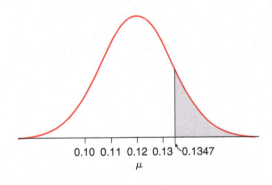

CONCLUSION   If the sample is 876 letters, the probability of getting a sample proportion of 0.1347 or larger, when the true proportion of Es in the population is 0.12, is about 9%.

TRY THIS!   Exercise 7.45

Estimating the Population Proportion with Confidence Intervals

An example of a real survey illustrates this situation. The Pew Research Center took a random sample of 446 registered Democrats in the United States in 2013. In this sample, 57% of the 446 people agreed with the statement that the news media spent too much time on unimportant stories. (Pew also asked the same question of Republicans and Independents.) However, this percentage just tells us about our sample. What percentage of the population—that is, what percentage of *all* Democrats in the United States—agree with this statement? How much larger or smaller than 57% might the percentage who agree be? Can we conclude that a majority (more than 50%) of Americans share this belief?

We don't know p, the population parameter. We do know $\hat{p}$ for this sample; it is equal to 0.57. Here's what else we know from the preceding sections:

1. Our estimator is unbiased, so even though our estimate of 57% may not be exactly equal to the population parameter, it's probably just a little higher or just a little lower.

2. The standard error can be estimated as

$$\sqrt{\frac{\hat{p}(1-\hat{p})}{n}} = \sqrt{\frac{0.57 \times 0.43}{446}} = 0.023, \text{ or about } 2.3\%$$

   Because the estimator is unbiased, the standard error tells us how much higher or lower our estimator might be from the population parameter.

3. Because the sample size is large, we also realize that the probability distribution of $\hat{p}$ is pretty close to being Normally distributed and is centered around the true population parameter value. Thus, there's about a 68% chance that $\hat{p}$ is closer than one standard error away from the population proportion, and a 95% chance that it is closer than two standard errors away. Also, there is almost a 100% chance (99.7%, actually) that the sample proportion is closer than three standard errors from the population proportion. Thus, we can feel very confident that the percentage of the population who agree with this statement is within three standard errors of 57%. Three standard errors is 3(2.3%) = 6.9%, so we can be almost certain that the value of the population parameter is within 6.9 percentage points of 57%.

In other words, we can be highly confident that the population parameter is between these two numbers:

$$57\% - 6.9\% \quad \text{to} \quad 57\% + 6.9\%, \text{ or}$$
$$50.1\% \quad\quad\quad \text{to} \quad\quad\quad 63.9\%$$

We have just calculated a **confidence interval**. Confidence intervals are often reported as the estimate plus or minus some amount:

$$57\% \text{ plus or minus } 6.9\%, \text{ or } 57\% \pm 6.9\%.$$

The "some amount," in this case the 6.9 percentage points, is called the **margin of error**. The margin of error tells how far from the population value our estimate can be.

A confidence interval provides two pieces of information: (1) a range of plausible values for our population parameter (50.1% to 63.9%), and (2) a **confidence level**, which expresses (no surprise here) our level of confidence in this interval. Our high confidence level of 99.7% assures us that we can be very confident that a majority of Democrats agree the news media spend too much time on unimportant stories because the smallest plausible level of agreement in the population is 50.1%, which is (just) bigger than a majority.

An analogy can help explain confidence intervals. Imagine a city park. In this park sit a mother and her daughter, a toddler. The mother sits in the same place every day, on a bench along a walkway, while her daughter wanders here and there. Most of the time, the child stays very close to her mother, as you would expect. In fact, our studies have revealed that 68% of the days we've looked, she is within 1 yard of her mother. Sometimes she strays a little bit farther, but on 95% of the days she is still within 2 yards of her mother. Only rarely does she move much farther; she is almost always within 3 yards of her mother.

One day the unimaginable happens, and the mother and the park bench become invisible. Fortunately, the child remains visible. The problem is to figure out where the mother is sitting.

Where is the mother? On 68% of the days, the child is within 1 yard of the mother, so at these times the mother must be within 1 yard of the child. If we think the mother is within 1 yard of the child on most days—that is, 68% of the days we observe—we will be right. But this also means we will be wrong on 32% of our visits. We could be more confident of being correct if we instead guessed that the mother is within 2 yards of the child. Then we would be wrong on only 5% of the days.

In this analogy, the mother is the population proportion. Like the mother, the population proportion never moves from its spot and never changes values. And just as we cannot see the invisible mother, we don't know where the parameter sits. The toddler is like our sample proportion, $\hat{p}$; we *do* know its value, and we know that it hangs out near the population proportion and moves around from sample to sample. Thus, even though we can't know exactly what the true population proportion is, we can infer that it is near the sample proportion.

Setting the Confidence Level

The confidence *level* tells us how often the estimation method is successful. Our method is to take a random sample and calculate a confidence interval to estimate the population proportion. If the method has a 100% confidence level, that method always works. If the method has a 10% confidence level, it works in 10% of surveys. We say the method works if the interval captures the true value of the population parameter. In this case, the interval works if the true population proportion is inside the interval.

Think of the confidence level as the capture rate; it tells us how often a confidence interval based on a random sample will capture the population proportion. Keep in mind that the population proportion, like the mother on the park bench, does not move—it is always the same. However, the confidence interval does change with every random sample collected. Thus, the confidence level measures the success rate of the *method*, not of any one particular interval.

 KEY POINT The confidence level measures the capture rate for our method of finding confidence intervals.

Figure 7.14 demonstrates what we mean by a 95% confidence level. Let's suppose that in the United States, 51% of all voters favor stricter laws with respect to buying and selling guns. We simulate taking a random sample of 1000 people. We calculate the percentage of the sample who favor stricter laws, and then we find the confidence interval that gives us a 95% confidence level. We do this again and keep repeating. Figure 7.14 shows 100 simulations.

Each blue point and each orange point represent a sample percentage. Note that the points are centered around the population percentage of 51%. The horizontal lines represent the confidence interval: the sample percentage plus or minus the margin of error. The

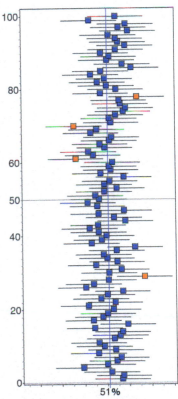

▲ **FIGURE 7.14** Results from 100 simulations in which we draw a random sample and then find and display a confidence interval with a 95% confidence level. The orange squares indicate "bad" intervals.

margin of error was chosen so that the confidence level is 95%. Notice that most of the lines cross the vertical line at 51%. These are successful confidence intervals that capture the population value of 51% (in blue). However, a few sample percentages miss the mark; these are indicated by orange points. In 100 trials, our method failed 4 times and was successful 96 times. In other words, it worked in 96% of the trials. When we use a 95% confidence level, our method works in 95% of all surveys we could conduct.

We can change the confidence level by changing the margin of error. The greater the margin of error, the higher our confidence level. For example, we can be 100% confident that the true percentage of Americans who favor stricter gun laws is between 0% and 100%. We're 100% confident in this interval because it can never be wrong. Of course, it will also never be useful. We really don't need to spend money on a survey to learn that the answer lies between 0% and 100%, do we?

It would be more helpful—more precise—to have a smaller margin of error than "plus or minus 50 percentage points." However, if the margin of error is too small, then we are more likely to be wrong. Think of the margin of error as a tennis racket. The bigger the racket, the more confident you are of hitting the ball. Choosing an interval that ranges from 0% to 100% is like using a racket that fills the entire court—you will definitely hit the ball, but not because you are a good tennis player. If the racket is too small, you are less confident of hitting the ball, so you don't want it too small. Somewhere between too big and too small is just right.

Selecting a Margin of Error

We select a margin of error that will produce the desired confidence level. For instance, how can we choose a margin of error with a confidence level of 95%? We already know that if we take a large enough random sample and find the sample proportion, then the CLT tells us that 95% of the time, the sample proportion is within two standard errors of the population proportion. This is what we learned from Example 6. It stands to reason, then, that if we choose a margin of error that is two standard errors, then we'll include the population proportion in 95% of our samples.

This means that

$$\hat{p} \pm 2SE$$

is a confidence interval with a 95% confidence level. More succinctly, we call this a 95% confidence interval.

Using the same logic, we understand that the interval

$$\hat{p} \pm 1SE$$

is a 68% confidence interval and that

$$\hat{p} \pm 3SE$$

is a 99.7% confidence interval.

Figure 7.15 shows the four margins of error with their corresponding confidence levels for a sample in which $\hat{p} = 0.50$.

▶ **FIGURE 7.15** Four confidence intervals with confidence levels ranging from 99.99% (plus or minus 4 standard errors — top) to 68% (plus or minus 1 standard error). Notice how the interval gets wider with increasing confidence level.

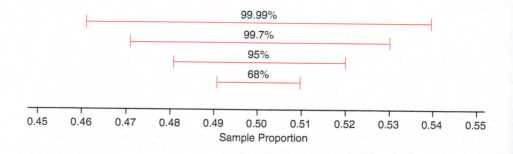

This figure illustrates one reason why a 95% confidence interval is so desirable. If we increase the margin of error from 2 standard errors to 3, we gain only a small amount of confidence; the level goes from 95% to 99.7%. However, if we decrease from 2 standard errors to 1, we lose a lot of confidence; the level falls from 95% to 68%. Thus, the choice of 2 standard errors is very economical.

The margin of error has this structure:

$$\text{Margin of error} = z^*SE$$

where z^* is a number that tells how many standard errors to include in the margin of error. If $z^* = 1$, the confidence level is 68%. If $z^* = 2$, the confidence level is 95%. Table 7.6 summarizes the margin of error for commonly used confidence levels.

Confidence Level	Margin of Error Is . . .
99.7%	3.0 standard errors
99%	2.58 standard errors
95%	1.96 (about 2) standard errors
90%	1.645 standard errors
80%	1.28 standard errors

◄ **TABLE 7.6** We can set the confidence level to the value we wish by choosing the appropriate margin of error.

Reality Check: Finding a Confidence Interval When *p* Is Not Known

As we have seen, a confidence interval for a population proportion has this structure:

$$\hat{p} \pm m$$

where m is the margin of error. Substituting for the margin of error, we can also write

$$\hat{p} \pm z^*SE$$

Finding the standard error requires us to know the value of p:

$$SE = \sqrt{\frac{p(1 - p)}{n}}$$

However, in real life, we don't know p. So instead, we substitute our sample proportion and use Formula 7.1b for the estimated standard error:

$$SE_{\text{est}} = \sqrt{\frac{\hat{p}(1 - \hat{p})}{n}}$$

The result is a confidence interval with a confidence level close to, but not exactly equal to, the correct level. This tends to be close enough for most practical purposes.

In real life, then, Formula 7.2 is the method we use to find approximate confidence intervals for a population proportion.

Formula 7.2: $\hat{p} \pm m$,   where $m = z^*SE_{\text{est}}$ and $SE_{\text{est}} = \sqrt{\frac{\hat{p}(1 - \hat{p})}{n}}$

where:
m is the margin of error;
$\hat{p}$ is the sample proportion of successes, or the proportion of people in the sample with the characteristic we are interested in;
n is the sample size;
z^* is a multiplier that is chosen to achieve the desired confidence level (as in Table 7.6); and SE_{est} is the estimated standard error.

EXAMPLE **11** Teachers and Digital Devices

Do a majority of teachers in the U.S. think that digital devices (smartphones, tablets, computers) are mostly helpful for students' education? In 2018, Gallup took a poll of 497 randomly selected adults who teach K–12 students and 42% of them said that such devices had "mostly helpful" effects on students' education. (Source: Busteed & Dugan, 2018, news.gallup.com.)

QUESTION   Estimate the standard error. Find a 95% confidence interval for the percentage of all K–12 teachers who believe that these devices have a mostly helpful effect on students' education. Is it plausible to conclude that as many as 50% or more teachers believe this?

SOLUTION   We first make sure the conditions of the Central Limit Theorem apply. We are told that Gallup took a random sample. We must assume their observations were independent (which makes sense since teachers' answers couldn't influence each other). We don't know whether the pollsters sampled with or without replacement, but because the population is very large—easily 10 times larger than the sample size—we don't need to worry about the replacement issue. (This confirms that conditions 1 and 3 apply.)

Next we need to check that the sample size is large enough to use the CLT. We do not know p, the proportion of all teachers who believe that digital devices have mostly helpful effects on students' education. We know only the value of the statistic $\hat{p}$ (equal to 0.42), which Gallup found on the basis of its sample. This means that our sample has at least 10 successes (teachers who agree with the statement) because $497 \times .42 = 208.74$, which is much larger than 10. Also, we know we have at least 10 failures (teachers who don't agree with the statement) because $497 \times (1 - 0.42)$ is even bigger than 208.74.

At this point we can go directly to technology, such as StatCrunch or Minitab, or we can continue to compute using a calculator. Figure 7.16 shows the output from StatCrunch.

▶ **FIGURE 7.16** StatCrunch output for a 95% confidence interval for the proportion of all teachers in the U.S. who believe that digital devices have a mostly helpful effect on students' education.

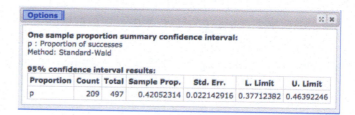

If we instead use a calculator, the next step is to estimate the standard error. Using Formula 7.1b yields

$$SE_{est} = \sqrt{\frac{0.42(1 - 0.42)}{497}} = 0.02214$$

We now use this result together with Formula 7.2 (using 2, rather than the slightly more accurate value 1.96, for our multiplier) to find the interval.

$$\hat{p} \pm 2SE_{est}$$

$$0.42 \pm 2(0.02214)$$

$$0.42 \pm 0.04428$$

or, if you prefer, after some rounding

$$42\% \pm 4\%$$

Expressing this as an interval, we get

$$42\% - 4\% = 38\%$$

$$42\% + 4\% = 46\%$$

The 95% confidence interval is 38% to 46%.

CONCLUSION The confidence interval tells us which values are plausible for the population percentage. We have to conclude that it is not plausible that more than 50% of teachers believe that digital devices are mostly helpful toward students' education because none of the plausible values are that large. The largest plausible value for this percentage is 46%.

TRY THIS!   Exercise 7.59

Interpreting Confidence Intervals

A confidence interval for a sample proportion gives a set of values that are plausible for the population proportion. If a value is not in the confidence interval, we conclude that it is implausible. It's not impossible that the population value is outside the interval, but it would be pretty surprising.

Suppose a candidate for political office conducts a poll and finds that a 95% confidence interval for the proportion of voters who will vote for him is 42% to 48%. He would be wise to conclude that he does *not* have 50% of the population voting for him. The reason is that the value 50% is not in the confidence interval, so it is implausible to believe that the population value is 50%.

There are many common misinterpretations of confidence intervals that you must avoid. The most common mistake that students (and, indeed, many others) make is trying to turn confidence intervals into some sort of probability statement. For example, if asked to interpret a 95% confidence interval of 45.9% to 53.1%, many people would mistakenly say, "This means there is a 95% *chance* that the population proportion is between 45.9% and 53.1%."

What's wrong with this statement? Remember that probabilities are long-run frequencies. This sentence claims that if we were to repeat this survey many times, then in 95% of the surveys the true population percentages would be a number between 45.9% and 53.1%. This claim is wrong because the true population percentage doesn't change. Either it is *always* between 45.9% and 53.1% or it is *never* between these two values. It can't be between these two numbers 95% of the time and somewhere else the rest of the time. In our story about the invisible mother, the mother, who represented the population proportion, *always* sat at the same place. Similarly, the population proportion (or percentage) is always the same value.

Another analogy will help make this clear. Suppose there is a skateboard factory. Say 95% of the skateboards produced by this factory are perfect, but 5% have no wheels. Once you buy a skateboard from this factory, you can't say that there is a 95% chance that it is a good board. Either it has wheels or it does not have wheels. It is *not* true that the board has wheels 95% of the time and, mysteriously, no wheels the other 5% of the time. A confidence interval is like one of these skateboards. Either it contains the true parameter (has wheels) or it does not. The "95% confidence" refers to the "factory" that "manufactures" confidence intervals: 95% of its products are good, and 5% are bad.

Our confidence is in the process, not in the product.

> **KEY POINT** Our confidence is in the process that produces confidence intervals, not in any particular interval. It is incorrect to say that a particular confidence interval has a 95% (or any other percent) chance of including the true population parameter. Instead, we say that the *process* that produces intervals captures the true population parameter with a 95% probability.

EXAMPLE 12 Buying or Renting?

After the Great Recession, the Pew Research Center noted there seemed to be a decline in households that rented their homes and were looking to purchase homes. However, they reported that in 2016 "a solid 72%" of renters reported that they wished to buy their own home. Pew reports that the "margin of error at 95% confidence level is plus-or-minus 5.4 points."

QUESTION State the confidence interval in interval form. How would you interpret this confidence interval? What does "95%" mean?

Data Moves: The complete data set can be downloaded from http://www.pewsocialtrends.org/dataset/2016-homeownership-survey/, although they are in a proprietary data format. Some open-source data analysis software can read this format, and the file homeowners.csv provides a cleaned-up, more friendly version of the survey data. Documentation for the data set is on the Pew website.
▶ **File:** homeowners.csv

CONCLUSION The margin of error, we are told is 5.4 percentage points. In interval form, then, the 95% confidence interval is

$$72\% - 5.4\% \text{ to } 72\% + 5.4\%, \text{or}$$

$$66.6\% \text{ to } 77.4\%$$

We interpret this to mean that we are 95% confident that the true percentage of renters who wish to buy their own homes is between approximately 67% and 77%. The 95% indicates that if we were to conduct not just this single survey, but infinitely many such surveys, then 95% of those surveys would result in confidence intervals that include the true population proportion.

TRY THIS Exercise 7.65

Example 13 demonstrates the use of confidence intervals to make decisions about population proportions.

EXAMPLE 13 Morse and Es

Recall from Example 10 that Morse believed the proportion of Es in the English language was 0.12 and that our sample showed 118 Es out of 876 randomly chosen letters from a modern-day book.

QUESTION Find a 95% confidence interval for the proportion of Es in the book. Is the proportion of Es in the book consistent with Morse's 0.12? Assume the conditions that allow us to interpret the confidence interval are satisfied. (The conditions were checked in Example 10.)

SOLUTION The best approach is to use technology. Figure 7.17 shows TI-84 output that gives a 95% confidence interval as

$$(0.112, 0.157) \quad \text{or} \quad (11.2\%, 15.7\%)$$

If you do not have access to statistical technology, then the first step is to find the sample proportion of Es: 118/876, or 0.1347.

The estimated standard error is

$$SE_{est} = \sqrt{\frac{\hat{p}(1 - \hat{p})}{n}} = \sqrt{\frac{0.1347(0.8653)}{876}} = \sqrt{0.00013305} = 0.0115349$$

Because we want a 95% confidence level, our margin of error is plus or minus 1.96 standard errors:

$$\text{Margin of error} = 1.96 SE_{est} = 1.96(0.0115349) = 0.022608$$

The interval boundaries are

$$\hat{p} \pm 1.96 SE_{est} = 0.1347 \pm 0.0226$$

Upper end of interval: $0.1347 + 0.0226 = 0.1573$
Lower end of interval: $0.1347 - 0.0226 = 0.1121$

This confirms the result we got through technology: A 95% confidence interval is (0.1121 to 0.1573). Note that this interval *does* include the value 0.12.

CONCLUSION We are 95% confident that the proportion of Es in the modern book is between 0.112 and 0.157. This interval captures 0.12. Thus, it is plausible that the population proportion of Es in the book is 0.12, as Morse suggested.

TRY THIS! Exercise 7.69

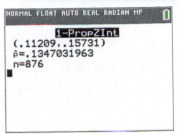

▲ FIGURE 7.17 TI-84 output for a confidence interval for the proportion of Es. The values differ slightly from the "by hand" calculation due to rounding.

Planning a Study: Finding the Sample Size

So far in this chapter, we've presented situations in which the data already exist and we find a confidence interval. But how do you know how much data to collect? For example, suppose we want to discover the percentage of children in the United States who have one or more sugary drinks per day.

We know we have to draw a random sample from this population (all children in the United States), but how many children do we need to survey? The first step to answering this question is to decide how precise you want your estimate to be. In other words, what margin of error would you like?

A common choice is a margin of error of 3 percentage points. That way, we could say something like "We are very confident that the percentage of children who drink one or more sugary drinks per day is ___, plus or minus three percentage points." Three points is a relatively small margin of error and is commonly chosen by researchers because it is often precise enough to be useful.

So how can we determine the correct sample size to get a margin of error that will be 3 percentage points or less (or whatever value we choose)?

We won't go through the algebra here. (If you're interested, see homework questions 7.107 and 7.108.) But our solution is based on the definition of a margin of error from Formula 7.2:

$$m = z^* SE_{est}$$

For example, if we choose a margin of error of three percentage points, then $0.03 = z^* SE_{est}$.

For a 95% confidence interval, we can approximate the multiplier $z^* = 1.96$ by using the value $z^* = 2.00$, or

$$m = 2 SE_{est}$$

Because $SE_{est} = \sqrt{\dfrac{\hat{p}(1 - \hat{p})}{n}}$

We can rewrite the margin of error in Formula 7.2 as

$$m = 2\sqrt{\frac{\hat{p}(1 - \hat{p})}{n}}$$

Our solution is based on solving for *n*. We also make use of the fact that the biggest margin of error occurs when $\hat{p} = 0.5$. And so if we replace the sample proportion with 0.5, we will be able to find the sample size for the "worst case" scenario regardless of what value the sample proportion actually turns out to be from our study. It takes a few steps of algebra (which we won't repeat) to find that Formula 7.3 gives us the sample size we need.

Formula 7.3: $n = \dfrac{1}{m^2}$

Example 14 illustrates how this formula can be applied.

EXAMPLE 14 Sugary Sodas

A recent study asked the question, "What percentage of children in California between the ages of 2 and 11 drink at least one sugary drink per day?" We wish to be able to calculate a 95% confidence interval for this estimate, and we want the margin of error to be 3 percentage points.

QUESTION How many children should we randomly sample to achieve a margin of error of 3 percentage points if we use a 95% confidence level?

SOLUTION We're given that $m = 0.03$, and so using Formula 7.3:

$$n = \frac{1}{0.03^2} = \frac{1}{0.0009} = 1111.111$$

CONCLUSION We'd need approximately 1111 children in our sample in order to achieve a 3 percentage point margin of error.

TRY THIS! Exercise 7.71

An actual study in California, based on a random sample of more than 1,111 children, estimated that a 95% confidence interval for the percentage of all children in California who drank at least one sugary drink per day was 28.3 to 33.5, which is a margin of error of 2.6 percentage points.

SECTION 7.5

Comparing Two Population Proportions with Confidence

Statistical questions that make comparisons can be very powerful. For example, has there been a change in the percentage of females graduating with a bachelor's degree? In particular, is there a difference in the percentage of women in the millennial generation (born between 1981 and 1996) and the Gen Xers (1965–1980)?

Data from a 2017 Pew poll found some evidence for a difference. Pew reported that 36% of millennial women graduated with a bachelor's degree compared to 40% of Gen Xer women. Since these statistics are based on random samples, it is possible that the percentages of *all* millennial women and all Gen X women are the same and only the sample percentages differ. How do we know if there was a real difference, in other words, that the population percentages differ?

What's the Difference?

What's the difference between Coke and Pepsi? When asked a question like this, you probably think of qualitative characteristics: flavor, color, and bubbliness. But when we ask about the difference between two *numbers*, we mean "How far apart are the two numbers?"

The answer to the question, "How far apart are two numbers?" is found by subtracting. How far apart are the sample percentages 58% and 43%?

$$58\% - 43\% = 15\%$$

The two sample percentages are 15 percentage points apart.

Much of our analysis in comparing two samples is based on subtraction. In this section, and in Section 8.4, our comparison of two population proportions will be based on the statistic

$$\hat{p}_1 - \hat{p}_2$$

This statistic will be used to estimate the difference between two population proportions:

$$p_1 - p_2$$

It might seem strange to you that we would have to go to so much trouble to determine whether two numbers are different from each other. After all, can't we just look and *see* that 0.43 does not equal 0.58?

This issue is subtle but important. Even when two proportions are equal in the *population*, their *sample* proportions can be different. This difference is caused by the fact that we see only *samples* of the populations and not the entire populations. This means that even if, say, 23% of all men and 21% of all women believe that embryonic stem cell research is wrong, a random sample of men and women might have different percentages of believers, perhaps 22% and 28%.

> **KEY POINT** Even if the proportions are equal for two populations, the sample proportions drawn from these populations are usually different.

Confidence intervals are one method for determining whether different sample proportions reflect "real" differences in the population. The basic approach is this:

First, we find a confidence interval, at the significance level we think best, for the difference in proportions $p_1 - p_2$.

Next, we check to see whether that interval includes 0. If it does, then this suggests that the two population proportions might be the same. Why? Because if $p_1 - p_2 = 0$, then $p_1 = p_2$ and the proportions are the same.

If the confidence interval does not contain 0, we also learn interesting things. As you will soon see, the confidence interval tells us how much greater one of the proportions might be than the other.

EXAMPLE 15 Do Men's and Women's Views Differ?

In a random sample of 2000 U.S. men and 2000 women the Pew Foundation found, in 2013, that 23% of the men in their sample believed that research using embryonic stem cells is "morally wrong." For the women, 21% believed it is morally wrong.

QUESTION Can we conclude, on the basis of these sample percentages *only*, that in the United States, a greater percentage of men than of women believe that embryonic stem cell research is morally wrong? Explain.

SOLUTION No, we cannot conclude this. Although a greater proportion of *the sample* of men than of *the sample* of women believe stem cell research is morally wrong, in the *population* of all men and women these proportions might be the same, might not be the same, or might even be reversed. We need a confidence interval to answer this question.

TRY THIS! Exercise 7.75a

Looking Back

Statistics
A statistic, as you learned in Section 7.1, is a number that is based on data and used to estimate a population parameter.

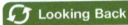
Looking Back

z-Scores
You've already seen subtraction used to compare numbers. The numerator of the z-score (the observed value minus the sample mean) is used to tell us the distance between a number and the mean of the sample.

Confidence Intervals for Two Population Proportions

The confidence interval for two proportions has the same structure as the confidence interval for one proportion, as it was presented in Formula 7.2:

$$\text{Statistic} \ \pm \ z^*SE_{\text{est}}$$

The statistic for two proportions is different; it is now $\hat{p}_1 - \hat{p}_2$. And so the standard error is also different:

$$SE_{\text{est}} = \sqrt{\frac{\hat{p}_1(1 - \hat{p}_1)}{n_1} + \frac{\hat{p}_2(1 - \hat{p}_2)}{n_2}}$$

The value for z^* is chosen to get the desired confidence level, exactly as we did for a one-proportion confidence interval. For a 95% confidence level, for example, use $z^* = 1.96$. The samples can be different sizes, so n_1 represents the size of the sample drawn from population 1, and n_2 represents the number of people or objects in the sample drawn from population 2.

Putting these together, we find that the confidence interval for the difference of two proportions is

Formula 7.4: $\hat{p}_1 - \hat{p}_2 \pm z^*\sqrt{\frac{\hat{p}_1(1 - \hat{p}_2)}{n_1} + \frac{\hat{p}_2(1 - \hat{p}_2)}{n_2}}$

Naturally, we recommend that you use technology to do this calculation whenever possible.

Figure 7.18 shows the calculations used in StatCrunch to compare the results from the Pew poll on views of men and women in the United States on embryonic stem cell research. We used the population of men as population 1 and the population of women as population 2. Both samples had 2000 people. The 95% confidence interval is -0.006 to $.0457$.

We'll discuss how to interpret intervals of differences of proportions later.

► **FIGURE 7.18** StatCrunch calculations for finding a 95% confidence interval for the difference between the proportions of men p_1 and women p_2 who believe stem cell research is morally wrong. After rounding, the lower limit of the interval is -0.006, and the upper limit is 0.0457. The confidence level is 95% only if conditions are met.

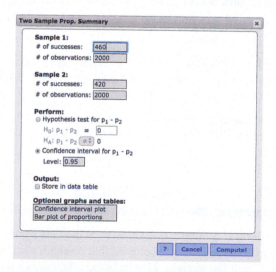

EXAMPLE 16 Estimating Changes in Graduation Rates

In a 2017 survey of 400 randomly selected women of the millennial generation and 400 women in the Gen X generation, 36% of the millennial women and 40% of the Gen X women graduated with a bachelor's degree. We wish to find a 95% confidence interval for the difference in the proportion of women with bachelor's degrees between the millennial and the Gen X generations.

QUESTION   Figure 7.19 shows the information that StatCrunch requires to calculate the 95% confidence interval for the difference in population proportions. Fill in the missing information. (Other statistical software packages require similar information.)

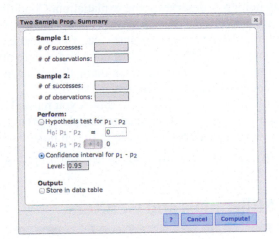

◀ **FIGURE 7.19** StatCrunch screenshot for a two-sample proportion test.

SOLUTION   First, we must decide which is sample 1 and which is sample 2. It doesn't matter which choice we make, but we must be consistent, and we must remember our choice when we interpret the interval. A common convention is to choose the more recent sample as sample 1, and so the millennials will be sample 1 and the Gen Xers sample 2.

Next we must determine the number of successes in each sample. There were 400 women in each sample. In sample 1, 36% received a bachelor's degree, and so there were $400 \times 0.36 = 144$ bachelor degrees in the sample.

Similarly, in sample 2 there were $400 \times 0.40 = 160$ successes.

We must also make sure that the "Confidence interval" button is checked and that the Level is set to 0.95.

CONCLUSION

Sample 1: number of successes, 144; number of observations, 400.
Sample 2: number of successes, 160; number of observations, 400.

TRY THIS!   Exercise 7.75c

Checking Conditions

These calculations "work" only when conditions are met. In a nutshell, the conditions that must exist in order for us to apply the Central Limit Theorem must hold for both samples, and one more condition must be met: the samples must be independent of each other.

To summarize, before you can interpret a confidence interval for two population proportions, you must check the following:

1. Random and Independent. Both samples are *randomly* drawn from their populations, and observations are independent of each other. (*Note:* An important exception to this is discussed at the end of this section.)

2. Large Samples. Both sample sizes are large enough that at least 10 successes and 10 failures can be expected in both samples.

3. Big Populations. If the samples are collected without replacement, then both population sizes must be at least 10 times bigger than their samples.

4. (New!) Independent Samples. The samples must be independent of each other.

The new condition, condition 4, deserves some explanation. Condition 4 requires that there be no relationship between the objects in one sample and the objects in another. This condition would have been violated, for instance, if Pew had interviewed *the same people* in 2002 and again in 2009. (Note that it is not a bad idea to interview the same people twice across time periods, if you can track them all down again. When that is done, however, the techniques presented here are not valid.)

Note that for condition 2 you now have four things you must check (1) at least 10 successes in sample 1, (2) at least 10 failures in sample 1, (3) at least 10 successes in sample 2, and (4) at least 10 failures in sample 2. In symbols,

$$n_1\hat{p}_1 \geq 10, \; n_1(1 - \hat{p}_1) \geq 10, \; n_2\hat{p}_2 \geq 10, \text{ and } n_2(1 - \hat{p}_2) \geq 10$$

> **Details**
>
> **Independent Samples**
> If the samples are not independent, then the standard error in the confidence interval will be the wrong value, and you'll have the wrong margin of error.

EXAMPLE 17 Millennials and GenXers

In Example 16, we did the preliminary steps for calculating the 95% confidence interval for the difference in the proportion of female millennials with bachelor's degrees and female Gen Xers with bachelor's degrees. The data came from a random sample of women from each generation. This interval turned out to be

$$-0.107 \text{ to } 0.027$$

or -10.7 percentage points to 2.7 percentage points.

QUESTION   Check that the conditions hold for interpreting this interval.

SOLUTION   Whether or not the conditions hold depends, to a great extent, on whether the researchers followed appropriate procedures. When in doubt, we will assume that they did. (This example is based on a study by the Pew Research Foundation, and its website goes to some length to convince us that it does follow appropriate procedures.)

Condition 1: Random and Independent. We are told that the samples are random, and we must assume that the observations are independent of each other in both samples.

Condition 2: Large samples. We check all four:

$$400 \times 0.36 = 144$$
$$400 \times (1 - 0.36) = 256$$
$$400 \times 0.40 = 160$$
$$400 \times (1 - 0.40) = 240$$

All values are bigger than 10, so the samples are large enough.

Condition 3. Big Populations. Clearly there are more than $10 \times 400 = 4000$ women in each of these populations.

Condition 4. Independent Samples. Because each sample was a random sample from a different population (the populations of all women born between 1981–1996 and all women born between 1965–1980), the samples are independent.

CONCLUSION   The conditions are satisfied.

TRY THIS!   Exercise 7.75b

Interpreting Confidence Intervals for Two Proportions

Having framed a comparison statistical question, chosen to analyze by finding a confidence interval and checked conditions, we're ready for the last phase of the Data Cycle: Interpretation. The basic interpretation of a confidence interval for the difference of two proportions is the same as for one proportion: we are 95% confident (or whatever our confidence level is) that the true population value is within the interval.

One important difference, though, is that the reason we are looking at the difference of two proportions, $p_1 - p_2$, is that we want to compare them. We want to know which proportion is larger than the other and how much larger it is, or whether they are the same. Therefore, in examining a confidence interval for two proportions, we ask these questions:

1. Is 0 included in the interval? If so, then we can't rule out the possibility that the population proportions are equal.

2. What does a positive value mean? A negative value? What is the greatest plausible difference for the population proportions? The smallest plausible difference? To answer these questions, we have to know which population was assigned to be population 1 and which to be population 2.

To understand how to interpret confidence intervals involving differences of proportions, consider another Pew study. This study asked whether attitudes towards stem cell research had changed over time. Based on a random sample of 1500 people in 2002, Pew estimated that 43% of the American public approved of stem cell research. Seven years later, a poll of the same size estimated that 58% of the public approved. A 95% confidence interval for the difference of proportions in the populations is 0.12 to 0.19. How do we interpret this interval?

All of the values in this interval are positive. What does a positive number mean? It means $p_1 - p_2 > 0$. This can happen only if $p_1 > p_2$—in other words, if the proportion of people who support embryonic stem cell research is greater in population 1 than in population 2.

Now we need to know which is population 1 and which is population 2. Looking back, we see that we defined the 2009 survey as population 1. This tells us that the proportion of people who support embryonic stem cell research was greater in 2009 than in 2002.

How much greater? We are confident that the increase was no fewer than 12 percentage points and no more than 19 percentage points.

On the other hand, the confidence interval for the difference in the proportion of men and women who believed that stem cell research is morally wrong was found to be −0.0006 to 0.0457. (See Figure 7.18). Because this interval contains 0, we should conclude that these data provide no evidence that the proportions of men and women in the population who believe this differ.

> **Details**
>
> **Choosing Populations**
> When comparing proportions from different years, researchers usually choose the most recent year as their "population 1." This makes it possible to interpret the difference in proportions as a change across time.

EXAMPLE 18   Interpreting CIs for Two Proportions

In Example 17, we found that a 95% confidence interval for the difference in the proportion of millennial women with bachelor's degrees and Gen X women with bachelor's degrees was −0.107 to 0.027. In percentages, this is −10.7 percentage points to 2.7 percentage points.

QUESTION   Interpret this confidence interval.

SOLUTION The interval contains 0. Thus, we can't rule out the possibility that the proportion of women with bachelor's degrees is the same in each generation.

A positive value means that the percentage of millennial women with bachelor's degrees is higher than the percentage of Gen X women with bachelor's degrees. We see that it is not implausible that the percentage of millennial women with bachelor's degrees is as much as 2.7 percentage points higher than the percentage of Gen X women with bachelor's degrees.

A negative value means that the proportion of millennial women with bachelor's degrees is smaller than it is for Gen X women. We see that it is not implausible that the percentage of millennial women with bachelor's degrees is as much as 10.7 points lower than for Gen X women.

CONCLUSION We find no evidence that the percentage of bachelor's degrees differs between millennial women and Gen X women.

TRY THIS! Exercise 7.75d

Looking Back

Controlled Experiments
An important feature of a well-designed controlled experiment is that subjects are assigned to treatment and control groups at random. If this doesn't happen, we cannot make cause-and-effect conclusions.

Looking Back

Random Assignment
You learned in Section 1.4 that researchers randomly assign subjects to treatment groups in order to determine whether there is a cause-and-effect relationship between the treatment and the response variable.

Random Assignment versus Random Sampling

There is an important exception to condition 1, the random sample condition, for confidence intervals of two proportions. Sometimes, a particular study is not concerned with generalizing to a larger population. Sometimes, the purpose is instead to determine whether there is a cause-and-effect relationship between two variables.

If the two samples are not random samples but, instead, objects are *randomly assigned* to groups, then if the other conditions are met, we can interpret a confidence interval for a difference in proportions. This is the situation in which we find ourselves when doing controlled experiments, as discussed in Chapter 1.

EXAMPLE **19** Crohn's Disease Proportions

In Example 11 of Chapter 1, you learned about a study to determine (among other things) which of two treatments for Crohn's disease, Inflix injections or Azath pills, was better. Patients were randomly assigned to receive either Inflix or Azath. One hundred sixty-nine patients received Inflix, and at the end of the study, 75 of them were in remission (a good outcome). One hundred seventy patients received Azath, and at the end of the study, 51 were in remission.

Let p_1 represent the proportion of Crohn's disease victims who would be in remission if they took Inflix, and let p_2 represent the proportion who would be in remission if they took Azath. A 95% confidence interval for the difference in sample proportions is 0.04 to 0.25.

QUESTION Assume that conditions 2, 3, and 4 are satisfied. Interpret the confidence interval. Which treatment is better?

SOLUTION Even though the two samples are not randomly selected from the population, the fact that patients were randomly assigned to one of the two treatments, together with the fact that the other three conditions hold, means we can interpret the confidence interval.

The interval does *not* include 0 (although it comes close!) This tells us that we are confident that one treatment is different from the other.

The values of the confidence interval are all positive. A positive value means the proportion of people in remission is greater for population 1, which consists of those who took Inflix.

CONCLUSION Inflix is the better treatment. The percentage of people who will go into remission is at least 4 percentage points greater with Inflix than with Azath, and it could be as much as 25 percentage points greater.

TRY THIS! Exercise 7.79

We will discuss the differences between random assignment and random sampling in greater depth in Chapter 12. In the meantime, keep these points in mind:

Random sampling allow us to make generalizations to the population from which the samples were taken.

Random assignment allow us to make cause-and-effect conclusions.

Applying this, Example 19 allows us to conclude that Inflix caused an improvement for these Crohns' sufferers compared to Azath, but based on this evidence alone, we can't generalize this finding beyond the sample.

SNAPSHOT ▶ Confidence Interval for the Difference of Two Sample Proportions

WHAT IS IT? ▶ The proportion of people/objects with a particular characteristic in one sample minus the proportion of people/objects with a particular characteristic in another sample.

WHAT IS IT USED FOR? ▶ To estimate the difference in proportions from two separate populations, for instance, control and treatment groups, Republicans and Democrats, or residents in 2015 with residents in 2013.

WHY DO WE USE IT? ▶ If the samples are independent of each other, and if both samples are randomly selected from their respective populations, then this statistic is an unbiased estimate of $p_1 - p_2$ and has standard error

$$\sqrt{\frac{p_1(1 - p_1)}{n_1} + \frac{p_2(1 - p_2)}{n_2}}$$

HOW IS IT USED? ▶ If, in addition to everything above, both sample sizes are fairly large, then the sampling distribution is approximately Normal, and we can use the Normal distribution to find probabilities for this statistic.

CASE STUDY REVISITED

Spring Break Fever: Just What the Doctors Ordered?

What was wrong with the American Medical Association's spring break survey? The AMA poll was actually based on an "online survey panel," which consists of a group of people who agree to take part in several different online surveys in exchange for a small payment. Marketing companies recruit people to join panels so that the marketers can investigate trends within various slices of the public. Such a sample may or may not be representative of the population we're interested in—we have no way of knowing. And because the sample is not chosen randomly, we also have no way of knowing how our estimate will behave from sample to sample.

For such a survey, it is impossible to find a confidence interval for the true proportion of women who "agree that spring break trips involve more or heavier drinking than occurs on college campuses" because (1) our estimate might be biased and (2) the true percentage might lie much farther from our estimate than two standard errors. For this reason, the AMA ended up removing the margin of error from its website and no longer claimed that the figures were a valid inference for all college women who participated in spring break.

DATAPROJECT ▸ Population Proportions

1 OVERVIEW

In order to estimate parameters from populations, we need data from a random sample that meets the conditions described in Chapter 7. But sometimes that is not enough. Often, the data need to be formatted in order to allow the software to compute the proper statistics. When working with survey data to estimate proportions, it is usually helpful to have the variable you're examining in a binary yes/no or 1/0 format. This project will show you how to prepare your data to estimate proportions.

2 GOALS

Be able to recode categorical variables and estimate population proportions.

3 RECODING MY TUNES

The file itunessample.csv contains a simple random sample of about 500 of the songs in one of the authors' iTunes library. (There are 37,000 songs in the entire library.) The app iTunes keeps the data it generates in a format called "xml," which stands for "extension markup language." While the name isn't too illuminating, what this means for us is that it is possible to retrieve this "library" and convert it to a comma-separated data set that we can analyze. We used the statistical programming language R to convert the library from xml format to a spreadsheet-like format, and the results are in the file itunessample.csv in StatCrunch.

Project Our statistical investigation question is, *What percentage of the songs in the author's library are in the Rock genre?* While you might not find this interesting (unless you are the author), understanding how to answer this question will illustrate a useful data wrangling skill called *recoding*. Recoding is the action of changing the values (almost always for a categorical variable, but it can be done to numerical values too) to something more useful.

Some principles to guide us:
1) We will always keep the original variable. The recoded values will be stored as a new variable. This way we can redo our work if we make a mistake.
2) Binary variables will be recoded with 1's and 0's, and the name of the variable will tell us what the 1 value represents. For example, if we had a variable named *emotion* with values "happy" and "sad," we'd create a variable named *happy* with values 1 and 0, and the 1 would indicate that the response was "happy" and 0 "not happy".

To begin, take a moment to acquaint yourself with the provided variables. The variable we wish to study is *Genre*. Make a bar plot of this variable. What do you notice?

Probably you notice that there are too many values for us to easily make sense of in a bar plot. A table might be better, but even it shows us that there are many different values. Note that the table takes us one step toward answering our question by giving us the proportion of songs in the sample that are in the Rock genre: $\hat{p} = 0.096$. But we still have work to do to be able to compute confidence intervals or hypothesis tests.

There are two approaches to recoding this variable provided by StatCrunch. Both are under the *Data* menu. The first is the "Indicator" option. Select this and then select *Genre* in the dialogue box. Click on "Compute!" What happens?

You'll see that you get one new variable for each value. We are only interested in the variable named *Genre = Rock*. We can either delete the others, or ignore them.

The next step is to rename this variable to *is.rock*, which will be less confusing to analyze than a variable name that looks like an equation. You can rename this variable (and any variable) by moving the mouse near the variable name at the top of the spreadsheet. A small triangle will appear and clicking on this brings up a drop-down menu. Choose "rename."

You can delete the unwanted columns using Edit > Columns > Delete.

The second approach is to select Data > Recode. This requires that you type in the new values. It's very important that before you hit "Compute!" you select "Create New Columns," as shown in Figure 7.20.

Once you select the *Genre* column and hit "Compute!" you'll get a dialogue box that shows you the

Recode Columns

Select column(s):
X1
Track ID
Name
Artist
Album

Where:
--optional--   **Build**

Column label:
--optional--

Options:
- ○ Replace current column(s)
- ⦿ Create new column(s)
- ○ Open in a new data table

To recode numeric ranges, first use Data > Bin, and then use this procedure on the binned columns.

?   Cancel   Compute!

▲ FIGURE 7.20

Recode Columns

Column to be recoded:
Genre

Original	Recoded
Alt. country	0
Alternative	0
Alternative & Punk	0
Christmas	0
Classical	0
Classical Music - Choral - Sacr...	0
Classical Music - Orchestral	0
Classical-vocal,early,chamber	0
Contemporary Classical	0
Country	0
Electronic	0
Electronica	0
Folk	Folk
Gothic	Gothic
Hip Hop/Rap	Hip Hop/Rap
Hip-Hop	Hip-Hop
Indie	Indie
Irish	Irish
Jazz	Jazz
Latin	Latin
Musical	Musical
NA	NA
NPR Music's The Austin 100 (2...	NPR Music's The Austin 100 (20
Opera	Opera
Podcast	Podcast
Pop	Pop
Punk	Punk
Rap	Rap
Rock	1

?   Cancel   Compute!

▲ FIGURE 7.21

values of the *Genre* variable. You'll need to change these one by one into 0's and change the *Rock* value to 1. You can advance from field to field with the tab key. Figure 7.21 shows a partially completed table.

You can now use the Stat > Proportion Stat > One Sample >With Data menu to find statistics.

You'll notice, perhaps, that StatCrunch, provides a way for you to compute one-proportion statistics from multivalued categorical variables without having to recode. But, as you'll soon see, recoding is a very useful data wrangling skill, and it is usually useful to create recoded variables that contain the information you need to use.

ASSIGNMENT

Find a 95% confidence interval for the proportion of Classical music in the author's library. Notice that some of the genre values seem to be either duplicates or particular categories that should probably be classified as "Classical." Use the Recode option to recode the values in Genre so that all of the values that you think should be considered Classical are grouped as Classical, and then find a 95% confidence level for the proportion of songs in the entire library that are Classical.

CHAPTER REVIEW

KEY TERMS

Population, *325*
Parameter, *325*
Census, *325*
Sample, *325*
Statistic, *325*
Estimator, *325*
Estimate, *325*

Statistical inference, *326*
Biased, *328*
Sampling bias, *328*
Measurement bias, *328*
Convenience sample, *330*
Simple random sample (SRS), *330*

With and without replacement, *330*
Nonresponse bias, *331*
Sampling distribution, *335*
Unbiased (accurate), *336*
Precision, *336*
Standard error (SE), *336*

Central Limit Theorem (CLT), *341*
Confidence interval, *348*
Margin of error, *348*
Confidence level, *348*

LEARNING OBJECTIVES

After reading this chapter and doing the assigned homework problems, you should

- Be able to estimate a population proportion from a sample proportion and quantify how far off the estimate is likely to be.

- Understand that random sampling reduces bias.

- Understand that large random samples produce more precise estimators.

- Understand when the Central Limit Theorem for sample proportions applies and know how to use it to find approximate probabilities for sample proportions.

- Understand how to find, interpret, and use confidence intervals for a single population proportion.

- Understand how to find, interpret and use confidence intervals that compare two proportions.

SUMMARY

The Central Limit Theorem (CLT) for sample proportions tells us that if we take a random sample from a population, and if the sample size is large and the population size much larger than the sample size, then the sampling distribution of $\hat{p}$ is approximately

$$N\left(p, \sqrt{\frac{p(1-p)}{n}}\right)$$

This result is used to infer the true value of a population proportion on the basis of the proportion in a random sample. The primary means for doing this is with a confidence interval:

Formula 7.2: $\hat{p} \pm m$ where $m = z^* \times SE_{est}$

and $SE_{est} = \sqrt{\dfrac{\hat{p}(1-\hat{p})}{n}}$

where
$\hat{p}$ is the sample proportion of successes, the proportion of people in the sample with the characteristic we are interested in;
m is the margin of error;
n is the sample size; and
z^* is a multiplier that is chosen to achieve the desired confidence level.

An important first step is to make sure that the sample size is large enough for the CLT to work. This means that we need the

sample size times the sample proportion to be at least 10 and that we need the sample size times (1 minus the sample proportion) to be at least 10.

A 95% confidence interval might or might not have the correct population value within it. However, we are confident that it does, because the method works for 95% of all samples.

Formula 7.4: $\hat{p}_1 - \hat{p}_2 \pm z^*\sqrt{\dfrac{\hat{p}_1(1-\hat{p}_1)}{n_1} + \dfrac{\hat{p}_2(1-\hat{p}_2)}{n_2}}$

where $\hat{p}_1$ is the sample proportion of successes in the first group and $\hat{p}_2$ is the sample proportion of successes in the second group. Here n_1 is the sample size of the first group and n_2 is the sample size of the second group. Also, z^* is the multiplier chosen to achieve the desired level of confidence.

We can compare two population proportions by finding a confidence interval for their difference (subtract one from the other). If the confidence interval contains 0, it means the population proportions could be equal. If it does not contain 0, then we are confident that the population proportions are not equal, and we should note whether the values in the interval are all positive (the first population proportion is greater than the second) or all negative (the first is less than the second).

SOURCES

Busteed, B., and Dugan, A. "U.S. Teachers See Digital Devices as Net Plus For Education," April 6, 2018, news.gallup.com.
Colombel et al. 2010. Infliximab, azathioprine, or combination therapy for Crohn's disease. *New England Journal of Medicine*, vol. 362: 1383–1395.
Cornish, A. 2007. *Do polls miss views of the young and mobile*? National Public Radio. October 1. http://www.npr.org.

Crowley, C. 2010. Soda tax or flat tax? Questions can influence poll results, Cornell expert says. *Albany Times Union*, February 5. http://www .timesunion.com.
Gallup. 2007. Shrunken majority now favors stricter gun laws. October 11. http://www.galluppoll.com.
The New York Times. 1993. 1 in 5 in new survey express some doubt about the Holocaust. April 20.

The New York Times. 1994. Pollster finds error on Holocaust doubts. May 20.

Pasternak, B., et al. 2013. Ondansetron in pregnancy and risk of adverse fetal outcomes. *New England Journal of Medicine*, vol. 368: 814–823.

Pew Research Center for the People & the Press. 2010. The growing gap between landline and dual frame election polls. November 12.

Rasmussen Reports. 2013. 62% say their home is worth more than what they still owe. http://www.rasmussenreports.com/public_content/.

Schweinhart, L., et al. 2005. Lifetime effects: The High/Scope Perry Preschool Study through age 40. *Monographs of the High/Scope Educational Research Foundation*, *14*. Ypsilanti, MI: High/Scope Press.

SECTION EXERCISES

SECTION 7.1

7.1 Population vs. Sample (Example 1) In 2017 the Gallup poll surveyed 1021 adults in the United States and found that 57% supported a ban on smoking in public places.

  a. Identify the population and the sample.

  b. What is the parameter of interest? What is the statistic?

7.2 Population vs. Sample In 2017 Pew Research Center polled 3930 adults in the United States and found that 43% reported playing video games often on some kind of electronic device.

  a. Identify the population and the sample.

  b. What is the parameter of interest? What is the statistic?

7.3 Parameter vs. Statistic (Example 2) Bob Ross hosted a weekly television show, *The Joy of Painting*, on PBS in which he taught viewers how to paint. During each episode, he produced a complete painting while teaching viewers how they could produce a similar painting. Ross completed 30,000 paintings in his lifetime. Although it was an art instruction show, PBS estimated that only 10% of viewers painted along with Ross during his show based on surveys of viewers. For each of the following, also identify the population and explain your choice.

  a. Is the number 30,000 a parameter or a statistic?

  b. Is the number 10% a parameter or a statistic?

7.4 Parameter vs. Statistic The website scholarshipstats.com collected data on all 5341 NCAA basketball players for the 2017 season and found a mean height of 77 inches. Is the number 77 a parameter or a statistic? Also identify the population and explain your choice.

7.5 x vs. μ Two symbols are used for the mean: μ and $\bar{x}$.

  a. Which represents a parameter, and which a statistic?

  b. In determining the mean age of all students at your school, you survey 30 students and find the mean of their ages. Is this mean $\bar{x}$ or μ?

7.6 σ vs. s Two symbols are used for the standard deviation: σ and s.

  a. Which represents a parameter, and which represents a statistic?

  b. To estimate the commute time for all students at a college, 100 students are asked to report their commute times in minutes. The standard deviation for these 100 commute times was 13.9 minutes. Is this standard deviation σ or s?

7.7 μ vs. $\bar{x}$ The mean weight of all professional NBA basketball players is 218.8 pounds. A sample of 50 professional basketball players has a mean weight of 217.6 pounds. Which number is μ, and which number is $\bar{x}$?

7.8 σ vs. s The standard deviation of all professional NBA basketball players is 29.9 pounds. A sample of 50 professional basketball players has a standard deviation of 26.7 pounds. which number is σ, and which number is s?

TRY 7.9 Notation (Example 3) The city of San Francisco provides an open data set of commercial building energy use. Each row of the data set represents a commercial building. A sample of 100 buildings from the data set had a mean floor area of 32,470 square feet. Of the sample, 28% were office buildings.

  a. What is the correct notation for the value 32,470?

  b. What is the correct notation for the value 28%?

7.10 Notation The city of Chicago provides an open data set of the number of WiFi sessions at all of its public libraries. For 2014, there were an average of 451,846.9 WiFi sessions per month at all Chicago public libraries. What is the correct notation for the value 451,846.9?

TRY 7.11 Samples and Populations (Example 4) Chapman University conducts an annual Survey of American Fears. One of the objectives of this survey is to collect annual data on the fears, worries, and concerns of Americans. In 2017 the survey sampled 1207 participants. One of the survey findings was that 16% believe that Bigfoot is a real creature. Identify the sample and population. Is the value 16% a parameter or a statistic? What symbol would be use for this value?

7.12 Samples and Populations The Centers for Disease Control and Prevention (CDC) conducts an annual Youth Risk Behavior Survey, surveying over 15,000 high school students. The 2015 survey reported that, while cigarette use among high school youth had declined to its lowest levels, 24% of those surveyed reported using e-cigarettes. Identify the sample and population. Is the value 24% a parameter or a statistic? What symbol would we use for this value?

7.13 Sample vs. Census You are receiving a large shipment of batteries and want to test their lifetimes. Explain why you would want to test a sample of batteries rather than the entire population.

7.14 Sampling GPAs Suppose you want to estimate the mean grade point average (GPA) of all students at your school. You set up a table in the library asking for volunteers to tell you their GPAs. Do you think you would get a representative sample? Why or why not?

7.15 Sampling with and without Replacement Explain the difference between sampling with replacement and sampling without replacement. Suppose you have the names of 10 students, each written on a 3-inch by 5-inch notecard, and want to select two names. Describe both procedures.

7.16 Simple Random Sampling Is simple random sampling usually done with or without replacement?

7.17 Finding a Random Sample (Example 5) You need to select a simple random sample of four from eight friends who will participate in a survey. Assume the friends are numbered 1, 2, 3, 4, 5, 6, 7, and 8.

Select four friends, using the two lines of numbers in the next column from a random number table.

Read off each digit, skipping any digit not assigned to one of the friends. The sampling is without replacement, meaning that you cannot select the same person twice. Write down the numbers chosen. The first person is number 7.

| 0 7 0 3 3 | 7 5 2 5 0 | 3 4 5 4 6 |
| 7 5 2 9 8 | 3 3 8 9 3 | 6 4 4 8 7 |

Which four friends are chosen?

7.18 Finding a Random Sample You need to select a simple random sample of two from six friends who will participate in a survey. Assume the friends are numbered 1, 2, 3, 4, 5, and 6.

Use technology to select your random sample. Indicate what numbers you obtained and how you interpreted them.

If technology is not available, use the line from a random number table that corresponds to the day of the month on which you were born. For example, if you were born on the fifth day of any month, you would use line 05. Show the digits in the line and explain how you interpreted them.

* 7.19 Random Sampling Assume your class has 30 students and you want a random sample of 10 of them. Describe how to randomly select 10 people from your class using the random number table.

7.20 Random Sampling with Coins Assume your class has 30 students and you want a random sample of 10 of them. A student suggests asking each student to flip a coin, and if the coin comes up heads, then he or she is in your sample. Explain why this is not a good method.

TRY 7.21 Survey Response (Example 6) A school district conducts a survey to determine whether voters favor passing a bond to fund school renovation projects. All registered voters are called. Of those called, 15% answer the survey call. Of those who respond, 62% say they favor passing the bond. Give a reason why the school district should be cautious about predicting that the bond will pass.

7.22 Survey Response To determine if patrons are satisfied with performance quality, a theater surveys patrons at an evening performance by placing a paper survey inside their programs. All patrons receive a program as they enter the theater. Completed surveys are placed in boxes at the theater exits. On the evening of the survey, 500 patrons saw the performance. One hundred surveys were completed, and 70% of these surveys indicated dissatisfaction with the performance. Should the theater conclude that patrons were dissatisfied with performance quality? Explain.

7.23 Views on Capital Punishment In carrying out a study of views on capital punishment, a student asked a question two ways:

1. With persuasion: "My brother has been accused of murder and he is innocent. If he is found guilty, he might suffer capital punishment. Now do you support or oppose capital punishment?"

2. Without persuasion: "Do you support or oppose capital punishment?"

Here is a breakdown of her actual data.

Men

	With persuasion	No persuasion
For capital punishment	6	13
Against capital punishment	9	2

Women

	With persuasion	No persuasion
For capital punishment	2	5
Against capital punishment	8	5

a. What percentage of those persuaded against it support capital punishment?

b. What percentage of those not persuaded against it support capital punishment?

c. Compare the percentages in parts a and b. Is this what you expected? Explain.

7.24 Views on Capital Punishment Use the data given in Exercise 7.23.

Make the two given tables into one table by combining men for capital punishment into one group, men opposing it into another, women for it into one group, and women opposing it into another. Show your two-way table.

The student who collected the data could have made the results misleading by trying persuasion more often on one gender than on the other, but she did not do this. She used persuasion on 10 of 20 women (50%) and on 15 of 30 men (50%).

a. What percentage of the men support capital punishment? What percentage of the women support it?

b. On the basis of these results, if you were on trial for murder and did not want to suffer capital punishment, would you want men or women on your jury?

SECTION 7.2

7.25 Targets: Bias or Lack of Precision?

a. If a rifleman's gunsight is adjusted incorrectly, he might shoot bullets consistently close to 2 feet left of the bull's-eye target. Draw a sketch of the target with the bullet holes. Does this show lack of precision or bias?

b. Draw a second sketch of the target if the shots are both unbiased and precise (have little variation).

The rifleman's aim is not perfect, so your sketches should show more than one bullet hole.

7.26 Targets: Bias or Lack of Precision, Again

a. If a rifleman's gunsight is adjusted correctly, but he has shaky arms, the bullets might be scattered widely around the bull's-eye target. Draw a sketch of the target with the bullet holes. Does this show variation (lack of precision) or bias?

b. Draw a second sketch of the target if the shots are unbiased and have precision (little variation).

The rifleman's aim is not perfect, so your sketches should show more than one bullet hole.

7.27 Bias? Suppose that, when taking a random sample of three students' GPAs, you get a sample mean of 3.90. This sample mean is far higher than the collegewide (population) mean. Does that prove that your sample is biased? Explain. What else could have caused this high mean?

7.28 Unbiased Sample? Suppose you attend a school that offers both traditional courses and online courses. You want to know the average age of all the students. You walk around campus asking those students that you meet how old they are. Would this result in an unbiased sample?

7.29 Proportion of Odd Digits A large collection of one-digit random numbers should have about 50% odd and 50% even digits, because five of the ten digits are odd (1, 3, 5, 7, and 9) and five are even (0, 2, 4, 6, and 8).

a. Find the proportion of odd-numbered digits in the following lines from a random number table. Count carefully.

57.283 pt	74834	81172
89281	48134	71185

b. Does the proportion found in part a represent $\hat{p}$ (the sample proportion) or p (the population proportion)?

c. Find the error in this estimate, the difference between $\hat{p}$ and p (or $\hat{p} - p$).

7.30 Proportion of Odd Digits 1, 3, 5, 7, and 9 are odd and 0, 2, 4, 6, and 8 are even. Consider a 30-digit line from a random number table.

a. How many of the 30 digits would you expect to be odd on average?

b. If you actually counted, would you get exactly the number you predicted in part a? Explain.

7.31 Driver's License (Example 7) According to *The Washington Post*, 72% of high school seniors have a driver's license. Suppose we take a random sample of 100 high school seniors and find the proportion who have a driver's license.

a. What value should we expect for our sample proportion?

b. What is the standard error?

c. Use your answers to parts a and b to complete this sentence: We expect _____% to have their driver's license, give or take _____%.

d. Suppose we increased the sample size from 100 to 500. What effect would this have on the standard error? Recalculate the standard error to see if your prediction was correct.

7.32 BA Attainment According to a 2017 Pew Research report, 40% of millennials have a BA degree. Suppose we take a random sample of 500 millennials and find the proportion who have a BA degree.

a. What value should we expect for our sample proportion?

b. What is the standard error?

c. Use your answers to parts a and b to complete this sentence: We expect _____% to have a BA degree give or take _____%.

d. Suppose we decreased the sample size from 500 to 100. What effect would this have on the standard error? Recalculate the standard error to see if your prediction was correct.

7.33 ESP A Zener deck of cards has cards that show one of five different shapes with equal representation, so that the probability of selecting any particular shape is 0.20. A card is selected randomly, and a person is asked to guess which card has been chosen. The graph below shows a computer simulation of experiments in which a "person" was asked to guess which card had been selected for a large number of trials. (If the person does not have ESP, then his or her proportion of successes should be about 0.20, give or take some amount.) Each dot in the dotplots represents the proportion of success for one person. For instance, the dot in Figure A farthest to the right represents a person with an 80% success rate. One dotplot represents an experiment in which each person had 10 trials; another shows 20 trials; and a third shows 40 trials.

Explain how you can tell, from the widths of the graphs, which has the largest sample ($n = 40$) and which has the smallest sample ($n = 10$).

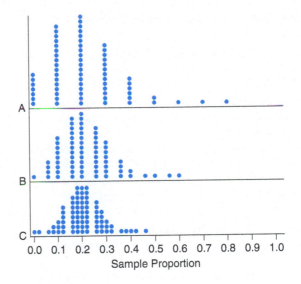

7.34 ESP Again In the graphs for Exercise 7.33, explain how you can tell from the *shape* of the graph which has the largest sample size and which has the smallest sample size.

7.35 Standard Error Which of the dotplots given in Exercise 7.33 has the largest standard error, and which has the smallest standard error?

7.36 Bias? Assuming that the true proportion of success for the trials shown in the graph for Exercise 7.33 is 0.2, explain whether any of the graphs shows bias.

7.37 Fair Coin? One of the graphs shows the proportion of heads from flipping a fair coin 10 times, repeatedly. The others do not. Which graph represents the coin flips? Explain how you know.

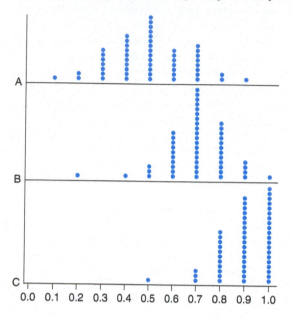

7.38 Far from Fair Which of the graphs in Exercise 7.37 is centered farthest from 0.50?

TRY 7.39 Variation in Sample Proportions (Example 7) Suppose it is known that 20% of students at a certain college participate in a textbook recycling program each semester.

a. If a random sample of 50 students is selected, do we expect that exactly 20% of the sample participates in the textbook recycling program? Why or why not?

b. Suppose we take a sample of 500 students and find the sample proportion participating in the recycling program. Which sample proportion do you think is more likely to be closer to 20%: the proportion from a sample size of 50 or the proportion from a sample size of 500? Explain your reasoning.

7.40 Variation in Sample Proportions Suppose it is known that 60% of employees at a company use a Flexible Spending Account (FSA) benefit.

a. If a random sample of 200 employees is selected, do we expect that exactly 60% of the sample uses an FSA? Why or why not?

b. Find the standard error for samples of size 200 drawn from this population. What adjustments could be made to the sampling method to produce a sample proportion that is more precise?

SECTION 7.3

TRY 7.41 Driver's License (Example 8) According to a 2017 article in *The Washington Post*, 72% of high school seniors have a driver's license. Suppose we take a random sample of 100 high school seniors and find the proportion who have a driver's license. Find the probability that more than 75% of the sample has a driver's license. Begin by verifying that the conditions for the Central Limit Theorem for Sample Proportions have been met.

7.42 BA Attainment According to a 2017 Pew Research report, 40% of millennials have a BA degree. Suppose we take a random sample of 500 millennials and find the proportion who have a driver's license. Find the probability that at most 35% of the sample has a BA degree. Begin by verifying that the conditions for the Central Limit Theorem for Sample Proportions have been met.

TRY 7.43 Stress (Example 9) According to a 2017 Gallup poll, 80% of Americans report being afflicted by stress. Suppose a random sample of 1000 Americans is selected.

a. What percentage of the sample would we expect to report being afflicted by stress?

b. Verify that the conditions for the Central Limit Theorem are met.

c. What is the standard error for this sample proportion?

d. According to the Empirical Rule, there is a 95% probability that the sample proportion will fall between what two values?

7.44 Print Books According to a 2018 Pew Research report, 40% of Americans read print books exclusively (rather than reading some digital books). Suppose a random sample of 500 Americans is selected.

a. What percentage of the sample would we expect to read print books exclusively?

b. Verify that the conditions for the Central Limit Theorem are met.

c. What is the standard error for this sample proportion?

d. Complete this sentence: We expect _____% of Americans to read print books exclusively, give or take _____%.

TRY 7.45 Streaming Services (Example 10) According to a 2017 Pew Research survey, 60% of young Americans aged 18 to 29 say the primary way they watch television is through streaming services on the Internet. Suppose a random sample of 200 Americans from this age group is selected.

a. What percentage of the sample would we expect to watch television primarily through streaming services?

b. Verify that the conditions for the Central Limit Theorem are met.

c. Would it be surprising to find that 125 people in the sample watched television primarily through streaming services? Why or why not?

d. Would it be surprising to find that more than 74% of the sample watched television primarily through streaming services? Why or why not?

7.46 Netflix Cheating According to a 2017 survey conducted by Netflix, 46% of couples have admitted to "cheating" on their significant other by streaming a TV show ahead of their partner. Suppose a random sample of 80 Netflix subscribers is selected.

a. What percentage of the sample would we expect have "cheated" on their partner?

b. Verify that the conditions for the Central Limit Theorem are met.

c. What is the standard error for this sample proportion?

d. Complete the sentence: We expect _____% of streaming couples to admit to Netflix "cheating," give or take _____%.

7.47 Voting According to a 2017 Pew Research Center report on voting issues, 59% of Americans feel that the everything should be done to make it easy for every citizen to vote. Suppose a random sample of 200 Americans is selected. We are interested in finding the probability that the proportion of the sample who feel with way is greater than 55%.

a. Without doing any calculations, determine whether this probability will be greater than 50% or less than 50%. Explain your reasoning.

b. Calculate the probability that the sample proportion is 55% or more.

7.48 Instagram According to a 2018 Pew Research Center report on social media use, 28% of American adults use Instagram. Suppose a sample of 150 American adults is randomly selected. We are interested in finding the probability that the proportion of the sample who use Instagram is greater than 30%.

a. Without doing any calculations, determine whether this probability will be greater than 50% or less than 50%. Explain your reasoning.

b. Calculate the probability that the sample proportion is 30% or more.

7.49 Super Bowl In 2018 it was estimated that approximately 45% of the American population watches the Super Bowl yearly. Suppose a sample of 120 Americans is randomly selected. After verifying the conditions for the Central Limit Theorem are met, find the probability that at the majority (more than 50%) watched the Super Bowl. (Source: vox.com)

7.50 College Enrollment According to data released in 2016, 69% of students in the United States enroll in college directly after high school graduation. Suppose a sample of 200 recent high school graduates is randomly selected. After verifying the conditions for the Central Limit Theorem are met, find the probability that at most 65% enrolled in college directly after high school graduation. (Source: nces.ed.gov)

7.51 Color Blindness While the majority of people who are color blind are male, the National Eye Institute reports that 0.5% of women of with Northern European ancestry have the common form of red-green color blindness. Suppose a random sample of 100 women with Northern European ancestry is selected. Can we find the probability that less than 0.3% of the sample is color blind? If so, find the probability. If not, explain why this probability cannot be calculated.

7.52 Blood Type Human blood is divided into 8 possible blood types. The rarest blood type is AB negative. Only 1% of the population has this blood type. Suppose a random sample of 50 people is selected. Can we find the probability that more than 3% of the sample has AB negative blood? If so, find the probability. If not, explain why this probability cannot be calculated.

SECTION 7.4

7.53 College Athletics A recent Monmouth University poll found that 675 out of 1008 randomly selected people in the United States felt that college and universities with big sports programs placed too much emphasis on athletics over academics. Assuming the conditions for using the CLT were met, use the Minitab output provided to answer these questions.

Descriptive Statistics

N	Event	Sample p	95% CI for p
1008	675	0.669643	(0.639648, 0.698643)

a. Complete this sentence: I am 95% confident that the population proportion believing that colleges and universities with big sports programs place too much emphasis on athletics over academics is between _____ and _____. Report each number as a percentage rounded to one decimal place.

b. Suppose a sports blogger wrote an article claiming that the majority of Americans believe that colleges and university with big sports programs place too much emphasis on athletics over academics. Does this confidence interval support the blogger's claim? Explain your reasoning.

7.54 Environment A 2017 Gallup poll found that 601 out of 1018 randomly selected adults in the United States said protection of the environment should be given priority over development of U.S. energy supplies such as coal, oil, and gas. Assuming the conditions for using the CLT were met, use the TI calculator output provided to answer the following.

a. Report a 95% confidence interval for the population proportion. Write a sentence that explains what this interval means.

b. Based on this interval, do you believe a majority of people in the United States believe protection of the environment should be given priority over development of U.S. energy supplies? Explain your reasoning.

7.55 Gun Control According to a 2017 Gallup Poll, 617 out of 1028 randomly selected adults living in the United States felt the laws covering the sale of firearms should be more strict.

a. What is the value of $\hat{p}$, the sample proportion who favor stricter gun laws?

b. Check the conditions to determine whether the CLT can be used to find a confidence interval.

c. Find a 95% confidence interval for the population proportion who favor stricter gun laws.

d. Based on your confidence interval, do a majority of Americans favor stricter gun laws?

7.56 Smokers According to a 2017 Gallup poll, 572 out of 1021 randomly selected smokers polled believed they are discriminated against in public life or in employment because of their smoking.

a. What percentage of the smokers polled believed they are discriminated against because of their smoking?

b. Check the conditions to determine whether the CLT can be used to find a confidence interval.

c. Find a 95% confidence interval for the population proportion of smokers who believe they are discriminated against because of their smoking.

d. Can this confidence interval be used to conclude the majority of Americans believe smokers are discriminated against because of their smoking? Why or why not?

7.57 Voting A random sample of likely voters showed that 55% planned to vote for Candidate X, with a margin of error of 2 percentage points and with a 95% confidence level.

a. Use a carefully worded sentence to report the 95% confidence interval for the percentage of voters who plan to vote for Candidate X.

b. Is there evidence that Candidate X could lose.

c. Suppose the survey was taken on the streets of New York City and the candidate was running for U.S. president. Explain how that would affect your conclusion.

7.58 Voting A random sample of likely voters showed that 49% planned to support Measure X. The margin of error is 3 percentage points with a 95% confidence level.

a. Using a carefully worded sentence, report the 95% confidence interval for the percentage of voters who plan to support Measure X.

b. Is there evidence that Measure X will fail?

c. Suppose the survey was taken on the streets of Miami and the measure was a Florida statewide measure. Explain how that would affect your conclusion.

TRY **7.59 Alien Life (Example 11)** The 2017 Chapman University Survey of American Fears asked a random sample of 1207 adults Americans if they believed that aliens had come to Earth in modern times, and 26% responded yes.

a. What is the standard error for this estimate of the percentage of all Americans who believe that aliens have come to Earth in modern times?

b. Find a 95% confidence interval for the proportion of all Americans who believe that aliens have come to Earth in modern times.

c. What is the margin of error for the 95% confidence interval?

d. A similar poll conducted in 2016 found that 24.7% of Americans believed aliens have come to Earth in modern times. Based on your confidence interval, can you conclude that the proportion of Americans who believe this has increased since 2016?

7.60 Diabetes In a simple random sample of 1200 Americans age 20 and over, the proportion with diabetes was found to be 0.115 (or 11.5%).

a. What is the standard error for the estimate of the proportion of all Americans age 20 and over with diabetes?

b. Find the margin of error, using a 95% confidence level, for estimating this proportion.

c. Report the 95% confidence interval for the proportion of all Americans age 20 and over with diabetes.

d. According to the Centers for Disease Control and Prevention, nationally, 10.7% of all Americans age 20 or over have diabetes. Does the confidence interval you found in part c support or refute this claim? Explain.

7.61 Marijuana Legalization A 2017 Gallup poll reported that 658 out of 1028 U.S. adults believe that marijuana should be legalized. When Gallup first polled U.S. adults about this subject in 1969, only 12% supported legalization. Assume the conditions for using the CLT are met.

a. Find and interpret a 99% confidence interval for the proportion of U.S. adults in 2017 that believe marijuana should be legalized.

b. Find and interpret a 95% confidence interval for this population parameter.

c. Find the margin of error for each of the confidence intervals found in parts a and b.

d. Without computing it, how would the margin of error of a 90% confidence interval compare with the margin of error for the 95% and 99% intervals? Construct the 90% confidence interval to see if your prediction was correct.

7.62 Organic Produce A 2016 Pew Research poll found that 61% of U.S. adults believe that organic produce is better for health than conventionally grown varieties. Assume the sample size was 1000 and that the conditions for using the CLT are met.

a. Find and interpret a 95% confidence interval for the proportion of U.S. adults to believe organic produce is better for health.

b. Find and interpret an 80% confidence interval for this population parameter.

c. Which interval is wider?

d. What happens to the width of a confidence interval as the confidence level decrease?

7.63 Democracy and Unpopular Views A 2017 survey of U.S. adults found that 74% believed that protecting the rights of those with unpopular views is a very important component of a strong democracy. Assume the sample size was 1000.

a. How many people in the sample felt this way?

b. Is the sample large enough to apply the Central Limit Theorem? Explain. Assume all other conditions are met.

c. Find a 95% confidence interval for the proportion of U.S. adults who believe that protecting the rights of those with unpopular views is a very important component of a strong democracy.

d. Find the width of the 95% confidence interval. Round your answer to the nearest tenth percent.

e. Now assume the sample size was 4000 and the percentage was still 74%. Find a 95% confidence interval and report the width of the interval.

f. What happened to the width of the confidence interval when the sample size was increased? Did it increase or decrease?

7.64 Democracy and Free Press A 2017 survey of U.S. adults found the 64% believed that freedom of news organization to criticize political leaders is essential to maintaining a strong democracy. Assume the sample size was 500.

a. How many people in the sample felt this way?

b. Is the sample large enough to apply the Central Limit Theorem? Explain. Assume all other conditions are met.

c. Find a 95% confidence interval for the proportion of U.S. adults who believe that freedom of news organizations to criticize political leaders is essential to maintaining a strong democracy.

d. Find the width of the 95% confidence interval. Round your answer to the nearest whole percent.

e. Now assume the sample size was increased to 4500 and the percentage was still 64%. Find a 95% confidence interval and report the width of the interval.

f. What happened to the width of the confidence interval when the sample size was increased. Did it increase or decrease?

TRY **7.65 Winter Olympics (Example 12)** According to a 2018 Rasmussen Poll, 40% of American adults were very likely to watch some of the Winter Olympic coverage on television. The survey polled 1000 American adults and had a margin of error of plus or minus 3 percentage points with a 95% level of confidence.

a. State the survey results in confidence interval form and interpret the interval.

b. If the Rasmussen Poll was to conduct 100 such surveys of 1000 American adults, how many of them would result in confidence intervals that included the true population proportion?

c. Suppose a student wrote this interpretation of the confidence interval: "We are 95% confident that the sample proportion is between 37% and 43%." What, if anything, is incorrect in this interpretation?

7.66 Marijuana Use The Gallup poll reported that 45% of Americans have tried marijuana. This was based on a survey of 1021 Americans and had a margin of error of plus or minus 5 percentage points with a 95% level of confidence.

a. State the survey results in confidence interval form and interpret the interval.

b. If the Gallup Poll was to conduct 100 such surveys of 1021 Americans, how many of them would result in confidence intervals that did not include the true population proportion?

c. Suppose a student wrote this interpretation of the interval: "We are 95% confident that the percentage of Americans who have tried marijuana is between 40% and 50%." What, if anything, is incorrect in this interpretation?

7.67 Past Presidential Vote In the 1960 presidential election, 34,226,731 people voted for Kennedy, 34,108,157 for Nixon, and 197,029 for third-party candidates (Source: www.uselectionatlas.org).

a. What percentage of voters chose Kennedy?

b. Would it be appropriate to find a confidence interval for the proportion of voters choosing Kennedy? Why or why not?

7.68 Batting Averages The website www.mlb.com compiles statistics on all professional baseball players. For the 2017 season, statistics were recorded for all 663 players. Of this population, the mean batting average was 0.236 with a standard deviation of 0.064. Would it be appropriate to use this data to construct a 95% confidence interval for the mean batting average of professional baseball players for the 2017 season? If so, construct the interval. If not, explain why it would be inappropriate to do so.

7.69 Picky Eaters (Example 13) In a 2017 Harris poll conducted for Uber Eats, 438 of 1019 U.S. adults polled said they were "picky eaters."

a. What proportion of the respondents said they were picky eaters?

b. Find a 95% confidence interval for the population proportion of U.S. adults who say they are picky eaters.

c. Would a 90% confidence interval based on this sample be wider or narrower than the 95% interval? Give a reason for your answer.

d. Construct the 90% confidence interval. Was your conclusion in part c correct?

7.70 Nutrition Labels Of 1019 U.S. adults responding to a 2017 Harris poll, 47% said they always or often read nutrition labels when grocery shopping.

a. Construct a 95% confidence interval for the population proportion of U.S. adults who always or often read nutrition labels when grocery shopping.

b. What is the width of the 95% confidence interval?

c. Name a confidence level that would produce an interval wider than the 95% confidence interval. Explain why you think this interval would be wider than a 95% confidence interval.

d. Construct the interval using the confidence level you proposed in part c and find the width of the interval. Is this interval wider than the 95% confidence interval?

7.71 Estimating Sample Size (Example 14) Pew Research conducted a study in 2018 to estimate the percentage of Americans who do not use the Internet.

a. If a 95% confidence level is used, how many people should be included in the survey if the researchers wanted to have a margin of error of 6%?

b. How would the sample size change if the researchers wanted to estimate the percentage with a margin of error of 4%?

c. What is the relationship between the size of the margin of error and the sample size?

7.72 Estimating Sample Size In the 2018 study Closing the STEM Gap, researchers wanted to estimate the percentage of middle school girls who planned to major in a STEM field.

a. If a 95% confidence level is used, how many people should be included in the survey if the researchers wanted to have a margin of error of 3%?

b. How could the researchers adjust their margin of error if they want to decrease the number of study participants?

SECTION 7.5

7.73 Happiness A Harris poll asked Americans in 2016 and 2017 if they were happy. In 2016, 31% reported being happy and in 2017, 33% reported being happy. Assume the sample size for each poll was 1000. A 95% confidence interval for the difference in these proportions $p_1 - p_2$ (where proportion 1 is proportion happy in 2016 and proportion 2 is the proportion happy in 2017) is $(-0.06, 0.02)$. Interpret this confidence interval. Does the interval contain 0? What does this tell us about happiness among American in 2016 and 2017?

7.74 Artificial Intelligence A Harris poll asked a sample of U.S. adults if they agreed with the statement "Artificial intelligence will widen the gap between the rich and poor in the U.S." Of those aged 18 to 35, 69% agreed with the statement. Of those aged 36 to 50, 60% agreed with the statement. A 95% confidence interval for $p_1 - p_2$ (where p_1 is the proportion of those aged 18–35 who agreed and p_2 is the proportion of those aged 36–50 who agreed) is $(0.034, 0.146)$. Does the interval contain 0? What does this tell us about the proportion of adults in these age groups who agree with the statement?

TRY 7.75 Democratic (Examples 15, 16, 17, 18) Voters and the FBI In 2003 and 2017 Gallup asked Democratic voters about their views on the FBI. In 2003, 44% thought the FBI did a good or excellent job. In 2017, 69% of Democratic voters felt this way. Assume these percentages are based on samples of 1200 Democratic voters.

a. Can we conclude, on the basis of these two percentages alone, that the proportion of Democratic voters who think the FBI is doing a good or excellent job has increase from 2003 to 2017? Why or why not?

b. Check that the conditions for using a two-proportion confidence interval hold. You can assume that the sample is a random sample.

c. Construct a 95% confidence interval for the difference in the proportions of Democratic voters who believe the FBI is doing a good or excellent job, $p_1 - p_2$. Let p_1 be the proportion of Democratic voters who felt this way in 2003 and p_2 be the proportion of Democratic voters who felt this way in 2017.

d. Interpret the interval you constructed in part c. Has the proportion of Democratic voters who feel this way increased? Explain.

7.76 Trust in Judiciary In 2016 and 2017 Gallup asked American adults about their amount of trust they had in the judicial branch of government. In 2016, 61% expressed a fair amount or great deal of trust in the judiciary. In 2017, 68% of Americans felt this way. These percentages are based on samples of 1022 American adults.

a. Explain why it would be inappropriate to conclude, based on these percentages alone, that the percentage of Americans who had a fair amount or great deal of trust in the judicial branch of government increased from 2016 to 2017.

b. Check that the conditions for using a two-proportion confidence interval hold. You can assume that the sample is a random sample.

c. Construct a 95% confidence interval for the difference in the proportions of Americans who expressed this level of trust in the judiciary, $p_1 - p_2$, where p_1 is the 2016 population proportion and p_2 is the 2017 population proportion.

d. Based on the confidence interval constructed in part c, can we conclude that proportion of Americans with this level of trust in the judiciary increased from 2016 to 2017? Explain.

g 7.77 Perry Preschool and Graduation from High School The Perry Preschool Project was created in the early 1960s by David Weikart in Ypsilanti, Michigan. In this project, 123 African American children were randomly assigned to one of two groups: One group enrolled in the Perry Preschool, and the other group did not. Follow-up studies were done for decades. One research question was whether attendance at preschool had an effect on high school graduation. The table shows whether the students graduated from regular high school or not and includes both boys and girls (Schweinhart et al. 2005). Find a 95% confidence interval for the difference in proportions, and interpret it.

	Preschool	No Preschool
Grad HS	37	29
No Grad HS	20	35

7.78 Preschool: Just the Boys Refer to Exercise 7.77 for information. This data set records results just for the boys.

	Preschool	No Preschool
Grad HS	16	21
No Grad HS	16	18

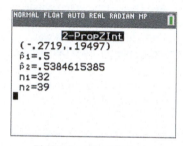

a. Find and compare the percentages that graduated for each group, descriptively. Does this suggest that preschool was linked with a higher graduation rate?

b. Verify that the conditions for a two-proportion confidence interval are satisfied.

c. Indicate which one of the following statements is correct.

   i. The interval does not capture 0, suggesting that it is plausible that the proportions are the same.

   ii. The interval does not capture 0, suggesting that it is not plausible that the proportions are the same.

   iii. The interval captures 0, suggesting that it is plausible that the population proportions are the same.

   iv. The interval captures 0, suggesting that it is not plausible that the population proportions are the same.

d. Would a 99% confidence interval be wider or narrower?

TRY 7.79 Fish Oil (Example 19) A double-blind study using random assignment was done of pregnant women in Denmark. Women were given fish oil or a placebo during pregnancy. Their children were followed during the first 5 years of life to see if they developed asthma. The results are summarized in the table. (Bisgaard et al., "Fish Oil-Derived Fatty Acids in Pregnancy and Wheeze and Asthma in Offspring," *New England Journal of Medicine*, vol. 375: 2530–2539. doi:10.1056/NEJMoa1503734)

Developed asthma	Fish Oil	Placebo
Yes	58	83
No	288	266

a. Calculate and compare the percentages of children who developed asthma in the fish oil group and in the placebo group.

b. Check that the conditions for using a two-population confidence interval hold.

c. Find the 95% confidence interval for the difference in the proportion of children who develop asthma in the two groups. Based on your confidence interval, can we conclude that there is a difference in the population proportions?

7.80 Sugar-Sweetened Beverages In 2017, the journal *Obesity* reported on trends in sugar-sweetened beverage (SSB) consumption. A random sample of youths aged 12 to 19 years old were asked to monitor all food and beverages consumed in a 24-hour period. The study was done in 2003 and repeated in 2014. The numbers who consumed a sugary beverage such as soda or fruit juice in a day are shown in the table. (Bleich et al., "Trends in Beverage Consumption among Children and Adults, 2003-2014," *Obesity*, vol. 26 [2018]: 432–441. doi:10.1002/oby.22056)

Consumed SSB	2003	2014
Yes	3416	2682
No	685	1419

a. Calculate and compare the percentages of youths in this age group who consumed an SSB during the recording period.

b. Check that the conditions for using a two-population confidence interval hold.

c. Find the 95% confidence interval for the difference in the proportion of youth consuming an SSB in 2003 and 2014. Based on your confidence interval, do you think there has been a change in sugar-sweetened beverage consumption among this age group? Explain.

7.81 Gender and Use of Turn Signals Statistics student Hector Porath wanted to find out whether gender and the use of turn signals when driving were independent. He made notes when driving in his truck for several weeks. He noted the gender of each person that he observed and whether he or she used the turn signal when turning or changing lanes. (In his state, the law says that you must use your turn signal when changing lanes, as well as when turning.) The data he collected are shown in the table.

	Men	Women
Turn signal	585	452
No signal	351	155
	936	607

a. What percentage of men used turn signals, and what percentage of women used them?

b. Assuming the conditions are met (although admittedly this was not a random selection), find a 95% confidence interval for the difference in *percentages*. State whether the interval captures 0, and explain whether this provides evidence that the proportions of men and women who use turn signals differ in the population.

c. Another student collected similar data with a smaller sample size:

	Men	Women
Turn Signal	59	45
No Signal	35	16
	94	61

First find the percentage of men and the percentage of women who used turn signals, and then, assuming the conditions are met, find a 95%

confidence interval for the difference in *percentages*. State whether the interval captures 0, and explain whether this provides evidence that the percentage of men who use turn signals differs from the percentage of women who do so.

d. Are the conclusions in parts b and c different? Explain.

7.82 Local TV News Pew Research reported that 46% of Americans surveyed in 2016 got their news from local television. A similar survey conducted in 2017 found that 37% of Americans got their news from local television. Assume the sample size for each poll was 1200.

a. Construct the 95% confidence interval for the difference in the proportions of Americans who get their news from local television in 2016 and 2017.

b. Based on your interval, do you think there has been a change in the proportion of Americans who get their news from local television? Explain.

7.83 Drug for Nausea Ondansetron (Zofran) is a drug used by some pregnant women for nausea. There was some concern that it might cause trouble with pregnancies. An observational study was done of women in Denmark (Pasternak et al. 2013). An analysis of 1849 exposed women and 7376 unexposed women showed that 2.1% of the women exposed to ondansetron and 5.8% of the unexposed women had miscarriages during weeks 7 to 22 of their pregnancies.

a. Was the rate of miscarriages higher with Zofron, as feared?

b. Which of the conditions for finding the confidence interval do not hold?

7.84 Preschool: Just the Girls The Perry Preschool Project was created in the early 1960s by David Weikart in Ypsilanti, Michigan. In this project, 123 African American children were randomly assigned to one of two groups: One group enrolled in the Perry Preschool, and one group did not enroll. Follow-up studies were done for decades. One research question was whether attendance at preschool had an effect on high school graduation. The table shows whether the students graduated from regular high school or not and includes *girls* only (Schweinhart et al. 2005).

	Preschool	No Preschool
HS Grad	21	8
No HS Grad	4	17

a. Find the percentages that graduated for both groups, and compare them descriptively. Does this suggest that preschool was associated with a higher graduation rate?

b. Which of the conditions fail so that we cannot use a confidence interval for the difference between proportions?

CHAPTER REVIEW EXERCISES

7.85 Organic Foods According to a Gallup poll, 45% of Americans actively seek out organic foods when shopping. Suppose a random sample of 500 Americans is selected and the proportion who actively seek out organic foods is recorded.

a. What value should we expect for the sample proportion?

b. What is the standard error?

c. Use your answers to parts a and b to complete this sentence: We expect _____% of Americans to actively seek out organic foods when shopping, give or take _____%.

d. Would it be surprising to find a sample proportion of 55%? Why or why not?

e. What effect would decreasing the sample size from 500 to 100 have on the standard error?

7.86 College Enrollment (Women) According to the Bureau of Labor Statistics, 71.9% of young women enroll in college directly after high school graduation. Suppose a random sample of 200 female high school graduates is selected and the proportion who enroll in college is obtained.

a. What value should we expect for the sample proportion?

b. What is the standard error?

c. Would it be surprising if only 68% of the sample enrolled in college? Why or why not?

d. What effect would increasing the sample size to 500 have on the standard error?

7.87 Environmental Satisfaction In 2018 Gallup reported that 52% of Americans are dissatisfied with the quality of the environment in the United States. This was based on a 95% confidence interval with a margin of error of 4 percentage points. Assume the conditions for constructing the confidence interval are met.

a. Report and interpret the confidence interval for the population proportion that are dissatisfied with the quality of the environment in the United States in 2018.

b. If the sample size were larger and the sample proportion stayed the same, would the resulting interval be wider or narrower than the one obtained in part a?

c. If the confidence level were 90% rather than 95% and the sample proportion stayed the same, would the interval be wider or narrower than the one obtained in part a?

d. In 2018 the population of the United States was roughly 327 million. If the population had been half that size, would this have changed any of the confidence intervals constructed in this problem? In other words, if the conditions for constructing a confidence interval are met, does the population size have any effect on the width of the interval?

7.88 Technology Anxiety In a 2018 survey conducted by Northeastern University, 28% of working adults with education levels less than a bachelor's degree worried that their job would be eliminated due to new technology or automation. This was based on a 95% confidence interval with a margin of error of 3 percentage points.

a. Report the confidence interval for the proportion of adults with education level less than a bachelor's degree who are worried about job loss due to new technology or automation.

b. If the sample size were smaller and the sample proportion stayed the same, would the resulting interval be wider or narrower than the one obtained in part a?

c. If the confidence level were 99% rather than 95% and the sample proportion stayed the same, would the interval be wider or narrower than the one obtained in part a?

d. In 2018 the population of the United States was roughly 327 million. If 50 million people were added to the population what effect, if any, would this have on the intervals obtained in this problem?

7.89 Sample Proportion A poll on a proposition showed that we are 95% confident that the population proportion of voters supporting it is between 40% and 48%. Find the sample proportion that supported the proposition.

7.90 Sample Proportion A poll on a proposition showed that we are 99% confident that the population proportion of voters supporting it is between 52% and 62%. Find the sample proportion that supported the proposition.

7.91 Margin of Error A poll on a proposition showed that we are 95% confident that the population proportion of voters supporting it is between 40% and 48%. Find the margin of error.

7.92 Margin of Error A poll on a proposition showed that we are 99% confident that the population proportion of voters supporting it is between 52% and 62%. Find the margin of error.

7.93 Dreaming in Color According to studies done in the 1940s, 29% of people dream in color. Assuming this is still true, find the probability that in a random sample of 200 independent people, 50% or more will report dreaming in color. Start by checking the conditions to see whether the Central Limit Theorem applies.

7.94 Hand Washing Ignaz Semmelweiss (1818–1865) was the doctor who first encouraged other doctors to wash their hands with disinfectant before touching patients. Before the new procedure was established, the rate of infection at Dr. Semmelweiss's hospital was about 10%. Afterward the rate dropped to about 1%. Assuming the population proportion of infections was 10%, find the probability that the sample proportion will be 1% or less, assuming a sample size of 200. Start by checking the conditions required for the Central Limit Theorem to apply.

7.95 Statistical Tie In the primaries leading up to the 2016 presidential election, the *Business Insider* reported that Bernie Sanders and Hilary Clinton were in a "statistical tie" in the polls leading up to the Vermont primary. Clinton led Sanders 43% to 35% in the polls, with a margin of error of 5.2%. Explain what this means to someone who may be unfamiliar with margin of error and confidence intervals.

7.96 School Bond Suppose a political consultant is hired to determine if a school bond is likely to pass in a local election. The consultant randomly samples 250 likely voters and finds that 52% of the sample supports passing the bond. Construct a 95% confidence interval for the proportion of voters who support the bond. Assume the conditions are met. Based on the confidence interval, should the consultant predict the bond will pass? Why or why not?

7.97 Trust in the Executive Branch Has trust in the executive branch of government declined? A Gallup poll asked U.S. adults if they trusted the executive branch of government in 2008 and again in 2017. The results are shown in the table.

	2008	2017
Yes	623	460
No	399	562
Total	1022	1022

a. Find and compare the sample proportion for those who trusted the executive branch in 2008 and in 2017.

b. Find the 95% confidence interval for the difference in the population proportions.

Assume the conditions for using the confidence interval are met. Based on the interval, has there been a change in the proportion of U.S. adults who trust the executive branch? Explain.

7.98 Trust in the Legislative Branch Has trust in the legislative branch of government declined? A Gallup poll asked U.S. adults if they trusted the legislative branch of government in 2008 and again in 2017. The results are shown in the table.

	2008	2017
Yes	399	358
No	623	664
Total	1022	1022

a. Find and compare the sample proportion for those who trusted the legislative branch in 2008 and in 2017.

b. Find the 95% confidence interval for the difference in the population proportions.

Assume the conditions for using the confidence interval are met. Based on the interval, has there been a change in the proportion of U.S. adults who trust the legislative branch? Explain.

* 7.99 Voters Poll: Sample Size A polling agency wants to determine the sample size required to get a margin of error of no more than 3 percentage points (0.03). Assume the pollsters are using a 95% confidence level. How large a sample should they take? See above for the .

* 7.100 Ratio of Sample Sizes Find the sample size required for a margin of error of 3 percentage points, and then find one for a margin of error of 1.5 percentage points; for both, use a 95% confidence level. Find the ratio of the larger sample size to the smaller sample size. To reduce the margin of error to half, by what do you need to multiply the sample size?

7.101 Criticize the Sampling Marco is interested in whether Proposition P will be passed in the next election. He goes to the university library and takes a poll of 100 students. Since 58% favor Proposition P, Marco believes it will pass. Explain what is wrong with his approach.

7.102 Criticize the Sampling Maria opposes capital punishment and wants to find out if a majority of voters in her state support it. She goes to a church picnic and asks everyone there for their opinion. Because most of them oppose capital punishment, she concludes that a vote in her state would go against it. Explain what is wrong with Maria's approach.

7.103 Random Sampling? If you walked around your school campus and asked people you met how many keys they were carrying, would you be obtaining a random sample? Explain.

7.104 Biased Sample? You want to find the mean weight of the students at your college. You calculate the mean weight of a sample of members of the football team. Is this method biased? If so, would the mean of the sample be larger or smaller than the true population mean for the whole school? Explain.

* 7.105 Bias? Suppose that, when taking a random sample of 4 from 123 women, you get a mean height of only 60 inches (5 feet). The procedure may have been biased. What else could have caused this small mean?

* 7.106 Bias? Four women selected from a photo of 123 were found to have a sample mean height of 71 inches (5 feet 11 inches). The population mean for all 123 women was 64.6 inches. Is this evidence that the sampling procedure was biased? Explain.

7.107 Sample Size Formula (Part 1) From Formula 7.2, an estimate for margin of error for a 95% confidence interval is

$$m = 2\sqrt{\frac{\hat{p}(1 - \hat{p})}{n}}$$ where n is the required sample size and $\hat{p}$ is the sample proportion. Since we do not know a value for $\hat{p}$, we use a conservative estimate of 0.50 for $\hat{p}$. Replace $\hat{p}$ with 0.50 in the formula and simplify.

7.108 Sample Size Formula (Part 2) Using your result from Exercise 7.107, solve for n by (1) dividing both sides of the equation by 2, (2) squaring both sides of the equation, (3) cross-multiplying, and (4) solving for n.

gUIDED EXERCISES

7.41 Driver's License (Example 8) According to a 2017 article in *The Washington Post*, 72% of high school seniors have a driver's license. Suppose we take a random sample of 100 high school seniors and find the proportion who have a driver's license.

QUESTION Find the probability that more than 75% of the sample has a driver's license. Begin by verifying that the conditions for the Central Limit Theorem for Sample Proportions have been met.

Step 1 ▶ Population proportion
The sample proportion is 0.75. What is the population proportion?

Step 2 ▶ Check assumptions
Because we are asked for an approximate probability, we might be able to use the Central Limit Theorem. In order to use the Central Limit Theorem for a proportion, we must check the assumptions.

a. Randomly sampled: Yes
b. Sample size: If a simple random sample of 100 high school seniors is selected, how many of them would you expect would have a driver's license, on the average? Calculate *n* times *p*. Also calculate how many you expect would not have a driver's license, *n* times $(1 - p)$. State whether both are greater than 10.
c. Assume the population size is at least 10 times the sample size, which would be at least 1000.

Step 3 ▶ Calculate
Part of the standardization follows. Finish it, showing all the numbers.
First find the standard error:

$$SE = \sqrt{\frac{p(1 - p)}{n}} = \sqrt{\frac{0.72(0.28)}{?}} = ?$$

Then standardize:

$$z = \frac{\hat{p} - p}{SE} = \frac{0.75 - 0.72}{?} = ? \text{ standard units}$$

Find the approximate probability that at least 0.67 pass by finding the area to the right of the *z*-value of 0.59 in the Normal curve. Show a well-labeled curve, starting with what is given in the figure below.

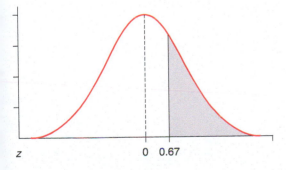

Step 4 ▶ Explain
Explain why the tail area in the accompanying figure represents the correct probability.

Step 5 ▶ Answer the question
Find the area of the shaded region in this figure.

g 7.77 Perry Preschool and Graduation from High School
The Perry Preschool Project was created in the early 1960s by David Weikart in Ypsilanti, Michigan. In this project, 123 African American children were randomly assigned to one of two groups: One group enrolled in the Perry Preschool and the other did not. Follow-up studies were done for decades. One research question was whether attendance at preschool had an effect on high school graduation. The table shows whether the students graduated from regular high school or not and includes both boys and girls (Schweinhart et al. 2005).

	Preschool	No Preschool
Grad HS	37	29
No Grad HS	20	35

QUESTION Find a 95% confidence interval for the difference in proportions, and interpret it. Follow the steps below.

Step 1 ▶ Calculate percentages
Looking at the children who went to preschool, 37/57, or 64.9%, graduated from high school. Looking at the children who did not go to preschool, what percent graduated from high school?

Step 2 ▶ Compare
In this sample, are the children who attend preschool more or less likely to graduate than the children who don't attend preschool?

Step 3 ▶ Verify conditions
Although we don't have a random sample of children, we do have random assignment to groups, and the two groups are independent.

We must verify that the sample sizes are large enough.

$$n_1\hat{p}_1 = 57(0.649) = 37$$
$$n_1(1 - \hat{p}_1) = 57(0.351) = 20$$
$$n_2\hat{p}_2 = 64(0.453) = ??$$
$$n_2\hat{p}_2 = ??$$

As you can see, because we are using the estimated values of $\hat{p}$ (p-hat), for our expected values we simply get the numbers in the table. Do the remaining two calculations showing that you get 29 and 35. Because all of these numbers are larger than 10, we can proceed.

Step 4 ► Calculate intervals

Refer to the TI-84 output, and fill in the blanks in this sentence:

I am 95% confident that the difference in proportions graduating (Preschool rate minus No Preschool rate) is between _____ and _____. You may round each number to three decimal digits.

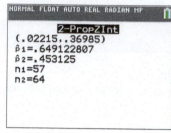

▲ TI-84 Output

Step 5 ► Draw conclusions

Indicate which one of the following statements is correct?

  i. The interval does not capture 0, suggesting that it is plausible that the proportions are the same.

  ii. The interval does not capture 0, suggesting that it is not plausible that the proportions are the same.

  iii. The interval captures 0, suggesting that it is plausible that the population proportions are the same.

  d. The interval captures 0, suggesting that it is not plausible that the population proportions are the same.

Step 6 ► Generalize

Can we generalize to a larger population from this data set? Why or why not?

Step 7 ► Determine causation

Can we conclude from this data set that preschool caused the difference? Why or why not?

TechTips

For generating random numbers, see the TechTips for Chapter 5.

EXAMPLE A: ONE-PROPORTION PROBABILITY USING NORMAL TECHNOLOGY ▶ A U.S. government survey in 2007 said that 87% of young Americans earn a high school diploma. If you took a simple random sample of 2000 young Americans, what is the approximate probability that 88% or more of the sample will have earned their high school diploma? To use the "Normal" steps from Chapter 6, we just need to evaluate the mean and standard error (the standard deviation). The population mean is 0.87. The standard error is

$$SE = \sqrt{\frac{p(1-p)}{n}} = \sqrt{\frac{0.87(1-0.87)}{2000}} = \sqrt{\frac{0.1131}{2000}} = 0.00752$$

Now you can use the steps (for TI-84, Minitab, Excel, or StatCrunch) from Chapter 6, TechTips, EXAMPLE A: NORMAL.

Figure 7a shows TI-84 output, and Figure 7b shows StatCrunch output.

Thus the probability is about 9%.

▲ **FIGURE 7a** TI-84 Normal Output

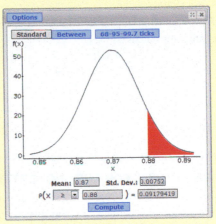

▲ **FIGURE 7b** StatCrunch Normal Output

EXAMPLE B: FINDING A CONFIDENCE INTERVAL FOR A PROPORTION ▶ Find a 95% confidence interval for a population proportion of heads when obtaining 22 heads in a sample of 50 tosses of a coin.

EXAMPLE C: FINDING A CONFIDENCE INTERVAL FOR (THE DIFFERENCE OF) TWO PROPORTIONS ▶ Find a 95% confidence interval for the difference between the two population proportions using the Perry Preschool Project results. The project reported that 37 of 57 children sampled who had attended preschool graduated from high school, whereas 29 of 64 children sampled who had not attended preschool graduated from high school.

EXAMPLE D: USING SIMULATIONS TO UNDERSTAND THE BEHAVIOR OF ESTIMATORS ▶ Use simulated sampling on a population of 1000, consisting of 250 cat lovers and 750 dog lovers, to demonstrate the effects of sample size on sample proportion as an estimator of population proportion.

TI-84

CONFIDENCE INTERVAL FOR ONE PROPORTION

1. Press **STAT**, choose **Tests, A: 1-PropZInt**.
2. Enter: **x: 22**; **n: 50**; **C-level, .95**.
3. Press **ENTER** when **Calculate** is highlighted.

Figure 7c shows the 95% confidence interval of (0.302, 0.578).

CONFIDENCE INTERVAL FOR TWO PROPORTIONS

1. Press **STAT**, choose **Tests, B: 2-PropZInt**.
2. Enter **x1: 37**; **n1: 57**; **x2: 29**; **n2: 64**; **C-Level: .95**.
3. Press **Enter** when **Calculate** is highlighted.
4. The output screen shows a 95% CI of (0.02215, 0.36985).

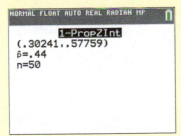

◀ **FIGURE 7c** TI-84 Output for a One-Proportion Z-interval

MINITAB

CONFIDENCE INTERVAL FOR ONE PROPORTION

1. **Stat > Basic Statistics > 1 Proportion**.
2. Choose **Summarized data**.

Enter: **Number of events, 22**; **Number of trials, 50**.

3. Click **Options** and, in the **Method:** box, select **Normal approximation**.

4. In the **Confidence level:** box, if you want a confidence level different from 95%, change it.

5. Click **OK**: click **OK**.

The relevant part of the output is

95% CI
(0.302411, 0.577589)

CONFIDENCE INTERVAL FOR TWO PROPORTIONS

1. **Stat > Basic Statistics > 2 Proportions**.

2. Choose **Summarized data**. Enter: **Sample 1, Number of events, 37, Number of trials, 57; Sample 2, Number of events, 29, Number of trials, 64**. Click **OK**.

3. The output shows a 95% CI for difference of (0.0221483, 0.369847).

EXCEL

The free version of XLSTAT does not have these capabilities.

CONFIDENCE INTERVAL FOR ONE PROPORTION

1. Click **XLSTAT, Parametric tests, Tests for one proportion**.

2. See Figure 7d. Enter: **Frequency, 22; Sample size, 50; Test Proportion, 0.5**. Leave **Frequency** and **z test** checked, but uncheck **Continuity correction**. Click **Options**, choose **Wald** and click **OK**. Click **OK**.

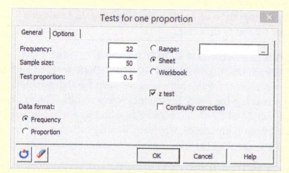

▲ **FIGURE 7d** XLSTAT Input for One Proportion

(If you wanted an interval other than 95%, you would click Options and change the significance level. For a 99% interval you would use 1. For a 90% interval you would use 10.)

The relevant part of the output is

95% confidence interval on the proportion (Wald):
(0.302, 0.578)

CONFIDENCE INTERVAL FOR TWO PROPORTIONS

1. Click **XLSTAT, Parametric tests, Test for two proportions**.

2. Enter: **Frequency 1: 37; Sample size 1: 57; Frequency 2: 29; Sample size 2: 64**.

3. Leave other checked options as given, and click **OK**.

4. The relevant part of the output is

95% confidence interval on the difference between the proportions:
(0.022, 0.370)

STATCRUNCH

CONFIDENCE INTERVAL FOR ONE PROPORTION

1. **Stat > Proportion Stats > One Sample > With Summary**

2. Enter: **# of successes, 22; # of observations, 50**.

3. Select the **Confidence interval for p** option.

4. Leave the default **Level: 0.95** for a 95% interval. For **Method**, leave the default **Standard-Wald**.

5. Click **Compute!**

The relevant part of the output is shown. "L. Limit" is the lower limit of the interval, and "U. Limit" is the upper limit of the interval.

L. Limit	U. Limit
0.30241108	0.5775889

CONFIDENCE INTERVAL FOR TWO PROPORTIONS

1. **Stat > Proportion Statistics > Two Sample > With Summary**.

2. Enter: **Sample 1: # of successes: 37; # of observations: 57; Sample 2: # of successes: 29; # of observations: 64**.

3. Select **Confidence interval for P₁ − P₂**, leave the **Level** at **0.95,** and click **Compute!**

4. The output will show **L. Limit** (for lower limit of confidence interval) **0.022148349**, and **U. Limit** (for upper limit of confidence interval) **0.36984727**.

SIMULATED SAMPLING

1. Enter the population data. This can be done either by hand or by loading from a data file. To enter by hand, you can "brute-force it" by entering **Cat** in first 250 rows of the first column, which is labeled **var1**, and then entering **Dog** in the next 750 rows. But *if you do it carefully*, it is quicker to do **Data > Compute > Expression** and then type the expression **conc at(rep("Cat",250),rep("Dog",750))**. Change the column label to **Lover**. Click **Compute!**

2. In this step you will generate 1000 samples of size 5 from the population, Lover; then you will count and list the number of cat lovers in each sample in a new column labeled Cfive. **Data > Sample**. **Select columns: Lover**. Enter **Sample size: 5, Number of samples: 1000**. Select **Compute statistic for each sample**. *Now be careful!* Enter **Expression: sum("Sample(Lover)" = "Cat")**. Enter **Column name: Cfive**. Click **Compute!**

3. In this step you will calculate the proportion of cat lovers in each sample and then list the 1000 sample proportions in a new column labeled p(n = 5). **Data > Compute > Expression**. Enter **Expression: Cfive/5**. Enter **Column label: p(n = 5)**. Click **Compute!**

4. To generate a column of 1000 sample proportions of sample size 10, repeat step 2, but substitute **Sample size: 10, Column name: Cten, Expression: Cten/10**, and repeat step 3, but substitute **Column label: p(n = 10)**.

5. To generate a column of 1000 sample proportions of sample size 100, repeat steps 2 and 3, but change to **Sample size: 100, Column name: Chundred, Expression: Chundred/100**, and **Column label: p(n = 100)**.

6. Now graph the results. **Graph > Histogram**. Select **p(n = 5), p(n = 10)**, and **p(n = 100)**; do this by holding the **Ctrl** key down while clicking on **p(n = 5), p(n = 10)**, and **p(n = 100)**. Scroll down and check **Horizontal lines** and **Use same X-axis**. Set **Rows per page** to **3**. Click **Compute!** Your output should look somewhat like Figure 7e.

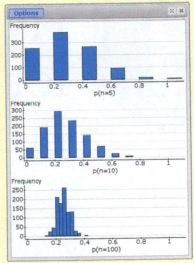

▲ **FIGURE 7e** StatCrunch Output

8 Hypothesis Testing for Population Proportions

THEME

In many scientific and business contexts, decisions must be made about the values of population parameters, even though our estimates of these parameters are uncertain. Hypothesis testing provides a method for making these decisions while controlling for the probability of making certain types of mistakes.

In science, business, and everyday life, we often have to make decisions on the basis of incomplete information. For example, a marketing company needs to determine which poster will attract the most movie-goers for an upcoming release. The company can't poll everyone in the country about which poster makes them most excited about the movie; they'll have to make their decision based on a small sample of people's opinions. An educational psychologist wonders whether kids who receive music training become more creative than other children. The test she uses to measure creativity does in fact show an increase, but might this increase be explained by chance alone? A sample of people who watched violent TV when they were children turn out to exhibit more violent behavior than a comparison sample made up of people who did not watch violent TV when young. Could this difference be due to chance, or is something else going on?

Hypothesis testing is a type of statistical inference. In Chapter 7, we used confidence intervals to estimate parameters and provide a margin of error for our estimates. In this chapter we make decisions on the basis of the information provided by our sample. If we knew everything about the population, we would definitely know what decision to make. But seeing only a sample from the population makes this decision harder, and mistakes are inevitable. Just as we measured our uncertainty in Chapter 7, our next task is to measure our mistake rate when testing hypotheses. In this chapter, we continue to work with population proportions. In the next chapter, we'll see how to find confidence intervals and perform hypothesis tests for means.

CASE STUDY

Dodging the Question

Have you ever become frustrated watching a politician in a debate, or maybe in an interview, completely dodge a difficult question? One way a politician might do this is by answering a different, easier question. Researchers have found that we are not always good at noticing when a question is dodged and that our attitudes toward the speaker and our situation help determine whether we are good "dodge detectors." In order to help viewers become better dodge detectors while watching political debates, some TV stations have started posting, at the bottom of the TV screen, the question that was asked. Does this practice really improve people's ability to notice question dodging?

Some researchers filmed hired actors to play politicians having a formal debate. In this video, one of the "politicians" is asked a question about health care—but he dodges the question by instead answering a question about illegal drugs. Test subjects were recruited to watch the video. Half of the subjects, randomly chosen, saw the video with the question about health care posted at the bottom of the screen. The other half saw nothing on the bottom of the screen. For those who saw the question posted, 88% noticed the dodge. For those who did not see the question posted, only 39% noticed that the politician evaded the question (Rogers and Norton 2011).

The difference between these two groups is 49 percentage points. Does this difference occur because we really and truly listen differently when the question is posted on the screen? Or might this difference be due merely to chance? After all, the subjects were randomly assigned either to see the post or not to see the post, so there was an element of chance involved. In other words, is it possible that even if posting the question had no effect at all, we would see a difference as large as 49 percentage points?

In this chapter we'll show you how to make this decision: Is the difference real, or could it be due to chance? At the end of the chapter, we'll return to this case study and see what decision the researchers made and why they made it.

SECTION 8.1

The Essential Ingredients of Hypothesis Testing

Football games and tennis matches begin with a coin toss to determine which team or player gets to play offense first. Coin tosses, in which the coin is flipped high into the air, are used for this because flipping a coin is believed to be fair. "Fair" means the coin is equally likely to land heads or tails, so both sides have an equal chance of winning.

But what if we spin the coin (on a hard, flat surface) rather than flipping it in the air? We claim that, because the "heads" side of a coin bulges out, the lack of symmetry will cause the spinning coin to land on one side more often than on the other. In other words, we believe a spun coin is not fair. Some people—maybe most—will find this claim to be outrageous and will insist it is false.

Suppose for the moment that, as evidence of our claim, we take a coin and spin it 20 times and that we get 20 heads. Reflect on what your reaction would be. We're betting that you'd be surprised. In your personal experience, this sort of thing, 20 heads in 20 spins, is rare when dealing with fair coins. You were expecting about 10 heads, give or take, because you didn't really believe that spinning the coin would matter. The fact that all 20 spins landed heads means that something surprising has happened, and you will probably think that we were right—that spun coins are biased.

If that describes your thought process, then you've just informally done a **hypothesis test**. Hypothesis testing is a procedure that enables us to choose between two claims when we have variability in our measurements. We will teach a particular procedure that we call formal hypothesis testing. We call this procedure formal because it is based on particular terminology and four well-specified steps. However, we hope to show you that this "formal" procedure is not too different from the common sense you just applied when thinking about getting 20 heads in 20 spins.

The formal procedure follows the Data cycle and looks like this:

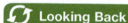 **Looking Back**

The Data Cycle
The Data Cycle, introduced in Chapter 1, includes the four stages of a statistical investigation: asking questions, considering data, analyzing data, and interpreting results.

- In the first step, pose a statistical question, but reformat this question into a pair of hypotheses. These two hypotheses should be competing claims about the world. In this case, our question of whether a spun coin is biased is turned into this pair of hypotheses: the spun coin is fair (our skeptical, neutral claim) and the spun coin is biased.

- The second step in the formal process is a preparation step in which we examine available data or collect new data. We determine how we'll use these data to make our decision and make sure we have enough data to minimize the probability of making mistakes.

- In the third step, we analyze our data by comparing them to our expectations. When you saw that the sample proportion of heads in 20 spins was 1.0 (20 heads out of 20 spins), you mentally compared this to your preconceived notion that the sample proportion was going to be close to 0.5. You might have even computed the probability of getting such an extreme outcome if, in fact, spinning a coin were fair. (The probability of getting either 20 heads or 0 heads for a fair coin is about 0.000002, about 2 in 1 million. That's why you would feel surprised if all 20 spins were heads and you were expecting a 50–50 outcome.)

Details

Pennies
In fact, spinning a coin *does* produce bias but, as far as we know, only for some coins. If you can find a 1962 penny, you might need to spin it about 50 times, but you should see reason to reject the null hypothesis.

- In the fourth and final step of a formal hypothesis test, we interpret our analysis by stating a conclusion: Do we believe the claim, or do we find that the claim doesn't have enough evidence to back it? If we did in fact get 20 heads in 20 spins, we would be willing to conclude that spinning a coin is biased.

The example we have just walked through has an extreme outcome: exactly 100% heads. But what about in-between outcomes? What if we saw 7 out of 20 heads (0.35)? This outcome is less than 0.5, but is it so much less that we would think spinning the coin is biased? Hypothesis testing helps us make decisions for in-between cases such as this.

Before going further, take out a coin. Spin it 20 times, and record the number of heads. We're going to illustrate the basic concepts of hypothesis testing using this "experiment." When *we* did this, we got 7 heads, and we'll use this outcome in the following discussion.

We're first going to show you some important concepts that are the essential ingredients in the recipe for preparing hypothesis tests: hypotheses, minimizing mistakes, a test statistic, and surprise. Then, in Section 8.2, we will show how these ingredients are combined into the four steps of a formal hypothesis test. Section 8.3 covers some details of calculations and some issues that help you better interpret the results of hypothesis tests. The final section shows how our template for hypothesis testing can be expanded to fit the new context of comparing two population proportions.

Main Ingredient: A Pair of Hypotheses

In a formal hypothesis test, statistical questions are converted to hypotheses. Hypotheses are always statements about population parameters. A hypothesis begins as a statement about the real world, but it must then be rephrased in terms of population parameters.

For example, in the coin-spinning example, our question about whether a spun coin is biased is actually a claim about the real world: a spun coin is not fair. We now must restate this in terms of a parameter.

The parameter we are making a claim about is the probability of the coin landing heads. Let's call that probability p. For a fair coin, $p = 0.5$. If the coin is not fair, then p is *not* equal to 0.5. In symbols, $p \neq 0.5$. These are statements about the population parameter p, the probability of getting heads.

The claim that the spun coin is fair is probably not that outlandish to you. Such a hypothesis is called a **null hypothesis**. Our null hypothesis is that $p = 0.50$.

On the other hand, the hypothesis that we hope to convince the world is true is called the **alternative hypothesis**. Ours is $p \neq 0.5$.

Hypotheses will come in pairs like the pair you've just seen:

The **null hypothesis**, which we write H_0 (and pronounce "H-naught" or simply "the null hypothesis"), is the neutral, status quo, skeptical statement about a population parameter. In the context of researching new ideas, the null hypothesis often represents "no change," "no effect," or "no difference." In this text, the null hypothesis will always have an $=$ sign.

The **alternative hypothesis**, H_a (pronounced "H-A"), is the research hypothesis. It is a statement about the value of a parameter that we intend to demonstrate is true.

For our coin-spinning experiment, we write the hypotheses as

$$H_0: p = 0.5$$
$$H_a: p \neq 0.5$$

We emphasize that in a formal hypothesis test, hypotheses are *always* statements about population parameters. In this case, our population consists of infinitely many spins of our coin, and p represents the probability of getting heads.

 Details

Definition of *Hypothesis*
Merriam-Webster's online dictionary defines *hypothesis* as "a tentative claim made in order to draw out and test its logical or empirical consequences."

 Looking Back

Statistic vs. Parameter
In Chapter 7 you learned that $\hat{p}$, a statistic, represents the proportion of successes in a sample, whereas p, a parameter, represents the proportion of successes in the population.

 Caution

Pronunciation
Do not say "Ho" or "Ha," as Santa Claus might.

 KEY POINT   Hypotheses are always statements about population parameters; they are never statements about sample statistics.

EXAMPLE 1 Marriage

Historically, about 70% of all U.S. adults were married. A sociologist who asks whether marriage rates in the United States have declined will take a random sample of U.S. adults and record whether or not they are married.

QUESTIONS

a. State the null and alternative hypotheses in words.

b. Let p represent the proportion of all adults in the United States who are married. State the null and alternative hypotheses in terms of this population parameter.

SOLUTIONS The alternative hypothesis is the claim that the sociologist wishes to make, and the null hypothesis is the skeptical claim that things have not changed.

a. Null hypothesis: The same proportion of adults are married now as in the past. Alternative hypothesis: The proportion of adults who are married now is *less than* it was in the past.

b. Because the historical proportion is $p = 0.70$, we have

$$H_0: p = 0.70$$

$$H_a: p < 0.70$$

TRY THIS! Exercise 8.3

The alternative hypothesis in Example 1 has a "less than" sign rather than the "not equal to" sign that we used in our coin-spinning experiment. This is because the alternative hypothesis must reflect the claim made about the real world. Here, the sociologist wants to establish that the proportion of married adults has *decreased* from its previous value. Therefore, he believes the proportion of married adults is less than the historical value of 0.70. Such hypotheses are called **one-sided hypotheses**. For the coin-spinning experiment, we wanted to show only that the spun coin is *not* fair; we didn't care whether it was biased toward heads or biased toward tails. A hypothesis with a "not equal to" sign is called a **two-sided hypothesis**.

There are three basic pairs of hypotheses, as Table 8.1 shows.

▶ **TABLE 8.1** Three pairs of hypotheses that can be used in a hypothesis test.

Two-Sided	One-Sided (Left)	One-Sided (Right)
$H_0: p = p_0$	$H_0: p = p_0$	$H_0: p = p_0$
$H_a: p \neq p_0$	$H_a: p < p_0$	$H_a: p > p_0$

EXAMPLE 2 Internet Advertising

An Internet retail business is trying to decide whether to pay a search engine company to upgrade its advertising. In the past, 15% of customers who visited the company's web page by clicking on the advertisement bought something (this is called a "click-through"). If the business decides to purchase premium advertising, then the search engine company will make that company's ad more prominent.

The search engine company offers to do an experiment: for one day, a random sample of customers will see the retail business's ad in the more prominent position. The retail business can then decide whether the advertising improves the percentage of click-throughs. The retailer agrees to the experiment, and when it is over, 17% of the customers have bought something.

A marketing executive wrote the following hypotheses:

$$H_0: \hat{p} = 0.15$$

$$H_a: \hat{p} = 0.17$$

where $\hat{p}$ represents the proportion of the sample that bought something from the website.

QUESTION   What is wrong with these hypotheses? Rewrite them so that they are correct.

SOLUTION   First, these hypotheses are written about the sample proportion, $\hat{p}$. We *know* that 17% of the sample bought something, so there is no need to make a hypothesis about it. What we don't know is what proportion of the entire population of people who will click on the advertisement will purchase something. The hypotheses should be written in terms of p, the proportion of the population that will purchase something.

A second problem is with the alternative hypothesis. The research question that the company wants to answer is not whether 17% of customers will purchase something. It wants to know whether the percentage of customers who do so has increased over what has happened in the past.

The correct hypotheses are

$$H_0: p = 0.15$$

$$H_a: p > 0.15$$

where p represents the proportion of all customers who click on the advertisement and purchase a product.

TRY THIS!   Exercise 8.9

Hypothesis tests are like criminal trials. In a criminal trial, two hypotheses are placed before the jury: the defendant either is not guilty or is guilty. These hypotheses are not given equal weight, however. The jury is told to assume the defendant is not guilty until the evidence overwhelmingly suggests this is not so. (Defendants charged with a crime in the United States must be found guilty "beyond all reasonable doubt.")

Hypothesis tests follow the same principle. The statistician plays the role of the prosecuting attorney, who hopes to show that the defendant is guilty. The hypothesis that the statistician or researcher hopes to establish plays the role of the prosecutor's charge of guilt. The null hypothesis is chosen to be a neutral, noncontroversial statement (such as the claim that the defendant is not guilty). Just as in a jury trial, where we ask the jury to believe that the defendant is not guilty unless the evidence against this belief is overwhelming, we will believe that the null hypothesis is true in the beginning. But once we examine the evidence, we may reject this belief if the evidence is overwhelmingly against it.

KEY POINT   The null hypothesis always gets the benefit of the doubt and is assumed to be true throughout the hypothesis-testing procedure. If we decide at the last step that the observed outcome is extremely unusual under this assumption, then *and only then* do we reject the null hypothesis.

The most important step of a formal hypothesis test is choosing the hypotheses. In fact, there are really only two steps of a formal hypothesis test that a computer cannot do, and this is one of those steps. (The other step is checking to make sure that the conditions necessary for the probability calculations to be valid are satisfied. Also, computers can't interpret the findings, as you will be asked to do.)

Add In: Making Mistakes

Mistakes are an inevitable part of the hypothesis-testing process. The trick is not to make them too often.

One mistake we might make is to reject the null hypothesis when it is true. For example, even a fair coin *can* fall heads in 20 out of 20 flips. If that happened, we might conclude that the coin was unfair when it really was fair. We can't prevent this mistake from happening, but we can try to make it happen infrequently.

The **significance level** is the name for a special probability: it is the probability of making the mistake of rejecting the null hypothesis when, in fact, the null hypothesis is true. The significance level is such an important probability that it has its own symbol, the Greek lowercase *alpha*: α.

In our experiment with spinning the coin, the significance level is the probability that we will conclude that spinning a coin is *not* fair when, in fact, it really *is* fair. In a criminal justice setting, the significance level is the probability that we conclude that the suspect is guilty when he is actually innocent.

EXAMPLE 3 Significance Level for Internet Advertising

In Example 2, an Internet retail business gave a pair of hypotheses about p, the proportion of customers who click on an advertisement and then purchase a product from the company. Recall that in the past, the proportion of customers who bought the product was 0.15, and the company hopes this proportion has increased. It intends to test these hypotheses with a significance level of 5%. In other words, $\alpha = 0.05$.

$$H_0\colon p = 0.15$$

$$H_a\colon p > 0.15$$

QUESTION   Describe the significance level in context.

SOLUTION   The significance level is the probability of rejecting H_0 when in fact it is true. In this context, this means that the probability is 5% that the company will conclude that the proportion of customers who will buy its product is bigger than 0.15 when, in fact, it is 0.15.

TRY THIS!   Exercise 8.11

Naturally, we want a procedure with a small significance level because we don't want to make mistakes too often. How small? Most researchers and statisticians use a significance level of 0.05. In some situations it makes sense to allow the significance level to be bigger, and some situations require a smaller significance level. But $\alpha = 0.05$ is a good place to start.

> **KEY POINT**   The significance level, α (Greek lowercase *alpha*), represents the probability of rejecting the null hypothesis when the null hypothesis is true. For many applications, $\alpha = 0.05$ is considered acceptably small, but 0.01 and 0.10 are also sometimes used.

Mix with: The Test Statistic

A **test statistic** compares the real world with the null hypothesis world. It compares our observed outcome with the outcome the null hypothesis says we should see.

For example, in our coin-spinning experiment, we saw 7 out of 20 heads, for a proportion of 0.35 heads. The null hypothesis tells us that we should expect half to be heads: 0.5. The test statistic tells us how far away our observation, 0.35, is from the null hypothesis value, 0.5.

To do this comparison, we use the **one-proportion z-test statistic**.

Formula 8.1: The one-proportion z-test statistic

$$z = \frac{\hat{p} - p_0}{SE}, \text{ where } SE = \sqrt{\frac{p_0(1-p_0)}{n}}$$

The symbol p_0 represents the value of p that the null hypothesis claims is true. For example, for the coin-spinning example, p_0 is 0.5. Most of the other test statistics you will see in this text have the same structure as Formula 8.1:

$$z = \frac{\text{observed value} - \text{null value}}{SE}$$

For our coin-spinning example, because we observed 0.35 heads in 20 spins, the observed value of our test statistic is

$$z_{\text{observed}} = \frac{0.35 - 0.50}{\sqrt{\dfrac{0.50(1 - 0.50)}{20}}} = \frac{-0.15}{\sqrt{0.0125}} = \frac{-0.15}{0.1118} = -1.34$$

> **↻ Looking Back**
>
> **Standard Error**
> The standard error of the sample proportion, given in Chapter 7, is
> $$SE = \sqrt{\frac{p(1 - p)}{n}}$$
> Formula 8.1 uses the symbol p_0 to remind us to use the value that the null hypothesis claims to be correct.

Why Is the z-Statistic Useful? The z-test statistic has the same structure as the z-score introduced in Chapter 3, and it serves the same purpose. By subtracting the value the null hypothesis expects from the observed value, $\hat{p} - p_0$, we learn how far away the actual sample value was from the expected value. A positive value means the outcome was greater than what was expected, and a negative value means it was smaller than what was expected.

If the test statistic value is 0, then the observed value and the expected value are the same. This means we have little reason to doubt the null hypothesis. The null hypothesis tells us that the test statistic should be 0, give or take some amount. If the value is far from 0, then we doubt the null hypothesis.

By dividing this distance by the standard error, we convert the distance into "standard error" units, and we learn how many standard errors away our outcome lies from what was expected. Our spun coin resulted in a z-statistic of -1.34. This tells us that we saw fewer heads than the null hypothesis expected and that our sample proportion was 1.34 standard errors below the null hypothesis proportion of 0.5.

> **↻ Looking Back**
>
> **z-Scores**
> A z-score has the structure
> $$\frac{\text{observed value} - \text{mean}}{\text{standard deviation}}$$

> **KEY POINT**   If the null hypothesis is true, then the z-statistic will be close to 0. Therefore, the farther the z-statistic is from 0, the more the null hypothesis is discredited.

EXAMPLE 4   Test Statistic for Homeownership

It is generally agreed among economists that 2005 was the year in which the greatest percentage of American households owned their own home. This percentage was 69%. This percentage plummeted during the "Great Recession" from 2007 to 2009. More recently, many economists wondered whether the housing market had "recovered" in the sense of whether the percentage who own their own homes had recovered since the Great Recession or was still lower than the historic value of 69%. The Pew Research

Foundation provided some data to help answer this question. Based on a random sample of 2000 households in 2016, they found that 63.5% owned their own homes (Fry and Brown 2016). While the sample statistic of 0.635 is lower than the 2005 population proportion of 0.69, this difference could very well be simply due to the fact that the 2016 value is based on a random sample of only 2000 households. (The U.S. Census says there are 126.2 million households in the United States, so 2000 is a small fraction of the total.) If we let p represent the proportion of households in the population that owned a house in 2016, then our hypotheses are

$$H_0: p = 0.69$$
$$H_a: p < 0.69$$

Or, in words, the null hypothesis says that the proportion of households that own their own home is the same as in 2005, and the alternative hypothesis says it was lower in 2016.

QUESTION   Calculate the observed value of the test statistic, and explain the value in context.

SOLUTION   The observed proportion of households in the sample who owned their home is 0.635. Therefore, the observed value of the test statistic is

$$z_{obs} = \frac{0.635 - 0.69}{\sqrt{\dfrac{0.69(0.31)}{2000}}} = \frac{-0.055}{0.0134166} = -5.32$$

CONCLUSION   The observed value of the test statistic is -5.32. Note that the negative sign indicates that the sample proportion was less than the null hypothesis proportion. This means that while the null hypothesis claims a proportion of 0.69, the observed proportion was much lower—5.32 standard errors lower—than what the null hypothesis claims.

TRY THIS!   Exercise 8.15

The Final Essential Ingredient: Surprise!

No, the final essential ingredient in hypothesis testing is not *a* surprise. The main ingredient is surprise itself.

Surprise happens when something unexpected occurs. The null hypothesis tells us what to expect when we look at our data. If we see something unexpected—that is, if we are surprised—then we should doubt the null hypothesis, and if we are really surprised, we should reject it altogether.

Figure 8.1 shows all possible outcomes of our coin-spinning experiment in terms of the test statistic and shows the "surprising" outcomes in red. If spinning a coin is fair, as the null hypothesis claims, we can compute the probability of every outcome. A fair coin spun 20 times will produce about 10 heads, give or take, which is associated with a test statistic of 0. If the null hypothesis is true, getting 5 or fewer heads OR 15 or more heads, is rare. (In fact, the probability that you will get 0 to 5 OR 15 to 20 heads—the outcomes shown in red in Figure 8.1—is less than 5%.) If we had spun a coin 20 times and saw one of these red outcomes, we would be surprised and would probably reject the null hypothesis that the spun coin was fair.

Because we are statisticians, we have a way of measuring our surprise. The **p-value** is a number that measures our surprise by reporting the probability that if the null hypothesis is true, a test statistic will have a value as extreme as or more extreme than the value we actually observe. Small p-values (closer to 0) mean we have received a surprise. Large p-values (closer to 1) mean no surprise: The outcome happens fairly often. Any of the outcomes in red in Figure 8.1 have a small p-value (all are less than 0.05).

H_0 says rarely happens ------>|<----------- H_0 says happens often ----------->|<------ H_0 says rarely happens

−4.5	−4.0	−3.6	−3.1	−2.7	−2.2	−1.8	−1.3	−0.9	−0.4	0.0	+0.5	+0.9	+1.3	+1.8	+2.2	+2.7	+3.1	+3.6	+4.0	+4.5
0	1	2	3	4	5	6	7	8	9	10	11	12	13	14	15	16	17	18	19	20

▲ **FIGURE 8.1** All possible outcomes for spinning a coin 20 times in terms of the test statistic and the number of heads. The red values are those that might be considered unusual and unexpected if the null hypothesis is true. If the coin is truly fair, "red" outcomes will happen less than 5% of the time. The "black" outcomes will happen a little more than 95% of the time. The lowest row shows the outcomes in terms of the number of heads. The row above it shows the outcomes in terms of the z-statistic.

EXAMPLE 5 Judging p-Values

Suppose you spin a coin 20 times and count the number of heads. The null hypothesis is that the coin is fair. The alternative hypothesis is that the coin is not fair.

QUESTION Which of these outcomes, 3 heads or 9 heads, has the smaller p-value? Why? (You might refer to Figure 8.1.)

SOLUTION If the null hypothesis is right, then we should get about half heads. This means we expect 10 heads, give or take. Because 9 heads is much closer to 10 heads than 3 heads is to 10, 3 is a more surprising outcome and will have a smaller p-value. We can see, by referring to Figure 8.1, that the test statistic value for 3 heads is −3.1 and that this is a more extreme outcome with a smaller p-value than 9 heads (with $z = -0.4$).

TRY THIS! Exercise 8.17

KEY POINT The p-value is a probability. Assuming that the null hypothesis is true, the p-value is the probability that if the experiment were repeated, you would get a test statistic as extreme as or more extreme than the one you actually got. A small p-value suggests that a surprising outcome has occurred and discredits the null hypothesis.

EXAMPLE 6 Fewer Homeowners

In Example 4 we were interested in whether the proportion of households who owned their own homes had risen back up to 0.69, the level before the Great Recession, or had remained below that level. A hypothesis test found that the observed value of the test statistic was −5.32. We can calculate that the p-value associated with this is very close to 0. (In the next section, you'll learn how to calculate the p-value.)

QUESTION Explain the meaning of the p-value in this context. If we believed that the null hypothesis was true, should we be surprised? (The null hypothesis was that $p = 0.69$, where p was the proportion of households that owned their own home.)

SOLUTION The p-value is very small (nearly 0). This tells us that if it is true that 69% of the population owns its own home, then getting a test statistic as extreme as or more extreme than −5.32 is highly unlikely. If you believed that the null hypothesis was true and $p = 0.69$, then you should be *very* surprised, because what you saw is nearly impossible.

TRY THIS! Exercise 8.19.

Hypothesis Testing and the Data Cycle: Asking Questions

It is with hypothesis testing that the Data Cycle really starts spinning! The alternative hypothesis is actually a very formalized type of statistical question. For example, a researcher might begin her research with the question, "I wonder if students who take notes by hand remember more of the lecture than those who take notes on a laptop?" And, after much thinking and preparation, fine-tune the question to a pair of hypotheses:

H_0: The mean score on a memory test about a lecture is the same for students who were required to take notes on laptops and for students required to take notes by hand.

H_a: The mean score was higher for students who took notes by hand than for students who took notes on a laptop.

In the next section you'll see that the Consider Data step is extremely important. Unless particular conditions were met when the data were collected, it might not be possible to analyze the data in a way that will allow us to answer the questions we wish to answer. If you are collecting the data yourself, then you need to collect the data carefully so that the conditions are met.

Hypothesis Testing in Four Steps

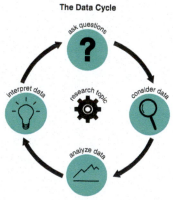

The Data Cycle

ask questions

research topic

consider data

interpret data

analyze data

▲ **FIGURE 8.2** The Data Cycle guides us to Hypothesize, Prepare, Compare, and Interpret.

Now that you know the main ingredients (hypotheses, minimizing the chance of mistakes, test statistic, and surprise), it's time to learn the recipe. Here, as with nearly every statistical investigation, our recipe is provided by the Data Cycle. (See Figure 8.2 for a reminder.) The Data Cycle can be translated into four steps that combine these ingredients into a useful, logical structure.

Step 1: Hypothesize.

In the Ask Questions phase, we turn our question into a pair of hypotheses about the population parameter.

Step 2: Prepare.

Choose a test statistic. Consider the data to check that they meet required conditions; state any assumptions about the data you must make if you cannot check the conditions.

Step 3: Compute to Compare.

To begin the Analyze Data phase, state a significance level and compute the observed value of the test statistic. The test statistic compares our observed value with the null hypothesis value. Find the p-value in order to measure your level of surprise.

Step 4: Interpret.

Answer the statistical question by deciding whether to reject or not reject the null hypothesis. What does this mean in the context of the data?

A Few Details

Step 1 we have covered in detail, but the other steps still require some explanation. After we've explained these steps, we'll show some examples of how these steps work together to carry out a test for a population proportion.

Detail for Step 2: Check Conditions to Find Probabilities This is the step where we Consider Data. To carry out and interpret our analysis, we need to know whether our test statistic is an unusual value when the null hypothesis is true. To find the p-value and significance level, which are both probabilities, we need to know the probability distribution of this test statistic. And to do this, we need to examine the conditions under which the data were collected.

If the conditions below are met *and* if the null hypothesis is true, then the sampling distribution for the one-proportion test statistic is, approximately, the standard Normal distribution. The conditions are the following:

1. *Random Sample.* The sample is collected randomly from the population.

2. *Large Sample Size.* The sample size, n, is large enough that the sample has 10 or more expected successes and 10 or more expected failures; in other words, $np_0 \geq 10$ and $n(1 - p_0) \geq 10$.

3. *Large Population.* If the sample is collected without replacement, then the population size is at least 10 times bigger than the sample size. (If the sample is drawn with replacement, then any size population will work.)

4. *Independence.* Each observation or measurement must have no influence on any others.

> **KEY POINT** Under the appropriate conditions, the sampling distribution of the z-statistic is approximately a standard Normal distribution, $N(0, 1)$.

It is very important that in step 2 ("Prepare") we make sure these conditions hold or that we can reasonably assume that they do. If not, then we can't find the p-value. If we can't find the p-value, there is no reason to proceed with the other steps.

EXAMPLE 7 Checking Conditions for Homeownership Test

The observed value of the test statistic for the hypothesis test of Example 4 was -5.32. If we know that the sampling distribution for the test statistic is a standard Normal distribution, $N(0,1)$, then we will know that this is an extremely rare outcome. The poll was based on a simple random sample of 2000 households in the United States, and the sample proportion of households that owned their home was 0.635. We were told that there were approximately 126.2 million households in the United States. The hypotheses were

$$H_0\text{: } p = 0.69$$
$$H_a\text{: } p < 0.69$$

QUESTION Check the conditions to show that the test statistic for this test approximately follows a standard Normal distribution.

SOLUTION We check each of the four conditions to see whether they are satisfied.

1. *Random sample.* We were told that the households were randomly selected from the population.
2. *Large Sample Size.* Because $p_0 = 0.69$, we expect $2000 \times 0.69 = 1380$ successes, which is bigger than 10. The remaining $1000 - 1380 = 620$ households must be "failures," and so we know we expect more than 10 failures. The sample size is large enough.

Looking Back

The Sampling Distribution
Recall from Chapter 7 that the probability distribution for a statistic is called the sampling distribution.

3. *Large Population.* The total number of households is easily more than 10 times the sample size (which is 20,000).

4. *Independence.* We assume the pollsters surveyed households in such a way that their responses were independent of each other.

CONCLUSION   The conditions are verified. We can use the Standard Normal distribution, $N(0, 10)$, to find probabilities for the test statistic.

TRY THIS!   Exercise 8.23

Detail for Step 3: Calculating the p-Value

After finding the observed value of the test statistic in step 3, we next must determine whether this value "surprises" the null hypothesis. The p-value measures this surprise. But to find the p-value, you need to know what the phrase *as extreme as or more extreme than* means when we say, "The p-value is the probability that, if the null hypothesis were true, we would get a statistic as extreme as or more extreme than the observed test statistic."

Think of our coin-spinning example. We expected a proportion of 0.50 heads, but we saw 0.35 heads. Our observed value of the test statistic was therefore found to be $z = -1.34$. We wish to find the p-value to answer this question: If we were to do this again, and if coin-spinning is really and truly fair, what's the probability that we would get a test statistic value *as extreme as or more extreme than* -1.34?

The meaning of this phrase depends on which of our three alternative hypotheses we're using.

If the alternative hypothesis is two-sided,

$$H_a: p \neq p_0 \text{ (The true value of } p \text{ is either bigger or smaller than}$$
what the null hypothesis claims.)

then *as extreme as or more extreme than* means "even farther away from 0 than the value you observed." This corresponds to finding the probability in both tails of the $N(0, 1)$ distribution. This is called a **two-tailed p-value**.

Our coin-spinning example will have a two-tailed p-value because the alternative is two-sided ($p \neq 0.5$).

There are two different types of one-sided hypotheses:

If the alternative hypothesis is

$$H_a: p < p_0 \text{ (The true value is less than the value claimed}$$
by the null hypothesis.)

then *as extreme as or more extreme than* means "less than or equal to the observed value." This corresponds to finding the probability in the left tail of $N(0, 1)$ and so is called a left-tailed p-value. Example 1, a test to see whether the marriage rate had decreased, uses a left-tailed p-value because the alternative hypothesis is $H_a: p < 0.70$.

Finally, if the alternative hypothesis is

$$H_a: p > p_0 \text{ (The true value is greater than the value}$$
claimed by the null hypothesis.)

then *as extreme as or more extreme than* means "greater than or equal to the observed value." This corresponds to finding the probability in the right tail of $N(0, 1)$. This p-value is called a right-tailed p-value. Example 2 ($H_a: p > 0.15$) and Example 7 ($H_a: p > 0.10$) both use right-tailed p-values.

Once we've determined which extremities to use, we can use Table 2 (Normal table) in Appendix A, a statistical calculator, or other technology. These three cases are illustrated in Figure 8.3, which uses an observed test statistic value of $z = 1.56$.

Details

Tails

Many statisticians use the terms *one-tailed hypotheses* and *one-sided hypothesis* interchangeably and likewise *two-sided* and *two-tailed*

(a)

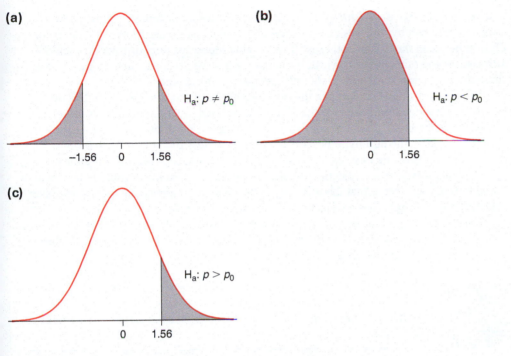

(b)

(c)

◄ **FIGURE 8.3** The shaded areas represent p-values for three different alternative hypotheses when the observed value of the test statistic is $z = 1.56$. **(a)** The p-value for a two-sided alternative hypothesis (0.119). **(b)** The p-value for a left-sided hypothesis (0.941). **(c)** The p-value for a right-sided hypothesis (0.059).

EXAMPLE 8 p-Value for Coin Spinning

We claimed that spinning a coin leads to a biased outcome. Our hypotheses were

$$H_0: p = 0.5$$

$$H_a: p \neq 0.5$$

We spun the coin 20 times and saw a sample proportion of $\hat{p} = 0.35$, which led to an observed test statistic of $z_{observed} = -1.34$. Assume the required conditions are satisfied so that the sampling distribution of the test statistic is approximately the standard Normal distribution.

QUESTION   Find the approximate p-value, and explain what it means.

SOLUTION   The alternative hypothesis is two-sided, so we will find a two-tailed p-value. The situation is similar to Figure 8.3a but with $z = -1.34$. We can use a statistical calculator, such as the Normal Calculator in StatCrunch or the TI-84, or Table 2 in Appendix A. Figure 8.4 shows the output from a TI-84.
   The p-value is 0.1797.

CONCLUSION   The p-value is about 0.18. This tells us that if spinning a coin is fair, then the probability of seeing an outcome as extreme as or more extreme than 0.35 heads is about 18%.

NOW TRY!   Exercise 8.25

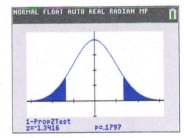

▲ **FIGURE 8.4** TI-84 output showing the probability that a test statistic will be as extreme or more extreme than −1.34 if the null hypothesis is true.

Compare the approximate p-value of 0.18 that we calculated in Example 8 with Figure 8.1. There we see that the test statistic value of −1.34 falls into the "black" region, which indicates an outcome that is not surprising if the coin is fair. Indeed, our p-value of 18% is relatively large: for fair coins, outcomes like this happen about 1 time in 5.

Detail for Step 4: Making a Decision   In the final step, you must make a decision between hypotheses and explain what the decision means. How, then, do we choose between the two hypotheses? The p-value measures our surprise (or lack of surprise) at the outcome of our test, but what should we do about this number? When is the outcome so unusual that we should reject the null hypothesis?

We apply a simple rule: reject the null hypothesis if the p-value is smaller than (or equal to) the value chosen for the significance level, α. If the p-value is larger than the significance level, do not reject the null hypothesis. For most applications, this means you reject the null hypothesis if the p-value is less than or equal to 0.05.

Following this rule ensures that our significance level is achieved. In other words, by following this rule and rejecting the null hypothesis when the p-value is less than or equal to 0.05, we ensure that there is only a 5% chance (at most) that we are mistakenly rejecting the null hypothesis (rejecting H_0 even though H_0 is true).

For our coin-spinning experiment, we found a p-value of about 0.18. If we wished to achieve a significance level of 0.05, then we would *not* reject the hypothesis that spinning a coin is fair, because 0.18 is greater than 0.05.

> **KEY POINT**   To achieve a significance level of α, reject the null hypothesis if the p-value is less than (or equal to) α. If the p-value is greater than α, do not reject the null hypothesis.

The Four-Step Approach

With those details taken care of, we're now ready to perform a hypothesis test. We'll work completely through one example to show you how the steps fit together.

In one Florida election, 47% of all registered voters voted. A researcher studying the behavior of academics asked this question: do political scientists turn out to vote in the same proportion as the general population? To answer this question, the researcher took a random sample, without replacement, of 54 political scientists who live in Florida and interviewed them to determine whether they voted in this election. As it turned out, 40 of the 54 political scientists in the sample voted, which is a sample proportion of 0.74 (Schwitzgebel and Rust 2010).

We wish to carry out a hypothesis test to determine whether the proportion of all political scientists who voted in this election differed from the proportion of the general public.

Step 1: Hypothesize   The null and alternative hypotheses are stated both in words and in symbols.

> H_0: Political scientists vote in the same proportion as the public, 0.47.
>
> H_a: Political scientists do not vote in the same proportion as the public.

$$H_0\colon p = 0.47$$
$$H_a\colon p \neq 0.47$$

The parameter p represents the proportion of all political scientists in Florida who voted in this election.

Step 2: Prepare   Choosing the correct test statistic is not a big deal at this point, because you have seen only one test statistic to choose from: the one-proportion z-test statistic. In Section 8.4 you'll study a new test statistic for comparing two population proportions. Later still, you will see test statistics for comparing means.

Details

Other Test Statistics
In Chapter 9 you will learn about the t-test statistic, used for testing means, and in Chapter 10 you will learn about the chi-square test statistic.

Next we must examine the data to make sure that conditions are met so that the distribution of the test statistic is approximately Normal. To do this, we must check the four conditions:

Random Sample. We are told that the data come from a random sample of 54 political scientists, and this satisfies the first condition.

Large Sample. We must next check that the sample size of 54 is large enough to produce at least 10 successes and 10 failures.

If the null hypothesis is true, the probability of success is $p_0 = 0.47$. Because $n = 54$,

$$np_0 = 54 \times 0.47 = 25.38, \text{ which is more than 10, and}$$

$$n(1 - p_0) = 54 \times (1 - 0.47) = 28.62, \text{ which is also more than 10.}$$

Large Population. The third condition is true if the population—all political scientists registered to vote in Florida—is more than 10 times bigger than the sample size; that is, if the population size is greater than $10 \times 54 = 540$. We confess that we do not know the number of political scientists in Florida. We cannot check this condition, but we will assume that it is true. (Remember, we need only check this if the sampling was done without replacement.)

Independence. The final condition to check is that the observations are independent. We assume that, because the researcher used a random sample of political scientists, their responses were independent of one another.

Because the conditions are verified, if the null hypothesis (that the proportion of all political scientists who voted is 0.47) is true, then the sampling distribution of the one-proportion z-test statistic is $N(0, 1)$.

Step 3: Compute to Compare

Before proceeding, we choose a significance level. We will use the standard value of $\alpha = 0.05$. This means that if it is true that political scientists voted at the same rate as the rest of the population of Florida, there is a 5% chance that we will mistakenly conclude that they did *not* vote at the same rate.

At this point, we are ready to turn to technology to complete step 3. For instance, Figure 8.5 shows the input that StatCrunch requires to carry out the hypothesis test. First we must enter the number of observed successes and the sample size ("# of observations"). Then we must enter the hypotheses by choosing the correct value for the null hypothesis (0.47) and the correct form for the alternative hypothesis (two-sided). Other statistical software packages are very similar.

The software will now calculate the observed value of the test statistic and the p-value. Before we look at the output, though, let's go over the calculations ourselves.

Find the observed value of the test statistic. Remember that our test statistic will compare the value of the statistic provided by the data, $\hat{p}$, with the value that the null hypothesis says we should see, p_0.

The researcher reports that in his sample of 54 political scientists, 40 of them voted. The sample proportion therefore is $\hat{p} = \dfrac{40}{54} = 0.7407$ (after rounding.) How far away, in terms of standard errors, is 0.7407 from 0.47? To answer, we find the standard error:

$$SE = \sqrt{\frac{p_0(1 - p_0)}{n}} = \sqrt{\frac{0.47 \times (1 - 0.47)}{54}} = 0.0679188$$

The observed value of our test statistic is

$$z_{observed} = \frac{\hat{p} - p_0}{SE} = \frac{0.740741 - 0.47}{0.0679188} = 3.99$$

▶ **FIGURE 8.5** The procedure for carrying out a hypothesis test in StatCrunch is similar to that in many statistical software packages. You must enter the data and the correct hypotheses. (StatCrunch will also compute the number of successes directly from the raw data if provided.)

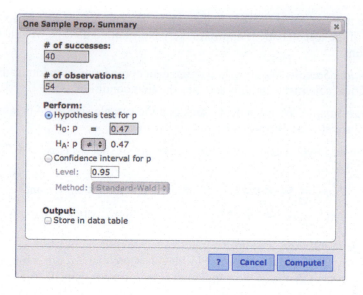

We see that the observed proportion is just less than 4 standard errors above the null hypothesis claim.

Find the p-value. Because the alternative hypothesis is two-sided ($p \neq 0.47$), we will find a two-tailed p-value. Using Table 2 in Appendix A or a statistical calculator, we find that the p-value is about 0.00006. This calculation is illustrated in Figure 8.6.

▶ **FIGURE 8.6** The p-value as a shaded area. This value is from a two-tailed test for which *z* is 3.99. The area has been enlarged a bit so that it can be seen readily.

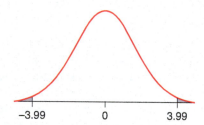

Our calculations confirm the StatCrunch output, shown in Figure 8.7.

▶ **FIGURE 8.7** The StatCrunch output gives us the test statistic value of 3.986 and tells us that the p-value is smaller than 0.0001.

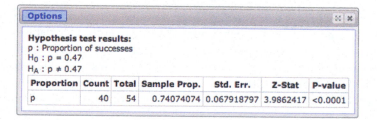

Hypothesis test results:
p : Proportion of successes
H_0 : p = 0.47
H_A : p ≠ 0.47

Proportion	Count	Total	Sample Prop.	Std. Err.	Z-Stat	P-value
p	40	54	0.74074074	0.067918797	3.9862417	<0.0001

Note that StatCrunch follows a common convention for very small p-values. Rather than stating the precise value, it simply states that the value is very small. StatCrunch does this for p-values less than 0.0001. However, to be consistent with other technologies, in this text we'll do it whenever the p-value is less than 0.001.

Step 4: Interpret Because the p-value is less than our stated significance of 0.05, we reject the null hypothesis. We conclude that political scientists did *not* vote in the same proportion as the rest of the population.

Often, technology is used to carry out a hypothesis test. Examples 9 and 10 illustrate how this is done using the statistical software package StatCrunch, but other packages are similar.

EXAMPLE 9 Men's Health

Health professionals are often concerned about our lifestyles and how they affect our well-being. A group of medical researchers knew from previous studies that in the past, about 39% of all men between the ages of 45 and 59 were regularly active. Because regular activity is good for our health, researchers were concerned that this percentage had declined over time. For this reason, they selected a random sample, without replacement, of 1927 men in this age group and interviewed them. Out of this sample, 680 said that they were regularly active (Elwood et al. 2013).

QUESTION Did the proportion of regularly active men decline? To begin to address this statistical question, carry out the first two steps of a hypothesis test. Figure 8.8 shows the input required by a statistical software package. Explain how you would fill in the required entries shown in Figure 8.8. Use a significance level of 5%.

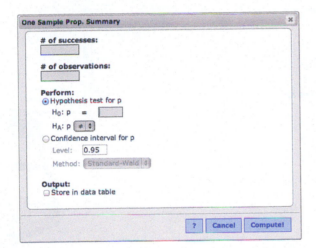

◀ **FIGURE 8.8** Most statistical software packages require input such as this in order to perform a hypothesis test based on summary statistics.

SOLUTION We will let p represent the proportion of all men in this age group who would say that they are regularly active.

Step 1: Hypothesize
In the past, the population proportion, p, was 0.39. The researchers wish to know whether this proportion has *decreased*, which means we have a left-sided alternative hypothesis. Thus,

$$H_0: p = 0.39$$
$$H_a: p < 0.39$$

Step 2: Prepare

We will do a one-proportion z-test.

We now check the conditions:

Random Sample. We are told the sample is random.

Large Sample. We have $p_0 = 0.39$. Thus, we expect $1927 \times 0.39 = 752$ successes, which is bigger than 10. Because $(1 - 0.39)$ times 1927 will lead to an even larger number of expected failures, the number of expected failures is also bigger than 10. And so the sample size is large enough.

Large Population. If sampling was done without replacement, we need a large population. The population of all men in this age group is certainly larger than 10×1927.

Independence. As long as the sample was random and the men were interviewed independently, this condition is satisfied.

The required input to compute the test statistic and p-value using StatCrunch are shown in Figure 8.9. Similar inputs are required by other statistical software packages.

► **FIGURE 8.9** Required input, using StatCrunch, to test the hypothesis that the proportion of regularly active men in this age group has declined.

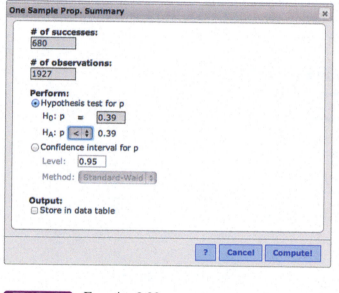

TRY THIS! Exercise 8.29

EXAMPLE **10** Men's Health, Continued

Using the output provided in Figure 8.10, carry out steps 3 and 4 of a hypothesis test to test whether the proportion of men aged 45 to 59 who say they are regularly active has declined from 0.39. Use a significance level of 5%.

► **FIGURE 8.10** StatCrunch output to test whether the proportion of regularly active men has declined from historical levels.

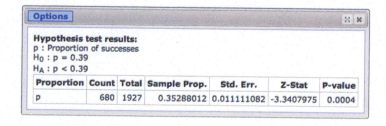

Hypothesis test results:
p : Proportion of successes
$H_0 : p = 0.39$
$H_A : p < 0.39$

Proportion	Count	Total	Sample Prop.	Std. Err.	Z-Stat	P-value
p	680	1927	0.35288012	0.011111082	-3.3407975	0.0004

Step 3: Compute to compare
We first note that the significance level was given to us: $\alpha = 0.05$. The observed value of the test statistic is -3.34, which tells us that our observed sample proportion was 3.34 standard errors below the value of 0.39. The p-value is 0.0004, which is quite small.

Step 4: Interpret
Because the p-value is less than 0.05, we reject the null hypothesis. We conclude that the proportion of all men in this age group who are regularly active was smaller in 2009 than in 1979.

TRY THIS!   Exercise 8.31

> **Details**
>
> **Small p-Values**
> When small p-values, such as 0.0001, occur, many software packages round off and report the p-value as "$p < 0.001$" (or use some other small value). We will follow that practice in this book.

SNAPSHOT ▸ One-Proportion *z*-Test

WHAT IS IT?	▸	A procedure for choosing between two hypotheses about the true value of a single population proportion. The test statistic is $$z = \frac{\hat{p} - p_0}{SE}, \text{ where } SE = \sqrt{\frac{p_0(1 - p_0)}{n}}$$
WHAT DOES IT DO?	▸	Because estimates of population parameters are uncertain, a hypothesis test gives us a way of making a decision while knowing the probability that we will incorrectly reject the null hypothesis.
HOW DOES IT DO IT?	▸	The test statistic z compares the sample proportion to the hypothesized population proportion. Large values of the test statistic tend to discredit the null hypothesis.
HOW IS IT USED?	▸	When proposing hypotheses about a single population proportion. The data must be from an independent, random sample, and the sample size must be sufficiently large.

SECTION 8.3

Hypothesis Tests in Detail

In this section, we cover a variety of concepts that are important in correctly using and interpreting hypothesis tests.

Xtreme Stats!

For many people, it seems a little odd that a small p-value leads to such a major action as rejecting the null hypothesis. But it is important to realize that when we see a small p-value, it means our test statistic is extreme. And an extreme test statistic means something unusual, and therefore unexpected, has happened.

Figure 8.11 illustrates how the p-value depends on the observed outcome of our coin-spinning study. Each graph represents the p-value for a different outcome, with the coin spun 20 times in each case. The null hypothesis in all cases is $p = 0.5$, and the alternative is the two-sided hypothesis that the probability of heads is not 0.5.

> **Caution**
>
> **So Many *p*'s!**
> p is the population proportion. p_0 is the value of the population proportion according to the null hypothesis. $\hat{p}$ is the sample proportion. The p-value is the probability that if the null hypothesis is true, our test statistic will be as extreme as or more extreme than the value we actually observed.

Note that the closer the number of heads is to 10, the closer the z-value is to 0 and the larger the p-value is. Also note that the p-value for an outcome of 11 heads is the same as for 9 heads, and the p-value for an outcome of 12 heads is the same as for 8 heads. This happens because the alternative hypothesis is two-sided and the Normal distribution is symmetric.

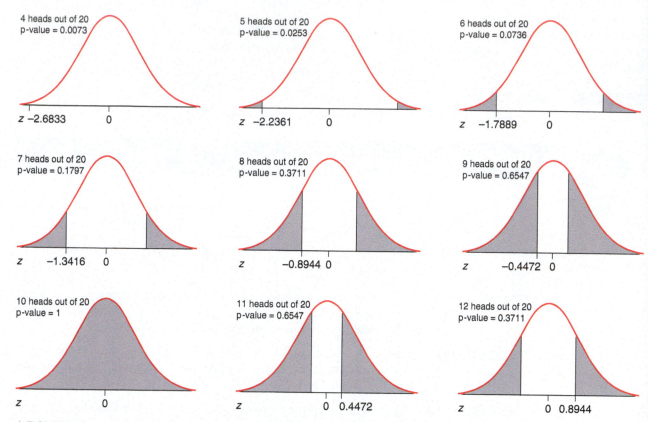

▲ **FIGURE 8.11** Each graph shows the p-value (shaded) for a different number of heads out of 20 spins of a coin, with the assumption that the coin is fair and using a two-sided alternative hypothesis. Note that the closer the number of heads is to 10 (out of 20), the closer the observed test statistic is to 0 and the larger the p-value is. Also, as the number of heads gets farther from 10, the observed test statistic gets farther from 0 and the p-value gets smaller.

EXAMPLE 11 p-Values for Coin Spinning

Two different students each did the coin-spinning experiment with a two-sided alternative hypothesis. Both students spun a coin 20 times. Their test statistics follow.

$$\text{Study 1: } z = 1.98$$

$$\text{Study 2: } z = -2.02$$

QUESTION Which of these test statistics has the smaller p-value, and why?

SOLUTION If the null hypothesis is correct, then the test statistic should be close to 0. Values farther from 0 are more surprising and so have smaller p-values. Because -2.02 is farther from 0 than is 1.98, the area under the Normal curve in the tails is smaller for -2.02 than for 1.98. Thus, -2.02 has the smaller p-value.

TRY THIS! Exercise 8.43

If Conditions Fail

If the conditions concerning the sampling distribution of the z-statistic fail to be met, then we cannot find the p-value using the Normal curve (using Table 2 in Appendix A or technology). However, other approaches often exist. (You'll learn about some of these approaches in Chapters 10 and 13.)

Sample Size Is Too Small　The Normal distribution is only an approximation to the true distribution of the z-statistic. If the sample size is large enough, then the approximation is very good. If the sample size is too small, then the approximation may not be good, but other tests can be used. (One such test that you will see in Section 10.4 is called Fisher's Exact Test.)

Samples Are Not Randomly Selected　If samples are not selected randomly, then it is not possible to make inferences about the populations the samples came from. That having been said, random samples are relatively rare in medical studies. For example, medical researchers cannot take a random sample of people with the medical condition they wish to study; they must rely on recruiting patients who come into hospitals. Psychologists at universities often study students, particularly students who are willing to submit to a study for a small amount of money or the chance to win a prize in a raffle. They cannot take a random sample of all people in the population they wish to study and fly them to the university to participate in their experiment.

These convenience samples are problematic for the statistical techniques in this text—and indeed for any statistical technique. Sometimes we get around this by assuming that the samples are random or, at the very least, representative of the population. However, we have no guarantee that the conclusions we make with this assumption are valid or useful. For instance, in the United States, there is growing recognition that women are not well represented in medical research studies, and this imbalance might negatively impact women's medical treatment, since evidence suggests that men and women experience many diseases differently (Kim, et al. 2010).

In many situations, however, researchers are not that interested in generalizing to the larger population. For example, in randomized controlled experiments, the emphasis is usually on understanding whether the treatment—maybe a new sleeping pill—works for anyone at all. A random sample of insomniacs is not available, but by randomizing the patients on hand to the treatment or a placebo, researchers can tell whether the pill is effective for *this* group. Research with other groups is still needed to see whether the results obtained are replicable, but the study is at least an encouraging start because it can inform researchers that the therapy works with some patients. You'll see an example of this reasoning in Section 8.4.

> **! Caution**
>
> **Cause and Effect**
> In Chapter 1 you learned that we can conclude that there is a cause-and-effect relationship between a treatment variable and a response variable only when we have a controlled experiment that uses random assignment, includes a placebo (or comparison) treatment, and is double-blind.

Balancing Two Types of Mistakes

One of the main ingredients in hypothesis testing is the probability of making a certain type of mistake. This type of mistake occurs when we reject the null hypothesis even though it is true. The probability of making this mistake is called the significance level. In step 2 of our hypothesis test, we deliberately set this probability to a small value, typically 5%.

If it is so important to have a small probability of making this mistake, why don't we choose an even smaller probability? Why not set it to 0?

The reason is that there is a trade-off. As the probability of mistakenly rejecting the null hypothesis is made smaller, the probability of making another type of mistake gets bigger. This other mistake is to *fail* to reject the null hypothesis even though it is wrong. For instance, we might conclude that there is no reason to think spinning a coin is biased, when, in fact, it *is* biased. Mistakes like this can be costly because we might fail to make an important discovery. For instance, medical researchers might fail to recognize that a new medical procedure is effective, and as a result, many people will not have this potential cure available to them.

To understand this trade-off, think about the criminal justice system. The null hypothesis, as the jury is told to believe, is that the defendant is innocent. The first type of mistake occurs when we convict an innocent person (mistakenly reject the null hypothesis). The probability of making this mistake is what we call the significance level. The second type of mistake occurs when we free a guilty person (fail to reject the null hypothesis even though it is false).

We can make the significance level (the probability of convicting an innocent man) 0 by following a simple rule: free every defendant. If everyone goes free, then it is impossible to convict an innocent person because we will convict no one. But now the probability of freeing a guilty person is 100%, since every guilty person will be set free.

Of course, we could lower the probability of freeing guilty people to 0% by simply convicting everyone. But now the significance level has gone to 100% as well, because every innocent person will be convicted.

There is only one way out if we want to lower the probability of *both* types of mistakes:

Increase the sample size. Increasing the sample size improves the precision of the test, and so we make mistakes less often.

> **KEY POINT**
> We cannot make the significance level arbitrarily small because doing so increases the probability that we will mistakenly fail to reject the null hypothesis.

Table 8.2 shows the two types of mistakes.

► **TABLE 8.2** The two types of mistakes. If the null hypothesis is true, we might reject it. If the null is false, we might fail to reject it.

Truth	Reject H$_0$	Fail to Reject H$_0$
H$_0$ True	Bad (The probability of doing this is called the significance level.)	Good
H$_0$ False	Good	Bad

EXAMPLE 12 Describing Mistakes

In Section 8.2, we considered whether political scientists vote in the same proportions as the public. Our hypotheses were

$$H_0: p = 0.47$$
$$H_a: p \neq 0.47$$

where p is the proportion of all political scientists in Florida who voted.

QUESTION Describe the two types of errors we might make in conducting this hypothesis test. Your descriptions should be in the context of this problem. Explain what it means to set the significance level to 5%.

SOLUTION The first type of mistake is to reject the null hypothesis when it is true. In the present context, this means concluding that political scientists vote in different proportions than the public even though they actually have the same voter turnout as the public. The second type of mistake is to fail to reject the null hypothesis when it is false. In the present context, this means concluding that there is no difference between political scientist voter turnout and the general turnout, even though there really is. The 5% significance level means that there is only a 5% chance that we will mistakenly conclude that the political scientists are different from the public when, in fact, they are not.

TRY THIS! Exercise 8.45

So What? Statistical Significance versus Practical Significance

Researchers call a result "statistically significant" when they reject the null hypothesis. This means that the difference between their data-estimated value for a parameter and the null hypothesis value for the parameter is so large that it cannot be convincingly explained by chance. However, just because a difference is statistically significant does not mean it is useful or meaningful.

A *practically* significant result is both statistically significant and meaningful. For example, suppose that the proportion of people who get a certain type of cancer is 1 in 10 million. However, a statistical analysis finds that those who talk on their cell phones every day have a statistically significant greater risk of getting that cancer and that the risk is doubled. It may be true that using your cell phone is therefore more dangerous than not using it, but would you really stop talking on the phone if your risk would change from 2 in 10 million to 1 in 10 million? That's a pretty big change of habit for a pretty small change in risk. Most people would conclude that the difference in risk is statistically significant but not practically significant.

 KEY POINT   Statistically significant findings do not necessarily mean that the results are useful.

Don't Change Hypotheses!

A researcher sets up a study to see whether caffeine affects our ability to concentrate. He has a large number of subjects, and he gives them a task to complete when they have not had any caffeine. The task takes some concentration to complete, and he records how long it takes them. Later, he asks them to complete the same task, only this time the subjects have had a dose of caffeine. Again he records the time, and he's interested in the proportion who take longer to complete the task with caffeine than without.

He isn't sure just what the effect of caffeine will be. It might help people concentrate, in which case only a small proportion of people will take longer. On the other hand, it might make people jittery so that a large proportion will take longer to complete the task. If caffeine has no effect, probably half will take the same amount of time or more, and half will take the same amount of time or less.

The researcher chooses a significance level of $\alpha = 0.05$ to test this pair of hypotheses:

$$H_0: p = 0.50$$

$$H_a: p \neq 0.50$$

The parameter p represents the proportion of all people who would take longer to complete the task with caffeine than without. His alternative hypothesis is two-sided because he does not know what the effect will be—that is, whether caffeine will increase or decrease concentration.

He collects his data and gets a z-statistic of -1.81. This leads to a p-value of 0.07—and to a moral dilemma! (Figure 8.12 illustrates this p-value.) The researcher needs a p-value less than or equal to 0.05 if he is to publish this paper, because no one wants to hear about an insignificant result.

However, it occurs to this researcher that if he had a different alternative hypothesis, his p-value would be different. Specifically, if he had used

$$H_a: p < 0.50$$

▶ **FIGURE 8.12** The shaded areas represent the p-value of 0.070 for a test statistic of $z = -1.81$ in a two-sided hypothesis test.

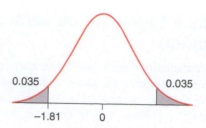

then the p-value would have been the area in just the lower tail. In that case, his p-value would be 0.035, and he would reject the null hypothesis.

What the researcher has thought of doing is the sort of thing that small children (and politicians) do in a contest: they change the rules midway through so that they can win. As they say in the western movies, "you gotta dance with the gal/guy that brung ya." You can't choose your hypotheses to fit the data. By doing so, you are increasing the true significance level of your test above the 5% "advertised" significance level.

Hypothesis-Testing Logic

Statisticians and scientists are rather touchy when it comes to talk about "proving" things. They often use softer words, such as "Our data demonstrate that . . ." or "Our data are consistent with the theory that . . ." One reason is that in mathematical and scientific circles, the word *prove* has a very precise and very definite meaning.

If something is proved, then it is absolutely, positively, and without any doubt true. However, in real life, and particularly in statistics and science (which we consider to be part of real life), you can never be completely certain. In fact, as you've seen, mistakes are built into the hypothesis-testing procedure. We design the procedure so that mistakes are rare, but we know that they happen. For this reason, we avoid saying, for example, that we have *proved* that drinking caffeine changes the proportion of people who take longer to complete a particular task.

On a similar note, it is improper (maybe even impolite!) to say that you have "accepted" the null hypothesis when your p-value is bigger than 0.05. Instead, we say, "We have failed to reject H_0," or "We cannot reject H_0." This is because several factors might make it difficult to determine whether the null hypothesis is false.

It could be that our sample size was too small for us to detect that the null hypothesis was wrong. Basically, the amount of variability in a test statistic based on a small sample is so large that it washes out our ability to see the relatively small difference between the null hypothesis value and the true value. With a larger sample size, we'd have less sampling variability and maybe see that the true population proportion was different from the null hypothesis's claim.

In our long-running coin-spinning experiment, we spun the coin 20 times, got 7 heads, and found a p-value of 0.18. We failed to reject the null hypothesis that the coin was fair, and so we conclude that there is no evidence that spinning a coin is biased. Could we make a stronger statement? Could we claim that spinning a coin is fair?

No. For instance, we might have had too small a sample size. That is, even though the sample size was large enough for the conditions of the Central Limit Theorem to give us a good approximation for the p-value, the sampling variability was still too great for us to see whether the probability of heads is different from 0.50 when spinning a coin.

KEY POINT Don't say you "proved" something with statistics. Say you "demonstrated" it or "showed" it. Similarly, don't say you "accept the null hypothesis"; say, rather, that you "cannot reject the null hypothesis" or that you "failed to reject the null hypothesis" or that "there is insufficient evidence to reject the null hypothesis."

EXAMPLE **13** Find the Flaws

Are public libraries in the United States an endangered species? In past years, suppose that it was believed that roughly 49% of Americans had visited a library. (In truth, the percentage was higher, but we're using this value to illustrate a point.) Imagine that a professional association of librarians wishes to know whether attendance is declining. They examine a Pew survey conducted in 2013, in which only 48% of those in a random sample had attended a library in the last year (Pew Research Center 2013). The librarians decide to use a strict significance level of $\alpha = 0.01$. They carry out a hypothesis test, and the result is shown in Figure 8.13. Based on this, they send out a press release that says, "We have proved that there is no decline in library attendance."

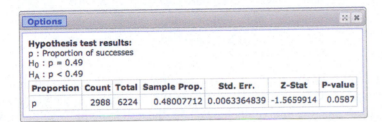

◄ FIGURE 8.13 StatCrunch output for the librarians' hypothesis test.

Hypothesis test results:
p : Proportion of successes
H_0 : p = 0.49
H_A : p < 0.49

Proportion	Count	Total	Sample Prop.	Std. Err.	Z-Stat	P-value
p	2988	6224	0.48007712	0.0063364839	-1.5659914	0.0587

QUESTION   What mistakes did this professional association make? How would you correct them?

SOLUTION   The professional association concluded that they had *proved* that attendance did not decline. However, they have *not* proved that the null hypothesis is true. At best, they have not found sufficient evidence to reject it, but this is very different from saying that they have found evidence that proves it is true.

The association should conclude that, using a 1% significance level, there is not enough evidence to conclude that library membership has declined.

TRY THIS!   Exercise 8.49

Confidence Intervals and Hypothesis Tests

Confidence intervals and hypothesis tests are closely related, even though they are used to answer (slightly) different questions. Confidence intervals are used to answer the question, "What is the value of this parameter?" For instance, suppose we spin a coin 20 times and get 7 heads. An approximate 95% confidence interval for the probability of getting heads is 0.14 to 0.56, as shown in Figure 8.14. This tells us that, on the basis of our data, we are highly confident that the true probability of getting heads is between 14% and 56%.

The hypothesis test answers a slightly different question: "Are the data consistent with the parameter being one particular value, or might the parameter be something else?" These hypothesis tests are a little more vague: we are not really asking what the value is; we simply want to know whether it is one thing or another. For instance, for the coin-spinning we ask, "Are the data consistent with the coin being fair? That is, $p = 0.50$? Or is the coin not fair?"

Even though they are designed to answer different questions, they are similar enough that you can often use a confidence interval to reach the same types of conclusions you would reach with a hypothesis test using a two-sided alternative hypothesis. In most situations, doing a hypothesis test with a two-sided alternative hypothesis and

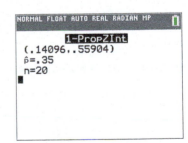

▲ FIGURE 8.14 TI-84 confidence interval output for 7 heads out of 20 spins.

Looking Back

Confidence Intervals for Proportions
You learned how to find a confidence interval for a population proportion in Section 7.4.

significance level α (in percentage points) will lead to the same conclusion as finding a confidence interval with a $(1 - \alpha)$ confidence level and rejecting the null hypothesis if its value is not captured by the interval. Table 8.3 shows this relationship between confidence level and significance level.

For instance, if we wish to test whether the spun coin is fair with a significance level of $\alpha = 0.05$, and our sample proportion after 20 spins is 0.35, then we find the $(1 - \alpha) = 1 - 0.05 = 0.95$, or 95%, confidence interval based on this outcome. The interval is 0.14 to 0.56, as shown in Figure 8.14. Because this interval includes 0.5, we would not reject the null hypothesis that the coin is fair.

▶ **TABLE 8.3** The table shows the relationship between confidence levels and significance levels for hypothesis tests. There are rare occasions when the conclusions of a hypothesis test would be different from what we might conclude from a confidence interval because of the different method of calculating the standard error.

Confidence Level = $(1 - \alpha)$	Alternative Hypothesis	Significance Level α = Conf. Level − α
90%	Two-sided ($\neq$)	10%
95%	Two-sided ($\neq$)	5%
99%	Two-sided ($\neq$)	1%

If we wished to do this test with a 10% significance level, we would find the 90% confidence interval. This turns out to be 0.17 to 0.53, as shown in the StatCrunch output in Figure 8.15. Again, the interval includes 0.50, the null hypothesis value, so we do not reject the null hypothesis.

Details

Rare Differences
It is possible to reach different conclusions using confidence intervals and hypothesis tests because the standard errors are computed using slightly different values for the population proportion p. The confidence interval estimates p with the sample proportion ($\hat{p}$), while the hypothesis test uses the value claimed by the null hypothesis, p_0.

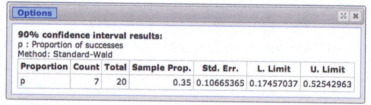

Options

90% confidence interval results:
p : Proportion of successes
Method: Standard-Wald

Proportion	Count	Total	Sample Prop.	Std. Err.	L. Limit	U. Limit
p	7	20	0.35	0.10665365	0.17457037	0.52542963

▲ **FIGURE 8.15** StatCrunch output for a 90% confidence interval for the proportion of heads for spinning a coin.

KEY POINT   A confidence interval answers the question, "What is the value of the parameter?" A hypothesis test is used to judge between two different claims about the value.

SECTION 8.4

Comparing Proportions from Two Populations

You have now seen how to carry out a hypothesis test for a single population proportion. With very few changes, this procedure can be altered to accommodate a more interesting situation: comparing proportions from two populations.

Consider as an example the opinion polls on embryonic stem cell research introduced in Section 7.5. This medical research shows great promise in the treatment of several major diseases but is controversial because it goes against many people's religious convictions. The Pew Forum on Religion & Public Life has, over time, conducted surveys to assess Americans' support for stem cell research. In 2002, 43% of Americans expressed support for stem cell research (Pew Forum 2008). Later, in 2009, 58% supported this research (Pew Forum 2009). Can we conclude that support has changed in the population of all Americans? Or could this difference be due to chance variation during the sampling procedure?

This problem involves two populations. One population consists of all Americans in 2009, and the second population consists of all Americans in 2002. Each population has a true proportion who support stem cell research, but in each case we cannot know this true value. Instead, we have a random sample taken from both populations, and we must estimate the two proportions from these two random samples.

Here are the changes we need to make to our "ingredients" in order to compare proportions from two populations.

Changes to Ingredients: The Hypotheses

Because we now have two population proportions to consider, we need some new notation. We'll let p_1 represent the proportion of Americans who supported stem cell research in 2009, and we'll let p_2 represent the proportion who supported it in 2002.

We are not interested in the actual numerical values of p_1 and p_2, as we were when dealing with just one population proportion. We are interested only in their relation to each other. The conservative, status quo, not-worth-a-headline hypothesis is that these proportions are the same. In other words, there has been no change in support. We write this as

$$H_0: p_1 = p_2$$

In words, the null hypothesis says that the proportion of Americans who support stem cell research was the same in 2009 as it was in 2002.

The alternative hypothesis is that the proportion of Americans who support stem cell research has changed. If so, the two proportions are no longer equal.

$$H_a: p_1 \neq p_2$$

One-sided hypotheses are also possible. Our research question might instead have been, "Has support for stem cell research decreased?" If that had been our question, then we would have used

$$H_a: p_1 < p_2$$

And if we had wished to answer the question, "Has support for stem cell research increased?" we would have used this alternative:

$$H_a: p_1 > p_2$$

These options lead to three pairs of hypotheses, as shown in Table 8.4. You choose the pair that corresponds to the research question your study hopes to answer. Note that the null hypothesis is always $p_1 = p_2$ because the neutral position is always that the two proportions are the same.

◄ **TABLE 8.4** Possible hypotheses for a two-proportion hypothesis test.

Hypothesis	Symbols	The Alternative in Words
Two-sided	$H_0: p_1 = p_2$ $H_a: p_1 \neq p_2$	The proportions are different in the two populations.
One-sided (left)	$H_0: p_1 = p_2$ $H_a: p_1 < p_2$	The proportion in population 1 is less than the proportion in population 2.
One-sided (right)	$H_0: p_1 = p_2$ $H_a: p_1 > p_2$	The proportion in population 1 is greater than the proportion in population 2.

Changes to Ingredients: The Test Statistic

We are interested in how p_1 and p_2 differ, so our test statistic is based on the difference between our sample proportions from the two populations. The test statistic we will use has the same structure as the one-sample z-statistic:

$$z = \frac{\text{estimator} - \text{null value}}{SE}$$

However, the estimator for the **two-proportion z-test** is $\hat{p}_1 - \hat{p}_2$ because we are estimating the difference $p_1 - p_2$. Here $\hat{p}_1$ and $\hat{p}_2$ are just the sample proportions for the different samples. In our case, $\hat{p}_1$ is the sample proportion for the people surveyed in 2009 (reported as 0.58), and $\hat{p}_2$ is the sample proportion for the people surveyed in 2002 (reported as 0.43).

The null value is 0 because the null hypothesis claims these proportions are the same, so $p_1 - p_2 = 0$.

The standard error, SE, is more complicated than in the one-sample case because the null hypothesis no longer tells us the value of the population proportion. All it tells us is that both populations have the same value. For this reason, when we estimate this single value, we pool the two samples together. Formula 8.2 shows you how to do this.

Formula 8.2: The two-proportion z-test statistic

$$z = \frac{\hat{p}_1 - \hat{p}_2 - 0}{SE}$$

where

$$SE = \sqrt{\hat{p}(1 - \hat{p})\left(\frac{1}{n_1} + \frac{1}{n_2}\right)}$$

n_1 = sample size in sample 1

n_2 = sample size in sample 2

$$\hat{p} = \frac{\text{number of successes in sample 1} + \text{number of successes in sample 2}}{n_1 + n_2}$$

$$\hat{p}_1 = \text{proportion of successes in sample 1} = \frac{\text{number of successes in sample 1}}{n_1}$$

$$\hat{p}_2 = \text{proportion of successes in sample 2} = \frac{\text{number of successes in sample 2}}{n_2}$$

Formula 8.2 is perhaps the most elaborate formula we have shown you so far. As usual, it is much more important to be able to use technology to perform this test than to apply the formula. Still, studying the formula does help us understand why the test statistic is useful.

EXAMPLE **14** Common Ground?

It might seem that Republicans and Democrats have become so divided that there is no common ground. The Pew Research Organization (2018) recently asked a sample of Republicans and Democrats if they believe it is "very important for the country that Republicans and Democrats work together on issues." They randomly sampled 2577 Democrats and 1978 Republicans to ask if they agree with this statement. The data are summarized in Table 8.5.

	Democrats	Republicans	Total
Agree	2036	1524	3560
Disagree	541	454	995
Total	2577	1978	4555

◀ **TABLE 8.5** Data for the Pew study.

QUESTION Find the observed value of the test statistic to test the hypotheses

$$H_0: p_1 = p_2$$
$$H_a: p_1 \neq p_2$$

where p_1 represents the proportion of Democrats who agree with this statement and p_2 represents the proportion of Republicans who agree.

SOLUTION We must bring all the pieces together and assemble them into the test statistic:

$$\hat{p}_1 = \frac{2036}{2577} = 0.79$$

$$\hat{p}_2 = \frac{1524}{1978} = 0.77$$

$$\hat{p} = \frac{2036 + 1524}{2577 + 1978} = \frac{3560}{4555} = 0.782 \quad \text{(a pooled estimate of the sample proportion)}$$

$$SE = \sqrt{0.782(1 - 0.782)\left(\frac{1}{2577} + \frac{1}{1978}\right)} = 0.0123$$

Now we assemble the pieces:

$$z_{obs} = \frac{0.79 - 0.77}{0.0123} = 1.63$$

TRY THIS! Exercise 8.65

> ### Looking Back
>
> **Two-way Tables**
> Two-way tables, such as the one used to summarize the data in Example 14, were first presented in Chapter 1.

> ### Details
>
> **The Sign of z**
> In a two-proportion test, whether z is positive or negative depends on which population you call "1" and which you call "2." It's important to pay attention to which proportion is subtracted from which!

Changes to Ingredients: Checking Conditions

The conditions that we need to check for a two-sample test of proportions are similar to those for a one-sample test but with some additional things to consider:

1. *Large Samples:* Both sample sizes must be large enough. Because we don't know the value of p_1 or p_2, we must use an estimate. The null hypothesis says that these two proportions are the same, so we use $\hat{p}$, the pooled sample proportion, to check this condition. Do not use $\hat{p}_1$ or $\hat{p}_2$. This means that we need

   a. $n_1\hat{p} \geq 10$ and $n_1(1 - \hat{p}) \geq 10$

   b. $n_2\hat{p} \geq 10$ and $n_2(1 - \hat{p}) \geq 10$

2. *Random Samples:* The samples are drawn randomly from the appropriate population. In practice, this condition is often impossible to check unless we were present when the data were collected. If we were not told explicitly the sample was randomly drawn, we may have to assume the condition is satisfied.

3. *Independent Samples:* The samples are independent of each other. This condition is violated if, for example, the same individuals are in both samples that we are comparing.

4. *Independent within Samples:* The observations within each sample must be independent of one another.

If these four conditions hold, then, if the null hypothesis is true, z follows a $N(0, 1)$ distribution.

EXAMPLE 15 Right of Way

Psychologists at the University of California, Berkeley, were interested in studying whether people driving "high-status" cars behaved differently than those driving other cars do. State law requires that cars come to a complete halt and not enter a crosswalk that contains a pedestrian. The researchers used an accomplice to step into a crosswalk on a busy street in California. The researchers took careful note of whether the driver illegally cut off the pedestrian accomplice by entering the crosswalk. They also rated the status of the car with a number between 1 and 5; a 1 meant "low" status, and a 5 meant "high" (such as a Rolls-Royce). A vehicle was recorded only if there were no cars in front of or behind it. This helped ensure that observations were independent. (For instance, if one car stops, a car behind it might be more likely to stop.) Although the actual study used an advanced statistical model to control for potential confounding factors, we can perform a simple analysis by combining the cars into two groups. The researchers observed 33 "low-status" cars (rated 1 or 2), and 119 "high-status" cars (rated 3 or higher). Of the low-status cars, 8 cars failed to stop. Of the high-status cars, 45 failed to stop (Piff et al. 2012).

QUESTION   Find and state the proportion of cars that failed to stop for both groups. Perform a four-step hypothesis test to test the hypothesis that high-status cars are more likely to cut off a pedestrian. Use a significance level of 0.10. For the purpose of this problem, assume the cars are random sample from the population of all cars that visit this intersection.

SOLUTION   The proportion of low-status cars that cut off the pedestrian was $8/33 = 0.242$. The proportion of high-status cars that cut off the pedestrian was $45/119 = 0.378$.

As always, our hypotheses are about populations. We witnessed only 33 low-status and 119 high-status cars, but we take these as representative of a population and hypothesize about *all* low-status and high-status cars that might pass through this crosswalk.

Let p_1 represent the proportion of all low-status cars that would fail to stop if they were passing through this crosswalk when a pedestrian stepped into the crosswalk. Let p_2 represent the proportion of all high-status cars that would fail to stop.

Step 1: Hypothesize
The null hypothesis is neutral; it states that both groups of cars are the same:

$$\text{H}_0\text{: } p_1 = p_2$$

The alternative hypothesis is that the high-status cars cut off pedestrians more often:

$$\text{H}_a\text{: } p_1 < p_2$$

Step 2: Prepare
We are comparing two population proportions, so our test statistic will be the two-proportion z-statistic. We must check to see whether conditions are satisfied so that we can use the Normal distribution as an approximate sampling distribution.

1. Are both samples large enough?
   We find that

$$\hat{p} = \frac{8 + 45}{33 + 119} = 0.34868$$

   First sample: $n_1 \times 0.34868 = 33 \times 0.34868 = 11.5$, which is greater than 10, and $33 \times (1 - 0.34868) = 21.5$, which is also greater than 10.

   Second sample: $n_2 \times 0.34868 = 119 \times 0.34868 = 41.5$ and $119 \times (1 - 0.34868) = 77.5$, both of which are greater than 10.

2. Are the samples drawn randomly from their respective populations?
   Frankly, probably not (but we hope they are representative). However, the problem statement allows us to make this assumption and proceed.

3. Are the samples independent of each other?
   Yes; there is no reason to suppose that the actions of low-status cars will affect high-status cars or the other way around.

4. Are observations within each sample independent?
   Yes; the researchers took care to make sure this was so by recording only cars that were not following other cars.

With these three conditions checked, we can proceed to step 3.

Step 3: Compute to compare
We use a significance level of $\alpha = 0.10$. We must find the individual pieces of Formula 8.2. We defined p_1 to be the proportion of low-status cars that cut off pedestrians. So let $\hat{p}_1$ be the proportion of low-status cars *in the sample* that cut off pedestrians. The sample size in this group was $n_1 = 33$. Earlier, we calculated

$$\hat{p}_1 = \frac{8}{33} = 0.2424$$

The sample proportion for the high-status cars was

$$\hat{p}_2 = \frac{45}{119} = 0.3782$$

To find the standard error, we need to use $\hat{p}$, the proportion of bad events that happen in the sample if we ignore the fact that the cars belong to two different groups. Earlier, we found that $\hat{p} = 0.34868$. The standard error is then

$$SE = \sqrt{0.34868(1 - 0.34868)\left(\frac{1}{33} + \frac{1}{119}\right)} = 0.0938$$

Putting it all together yields

$$z_{observed} = \frac{0.2424 - 0.3782}{0.0938} = -1.45$$

Now that we know the observed value, we must measure our surprise. The null hypothesis assumes that the two population proportions are the same and that our sample proportions differed only by chance. The p-value will measure the probability of getting an outcome as extreme as or more extreme than -1.45, assuming that the population proportions are the same.

The p-value is calculated exactly the same way as with a one-proportion z-test. Our alternative hypothesis is a left-sided hypothesis, so we need to find the probability of getting a value less than the observed value (Figure 8.16).

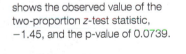

▲ FIGURE 8.16 The shaded area represents the probability to the left of a test statistic value of -1.45.

◀ FIGURE 8.17 StatCrunch output shows the observed value of the two-proportion z-test statistic, -1.45, and the p-value of 0.0739.

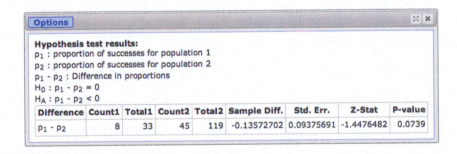

Hypothesis test results:
p_1 : proportion of successes for population 1
p_2 : proportion of successes for population 2
$p_1 - p_2$: Difference in proportions
$H_0 : p_1 - p_2 = 0$
$H_A : p_1 - p_2 < 0$

Difference	Count1	Total1	Count2	Total2	Sample Diff.	Std. Err.	Z-Stat	P-value
$p_1 - p_2$	8	33	45	119	-0.13572702	0.09375691	-1.4476482	0.0739

Figure 8.17 provides the z-value and the p-value from StatCrunch:

$$\text{p-value} = 0.074$$

Step 4: Interpret

The p-value is less than our stated significance level of 0.10, so we reject the null hypothesis. We conclude that the high-status cars were more likely to cut off the pedestrian.

Note that because this was an observational study, we can't conclude cause and effect, which means that we can't conclude that driving a high-status car makes a person less concerned about pedestrians. Also, our sample was not randomly selected from a larger population, so perhaps the results would differ in different parts of the country. The p-value was relatively high; if we had used a significance level of 0.05, we would *not* have rejected the null hypothesis. The original researchers used more sophisticated methods that enabled them to control for potential confounders and allowed them to see how the proportion of cutting off a pedestrian varied for each of the five status levels. With these more refined methods, they got a smaller p-value.

TRY THIS!   Exercise 8.69

As explained at the start of this section, to compare changing attitudes towards stem cell research, the Pew Foundation's researchers surveyed a sample of 1500 Americans in 2002 and a new sample of the same size in 2009. In 2002, 43% of Americans supported stem cell research and in 2009 this had increased to 58%. We asked earlier if this change might be explained by chance variation, since there is a chance element in who gets included in a random sample.

To answer this question, we can carry out a hypothesis test in which the null hypothesis is that the population proportions of the U.S. who support stem cell research is the same in 2002 and 2009. The alternative is that they are different. The test statistic, after some calculation, is $z = 8.22$. You should not need a table or calculator to know that the associated p-value must be very small, since very rarely do we see values in a Normal distribution smaller than -3 or bigger than $+3$. So we can conclude that the change was not due to chance, and there was a real shift in support over this time period.

SNAPSHOT ▶ Two-Proportion *z*-Test

WHAT IS IT? ▶	A hypothesis test.
WHAT DOES IT DO? ▶	Provides a procedure for comparing two population proportions. The null hypothesis is always that the proportions are the same, and this procedure gives us a way to reject or fail to reject that hypothesis.
HOW DOES IT DO IT? ▶	The test statistic *z* compares the differences between the sample proportions and the value 0 (which is what the null hypothesis says this difference should be):

$$z = \frac{\hat{p}_1 - \hat{p}_2 - 0}{SE}, \text{ where } SE = \sqrt{\hat{p}(1 - \hat{p})\left(\frac{1}{n_1} + \frac{1}{n_2}\right)}$$

Values of *z* that are far from 0 tend to discredit the null hypothesis.

HOW IS IT USED? ▶ When comparing two proportions, each from a different population. The data must come from two independent, random samples, and each sample must be sufficiently large. Then $N(0, 1)$ can be used to compute the p-value for the observed test statistic.

CASE STUDY REVISITED

Dodging the Question

During political debates, candidates sometimes dodge questions by answering a question different from the question asked. Can posting the correct question at the bottom of the TV screen make it more likely that viewers will notice this evasion? Two researchers randomly assigned viewers to one of two conditions. In both conditions, viewers watched actors in a debate in which one of the actors dodged the question. Under one condition, however, the question was posted on the screen while the viewers watched. In this group, 88% of viewers noticed the dodge. The other group saw nothing posted, and only 39% of them noticed the dodge. Can this difference of 88% − 39% = 49 percentage points be due to chance? Or does it suggest that we can become better dodge detectors if we are reminded of the question?

To find out, the researchers carried out a hypothesis test. The data are summarized in Table 8.6 and Figure 8.18.

	Question Posted	No Question Posted	Total
Detected	63	28	91
Not Detected	9	43	52
Total	72	71	143

▲ **TABLE 8.6** Contingency table for the relationship between posting of a question and dodge detection.

We carried out a two-proportion z-test to see whether posting questions improves dodge detection. We'll used a significance level of 0.05. Say we let p_1 represent the proportion of people who will notice the dodge when the question is posted during the debate, and we let p_2 represent the proportion of people who will notice the dodge when no question is posted. Then our hypotheses are

H_0: $p_1 = p_2$ (The same percentage in both groups will recognize the dodge.)

H_a: $p_1 > p_2$ (A greater percentage will recognize the dodge when the question is posted.)

► **FIGURE 8.18** Relationship between detecting a dodge and whether or not the viewer saw the question on his or her TV screen.

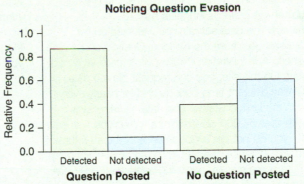

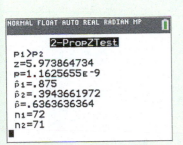

▲ **FIGURE 8.19** TI-84 output for a two-sample z-test using a two-sided alternative hypothesis. The p-value is 0.00000000116.

A quick check shows that the sample sizes are large enough and that the other necessary conditions hold. The z-statistic, calculated with technology (see Figure 8.19), is 5.97. We know from the Empirical Rule that the p-value will be very small since z-statistics are almost never that far from 0. And indeed, if we were to calculate the p-value, we would find that if the null hypothesis were true, then the probability of getting a test statistic as large as 5.97 or larger would be 0.000000001.

With such a small p-value, we reject the null hypothesis and conclude that posting the question on the screen really did help viewers notice the "candidate's" evasion of the question.

DATAPROJECT ▶ Dates as Data

1   OVERVIEW

Many data sets include information about the date of the observations. Dates are interesting values because they are not quite numerical and not quite categorical, and so special tools are often needed to work with dates.

2   GOALS

Learn to identify date formats and create useful new variables from dates. Use the data move *sort* to examine the structure of a data set.

3   DATA MOVES AROUND THE CALENDAR

How much rain falls in Los Angeles, California, in a year? This question is of great relevance to many southern Californians, since their state, like many in the West and Southwest of the United States, is vulnerable to drought. As you'll see, answering this question is not as straightforward as it might at first seem. In fact, we won't have an answer until Chapter 9. In this chapter, we'll do some important first steps.

Rainfall amounts are accessible through data.gov, a data "clearinghouse" maintained by the federal government. Rainfall is measured with a rain gauge, a standardized tube with markings on it. Weather stations all over the world have rain gauges, and each day they are read to see how much rain fell. There are thousands of weather stations in the United States; the greater metropolitan area of Los Angeles has 70.

Project The data set rain.csv was created by automated request from the National Oceanic and Atmospheric Administration (NOAA) via https://www.ncdc.noaa.gov/. On this site, you can request climate data across many different time periods and for nearly every location in the United States. The data set rain.csv is large, with over 54,000 observations. When you first view the data, you do not get a good sense of the structure of the data. For instance, the first 45 or so observations are all from the same station, named "Pasadena 2.0 SE, CA US," and we are given the latitude, longitude and elevation of this station. (Elevation is in meters; 225 meters is about 735 feet.) However, we don't know from these first few rows of the data set whether any other stations are included and, if so, how many.

We also see a date variable. For instance, the first date we see is February 1, 2016, written as "2016-02-01." We also notice that, as we read down the column, the date increases by one day per row, suggestion that we have daily readings of the amount of rain. What

we do not know yet is which years are available in this data set.

Dates are slightly more difficult to work with than numbers or categories. First, notice that there are many different ways we could represent a particular date. We might write "Feb 1, 2016" or "02-01-16" or "2/1/16" or, well, you get the idea. The computer sees dates such as these as a funny mix of letters, symbols, and numbers, and so many computer programs will first think that dates are categorical values. Most software, such as StatCrunch, won't let you find means or standard deviations of dates.

And so a simple question such as "Which years are represented in my data?" can be a challenge to answer.

For these reasons, the first step when working with dates is often to create new variables that extract the information we wish. The **Data > Compute > Expression** command will allow you to create new variables (see the Data Project for Chapter 4). Click on **Build** and you'll see a dialogue box like the one in Figure 8.20.

▲ FIGURE 8.20

The expression shown will create a new variable whose values are the year the observation was recorded.

This function requires that you give it the *format* of the dates. The string "yyyy-MM-dd" tells StatCrunch the format, and means "the year comes first and is represented with four digits; next comes a dash and then a two-digit month, then another dash and a two-digit day." If, instead, our dates looked like "2/1/16" we would have indicated the format with "m/d/yy." A complete list of the formats is in the help section of StatCrunch: https://www.statcrunch.com/app/helpContent.php.

Once you click on **Okay**, you'll be given the option of naming the variable (see Figure 8.21). Name it "Year" and click **Compute!**, and you'll have a new column that gives the year of each observation.

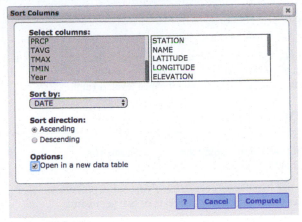

▲ FIGURE 8.22

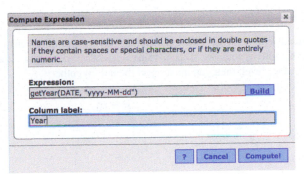

▲ FIGURE 8.21

Assignment

Once we've created new columns from the *Date* variable, we're going to use the *Sort* function to explore the data. We can learn quite a bit simply through sorting by a variable.

1. Make a frequency table of the *Year* variable and answer the question, "Which years are represented in the data set?"
2. What are the first and last dates? To see the first date, select **Data > Sort** and fill out the dialogue box as in Figure 8.22. It's very important that you select *all* of the columns to be sorted.

   Sorting data is a common "data move" for quickly investigating the extremes of a data set. How can you use the *Sort* function to determine which date is the last? StatCrunch lets you sort on many columns by selecting **Add Sort by Column** in the Sort Columns dialogue box. How can you use *Sort* to determine the largest amount of rainfall at the Woodland Hills Pierce College location?

   Notice that now you've sorted the data by date, you can see that there are multiple weather stations included in the data set. How many stations are included?

3. Use the *getDay(x, format)* function to create a column for the day of the week. (The result will have the values 1, 2, ... 7, where 1 represents Sunday and 7 represents Saturday.). Use *getMonth(x, format)* to create a column for the month. Create graphs that can be used to answer the questions: Which day of the week has the most observations? Which month?

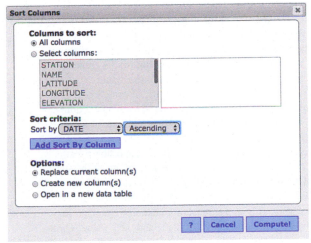

▲ FIGURE 8.23

CHAPTER REVIEW

KEY TERMS

Hypothesis testing, *384*
Null hypothesis, H₀, *385*
Alternative hypothesis, Hₐ, *385*

one-sided hypothesis, *386*
two-sided hypothesis, *386*
significance level, α (alpha), *388*

test statistic, *389*
one-proportion z-test statistic,
   389

p-value, *390*
two-tailed p-value, *394*
two-proportion z-test, *410*

LEARNING OBJECTIVES

After reading this chapter and doing the assigned homework problems, you should

- Know how to test hypotheses concerning a population proportion and hypotheses concerning the comparison of two population proportions.

- Understand the meaning of p-value and how it is used.

- Understand the meaning of significance level and how it is used.

- Know the conditions required for calculating a p-value and significance level.

SUMMARY

Hypothesis tests are performed in the following four steps.

Step 1: Hypothesize.
Step 2: Prepare.
Step 3: Compute to compare.
Step 4: Interpret.

Step 1 is the most important, because it establishes the entire procedure. Hypotheses are *always* statements about parameters. The alternative hypothesis is the hypothesis that the researcher wishes to convince the public is true. The null hypothesis is the skeptical, neutral hypothesis. Each step of the hypothesis test is carried out assuming that the null hypothesis is true. For all tests in this book, the null hypothesis will always contain an equals (=) sign, whereas the alternative hypothesis can contain the symbol for "is greater than" (>), the symbol for "is less than" (<), or the symbol for "is not equal to" (≠).

Step 2 requires that you decide what type of test you are doing. In this chapter, this means you are either testing the value of a proportion from a single population (one-proportion z-test) or comparing two proportions from different populations (two-proportion z-test). You must also check that the conditions necessary for using the standard Normal distribution as the sampling distribution are all met.

Step 3 is where the observed value of the statistics is compared to the null hypothesis. This step is most often handled by technology, which will compute a value of the test statistic and the p-value. These values are valid only if the conditions in step 2 are satisfied.

Step 4 requires you to compare the p-value, which measures our surprise at the outcome if the null hypothesis is true, to the

significance level, which is the probability that we will mistakenly reject the null hypothesis. If the p-value is less than (or equal to) the significance level, then you must reject the null hypothesis.

For a one-proportion z-test,

$$\textbf{Formula 8.1: } z = \frac{\hat{p} - p_0}{SE}$$

where

$$SE = \sqrt{\frac{p_0(1 - p_0)}{n}}$$

p_0 is the proposed population proportion
$\hat{p}$ (p-hat) is the sample proportion, x/n
n is the sample size
For a two-proportion z-test,

$$\textbf{Formula 8.2: } z = \frac{\hat{p}_1 - \hat{p}_2 - 0}{SE}$$

where

$$SE = \sqrt{\hat{p}(1 - \hat{p})\left(\frac{1}{n_1} + \frac{1}{n_2}\right)}$$

$$\hat{p} = \frac{\text{number of successes in both samples}}{n_1 + n_2}$$

$\hat{p}_1$ is the proportion of successes in the first sample, and $\hat{p}_2$ is the proportion in the second sample

Calculating the p-value depends on which alternative hypothesis you are using. Figure 8.24 shows, from left to right, a two-tailed p-value, one-tailed (right-tailed) p-value, and a one-tailed (left-tailed) p-value.

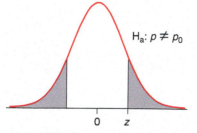
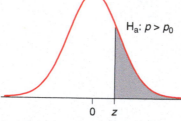
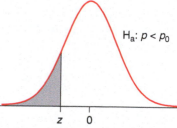

▲ **FIGURE 8.24** Representations of possible p-values for three different alternative hypotheses. The area of the shaded regions represents the p-value.

SOURCES

Einstein-PE Investigators. 2012. Oral rivaroxaban for the treatment of symtomatic pulmonary embolism. *New England Journal of Medicine*, vol. 336 (April 5): 12.87–12.97.

Elwood, P., et al. 2013. Healthy lifestyles reduce the incidence of chronic diseases and dementia: Evidence from the Caerphilly Cohort Study. *PLoS ONE*, vol. 8: e81877. doi:10.137/journal.pone.0081877

Feder, L., and L. Dugan. 2002. A test of the efficacy of court-mandated counseling for domestic violence offenders: The Broward Experiment. *Justice Quarterly*, vol. 19: 343–375.

Fry, R. and A. Brown. 2016. In a recovering market, home ownership rates are down sharply for blacks, young adults. Pew Research Center, December. www.pewresearch.org.

Furman, R., et al. 2014. Idelalisib and rituximab in relapsed chronic lymphocytic leukemia. *New England Journal of Medicine*, January 22. doi:10.1056/NEJMoa1315226

Kim, A., et al. 2010. Sex bias in trials and treatments must end. *Nature*, June 10 2010.

Leschied, A., and A. Cunningham. 2002. *Seeking effective interventions for serious young offenders: Interim results of a four-year randomized study of multisystemic therapy in Ontario, Canada*. London, Canada: London Family Court Clinic.

Pew Forum. 2008. Declining majority of Americans favor embryonic stem cell research. http://www.pewforum.org/2008/07/17/declining-majority-of-americans-favor-embryonic-stem-cell-research/.

Pew Forum. 2009. Support for health care overhaul, but it's not 1993: Stable views of stem cell research. The Pew Forum on Religion & Public Life Issues. http://www.people-press.org/2009/03/19/support-for-health-care-overhaul-but-its-not-1993/.

Pew Research Center. 2013. How Americans value public libraries in their communities. http://libraries.pewinternet.org/2013/12/11/libraries-in-communities/

Pew Research Center. 2018. The public the political system and American democracy, April 26. http://www.people-press.org/2018/04/26/the-public-the-political-system-and-american-democracy/

Piff, P., et al. 2012. Higher social class predicts increased unethical behavior. *Proceedings of the National Academy of Sciences*, vol. 109 (11): 4086–4091.

Rogers, T. and M. Norton. 2011. The artful dodger: Answering the wrong question the right way. *Journal of Experimental Psychology: Applied*, vol. 17 (2): 139–147.

Schwitzgebel, E., and J. Rust. 2010. Do ethicists and political philosophers vote more often than other professors? *Review of Philosophy and Psychology*, vol. 1, 189–199.

Shankaran, S., et al. 2012. Childhood outcomes after hypothermia for neonatal encephalopathy. *New England Journal of Medicine*, vol. 366: 2085–2092.

SECTION EXERCISES

SECTION 8.1

8.1 Choose one of the answers given. The null hypothesis is always a statement about a _____ (sample statistic or population parameter).

8.2 Choose one of the answers in each case. In statistical inference, measurements are made on a _____ (sample or population), and generalizations are made to a _____ (sample or population).

8.3 Vegetarians (Example 1) In 2016 a Harris poll estimated that 3.3% of American adults are vegetarian. A nutritionist thinks this rate has increased and will take a random sample of American adults and record whether or not they are vegetarian. State the null and alternative hypotheses in words and in symbols.

8.4 Embedded Tutors A college chemistry instructor thinks the use of embedded tutors (tutors who work with students during regular class meeting times) will improve the success rate in introductory chemistry courses. The passing rate for introductory chemistry is 62%. The instructor will use embedded tutors in all sections of introductory chemistry and record the percentage of students passing the course. State the null and alternative hypotheses in words and in symbols. Use the symbol p to represent the passing rate for all introductory chemistry courses that use embedded tutors.

8.5 Teen Drivers According to a 2015 University of Michigan poll, 71.5% of high school seniors in the United States had a driver's license. A sociologist thinks this rate has declined. The sociologist surveys 500 randomly selected high school seniors and finds that 350 have a driver's license.

  a. Pick the correct null hypothesis.
   i. $p = 0.715$ ii. $p = 0.70$ iii. $\hat{p} = 0.715$ iv. $\hat{p} = 0.70$
  b. Pick the correct alternative hypothesis.
   i. $p > 0.715$ ii. $p < 0.715$ iii. $\hat{p} < 0.715$ iv. $p \neq 0.715$
  c. In this context, the symbol p represents (choose one)
   i. the proportion of high school seniors in the entire United States that have a driver's license.
   ii. the proportion of high school seniors in the sociologist's random sample that have a driver's license.

8.6 Water A friend is tested to see whether he can tell bottled water from tap water. There are 30 trials (half with bottled water and half with tap water), and he gets 18 right.

  a. Pick the correct null hypothesis:
   i. $\hat{p} = 0.50$ ii. $\hat{p} = 0.60$ iii. $p = 0.50$ iv. $p = 0.60$
  b. Pick the correct alternative hypothesis:
   i. $\hat{p} \neq 0.50$ ii. $\hat{p} = 0.875$ iii. $p > 0.50$ iv. $p \neq 0.875$

8.7 Flu Vaccine In 2016, the Centers for Disease Control and Prevention estimated that the flu vaccine was 73% effective against the influenza B virus. An immunologist suspects that the current flu vaccine is less effective against this virus. Pick the correct pair of hypotheses the immunologist could use to test this claim.

   i. $H_0: p > 0.73$
      $H_a: p < 0.73$

   ii. $H_0: p = 0.73$
      $H_a: p < 0.73$

   iii. $H_0: p = 0.73$
      $H_a: p > 0.73$

   iv. $H_0: p = 0.73$
      $H_a: p \neq 0.73$

8.8 Law School Grad Employment The National Association for Law Placement estimated that 86.7% of law school graduates in 2015 found employment. An economist thinks the current employment rate for law school graduates is different from the 2015 rate. Pick the correct pair of hypotheses the economist could use to test this claim.

i. $H_0: p \neq 0.867$
$H_a: p = 0.867$

ii. $H_0: p = 0.867$
$H_a: p > 0.867$

iii. $H_0: p = 0.867$
$H_a: p < 0.867$

iv. $H_0: p = 0.867$
$H_a: p \neq 0.867$

TRY 8.9 Soda Orders (Example 2) A manager at a casual dining restaurant noted that 15% of customers ordered soda with their meal. In an effort to increase soda sales, the restaurant begins offering free refills with every soda order for a two-week trial period. During this trial period, 17% of customers ordered soda with their meal. To test if the promotion was successful in increasing soda orders, the manager wrote the following hypotheses: $H_0: p = 0.15$ and $H_a: \hat{p} = 0.17$, where $\hat{p}$ represents the proportion of customers who ordered soda with their meal during promotion. Are these hypotheses written correctly? Correct any mistakes as needed.

8.10 Mixed Nuts The label on a can of mixed nuts says that the mixture contains 40% peanuts. After opening a can of nuts and finding 22 peanuts in a can of 50 nuts, a consumer thinks the proportion of peanuts in the mixture differs from 40%. The consumer writes these hypotheses: $H_0: p \neq 0.40$ and $H_a: p = 0.44$ where p represents the proportion of peanuts in all cans of mixed nuts from this company. Are these hypotheses written correctly? Correct any mistakes as needed.

TRY 8.11 Flu Vaccine (Example 3) An immunologist is testing the hypothesis that the current flu vaccine is less than 73% effective against the flu virus. The immunologist is using a 1% significance level and these hypotheses: $H_0: p = 0.73$ and $H_a: p < 0.73$. Explain what the 1% significance level means in context.

8.12 Law School Grad Employment An economist is testing the hypothesis that the employment rate for law school graduates is different from 86.7%. The economist is using a 5% significance level and these hypotheses: $H_0: p = 0.867$ and $H_a: p \neq 0.867$. Explain what the 5% significance level means in context.

8.13 Student Loans According to a 2017 Pew Research Center report, 37% of adults aged 18 to 29 had student loan debt. Suppose in a random sample of adults from this age group 48 out of 120 had student loan debt.

a. Give the null and alternative hypotheses to test that the student loan rate is not 37%.

b. Report the test statistic (z) from the output given.

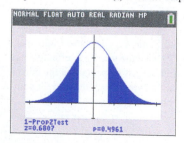

8.14 Hybrid Car Sales According to Green Car Reports, 4.4% of cars sold in California in 2017 were hybrid cars. Suppose in a random sample of 500 recently sold cars in California 18 were hybrid.

a. Write the null and alternative hypotheses to test that hybrid car sales in California have declined.

b. Report the value of the test statistic (z) from the figure.

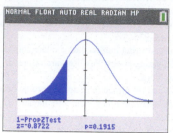

TRY 8.15 Vegetarians (Example 4) In 2016 the Harris poll estimated that 3.3% of American adults are vegetarian. A nutritionist thinks this rate has increased. The nutritionist samples 150 American adults and finds that 11 are vegetarian.

a. What is $\hat{p}$, the sample proportion of vegetarians?

b. What is p_0, the hypothetical proportion of vegetarians?

c. Find the value of the test statistic. Explain the test statistic in context.

8.16 Embedded Tutors A college chemistry instructor thinks the use of embedded tutors will improve the success rate in introductory chemistry courses. The passing rate for introductory chemistry is 62%. During one semester, 200 students were enrolled in introductory chemistry courses with an embedded tutor. Of these 200 students, 140 passed the course.

a. What is $\hat{p}$, the sample proportion of students who passed introductory chemistry.

b. What is p_0, the proportion of students who pass introductory chemistry if the null hypothesis is true?

c. Find the value of the test statistic. Explain the test statistic in context.

TRY 8.17 Coke versus Pepsi (Example 5) Suppose you are testing someone to see whether she or he can tell Coke from Pepsi, and you are using 20 trials, half with Coke and half with Pepsi. The null hypothesis is that the person is guessing.

a. About how many should you expect the person to get right under the null hypothesis that the person is guessing?

b. Suppose person A gets 13 right out of 20, and person B gets 18 right out of 20. Which will have a smaller p-value, and why?

8.18 St. Louis Jury Pool St. Louis County is 24% African American. Suppose you are looking at jury pools, each with 200 members, in St. Louis County. The null hypothesis is that the probability of an African American being selected into the jury pool is 24%.

a. How many African Americans would you expect on a jury pool of 200 people if the null hypothesis is true?

b. Suppose pool A contains 40 African American people out of 200, and pool B contains 26 African American people out of 200. Which will have a smaller p-value and why?

8.19 Vegetarians (Example 6) In problem 8.15 the nutritionist was interested in knowing if the rate of vegetarianism in American adults has increased. She carried out a hypothesis test and found that the observed value of the test statistic was 2.77. We can calculate that the p-value associated with this is 0.0028, which is very close to 0. Explain the meaning of the p-value in this context. Based on this result, should the nutritionist believe the null hypothesis is true?

8.20 Embedded Tutors In problem 8.16, a college chemistry instructor thinks the use of embedded tutors will improve the success rate in introductory chemistry courses. The instructor carried out a hypothesis test and found that the observed value of the test statistic was 2.33. The p-value associated with this test statistic is 0.0099. Explain the meaning of the p-value in this context. Based on this result, should the instructor believe the success rate has improved?

8.21 Hospital Readmission A hospital readmission is an episode when a patient who has been discharged from a hospital is readmitted again within a certain period. Nationally the readmission rate for patients with pneumonia is 17%. A hospital was interested in knowing whether their readmission rate for pneumonia was less than the national percentage. They found 11 patients out of 70 treated for pneumonia in a two-month period were readmitted.

a. What is $\hat{p}$, the sample proportion of readmission?

b. Write the null and alternative hypotheses.

c. Find the value of the test statistic and explain it in context.

d. The p-value associated with this test statistic is 0.39. Explain the meaning of the p-value in this context. Based on this result, does the p-value indicate the null hypothesis should be doubted?

8.22 Guessing A 20-question multiple choice quiz has five choices for each question. Suppose that a student just guesses, hoping to get a high score. The teacher carries out a hypothesis test to determine whether the student was just guessing. The null hypothesis is $p = 0.20$, where p is the probability of a correct answer.

a. Which of the following describes the value of the z-test statistic that is likely to result? Explain your choice.

   i. The z-test statistic will be close to 0.

   ii. the z-test statistic will be far from 0.

b. Which of the following describes the p-value that is likely to result? Explain your choice.

   i. The p-value will be small.

   ii. The p-value will not be small.

SECTION 8.2

8.23 Dreaming (Example 7) A 2003 study of dreaming published in the journal *Perceptual and Motor Skills* found that out of a random sample of 113 people, 92 reported dreaming in color. However, the proportion of people who reported dreaming in color that was established in the 1940s was 0.29 (Schwitzgebel 2003). Check to see whether the conditions for using a one-proportion z-test are met assuming the researcher wanted to see whether the proportion dreaming in color had changed since the 1940s.

8.24 Age Discrimination About 30% of the population in Silicon Valley, a region in California, are between the ages of 40 and 65, according to the U.S. Census. However, only 2% of the 2100 employees at a laid-off man's former Silicon Valley company are between the ages of 40 and 65. Lawyers might argue that if the company hired people regardless of their age, the distribution of

ages would be the same as though they had hired people at random from the surrounding population. Check whether the conditions for using the one-proportion z-test are met.

TRY **8.25 Self-Driving Cars (Example 8)** In a Northeastern University/Gallup poll of 461 young Americans aged 18 to 35, 152 reported they would be comfortable riding in a self-driving car. Suppose we are testing the hypothesis that more than 30% of Americans in this age group would be comfortable riding in a self-driving car, using a significance level of 0.05. Which of the following figures correctly matches the alternative hypothesis $p > 0.30$. Report and interpret the correct p-value.

(A)
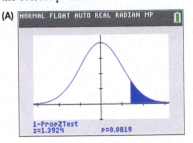

(B)
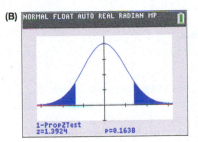

8.26 Diabetes According to a Gallup poll, 11.55% of American adults have diabetes. Suppose a researcher wonders if the diabetes rate in her area is higher than the national rate. She surveys 150 adults in her area and finds that 21 of them have diabetes.

a. If the region had the same rate of diabetes as the rest of the country, how many would we expect have diabetes?

b. Suppose you are testing the hypothesis that the diabetes rate in this area differs from the national rate, using a 0.05 significance level. Choose the correct figure and interpret the p-value.

(A)
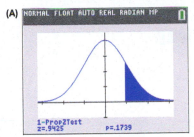

(B)
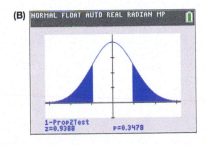

8.27 Coke versus Pepsi A taste test is done to see whether a person can tell Coke from Pepsi. In each case, 20 random and independent trials are done (half with Pepsi and half with Coke) in which the person determines whether she or he is drinking Coke or Pepsi. One person gets 13 right out of 20 trials. Which of the following is the correct figure to test the hypothesis that the person can tell the difference? Explain your choice.

(A)
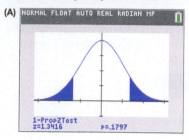

(B)
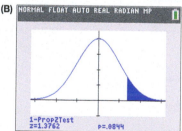

8.28 Seat Belts Suppose we are testing people to see whether the rate of use of seat belts has changed from a previous value of 88%. Suppose that in our random sample of 500 people we see that 450 have the seat belt fastened. Which of the following figures has the correct p-value for testing the hypothesis that the proportion who use seat belts has changed? Explain your choice.

(A)
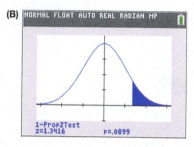

(B)
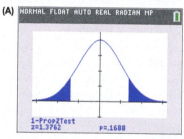

TRY 8.29 Working Out (Example 9) According to a 2018 survey by Timex reported in *Shape* magazine, 73% of Americans report working out one or more times each week. A nutritionist is interested in whether this percentage has increased. A random sample of 200 Americans found 160 reported working out one or more times each week. Carry out the first two steps of a hypothesis test to determine whether the proportion has increased. Explain how you would fill in the required TI calculator entries for p_0, x, and n.

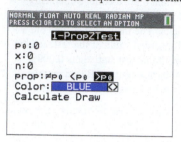

8.30 Vacations According to a 2017 AAA survey, 35% of Americans planned to take a family vacation (a vacation more than 50 miles from home involving two or more immediate family members. Suppose a recent survey of 300 Americans found that 115 planned on taking a family vacation. Carry out the first two steps of a hypothesis test to determine if the proportion of Americans planning a family vacation has changed. Explain how you would fill in the required entries in the figure for # of success, # of observations, and the value in H_0.

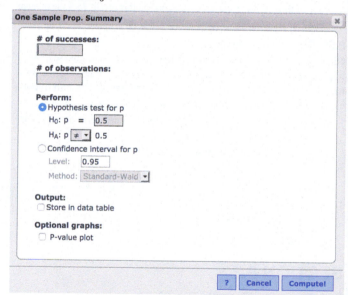

TRY 8.31 Working Out (Example 10) According to a 2018 survey by Timex reported in *Shape* magazine, 73% of Americans report working out one or more times each week. A nutritionist is interested in whether this percentage has increased. A random sample of

200 Americans found 160 reported working out one or more times each week. The first two steps were asked for in Exercise 8.29. Use the output provided to carry out the third and fourth steps of a hypothesis test that will test whether the proportion of Americans who work out one or more times per week has increased. Use a significance level of 0.05.

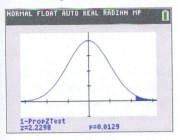

8.32 Vacations According to a 2017 AAA survey, 35% of Americans planned to take a family vacation (a vacation more than 50 miles from home involving two or more immediate family members). Suppose a recent survey of 300 Americans found that 115 planned on taking a family vacation. In Exercise 8.30 you carried out the first two steps of a hypothesis test that will test whether the proportion has changed. Use the output provided to carry out the third and fourth steps of the hypothesis test. Use a significance level of 0.05.

One sample proportion summary hypothesis test:
p : Proportion of successes
H_0 : p = 0.35
H_A : p ≠ 0.35

Hypothesis test results:

Proportion	Count	Total	Sample Prop.	Std. Err.	Z-Stat	P-value
p	115	300	0.38333333	0.027537853	1.2104551	0.2261

8.33 p-Values For each graph, indicate whether the shaded area could represent a p-value. Explain why or why not. If yes, state whether the area could represent the p-value for a one-sided or a two-sided alternative hypothesis.

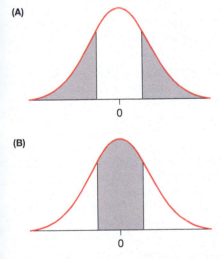

8.34 p-Values For each graph, state whether the shaded area could represent a p-value. Explain why or why not. If yes, state whether the area could represent the p-value for a one-sided or a two-sided alternative hypothesis.

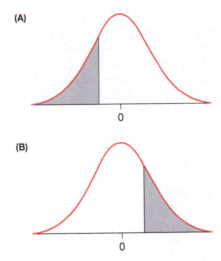

g 8.35 News on Facebook A 2018 Gallup poll of 3635 randomly selected Facebook users found that 2472 get most of their news about world events on Facebook. Research done in 2013 found that only 47% of all Facebook users reported getting their news about world events on Facebook. *See page 430 for guidance.*

a. Does this sample give evidence that the proportion of Facebook users who get their world news on Facebook has changed since 2013? Carry out a hypothesis test and use a 0.05 significance level.

b. After conducting the hypothesis test, a further question one might ask is what proportion of all Facebook users got most of their news about world events on Facebook in 2018. Use the sample data to construct a 90% confidence interval for the population proportion. How does your confidence interval support your hypothesis test conclusion?

8.36 Olympic Viewing A 2018 Gallup poll of 2228 randomly selected U.S. adults found that 39% planned to watch at least a "fair amount" of the 2018 Winter Olympics. In 2014, 46% of U.S. adults reported planning to watch at least a "fair amount."

a. Does this sample give evidence that the proportion of U.S. adults who planned to watch the 2018 Winter Olympics was less than the proportion who planned to do so in 2014? Use a 0.05 significance level.

b. After conducting the hypothesis test, a further question one might ask is what proportion of all U.S. adults planned to watch at least a "fair amount" of the 2018 Winter Olympics. Use the sample data to construct a 90% confidence interval for the population proportion. How does your confidence interval support your hypothesis test conclusion?

8.37 Global Warming Historically (from about 2001 to 2014), 57% of Americans believed that global warming is caused by human activities. A March 2017 Gallup poll of a random sample of 1018 Americans found that 692 believed that global warming is caused by human activities.

a. What percentage of the sample believed global warming was caused by human activities?

b. Test the hypothesis that the proportion of Americans who believe global warming is caused by human activities has changed from the historical value of 57%. Use a significance level of 0.01.

c. Choose the correct interpretation:

   i. In 2017, the percentage of Americans who believe global warming is caused by human activities is not significantly different from 57%.

   ii. In 2017, the percentage of Americans who believe global warming is caused by human activities has changed from the historical level of 57%.

8.38 Plane Crashes According to one source, 50% of plane crashes are due at least in part to pilot error (http://www.planecrashinfo.com). Suppose that in a random sample of 100 separate airplane accidents, 62 of them were due to pilot error (at least in part.)

a. Test the null hypothesis that the proportion of airplane accidents due to pilot error is not 0.50. Use a significance level of 0.05.

b. Choose the correct interpretation:

i. The percentage of plane crashes due to pilot error is not significantly different from 50%.

ii. The percentage of plane crashes due to pilot error is significantly different from 50%.

8.39 Mercury in Freshwater Fish Some experts believe that 20% of all freshwater fish in the United States have such high levels of mercury that they are dangerous to eat. Suppose a fish market has 250 fish tested, and 60 of them have dangerous levels of mercury. Test the hypothesis that this sample is *not* from a population with 20% dangerous fish. Use a significance level of 0.05.

Comment on your conclusion: Are you saying that the percentage of dangerous fish is definitely 20%? Explain.

8.40 Twitter Suppose a poll is taken that shows 220 out of 400 randomly selected Twitter users feel that Twitter should do more to decrease hateful and abusive content on the site. Test the hypothesis that the majority (more than 50%) of Twitter users feel the site should do more to decrease hateful and abusive content on the site. Use a significance level of 0.01.

8.41 Morse's Proportion of *t*'s Samuel Morse determined that the percentage of *t*'s in the English language in the 1800s was 9%. A random sample of 600 letters from a current newspaper contained 48 *t*'s. Using the 0.10 level of significance, test the hypothesis that the proportion of *t*'s in this modern newspaper is 0.09.

8.42 Morse's Proportion of *a*'s Samuel Morse determined that the percentage of *a*'s in the English language in the 1800s was 8%. A random sample of 600 letters from a current newspaper contained 60 *a*'s. Using the 0.10 level of significance, test the hypothesis that the proportion of *a*'s in this modern newspaper is 0.09.

SECTION 8.3

TRY **8.43 p-Values (Example 11)** A researcher carried out a hypothesis test using a two-sided alternative hypothesis. Which of the following z-scores is associated with the smallest p-value? Explain.

i. $z = 0.50$ ii. $z = 1.00$ iii. $z = 2.00$ iv. $z = 3.00$

8.44 Coin Flips A test is conducted in which a coin is flipped 30 times to test whether the coin is unbiased. The null hypothesis is that the coin is fair. The alternative is that the coin is not fair. One of the accompanying figures represents the p-value after getting 16 heads out of 30 flips, and the other represents the p-value after getting 18 heads out of 30 flips. Which is which, and how do you know?

(A)
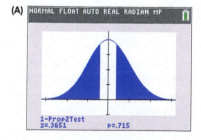

(B)
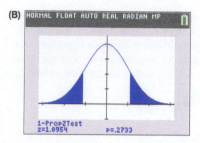

TRY **8.45 Young Voters (Example 12)** According to the Brookings Institution, 50% of eligible 18- to 29-year-old voters voted in the 2016 election. Suppose we were interested in whether the proportion of voters in this age group who voted in the 2018 election was higher. Describe the two types of errors we might make in conducting this hypothesis test.

8.46 Errors with Toast Suppose you are testing someone to see whether he or she can tell butter from margarine when it is spread on toast. You use many bite-sized pieces selected randomly, half from buttered toast and half from toast with margarine. The taster is blindfolded. The null hypothesis is that the taster is just guessing and should get about half right. When you reject the null hypothesis when it is actually true, that is often called the first kind of error. The second kind of error is when the null is false and you fail to reject. Report the first kind of error and the second kind of error.

8.47 Blackstone on Errors in Trials Sir William Blackstone (1723–1780) wrote influential books on common law. He made this statement: "All presumptive evidence of felony should be admitted cautiously; for the law holds it better that ten guilty persons escape, than that one innocent party suffer."

Keep in mind that the null hypothesis in criminal trials is that the defendant is not guilty. State which of these errors (in blue) is the first type of error (rejecting the null hypothesis when it is actually true) and which is the second type of error.

8.48 Alpha By establishing a small value for the significance level, are we guarding against the first type of error (rejecting the null hypothesis when it is true) or guarding against the second type of error?

TRY **8.49 Superpower (Example 13)** What superpower do Americans want most? In past years, 10% of Americans chose invisibility as the most desired superpower, based on the Marist poll. Assume this is an accurate representation of all Americans. A group of futurists examines a more recent 2018 Marist poll that found that 12% of those sampled picked invisibility as their desired superpower. The futurists carry out a hypothesis test using a significance level of 0.01, and the result is shown in the StatCrunch output. Based on this, can they conclude that the percentage of all Americans who would pick invisibility as their superpower is still 10%? If not, what conclusion would be appropriate based on these sample data?

One sample proportion summary hypothesis test:
p : Proportion of successes
$H_0 : p = 0.1$
$H_A : p > 0.1$

Hypothesis test results:

Proportion	Count	Total	Sample Prop.	Std. Err.	Z-Stat	P-value
p	123	1033	0.11907067	0.0093340709	2.0431244	0.0205

8.50 Flaws The null hypothesis on true/false tests is that the student is guessing, and the proportion of right answers is 0.50. A student taking a five-question true/false quiz gets 4 right out of 5. She says that this shows that she knows the material, because the one-tailed p-value from the one-proportion z-test is 0.090, and she is using a significance level of 0.10. What is wrong with her approach?

8.51 Which Method? A proponent of a new proposition on a ballot wants to know whether the proposition is likely to pass. Suppose a poll is taken, and 580 out of 1000 randomly selected people support the proposition. Should the proponent use a hypothesis test or a confidence interval to answer this question? Explain. If it is a hypothesis test, state the hypotheses and find the test statistic, p-value, and conclusion. If a confidence interval is appropriate, find the approximate 95% confidence interval. In both cases, assume that the necessary conditions have been met.

8.52 Which Method? A proponent of a new proposition on a ballot wants to know the population percentage of people who support the bill. Suppose a poll is taken, and 580 out of 1000 randomly selected people support the proposition. Should the proponent use a hypothesis test or a confidence interval to answer this question? Explain. If it is a hypothesis test, state the hypotheses and find the test statistic, p-value, and conclusion. Use a 5% significance level. If a confidence interval is appropriate, find the approximate 95% confidence interval. In both cases, assume that the necessary conditions have been met.

★ 8.53 Effectiveness of Financial Incentives A psychologist is interested in testing whether offering students a financial incentive improves their video-game-playing skills. She collects data and performs a hypothesis test to test whether the probability of getting to the highest level of a video game is greater with a financial incentive than without. Her null hypothesis is that the probability of getting to this level is the same with or without a financial incentive. The alternative is that this probability is greater. She gets a p-value from her hypothesis test of 0.003. Which of the following is the best interpretation of the p-value?

i. The p-value is the probability that financial incentives are *not* effective in this context.

ii. The p-value is the probability of getting exactly the result obtained, assuming that financial incentives are *not* effective in this context.

iii. The p-value is the probability of getting a result as extreme as or more extreme than the one obtained, assuming that financial incentives are *not* effective in this context.

iv. The p-value is the probability of getting exactly the result obtained, assuming that financial incentives *are* effective in this context.

v. The p-value is the probability of getting a result as extreme as or more extreme than the one obtained, assuming that financial incentives *are* effective in this context.

8.54 Is it acceptable practice to look at your research results, note the direction of the difference, and then make the alternative hypothesis one-sided in order to achieve a significant difference? Explain.

8.55 If we reject the null hypothesis, can we claim to have *proved* that the null hypothesis is false? Why or why not?

8.56 If we do not reject the null hypothesis, is it valid to say that we *accept* the null hypothesis? Why or why not?

8.57 When a person stands trial for murder, the jury is instructed to assume that the defendant is innocent. Is this claim of innocence an example of a null hypothesis, or is it an example of an alternative hypothesis?

8.58 When, in a criminal court, a defendant is found "not guilty," is the court saying with certainty that he or she is innocent? Explain.

★ 8.59 Arthritis A magazine advertisement claims that wearing a magnetized bracelet will reduce arthritis pain in those who suffer from arthritis. A medical researcher tests this claim with 233 arthritis sufferers randomly assigned either to wear a magnetized bracelet or to wear a placebo bracelet. The researcher records the proportion of each group who report relief from arthritis pain after 6 weeks. After analyzing the data, he fails to reject the null hypothesis. Which of the following are valid interpretations of his findings? There may be more than one correct answer.

a. The magnetized bracelets are not effective at reducing arthritis pain.

b. There's insufficient evidence that the magnetized bracelets are effective at reducing arthritis pain.

c. The magnetized bracelets had exactly the same effect as the placebo in reducing arthritis pain.

d. There were no statistically significant differences between the magnetized bracelets and the placebos in reducing arthritis pain.

★ 8.60 No-Carb Diet A weight-loss diet claims that it causes weight loss by eliminating carbohydrates (breads and starches) from the diet. To test this claim, researchers randomly assign overweight subjects to two groups. Both groups eat the same amount of calories, but one group eats almost no carbs, and the other group includes carbs in their meals. After 2 months, the researchers test the claim that the no-carb diet is better than the usual diet. They record the proportion of each group that lost more than 5% of their initial weight. They then announce that they failed to reject the null hypothesis. Which of the following are valid interpretations of the researchers' findings?

a. There were no significant differences in effectiveness between the no-carb diet and the carb diet.

b. The no-carb diet and the carb diet were equally effective.

c. The researchers did not see enough evidence to conclude that the no-carb diet was more effective.

d. The no-carb diet was less effective than the carb diet.

SECTION 8.4

8.61 When comparing two sample proportions with a two-sided alternative hypothesis, all other factors being equal, will you get a smaller p-value if the sample proportions are close together or if they are far apart? Explain.

8.62 When comparing two sample proportions with a two-sided alternative hypothesis, all other factors being equal, will you get a smaller p-value with a larger sample size or a smaller sample size? Explain.

g 8.63 Treatment for HIV-1 In a 2018 study reported in *The Lancet*, Molina et al. reported on a study for treatment of patients with HIV-1. The study was a randomized, controlled, double-blind study that compared the effectiveness of ritonavir-boosted daruna-vir (rbd), the drug currently used to treat HIV-1, with dorovirine, a newly developed drug. Of the 382 subjects taking ritonavir-boosted darunavir, 306 achieved a positive result. Of the 382 subjects taking dorovirine, 321 achieved a positive outcome. *See page 430 for guidance.*

a. Find the sample percentage of subjects who achieved a positive outcome in each group.

b. Perform a hypothesis test to test whether the proportion of patients who achieve a positive outcome with the current treatment (ritonavir-boosted darunavir) is different from the proportion of patients who achieve a positive outcome with the new treatment (dorovirine). Use a significance level of 0.01. Based on this study, do you think dorovirine might be a more effective treatment option for HIV-1 than ritonavir-boosted darunavir? Why or why not?

8.64 Smoking Cessation in HIV Patients In a 2018 study reported in *The Lancet*, Mercie et al. investigated the efficacy and safety of varenicline for smoking cessation in people living with HIV. The study was a randomized, double-blind, placebo-controlled trial. Of the 123 subjects treated with varenicline, 18 abstained from smoking for the entire 48-week study period. Of the 124 subjects assigned to the placebo group, 8 abstained from smoking for the entire study period.

a. Find the sample percentage of subjects in each group who abstained from smoking for the entire study period.

b. Determine whether varenicline is effective in reducing smoking among HIV patients. Note that this means we should test if the proportion of the varenicline group who abstained from smoking for the entire study period is significantly greater than that of the placebo group. Use a significance level of 0.05.

TRY **8.65 Reading (Example 14)** The researchers in a Pew study interviewed two random samples, one in 2015 and one in 2018. Both samples were asked, "Have you read a print book in the last year?" The results are shown in the table below.

Read a print book	2015	2018	Total
Yes	1201	1341	2542
No	705	661	1366
Total	1906	2002	

a. Find and compare the sample proportions that had read a print book for these two groups.

b. Find a pooled estimate of the sample proportion.

c. Has the proportion who read print books increased? Find the observed value of the test statistic to test the hypotheses H_0: $p_{2015} = p_{2018}$ and H_a: $p_{2015} < p_{2018}$ assuming the conditions for a two-proportion z-test hold.

8.66 Audio Books Pew Research published survey results from two random samples. Both samples were asked, "Have you listened to an audio book in the last year?" The results are shown in the table below.

Listened to an audio book	2015	2018	Total
Yes	229	360	589
No	1677	1642	3319
Total	1906	2002	

a. Find and compare the sample proportions that had listened to an audio book for these two groups.

b. Are a greater proportion listening to audio books in 2018 compared to 2015? Test the hypothesis that a greater proportion of people listened to an audio book in 2018 than in 2015. Use a 0.05 significance level.

8.67 Freedom of the Press A Gallup poll asked college students in 2016 and again in 2017 whether they believed the First Amendment guarantee of freedom of the press was secure or threatened in the country today. In 2016, 2489 out of 3072 students surveyed said that freedom of the press was secure or very secure. In 2017, 1808 out of 2014 students surveyed felt this way.

a. Determine whether the proportion of college students who believe that freedom of the press is secure or very secure in the country changed from 2016. Use a significance level of 0.05.

b. Use the sample data to construct a 95% confidence interval for the difference in the proportions of college students in 2016 and 2017 who felt freedom of the press was secure or very secure. How does your confidence interval support your hypothesis test conclusion?

8.68 Freedom of Religion A Gallup poll asked college students in 2016 and again in 2017 whether they believed the First Amendment guarantee of freedom of religion was secure or threatened in the country today. In 2016, 2089 out of 3072 students surveyed said that freedom of religion was secure or very secure. In 2017, 1929 out of 3014 students felt this way.

a. Determine whether the proportion of college students who believe that freedom of religion is secure or very secure in this country has changed from 2016. Use a significance level of 0.05.

b. Use the sample data to construct a 95% confidence interval for the difference in the proportions of college students in 2016 and 2017 who felt freedom of religion was secure or very secure. How does your confidence interval support your hypothesis test conclusion?

TRY **8.69 Environmental Quality (Example 15)** A Gallup poll asked a random samples of Americans in 2016 and 2018 if they were satisfied with the quality of the environment. In 2016, 543 were satisfied with the quality of the environment and 440 were dissatisfied. In 2018, 461 were satisfied and 532 were dissatisfied. Determine whether the proportion of Americans who are satisfied with the quality of the environment has declined. Use a 0.05 significance level.

8.70 Caregiving Responsibilities In 2017 the Pew Research Center conducted a survey on family-leave practices and attitudes. Respondents were asked to complete this sentence: "When a family member has a serious health condition, caregiver responsibilities . . ." with choices being "mainly on women," "mainly on men," or "on both men and women equally." The percentages for each response are shown in the table below. For these age groups, the responses fell only into the two categories shown in the table. Assume a sample size of 1200 for each age group.

Age	Fall mainly on women	Fall equally on men and women
30–49	60%	40%
50–64	62%	38%

Can we conclude that there is a difference in the proportion of people aged 30 to 49 and aged 50 to 64 who feel the primary caregiver responsibility falls on women? Use a significance level of 0.05.

CHAPTER REVIEW EXERCISES

8.71 Choosing a Test and Naming the Population(s)
For each of the following, state whether a one-proportion z-test or a two-proportion z-test would be appropriate, and name the population(s).

a. A polling agency takes a random sample of voters in California to determine if a ballot proposition will pass.

b. A researcher asks a random sample of residents from coastal states and a random sample of residents of non-coastal states whether they favor increased offshore oil drilling. The researcher wants to determine if there is a difference in the proportion of residents who support off-shore drilling in the two regions.

8.72 Choosing a Test and Naming the Population(s)
For each of the following, state whether a one-proportion z-test or a two-proportion z-test would be appropriate, and name the population(s).

a. A researcher takes a random sample of voters in western states and voters in southern states to determine if there is a difference in the proportion of voters in these regions who support the death penalty.

b. A sociologist takes a random sample of voters to determine if support for the death penalty has changed since 2015.

8.73 Choosing a Test and Giving the Hypotheses Give the null and alternative hypotheses for each test, and state whether a one-proportion z-test or a two-proportion z-test would be appropriate.

a. You test a person to see whether he can tell tap water from bottled water. You give him 20 sips selected randomly (half from tap water and half from bottled water) and record the proportion he gets correct to test the hypothesis.

b. You test a random sample of students at your college who stand on one foot with their eyes closed and determine who can stand for at least 10 seconds, comparing athletes and nonathletes.

8.74 Choosing a Test and Naming the Population(s)
In each case, choose whether the appropriate test is a one-proportion z-test or a two-proportion z-test. Name the population(s).

a. A researcher takes a random sample of 4-year-olds to find out whether girls or boys are more likely to know the alphabet.

b. A pollster takes a random sample of all U.S. adult voters to see whether more than 50% approve of the performance of the current U.S. president.

c. A researcher wants to know whether a new heart medicine reduces the rate of heart attacks compared to an old medicine.

d. A pollster takes a poll in Wyoming about homeschooling to find out whether the approval rate for men is equal to the approval rate for women.

e. A person is studied to see whether he or she can predict the results of coin flips better than chance alone.

8.75 Water Taste Test A student who claims that he can tell tap water from bottled water is blindly tested with 20 trials. At each trial, tap water or bottled water is randomly chosen and presented to the student who much correctly identify the type of water. The experiment is designed so that the student will have exactly 10 sips from each type of water. He gets 13 identifications right out of 20. Can the student tell tap water from bottled water at a 0.05 level of significance? Explain.

8.76 Butter Taste Test A student is tested to determine whether she can tell butter from margarine. She is blindfolded and given small bites of toast that has been spread with either butter or margarine that have been randomly chosen. The experiment is designed so that she will have exactly 15 bites with butter and 15 bites with margarine. She gets 20 right out of 30 trials. Can she tell butter from margarine at a 0.05 level of significance? Explain.

* **8.77 Biased Coin?** A study is done to see whether a coin is biased. The alternative hypothesis used is two-sided, and the obtained z-value is 2. Assuming that the sample size is sufficiently large and that the other conditions are also satisfied, use the Empirical Rule to approximate the p-value.

* **8.78 Biased Coin?** A study is done to see whether a coin is biased. The alternative hypothesis used is two-sided, and the obtained z-value is 1. Assuming that the sample size is sufficiently large and that the other conditions are also satisfied, use the Empirical Rule to approximate the p-value.

8.79 ESP A researcher studying extrasensory perception (ESP) tests 300 students. Each student is asked to predict the outcome of a large number of coin flips. For each student, a hypothesis test using a 5% significance level is performed. If the p-value is less than or equal to 0.05, the researcher concludes that the student has ESP. Assuming that none of the 300 students actually have ESP, about how many would you expect the researcher to conclude *do* have ESP? Explain.

8.80 Coin Flips Suppose you tested 50 coins by flipping each of them many times. For each coin, you perform a significance test with a significance level of 0.05 to determine whether the coin is biased. Assuming that none of the coins is biased, about how many of the 50 coins would you expect to appear biased when this procedure is applied?

8.81 Student Age A community college used enrollment records of all students and reported that that the percentage of the student population identifying as female in 2010 was 54% whereas the proportion identifying as female in 2018 was 52%. Would it be appropriate to use this information for a hypothesis test to determine if the proportion of students identifying as female at this college had declined? Explain.

8.82 Taste Test A student was tested to see if he could tell the difference between two different brands of cola. He was presented with 20 samples of cola and correctly identified 12 of the samples. Since he was correct 60 percent of the time, can we conclude that he can correctly tell the difference between two brands of cola based on this sample alone?

8.83 Facebook Pew Research conducts polls on social media use. In 2012, 66% of those surveyed reported using Facebook. In 2018, 76% reported using Facebook.

a. Assume that both polls used samples of 100 people. Do a test to see whether the proportion of people who reported using Facebook was significantly different in 2012 and 2018 using a 0.01 significance level.

b. Repeat the problem, now assuming the sample sizes were both 1500. (The actual survey size in 2018 was 1785.)

c. Comment on the effect of different sample sizes on the p-value and on the conclusion.

8.84 Television In the Pew Research social media survey, television viewers were asked if it would be very hard to give up watching television. In 2002, 38% responded yes. In 2018, 31% said it would be very hard to give up watching television.

a. Assume that both polls used samples of 200 people. Do a test to see whether the proportion of people who reported it would be very hard to give up watching television was significantly different in 2002 and 2018 using a 0.05 significance level.

b. Repeat the problem, now assuming the sample sizes were both 2000. (The actual sample size in 2018 was 2002.)

c. Comment on the effect of different sample sizes on the p-value and on the conclusion.

8.85 Presidential Election Pew Research reported that in the 2016 presidential election, 53% of all male voters voted for Trump and 41% voted for Clinton. Among all women voters, 42% voted for Trump and 54% voted for Clinton. Would it be appropriate to do a two-proportion z-test to determine whether the proportions of men and women who voted for Trump were significantly different (assuming we knew the number of men and women who voted)? Explain.

8.86 Educational Attainment According to a 2016 report by the Census Bureau, 60.1% of women and 57.6% of men have completed some college education or higher. Would it be appropriate to do a two-proportion z-test to determine whether the proportions of men and women who had completed some college education or higher were different (assuming we knew the total number of men and women)? Why or why not?

8.87 Self-Employment According to the Bureau of Labor Statistics, 10.1% of Americans are self-employed. A researcher wants to determine if the self-employment rate in a certain area is different. She takes a random sample of 500 working residents from the area and finds that 62 are self-employed.

a. Test the hypothesis that the proportion of self-employed workers in this area is different from 10.1%. Use a 0.05 significance level.

b. After conducting the hypothesis test, a further question one might ask, "What proportion of workers in this area are self-employed?" Use the sample data to find a 95% confidence interval for the proportion of workers who are self-employed in the area from which the sample was drawn. How does this confidence interval support the hypothesis test conclusion?

8.88 Student Loans According to a 2016 report from the Institute for College Access and Success 66% of all graduates from public colleges and universities had student loans. A public college surveyed a random sample of 400 graduates and found that 62% had student loans.

a. Test the hypothesis that the percentage of graduates with student loans from this college is different from the national percentage. Use a significance level of 0.05.

b. After conducting the hypothesis test, a further question one might ask is what proportion of graduates from this college have student loans? Use the sample data to find a 95% confidence interval for the proportion of graduates from the college who have student loans. How does this confidence interval support the hypothesis test conclusion?

8.89 Gun Control A Quinnipiac poll conducted on February 20, 2018, found that 824 people out of 1249 surveyed favored stricter gun control laws. A survey conducted one week later on February 28, 2018, by National Public Radio found that 754 out of 1005 people surveyed favored stricter gun control laws.

a. Find both sample proportions and compare them.

b. Test the hypothesis that the population proportions are not equal at the 0.05 significance level.

c. After conducting the hypothesis test, a further question one might ask is what is the difference between the two population proportions? Find a 95% confidence interval for the difference between the two proportions and interpret it. How does the confidence interval support the hypothesis test conclusion?

8.90 Gay Marriage A Gallup poll conducted in 2017 found that 648 out of 1011 people surveyed supported same-sex marriage. An NBC News/*Wall Street Journal* poll conducted the same year surveyed 1200 people and found 720 supported same-sex marriage.

a. Find both sample proportions and compare them.

b. Test the hypothesis that the population proportions are not equal at the 0.05 significance level.

* **8.91 Three-Strikes Law** California's controversial "three-strikes law" requires judges to sentence anyone convicted of three felony offenses to life in prison. Supporters say that this decreases crime both because it is a strong deterrent and because career criminals are removed from the streets. Opponents argue (among other things) that people serving life sentences have nothing to lose, so violence within the prison system increases. To test the opponents' claim, researchers examined data starting from the mid-1990s from the California Department of Corrections. "Three Strikes: Yes" means the person had committed three or more felony offenses and was probably serving a life sentence. "Three Strikes: No" means the person had committed no more than two offenses. "Misconduct" includes serious offenses (such as assaulting an officer) and minor offenses (such as not standing for a count). "No Misconduct" means the offender had not committed any offenses in prison.

a. Compare the proportions of misconduct in these samples. Which proportion is higher, the proportion of misconduct for those who had three strikes or that for those who did not have three strikes? Explain.

b. Treat this as though it were a random sample, and determine whether those with three strikes tend to have more offenses than those who do not. Use a 0.05 significance level.

	Three Strikes	
	Yes	**No**
Misconduct	163	974
No Misconduct	571	2214
Totals	734	3188

8.92 Cardiovascular Disease and Gout Patients with gout also have an increased risk of cardiovascular disease. A 2018 double-blind study by White et al. reported in *The New England Journal of Medicine*, patients with cardiovascular disease and gout were randomly assigned to receive one of two drugs: febuxostat or allopurinol. The number of patients in each group who experienced an adverse cardiovascular event was recorded and the data is shown in the table. Test the hypothesis that the proportion of patients who will experience adverse cardiovascular events differs for these two drugs. Use a 0.01 significance level.

	Adverse cardiovascular event	
Drug	Yes	No
Febuxostat	335	2767
Allopurinol	321	2766

8.93 Environment In 2015 a Gallup poll reported that 52% of Americans were satisfied with the quality of the environment. In 2018, a survey of 1024 Americans found that 461 were satisfied with the quality of the environment. Does this survey provide evidence that satisfaction with the quality of the environment among Americans has decreased? Use a 0.05 significance level.

8.94 Cloning Dolly the Sheep, the world's first mammal to be cloned, was introduced to the public in 1997. In a Pew Research poll taken soon after Dolly's debut, 63% of Americans were opposed to the cloning of animals. In a Pew Research poll taken 20 years after Dolly, 60% of those surveyed were opposed to animal cloning. Assume this was based on a random sample of 1100 Americans. Does this survey indicate that opposition to animal cloning has declined since 1997? Use a 0.05 significance level.

8.95 A friend claims he can predict the suit of a card drawn from a standard deck of 52 cards. There are four suits and equal numbers of cards in each suit. The parameter, p, is the probability of success, and the null hypothesis is that the friend is just guessing.

a. Which is the correct null hypothesis?
   i. $p = 1/4$ ii. $p = 1/13$ iii. $p > 1/4$ iv. $p > 1/13$

b. Which hypothesis best fits the friend's claim? (This is the alternative hypothesis.)
   i. $p = 1/4$ ii. $p = 1/13$ iii. $p > 1/4$ iv. $p > 1/13$

8.96 A friend claims he can predict how a six-sided die will land. The parameter, p, is the long-run likelihood of success, and the null hypothesis is that the friend is guessing.

a. Pick the correct null hypothesis.
   i. $p = 1/6$ ii. $p > 1/6$ iii. $p < 1/6$ iv. $p > 1/2$

b. Which hypothesis best fits the friend's claim? (This is the alternative hypothesis.)
   i. $p = 1/6$ ii. $p > 1/6$ iii. $p < 1/6$ iv. $p > 1/2$

8.97 Votes for Independents Judging on the basis of experience, a politician claims that 50% of voters in Pennsylvania have voted for an independent candidate in past elections. Suppose you surveyed 20 randomly selected people in Pennsylvania, and 12 of them reported having voted for an independent candidate. The null hypothesis is that the overall proportion of voters in Pennsylvania that have voted for an independent candidate is 50%. What value of the test statistic should you report?

8.98 Votes for Independents Refer to Exercise 8.97. Suppose 14 out of 20 voters in Pennsylvania report having voted for an independent candidate. The null hypothesis is that the population proportion is 0.50. What value of the test statistic should you report?

* 8.99 Texting While Driving The mother of a teenager has heard a claim that 25% of teenagers who drive and use a cell phone reported texting while driving. She thinks that this rate is too high and wants to test the hypothesis that fewer than 25% of these drivers have texted while driving. Her alternative hypothesis is that the percentage of teenagers who have texted when driving is less than 25%.

$$H_0: p = 0.25$$
$$H_a: p < 0.25$$

She polls 40 randomly selected teenagers, and 5 of them report having texted while driving, a proportion of 0.125. The p-value is 0.034. Explain the meaning of the p-value in the context of this question.

* 8.100 True/False Test A teacher giving a true/false test wants to make sure her students do better than they would if they were simply guessing, so she forms a hypothesis to test this. Her null hypothesis is that a student will get 50% of the questions on the exam correct. The alternative hypothesis is that the student is not guessing and should get more than 50% in the long run.

$$H_0: p = 0.50$$
$$H_a: p > 0.50$$

A student gets 30 out of 50 questions, or 60%, correct. The p-value is 0.079. Explain the meaning of the p-value in the context of this question.

8.101 ESP Suppose a friend says he can predict whether a coin flip will result in heads or tails. You test him, and he gets 10 right out of 20. Do you think he can predict the coin flip (or has a way of cheating)? Or could this just be something that occurs by chance? Explain without doing any calculations.

8.102 ESP Again Suppose a friend says he can predict whether a coin flip will result in heads or tails. You test him, and he gets 20 right out of 20. Do you think he can predict the coin flip (or has a way of cheating)? Or could this just be something that is likely to occur by chance? Explain without performing any calculations.

8.103 Does Hand Washing Save Lives? In the mid-1800s, Dr. Ignaz Semmelweiss decided to make doctors wash their hands with a strong disinfectant between patients at a clinic with a death rate of 9.9%. Semmelweiss wanted to test the hypothesis that the death rate would go down after the new handwashing procedure was used. What null and alternative hypotheses should he have used? Explain, using both words and symbols. Explain the meaning of any symbols you use.

8.104 Opioid Crisis Suppose you wanted to test the claim that the majority of U.S. voters are satisfied with the government response to the opioid crisis. State the null and alternative hypotheses you would use in both words and symbols.

8.105 Guessing on a True/False Test A true/false test has 50 questions. Suppose a passing grade is 35 or more correct answers. Test the claim that a student knows more than half of the answers and is not just guessing. Assume the student gets 35 answers correct out of 50. Use a significance level of 0.05. Steps 1 and 2 of a hypothesis test procedure are given. Show steps 3 and 4, and be sure to write a clear conclusion.

Step 1: $H_0: p = 0.50$
          $H_a: p > 0.50$

Step 2: Choose the one-proportion z-test. Sample size is large enough, because np_0 is $50(0.5) = 25$ and $n(1 - p_0) = 50(0.50) = 25$, and both are more than 10. Assume the sample is random and $\alpha = 0.05$.

8.106 Guessing on a Multiple-Choice Test A multiple-choice test has 50 questions with four possible options for each question. For each question, only one of the four options is correct. A passing grade is 35 or more correct answers.

a. What is the probability that a person will guess correctly on one multiple-choice question?

b. Test the hypothesis that a person who got 35 right out of 50 is not just guessing, using an alpha of 0.05. Steps 1 and 2 of the hypothesis testing procedure are given. Finish the question by doing steps 3 and 4.

Step 1: $H_0: p = 0.25$
          $H_a: p > 0.25$

Step 2: Choose the one-proportion z-test. n times p is 50 times 0.25, which is 12.5. This is more than 10, and 50 times 0.75 is also more than 10. Assume a random sample.

gUIDED EXERCISES

g 8.35 News on Facebook A 2018 Gallup poll of 3635 randomly selected Facebook users found that 2472 get most of their news about world events on Facebook. Research done in 2013 found that only 47% of all Facebook users reported getting their news about world events on Facebook.

QUESTION Does this sample give evidence that the proportion of Facebook users who get their world news on Facebook has changed since 2013? Carry out a hypothesis test and use a 0.05 significance level.

Step 1 ▶ Hypothesize

H_0: The population proportion that got their news about world events from Facebook in 2013 was 0.47, $p =$ _____
H_a: p_____

Step 2 ▶ Prepare

Choose the one-proportion z-test.
Random and independent sample: Yes
Sample size: $np_0 = 3635(0.47) =$ about 1708, which is more than 10, and $n(1 - p_0) =$ about_____, which is more than _____.
Population size is more than 10 times 3635.

Step 3 ▶ Compute to compare

$\alpha = 0.05$
$\hat{p} =$ _____

$$SE = \sqrt{\frac{p_0(1 - p_0)}{n}} = \sqrt{\frac{0.47(\_\_)}{3635}} = \_\_\_\_\_$$

$$z = \frac{\hat{p} - p_0}{SE} = \frac{0.6801 - \_\_\_}{\_\_\_\_} = \_\_\_\_\_$$

Report your p-value with three decimal digits.
   Check your answers with the accompanying figure. Do not worry if the last digits are a bit different (this can occur due to rounding).

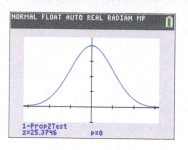

Step 4 ▶ Interpret

Reject H_0 (if the p-value is 0.05 or less) or do not reject H_0 and choose one of the following conclusions:

i. The proportion is not significantly different from 47%. (A significance difference is one for which the p-value is less than or equal to 0.05.)

ii. The proportion is significantly different from 47%.

8.63 Treatment for HIV-1 Molina et al. reported on a study for treatment of patients with HIV-1. The study was a randomized, controlled, double-blind study that compared the effectiveness of ritonavir-boosted darunavir (rbd), the drug currently used to treat HIV-1 with dorovirine, a newly developed drug. Of the 382 subjects taking ritonavir-boosted darunavir, 306 achieved a positive result. Of the 382 subjects taking dorovirine, 321 achieved a positive outcome.

a. Find the sample percentage of subjects who achieved a positive outcome in each group.

b. Perform a hypothesis test to test whether the proportion of patients who achieve a positive outcome with the current treatment (ritonavir-boosted darunavir) is different from the proportion of patients who achieve a positive outcome with the new treatment (dorovirine). Use a significance level of 0.01.

Based on this study, do you think dorovirine might be a more effective treatment option for HIV-1 than ritonavir-boosted darunavir? Why or why not?

QUESTION Find and compare the sample percentage of subjects with a positive outcome in each group. Then perform a hypothesis test to determine if the population proportions are different.

Percentage of positive outcomes for ritonavir-boosted darunavir

$$(rbd) = \frac{306}{382} = 0.801,$$ percentage of positive outcomes for doro-

virine $$(dor) = \frac{321}{382} = 0.840.$$

Step 1 ▶ Hypothesize

Let p_{rbd} be the proportion of those taking ritonavir-boosted darunavir who achieved a positive outcome and let p_{dor} be the proportion of those taking dorovirine who had a positive outcome.

H_0:_____

H_a: $p_{rbd} \neq p_{dor}$

Step 2 ▶ Prepare

Choose the two-proportion z-test. Although we don't have a random sample, we have random assignment to two independent groups. The pooled proportion of positive outcomes is

$$\hat{p} = \frac{306 + 321}{382 + 382} = \frac{627}{764} = 0.8207$$

We must check the following products to make sure none is below 10:

$$n_1\hat{p} = 382(0.8207) = 313.5$$
$$n_1(1 - \hat{p}) = \underline{\quad}(0.1793) = \underline{\quad}$$

Note: In this problem, since $n_1 = n_2$ we do not have to repeat these calculations for n_2 since the resulting products would be the same.

Step 3 ▶ Compute to compare

$\alpha = 0.05$

Refer to the figure:

$z = \underline{\quad}$

$p - value = \underline{\quad}$

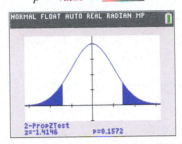

Step 4 ▶ Interpret

Reject or do not reject the null hypothesis and choose i or ii.

i. There is no evidence that there is a difference in the proportion of patients who achieve a positive outcome between the two treatments.

ii. There is a significant different in the proportion of patients who achieve a positive outcome between the two treatment.

TechTips

General Instructions for All Technology

All technologies will use the examples that follow.

EXAMPLE A ▶ Do a one-proportion z-test to determine whether you can reject the hypothesis that a coin is a fair coin if 10 heads are obtained from 30 flips of the coin. Find z and the p-value.

EXAMPLE B ▶ Do a two-proportion z-test: Find the observed value of the test statistic and the p-value that tests whether the proportion of people who support stem cell research changed from 2002 to 2007. In both years, the researchers sampled 1500 people. In 2002, 645 people expressed support. In 2007, 765 people expressed support.

TI-84

One-Proportion z-Test

1. Press **STAT**, choose **TESTS**, and choose **5: 1-PropZTest**.
2. See Figure 8a.

Enter: p_0, .5; x, 10; n, 30.
Leave the default $\neq p_0$
Scroll down to **Calculate** and press **ENTER**.

▲ **FIGURE 8a** TI-84 Input for One-Proportion z-Test

You should get a screen like Figure 8b. If you choose **Draw** instead of **Calculate**, you can see the shading of the Normal curve, as shown in Figure 8.3.

▲ **FIGURE 8b** TI-84 Output for One-Proportion z-Test

Two-Proportion z-Test

1. Press **STAT**, choose **TESTS**, and choose **6: 2-PropZTest**.
2. See Figure 8c.

Enter: **x1, 645; n1, 1500; x2, 765; n2, 1500**.
Leave the default **p1 ≠ p2**
Scroll down to **Calculate** (or **Draw**) and press **ENTER**.

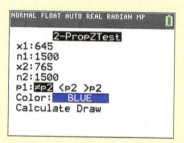

▲ **FIGURE 8c** TI-84 Input for Two-Proportion z-Test

You should get a screen like Figure 8d.

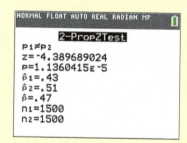

▲ **FIGURE 8d** TI-84 Output for Two-Proportion z-Test

Caution! Beware of p-values that appear at first glance to be larger than 1. In Figure 8d, the p-value is 1.1 times 10 to the negative fifth power (1.1×10^{-5}) or 0.000011.

MINITAB

One-Proportion z-Test

1. **Stat > Basic Statistics > 1-Proportion**.
2. Refer to Figure 8e. Select **Summarized data**. Enter **number of events, 10**; **Number of trials, 30**; check **Perform hypothesis test**; and enter **hypothesized proportion, .5**
3. Click **Options** and for **Method**, select **either Exact or Normal approximation**. (If you wanted to change the alternative hypothesis to one-sided, you would do that also through **Options**.) Click **OK**; click **OK**.

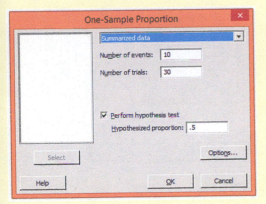

▲ FIGURE 8e Minitab Input for One-Proportion *z*-Test

You should get output that looks like Figure 8f. Note that you also get a 95% confidence interval (95% CI) for the proportion.

Test and CI for One Proportion

Method

p: event proportion
Exact method is used for this analysis.

Descriptive Statistics

N	Event	Sample p	95% CI for p
30	10	0.333333	(0.172874, 0.528120)

Test

Null hypothesis H_0: p = 0.5
Alternative hypothesis H_1: p ≠ 0.5

P-Value
0.099

▲ FIGURE 8f Minitab Output for One-Proportion *z*-Test

Two-Proportion *z*-Test

1. **Stat > Basic Statistics > 2 Proportions**
2. See Figure 8g.
   Select **Summarized data**. Enter **Sample 1 Number of events: 645, Number of trials: 1500; Sample 2 Number of events: 765, Number of trials: 1500.**

3. Click **OK**.

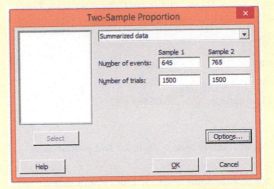

▲ FIGURE 8g Minitab Input for Two-Proportion *z*-test

Your output should look like Figure 8h. Note that it includes a 95% confidence interval for the difference between the proportions, the $z_{observed}$, and the p-value for the normal approximation. The exact p-value based on binomial distribution is also included.

Test and CI for Two Proportions

Method

p_1: proportion where Sample 1 = Event
p_2: proportion where Sample 2 = Event
Difference: p_1 - p_2

Descriptive Statistics

Sample	N	Event	Sample p
Sample 1	1500	645	0.430000
Sample 2	1500	765	0.510000

Estimation for Difference

Difference	95% CI for Difference
-0.08	(-0.115605, -0.044395)

CI based on normal approximation

Test

Null hypothesis H_0: p_1 - p_2 = 0
Alternative hypothesis H_1: p_1 - p_2 ≠ 0

Method	Z-Value	P-Value
Normal approximation	-4.40	0.000
Fisher's exact		0.000

▲ FIGURE 8h Minitab Output for Two-Proportion *z*-Test and Interval

---

EXCEL

One-Proportion *z*-Test

1. **XLSTAT > Parametric tests > Tests for one proportion**.
2. See Figure 8i.

Enter: **Frequency, 10; Sample size, 30; Test proportion, .5.** Click to uncheck the continuity correction. (If you wanted a one-sided hypothesis, you would click **Options**.)
Click **OK**.

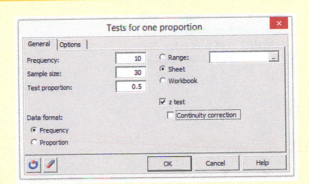

▲ FIGURE 8i XLSTAT input for One Proportion *z*-Test

When the output appears, you may need to change the column width to see the answers. Click **Home**, and in the **Cells** group click **Format** and **AutoFit Column Width**. The relevant parts of the output are shown here.

Difference	−0.167
z (Observed value)	−1.826
p-value (Two-tailed)	0.068

Two-Proportion z-Test

1. **XLSTAT > Parametric tests > Tests for two proportions**.
2. See Figure 8j.

Enter: **Frequency 1, 645; Sample size 1, 1500; Frequency 2, 765; Sample size 2, 1500**.
(If you wanted a one-sided alternative, you would click **Options**.)
Click **OK**.

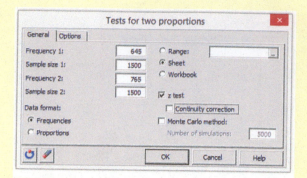

▲ **FIGURE 8j** XLSTAT Input for Two-Proportion z-Test

The relevant parts of the output are shown.

Difference	−0.080
z (Observed value)	−4.404
p-value (Two-tailed)	<0.0001

STATCRUNCH

One-Proportion z-Test and Confidence Interval

1. **Stat > Proportion Stats > One Sample > With Summary**
2. Enter: **# of successes, 10; # of observations, 30**.
   Select the **Hypothesis Test** or **Confidence interval** option.
   a. Leave Hypothesis test checked. Enter: **Ho: p=, 0.5**. Leave the **Alternative 2-tailed**, $H_A: P \neq 0.5$, which is the default.
   b. For a Confidence interval, leave the **Level, 0.95** (the default). For **Method**, leave *Standard-Wald*, the default.
3. Click **Compute!**

Figure 8k shows the output for the hypothesis test.

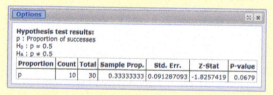

▲ **FIGURE 8k** StatCrunch Output for One-Proportion z-Test

Two-Proportion z-Test

1. **Stat > Proportion Stats > Two Sample > With Summary**
2. Refer to Figure 8l.

Enter: **Sample 1:# of successes, 645;**
              # of observations, 1500;
   Sample 2:# of successes, 765;
              # of observations, 1500.

3. You may want to change the alternative hypothesis from the two-sided default (or want a confidence interval). Otherwise, click **Compute!**

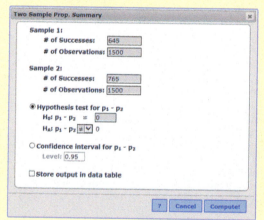

▲ **FIGURE 8l** StatCrunch Input for Two-Proportion z-Test

Figure 8m shows the StatCrunch output.

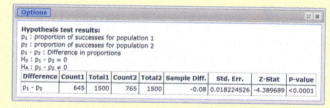

▲ **FIGURE 8m** StatCrunch Output for Two-Proportion z-Test

9 Inferring Population Means

THEME

The sample mean can be used to estimate the population mean. To understand how to form confidence intervals and carry out hypothesis tests, we need to understand the precision and accuracy of the sample mean as an estimate, and we need to know its probability distribution. To compare two means, we calculate the difference of the sample means, and we also need to know the accuracy, precision, and probability distribution for the difference of two samples means.

Brewing beer is a tricky business. Beer has only four main ingredients: malted barley, hops, yeast, and water. However, these four ingredients must be mixed in precise quantities at precise temperatures. Having some experience in trying to get this mix just right, in the late 1800s the Guinness brewery in Dublin, Ireland, began to hire the best and brightest science graduates to help them perfect the brewing process. One of these graduates, hired in 1899 at the age of 23, was William Sealy Gosset (1876–1937), who had majored in chemistry and mathematics. One of Gosset's jobs was to measure the amount of yeast in a small sample of beer. On the basis of this sample, he was to estimate the mean amount of yeast in the beers produced. If this yeast amount was too high or too low, then something was wrong, and the process would have to be fixed.

Naturally, uncertainty played a role. Suppose the average yeast count in his sample was too high or too low. Did this indicate that the mean yeast count in the entire factory was off? Or was his sample different just because of chance? The statistical science of the day knew how to answer this question if the number of samples was large, but Gosset worked in a context in which large samples were just too expensive and time-consuming to collect. A decision had to be based on a small sample. Gosset solved the problem, and his approach (which is called the *t*-test) is now one of the most widely used techniques in statistics.

Estimating one or more population means is still an important part of science and public policy. How do we compare the effects of different drugs on epilepsy? How do commuting times vary between cities? Does our sense of smell differ when we are sitting upright compared to lying down? To answer these questions, we need good estimates of the population means, and these estimates must be based on reliable data from small samples.

In Chapters 7 and 8 you learned two important techniques for statistical inference: the confidence interval and the hypothesis test. In those chapters, we applied statistical inference for population proportions. In this chapter, we use the same two techniques for making inferences about means of populations. We begin with inference for one population and conclude with inferring the difference between the means of two populations.

CASE STUDY

You Look Sick! Are You Sick?

Have you ever looked at a person, perhaps a stranger passing on the street, and thought "Uh oh! He's got the flu." Perhaps it was because of obvious symptoms, such as sneezing or coughing. But maybe it was nothing you could put a finger on; you just felt that they were sick. Being able to avoid sick people is one strategy for avoiding illness yourself. Do humans have the ability to identify sick people simply by looking at them?

Researchers in Sweden decided to find out. They took pictures of a variety of people in two states: healthy and sick. For the sick state, they injected the people with bacteria, "lipopolysaccharide" (LPS), which would cause short-term symptoms of a cold or flu. They took a picture before this injection, when the person was healthy, and then again two hours after this injection. A sample of 60 raters were shown these pictures, but were not told whether the person was healthy or sick. (The pictures were shown in a random order, except that a picture of the same person was never shown sequentially, to prevent the raters from doing a side-by-side comparison.) The raters were asked to rate how sick the person looked on a scale of 1 to 7, with 7 being the sickest (Axelsson, et. al 2018).

The researchers concluded that, compared to the placebo, the injection made people look sicker and provided a confidence interval of 0.41 to 0.55 for the mean difference, with units corresponding to the sickness rating scale of 1 to 7. They also concluded that the effect was significant with "$P < 0.001$".

In this chapter we discuss how confidence intervals can be used to estimate characteristics of a population and to estimate the size of an effect in a randomized experiment. We'll also see that confidence intervals and hypothesis tests can be used together to gain information about a study. At the end of this chapter, we will return to this study and see if we can better understand its conclusions.

SECTION 9.1

Sample Means of Random Samples

As you learned in Chapter 7, we estimate population parameters by collecting a random sample from that population. We use the collected data to calculate a statistic, and this statistic is used to estimate the parameter. Whether we are using the statistic $\hat{p}$ to estimate the parameter p or are using $\bar{x}$ to estimate μ, if we want to know how close our estimate is to the truth, we need to know how far away that statistic is, typically, from the parameter.

Just as we did in Chapter 7 with $\hat{p}$, we now examine three characteristics of the behavior of the sample mean: its accuracy, its precision, and its probability distribution. By understanding these characteristics, we'll be able to measure how well our estimate performs and thus make better decisions.

As a reminder, Table 9.1 shows some commonly used statistics and the parameters they estimate. (This table originally appeared as Table 7.2.)

Statistic (based on data)		Parameter (typically unknown)	
Sample mean	$\bar{x}$	Population mean	μ (mu)
Sample standard deviation	s	Population standard deviation	σ (sigma)
Sample variance	s^2	Population variance	σ^2
Sample proportion	$\hat{p}$ (p-hat)	Population proportion	p

◀ **TABLE 9.1** Commonly used statistics and their corresponding parameters.

This chapter uses much of the vocabulary introduced in Chapters 7 and 8, but we'll remind you of important terms as we proceed. To help you visualize how a sample mean based on randomly sampled data behaves, we'll make use of the by-now-familiar technique of simulation. Our simulation is slightly artificial, because to do a simulation, we need to know the population. However, after using a simulation to understand how the sample mean behaves in this artificial situation, we will discuss what we do in the real world when we do not know very much about the population.

Accuracy and Precision of a Sample Mean

The reason why the sample mean is a useful estimator for the population mean is that the sample mean is accurate and, with a sufficiently large sample size, very precise. The accuracy of an estimator, you'll recall, is measured by the **bias**, and the precision is measured by the standard error. You will see the following in this simulation:

1. The sample mean is unbiased when estimating the population mean—that is, on average, the sample mean is the same as the population mean.

2. The **precision** of the sample mean depends on the variability in the population, but the more observations we collect, the more precise the sample mean becomes.

For our simulation, we'll use the population that consists of all emergency calls to the Los Angeles Fire Department between 2013 and 2016. We're going to examine the *response time*, the time it takes for the first emergency vehicle to arrive on scene from the time the emergency is called in. To keep things simple, we're including only response times of less than one hour (60 minutes). (Keep in mind that some events often result in multiple call-ins over a very long time.) Figure 9.1 shows a sketch of the population distribution, which consists of over 1.5 million arrivals. It's right-skewed, with a mean response time of 6.3 minutes and a standard deviation of 2.8 minutes.

Details

Mu-sings
The mean of a population, represented by the Greek character μ, is pronounced "mu" as in *music*.

Data Moves ▶
The data come from https://data.lacity.org and have been cleaned and then filtered so that only positive-valued response times less than 60 minutes are included. For each event, only the first arriving vehicle is provided. The response is the difference between "Incident Time" and "On Scene Time." The file is too large for some systems, and so we provide a reduced version.

Data ▶ lafdsmall.csv

The population parameters are, in symbols,

$$\mu = 6.29 \text{ minutes}$$

$$\sigma = 16 \text{ minutes}$$

▶ FIGURE 9.1 The population distribution for the response times for the Los Angeles Fire Department, with a mean of 6.3 minutes and a standard deviation of 2.8 minutes.

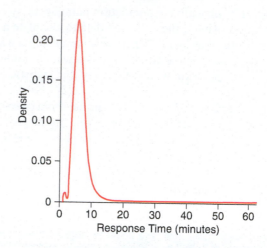

Imagine taking a random sample of 50 rows from this data set. You'd end up with 50 response times. What would the mean be of these 50 numbers be? Would it likely be within 1 minute of the population mean of 6.3 minutes? Within 30 seconds? Would it be unusual for it to be more than 2 minutes above or below the population mean? (That is, are response times smaller than 4.3 minutes or greater than 8.3 minutes unusual?)

We're going to carry out a simulation in which we randomly sample 50 observations from this data set and calculate the average response time. We'll then repeat this action many, many times. (The exact number of repetitions we perform isn't important for this discussion.) We are interested in two question: (1) What is the typical value of the sample mean in this simulation? If it is 6.3—the value of the population mean—then the sample mean is unbiased. (2) If the estimator is unbiased, then how far will it be from 6.3? In other words, how much spread is there in the distribution of sample means? This spread helps us measure the precision of the sample mean as an estimator of the population mean.

For example, our very first sample of 50 response times had a mean time of 5.9 minutes. We plotted this sample mean, as well as the means of the many other samples we took, in Figure 9.2a. The plot is on the same scale as in Figure 9.1 so that you can see how much narrower this distribution of sample means is compared to the population distribution. In fact, it is so narrow it is a little difficult to see, and so Figure 9.2b expands the scale so that we can see more detail. The triangle marks the location of the mean of this distribution of sample means.

▶ FIGURE 9.2 Each dot represents the sample mean of 50 observations randomly selected from the population whose distribution is shown on Figure 9.1. Note that the spread of this distribution as shown in the original x-axis scale, Figure **(a)**, is quite narrow. Figure **(b)** expands the scale somewhat so we can get a better sense of the shape and spread of the distribution of sample means.

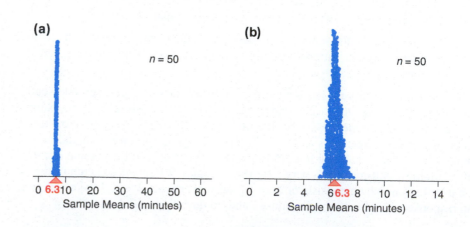

Figure 9.2 is a very approximate picture of the **sampling distribution** of the sample mean for samples of size 50. Recall that a sampling distribution is a probability distribution of a statistic; in this case, the statistic is the sample mean. You can think of the sampling distribution as the distribution of *all* possible sample means that would result from drawing repeated random samples of a certain size from the population.

When the mean of the sampling distribution is the same value as the population mean, we say that the statistic is an **unbiased estimator**. This appears to be the case here, because both the mean of the sampling distribution in Figure 9.2 and the population mean are both about 6.3 minutes.

The standard deviation of the sampling distribution is what we call the **standard error**. The standard error measures the precision of an estimator by telling us how much the statistic varies from sample to sample. For the sample mean, the standard error is smaller than the population standard deviation. We can see this by comparing the spread of the population distribution, shown in Figure 9.1, with the spread in Figure 9.2, which shows an approximate sampling distribution for the mean based on a sample size of 50. Soon you'll see how to calculate the standard error.

What happens to the center and spread of the sampling distribution if we increase the sample size? Let's start all over with the simulation. This time, we take a random sample of 100 emergencies—twice as many as last time—and calculate the mean. We then repeat this action many times. Figure 9.3 shows the results for this simulation and also for simulations in which each sample mean is based on 400 and then 900 emergencies. The scale of the *x*-axis is the same. Note that the spread of the distributions decreases as the sample size gets larger.

🔄 Looking Back

Sampling Distribution
In Chapter 7 we introduced the sampling distribution of sample proportions. The sampling distribution of sample means is the same concept: It is a distribution that gives us probabilities for sample means drawn from a population. The sampling distribution is the distribution of all possible sample means.

What Have We Demonstrated with These Simulations?

Because the sampling distributions are always centered at the population mean, we have demonstrated that the sample mean is an unbiased estimator of the population mean. We saw this for only one type of population distribution: the Normal distribution. But in fact, this is the case for any population distribution.

(a)
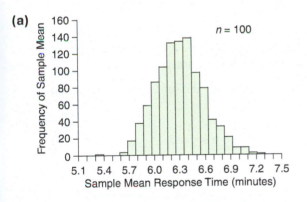

(b)
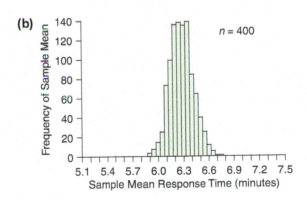

(c)
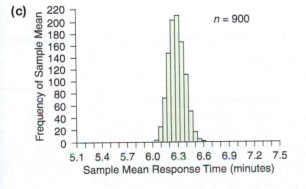

◄ FIGURE 9.3 **(a)** Histogram of a large number of sample means. Each sample mean is based on a sample of 100 randomly selected emergencies. **(b)** Sample means based on samples of 400 emergencies. **(c)** Sample means based on samples of 900 emergencies. The sampling distribution is narrower for larger sample sizes, reflecting a smaller standard error.

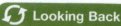 **Looking Back**

Sample Proportions from Random Samples
Compare the properties of the sample mean to those of the sample proportion, $\hat{p}$, given in Chapter 7. The sample proportion is also an unbiased estimator (for estimating the population proportion, p). It has standard error
$$\sqrt{\frac{p(1 - p)}{n}}.$$

We have demonstrated that the standard deviation of the sampling distribution, which is called the standard error of the sample mean, is smaller when based on a larger sample size. This is true for any population distribution.

We can be more precise. If the symbol μ represents the mean of the population and if σ represents the standard deviation of the population, then

1. The mean of the sampling distribution is also μ (which tells us that the sample mean is unbiased when estimating the population mean).

2. The standard error is $\dfrac{\sigma}{\sqrt{n}}$ (which tells us that the standard error depends on the population distribution and is smaller for larger samples).

KEY POINT For all populations, the sample mean, if based on a random sample, is unbiased when estimating the population mean. The standard error of the sample mean is $\dfrac{\sigma}{\sqrt{n}}$, so the sample mean is more precise for larger sample sizes.

EXAMPLE 1 iTunes Music Library Statistics

A student's iTunes music library has a very large number of songs. The mean length of the songs is 243 seconds, and the standard deviation is 93 seconds. The distribution of song lengths is right-skewed. Using his digital music player, this student will create a playlist that consists of 25 randomly selected songs.

QUESTIONS

a. Is the mean value of 243 minutes an example of a parameter or a statistic? Explain.

b. What should the student expect the average song length to be for his playlist?

c. What is the standard error for the sample mean song length of 25 randomly selected songs?

SOLUTIONS

a. The mean of 243 is an example of a parameter, because it is the mean of the population that consists of all of the songs in the student's library.

b. The sample mean length can vary, but it is typically the same as the population mean: 243 seconds.

c. The standard error is $\dfrac{\sigma}{\sqrt{n}} = \dfrac{93}{\sqrt{25}} = \dfrac{93}{5} = 18.6$ seconds.

TRY THIS! Exercise 9.9

SECTION 9.2

The Central Limit Theorem for Sample Means

In our simulations of drawing random emergencies from the Los Angeles Fire Department data set, you might have noticed a surprising fact: Although the shape of the population was right-skewed and a little unusual, the shape of the distribution of sample means was symmetric. Your eyes weren't deceiving you; in fact, the distributions of sample means were very close to being Normal distributions.

This was no accident. The shape of the sampling distribution of means is always close to Normal, regardless of the shape of the population distribution. Just how "close" to Normal it gets depends on the sample size. For small sample sizes, the sampling distribution of means can still be pretty far from Normal looking. But for larger sample sizes, in many contexts 25 or more observations will do, the approximation is quite good. This is the conclusion of the Central Limit Theorem, an important mathematical theorem that tells us that as long as the sample size is large, we can use the Normal distribution to perform statistical inference with the mean, regardless of the population the data are sampled from.

The **Central Limit Theorem (CLT)** assures us that no matter what the shape of the population distribution, if a sample is selected such that the following conditions are met, then the distribution of sample means follows an approximately Normal distribution. The mean of this distribution is the same as the population mean. The standard deviation (also called the standard error) of this distribution is the population standard deviation divided by the square root of the sample size. As a rule of thumb, sample sizes of 25 or more may be considered "large."

When determining whether you can apply the Central Limit Theorem to analyze data, there are three conditions to consider:

Condition 1: *Random Sample and Independence.* Each observation is collected randomly from the population, and observations are independent of each other. The sample can be collected either with or without replacement.

Condition 2: *Large Sample.* Either the population distribution is Normal or the sample size is large.

Condition 3: *Big Population.* If the sample is collected without replacement (as is done in an Simple Random Sample (SRS)), then the population must be at least 10 times larger than the sample size.

Each of these conditions is important. Consider first the random sample and independence condition. If this condition is met, then the sample mean will be unbiased. In other words, the mean of the sampling distribution will be μ, the same as the mean of the population. Condition 1 tells us about the location of the center of the sampling distribution.

If both the Random Sample and Independence condition *and* the Big Population condition are satisfied, then the standard error is $\sigma / \sqrt{n}$. In other words, the standard deviation of the sampling distribution is the population standard deviation divided by the square root of the sample size. Conditions 1 and 3 give us the value of the spread of the sampling distribution.

If all three conditions are true, then we also know the shape, approximately. For a large sample size, the shape will be approximately Normal. But to know the shape, all three conditions must be true.

Looking Back

CLT for Proportions
In Chapter 7, you saw that the Central Limit Theorem applies to sample proportions. Here you'll see that it also applies to sample means.

> **KEY POINT**
>
> The sampling distribution of $\bar{x}$ is approximately $N\left(\mu, \dfrac{\sigma}{\sqrt{n}} \right)$, where μ is the mean of the population and σ is the standard deviation of the population. The larger the sample size, n, the better the approximation. If the population is Normal to begin with, then the sampling distribution is exactly a Normal distribution, regardless of the sample size.

EXAMPLE **2** Describing Sampling Distributions

The population distribution of all emergency resopnse times from the Los Angeles Fire Department is right-skewed. Suppose we take a random sample of a certain size from this population. We calculate the mean response time for this sample and record the value. We then repeat this many thousands of times. Recall that the mean of the population is 6.3 minutes and the standard deviation is 2.8 minutes.

QUESTIONS

a. Describe the sampling distribution if the sample size is 9. Be sure to describe the shape, center, and spread.

b. Describe the sampling distribution if the sample size is 81.

SOLUTIONS

a. A sample size of 9 is too small for the Central Limit Theorem to "kick in". So the shape of the distribution will likely still be right-skewed, since the population distribution is right-skewed. Because the sample mean is an unbiased estimator of the population mean, the center of the sampling distribution will be 6.3 minutes. The spread of the sampling distribution is measured by the standard error:

$$\sigma/\sqrt{n} = 2.8/\sqrt{9} = 2.8/3 = 0.93 \text{ minutes}$$

b. Now the sample size is large enough for us to expect the sampling distribution to be very close to a Normal distribution. The mean will still be 6.3 minutes, but the spread is now smaller:

$$\sigma/\sqrt{n} = 2.8/\sqrt{9} = 2.8/3 = 0.93 \text{ minutes}$$

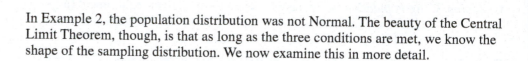

TRY THIS!   Exercise 9.11

In Example 2, the population distribution was not Normal. The beauty of the Central Limit Theorem, though, is that as long as the three conditions are met, we know the shape of the sampling distribution. We now examine this in more detail.

Visualizing Distributions of Sample Means

The sketch in Figure 9.4 shows the distribution of in-state tuition and fees for all two-year colleges in the United States for the 2014–2015 academic year (Integrated Postsecondary Education Data System IPEDS, U.S. Dept. of Education). Note that the distribution looks nothing at all like a Normal distribution. It is skewed and multimodal. (The mode around $1000 is due largely to the cost of two-year colleges in California.)

▶ **FIGURE 9.4** Population distribution of annual tuitions and fees at all two-year colleges in the United States for the 2014–2015 academic year.

Data Moves ▶ The IPEDS website provides a tool to download a wide variety of data in ".csv format" on all institutions of higher learning in the United States. We cleaned the data by providing friendly variable names, keeping only data on two-year colleges, and removing "administrative units" and "for-profits."

Data ▶ twoyears.csv

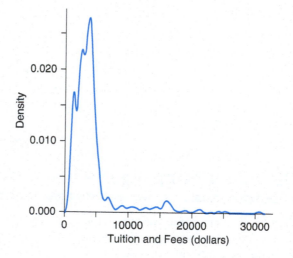

This distribution represents the distribution of a population, because it includes *all* two-year colleges. The mean of this population—the "typical" tuition of all two-year colleges—is $4274.

Using this distribution, we now show the results of a simulation that should be starting to feel familiar. First, we take a random sample of 30 colleges. The distribution of this sample is shown in Figure 9.5. We find the mean tuition of the 30 colleges in the sample and record this figure; for example, the sample mean for the sample shown in Figure 9.5 is about $3425.

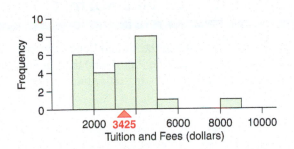

Tuition and Fees (dollars)

◄ **FIGURE 9.5** Distribution of one sample of 30 colleges taken from the population of all colleges. The mean of this sample, $3425, is indicated.

We repeat this activity (that is, we sample another 30 colleges from the population of all colleges and record the mean tuition of the sample) 200 times. When we are finished, we have 200 sample mean tuitions, each sample mean based on a sample of 30 colleges. Figure 9.6a shows this distribution. Figure 9.6b shows the distribution of averages when, instead of sampling 30 colleges, we triple the number and sample 90 colleges. What differences do you see among the population distribution (Figure 9.4), the distribution of one sample (Figure 9.5), the sampling distribution when the sample size is 30 (Figure 9.6a), and the sampling distribution when the sample size is 90 (Figure 9.6b)?

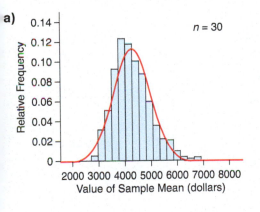

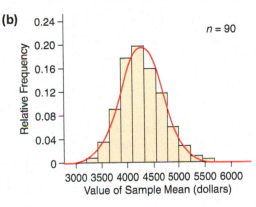

◄ **FIGURE 9.6 (a)** Distribution of sample means, where each sample mean is based on a sample size of $n = 30$ college tuitions and is drawn from the population shown in Figure 9.4. This is (approximately) the sampling distribution of $\bar{x}$ when $n = 30$. A Normal curve is superimposed. **(b)** The approximate sampling distribution of $\bar{x}$ when $n = 90$.

Both of the sampling distributions in Figures 9.6a and 9.6b show us the values and relative frequencies for $\bar{x}$, but they are based on different sample sizes. We see that even though the *population* distribution has an unusual shape (Figure 9.4), the sampling distributions for $\bar{x}$ are fairly symmetric and unimodal. Although the Normal curve that is superimposed doesn't match the histogram too closely when $n = 30$, the match is pretty good for $n = 90$. This is an example of one of those fairly rare times in which our rule of thumb doesn't work. Even though the sample size is bigger than 25, the sampling distribution is not quite a Normal distribution. (But it's close!)

This is exactly what the CLT predicts. When the sample size is large enough, we can use the Normal distribution to find approximate probabilities for the values of $\bar{x}$ when we take a random sample from the population.

How large is large enough? Unlike in Chapter 7, where we worked with sample proportions, we can't provide a hard-and-fast rule for sample size. For nearly all examples in this text, though not always in real life, 25 is large enough.

🔄 Looking Back

Distribution of a Sample vs. Sampling Distribution
Remember that these are two different concepts. The *distribution of a sample*, from Chapter 3, is the distribution of one single sample of data (Figure 9.5). The *sampling distribution*, on the other hand, is the probability distribution of an estimator or statistic such as the sample mean (Figures 9.6a and 9.6b).

Applying the Central Limit Theorem

The Central Limit Theorem helps us find probabilities for sample means when those means are based on a random sample from a population. Example 3 demonstrates how we can answer probability questions about the sample mean even if we can't answer probability questions about individual outcomes.

EXAMPLE 3 Pulse Rates Are Not Normal

According to one very large study done in the United States, the mean resting pulse rate of adult women is about 74 beats per minute (bpm), and the standard deviation of this population is 13 bpm. The distribution of resting pulse rates is known to be skewed right.

QUESTIONS

a. Suppose we take a random sample of 36 women from this population. What is the approximate probability that the average pulse rate of this sample will be below 71 or above 77 bpm? (In other words, what is the probability that it will be more than 3 bpm away from the population mean of 74 bpm?)

b. Can you find the probability that a single adult woman, randomly selected from this population, will have a resting pulse rate more than 3 bpm away from the mean value of 74?

SOLUTIONS

a. It doesn't matter that the population distribution is not Normal. Because the sample size of 36 women is relatively large, the distribution of sample means will be approximately (though not exactly) Normal.

The mean of this Normal distribution will be the same as the population mean: $\mu = 74$ bpm. The standard deviation of this distribution is the standard error:

$$SE = \frac{\sigma}{\sqrt{n}} = \frac{13}{\sqrt{36}} = \frac{13}{6} = 2.167$$

To use the Normal table to find probabilities requires that the values of 71 bpm and 77 bpm be converted to standard units. First, we convert 71 bmp to standard units:

$$z = \frac{\bar{x} - \mu}{SE} = \frac{71 - 74}{2.167} = \frac{-3}{2.167} = -1.38$$

Similarly, we find that 77 bmp converts to $+1.38$ standard units. Figure 9.7 shows that the area that corresponds to the probability that the sample mean pulse rate will be more than 1.38 standard errors away from the population mean pulse rate. This probability is calculated to be about 17%.

▶ **FIGURE 9.7** Area of the Normal curve outside of z-scores of −1.38 and 1.38.

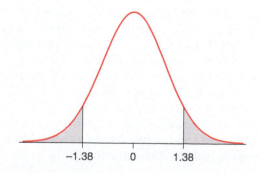

CONCLUSIONS

a. The approximate probability that the average pulse of 36 adult women will be more than 3 bpm away from 74 bpm is about 17%.

b. We cannot find the probability for a single woman because we do not know the probability distribution. We know only that it is "right-skewed," which is not enough information to find actual probabilities.

TRY THIS!   Exercise 9.13

Many Distributions

It's natural at this point to feel that you have seen a confusingly large number of types of distributions, but it's important that you keep them straight. The *population distribution* is the distribution of values from the population. Figure 9.4 (two-year college tuitions) is an example of a population distribution because it shows the distribution of *all* two-year colleges. Figure 9.1 (response times for all emergency calls) is another example of a population distribution. For some populations, we don't know precisely what this distribution is. Sometimes we assume (or know) it is Normal, sometimes we know it is skewed in one direction or the other, and sometimes we know almost nothing.

From the population we draw a random sample of *n* observations. We can make a histogram of these data. This histogram gives us a picture of the *distribution of the sample*. If the sample size is large, and if the sample is random, then the sample will be representative of the population, and the distribution of the sample will look similar to (but not the same as!) the population distribution. Figure 9.5 is an example of the distribution of a sample of size $n = 30$ taken from the population of two-year college tuitions.

The *sampling distribution* is more abstract. If we take a random sample of data and calculate the sample mean (the center of the distribution of the sample) and then repeat this many, many times, we will get an idea of what the sampling distribution looks like. Figures 9.6a and 9.6b are examples of approximate sampling distributions for the sample mean, based on samples from the two-year college tuition data. Note that these do not share the same shape as the population or the sample; they are both approximately Normal.

> **! Caution**
>
> **CLT Not Universal**
> The CLT does not apply to all statistics you run across. It does not apply to the sample median, for example. No matter how large the sample size, you cannot use the Normal distribution to find a probability for the median value. It also does not apply to the sample standard deviation.

EXAMPLE 4   Identify the Distribution

Figure 9.8 shows three distributions. One distribution is a population. The other two distributions are (approximate) sampling distributions of sample means randomly sampled from that population. One sampling distribution is based on sample means of size 10, and the other is based on sample means of size 25.

QUESTION   Which graph (a, b, or c) is the population distribution? Which shows the sampling distribution for the mean with $n = 10$? Which with $n = 25$?

SOLUTION   The Central Limit Theorem tells us that sampling distributions for means are approximately Normal. This implies that Figure 9.8b is not a sampling distribution, so it must be the population distribution from which the samples were taken. We know that the sample mean is more precise for larger samples, and because Figure 9.8a has the larger standard error (is wider), it must be the graph associated with $n = 10$. This means that Figure 9.8c is the sampling distribution of means with $n = 25$.

(a)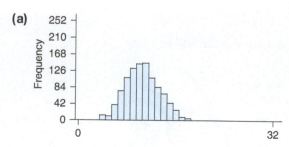

(b)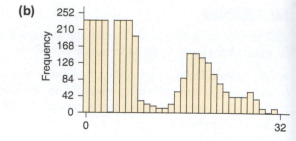

(c)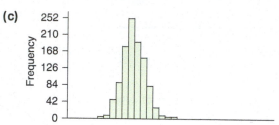

◀ **FIGURE 9.8** Three distributions, all on the same scale. One is a population distribution, and the other two are sampling distributions for means sampled from the population.

TRY THIS! Exercise 9.15

SNAPSHOT ▶ The Sample Mean ($\bar{x}$)

WHAT IS IT? ▶ The arithmetic average of a sample of data.

WHAT DOES IT DO? ▶ Estimates the mean value of a population, μ. The mean is used as a measure of what is "typical" for a population.

HOW DOES IT DO IT? ▶ If the sample was a random sample, then the sample mean is unbiased, and we can make the precision of the estimator as good as we want by taking a large enough sample size.

HOW IS IT USED? ▶ If the sample size is large enough (or the population is Normal), we can use the Normal distribution to find the probability that the sample mean will take on a value in any given range. This lets us know how wrong our estimate could be.

The *t*-Distribution

The hypothesis tests and confidence intervals that we will use for estimating and testing the mean are based on a statistic called the **t-statistic**:

$$t = \frac{\bar{x} - \mu}{SE_{est}}$$

$$\text{where } SE_{est} = \frac{s}{\sqrt{n}}$$

The *t*-statistic is very similar to a *z*-score for the sample mean. In the numerator, we subtract the population mean from the sample mean. Then we divide not by the standard error, but instead by an *estimate* of the standard error.

Why do we use an estimate, $s/\sqrt{n}$, and not the actual standard error $\sigma/\sqrt{n}$? The reason is that in real life, we almost never know the value of σ, the population standard deviation. So instead, we replace it with an estimate: the sample standard deviation, *s*.

🔄 Looking Back

Sample Standard Deviation
In Chapter 3 we gave the formula for the sample standard deviation:

$$s = \sqrt{\frac{\Sigma(x - \bar{x})^2}{n - 1}}$$

If we find ourselves in a situation in which we *do* know the population standard deviation then we can use the actual standard error in our test statistic. In this case, the resulting statistic is called a *z*-statistic because it is simply a *z*-score:

$$z = \frac{\bar{x} - \mu}{\left(\dfrac{\sigma}{\sqrt{n}}\right)}$$

The *z*-statistic follows an approximately Normal distribution if the sample size is large enough, for exactly the same reasons as the *z*-statistic provided in Formula 8.1 in Chapter 8.

However, we rarely get to use a *z*-statistic and so must instead use the *t*-statistic. The *t*-statistic does *not* follow the Normal distribution. One reason for this is that the denominator changes with every sample. For this reason, the *t*-statistic is more variable than the *z*-statistic (whose denominator is the same in each sample of the same size). Instead, if the three conditions for using the Central Limit Theorem hold, the *t*-statistic follows a distribution called—surprise!—the ***t*-distribution**. This was Gosset's great discovery at the Guinness brewery. When small sample sizes were used to make inferences about the mean, even if the population was Normal, the Normal distribution just didn't fit the results that well. Gosset discovered a new distribution, which he, along with his collaborator Ronald Fisher, called the *t*-distribution. The *t*-distribution turned out to be a better model than the Normal for the sampling distribution of $\bar{x}$ when σ is not known.

The *t*-distribution shares many characteristics with the $N(0, 1)$ distribution. Both are symmetric, are unimodal, and might be described as "bell-shaped." However, the *t*-distribution has thicker tails. This means that in a *t*-distribution, it is more likely that we will see extreme values (values far from 0) than it is in a standard Normal distribution.

The *t*-distribution's shape depends on only one parameter, called the **degrees of freedom (df)**. The number of degrees of freedom is (usually) an integer: 1, 2, 3, and so on. If the df is small, then the *t*-distribution has very thick tails. As the degrees of freedom get larger, the tails get thinner. Ultimately, when the df is infinitely large, the *t*-distribution is exactly the same as the $N(0, 1)$ distribution.

Figure 9.9 shows *t*-distributions with 1, 10, and 40 degrees of freedom. In each case, the *t*-distribution is shown with a $N(0, 1)$ curve so that you can compare them. (We compare to the $N(0, 1)$ because it is familiar and because, as you can see, the *t*-distribution and the Normal distribution are very similar.) The *t*-distribution is the one whose tails are "higher" at the extremes. Note that by the time the degrees of freedom reaches 40 (Figure 9.9c), the *t*-distribution and the $N(0, 1)$ distribution are too close to tell apart (on this scale).

> **Details**
>
> **Degrees of Freedom**
> Degrees of freedom are related to the sample size: Generally, the larger the sample size, the larger the degrees of freedom. When estimating a single mean, as we are doing here, the number of degrees of freedom is equal to the sample size minus one.
> $$df = n - 1$$

(a)

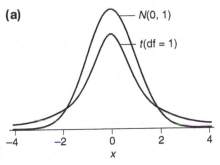

(b)

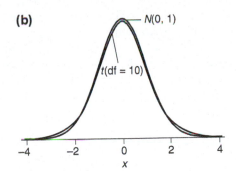

(c)

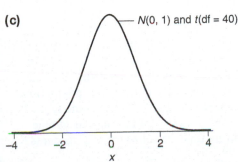

◀ **FIGURE 9.9 (a)** A *t*-distribution with 1 degree of freedom, along with a $N(0, 1)$ distribution. The *t*-distribution has much thicker tails. **(b)** The degrees of freedom are now equal to 10, and the tails are only slightly thicker in the *t*-distribution. **(c)** The degrees of freedom are now equal to 40, and the two distributions are visually indistinguishable.

Answering Questions about the Mean of a Population

Do you commute to work? How long does it take you to get there? Is this amount of time typical for others in your state? Which state has the greatest commuting times? This information is important not just to those of us who must fight traffic every day but also to business leaders and politicians who make decisions about quality of living and the cost of doing business. The U.S. Census performs periodic surveys that determine, among other things, commuting times around the country. In 2012, the state of Maryland had the greatest mean commuting time, which was 31.9 minutes. South Dakota was lowest at 16.7 minutes.

These means are not population means. They are *estimates* of the mean commuting time for all residents in these states who work outside of their homes. We can learn the true mean commuting time only by asking all residents, which is clearly too time-consuming to do very often. Instead, the U.S. Census takes a random sample of U.S. residents to estimate these values.

In this section we present two techniques for answering questions about the population mean. Confidence intervals are used for estimating parameter values. Hypothesis tests are used for deciding whether a parameter's value is one thing or another. These are the same methods that were introduced in Chapter 7 (confidence intervals) and Chapter 8 (hypothesis tests) for population proportions, but here you'll see how they are modified to work with means.

Estimation with Confidence Intervals

Confidence intervals are a technique for communicating an estimate of the mean, along with a measure of the uncertainty in our estimate. The job of a confidence interval is to provide us with a range of plausible values that, according to the data, are highly plausible values for the unknown population mean. For instance, the range of plausible values for the mean commuting time for all South Dakota residents is 16.3 to 17.1 minutes.

Not all confidence intervals do an equally good job; the "job performance" of a confidence interval is therefore measured with something called the **confidence level**. The higher this level, the better the confidence interval performs. The confidence level for mean South Dakota commuting times is 90%, which means we can be confident that this interval contains the true mean.

Sometimes, you will be in a situation in which you will know only the sample mean and sample standard deviation. In these situations, you can use a calculator to find the confidence interval. However, if you have access to the actual data, you are much better off using statistical software to do all the calculations for you. We will show you how to respond to both situations.

No matter which situation you are in, you will need to judge whether a confidence interval is appropriate for the situation, and you will need to interpret the confidence interval. Therefore, we will discuss these essential skills before demonstrating the calculations.

When Are Confidence Intervals Useful? A confidence interval is a useful answer to the following statistical questions "What's the typical value in this large group of objects or people? And how far away from the truth might this estimate of the typical value be?" You should provide a confidence interval whenever you are estimating the value of a population parameter on the basis of a random sample from that population. For example, judging on the basis of a random sample of 30 adults, what's the typical body temperature of all healthy adults? On the basis of a survey of a random sample of Maryland residents, what's the typical commuting time for all Maryland residents?

A confidence interval is useful for answering questions such as these because it communicates the uncertainty in our estimate and provides a range of plausible values.

A confidence interval is not appropriate if there is no uncertainty in your estimate. This would be the case if your "sample" were actually the entire population. For example, it is not necessary to find a confidence interval for the mean score on your class's statistics exam. The population is your class, and all the scores are known. Thus, the population mean is known, and there is no need to estimate it.

Checking Conditions In order for us to measure the correct confidence level, these conditions must hold:

Condition 1: *Random Sample and Independence*. The data must be collected randomly, and each observation must be independent of the others.

Condition 2: *Large Sample*. Either the population must be Normally distributed or the sample size must be fairly large (at least 25).

Condition 3: *Big Population*. If the sample is collected without replacement, then the population must be at least 10 times larger than the sample size.

If these conditions do not hold, then we cannot measure the job performance of the interval; the confidence level may be incorrect. This means that we may advertise a 95% confidence level when, in fact, the true performance is much worse than this.

To check the first condition, you must know how the data were collected. This is not always possible, so rather than checking these conditions, you must simply assume that they hold. If they do not, your interval will not be valid.

The requirement for independence means that measurement of one object in the sample does not affect any other. Essentially, if we know the value of any one observation, this knowledge should tell us nothing about the values of other observations. This condition might be violated if, say, we randomly sampled several schools and gave all the students math exams. The individual math scores would not be independent because we would expect that students within the same school might have similar scores.

The second condition is due to the Central Limit Theorem. If the population distribution is Normal (or very close to it), then we have nothing to worry about. But if it is non-Normal, then we need a large enough sample size so that the sampling distribution of sample means is approximately Normal. For many applications, a sample size of 25 is large enough, but for extremely skewed distributions, you might need an even larger sample size.

Throughout this chapter, we will assume that the population is large enough to satisfy the third condition, unless stated otherwise.

EXAMPLE 5 Is the Cost of College Rising?

Many cities and states are finding it more difficult to offer low-cost college educations. In the 2011–2012 academic year, the mean cost of attending two year colleges in the United States was $3831. Has this increased over time? Three years later, during the 2014–2015 academic year, some researchers took a random sample of 35 two-year colleges in the U.S. and found that the average tuition charged in 2014–2015 was $4173, with a standard deviation of $2589.80. Figure 9.10 provides the Minitab output, which shows that a 90% confidence interval for the mean cost of attending two-year colleges in 2014–2015 was $3433 to $4914.

◄ FIGURE 9.10 Minitab output for a 90% confidence interval of mean in-state tuition and fees of all two-year colleges in the United States during the 2014–2015 academic year.

One-Sample T

N	Mean	StDev	SE Mean	90% CI
35	4173.3	2589.8	437.8	(3433.1, 4913.6)

QUESTIONS

a. Describe the population. Is the number $4173 an example of a parameter or a statistic?

b. Verify that the conditions for a valid confidence interval are met.

SOLUTIONS

a. The population consists of all two-year college tuitions (for in-state residents) in the academic year 2014–2015. (There are more than 1000 two-year colleges in the United States.) The number $4173 is the mean of a sample of only 35 colleges. Because it is the mean of a *sample* (and not of a population), it is a statistic.

b. The first condition is that the data represent a random sample of independent observations. We are told the sample was collected randomly, so we assume this is true. Independence also holds, because knowledge about any one school's tuition tells us nothing about other schools in the sample. The second condition requires that the population be roughly Normally distributed or the sample size be equal to or larger than 25. We do not know the distribution of the population, but because the sample size is large enough (greater than 25), this condition is satisfied.

 Exercise 9.17

Interpreting Confidence Intervals To understand confidence intervals, you must know how to interpret a confidence interval and how to interpret a confidence level.

A confidence *interval* can be interpreted as a range of plausible values for the population parameter. In other words, in the case of population means, we can be confident that if we were to someday learn the true value of the population mean, it would be within the range of values given by our confidence interval. For example, the U.S. Census estimates that the mean commuting time for South Dakota residents is 16.3 minutes to 17.1 minutes, with a 90% confidence level. We interpret this to mean that we can be fairly confident that the true mean commute for *all* South Dakota residents is between 16.3 and 17.1 minutes. Yes, we could be wrong. The mean might actually be less than 16.3 minutes, or it might be more than 17.1 minutes. However, we would be rather surprised to find this was the case; we are highly confident that the mean is within this interval.

> **KEY POINT**   A confidence interval can be interpreted as a range of plausible values for the population parameter.

EXAMPLE **6** Evidence for Changing College Costs

Based on a random sample of 35 two-year colleges, a 90% confidence interval for the mean tuition at two-year colleges for the 2014–2015 academic year is $3433 to $4914. When we examined the data for *all* two-year colleges in 2011–2012, we learned that the population mean tuition in that year was $3831.

QUESTION   Does the confidence interval for mean tuitions in 2014–2015 provide evidence that the mean tuition has changed since the 2011–2012 academic year?

SOLUTION   No, it does not. Although, on the basis of this random sample of 35 colleges, we cannot know the population mean tuition in 2014–2015 with certainty, we are highly confident it is between $3433 and $4914. Because this range includes the value of $3831, which was the mean in 2011–2012, we have no evidence that the mean cost has changed.

TRY THIS!   Exercise 9.19

Measuring Performance with the Confidence Level

The confidence *level*, which in the case of both the intervals for mean commuting times and for mean tuition costs was 90%, tells us about the method used to find the interval. A value for the level of 90% tells us that the U.S. Census used a method that works in 90% of all samples. In other words, if we were to take many same-sized samples of commuters, and for each sample calculate a 90% confidence interval, then 90% of those intervals would contain the population mean.

The confidence level does *not* tell us whether the interval 16.3 to 17.1 contains or does not contain the population mean. The "90%" just tells us that the method that produced this interval is a pretty good method.

Suppose you decided to purchase a new phone online. You have your choice of several manufacturers, and they are rated in terms of their performance level. One manufacturer has a 90% performance level, which means that 90% of the phones it produces are good and 10% are defective. Some other manufacturers have lower levels: 80%, 60%, and worse. From whom do you buy? You choose to buy from the manufacturer with the 90% level, because you can be very confident that the phone it sends you will be good. Of course, once the phone arrives at your home, the confidence level isn't too useful. Your phone either works or does not work; there's no 90% about it.

Confidence levels work the same way. We prefer confidence intervals that have 90% or higher confidence levels, because then we know that the process that produced these levels is a good process, and therefore, we are confident in any decisions or conclusions we reach. But the level doesn't tell us whether this one particular interval sitting in front of us is good or bad. In fact, we shall never know that unless we someday gain access to the entire population.

> **KEY POINT**   The confidence level is a measure of how well the method used to produce the confidence interval performs. We can interpret the confidence level to mean that if we were to take many random samples of the same size from the same population, and for each random sample calculate a confidence interval, then the confidence level is the proportion of intervals that "work"—the proportion that contain the population parameter.

Figure 9.11 illustrates this interpretation of confidence levels. From the population of all U.S. movies that made over 100 million dollars (adjusted for inflation; http://www.thenumbers.com/), we took a random sample (with replacement) of 30 movies and calculated the mean revenue in this sample (in millions of dollars). Because the samples were random, each sample produced a different sample mean. For each sample we also calculated a 95% confidence interval. We repeated this process 100 times, and each time we made a plot of the confidence interval. Figure 9.11a shows the results from the first 10 samples of 30 randomly selected movies. Nine of the ten intervals were "good"—intervals that contained the true population mean of $172 million. Figure 9.10b shows what happened after we collected 100 different 95% confidence intervals. With a 95% confidence interval, we would expect about 95% of the intervals to be good and 5% to be bad. And in fact, six intervals (shown in red) were bad.

Details

Making Mistakes
Because the U.S. Census provides 90% confidence intervals for the mean commuting times in all 50 states, we can expect about 5 of these intervals (roughly 10% of them) to be wrong.

! Caution

Confidence Levels Are Not Probabilities
A confidence level, such as 90%, is not a probability. Saying we are 90% confident the mean is between 21.1 minutes and 21.3 minutes does *not* mean that there is a 90% chance that the mean is between these two values. It either is or isn't. There's no probability about it.

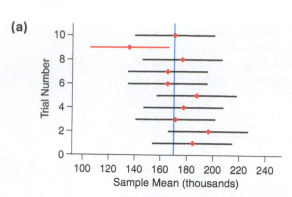

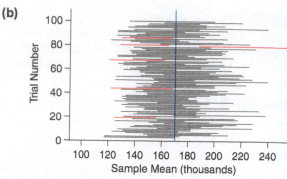

▲ FIGURE 9.11 **(a)** Ten different 95% confidence intervals, each based on a separate random sample of 30 movies. The population mean of $172 million is shown with a vertical bar. Nine of the ten intervals are good because they include this population mean. **(b)** One hundred confidence intervals, each based on a random sample of 30 movies. Because we are using a 95% confidence level, we expect about 95% of the intervals to be good. In fact, 94 of the 100 turned out to be good, this time. The red intervals are "bad" intervals that do not contain the population mean.

EXAMPLE 7 iPad Batteries

A consumer group wishes to test Apple's claim that the iPad has a 10-hour battery life. With a random sample of 5 iPads, and running them under identical conditions, the group finds a 95% confidence interval for the mean battery life of an iPad to be 9.5 hours to 12.5 hours. One of the following statements is a correct interpretation of the confidence *level*. The other is a correct interpretation of the confidence *interval*.

  (i) We are very confident that the mean battery life of all iPads is between 9.5 and 12.5 hours.

  (ii) In about 95% of all samples of 5 iPads, the resulting confidence interval will contain the mean battery life of all iPads.

QUESTION   Which of these statements is a valid interpretation of a confidence interval? Which of these statements is a valid interpretation of a confidence level?

SOLUTION   Statement (i) interprets a confidence *interval* (9.5, 12.5). Statement (ii) tells us the meaning of the 95% confidence *level*.

TRY THIS!   Exercise 9.23

Calculating the Confidence Interval You will be calculating confidence intervals in two situations. In the first situation, you'll have only summary statistics from the sample: the sample mean, the standard deviation, and the sample size. In this situation, you can often find the confidence interval using a calculator, although a computer will generally give a more accurate interval. In the second situation, you'll have the actual data. In this case, you should definitely use a computer. In both situations, it is very important to know the general structure of the formula in order to understand how confidence intervals are interpreted.

Confidence intervals for means have the same basic structure as they did for proportions:

$$\text{Estimator} \pm \text{margin of error}$$

As in Chapter 7, the margin of error has the structure

$$\text{Margin of error} = (\text{multiplier}) \times SE$$

The standard error (*SE*) is $SE = \dfrac{\sigma}{\sqrt{n}}$. Because we usually do not know the standard deviation of the population and hence the *standard error*, we replace *SE* with its estimate. This leads to a formula similar in structure, but slightly different in details, from the one you learned for proportions.

Formula 9.1: One-Sample *t*-Interval

$$\bar{x} \pm m$$

$$\text{where} \qquad m = t^*SE_{\text{est}} \quad \text{and} \quad SE_{\text{est}} = \dfrac{s}{\sqrt{n}}$$

The multiplier *t** is a constant that is used to fine-tune the margin of error so that it has the level of confidence we want. This multiplier is found using a *t*-distribution with $n - 1$ degrees of freedom. (The degrees of freedom determine the shape of the *t*-distribution.) SE_{est} is the estimated standard error.

To compute a confidence interval for the mean, you first need to choose the level of confidence. After that, you need either the original data or these four pieces of information:

1. The sample average, $\bar{x}$, which you calculate from the data

2. The sample standard deviation, *s*, which you calculate from the data

3. The sample size, *n*, which you know from looking at the data

4. The multiplier, *t**, which you look up in a table (or use technology) and which is determined by your confidence level and the sample size *n*. The value of *t** tells us how wide the margin of error is, in terms of standard errors. For example, if *t** is 2, then our margin of error is two standard errors wide.

The structure of the confidence interval for the mean is the same as for the proportion: the estimated value plus-or-minus the margin of error. One difference is that for the proportion, the multiplier in the margin of error depends only on our desired confidence level. If we want a 95% confidence level, we always use 1.96 as the multiplier value z^* (and we sometimes round to 2). If we want a 90% confidence level, we always use 1.64 for z^*. For means, however, the multiplier is also determined by the sample size.

The reason for this is that when we are finding a confidence interval for a proportion, our confidence level is determined by the Normal distribution. But when we are working with the mean, the level is determined by the *t*-distribution based on $n - 1$ degrees of freedom. The correct values can be found in Table 4 in Appendix A, or you can use technology. Table 4 is organized such that each row represents possible values of *t** for each degree of freedom. The columns contain the values of *t** for a given confidence level. For example, for a 95% confidence level and a sample size of $n = 30$, we use $t^* = 2.045$. We find this in the table by looking in the row with df $= n - 1 = 30 - 1 = 29$ and using the column for a 95% confidence level. Refer to Table 9.2, which is excerpted from the table in Appendix A.

Example 8 shows how to use Table 4 to find the multiplier, a technique that is useful if you do not have access to a statistical calculator.

EXAMPLE 8 Finding the Multiplier *t**

A study to test the life of iPad batteries reports that in a random sample of 30 iPads, the mean battery life was 9.7 hours, and the standard deviation was 1.2 hours. The raw data were not made available to the public.

QUESTION Using Table 9.2, which is from the table in Appendix A, find t^* for a 90% confidence interval when $n = 30$.

▶ **TABLE 9.2** Critical values of t.

DF	Confidence Level			
	90%	95%	98%	99%
28	1.701	2.048	2.467	2.763
29	1.699	2.045	2.462	2.756
30	1.697	2.042	2.457	2.750
34	1.691	2.032	2.441	2.728

SOLUTION We find the number of degrees of freedom from rd the sample size:

$$df = n - 1 = 30 - 1 = 29$$

And so we find, from Table 9.2, $t^* = 1.699$ (shown underlined).

TRY THIS! Exercise 9.25

> **🔄 Looking Back**
>
> **Why Not 100%?**
> In Chapter 7, you learned that one reason why a 95% confidence level is popular is that increasing the confidence level only a small amount beyond 95% requires a much larger margin of error.

It is best to use technology to find the multiplier, because most tables stop at 35 or 40 degrees of freedom. For a 95% confidence level, if you do not have access to technology and the sample size is bigger than 40, it is usually safe to use $t^* = 1.96$—the same multiplier that we used for confidence intervals for sample proportions (for 95% confidence). The precise value, if we used a computer, is 2.02, but this is only 0.06 unit away from 1.96, so the result is probably not going to be affected in a big way.

Example 9 illustrates the use of statistical software to find a confidence interval when only summary statistics are provided.

EXAMPLE 9 Pizza Size

Eagle Boys, an Australian chain of pizza stores, published data on the size of its pizzas to convince the public of their value. Using a random sample of 125 pizzas, the store found that the mean diameter was 11.5 inches with a standard deviation of 0.25 inch. Figure 9.12a shows a screenshot from StatCrunch for calculating a confidence interval using summary statistics. Figure 9.12b shows the results (Dunn 2012).

QUESTION For each field in StatCrunch, provide the value or setting required to calculate a 95% confidence interval for the mean diameter of all pizzas produced by this store. State and interpret the confidence interval.

▶ **FIGURE 9.12 (a)** StatCrunch fields for calculating a confidence interval for the population mean when only summary statistics are available. **(b)** Result of the calculation.

(a)

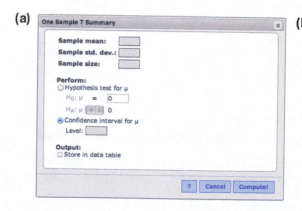

(b)

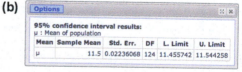

SOLUTION Sample mean: 11.5
Sample standard deviation: 0.25
Sample size: 125
Confidence interval level for μ: 0.95

The 95% confidence level for the mean pizza diameter is 11.46 to 11.54 inches. We are 95% confident that the mean diameter of all pizzas produced by this company is between 11.46 and 11.54 inches.

TRY THIS! Exercise 9.27

Example 10 shows that in order to have a higher level of confidence, we need a larger margin of error. This larger margin of error means that the confidence interval is wider, so our estimate is less precise.

EXAMPLE 10 College Tuition Costs

A random sample of 35 two-year colleges in 2014–2015 had a mean tuition (for in-state students) of $4173, with a standard deviation of $2590.

QUESTION Find a 90% confidence interval and a 95% confidence interval for the mean in-state tuition of all two-year colleges in 2010–2011. Interpret the intervals. First, verify that the necessary conditions hold.

SOLUTION This time, we will show how to find the confidence interval using the formula, which is what you must do if you do not have a statistical calculator or statistical software.

We are told that the sample is random and that the sample size is larger than 25, so the necessary conditions hold.

We are given the desired confidence levels, the standard deviation, and the sample mean. Formula 9.1 gives us the form of the confidence interval:

$$\bar{x} \pm m$$

The next step is to calculate the margin of error:

$$m = t^* \, SE_{est}$$

First we find the estimated standard error:

$$SE_{est} = \frac{2590}{\sqrt{35}} = 437.7899$$

We find the appropriate values of t^* from Table 9.2:

$$t^* = 1.691 \text{ (for 90\% confidence level)}$$
$$t^* = 2.032 \text{ (for 95\% confidence level).}$$

For the 90% confidence interval,

$$\bar{x} \pm t^* \, SE_{est} \quad \text{becomes}$$
$$4173 \pm (1.691 \times 437.7899),$$
$$\text{or} \quad 4137 \pm 740.3027$$
$$\text{Lower limit: } 4137 - 740.3027 = 3432.70$$
$$\text{Upper limit: } 4173 + 740.3027 = 4913.30$$

A 90% confidence interval for the mean tuition of all two-year colleges in the 2014–2015 academic year is ($3433, $4913).

For the 95% confidence interval,

$$\bar{x} \pm t^* SE_{est} \text{ becomes}$$

$$4173 \pm (2.032 \times 437.7899)$$

$$\text{or } 4173 \pm 889.5890$$

$$\text{Lower Limit: } 4173 - 889.5890 = 3283.4109$$

$$\text{Upper Limit: } 4173 + 889.5890 = 5062.5890$$

CONCLUSION The 90% confidence interval is ($3433, $4913). The 95% confidence interval is ($3283, $5063), which is wider. We are 90% confident that the mean tuition of all two-year colleges is between $3433 and $4913. We are 95% confident that the mean tuition of all two-year colleges is between $3283 and $5062.

TRY THIS! Exercise 9.29

 Tech If you have access to the original data (and not just to the summary statistics, as we were given in Example 10), then it is always best to use a computer to find the confidence interval for you. Figure 9.13a shows the information required by StatCrunch to calculate a 95% confidence interval for the mean tuition of two-year colleges in the 2014–2015 academic year using the data themselves (which you'll recall are drawn from a random sample of 35 two-year colleges). The output in Figure 9.13b shows the estimated mean ($4173.3429), the standard error ($437.75842), the degrees of freedom (34), the lower limit of the confidence interval ($3283.7107), and the upper limit ($5062.975).

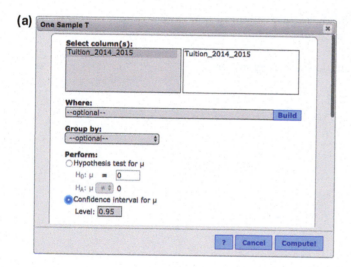

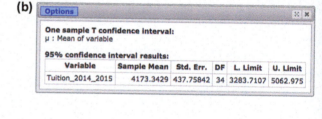

◄ **FIGURE 9.13** Screenshot showing **(a)** required StatCrunch input and **(b)** the resulting output for a 95% confidence interval for the mean in-state tuition at two-year colleges in the 2014–2015 academic year.

Reporting and Reading Confidence Intervals

There are two ways of reporting confidence intervals. Professional statisticians tend to report (lower boundary, upper boundary). This is what we've done so far in this chapter. Thus, in Example 10 we reported the 95% confidence interval for the mean of two-year college tuitions in 2014–2015 as ($3283, $5063).

In the press, however, and in some scholarly publications, you'll also see confidence intervals reported as

$$\text{Estimate} \pm \text{margin of error}$$

For the two-year college tuitions, we calculated the margin of error to be $889.5890 for 95% confidence. Thus, we could also report the confidence interval as

$$\$4173 \pm \$890$$

This form is useful because it shows our estimate for the mean ($4173) as well as our uncertainty (the mean could plausibly be $890 lower or $890 more).

You're welcome to choose whichever you think best, although you should be familiar with both forms.

Understanding Confidence Intervals   As shown in Chapter 7 and illustrated in Example 10, a wider interval results in a higher level of confidence. Imagine that the population mean is a tennis ball and the confidence interval is a tennis racquet. Which would make you more confident of hitting the ball: using a (small) ping-pong-paddle-sized racquet or using a (larger) tennis racquet? The larger racquet fills more space, so you should feel more confident that you'll connect with the ball. Using a wider confidence interval gives us a higher level of confidence that we'll "connect" with the true population mean.

Wider intervals are not always desirable, however, because they mean that we have less precision. For example, we could have a 100% confidence interval for the mean tuition at two-year colleges: $0 to infinity dollars. But this interval is so imprecise that it is useless. The 95% confidence interval offers less than 100% confidence, but it is much more precise.

Because the margin of error depends on the standard error, and because the standard error depends on the sample size, we can make our interval more precise by collecting more data. A larger sample size provides a smaller standard error, and this means a smaller margin of error at the same level of confidence.

EXAMPLE **11**   Confidence in Growing Enrollment

The IPEDS website provides much data on colleges and universities in the United States. We took a random sample of 35 two-year colleges from this data set and calculated two confidence intervals for the mean total enrollment of all two-year colleges in the United States. Both intervals are based on the same sample, but have different confidence levels. One of the intervals is 4948 students to 16,170 students. The other is 3815 students to 17,303 students. (IPEDS stands for Integrated Postsecondary Education Data System.)

QUESTIONS

a. Which interval has the higher confidence level, and why?

b. What will be the effect of taking a larger sample on the width of the interval?

SOLUTIONS

a. The interval (3815, 17303) has the higher level of confidence because it is the wider interval. This interval has width $17303 - 3815 = 13{,}488$ students and is wider than the other interval, which is 11,222 students wide.

b. If we take a larger sample, the standard error for our estimator will be smaller. This means the margin of error will be smaller, so both intervals will be narrower.

NOW TRY!   Exercise 9.31

Hypothesis Testing for Means

In Chapter 8 we laid out the foundations of hypothesis testing. Here, you'll see that the same four steps can be used to test hypotheses about means of populations. These four steps are

Step 1: Hypothesize

Frame your statistical question as a hypothesis about the population parameter.

Step 2: Prepare.

Choose a test statistic and consider the data: Are the necessary conditions met to allow inference with your chosen test statistic? State any assumptions you must make if you are to proceed.

Step 3: Compute to Compare.

Analyze the data by stating a significance level, computing the observed test statistic and finding the p-value. The observed value of the test statistic compares our observed statistic with the value of the population parameter as claimed by the null hypothesis.

Step 4: Interpret.

Answer the statistical question by deciding whether to reject or not to reject the null hypothesis. What does this mean in the context of the data?

As an example of testing a mean, consider this "study" one of the authors did. McDonald's advertises that its ice cream cones have a mean weight of 3.2 ounces ($\mu = 3.2$). A human server starts and stops the machine that dispenses the ice cream, so we might expect some variation in the amount. Some cones might weight slightly more, some cones slightly less. If we weighed all the McDonald's ice cream cones at a particular store, would the mean be 3.2 ounces, as the company claims?

One of the authors collected a sample of five ice cream cones (all in the name of science) and weighed them on a food scale. The weights were (in ounces)

$$4.2, 3.6, 3.9, 3.4, \text{ and } 3.3$$

We summarize these data as

$$\bar{x} = 3.68 \text{ ounces}, s = 0.3701 \text{ ounce}$$

Do these observations support the claim that the mean weight is 3.2 ounces? Or is the mean a different value? We'll apply the four steps of the hypothesis test to make a decision.

Step 1: Hypothesize

We ask the statistical question: does the true population mean weight of ice cream cones differ from the advertised weight? We restate this as a pair of hypotheses that, as always, are statements about the population. In this case, the population consists of all ice cream cones that have been, will be, or could be dispensed from a particular McDonald's. In this chapter, our hypotheses are about the mean values of populations.

The null hypothesis is the status quo position, which is the claim that McDonald's makes. An individual cone might weigh a little more than 3.2 ounces, or a little less, but after looking at a great many cones, we would find that McDonald's is right and that the mean weight is 3.2 ounces.

We state the null hypothesis as

$$H_0: \mu = 3.2$$

Recall that the null hypothesis always contains an equals sign.

The alternative hypothesis, on the other hand, says that the mean weight is different from 3.2 ounces:

$$H_a: \mu \neq 3.2$$

This is an example of a two-sided hypothesis. We will reject the null hypothesis if the average of our sample cones is very big (suggesting that the population mean is greater than 3.2) or very small (suggesting that the population mean is less than 3.2). It is also possible to have one-sided hypotheses, as you will see later in this chapter.

KEY POINT   Hypotheses are always statements about population parameters. For the test you are about to learn, this parameter is always μ, the mean of the population.

Step 2: Prepare

The test statistic, called the one-sample t-test, is very similar in structure to the test for one proportion and is based—not surprisingly, given the name of the test—on the t-statistic introduced in Section 9.2. The idea is simple: Compare the observed value of the sample mean, $\bar{x}$, to the value claimed by the null hypothesis, μ_0.

Formula 9.2: Test Statistic for the One-Sample t-Test

$$t = \frac{\bar{x} - \mu_0}{SE_{est}}, \qquad \text{where} \qquad SE_{est} = \frac{s}{\sqrt{n}}$$

If conditions hold, the test statistic follows a t-distribution with df $= n - 1$

This test statistic works because it compares the value of the parameter that the null hypothesis says is true, μ_0, to the estimate of that value that we actually observed in our data. If the estimate is close to the null hypothesis value, then the t-statistic is close to 0. But if the estimate is far from the null hypothesis value, then the t-statistic is far from 0. The farther t is from 0, the worse things look for the null hypothesis.

Anyone can make a decision, but only a statistician can measure the probability that the decision is right or wrong. To do this, we need to know the sampling distribution of our test statistic.

The sampling distribution will follow the t-distribution under these conditions:

Condition 1: *Random Sample and Independence.* The data must be a random sample from a population, and observations must be independent of one another.

Condition 2: *Large Sample.* The population distribution must be Normal, or the sample size must be large. For most situations, 25 is large enough.

Now let's apply this to our ice cream problem. The population for testing the mean ice cream cone weight is somewhat abstract, because a constant stream of ice cream cones is being produced by McDonald's. However, it seems logical that if some cones weigh slightly more than 3.2 ounces and some weigh slightly less, then this distribution of weights should be symmetric and not too different from a Normal distribution. Because our population distribution is Normal, the fact that we have a small sample size, $n = 5$, is not a problem here.

Looking Back

The z-Test
Compare this to the z-test statistic for one proportion in Chapter 8, which has a very similar structure:

$$z = \frac{\hat{p} - p_0}{SE}$$

The ice cream cone weights are independent of each other because we were careful, when weighing, to recalibrate the scale, and each cone was obtained on a different day. The cones were not, strictly speaking, randomly sampled, although because the cones were collected on different days and at different times, we will assume that they behave as though they come from a random sample. (But if we're wrong, our conclusions could be *very* wrong!)

Step 3: Compute to compare

The first step is to set the significance level α, as we discussed in Chapter 8. The significance level is a performance measure that helps us evaluate the quality of our test procedure. It is the probability of making the mistake of rejecting the null hypothesis when, in fact, the null hypothesis is true. In this case, this is the probability that we will say McDonald's cones do not weigh an average of 3.2 ounces when, in fact, they really do. We will use $\alpha = 0.05$.

The conditions of our data tell us that our test statistic should follow a *t*-distribution with $n - 1$ degrees of freedom. Therefore, we proceed to do the calculations necessary to compare our observed sample mean to the hypothesized value of the population mean and to measure our surprise.

To find the observed value of our *t*-statistic, we need to find the sample mean and the standard deviation of our sample. These values were given earlier, but you can easily calculate them from the data.

$$SE_{est} = \frac{0.3701}{\sqrt{5}} = 0.1655$$

$$t = \frac{3.68 - 3.2}{0.1655} = \frac{0.48}{0.1655} = 2.90$$

The observed sample mean was 2.90—almost 3 standard errors above the value expected by the null hypothesis.

> **KEY POINT** The *t*-statistic measures how far (how many standard errors) away our observed mean, $\bar{x}$, lies from the hypothesized value, μ_0. Values far from 0 tend to discredit the null hypothesis.

How unusual is such a value, according to the null hypothesis? The p-value tells us exactly that—the probability of our getting a *t*-statistic as extreme as or more extreme than what we observed, if in fact the mean is 3.2 ounces.

Because our alternative hypothesis says we should be on the lookout for *t*-statistic values that are much bigger or smaller than 0, we must find the probability in both tails of the *t*-distribution. The p-value is shown in the small shaded tails of Figure 9.14. Our sample size is $n = 5$, so our degrees of freedom are $n - 1 = 5 - 1 = 4$.

The p-value of 0.044 tells us that if the typical cone really weighs 3.2 ounces, our observations are somewhat unusual. We should be surprised.

Step 4: Interpret

The last step is to compare the p-value to the significance level and decide whether to reject the null hypothesis. If we follow a rule that says we will reject whenever the p-value is less than or equal to the significance level, then we know that the probability of mistakenly rejecting the null hypothesis will be the value of α.

Our p-value (0.044) is less than the significance level we chose (0.05), so we should reject the null hypothesis and conclude that at this particular McDonald's, cones do *not* weigh, on average, 3.2 ounces.

This result makes some sense from a public relations standpoint. If the mean were 3.2 ounces, about half of the customers would be getting cones that weighed too little. By setting the mean weight a little higher than what is advertised, McDonald's can give everyone more than they thought they were getting.

Details

What Value for α?
For most situations, using a significance level of 0.05 is a good choice and is recommended by many scientific journals. Values of 0.01 and 0.10 are also commonly used.

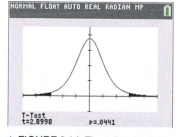

▲ **FIGURE 9.14** The tail areas above 2.90 and below −2.90 are shown as the small shaded areas on both sides. The p-value is 0.0441, the probability that if $\mu = 3.2$, a test statistic will be bigger than 2.90 or smaller than −2.90. The distribution shown is a *t*-distribution with 4 degrees of freedom.

Looking Back

p-Values
In Chapter 8 you learned that the p-value is the probability that when the null hypothesis is true, we will get a test statistic as extreme as or more extreme than what we actually saw. (What is meant by "extreme" depends on the alternative hypothesis.) The p-value measures our surprise at the outcome.

One- and Two-sided Alternative Hypotheses

The alternative hypothesis in the ice cream cone test was two-sided. As you learned in Chapter 8, alternative hypotheses can also be one-sided. The exact form of the alternative hypothesis depends on the research question. In turn, the form of the alternative hypothesis tells us how to find the p-value. Two-sided hypotheses will require two-tailed p-value calculations, and one-sided hypotheses will require one-tailed p-value calculations.

You will always use one of the following three pairs of hypotheses for the one-sample *t*-test:

Two-Sided	One-Sided (Left)	One-Sided (Right)
$H_0: \mu = \mu_0$	$H_0: \mu = \mu_0$	$H_0: \mu = \mu_0$
$H_a: \mu \neq \mu_0$	$H_a: \mu < \mu_0$	$H_a: \mu > \mu_0$

You choose the pair of hypotheses on the basis of your research question. For the ice cream cone example, we asked if the mean weight is *different* from the value advertised, so we used a two-sided alternative hypothesis. Had we instead wanted to know whether the mean weight was less than 3.2 ounces, we would have used a one-sided (left) hypothesis.

Your choice of alternative hypothesis determines how you calculate the p-value. Figure 9.15 shows how to find the p-value for each alternative hypothesis, all using the same *t*-statistic value of $t = 2.1$ and the same sample size of $n = 30$.

Note that the p-value is always an "extreme" probability; it's always the probability of the tails (even if the tail is pretty big, as it is in Figure 9.15b).

Looking Back

What Does "as extreme as or more extreme than" Mean?
See Chapter 8 for a detailed discussion of how the p-value depends on the alternative hypothesis.

(a)

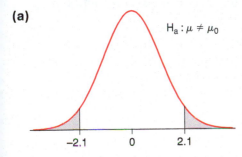

(b)

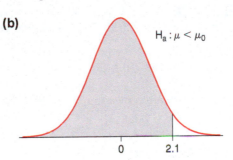

(c)

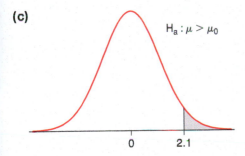

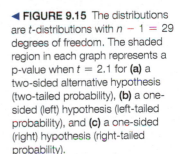

◀ **FIGURE 9.15** The distributions are *t*-distributions with $n - 1 = 29$ degrees of freedom. The shaded region in each graph represents a p-value when $t = 2.1$ for **(a)** a two-sided alternative hypothesis (two-tailed probability), **(b)** a one-sided (left) hypothesis (left-tailed probability), and **(c)** a one-sided (right) hypothesis (right-tailed probability).

EXAMPLE 12 College Costs

In Example 5 we asked whether the mean cost of two-year colleges had increased since the 2011–2012 academic year. In 2011–2012 the mean cost of all two-year colleges (for in-state tuition and fees) was $3831. We will now examine the same question using a hypothesis test. Our data are a random sample of tuition prices at 35 two-year colleges during 2014–2015 the academic year. Figure 9.16a shows summary statistics for this sample, and Figure 9.16b shows the distribution of the sample.

(a)

(b)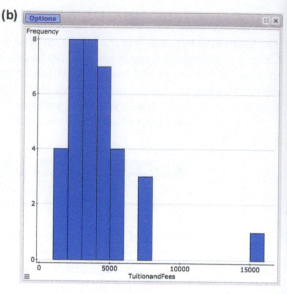

▲ **FIGURE 9.16** Summary statistics for tuitions at 35 two-year colleges. **(a)** The sample mean tuition for these 35 colleges was $4173, with a standard deviation of $2590. **(b)** The distribution of the sample is somewhat right-skewed.

QUESTION Test the hypothesis that the mean tuition cost at two-year colleges has increased since the 2011–2012 academic year.

SOLUTION First, we note that if we had the data for *all* two-year colleges in 2014–2015 (as we did for the 2011–2012 academic year), we would not need to do a hypothesis test. We could simply compute the population mean and see whether it was greater than the population mean in 2011–2012. But because we have only a sample of all the two-year colleges, we will use a hypothesis test.

Step 1: Hypothesize
Let μ represent the mean tuition at all two-year colleges in 2014–2015. We are asked to compare the population mean in 2014–2015 with the reported population mean of $3831 from 2011–2012.

$$H_0 : \mu = 3831$$

$$H_a : \mu > 3831$$

The null hypothesis says the mean is the same as in 2011–2012. The alternative hypothesis says that the mean is higher now than it was then. This differs from the ice cream cone example, where we only wanted to know whether or not the mean weight was 3.2. Here we care about the direction: Is the new mean greater than the old mean?

Step 2: Prepare
We need to check the conditions to see whether the *t*-statistic will follow a *t*-distribution (at least approximately).

Condition 1: *Random Sample and Independence.* The colleges in this study were selected randomly from the population of all two-year colleges. Our sampling procedure ensured that observations were independent of each other.

Condition 2: *Large Sample.* The distribution of the sample is not necessarily Normal (although it is not too different from Normal). But because the sample size is larger than 25, this sampling distribution will be approximately a *t*-distribution with $n - 1 = 34$ degrees of freedom.

Step 3: Compute to compare
We will test using a 5% significance level. This calculation is best done using technology, and StatCrunch output is shown in Figure 9.17a. We will use Formula 9.2 to show how it is applied in this situation.

$$SE_{est} = \frac{2589.8137}{\sqrt{35}} = 437.7584$$

$$t = \frac{\bar{x} - \mu_0}{SE_{est}} = \frac{4173.3429 - 3831}{437.7584} = \frac{342.3429}{437.7584} = 0.7820$$

(a)

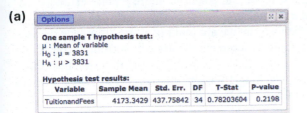

One sample T hypothesis test:
μ : Mean of variable
$H_0 : \mu = 3831$
$H_A : \mu > 3831$

Hypothesis test results:

Variable	Sample Mean	Std. Err.	DF	T-Stat	P-value
TuitionandFees	4173.3429	437.75842	34	0.78203604	0.2198

(b)

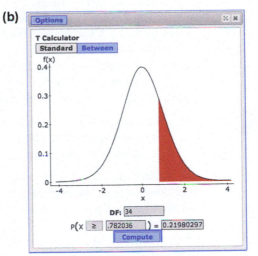

▲ **FIGURE 9.17** StatCrunch was used to perform the calculations to enable us to comapre the null hypothesis value of 3831 to our observed sample mean. **(a)** The *t*-statistic value is 0.7820 with a p-value of 0.2198. **(b)** The p-value is a right-tailed probability, using a *t*-distribution with 34 degrees of freedom.

This tells us that our observed mean was only 0.78 standard errors above where we would expect it to be if the null hypothesis were true.

Our p-value is a right-tailed probability because the alternative hypothesis cares only whether we see values greater than expected. (Remember, we are using a one-sided hypothesis.) Using technology, we find the p-value to be 0.23198, which is illustrated in Figure 9.17b.

Step 4: Interpret
The p-value is greater than 0.05, so we conclude that this sample provides no evidence that the mean tuition was higher in 2014–2015 than in the earlier year.

TRY THIS!   Exercise 9.37

SNAPSHOT ▶ One Sample *t*-Test

WHAT IS IT ▶	$t = \dfrac{\bar{x} - \mu_0}{SE_{est}}$, where $SE_{est} = \dfrac{s}{\sqrt{n}}$
WHAT DOES IT DO? ▶	It tests hypotheses about a mean of a single population.
HOW DOES IT DO IT? ▶	If the estimated mean differs from the hypothesized value, then the test statistic will be far from 0. Thus, values of the *t*-statistic that are unexpectedly far from 0 (in one direction or the other) discredit the null hypothesis.
HOW IS IT USED? ▶	When proposing values about a single population mean. The observed value of the test statistic is compared to a *t*-distribution with $n - 1$ degrees of freedom to calculate the p-value. If the p-value is small, you should be surprised by the outcome and reject the null hypothesis.

Comparing Two Population Means

Do people comprehend better when they read on paper rather than on a computer screen? Do those who identified as male spend less time doing laundry than those who identified as female? If so, how much less? These questions can be answered, in part, by comparing the means of two populations. Although we could construct separate confidence intervals to estimate each mean, we can construct more precise estimates by focusing on the difference between the two means.

In Section 7.5 you learned that when we say we are looking at the difference between two proportions, we mean that we are going to subtract one proportion from the other. The same is true when examining differences of means. For instance, we might subtract the mean amount of time Americans spend watching TV every day from the mean amount of time they spend exercising. From subtraction, we learn that:

if the result is positive, the first mean is greater than the second.
if the result is negative, the first mean is less than the second.
if the result is 0, the means are equal.

When comparing two populations, it is important to pay attention to whether the data sampled from the populations are two **independent samples** or are, in fact, one sample of related pairs (paired samples). With **paired (dependent) samples**, if you know the value that a subject has in one group, then you know something about the other group, too. In such a case, you have somewhat less information than you might have if the samples were independent. We begin with some examples to help you see which is which.

Usually, dependence occurs when each object or person in your sample is measured twice (as is common in "before and after" comparisons) or when the objects are related somehow (for example, if you are comparing twins, siblings, or spouses) or when the experimenters have deliberately matched subjects in the groups to have similar characteristics.

> **! Caution**
>
> **Paired (Dependent) vs. Independent Samples**
> One indication that you have paired samples is that each observation in one group is coupled with *one particular observation* in the other group. In this case, the two groups will have the same sample size (assuming no observations are missing).

EXAMPLE 13 Independent or Dependent Samples?

Here are four descriptions of research studies:

a. People chosen in a random sample were asked how many minutes they had spent the day before watching television and how many minutes they had spent exercising. Researchers want to know how different the mean amounts of times are for these two activities.

b. Men and women each had their sense of smell measured. Researchers want to know whether, typically, men and women differ in their ability to sense smells.

c. Researchers randomly assigned overweight people to one of two diets: Weight Watchers and Atkins. Researchers want to know whether the mean weight loss on Weight Watchers was different from that on Atkins.

d. The numbers of years of education for husbands and wives are compared to see whether the means are different.

QUESTION For each study, state whether it involves two independent samples or paired (that is, dependent) samples.

SOLUTION

a. This study is based on a single sample of people who are measured twice. One population consists of people watching TV. The second population consists of the same people exercising. These are *paired* (dependent) samples.

b. The two populations consist of men in one and women in the other. As long as the people are not related, knowledge about a measurement of a man could not tell us anything about any of the women. These are *independent* samples.

c. The two populations are people on the Weight Watchers diet and people on the Atkins diet. We are told that the two samples consist of different people; subjects are randomly assigned to one diet but not to the other. These are *independent* samples.

d. The populations are matched. Each husband is coupled with one particular wife, so the samples are *paired* (or *dependent*).

TRY THIS!   Exercise 9.53

As you shall see, we analyze paired data differently from data that come from two independent samples. Paired data are turned into "difference" scores: We simply subtract one value in each pair from the other. We now have just a single variable, and we can analyze it using the one-sample techniques discussed in Sections 9.3 and 9.4.

Estimating the Difference of Means with Confidence Intervals (Independent Samples)

Researchers at Princeton and the University of California, Los Angeles were curious about the best way for students to take notes. Is it better to take notes by hand (pen and paper) or with a computer (Mueller & Oppenheimer 2014). The analysis we show here simplifes and summarizes the researchers' more thorough analysis.

The researchers randomly assigned student volunteers to watch videos of lectures and take notes either by hand ("Longhand") or with a laptop ("Laptop"). The boxplots in Figure 9.18 show summary statistics for their scores on a test taken after the lecture that measures their conceptual understanding of the lecture.

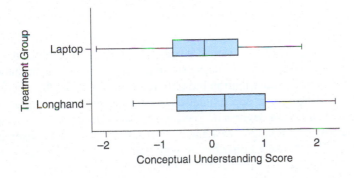

◄ **FIGURE 9.18** Scores on a test of conceptual understanding of a lecture. One group took notes on laptops, and the other took notes by hand. Scores range from −3 to 3 with higher scores reflecting greater conceptual understanding.

The mean score of the longhand group was 0.28 and for the laptop group −0.15. Higher scores indicate stronger conceptual understanding.

Is this evidence that taking notes by is better for conceptual understanding and, if so, by how much?

To guarantee a particular confidence level—for example, 95%—requires that certain conditions hold:

Condition 1: *Random Samples and Independence.* Both samples are randomly taken from their populations, or subjects are randomly assigned to one of the two groups, and each observation is independent of any other.

Condition 2: *Independent Samples.* The two samples are independent of each other (not paired).

Condition 3: *Large Samples.* The populations are approximately Normal, or the sample size in each sample is 25 or more. (In special cases, you might need even larger sample sizes.)

>  **Looking Back**
>
> **Boxplots**
> In Section 3.5, we explained that the middle line in the boxplot represents the median values, and the box includes the middle 50% of values.

If these conditions hold, we can use the following procedure to find an interval with a 95% confidence level.

The formula for a confidence interval comparing two means, when the data are from independent samples, is the same structure as before:

$$\text{(Estimate)} \pm \text{margin of error}$$

which is

$$\text{(Estimate of difference)} \pm t^*(SE_{\text{estimate of difference}})$$

We estimate the difference with

$$\text{(Mean of first sample)} - \text{(mean of second sample)}$$

It doesn't matter which sample you use as the "first" and which as the "second." But it *is* important that you remember which is which.

The standard error of this estimator depends on the sample sizes of both samples and on the standard deviations of both samples:

$$SE_{\text{est}} = \sqrt{\frac{s_1^2}{n_1} + \frac{s_2^2}{n_2}}$$

We can put these together into a confidence interval:

Formula 9.3: Two-Sample *t*-Interval

$$(\bar{x}_1 - \bar{x}_2) \pm t^* \sqrt{\frac{s_1^2}{n_1} + \frac{s_2^2}{n_2}}$$

The multiplier t^* is based on an approximate *t*-distribution. If a computer is not available, you can conservatively calculate the degrees of freedom for the t^* multiplier as the smaller of $n_1 - 1$ and $n_2 - 1$, but a computer provides a more accurate value.

Choosing the value of t^* (the critical value of t) by hand to get your desired level of confidence is tricky. For reasons involving some pretty advanced mathematics, the sampling distribution is not a *t*-distribution but only approximately a *t*-distribution. To make matters worse, to get the approximation to be good requires using a rather complex formula to find the degrees of freedom. If you must do these calculations by hand, we recommend taking a "fast and easy" (but also safe and conservative) approach instead. For t^*, use a *t*-distribution with degrees of freedom equal to the smaller of $n_1 - 1$ and $n_2 - 1$. That is, use the smaller of the two samples, and subtract 1. For a 95% confidence level, if both samples contain 40 or more observations, you can use 1.96 for the multiplier.

Interpreting Confidence Intervals of Differences

The most important thing to look for is whether or not the interval includes 0. If it does *not*, then we have evidence that the two population means are different from each other. In this case, check to see whether the interval contains all positive values. If so, then we are confident that the first population mean is greater than the second population mean. If the interval contains all negative values, we are confident that the first mean is less than the second mean. (Remember, you get to choose which mean is "first" and which is "second" when you do the subtraction.)

For instance, did the students who took notes by longhand have a different mean score on the conceptual understanding test than the students who took notes with a laptop? Let's use the "longhand" group as the first group and the laptop group as the second. After verifying that the conditions are met, a 95% confidence interval for the mean difference in scores between the two groups is 0.05 to 0.80. Because this interval does not include 0, we are confident that the differences in the mean scores of the two groups is different. And because the interval contains all positive values, we're confident that the longhand group (group 1) did better than the laptop group (group 2).

You should also pay attention to how great or how small the difference between the two means could be. This is particularly important if the interval includes 0. In this case, the interval will contain both negative and positive values.

Example 14 shows you how to calculate and interpret a confidence interval for the difference of two means.

EXAMPLE **14** Sleeping In

Do people in the United States sleep more on holidays and weekends than on weekdays? The Bureau of Labor Statistics carries out a "time use" survey, in which randomly chosen people are asked to record every activity they do on a randomly chosen day of the year. For instance, you might be chosen to take part in the survey on Tuesday, April 18, while someone else will be chosen to take part on Sunday, December 5.

Because we have two separate groups of people reporting their amount of sleep—one group that reported only on weekends and holidays, and another group that reported only on weekdays—these data are two independent samples. The summary statistics follow.

Weekday: $\bar{x} = 499.7$ minutes (about 8.3 hours), $s = 126.9$ minutes, $n = 6007$
Weekend/Holiday: $\bar{x} = 555.9$ minutes (about 9.3 hours), $s = 140.9$ minutes, $n = 6436$

Boxplots are shown in Figure 9.19.

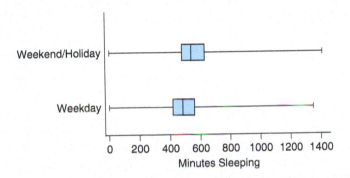

Data Moves ▶ The American Time Use Survey (https://www.bls.gov/tus/) provides data based on interviews with a random sample of Americans on a random sample of days. The data tell us how many minutes people spend doing various activities. We used the provided date variable to create a new variable to distinguish holidays and weekends from weekdays.

Data ▶ atus.csv

◀ **FIGURE 9.19** Distribution of minutes spent sleeping on weekends and holidays and on weekdays. The sample median sleeping time is slightly greater on weekends and holidays.

QUESTION   Verify that the necessary conditions hold to guarantee that our 95% confidence level will indeed be 95%. Find a 95% confidence interval for the mean difference in time spent sleeping on a weekend/holiday compared to time spent sleeping on a weekday. Interpret this interval.

SOLUTION   We are told that people are selected randomly, and it also seems reasonable that the amount of sleep they get is independent. The question explained that the samples were themselves independent of each other (because different people appear in each sample). The boxplots suggest that the distributions are roughly symmetric; it is difficult to know for sure with boxplots. However, with such large sample sizes, we do not have to worry and can proceed.

When we have the raw data at hand, the best approach (and, for a sample as large as this, the only approach) is to use a computer. Figure 9.20a shows the input required to get StatCrunch to compute the interval for us. Figure 9.20b shows the output.

The 95% confidence interval is (51.5 to 60.9) minutes. We are 95% confident that the true mean difference in amount of time spent sleeping on weekends/holidays and amount of time spent sleeping on weekdays is between 51.5 minutes and 60.9 minutes. The interval contains all positive values, so we are confident that people typically sleep

(a)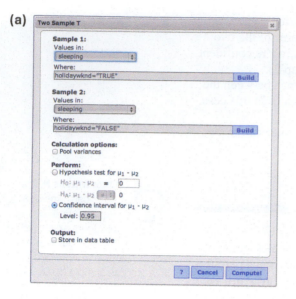

(b)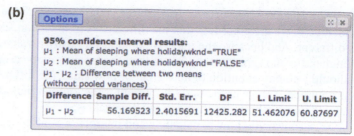

◀ **FIGURE 9.20** StatCrunch input **(a)** and results **(b)** for the mean difference in time spent sleeping on weekends/holidays and on weekdays. Note that the "Pool variances" option should be unchecked.

longer on weekends and holidays than they do on weekdays. The difference may be as little as 51.5 minutes or as much as 60.9 minutes. Generally, it looks like people tend to sleep about an hour later when they do not have to work or go to school.

Notice that the degrees of freedom are a decimal: 12,425.282. Statistical software packages apply a fairly complex formula to approximate the degrees of freedom. If you are doing this "by hand," we recommend that you use the simpler method of taking the smallest sample size minus 1, as demonstrated next. Although for sample sizes this large, it makes very little difference.

TRY THIS!   Exercise 9.55

Tech   If we do not have the original data and are provided only summary statistics, then we might use a calculator to find the interval for the difference between the mean amount of time spent sleeping for the people who reported on weekends and holidays and the mean amount of time spend sleeping for the people who reported on weekdays.

How would this work for finding a confidence interval for the difference in mean sleeping times in Example 14? First, we must find the multiplier t^*. The sample sizes of both groups are quite large. Our rule of thumb for finding an approximate number of degrees of freedom for this test is to use the smaller of the two sample sizes and subtract 1. The smaller sample size is 6007, so we will use 6006 as the degrees of freedom. Because this is much greater than 40, we simply use 1.96 as a conservative approximation to t^*. (If the degrees of freedom are greater than 40, then for t^* use the values for the Normal distribution: 1.96 for 95% and 1.64 for 90%.)

We'll use the Weekend/Holiday group as our sample 1 and the Weekday group as sample 2.

Estimate of difference: $555.9 - 499.7 = 56.2$ minutes

$$m = t^*\sqrt{\frac{s_1^2}{n_1} + \frac{s_2^2}{n_2}} = t^*\sqrt{\frac{140.9^2}{6436} + \frac{126.9^2}{6007}} = t^*2.401137$$

So using $t^* = 1.96$, we have $m = 1.96 \times 2.401137 = 4.706229$ minutes. Therefore, a 95% confidence interval is

$$56.2 \pm 4.706229, \text{ or about } (51.5, 60.9) \text{ minutes.}$$

This is very close to what the software found.

Testing Hypotheses about Two Means

Hypothesis tests to compare two means from independent samples follow the same structure we discussed in Chapter 8, although the details change slightly because we now are comparing means rather than proportions. We show this structure by revisiting our comparison of sleeping times between weekends and weekdays. In Example 14 you found a confidence interval for the mean difference. Here we approach the same question with a hypothesis test.

In Example 14 we used boxplots to investigate the shape of the distribution of amount of sleep. Here, we examine histograms (Figure 9.21), which show a more detailed picture of the distributions. Both distributions have roughly the same amount of spread, and the histograms seem fairly symmetric.

We call weekend and holiday sleepers "population 1" and weekday sleepers "population 2." Then the symbol μ_1 represents the mean amount of sleep for all people in the United States on weekends and holidays, and μ_2 represents the mean amount of sleep for all people in the United States on weekdays.

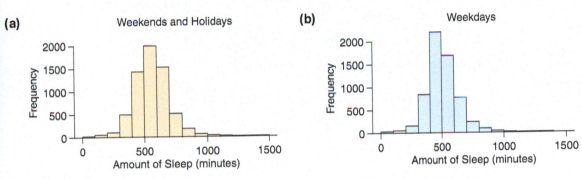

▲ **FIGURE 9.21** Amount of sleep reported on **(a)** weekends and holidays and **(b)** weekdays. Note that 500 minutes is about 8 hours 18 minutes.

Step 1: Hypothesize

H_0: $\mu_1 = \mu_2$ (typical amount of sleep is the same on weekends/
            holidays as weekdays)

H_a: $\mu_1 \neq \mu_2$ (typical amount of sleep time is different on weekends/
            holidays than on weekdays)

Step 2: Prepare

The conditions for testing two means are not very different from those for testing one mean and are identical to those for finding confidence intervals of the difference of two means.

Condition 1: *Random Samples and Independent Observations.* Observations are taken at random from two populations, producing two samples, or all subjects are randomly assigned to one of the two groups. Observations within a sample are independent of one another, which means that knowledge of one value tells us nothing about other observed values in that sample.

Condition 2: *Independent Samples.* The samples are independent of each other. Knowledge about a value in one sample does not tell us anything about any value in the other sample.

Condition 3: *Large Sample*. Both populations are approximately Normal, or both sample sizes are 25 or more. (In extreme situations, larger sample sizes may be required.)

Condition 1 holds because we are told that the people were selected randomly and independently. And because the two groups consist of different people, condition 2 holds. The distributions of the samples both look reasonably symmetric, but with such large sample sizes (over 6000 per group), condition 3 also holds.

Step 3: Compute to compare

First, we choose a significance level. It is common to use $\alpha = 0.05$, and we will do so for this example. The test statistic used to test this hypothesis is based on the difference between the sample means. Basically, the test statistic measures how far away the observed difference in sample means is from the hypothesized difference in population means. Yes, you guessed it: The distance is measured in terms of standard errors.

$$t = \frac{(\text{difference in sample means} - \text{what null hypothesis says the difference is})}{SE_{est}}$$

Using the test statistic is made easier by the fact that the null hypothesis almost always says that the difference is 0.

$$\text{Difference in sample means: } \bar{x}_1 - \bar{x}_2 = 555.9 - 499.7 = 56.2$$

(The summary statistics are given in Example 14.)

$$SE_{est} = \sqrt{\frac{s_1^2}{n_1} + \frac{s_2^2}{n_2}} = \sqrt{\frac{140.9^2}{6436} + \frac{126.9^2}{6007}} = 2.401137$$

$$t = \frac{56.2}{2.401137} = 23.4$$

Formula 9.4: Two-Sample *t*-Test

$$t = \frac{\bar{x}_1 - \bar{x}_2 - 0}{SE_{est}}, \text{ where } SE_{est} = \sqrt{\frac{s_1^2}{n_1} + \frac{s_2^2}{n_2}}$$

If all the conditions are met, the test statistic follows an approximate *t*-distribution, where the degrees of freedom are conservatively estimated to be the smaller of $n_1 - 1$ and $n_2 - 1$.

If there is no difference in the mean amounts of sleep in the United States, then the sample means should be nearly equal, and their difference should be close to 0. Our *t*-statistic tells us that the difference in means is 23.4 standard errors away from where the null hypothesis expects it to be.

Intuitively, you should understand that this *t*-statistic is extremely large and therefore casts quite a bit of doubt on our null hypothesis. But let's measure how surprising this is. To do so, we need to know the sampling distribution of the test statistic *t*, because we measure our surprise by finding the probability that if the null hypothesis were true, we would see a value as extreme as or more extreme than the value we observed. In other words, we need to find the p-value.

If the conditions listed in the Prepare step hold, then t follows, approximately, a t-distribution with minimum $(n_1 - 1, n_2 - 1)$ degrees of freedom. This approximation can be made even better by adjusting the degrees of freedom, but this adjustment is, for most cases, too complex for a "by hand" calculation. For this reason, we recommend using technology for two-sample hypothesis tests, because you will get more accurate p-values.

The smallest of the sample sizes is 6007, so our conservative number of degrees of freedom is $6007 - 1 = 6006$.

Our alternative hypothesis is two-sided, and it says that the true difference is either much bigger than 0 or much smaller than 0, so we use a two-tailed area under the t-distribution. Figure 9.22 shows this calculation using a statistical calculator. As you might have guessed, the value of 23.4 is so extreme that the area is invisible.

The p-value is nearly 0.

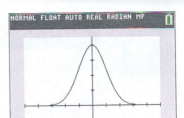

▲ FIGURE 9.22 The TI-84 output shows us that the p-value is nearly 0, because it is extremely unlikely the t statistic will be more than 23.4 standard errors away from 0 when the null hypothesis is true.

Step 4: Interpret
Again, we compare the p-value to the significance level, α. If the p-value is less than or equal to α, we reject the null hypothesis. In this example, the p-value is nearly 0, so it is certainly much less than 0.05. If people do tend to sleep the same amount of time on weekends as on weekdays, then this outcome is extremely surprising. Nearly impossible, in fact. Therefore, we reject the null hypothesis and conclude that people do tend to sleep a different amount on weekends and holidays than on weekdays.

The previous analysis was done using only the summary statistics provided. If you have the raw data, then you should use computer software to do the analysis. You will get more accurate values and save yourself lots of time. Figure 9.23 shows StatCrunch output for testing whether the mean sleeping time on weekends and holidays is different from the mean sleeping time on weekdays.

> **! Caution**
>
> **Don't Accept!**
> Remember from Chapter 8 that we do not "accept" the null hypothesis. It is possible that the sample size is too small (the test has low power) to detect the real difference that exists. Instead, we say that there is not enough evidence for us to reject the null.

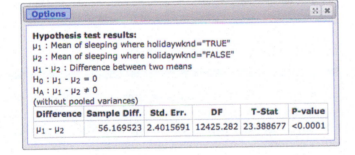

◀ FIGURE 9.23 StatCrunch output to test whether people typically sleep a different amount of time on weekends and holidays than they do on weekdays.

Into the Pool Some software packages, and some textbooks too, provide for another version of this t-test called the "pooled two-sample t-test." We have presented the unpooled version (you can see this in the StatCrunch output above the table, where it says "without pooled variances"). The unpooled version is preferred over the other version because the pooled version works only in special circumstances (when the population standard deviations are equal). The unpooled version works reasonably well in all situations, as long as the listed conditions hold.

> **! Caution**
>
> **Don't Pool**
> When using software to do a two-sample t-test, make sure it does the unpooled version. You might have to tell the software explicitly. The unpooled version is more accurate in more situations than the pooled version.

SNAPSHOT ▶ Two Sample *t*-Test (From Independent Samples)

WHAT IS IT? ▶	A procedure for deciding whether two means, estimated from independent samples, are different. The test statistic used is $$t = \frac{\bar{x}_1 - \bar{x}_2 - 0}{SE_{est}}, \text{ where } SE_{est} = \sqrt{\frac{s_1^2}{n_1} + \frac{s_2^2}{n_2}}$$
WHAT DOES IT DO? ▶	Provides us with a decision on whether to reject the null hypothesis that the two means are the same and lets us do so knowing the probability that we are making a mistake.
HOW DOES IT DO IT? ▶	Compares the observed difference in sample means to 0, the value we expect if the population means are equal.
HOW IS IT USED? ▶	The observed value of the test statistic can be compared to a *t*-distribution.

Hypotheses: Choosing Sides So far, we've presented the hypotheses with one mean on the left side and one on the right, like this: H_0: $\mu_1 = \mu_2$. But you will generally see hypotheses written as differences: H_0: $\mu_1 - \mu_2 = 0$. These two hypotheses mean exactly the same thing, because if the means are equal, then subtracting one from the other will result in 0.

As with the other hypothesis tests you've seen, there are three different alternative hypotheses we can use. Here are your choices:

Two-Sided	One-Sided (Left)	One-Sided (Right)
H_0: $\mu_1 - \mu_2 = 0$	H_0: $\mu_1 - \mu_2 = 0$	H_0: $\mu_1 - \mu_2 = 0$
H_a: $\mu_1 - \mu_2 \neq 0$	H_a: $\mu_1 - \mu_2 < 0$	H_a: $\mu_1 - \mu_2 > 0$

EXAMPLE 15 Reading Electronics

More and more often, people are reading on computer screens or other electronic "e-readers." Do we read differently when we read on a computer screen than we do when we read material on ordinary paper? Researchers in Norway carried out a study to determine whether children (of high school age) read differently when reading material from a PDF on a computer screen than when reading a printed copy. Specifically, they measured whether reading comprehension differed between the two types of material.

To carry out this study, 72 tenth-grade students were randomly assigned to one of two groups, which we'll call "electronic" and "paper." All students were asked to read two texts, both of roughly equal length. However, the students in the "electronic" group read the texts on computer screens, and the "paper" group read them on paper. The texts were formatted so that they appeared the same both on the computer screen and on paper. After reading, all students took the same reading comprehension test (Mangen et al. 2013).

Figure 9.24 provides summary statistics for the reading comprehension scores for the two groups, and Figure 9.25 shows their histograms. The typical reading comprehension score is larger for the students who read on paper, which indicates that the students in this sample typically had a greater level of understanding of what they had read.

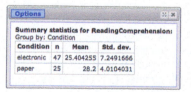

▲ **FIGURE 9.24** Summary statistics produced by StatCrunch.

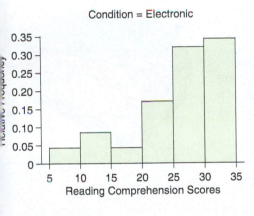

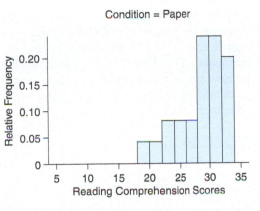

◀ **FIGURE 9.25** The distribution of reading comprehension scores for tenth-grade students. One group read the texts on a computer screen, the other on paper.

QUESTION Carry out the four steps of a hypothesis test to test whether students who read on paper have a different level of comprehension than those who read on computer screens. Use a significance level of 5%. If a required condition (in step 2) doesn't hold, explain the consequences and state any assumptions you must make in order to continue with the test. In step 3, refer to Figure 9.26, which displays an excerpt of StatCrunch output for this hypothesis test.

SOLUTION

Step 1: Hypothesize

We will let μ_1 represent the mean reading comprehension score of all tenth-grade students in Norway who might read these texts in *paper* format, and we'll let μ_2 represent the mean reading comprehension scores of all tenth-grade students in Norway who might read these texts on a *computer*.

$$H_0: \mu_1 = \mu_2$$
$$H_a: \mu_1 \neq \mu_2$$

(You might also write these as $H_0: \mu_1 - \mu_2 = 0$ and $H_a: \mu_1 - \mu_2 \neq 0$.)

Step 2: Prepare

We do not have a random sample of tenth-grade Norwegians. This means we cannot generalize our results to all students of this age from Norway. However, because random assignment was used, we can conclude that, if we find a statistically significant difference, it is due to the reading condition and not to a (possibly unidentified) confounding factor. The second condition holds because we are told that different children are in the different groups. Finally, although the distributions are left-skewed, the sample sizes are large, so we can use the *t*-distribution. (But note that the paper group is just barely large enough, $n = 25$. It is possible that our p-value might be more approximate than we would like.)

Step 3: Compute to compare

We'll use a significance level of 5%, as directed. Referring to Figure 9.26, we see that the value of the *t*-statistic is 2.11, which tells us that the observed difference in means was 2.1 standard errors above the value we would expect if the null hypothesis were true. The p-value is 0.0388.

Difference	Sample Diff.	Std. Err.	DF	T-Stat	P-value
$\mu_1 - \mu_2$	2.7957447	1.3271877	69.844774	2.1065179	0.0388

◀ **FIGURE 9.26** Excerpt from statistical software to test whether mean reading comprehension for students reading on paper (group 1) was different from reading comprehension for students reading on a computer (group 2).

Step 4: Interpret
Because the p-value of 0.0388 is less than our significance level of 0.05, we reject the null hypothesis. For these students, mean reading comprehension for those reading on paper was different from reading comprehension for those reading on a computer screen.

TRY THIS!   Exercise 9.57

The researchers themselves did a different analysis from the one we have presented. Their analysis took into account additional factors that we chose to leave out so that we could focus on the structure of the hypothesis test. The researchers' conclusion was that reading comprehension was higher for students reading on paper than for students reading on the computer.

CI for the Mean of a Difference: Dependent Samples

With paired samples, we turn two samples into one. We do this by finding the differences in each pair. For example, researchers wanted to know whether our sensitivity to smells is different when we were sitting up compared to when we are lying down (Lundström et al. 2006). They devised a measure of the ability of a person to detect smells, and then they measured a sample of people twice: once when they were lying down and once when they were sitting upright.

How much might this smell-sensitivity score change, if at all? One way to answer this question is with a confidence interval for the mean difference in scores. These data differ from the examples you've seen with two independent samples in that even though there are two groups (lying down, sitting upright), each person in the sample is measured twice. For this reason, these are *dependent*, or *paired*, samples.

When we are dealing with paired samples, our approach is to transform the original data from two variables (lying down and sitting upright, or, if you prefer, group 1 and group 2), into a single variable that contains the difference between the scores in group 1 and group 2. As before, it doesn't matter which is group 1 and which is group 2, but we do have to remember our choice. Once that is done, we have a single sample of difference scores, and we apply our one-sample confidence interval from Formula 9.1.

For the "smell" study, we create the new difference variable by subtracting each person's score when lying down from her or his score when sitting.

The first few lines of the original data are shown in Table 9.3a.

▶ **TABLE 9.3a** Smelling ability for the first four people, sitting and lying down.

Subject Number	Sex	Sitting	Lying Down
1	Woman	13.5	13.25
2	Woman	13.5	13
3	Woman	12.75	11.5
4	Man	12.5	12.5

We create a new variable, call it *Difference*, and define it to be the difference between smelling ability sitting upright and smelling ability lying down. We show this new variable in Table 9.3b.

Subject Number	Sex	Sitting	Lying Down	Difference
1	Woman	13.5	13.25	0.25
2	Woman	13.5	13	0.50
3	Woman	12.75	11.5	1.25
4	Man	12.5	12.5	0

◄ TABLE 9.3b Difference between smelling ability while sitting upright and smelling ability while lying down.

Here are summary statistics for the *Sitting, Lying,* and *Difference* variables:

Variable	n	Sample Mean	Sample Standard Deviation
Sitting	36	11.47	3.26
Lying	36	10.60	3.06
Difference	36	0.87	2.39

After verifying that the necessary conditions hold (they do), we can find a 95% CI for the mean difference using Formula 9.1. Because we have 36 observations, the degrees of freedom for t^* are $36 - 1 = 35$. The table in Appendix A tells us that $t^* = 2.03$.

Thus, a 95% confidence interval for the mean difference in smelling scores is

$$\bar{x} \pm t^* \frac{s}{\sqrt{n}},$$

$$\text{or } 0.87 \pm 2.03 \frac{2.39}{\sqrt{35}},$$

which works out (after rounding) to (0.05, 1.69).

Because these values are all positive, we conclude that the mean for the first group (sitting upright) is higher than the mean for the second group (lying down). For this reason, we are confident that smelling sensitivity is greater when sitting upright than when lying down. This difference could be fairly small, 0.05 unit, or as large as 1.7 units.

Many statistical software packages allow you to compute the confidence interval for two means in a paired sample by selecting a "paired two-sample" option. Example 16 shows how statistical software can be used to find a confidence interval for the difference of two means when the data are from paired samples.

EXAMPLE 16 Dieting

Americans who want to lose weight can choose from many different diets. In one study (Dansinger et al. 2005), researchers compared results from four different diets. In this example, though, we look only at a small part of these data and examine only the 40 subjects who were randomly assigned to the Weight Watchers diet. These subjects were measured twice: at the start of the study and then 2 months later. The data consist of two variables: weight (in kilograms) at 0 months and weight at 2 months. One question we can ask is, How much weight will the typical person lose on the Weight Watchers diet after 2 months?

Figure 9.27 shows two different 90% confidence intervals for these data. Figure 9.27a shows a confidence interval treating the data as if they were independent samples, and Figure 9.27b shows a confidence interval that treats the data as paired samples. Note that these data are not a random sample, but random assignment was used.

(a)
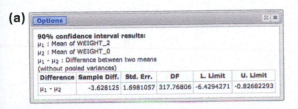

(b)
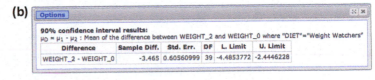

▲ FIGURE 9.27 StatCrunch output showing 90% confidence intervals **(a)** as if the data were independent samples and **(b)** as if they were paired samples.

> **QUESTION**   State and interpret the correct interval.

> **SOLUTION**   Because the same subjects are in both samples (because each subject was measured twice), the data are paired samples. For this reason, Figure 9.27b shows the correct output.

The 90% confidence interval for the mean difference in weight is -4.5 kilograms to -2.4 kilograms. The fact that the interval contains only negative values means that the mean weight in the first measurement (at 2 months) is less than the mean weight in the second measurement (at 0 months). We are therefore confident that the typical subject lost weight: as much as 4.5 kilograms (about 10 pounds) or as little as 2.4 kilograms (about 5 pounds).

> **TRY THIS!**   Exercise 9.67

Test of Two Means: Dependent Samples

In Example 16, we asked a question about amounts: *How much* weight did the typical Weight Watchers dieter lose? Sometimes researchers aren't as interested in "How much?" as in answering the question, "Did anything change at all?"

Questions such as this can be answered with hypothesis tests about paired data. In this case, as with confidence intervals for paired data, we convert the two variables into a difference variable, and our hypotheses are now not about the individual groups but about the difference.

To illustrate, let's consider another subgroup of the dieters' data. The diet program known as The Zone promises that you'll lose weight, burn fat, and not feel hungry. The diet requires that you eat 30% protein, 30% fat, and 40% carbs, and it also imposes restrictions on the times at which you eat your meals and snacks. Can people lose weight on The Zone?

In words, our null hypothesis is that after 2 months of dieting, the mean weight of people on The Zone diet is the same as the mean weight before dieting. Our alternative hypothesis is that the mean weight is *less* after 2 months of dieting (a one-sided hypothesis).

For the data we are analyzing, subjects were randomly assigned to one of four diets, although we will consider only those on The Zone. For each subject, weight was measured at 0 months and at 2 months. Because the same subjects appear in both groups, the data are *paired*. Instead of considering weight at 0 months and weight at 2 months as separate variables, we will calculate the *change* in weight and name this variable *Difference*. For each subject,

$$\text{Difference} = (\text{weight at 2 months}) - (\text{weight at 0 months})$$

Our hypotheses are now about just one mean, the mean of *Difference*:

$$H_0: \mu_{\text{difference}} = 0 \quad (\text{or } \mu_{\text{2months}} = \mu_{\text{0months}})$$

$$H_a: \mu_{\text{difference}} < 0 \quad (\text{or } \mu_{\text{2months}} < \mu_{\text{0months}})$$

Our test statistic is the same as for the one-sample t-test:

$$t = \frac{\bar{x}_{\text{difference}} - 0}{SE_{\text{difference}}}, \quad \text{where } SE_{\text{difference}} = \frac{s_{\text{difference}}}{\sqrt{n}}$$

We find $\bar{x}$ by averaging the difference variable: $\bar{x} = -3.795$ kilograms.

We find $s_{\text{difference}}$ by finding the standard deviation of the difference variable: $s = 3.5903$ kilograms.

There were 40 subjects, so

$$SE = \frac{3.5903}{\sqrt{40}} = 0.5677$$

and then

$$t = \frac{-3.795}{0.5677} = -6.685$$

To find the p-value, we use a *t*-distribution (assuming the conditions for a one-sample *t*-test hold) with $n - 1$ degrees of freedom, where *n* is the number of data pairs. Because there are 40 pairs, one for each subject, the degrees of freedom are 39. The alternative hypothesis is left-sided, so we find the area to the left of -6.685. The test statistic tells us that the mean difference was 6.685 standard errors below where the null hypothesis expected it to be. We should expect the p-value to be very small.

In fact, the area to the left of -6.685 is too small to see in the statistical calculator in Figure 9.28. But the calculator does verify that the p-value is nearly 0.

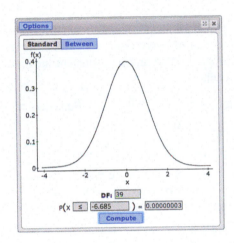

◄ **FIGURE 9.28** StatCrunch statistical calculator output, showing that the probability below -6.685 in a *t*-distribution with 39 degrees of freedom is extremely small.

Because the p-value is so small, we reject the null hypothesis and conclude that the typical subject in the study really did lose weight on The Zone diet.

EXAMPLE 17 More Rising College Costs

Earlier we questioned whether costs at two-year colleges increased by comparing a sample of schools in one academic year with the population mean from the past. However, by drawing a sample of schools and examining their current tuition with their own previous tuition, we can gain a more precise answer to the question: Did schools, on average, increase their tuition and fees? To answer this question, we took a random sample of 35 schools and compare their tuition and fees in 2011–2012 with their tuition and fees in 2014–2015. (Note that this is different random sample from the sample used in Examples 5 and 6.)

The first few rows look like this:

Institution Name	Tuition 2011–2012	Tuition 2014–2015
Mohawk Valley Community College	4010	4415
George C Wallace State Community College–Hanceville	4080	4260
College of the Redwoods	888	1182

The summary statistics show us that the sample mean is higher in 2014–2015.

Tuition and Fees	n	Sample Mean	Sample SD
2011–2012	35	4742.74	4970.37
2014–2015	35	5380.23	5512.00

Paired studies such as these can sometimes give more insight into change, since they have less sampling variability than a single sample design as in Example 5. The observed value of the test statistic from a paired t-test was 5.9, which corresponds to a p-value that is less than 0.0001. (We used the more recent academic year as group 1 in the paired t-test, so a positive difference for a school means that the school charged more for tuition in 2014–2015 than it did in 2011–2012.)

QUESTION   Carry out a hypothesis test at a 5% level to answer the question of whether schools have increased their tuition and fees from 2011–2012 to 2014–2015.

SOLUTION

Step 1: Hypothesize

To restate the question as a pair of hypotheses, we'll let μ_1 represent the mean tuition of all two-year colleges in 2014–2015 and μ_2 represent the mean tuition of all two-year colleges in 2011–2012. Then

$$H_0: \mu_{\text{difference}} = 0 \quad \text{where} \quad \mu_{\text{difference}} = \mu_1 - \mu_2$$
$$H_a: \mu_{\text{difference}} > 0$$

Step 2: Prepare

Because these are a random sample, the first condition is satisfied. The same colleges appear in both groups, so we have paired data. The number of pairs (colleges) is greater than 25, so the Large Sample condition is satisfied. We can proceed.

Step 3: Calculate to Compare

We were told to use a 5% significance level. The calculations were also provided for us: $t = 5.9$. The p-value is very small.

Step 4: Interpret

Because the p-value is less than the significance level, we reject the null hypothesis and conclude that, on average, two-year colleges have raised their tuition and fees.

TRY THIS!   Exercise 9.69

SNAPSHOT ▸ Paired *t*-Test (Dependent Samples)

WHAT IS IT? ▸ A procedure for deciding whether two dependent (paired) samples have different means. Each pair is converted to a difference. The test statistic is the same as for the one-sample *t*-test, except that the null hypothesis value is 0:

$$t = \frac{\bar{x}_{\text{difference}} - 0}{SE_{\text{difference}}}$$

WHAT DOES IT DO? ▸ Lets us make decisions about whether the means are different while knowing the probability that we are making a mistake.

HOW DOES IT DO IT? ▸ The test statistic compares the observed average difference, $\bar{x}_{\text{difference}}$, with the average difference we would expect if the means were the same: 0. Values far from 0 discredit the null hypothesis.

HOW IS IT USED? ▸ If the required conditions hold, the value of the observed test statistic can be compared to a *t*-distribution with $n - 1$ degrees of freedom.

SECTION 9.6

Overview of Analyzing Means

We hope you've been noticing a lot of repetition. The hypothesis test for two means is very similar to the test for one mean, and the hypothesis test for paired data is really a special case of the one-sample *t*-test. Also, the hypothesis tests use almost the same calculations as the confidence intervals, and they impose the same conditions, arranged slightly differently.

All the test statistics (for one proportion, for one mean, for two means, and for two proportions) have this structure:

$$\text{Test statistic} = \frac{(\text{estimated value}) - (\text{null hypothesis value})}{SE}$$

All the confidence intervals have this form:

$$\text{Estimated value} \pm (\text{multiplier}) \, SE_{\text{EST}}$$

Not all confidence intervals used in statistics have this structure, but most that you will encounter do.

The method for computing a p-value is the same for all tests, although different distributions are used for different situations. The important point is to pay attention to the alternative hypothesis, which tells you whether you are finding a two-tailed or a one-tailed p-value.

Don't Accept the Null Hypothesis

If the p-value is larger than the significance level, then we do not reject the null hypothesis. But this doesn't mean we "accept" it. In other words, this doesn't mean we think the null hypothesis is true.

In Example 14 and the discussion that followed, we concluded that people tend to sleep longer on weekends and holidays than on weekdays—roughly an hour more on average. We reached this conclusion on the basis of a random sample of over 12,000 people. But let's consider what might have happened if we had taken a random sample of only 30 people from each group.

The following summary statistics are based on a random sample of just 30 people who reported their sleeping amounts on weekdays and another 30 people who reported their sleeping amounts on weekends and holidays.

$$\text{Weekday:} \qquad \bar{x} = 532 \text{ minutes, } s = 138 \text{ minutes}$$

$$\text{Weekend/holiday: } \bar{x} = 585 \text{ minutes, } s = 155 \text{ minutes}$$

If we now test the hypothesis that mean hours of sleep are different on weekends and holidays, we will get different results from our previous calculations. Now we find that our test statistic is $t = 1.40$ and the p-value is 0.167. If we believe the mean hours of sleep in the population are really the same, then the t-statistic value was not a surprise to us. And so we do not reject the null hypothesis.

But even though we do not reject the null, we do not necessarily believe the null hypothesis is true. The variation in amount of sleep is quite large: more than two hours. When our sample size is small, it is unlikely that we'll be able to tell whether two population means are truly different, because there is so much variability in our test statistic.

This is one reason why we never "accept" the null hypothesis. With a larger sample size, our test statistic will be more precise, and if the population means are truly different, we will have a better chance of correctly rejecting the null. In fact, when we did this test with more than 6000 people in each group, we determined that the means are different.

Confidence Level	Equivalent α (Two-Sided)
99%	0.01
95%	0.05
90%	0.10

▲ TABLE 9.4 Equivalences between confidence intervals and tests with two-sided alternative hypotheses.

Confidence Intervals and Hypothesis Tests

If the alternative hypothesis is two-sided, then a confidence interval can be used instead of a hypothesis test. In fact, these two choices will always reach the same conclusion:

Choice 1: Perform the two-sided hypothesis test with significance level α.

Choice 2: Find a $(1 - \alpha) \times 100\%$ confidence interval (using methods given earlier). Reject the null hypothesis if the value for the null hypothesis does *not* appear in the interval.

KEY POINT   A 95% confidence interval is equivalent to a test with a two-sided alternative with a significance level of 0.05. Table 9.4 shows some other equivalences. All are true only for *two-sided* alternative hypotheses.

EXAMPLE 18   Calcium Levels in the Elderly

The boxplots in Figure 9.29 show the results of a study to determine whether calcium levels differ substantially between senior men and senior women (all older than 65 years). Calcium is associated with strong bones, and people with low calcium levels are believed to be more susceptible to bone fractures. The researchers carried out a hypothesis test to see whether the mean calcium levels for men and women were the same. Figure 9.30 shows the results of the analysis. Calcium levels (the variable *cammol*) are measured in millimoles per liter (mmol/L).

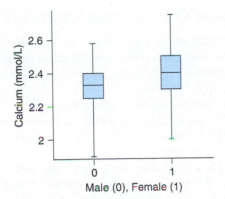

◀ **FIGURE 9.29** Boxplots of calcium levels (mmol/L) for males and females.

○ ○ ○   Two sample T statistics with data

Options

Hypothesis test results:

μ_1 : mean of cammol where female=1

μ_2 : mean of cammol where female=0

$\mu_1 - \mu_2$: mean difference

$H_0 : \mu_1 - \mu_2 = 0$

$H_A : \mu_1 - \mu_2 \neq 0$

(without pooled variances)

Difference	Sample Mean	Std. Err.	DF	T-Stat	P-value
$\mu_1 - \mu_2$	0.07570534	0.019790715	168.45009	3.825296	0.0002

◀ **FIGURE 9.30** StatCrunch output for testing whether mean calcium levels in men and women differ. The difference estimated is mean of females minus mean of males.

QUESTIONS

a. Assuming that all conditions necessary for carrying out *t*-tests and finding confidence intervals hold, what conclusion should the researchers make on the basis of this output? Use a significance level of 0.05.

b. Suppose the researchers calculate a confidence interval for the difference of the two means. Will this interval include the value 0? If not, will it include all negative values or all positive values? Explain.

SOLUTIONS

a. The p-value, 0.0002, is less than 0.05, so the researchers should reject the null hypothesis and conclude that men and women have different calcium levels.

b. Because we rejected the null hypothesis, we know that the confidence interval cannot include the value 0. If it did, then 0 would be a plausible difference between the means, and our hypothesis test concluded that 0 is not plausible. The estimated difference between the two means is, from the output, 0.0757. Because this value is positive, and because the interval cannot include 0, all values in the interval must be positive, showing a higher mean level of calcium in women than in men.

TRY THIS!   Exercise 9.75

Hypothesis Test or Confidence Interval?

If you can use either a confidence interval or a hypothesis test, how do you choose? First of all, remember that these two techniques produce the same results only when you have a two-sided alternative hypothesis, so you need to make the choice only when you have a two-sided alternative.

These two approaches answer slightly different questions. The confidence interval answers the questions, "What's the estimated value? And how much uncertainty do you have in this estimate?" Hypothesis tests are designed to answer the question, "Is the parameter's value one thing, or another?"

For many situations, the confidence interval provides much more information than the hypothesis test. It not only tells us whether or not we should reject the null hypothesis but also gives us a plausible range for the population value. The hypothesis test, on the other hand, simply tells us whether to reject or not (although it does give us the p-value, which helps us see just how unusual our result is if the null hypothesis is true).

For example, in our hypothesis test of whether people tend to sleep a different amount on weekend and holidays than on weekdays, we rejected the null hypothesis and concluded that, on average, they did sleep different amounts. But that is all we can say with the hypothesis test. We cannot say, with the same significance level, whether they slept more on weekends or on weekdays, and we cannot say how much more. However, by finding that the 95% confidence interval was about 51 minutes to 61 minutes, we know now that, typically, people sleep about an hour longer on weekends and holidays than on weekdays. (Because group 1 was "weekend/holiday," the positive values mean that this mean was greater than the mean for the "weekday" group.) Because confidence intervals provide so much more information than hypothesis tests, there is a growing trend in scientific journals to require researchers to provide confidence intervals either in place of, or in support of, hypothesis tests.

CASE STUDY REVISITED

You look sick! Are you sick?

Can we tell who is sick just by looking at them? Researchers showed people pictures of volunteers, some of whom were healthy and others sick. The people rated the pictures on how healthy they thought the person in the picture appeared. To analyze the data, the researchers used some advanced statistical techniques to supplement a complex study design. However, we can simplify somewhat to apply the techniques in this chapter and reach a pretty good understanding of the researchers' findings.

The researchers actually considered a number of measures of health and appearance. Here, we'll focus on just one: the perceived sickness level of the people in the photos. Recall that raters were shown several pictures of people who were healthy and several who had been injected with bacteria. To simplify the data, for each rater, we computed the average sickness rating they awarded to the "healthy" pictures and to the "sick" pictures. (The raters didn't know which were healthy and which were sick, of course.)

The boxplot in Figure 9.31 shows that the raters did typically rate the sick photos as looking less healthy than the photos of healthy people. But this boxplot hides an important feature of the data. The data are paired, since each rater provided ratings for both groups of photos.

▶ FIGURE 9.31 The photos of sick people tended to be rated higher on the sickness rating scale, although there was some overlap.

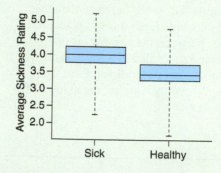

To get a more accurate picture of the data, we computed the differences in ratings for each rater by subtracting the score they assigned to healthy people from the score they assigned to sick people. A positive difference indicates that they tended to rate the photos of sick people as looking sicker than the healthy photos. A negative score means they tended to rate the sick photos as looking healthier than the healthy photos.

Figure 9.32 shows the distribution of the sample of difference scores. We note that very few raters had negative difference scores. The mean difference score is 0.48 and the standard deviation is 0.36.

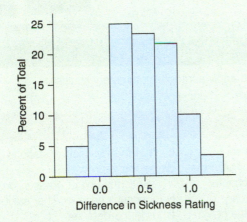

◄ FIGURE 9.32 The distribution of ratings is fairly symmetric, and centered at 0.48, which suggests that, on average, raters could tell the difference between healthy and sick photos.

This positive mean could have resulted from chance. Since the raters were shown photos in a random order, it is conceivable that, had the order been changed, the ratings would have come out differently. To test this hypothesis, we carried out a paired t-test, comparing the null hypothesis that the mean difference score is 0 against the alternative that the mean difference is positive. The observed test statistic was 10.245 with a p-value close to 0, and so we reject the null hypothesis and conclude that these raters really did see that the sick people were sicker than the healthy people.

A 95% confidence interval confirms this story. The resulting interval is (0.39, 0.58). Because these values are all positive, we conclude that the mean difference is positive and so, again, conclude that these raters really could discern the difference between photos of sick people and photos of healthy people. We note that this doesn't match precisely the interval shown at the start of the chapter due to the fact that the researchers used more precise statistical analyses than we did here. However, the conclusions are the same.

Also note that the report used a different format for reporting p-values than we've used: "P < 0.001". Different journals use different formatting standards, but be aware that this means the same thing as "p-value < 0.001".

DATAPROJECT ▸ Data Structures

1 OVERVIEW

We continue the work from Chapter 8 to answer the question, How much rain falls in Los Angeles in a year?

2 GOALS

Identify data sets with a "nested" structure and apply appropriate summary statistics.

3 ANALYZING ACROSS GROUPS

We're going to continue our work with the data set *rain.csv*, which contains daily rainfall amounts for all of the 70 weather stations in Los Angeles. In the last chapter we learned that the data include the dates February 1, 2016, through May 25, 2018. This means that the only complete year is 2017. For this reason, we'll focus on that year. You'll need to use the data set you saved from your work on the project in Chapter 8, or you'll need to re-create that work so that you have a variable that indicates the year in which the observation was recorded.

Project

We wish to know how much rain fell in Los Angeles in 2017. Your first impulse might be to simply add up all of the rainfall amounts contained in the *PRCP* variable. (PRCP means "precipitation.") But this approach doesn't quite make sense. Imagine that you have two rain gauges sitting next to each other. After a rainy day, one reads 1.25 inches and the other also reads 1.25 inches. But this doesn't mean you had a total of 1.25 + 1.25 = 2.50 inches of rain. It's not appropriate to simply add rain amounts from the same location on the same day. You don't want the amount of rainfall you report to depend on the number of rain gauges you're using to catch the rainfall.

On the other hand, if it rained 1.25 inches on Monday, 1.00 inch on Tuesday, and no rain for the rest of the week, it is legitimate to say that there were 1.25 + 1.00 = 2.25 inches of rain that week.

This data set has a fairly complex structure, which is typical of data sets that include many observations across time. Remember that one of the important questions we should consider when seeing a data set for the first time is, "What's the primary unit of observation?" This question is difficult to answer in this case. Would we say that it is a day of the year? If so, then we would expect each row to uniquely represent one day of the year. But as you saw in the last chapter, each day of the year appears multiple times because

each of the 70 weather stations had a record on a given day of the year.

So is the unit of observation the weather station? No, because each weather station appears multiple times in the data set.

In fact, the unit of observation is a strange abstraction: the station-date. In other words, Alameda-July-1-2017 is one unique observation. Alameda-July-2-2017 is another, as is Pasadena-July-2-2017. Data sets with this structure are sometimes called "nested" or "hierarchical." The reason is that observations for each date are "nested" inside each station. Pick any station and it has all (or almost all) the dates recorded for it.

In some data sets, and this one is a good example, the nesting can work both ways. For example, pick any date, and it has all the stations recorded for it. So we can think of the data as either consisting of stations with daily observations nested within, or as dates with observations from all of the stations nested within.

This fact will allow us to ask and answer slightly different sets of questions.

First, if the primary unit is the station and days are nested within stations, then we can ask: How much total rainfall fell *at each* station during 2017? How do these amounts vary across the greater Los Angeles area? What was the average annual rainfall in Los Angeles in 2017?

Second, if the primary unit is the day, and stations are nested within days, we can ask: What was the daily average rainfall in Los Angeles (averaged across all stations)? How did this average vary throughout the year?

Assignment

To answer the first set of questions, we need to tell StatCrunch to perform operations separately for each weather station. To do this we must "group by" station. We choose Stat > Summary Stats > Columns and fill out the dialogue box as shown in Figure 9.32. Notice that we restrict our analysis to 2017 in order to simplify our calculations. (2017 is the only complete year of observations.) The "group by" option means that we will compute statistics separately for each group. Because we choose NAME as our groups, we'll get a separate statistic for each station. (You could also choose STATION and get the same result.)

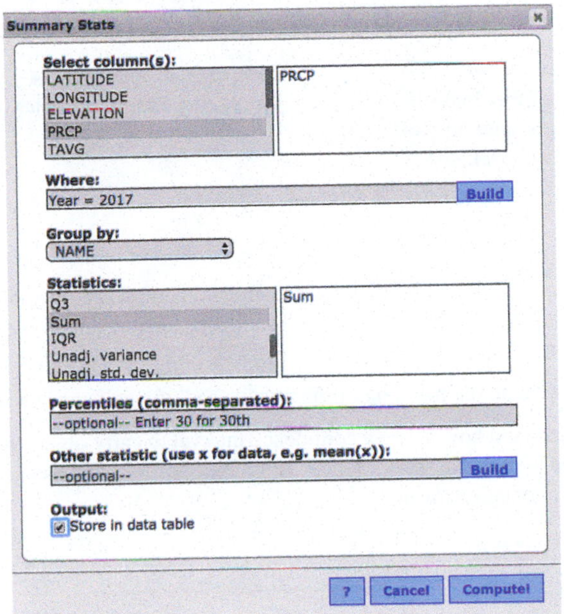

Summary Stats

Select column(s):
LATITUDE
LONGITUDE
ELEVATION
PRCP
TAVG

PRCP

Where:
Year = 2017   Build

Group by:
NAME

Statistics:
Q3
Sum
IQR
Unadj. variance
Unadj. std. dev.

Sum

Percentiles (comma-separated):
--optional-- Enter 30 for 30th

Other statistic (use x for data, e.g. mean(x)):
--optional--   Build

Output:
☑ Store in data table

?   Cancel   Compute!

▲ FIGURE 9.32

Because we want to know the total at the station, we select *Sum* as our statistic. This will simply add the precipitation amounts at each station.

Selecting the "store in data table" option lets us access the data for future analysis.

Once you've carried out these steps, create a visualization of the distribution of station totals and write a description of the distribution. Finally, what was the mean total rainfall across all stations in 2017? (You might be surprised that there's quite a bit of variability. One reason for this is that the Los Angeles area has a very diverse geography, ranging from mountains to the ocean to many hills that act as rain "shade" for other parts of the city.)

To answer the second set of questions, group by "DATE" and calculate the mean daily rainfall for each day. Explain why it would not be appropriate to calculate the total daily rainfall by using the sum instead of the mean. Describe the distribution of average daily rainfall for 2017. How does the mean of the average daily rainfalls here compare to the mean total rainfall from the previous set of questions?

Create a graphic that helps answer the question: How did rainfall vary in Los Angeles in 2017? Interpret your graph.

CHAPTER REVIEW

KEY TERMS

You may want to review the following terms, which were introduced in Chapters 7 and 8.
     Chapter 7: statistic, estimator, bias, precision, sampling distribution, standard error, confidence interval, confidence level, margin of error
     Chapter 8: null hypothesis, alternative hypothesis, significance level, test statistic, p-value, one-sided hypothesis, two-sided hypothesis

Bias, *437*
Precision, *437*
Sampling distribution, *439*
Unbiased estimator, *439*

Standard error, *439*
Central Limit Theorem
  (CLT), *441*
t-Statistic, *446*

t-Distribution, *447*
Degrees of freedom (df), *447*
Confidence intervals, *448*
Confidence level, *448*

Independent samples, *464*
Paired (dependent) samples, *464*

LEARNING OBJECTIVES

After reading this chapter and doing the assigned homework problems, you should

- Understand when the Central Limit Theorem for sample means applies and know how to use it to find approximate probabilities for sample means.

- Know how to test hypotheses concerning a population mean and concerning the comparison of two population means.

- Understand how to find, interpret, and use confidence intervals for a single population mean and for the difference of two population means.

- Understand the meaning of the p-value and of significance levels.

- Understand how to use a confidence interval to carry out a two-sided hypothesis test for a population mean or for a difference of two population means.

SUMMARY

The sample mean gives us an unbiased estimator of the population mean, provided that the observations are sampled randomly from a population and are independent of each other. The precision of this estimator, measured by the standard error (the standard deviation of the sampling distribution), improves as the sample size increases. If the population distribution is Normal, then the sampling distribution is Normal also. Otherwise, according to the Central Limit Theorem, the sampling distribution is approximately Normal, although for small sample sizes the approximation can be very bad. If the population distribution is not Normal, we recommend that you use a sample size of 25 or more.

    Although we did not go into the (fairly complex) mathematics, our ability to measure confidence levels (which tell us how well confidence intervals perform), p-values, and significance levels (α) depends on the Central Limit Theorem (CLT). If the conditions for applying the CLT are not satisfied, then our reported values for these performance measures may be wrong.

    Confidence intervals are used to provide estimates of parameters, along with a measure of our uncertainty in that estimate. The confidence intervals in this chapter differ only a little from those for proportions (Chapter 7). All are of the form

$$\text{Estimate} \pm \text{margin of error}$$

    One thing that is different is that now you must decide whether your two samples are independent or paired before performing your analysis.

    Another difference between confidence intervals for means and those for proportions is that the multiplier in the margin of error is based on the *t*-distribution, not on the Normal distribution.

    By this point you have learned several different hypothesis tests, including the z-test for one-sample proportion and for two-sample proportions and the *t*-test for one-sample mean, for two means from independent samples, and for two means from dependent samples. (You may also find pooled and unpooled versions of

the two-mean independent-samples *t*-test, but you should always use the unpooled version.) It is important to learn which test to choose for the data you wish to analyze.

    Hypothesis tests follow the structure described in Chapter 8, and, just as for confidence intervals, you must decide whether you have independent samples or paired samples. Tests of two means based on independent samples are based on the difference between the means. The null hypothesis is (almost) always that the difference is 0. The alternative hypothesis depends on the research question.

    To find the p-value for a test of two means, use the *t*-distribution. To use the *t*-distribution, you must know the degrees of freedom (df), and this depends on whether you are doing a test for one mean (df = $n - 1$); two means from independent samples (use your computer, or, if working by hand, use the df for the smaller of $n_1 - 1$, $n_2 - 1$); or two means from paired data (number of pairs -1).

Formulas

Samples are selected randomly from each population and are independent. Population distributions are Normal, or if not, sample sizes need to be 25 or bigger for each sample.

Formula 9.1: One-Sample Confidence Interval for Mean

$$\bar{x} \pm m$$

where $m = t^* SE_{\text{est}}$ and $SE_{\text{est}} = \dfrac{s}{\sqrt{n}}$

    The multiplier t^* is a constant that is used to fine-tune the margin of error so that it has the level of confidence we want. It is chosen on the basis of a *t*-distribution with $n - 1$ degrees of freedom. SE_{est} is the estimated standard error.

      Paired: $\bar{x}_{\text{difference}} \pm m$,   where $m = t^* SE_{\text{est}}$ and

$$SE_{\text{est}} = \dfrac{s_{\text{difference}}}{\sqrt{n}}$$

(where $\bar{x}_{\text{difference}}$ is the average difference, $s_{\text{difference}}$ is the standard deviation of the differences, and n is the number of data pairs)

Formula 9.2: The One-Sample t-Test for Mean

$t = \dfrac{\bar{x} - \mu}{SE_{\text{est}}}$, where $SE_{\text{est}} = \dfrac{s}{\sqrt{n}}$ and, if conditions hold,

t follows a t-distribution with df $= n - 1$

$$\textbf{Paired: } t = \frac{\bar{x}_{\text{difference}} - 0}{SE_{\text{difference}}}, \text{ where } SE_{\text{difference}} = \frac{s_{\text{difference}}}{\sqrt{n}}$$

(where $\bar{x}_{\text{difference}}$ is the average difference, $s_{\text{difference}}$ is the standard deviation of the differences, and n is the number of data pairs)

If conditions hold, t follows a t-distribution with degrees of freedom df $= n - 1$ (where n is the number of data pairs)

Formula 9.3: Two-Sample Confidence Interval

$$(\bar{x}_1 - \bar{x}_2) \pm t^* \sqrt{\frac{s_1^2}{n_1} + \frac{s_2^2}{n_2}}$$

If conditions hold, t^* is based on a t-distribution. If no computer is available, the degrees of freedom are conservatively estimated as the smaller of $n_1 - 1$ and $n_2 - 1$.

Formula 9.4: Two-Sample t-Test (Unpooled)

$$t = \frac{\bar{x}_1 - \bar{x}_2}{SE_{\text{est}}}, \text{ where } SE_{\text{est}} = \sqrt{\frac{s_1^2}{n_1} + \frac{s_2^2}{n_2}}$$

If conditions hold, t follows an approximate t-distribution. If no computer is available, the degrees of freedom, df, are conservatively estimated as the smaller of $n_1 - 1$ and $n_2 - 1$ (Do not use the pooled version.)

SOURCES

Axelsson, J., T. Sundelin, M.J. Olsson, K. Sorjonen, C. Axelsson, J. Lasselin, and M. Lekander. 2018. Identification of acutely sick people and facial cues of sickness. *Proc. R. Soc. B* 285: 20172430. http://dx.doi.org/10.1098/rspb.2017.2430

Bureau of Labor Statistics. American Time Use Survey. http://www.bls.gov/tus/ (accessed January 2014).

Dansinger, M., J. Gleason, J. Griffith, H. Selker, and E. Schaefer. 2005. Comparison of the Atkins, Ornish, Weight Watchers, and Zone diets for weight loss and heart disease risk reduction: A randomized trial. *Journal of the American Medical Association* 293(1), 43–53.

Dunn, P. 2012. Assessing claims made by a pizza chain. *Journal of Statistics Education* 20(1). http://www.amstat.org/publications/jse/v20n1/dunn.pdf

Integrated Postsecondary Data System. http://nces.ed.gov/ipeds/ (accessed December 2013).

Lundström, J., J. Boyle, and M. Jones-Gotman. 2006. Sit up and smell the roses better: Olfactory sensitivity to phenyl ethyl alcohol is dependent on body position. *Chemical Senses* 31(3), 249–252. doi:10.1093/chemse/bjj025.

Mangen, A., B. Walgermo, and K. Bronnick. 2013. Reading linear texts on paper versus computer screens: Effects on reading comprehension. *International Journal of Educational Research* 58, 61–68. http://dx.doi.org/10.1016/j.ijer.2012.12.002

Meador, K. J., et al. 2009. Cognitive function at 3 years of age after fetal exposure to antiepileptic drugs. *New England Journal of Medicine* 360(16), 1597–1605.

Mueller, P. and D. Oppenheimer. 2014. The pen is mightier than the keyboard: advantages of longhand over laptop note taking. *Psychological Science* 25(6), 1159–1168. doi:10.1177/0956797614524581

National Health and Nutrition Examination Survey (NHANES). Centers for Disease Control and Prevention (CDC). National Center for Health Statistics (NCHS). National Health and Nutrition Examination Survey Data. Hyattsville, MD: U.S. Department of Health and Human Services, Centers for Disease Control and Prevention, 2003–2004.

SECTION EXERCISES

SECTION 9.1

9.1 Ages A study of all the students at a small college showed a mean age of 20.7 and a standard deviation of 2.5 years.

a. Are these numbers statistics or parameters? Explain.

b. Label both numbers with their appropriate symbol (such as $\bar{x}$, μ, s, or σ).

9.2 Units A survey of 100 random full-time students at a large university showed the mean number of semester units that students were enrolled in was 15.2 with a standard deviation of 1.5 units.

a. Are these numbers statistics or parameters? Explain.

b. Label both numbers with their appropriate symbol (such as $\bar{x}$, μ, s, or σ).

9.3 Exam Scores The distribution of the scores on a certain exam is $N(80, 5)$ which means that the exam scores are Normally distributed with a mean of 80 and a standard deviation of 5.

a. Sketch or use technology to create the curve and label on the x-axis the position of the mean, the mean plus or minus one standard deviation, the mean plus or minus two standard deviations, and the mean plus or minus three standard deviations.

b. Find the probability that a randomly selected score will be greater than 90. Shade the region under the Normal curve whose area corresponds to this probability.

9.4 Exam Scores The distribution of the scores on a certain exam is $N(100, 10)$ which means that the exam scores are Normally distributed with a mean of 100 and a standard deviation of 10.

a. Sketch or use technology to create the curve and label on the x-axis the position of the mean, the mean plus or minus one standard deviation, the mean plus or minus two standard deviations, and the mean plus or minus three standard deviations.

b. Find the probability that a randomly selected score is between 90 and 110. Shade the region under the Normal curve whose area corresponds to this probability.

9.5 Showers According to home-water-works.org, the average shower in the United States lasts 8.2 minutes. Assume this is correct, and assume the standard deviation of 2 minutes.

a. Do you expect the shape of the distribution of shower lengths to be Normal, right-skewed, or left-skewed? Explain.

b. Suppose that we survey a random sample of 100 people to find the length of their last shower. We calculate the mean length from the sample and record the value. We repeat this 500 times. What will be the shape of the distribution of these sample means?

c. Refer to part b. What will be the mean and the standard deviation of the distribution of these sample means?

9.6 Smartphones According to a 2017 report by ComScore .com, the mean time spent on smartphones daily by the American adults is 2.85 hours. Assume this is correct and assume the standard deviation is 1.4 hours.

a. Suppose 150 American adults are randomly surveyed and asked how long they spend on their smartphones daily. The mean of the sample is recorded. Then we repeat this process, taking 1000 surveys of 150 American adults and recording the sample means. What will be the shape of the distribution of these sample means?

b. Refer to part (a). What will be the mean and the standard deviation of the distribution of these sample means?

9.7 Retirement Income Several times during the year, the U.S. Census Bureau takes random samples from the population. One such survey is the American Community Survey. The most recent such survey, based on a large (several thousand) sample of randomly selected households, estimates the mean retirement income in the United States to be $21,201 per year. Suppose we were to make a histogram of all of the retirement incomes from this sample. Would the histogram be a display of the population distribution, the distribution of a sample, or the sampling distribution of means?

9.8 Time Employed A human resources manager for a large company takes a random sample of 50 employees from the company database. She calculates the mean time that they have been employed. She records this value and then repeats the process: She takes another random sample of 50 names and calculates the mean employment time. After she has done this 1000 times, she makes a histogram of the mean employment times. Is this histogram a display of the population distribution, the distribution of a sample, or the sampling distribution of means?

TRY **9.9 Driving (Example 1)** Drivers in Wyoming drive more miles yearly than motorists in any other state. The annual number of miles driven per licensed driver in Wyoming is 22,306 miles. Assume the standard deviation is 5500 miles. A random sample of 200 licensed drivers in Wyoming is selected and the mean number of miles driven yearly for the sample is calculated. (Source: *2017 World Almanac and Book of Facts*)

a. What value would we expect for the sample mean?

b. What is the standard error for the sample mean?

9.10 Driving Drivers in Alaska drive fewer miles yearly than motorists in any other state. The annual number of miles driven per licensed driver in Alaska is 9134 miles. Assume the standard deviation is 3200 miles. A random sample of 100 licensed drivers in Alaska is selected and the mean number of miles driven yearly for the sample is calculated. (Source: *2017 World Almanac and Book of Facts*)

a. What value would we expect for the sample mean?

b. What is the standard error for the sample mean?

SECTION 9.2

TRY **9.11 Babies Weights (Example 2)** Some sources report that the weights of full-term newborn babies have a mean of 7 pounds and a standard deviation of 0.6 pound and are Normally distributed.

a. What is the probability that one newborn baby will have a weight within 0.6 pound of the mean—that is, between 6.4 and 7.6 pounds, or within one standard deviation of the mean?

b. What is the probability the average of four babies' weights will be within 0.6 pound of the mean—that is, between 6.4 and 7.6 pounds?

c. Explain the difference between a and b.

9.12 Babies' Weights, Again Some sources report that the weights of full-term newborn babies have a mean of 7 pounds and a standard deviation of 0.6 pound and are Normally distributed. In the given outputs, the shaded areas (reported as p=) represent the probability that the mean will be larger than 7.6 or smaller than 6.4. One of the outputs uses a sample size of 4, and one uses a sample size of 9.

a. Which is which, and how do you know?

b. These graphs are made so that they spread out to occupy the room on the face of the calculator. If they had the same horizontal axis, one would be taller and narrower than the other. Which one would that be, and why?

(A)
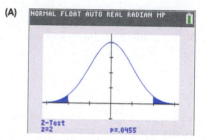

(B)
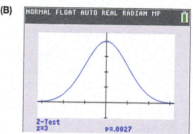

TRY **9.13 (Example 3) Income in Maryland** According to a 2018 *Money* magazine article, Maryland has one of the highest per capita incomes in the United States, with an average income of $75,847. Suppose the standard deviation is $32,000 and the distribution is right-skewed. A random sample of 100 Maryland residents is taken.

a. Is the sample size large enough to use the Central Limit Theorem for means? Explain.

b. What would the mean and standard error for the sampling distribution?

c. What is the probability that the sample mean will be more than $3200 away from the population mean?

9.14 Income in Kansas According to a 2018 *Money* magazine article, the average income in Kansas is $53,906. Suppose the standard deviation is $3000 and the distribution of income is right-skewed. Repeated random samples of 400 Kansas residents are taken, and the sample mean of incomes is calculated for each sample.

a. The population distribution is right-skewed. Will the distribution of sample means be Normal? Why or why not?

b. Find and interpret a *z*-score that corresponds with a sample mean of $53,606.

c. Would it be unusual to find a sample mean of $54,500? Why or why not?

TRY **9.15 CLT Shapes (Example 4)** One of the histograms is a histogram of a sample (from a population with a skewed distribution) one is the distribution of many means of repeated random samples of size 5, and one is the distribution of repeated means of random samples of size 25; all the samples are from the same population. State which is which and how you know.

(A)

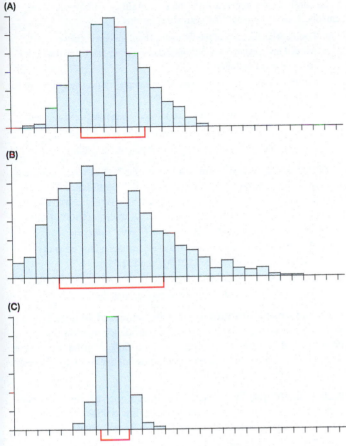

(B)

(C)

(C)

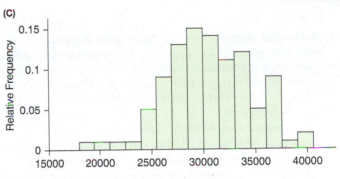

(D)

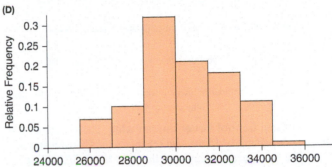

9.16 Used Van Costs One histogram shows the distribution of costs for all used Toyota Sienna vans for sale within a 100 mile radius of San Leandro, CA, for a day in 2018. The other three graphs show distributions of means from random samples taken from this population based on samples of 2 vans, 5 vans, and 15 vans. Each graph based on means was done with many repetitions. Which distribution is which, and why?

(A)

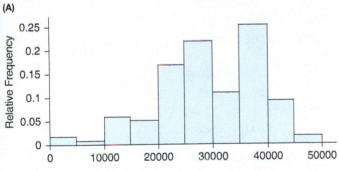

(B)

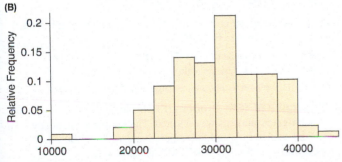

TRY **9.17 (Example 5) Age of Used Vans** The mean age of all 118 used Toyota vans for sale (see exercise 9.16) was 3.1 years with a standard deviation of 2.7 years. The distribution of ages is right-skewed. For a statistics project, a student randomly selects 35 vans from this data set and finds the mean of the sample is 2.7 years with a standard deviation of 2.1 years.

a. Find each of these values: $\mu = ?$ $\sigma = ?$ $\bar{x} = ?$ $s = ?$

b. Which of the values listed in part a are parameters? Which are statistics?

c. Are the conditions for using the CLT fulfilled? What would be the shape of the approximate sampling distribution of a large number of means, each from a sample of 35 vans?

9.18 Student Ages The mean age of all 2550 students at a small college is 22.8 years with a standard deviation is 3.2 years, and the distribution is right-skewed. A random sample of 4 students' ages is obtained, and the mean is 23.2 with a standard deviation of 2.4 years.

a. $\mu = ?$ $\sigma = ?$ $\bar{x} = ?$ $s = ?$

b. Is μ a parameter or a statistic?

c. Are the conditions for using the CLT fulfilled? What would be the shape of the approximate sampling distribution of many means, each from a sample of 4 students? Would the shape be right-skewed, Normal, or left-skewed?

SECTION 9.3

TRY **9.19 Medical School Undergraduate GPAs (Example 6)** The undergraduate grade point average (GPA) for students accepted at a random sample of 10 medical schools in the United States was taken. The mean GPA for these accepted students was 3.75 with a standard error of 0.06. The distribution of undergraduate GPAs is Normal. (Source: Accepted.com)

a. Decide whether each of the following statements is worded correctly for the confidence interval. Fill in the blanks for the correctly worded one(s). Explain the error for the ones that are incorrectly worded.

  i. We are 95% confident that the sample mean is between ___ and ___.

  ii. We are 95% confident that the population mean is between ___ and ___.

  iii. There is a 95% probability that the population mean is between ___ and ___.

b. Based on your confidence interval, would you believe that the population mean GPA is 3.80? Why or why not?

9.20 Medical School Acceptance Rates The acceptance rate for a random sample of 15 medical schools in the United States was taken. The mean acceptance rate for this sample was 5.77 with a standard error of 0.56. Assume the distribution of acceptance rates is Normal. (Source: Accepted.com)

a. Decide whether each of the following statements is worded correctly for the confidence interval. Fill in the blanks for the correctly worded one(s). Explain the error for the ones that are incorrectly worded.

  i. We are 95% confident that the sample mean is between ___ and ___.

  ii. We are 95% confident that the population mean is between ___ and ___.

  iii. There is a 95% probability that the population mean is between ___ and ___.

b. Based on your confidence interval, would you believe the average acceptance rate for medical schools is 6.5? Explain.

9.21 Oranges A statistics instructor randomly selected four bags of oranges, each bag labeled 10 pounds, and weighed the bags. They weighed 10.2, 10.5, 10.3, and 10.3 pounds. Assume that the distribution of weights is Normal. Find a 95% confidence interval for the mean weight of all bags of oranges. Use technology for your calculations.

a. Decide whether each of the following three statements is a correctly worded interpretation of the confidence interval, and fill in the blanks for the correct option(s).

  i. I am 95% confident that the population mean is between ___ and ___.

  ii. There is a 95% chance that all intervals will be between ___ and ___.

  iii. I am 95% confident that the sample mean is between ___ and ___.

b. Does the interval capture 10 pounds? Is there enough evidence to reject the null hypothesis that the population mean weight is 10 pounds? Explain your answer.

9.22 Carrots The weights of four randomly chosen bags of horse carrots, each bag labeled 20 pounds, were 20.5, 19.8, 20.8, and 20.0 pounds. Assume that the distribution of weights is Normal. Find a

95% confidence interval for the mean weight of all bags of horse carrots. Use technology for your calculations.

a. Decide whether each of the following three statements is a correctly worded interpretation of the confidence interval, and fill in the blanks for the correct option(s).

  i. 95% of all sample means based on samples of the same size will be between ___ and ___.

  ii. I am 95% confident that the population mean is between ___ and ___.

  iii. We are 95% confident that the boundaries are ___ and ___.

b. Can you reject a population mean of 20 pounds? Explain.

TRY **9.23 Private University Tuition (Example 7)** A random sample of 25 private universities was selected. A 95% confidence interval for the mean in-state tuition costs at private universities was (22,501, 32,664). Which of the following is a correct interpretation of the confidence level? (Source: *Chronicle of Higher Education*)

a. There is a 95% probability that the mean in-state tuition costs at a private university is between $22,501 and $32,664.

b. In about 95% of the samples of 25 private universities, the confidence interval will contain the population mean in-state tuition.

9.24 Random Numbers If you take samples of 40 lines from a random number table and find that the confidence interval for the proportion of odd-numbered digits captures 50% 37 times out of the 40 lines, is it the confidence *interval* or confidence *level* you are estimating with the 37 out of 40?

9.25 t* (Example 8) A researcher collects one sample of 27 measurements from a population and wants to find a 95% confidence interval. What value should he use for t^*? (Recall that df $= n - 1$ for a one-sample t-interval.)

	C-level		
df	**90%**	**95%**	**99%**
24	1.711	2.064	2.797
25	1.708	2.060	2.787
26	1.706	2.056	2.779
27	1.703	2.052	2.771
28	1.701	2.048	2.763

	C-level		
df	**90%**	**95%**	**99%**
28	1.701	2.048	2.763
29	1.699	2.045	2.756
30	1.697	2.042	2.750
34	1.691	2.032	2.728

9.26 t* A researcher collects a sample of 25 measurements from a population and wants to find a 99% confidence interval.

a. What value should he use for t^*? (Recall that df $= n - 1$ for a one-sample t-interval.) Use the table given for Exercise 9.25.

b. Why is the answer to this question larger than the answer to Exercise 9.25?

TRY **9.27 Heights of 12th Graders (Example 9)** A random sample of 30 12th-grade students was selected. The sample mean height was 170.7 centimeters, and the sample standard deviation was 11.5 centimeters. (Source: AMSTAT Census at School)

a. State how you would fill in the numbers below to do the calculation with Minitab Express.

b. Report the confidence interval in a carefully worded sentence, using the Minitab Express output provided.

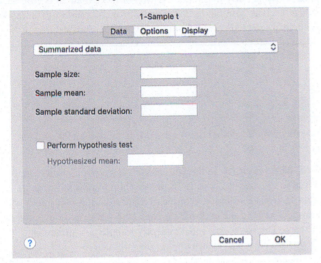

1-Sample t

Descriptive Statistics

N	Mean	StDev	SE Mean	95% CI for μ
30	170.700	11.500	2.100	(166.406, 174.994)

μ:mean of Sample

9.28 Drinks A fast-food chain sells drinks that it calls HUGE. When we take a sample of 25 drinks and weigh them, we find that the mean is 36.3 ounces with a standard deviation of 1.5 ounces.

a. State how you would fill in the numbers below to do the calculation with a TI-84.

b. Report the confidence interval in a carefully worded sentence.

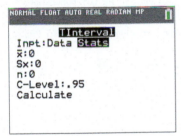

TI-84 Input

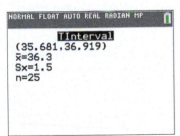

TI-84 Output

9.29 Men's Pulse Rates (Example 10) A random sample of 25 men's resting pulse rates shows a mean of 72 beats per minute and a standard deviation of 13.

a. Find a 95% confidence interval for the population mean pulse rate for men, and report it in a sentence. You may use the table given for Exercise 9.25.

b. Find a 99% confidence interval.

c. Which interval is wider, and why?

9.30 Travel Time to School A random sample of 50 12th-grade students was asked how long it took to get to school. The sample mean was 16.2 minutes, and the sample standard deviation was 12.3 minutes. (Source: AMSTAT Census at School)

a. Find a 95% confidence interval for the population mean time it takes 12th-grade students to get to school.

b. Would a 90% confidence interval based on this sample data be wider or narrower than the 95% confidence interval? Explain. Check your answer by constructing a 90% confidence interval and comparing this width of the interval with the width of the 95% confidence interval you found in part a.

TRY **9.31 RBIs (Example 11)** A random sample of 25 baseball players from the 2017 Major League Baseball season was taken and the sample data was used to construct two confidence intervals for the population mean. One interval was (22.0, 42.8). The other interval was (19.9, 44.0). (Source: mlb.com)

a. One interval is a 95% interval, and one is a 90% interval. Which is which, and how do you know?

b. If a larger sample size was used, for example, 40 instead of 25, how would this affect the width of the intervals? Explain.

9.32 GPAs, Again In exercise 9.31, two intervals were given for the same data, one for 95% confidence and one for 90% confidence.

a. How would a 99% interval compare? Would it be narrower than both, wider than both, or between the two in width. Explain.

b. If we wanted to use a 99% confidence level and get a narrower width, how could we change our data collection?

9.33 Confidence Interval Changes State whether each of the following changes would make a confidence interval wider or narrower. (Assume that nothing else changes.)

a. Changing from a 90% confidence level to a 99% confidence level

b. Changing from a sample size of 30 to a sample size of 200

c. Changing from a standard deviation of 20 pounds to a standard deviation of 25 pounds

9.34 Confidence Interval Changes State whether each of the following changes would make a confidence interval wider or narrower. (Assume that nothing else changes.)

a. Changing from a 95% level of confidence to a 90% level of confidence

b. Changing from a sample size of 30 to a sample size of 20

c. Changing from a standard deviation of 3 inches to a standard deviation of 2.5 inches

9.35 Potatoes The weights of four randomly and independently selected bags of potatoes labeled 20 pounds were found to be 21.0, 21.5, 20.5, and 21.2 pounds. Assume Normality.

a. Find a 95% confidence interval for the mean weight of all bags of potatoes.

b. Does the interval capture 20.0 pounds? Is there enough evidence to reject a mean weight of 20 pounds?

9.36 Tomatoes The weights of four randomly and independently selected bags of tomatoes labeled 5 pounds were found to be 5.1, 5.0, 5.3, and 5.1 pounds. Assume Normality.

a. Find a 95% confidence interval for the mean weight of all bags of tomatoes.

b. Does the interval capture 5.0 pounds? Is there enough evidence to reject a mean weight of 5.0 pounds?

SECTION 9.4

TRY 9.37 Human Body Temperatures (Example 12) A random sample of 10 independent healthy people showed the following body temperatures (in degrees Fahrenheit):

98.5, 98.2 99.0, 96.3, 98.3, 98.7, 97.2, 99.1, 98.7, 97.2

Test the hypothesis that the population mean is not 98.6 °F, using a significance level of 0.05. *See page 500 for guidance.*

9.38 Reaction Distance Data on the disk and website show reaction distances in centimeters for the dominant hand for a random sample of 40 independently chosen college students. Smaller distances indicate quicker reactions.

a. Make a graph of the distribution of the sample, and describe its shape.

b. Find, report, and interpret a 95% confidence interval for the population mean.

c. Suppose a professor said that the population mean should be 10 centimeters. Test the hypothesis that the population mean is not 10 centimeters, using the four-step procedure, with a significance level of 0.05.

9.39 Potatoes Use the data from exercise 9.35.

a. If you use the four-step procedure with a two-sided alternative hypothesis, should you be able to reject the hypothesis that the population mean is 20 pounds using a significance level of 0.05? Why or why not? The confidence interval is reported here: I am 95% confident that the population mean is between 20.4 and 21.7 pounds.

b. Now test the hypothesis that the population mean is not 20 pounds using the four-step procedure. Use a significance level of 0.05.

c. Choose one of the following conclusions:

  i. We cannot reject a population mean of 20 pounds.

  ii. We can reject a population mean of 20 pounds.

  iii. The population mean is 21.05 pounds.

9.40 Tomatoes Use the data from exercise 9.36.

a. Using the four-step procedure with a two-sided alternative hypothesis, should you be able to reject the hypothesis that the population mean is 5 pounds using a significance level of 0.05? Why or why not? The confidence interval is reported here: I am 95% confident the population mean is between 4.9 and 5.3 pounds.

b. Now test the hypothesis that the population mean is not 5 pounds using the four-step procedure. Use a significance level of 0.05 and number your steps.

9.41 Cholesterol In the U.S. Department of Health has suggested that a healthy total cholesterol measurement should be 200 mg/dL or less. Records from 50 randomly and independently selected people from the NHANES study showed the results in the Minitab output given:

One-Sample T

Test of μ = 200 vs > 200

N	Mean	StDev	SE Mean	T	P
50	208.26	40.67	5.75	1.44	0.079

Test the hypothesis that the mean cholesterol level is more than 200 using a significance level of 0.05. Assume that conditions are met.

9.42 BMI A body mass index (BMI) of more than 25 is considered unhealthy. The Minitab output given is from 50 randomly and independently selected people from the NHANES study.

One-Sample T

Test of μ = 25 vs > 25

N	Mean	StDev	SE Mean	T	P
50	27.874	6.783	0.959	3.00	0.002

Test the hypothesis that the mean BMI is more than 25 using a significance level of 0.05. Assume that conditions are met.

9.43 Male Height In the United States, the population mean height for 3-year-old boys is 38 inches (http://www.kidsgrowth.com). Suppose a random sample of 15 non-U.S. 3-year-old boys showed a sample mean of 37.2 inches with a standard deviation of 3 inches. The boys were independently sampled. Assume that heights are Normally distributed in the population.

a. Determine whether the population mean for non-U.S. boys is significantly different from the U.S. population mean. Use a significance level of 0.05.

b. Now suppose the sample consists of 30 boys instead of 15, and repeat the test.

c. Explain why the *t*-values and p-values for parts a and b are different.

9.44 Female Height In the United States, the population mean height for 10-year-old girls is 54.5 inches. Suppose a random sample of 15 10-year-old girls from Brazil is taken and that these girls had a sample mean height of 53.2 inches with a standard deviation of 2.5 inches. Assume that heights are Normally distributed. (Source: cdc.gov)

a. Determine whether the population mean for height for 10-year-old girls from Brazil is significantly different from the U.S. population mean. Use a significance level of 0.05.

b. Now suppose the sample consists of 40 girls instead of 15. Repeat the test.

c. Explain why the *t*-values and p-value for parts a and b are different.

9.45 Deflated Footballs? Patriots In the 2015 AFC Championship game, there was a charge the New England Patriots deflated their footballs for an advantage. The balls should be inflated to between 12.5 and 13.5 pounds per square inch. The measurements were 11.50, 10.85, 11.15, 10.70, 11.10, 11.60, 11.85, 11.10, 10.95, 10.50, and 10.90 psi (pounds per square inch). (Source: http://online.wsj.com/public/resources/documents/Deflategate.pdf)

a. Test the hypothesis that the population mean is less than 12.5 psi using a significance level of 0.05. State clearly whether the Patriots' balls are deflated or not. Assume the conditions for a hypothesis test are satisfied.

b. Use the data to construct a 95% confidence interval for the mean psi for the Patriots' footballs. How does this confidence interval support your conclusion in part a?

9.46 Deflated Footballs? Colts In the 2015 AFC Championship game, there was a charge the New England Patriots deflated their footballs for an advantage. The Patriots' opponents during the championship game were the Indianapolis Colts. Measurements of the

Colts footballs were taken. The balls should be inflated to between 12.5 and 13.5 pounds per square inch (psi). The measurements were 12.70, 12.75, 12.50, 12.55, 12.35, 12.30, 12.95, and 12.15 psi. (Source: http://online.wsj.com/public/resources/documents/Deflategate.pdf)

a. Test the hypothesis that the population mean is less than 12.5 psi using a significance level of 0.05. State clearly whether the Colts' balls are deflated or not. Assume the conditions for a hypothesis test are satisfied.

b. Use the data to construct a 95% confidence interval for the mean psi for the Colts' footballs. How does this confidence interval support your conclusion in part a?

9.47 Movie Ticket Prices According to Deadline.com, the average price for a movie ticket in 2018 was $8.97. A random sample of movie prices in the San Francisco Bay Area 25 movie ticket prices had a sample mean of $12.27 with a standard deviation of $3.36.

a. Do we have evidence that the price of a movie ticket in the San Francisco Bay Area is different from the national average? Use a significance level of 0.05.

b. Construct a 95% confidence interval for the price of a movie ticket in the San Francisco Bay Area. How does your confidence interval support your conclusion in part a?

9.48 Broadway Ticket Prices According to Statista.com, the average price of a ticket to a Broadway show in 2017 was $109.21. A random sample of 25 Broadway ticket prices in 2018 had a sample mean of $114.7 with a standard deviation of $43.3.

a. Do we have evidence that Broadway ticket prices changed from 2017 prices? Use a significance level of 0.05.

b. Construct a 95% confidence interval for the price of a Broadway ticket. How does your confidence interval support your conclusion in part a?

9.49 Atkins Diet Difference Ten people went on an Atkins diet for a month. The weight losses experienced (in pounds) were

3, 8, 10, 0, 4, 6, 6, 4, 2, and −2

The negative weight loss is a weight gain. Test the hypothesis that the mean weight loss was more than 0, using a significance level of 0.05. Assume the population distribution is Normal.

9.50 Pulse Difference The following numbers are the differences in pulse rate (beats per minute) before and after running for 12 randomly selected people.

24, 12, 14, 12, 16, 10, 0, 4, 13, 42, 4, and 16

Positive numbers mean the pulse rate went up. Test the hypothesis that the mean difference in pulse rate was more than 0, using a significance level of 0.05. Assume the population distribution is Normal.

9.51 Student Ages Suppose that 200 statistics students each took a random sample (with replacement) of 50 students at their college and recorded the ages of the students in their sample. Then each student used his or her data to calculate a 95% confidence interval for the mean age of all students at the college. How many of the 200 intervals would you expect to capture the true population mean age, and how many would you expect not to capture the true population mean? Explain by showing your calculation.

9.52 Presidents' Ages at Inauguration A 95% confidence interval for the ages of the first six presidents at their inaugurations

is (56.2, 59.5). Either interpret the interval or explain why it should not be interpreted.

SECTION 9.5

TRY **9.53 Independent or Paired (Example 13)** State whether each situation has independent or paired (dependent) samples.

a. A researcher wants to know whether pulse rates of people go down after brief meditation. She collects the pulse rates of a random sample of people before meditation and then collects their pulse rates after meditation.

b. A researcher wants to know whether professors with tenure have fewer posted office hours than professors without tenure do. She observes the number of office hours posted on the doors of tenured and untenured professors.

9.54 Independent or Paired State whether each situation has independent or paired (dependent) samples.

a. A researcher wants to compare food prices at two grocery stores. She purchases 20 items at Store A and finds the mean and the standard deviation for the cost of the items. She then purchases 20 items at Store B and again finds the mean and the standard deviation for the cost of the items.

b. A student wants to compare textbook prices at two bookstores. She has a list of textbooks and finds the price of the text at each of the two bookstores.

TRY **9.55 Televisions: CI (Example 14)** Minitab output is shown for a two-sample t-interval for the number of televisions owned in households of random samples of students at two different community colleges. Each individual was randomly chosen independently of the others. One of the schools is in a wealthy community (MC), and the other (OC) is in a less wealthy community.

Two-Sample T: CI

Sample	N	Mean	StDev	SE Mean
OCTVs	30	3.70	1.49	0.27
MCTVs	30	3.33	1.49	0.27

Difference = $\mu(1) - \mu(2)$
Estimate for difference: 0.370
95% CI for difference: (−0.400, 1.140)

a. Are the conditions for using a confidence interval for the difference between two means met?

b. State the interval in a clear and correct sentence.

c. Does the interval capture 0? Explain what that shows.

9.56 Pulse and Gender: CI Using data from NHANES, we looked at the pulse rate for nearly 800 people to see whether it is plausible that men and women have the same population mean. NHANES data are random and independent. Minitab output follows.

Two-Sample T: CI

Sample	N	Mean	StDev	SE Mean
Women	384	76.3	12.8	0.65
Men	372	72.1	13.0	0.67

Difference = $\mu(1) - \mu(2)$
Estimate for difference: 4.200
95% CI for difference: (2.357, 6.043)

a. Are the conditions for using a confidence interval for the difference between two means met?

b. State the interval in a clear and correct sentence.

c. Does the interval capture 0? Explain what that shows.

TRY 9.57 Televisions (Example 15) The table shows the Minitab output for a two-sample t-test for the number of televisions owned in households of random samples of students at two different community colleges. Each individual was randomly chosen independently of the others; the students were not chosen as pairs or in groups. One of the schools is in a wealthy community (MC), and the other (OC) is in a less wealthy community. Test the hypothesis that the population means are not the same, using a significance level of 0.05. *See page 501 for guidance.*

Two-Sample T-Test and CI: OCTV, MCTV

Two-sample T for OCTV cs MCTV

	N	Mean	StDev	SE Mean
OCTV	30	3.70	1.49	0.27
MCTV	30	3.33	1.49	0.27

Difference = μ(OCTV) $-$ μ(MCTV)
Estimate for difference: 0.367
95% CI for difference: (-0.404, 1.138)
T-Test of difference = 0 (vs $\neq$):   T-Value = 0.95
                                        P-Value = 0.345

9.58 Pulse Rates Using data from NHANES, we looked at the pulse rates of nearly 800 people to see whether men or women tended to have higher pulse rates. Refer to the Minitab output provided.

a. Report the sample means, and state which group had the higher sample mean pulse rate.

b. Use the Minitab output to test the hypothesis that pulse rates for men and women are not equal, using a significance level of 0.05. The samples are large enough so that Normality is not an issue.

Two-Sample T-Test and CI: Pulse, Sex

Two-sample T for Pulse

Sex	N	Mean	StDev	SE Mean
Female	384	76.3	12.8	0.65
Male	372	72.1	13.0	0.67

Difference = μ(Female) $-$ μ(Male)
Estimate for difference: 4.248
95% CI for difference: (2.406, 6.090)
T-Test of difference = 0 (vs $\neq$):   T-Value = 4.53
                                        P-Value = 0.000
                                        DF = 752

9.59 Triglycerides Triglycerides are a form of fat found in the body. Using data from NHANES, we looked at whether men have higher triglyceride levels than women.

a. Report the sample means, and state which group had the higher sample mean triglyceride level. Refer to the Minitab output in figure (A).

b. Carry out a hypothesis test to determine whether men have a higher mean triglyceride level than women. Refer to the Minitab output provided in figure (A). Output for three different alternative hypotheses is provided—see figures (B), (C), and (D)—and you must choose and state the most appropriate output. Use a significance level of 0.05.

(A)

Two-Sample T-Test and CI: Triglycerides, Gender

Two-sample T for Triglycerides

Gender	N	Mean	StDev	SE Mean
Female	44	84.4	40.2	6.1
Male	48	139.5	85.3	12

Difference = μ(Female) $-$ μ(Male)
Estimate for difference: -55.1
95% CI for difference: (-82.5, -27.7)

(B)

B: T-Test of difference = 0 (vs <):
                           T-Value = -4.02
                           P-Value = 0.000

(C)

C: T-Test of difference = 0 (vs >):
                           T-Value = -4.02
                           P-Value = 1.000

(D)

D: T-Test of difference = 0 (vs $\neq$):
                           T-Value = -4.02
                           P-Value = 0.000

9.60 Systolic Blood Pressures When you have your blood pressure taken, the larger number is the systolic blood pressure. Using data from NHANES, we looked at whether men and women have different systolic blood pressure levels.

a. Report the two sample means, and state which group had the higher sample mean systolic blood pressure. Refer to the Minitab output in figure (A).

b. Refer to the Minitab output given in figure (A) to test the hypothesis that the mean systolic blood pressures for men and women are not equal, using a significance level of 0.05. Although the distribution blood pressures in the population are right-skewed, the sample size is large enough for us to use t-tests. Choose from figures (B), (C), and (D) for your t- and p-values, and explain.

(A)

Two-sample T for BPSys

Gender	N	Mean	StDev	SE Mean
Female	404	116.8	22.7	1.1
Male	410	118.7	18.0	0.89

Difference = μ(Female) $-$ μ(Male)
Estimate for difference: -1.93
95% CI for difference: (-4.75, 0.89)

(B)

```
B: T-Test of difference =   0 (vs ≠):
                          T-Value = -1.34
                          P-Value = 0.180
```

(C)

```
C: T-Test of difference = 0 (vs >): T-Value = -1.34  P-Value = 0.910
```

(D)

```
D: T-Test of difference = 0 (vs <): T-Value = -1.34  P-Value = 0.090
```

9.61 Triglycerides, Again Report and interpret the 95% confidence interval for the difference in mean triglyceride levels for men and women (refer to the Minitab output in exercise 9.59). Does this support the hypothesis that men and women differ in mean triglyceride level? Explain.

9.62 Blood Pressures, Again Report and interpret the 95% confidence interval for the difference in mean systolic blood pressure for men and women (refer to the Minitab output in exercise 9.60). Does this support the hypothesis that men and women differ in mean systolic blood pressure? Explain.

9.63 Baseball Salaries A random sample of 40 professional baseball salaries from 1985 through 2015 was selected. The league of the player (American or National) was also recorded. Salary (in hundred thousand dollars) and league are shown in the table. Test the hypothesis that there is a difference in the mean salary of players in each league. Assume the distributions are Normal enough to use the *t*-test. Use a significance level of 0.05. (Source: sampled from All MLB Salaries, accessed via StatCrunch. Owner: StatCrunch featured.)

Salary	League	Salary	League
275	American	1600	National
3400	National	5000	American
5725	National	385	National
1500	American	400	American
558.3	National	380	National
320	American	175	American
1791.7	National	400	American
1200	American	900	American
1500	National	62.5	American
3400	National	3200	American
240	National	2500	National
481	National	316.5	American
5050	American	4175	American
491.7	American	500	National
1750	American	1500	American

9.64 College Athletes' Weights A random sample of male college baseball players and a random sample of male college soccer players were obtained independently and weighed. The following table shows the unstacked weights (in pounds). The distributions of both data sets suggest that the population distributions of both data sets suggest that the population distributions of both data sets suggest that the population distributions of both data sets suggest that the population distributions of

Normal. Determine whether the difference in means is significant, using a significance level of 0.05.

Baseball	Soccer	Baseball	Soccer
190	165	186	156
200	190	210	168
187	185	198	173
182	187	180	158
192	183	182	150
205	189	193	172
185	170	200	180
177	182	195	184
207	172	182	174
225	180	193	190
230	167	190	156
195	190	186	163
169	185		

9.65 Baseball Salaries In exercise 9.63 you could not reject the null hypothesis that the mean salary for the two leagues are the same, using a two-tailed test with a significance level of 0.05.

a. If you found a 95% confidence interval for the difference between means, would it capture 0? Explain.

b. Using the data in exercise 9.63, find a 95% confidence interval for the difference in the population means and explain what it shows.

*9.66 College Athletes' Weights In exercise 9.64, you could reject the null hypothesis that the mean weights of soccer and baseball players were equal using a two-tailed test with a significance level of 0.05.

a. If you found a 95% confidence interval for the difference between means, would it capture 0? Explain.

b. If you found a 90% interval, would it capture 0? Explain.

c. Now go back to exercise 9.64. Find a 95% confidence interval for the difference between means, and explain what it shows.

TRY **9.67 Textbook Prices, UCSB vs. CSUN (Example 16)** The prices of a sample of books at University of California at Santa Barbara (UCSB) were obtained by statistics students Ricky Hernandez and Elizabeth Alamillo. Then the cost of books for the same subjects (at the same level) were obtained for California State University at Northridge (CSUN). Assume that the distribution of differences is Normal enough to proceed, and assume that the sampling was random. The data are at this text's website.

a. First find both sample means and compare them.

b. Test the hypothesis that the population means are different, using a significance level of 0.05.

9.68 Textbook Prices. OC vs. CSUN The prices of a random sample of comparable (matched) textbooks from two schools were recorded. We are comparing the prices at OC (Oxnard Community College) and CSUN (California State University at Northridge). Assume that the population distribution of differences is approximately Normal. Each book was priced separately; there were no books "bundled" together.

a. Compare the sample means.

b. Determine whether the mean prices of all books are significantly different. Use a significance level of 0.05.

g 9.69 Females' Pulse Rates before and after a Fright (Example 17) In a statistics class taught by one of the authors, students took their pulses before and after being frightened. The frightening event was having the teacher scream and run from one side of the room to the other. The pulse rates (beats per minute) of the women before and after the scream were obtained separately and are shown in the table. Treat this as though it were a random sample of female community college students. Test the hypothesis that the mean of college women's pulse rates is higher after a fright, using a significance level of 0.05. *See page 501 for guidance.*

Women		Women	
Pulse Before	**Pulse After**	**Pulse Before**	**Pulse After**
64	68	84	88
100	112	80	80
80	84	68	92
60	68	60	76
92	104	68	72
80	92	68	80
68	72		

9.70 Males' Pulse Rates before and after a Fright Follow the instructions for exercise 9.69, but use the data for the men in the class. Test the hypothesis that the mean of college men's pulse rates is higher after a fright, using a significance level of 0.05.

Men		Men	
Pulse Before	**Pulse After**	**Pulse Before**	**Pulse After**
50	64	64	68
84	72	88	100
96	88	84	80
80	72	76	80
80	88		

9.71 Organic Food A student compared organic food prices at Target and Whole Foods. The same items were priced at each store. The first three items are shown in Figure A. (Source: StatCrunch Organic food price comparison fall 2011. Owner: kerrypaulson) Choose the correct output (B or C) for the appropriate test, explaining why you chose that output. Then test the hypothesis that the population means are not equal using a significance level of 0.05.

Food	Target	Whole
Bananas/1 lb	0.79	0.99
Grape tomatoes/1 lb	4.49	3.99
Russet potato/5 lb	4.49	4.99

Figure A

Paired T-Test and CI: Target, Whole
```
Paired T for Target - Whole

                N      Mean     StDev    SE Mean
Target         30     2.879     1.197      0.219
Whole          30     3.144     1.367      0.250
Difference     30    -0.265     1.152      0.210

95% CI for mean difference: (-0.695, 0.165)
T-Test of mean difference = 0 (vs ≠ 0):  T-Value = -1.26
                                          P-Value = 0.217
```

Figure B

Two-Sample T-Test and CI: Target, Whole
```
Two-sample T for Target vs Whole

                N      Mean     StDev    SE Mean
Target         30     2.88      1.20       0.22
Whole          30     3.14      1.37       0.25

Difference = (Target) - μ (Whole)
Estimate for difference: -0.265
95% CI for difference: (-0.930, 0.399)
T-Test of difference = 0 (vs ≠ 0):  T-Value = -0.80
                                     P-Value = 0.427
```

Figure C

9.72 Body Temperature The body temperatures of 65 men and 65 women were compared. The results of a hypothesis test are shown. Assume the conditions for using a *t*-test are satisfied. (Source: Body Temperature, accessed via StatCrunch. Owner: StatCrunch featured.)

a. Why was a two-sample *t*-test used instead of a paired *t*-test?

b. Which sample mean is larger? How do you know?

c. Write the conclusion for the hypothesis test.

Two sample T hypothesis test:
μ_1 : Mean of Body Temp where Gender="Female"
μ_2 : Mean of Body Temp where Gender="Male"
$\mu_1 - \mu_2$: Difference between two means
$H_0 : \mu_1 - \mu_2 = 0$
$H_A : \mu_1 - \mu_2 \neq 0$
(without pooled variances)

Hypothesis test results:

Difference	Sample Diff.	Std. Err.	DF	T-Stat	P-value
$\mu_1 - \mu_2$	0.28923077	0.12655395	127.5103	2.2854345	0.0239

9.73 Ales vs. IPAs Data were collected on calorie content in ales and IPAs and is summarized in the following table. (Source: efficientdrinker.com)

a. Does the sample data provide evidence that there is a difference in calorie content for ales and IPAs? Assume the conditions for using a *t*-test are satisfied. Use a significance level of 0.05 for the hypothesis test.

b. If the test had been to determine whether the mean for IPAs is significantly larger than the mean for Ales, how would this change the alternative hypothesis and p-value? Test this hypothesis write the conclusion for the test.

Type	Ale	IPA
N	43	12
Sample mean	173.7	230.8
Sample standard deviation	24.8	75.9

For questions 9.77 and 9.78 the data set is given at this text's website. Assume that the data sets are from random samples and the distributions are Normal.

9.74 Surfers Surfers and statistics students Rex Robinson and Sandy Hudson collected data on the number of days on which surfers surfed in the last month for 30 longboard (L) users and 30 shortboard (S) users. Treat these data as though they were from two independent random samples. Test the hypothesis that the mean days surfed for all longboarders is larger than the mean days surfed for all shortboarders (because longboards can go out in many different surfing conditions). Use a level of significance of 0.05.

Longboard: 4, 9, 8, 4, 8, 8, 7, 9, 6, 7, 10, 12, 12, 10, 14, 12, 15, 13, 10, 11, 19, 19, 14, 11, 16, 19, 20, 22, 20, 22

Shortboard: 6, 4, 4, 6, 8, 8, 7, 9, 4, 7, 8, 5, 9, 8, 4, 15, 12, 10, 11, 12, 12, 11, 14, 10, 11, 13, 15, 10, 20, 20

9.75 Self-Reported Heights of Men (Example 18) A random sample of students at Oxnard College reported what they believed to be their heights in inches. Then the students measured each others' heights in centimeters, without shoes. The data shown are for the men. Assume that the conditions for t-tests hold.

a. Convert heights in inches to centimeters by multiplying inches by 2.54. Find a 95% confidence interval for the mean difference as measured in centimeters. Does it capture 0? What does that show?

b. Perform a t-test to test the hypothesis that the means are not the same. Use a significance level of 0.05, and show all four steps.

Men		Men	
Centimeters	Inches	Centimeters	Inches
166	66	178	70
172	68	177	69
184	73	181	71
166	67	175	69
191	76	171	67
173	68	170	67
174	69	184	72
191	76		

9.76 Eating Out Jacqueline Loya, a statistics student, asked students with jobs how many times they went out to eat in the last week. There were 25 students who had part-time jobs, and 25 students who had full-time jobs. Carry out a hypothesis test to determine whether the mean number of meals out per week for students with full-time jobs is greater than that for those with part-time jobs. Use a significance level of 0.05. Assume that the conditions for a two-sample t-test hold.

Full-time jobs: 5, 3, 4, 4, 4, 2, 1, 5, 6, 5, 6, 3, 3, 2, 4, 5, 2, 3, 7, 5, 5, 1, 4, 6, 7

Part-time jobs: 1, 1, 5, 1, 4, 2, 2, 3, 3, 2, 3, 2, 4, 2, 1, 2, 3, 2, 1, 3, 3, 2, 4, 2, 1

9.77 Heart Rate Heart rate data for a random sample of males and females was obtained. The data are available at the text's website. Use the data to answer these questions. Data were originally published in the *Journal of Statistics Education* online data archive and accessed through StatCrunch.

a. Find a 95% confidence interval for the difference between the mean heart rates of males and females. Based on your confidence interval, is there a significant difference in the mean heart rates of females and males? Explain.

b. Use a significance level of 0.05, test the hypothesis that the mean heart rates of males and females are significantly different.

9.78 Self-Driving Cars A survey of asked respondents how long (in years) they thought it would be before self-driving cars were the majority of vehicles on the road, and a random sample from this survey was selected. The data are available at the text's website. (Source: StatCrunch Survey: Responses to self-driving cars. Owner: scsurvey)

a. Find a 95% confidence interval for the difference in mean lengths of time before self-driving cars are the majority of vehicles on the road between males and females.

b. Use a significance level of 0.05, test the hypothesis that the mean lengths of time for males and females are significantly different.

CHAPTER REVIEW EXERCISES

9.79 Women's Heights Assume women's heights are approximately Normally distributed with a mean of 65 inches and a standard deviation of 2.5 inches. Which of the following questions can be answered using the Central Limit Theorem for sample means as needed? If the question can be answered, do so. If the question cannot be answered, explain why the Central Limit Theorem cannot be applied.

a. Find the probability that a randomly selected woman is less than 63 inches tall.

b. If five women are randomly selected, find the probability that the mean height of the sample is less than 63 inches.

c. If 30 women are randomly selected, find the probability that the mean height of the sample is less than 63 inches.

9.80 Showers According to home-water-works.org, the average shower in the United States lasts 8.2 minutes. Assume the standard deviation of shower times is 2 minutes and the distribution of shower times is right-skewed.

Which of the following questions can be answered using the Central Limit Theorem for sample means as needed? If the question can be answered, do so. If the question cannot be answered, explain why the Central Limit Theorem cannot be applied.

a. Find the probability that a randomly selected shower lasts more than 9 minutes.

b. If five showers are randomly selected, find the probability that the mean length of the sample is more than 9 minutes.

c. If 50 showers are randomly selected, find the probability that the mean length of the sample is more than 9 minutes.

9.81 Choose a test for each situation: one-sample t-test, two-sample t-test, paired t-test, and no t-test.

a. A random sample of students who transfered to a 4-year university from community colleges are asked their GPAs. Our goal is to determine whether the mean GPA for transfer students is significantly different from the population mean GPA for all students at the university.

b. Students observe the number of office hours posted for a random sample of tenured and a random sample of untenured professors.

c. A researcher goes to the parking lot at a large grocery chain and observes whether each person is male or female and whether they return the cart to the correct spot before leaving (yes or no).

9.82 Choose a t-test for each situation: one-sample t-test, two-sample t-test, paired t-test, and no t-test.

a. A random sample of car dealerships is obtained. Then a student walks onto each dealer's lot wearing old clothes and finds out how long it takes (in seconds) for a salesperson to approach the student. Later the student goes onto the same lot dressed very nicely and finds out how long it takes for a salesperson to approach.

b. A researcher at a preschool selects a random sample of 4-year-olds, determines whether they know the alphabet (yes or no), and records gender.

c. A researcher calls the office phone for a random sample of faculty at a college late at night, measures the length of the outgoing message, and records gender.

9.83 Cones: 3 Tests A McDonald's fact sheet says its cones should weigh 3.18 ounces (converted from grams). Suppose you take a random sample of four cones, and the weights are 4.2, 3.4, 3.9, and 4.4 ounces. Assume that the population distribution is Normal, and, for all three parts, report the alternative hypothesis, the t-value, the p-value, and your conclusion. The null hypothesis in all three cases is that the population mean is 3.18 ounces.

a. Test the hypothesis that the cones do not have a population mean of 3.18 ounces.

b. Test the hypothesis that the cones have a population mean less than 3.18 ounces.

c. Test the hypothesis that the cones have a population mean greater than 3.18 ounces.

9.84 French Fries A fast-food chain advertises that the medium-size serving of French fries weighs 135 grams. A reporter took a random sample of 10 medium orders of French fries and weighed each order. The weights (in grams) were 111, 124, 125, 156, 127, 134,

135, 136, 139, 141. Assume the population distribution is Normal. (Source: soranews24.com)

a. Test the hypothesis that the medium servings have a population mean different from 135 grams. Use a significance level of 0.05.

b. Construct a 95% confidence interval for the population mean. How does your confidence interval support your conclusion in part a? Do you think the consumers are being misled about the serving size? Explain.

9.85 Brain Size Brain size for 20 random women and 20 random men was obtained and is reported in the table (in hundreds of thousands of pixels shown in an MRI). Test the hypothesis that men tend to have larger brains than women at the 0.05 level. (Source: Willerman, L., Schultz, R., Rutledge, J. N., and Bigler, E., "In Vivo Brain Size and Intelligence," *Intelligence*, 15(1991):223–228.)

Brain Size (100,000s of pixels)		Brain Size (100,000s of pixels)	
Female	Male	Female	Male
8.2	10.0	8.1	9.1
9.5	10.4	7.9	9.6
9.3	9.7	8.3	9.4
9.9	9.0	8.0	10.6
8.5	9.6	7.9	9.5
8.3	10.8	8.7	10.0
8.6	9.2	8.6	8.8
8.8	9.5	8.3	9.5
8.7	8.9	9.5	9.3
8.5	8.9	8.9	9.4

9.86 Reducing Pollution A random sample of 12th-grade students were asked to rate the importance of reducing pollution on a scale from 0 to 1000. Responses were recorded by gender. The results of a hypothesis test are shown. Assume the conditions for using a two-sample t-test are met.

a. Which group has a higher mean, and how do you know?

b. Test the hypothesis that the mean rating for males and females are different, using a significance level of 0.05.

Two sample T hypothesis test:
μ_1 : Mean of Importance_reducing_pollution where Gender="Male"
μ_2 : Mean of Importance_reducing_pollution where Gender="Female"
$\mu_1 - \mu_2$: Difference between two means
$H_0 : \mu_1 - \mu_2 = 0$
$H_A : \mu_1 - \mu_2 \neq 0$
(without pooled variances)

Hypothesis test results:

Difference	Sample Diff.	Std. Err.	DF	T-Stat	P-value
$\mu_1 - \mu_2$	-46.652174	65.836748	93.638114	-0.70860386	0.4803

9.87 Heart Rate before and after Coffee Elena Lucin, a statistics student, collected the data in the table showing heart rate (beats per minute) for a random sample of coffee drinkers before and 15 minutes after they drank coffee. Carry out a complete analysis, using the techniques you learned in this chapter. Use a 5% significance level to test whether coffee increases heart rate. The same amount of caffeinated coffee was served to each person, and you may assume that conditions for a *t*-test hold.

Before	After
90	92
84	88
102	102
84	96
74	96
88	100
80	84
68	68

Before	After
74	78
72	82
72	76
92	96
86	88
90	92
80	74

*9.88 Exam Grades The final exam grades for a sample of day-time statistics students and evening statistics students at one college are reported. The classes had the same instructor, covered the same material, and had similar exams. Using graphical and numerical summaries, write a brief description about how grades differ for these two groups. Then carry out a hypothesis test to determine whether the mean grades are significantly different for evening and daytime students. Assume that conditions for a *t*-test hold. Select your significance level.

Daytime grades: 100, 100, 93, 76, 86, 72.5, 82, 63, 59.5, 53, 79.5, 67, 48, 42.5, 39

Evening grades: 100, 98, 95, 91.5, 104.5, 94, 86, 84.5, 73, 92.5, 86.5, 73.5, 87, 72.5, 82, 68.5, 64.5, 90.75, 66.5

9.89 Hours of Television Viewing The number of hours per week of television viewing for random samples of fifth-grade boys and fifth-grade girls were obtained. Each student logged his or her hours for one Monday-through-Friday period. Assume that the students were independent; for example, there were no pairs of siblings who watched the same shows. The data are available at the text's website.

Using graphical and numerical summaries, write a brief description of how the hours differed for the boys and girls. Then carry out a hypothesis test to determine whether the mean hours of television viewing are different for boys and girls. Evaluate whether the conditions for a *t*-test are met, and state any assumptions you must make in order to carry out a *t*-test.

9.90 Reaction Distances Reaction distances in centimeters for a random sample of 40 college students were obtained. Shorter distances indicate quicker reactions. Each student tried catching a meter stick with his or her dominant hand and nondominant hand. The subjects all started with their dominant hand. The data are available at the text's website.

Examine summary statistics, and explain what we can learn from them. Then do an appropriate test to see whether the mean reaction distance is shorter for the dominant hand. Use a significance level of 0.05.

9.91 Ales vs. Lagers Data were collected on calorie content for a random sample of ales and lagers and are summarized in the table. Assume the conditions for using a *t*-test are satisfied. (Source: efficientdrinker.com)

Summary statistics for Calories:
Where: Type = "L" or Type = "A"
Group by: Type

Type	n	Mean	Variance	Std. dev.	Std. err.	Median	Range	Min	Max	Q1	Q3
A	43	173.72093	614.49169	24.788943	3.7802784	165	106	125	231	158	190
L	20	143.9	1248.9368	35.340301	7.9023314	150	136	64	200	119	172

a. If we wanted to use a hypothesis test to determine if there is a difference in the calorie content of ales and IPAs, would we use a paired *t*-test or a two-sample *t*-test? Explain.

b. Construct a 95% confidence interval for the difference in the mean calorie content of ales and lagers. Assume that the populations are approximately Normal. Based on your confidence interval, is there a difference in the mean calorie content for ales and lagers? Explain.

9.92 Weights of Hockey and Baseball Players Data were collected on the weights of random samples of professional hockey players and professional baseball players. (Source: NHL.com, MLB.com)

Summary statistics:

Column	n	Mean	Variance	Std. dev.	Std. err.	Median	Range	Min	Max	Q1	Q3
Hockey	40	185.675	231.60962	15.218726	2.4062918	186.6	67	149	216	176.5	196
Baseball	26	206.92308	582.87385	24.14278	4.7347888	210	85	155	240	185	225

a. Using these descriptive statistics, can a two sample *t*-test be used to determine if there is a difference in the mean weights of professional hockey and baseball players? If so, do the test using a significance level of 0.05. If not, explain why the test cannot be done given the information provided.

b. Construct a 95% confidence interval for the difference in the mean weights of professional hockey and baseball players. Based on your confidence interval, is there a difference in the weights of players in these two sports? Explain.

9.93 Grocery Delivery The table shows the prices for identical groceries at two online grocery delivery services, Amazon and Walmart. Assume the sampling is random and the populations are Normal.

item	Amazon	Walmart
Strawberry (basket)	2.5	2.98
Avocado (large)	0.99	0.78
Nutella	3	3.58
Honeynut Cheerios	3	3.64
Cage-free eggs	2.49	2.98
Butter	3.49	3.24
Bread	3.49	3.42
Tortillas	2.29	2.28
Quaker Oats	3.79	3.98
Pretzel Sticks	3	2.98
Frozen Pizza	3.69	3.5
Chips Ahoy	2.5	2.56

a. Give the data, would a two-sample *t*-test or a paired *t*-test be appropriate to test the hypothesis that the mean prices at the two delivery services are different? Explain.

b. Use the appropriate method to test the hypothesis. Use a significance level of 0.05.

9.94 Parents The following table shows the heights (in inches) of a random sample of students and their parent of the same gender. Test the hypothesis that the mean for the students is more than the mean for the parents, at the 0.05 level. Assume the data are Normal.

a. Use the paired *t*-test that is appropriate.

b. Use the two-sample *t*-test, even though it is not appropriate.

c. Compare the results. The numerator of both *t*-values is the difference in sample means, which is 1.12 inches. What must be causing the different *t*-values if the numerators are the same?

Height	
Student	Parent
70	65
71	71
61	60
63	65
65	67
68	66
70	72
63	63
65	64

Height	
Student	Parent
63	62
65	65
73	70
68	64
72	70
71	69
68	65
69	68

*9.95 Why Is $n - 1$ **in the Sample Standard Deviation?** Why do we calculate *s* by dividing by $n - 1$, rather than just *n*?

$$s^2 = \frac{\sum(x - \bar{x})^2}{n - 1}$$

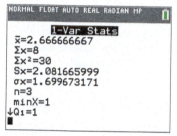

TI-84 Output

The reason is that if we divide by $n - 1$, then s^2 is an unbiased estimator of σ^2, the population variance.

We want to show that s^2 is an unbiased estimator of σ^2, sigma squared. The mathematical proof that this is true is beyond the scope of an introductory statistics course, but we can use an example to demonstrate that it is.

First we will use a very small population that consists only of these three numbers: 1, 2, and 5.

You can determine that the population standard deviation, σ, for this population is 1.699673 (or about 1.70), as shown in the TI-84 output. So the population variance, sigma squared, σ^2, is 2.888889 (or about 2.89).

Now take all possible samples, with replacement, of size 2 from the population, and find the sample variance, s^2, for each sample. This process is started for you in the table. Average these sample variances (s^2), and you should get approximately 2.88889. If you do, then you have demonstrated that s^2 is an unbiased estimator of σ^2, sigma squared.

Sample	*s*	s^2		Sample	*s*	s^2
1, 1	0	0		2, 5		
1, 2	0.7071	0.5		5, 1		
1, 5	2.8284	8.0		5, 2		
2, 1				5, 5		
2, 2						

Show your work by filling in the accompanying table and show the average of s^2.

9.96 Basketball Players The table shows the heights (in inches) of all the basketball players on the 2017–2018 Santa Clara University basketball team. Explain why it would be inappropriate to do a *t*-test with these data.

Center/Guards	83	75	73	73	74	71	73	82
Forwards	83	78	80	76	78	77		

*9.97 Construct two sets of body temperatures (in degrees Fahrenheit, such as 96.2 °F), one for men and one for women, such that the sample means are different but the hypothesis test shows the population means are not different. Each set should have three numbers in it.

9.98 Construct heights for 3 or more sets of twins (6 or more people). Make the twins similar, but not exactly the same, in height. Put all of the shorter twins in set A and all of the taller twins in set B. Create the numbers such that a two-sample *t*-test will *not* show a significant difference in the mean heights of the shortest of each pair, and the mean heights of the tallest of each pair, but the paired *t*-test *does* show a significant difference. (*Hint:* Make one of the pairs really tall, one of the pairs really short, and one of the pairs in between.) Report all the numbers and the *t*- and p-values for the tests. Explain why the paired *t*-test shows a difference and the two-sample *t*-test does not show a difference. Remember that 5 feet is 60 inches and that 6 feet is 72 inches.

gUIDED EXERCISES

g 9.37 Human Body Temperatures A random sample of 10 independent healthy people showed body temperatures (in degrees Fahrenheit) as follows:

98.5, 98.2, 99.0, 96.3, 98.3, 98.7, 97.2, 99.1, 98.7, 97.2

The Minitab output of the results of a one-sample *t*-test is shown.

```
One-Sample T: Sample of 10
Test of mu = 98.6 vs not = 98.6

Variable   N    Mean    StDev   SE Mean     95% CI              T      P
sam10      10   98.1200  0.9187  0.2905   (97.4628, 98.7772)  -1.65  0.133
```

QUESTION Test the hypothesis that the population mean is not 98.6 °F, using a significance level of 0.05. Write out the steps given, filling in the blanks.

Step 1 ▶ Hypothesize

H₀: $\mu = 98.6$
Hₐ: _____

Step 2 ▶ Prepare

A stemplot is shown that is not strongly skewed, suggesting that the distribution of the population is also approximately Normal.

(A histogram would also be appropriate.) Comment on the data collection, and state the test to be used. State the significance level.

```
96  3
97  22
98  23577
99  01
```

Step 3 ► Compute to compare

$$t = \underline{\hspace{2cm}}$$
$$\text{p-value} = \underline{\hspace{2cm}}$$

Step 4 ► Interpret

Reject or do not reject H_0, and choose interpretation i, ii, or iii:

i. We cannot reject 98.6 as the population mean from these data at the 0.05 level.

ii. The population mean is definitely 98.6 on the basis of these data at the 0.05 level.

iii. We can reject the null hypothesis on the basis of these data, at the 0.05 level. The population mean is not 98.6.

9.57 Televisions Minitab output is shown for a two-sample t-test for the number of televisions owned in households of random samples of students at two different community colleges. Each individual was randomly chosen independently of the others; the students were not chosen as pairs or in groups. One of the schools is in a wealthy community (MC), and the other (OC) is in a less wealthy community.

QUESTION Complete the steps to test the hypothesis that the mean number of televisions per household is different in the two communities, using a significance level of 0.05.

```
Two-Sample T-Test and CI: OCTV, MCTV

        N    Mean    StDev    SE Mean
OCTV   30    3.70    1.49     0.27
MCTV   30    3.33    1.49     0.27

Differen ce = mu OCTV - mu MCTV
Estimate for differe nce: 0.367
95% CI for difference: (-0.404, 1.138)
T-Test of diff erence = 0 (vs not =): T-Value = 0.95  P-Value = 0.345
```

Step 1 ► Hypothesize

Let μ_{oc} be the population mean number of televisions owned by families of students in the less wealthy community (OC), and let μ_{mc} be the population mean number of televisions owned by families of students in the wealthier community (MC).

$$H_0: \mu_{oc} = \mu_{mc}$$
$$H_a: \underline{\hspace{2cm}}$$

Step 2 ► Prepare

Choose an appropriate t-test. Because the sample sizes are 30, the Normality condition of the t-test is satisfied. State the other conditions, indicate whether they hold, and state the significance level that will be used.

Step 3 ► Compute to compare

$$t = \underline{\hspace{2cm}}$$
$$\text{p-value} = \underline{\hspace{2cm}}$$

Step 4 ► Interpret

Reject or do not reject the null hypothesis. Then choose the correct interpretation:

i. At the 5% significance level, we cannot reject the hypothesis that the mean number of televisions of all students in the wealthier community is the same as the mean number of televisions of all students in the less wealthy community.

ii. At the 5% significance level, we conclude that the mean number of televisions of all students in the wealthier community is different from the mean number of televisions of all students in the less wealthy community.

Confidence Interval

Report the confidence interval for the difference in means given by Minitab, and state whether it captures 0 and what that shows.

g 9.69 Females' Pulse Rates before and after a Fright In a statistics class taught by one of the authors, students took their pulses before and after being frightened. The frightening event was having the teacher scream and run from one side of the room to the other. The pulse rates (beats per minute) of the women before and after the scream were obtained separately and are shown in the table. Treat this as though it were a random sample of female community college students.

QUESTION Test the hypothesis that the mean of college women's pulse rates is higher after a fright, using a significance level of 0.05, by following the steps below.

Women		Women	
Pulse Before	Pulse After	Pulse Before	Pulse After
64	68	84	88
100	112	80	80
80	84	68	92
60	68	60	76
92	104	68	72
80	92	68	80
68	72		

Step 1 ► Hypothesize

μ is the mean number of beats per minute.

$$H_0: \mu_{\text{before}} = \mu_{\text{after}}$$
$$H_a: \mu_{\text{before}} \underline{\hspace{2cm}} \mu_{\text{after}}$$

Step 2 ► Prepare

Choose a test: Should it be a paired t-test or a two-sample t-test? Why? Assume that the sample was random and that the distribution of differences is sufficiently Normal. Mention the level of significance.

Step 3 ► Compute to compare

$$t = \underline{\hspace{2cm}}$$
$$\text{p-value} = \underline{\hspace{2cm}}$$

Step 4 ► Interpret

Reject or do not reject H_0. Then write a sentence that includes *significant* or *significantly* in it. Report the sample mean pulse rate before the scream and the sample mean pulse rate after the scream.

TechTips

General Instructions for All Technology

Because of the limitations of the algorithms, precision, and rounding involved in the various technologies, there can be slight differences in the outputs. These differences can be noticeable, especially for the calculated p-values involving t-distributions. It is suggested that the technology that was used be reported along with the p-value, especially for two-sample t-tests.

EXAMPLE A: ONE-SAMPLE t-TEST AND CONFIDENCE INTERVAL ▶ McDonald's sells ice cream cones, and the company's fact sheet says that these cones weigh 3.2 ounces. A random sample of 5 cones was obtained, and the weights were 4.2, 3.6, 3.9, 3.4, and 3.3 ounces. Test the hypothesis that the population mean is 3.2 ounces. Report the t- and p-values. Also find a 95% confidence interval for the population mean.

EXAMPLE B: TWO-SAMPLE t-TEST AND CONFIDENCE INTERVAL ▶ Below are the GPAs for random samples of male and female.

Male: 3.0, 2.8, 3.5

Female: 2.2, 3.9, 3.0

Perform a two sample t-test to determine whether you can reject the hypothesis that the population means are equal. Find the t- and p-values. Also find a 95% confidence interval for the difference in means.

EXAMPLE C: PAIRED t-TEST ▶ Here are the pulse rates (in beats per minute) before and after exercise for three randomly selected people.

Person	Before	After
A	60	75
B	72	80
C	80	92

Determine whether you can reject the hypothesis that the population mean change is 0 (in other words, that the two population means are equal). Find the t- and p-values.

TI-84

One-Sample t-Test

1. Press **STAT** and choose **EDIT**, and type the data into **L1** (list one).
2. Press **STAT**, choose **TESTS**, and choose **2: T-Test**.
3. Note this but don't do it: If you did not have the data in the list and wanted to enter summary statistics such as $\bar{x}$, s, and n, you would put the cursor over **Stats** and press **ENTER** and put in the required numbers.
4. See Figure 9a. Because you have raw data, put the cursor over **Data** and press **ENTER**.

Enter: μ_0, **3.2**; **List**, **L1**; **Freq: 1**; put the cursor over $\neq$ μ_0 and press **ENTER**; scroll down to **Calculate** and press **ENTER**.

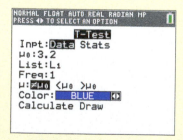

▲ FIGURE 9a TI-84 Input for One-Sample t-Test

Your output should look like Figure 9b.

▲ FIGURE 9b TI-84 Output for One-sample t-Test

One-Sample t-Interval

1. Press **STAT**, choose **TESTS**, and choose **8: TInterval**.
2. Choose **Data** because you have raw data. (If you had summary statistics, you would choose **Stats**.) Choose the correct **List** (to select **L1**, press **2nd** and **1**) and **C-Level**, here 0.95. Leave **Freq:1**, which is the default. Scroll down to **Calculate** and press **ENTER**.

The 95% confidence interval reported for the mean weight of the cones (ounces) is (3.2204, 4.1396).

Two-Sample t-Test

1. Press **STAT**, choose **EDIT**, and put your data (GPAs) in two separate lists (unstacked). We put the men's GPAs into **L1** and the women's GPAs into **L2**.
2. Press **STAT**, choose **TESTS**, and choose **4:2-Samp TTest**.
3. For **Inpt**, choose **Data** because we put the data into the lists. (If you had summary statistics, you would choose **Stats** and put in the required numbers.)

4. In choosing your options, be sure the lists chosen are the ones containing the data; leave the **Freq**s as 1, choose $\neq \boldsymbol{\mu_0}$ as the alternative, and choose **Pooled No** (which is the default). Scroll down to **Calculate** and press **ENTER**.

You should get the output shown in Figure 9c. The arrow down on the left-hand side means that you can scroll down to see more of the output.

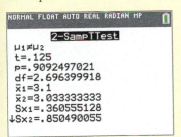

▲ **FIGURE 9c** TI-84 Output for Two-Sample *t*-Test

Two-Sample *t*-Interval

1. After entering your data into two lists, press **STAT**, choose **TESTS**, and choose **0:2-SampTInt**.
2. Choose **Data** because you have raw data. (If you had summary statistics, you would choose **Stats**.) Make sure the lists chosen are the ones with your data. Leave the default for **Freq1:1**, **Freq2:1** and **Pooled No**. Be sure the **C-Level** is **0.95**. Scroll down to **Calculate** and press **ENTER**.

The interval for the GPA example will be $(-1.744, 1.8774)$ if the men's data correspond to **L1** and the women's to **L2**.

Paired *t*-Test

1. Enter the data given in Example C into **L1** and **L2** as shown in Figure 9d.

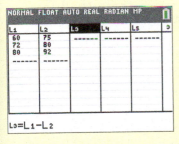

▲ **FIGURE 9d** Obtaining the List of Differences for the TI-84

2. See Figure 9d. Use your arrows to move the cursor to the top of **L3** so that you are in the label region. Then press **2ND L1 − 2ND L2**. For the minus sign, be sure to use the button above the plus button. Then press **ENTER**, and you should see all the differences in **L3**.
3. Press **STAT**, choose **TESTS**, and choose **2: T-Test**.
4. See Figure 9e. For **Inpt** choose **Data**. Be sure μ_0 is **0** because we are testing to see whether the mean difference is 0. Also be sure to choose **L3**, if that is where the differences are. Scroll down to **Calculate** and press **ENTER**.

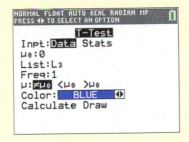

▲ **FIGURE 9e** TI-84 Input for Paired *t*-Test

Your output should look like Figure 9f.

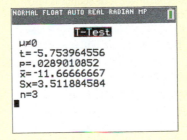

▲ **FIGURE 9f** TI-84 Output for Paired *t*-Test

MINITAB

One-Sample *t*-Test and Confidence Interval

1. Type the weight of the cones in **C1** (column 1).
2. **Stat > Basic Statistics > 1-Sample t**
3. See Figure 9g. Click in the empty white box below **One or more samples…**. A list of columns containing data will appear to the left. Double click **C1** to choose it. Check **Perform hypothesis test**, and put in **3.2** as the **Hypothesized mean**. (If you wanted a one-sided test or a confidence level other than 95%, you would use **Options**.)
4. Click **OK**.

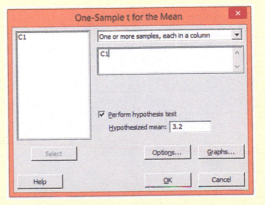

▲ **FIGURE 9g** Minitab Input Screen for One-Sample *t*-Test

The output is shown in Figure 9h.

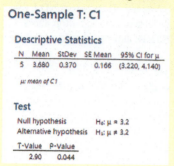

One-Sample T: C1

Descriptive Statistics

N	Mean	StDev	SE Mean	95% CI for μ
5	3.680	0.370	0.166	(3.220, 4.140)

μ: mean of C1

Test

Null hypothesis	H₀: μ = 3.2
Alternative hypothesis	H₁: μ ≠ 3.2

T-Value	P-Value
2.90	0.044

▲ **FIGURE 9h** Minitab Output for One-Sample *t*-Test and Confidence Interval

Note that the Minitab output (Figure 9h) includes the 95% confidence interval for the mean weight (3.22, 4.14) as well as the *t*-test.

Two-Sample *t*-Test and Confidence Interval

Use stacked data with all the GPAs in one column. The second column will contain the categorical variable that designates groups: male or female.

1. Upload the data from the disk or use the following procedure: Enter the GPAs in the first column. In the second column put the corresponding **m** or **f**. (Complete words or coding is also allowed for the second column, but you must decide on a system for one data set and stick to it. For example, using F one time and f the other times within one data set will create problems.) Use headers for the columns: **GPA** and **Gender.**

2. **Stat > Basic Statistics > 2-Sample t**

3. Refer to Figure 9i. Choose **Both samples are in one column,** because we have stacked data. Click in the small box to the right of **Samples** at the top to activate the box. Then double-click **GPA** and then double click **Gender** to get it into the **Sample IDs** box. If you wanted to do a one-sided test or to use a confidence level other than 95%, you would click **Options.**

   (If you had unstacked data, you would choose **Each sample is in its own column**, and choose both columns of data.)

4. Click **OK**.

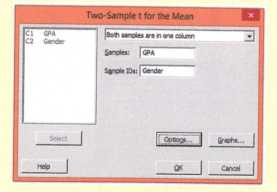

▲ **FIGURE 9i** Minitab Input Screen for Two-Sample *t*-Test and Confidence Interval

The output is shown in Figure 9j. Note that the confidence interval is included.

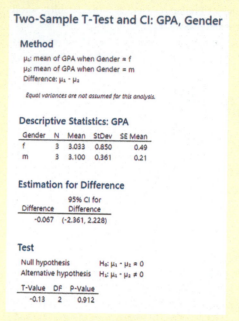

Two-Sample T-Test and CI: GPA, Gender

Method

μ₁: mean of GPA when Gender = f
μ₂: mean of GPA when Gender = m
Difference: μ₁ - μ₂

Equal variances are not assumed for this analysis.

Descriptive Statistics: GPA

Gender	N	Mean	StDev	SE Mean
f	3	3.033	0.850	0.49
m	3	3.100	0.361	0.21

Estimation for Difference

Difference	95% CI for Difference
-0.067	(-2.361, 2.228)

Test

Null hypothesis	H₀: μ₁ - μ₂ = 0
Alternative hypothesis	H₁: μ₁ - μ₂ ≠ 0

T-Value	DF	P-Value
-0.13	2	0.912

▲ **FIGURE 9j** Minitab Output for Two-Sample *t*-Test and Confidence Interval

Paired *t*-Test and Confidence Interval

1. Type the numbers (pulse rates) in two columns. Label the first column "**Before**" and the second column "**After**."

2. **Stat > Basic Statistics > Paired t**

3. Click in the small box to the right of **Sample 1** to activate the box. Then double-click **Before** and double-click **After** to get it in the **Sample 2** box. (If you wanted a one-sided test or a confidence level other than 95%, you would click **Options.**)

4. Click **OK**.

Figure 9k shows the output.

Paired T-Test and CI: Before, After

Descriptive Statistics

Sample	N	Mean	StDev	SE Mean
Before	3	70.67	10.07	5.81
After	3	82.33	8.74	5.04

Estimation for Paired Difference

Mean	StDev	SE Mean	95% CI for μ_difference
-11.67	3.51	2.03	(-20.39, -2.94)

μ_difference: mean of (Before - After)

Test

Null hypothesis	H₀: μ_difference = 0
Alternative hypothesis	H₁: μ_difference ≠ 0

T-Value	P-Value
-5.75	0.029

▲ **FIGURE 9k** Minitab Output for Paired *t*-Test and Confidence Interval

We are saving the one-sample *t*-test for last, because we have to treat it strangely.

Two-Sample *t*-Test

1. Type the GPAs in two columns side by side.
2. Click on **Data** and **Data Analysis,** and then scroll down to **t-Test: Two-Sample Assuming Unequal Variances** and double-click it.
3. See Figure 9l. For the **Variable 1 Range** select one column of numbers (don't include any labels). Then click inside the box for **Variable 2 Range**, and select the other column of numbers.

    You may leave the hypothesized mean difference empty, because the default value is 0, and that is what you want.
4. Click **OK**.

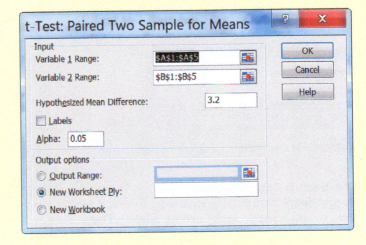

▲ FIGURE 9l Excel Input for Two-Sample *t*-Test

To see all of the output, you may have to click **Home, Format** (in the **Cells** group), and **AutoFit Column Width**.

Figure 9m shows the relevant part of the output.

df	3
t Stat	0.125
P(T<=t) one-tail	0.45421
t Critical one-tail	2.35336
P(T<=t) two-tail	0.90843
t Critical two-tail	3.18245

▲ FIGURE 9m Part of the Excel Output for Two-Sample *t*-Test

Caution: For a one-sided alternative hypothesis, Excel always reports one-half the p-value for the two-sided hypothesis. This is the correct p-value only when the value of the test statistic is consistent with the direction of the alternative hypothesis. (In other words, if the alternative hypothesis is ">," then the test statistic must be positive; if "<," then the value must be negative.) If this is not the case, to find the correct p-value, calculate 1 minus the reported one-tailed p-value.

In our example here, if the alternative hypothesis had been *less than* (<), the correct p-value would be $1 - 0.4542 = 0.5458$ since the calculated test statistic is *positive* (0.125).

Paired *t*-Test

1. Type the data into two columns.
2. Click on **Data, Data Analysis,** and **t-Test: Paired Two Sample for Means,** and follow the same procedure as for the two-sample *t*-Test.

One-Sample *t*-Test

1. You need to use a trick to force Excel to do this test. Enter the weights of the cones in column A, and put zeros in column B so that the columns are equal in length, as shown in Figure 9n.

A	B	
4.2	0	
3.6	0	
3.9	0	
3.4	0	
3.3	0	

▲ FIGURE 9n Excel Data Entry for One-Sample *t*-Test

2. Click **Data**, **Data Analysis**, and **t-Test: Paired Two Sample for Means**.
3. See Figure 9o. After selecting the two groups of data, you need to put the hypothesized mean in the box labeled **Hypothesized Mean Difference**. For the way we set up this example, it is **3.2**. (If you had entered 3.2s in column B, you would put in 0 for the Hypothesized Mean Difference.)
4. Click **OK**.

▲ FIGURE 9o Excel Input for One-Sample *t*-Test

Figure 9p shows the relevant part of the Excel output.

Hypothesized Mean Difference	3.2
df	4
t Stat	2.89979
P(T<=t) one-tail	0.02206
t Critical one-tail	2.13185
P(T<=t) two-tail	0.04413
t Critical two-tail	2.77645

▲ FIGURE 9p Relevant Part of the Excel Output for One-Sample *t*-Test

Again, the p-value reported in Figure 9p for the one-sided alternative hypothesis is consistent with the alternative hypothesis that the mean weight is *more than* (>) 3.2 ounces because the test statistic has a *positive* value.

One-Sample *t*-Test

1. Type the weights of the cones into the first column.
2. **Stat > T Stats > One Sample > With Data**
   (If you had summary statistics, then after **One Sample**, you would choose **With Summary**.)
3. See Figure 9q. Click on the column containing the data, **var1**. Put in the H_0: μ, which is **3.2**. Leave the default not equal for the H_A: μ, and click **Compute!**

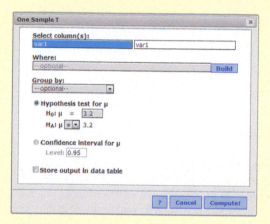

▲ **FIGURE 9q** StatCrunch Input for One-Sample *t*-Test

You will get the output shown in Figure 9r.

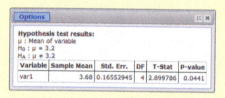

▲ **FIGURE 9r** StatCrunch Output for One-Sample *t*-Test

One-Sample Confidence Interval
Go back and perform the same steps as for the one-sample *t*-test, but when you get to step 3, check **Confidence interval for** μ. You may change the confidence level from the default 0.95 if you want. See Figure 9q.

Two-Sample *t*-Test
Use stacked data with all the GPAs in one column. The second column will contain the categorical variable that designates groups: male or female.

1. Upload the data from the disk or follow these steps: Enter the GPAs in column 1 (**var1**). Put **m** or **f** in column 2 (**var2**) as appropriate. (Complete words or coding for column 2 is also allowed, but whatever system you use must be maintained within the data set.) Put labels at the top of the columns; change **var1** to **GPA** and **var2** to **Gender**.
2. **Stat > T Stats > Two Sample > With Data**
3. See Figure 9s. For **Sample 1: Values in:** choose **GPA**, and for **Where:** put **Gender=m**. For **Sample 2:** choose **GPA** and for

Where: put **Gender=f**. Click off **Pool variances**. (If you had unstacked data, you would choose the two lists for the two samples and not use the **Where:** boxes.)

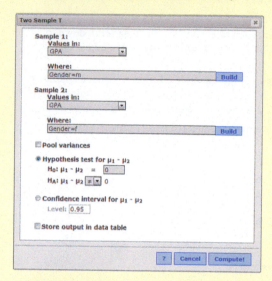

▲ **FIGURE 9s** StatCrunch Input for Two-Sample *t*-Test (Stacked Data)

4. Click **Compute!**

Figure 9t shows the output.

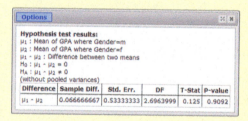

▲ **FIGURE 9t** StatCrunch Output for Two-Sample *t*-Test

Two-Sample Confidence Interval
1. Go back and do the preceding first three numbered steps.
2. Check **Confidence interval for** $\mu_1 - \mu_2$.
3. Click **Compute!**

Paired *t*-Test
1. Type the pulse rates Before in column 1 and the pulse rates After in column 2. The headings for the columns are not necessary.
2. **Stat > T Stats > Paired**
3. Select the two columns. Ignore the **Where:** and **Group by:** boxes.
4. Click **Compute!**

Figure 9u shows the output.

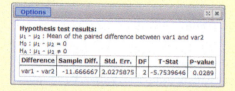

▲ **FIGURE 9u** StatCrunch Output for Paired *t*-Test

10 Associations between Categorical Variables

THEME

Distributions of categorical variables are often summarized in two-way tables. We can make inferences about these distributions by calculating how many observations we would expect in each cell if the null hypothesis were true and then comparing this to the actual counts.

As we have shown in the last few chapters, statistical tests are comparisons between two different views. In one view, if two samples have different means, then it might be because the populations are truly different. In the other view, randomness rules, and this difference is due merely to chance. A sample of people who had watched violent TV as children may behave more aggressively toward their partners when adults. But this might just be due to chance. Changing the placement of an advertisement on an Internet search results page might lead to a greater number of "click-through" visits to the company that placed the ad. Or the difference could be due to chance—that is, if we were to repeat the study, we might see a very different outcome.

In this chapter we ask the same question as before: Are the results we see due to chance, or might something else be going on? In Chapter 8, we asked this question about one-sample and two-sample proportions. We now ask the same question with respect to categorical variables with multiple categories. For example, we might compare two categorical variables: the results of popping a batch of popcorn ("good" and "bad") for different amounts of oil ("no oil," "medium amount of oil," and "maximum oil"). Or we might compare a single categorical variable with a proposed model: Do the proportions of people hospitalized for swine flu for different age groups match the proportions of people in those age groups in the general population? If yes, this suggests that swine flu affects all age groups the same.

One of the nice surprises in this chapter is that only one test statistic is needed to analyze data collected under several different situations. This test statistic can be used to compare distributions of categorical variables, to compare population proportions, and to test whether two categorical variables are independent. This chapter also introduces a hypothesis test that requires no assumptions about the distribution of the population or the sample.

CASE STUDY

Popping Better Popcorn

You're planning a movie night and decide to cook up some popcorn. Many factors might affect how good the popcorn tastes: the brand, for example, or how much oil you use, or how long you let the popcorn pop before stopping. Some researchers (Kukuyeva et al. 2008) investigated factors that might determine how to pop the perfect batch of popcorn. They decided that if more than half of the kernels in a bag were popped in the first 75 seconds, then the batch was a "success." If fewer than half, then it was a "failure." They popped 36 bags under three different treatments: no oil, "medium" amount of oil (1/2 tsp), and "maximum" oil (1 tsp). The bags were randomly assigned to an oil group. Each bag had 50 kernels. The outcome is shown in the accompanying table. From the table, it looks as though it is bad to use too much oil. But might this just be due to chance? In other words, if other investigators were to do this experiment the same way, would they also find so few successful bags with the maximum amount of oil?

Result	Oil Amount		
	No Oil	Medium Oil	Maximum Oil
Failure	23	22	33
Success	13	14	3

This question—Is the outcome due to chance?—is one we've asked before. In this study, the two variables are the amount of oil (with three values: none, medium, and maximum) and the result (success or failure). Both *Oil* and *Result* are categorical variables. In this chapter, we will see how to test the hypothesis that the amount of oil had an effect on the outcome.

SECTION 10.1

The Basic Ingredients for Testing with Categorical Variables

The statistical questions we ask when working with categorical variables are similar in many ways to those we ask of numerical variables: What's the typical outcome? Are two variables associated or not?

You'll also see that the basic structure of our analyses are similar. For example, hypothesis tests follow the same four steps you learned in Chapters 8 and 9. And we're still concerned with whether the observed data match the values we expected under our probability model. However, you'll see that we use slightly different ingredients: expected counts, the chi-square statistics, and the chi-square distribution.

To provide a context for our discussion, we're going to consider a particular six-sided die and ask the statistical question, Is this die "fair"? In other words, is each side of a thrown die equally likely to end up on top? We might think so, because of the symmetry of the die. However, the manufacturing process must be very precise to achieve this. Surely, the dice we buy have a few slight defects here and there that would lead to their being biased.

We took a die from our favorite board game and decided to test whether it was fair. Our null hypothesis was that the die is balanced so that each side is equally likely to land face-up. The alternative hypothesis was that one or more sides are favored. To test whether this die was fair, we rolled it 60 times. For each outcome, we counted the number of dots showing on top of the die.

Before we look at our data, let's take a moment to think about what the outcome would look like in a perfect world if the die really were fair. If we rolled a perfect die infinitely many times, the outcome should look like Figure 10.1—uniform distribution, because each outcome is equally likely.

> **Details**
>
> **Pronunciation**
> *Chi* is pronounced "kie" (rhymes with "pie"), with a silent *h*.

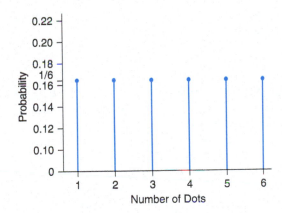

◀ **FIGURE 10.1** Probability distribution for an ideal fair die. Each outcome is equally likely (with probability 1/6).

How did our real die compare to the perfect, ideal die? Table 10.1 provides a summary of the data.

> **Details**
>
> **Raw Data**
> The raw data look like this: "three dots," "two dots," "six dots," "two dots" . . . There are 60 entries.

◀ **TABLE 10.1** Summary of outcomes with a six-sided die.

Outcome:	One dot	Two dots	Three dots	Four dots	Five dots	Six dots
Frequency:	12	8	10	11	9	10

Figure 10.2 shows the graph of the results with this real die, tossed 60 times.

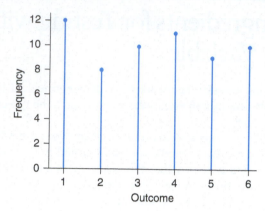

You can see more clearly from the graph than from the table that we did not get a perfectly uniform distribution. The question now is whether these deviations from the "perfect" model in Figure 10.1 occur because the die is not fair or are due merely to chance.

> **KEY POINT**   Comparing real outcomes to expected outcomes is a fundamental analysis technique.

Our approach to solving this problem will be to compare our actual outcomes with the outcomes we would expect to get if the die were behaving as dice are supposed to—giving us roughly equal proportions of outcomes. To compare the real outcomes with the expected outcomes, we'll make use of a statistic, called the chi-square statistic, that measures how far apart reality is from expectation. Next, we'll see that this statistic's sampling distribution can be used to find p-values, which will tell us whether we should be suspicious that our outcomes aren't conforming to our expectations.

The basics of this approach should sound familiar. It's the same outline we've followed with every other hypothesis test. Some of the details are different, of course, but the approach is the same.

1. The Data

Recall that categorical variables are those whose values are categories, as opposed to numbers. The outcome of the roll of a die can be thought of as a category, particularly if we don't care about the numerical values on the die.

When two categorical variables are analyzed, as we are doing in this chapter, they are often displayed in a **contingency table** (also called a **two-way table**), a summary table that displays frequencies for the outcomes. Such tables can give the impression that the data are numerical, because you are seeing numbers in the table. But it's important to keep in mind that the numbers are *summaries* of variables whose values are *categories*.

For instance, the General Social Survey (GSS) asked respondents, "There are always some people whose ideas are considered bad or dangerous by other people. For instance, somebody who is against churches and religion. If such a person wanted to make a speech in your (city/town/community) against churches and religion, should he be allowed to speak, or not?" These respondents were also asked about their income, which was recorded as one of four annual income levels: $0–$20K, $20–$40K, $40–$70K, and $70K and more. The first four lines of the actual data would look something like Table 10.2. A value on the boundary would be assigned to the group with the smaller incomes, so an income of $40,000 would be in the $20–$40K group.

Observation ID	Response	Income Level
1	allowed	20–40
2	allowed	20–40
3	not allowed	70 and over
4	allowed	0–20
. . .	. . .	. . .

◀ **TABLE 10.2** The first four lines of the raw data (based on an actual data set) showing responses to two questions on the General Social Survey.

We have chosen to display another variable, *Observation ID*, which is useful simply for keeping track of the order in which the observations are stored.

A two-way table summarizes the association between the two variables *Response* and *Income Level* as in Table 10.3.

Response	Income			
	0–20	20–40	40–70	70+
Allowed	248	311	223	229
Not Allowed	100	105	55	21

◀ **TABLE 10.3** Two-way table summarizing the association between responses to the question about whether speeches opposed to church and religion should be allowed, and income level. (Source: GSS 2012–2014, http://www.teachingwithdata.org)

2. Expected Counts

The **expected counts** are the numbers of observations we would see in each cell of the summary table if the null hypothesis were true. A little later, we will set up some formal hypothesis tests, and you will see more precisely how to calculate the expected counts.

For example, Table 10.1 shows the actual counts we observed when we rolled a six-sided die 60 times. If the die were truly fair, then because each outcome would have the same probability, we should see the same number of outcomes in each category.

When the die is rolled 60 times, the expected counts are therefore 10 in each outcome category. Because the probability for each category is $1/6$ we expect $1/6$ of the total outcomes to fall in each category: $(1/6) \times 60 = 10$. Table 10.4 compares these expected counts with our actual counts.

Details

Expectations
Expected counts are actually long-run averages. When we say that we "expect" 10 observations in a cell of a table, we mean that if the null hypothesis were true and we were to repeat this data collection many, many times, then, on average, we would see 10 observations in that cell.

Outcome	One Dot	Two Dots	Three Dots	Four Dots	Five Dots	Six Dots
Expected Counts	10	10	10	10	10	10
Observed Counts	12	8	10	11	9	10

◀ **TABLE 10.4** The expected (top) and observed (bottom) counts for rolling a fair die 60 times.

For a slightly more complex (but still quite typical) example, consider a study of the link between TV viewing and violent behavior. Researchers examined TV viewing habits for children, comparing children who watched TV programs with a high level of violence with those who did not. Many years later, they interviewed the same respondents (now adults) about violent behavior in their personal lives (Husemann et al. 2003). Table 10.5 summarizes the results.

	High TV Violence	Low TV Violence	Total
Yes, Physical Abuse	25	57	82
No Physical Abuse	41	206	247
Total	66	263	329

◀ **TABLE 10.5** A two-way summary of TV violence and later abusive behavior.

There are two categorical variables: *TV Violence* ("High" or "Low") and *Physical Abuse* ("Yes" or "No"). (The subjects were asked whether they had pushed, grabbed, or shoved their partner.) We wish to know whether these variables are associated. The null hypothesis says that they are not—that these variables are independent. In other words, any patterns you might see are due purely to chance.

What counts should we expect if these variables are truly not related to each other? That is, if the null hypothesis is true, what would we expect the table to look like? There are two ways of answering this question, and they both lead to the same answer. Let's look at them both.

Starting with the Physical Abuse *Variable*

Starting with the Physical Abuse *Variable* We notice that out of the entire sample, 82/329 (a proportion of 0.249240, or about 24.92%) said that they had physically abused their partners. If abuse is independent of TV watching—that is, if there is no relationship between these two variables—then we should expect to find the same percentages of violent abuse in those who watched high TV violence and in those who watched low TV violence.

So when we consider the 66 people who watched high TV violence as children, we expect 24.92% of them to have abused their partners. This translates to $(0.249240 \times 66) = 16.4498$ people. When we consider the 263 people who watched low TV violence, we expect to find that 24.92%, or $0.249240 \times 263 = 65.5501$, of them have abused their partners.

We can use similar reasoning to find the other expected counts. We know that if a proportion of 0.249240 have abused their partners, then $1 - 0.249240 = 0.750760$ have not. This proportion should be the same no matter which level of TV violence was watched. Thus, we expect the following:

In the "no physical abuse and high TV violence" group,

$$0.750760 \times 66 = 49.5502$$

In the "no physical abuse and low TV violence" group,

$$0.750760 \times 263 = 197.4499$$

(We are avoiding rounding in these intermediate steps so that our answers for the expected counts will be as accurate as possible.)

We summarize these calculations in Table 10.6, which shows the expected counts in parentheses. We include the actual counts in the same table so that we can compare. Note that we rounded to two decimal digits for ease of presentation.

▶ TABLE 10.6 TV violence and later physical abuse: a summary including expected counts (in parentheses).

	High TV Violence	Low TV Violence	Total
Yes, Physical Abuse	25 (16.45)	57 (65.55)	82
No Physical Abuse	41 (49.55)	206 (197.45)	247
Total	66	263	329

The values 82 and 247 are examples of what we call row totals (for obvious reasons, we hope!), 66 and 263 are the column totals, and 329 is the grand total. We can generate a formula for automatically are the column expected counts for each cell. This formula is rarely needed, for two reasons. First, you can and should always think through the calculations as we did here. Second, most software will do this automatically for you.

Formula 10.1:   Expected count for a cell $= \dfrac{(\text{row total}) \times (\text{column total})}{\text{grand total}}$

Starting with the **TV Violence** *Variable* The other way of finding the expected counts is to begin by considering the *TV Violence* variable rather than beginning with the *Physical Abuse* variable. We see that $66/329 = 0.200608$, or 20.06%, watched high TV violence. The rest, $263/329 = 0.799392$, or 79.94%, watched low TV violence.

If these variables are not related, then when we look at the 82 people who committed physical abuse, we should expect about 20.06% of them to fall in the High TV Violence category and the rest to fall in the Low TV Violence category.

Also, when we look at the 247 who did not commit physical abuse, we should expect 20.06% of them to have viewed high TV violence.

Among the abusive, the expected number with high TV violence is $0.200608 \times 82 = 16.45$. Among the nonabusive, the expected number with high TV violence is $0.200608 \times 247 = 49.55$.

You see that you will get the same result no matter which variable you consider first.

> **Details**
>
> **Fractions of People**
> Does it bother you that we have fractions of people in each category? It is a little strange until you think about this in terms of an ideal model. These expected counts are like averages. We say the average family has 2.4 children, and we know very well that there is no single family with a 0.4 child. This number 2.4 is a description of the collection of all families. Our claim that we expect 16.45 people (in this group) who have seen high TV violence to be abusive as adults is a similar idealization.

EXAMPLE 1 Gender and Opinion on Same-Sex Marriage

Do men and women feel differently about the issue of same-sex marriage? In 2014, the General Social Survey took a random sample of 1690 adult Americans, recording their gender and their level of agreement with the statement "Homosexuals should have the right to marry." The results are summarized in Table 10.7.

Opinion	Female	Male	Total
Strongly Agree	261	271	**532**
Agree	184	239	**423**
Neutral	76	109	**185**
Disagree	90	133	**223**
Strongly Disagree	130	197	**327**
Total	**741**	**949**	**1690**

◄ **TABLE 10.7** Summary of responses from the 2014 General Social Survey

Data Moves ► These data come from the General Social Survey, which is a long-running survey that polls Americans on various opinions and perspectives (gss.norc.org). Its website provides a "Data Explorer" interface that, once you've created a free account, lets you select and download variables you are interested in examining. The data, unfortunately, do not come in a simple comma-separated file, and require that you have the software packages R, SPSS, Excel, or SAS available. Once they are downloaded, you might have to do some work, as we did with the data in this example, because the responses are sometimes coded as numbers and need to be translated back to words in order for the data to be interpretable.

File: gaymarriagefull.csv includes results from this survey on this question and two others for several years prior to 2016.

QUESTIONS Assuming that the two variables *Opinion* and *Gender* are *not* associated, find the number of males who would be expected to agree strongly and the number of females who would be expected to agree strongly.

SOLUTION We consider *Gender* first. The percentage of men in the sample is $(949/1690) \times 100\% = 56.1539\%$. The percentage of women is therefore $100\% - 56.1539\% = 43.8462\%$. If *Gender* is not associated with *Opinion*, then we should see that about 56% of those who strongly agree are male, and about 44% of those who strongly agree are female. In other words:

CONCLUSION Expected count of males who strongly agree $= 532 \times 0.561539 = 298.7387$ or about 298.7.

Expected count of females who strongly agree $532 \times 0.438462 = 233.2615$ or about 233.3.

TRY THIS! Exercises 10.9c, 10.9d

3. The Chi-Square Statistic

Let's examine again our table of the relation of expected die tosses to our actual die tosses (Table 10.8).

▶ **TABLE 10.8** Sixty rolls of a six-sided die.

Outcome:	One Dot	Two Dots	Three Dots	Four Dots	Five Dots	Six Dots
Expected Counts (E)	10	10	10	10	10	10
Observed Counts (O)	12	8	10	11	9	10

We note that in the first category, we saw two more "aces" ("One Dot") than expected. On the other hand, we saw two fewer "Two Dots" and exactly the expected number of "Three Dots" and "Six Dots." Are these differences big or small? If they are small, then we can believe that the deviations between actual and expected counts are just due to chance. But if they are big, then maybe the die is not fair.

The **chi-square statistic** is a statistic that measures the amount that our expected counts differ from our observed counts. This statistic is shown in Formula 10.2.

$$\text{Formula 10.2:} \quad X^2 = \sum_{\text{cells}} \frac{(O - E)^2}{E}$$

where

O is the observed count in each cell
E is the expected count in each cell
Σ means add the results from each cell

Why does this statistic work? The term $(O - E)$ is the difference between what we observe and what we expect under the null hypothesis. To measure the total amount of deviation between Observed and Expected, it is tempting to just add together the individual differences. But this doesn't work, because the expected counts and the observed counts always add to the same value; if we sum the differences, they will always add to 0.

You can see that the differences between Observed and Expected add to 0 in Table 10.9, where we've added a row of differences (Observed minus Expected). You'll notice the sum of the differences is 0: $2 - 2 + 0 + 1 - 1 + 0 = 0$. In fact, the sum is always zero, and for that reason, we can't measure the distance between expected values and observed values simply by averaging the differences because the average will always be 0.

Outcome:	One Dot	Two Dots	Three Dots	Four Dots	Five Dots	Six Dots
Expected Counts	10	10	10	10	10	10
Observed Counts	12	8	10	11	9	10
Observed minus Expected	2	–2	0	1	–1	0

▲ **TABLE 10.9** Sixty rolls of a six-sided die, emphasizing the observed minus expected counts.

One reason why the chi-square statistic uses squared differences is that by squaring the differences, we always get a positive value, because negative numbers multiplied by themselves result in positive numbers:

$$2^2 + (-2)^2 + 0^2 + 1^2 + (-1)^2 + 0^2 = 4 + 4 + 0 + 1 + 1 + 0 = 10$$

Why divide by the expected count? The reason is that a difference between the expected and actual counts of, say, 2 is a small difference if we were expecting 1000 counts. But if we were expecting only 5 counts, then this difference of 2 is substantial. By dividing by the expected count, we're controlling for the size of the expected count. Basically, for each cell, we are finding what proportion of the expected count the squared difference is.

If we apply this formula to our test of whether the die is unbalanced, we get $X^2 = 1.0$. We must still decide whether this value discredits the null hypothesis that the die is fair. Keep reading.

EXAMPLE 2 Viewing Violent TV as a Child and Abusiveness as an Adult

Table 10.10 shows summary statistics from a study that asked whether there was an association between watching violent TV as a child and aggressive behavior toward one's spouse later in life. The table shows both actual counts and expected counts (in parentheses).

	High TV Violence	Low TV Violence	Total
Yes, Physical Abuse	25 (16.45)	57 (65.55)	82
No Physical Abuse	41 (49.55)	206 (197.45)	247
Total	66	263	329

◀ **TABLE 10.10** A two-way summary of the relationship between viewing TV violence and later abusiveness (expected counts are shown in parentheses).

QUESTION Find the chi-square statistic to measure the difference between the observed counts and expected counts for the study of the relationship between violent TV viewing and future behavior.

SOLUTION We use Formula 10.2 with the values for O and E taken from Table 10.10.

$$X^2 = \sum \frac{(O - E)^2}{E}$$

$$= \frac{(25 - 16.45)^2}{16.45} + \frac{(57 - 65.55)^2}{65.55} + \frac{(41 - 49.55)^2}{49.55} + \frac{(206 - 197.45)^2}{197.45} = 7.4047$$

CONCLUSION

$$X^2 = 7.40$$

Later we will see whether this is an unusually large value for two independent variables.

TRY THIS! Exercise 10.9e

As you might expect, for tables with many cells, these calculations can quickly become tiresome. Fortunately, technology comes to our rescue. Most statistical software will calculate the chi-square statistic for you, given data summarized in a two-way table or presented as raw data as in Table 10.2, and some software will even display the expected counts alongside the observed counts. Figure 10.3 shows the output from StatCrunch for the TV violence study.

What happens to the chi-square statistic when the expected counts are exactly the same as the observed counts for every cell of a table? In our test to see whether a die was fair, we rolled the die 60 times. We expected 10 outcomes in each category. If we had gotten exactly 10 (for each cell), then our observations would have matched our expectations perfectly. In that case, the chi-square statistic would equal 0, because

$$(\text{Observed} - \text{Expected})^2 = (10 - 10)^2 = 0 \text{ in each cell}$$

► **FIGURE 10.3** StatCrunch output for TV violence and abusiveness. The expected counts are below the observed values.

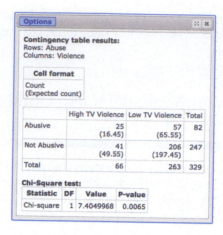

	High TV Violence	Low TV Violence	Total
Abusive	25 (16.45)	57 (65.55)	82
Not Abusive	41 (49.55)	206 (197.45)	247
Total	66	263	329

Chi-Square test:

Statistic	DF	Value	P-value
Chi-square	1	7.4049968	0.0065

Thus, when expectations and reality are exactly the same, the chi-square statistic is 0.

Even when the null hypothesis is true, our real-life observations will differ slightly from the expected counts just by chance. When this happens, the chi-square statistic will be a small value.

If reality is very different from what our null hypothesis claims, then our observed counts should differ substantially from the expected counts. When that happens, the chi-square statistic is a big value.

The trick, then, is to decide what values of the chi-square statistic are "big." Big values discredit the null hypothesis. To determine whether an observed value is big, we need to know its probability distribution when the null hypothesis is true.

> **KEY POINT** If the data conform to the null hypothesis, then the value of the chi-square statistic will be small. For this reason, large values of the chi-square statistic make us suspicious of the null hypothesis.

 Looking Back

The p-Value
You learned in Chapters 8 and 9 that the p-value measures our surprise, if the null hypothesis is believed.

4. Finding the p-Value for the Chi-Square Statistic

We are wondering whether our die is fair or biased. If it's fair, then the deviations we see from our expected counts should be small, and the chi-square statistic should be small. We found that $X^2 = 1.0$ for the die. Is 1.0 a small value, or is it surprisingly large for a fair die?

We are wondering whether television-viewing habits as a child are associated with violent behavior as an adult. If there is *no* association, then the observed counts in Table 10.10 should be close to the expected counts, and our chi-square statistic should be small. We found that $X^2 = 7.40$. Is this small or surprisingly large if there's no association?

The p-value is the probability, assuming that the null hypothesis is true, that X^2 will be as large as or larger than the value observed. If the die is truly fair, the p-value is the probability of getting a statistic as large as 1.0 or bigger. If there is no association between violence and viewing behavior, the p-value is the probability of getting a value as large as or larger than 7.40. But to find this probability, we need to know the sampling distribution of X^2.

If the sample size is large enough, there is a probability distribution that gives a fairly good approximation to the sampling distribution. Not surprisingly, this approximate distribution is called the **chi-square distribution**. The chi-square distribution is often represented with the Greek lowercase letter *chi* (χ) raised to the power of 2—that is, χ^2.

Unlike the normal distribution and the *t*-distribution, the χ^2 distribution allows for only positive values. It also differs from the other sampling distributions you've seen in that it is (usually) not symmetric and is instead right-skewed.

Like the shape of the *t*-distribution, the shape of the chi-square distribution depends on a parameter called the **degrees of freedom**. The lower the degrees of freedom, the more skewed the shape of the chi-square distribution. Figure 10.4 shows the chi-square distribution for several different values of the degrees of freedom. Sometimes the degrees of freedom are indicated using this notation: χ^2_{df}. For example, χ^2_6 represents a chi-square distribution with 6 degrees of freedom, as in Figure 10.4b.

> ! **Caution**
>
> **Symbols**
> The symbol X^2 is used to represent the chi-square *statistic*. The symbol χ^2 is used to represent the chi-square *distribution*. Do not confuse the two. One is a statistic whose value is based on data; the other provides (approximate) probabilities for the values of that statistic.

(a)

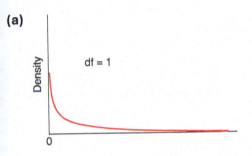

(b)

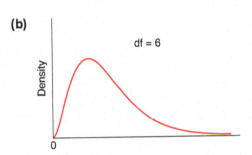

(c)

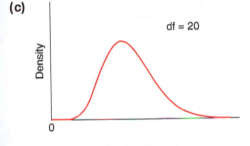

◀ **FIGURE 10.4** Three chi-square distributions for **(a)** df = 1, **(b)** df = 6, and **(c)** df = 20. Note that the shape becomes more symmetric as the degrees of freedom (df) increase. No negative values are possible with the chi-square distribution; the smallest possible value is 0.

The degrees-of-freedom parameter is different for different tests, but in general it depends on the number of categories in the summary table.

The chi-square distribution χ^2 is only an approximation to the true sampling distribution of the statistic X^2. The approximation is usually quite good if all the expected counts are 5 or higher.

> **Details**
>
> **Degrees of Freedom (df)**
> The degrees of freedom for the chi-square distribution are determined by the number of categories, not by the number of observations. For example, when a six-sided die is tossed, there are six categories.

> **KEY POINT**   The chi-square distribution provides a good approximation to the sampling distribution of the chi-square statistic only if the sample size is large. For many applications, the sample size is large enough if each expected count is 5 or higher.

At the start of this section, we asked whether the value we saw for the chi-square statistic when comparing an ideal die to our real die was big or small. Recall that the value was $X^2 = 1.0$. As you will see in Section 10.2, df = 5 for this problem. Figure 10.5 shows that 1.0 is not a very big value. Specifically, about 96% of the total area of the chi-square distribution is above 1.0. This tells us that the observed outcomes come close to what we would expect of a fair die, so we should not conclude that the die is biased.

▶ **FIGURE 10.5** A chi-square distribution with 5 degrees of freedom. The shaded area is the area above the value 1.0. This suggests that 1.0 is not a very large value for this distribution; most values are bigger than 1.0.

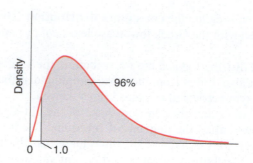

In Example 1 we calculated how many men and how many women we would expect to "strongly agree" with same-sex marriage if men and women had the same levels of agreement. The chi-square statistic that compares the expected proportions of support if men and women were the same against the actual observed levels of support can be calculated to be 9.793 with 4 degrees of freedom. Figure 10.6 illustrates that statistical calculators can also be useful for finding p-values for these situations. Here, 9.793 is a relatively large value if men and women have the same levels of support for same-sex marriage. This tells us that our observations are far away from the hypothesis that support levels are the same, so it would be reasonable to conclude that men and women differ in their levels of support.

▶ **FIGURE 10.6** StatCrunch calculator showing that the probability of getting a chi-square statistic of 9.793 or larger (when there are 4 degrees of freedom) is 0.044 or about 4.4%.

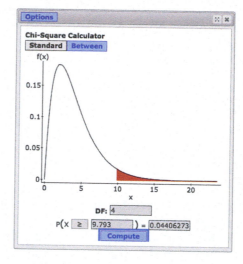

SECTION 10.2

The Chi-Square Test for Goodness of Fit

We now show how these ingredients can be combined to do a hypothesis test to determine whether the distribution of a categorical variable is following a proposed distribution. These tests are called **goodness-of-fit tests**. Our experiment with the potentially unfair die was a goodness-of-fit test. We had a proposed distribution (each outcome is equally likely and happens with probability 1/6) and an experimental distribution that we got from rolling a real die.

The goodness-of-fit test is a hypothesis test, and it follows the same four-step procedure as the tests we presented in Chapters 8 and 9; only the details have changed. The goodness-of-fit test relies on the chi-square statistic and can be applied when you are comparing the distribution of counts of a categorical variable (*one* categorical variable) with a distribution of expected counts. This hypothesis test is used to answer the question: Does the distribution I'm seeing match a particular theoretical distribution?

Goodness of Fit

Step 1: Hypothesize

The hypotheses for a goodness-of-fit test are always the same:

> H_0: The population distribution of the variable is the same as the proposed distribution.
> H_a: The distributions are different.

For our die-testing example, we began with the question "Is this die fair?" We turn this into a statistical question by rephrasing as "Is each of the six sides equally likely to land face up?" This question, in turn, becomes the hypotheses:

> H_0: The die is fair (each outcome is equally likely).
> H_a: The die is not fair.

Step 2: Prepare

We plan to use the chi-square goodness-of-fit test statistic to compare the observed number of outcomes in each category (O) with the expected number of outcomes (E).

Recall that an important feature of a hypothesis test is being able to measure how often our conclusions will be wrong. To do this, we need to know the probability distribution of our test statistic so that we can find the p-value and significance level for our test. Under the following conditions, the chi-square distribution with $k - 1$ degrees of freedom gives good approximations for the p-value and significance level. The letter k is used to represent the number of categories for our variable. In the die example $k = 6$, so we had $6 - 1 = 5$ degrees of freedom.

> Conditions:
>
> 1. *Random Sample.* The sample was collected randomly.
>
> 2. *Independent Measurements.* Each measurement on an individual is independent of all other measurements.
>
> 3. *Large Sample.* The expected count is at least 5 in each cell.

The first condition shouldn't surprise you. In order for us to learn about the population, our sample must be representative of that population, so it must be collected randomly.

The second condition is there to make sure that each observation provides a useful additional bit of information. This condition would be violated if, for example, you had randomly sampled married couples, instead of individuals, and recorded their individual marital statuses. Then, if you knew the marital status of one person in your sample, you would definitely know the marital status of another person. The "measurement" of marital status would not be independent for these two people.

These first two conditions are difficult to check unless you were present when the data were collected or have access to very detailed notes about how they were collected. However, we often know enough to reasonably *assume* that these conditions hold.

Unlike the first two conditions, the third condition can be checked by looking at the data. This third condition is another way of saying that a large sample is required. It must be large enough for us to *expect* at least 5 observations in each cell (each category). Note that we don't need to actually get 5 or more observations in each cell. We just need to *expect* 5 or more.

We use the chi-square statistic (Formula 10.2) for the test statistic:

$$X^2 = \sum_{cells} \frac{(O - E)^2}{E}$$

If the expected counts are large enough, the chi-square distribution with degrees of freedom = (number of categories − 1) is a good approximation for the distribution of X^2.

To complete the "Prepare" step, you need to calculate the expected counts and verify that all are greater than or equal to 5. This can be time-consuming, and we recommend that you use technology.

> **! Caution**
>
> **Counts, Not Proportions**
> Sometimes, two-way tables present proportions (or percentages), not counts. Before carrying out these tests, you will need to convert the proportions to counts (if possible).

Step 3: Compute to compare

The chi-square statistic compares the observed counts to the expected counts. Calculating the value of this statistic is best done with technology, although for tables with only a few cells, it is not too tedious to do it by hand. Once we have the observed value of the chi-square statistic, we compute the p-value, which measures whether we should be surprised by how large the chi-square statistic is (if the null hypothesis is true).

The p-value is the probability that, if the null hypothesis is true, a chi-square statistic will be as big as or bigger than the observed value. In other words, it is always the area under the χ^2 probability curve for values to the right of the one you observed. Figure 10.5 is an example of this. The shaded area represents the p-value for our die-rolling experiment. If the die really is fair (so that each outcome is equally likely), then the probability of getting a chi-square statistic as big as or bigger than 1.0 is about 96%.

As always, we must set a significance level, and 5% is a common choice.

Step 4: Interpret

This step is the same, no matter which test we use. If the p-value is less than or equal to the significance level, α (alpha), then we reject the null hypothesis. If the p-value is bigger than α, then we do not reject.

EXAMPLE 3 Feeling Sleepy?

Sleep deprivation can be a big problem in our lives. Does our ability to get a good night's sleep depend on our age? Table 10.12 shows the sample distribution of ages, by decade, for those in a random sample of people who reported to a health professional that they were having sleeping troubles. (The data come from the National Health and Nutrition Examination Survey (NHANES), conducted by the Center for National Health Statistics since the 1960s. These data have been modified slightly to adjust for the sampling scheme.) (Source: Pruim, 2015 and www.cedc.gov/nchs/nhanes.htm.) A graph of the distribution of ages for a sample of those with sleeping trouble is shown in Figure 10.7a. There were 1893 people who reported sleeping troubles.

The Expected Frequency column shows the number of people we would expect in each age category if we simply took a random sample of 1893 people from the same population. If these two distributions are nearly the same, it suggests that sleep troubles are unrelated to age. However, if they are very different, then it might mean that some age groups have greater or lower risk of sleep trouble than other age groups. The population distribution of ages, by decade, in the general U.S. population is shown in Figure 10.7b. Recall that although the two distributions look different, the distribution in Figure 10.7a might still be a sample from the population shown in 10.7b, and might differ from it only because of the chance that results from the random sampling. This hypothesis test will help us determine if the difference is too great to be blamed on chance.

▶ **TABLE 10.12** The observed number of people in a random sample who reported sleep troubles to a health professional in the last year. The Expected Frequency shows what this count would be if the distribution of those reporting sleep troubles followed the same distribution as the general U.S. population.

Age Group (years)	Sleep Troubles	Expected Frequency
10–19	63	301.52
20–29	262	301.30
30–39	269	283.33
40–49	387	307.74
50–59	444	296.19
60–69	286	206.48
70+	182	196.45

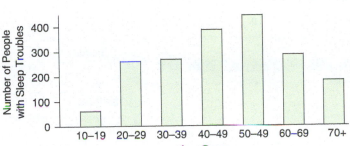

◀ **FIGURE 10.7a** Sample distribution of sleep troubles by age group.

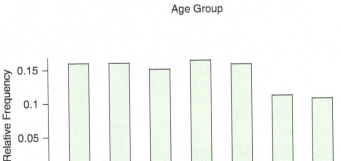

◀ **FIGURE 10.7b** Population distribution of the ages for the U.S. population. (Source: CIA Factbook and Wikipedia.)

QUESTION   Test whether the observed distribution fits the distribution of ages in the United States using a significance level of 0.05. (In other words, is the distribution in Figure 10.7a the type of sample distribution we get when sampling from the population whose distribution is shown in Figure 10.7b?)

SOLUTION   The *Age Group* variable has 7 categories. By summing the counts in Sleep Troubles, we see that there are 1893 people who reported sleep trouble.

Step 1: Hypothesize
We convert our question into a pair of hypotheses:

   H_0: the distribution of ages of those reporting sleeping problems follows that of the
      general U.S. population
   H_a: the distribution of ages of those reporting sleeping problems does not follow the
      general U.S. population.

Step 2: Prepare
Our goal is to compare an observed distribution (the distribution of ages based on those who reported sleeping troubles) with a theoretical distribution (the distribution of ages of all people in the U.S. population who are older than 10). Therefore, the chi-square goodness of fit test is appropriate. However, we must check the conditions to make sure that we can find the sampling distribution of the test statistic.

   In order for the chi-square distribution to be a good approximation to the true sampling distribution of the chi-square statistic, we need to have a random sample (which we are told we have), independent observations (which is a highly plausible assumption), and a large enough sample size. Because Table 10.12 assures us we have more than 5 expected counts in each age category, we have a large enough sample size.

Step 3: Compute to compare
As directed, we will use a 5% significance level. To carry out the analysis step, we use technology. Figure 10.8 shows both the data as entered in StatCrunch and the output from the goodness-of-fit test.

   We see that the chi-square statistic is 320.3922. This is a very large value. The small p-value (less than 0.0001) tells us that if it were true that the age distribution of people with sleep troubles is the same as the general population, then seeing a sample

 Looking Back

p-Value
Recall that the p-value is the probability of getting a test statistic as extreme as or more extreme than what was observed, assuming that the null hypothesis is true. For the chi-square statistic, "as extreme as or more extreme than" means "as big as or bigger than."

▶ **FIGURE 10.8** Data to calculate the goodness-of-fit test, and also the output from that test, using StatCrunch.

Data Moves ▶ The data come from the "NHANES" package within the statistical software R (Pruim, 2015). However, this dataset provided the ages as integers, not categories (e.g., 34 instead of "30–39"). Also, it did not provide data about the age distribution in the general U.S. population. For this reason, we "scraped" the age distribution from Wikipedia using the *StatCrunch This* feature. There, the ages were reported by decade, as they are in this example. We then created a new variable for the NHANES data that grouped the ages by decades. This allowed us to merge the two data sets together based on the *age.group* variable. Merging is a useful tool for enhancing the context of a given data set.

Data ▶ example10.3.csv

Row	Age group	Sleep troubles	Expected frequency	var4	var5	var6	var7	var8	var9
1	10-19	63	301.52						
2	20-29	262	301.3						
3	30-39	269	283.33						
4	40-49	387	307.74						
5	50-59	444	296.19						
6	60-69	286	206.48						
7	70+	182	196.45						

Options

Chi-Square goodness-of-fit results:
Observed: Sleep troubles
Expected: Expected frequency

N	DF	Chi-Square	P-value
1693	6	320.40023	<0.0001

Observed	Expected
63	301.51841
262	301.29841
269	283.3285
387	307.73837
444	296.18844
286	206.47891
182	196.44896

distributed like the one we got is extremely rare. Put differently, if the null hypothesis is correct, then we should be very surprised to see a distribution such as this.

Note that the number of degrees of freedom (labeled "DF") in Figure 10.8 is the number of categories, 7, minus 1.

Step 4: Interpret
Because of the small p-value (much less than our significance level of 0.05), we reject the null hypothesis.

CONCLUSION The distribution of ages of those with sleeping trouble differs from the distribution of ages of the general population. There is likely an association between age and having trouble sleeping.

TRY THIS! Exercise 10.17

SNAPSHOT ▶ Chi-Square Test for Goodness of Fit

WHAT IS IT? ▶ A test of whether the distribution of a single categorical variable follows a proposed distribution.

WHAT DOES IT DO? ▶ Using the chi-square statistic, it compares the observed counts with the counts we would expect if the proposed distribution were the true distribution.

HOW DOES IT DO IT? ▶ If the sample size is large enough and basic conditions are met, the chi-square statistic follows a chi-square distribution with df = (number of categories − 1). The p-value is the probability of getting a value as large as or larger than the observed chi-square statistic value, using the chi-square distribution.

HOW IS IT USED? ▶ To compare a single categorical variable to a theoretical distribution.

Note: The chi-square statistic is designed so that it ranges from 0 to potentially very large positive values. If the null hypothesis is true, the chi-square statistic will be close to 0. If the alternative hypothesis is true, the chi-square statistic should be big. For these reasons, when finding the p-value, we always use the area to the right of the observed test statistic.

SECTION 10.3

Chi-Square Tests for Associations between Categorical Variables

There are two tests to determine whether two categorical variables are associated. Which test you use depends on how the data were collected.

We often summarize two categorical variables in a two-way table, in part because it helps us see whether associations exist between these variables. When we examine a two-way table (or any data summary, for that matter), one aspect that gets hidden is the method used to collect the data. We can collect data that might appear in a two-way table in either of two ways.

The first method is to collect two or more distinct, independent samples, one from each population. Each object sampled has a categorical value that we record. For example, we could collect a random sample of men and a distinct random sample of women. We could then ask them to what extent they agree with the statement that same-sex marriage should be allowed: Strongly Agree, Agree, Neutral, Disagree, or Strongly Disagree. We now have one categorical response variable, *Opinion*. We also have another categorical variable, *Gender*, that keeps track of which population the response belongs to. Hence we have two samples, one categorical response variable, and one categorical grouping variable.

The second method is to collect just one sample. For the objects in this sample, we record two categorical response variables. For example, we might collect a large sample of people and record their marital status (single, married, divorced, or widowed) and their educational level (high school, college, graduate school). From this one sample, we get two categorical variables: *Marital Status* and *Educational Level*.

In both data collection methods, we are interested in knowing whether the two categorical variables are related or unrelated. However, because the data collection methods are different, the ways in which we test the relationship between variables differ also. That's the bad news. The good news is that this difference is all behind the scenes. The result of careful calculation shows us that no matter which method we use to collect data, we can use the same chi-square statistic and the same chi-square distribution to test the relation between variables.

These two methods have different names. If we test the association, based on two samples, between the grouping variable and the categorical response variable (the first method), the test is called a test of **homogeneity**. If we base the test on one sample (the second method), the test is called a test of **independence**. Two different data collection methods, two different names—but, fortunately, the same test! The Consider Data phase of the Data Cycle is important for correctly identifying the hypothesis test. It's also important, of course, for verifying that the conditions of the test are satisfied.

 Details

Homogeneity
The word *homogeneity* is based on the word *homogeneous*, which means "of the same, or similar, kind or nature."

> **KEY POINT** There are two tests to determine whether two categorical variables are associated. For two or more samples and one categorical response variable, we use a test of homogeneity. For one sample and two categorical response variables, we use a test of independence.

EXAMPLE 4 Independence or Homogeneity?

Perhaps you've taken an online class or two? What are your feelings about the value of online classes? The Pew Foundation surveyed two distinct groups: the general American public and presidents of colleges and universities. About 29% of the sample from the public said that online courses "offer an equal value compared with courses taken in the classroom." By comparison, over half of the sample of college presidents felt that online courses and classroom courses had the same value. The Pew researchers carried out a hypothesis test to determine whether the variables *Online Course Value* (which recorded whether a respondent agreed or disagreed that online courses were equivalent to classroom courses) and *President* (which recorded whether or not a respondent was the president of a college or university) were associated).

QUESTION Will this be a test of independence or of homogeneity?

SOLUTION The Pew Foundation took two distinct samples of people: presidents and the public. The variable *President* simply tells us which group a person belongs to. There is only one response variable: *Online Course Value*.

CONCLUSION This is a test of a homogeneity.

TRY THIS! Exercise 10.27

EXAMPLE 5 Independence or Homogeneity?

Do movie critics' opinions align with the public's? Even though the answer may be an obvious "no" to you, it is nice to know that we can examine data to answer this question. We took a random sample of about 200 movies from the Rotten Tomatoes website. Rotten Tomatoes keeps track of movie reviews and summarizes the general opinion about each movie. The critics' summary opinion is given as (from Best to Worst) "Certified Fresh," "Fresh," or "Rotten." The audience general opinions are classified as "Upright" (which is good) or "Spilled" (which is bad). Thus, for each movie in our sample, we have two variables: *Critics' Opinion* and *Audience Opinion*.

QUESTIONS If we test whether there is an association between *Critics' Opinion* and *Audience Opinion*, is this a test of homogeneity or of independence?

CONCLUSION There is one sample, consisting of about 200 movies. Each movie provides two responses: a critics' opinion and an audience opinion. This is a test of independence.

TRY THIS! Exercise 10.29

Tests of Independence and Homogeneity

Again, the tests follow the four-step procedure of all hypothesis tests. We'll give you an overview and then fill in the details with an example.

Step 1: Hypothesize
The hypotheses are always the same.

> H_0: There is *no* association between the two variables (the variables are independent).
>
> H_a: There is an association between the two variables (the variables are not independent).

Although the hypotheses are always the same, you should phrase these hypotheses in the context of the problem. For example, our null hypothesis for the movie ratings is that there is no association between critics' opinions and the public's opinion. The alternative hypothesis is that there is an association.

Step 2: Prepare

Whether you are testing independence or homogeneity, the test statistic you should use to compare counts is the chi-square statistic, shown in Formula 10.2 and repeated here.

$$X^2 = \sum_{\text{cells}} \frac{(O - E)^2}{E}$$

If the conditions are right, then this statistic follows, approximately, a chi-square distribution with

$$df = (\text{number of rows} - 1)(\text{number of columns} - 1)$$

Conditions:

1. *Random Samples.* All samples were collected randomly.

2. *Independent Samples and Observations.* All samples are independent of each other. Always, the observations within a sample must be independent of each other.

3. *Large Samples.* The expected count must be 5 or more in each cell.

Note that in a test of independence, there is always only one sample. But a test of homogeneity might have several independent samples.

Step 3: Compute to compare

Begin by stating a significance level. The rest is best done with technology. The p-value is the probability that, assuming the null hypothesis is true, we could get a value as large as or larger than the observed chi-square statistic. In other words, the p-value is the probability that, if the variables really are not associated, we would see a test statistic as large or larger than the one observed. A small p-value therefore means a large test statistic, which casts doubt on the hypothesis that the variables are not associated. Using technology, we found that the chi-square statistic for the movie data was 69.8. The p-value was less than 0.001.

Step 4: Interpret

If the p-value is less than or equal to the stated significance level, we reject the null hypothesis and conclude that the variables are associated. Because the p-value for the movie data is so small, we would reject the null hypothesis and conclude that there was an association between critics and the public's opinions. You might enjoy looking at the data, provided on this text's website, to see whether the association is negative (critics tend to disagree with the audience) or positive (critics and audience agree).

EXAMPLE 6   Confidence and Climate Change?

Many, perhaps all, people in the United States have opinions about whether climate change is happening. But people also differ in how strongly they hold this opinion. Researchers interested in how people thought about climate change randomly sampled almost 2000 U.S. adults and asked them their opinions about a variety of climate-related issues, including their opinion on climate change ("Climate change is happening"and "Climate change is not happening") and how certain they were of this opinion ("Extremely Sure," "Not at All Sure," "Somewhat Sure," "Very Sure"). (Bloodhart 2015)

QUESTIONS   Is this a test of homogeneity or independence? Is confidence in opinion associated with the actual opinion?

SOLUTION   Because there is one sample with two response variables, this is a test of independence.

Step 1: Hypothesize

Our statistical question can be rephrased in terms of these hypotheses:

H_0: Among all U.S. residents, confidence about climate change is independent of the opinion itself.

H_a: Confidence in opinion about climate change depends on the opinion.

Step 2: Prepare

Examining the data, we can see from the output that all expected counts are 5 or greater. (The smallest expected count is 16.58.) This means that the chi-square distribution will provide a good approximation for the p-value. The data are summarized in Figure 10.9. This table displays the actual counts for this random sample, the expected counts if the null hypothesis is true, and the calculated values for the chi-square statistic and p-value.

▶ **FIGURE 10.9** StatCrunch output showing the summary table with expected counts, chi-square statistic, and p-value to test whether opinion in climate change is associated with confidence in that opinion.

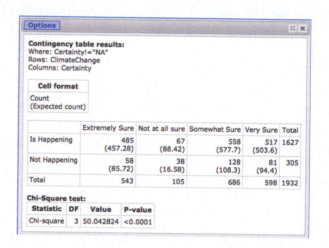

Step 3:

We will use a significance level of 0.05.

The chi-square statistic is $X^2 = 50.04$ (rounding to two decimal digits.) The chi-square distribution has 3 degrees of freedom, and the p-value is reported as smaller than 0.0001.

Step 4: Interpret

Because the p-value is less than the significance level of 0.05, we reject the null hypothesis.

CONCLUSION Opinion and Confidence in the opinion are associated.

TRY THIS! Exercise 10.33

Details

Random Assignment
Experiments that randomly assign subjects to treatment groups result in distinct, independent samples, so if other conditions are satisfied, they can be analyzed with tests of homogeneity.

EXAMPLE 7 Hungry Monkeys

Research in the past has suggested that mice and rats that are fed less food live longer and healthier lives. More recently, a study of rhesus monkeys was carried out that involved caloric restriction (less food). It is believed that monkeys have many similarities to humans, which is what makes this study so interesting.

Seventy-six rhesus monkeys, all young adults, were randomly divided into two groups. Half of the monkeys (38) were assigned to caloric restriction. Their food was decreased about 10% per month for three months.

For those on the normal diet, 14 out of 38 had died of age-related causes by the time the article was written. For those on caloric restriction, only 5 out of 38 had died of age-related causes (Colman et al. 2009).

QUESTION Because this is a randomized study, the hypothesis that diet is associated with aging can be stated as a cause-and-effect hypothesis. Therefore, test the hypothesis that diet causes differences in aging. Will this be a test of homogeneity or of independence? Minitab output is shown in Figure 10.10.

SOLUTION There are two samples (monkeys with caloric restriction and monkeys without) and one outcome variable: whether the monkey died of age-related causes. Therefore, this is a test of homogeneity.

Step 1: Hypothesize
H_0: For these monkeys, the amount of calories in the diet is independent of aging.
H_a: For these monkeys, the amount of calories in the diet causes differences in aging.

Because we do not have a random sample, our results do not generalize beyond this group of monkeys. But because we have randomized assignment to treatment groups, we are able to conclude that any differences we see in the aging process are caused by the diet.

Step 2: Prepare
Because we wish to use the chi-square test, we must confirm that the expected counts are all 5 or more. This is the case here, as Figure 10.10 confirms.

Step 3: Compute to compare
From the Minitab output, Figure 10.10, you can see that the chi-square value is 5.68 and the p-value is 0.017.

Step 4: Interpret
Because the p-value is less than 0.05, we can reject the null hypothesis.

CONCLUSION Changing the calorie content of these monkeys' diets caused them to live longer.

TRY THIS! Exercise 10.39

Tabulated Statistics: Died, Diet

Rows: Died   Columns: Diet

	Normal	Restricted	All
Yes	14	5	19
	9.500	9.500	
No	24	33	57
	28.500	28.500	
All	38	38	76

Cell Contents
   Count
   Expected count

Chi-Square Test

	Chi-Square	DF	P-Value
Pearson	5.684	1	0.017

▲ **FIGURE 10.10** Minitab output showing *Diet* (normal or caloric) and *Aging* (died from age-related causes or not) for 76 monkeys.

> 🔁 **Looking Back**
>
> **Comparing Two Proportions**
> In Section 8.4, you learned to test two population proportions based on data from independent samples using the z-test. The test of diet in monkeys could have been done using this test. The chi-square test provides an alternative, but equivalent, approach.

The research article contains other information that suggests that monkeys on a restricted diet are generally healthier than monkeys on a normal diet.

Random Samples and Randomized Assignment

You have now seen randomization used in two different ways. Random sampling is the practice of selecting objects in our sample by choosing them at random from the population, as is done in many surveys. We can make generalizations about the population only if the sample is selected randomly, because this is the only way of ensuring that the sample is representative of the population. The survey in Example 6 is a study based on random sampling. When we conclude, as we did in Example 6, that opinions about climate change and confidence in option are associated, we are stating a conclusion about the entire population. We are confident that these variables are associated in the population because our sample was selected at random.

In Example 7, on the other hand, there was no random sample. However, the monkeys were randomly assigned to a treatment group (low-calorie diet) or the control group (normal diet). Because the monkeys were not selected randomly, we have no means of generalizing about the population as a whole, statistically speaking. (There might be a biological argument, or an assumption, that a diet that works on one group of monkeys would work on any other group, but as statisticians, we have no data to support this assumption.) However, the researchers performed this study because they were interested in a cause-and-effect relationship: Does changing the calories in a monkey's diet change the monkey's health and longevity?

Because researchers controlled which monkeys got which diet, this is a controlled experiment. And because they used randomized assignment and we rejected the null hypothesis that diet and health were independent, we can conclude that in fact the caloric restriction *did* affect the monkeys' health.

Pay attention to how the data were collected. If they were randomly sampled, we can make generalizations about the population. If they were obtained by randomly assigning subjects to treatment groups, then we may be able to draw cause-and-effect conclusions.

SNAPSHOT ▶ Chi-Square Test of Independence and Homogeneity

WHAT IS IT? ▶	A test of whether two categorical variables are associated.
WHAT DOES IT DO? ▶	Using the chi-square statistic, we compare the observed counts in each outcome category with the counts we would expect if the variables were *not* related. If observations are too different from expectations, then the assumption that there is no association between the categorical variables looks suspicious.
HOW DOES IT DO IT? ▶	If the sample size is large enough and basic conditions are met, then the chi-square statistic follows, approximately, a chi-square distribution with df = (number of rows − 1) × (number of columns − 1). The p-value is the probability of getting a value as large as or larger than the observed chi-square statistic, using the chi-square distribution.
HOW IS IT USED? ▶	To compare distributions of two categorical variables.

Relation to Tests of Proportions

In the special case in which both categorical variables have exactly two categories, the test of homogeneity is identical to a z-test of two proportions, using a two-sided alternative hypothesis. The following analysis illustrates this.

In a landmark study of a potential AIDS vaccine, published in 2009, researchers from the U.S. Army and the Thai Ministry of Health randomly assigned about 8200 volunteers to receive a vaccine against AIDS and another 8200 to receive a placebo. (We rounded the numbers slightly to make this discussion easier.) Both groups received counseling on AIDS prevention measures and were promised lifetime treatment should they contract AIDS. Of those who received the vaccine, 51 had AIDS at the end of the study (three years later). Of those that received the placebo, 74 had AIDS (http://www.hivresearch.org/, accessed September 29, 2009). We will show two ways of testing whether an association existed between receiving the vaccine and getting AIDS. The data are summarized in Table 10.13.

▶ **TABLE 10.13** Summary of data from AIDS vaccine experiment.

	Vaccine	No Vaccine	Total
AIDS	51	74	125
No AIDS	8,149	8,126	16,275
Total	8,200	8,200	16,400

If we use the approach of this chapter, we would recognize that this is a test of homogeneity, because there are two samples (*Vaccine* and *Placebo*) and one outcome variable (*AIDS*). Although we cannot generalize to a larger population (because the volunteers were not randomly selected), we can make a cause-and-effect conclusion about whether differences in AIDS rates are due to the vaccine, because this is a controlled, randomized study.

As a first step, we calculate the expected counts, under the assumption that the two variables are not associated.

Because the proportion of those who got AIDS was $125/16400 = 0.007622$, if the risk of getting AIDS had nothing to do with the vaccine, then we should see about the same proportion of those getting AIDS in both groups. If the proportion of people who got AIDS in the *Vaccine* group was 0.007622, then we would expect $8200 \times 0.007622 = 62.5$ people to get AIDS in the *Vaccine* group.

Both groups are the same size, so we would expect the same number of AIDS victims in the *Placebo* group. This means that in both groups, we would expect $8200 - 62.5 = 8137.5$ not to get AIDS.

The results, with expected counts in parentheses to the right of the observed counts, are shown in Table 10.14.

	Vaccine	No Vaccine	Total
AIDS	51 (62.5)	74 (62.5)	125
No AIDS	8,149 (8137.5)	8,126 (8137.5)	16,275
Total	8,200	8,200	16,400

◀ **TABLE 10.14** Expected counts, assuming no association between variables, are shown in parentheses.

Note that all expected counts are much greater than 5. The chi-square statistic is not difficult to calculate:

$$X^2 = \sum \frac{(\text{Observed} - \text{Expected})^2}{\text{Expected}}$$

$$= \frac{(51 - 62.5)^2}{62.5} + \frac{(74 - 62.5)^2}{62.5} + \frac{(8149 - 8137.5)^2}{8137.5} + \frac{(8126 - 8137.5)^2}{8137.5}$$

$$= 4.26$$

The degrees of freedom of the corresponding chi-square distribution is

(Number of rows $- 1$)(number of columns $- 1$) $= (2 - 1)(2 - 1) = 1 \times 1 = 1$

The p-value is illustrated in Figure 10.11. It is the area under a chi-square distribution with 1 degree of freedom and to the right of 4.26. The p-value turns out to be 0.039. We therefore reject the null hypothesis and conclude that there is an association between getting the vaccination and contracting AIDS. The difference in the numbers of AIDS victims was caused by the vaccine.

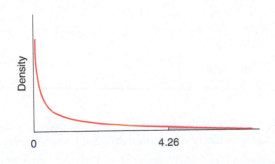

◀ **FIGURE 10.11** The area to the right of 4.26 represents the p-value to test whether there is an association between receiving the AIDS vaccine and contracting AIDS. The distribution is a chi-square distribution with 1 degree of freedom. The p-value is 0.039.

One thing that's disappointing about this conclusion is that the alternative hypothesis states only that the variables are associated. That's nice, but what we really want to know is *how* they are associated. Did the vaccine decrease the number of people who got AIDS? That's what the researchers wanted to know. They didn't want to know merely whether there was an association. They had a very specific direction in mind for this association.

One drawback with chi-square tests is that they reveal only whether two variables are associated, not *how* they are associated. Fortunately, when both categorical variables have only two categories, we can instead do a two-proportion *z*-test.

By doing a two-proportion *z*-test, we can test for the *direction* of the effect: whether the vaccine improved AIDS infection rates. We'll use Figure 10.12, which shows Stat-Crunch input and output, to test this hypothesis using a two-proportion *z*-test.

To begin, we define p_1 to be the proportion of those who received the vaccine and developed AIDS and p_2 to be the proportion of those who received the placebo and developed AIDS.

► FIGURE 10.12 StatCrunch enables us to carry out a two-proportion *z*-test by entering the summary information as shown in part **(a)**. The resulting output is shown in part **(b)**.

(a)

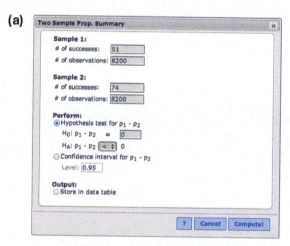

(b)

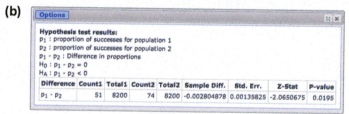

Step 1: Hypothesize

$$H_0: p_1 - p_2 = 0 \ (or \ \ p_1 = p_2)$$
$$H_a: p_1 - p_2 < 0 \ (or \ \ p_1 < p_2)$$

In words, our null hypothesis is that the proportion of AIDS victims will be the same in the vaccine and the placebo groups. The alternative hypothesis states that the proportion of AIDS victims will be lower in the vaccine group.

Step 2: Prepare

The conditions are satisfied for a two-sample proportion test. (We'll leave it to you to confirm this, but it is important to note that because of the random assignment—the vaccine was randomly assigned to some subjects and a placebo to others—the two samples are independent of each other.)

Step 3: Compute to compare

The observed value of the *z*-statistic is -2.07, with a p-value of 0.0195.

Step 4: Interpret

Based on the p-value, we reject the null hypothesis and conclude that people who receive the vaccine are less likely to contract AIDS than those who do not. Because this was a randomized, controlled study, we can conclude that the vaccine caused a decrease in the probability of people getting AIDS.

Note that if we had doubled the p-value—that is, if we had instead used a two-sided alternative hypothesis—we would have got $2 \times 0.019458 = 0.039$, exactly what we got for the chi-square test.

Although medical researchers were very excited about this study—not too many years before, it was thought to be impossible to develop an AIDS vaccine—they caution that this vaccine lowers risk by only about 30%. Many medical professionals consider a vaccine to be useful only if it lowers risk by at least 60%. There was also some controversy over this study: Some argued that a few of the subjects who were dropped from the initial analysis (because of preexisting medical conditions) should have been included. If these subjects had been included, then the vaccine would no longer have been judged effective, statistically speaking.

> **KEY POINT**
> For a 2-by-2 contingency table of counts, a two-proportion z-test with a two-sided alternative is equivalent to a test of homogeneity.

When should you use the two proportion z-test, and when should you use the test for homogeneity? If you need to use a one-sided alternative hypothesis, then you should use the z-test. However, if you plan to use a two-sided alternative hypothesis, then it doesn't matter which test you use.

SECTION 10.4

Hypothesis Tests When Sample Sizes Are Small

We demonstrate two approaches for dealing with data in which the expected counts are less than 5. One approach is to combine categories so that each of the new, larger categories has an expected count of 5 or more. The other approach is to use a test called Fisher's Exact Test. Unlike the chi-square approach, which gives us approximate p-values, Fisher's Exact Test gives the actual p-value.

Combining Categories

The next example illustrates how to combine categories with small expected counts so that you can do a chi-square test.

EXAMPLE 8 Swine Flu Hospitalizations

In 2009, the United States (and much of the rest of the world) was hit by an epidemic of "swine flu" (the H1N1 virus). One characteristic of swine flu was that older people were less susceptible than younger people, possibly because of immunity they gained during previous flu seasons with related viruses. But older people, though less likely to catch the virus, can experience more severe symptoms. To gain some insight as to whether the severity of symptoms from the swine flu varied with age, we look at rates of hospitalization for swine flu in different age categories (http://www.cdc.gov/mmwr). Different hospitalization rates might suggest that the severity of the flu is different by age group.

Table 10.15 looks at the relation between the variables *Age Category* and *Hospitalized* for all swine flu victims in the United States (before June 11, 2009). Presumably, hospitalization occurs when the victim has a more severe form of the illness.

▶ **TABLE 10.15** Summary of swine flu cases in the United States from May 1, 2009, to June 11, 2009.

		Age Category						
		Under 5	5–14	15–29	30–44	45–60	over 60	Totals
Hospitalized?	**Yes**	7	9	9	9	1	0	35
	No	44	195	241	59	35	10	584
	Totals	51	204	250	68	36	10	619

QUESTION Test the hypothesis that hospitalization for swine flu is associated with age category, using a significance level of 0.05. Use the Minitab output provided in Figure 10.13.

SOLUTION This is a test of independence. There are one sample (swine flu victims) and two response variables (*Age Category* and *Hospitalized*).

Step 1: Hypothesize

H₀: Age and whether a swine flu victim is hospitalized are independent.
Hₐ: Age and whether a swine flu victim is hospitalized are associated.

Step 2: Prepare

Our first choice would be a chi-square test of independence. However, several of the expected values shown in Figure 10.13 are less than 5. Minitab even alerts us to this fact (as most, but not all, packages do). For emphasis, we highlighted this warning and the low expected values in red ink.

▶ **FIGURE 10.13** Minitab output for swine flu hospitalizations, showing low expected counts. "Yes" means the people were hospitalized, and "No" means they were not hospitalized. (Note that Minitab also warns us if the expected count is less than 1.)

Chi-Square Test for Association: Hospitalized, Age Category

Rows: Hospitalized   Columns: Age Category

	Under 5	5 - 14	15 - 29	30 - 44	45 - 60	Over 60	All
Yes	7	9	9	9	1	0	35
	2.88	11.53	14.14	3.84	2.04	0.57	
No	44	195	241	59	35	10	584
	48.12	192.47	235.86	64.16	33.96	9.43	
All	51	204	250	68	36	10	619

Cell Contents
  Count
  Expected count

Chi-Square Test

	Chi-Square	DF
Pearson	17.280	5
Likelihood Ratio	14.707	5

1 cell(s) with expected counts less than 1.
Chi-Square approximation probably invalid.
4 cell(s) with expected counts less than 5.

One solution, which can be a little unsatisfying, is to combine categories so that each expected count is 5 or more. For example, you can see that if we combine the "Under 5" and "5–14" categories, we get a new category, "Under 15," with expected count $2.88 + 11.53 = 14.41$. Similarly, we can combine the upper three categories into a new category of "30 and older," with expected count $3.84 + 2.04 + 0.57 = 6.45$.

The result is shown in Figure 10.14. We now have only three age categories, but the expected counts are large enough for us to use the chi-square distribution to find a good approximation of the p-value.

Chi-Square Test for Association: Hospitalized, Age Category

Rows: Hospitalized   Columns: Age Category

	Under 15	15 - 29	30 and older	All
Yes	16	9	10	35
	14.42	14.14	6.45	
No	239	241	104	584
	240.58	235.86	107.55	
All	255	250	114	619

Cell Contents
   Count
   Expected count

Chi-Square Test

	Chi-Square	DF	P-Value
Pearson	4.239	2	0.120
Likelihood Ratio	4.227	2	0.121

◄ **FIGURE 10.14** Minitab output with groups combined so that none of the expected counts is less than 5. "Yes" means the people were hospitalized, and "No" means they were not hospitalized.

Step 3: Compute to compare
Chi-square = 4.24, p-value = 0.120

Step 4: Interpret
Do not reject H_0.

CONCLUSION   With these limited number of categories, there is not enough evidence to conclude that there is an association between age and hospitalization.

TRY THIS!   Exercise 10.49

Advantages and Disadvantages of Combining Categories   The biggest advantage of combining categories is that you are able to use a procedure, the chi-square test, that otherwise would give you very inaccurate p-values and might therefore lead to the wrong conclusion. Had we not combined cells, the p-value for the chi-square test shown in Figure 10.14 would have been 0.004. We would have rejected the null hypothesis and concluded that there was an association. But this conclusion would have been based on a potentially very inaccurate p-value.

The disadvantage is that our knowledge becomes more crude. Having an age category of "Under 15" is somewhat disappointing, because we know that children change a lot between birth and their early teenage years. The same can be said of those "30 and older." Thirty-year-olds should have very different immune systems than do 90-year-olds. Unfortunately, using the Chi-square approach, we cannot say more without more data. (However, given that "more data" means more swine flu victims, it is not necessarily a bad thing that we can't tell more about this association.)

Fisher's Exact Test

Another approach that works in some cases is **Fisher's Exact Test**. This test is called exact because in many situations we can find the exact p-value, not the approximate p-value found by the chi-square test. One price we pay is that this method is not widely implemented in statistical software packages. Even when it is, it is often implemented in software only for two-way contingency tables—that is, when there are only two outcomes for both variables.

To help you visualize Fisher's Exact Test, we've chosen a study based on a small sample size. Children who were suffering from (nonlethal) scorpion stings were randomly assigned to receive either an antivenom or a placebo. After several hours, the investigators recorded whether the children showed improvement. (The antivenom was new and was not known to be effective.) As you might expect, there were not many such children coming through the emergency room during the time in which this study was conducted. Table 10.16 summarizes the results for the 15 children who participated in the study.

▶ **TABLE 10.16** Summary of data on children with scorpion stings.

	Antivenom	Placebo	Total
No Improvement	1	6	7
Improvement	7	1	8
Total	8	7	15

At first glance, the outcome looks favorable for the antivenom: 7 out of 8 of those who received the antivenom improved, and only 1 out of 7 who received the placebo improved. However, with such small sample sizes, might such a seemingly favorable outcome be due to chance?

Fisher's Exact Test asks you to imagine a parallel world similar to, but different from, the world the researchers observed. This parallel world is similar in that it also has 15 children with scorpion stings. It is also similar in that 8 of the children received the antivenom, and 8 children improved. The parallel world is different, though, in that there is *no* association between the treatment and the outcome: The results are determined purely by chance. It is (almost) as though you were blindfolded and throwing darts at Table 10.16. Say you throw 8 darts at the Antivenom side of the table, and some of them land randomly in the "No Improvement" space, and some of them land randomly in the "Improvement" space.

If you happened to throw 4 darts into the "No Improvement" space (and so 4 into the "Improvement" space), the outcome might look like this:

	Antivenom	Placebo	Total
No Improvement	4		7
Improvement	4		8
Total	8	7	15

We have left the totals in the rows and columns unchanged because, in this parallel world, the numbers here are the same as in the real world. But note that once you've thrown your darts at the Antivenom squares, the rest of the table can be completed in only one way:

	Antivenom	Placebo	Total
No Improvement	4	3	7
Improvement	4	4	8
Total	8	7	15

This particular outcome doesn't look so good for the antivenom: Only half of the children who received it improved.

Another possible outcome in the parallel world is this:

	Antivenom	Placebo	Total
No Improvement	6	1	7
Improvement	2	6	8
Total	8	7	15

This outcome looks even worse for the antivenom, whereas the following looks quite good for the antivenom:

	Antivenom	Placebo	Total
No Improvement	0	7	7
Improvement	8	0	8
Total	8	7	15

In short, there are many different tables we might see in this parallel world, just by chance. Of course, some are more likely than others. Some of them suggest that the antivenom may be successful; some of them do not. Mathematically, we can calculate the p-value by figuring out the probability of each table in the parallel world, and then calculating the probability of getting an outcome as extreme as or more extreme than the 7/8 improved that the researchers actually saw. (This p-value is found using something called the hypergeometric distribution.) For larger sample sizes, this probability calculation can be approximated with a simulation in which the computer randomly "throws" 8 "darts" at the antivenom side of the table and then repeats this a great many times to see how often the outcome is 7/8 or 8/8.

Figure 10.15 shows StatCrunch input and output for the Fisher's Exact Test and the equivalent Minitab output. Each gives us the p-value of 0.0101. Based on this, we would reject the null hypothesis and conclude that the antivenom was effective.

(a)

Row	Antivenom	Placebo	Outcome	var4	var5
1	1		6 No Improvement		
2	7		1 Improvement		
3					

Options

Contingency table results:
Rows: Outcome
Columns: None

	Antivenom	Placebo	Total
No Improvement	1	6	7
Improvement	7	1	8
Total	8	7	15

Fisher's exact test:
P-value = 0.0101

(b)   Tabulated Statistics: Response, Treatment

Rows: Response   Columns: Treatment

	Antivenom	Placebo	All
No Improvement	1	6	7
Improvement	7	1	8
All	8	7	15

Cell Contents
   Count

Fisher's Exact Test

P-Value
0.0101010

◄ **FIGURE 10.15** There are several statistical packages that will do a basic Fisher's Exact Test, based either on the original data or on a contingency table. The chi-square test provides only an approximate p-value, but Fisher's value is exact. **(a)** StatCrunch summary data and output and **(b)** Minitab output.

If we were to perform this same test as a chi-square test, then many software packages would give us a warning that our sample sizes were too small and that, therefore, the p-value may be inaccurate. If you ignore the warning and carry out a chi-square test anyways, then for these data, the resulting p-value is too small: 0.0046. This too-small value might give us a misleading sense of confidence in our results.

> **Details**
>
> **Larger Tables**
> Fisher's Exact Text can be used for tables with more than two rows or columns, although there are some technical considerations that are beyond an introductory course.

We can, of course, use Fisher's Exact Test even when the sample sizes are large enough to satisfy the chi-square test conditions. For instance, we can carry out a Fisher's Exact Test for the monkey data to determine whether there is an association between the monkeys' diets (normal or restricted) and their survival. In Example 7 we found that an approximate p-value based on the chi-square test was 0.017. Although the expected counts were all greater than 5, they were not that much greater (the smallest was 9.5). Fisher's Exact Test gives the exact p-value as 0.0324, which tells us that, even with the chi-square conditions satisfied, the approximate p-value can still be too small.

CASE STUDY REVISITED

Popping Better Popcorn

To better understand the ideal conditions for popping corn, experimenters designed a randomized, controlled study to observe how well the popcorn popped under different conditions. Of interest here was how the amount of oil affected the outcome. A summary of results was shown in the case study at the beginning of the chapter, where we observed that it looked as though using the maximum amount of oil did not work well, because so few popcorn bags were successful by the criteria adopted for success.

Did the amount of oil affect the resulting quality of the popcorn? To test this using the methods of this chapter, we carry out a test of homogeneity, because we have three independent samples (no oil, medium oil, and maximum oil) and one categorical response variable (*Result*: success or failure). We have three independent samples, because bags were randomly assigned to one of these three groups. Each group was assigned 36 bags of popcorn, and each bag had 50 kernels.

The hypotheses follow:

H_0: The quality of popcorn and the amount of oil are independent.
H_a: The amount of oil affects the quality of the popcorn.

The results (and some of the raw data) are shown in Figure 10.16 in output generated by StatCrunch. The output shows that the expected counts are all greater than 5, so our sample sizes are large enough for the chi-square distribution (with 2 degrees of freedom) to produce a good approximation to the p-value.

▶ **FIGURE 10.16** StatCrunch output shows the results of the analysis in the foreground and the raw data in the background. The expected counts in the table are given below the observed values.

popcornexperiment.txt

Row	brand	time	oil	container	good	result	var7	var8
1	1	0	maximum	1	8	Failure		
2	0	1	medium	0	26	Success		
3	0	0	maximum	1	5	Failure		
4	1	0	none	1	8	Failure		
5	1	0	none	0	7	Failure		
6	1	1	medium	1	20	Failure		
7	1							
8	0							
9	1							
10	0							
11	1							
12	0							
13	0							
14	1							
15	0							
16	0							

Description:
popcorn popped under different randomly
more than 25, then result was deemed a

Source:
Irina A. Kukuyeva and Jean Wang and Yu

Tags:
anova, chi-square, design

Contingency Table with data

Options

Contingency table results:
Rows: result
Columns: oil

Cell format
Count
Expected count

	maximum	medium	none	Total
Failure	33	22	23	78
	26	26	26	
Success	3	14	13	30
	10	10	10	
Total	36	36	36	108

Statistic	DF	Value	P-value
Chi-square	2	10.246154	0.006

From the output, we see that the test statistic has a value of 10.25, with a p-value of 0.006. This is quite small. Certainly it is less than 0.05, so at the 5% significance level we reject the null hypothesis. We conclude that quality is affected by the amount of oil used. (At least, it is if you believe that quality is measured by the number of kernels popped after 75 seconds.)

DATAPROJECT ▸ Stacking Data

OVERVIEW

One of the first things we discussed about data structures, back in Chapter 1, was the difference between stacked and unstacked data. Which method you prefer often depends on your software, which often has very definite preferences! For example, most calculators prefer the unstacked version for data. In this project, we'll explore the Stack and Split data moves, which allow us to move between stacked and unstacked data.

GOAL

2

Convert stacked data sets to unstacked, and vice versa.

PROJECT

3

These data come from a study that attempted to determine whether our sense of smell differs by gender and by our body's position (sitting up vs. lying down; Lundstrom et al. 2006). The table you see below shows you the data allowing us to compare men and women while they were sitting. (The details of the study aren't important for this activity, but it is helpful to know that the larger the value, the stronger the sense of smell.)

Row	man Sitting Up	woman Sitting Up
1	12.5	13.5
2	9.25	13.5
3	5	12.75
4	8.75	12.5
5	12.75	14
6	9	12.75
7	7.5	13.75
8	10	10
9	4.5	12.25
10	12.75	13.5
11	13.5	12.75
12	15.5	13
13	15.5	12.5
14	7.5	9
15	16.75	3.75
16	13	14.75
17	8.25	7.75
18	13	15.75
19		

The advantages of the unstacked format is that we quickly see that the data contain values for both men and women, and we see that there are equal numbers of men and women (18). The disadvantage is that this format "hides" the fact that we really have two variables. One variable is the numerical variable that measures their sense of smell, and the other is a categorical variable that measures their gender. There's another, hidden variable, which records whether the

measurement was taken when they were sitting up or lying down. More on that in the assignment!

Project:

A statistical question we would like to answer with data such as these is, "Do men and women differ in their sense of smell when sitting up?" If your software is capable of finding a confidence interval for the difference of two means, or carrying out the necessary hypothesis tests, then you might be content to keep the data unstacked. However, many packages require stacked data.

Fortunately, stacking is quite simple. Select **Data > Arrange > Stack** and select both columns. Be sure to check the box for "Open in a new data table." See the following figure.

Stack Columns

Select columns:
man Sitting Up
woman Sitting Up

man Sitting Up
woman Sitting Up

ID column:
--optional--

Store labels in:
--optional--

Store values in:
--optional--

Store IDs in:
--optional--

Options:
☑ Open in a new data table

? | Cancel | Compute!

Assignment:

Now is a chance to practice and to see what stacking/splitting can and cannot do:

1. Notice that StatCrunch did not name the new, stacked variables. How can you assign them names using the Stack dialogue box?

2. Now unstack these data by using **Data > Arrange > Split**. What type of variable is required in the "Select column" field: numerical or categorical? How about the "Split by" field?

3. A more complete, stacked dataset is contained in smellsense.csv. Try converting this data set to unstacked. What is lost? (This loss is the reason we prefer stacked data in this textbook.)

CHAPTER REVIEW

KEY TERMS

Contingency table (Two-way table), *510*
Expected counts, *511*

Chi-square statistic, *514*
Chi-square distribution, *517*
Degrees of freedom, *517*

Goodness-of-fit tests, *518*
Homogeneity, *523*

Independence, *523*
Fisher's Exact Test, *533*

LEARNING OBJECTIVES

After reading this chapter and doing the assigned homework problems, you should

- Understand when a goodness-of-fit test is needed and appropriate and know how to perform the test and interpret results.

- Distinguish between tests of homogeneity and tests of independence.

- Understand when it is appropriate to use a chi-square statistic to test whether two categorical variables are associated and know how to perform this test and interpret the results.

- Know when and how to perform Fisher's Exact Test.

SUMMARY

We presented three types of hypotheses we can test when analyzing categorical variables. The goodness-of-fit test is used to compare a proposed distribution for one categorical variable with the observed sample distribution. Although the test of homogeneity is conceptually different from the test of independence, they are exactly the same in terms of the calculations required. Both of these tests attempt to determine whether two categorical variables are associated. The only difference is in the way the data for the study were collected. When researchers collect two or more independent samples and measure one categorical response variable, they are performing a test of homogeneity. When instead they collect one sample and measure two categorical response variables, they are performing a test of independence.

All three tests rely on the chi-square statistic. For each cell of a two-way summary table, we compare the observed count with the count we would expect if the null hypothesis were true. If the chi-square statistic is big, it means that these two counts don't agree, and it discredits the null hypothesis.

An approximate p-value is calculated by finding the area to the right of the observed chi-square statistic using a chi-square distribution. To do this, you need to know the degrees of freedom for the chi-square distribution, and this depends on which test you are using.

If the sample size is too small to use the chi-square test, you might consider combining categories. Fisher's Exact Test is another alternative you can try.

Formulas

Expected Counts

Formula 10.1: Expected count for a cell $= \dfrac{\text{(row total)} \times \text{(column total)}}{\text{grand total}}$

Chi-Square Statistic

Formula 10.2: $X^2 = \sum\limits_{\text{cells}} \dfrac{(O - E)^2}{E}$

Goodness-of-Fit Test

Hypotheses

H_0: The true distribution is the same as the proposed distribution.
H_a: The true distribution is different from the proposed distribution.

Conditions

1. *Random Sample.* The sample was collected randomly.
2. *Independent Measurements.* Each measurement on an individual is independent of all other measurements.
3. *Large Sample.* The expected count is at least 5 in each category.

Sampling Distribution

If conditions hold, then the sampling distribution of this statistic follows a chi-square distribution with degrees of freedom = (number of categories − 1).

Test of Homogeneity and Independence

Hypotheses

H_0: The variables are independent.
H_a: The variables are associated.

Conditions (Homogeneity)

1. *Random Samples.* Two or more samples that are all sampled randomly.
2. *Independent Samples and Observations.* Samples are independent of each other. The observations within each sample are independent.
3. *Large Samples.* At least 5 expected counts in each cell of the summary table.

Conditions (Independence)

1. *Random Sample.* One sample, selected randomly.
2. *Independent Observations.* Observations are independent of each other.
3. *Large Sample.* There are at least 5 expected counts in each cell of the summary table.

Sampling Distribution

If conditions hold, the sampling distribution follows a chi-square distribution with degrees of freedom = (number of rows − 1) × (number of columns − 1).

SOURCES

Bloodhart, B., et. al. 2015. Local climate experts: The influence of local TV weather information on climate change perceptions. *PLoS ONE* 10(11): e0141526. doi:10.1371/journal.pone.0141526

Boyer, L. V., et al. 2009. Antivenom for critically ill children with neurotoxicity from scorpion stings. *New England Journal of Medicine* 360: 2090–2098.

Cho, J., et al. 2015. Complementary relationships between traditional media and health apps among American college students. *Journal of American College Health* 63(4): 248–257. doi:10.1080/07448481.215.1015025.

Colman, R. J., et al. 2009. Caloric restriction delays disease onset and mortality in Rhesus monkeys. *Science* 325: 201.

Du Toit, G., et al. 2015. Randomized trial of peanut consumption in infants at risk for peanut allergy. *New England Journal of Medicine* 372 (February 26): 803–813. doi:10.1056/NEJMoa1414850.

Halloy, J., et al. 2007. Social integration of robots into groups of cockroaches to control self-organized choices. *Science Magazine* 318 (5853): 1155–1158.

Husemann, L. R., J. Moise-Titus, C. Podolski, and L. D. Eron. 2003. Longitudinal relations between children's exposure to TV violence and their aggressive and violent behavior in young adulthood: 1977–1992. *Developmental Psychology* 39(2): 201–221.

Kappos, L., et al. 2018. Siponimod versus placebo in secondary progressive multiple sclerosis (EXPAND): A double-blind, randomised, phase 3 study. *The Lancet* 391 (10127): P1263– P1273. https://doi.org/10.1016/S0140-6736(18)30475-6

Kukuyeva, I. A., J. Wang, and Y. Yaglovskaya. 2008. Popcorn popping yield: An investigation presented at the Joint Statistics Meetings.

Lundstrom, J., Boyle, J., and Jones-Gotman, M. 2006. Sit up and smell the roses better: Olfactory sensitivity to phenyl ethyl alcohol is dependent on body position. *Chemistry Senses* 31(3): 249–252. doi:10.1093/chemse/bjj025.

Pruim, R., the NHANES R Package, 2015. Weigh and weigh-related behaviors among 2-year college students. *Journal of American College Health* 63(4): 221– 229. doi:10.1080>07448481.2015.1015022.

Schweinhart, L. J., et al. 2005. *Lifetime effects: the High/Scope Perry Preschool Study through age 40.* Monographs of the High/Scope Educational Research Foundation, 14. Ypsilanti, MI: High/Scope Press.

Sprigg, N., et al. 2018. Tranexamic acid for hyperacute primary IntraCerebral Haemorrhage (TICH-2): An international randomised, placebo-controlled, phase 3 superiority trial. *The Lancet*, published ahead of print, May 16, 2018. https//doi.org/10.1016/S0140-6736(18)31033-X.

Trembley, R. E., et al. 1996. From childhood physical aggression to adolescent maladjustment: The Montreal prevention experiment. In *Preventing childhood disorders, substance use and delinquency,* edited by R. D. Peters and R. J. McMahon, 269–298. Thousand Oaks, CA: Sage Publications.

SECTION EXERCISES

SECTION 10.1

10.1 Tests

a. In Chapter 8, you learned some tests of proportions. Are tests of proportions used for categorical or numerical data?

b. In this chapter, you are learning to use chi-square tests. Do these tests apply to categorical or numerical data?

10.2 In Chapter 9, you learned some tests of means. Are tests of means used for numerical or categorical data?

10.3 Crime and Gender A statistics student conducted a study in Ventura County, California, that looked at criminals on probation who were under 15 years of age to see whether there was an association between the type of crime (violent or nonviolent) and gender. Violent crimes involve physical contact such as hitting or fighting; nonviolent crimes include vandalism, robbery, and verbal assault. The raw data are shown in the accompanying table; v stands for violent, n for nonviolent, b for boy, and g for girl.

Gen	Viol?	Gen	Viol?	Gen	Viol?
b	n	b	n	g	n
b	n	b	n	g	n
b	n	b	n	g	n
b	n	b	n	g	v
b	n	b	v	g	v
b	n	b	v	g	v
b	n	b	v	g	v
b	n	b	v	g	v
b	n	b	v	g	v
b	n	b	v	g	v
b	n	b	v	g	v
b	n	b	v	g	v
b	n	b	v	g	v
b	n	b	v	g	v
b	n	g	n		

Create a two-way table to summarize these data. Notice that the two variables are categorical, as can be seen from the raw data. If you are doing this by hand, create a table with two rows and two columns. Label the columns Boy and Girl (across the top). Label the rows Violent and Nonviolent. Begin with a big table, making a tally mark in one of the four cells for each observation, and then summarize the tally marks as counts.

10.4 Red Cars and Stop Signs The table shows the raw data for the results of a student survey of 22 cars and whether they stopped completely at a stop sign or not. In the Color column, "Red" means the car was red, and "No" means the car was not red. In the Stop column, "Stop" means the car stopped, and "No" means the car did not stop fully.

Create a two-way table to summarize these data. Use Red and No for the columns (across the top) and Stop and No for the rows. (We gave you an orientation of the table so that your answers would be easy to compare.)

Are the two variables categorical or numerical?

Color	Stop	Color	Stop
Red	Stop	No	No
Red	Stop	Red	Stop
Red	No	Red	No
Red	No	Red	No
Red	No	No	Stop
No	Stop	No	Stop
No	Stop	No	Stop
No	Stop	No	Stop
No	No	Red	Stop
Red	Stop	Red	No
Red	No	Red	No

10.5 The table summarizes the outcomes of a study that students carried out to determine whether humanities students had a higher mean grade point average (GPA) than science students. Identify both of the variables, and state whether they are numerical or categorical. If numerical, state whether they are continuous or discrete.

	Mean GPA
Science	3.4
Humanities	3.5

10.6 Finger Length There is a theory that relative finger length depends on testosterone level. The table shows a summary of the outcomes of an observational study that one of the authors carried out to determine whether men or women were more likely to have a ring finger that appeared longer than their index finger. Identify both of the variables, and state whether they are numerical or categorical. If numerical, state whether they are continuous or discrete.

	Men	Women
Ring Finger Longer	23	13
Ring Finger Not Longer	4	14

10.7 Student Loans—California According to a 2017 report, 53% of college graduates in California had student loans. Suppose a random sample of 120 college graduates in California shows that 72 had college loans. (Source: Lendedu.com)

a. What is the observed frequency of college graduates in the sample who had student loans?

b. What is the observed proportion of college graduates in the sample who had student loans?

c. What is the expected number of college graduates in the sample to have student loans if 53% is the correct rate? Do not round off.

10.8 Student Loans—Illinois According to a 2017 report, 64% of college graduate in Illinois had student loans. Suppose a random sample of 80 college graduates in Illinois is selected and 48 of them had student loans. (Source: Lendedu.com)

a. What is the observed frequency of college graduates in the sample who had student loans?

b. What is the observed proportion of college graduates in the sample who had student loans?

c. What is the expected number of college graduates in the sample to have student loans if 64% is the correct rate? Do not round off.

TRY 10.9 Breakfast Habits (Example 1 & 2) In a 2015 study by Nanney et al. and published in the *Journal of American College Health,* a random sample of community college students was asked whether they ate breakfast 3 or more times weekly. The data are reported by gender in the table.

Eat breakfast at least 3× weekly	Females	Males
Yes	206	94
No	92	49

a. Find the row, column, and grand totals, and prepare a table showing these values as well as the counts given.

b. Find the percentage of students overall who eat breakfast at least three times weekly. Round off to one decimal place.

c. Find the expected number who eat breakfast at least three times weekly for each gender. Round to two decimal places as needed.

d. Find the expected number who did not eat breakfast at least three times weekly for each gender. Round to two decimal places as needed.

e. Calculate the observed value of the chi-square statistic.

10.10 Fast Food Habits In the study referenced in exercise 10.8, researchers also asked whether or not students bought fast food at least one to two times per week. The data are reported by gender in the table.

Buy fast food at least 1–2 times weekly	Females	Males
Yes	138	85
No	160	58

a. Find the row, column, and grand totals, and prepare a table showing these values as well as the counts given.

b. Find the percentage of students overall who buy fast food at least 1 or 2 times weekly. Round off to one decimal place.

c. Find the expected number who buy fast food at least 1 or 2 times weekly for each gender. Round to two decimal places as needed.

d. Find the expected number who did not buy fast food at least 1 or 2 times weekly for each gender. Round to two decimal places as needed.

e. Calculate the observed value of the chi-square statistic.

10.11 Mummies with Heart Disease

According to the website MedicalNewsToday.com, coronary artery disease accounts for about 40% of deaths in the United States. Many people believe this is due to modern-day factors such as high-calorie fast food and lack of exercise. However, a study published in the *Journal of the American Medical Association* in November 2009 (www.medicalnewstoday.com) reported on 16 mummies from the Egyptian National Museum of Antiquities in Cairo. The mummies were examined, and 9 of them had hardening of the arteries, which seems to suggest that hardening of the arteries is not a new problem.

a. Calculate the expected number of mummies with artery disease (assuming the rate is the same as in the modern day). Then calculate the expected number of mummies without artery disease (the rest).

b. Calculate the observed value of the chi-square statistic for these mummies.

10.12 Texting While Driving

According to the 2015 High School Youth Risk Behavior Survey, 41.5% of high school students reported they had texted or emailed while driving a car or other vehicle. Suppose you randomly sample 80 high school students and ask if they have texted or emailed while driving. Suppose 38 say yes and 42 say no. Calculate the observed value of the chi-square statistic for testing the hypothesis that 41.5% of high school students engage in this behavior.

10.13 Violins

Stradivarius violins, made in the 1700s by a man of the same name, are worth millions of dollars. They are prized by music lovers for their uniquely rich, full sound. In September 2009, an audience of experts took part in a blind test of violins, one of which was a Stradivarius. There were four other violins (modern-day instruments) made of specially treated wood. When asked to pick the Stradivarius after listening to all five violins, 39 got it right and 113 got it wrong (*Time*, November 23, 2009).

a. If this group were just guessing, how many people (out of the 152) would be expected to guess correctly? And how many would be expected to guess incorrectly?

b. Calculate the observed value of the chi-square statistic showing each step of the calculation.

10.14 Coin Flips

You flip a coin 100 times and get 58 heads and 42 tails. Calculate the chi-square statistic by hand, showing your work, assuming the coin is fair.

SECTION 10.2

10.15

Fill in the blank by choosing one of the options given: Chi-square goodness-of-fit tests are applicable if the data consist of _____ (one categorical variable, two categorical variables, one numerical variable, or two numerical variables).

10.16

Fill in the blank by choosing one of the options given: Chi-square goodness-of-fit data are often summarized with _____ (one row or one column of observed counts—but not both, or at least two rows and at least two columns of observed counts).

TRY 10.17 Are Humans Like Random Number Generators?

g (Example 3) One of the authors collected data from a class to see whether humans made selections randomly, as a random number generator would. Each of 38 students had to pick an integer from one to five. The data are summarized in the table.

Integer:	One	Two	Three	Four	Five
Times Chosen:	3	5	14	11	5

A true random number generator would create roughly equal numbers of all five integers. Do a goodness-of-fit analysis to test the hypothesis that humans are not like random number generators. Use a significance level of 0.05, and assume these data were from a random sample of students. *See page 552 for guidance.*

10.18 Is the Random Number Table Really Random?

We counted ones, twos, threes, fours, and fives from a few lines of a random number table, and we should expect to get equal numbers of each. (We ignored the sixes, sevens, eights, nines, and zeros.) There were 14 ones, 12 twos, 16 threes, 11 fours, and 8 fives, which is 61 numbers in the categories selected. First, find the expected counts, which should all be the same.

▲ TI-84 GOF Output

Then test the hypothesis that the random number table does not generate equal proportions of ones, twos, threes, fours, and fives, using a significance level of 0.05. Refer to the goodness-of-fit (GOF) output shown.

10.19 Coin Spins

A penny was spun on a hard, flat surface 50 times, and the result was 15 heads and 35 tails. Using a chi-square test for goodness of fit, test the hypothesis that the coin is biased, using a 0.05 level of significance.

10.20 Internet Usage

In 2018 Pew Research reported that 11% of Americans do not use the Internet. Suppose in a random sample of 200 Americans, 26 reported not using the Internet. Using a chi-square test for goodness-of-fit, test the hypothesis that the proportion of Americans who do not use the Internet is different from 11%. Use a significance level of 0.05.

10.21 Dreidel Spinning

When playing Dreidel, (see photo) you sit in a circle with friends or relatives and take turns spinning a wobbly top (the dreidel). In the center of the circle is a pot of several foil-wrapped chocolate coins. If the four-sided top lands on the Hebrew letter *gimmel*, you take the whole pot and everyone needs to contribute to the pot again. If it lands on *hey*, you take half the pot. If it lands on *nun*, nothing happens. If it lands on *shin*, you put a coin in. Then the next player takes a turn.

Each of the four outcomes is believed to be equally likely. One of the author's families got the following outcomes while playing with a wooden dreidel during Hanukah.

Determine whether the outcomes allow us to conclude that the dreidel is biased (the four outcomes are not equally likely). Use a significance level of 0.05.

gimmel	hey	nun	shin
5	1	7	27

10.22 Plastic Dreidel See exercise 10.21 for an explanation of playing with the dreidel. This time the family used a plastic dreidel and got the following outcomes. The four outcomes are believed to be equally likely (that is, has a uniform probability distribution). Determine whether the plastic dreidel does not follow the uniform distribution using a significance level of 0.05.

gimmel	hey	nun	shin
11	9	11	9

10.23 Is the Six-Sided Die Fair? The table shows the results of rolling a six-sided die 120 times.

Outcome on Die	Frequency
1	27
2	20
3	22
4	23
5	19
6	9

Test the hypothesis that the die is not fair. A fair die should produce equal numbers of each outcome.

Use the four-step procedure with a significance level of 0.05, and state your conclusion clearly.

10.24 Is the Six-Sided Die Fair? Repeat the chi-square test (all four steps) from exercise 10.23, but this time assume that you got exactly 20 outcomes in each of the six categories. Refer to the figure. Explain.

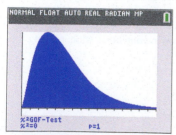

▲ TI-84 GOF Output for "Draw"

* **10.25 Violins** Professional musicians listened to five violins being played, without seeing the instruments. One violin was a Stradivarius, and the other four were modern-day violins. When asked to pick the Stradivarius (after listening to all five), 39 got it right and 113 got it wrong.

    a. Use the chi-square goodness-of-fit test to test the hypothesis that the experts are not simply guessing. Use a significance level of 0.05.

    b. Perform a one-proportion z-test with the same data, using a one-tailed alternative that the experts should get more than 20% correct. Use a significance level of 0.05.

    c. Compare your p-values and conclusions.

10.26 Mummies with Heart Disease Exercise 10.11 on artery disease in mummies indicated that 9 out of 16 mummies showed heart disease (hardening of the arteries).

    Test the hypothesis that the population proportion of mummies with hardening of the arteries is not the same as in the modern United States (that it is not 40%). Use a significance level of 0.05.

SECTION 10.3

TRY **10.27 Party and Right Direction (Example 4)** Suppose a polling organization asks a random sample of people if they are Democrat, Republican, or Other and asks them if they think the country is headed in the right direction or the wrong direction. If we wanted to test whether party affiliation and answer to the question were associated, would this be a test of homogeneity or a test of independence? Explain.

10.28 Antibiotic or Placebo A large number of surgery patients get infections after surgery, which can sometimes be quite serious. Researchers randomly assigned some surgery patients to receive a simple antibiotic ointment after surgery, others to receive a placebo, and others to receive just cleansing with soap. If we wanted to test the association between treatment and whether patients get an infection after surgery, would this be a test of homogeneity or of independence? Explain. (Source: Hospitals Could Stop Infections by Tackling Bacteria Patients Bring In, Study Finds, *New York Times*, January 6, 2010.)

TRY **10.29 Relevant Education (Example 5)** A 2018 Gallup poll asked college graduates if they agreed that the courses they took in college were relevant to their work and daily lives. The respondents were also classified by their field of study. If we wanted to test whether there was an association between response to the question and the field of study of the respondent, should we do a test of independence or homogeneity?

10.30 Diabetes Treatment In a 2018 study by Zhu et al. reported in *The Lancet*, researchers conducted an experiment to determine the efficacy and safety of the drug dorzagliatin in the treatment of patients with Type 2 diabetes. In this double-blind study, patients were randomly assigned to one of two treatment groups (drug or placebo) and the glucose levels of the two groups were compared after 12 weeks. If we test whether treatment group is associated with glucose level, are we doing a test of homogeneity or a test of independence?

10.31 Fitness The table shows the percentage of all men and women in the United States aged 18 to 44 who meet aerobic fitness guidelines. Give two reasons why a chi-square test is not appropriate for this data. (Source: *2017 World Almanac and Book of Facts*)

Percentage Meeting Fitness Guidelines		
Year	Men	Women
2005	50.0	43.1
2010	59.0	48.5
2014	60.8	52.7

10.32 Food Security The table shows the percentage of all U.S. households who are food secure, have low food security, or who have very low food security. The data are reported by area of residence. Give two reasons why it would be inappropriate to do a chi-square test to determine if there is an association between food security and area of residence. (Source: *2017 World Almanac and Book of Facts*)

	Food secure	Low food security	Very low food security
In metropolitan areas	87.8	7.3	4.9
Outside metropolitan area	84.6	9.3	6.1

10.33 Fitness App Use and Gender (Example 6) In a 2015 study reported in the *Journal of American College Health*, Cho et al. surveyed college students on their use of apps to monitor their exercise and fitness. The data are reported in the table. Test the hypothesis that fitness app use and gender are associated. Use a 0.05 significance level. *See page 552 for guidance.*

Use	Male	Female
Yes	84	268
No	9	57

10.34 Diet App Use and Gender In the study referenced in exercise 10.33, researchers also collected data on use of apps to monitor diet and calorie intake. The data are reported in the table. Test the hypothesis that diet app use and gender are associated. Use a 0.05 significance level.

Use	Male	Female
Yes	43	241
No	50	84

10.35 HPV Vaccination Rates A vaccine is available to prevent the contraction of human papillomavirus (HPV). The Centers for Disease Control and Prevention recommends this vaccination for all young girls in two doses. In a 2015 study reported in the *Journal of American College Health*, Lee et al. studied vaccination rates among Asian American and Pacific Islander (AAPI) women and non-Latina white women. Data are shown in the table. Test the hypothesis that vaccination rates and race are associated. Use a 0.05 significance level.

Completed HPV vaccinations	AAPI	White
Yes	136	1170
No	216	759

10.36 HPV Vaccine Literacy Rates In the study described in 10.35 researchers also asked survey respondents if they had heard of the HPV vaccine. Data are shown in the table. Test the hypothesis that knowledge of the vaccine and race are associated. Use a 0.05 significance level.

Heard of HPV vaccine	AAPI	White
Yes	248	1737
No	103	193

10.37 Gender and Happiness of Marriage The table shows the results of a two-way table of gender and whether a person is happy in his or her marriage, according to data obtained from a General Social Survey.

Happiness of Marriage × Respondent's Sex Cross Tabulation Count

		Respondent's Sex		
		Male	Female	Total
Happiness of Marriage	Very Happy	278	311	589
	Pretty Happy	128	154	282
	Not Too Happy	4	22	26
Total		410	487	897

a. If we carry out a test to determine whether these variables are associated, is this a test of independence, homogeneity, or goodness of fit?

b. Do a chi-square test with a significance level of 0.05 to determine whether gender and happiness of marriage are associated.

c. Does this suggest that women and men tend to have different levels of happiness or that their rates of happiness in marriage are about equal?

10.38 Is Smiling Independent of Age? Randomly chosen people were observed for about 10 seconds in several public places, such as malls and restaurants, to see whether they smiled during that time. The table shows the results for different age groups.

	Age Group				
	0–10	11–20	21–40	41–60	61+
Smile	1131	1748	1608	937	522
No Smile	1187	2020	3038	2124	1509

(Source: M. S. Chapell, Frequency of Public Smiling over the Life Span, *Perceptual and Motor Skills* 85[1997]: 1326.)

a. Find the percentage of each age group that were observed smiling, and compare these percentages.

b. Treat this as a single random sample of people, and test whether smiling and age group are associated, using a significance level of 0.05. Comment on the results.

10.39 Preschool Attendance and High School Graduation Rates (Example 7) The Perry Preschool Project was created in the early 1960s by David Weikart in Ypsilanti, Michigan. One hundred twenty three African American children were randomly assigned to one of two groups: One group enrolled in the Perry Preschool, and one did not enroll. Follow-up studies were done for decades to answer the research question of whether attendance at preschool had an effect on high school graduation. The table shows whether the students graduated from regular high school or not. Students who received GEDs were counted as not graduating from high school. This table includes 121 of the original 123. This is a test of homogeneity, because the students were randomized into two distinct samples. (Schweinhart et al. 2005)

	Preschool	No Preschool
HS Grad	37	29
No HS Grad	20	35

a. For those who attended preschool, the high school graduation rate was 37/57, or 64.9%. Find the high school graduation rate for those not attending preschool, and compare the two. Comment on what the rates show for *these* subjects.

b. Are attendance at preschool and high school graduation associated? Use a 0.05 level of significance.

10.40 Preschool Attendance and High School Graduation Rates for Females

The Perry Preschool Project data presented in exercise 10.39 (Schweinhart et al. 2005) can be divided to see whether the preschool attendance effect is different for males and females. The table shows a summary of the data for females, and the figure shows Minitab output that you may use.

```
Chi-Square Test: Preschool, No Preschool: Girls
Expected counts are printed below observed counts

                        No
          Preschool  Preschool   Total

Grad         21           8         29
           14.50       14.50

No Grad       4          17         21
           10.50       10.50

Total        25          25         50

Chi-Sq = 13.875, DF = 1, P-Value = 0.000
```

	Preschool	No Preschool
HS Grad	21	8
HS Grad No	4	17

a. Find the graduation rate for those females who went to preschool, and compare it with the graduation rate for females who did not go to preschool.

b. Test the hypothesis that preschool and graduation rate are associated, using a significance level of 0.05.

10.41 Preschool Attendance and High School Graduation Rates for Males

The Perry Preschool Project data presented in exercise 10.39 can be divided to see whether there are different effects for males and females. The table shows a summary of the data for males (Schweinhart et al. 2005).

	Preschool	No Preschool
HS Grad	16	21
HS Grad No	16	18

a. Find the graduation rate for males who went to preschool, and compare it with the graduation rate for males who did not go to preschool.

b. Test the hypothesis that preschool and graduation are associated, using a significance level of 0.05.

c. Exercise 10.40 showed an association between preschool and graduation for just the females in this study. Write a sentence or two giving your advice to parents with preschool-eligible children about whether attending preschool is good for their children's future academic success, based on this data set.

10.42 Same-Sex Marriage

A 2018 Gallup poll asked respondents if they supported same-sex marriage. Results are reported by political party in the StatCrunch output that follows.

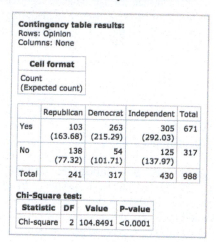

Contingency table results:
Rows: Opinion
Columns: None

Cell format
Count
(Expected count)

	Republican	Democrat	Independent	Total
Yes	103 (163.68)	263 (215.29)	305 (292.03)	671
No	138 (77.32)	54 (101.71)	125 (137.97)	317
Total	241	317	430	988

Chi-Square test:

Statistic	DF	Value	P-value
Chi-square	2	104.8491	<0.0001

a. Find the percentage in each political party who support same-sex marriage. Round off to one decimal place as needed.

b. Test the hypothesis that support of same-sex marriage and political party affiliation are independent using a significance level of 0.05.

c. Does this suggest that political parties differ significantly in their support of same-sex marriage?

10.43 Marijuana Legalization

A 2018 Pew Research poll asked a random sample of Millennials and GenXers if they supported legalization of marijuana. Survey results found 70% of Millennials and 66% of GenXers supported marijuana legalization.

a. Use these results to fill in the following two-way table with the counts in each category. Assume the sample size for each group was 200.

b. Test the hypothesis that support of marijuana legalization is independent of generation for these two groups using a significance level of 0.05.

c. Does this suggest that these generations differ significantly in their support of marijuana legalization?

	Supports Marijuana Legalization	
Generation	**Yes**	**No**
Millennial		
GenX		
Total		

10.44 Marijuana Legalization

The 2018 Pew Research poll in exercise 10.43 also reported responses by political party. Survey results found 45% of Republicans and 69% of Democrats supported marijuana legalization.

a. Use these results to fill in the following two-way table with the counts in each category. Assume the sample size for each group was 200.

b. Test the hypothesis that support of marijuana legalization is independent of political party for these two groups using a significance level of 0.05.

c. Does this suggest that these political parties differ significantly in their support of marijuana legalization?

Political Party	Supports Marijuana Legalization	
	Yes	No
Republican		
Democrat		

10.45 Brain Bleed Treatment In a 2018 article published in *The Lancet*, Sprigg et al. studied the effect of tranexamic acid in treating patients with intracerebral hemorrhages using a randomized, placebo-controlled trial. Of the 1161 subjects treated with tranexamic acid, 383 suffered an adverse outcome after 2 days. Of the 1164 subjects given a placebo, 419 suffered an adverse outcome after 2 days.

a. Find the percentage in each group that suffered an adverse outcome. Round off to one decimal place as needed.

b. Create a two-way table with the treatment labels (drug/placebo) across the top.

c. Test the hypothesis that treatment and adverse outcome are associated using a significance level of 0.05.

10.46 Multiple Sclerosis Treatment In a 2018 article published in *The Lancet*, Kappos et al. studied the effect of the drug siponimod in treating patients with secondary progressive multiple sclerosis (SPMS) using a double-blind, randomized, controlled study. Of the 1099 patients given the drug, 198 experienced a severe adverse outcome. Of the 546 patients given the placebo, 82 experienced a severe adverse outcome.

a. Find the percentage in each group that suffered a severe adverse outcome.

b. Create a two-way table with the treatment labels (drug/placebo) across the top.

c. Test the hypothesis that treatment and severe adverse outcome are associated using a significance level of 0.05.

10.47 Political Party Affiliation and Education A 2018 Pew Research poll recorded respondents' political affiliation and education attainment. A summary of the data is shown in the following table. Test the hypothesis that political party affiliation and educational attainment are associated at the 0.05 level.

	Political Party Affiliation	
Educational Attainment	Democrat/Lean Democrat	Republican/Lean Republican
High school or less	144	150
Some college	132	126
College graduate	135	98
Postgraduate degree	95	47

10.48 Political Party Affiliation and Generation A 2018 Pew Research poll recorded respondents' political affiliation and generation. A summary of the results for Millennials and GenXers are

shown in the following table, assuming a sample size of 200. Test the hypothesis that political party affiliation and generation are associated at the 0.05 level for these generations.

	Political Party Affiliation	
Generation	Democrat	Republican
Millennials	118	64
GenX	98	86

SECTION 10.4

TRY **10.49 Alcohol Use (Example 8)** In a 2016 article published in the *Journal of American College Health*, Heller et al. surveyed a sample of students at an urban community college. Students' ages and frequency of alcohol use per month are recorded in the following table. Because some of the expected counts are less than 5, we should combine some groups. For this question, combine the frequencies *10–29 days* and *Every day* into one group. Label this group *10+ days* and show your new table. Then test the new table to see whether there is an association between age group and alcohol use using a significance level of 0.05. Assume this is a random sample of students from this college.

	Alcohol Use			
Age	None	1–9 days	10–29 days	Every day
18–20	182	100	27	4
21–24	142	109	35	4
25–29	49	41	5	2
30+	76	32	8	2

10.50 Night Shifts A random sample of nurses working night shifts were asked whether they were fatigued (no or yes) and what form of sleep aid they used. Assume that each person used one sleep aid. The table shows the results.

a. Why is a chi-square test for independence inappropriate for this data set?

b. Combine some groups. Put those who used a Dark Room or White Noise (natural) in one group, and in the other group (chemical) include all the others. Show the new table.

c. Using the table for part b, test the hypothesis that fatigue and form of sleep aid are independent at the 0.05 level.

d. Does this show that sleep aids cause fatigue?

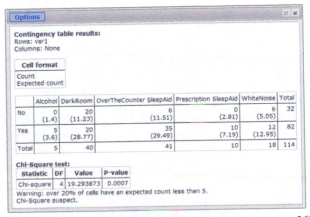

(Source: StatCrunch: Night Shift Fatigue. Owner: armyman36)

10.51 Gender and Political Party Affiliation The data in the table are from a General Social Survey and concern gender and political party.

a. Find the expected counts and report the smallest. Could we use the table as is, without combining categories?

b. Create a new table, using the data from the table shown, with fewer categories. Merge the Strong Democrat and the Not Strong Democrat categories into one group called Democrats. Merge the Strong Republican and the Not Strong Republican categories into one group called Republicans. Put all the Independent categories and the Other category into one group called Other.

c. What percentage of the women and what percentage of the men are Democrats? Compare these percentages.

d. Assume that conditions are met, and do a chi-square test with a significance level of 0.05 to see whether the variables *Gender* and *Political Party* are associated using your new grouping.

Political Party Affiliation × Respondent's Sex Cross Tabulation
Count

		Respondent's Sex		
		Male	Female	Total
Political Party Affiliation	Strong democrat	142	214	356
	Not str democrat	136	207	343
	Ind, near dem	127	108	235
	Independent	147	226	373
	Ind, near rep	93	64	157
	Not str republican	115	135	250
	Strong republican	83	109	192
	Other party	36	18	54
Total		879	1081	1960

10.52 Children and Happiness The data in the table come from a General Social Survey. The top row is the number of children reported for the respondents. The respondents also reported their level of happiness; Very H means Very Happy, and so on. The counts are shown in the table. Is happiness associated with having at least one child?

General Happiness × Number of Children Cross Tabulation
Count

		Number of Children									
		0	1	2	3	4	5	6	7	Eight or More	Total
General happiness	Very Happy	143	88	187	85	48	16	13	6	6	592
	Pretty happy	316	153	305	162	97	28	16	4	12	1093
	Not too happy	76	33	73	51	22	12	3	2	5	277
Total		535	274	565	298	167	56	32	12	23	1962

a. Merge all the Number of Children categories into two groups: those who have 0 children and those who have at least 1 child. For the rows, merge the Very Happy and Pretty Happy into one group called Happy. Rename the Not Too Happy group Not Happy. Report the new table, which should have two rows and two columns.

b. We wish to test whether happiness is associated with having children. Why was it necessary to merge categories?

c. With the merged data, determine whether there is an association between happiness and whether a person has at least one child. Use a significance level of 0.05.

* **10.53 Scorpion Antivenom** In a 2009 study reported in the New England Journal of Medicine, Boyer et al. randomly assigned children aged 6 months to 18 years who had nonlethal scorpion stings to receive an experimental antivenom or a placebo. "Good" results were no symptoms after four hours and no detectable plasma venom.

	Antivenom	Placebo	Total
No Improvement	1	6	7
Improvement	7	1	8
Total	8	7	15

The alternative hypothesis is that the antivenom leads to improvement. The p-value for a one-tailed Fisher's Exact Test with these data is 0.009.

a. Suppose the study had turned out differently, as in the following table.

	Antivenom	Placebo
Bad	0	7
Good	8	0

Would Fisher's Exact Test have led to a p-value larger or smaller than 0.009? Explain.

b. Suppose the study had turned out differently, as in the following table.

	Antivenom	Placebo
Bad	2	5
Good	6	2

Would Fisher's Exact Test have led to a p-value larger or smaller than 0.009? Explain.

c. Try the two tests, and report the p-values. Were you right? Search for a Fisher's Exact Test calculator on the Internet, and use it.

10.54 Nice Rats Rats had a choice of freeing another rat or eating chocolate by themselves. Most of the rats freed the other rat and then shared the chocolate with it. The table shows the data concerning the gender of the rat in control.

	Male	Female
Freed Rat	17	6
Did not	7	0

a. Can a chi-square test for homogeneity or independence be performed with this data set? Why or why not?

b. Determine whether the sex of a rat influences whether or not it frees another rat using a significance level of 0.05.

(Source: P. Mason, *Ventura County Star*, December 11, 2011, http://video.sciencemag.org/VideoLab/1310979895001/1.)

10.55 Peanut Allergies In a 2015 study reported in the *New England Journal of Medicine*, Du Toit et al. randomly assigned infants who were likely to develop a peanut allergy (as measured by having eczema, egg allergies, or both) to either consume or avoid peanuts until 60 months of age. The infants in this cohort did not previously show any preexisting sensitivity to peanut extract. The numbers in each group developing a peanut allergy by 60 months of age are shown in the following table.

Peanut allergy at age 60 mos.	Treatment Group	
	Consume peanuts	Avoid peanuts
Yes	5	37
No	267	233

a. Compare the percentages in each group that developed a peanut allergy by age 60 months.

b. Test the hypothesis that treatment group and peanut allergy are associated using the chi-square statistic. Use a significance level of 0.05.

c. Do a Fisher's Exact Test for the data with the same significance level. Report the two-tailed p-value and your conclusion. (Use technology to run the test.)

d. Compare the p-values for parts b and c. Which is more accurate? Explain.

10.56 Peanut Allergies In the study described in exercise 10.55, researchers (Du Toits et al., 2015) also studied infants with eczema, egg allergies, or both who also had a preexisting sensitivity to peanut extract. These infants were also randomly assigned to either consume or avoid peanuts until 60 months of age. The numbers in each group developing a peanut allergy by 60 months of age are shown in the following table.

Peanut allergy at age 60 mos.	Treatment Group	
	Consume peanuts	Avoid peanuts
Yes	5	18
No	43	33

a. Compare the percentages in each group that developed a peanut allergy by age 60 months.

b. Test the hypothesis that treatment group and peanut allergy are associated using the chi-square statistic. Use a significance level of 0.05.

c. Do a Fisher's Exact Test for the data with the same significance level. Report the two-tailed p-value and your conclusion. (Use technology to run the test.)

d. Compare the p-values for parts b and c. Which is more accurate? Explain.

CHAPTER REVIEW EXERCISES

In exercises 10.57 to 10.64, choose an appropriate test from those listed below. Unless otherwise stated, assume large expected counts.

Chi-square test of goodness of fit
Chi-square test of independence
Chi-square test of homogeneity
Fisher's Exact Test
No test because the data represent a population, not a sample

10.57 Suppose you have a random sample of firefighters in Illinois and want to determine if the racial distribution of the firefighters is different from the racial distribution of the state as a whole.

10.58 Suppose you know the class (first class, second class, third class, or steerage) and whether each person on the *Titanic* survived or died, for *all* passengers on the *Titanic*.

10.59 Suppose you take a random sample of students at a college, asking them their gender and in which of the following areas they are majoring: Fine Arts, Liberal Arts, Business, STEM (Science, Technology, Engineering, and Mathematics), Social Science or Other.

10.60 Suppose you randomly assign some parolees to check in once a week with their parole officers and others to check in once a month, and observe whether they are arrested within 6 months of starting parole.

10.61 Based on a random sample of residents in a large city, you wish to determine if there is an association between income level (low, middle, or high) and whether the residents own or rent their home.

10.62 Suppose you are testing two different injections by randomly assigning them to children who react badly to bee stings and go to the emergency room. You observe whether the children are substantially improved within an hour after the injection. However, one of the expected counts is less than 5.

10.63 Suppose you are interested in whether more than 50% of voters in California support a proposition (Prop X). After the vote, you find the total number that support it and the total number that oppose it.

10.64 Suppose there is a theory that 90% of the people in the United States dream in color. You survey a random sample of 200 people; 198 report that they dream in color, and 2 report that they do not. You wish to verify the claim made in the theory.

* **10.65 Perry Preschool Arrests** The Perry Preschool Project discussed in exercises 10.39 to 10.41 found that 8 of the 58 students who attended preschool had at least one felony arrest by age 40 and that 31 of the 65 students who did not attend preschool had at least one felony arrest (Schweinhart et al. 2005).

a. Compare the percentages descriptively. What does this comparison suggest?

b. Create a two-way table from the data and do a chi-square test on it, using a significance level of 0.05. Test the hypothesis that preschool attendance is associated with being arrested.

c. Do a two-proportion z-test. Your alternative hypothesis should be that preschool attendance lowers the chances of arrest.

d. What advantage does the two-proportion z-test have over the chi-square test?

* **10.66 Parental Training and Criminal Behavior of Children** In Montreal, Canada, an experiment was done with parents of children who were thought to have a high risk of committing crimes when they became teenagers (Tremblay et al., 1996). Some of the families were randomly assigned to receive parental training, and the others were not. Out of 43 children whose parents were randomly assigned to the parental training group, 6 had been arrested by the age of 15. Out of 123 children whose parents were not in the parental training group, 37 had been arrested by age 15.

a. Find and compare the percentages of children arrested by age 15. Is this what researchers might have hoped?

b. Create a two-way table from the data, and test whether the treatment program is associated with arrests. Use a significance level of 0.05.

c. Do a two-proportion z-test, testing whether the parental training lowers the rate of bad results. Use a significance level of 0.05.

d. Explain the difference in the results of the chi-square test and the two-proportion z-test.

e. Can you conclude that the treatment causes the better result? Why or why not?

10.67 Active Military Demographics In 2017 the Pew Research Center published a report on the demographics of the U.S. military. The following table shows the ethnic breakdown of active-duty U.S. military services and the ethnic breakdown of the U.S. population. Would it be appropriate to use the data in the table to conduct a chi-square goodness of fit test to determine if the ethnic breakdown of the military matches the ethnic distribution of the U.S. population? If so, do the test. If not, explain why it would be inappropriate to do so.

Ethnicity	% Active-Duty Military	% U.S. Population
White	60	61
Black	17	12
Hispanic	12	18
Asian	4	6
Other	7	3

10.68 Vehicle Sales The following table shows the average number of vehicles sold in the United States monthly (in millions) for the years 2001 through 2018. Data on all monthly vehicle sales for these years were obtained and the average number per month was calculated. Would it be appropriate to do a chi-square analysis of this data set to see if vehicle sales are distributed equally among the months of the year? If so, do the analysis. If not, explain why it would be inappropriate to do so. (Source: www.fred.stlouisfed.org)

Month	Avg Sales per Month (in millions)
Jan	15.7
Feb	15.7
Mar	15.8
Apr	15.8
May	15.8
June	15.7
July	16.1
Aug	16.1
Sept	15.8
Oct	15.9
Nov	15.9
Dec	15.9

10.69 Harassment in the Workplace An 2017 NPR/Marist poll asked a random sample of Americans if they had personally experienced sexual harassment in the workplace. The results are shown in the following table.

	Personally Experienced Sexual Harassment in the Workplace	
	Yes	No
Men	120	380
Women	182	337

a. Find the percentage in each group who had personally experienced sexual harassment in the workplace.

b. Test the hypothesis that experience of sexual harassment in the workplace is associated with gender at the 0.05 significance level.

10.70 CPR in Sweden Three million people in Sweden are trained in CPR, which is more than 30% of the population. The data set is summarized below. A 2015 study by Hasselqvist-Ax et. al reported in the *New England Journal of Medicine* examined the relationship between bystander CPR (Cardio Pulmonary Resuscitation) and positive outcomes for out-of-hospital cardiac arrest cases. Researchers collected data on whether or not the victim received CPR from a bystander and whether or not the victim survived. The results are shown in the following table.

	CPR	No CPR
Survived	1,629	595
Died	13,883	14,274

a. Find the survival percentage for both groups and compare them.

b. Test the hypothesis that bystander CPR is independent of survival using a 0.05 significance level.

* **10.71 Robot Cockroaches** Cockroaches tend to rest in groups and prefer dark areas. In a study by Halloy et al. published in *Science Magazine* in November 2007, cockroaches were introduced to a brightly lit, enclosed area with two different available shelters, one darker than the other. Each time a group of cockroaches was put into the brightly lit area will be called a trial. When groups of 16 real cockroaches were put in a brightly lit area, in 22 out of 30 trials, all the cockroaches went under the same shelter. In the other 8 trials, some of the cockroaches went under one shelter, and some under the other one.

Another group consisted of a mixture of real cockroaches and robot cockroaches (4 robots and 12 real cockroaches). The robots did not look like cockroaches but had the odor of male cockroaches, and they were programmed to prefer groups (and brighter shelters). There were 30 trials. In 28 of the trials, all the cockroaches and robots rested under the same shelter, and in 2 of the trials they split up.

	Cockroaches Only	Robots Also
One Shelter Used	22	28
Both Shelters Used	8	2

Is the inclusion of robots associated with whether they all went under the same shelter? To answer the following questions, assume the cockroaches are a random sample of all cockroaches.

a. Use a chi-square test for homogeneity with a significance level of 0.05 to see whether the presence of robots is associated with whether roaches went into one shelter or two.

b. Repeat the question using Fisher's Exact Test. (If your software will not perform the test for you, search for Fisher's Exact Test on the Internet to do the calculations.) Conduct a two-sided hypothesis test so that the test is consistent with the test in part a.

c. Compare the p-values and conclusions from part a and part b. Which statistical test do you think is the better procedure in this case? Why?

∗ 10.72 Robot Cockroaches Refer to the description in exercise 10.71. There were 22 trials with only cockroaches (no robots) that went under one shelter. In 16 of these 22 trials, the group chose the darker shelter, and in 6 of the 22 the group chose the lighter shelter. There were 28 trials with a mixture of real cockroaches and robots that all went under one shelter. In 11 of these trials, the group chose the darker shelter, and in 17 the group chose the lighter shelter. The robot cockroaches were programmed to choose the lighter shelter (as well as preferring groups; Halloy et al. 2007)

All under One Shelter		
	Cockroaches Only	Robots Also
Darker	16	11
Lighter	6	17

Is the introduction of robot cockroaches associated with the type of shelter when the group went under one shelter? Assume cockroaches were randomly sampled from some meaningful population of cockroaches.

a. Use the chi-square test to see whether the presence or absence of robots is associated with whether they went under the darker or the brighter shelter. Use a significance level of 0.05

b. Do Fisher's Exact Test with the data. If your software does not do Fisher's Exact Test, search the Internet for a Fisher's Exact Test calculator and use it. Report the p-value and your conclusion.

c. Compare the p-values for parts a and b. Which do you think is the more accurate procedure? The p-values that result from the two methods in this question are closer than the p-values in the previous question. Why do you think that is?

10.73 Stand Your Ground: Race of Defendant In July 2013, Jeff Witmer obtained a data set from the *Tampa Bay Times* after the Zimmerman case was decided. Zimmerman shot and killed Trayvon Martin (an unarmed black teenager) and was acquitted. The data set concerns "stand your ground" cases with male defendants. Some of these were fatal attacks, and some were not fatal. Many of those charged used guns, but some used various kinds of knives or other methods.

	Accused		
	Nonwhite	White	All
Not Convicted	48	80	128
Convicted	19	38	57
All	67	118	185

a. What percentage of the nonwhite defendants were convicted?

b. What percentage of the white defendants were convicted?

c. Test the hypothesis that conviction is independent of race at the 0.05 level. Assume you have a random sample.

∗ 10.74 Conviction Rate with Opposite Race Here are the conviction rates with the "stand your ground" data mentioned in the previous exercise. "White shooter on nonwhite" means that a white assailant shot a minority victim.

	Conviction Rate
White shooter on white	$35/97 = 36.1\%$
White shooter on nonwhite	$3/21 = 14.3\%$
Nonwhite shooter on white	$8/22 = 36.4\%$
Nonwhite shooter on nonwhite	$11/45 = 24.4\%$

a. Which has the higher conviction rate: white shooter on nonwhite or nonwhite shooter on white?

b. Create a two-way table using White Shooter on Nonwhite and Non-White Shooter on White across the top and Convicted and Not Convicted on the side.

c. Test the hypothesis that race and conviction rate (for these two groups) are independent at the 0.05 level.

d. Because some of the expected counts are pretty low, try a Fisher's Exact Test with the data, reporting the p-value (two-tailed) and the conclusion.

gUIDED EXERCISES

g 10.17 Are Humans Like Random Number Generators?

One of the authors collected data from a class to see whether humans made selections randomly like a random number generator would. Each of 38 students had to pick an integer from one to five. The data are summarized in the table.

Integer:	One	Two	Three	Four	Five
Times Chosen:	3	5	14	11	5

A true random number generator would create roughly equal numbers of all five integers.

QUESTION Test the hypothesis that humans are not like random number generators by following the steps below. Use a significance level of 0.05, and assume the data were collected from a random sample of students.

Step 1 ▶ Hypothesize

H_0: Humans are like random number generators and produce numbers in equal quantities.

H_a: ?

▲ TI-84 GOF Output

Step 2 ▶ Prepare

Choose chi-square goodness of fit (GOF). We have only one variable: *Integer Chosen*. Use a significance level of 0.05. We are assuming randomness. We must check to see that all the expected counts are 5 or more. There were 38 students. Explain why all the expected counts are 7.6.

Step 3 ▶ Compute to compare

$$X^2 = \frac{(3 - 7.6)^2}{7.6} + \frac{(5 - 7.6)^2}{7.6}$$
$$+ \underline{\quad} + \underline{\quad} + \underline{\quad} = \underline{\quad}$$

Complete the calculation of the chi-square statistic. Then check your calculated chi-square with the output. Obtain the p-value from the TI-84 output.

Step 4 ▶ Interpret

Reject or do not reject the null hypothesis, and pick an interpretation from those below.

i. Humans have not been shown to be different from random number generators.

ii. Humans have been shown to be different from random number generators.

10.33 Fitness App Use and Gender

In a 2015 study reported in the *Journal of American College Health*, Cho et al. surveyed college students on their use of apps to monitor their exercise and fitness. The data are reported in the table.

Use	Male	Female
Yes	84	268
No	9	57

QUESTION Test the hypothesis that fitness app use and gender are associated. Use a 0.05 significance level. The steps will guide you through the process. StatCrunch output is provided.

Step 1 ▶ Hypothesize

H_0: Fitness app use and gender are independent (not associated).

H_a: ?

Contingency table results:
Rows: Use
Columns: None

Cell format
Count
(Expected count)

	Male	Female	Total
Yes	84 (78.32)	268 (273.68)	352
No	9 (14.68)	57 (51.32)	66
Total	93	325	418

Chi-Square test:

Statistic	DF	Value	P-value
Chi-square	1	3.3605955	0.0668

Step 2 ▶ Prepare

We choose the chi-square test of independence because the data were from *one* random sample in which the people were classified two different ways. The expected count for each cell has been calculated, and these counts are shown in the StatCrunch output in parentheses. Check that each expected count is greater than 5.

Step 3 ▶ Compute to compare.

Report the significance level. Refer to the output given.

$$X^2 = \underline{\qquad}$$

$$\text{p-value} = \underline{\qquad}$$

Step 4 ▶ Interpret

Reject or do not reject the null hypothesis and state what this means.

Techtips

General Instructions for All Technology

EXAMPLE A (CHI-SQUARE TEST FOR TWO-WAY TABLES): PERRY PRESCHOOL AND GRADUATION FROM HIGH SCHOOL ▶ In the 1960s an experiment was started in which a group of children were randomly assigned to attend preschool or not to attend preschool. They were studied for years, and whether they graduated from high school is shown in Table A.

We will show the chi-square test for two-way tables to see whether the factors are independent or not. For Minitab and StatCrunch, we also show Fisher's Exact Test.

	Preschool	No Preschool
Grad HS	37	29
No Grad HS	20	35

▲ **TABLE A** Two-way table summary for preschool and graduation from high school

Discussion of Data

Much of technology is set up so that you can use either a table summary (such as Table A) or a spreadsheet table containing the raw data. Table B shows the beginning of a spreadsheet table of the raw data, for which there would be 121 rows for the 121 children.

Preschool	Graduate HS
Yes	No
Yes	Yes
No	Yes
No	Yes
Yes	Yes
No	No

▲ **TABLE B** Some raw data

EXAMPLE B CHI-SQUARE FOR GOODNESS OF FIT: ARE HUMANS LIKE RANDOM NUMBER GENERATORS? ▶ One of the authors collected data from a class to see whether humans made selections randomly, as a random number generator would. Each of 38 students had to pick one of the integers one, two, three, four, or five. Table C summarizes the collected (*Observed*) data. If the students picked randomly, we would expect each number (one through five) to be chosen an equal number of times (uniform distribution). Because there were 38 students and 5 choices, each *expected* count is $38 \times (1/5)$, or 7.6, as listed in Table C.

Integer:	One	Two	Three	Four	Five
Observed Counts (O):	3	5	14	11	5
Expected Counts (E):	7.6	7.6	7.6	7.6	7.6

▲ **TABLE C** Goodness-of-fit data

TI-84

Chi-Square Test for Two-way Tables

You will not put the data into the lists. You will use a matrix (table), and the input must be in the form of a summary such as Table A.

1. Press **2ND** and **MATRIX** (or **MATRX**).
2. Scroll over to **EDIT** and press **ENTER** when **1:** is highlighted.
3. See Figure 10a. Put in the dimensions. Because the table has two rows and two columns, press **2, ENTER, 2, ENTER**. (The first number is the number of rows, and the second number is the number of columns.)

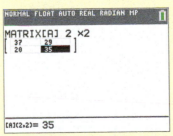

▲ **FIGURE 10a** TI-84 Input for Two-way Table

4. Enter each of the four numbers in the table, as shown in Figure 10a. Press **ENTER** after typing each number.

5. Press **STAT**, and scroll over to **TESTS**.

6. Scroll down (or up) to **C: χ^2-Test** and press **ENTER**.

7. Leave the **Observed** as **A** and the **Expected** as **B**. Scroll down to **Calculate** and press **ENTER**.

You should get the output shown in Figure 10b.

▲ **FIGURE 10b** TI-84 Output for a Chi-Square Test

8. To see the expected counts, click **2ND**, **MATRIX**, scroll over to **EDIT**, scroll down to **2: [B]**, and press **ENTER**. You may have to scroll to the right to see some of the numbers. They will be arranged in the same order as the table of observed values. Check these numbers for the required minimum value of 5.

Goodness-of-Fit Test

(The TI-83 does not provide the goodness-of-fit test.)

1. Put the observed counts in one list, such as **L1**, and put the expected counts in another list, such as **L2**, using the same order.

2. Then choose **Stat**, **Tests**, and **D: X^2GOF-Test**.

3. Make sure the **Observed** list is the one in which you have the observed counts and that the **Expected** list is where you have the expected counts. You will have to put in the value for **df**, which is the number of categories minus 1. In Example B there were five numbers to choose from, so df is 4 because $5 - 1 = 4$.

4. Then choose **Calculate** (or **Draw**).

5. The chi-square value comes out to be 11.47368, and the p-value is 0.0217258

Chi-Square Test for Two-way Tables

For Minitab you may have your data as a table summary (as shown in Table A) or as raw data (as shown in Table B).

TABLE SUMMARY

1. Type in the data summary, together with the optional row and column labels, as shown below.

↓	C1-T	C2	C3
		Preschool	No Preschool
1	GradHS	37	29
2	NoGradHS	20	35

2. **Stat > Tables > Cross Tabulation and Chi-Square**

3. See Figure 10c. Choose **Summarized data in a two-way table**. For **Columns containing the table:**, select both columns (by double clicking them). For **Rows:** select **C1**.

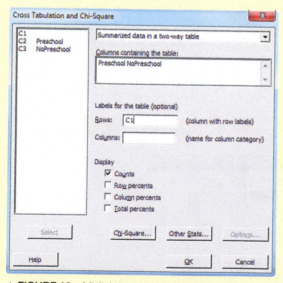

▲ **FIGURE 10c** Minitab Input for Two-way Tables

4. Click **Chi-square . . .**, check **Chi-square test** and **Expected cell counts**; click **OK**. Click **Other Stats . . .** and check **Fisher's exact test**; click **OK**. Click **OK**.

Figure 10d shows the output.

	Preschool	NoPreschool	All
GradHS	37	29	66
	31.09	34.91	
NoGradHS	20	35	55
	25.91	29.09	
All	57	64	121

Cell Contents
   Count
   Expected count

Chi-Square Test

	Chi-Square	DF	P-Value
Pearson	4.671	1	0.031
Likelihood Ratio	4.710	1	0.030

Fisher's Exact Test

P-Value
0.0439073

▲ **FIGURE 10d** Minitab Output for Two-way Tables.

RAW DATA

1. Make sure your raw data are in the columns. See Table B.

2. **Stat > Tables > Cross Tabulation and Chi-square**.

3. Choose **Raw data (categorical variables)**. Click C1 **For rows** and C2 **For columns** (or vice versa).

4. Do step 4 above.

Goodness-of-Fit Test

1. Type your observed counts in one column, such as **C1**.

2. If you have expected counts that are not equal, type them in another column, such as **C2**. (If all the expected counts are the same, as in our Example B, they don't have to be typed in.)

3. **Stat > Tables > Chi-Square Goodness-of-Fit Test (One Variable)**

4. See Figure 10e. Click in the box next to **Observed Counts**, and then double-click the column with the observed counts.

5. If there are **Equal proportions**, leave it selected. If you have a column of expected counts, click **Proportions specified by historical counts** and then choose the column containing the expected counts.

   Your results will show bar charts (as well as chi-square output) unless you click on **Graphs** and turn the graphs off.

6. Click **OK**.

   You should get a chi-square value of 11.4737and a p-value of 0.022.

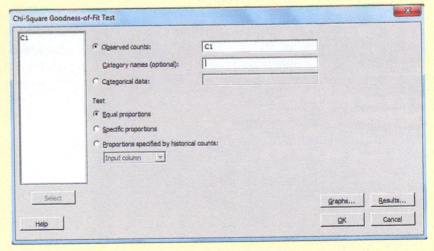

▲ **FIGURE 10e** Minitab Goodness-of-Fit Options

Chi-Square Test for Two-way Tables

TABLE SUMMARY

1. Type a summary of your data into two (or more) columns, as shown in columns A and B in Figure 10f.

	B2			f_x	35		
	A	B	C	D	E	F	G
1	37	29					
2	20	35					
3							
4							

▲ **FIGURE 10f** Excel Input for Two-way Table

2. To get the row total of 66 (from 37 + 29), click in the cell to the right of 29, C1, then click *fx*, double click **SUM**, and click **OK**. You can do the same thing for the other totals. Save the grand total for last. Your table with the totals should look like columns A, B, and C in Figure 10g.

3. To get the first expected count, 31.09091, click in the cell E1. Then type = and click on the **66** in the table, type * (for multiplication), click on the **57**, type / (for division), click on the **121**, and press **Enter**. The input for getting the expected count of 25.909 in cell E2 is shown in the formula bar at the top of Figure 10g. For each of the expected counts, you start from the cell you want filled, and you click on the row total, * (for multiply), the column total, / (for divide), and the grand total and press **Enter**.

	E2			f_x	=A3*C2/C3		
	A	B	C	D	E	F	G
1	37	29	66		31.091	34.909	66
2	20	35	55		25.909	29.091	55
3	57	64	121		57	64	121
4							

▲ **FIGURE 10g** Excel, Including Totals and Expected Counts

4. Click in an empty cell, J1. Click *fx*.

5. Select a category: **Statistical** or **All**.

6. Choose **CHISQ.TEST**. For the **Actual_range**, highlight the cells containing the observed counts, A1:B2; do *not* include the row and column totals or the grand total. For the **Expected range**, highlight the cells with the expected counts, E1:F2.

   You will see the p-value (0.030671). Press **OK** and it will show up in the active cell in the worksheet.

   The previous steps for Excel will give you the p-value but not the value for chi-square. If you want the numerical value for chi-square, continue with the steps that follow.

7. Click in an empty cell, J2, then click *fx*.

8. Select a category: **Statistical** or **All**.

9. Choose **CHISQ.INV.RT** (for inverse, right tail).

10. For the **Probability**, click on the cell J1 that shows the p-value of **0.030671**. For **Deg_freedom**, put in the degrees of freedom (df). For two-way tables,

    df = (number of rows − 1)(number of columns − 1).

    For Example A, df is 1. Click **OK**. You should get a chi-square of 4.67.

RAW DATA

1. Input the raw data, including the column labels. See Table B.

2. **Insert > PivotTable**

3. In the Create PivotTable dialog box, for **Table/Range**, click on the column heading A and drag across B. The entry will read **$A:$B**. For Existing Worksheet Location, click on cell **E3**. Click **OK**.

4. See Figure 10h, Excel PivotTable Fields. Choose both **Preschool** and **Graduate HS**; both will appear below in the **Rows** Box. Drag the **Preschool** label from the **Rows** Box into the **Columns** Box. Now go back to the upper box and drag **Graduate HS** down into the ∑ **Values** Box.

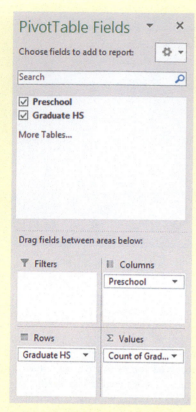

5. Go to step 3 in the TABLE SUMMARY section above, but use new cell locations as needed.

Goodness-of-Fit Test

1. Type your observed counts in one column and your expected counts in another column, in the same order.

2. Follow steps 4, 5, and 6 above.

   You will get a p-value of 0.021726.

   If you want the numerical value of chi-square, follow the instructions in steps 7–11 above. However, in step 11, for goodness-of-fit,

$$\text{df} = \text{number of categories} - 1$$

   For Example B, the number of categories is 5, so df is 4.

   You should get a chi-square value of 11.47.

▲ **FIGURE 10h** StatCrunch Input for Two-way Table

Chi-Square Test for Two-way Tables

TABLE SUMMARY

1. Enter your data summary as shown in Figure 10i. Note that you can have column labels (Preschool or NoPreschool) and also row labels (GradHS or NoGrad).

Row	var1	Preschool	NoPreschool	var
1	GradHS	37	29	
2	NoGrad	20	35	
3				

▲ **FIGURE 10i** StatCrunch Input for Two-way Table

2. **Stat > Tables > Contingency > With Summary**

3. See Figure 10j. Select the columns that contain the summary counts (press keyboard **Ctrl** when selecting the second column), and then select the column that contains the **Row labels**, here **var1**.

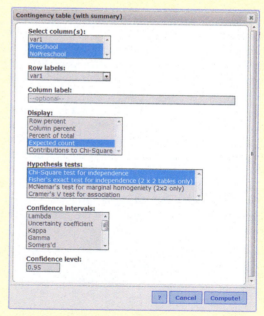

▲ **FIGURE 10j** StatCrunch Two-way-Table Options

4. For **Display**, select **Expected Count**. If Fisher's Exact Test is also wanted, press **Ctrl** on keyboard while selecting **Fisher's exact test for independence** . . .

5. Click **Compute!**

Figure 10k shows the well-labeled output.

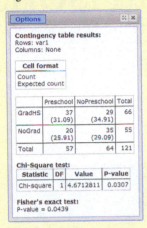

▲ FIGURE 10k StatCrunch Output for Two-way Table

RAW DATA

1. Be sure you have raw data in the columns; see Table B.
2. **Stat > Tables > Contingency > With Data**
3. Select both columns.
4. Select **Expected Count**.
5. Click **Compute!**

Goodness-of-Fit Test

1. Enter the observed counts in one column and the expected counts in another column, in the same order.
2. **Stat > Goodness-of-fit > Chi-Square test**
3. Select the column for **Observed**, and select the column for **Expected**.
4. Click **Compute!**

With the data on integer selection you should get a chi-square of 11.47 and a p-value of 0.0217.

11 Multiple Comparisons and Analysis of Variance

THEME

Comparing the means of more than two groups is a method for examining whether a numerical variable is associated with a categorical variable. Although it is tempting to compare groups pair-by-pair, this leads to problems. ANOVA is a procedure that compares means across groups while avoiding these problems.

n Chapter 9, you learned how to compare the means of two groups using confidence intervals and hypothesis tests. There is nothing magical about the number 2: We often need to compare three, four, or even more groups with each other. Analysis of variance, or ANOVA, is a method for doing just that. The name *analysis of variance* is slightly misleading, because although we will indeed use the variance (recall that the variance is just the standard deviation squared), it is really the *means* we are interested in.

ANOVA is a method for testing whether there is an association between a categorical variable that identifies different groups, and a numerical variable. For example, do you want to know which diet (the categorical variable) is best for losing weight (the numerical variable)? One approach is to randomly assign people to one of several diets and then compare the mean weight loss for each type of diet. If there is an association between the categorical variable *Diet Type* and the numerical variable

Weight Loss, then it may be that the type of diet you choose determines the amount of weight you could lose. Is success in sports determined in part by the color of the team uniform? To examine this question, we could compare the success rates of several different teams after grouping the teams by the categorical variable *Uniform Color*.

We begin this chapter by introducing a famous difficulty in statistical inference: the problem of multiple comparisons. Comparing several groups requires some care. We then discuss a basic mindset for thinking about the variation within and between groups, and we show how this leads to a test statistic that can compare the means of several different groups without raising the problem of multiple comparisons. This test statistic works best under certain conditions, and we discuss these in detail. Finally, we consider post hoc tests, which allow us to explore the data at a greater level of detail.

CASE STUDY

Seeing Red

Does the color of your team uniform affect your performance? That's what researchers in England claimed, after studying data from 68 soccer teams over 56 seasons. On average, teams wearing red won 53% of their home games. Teams wearing blue won, on average, 51%, and teams wearing yellow/orange won only 50%. The researchers compared the performance of teams wearing four different color groups (blue, orange or yellow, red, white) and found significant differences in performance, as measured by the percentage of home games won. (Teams wear their team colors only for home games.) Further investigation showed that teams wearing red did the best of all (Attrill et al. 2008).

You have seen two groups compared before, but how do we compare means across *four* different groups? This chapter covers techniques for doing so. Then we will return to this case study at the end of the chapter and see whether red is, statistically speaking, the "winningest" color.

Multiple Comparisons

In earlier chapters, we compared two groups to each other, to answer questions such as "Is reading comprehension higher on a e-reader than on traditional paper?" or "Do people sleep longer on weekends and holidays than during the week?" In both these situations, there were two groups (e-reader/paper, weekend/weekday), and we used *t*-tests or confidence intervals to compare the differences between the means. We now consider situations that involve more than two gropus. As it turns out, special care is needed, and in this section we will show you why. We begin by showing what the data you encounter might look like for these situations.

Data for Multiple Comparisons

How much oil—none, a medium amount, or a lot—will produce the most popped popcorn kernels, on average? Is it better, if you want to remember what was said during a lecture, to takes notes with paper and pencil, with a laptop, or with a laptop but summarizing rather than transcribing the lecture? These statistical questions involve comparing a numerical response variable—the number of kernels popped and the score on a memory test—with a categorical variable that includes more than two groups.

Table 11.1 shows the first six rows of data from the popcorn-popping experiment in the Case Study of Chapter 10. One factor that the experimenters examined was the amount of oil. They tried three different amounts of oil: none, 1/2 teaspoon, and 1 teaspoon. For our purposes, we call these groups None, Medium, and Maximum. Then they popped 36 bags using no oil, another 36 using a medium amount of oil, and another 36 using the maximum amount. Each bag contained exactly 50 kernels of popcorn. After 75 seconds, the bags were removed from heat, and the experimenters counted and recorded the number of kernels popped. Does the amount of oil that is used affect the number of kernels that pop?

The data set has two variables: a categorical variable (*Amount of Oil*) and a numerical variable (*Number of Kernels Popped*).

We might summarize these data by showing the mean and standard deviation of the number of kernels popped in each group. Graphically, we might show boxplots to summarize the shape of the distributions. Table 11.2 shows the summary statistics. Figure 11.1 shows boxplots of these data.

Amount of Oil	Number of Kernels Popped
Maximum	8
Medium	26
Maximum	5
None	8
None	7
Medium	20

▲ **TABLE 11.1** The first six lines of the popcorn data set.

▶ **TABLE 11.2** The mean and standard deviation (in parentheses) for the number of kernels popped for each amount of oil.

Group:	None	Medium	Maximum
Mean (SD)	19.75 (11.76)	19.75 (11.83)	13.47 (9.30)

▶ **FIGURE 11.1** Boxplots showing the five-number summary of the distribution of number of kernels popped. The group in which the maximum amount of oil was used had a lower median number of good kernels than the other two groups.

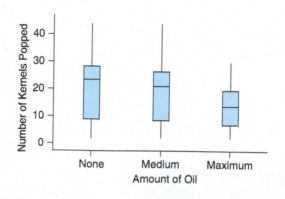

The Problem of Multiple Comparisons

Figure 11.2 shows another perspective on the popcorn data. Each dot represents a bag of popcorn and shows us how many kernels were popped. The mean number of popped kernels in each group is indicated by a red dot. We want to know whether the amount of oil affects the number of kernels popped. As you can see, this varies from bag to bag. But are the population means different? Or are observed differences of sample means due to chance? That is, if we were to pop a very large number of bags under these three different conditions, might the mean number of kernels popped all be the same?

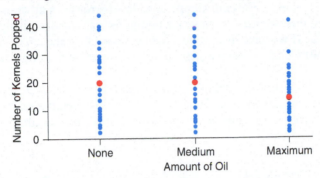

◄ FIGURE 11.2 Each dot represents a bag with 50 popcorn kernels in it. The bags were heated for 75 seconds, and the graph shows the number of kernels that had popped after this time. Different amounts of oil were used. The large red dots show the mean number of kernels popped under each treatment condition.

In Chapter 9, we would have answered these questions by doing a two-sample *t*-test. But as the name implies, two-sample *t*-tests are used for comparing only two groups. Here, we need to compare three. We might try doing three separate two-sample *t*-tests. For example, we might test

Hypothesis 1
    H_0: Mean with no oil = mean with medium amount of oil
    H_a: Mean with no oil ≠ mean with medium amount of oil

Hypothesis 2
    H_0: Mean with no oil = mean with maximum oil
    H_a: Mean with no oil ≠ mean with maximum oil

Hypothesis 3
    H_0: Mean with medium oil = mean with maximum oil
    H_a: Mean with medium oil ≠ mean with maximum oil

To compare all three groups, we need three different hypothesis tests. This is called a **multiple comparison**, because we are comparing multiple pairs of means. But this causes a problem.

To understand the problem, recall that whenever we do a significance test, there is a small probability that we will make the mistake of rejecting the null hypothesis when it is true. This probability is called the significance level, and we usually set it to be 0.05, or 5%. Here, for instance, there is a 5% chance that if the mean number of kernels is the same with no oil as with medium oil, we will mistakenly conclude that the amount is different.

The basic idea with multiple comparisons is that even though the probability of something going wrong on one occasion is small, if we keep repeating the experiment, eventually something *will* go wrong. Essentially, by doing multiple tests, we are creating more opportunities to mistakenly reject the null hypothesis. The more tests we do, the greater the probability that we'll mistakenly reject a null hypothesis at least once.

The probability that we will mistakenly reject a null hypothesis at least once after doing several hypothesis tests is called the **overall significance level**. Making "at least one" such mistake means that we make this error for one test, for two tests, or for all tests. When three hypothesis tests are performed, each with a significance level of 5%, the overall significance level is about 14%. In other words, the probability that we will conclude that at least one amount of oil is more effective than another, when the truth is that all amounts are equally effective, is 14%. This is not terribly high, but we were shooting for a 5% error rate, and 14% is quite a bit higher.

> **KEY POINT** The overall significance level is the probability that you will mistakenly reject a null hypothesis (that is, reject a null hypothesis when it is true) in at least one of several hypothesis tests. The overall significance level is always larger than the significance level for any one of the individual tests.

Details

Six Comparisons from Four Groups

The groups are A, B, C and D; the comparisons are AB, AC, AD, BC, BD, and CD.

When we have only three groups to compare, we need to make only three comparisons, and the overall significance level goes from 5% (for one comparison) to 14%. But if we have more groups, the overall significance level error goes up dramatically. If we have four groups, then we need to make six comparisons. The overall significance level is now about 26%. If we have five groups, then we need to make ten comparisons, and there is about a 40% chance that we will mistakenly reject at least one of the ten null hypotheses.

Figure 11.3 shows how the overall significance level increases as the number of groups increases. You can see that if you have ten groups, then the probability that you will mistakenly reject at least one null hypothesis is 90%. You are almost certain to make a mistake!

▶ **FIGURE 11.3** The overall significance level—the probability that at least once we'll reject the null hypothesis when in fact it is true—increases dramatically as the number of groups increases. If there are ten groups and all the means are equal, then the probability that we will mistakenly conclude that at least one is different from the others is about 90%.

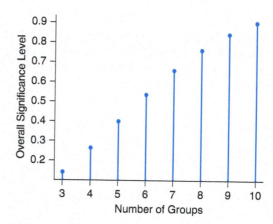

One Solution: The Bonferroni Correction

Through the years, statisticians have worked out different ways of solving the problem of multiple comparisons. Some methods are fairly complex (both mathematically and computationally) and are best left to a computer. We'll talk about one of these in Section 11.4. But one method, the Bonferroni Correction, is fairly straightforward.

The **Bonferroni Correction** is done by using a smaller significance level for each individual test. If you want an overall significance level of α, and you are doing m comparisons, then you should do each individual comparison with significance level α/m. If you do this, then the overall significance level will be α or smaller. (Usually, it will be smaller.)

Finding the Number of Comparisons The number of comparisons required depends on the number of groups that you are comparing. In the popcorn experiment, there were three groups and three comparisons. The fact that in this case the number of comparisons is the same as the number of groups is a bit of a coincidence, and it happens only when there are three groups. In general, the number of comparisons is the same as the number of different ways you can select two objects from a group of objects. Formula 11.1 shows how to find this number.

Formula 11.1: Number of comparisons $= \dfrac{k(k-1)}{2}$

where k is the number of groups you are comparing.

EXAMPLE 1 Health of Sea Lions

Biologists collected data on the health of sea lions living in four distinct regions on the northern Pacific Coast. They took measurements that reflect the overall health of sea lion colonies living in these regions. They wish to test the hypothesis that the mean health is the same for each region.

QUESTIONS

a. If they choose to do t-tests comparing each region to every other region, how many comparisons must they make?

b. If they wish to have an overall significance level of $\alpha = 0.05$, applying the Bonferroni Correction, what significance level should they use for each individual t-test?

SOLUTIONS

a. We have $k = 4$ groups, so we need

$$\text{Number of comparisons} = \frac{k(k-1)}{2} = \frac{4(3)}{2} = \frac{12}{2} = 6$$

b. The Bonferroni Correction says to divide the overall significance by the number of comparisons, so we should use $\alpha = (0.05/6)$, which is about 0.0083, for each individual test.

TRY THIS! Exercise 11.3

KEY POINT

When comparing more than two groups with hypothesis tests, in order to keep the overall significance level at or below the given value of α (usually 0.05), the Bonferroni Correction divides α by the number of comparisons, not the number of groups.

EXAMPLE 2 Popping Corn in Oil

The following are results from three t-tests that compare all possible pairs of groups to test which amount of oil works best for popping corn. Each group had 36 bags of popcorn, each with the same number of kernels. The researchers recorded the number of kernels popped after a fixed amount of time and then tested whether the mean amount differed for different amounts of oil.

Hypothesis 1
$H_0: \mu_{none} = \mu_{med}$
$H_a: \mu_{none} \neq \mu_{med}$
$t = 0$, p-value $= 1.000$

Hypothesis 2
$H_0: \mu_{none} = \mu_{max}$
$H_a: \mu_{none} \neq \mu_{max}$
$t = 2.51$, p-value $= 0.014$

Hypothesis 3
$H_0: \mu_{med} = \mu_{max}$
$H_a: \mu_{med} \neq \mu_{max}$
$t = 2.50$, p-value $= 0.015$

QUESTION Apply the Bonferroni Correction to test whether the mean number of kernels popped for each group is different from those for the other groups. Use an overall significance level of $\alpha = 0.05$.

SOLUTION There are three groups, so we are making three comparisons. This means that to have an overall significance level of 0.05, we conduct each individual test with a significance level of $0.05/3 = 0.0167$. Thus we reject the null hypothesis if the p-value is less than or equal to 0.0167.

The p-value for Hypothesis 1 (1.000) is clearly bigger than 0.0167, so we do not reject the null hypothesis.

The p-value for Hypothesis 2 (0.014) is smaller than 0.0167, so we can reject the null hypothesis.

The p-value for Hypothesis 3 (0.015) is also smaller than 0.0167, so we can reject the null hypothesis.

CONCLUSION We find that the mean for no oil is significantly different from the mean for maximum oil and also that the mean for medium oil is significantly different from the mean for maximum oil. However, there is no significant difference between the means for medium oil and no oil.

In practice, you should remember to check whether the conditions necessary for using the *t*-test hold before carrying out these tests. For this example, we skipped these details to focus on the main idea: When doing multiple comparisons, we need to make adjustments to keep the overall significance level to a reasonable value.

TRY THIS! Exercise 11.5

> **Details**
>
> **Statistical Power**
> Statistical power is the name for the probability that a hypothesis will correctly reject the null hypothesis when the null is wrong.

One drawback of the Bonferroni Correction is that it is very conservative. This can be good, because it means that the true overall significance level is even lower than α. The bad news is that using the Bonferroni Correction will lower the probability that we will *correctly* determine that two groups have different means.

Bonferroni Confidence Intervals In Chapter 9 you learned that you can compare the means of two independent groups either by using a two-sample *t*-test or by constructing a two-sample confidence interval for the difference between means. Confidence intervals are useful because they give more information than a *t*-test, such as estimates about the size of the difference between the means.

When examining several groups, we can also construct confidence intervals for each pair of groups. However, we have the same problem that we have with the hypothesis test. If we construct a 95% confidence interval for each pair of groups, it is true that we're 95% confident that each interval includes the true difference. This means that if we were to repeat this procedure with new data, there would be only a 5% chance that any given interval would *not* include the true difference in means. Unfortunately, there would be more than a 5% chance that *at least one* of the confidence intervals we constructed would not include the true difference in means.

The Bonferroni approach can fix this. The idea is that we make each individual confidence interval slightly wider (use a larger margin of error) so that we achieve the desired overall confidence level.

To see how to do this, it helps to use a special notation and write the confidence level like this:

$$L = (1 - \alpha)100\%$$

For example, if $\alpha = 0.05$, then the confidence level is

$$L = (1 - 0.05)100\% = 0.95 \times 100\% = 95\%$$

The Bonferroni Correction says that if you want an overall confidence level of L, then you must make each individual confidence interval with this level:

$$(1 - \alpha/m)100\%$$

where m is the number of intervals you are making.

For example, suppose we have $m = 3$ comparisons, and so three intervals are needed. If we want an overall confidence level of 95%, then $\alpha = 0.05$ (because $95\% = 100\% - 5\%$, so $\alpha = 0.05$). The Bonferroni Correction says to make each interval with confidence level

$$(1 - 0.05/3)100\% = (1 - 0.0167)100\% = 98.33\%$$

Below you can see the output for Bonferroni intervals for the popcorn example. Because we have three comparisons and are finding three intervals, we use $m = 3$. Each is a 98.33% confidence interval. Together, these are called *95% **simultaneous confidence intervals**,* because if we follow the Bonferroni approach, we can be 95% confident that *all three* of the intervals include the true difference in means. Note that we come to the same conclusion as we did with the *t*-tests. These intervals are illustrated in Figure 11.4.

	Lower Boundary	Upper Boundary	Conclusion
No oil − medium oil	−6.82	6.82	Not different
No oil − maximum oil	0.14	12.41	No oil pops more, on average, than maximum oil
Medium oil − maximum oil	0.12	12.44	Medium oil pops more, on average, than maximum oil

Tech

◀ FIGURE 11.4 Visual representation of three Bonferroni-corrected confidence intervals. We see that only the lower interval includes 0, which tells us that there is no difference in the mean number of good kernels popped for no oil and medium oil. The other two intervals just barely avoid 0, so we learn that using maximum oil produces fewer popped kernels than using either medium or no oil.

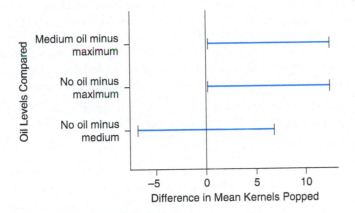

To interpret these intervals, we use the same approaches as in previous chapters. If the interval includes the value 0, then it is plausible that the true difference in means is 0. In this case, we would say that there's not enough evidence to conclude that the means are different.

If, on the other hand, the interval does not include 0, we can conclude (with a 95% overall level of confidence) that the means are different, because their difference is not plausibly equal to 0.

The interval for the difference in the mean number of kernels popped with no oil and the mean number popped with medium amounts of oil captures 0, so there is no evidence that you will get different results using no oil and using a medium amount of oil. On the other hand, the other two intervals do not include 0 and include all positive

---

Looking Back

Confidence Level
A confidence level of L tells us that if we were to take a great many different samples, and for each compute a confidence interval with confidence level L, then close to L% of those intervals would include the population parameter, and $100\% - L$ would not.

values. So we can be confident that using no oil produces, on average, more popped kernels than using the maximum amount of oil, and using medium amounts of oil results in a greater number of popped kernels, on average, than using the maximum amount of oil.

> **KEY POINT**
>
> We can use the Bonferroni Correction to produce a collection of simultaneous confidence intervals so that we can be 95% confident that all of the intervals include the population parameters.

The ANOVA procedure (which we introduce next), like the Bonferroni approach, "solves" the problem of multiple comparisons for some contexts. ANOVA has the advantage of providing a higher power than the Bonferroni approach.

The Analysis of Variance

 Details

Pronunciation
ANOVA is pronounced with the emphasis on the second syllable, an-**OH**-va.

 **Looking Back**

Variance vs. Variation
Both *standard deviation* and *variance* have specific mathematical definitions. Standard deviation and variance are methods for measuring *variation*, which is itself a more general term and has no mathematical definition.

The **analysis of variance**, also known as **ANOVA**, can be used to answer a very specific type of question: "Is this numerical variable associated with this categorical variable?" This type of question is answered by refining it into a more statistical question along the lines of "Do the population means of a response variable differ across these categories?" ANOVA is a procedure for comparing the means of several groups.

ANOVA is a more powerful approach than the Bonferroni Correction described in the last section. What we mean by this is that if there really is a difference in the means, then ANOVA has a better chance of finding it. ANOVA is also quite a bit less tedious than doing a series of *t*-tests.

When we are comparing means of one variable across several different groups (one categorical variable and one numerical variable), we use a **one-way analysis of variance**. It is also possible to have two-way ANOVAs, which are used to compare means when there are two categorical variables and one numerical response variable. Even "higher-way" ANOVAs are possible. These are all usually covered in later statistics courses.

The null and alternative hypotheses are the same for every ANOVA:

H_0: There is no association between the categorical variable and the numerical variable; the population means are the same for all categories.

H_a: There is an association; the population means of the categories differ.

Another way of stating this is to consider the different groups of the categorical variable. Suppose we have k different groups. For example, for the popcorn experiment, $k = 3$: no oil, medium oil, maximum oil. Then the hypotheses are

H_0: $\mu_1 = \mu_2 = \cdots = \mu_k$

H_a: At least one of the means is not equal to another.

In the context of popping corn, the null hypothesis says that no matter what amount of oil you use, you'll get the same mean number of kernels popped.

The alternative hypothesis claims that the two variables are associated and that, therefore, at least one mean is different from another. Different amounts of oil are associated with differing numbers of kernels popped. In a controlled study such as this one, we can claim causality: changing the amount of oil results in a different mean number of popped kernels

Rejecting the null hypothesis for an ANOVA is usually unfulfilling. We learn that at least one of the means is different from another, but we do not learn which one.

We do not learn whether there is more than one. We don't learn which is the biggest, which is the smallest, or really any information other than that the null hypothesis is not true. For this reason, an ANOVA test is often followed up with an exploration to look further at the relationship of the groups to one another. We'll give some approaches for this exploration, which is often called post hoc analysis, in Section 11.4.

> **KEY POINT**
>
> ANOVA tests whether a categorical variable is associated with a numerical variable. This is the same as testing whether the mean value of a numerical variable is different in different groups.

The ANOVA is based on comparing the amount of variation that exists between groups with the variation within each group. To help you understand what this means, let's think about our analysis visually.

Visualizing It

As the name suggests, the analysis of variance determines whether the means from the groups are the same or different by comparing the variances of the groups. Basically, if at least one of the means is very far away from another, we want to reject the null hypothesis. Here *very far* means "very far, relative to the amount of variation within the data."

Figure 11.5a shows a plot of means from four groups. (The data are simulated.) Do the means look similar or different to you?

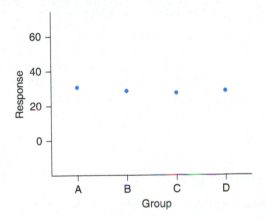

◄ **FIGURE 11.5a** The means of four different groups. The means are slightly different, but are they close enough together for the observed differences to be due to chance? Or are they far enough apart that we conclude they're actually different?

Figure 11.5b shows the same four means, but now we also show the data points themselves. You can see that each group has lots of variability. We would say that these means are actually fairly close together, because the means are more similar to each other than are individual values within any of the groups. In this case, we would expect

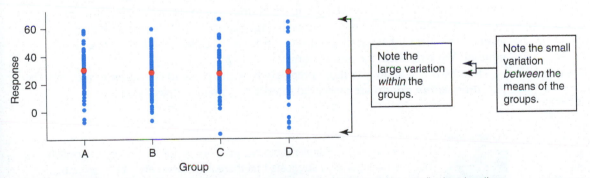

▲ **FIGURE 11.5b** With the actual data points displayed, we now see that the means are actually close together. The distances between means are much less than the distances between points within any of the groups.

our ANOVA *not* to reject the null hypothesis. There is not enough evidence to convince us that the means are different from each other. The small differences we see could easily be explained by chance.

Figure 11.5c shows another scenario. The means *are exactly the same as in Figures 11.5a and 11.5b*, but in this new data set, the variability is so small that the group means look very different from each other. The standard deviation *between* the group means is unchanged. But now the standard deviation within each group is much smaller. The amount of variation between the group means is much greater than the variation within each group, so the means are more different from each other. In this case, we would expect our ANOVA procedure to reject the null hypothesis.

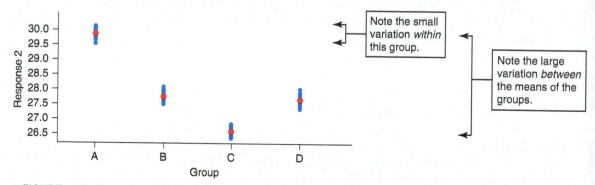

▲ **FIGURE 11.5c** These data have the same means as Figures 11.5a and 11.5b, but there is much less variation *within* each group. In this context, the means are very different from each other.

Although ANOVA focuses on means, boxplots are often useful as a first step for visualizing ANOVA. This is true even though, as you'll recall, boxplots display the median, not the mean, of a sample of data. Particularly if the distributions of the groups are roughly symmetric, then the mean and the median are nearly equal, so in this case the boxplots help us see how different the means are. Figure 11.6 shows boxplots for the four groups in Figure 11.5c.

Putting a Number on It

To perform a hypothesis test, we need a test statistic. This means that these informal ideas—variation between groups and variation within groups—need to be turned into a number. The *F*-statistic, the test statistic we will use for ANOVA, does just that.

$$F = \frac{\text{Variation between groups}}{\text{Variation within groups}}$$

The *F*-statistic compares the variation between groups to the variation within groups. If the variation between groups is big relative to the variation within groups (as in Figure 11.5c), then *F* will tend to be a large number. If the variation between groups is small relative to that within groups (as in Figure 11.5b), then *F* will be a small number. We will reject the null hypothesis when *F* is too big.

Soon we will show you more precisely how the *variation between groups* and the *variation within groups* are measured in order to calculate the *F*-statistic.

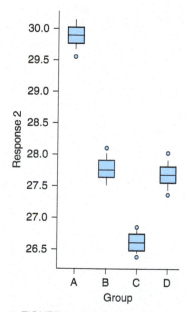

▲ **FIGURE 11.6** These boxplots show us mostly the same information as in Figure 11.5c. We see that the medians of the groups are very different with respect to the variation within each group. Because the distributions are symmetric, the medians (represented by the horizontal lines in the middle of the boxes) are very close to the means.

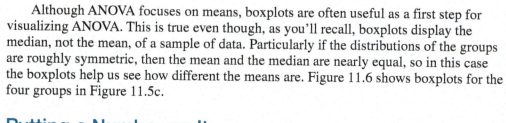

KEY POINT Large values of the *F*-statistic discredit the null hypothesis of no association, because a large value suggests that there is more variation between the groups than within each group.

EXAMPLE 3 Large and Small *F*-Values

We calculated the *F*-statistics for Figures 11.5b and 11.5c. One of these had an *F*-value of 1.03, and the other had an *F*-value of 17344.81.

QUESTION   Which observed *F*-statistic belongs to which figure, and why?

CONCLUSION   The null hypothesis is that the means are the same. Large values of the *F*-statistic discredit the null hypothesis. Although the means in Figures 11.5b and 11.5c have the same numerical value, the variation between groups is much greater in Figure 11.5c when compared to the variation within each group. For this reason, the large *F*-statistic of 17344.81 must belong to Figure 11.5c.

TRY THIS!   Exercise 11.15

Most statistical software packages produce something called an **ANOVA table**. This table includes lots of information (some of it redundant), including the value of the *F*-statistic. Example 4 shows an example of an ANOVA table.

EXAMPLE 4 Sleep and Economic Group

The American Time Use Survey (www.bls.gov/tus/) asked a random sample of Americans to report the number of hours of sleep they got, their weekly income, and other demographic questions, as well as to answer questions about how they spent their time on a randomly chosen day. We divided the sample into three economic groups, because we wondered whether the amount of sleep (in minutes) we get each night is associated with how much money we make.

The economic groups are called "lower" (those in the lower third of reported weekly income), "middle," and "upper" (those in the top third). The samples are large; each group has about 2100 people in it. Figure 11.7a summarizes the amounts of sleep for these groups. Although the typical amounts of sleep seem similar in these groups, the large sample size means that if, in fact, the mean amount of sleep truly differs in the population, we'll have a good chance of detecting this difference.

(a)

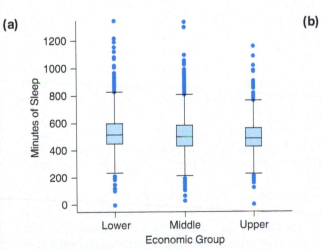

(b)

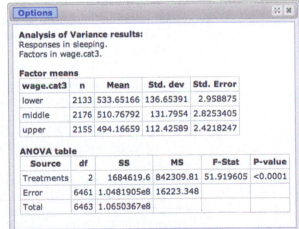

Options

Analysis of Variance results:
Responses in sleeping.
Factors in wage.cat3.

Factor means

wage.cat3	n	Mean	Std. dev	Std. Error
lower	2133	533.65166	136.65391	2.958875
middle	2176	510.76792	131.7954	2.8253405
upper	2155	494.16659	112.42589	2.4218247

ANOVA table

Source	df	SS	MS	F-Stat	P-value
Treatments	2	1684619.6	842309.81	51.919605	<0.0001
Error	6461	1.0481905e8	16223.348		
Total	6463	1.0650367e8			

▲ **FIGURE 11.7 (a)** Boxplots comparing the three economic groups and the amount of reported sleep (minutes). **(b)** ANOVA output provided by StatCrunch.

QUESTION   State the null and alternative hypotheses. Give the value of the *F*-statistic. What does this value tell us about the amount of variability between the three groups compared to the amount of variability within each group?

SOLUTION   The null hypothesis is that the mean amount of sleep is the same for all three economic groups. The alternative is that at least one group sleeps for a different mean length of time than another.

The *F*-value of 51.9 tells us that the variability between the groups was almost 52 times greater than the variability within groups.

TRY THIS!   Exercise 11.17

ANOVA in Context: A Tour of the ANOVA Table

What's the best way to take notes when attending a lecture? Researchers at Yale University and the University of California, Los Angeles, wanted to know (Mueller and Oppenheimer 2014). One hundred fifty-one student volunteers were shown a video of a lecture on an arbitrary topic. Students were randomly assigned to one of three groups. One group was asked to take notes using longhand (that is, writing them out with pencil and paper). Another group was asked to take notes using a laptop. The third group used a laptop but were asked to not literally transcribe the lecture but, as much as possible, to summarize what is said in their own words. One reason for this third condition is that the researchers had seen from a previous study that students who took notes by laptop were more likely to attempt to transcribe the lecturer verbatim than were students taking notes longhand.

After the lecture, students were given tests of their recall of the factual content of the lecture and of their conceptual understanding of the lecture. We'll focus on their conceptual understanding. Students' scores on conceptual understanding varied a fair amount. What is the explanation for the varying levels of conceptual understanding?

Explained Variation + Unexpected Variation = Total Variation   You can probably think of lots of reasons why the scores on the conceptual test varied. Maybe prior knowledge mattered. Maybe hearing difficulties mattered. One possible explanation, the one the researchers were most interested in, was whether the *method* that students used to take notes mattered.

Most statistical software packages produce an ANOVA table. This table breaks the total variation (measured by the **total sum of squares**) into two pieces: the **explained variation** and the **unexplained variation**. The explained variation is just the **variation between groups**. The farther apart the means of the groups are, the bigger the explained variation will be. The explained variation is sometimes called the **treatment variation**, because it is the amount of variation that is explained by the treatment (in this case, the diet to which the subjects were assigned).

The unexplained variation is the **variation within groups**. This is sometimes called the **error variation** or the **residual variation**. It is called residual because it is the variation that is "left over" after we have explained some of the variation with the *Treatment.*

Figure 11.8 shows the total variation in conceptual understanding scores (144.7527) in the row labeled Total under the column labeled SS (for *s*um of *s*quares). You can see that if you add the variation due to treatments (labeled with the variable name "condition"), 139.38244, to the variation due to error, 5.3702693, you get the total variation. However, in order to compare within-group and between-group variation, we need to rely instead on the average, or mean, amount of variation.

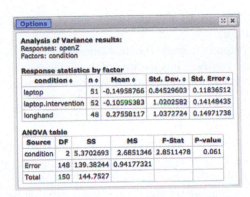

◀ FIGURE 11.8 ANOVA output for conceptual understanding scores for three groups of notetakers. The explained variation is caused by the different methods used to take notes (the *condition*) and is listed here as "condition." The mean variation due to condition is 2.69 units.

The Mean Sum of Squares The column labeled MS in the ANOVA table contains the **mean sum of squares**. This number is simply the SS value divided by the df value. We need to divide the SS by something in order to compare the "treatment" variation and the "error" variation. This is because the SS Treatment gets bigger as the number of groups gets bigger, and the SS Error gets bigger as the sample size gets bigger. For this reason, we need to adjust the SS due to treatments for the number of groups and adjust the SS due to error by the number of observations.

> **KEY POINT** Total variation as measured by the sum of squares can be broken into two pieces: explained (Treatment) variation and unexplained (Error) variation. If the explained variation is large compared to the unexplained variation, it suggests that the means of the groups are very different.

The df value, the degrees of freedom, does the adjustment. The df for SS Treatment is the number of groups minus 1. The df for SS Error is the sample size minus the number of groups. By dividing the SS by their degrees of freedom, we get a sort of average sum of squares. This is why it is called the mean sum of squares.

For instance, in Figure 11.8, the sum of squares due to the treatments (the methods for notetaking) is 5.3702693. Because there are 2 degrees of freedom, the mean sum of squares, abbreviated MS, is $5.3702693/2 = 2.6851346$. The mean sum of squares due to the error (the unexplained variation) is $139.38244/148 = 0.94177321$.

The *F*-statistic is calculated with the ratio of these two values:

$$F = \frac{\text{Variation between groups}}{\text{Variation within groups}} = \frac{MS_{\text{between}}}{MS_{\text{within}}} = \frac{2.6851346}{.94177321} = 2.8511478$$

The value for the *F*-statistic tells us that the variation between groups is close to three times bigger than the variation within groups. The situation is more similar to Figure 11.5c than to Figure 11.5b.

In Other Words There are many different phrases used in ANOVA to describe the same thing.

Variation between groups is also called *variation due to treatment, explained variation, and variation due to factors.*

Variation within groups is also called *variation due to error, unexplained variation,* and *residual variation.*

Which term is used depends on the software you are using, but many authors use all of these terms interchangeably. Different software label the columns and rows of an ANOVA table with slightly different labels.

Relation of Total Sum of Squares to Variance ANOVA stands for the analysis of *variance*. We have discussed measuring variation and have shown that the ANOVA table measures variation with the sum of squares and with the mean sum of squares. What does this have to do with the variance and standard deviation that we used earlier to measure variation?

In fact, they are very closely related. The total sum of squares is simply the numerator in the formula for finding the variance. It is the sum of the squared deviations from the mean.

Recall that the variance is the square of the standard deviation:

$$s^2 = \frac{\sum(x - \bar{x})^2}{n - 1} = \frac{\sum(\text{deviation})^2}{n - 1} = \frac{SS}{n - 1}$$

This means that we can get the variance of the conceptual scores for all of the participants by dividing the total sum of squares (144.7527) by the total sample size minus 1, which for our note takers is 150 (because n is 151 and $151 - 1 = 150$). There is no need to do this calculation, but we mention it to emphasize that essentially, the ANOVA table uses the same measure of variation that you learned in Chapter 3.

EXAMPLE 5 Fuel Economy

Does fuel economy on the highway (measured in the United States by the number of miles a car typically travels on one gallon of fuel) vary by the number of cylinders in a car's engine? Data from the U.S. Department of Energy are used to perform an ANOVA. The sum of squares due to treatments has been hidden in Figure 11.9.

◀ **FIGURE 11.9** ANOVA table to test whether mean mileage differs for different types of cars. Produced by StatCrunch.

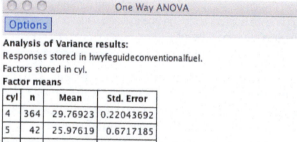

One Way ANOVA

Options

Analysis of Variance results:
Responses stored in hwyfeguideconventionalfuel.
Factors stored in cyl.
Factor means

cyl	n	Mean	Std. Error
4	364	29.76923	0.22043692
5	42	25.97619	0.6717185
6	400	24.315	0.13559277
8	243	20.074074	0.15366969

ANOVA table

Source	df	SS	MS	F-Stat	P-value
Treatments	3	▬▬▬	▬▬▬	432.80005	<0.0001
Error	1045	11520.568	11.024467		
Total	1048	25834.738			

QUESTION   What is the value of the sum of squares due to treatments? What is the value of the mean sum of squares due to treatments?

SOLUTION   Total SS = SS due to treatments + SS due to error

Therefore,

SS due to treatments = total SS − SS due to error

SS due to treatments = 25834.738 − 11520.568 = 14314.17

To find the MS due to treatments, we divide the SS due to treatments by the degrees of freedom:

MS due to treatments = 14314.17/3 = 4771.4

CONCLUSION   The sum of squares due to treatments is 14314.17.

TRY THIS!   Exercise 11.19

The ANOVA table actually has many redundant parts. Table 11.3 shows a generic ANOVA table and helps us see how the different parts are related.

Source	df	SS	MS	F-Statistic	p-value
Treatment/Factor/Explained	$k - 1$	_____	$SS_{tmt}/(k - 1)$	MS_{tmt}/MS_{error}	xxxx
Error/Residual/Unexplained	$N - k$	_____	$SS_{error}/(N - k)$		
Total	$N - 1$	$\sum (x - \bar{\bar{x}})^2$			

◀ **TABLE 11.3** Generic ANOVA table.

The x with the double bar, $\bar{\bar{x}}$, is used to represent the average of all the observations pooled together; N is the total number of observed values; and k is the number of groups. We do not give the formulas for the SS because these require introducing a great deal of special notation, and the details are more than you need at this point.

SECTION 11.3

The ANOVA Test

You've seen that if the null hypothesis is true and the mean value is the same for all categories, then the F-statistic will be fairly small. Large values of the F-statistic discredit the null hypothesis, but how do we tell whether a particular F-statistic is large because the null hypothesis is wrong and the means are different, or is it large just because of chance?

Finding the p-Value

The p-value measures our surprise at the outcome and is used to determine whether the F-statistic is "large." Recall that the p-value is the probability of getting a test statistic as extreme as or more extreme than the observed value, assuming the null hypothesis is true. For ANOVA, the p-value is the probability that you get an F-statistic equal to or larger than the observed value, assuming all the means are equal.

To find this probability, we need to know the sampling distribution of the F-statistic. This requires that certain conditions hold:

1. *Random Sample and Independent Measurements.* Observations can be thought of as a random sample from a population, and individual measurements within a group are independent of each other.

2. *Independent Groups.* The groups are independent of each other.

3. *Same Variance.* The population variances (or, if you prefer, the standard deviations) of the groups are the same.

4. *Normal Distribution or Large Sample.* The distribution of observations is Normal in each group's population or the sample size is large (at least 25 in each group).

Under these conditions, the distribution of the F-statistic follows something called the F-distribution (no surprise here). The F-distribution has two parameters, and both are called the degrees of freedom. These are found in the ANOVA table. The first df parameter is the one given for treatment/explained variation, and the second is the one given for the error/residual/unexplained variation.

The p-value is usually given in the ANOVA table, as you can see from Figures 11.8 and 11.9. Still, it's good to keep a visual in mind so that you understand where the p-value comes from. Figure 11.10 shows the F-distribution with 2 and 148 degrees of freedom, to test the hypothesis that the three note-taking groups all have the same means. The shaded area represents the p-value, which is 0.061.

▶ **FIGURE 11.10** An *F*-distribution with 2 and 148 degrees of freedom. The shaded area is the probability that an *F*-statistic will be larger than 2.85. This value is about 6.1%. This is the p-value calculated for the ANOVA shown in Figure 11.8 to test whether the note-taking methods differed in conceptual understanding.

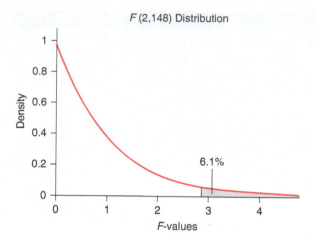

If we use a significance level of 0.05, then this p-value is too large to reject the null hypothesis. We conclude that these data provide no evidence that method of note taking affects conceptual understanding. However, the p-value is quite close to 0.05. In fact, the authors of this study went beyond what's described here, and were able to demonstrate that students who take notes by hand and get a chance to study their notes, do better than those who take notes by laptop and study their notes.

Of the four conditions required for finding the p-value, the first two (independent groups and independent observations within groups) are the most important. Unfortunately, these two are also the hardest to check. Independence of groups and independence of observations within groups are results of the data collection method, so if this is well documented, we can gain some idea whether these conditions hold. Sometimes we just have to assume that they do, understanding that if we are wrong, our results could be very wrong.

The first condition, random sample and independent measurements, is similar to the requirement for the one-sample *t*-test of Chapter 9. Having objects within a group that are associated with each other violates this condition. For example, if we randomly sample family members within each group, then for many purposes these measurements would be associated with each other, not independent.

The second condition, independent groups, would be violated if, say, the same objects were measured in each group. This is a fairly common data structure, particularly in medical research. For example, subjects might receive three different medications, one every six months. Researchers measure the subjects' reaction to the medications. The "groups" are the different medications, but because the same subjects appear in each group, the groups are not independent. A procedure called repeated measures ANOVA exists for analyzing data such as these, but it is beyond the scope of this book.

The third condition, that the variances or standard deviations of the groups be equal, is sometimes called the homoskedastic condition. **Homoskedasticity** is a fancy word for "having the same variance." One rule of thumb for checking this condition is that the largest standard deviation must be no more than twice the smallest.

For example, the standard deviations of the note-taking groups are

Longhand	0.85
Laptop	1.02
Laptop Intervention	1.04

The smallest SD is 0.85 units so as long as the largest SD is smaller than $2 \times 0.85 = 1.7$ we may assume this condition holds. Another way to see this is to divide the largest SD by the smallest: $1.04/0.85 = 1.22$, which is smaller than 2.

✗ **Details**

Same Variance

One advantage of doing *t*-tests for all comparisons, and applying a correction such as the Bonferroni correction, is that we do not need to require equal variances for the groups.

In other books, you may come across other rules of thumb for checking this condition. The ANOVA procedure is fairly resistant to slight deviations from homoskedasticity, which is why some rules of thumb are slightly different.

The final condition is that the observations in each group be sampled from a population that follows a Normal distribution. This can be checked by examining a histogram for each group. However, unless the sample sizes are fairly large in each group, this can be hard to check because histograms from Normal populations don't always look Normal when the sample size is small. Another approach requires more work but gives better insight: Subtract each observation from the average of its group. Then make a histogram of all these differences (called **residuals**). If the distribution of these residuals looks Normal, the condition holds. The *F*-test is more sensitive to departures from Normality than other procedures we've studied up to this point, so some caution is needed. There are other, more sophisticated methods for checking the Normality condition, though this book will not cover them in detail. For example, the Data Project in Chapter 6 mentions QQ plots as one useful method.

What If Conditions Are Not Satisfied? If either of the assumptions about independence is violated, then a one-way ANOVA is not the correct procedure to use. Other procedures exist for different situations, and these are usually covered in more advanced statistics courses. In Chapter 13, we discuss some techniques (called nonparametric procedures) that are sometimes useful if the same variance condition fails.

EXAMPLE **6** Checking Conditions

Atmospheric scientists often use *anomaly temperatures* to study global climate change. These temperatures are the difference between the readings at a particular location and the average global temperature for the entire twentieth century (1901–2000). A negative anomaly temperature is lower than the reference point, and a positive value means it is above that reference point. These are also called "residual" temperatures. All temperatures are measured in degrees Celsius (°C). The boxplots in Figure 11.11 on the next page show global surface temperatures grouped by "double-decades" for the last century (that is, 1880–1899, 1900–1919, and so on). The data are global monthly averages—the average temperature on land around the world for a particular month. The temperatures are collected by a network of sensors at fixed locations around the planet. Figure 11.12 on the next page shows a histogram of the residuals from an ANOVA. It is tempting to do an ANOVA, because we have a numerical response variable (*Temperature Anomaly*) and a categorical predictor variable (*Double-Decade*). However, the conditions do not hold.

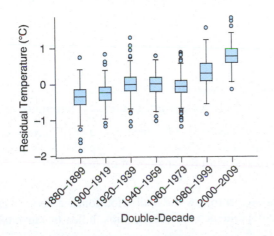

◀ FIGURE 11.11 Mean global surface temperatures for each "double-decade." Temperatures were measured relative to a reference temperature, which was the mean surface temperature for the twentieth century. Temperatures are in degrees Celsius.

▶ **FIGURE 11.12** Residual temperatures. Each observation is the temperature minus the average temperature for its double-decade.

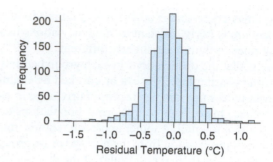

QUESTION   Which conditions are not satisfied for using ANOVA to test whether the mean global surface temperature is different for the given time periods?

CONCLUSION   Judging on the basis of the boxplots, the amount of variation in each group seems about the same, so we would expect the standard deviations to be about the same. This means that the third condition—same variance—is met. The histogram of residuals looks roughly Normal, so the fourth condition is met. However, the first two conditions are not met. From month to month, temperatures are not independent. One unusually warm month is likely to be followed by another unusually warm month, so observations within groups might be dependent. Also, because the same objects were measured in each group, the groups are not independent. A method other than ANOVA must be used to decide whether differences in mean temperatures are due solely to chance variation.

TRY THIS!   Exercise 11.27

Carrying Out an ANOVA Test   ANOVA is just another hypothesis test, like all of the others. Example 7 shows you how to do a complete ANOVA by applying the same four-step procedure used for other hypothesis tests.

EXAMPLE 7 Testing Popped Popcorn

Does the amount of oil used affect the number of kernels popped when one is making popcorn? Recall that researchers randomly assigned bags with 50 unpopped kernels to be popped with no oil, a medium amount of oil (1/2 tsp), or the maximum amount of oil (1 tsp). Thirty-six bags were assigned to each group. After 75 seconds, the popped kernels were counted.

QUESTION   Using a significance level of 5%, carry out an ANOVA test, showing each of the four steps of a hypothesis test.

SOLUTION

Step 1: Hypothesize

   H_0: $\mu_{none} = \mu_{medium} = \mu_{max}$
   H_a: The mean number of kernels popped differs by amount of oil used.

In words, the null hypothesis says that the mean number of kernels popped has nothing to do with the amount of oil used. The alternative hypothesis says that the number of kernels popped is associated with the amount of oil used.

Step 2: Prepare
In order for the F-statistic to follow the F-distribution, we need to check conditions. The first condition, independence between groups, holds because bags were randomly

Looking Back

Four Steps
The four steps of a hypothesis test are (1) hypothesize, (2) prepare, (3) compute to compare, and (4) interpret.

assigned to each group. You must assume that the second condition, independence within groups, holds. This condition would fail if, for example, the oven was still heating up while some bags in a group were tested. For the third condition, equal variances, refer to Figure 11.1, which shows boxplots of the three groups. The interquartile ranges are about the same, so you can expect the standard deviations to be roughly equal as well. Using technology, you could check this by computing the standard deviation for each group, but many statistical packages make this difficult to do. The final condition, Normal distribution, is also difficult to check with most statistical software packages. We note here that the response variable consists of counts, which are not Normally distributed because the Normal distribution has continuous, not discrete, values. However, because the counts have a large range (from 0 to 50), these distributions might be roughly Normal. You might not have the tools to check, so you'll have to make an assumption that the distributions are close enough to Normal for you to continue.

Step 3: Compute to compare
We used Minitab to do the ANOVA, and the output is shown in Figure 11.13.

One-way ANOVA: good versus oil

Analysis of Variance

Source	DF	Adj SS	Adj MS	F-Value	P-Value
oil	2	945.9	472.9	3.89	0.023
Error	105	12762.5	121.5		
Total	107	13708.3			

Means

oil	N	Mean	StDev	95% CI
maximum	36	13.47	9.30	(9.83, 17.12)
medium	36	19.75	11.83	(16.11, 23.39)
none	36	19.75	11.76	(16.11, 23.39)

Pooled StDev = 11.0249

◀ **FIGURE 11.13** ANOVA output (Minitab) for the popcorn experiment.

The *F*-statistic compares the between-groups variation (or the Treatments variation, labeled "oil") to the within-groups variation (Error). For these data, the between-groups variation is almost four times greater than the within-groups variation, as measured by the mean sum of squares:

$$F = 3.89$$

If the means are equal, then the probability of getting an *F*-statistic this large or larger is 0.023.

$$p\text{-value} = 0.023$$

Step 4: Interpret
Because the p-value is small (smaller than our significance level of 0.05), we reject the null hypothesis and conclude that the mean amount of kernels popped differs by amount of oil used.

CONCLUSION   Because we reject the null hypothesis, we conclude that the amount of oil used affects the number of kernels that get popped. We can make a causal conclusion here—that the amount of oil caused the differences we observed—because the bags were randomly assigned to the groups.

TRY THIS!   Exercise 11.31

SNAPSHOT ▸ Analysis of Variance (ANOVA)

WHAT IS IT? ▸ ANOVA is a procedure for testing whether the population means for several groups differ.

WHAT DOES IT DO? ▸ It compares the amount of variation between groups to the amount of variation within groups.

HOW DOES IT DO IT? ▸ If the population means of the groups are different, the variation between groups will be larger than the variation within groups.

HOW IS IT USED? ▸ The F-statistic is a ratio of the variation between groups (measured with the mean sum of squares) to the variation within groups (also measured with its mean sum of squares). If the F-statistic is large, reject the null hypothesis and conclude that at least one mean is different from another.

SECTION 11.4

Post Hoc Procedures

As we noted earlier, rejecting the null hypothesis of an ANOVA can be anticlimactic. We learn that oil affects the results of popping popcorn, but we don't know which amount of oil works best. For this reason, after rejecting the null hypothesis of an ANOVA, we next do a **post hoc analysis**. *Post hoc* is Latin for "after this." The phrase reminds us that we do this analysis after we have looked at the data and determined (with a chance of making an error) that at least one of the means is too different from another to be convincingly explained by chance. If we do not reject the null hypothesis, then there is no need to do a post hoc procedure, because we do not have enough evidence to conclude that the means differ.

The goal of a post hoc analysis is to determine which groups have different means from the others, which groups have the highest means, which the lowest means, and so on. The spirit of post hoc analysis is that we go into it knowing nothing other than that at least one of the means is different from another. As you saw in Section 11.1, we must be cautious when doing multiple comparisons of groups.

A popular approach for post hoc analysis is based on the Bonferroni Correction. However, in this section, we will use an approach similar to Bonferroni but more powerful, called the Tukey Honestly Significant Difference (Tukey HSD) approach. "More powerful" means that with Tukey HSD, you are more likely to correctly conclude that two means are different than you are with the Bonferroni Correction. Also, most software packages automatically produce the Tukey HSD intervals by default, or at least they make these an easy option to choose.

Rather than doing hypothesis tests for our post hoc analysis, we will instead find confidence intervals for the differences between all possible pairs of means. This is because we learn more by looking at confidence intervals. Not only do we learn that one mean is higher than another, but we also get an estimate of how much higher it might be.

A post hoc analysis is performed after rejecting the null hypothesis from an ANOVA and concluding that at least one of the group means is different from another. The goal is to determine *which* means are different by comparing all possible pairs of means. An appropriate correction must be made in order to keep the overall significance level at a specified value.

Tukey Honestly Significant Difference Confidence Intervals The **Tukey Honestly Significant Difference (HSD)** approach provides a set of simultaneous 95% confidence intervals similar to the Bonferroni simultaneous intervals in Section 9.1. Statisticians prefer the Tukey HSD to the Bonferroni Correction, because in many situations, the Tukey HSD intervals tend to be narrower—and therefore more precise— estimates. In a post hoc analysis, the Tukey HSD intervals are used to determine which means are greater than the others.

Figure 11.14 shows an extended ANOVA output for examining the popcorn experiment. Two additional tables provide confidence intervals for the difference of means using the Tukey HSD approach. The first of these tables gives us confidence intervals for comparing these mean differences:

- (Medium oil) minus (maximum oil)

- (No oil) minus (maximum oil)

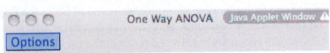

◀ **FIGURE 11.14** ANOVA Table for popcorn data with Tukey HSD intervals included (StatCrunch).

Tech

One Way ANOVA   Java Applet Window

Options

Analysis of Variance results:
Responses stored in good.
Factors stored in oil.

Factor means

oil	n	Mean	Std. Error
maximum	36	13.472222	1.5494424
medium	36	19.75	1.971775
none	36	19.75	1.9596627

ANOVA table

Source	df	SS	MS	F-Stat	P-value
Treatments	2	945.85187	472.92593	3.890878	0.0234
Error	105	12762.473	121.547356		
Total	107	13708.324			

Tukey 95% Simultaneous Confidence Intervals

maximum subtracted from

	Lower	Upper
medium	0.09988753	12.455668
none	0.09988753	12.455668

medium subtracted from

	Lower	Upper
none	-6.1778903	6.1778903

The next table tells us the mean difference for

- (No oil) minus (medium oil)

For example, the confidence interval for (none minus maximum oil) is (0.0999 to 12.4557). Because this interval includes only positive numbers, we can be confident that the mean number of kernels popped is higher with no oil than with the maximum amount of oil.

Interestingly, the mean number of kernels popped was the same for both the None and the Medium Oil groups, and for this reason, the confidence intervals are the same.

The fact that the confidence interval that compares using medium amounts of oil to using no oil includes 0 tells us that no significant differences appear between using no oil and using medium amounts of oil.

From the confidence intervals, we conclude that using medium oil or no oil is better than using the maximum amount of oil. Also, it does not matter whether you use no oil or medium amounts of oil.

EXAMPLE 8 Creepy Computers

Why do some robots seem cute and cuddly and others give us the creeps? Some researchers hypothesized that people do not like machines if the machines seem too human-like. In particular, they suggested that when machines seem to have "the capacity to feel and to sense," then we humans find this "unnerving" (Gray and Wegner 2012). To test this hunch, they asked a sample of humans about what they claimed was a new type of computer. Some of the humans were randomly selected to be in the control group, and they were told that this new computer was just like a regular computer, only "much more powerful." A second group were selected to be in the "with-agency" group, and they were told that this new computer had the ability to plan ahead and carry out actions on its own. The third group, the "with-experience" group, were told the computer could feel some form of "hunger, fear, and other emotions." After being told about this "new" computer, everyone was given a questionnaire, and from the respondents' answers to this questionnaire, the researchers extracted a scale that measured how "unnerving" they found the machine to be. A high score suggests that they were very disturbed by this potential computer.

To illustrate this study, we simulated data that closely match the descriptions provided by the researchers. Figure 11.15 shows a boxplot of the "unnerving" scores for the three groups. This suggests that the mean scores were different for the different groups, but might this difference be due to chance? The statistical software output in Figure 11.16 helps us answer this question and also provides a post hoc analysis.

► FIGURE 11.15 Boxplots show that in this sample, people typically felt that computers that could sense and feel (with-experience) were more unnerving than merely very fast computers (Control) or computers that could act on their own (with-agency).

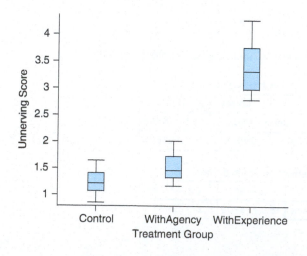

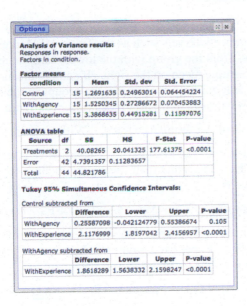

Options

Analysis of Variance results:
Responses in response.
Factors in condition.

Factor means

condition	n	Mean	Std. dev	Std. Error
Control	15	1.2691635	0.24963014	0.064454224
WithAgency	15	1.5250345	0.27286672	0.070453883
WithExperience	15	3.3868635	0.44915281	0.11597076

ANOVA table

Source	df	SS	MS	F-Stat	P-value
Treatments	2	40.08265	20.041325	177.61375	<0.0001
Error	42	4.7391357	0.11283657		
Total	44	44.821786			

Tukey 95% Simultaneous Confidence Intervals:

Control subtracted from

	Difference	Lower	Upper	P-value
WithAgency	0.25587098	-0.042124779	0.55386674	0.105
WithExperience	2.1176999	1.8197042	2.4156957	<0.0001

WithAgency subtracted from

	Difference	Lower	Upper	P-value
WithExperience	1.8618289	1.5638332	2.1598247	<0.0001

◀ **FIGURE 11.16** Statistical software output for a one-way ANOVA, and Tukey HSD intervals for these data (StatCrunch).

QUESTIONS

a. Carry out an ANOVA. The required conditions for an ANOVA all hold, but you might want to go through the checklist yourself to verify this.

b. Perform a post hoc analysis to determine which groups are different from the others. How are they different?

c. Is there evidence to support the researchers' claim that the "with-experience" group will be the most unnerved?

SOLUTIONS

a. The null hypothesis is that the mean "unnerving" score is the same for all three groups. The alternative hypothesis is that it differs for at least one of the three groups. The F-statistic is 177.6 and the p-value is small; it is less than 0.0001. For this reason, we reject the null hypothesis and conclude that at least one of the three groups is different from another.

b. The first Tukey Simultaneous Confidence interval, "*Control* subtracted from *WithAgency*" includes 0, so we conclude that there is insufficient evidence that the control group and the "with-agency" group differ in their feelings about these types of computers. However, the "*Control* subtracted from . . . *WithExperience*" interval contains all positive values and does not include 0. This means we are confident that the "with-experience" group had higher mean unnerving scores than the control group. On average, people found computers that could sense and feel to be more unnerving than computers that were merely very fast.

   The final interval, "*WithAgency* subtracted from *WithExperience*" also contains all positive values. From this, we conclude that the people felt the computers that could sense and feel (the with-experience group) were more unnerving than those that could act independently (with agency).

c. Yes, there is evidence that people find feeling computers to be the most unnerving. The people who were told about such a computer were more unnerved, on average, than those who were told about a computer that could act on its own (with agency) and were more unnerved than those who were told about a computer that was merely very fast.

TRY THIS!   Exercise 11.43

Visualizing Simultaneous Confidence Intervals A useful trick for sorting out simultaneous confidence intervals such as the Tukey HSD intervals is first to write the names of the groups, from the group with the lowest sample mean on the left to the group with the highest sample mean on the right. (If sample means are equal, you can put them in any convenient order.) For example, the groups in the computer study are as follows, sorted from lowest mean to highest mean (refer to Figure 11.16).

Control   With-Agency   With-Experience

Next we draw lines underneath any groups whose confidence interval for the difference in means includes 0. (This tells us that their means are so close together that we can't tell the population means apart.)

<u>Control   With-Agency</u>   With-Experience

This helps us see that the control group and the "with-agency" group have about the same mean level and that the With-Experience group was higher than both of the other groups.

We can apply this technique to the popcorn experiment too:

Max Oil   <u>Medium Oil   No Oil</u>

EXAMPLE **9** Which Is Best Way to Take Notes?

In Section 11.2 you learned about a study to determine the best way of taking notes. The researchers randomly assigned participants to a Longhand group (watch a lecture and take notes by hand), a Laptop group (take notes by laptop) and a Laptop Intervention group (take notes with a laptop but are instructed to summarize the lecture in their own words.) Afterward, the participants were given a test of their conceptual understanding of the lecture. We carried out an ANOVA, using a significance level of 10% ($\alpha = 0.10$), which suggests that the three methods result in different mean scores on the conceptual understanding test. But which method scores highest? Figure 11.17 shows aa post hoc analysis.

▶ **FIGURE 11.17** ANOVA and post hoc analysis comparing the conceptual understanding scores for three methods of taking notes.

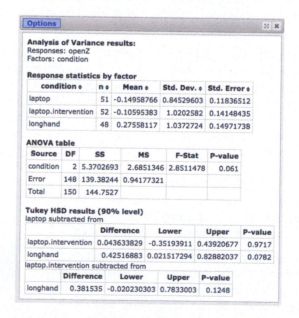

QUESTION Write down the groups in order of the lowest mean score (left) to highest mean score (right). Underline groups for which the difference in means is statistically *in*significant using a 10% significance level. (Note that the Tukey HSD intervals are at a 90% confidence level.) What can we conclude?

Because the confidence interval for the difference between the laptop method and the laptop intervention method includes 0, we connect them with a line. The same is true for the laptop intervention group and the longhand group.

<u>Laptop</u>   <u>Laptop.intervention</u>   <u>Longhand</u>

However, we do not draw a line to connect the Laptop and Longhand group, because the confidence interval for their difference does not include 0.

CONCLUSION   Taking notes by longhand is better than taking notes by laptop, although there's no evidence it is better than taking notes by laptop if you summarize the lecture in your own words.

TRY THIS!   Exercise 11.47

It might seem strange that Longhand is "the same" as the Laptop Intervention, and Laptop Intervention is the same as Laptop, but Longhand is NOT the same as Laptop Intervention. The reason for this result is that the mean score for the Longhand group was far enough above the mean of the Laptop group that we could see a significant difference. But the differences between the other two groups were too small. We might be able to clarify the situation had we the ability to repeat the study with a larger sample size. In fact, the researchers used a somewhat more sophisticated analysis that took into account the content of the lectures that the participants viewed.

If you do *not* reject the null hypothesis for an ANOVA, then you do not need to carry out a post hoc test. If we fail to reject the null, we are saying that we do not have evidence that the means of the groups are different. Because we have no evidence of difference, there is no need to do a post hoc analysis to see which groups are different.

Table 11.4 shows the output for a test to determine whether the three note-taking groups differed on their scores to test their factual recall of material. We see that the F-statistic is 1.007, with a p-value of 0.3677. Because the p-value is larger than a significance level of 0.10, we fail to reject the null hypothesis. We conclude that there is not enough evidence to suggest that the ability to recall facts from a lecture depends on the method of taking notes. Because we did not reject the null, it does not make sense to perform a post hoc analysis.

Source	df	SS	MS	F-Stat	p-value
Treatments (condition)	2	1.96240	0.98120	1.007	0.3677
Error	148	144.16455	0.97408		
Total	150	146.12696			

◄ **TABLE 11.4** ANOVA table to compare factual recall scores across three note-taking styles.

KEY POINT   Do not carry out post hoc tests if you failed to reject the null hypothesis in your F-test.

Final Thoughts on Post Hoc Analyses

Different software packages provide very different options for performing the post hoc analysis for an ANOVA. Most do some version of what we've presented: They give you simultaneous confidence intervals to compare pairs of means. A few, such as Minitab, default to showing you confidence intervals for *individual* groups, not for the *differences* between groups. We recommend that you ignore these individual intervals, because they fail to take into account the necessary post hoc adjustments and because, unless you take great care, you can easily reach the wrong conclusion. Instead, ask the software to find simultaneous confidence intervals for *differences* between all pairs of means.

The post hoc analyses we've discussed are intended to show us all possible comparisons between groups. This is appropriate if, when you began the ANOVA, you did not know which groups would be bigger or smaller than the others. Sometimes, researchers have very definite hypotheses about the order of groups. For example, medical researchers testing a drug might be interested in studying whether low-dose, medium-dose, or high-dose patients recover more quickly from their illness. They suspect that the high-dose patients will recover quickest, followed by medium-dose patients, followed by low-dose patients. In more advanced books on ANOVA, you will learn that we can do more focused analyses that concentrate on comparing particular groups, not all possible pairs of groups.

CASE STUDY REVISITED

Seeing Red

The researchers were interested in understanding whether the color of a soccer team's uniform affects its performance. They cited a history of research that suggests that some colors are more successful than others. Some researchers explain this by citing biological or evolutionary factors (many animals display red as a sign of dominance, and humans associate red with the emotions of anger and aggression). The researchers examined 55 years of soccer results in the United Kingdom. Part of their analysis focused on how the percentage of games won varied between color groups. A boxplot of the data, Figure 11.18, shows that the median values differ and supports the hypothesis that teams with red jerseys are slightly more successful than other teams. The boxplot shows that the same-variance requirement is satisfied, but the number of outliers and a suggestion of skewness for certain groups suggest that the Normal condition might not be satisfied. Also, the data might not really be independent, because the teams play each other. (If one team wins a lot, the other teams must be winning less.) Because of these possible problems, the researchers decided to use a different approach, examples of which you'll see in Chapter 13. But it's instructive first to consider what we might learn from the ANOVA approach in this chapter. The output is shown in Figure 11.19 on the next page.

▶ **FIGURE 11.18** The percentages of home game wins for teams with jerseys of different colors: blue (B), red (R), white (W), and yellow or orange (YO). Teams wearing red shirts seem to have a slightly higher winning percentage than other teams, on average.

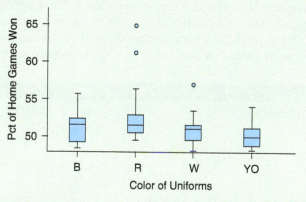

The p-value of 0.0361 means that, with a significance level of 5%, we reject the null hypothesis and conclude that the mean number of wins differs by color of the team jersey.

We visualize the post hoc analyses by writing the groups in order of their mean percentage of wins and underlining groups that have (statistically speaking) the same mean.

$$\underline{YO \quad B \quad W \quad R}$$

This visualization shows us that the only confidence interval that does not include 0 is the one that compares red teams with yellow/orange teams. From this, we are confident that teams with red uniforms win a greater percentage of home games, on average, than teams with yellow/orange uniforms.

Although not all of the conditions for an ANOVA were satisfied, the conclusions are supported by more careful analyses that the authors of this study performed in addition to the basic ANOVA.

► **FIGURE 11.19** Output from ANOVA to test whether teams with uniforms of different colors have different winning percentages, on average.

○ ○ ○   One Way ANOVA

Options

Analysis of Variance results:

Responses stored in percentwin.

Factors stored in color.

Factor means

color	n	Mean	Std. Error
B	23	51.239132	0.42154145
R	16	53.025	1.066634
W	11	51.29091	0.71343654
YO	18	50.283333	0.37593022

ANOVA table

Source	df	SS	MS	F–Stat	P–value
Treatments	3	65.40642	21.80214	3.0189104	0.0361
Error	64	462.19888	7.2218575		
Total	67	527.6053			

Tukey 95% Simultaneous Confidence Intervals

B subtracted from

	Lower	Upper
R	-0.5218412	4.0935802
W	-2.5468976	2.650455
YO	-3.186618	1.275024

R subtracted from

	Lower	Upper
W	-4.5105963	1.0424145
YO	-5.1773243	-0.30600926

W subtracted from

	Lower	Upper
YO	-3.7205107	1.7053592

DATAPROJECT ▸ Where to Begin

1 OVERVIEW

Often, when considering a new data set, it's hard to know where to begin. These next three projects involve the same data set and will give you some general tips and strategies to apply when working with new data sets.

2 GOAL

Identify strategies for analyzing data.

3 GETTING TO KNOW YOU

We begin our spin around the Data Cycle at the Consider Data phase. Often, you find yourself looking at a dataset because it was assigned to you (as happens both in the classroom and the workplace) or because some general curiosity or other research interest has made you wonder what's inside the data. For example, you might want to know something about crime in your city. A logical place to look to answer questions about crime is the city's open data portal. As we discussed earlier, not every city has one, but we're going to consider a random sample of 35000 entries from the City of Los Angeles's Crime data set.

Sometimes it can be quite intimidating to be faced with a large data set, and so our intent is to give you some advice for starting out.

The primary piece of advice is to get to know your data. The title of this section is a quote from a song from the musical *The King And I*. The lyrics are "Getting to know you, getting to know all about you." Your first step is to know all about your data.

In classroom settings, particularly with homework, we often neglect this step. This can give you the impression that this step either isn't important or can be done quickly. In fact, this is one of the most important parts of your statistical investigation, and you should plan to spend a good chunk of time on this step.

Project: Upload the sampleofcrime.csv datafile into StatCrunch. These data are a record of crimes reported in the City of Los Angeles. The complete data set has all crimes, but this data set is too large for StatCrunch, and so a smaller, random sample is provided instead. Even though it is smaller, it is still quite large.

We have two main goals. The first is to understand the structure of the data. The second is to understand its meaning.

By structure, we mean basic things, such as how many observations are there? How many variables? What type of variables do we have (for example, categorical, numerical, dates, locations, text, etc.)?

By meaning, we mean What do the variables represent? Some of them you might be able to guess at. For example, "Date Reported." Others, might be more mysterious? What does "DR" stand for? "MO"? What are the weapon codes?

Assignment: Use the Validate command, under the Data menu, to examine the data. How many variables and how many observations? Which variables are numeric, which are nonnumeric, and which are both? Why are some variables both? To answer these and the following questions, you might wish to refer to the documentation at https://data.lacity.org/A-Safe-City/Crime-Data-from-2010-to-Present/y8tr-7khq.

Write a brief report that answers the preceding questions and these additional questions:

1. Who collected these data?
2. How were they collected?
3. What is the scope of the data? They are crimes, but are any types of crimes included? In what city? Over what period?
4. Choose a value in the *Crime Code* variable and explain what it means, how it is related to the *Crime Code Description* variable and how it is related to the MO Code variable.
5. How many reporting districts are there? Which reports the most crimes?
6. What do the codes for *Victim Sex* mean?

CHAPTER REVIEW

KEY TERMS

Multiple comparisons, *561*
Overall significance
  level, *561*
Bonferroni Correction, *562*
Simultaneous confidence
  intervals, *565*

One-way analysis of variance
  (ANOVA), *566*
ANOVA table, *569*
Total sum of squares, *570*
Explained variation (variation
  between groups, treatment
  variation), *570*

Unexplained variation (variation
  within groups, error variation,
  residual variation), *570*
Mean sum of squares, *571*
Homoscedasticity, *574*
Residuals, *575*

Post hoc analysis, *578*
Tukey Honestly Significant
  Difference (HSD), *579*

LEARNING OBJECTIVES

After reading this chapter and doing the assigned homework problems, you should

- Understand the problem of multiple comparisons and the overall significance level.

- Know how to apply the Bonferroni Correction to multiple hypothesis tests and confidence intervals.

- Understand how the *F*-statistic allows us to compare means of multiple groups by comparing the variance within a group to the variance between groups.

- Be able to perform and interpret an ANOVA procedure to analyze potential associations between a categorical predictor variable and a numerical response variable.

SUMMARY

When comparing means from more than two groups, we need to be aware that with each comparison, there is a probability that we will mistakenly conclude the means are different when in fact they are not. In order to keep the probability of making such an error across all comparisons less than or equal to α (alpha), usually 0.05, we need to make an adjustment.

The Bonferroni Correction is one method that adjusts for multiple comparisons by using a smaller significance level for each individual comparison. To achieve an overall significance level of α, each comparison is made with a significance level of $\alpha/$(number of comparisons).

ANOVA (one-way) is an overall test to determine whether the means of three or more groups differ. If you reject the null hypothesis with ANOVA, you may go on to do post hoc tests to judge which means are significantly different from which others. If you cannot reject the null hypothesis that all the means are the same, you should not go on to do post hoc tests.

A variety of different post hoc procedures are available, and you will be limited, to some extent, by which procedures your software provides. Most software provides something similar to the Tukey HSD method, which finds confidence intervals for the difference in means for all possible pairs of groups.

SOURCES

Attrill, M. J., K. A. Gresty, R. A. Hill, and R. A. Barton. 2008. Red shirt colour is associated with long-term team success in English football. *Journal of Sports Science* 26 (6): 577–582.

Gray, K., and D. Wegner. 2012. Feeling robots and human zombies: Mind perception and the uncanny valley. *Cognition* 125, 125–130.

Mueller, P., Oppenheimer, D., 2014. The Pen Is Mightier Than the Keyboard: Advantages of Longhand over Laptop Note Taking. *Psychological Science* 25 (6): 1159–1168.

SECTION EXERCISES

SECTION 11.1

For all t-tests in this section, do not assume equal variances.

In Exercises 11.1 and 11.2, for each situation, choose the appropriate test: one-sample t-test, two-sample t-test, ANOVA, or chi-square test.

11.1 Choosing a Test

a. You want to test whether there is an association between a categorical variable and a numerical variable. For example, you want to test whether there is an association between type of apartment (studio, one-bedroom, or two-bedroom) and monthly rent.

b. You want to test whether the means of a numerical variable are different for two possible values of a categorical variable. For example, you want to test whether the mean income of paramedics is the same for male and female paramedics.

11.2 Choosing a Test

a. You want to test whether an association exists between two categorical variables. For example, you want to test whether there is an association between belief in global warming and political party affiliation.

b. You want to test whether the sample mean of a numerical variable is different from a known population mean. For example, *The New York Times* reported that the average Broadway show ticket was $109 in 2017. You take a recent random sample of 30 Broadway show ticket prices and want to determine if the mean price of a Broadway show ticket has changed from 2017.

TRY **11.3 Bonferroni Correction (Example 1)** Suppose you have five groups of observations, and you do hypothesis tests (*t*-tests) to compare all possible pairs of means.

a. How many pairwise comparisons can be done with five groups? List all comparisons with five groups labeled A, B, C, D, and E, starting with AB, AC, and so on.

b. Using the Bonferroni Correction, what significance level should you use for each hypothesis test if you want an overall significance level of 0.05?

11.4 Bonferroni Correction
Suppose you have four groups of data, and you want to do hypothesis tests (*t*-tests) to compare all possible pairs of means.

a. How many pairwise comparisons can be done with four groups called A, B, C, and D? Show all possible pairs, starting with AB.

b. Using the Bonferroni Correction, which significance level should you use for each comparisons if you want an overall significance level of 0.05?

g **11.5 Apartment Rents** Random samples of rents for 1-bedroom 1-bath apartments in Seattle, San Francisco, and Santa Monica were selected and shown in the following table. Use three two-sample *t*-tests, applying the appropriate Bonferroni Correction to achieve an overall significance level of 0.05, to compare all possible pairs of means. Instead of using the four-step method, summarize your findings using a table that reports, for each pair of cities, the observed *t*-test statistics, p-values, and conclusion. Assume the conditions for two-sample *t*-tests are met. *See page 599 for guidance.*

Seattle Rent (dollars)	San Francisco Rent (dollars)	Santa Monica Rent (dollars)
1165	2750	1899
1854	2595	2700
1750	3900	1799
1474	2445	3284
2495	4515	2300
1425	3800	2300
2010	3400	3104
1615	2795	2895

11.6 Commuting Times Mark Bates, a statistics professor at Oxnard College, recorded his commuting times using three different routes from home to work. The routes are named for the streets on which he traveled, and the times are in seconds.

a. For the boxplots given, compare the medians, interquartile ranges, and shapes, and mention any potential outliers.

Oxnard	Rose	Rice
732	869	694
842	648	629
736	1045	863
732	674	748
736	821	767
833	708	574
655	840	628
688	1029	637
727	735	620
721	745	752
695	794	608
707	652	983
843	552	765
852	732	666
789	578	727
	661	729
	657	605
	869	717
		679

b. Assuming that the overall significance level is 0.05, what is the Bonferroni-corrected level of significance for each pair of comparisons?

c. Carry out *t*-tests to compare the means of all pairs, and summarize your findings by reporting *t*-statistics, p-values, and conclusions. Assume that the conditions for using two-sample *t*-tests are met.

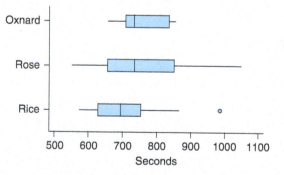

11.7 Gas Prices The website Gasbuddy.com reports the least expensive gas prices in some cities on a daily basis. The following table shows the least expensive gas prices for three cities on June 1, 2018.

Los Angeles (dollars)	Chicago (dollars)	Boston (dollars)
3.43	3.04	2.85
3.45	3.09	2.85
3.45	3.12	2.93
3.45	3.15	2.95
3.47	3.15	2.97
3.47	3.16	2.97
3.47	3.16	2.97
3.48	3.16	2.99

a. Assuming the overall level of significance is 0.05, what is the Bonferroni-corrected level of significance for each of the separate tests?

b. Report all three sample means. Which two means are closest to each other?

c. Carry out two-sample t-tests for all three pairs. Instead of using the four-step method, summarize your findings using a table that reports t-test statistics, p-values, and conclusions. Base your conclusions on the Bonferroni-corrected level of significance. Do not assume equal variances. Assume all conditions for two-sample t-tests are met.

11.8 More Gas Prices The following table shows the least expensive gas prices for three cities on June 1, 2018, as reported by Gasbuddy.com.

Seattle (dollars)	Austin (dollars)	Pittsburgh (dollars)
3.19	2.53	3.05
3.25	2.55	3.09
3.26	2.58	3.09
3.29	2.59	3.13
3.29	2.59	3.15
3.29	2.59	3.17
3.29	2.59	3.17
3.31	2.59	3.17

a. Assuming the overall level of significance is 0.05, what is the Bonferroni-correct level of significance for the three pairs of cities?

b. Report the sample mean for each city. Which two means are closest to each other?

c. Carry out two-sample t-tests for all three pairs. Instead of using the four-step method, summarize your findings using a table that reports t-test statistics, p-values, and conclusions. Base your conclusions on the Bonferroni-corrected level of significance. Do not assume equal variances. Assume all conditions for two-sample t-tests are met.

11.9 Bonferroni Intervals Assume you have three groups to compare through hypothesis tests and confidence intervals, and you want the overall level of significance to be 0.05 for the hypothesis tests (which is the same as a 95% confidence level for the confidence intervals).

a. How many possible comparisons are there?

b. What is the Bonferroni-corrected value of the significance level for each hypothesis test?

c. What is the Bonferroni-corrected confidence level for each interval? Report the percentage rounded to two decimal digits, and show your calculations.

11.10 Bonferroni Intervals Assume you have four groups to compare through hypothesis tests and confidence intervals, and you want the overall level of significance to be 0.05.

a. How many possible comparisons are there?

b. What is the Bonferroni-corrected value of the significance level for each comparison?

c. What is the Bonferroni-corrected confidence level for each interval assuming an overall significance level of 0.05? Report the percentage rounded to two decimal digits, and show your calculations.

11.11 Gas Price Intervals Use the data from exercise 11.7 and find Bonferroni-corrected intervals for all three comparisons assuming an overall confidence level of 95%, that is, an individual confidence level of 98.33%. Then state whether the means are significantly different based on whether the intervals capture 0 or not. Compare your conclusions with the conclusions in exercise 11.7.

11.12 Gas Price Intervals Use the data from exercise 11.8 and find Bonferroni-corrected intervals for all three comparisons assuming an overall confidence level of 95%, using an individual confidence level of 98.33%. Then state whether the means are significantly different based on whether the intervals capture 0 or not. Compare your conclusions with the conclusions in exercise 11.8.

11.13 Baseball Position and Hits The following table shows the number of hits for a random sample of Major League Baseball players two months into the season. The table shows data for samples of three positions: shortstop, left field, and first base.

Shortstop	Left Field	First Base
47	49	63
47	50	59
48	67	44
51	58	58
39	44	59
53	39	48
69	50	52
48	32	57
59	62	42
45	62	41

a. Compare the sample mean number of hits for the three positions.

b. Find the number of pairwise comparisons. Report the Bonferroni corrected significance level required to carry out two-sample t-tests to see whether each pair of sample means is different.

c. Assume the conditions for two-sample t-tests are met. Use an overall significance level of 0.05 and, for each pair, test whether the means are different. Report the value of the t-test statistic, the p-value, and the conclusion for each test. Are position and number of hits associated?

11.14 Baseball Position and Hits Use the data in the previous question to find all the pairwise confidence intervals for the difference in the population means. How do these confidence intervals support your conclusion in the previous problem?

SECTION 11.2

TRY **11.15 Comparing *F*-Values from Boxplots (Example 3)**
Refer to the figure. Assume that all distributions are symmetric (therefore the sample mean and median are approximately equal) and that all the samples are the same size. Imagine carrying out two ANOVAs. The first compares the means based on samples A, B, and C (above the horizontal line), and the second is based on samples L, M, and N (below the horizontal line). One of the calculated values of the *F*-statistic is 9.38, and the other is 150.00. Which value is which? Explain.

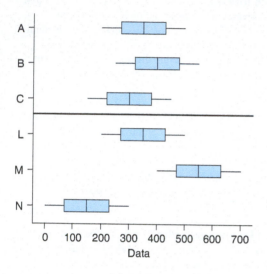

11.16 Comparing *F*-Values from Boxplots Refer to the following figure. Assume that all data sets are symmetric and that all the samples are the same size. Imagine carrying out two ANOVAs. The first compares the means based on samples A, B, and C (above the horizontal line), and the other is based on samples G, H, and K (below the horizontal line). One of the calculated values of the *F*-statistic is 9.38, and the other is 25.00. Which value is which? Explain.

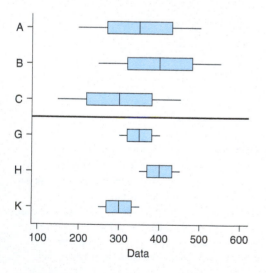

TRY **11.17 Marital Status and Cholesterol (Example 4)** Refer to the StatCrunch output from National Health and Nutrition Examination Survey (NHANES) data, which shows the association between marital status and cholesterol. Assume the population distributions are close enough to Normal to justify using ANOVA.

Analysis of Variance results:
Responses in Cholest.
Factors in Marital.

Factor means

Marital	n	Mean	Std. dev	Std. Error
Divorced	78	212.52564	51.753844	5.8599689
Living with partner	61	198.29508	40.437748	5.1775231
Married	507	208.13412	41.703425	1.8521141
Never married	184	184.49457	46.456631	3.4248283
Separated	20	208	42.631839	9.532769
Widowed	90	209.58889	48.150955	5.0755563

ANOVA table

Source	df	SS	MS	F-Stat	P-value
Treatments	5	89082.105	17816.421	9.1420934	<0.0001
Error	934	1820210.8	1948.8338		
Total	939	1909292.9			

a. Write the null and alternative hypotheses for the association between marital status and cholesterol.

b. Identify the *F*-statistic from the StatCrunch output.

c. Which marital status had the largest sample mean and which had the smallest sample mean?

d. Assuming that you did find an association between marital status and cholesterol levels, would this association mean that marital status caused different cholesterol levels? Can you think of a confounding factor?

11.18 Marital Status and Blood Pressure Test the hypothesis that people with different marital statuses differ in mean systolic blood pressure, using a significance level of 0.05. Refer to the StatCrunch output from NHANES data.

a. Write the null and alternative hypotheses for the association between marital status and blood pressure.

b. Identify the *F*-value from the output.

c. Which group had the largest sample mean, and which had the smallest?

d. Assuming that you did find an association between marital status and systolic blood pressure, would this association show that marital status caused different systolic blood pressures? Can you think of a confounding factor?

Analysis of Variance results:
Responses in Systolic.
Factors in Marital.

Factor means

Marital	n	Mean	Std. dev	Std. Error
Divorced	73	128.9726	23.565495	2.7581326
Living with partner	63	115.36508	15.664117	1.9734933
Married	503	124.84294	18.998249	0.84709004
Never married	191	116.94764	15.961051	1.1549011
Separated	20	121.45	17.500301	3.9131862
Widowed	82	146.13415	26.261159	2.9000598

ANOVA table

Source	df	SS	MS	F-Stat	P-value
Treatments	5	56195.526	11239.105	30.038499	<0.0001
Error	926	346469.09	374.15669		
Total	931	402664.62			

11.19 Schoolwork and Class (Example 5) A random survey was done at a small Lutheran college, and the students were asked how many hours a week they spent studying outside of class time. They were also asked what class they were in (1 = Freshman, 2 = Sophomore, 3 = Junior, and 4 = Senior).

a. Figure out the missing SS (sum of squares).

b. Find MS Error by dividing SS Error (from part a) by DF Error, and compare it with 65.1.

c. Divide MS class by MS Error (calculated in part b), and compare the result with the *F*-value.

d. When MS factor (in this case MS class) is more than MS Error, what does that show about the *F*-value? Will it be more or less than 1?

```
One-way ANOVA: SchoolWork versus class

Source   DF     SS     MS     F      P
class    3     893.5  297.8  4.58  0.005
Error    106   ?????   65.1
Total    109  7790.5


Level    N     Mean
1        28   15.679
2        24   13.583
3        33   12.727
4        25    7.680
```

11.20 TV Hours A random survey was done at a small Lutheran college, and the students were asked how many hours a week they spent watching TV. They were also asked what class they were in (1 = Freshman, 2 = Sophomore, 3 = Junior, 4 = Senior). The survey was given only to psychology students. Minitab output is shown.

a. Figure out the missing SS (sum of squares).

b. Figure out MS class by dividing SS class (from part a) by DF class, and compare it with 8.9.

c. Check the *F*-value by dividing MS class by MS Error.

d. When MS factor (in this case MS class) is smaller than MS Error, what does that show about the *F*-value? Will it be more than 1 or less than 1?

```
One-way ANOVA: TVHours versus class

Source   DF     SS     MS     F      P
class    3     ????    8.9   0.29  0.833
Error    106  3247.2  30.6
Total    109  3273.8


Level    N     Mean
1        27    5.537
2        23    6.870
3        33    5.909
4        27    5.648
```

11.21 Schoolwork and Class Use the information for exercise 11.19.

a. Which class had the highest sample mean, and which class had the lowest sample mean? (1 is for freshman, 2 is for sophomore, and so on.)

b. Write out the null and alternative hypotheses for the effect of class on schoolwork.

c. Identify the *F*-value from the output.

d. Assuming that you found an association between class and schoolwork would that show that class caused the different levels of schoolwork? Explain.

11.22 TV Hours Use the information for exercise 11.20.

a. Which class had the highest sample mean number of TV hours, and which class had the lowest sample mean? (1 is for freshman, 2 is for sophomore, and so on.)

b. Write out the null and alternative hypotheses for the effect of class on TV hours.

c. Identify the *F*-value from the output.

d. Assuming that you found an association between class and TV hours, would that show that class caused the different levels of TV hours? Explain.

SECTION 11.3

11.23 Stacking and Coding Some software (such as SPSS) requires that ANOVA data be stacked and coded. Some software works with both stacked and unstacked data, and some (such as the TI-84) requires unstacked data. Go back to the information given in exercise 11.8. Stack and code the data. For codes, use 1 for Denver, 2 for Houston, and 3 for Cleveland.

11.24 Stacking and Coding Some software (such as SPSS) requires that ANOVA data be stacked and coded. Some software works with both stacked and unstacked data, and some (such as the TI-84) requires unstacked data. Go back to the information given in exercise 11.7. Stack and code the data. For codes, use 1 for Chicago, 2 for Miami, and 3 for New York.

11.25 Schoolwork Again Go back to the information in exercise 11.19. Assuming the conditions for ANOVA are met, test the hypothesis that the mean number of hours of schoolwork varies by class, reporting the p-value and conclusion. Use the 0.05 level of significance. State your conclusion in the context of the data.

11.26 TV Hours Again Go back to the information in exercise 11.20. Assuming the conditions for ANOVA are met, test the hypothesis that the mean number of hours of TV varies by class, reporting the p-value and conclusion. Use the 0.05 level of significance. State your conclusion in the context of the data.

TRY **11.27 Pulse Rates (Example 6)** Pulse rates were taken for five people, each in three different situations: sitting, after meditation, and after exercise. Explain why it would not be appropriate to use one-way ANOVA to test whether the population mean pulse rates were associated with activity.

Person	Sitting	Meditation	Exercise
A	84	72	96
B	76	72	84
C	68	64	76
D	68	68	76
E	76	84	80

11.28 UCLA Music Survey The figure shows side-by-side box-plots of the number of hours per week that University of California, Los Angeles (UCLA) students spent listening to music. Minitab output for ANOVA is also shown. Check whether the conditions for ANOVA hold. If not, state which ones fail and why.

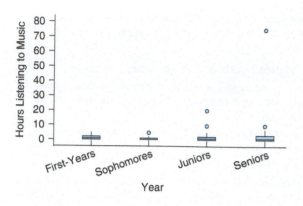

Analysis of Variance results:
Data stored in separate columns.

Column statistics

Column ◆	n ◆	Mean ◆	Std. Dev. ◆	Std. Error ◆
Seattle	8	3.27125	0.037961447	0.013421398
Austin	8	2.57625	0.023260942	0.008223985
Pittsburgh	8	3.1275	0.045903626	0.016229383
Los Angeles	8	3.45875	0.016420806	0.0058056315
Boston	8	2.935	0.055291436	0.019548475

ANOVA table

Source	DF	SS	MS	F-Stat	P-value
Columns	4	3.655025	0.91375625	616.06489	<0.0001
Error	35	0.0519125	0.0014832143		
Total	39	3.7069375			

One-way ANOVA: Hours versus Year

```
Source    DF      SS     MS     F      P
YR         3    116.9   39.0   0.76   0.520
Error    117   6017.1   51.4
Total    120   6134.0


S = 7.171    R-Sq = 1.91%    R-Sq(adj) = 0.00%


Level        N    Mean    StDev
Freshman     6   1.333    1.862
Sophomore   20   0.900    1.165
Junior      47   1.596    3.295
Senior      48   3.333   10.793
```

11.29 Commute Times by Method A survey was given to StatCrunch users on the length of time for commuting and the method of commuting. Assume this is a random sample. Minitab output for one-way ANOVA is given, along with the means and standard deviations. Divide the largest standard deviation (StDev) by the smallest, and explain why you should not use ANOVA on this data set. Assume Normality.

One-way ANOVA: TimeCom versus How

```
Source    DF     SS     MS     F      P
How        4   9439   2360   4.28   0.003
Error    130  71722    552
Total    134  81162

Level         N    Mean   StDev
Bus          11   30.91   22.34
Car/Truck    94   25.94   22.14
Foot         14   18.93   24.51
Other         4   38.75   14.36
Train        12   52.75   34.02
```

(Source: StatCrunch Responses to Commuting survey. Owner: Webster West)

11.30 Gas Price ANOVA Based on the following output, would it be appropriate to use ANOVA to determine if the mean gas prices for one of these five cities is significantly different from that of the other cities? If so, perform the ANOVA. If not, explain why it is not appropriate. Assume we have random samples from each city and Normality.

TRY 11.31 Apartment Rents (Example 7) Samples of rents for one-bedroom, one-bath apartments in three cities were obtained. Assume the distribution of each population is Normal to satisfy the conditions for using ANOVA. Test the hypothesis that the mean rent for one-bedroom apartments in at least one of these cities differs from the mean rent in the other, using a significance level of 0.05. *See page 599 for guidance.*

Seattle	San Francisco	Santa Monica
1165	2750	1899
1854	2595	2700
1750	3900	1799
1474	2445	3284
2495	4515	2300
1425	3800	2300
2010	3400	3104
1615	2795	2895

11.32 Study Hours by Major Three independent random samples of full-time college students were asked how many hours per week they studied outside of class. Their responses and their majors are shown in the table. Test the hypothesis that the mean number of hours studying varies by major by reporting the *F*-statistic, the p-value, and the conclusion. Assume the conditions for ANOVA are met.

Math	SocSci	English		Math	SocSci	English
15	10	14		10	3	5
20	12	12		8	5	5
14	8	12		7	7	4
15	7	10		7	6	8
14	7	10		5	3	9
10	10	10		5	3	10
12	6	8		5	3	4
9	8	8		15	2	3
10	5	5		6	2	3
11	4	7		5	1	3
9	4	6		3	5	6
8	4	4		5	4	4
8	8	6				

11.33 Salary by Type of College Information was gathered on the starting median salary for students who attended four different types of colleges. Assume the samples are random and Normal. Test the hypothesis that the population means are equal for all the types of colleges. Show all four steps for ANOVA. Do not do post hoc comparisons. Use a significance level of 0.05.

```
One-way ANOVA: Ivy League, Liberal, Party, State

Source      DF        SS        MS       F      P
Factor       3  2084185033  694728344  39.02  0.000
Error      246  4379756607   17803889
Total      249  6463941640

Level           N    Mean   StDev
Ivy League      8   60475    3219
Liberal Arts   47   45747    4369
Party          20   45715    3686
State         175   44126    4269
```

(Source: http://online.wsj.com/public/resources/documents/info-Salaries_for_Colleges_by_Type-sort.html, accessed via StatCrunch)

11.34 Draft Lottery When the draft lottery for military service in the Vietnam War was conducted, officials "randomly" selected birthdays. For example, September 14 was selected first, and that date was assigned the rank of 1. If March 7 were selected second, it would be assigned the rank of 2. This meant that all eligible men with birthdays on September 14 were drafted first, and men with birthdays on March 7 were selected next. If the birthdays were selected truly at random, the mean rank for each month should be about the same as that of any other month. The output shown below compares the mean ranks for each month. Small values mean the people born in that month were more likely to be called up.

```
One-way ANOVA: Jan, Feb, Mar, Apr, May, Jun,
               Jul, Aug, Sep, Oct, Nov, Dec

Source      DF       SS       MS      F      P
Factor      11   290507    26410   2.46  0.006
Error      354  3795120    10721
Total      365  4085628

Level    N    Mean   StDev
Jan     31   201.2    99.7
Feb     29   203.0   104.0
Mar     31   225.8    95.8
Apr     30   203.7   109.4
May     31   208.0   115.0
Jun     30   195.7   117.9
Jul     31   181.5   109.6
Aug     31   173.5   112.7
Sep     30   157.3    87.2
Oct     31   182.5    96.8
Nov     30   148.7    94.4
Dec     31   121.5    95.1
```

a. Which month had the smallest mean?

b. Test the hypothesis that the population means are equal for all twelve months. Show all four steps for ANOVA. Do not do post hoc comparisons. Use a significance level of 0.05.

(Source: StatCrunch DraftLottery.xls Owner: psuskp)

11.35 Reaction Times for Athletes A random sample of people were asked whether they were athletic, moderately athletic (Mod), or not athletic (NotAth). Then they were tested for reaction speed. Reaction speed was measured indirectly, through reaction distance, as follows: A vertical meter stick was dropped, and they caught it. The distance (in centimeters) that the stick fell is the reaction distance, and shorter distances correspond to faster reaction times. The data are shown in the following table.

NotAth	Mod	Athletic
16.0	11.7	24.7
25.3	22.3	15.7
19.3	12.7	17.7
19.0	21.3	12.7
14.3	16.0	21.0
34.3	14.0	21.0
17.7	16.3	19.3
	28.3	25.0
	30.7	27.0
	15.0	10.3
	26.7	
	26.7	

a. Interpret the boxplots given. Compare the medians, interquartile ranges, and shapes, and mention any potential outliers.

b. Test the hypothesis that people with different levels of athletic ability (self-described) have different mean reaction distances, reporting the F-statistic, p-value, and conclusion. Assume that the distribution of each population is close enough to Normal to satisfy the Normal condition of an ANOVA and that the sample is randomly selected. Do not do post hoc tests.

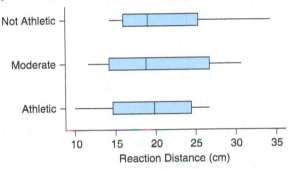

11.36 Tomato Plants and Colored Light Jennifer Brogan, a biology student who was taking a statistics class, exposed similar tomato plants to different colors of light. The average growth rates (in millimeters per week) are given in the table.

a. Interpret the boxplots given. Compare medians and interquartile ranges, and mention potential outliers.

b. Test the hypothesis that the color of light affects growth rate. In other words, test the hypothesis that the population mean growth rates differ by color, reporting the F-statistic, p-value, and conclusion. Assume that conditions for ANOVA are met. Use a significance level of 0.05. Do not do post hoc tests.

Blue	Red	Yellow	Green
5.34	13.67	4.61	2.72
7.45	13.04	6.63	1.08
7.15	10.16	5.29	3.97
5.53	13.12	5.29	2.66
6.34	11.06	4.76	3.69
7.16	11.43	5.57	1.96
7.77	13.98	6.57	3.38
5.09	13.49	5.25	1.87

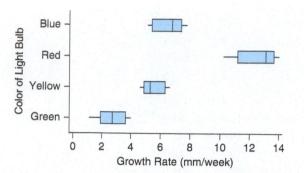

11.37 GPAs by Seating Choice A random sample of students was studied. Whether the student chose to sit in the front, middle, or back row in a college class was observed, along with the person's GPA. Test the hypothesis that GPA is associated with row using a significance level of 0.05. Assume Normality. Show all four steps, and use a significance level of 0.05. Do not do post hoc tests.

Front	Middle	Back	Front	Middle	Back
3.062	2.859	2.583	2.690	3.080	2.598
3.894	2.639	2.653	3.523	2.937	2.879
2.966	3.634	3.090	3.332	3.091	2.926
3.575	3.564	3.060	3.885	2.655	3.221
4.000	2.115	2.463	3.559	2.526	2.646

(Source: StatCrunch: Seating Choice versus GPA [For 3 rows, with Text and Indicator Columns]. Owner: bartonpoulson)

11.38 Reading Comprehension Sixty-six reading students were randomly assigned to be taught using one of three teaching methods. One method is called Basal and uses textbooks that are organized to develop reading skills. Another method is called DRTA (Directed Reading Thinking Activity) and is a comprehension strategy that guides students in asking questions about a text, making predictions, and then reading to confirm or refute their predictions. The third method, which is called Strat, uses different strategies. The test scores are given in the box. Test the hypothesis that reading method affects test scores using a significance level of 0.05. Show all four steps. Do not do post hoc tests. Assume Normality.

Basal	DRTA	Strat	Basal	DRTA	Strat
41	31	53	45	55	42
41	40	47	39	57	34
43	48	41	44	53	48
46	30	49	36	37	51
46	42	43	49	50	33
45	48	45	40	54	44
45	49	50	35	41	48
32	53	48	36	49	49
33	48	49	40	47	33
39	43	42	54	49	45
42	55	38	32	49	42

(Source: Moore, David S., and George P. McCabe (1989). Introduction to the Practice of Statistics. Original source: study conducted by Jim Baumann and Leah Jones of the Purdue University Education Department, accessed via StatCrunch)

11.39 Hours of Sleep and Health Status In a study done on a random sample of employees at a company, the employees wrote down how many hours they slept and their health status. StatCrunch output for an ANOVA is shown. Test the hypothesis that health status and number of hours of sleep are associated. Use a significance level of 0.05. Show all four steps. Do not do post hoc tests.

Options

Analysis of Variance results:
Responses in Hours Sleep.
Factors in Health Status.

Factor means

Health Status	n	Mean	Std. dev	Std. Error
excellent	32	6.78125	1.3733494	0.24277618
fair	18	6.0277778	1.7861577	0.42100142
good	72	6.4201389	1.3756089	0.16211707
poor	5	4.4	0.89442719	0.4

ANOVA table

Source	df	SS	MS	F-Stat	P-value
Treatments	3	27.101092	9.0336974	4.4399942	0.0053
Error	123	250.25816	2.0346192		
Total	126	277.35925			

(Source: StatCrunch: survey results. Group Data.xlsx owner: jib4wolf)

11.40 Happiness and Age Category StatCrunch surveyed users on happiness. Each respondent scored herself or himself between 1 (least happy) and 100 (most happy). We would like to determine whether age category has an effect on happiness for users of StatCrunch. Assume the data are from a random sample of StatCrunch users. Here 1 denotes people between 10 and 19 years old, 2 denotes people between 20 and 29, and so on.

a. Based on the given sample means, which group reported themselves as happiest, and which group reported the least happiness?

b. Test the hypothesis that age category is associated with happiness for users of StatCrunch. Use a significance level of 0.05. Assume Normality. Do not do post hoc tests.

```
Source    DF      SS       MS      F      P
Factor     6    5746      959    1.36  0.230
Error    673  475803      707
Total    679  481559

Level       N     Mean    StDev
Happy_1    108    68.16   25.45
Happy_2    239    66.49   28.38
Happy_3    143    68.19   27.81
Happy_4    100    74.06   23.63
Happy_5     66    69.73   23.36
Happy_6     20    75.00   25.73
Happy_7      4    54.75   23.41
```

SECTION 11.4

For all post hoc *t*-tests, pool variances.

11.41 Travel Time to School Random samples of 12th-grade students from California, Utah, and New York were asked how long it took them to get to school (in minutes). StatCrunch output for an ANOVA is shown, along with the Tukey HSD confidence intervals. Using the confidence intervals, is it possible to determine which state has the longest travel time? Is there evidence that the states differ in their mean travel times? Explain.

Analysis of Variance results:
Data stored in separate columns.

Column statistics

Column	n	Mean	Std. Dev.	Std. Error
CA	47	23.021277	17.624698	2.5708264
Utah	46	12.5	14.360053	2.1172742
NY	46	25.847826	22.685255	3.3447583

ANOVA table

Source	DF	SS	MS	F-Stat	P-value
Columns	2	4558.2484	2279.1242	6.6335262	0.0018
Error	136	46726.414	343.57657		
Total	138	51284.662			

Tukey HSD results (95% level)
CA subtracted from

	Difference	Lower	Upper	P-value
Utah	-10.521277	-19.6311	-1.4114535	0.0191
NY	2.8265495	-6.2832736	11.936373	0.743

Utah subtracted from

	Difference	Lower	Upper	P-value
NY	13.347826	4.1891564	22.506496	0.0021

11.42 House Prices Tukey HSD confidence intervals (with an overall significance level of 0.05) were calculated for the mean housing prices in three southern California neighborhoods: Agoura, Ventura, and Oxnard. The sample mean was largest for Agoura and smallest for Ventura. Arrange the towns from smallest sample mean on the left to largest sample mean on the right, underlining pairs for which the sample means are not significantly different. Assume that conditions for ANOVA are met. Then write a sentence or two interpreting your results.

```
Agoura subtracted from:

          Lower   Center  Upper   ---+---------+---------+---------+------
Ventura  -264.21  -176.78  -89.34     (----------*----------)
Oxnard   -259.09  -172.50  -85.91      (----------*-----------)
                                    ---+---------+---------+---------+------
                                     -240      -160      -80        0

Ventura subtracted from:

         Lower  Center  Upper   ---+---------+---------+---------+------
Oxnard  -43.17   4.28   51.73                              (-----*----)
                                ---+---------+---------+---------+------
                                 -240      -160      -80        0
```

11.43 GPA and Row (Example 8) A random sample of students was studied. Whether the student chose to sit in the front, middle, or back row in a college class was recorded, along with the person's grade point average (GPA). Test the hypothesis that GPA is associated with choice of row using a significance level of 0.05. Assume Normality. Assume the conditions for ANOVA are met.

Show the F- and p-values and the conclusion for the ANOVA using an overall significance level of 0.05. Do post hoc tests using confidence intervals (98.33% intervals for three groups) if post hoc

tests are appropriate. Assume equal variances. Use the underlining method, listing the category with the lowest mean on the left and the category with the highest mean on the right, with any pairs (or groups of three) that are not significantly different underlined. Finally, write your conclusions in complete sentences.

Front	Middle	Back
3.062	2.859	2.583
3.894	2.639	2.653
2.966	3.634	3.090
3.575	3.564	3.060
4.000	2.115	2.463

Front	Middle	Back
2.690	3.080	2.598
3.523	2.937	2.879
3.332	3.091	2.926
3.885	2.655	3.221
3.559	2.526	2.646

(Source: StatCrunch: Seating Choice versus GPA [For 3 rows, with Text and Indicator Columns]. Owner: bartonpoulson)

11.44 Reading Scores by Teaching Method Refer to exercise 11.38 to find the data. Follow the instructions given in exercise 11.43.

11.45 Reaction Distances Use the data given in exercise 11.35. Follow the instructions given in exercise 11.43.

11.46 Study Hours Use the data given in exercise 11.32. Follow the instructions given in exercise 11.43.

TRY **11.47 Apartment Rents (Example 9)** Use the Minitab output (at the bottom of the page) showing the Tukey HSD post hoc analysis for exercise 11.31 to order the mean rents for these three cities from least expensive to most expensive.

11.48 Tomatoes Use the data given in exercise 11.36. Follow the instructions given in exercise 11.43.

11.49 Concern over Nuclear Power Following the large-scale nuclear power plant failure in Japan, a StatCrunch survey was conducted in which respondents were asked about their level of concern over nuclear power and their political party. The data were coded so that 1 represented the lowest level of concern and 100 the greatest level of concern. See the following output. Do a complete analysis using ANOVA with a significance level of 0.05. Do post hoc tests based on the intervals given. Remember that an interval for a difference that captures 0 shows that there could be no difference in population means. Use the underlining method, listing the parties from the party with the lowest mean on the left to the party with the highest mean on the right, and underline any pairs or groups of three that are not significantly different. Finally, write your conclusions in sentences. For the purpose of this exercise, treat the respondents as a random sample of all StatCrunch users.

Tukey Simultaneous Tests for Differences of Means

Difference of Levels	Difference of Means	SE of Difference	95% CI	T-Value	Adjusted P-Value
SF-Seattle	1551.5	291.7	(817.3, 2285.7)	5.32	<0.0001
Santa Monica-Seattle	811.6	291.7	(77.4, 1545.8)	2.78	0.0288
Santa Monica-SF	-739.9	291.7	(-1474.1, -5.7)	-2.54	0.0484

Options

Analysis of Variance results:
Responses in Concern.
Factors in Party.

Factor means

Party	n	Mean	Std. dev	Std. Error
Democrat	130	63.3	33.674145	2.9534179
Independent	96	58.385417	31.960953	3.2620011
Other	77	57.454545	36.335261	4.1407873
Republican	85	43.505882	34.661157	3.7595303

ANOVA table

Source	df	SS	MS	F-Stat	P-value
Treatments	3	20771.643	6923.881	5.980431	0.0005
Error	384	444578.38	1157.7562		
Total	387	465350.02			

Tukey 95% Simultaneous Confidence Intervals:

Democrat subtracted from

	Difference	Lower	Upper	P-value
Independent	-4.9145833	-16.729436	6.9002697	0.706
Other	-5.8454545	-18.470982	6.7800729	0.6306
Republican	-19.794118	-32.040829	-7.5474062	0.0002

Independent subtracted from

	Difference	Lower	Upper	P-value
Other	-0.93087121	-14.362339	12.500596	0.998
Republican	-14.879534	-27.955555	-1.8035135	0.0184

Other subtracted from

	Difference	Lower	Upper	P-value
Republican	-13.948663	-27.761538	-0.13578823	0.0468

(Source: StatCrunch Responses to Nuclear Power Survey Owner: scsurvey)

11.50 Social Media Use A StatCrunch survey asked respondents how much time they spent daily on various social media sites. Is there a difference in the mean time spent on these social media sites? Use the following ANOVA output. Do a complete ANOVA analysis with a significance level of 0.05, including post hoc tests using the intervals given in the output. Assume that the Normality condition for the ANOVA is met, and for the purposes of this exercise treat the respondents as a random sample of all StatCrunch users. (Source: scsurvey)

Analysis of Variance results:
Responses: Time spent
Factors: Media app

Response statistics by factor

Media app ⬍	n ⬍	Mean ⬍	Std. Dev. ⬍	Std. Error ⬍
Facebook	1735	86.017291	136.61756	3.2798711
Instagram	1054	109.68406	144.08148	4.4380056
Other	464	83.860517	168.89873	7.8409268
Pinterest	141	119.20567	234.97615	19.788562
Snapchat	1268	145.78572	175.94957	4.9411563
Twitter	498	128.14056	158.04099	7.081986

ANOVA table

Source	DF	SS	MS	F-Stat	P-value
Media app	5	3114199	622839.81	25.316535	<0.0001
Error	5154	1.267992e8	24602.096		
Total	5159	1.299134e8			

Tukey HSD results (95% level)
Facebook subtracted from

	Difference	Lower	Upper	P-value
Instagram	23.66677	6.2043675	41.129172	0.0016
Other	-2.1567738	-25.526502	21.212954	0.9998
Pinterest	33.188383	-5.9683904	72.345156	0.1507
Snapchat	59.768427	43.248121	76.288732	<0.0001
Twitter	42.123271	19.391686	64.854856	<0.0001

Instagram subtracted from

	Difference	Lower	Upper	P-value
Other	-25.823543	-50.735415	-0.9116715	0.037
Pinterest	9.521613	-30.574647	49.617873	0.9845
Snapchat	36.101657	17.463587	54.739727	<0.0001
Twitter	18.456502	-5.857736	42.770739	0.255

Other subtracted from

	Difference	Lower	Upper	P-value
Pinterest	35.345157	-7.6538988	78.344212	0.177
Snapchat	61.9252	37.664401	86.185999	<0.0001
Twitter	44.280045	15.428861	73.131229	0.0002

Pinterest subtracted from

	Difference	Lower	Upper	P-value
Snapchat	26.580044	-13.114982	66.27507	0.3967
Twitter	8.9348885	-33.720704	51.590481	0.9913

Snapchat subtracted from

	Difference	Lower	Upper	P-value
Twitter	-17.645155	-41.291871	6.00156	0.2733

CHAPTER REVIEW EXERCISES

11.51 Happiness and Age Consider the data from the happiness survey (see exercise 11.40). If you do a series of two-sample *t*-tests comparing mean happiness of all pairs of the seven age groups, you will find that the p-value that compares happiness of those in their forties with those in their twenties, is 0.012. Would it be correct to say that the difference in mean happiness is statistically significant for those in their twenties and those in their forties? Explain.

11.52 GPA and Row Number Suppose you collect data on GPAs by classroom row in which the student chose to sit, and that there are four rows. Suppose you do multiple two-sample *t*-tests to compare the mean GPA of the rows, and discover that the p-value comparing the means of Row 1 and Row 4 is 0.025. Would it be correct to say that the difference in mean GPA is statistically significant for Row 1 and Row 4? Explain.

11.53 Contacting Mother Professors of ethics (Eth), professors of philosophy (Phil), and professors in fields other than philosophy or ethics (Other) were asked how many days it had been since they had last been in contact with their mothers. Contact was defined as face-to-face or telephone contact. Professors whose mothers had died were not included. Random samples of 30 from each group were taken.

Because the standard deviations are too different to use ANOVA, compare each pair of means using Bonferroni intervals without pooling the standard deviations. Use a Bonferroni-corrected confidence level of 98.33% in order to achieve an overall confidence of 95%. Write a sentence of two interpreting your findings.

Eth	Phil	Other	Eth	Phil	Other
28	14	3	5	2	0
4	1	9	5	2	1
7	3	1	14	3	7
6	4	1	14	1	1
100	2	0	1	7	0
1	2	0	10	2	0
1	1	1	5	3	1
70	40	1	2	9	3
5	2	3	5	1	1
7	2	6	4	3	12
10	7	0	3	0	20
2	10	4	2	10	4
2	1	1	60	5	1
1	6	4	1	4	7
3	1	1	4	45	3

(Source: Eric Schwitzgebel)

11.54 Ideal Percentage to Charity Professors of ethics (Eth), professors of philosophy (Phil), and professors in fields other than philosophy or ethics (Other) were asked what percentage of their income professors *should* donate to charity. Assume the professors are randomly sampled from the population of professors. Determine whether there are significant differences among these three groups. Do post hoc tests, if they are warranted, using an overall significance level of 0.05. Arrange the means from lowest on the left to highest on the right, underlining any pairs that are not significantly different. Then write out your conclusions in complete sentences. Minitab output is given.

One-way ANOVA:

```
Source   DF        SS       MS      F      P
Factor    2     650.4    325.2   7.96  0.000
Error   498   20344.7     40.9
Total   500   20995.1

S = 6.392     R-Sq = 3.10%     R-Sq(adj) = 2.71%

Level               N    Mean    StDev
CharShouldEth     177   7.023    8.381
CharShouldPhil    186   4.634    4.560
CharShouldOther   138   4.645    5.494
```

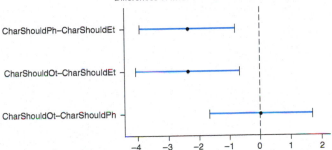

Tukey Simultaneous 95% CIs
Differences of Means for CharShouldEt, CharShouldPh, ...

(Source: Eric Schwitzgebel)

11.55 Actual Percentage to Charity Professors of ethics (Eth), professors of philosophy (Phil), and professors in fields other than philosophy or ethics (Other) were asked what percentage of income they actually donated to charity. Assume the professors are randomly sampled. The data are at this text's website. Ethicists tend to believe that a greater amount of money *should* be given to charity than do the other two groups. And so we wish to know whether ethicists tend to contribute more money to charity than professors in other fields. As a first step, determine whether the mean contribution level differs among these three groups. Do post hoc tests, if they are warranted, with an overall significance level of 0.05. Arrange the means from lowest on the left to highest on the right, underlining any pairs that are not significantly different. Then write out your conclusions in sentences.

```
One-way ANOVA: EthChar, PhilChar, OtherChar

Source    DF      SS     MS     F     P
Factor     2    307.1  153.5  7.33  0.001
Error    523  10955.3   20.9
Total    525  11262.3

S = 4.577   R-Sq = 2.73%   R-Sq(adj) = 2.35%

Pooled StDev = 4.577

Grouping Information Using Tukey Method

            N   Mean  Grouping
EthChar    181  5.296    A
OtherChar  156  5.059    A
PhilChar   189  3.606    B

Means that do not share a letter are significantly different.
```

(Source: Eric Schwitzgebel)

11.56 Hours of Television by Age Group The StatCrunch output shows the ANOVA results for testing whether there is an association between the number of hours of TV watched per week and age group: 50 and over (AdultTV), college students (TeenTV), and grade school students (ChildTV).

a. Test the hypothesis that people in different age groups spend different amounts of time, on average, watching television. Use a significance level of 0.05. Assume that the conditions for ANOVA are met.

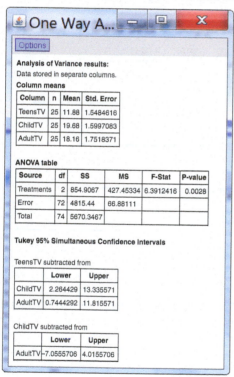

b. Using the output provided, determine which sample means are significantly different from each other. Report all the given confidence intervals for the difference between means, and tell what they show. Then arrange the groups with lowest mean on the left and highest on the right, with underlines connecting groups that do not have significantly different means. Finally, write a sentence or two explaining which group means are different and how they differ.

11.57 Triglycerides and Gender Using the NHANES data, we performed an ANOVA to test whether gender is associated with level of triglycerides, a form of fat, in the blood. ANOVA and *t*-test output from a two-sample *t*-test is shown. In both cases, we are testing the hypothesis that the mean triglyceride levels for men and women are different. Assume that conditions for both tests have been met.

```
Source    DF        SS       MS      F      P
Sex        1    132611   132611   3.16  0.076
Error    940  39476146    41996
Total    941  39608756
```

```
Differen ce = mu (Female) - mu (Male)
Estimate for differ ence: -23.8
95% CI for difference: (-50.0, 2.5)
T-Test of diff erence = 0 (vs not =): T-Value = -1.78  P-Value = 0.076
```

Compare the output by comparing p-values. Square the *t*-value to see what you get for comparison. Explain.

11.58 Cholesterol and Gender Using NHANES data, we performed one-way ANOVA and a two-sample *t*-test. In both cases we were testing the hypothesis that the mean cholesterol values for men and women are different. Compare the output of ANOVA and the two-sample *t*-test by looking at the *t*-statistics, *F*-statistics, and p-values.

Difference	Sample Mean	Std. Err.	DF	T-Stat	P-value
$\mu_1 - \mu_2$	7.5490246	2.93151	940	2.5751317	0.0102

ANOVA table

Source	df	SS	MS	F-Stat	P-value
Treatments	1	13393.94	13393.94	6.631303	0.0102
Error	940	1898616.9	2019.8053		
Total	941	1912010.9			

GUIDED EXERCISES

11.5 Apartment Rents Random samples of rents for 1-bedroom 1-bath apartments in Seattle, San Francisco, and Santa Monica were selected and shown in the following table. Assume the conditions for two-sample *t*-tests are met.

Seattle Rent (dollars)	San Francisco Rent (dollars)	Santa Monica Rent (dollars)
1165	2750	1899
1854	2595	2700
1750	3900	1799
1474	2445	3284
2495	4515	2300
1425	3800	2300
2010	3400	3104
1615	2795	2895

QUESTION Use three two-sample *t*-tests, applying the appropriate Bonferroni Correction to achieve an overall significance level of 0.05, to compare all possible pairs of means. Instead of using the four-step method, summarize your findings using a table that reports, for each pair of cities, the observed *t*-test statistics, p-values, and conclusion.

Step 1 ▶ Means and standard deviations
Fill out the table, reporting all the sample means and standard deviations.

	Mean	SD
Seattle	1723.50	408.84
San Francisco	3275.00	
Santa Monica		

Step 2 ▶ List comparisons
Finish the list of all three possible comparisons.

1. Seattle compared to San Francisco
2. _____
3. _____

Step 3 ▶ Find the corrected alpha
Find the corrected value for the significance level by dividing 0.05 by the number of comparisons.

Step 4 ▶ Compute to compare
We have assumed that the conditions for two-sample *t*-tests are met. For all tests, the null hypothesis is that the two population means are the same, and the alternative hypothesis is that the two population means are different.

Complete the following table. For a significant difference, the p-value must be less than the Bonferroni-corrected value for the significance level.

	t	p-Value	Conclusion
Seattle-San Francisco	5.17	0.0003	
Seattle-Santa Monica	3.36		
San Francisco-Santa Monica	2.27		

CONCLUSION Write a clear conclusion based on what you found. Which groups have sample means that are significantly different, and how do they differ?

11.31 Apartment Rents Samples of rents for one-bedroom, one-bath apartments in three cities were obtained. Assume the distribution of each population is Normal to satisfy the conditions for using ANOVA.

QUESTION Test the hypothesis that the mean rent for one-bedroom apartments in at least one of these cities differs from the mean rent in the other, using a significance level of 0.05. (Do not do post hoc tests.) Use the following numbered steps for guidance.

Seattle	San Francisco	Santa Monica
1165	2750	1899
1854	2595	2700
1750	3900	1799
1474	2445	3284
2495	4515	2300
1425	3800	2300
2010	3400	3104
1615	2795	2895

Step 1 ▶ Hypothesize
H_0: $\mu_{Seattle} = \mu_{SF} = ?$
H_a: At least one population mean is different from another

Step 2 ▶ Prepare
Choose one-way ANOVA.
Conditions:
Random sample and observations independent of each other.
Independent groups.
Sample variance: The ration of the largest standard deviation to the smallest is 744.2/408.8 = 1.82 which is less than 2.
Assume that the distribution of each population is close enough to Normal to satisfy the Normal condition for using ANOVA.

Step 3 ▶ Compute to compare.
$F = ?$
p-value $= ?$

Step 4 ▶ Interpret
Reject the null hypothesis if the p-value is less than or equal to 0.05. Explain in words what the conclusion implies about the three means.

11.47 Apartment A Minitab output showing the Tukey HSD post hoc analysis for exercise 11.31 is provided.

QUESTION Use the Minitab output below to order the mean rents for these three cities from least expensive to most expensive.

Step 1 ▶ Arrange the names of the three cities, placing the city with the lowest rent on the left and the city with the highest rent on the right. The city with the highest rent is filled in for you.

_____ _____ <u>San Francisco</u>

Step 2 ▶ Underline any pairs that are not significantly different. You can do this by examining the three confidence intervals for the differences to see whether each interval captures 0.

Step 3 ▶ Pick the correct conclusion from the options that follow.

i. There are no significant differences in the means.

ii. All the means are significantly different from each other, with San Francisco having the highest rents and Seattle having the lowest rents.

iii. San Francisco has the highest rents. There is no significant difference in rents between Seattle and Santa Monica.

iv. Seattle has the lowest rents. There is no significant difference in rents between San Francisco and Santa Monica.

Tukey Simultaneous Tests for Differences of Means

Difference of Levels	Difference of Means	SE of Difference	95% CI	T-Value	Adjusted P-Value
SF-Seattle	1551.5	291.7	(817.3, 2285.7)	5.32	<0.0001
Santa Monica-Seattle	811.6	291.7	(77.4, 1545.8)	2.78	0.0288
Santa Monica-SF	-739.9	291.7	(-1474.1, -5.7)	-2.54	0.0484

TechTips

General Instructions for All Technology

All the technology will use the following data set.

EXAMPLE (ANALYSIS OF VARIANCE) ▶ The two tables show the data we will use in two different forms. Table 11A consists of unstacked data, and Table 11B gives stacked data. Some technologies require one form and some the other form. The data were constructed for simplicity. To have context, imagine that each number represents the number of children in a random sample of families with children in fictitious countries C, E, and G.

C	G	E
1	11	2
2	13	3

▲ **TABLE 11A** Unstacked Data

Children	Country
1	C
2	C
11	G
13	G
2	E
3	E

▲ **TABLE 11B** Stacked Data

TI-84

1. Press **STAT**, choose **EDIT**, and enter the unstacked numbers into **L1**, **L2**, and **L3**.
2. Press **STAT**, choose **TESTS**, and scroll up to **ANOVA** and press **ENTER**.
3. See Figure 11a. When you see **ANOVA** (enter the three lists' names separated from each other by commas (the comma button is above **7**). To get **L1**, for example, press **2ND** and **1**. Enter: **L1, L2, L3**)

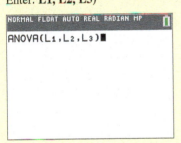

▲ **FIGURE 11a** TI-84 ANOVA Input

4. Press **ENTER**.
   You should get the output shown in Figure 11b.

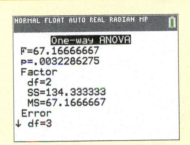

▲ **FIGURE 11b** TI-84 ANOVA Output

Post Hoc Tests (Bonferroni Correction)

To do post hoc tests, you will need to use two sample t-intervals (or t-tests), which are shown at the end of Chapter 9. Use "Pooled Yes" for the t-test if you previously performed ANOVA with the data. If you are using intervals and the overall significance level is 0.05:

> Use 98.33% intervals for three groups, because $(0.05/3 = 0.1667$ and $1 - 0.1667 = 0.9833)$.
> Use 99.17% intervals for four groups, six comparisons.

MINITAB

With Minitab, it is possible to use either stacked data or unstacked data, but using stacked data is the preferred convention.

1. See Table 11B. Enter the numbers in column C1 and the category code values in column C2. Include the labels at the top.
2. **Stat > ANOVA > One-Way**
3. See Figure 11c. Choose **Response data are in one column for all factor levels**. Be sure that the variable with the numbers (here, **C1 Children**) goes in the **Response** box and that the variable with the code (here, **C2 Country**) goes in the **Factor** box.
4. For post hoc tests, click **Comparisons** then choose, **Tukey, Interval plot**, and **Grouping information**. Click **OK**. **Error rate for comparisons: 5** (for a significance level of 0.05)

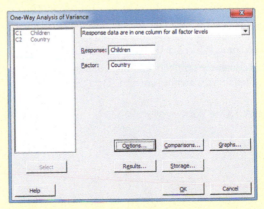

▲ **FIGURE 11c** Minitab ANOVA Input

5. Click **Graphs** and uncheck **Interval plots**. Click **OK** and **OK** again.

Part of the output is shown in Figure 11d.

One-way ANOVA: Children versus Country

Analysis of Variance

Source	DF	Adj SS	Adj MS	F-Value	P-Value
Country	2	134.333	67.167	67.17	0.003
Error	3	3.000	1.000		
Total	5	137.333			

Tukey Pairwise Comparisons

Grouping Information Using the Tukey Method and 95% Confidence

Country	N	Mean	Grouping
G	2	12.00	A
E	2	2.500	B
C	2	1.500	B

Means that do not share a letter are significantly different.

Tukey Simultaneous 95% CIs

▲ **FIGURE 11d** Minitab ANOVA Output

6. See Figure 11d. Click on **Tukey Simultaneous 95% CIs** to display the graph shown in Figure 11e.

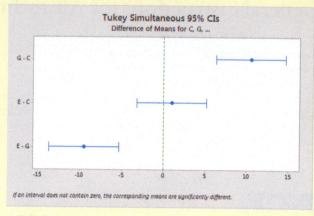

If an interval does not contain zero, the corresponding means are significantly different.

▲ **FIGURE 11E** Minitab Tukey Output

EXCEL

1. See Figure 11f: Enter unstacked data with labels into columns A, B, and C.

▲ **FIGURE 11f** Excel ANOVA Input

2. Click **Data, Data Analysis** and choose **Anova: Single Factor**.

3. For **Input Range**, drag the cursor over all the numbers; do not include the labels. Click **OK**.

Figure 11g shows part of the output.

ANOVA					
Source of Variation	SS	df	MS	F	P-value
Between Groups	134.3333	2	67.16667	67.16667	0.003229
Within Groups	3	3	1		
Total	137.3333	5			

▲ **FIGURE 11g** Excel ANOVA Output

Post Hoc Tests (Bonferroni Correction)

To do post hoc tests, use two sample *t*-tests, which are described in Chapter 9, but use *t*-**Test Two-Sample Assuming Equal Variances** and reset **Alpha** to **0.00167**.

STATCRUNCH

You may use stacked or unstacked data. We prefer stacked data.

1. See Table 11B. Enter the numbers into one column and your category code letters into another column. Type the labels at the top.

2. **Stat > ANOVA > One Way.**

3. See Figure 11h. Click **Compare: Values in a single column**. For **Responses in**: Click the numerical variable (here, **Children**). For **Factors in**: Click the variable with the code (here, **Country**). (If you had unstacked data, you would click **Compare: Selected columns**.)

4. To include post hoc tests, click **Tukey HSD: Compute at level:** and leave **0.95** if alpha is 0.05.

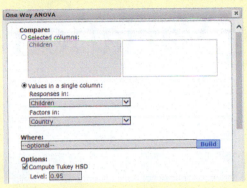

▲ **FIGURE 11h** StatCrunch ANOVA Input

5. Click **Compute!**

The output is shown in Figure 11i.

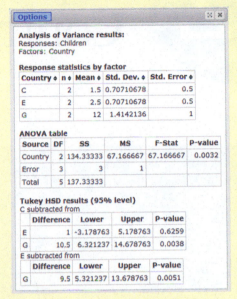

Analysis of Variance results:
Responses: Children
Factors: Country

Response statistics by factor

Country ◆	n ◆	Mean ◆	Std. Dev. ◆	Std. Error ◆
C	2	1.5	0.70710678	0.5
E	2	2.5	0.70710678	0.5
G	2	12	1.4142136	1

ANOVA table

Source	DF	SS	MS	F-Stat	P-value
Country	2	134.33333	67.166667	67.166667	0.0032
Error	3	3	1		
Total	5	137.33333			

Tukey HSD results (95% level)
C subtracted from

	Difference	Lower	Upper	P-value
E	1	-3.178763	5.178763	0.6259
G	10.5	6.321237	14.678763	0.0038

E subtracted from

	Difference	Lower	Upper	P-value
G	9.5	5.321237	13.678763	0.0051

▲ **FIGURE 11i** StatCrunch ANOVA Output

12 Experimental Design: Controlling Variation

THEME

Researchers sometimes wish to detect subtle differences between experimental groups. This task can be difficult, because variation in the data may be caused by factors other than the ones that researchers are studying. We can increase the probability of seeing small differences between groups by controlling variation.

We began our study of statistics in this text by talking about why and how we collect data. In Chapter 1, you saw that if the data are collected through a randomized, controlled experiment, then we can draw cause-and-effect conclusions. In Chapter 7 you learned that if the data are chosen at random from the population, then we can generalize our conclusions to that population. One lesson of Chapters 7 through 11 is that the method of data collection affects how we study and analyze the data. For example, if the data consist of paired samples and we wish to compare means, then we apply the paired *t*-test; if the data consist of two independent samples, then we apply the independent two-sample *t*-test.

In this chapter, we focus on the role that surveys and controlled experiments play in helping us understand the world by examining data. Chapter 7 introduced surveys based on simple random samples. But in many realistic situations—for example, trying to determine the number of female smokers in China or the proportion of trees in a forest infected by the dangerous bark beetle—a simple random sample is very difficult to collect. In this chapter, we examine some alternative survey designs that are more practical and effective for collecting and analyzing data.

Controlled experiments can be used to determine cause-and-effect associations. Can you do better on exams by writing about your anxieties just before the exam? Does stretching before a race make you run faster? These are cause-and-effect questions that can be difficult to answer, partly because there is much variation between people when they are taking exams or running races.

Most of us will carry out very few experiments in our lifetimes, but we will read about them. The results of experiments are reported in the nightly news, on blogs, and in newspapers. Sometimes the results seem highly counterintuitive or even contradict what we heard just last week. This chapter will include some basic suggestions for reading and interpreting published scientific studies.

CASE STUDY

Does Stretching Improve Athletic Performance?

Most athletes take stretching seriously. Runners in particular are taught that stretching before and after running is important not just to prevent injuries but also to maximize performance. However, a growing body of research is questioning the effectiveness of stretching. For example, Nelson et al. (2005) measured the speed of 16 NCAA track athletes who ran after stretching and after not stretching. On different days, each athlete followed these protocols: They stretched both legs, or only their forward leg positioned as at the start of a race, or only their back leg positioned as at the start of a race, or neither leg. All athletes did the same stretches for the same amount of time. The researchers found that the athletes ran the 20-meter sprint slower after stretching.

What does this study tell us about preparing competitive athletes? What does it tell us about "weekend" athletes? Is this evidence that you shouldn't stretch if you plan on running a 10K race? Are the results believable? Why was there no control group that did no stretches at all? Does it matter that all the subjects did all the stretching regimens? In this chapter we examine the characteristics of controlled experiments so that we can understand whether we can generalize conclusions of studies to other groups. At the end of the chapter, we will return to this case study and consider the answers to these questions in light of what we have learned.

SECTION 12.1

Variation Out of Control

In this chapter we return our focus to the Consider Data phase of the Data Cycle (see Figure 12.1). For much of this text, the data have been provided with only a little discussion about how they were collected. But often, much thought and effort went into collecting the data, and in this chapter we go into greater detail on how this can be done.

▶ **FIGURE 12.1** The Consider Data phase includes collecting data in a way that's appropriate for the questions we wish to answer.

The Data Cycle

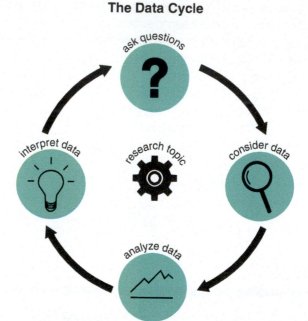

Consider the March 2014 outbreak of the Ebola virus, a deadly and contagious disease. By August 2014, according to the World Health Organization, about 3000 people had become infected and slightly more than half had died (www.who.int, viewed September 3, 2014). In the United States, there was a vaccine in very early testing stages that had never been used on humans. When two American medical workers became infected, they were given the vaccine. Both recovered. But did they recover because of the vaccine? About half of people recovered without the vaccine, and so they might have been lucky. Or it might have been the very high quality of care they received at the hospital. Again, natural variability makes it challenging to determine how effective a vaccine actually is at preventing the spread of a disease. But the results were promising enough that a full scale study of the effectiveness and safety of the vaccine began at the end of August 2014, and in December 2016 the World Health Organization was able to claim that the vaccine provided "high protection" against the disease (http://www.who.int/news-room/ viewed June 2018).

For another example of the importance of carefully collecting data, consider the Mediterranean diet, which emphasizes fruits, olive oil, vegetables, and fish, which was found to be effective against heart disease based on a controlled, randomized trial published in 2013. This study had over 7000 participants, and they were randomly assigned to one of three groups: the Mediterranean diet with one extra servings of olive oil per day, the Mediterranean diet but with an extra serving of nuts instead of the olive oil, and a traditional low-fat diet. After five years, the researchers reported that the groups on the Mediterranean diet had a lower rate of heart disease than did those on the low-fat diet.

But this study was recently retracted, when it was discovered that the randomized assignment was faulty. This fault has led some researchers to doubt the conclusions of the original study (Kolata 2018).

Statisticians and researchers have developed several methods for controlling variability in order to find answers to important questions such as whether vaccines are effective and whether diets can prolong our lives. In Chapter 1 we explained why questions such as these are best answered using data collected through controlled experiments. In this chapter, we show you how it's done.

Review of Experimental Basics

Before getting to the heart of the chapter, let's review a few of the important concepts covered in Chapter 1.

The phrase *cause and effect* has a common English meaning, but nonetheless, we need to be clear about its meaning in this context. For our purposes, we say that two variables X and Y have a cause-and-effect relationship if, whenever we change the value of X for a person (or object), then the mean value of Y changes in a predictable way. For example, if napping causes test scores to improve, then if you changed your test preparation from "no nap" to "nap," you would see an increase in your average test scores. If we gave you a flu vaccine, we would change your status from "not vaccinated" to "vaccinated," and we would hope to see your risk of getting the flu go down as a result.

Note that we were careful to say that napping would improve your "average" performance. The reason for this is that we are concerned with situations in which the response to the treatment variable has variability. So napping today might not raise your test score; it might even go down. But if you continue to nap, and if your test scores go up on average, then we will say that napping causes test scores to change.

We call the variable that we change—the "cause"—the **treatment variable**. The variable that we are interested in seeing changed—the "effect"—is called the **response variable**. The treatment variable records, for example, whether a subject received the vaccine or did not receive the vaccine, and the response variable records whether or not that person later got the flu.

The trick is to determine whether changes that we see in the response variable are due to changes in the treatment variable; are due to changes in some other, unseen variables; or are due to chance. Unseen variables that might affect our conclusions about causality are called **confounding variables** (or confounders).

For example, in an evaluation of an SAT preparation company, researchers found that students who completed the SAT preparation course did better on the SAT than students who did not. The treatment variable here records whether or not a student took the preparation class, and the response variable records the student's SAT scores. One problem with the study was that the preparation course was expensive, and a student's decision to take the course depended somewhat on the student's ability to pay (or the family's ability to pay)—a possible confounding variable.

To be a confounding variable, a variable must affect *both* the treatment *and* the response. The student's ability to pay meets this criterion, because it affects both the treatment (attend or not attend the SAT preparation course) and the response (score on SAT). A student's ability to pay for SAT preparation affects the treatment variable because, on average, students who take the SAT preparation class are better off financially than students in the group who do not take the class. And this ability to pay means that the students might, typically, have increased access to other resources that help them do better on the SAT (for example, they might not need jobs and so would have more time to prepare); for this reason, the response variable is also affected.

Another possible confounding variable is motivation. Students who are motivated enough to spend money on a prep course, to go through the trouble of finding a prep course, and to put in the extra time attending the course might also be motivated enough to study in other ways. So even if the prep course is not effective, the students' extra motivation might be the reason for their higher SAT scores (Briggs 2002).

Maybe you can think of other confounding variables as well. The presence of confounding variables means we cannot conclude that the prep course caused the higher SAT scores. To be able to conclude that the prep course caused the change, we would have to eliminate all other explanations.

To eliminate other explanations, the experimenter must make sure that the subjects in the experiment are alike in every way except that they have different values for the treatment variable. To achieve this, experimenters can take control and assign subjects to a treatment group. When this is done, we call the experiment a **controlled experiment**.

> **Details**
>
> **Lurking Variables**
> Another term for a confounding variable is a lurking variable.

Still, making sure that the subjects in the different treatment groups are all alike is difficult, because people (and animals and even some inanimate objects) are not all alike. This variability means that if we're not careful, some differences between the groups will arise that might be confounding variables and prevent us from concluding that the treatment caused any observed changes in the response variable.

Researchers use **random assignment** to ensure that different treatment groups are as similar as possible. For example, in the study of the effectiveness of the Mediterranean diet, researchers could have randomly assigned individuals to one of the three diets (the Mediterranean diet with extra olive oil, the Mediterranean diet with nuts instead of olive oil, and a traditional low-fat diet.) Had they done so, the three groups would have had similar distributions on potential confounding factors: similar percentages of genders, similar distributions of weights, similar predispositions for health conditions. Instead, the researchers assigned a few entire families to the same diet. This meant that perhaps the family was saved from heart disease by the diet, or perhaps they were saved because of other lifestyle or genetic factors they shared. There was no way to tell. Random assignment ensures that other variables, even some we don't know about, will be distributed similarly among the groups.

Observational studies are quite common, in part because it is not always possible to do a controlled experiment. How could researchers decide whether using a cell phone over many years causes brain cancer in a randomized study? First, such a study would be unethical. Second, even if they ignored ethical considerations, how could researchers make sure that subjects who were assigned to the treatment group used their phones often enough over many years? And how could they be sure that subjects in the control group did *not* use cell phones for many years?

A major drawback with observational studies is that they can establish only an association between the treatment and response variables; they cannot establish a cause-and-effect relationship. The news media often ignore this fact and report the results of observational studies as though they were controlled experiments. For example, Discovery News published this headline: "Dogs Walked by Men Are More Aggressive" (via http://jfmueller .faculty.noctrl.edu/). The reported study was an observational study, based on watching how dogs and their owners interact with others. Therefore, we can't conclude that a dog owned by a man could be made less aggressive by having a woman take it for walks.

 Looking Back

Observational Studies and Confounding
Recall that in an observational study, we can never eliminate the possibility that a confounding variable exists. This means we cannot reach cause-and-effect conclusions on the basis of a single observational study.

KEY POINT	Random assignment to treatment groups ensures that the subjects in all groups are as similar as possible so that the only real difference between groups is the treatment they receive.

EXAMPLE 1 Nonsense Math

Does including nonsensical mathematical equations in a research abstract make people think the research is more believable? A research abstract is a brief summary of a research paper that is intended to give the reader a general understanding of the results of the paper. A psychologist recruited 200 American adults and asked them to read a scientific abstract and then rate what they felt was the "general quality" of the research. A score of 0 meant that they felt the research had no quality, and a score of 100 meant that it was of the highest possible quality. The abstract was a real one that had appeared in a high-quality research journal. The psychologist chose it because it was on a topic that was easily understood by the general public. Half of the subjects were randomly assigned to read this abstract as it originally appeared in its published form. The other half read it with one added sentence of nonsensical mathematics. The researcher found that the group that saw the nonsensical mathematics rated the paper of higher quality, on average, than did the group that read the abstract in its original form (Eriksson 2012).

 For this controlled experiment, identify the treatment and response variables. Restate the conclusion of the study in terms of a cause-and-effect conclusion.

SOLUTION   The treatment variable records the treatment group that each subject was assigned to: nonsensical math or no math. The response variable is the subjects' rating of the quality of the abstract. The conclusion could be stated as "Nonsense math improves perceived quality of research abstracts."

TRY THIS!   Exercise 12.9

Statistical Power

Random assignment helps researchers eliminate (or at least minimize) the effects of possible confounding variables. But there is always the possibility that observed changes in the response variable are simply due to chance rather than due to the treatment the subjects received.

Because statistical tests involve chance, even when the two groups we're comparing are truly different, there is a probability that we won't see this difference. On the other hand, there is a probability that we will correctly find that the groups are truly different. This probability is called the **power**. The power is the probability of detecting differences that really exist between groups.

Obviously, we want this probability to be big. Many researchers like to have a power of at least 80%. Higher is even better.

The power depends primarily on three things: the sample size, the size of the true difference between the groups, and the natural variability within the population. Of these three factors, researchers and statisticians have direct control only over sample size. Still, by identifying sources of natural variability in the population, researchers can design experiments that can improve the power by controlling for the variation.

Variability that exists within a population can arise from several different sources. One source is natural variability. People are not all the same height, or the same weight, for example. Another important source of variability is measurement error. This is the variability caused by the devices or methods used to measure something. For example, if you weigh yourself twice in a row, you don't always get exactly the same weight the second time. Figure 12.2 shows the distribution of weights of two thousand 1-euro coins collected by statistician Herman Callaert in Belgium. You might expect the coins would weigh exactly the same, because heavier coins might be perceived as more valuable. Still, whether as a consequence of measurement error or of variations in the coin-minting process, we still see variability.

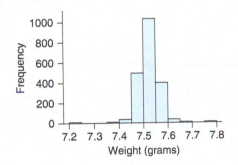

◀ **FIGURE 12.2** Even though the minting of coins is very carefully controlled, there is still variability, as this distribution of the weights of two thousand 1-euro coins shows (data from Shkedy, et al. 2006).

The best way to control for this variation when conducting a research study is to hold a variable fixed at a particular value. This is called **controlling for a variable**. If we are performing a study to see how naps affect memory, and we believe that caffeine intake affects naps, then we might control for caffeine amounts by requiring everyone to be caffeine-free. Or we might even give all of the groups the same amount of caffeine, if we were interested in seeing how our conclusions were affected by caffeine.

Blocking

For factors that the researchers cannot control, they can apply blocking. **Blocking** is a technique in which researchers group similar objects into "blocks" and then assign treatments randomly within each block. Blocking helps achieve two purposes. It reduces bias (and so is particularly useful when the sample sizes are small), and it increases statistical power.

In many medical studies, for example, different age groups respond differently to treatments. Researchers cannot control a person's age, but they can create blocks of subjects who are the same age (or nearly the same age) and in that way increase the statistical power of their study.

The basic idea is that once we've grouped similar subjects or objects together, we hope that subjects within a block will respond similarly to the treatment. If so, then we don't need to measure quite so many people, because there is less variability than we originally had.

> **KEY POINT**   A block consists of subjects or objects that are similar on one or more variables. Objects *within* a block are then randomly assigned to treatment groups.

High-anxiety students   Low-anxiety students

▲ **FIGURE 12.3** Each letter represents a student in the study. Students within each block are randomly assigned either to the Expressive Writing treatment (T) or to the control group (C).

Creating Blocks   To create blocks, ask whether you know of any variables that might affect the outcome and cannot be controlled. If so, create the blocks by grouping together objects that are similar to one another on those variables.

Once you have created the blocks, you then randomly assign objects within each block either to the control group or to the treatment group (or to the many different treatment groups, if the study is considering several different treatments). Each block should have objects in all treatment and control groups.

For example, psychologists at the University of Chicago had evidence that "expressive writing"—writing about emotions—might help prevent students from "choking" on exams (Ramirez and Beilock 2011). To test this, they recruited ninth-grade biology students at a large high school. Students were evaluated about their upcoming final exam. Some students were very anxious, but others were much less anxious. Because researchers suspected that one source of variation in the effects of the treatment might be the student's anxiety level, they created two blocks. One block consisted of students who were the most anxious about the final exam. The other block consisted of the least anxious students. Ten minutes before the exam, students in each block were randomly assigned to do one of two activities. Half of the students were told to write about their anxieties concerning the test; this expressive writing formed the treatment. The other half of the students were told to write about any topic that was not covered by the exam; these students formed the control group. Figure 12.3 illustrates this blocking design.

At the end of the study, researchers found that the high-anxiety students who did expressive writing outperformed the high-anxiety students in the control group. On the other hand, low-anxiety students who did expressive writing did not perform any differently than their control-group counterparts. The blocking structure of the experiment helped the researcher tease out some complex relations among anxiety, the treatment, and exam performance.

Blocking is often used in agricultural and biological studies done "in the field." The reason is that when you are studying plants in the great outdoors, the amount of sunlight, water, or soil nutrients that a plant might receive can vary widely just a few feet in any direction. Figure 12.4 shows an illustration of two blocks used to study the growth of Mojave yucca. Researchers, concerned about campers and outdoor recreationists

Block   Block

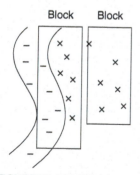

▲ **FIGURE 12.4** The Xs represent the locations of yucca plants. The plants near the (usually dry) riverbed have different soil types and moisture than the other plants, so it makes sense to create blocks grouping together the plants that are in a similar environment.

damaging the plants, are wondering whether the plants can be made healthier by protecting some with small fences.

The illustration shows that some of the plants are near a riverbed, which fills with water after a rainstorm. Plants near the riverbed receive more moisture, so if a study is being done on these plants, then the plants near the river should be put in one block and the other plants in another. Then, within each block, some plants are randomly chosen to be surrounded by fences, and others are randomly chosen to have no protection.

EXAMPLE 2 Wash That Statistician Right Out of Your Hair

A cosmetics company has created a new dandruff shampoo, which it believes will work better than its older shampoo formula. Company investigators believe that the effects of the shampoo depend on hair color. For this reason, they plan to block on hair color. To do this, they create blocks that have people with blond hair, brown hair, black hair, red hair, and gray hair, respectively. They then randomly select three of the blocks to receive the new shampoo and the other two blocks to receive the old shampoo.

QUESTION   Is this an effective design for the study? If not, describe an improvement.

SOLUTION   No. The researchers randomly assigned entire blocks to treatment groups. Instead, they should randomize *within* blocks. If an entire block is assigned to the same treatment—for example, if all the black-hair subjects are assigned to the old shampoo—then researchers won't be able to determine whether the new shampoo works on black hair.

To properly conduct this study, they should randomly assign half of the blond-hair group to receive the new formula and half to receive the old. Then they should randomly assign half of the brown-hair subjects to receive the new formula and half to receive the old, and so on.

TRY THIS!   Exercise 12.17

EXAMPLE 3 Music and Pain

Does listening to music help us deal with pain? Researchers wanted to determine whether "music therapy" could help patients recovering from surgery to handle pain better. The patients were given the ability to control the amount of morphine (a very powerful pain reliever) that they received during postsurgical recovery. The researchers planned to randomly assign some of the patients to music therapy, in which music would be played in their hospital room. The other patients would be assigned to a room with no music playing. The researchers would measure the amount of morphine the patients used during the first day of recovery.

However, the researchers believed that some patients were better able to tolerate pain than others. These patients might find the music to be less helpful than either patients with typical pain tolerance abilities or patients with low pain tolerance. For this reason, the researchers wished to create blocks based on pain sensitivity. So before surgery, the patients were given a questionnaire that asked them about how they dealt with pain. Based on the answers to this questionnaire, the researchers created three groups: low pain tolerance, typical pain tolerance, and high pain tolerance.

QUESTION   Table 12.1 shows the pain tolerance for the twelve patients who will participate in this study. Explain how to create blocks to test whether music therapy is effective for relieving pain.

▶ **TABLE 12.1** Twelve patients and their levels of pain tolerance.

Patient ID Number	Pain Tolerance
34	Low
35	High
36	Low
37	Medium
38	Medium
39	High
40	Low
41	Low
42	Medium
43	High
44	High
45	Medium

SOLUTION The researchers should create three blocks. One block consists of the patients with low pain tolerance: patients with IDs 34, 36, 40, and 41. Another block consists of those with medium pain tolerance (IDs 37, 38, 42, and 45), and the last block consists of those with high pain tolerance (IDs 35, 39, 43, and 44). Next, the researchers must randomly assign half of the patients within each block to Music Therapy and the other half to No Music. This can be done, for example, by putting slips of paper with the patients' numbers on them into a box and then randomly selecting two to be in the Music Therapy group. For example, in the Low Tolerance block, researchers put four slips of paper with the numbers 34, 36, 40, and 41 on them into a box. They shake the box and then draw out two slips of paper without looking. The patients' whose numbers are chosen will receive Music Therapy, and the other two will be in the No Music group. This process is then repeated for each block.

TRY THIS! Exercise 12.19

Blocking is useful because if you have correctly identified a variable that leads to different outcomes in your study, then by grouping subjects together on the basis of that variable, you have decreased the variability in the study—and decreasing the variability leads to greater statistical power.

Studies with small sample sizes can be greatly improved by blocking. Suppose that you wish to replicate the study about the effects of expressive writing on students in your own class. Imagine that there are 20 students, and most—say, 15 students—are not anxious. If you do not create blocks before you randomly assign 10 students to the treatment group and the other 10 to control, then it is possible that just by chance, all 5 of the anxious students will be in the control group. If this happens, then you will not be able to see the effect of the treatment, because the people who would be best helped by the treatment will not receive it.

If you created blocks instead, then one block would consist of the 5 anxious students and about half (2 or 3) would be guaranteed to be assigned to the treatment. Figure 12.5 compares one possible randomization with no blocks to a blocking design.

(a)

T	C
LA	HA
LA	LA
LA	LA
LA	HA
LA	LA
LA	HA
LA	HA
LA	HA
LA	LA
LA	LA

(b)

◀ **FIGURE 12.5 (a)** If there are no blocks, it is possible that all 5 of the higher-anxiety students (HA) will be assigned to the control group. This would result in one group consisting solely of lower-anxiety students (LA). **(b)** Creating blocks based on anxiety level ensures that roughly half of the higher-anxiety students (that is, 2 or 3 students) will be assigned to receive the treatment (T) and half the control (C).

Blocking and Matching As we have just discussed, blocking combines units that are similar in order to improve the power of an experiment. What unit can be more similar than matching a person with himself or herself? Researchers can sometimes exploit the fact that measuring the same person twice (usually before and then after applying a treatment) results in positively correlated values. Studies that use this structure illustrate the **matched-pairs design**. In Chapter 9 you saw examples of such paired samples, and you learned how to apply the paired *t*-test to test hypotheses about differences between two groups of paired data. Pairs of data are, in fact, a form of blocking.

For example, in a study to detect the effects of the gastric band on weight loss (O'Brien et al. 2010), patients had their weight measured without the band (before it was surgically put in place) and some time after the band was in place. In such a study, each patient is basically serving as his or her own control. The control group consists of the patients before the treatment. The treatment group consists of the *same* patients as in the control group but after they've received the treatment. This sort of study design is often more effective at measuring the effects of a treatment than one in which we have two groups of people (one group with no gastric band and the other with one) and one numeric treatment variable (weight).

> **↻ Looking Back**
>
> **Paired Samples**
> Paired samples are those in which each object in one group is associated with one particular object in the other group. They are also sometimes called dependent samples or dependent observations.

EXAMPLE 4 Comparing Tuitions

In Example 9.17, we concluded that, based on a random sample of 35 two-year colleges, that tuitions increased between the 2011–2012 academic year and the 2014–2015 academic year. The first few rows of the data are here:

Institution Name	Tuition 2011–2012	Tuition 2014–2015
Mohawk Valley Community College	$4010	$4415
George C Wallace State Community College-Hanceville	$4080	$4260
College of the Redwoods	$888	$1182

Our conclusion was based on a paired *t*-test in which the alternative hypotheses was that the mean tuition of all two-year schools was greater in 2014–2015. The value of the *t*-statistic was 5.9 with a p-value less than 0.0001.

QUESTIONS

a. Perform a two-sample *t*-test (do not treat the data as paired data) to test whether the mean tuition increased in 2014–2015 compared to 2011–2012. Use a 5% significance level.

b. Which approach is better? Why are the conclusions different?

SOLUTIONS

a. We will skip over the steps of the hypothesis test so that we can emphasize the results of the two different procedures. Figure 12.6 shows the output of StatCrunch, which displays the null and alternative hypotheses, the observed value of the *t*-statistic, and the p-value.

► **FIGURE 12.6** Output from two-sample *t*-test, no pairing.

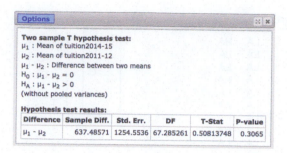

From this we see that the value of the *t*-statistic is 0.508 and the p-value is 0.3065. Because the p-value is bigger than 0.05, we do not reject the null hypothesis. Our conclusion contradicts the earlier conclusion that the mean tuition had increased.

b. The data in this sample are paired; each college contributes two data values to the sample. Because the values within pairs are positively correlated, the paired *t*-test has greater power than the two-sample *t*-test to detect differences in the population. The paired *t*-test is the more appropriate test for this situation. Note that the standard error of the two-sample *t*-test is 1254.5. This reflects how much variation there is in this test statistic from sample to sample. By way of contrast, the standard error for the paired *t*-test can be calculated as 67.3 (not shown). This smaller value reflects the fact that with this study design, correctly using the paired test statistic results in a more precise estimator.

Note that in both tests, the difference in the *sample* means is the same; $637.49. But our hypothesis is not about the sample; it is about the population. The paired *t*-test allowed us to see that the *population* means are different.

TRY THIS!   Exercise 12.25

SECTION 12.2

Controlling Variation in Surveys

In Chapter 7 we discussed a very basic method for collecting survey data called the simple random sample (SRS). In an SRS, we first create a list of all people in our population. Then we randomly choose people without replacement to participate in our study.

This basic method creates a sample that, if large enough, is representative of the population. Because it is based on a random scheme, we can compute a margin of error that measures how far away our estimate is likely to be from the population value, even when we don't know what the population value is.

In practice, however, simple random samples are difficult to carry out. In very large populations such as the United States (or any other country, for that matter), no single list of names exists for researchers to use. Also, this method can be imprecise: The margin of error can be big. In this section, we give an overview of other random sampling techniques that are more practical than simple random sampling.

Review of Sampling Basics

Before talking about more complex sampling schemes, let's take a moment to review some of the basics we introduced in Chapter 7.

One goal of a survey is to estimate the value of a population parameter on the basis of a sample. *Parameters* are numbers that characterize populations. The proportion of Americans who bought a new car in the last two years is a parameter, as is the mean weight of all eighth graders in your town. If we collect a representative sample from the population of interest, then a statistic from that sample can be used to estimate the parameter. A *statistic* is a number based on a sample of data. The average weight of 20 eighth graders chosen from your town is a statistic, as is the proportion of 100 people selected from the United States who say they bought a new car in the last two years.

Two features that measure the quality of an estimator are bias and precision. *Bias* measures how far away an estimator tends to be from the correct population value. If, on average, the estimator tends to get the value just right, then we say that the estimator is unbiased. *Precision* measures how the estimator varies from sample to sample. Different researchers using different random samples should get similar estimates when using a precise estimator. Precision is measured by the standard error, which is just the standard deviation the estimator would have if we were to repeat the survey many times. Ideally, an estimator is unbiased and has a small standard error.

Estimators can be biased for many different reasons. One common reason is that the sample is not representative of the population. For example, if we were to survey customers at a car dealership and ask whether they bought a new car in the last two years, these people would probably not be representative of the population as a whole. (They are probably less likely than the general population to have a new car at home if they're out shopping for a car.)

One way in which a nonrepresentative sample often arises is in situations where the sample chooses itself by volunteering to participate. For example, a political website featured a poll asking this question: "Do unwed fathers have a legal right to be in the delivery room to witness the birth of their child?" Of those responding, 49% said "No," and 36% said "Yes"; the rest were "Not Sure" (www.foxnews.com, viewed March 14, 2014). Is this representative of the entire U.S. population? Probably not. People chose whether to click on the link, and this decision is quite likely to be based on the fact that they have strong opinions on the topic, which they wish to share.

This situation is similar to an observational study. People choose whether or not to participate in the poll, so we don't know whether our estimate of the population parameter is representative only of the sort of person who would answer this poll or is representative of the population as a whole. (And visitors to a political website are likely to be more similar in their opinions than we would see in the entire country. If the poll were on a liberal news site, the results might very well go in the opposite direction.)

The only way to avoid this potential bias when collecting data is to select the sample randomly. A random sample has another advantage: It gives us a way to measure our confidence in the estimate. This measure takes the form of a confidence interval. A confidence interval enables us to say we're highly confident that the true population proportion is within, say, 3 percentage points of our estimate.

The precision of a survey estimator depends on the size of the sample and on the variability in the population. The good news is that the precision can always be improved by taking a larger sample (assuming you can afford to collect more data, that is). If everyone in the population answers the same way—in other words, if there is no variability—then the estimator will be extremely precise. For example, if everyone believes unwed fathers have a legal right to be in the delivery room, then $p = 1$ and everyone in the random sample will also believe this, and so the estimator will have perfect precision since there is no variability.

On the other hand, the worst-case scenario as far as precision is concerned is when the population is evenly split. If 50% of the people in the population would answer yes to the Fox poll, in other words, if $p = 0.50$, then the precision will be as bad as it can be.

For example, if $p = 0.50$ and we take a simple random sample of 100, then the standard error is

$$\sqrt{\frac{p(1-p)}{100}} = \sqrt{\frac{0.5 \times 0.5}{100}} = 0.05$$

> **! Caution**
>
> **Measurement Bias**
> Do not overlook the fact that estimates of the population proportion can also be biased if the questions are poorly worded.

A population with less diversity, say, $p = 0.90$, would produce more precise estimators with a smaller standard error:

$$\sqrt{\frac{p(1-p)}{100}} = \sqrt{\frac{0.9 \times 0.1}{100}} = 0.03$$

Several approaches to collecting samples get around the difficulties posed by SRSs. These approaches either make it easier to access people to survey or help by increasing the precision without the costs associated with a larger sample size.

Systematic Sampling

Systematic sampling is one technique that can be used, in some situations, to obtain a random sample. In a systematic sample, objects from the population are sampled at regular intervals, such as taking every fifth person. One of the most common examples of systematic sampling occurs in exit polls held during elections.

The purpose of an exit poll is to estimate the winner of an election before the votes are counted. They are called exit polls because people are surveyed as they exit the polling place. The pollsters—the people collecting the data—are instructed to stop people at specific intervals. For example, they might stop every fifth person (or every tenth person) throughout the day.

Systematic sampling works well when you have subjects coming at you in some sort of sequence and when you have good reason to believe that the opinions will be randomly mixed during the time period when you collect that data. Exit polls would fail if, for example, pollsters collected data only first thing in the morning or only in the middle of the afternoon. This is because it is possible that a certain type of voter goes to the polls early in the morning, and if so, then a systematic sample taken only in the morning will be biased.

Systematic samples are also used in quality control studies at manufacturing plants. Products come off an assembly line, and every tenth product might be inspected for defects. At the end of the day, this method yields a good estimate of the proportion of all products produced that day that have defects, as long as there is no particular pattern to the order in which defects arrive.

Stratified Sampling

Stratified sampling is a method for collecting data with increased precision. To perform stratified sampling, researchers must first identify **strata**. Strata are similar to blocks; they are collections of people who are very similar to each other. Stratified sampling is based on a very simple principle: If we know that everyone in some group will answer the question the same way, then there's no need to ask everyone.

To create strata, choose a variable that identifies people who hold similar beliefs that are associated with the variable you are studying. For example, suppose you were designing a survey to determine the proportion of the population that approves of allowing gay and lesbian citizens to serve in the military. According to Gallup polls and other research, approval for this idea depends on age: Younger people are more supportive than older people. Thus, even though we could take a large sample of 1000 people to get an approximately 3-point margin of error for a 95% confidence interval, we could also stratify on age and take a smaller sample to get the same precision.

A stratified sample would divide the population into strata based on age group. One stratum (the singular form of *strata*) might consist of 18- to 25-year-olds, another of 26- to 40-year-olds, and so on. We would then randomly select people (using a simple random sample) from each stratum and ask their opinion.

Stratified sampling works because it creates "mini-populations" that have lower variability than the population as a whole. If the strata are chosen carefully, then within each stratum, there is less variability than in the population as a whole. This means that within each stratum, we can produce a more precise estimator.

The result is that we have mini-estimates from each of the mini-populations. These estimates are very precise, because the strata were chosen to have low variability within them. We then combine the mini-estimates into one estimate for the entire population. We can show mathematically that in many cases, this combined estimator is more precise.

> **KEY POINT**
>
> Stratified sampling improves the precision of estimators by sampling from within strata. Each stratum is created in such a way that the people (or objects) within the stratum are similar to one another.

Age is one variable commonly used for creating strata. Political party (Republican or Democrat) is another. Other possibilities include gender, educational level, and geographic region. For example, if we were to take an opinion poll about increasing taxes to improve funding of two-year colleges, we could probably improve the precision of our survey by stratifying on education level. The strata might be "high school diploma or less," "two-year degree," "four-year degree," "graduate degree." From each of these groups, we would select a random sample for our survey.

If the variable chosen to create the strata does not actually end up creating groups with similar-thinking members, then stratification will not offer any advantage over SRS. In fact, in some very unusual cases, it can make the precision worse! For this reason, it's important to choose the right stratification plan.

> **! Caution**
>
> **Stratified Sampling and Blocking**
> Stratified sampling and blocking are similar. Both put subjects into groups (strata or blocks) so that subjects are similar within the group. However, stratified sampling is about ways of selecting subjects, and blocking is about a way of randomly assigning already selected subjects to treatment groups.

Cluster Sampling

Cluster sampling is a method that makes it easier to access very large populations. To perform a cluster sample, researchers divide the population into distinct groups, or **clusters**, using some sort of natural or convenient distinction. Next, they select a random sample (using SRS) of clusters and survey *everyone* within the clusters selected. For example, suppose you are interested in determining the number of children living in rental apartments in a city. Each apartment building can be a cluster, and you can then randomly select some number of apartment buildings and survey everyone within those buildings to find out how many children are living in the apartments. (Surveys such as this are sometimes done to look for evidence that landlords are discriminating against families with children.)

Clustering can happen over several stages. Randomly sampling, say, ten states in the United States and surveying everyone in these ten states is still a gigantic undertaking. Instead, after the states are selected, we might define a second stage of clusters consisting of cities, or maybe counties, zip codes, or the like. We would then take a random sample of these smaller clusters and survey everyone inside the clusters selected.

Foresters have used cluster sampling to estimate the extent to which bark beetles have infested forests. Bark beetles are a serious threat to forests, so being able to detect the proportion of damaged trees in areas in which there is not yet much damage is crucial to deterring their attack. By using satellite images, researchers can randomly select parcels of land to serve as clusters and then investigate every tree within those parcels (Coggins, Coops, and Wulder 2010). Cluster sampling is also used in humanitarian relief efforts to understand the extent to which people have been injured or killed by natural disasters. Researchers randomly select regions and then go door to door within them to collect damage reports.

Cluster sampling is sometimes confused with stratified sampling because both techniques break the population into groups. However, in some ways, cluster sampling is the opposite of stratified sampling. In stratified sampling, the strata contain people who are all as similar as possible (for example, all Republicans in one stratum and all Democrats in another). Cluster sampling, on the other hand, works best when each cluster by itself represents the entire population. This is rarely achieved, but when it *is* achieved, it ensures that estimates are as precise as possible.

> **KEY POINT**
>
> Cluster sampling can be used to access very large populations. A cluster consists of a group of objects or people. To perform a cluster sample, we take a random sample of clusters and survey everyone within the selected clusters.

EXAMPLE 5 Counting Young Female Smokers

Health experts are concerned about the large number of smokers among young Chinese women. However, experts do not know how many female smokers there are between the ages of 14 and 24 in China. To estimate the number of young, female smokers in China, researchers in California and China conducted a survey in which they randomly selected schools in China and then randomly selected classrooms in the selected schools. They surveyed all girls and young women in the selected classrooms and concluded that about 20% of all Chinese females in this age group were smokers (Ho et al. 2009).

QUESTION Is this a systematic sample, a stratified sample, or a cluster sample?

SOLUTION This is a cluster sample. More precisely, this is a two-stage cluster sample. The clusters are schools at one stage and then classrooms at the second stage.

TRY THIS! Exercise 12.27

SECTION 12.3

Reading Research Papers

One goal of this book is to teach you enough of the basic concepts of statistics that you can critically evaluate published research. The medical literature, in particular, records many findings that can have major consequences in our lives. When we rely on the popular media to interpret these findings, we often get contradictory messages. However, you now know enough statistics so that, with a few guiding principles, you can often make sense of research results yourself.

Before discussing ways of evaluating individual research articles, we offer a few overarching, guiding principles:

1. *Pay attention to how randomness is used.* Random sampling is used to obtain a representative sample, so that we can make inferences about a larger population. Random assignment is used to test causal associations so that we can conclude that the treatment was truly effective (or was not) on a particular sample of subjects. Many medical studies use random assignment but do not use random sampling. This means the results are not necessarily applicable to the entire population. Table 12.2, summarizes these possibilities.

▶ **TABLE 12.2** Four different study designs and the inferences possible.

	Sample Selected Randomly	No Random Sample
Random Assignment	You can make a causal conclusion and conclude that the entire population would be affected similarly.	You can make a causal conclusion, but you do not know whether everyone would respond similarly.
No Random Assignment	You can assume that an association between the variables exists in the population, but you cannot conclude that it is a causal relationship.	You can conclude that an association exists within the sample but not in the entire population, and you cannot conclude that there is a causal relationship.

2. *Don't rely solely on the conclusions of any single paper.* Research advances in small steps. A single research study, even when the conclusions are grand and ambitious, can tell us only a small part of the real story. For example, the *Los Angeles Times* reported, on the basis of a few published studies, that many people considered vitamin D to be useful for preventing cancer, cardiovascular disease, depression, and other maladies. But a panel of medical experts concluded that these beliefs were based on preliminary studies. The body of medical research, they concluded, was in fact quite mixed in its view of just how effective vitamin D really is (Healy 2010).

3. *Extraordinary claims require extraordinary evidence.* This advice was a favorite piece of wisdom of magician and professional skeptic (The Amazing) James Randi. If someone claims to have done something that had been believed impossible, don't believe it until you've seen some very compelling evidence.

4. *Be wary of conclusions based on very complex statistical or mathematical models.* Complex statistics often require complex assumptions, and one consequence of this is that the findings might be correct and true but only in a limited set of circumstances. Also, some research studies are essentially "what if" studies: "We're studying *what* will happen *if* these assumptions hold." It is important to remember that those assumptions may not be practical—or even possible.

5. *Stick to peer-reviewed journals.* **Peer review** means that papers were read by two or three (and sometimes more) knowledgeable and experienced researchers in the same field. These reviewers can prevent papers from being published if they do not think the methods employed were sound or if they find too many mistakes. They can also demand that the authors make changes and submit the paper again. Many published papers have gone through several rounds of reviews. But be careful: Not all peer-reviewed journals are equal. The more prestigious journals have greater resources to check papers for mistakes, and they have much more careful and knowledgeable reviewers on hand to find sometimes subtle errors. Good journals have editorial boards that reflect the general practice of the community and are not restricted to people who reflect a minority point of view.

The first of these principles needs more explanation. Some students get confused about the differences between random sampling and random assignment. One way to ease the confusion is to ignore the word *random*. We are left with understanding the difference between *sampling* and *assignment*.

When we *sample* people or objects, we select them from a larger group to participate in our study or survey. This larger group is what we've called the population. Random sampling means that we make this selection at random. If the investigators in a study or survey say that they conducted sampling, then you know that they did not use the entire population for their study; instead, they chose people from that population. Most medical studies are based on samples. However, most medical studies do *not* choose these samples randomly, because it is much too difficult to select, say, a random sample of emergency room patients who are willing to fly to a clinic in Minneapolis for an experimental treatment.

Figure 12.7 shows a random sample of ten people selected from a larger population of 50 people. Those selected randomly to be in the sample are shown in red.

In a controlled experiment, once we have a sample, we need to *assign* them to a treatment group. For instance, in our sample of ten people (numbered 03, 06, 17, 19, 21, 22, 25, 30, 31 and 33) five people will receive aspirin and five will receive a placebo. To do this randomly, we can have a computer or calculator randomly select half of these numbers, and then assign those who are selected to the aspirin group. The rest will receive a placebo. Or, if we wanted a more low-tech approach, we might write these ten numbers on slips of paper, put them in a box, mix the slips up, and then randomly select

```
00 01 02 03 04 05 06 07 08 09
10 11 12 13 14 15 16 17 18 19
20 21 22 23 24 25 26 27 28 29
30 31 32 33 34 35 36 37 38 39
40 41 42 43 44 45 46 47 48 49
```

▲ **FIGURE 12.7** Each number represents a person. The red numbers represent people who were randomly selected to appear in our sample.

half of them. The ones we select will be assigned to the aspirin group. For instance, if we selected the numbers 3, 21, 22, 25, and 33, then our assignment would be

Aspirin: 3, 21, 22, 25, 33
Placebo: 6, 17, 19, 30, 31

Note that in this random assignment, we have used everyone in the sample. The first five numbers selected were assigned to the aspirin group, and the five remaining numbers have been assigned to a different group (the placebo group).

EXAMPLE 6 Improving Tips

A sociologist wonders whether the tips for servers at restaurants are affected by whether or not the server touches the customers. Below are two study designs that the sociologist might use. Read these, and then answer the questions that follow.

Design A: The sociologist chooses four large restaurants in his city and gets all servers at these restaurants to agree to participate in his study. On several nights during the next few weeks, the sociologist visits the restaurants and records the number of times all servers touch their customers on the customer's back or shoulder. He also records the total amount of tips earned by each server. He finds that servers who sometimes touched their customers earned larger tips, on average, than those who did not.

Design B: The sociologist chooses four large restaurants in his city and gets all servers at these restaurants to agree to participate in his study. At each restaurant, half of the servers are randomly assigned to either the "touch" group or the "no touch" group. The "touch" group is instructed to lightly touch each customer on the back or shoulder two or three times during the meal. The "no touch" group is instructed not to touch customers at all. At the end of the study, the sociologist finds that servers in the "touch" group earned larger tips, on average, than those in the "no touch" group.

QUESTION For each study design, state whether it is possible to generalize the results to a larger population and whether the researcher can make a cause-and-effect conclusion.

SOLUTION For Design A, the servers (or restaurants) were not randomly selected from a larger population, so we cannot generalize the results beyond the sample. This was an observational study: The servers themselves chose whether to be "touchers" or "no touchers." Because random assignment was not used, we cannot conclude that the difference in tips was caused by touching. A possible confounding effect is the restaurant itself, which might encourage touching and which might have a clientele that tends to tip more generously than the clientele at other restaurants.

For Design B also, we cannot generalize to a larger population. But here random assignment was used. The researcher blocks on restaurant, so he also controls for the environment of the restaurant. Because random assignment was used, we can conclude that the touching caused the increased amount of tips.

TRY THIS! Exercise 12.35

Reading Abstracts

An **abstract** is a short paragraph at the beginning of a research article that describes the basic findings. If you look up science papers using Google Scholar, say, clicking on the link will usually take you to an abstract. Often, you will be able to read the abstract at no charge. (Reading the papers themselves often requires that you subscribe to the journal or that you read from the computers at a library that has purchased a subscription.)

In addition, the websites of many journals display the abstracts of their published papers, even if they do not provide access to the papers themselves.

For example, an article published in *The New England Journal of Medicine* (Poordad et al. 2011) reported the findings of a study on the effectiveness of a new treatment for chronic hepatitis C virus (HCV). Currently, the standard treatment, peginterferon-ribavirin, has a relatively low success rate. A success, in treating HCV, is called a "sustained virologic response." In a sustained virologic response, the virus is not eliminated from the body but is undetectable for a long period. Researchers hoped that adding a new medicine, boceprevir, to the standard treatment would lead to a great proportion of patients achieving a sustained virologic response. An excerpt from the abstract follows:

Methods: We conducted a double-blind study in which previously untreated adults with HCV genotype 1 infection were randomly assigned to one of three groups. In all three groups, peginterferon alfa-2b and ribavirin were administered for 4 weeks (the lead-in period). Subsequently, group 1 (the control group) received a placebo plus peginterferon-ribavirin for 44 weeks; group 2 received boceprevir plus peginterferon-ribavirin for 24 weeks, and those with a detectable HCV RNA level between weeks 8 and 24 received a placebo plus peginterferon-ribavirin for an additional 20 weeks; and group 3 received boceprevir plus peginterferon-ribavirin for 44 weeks. Nonblack patients and black patients were enrolled and analyzed separately.

Results: A total of 938 nonblack and 159 black patients were treated. In the nonblack cohort, a sustained virologic response was achieved in 125 of the 311 patients (40%) in group 1, in 211 of the 316 patients (67%) in group 2 ($P < 0.001$), and in 213 of the 311 patients (68%) in group 3 ($P < 0.001$). In the black cohort, a sustained virologic response was achieved in 12 of the 52 patients (23%) in group 1, in 22 of the 52 patients (42%) in group 2 ($P = 0.04$), and in 29 of the 55 patients (53%) in group 3 ($P = 0.004$). In group 2, a total of 44% of patients received peginterferon-ribavirin for 28 weeks. Anemia led to dose reductions in 13% of controls and 21% of boceprevir recipients, with discontinuations in 1% and 2%, respectively.

Conclusions: The addition of boceprevir to standard therapy with peginterferon-ribavirin, as compared with standard therapy alone, significantly increased the rates of sustained virologic response in previously untreated adults with chronic HCV genotype 1 infection. The rates were similar with 24 weeks and 44 weeks of boceprevir. (Funded by Schering-Plough [now Merck]; SPRINT-2 ClinicalTrials.gov, number NCT00705432.)

Note that the p-values here are denoted differently than they are in this text. "$P = 0.004$" is the same thing as "p-value $= 0.004$". You'll see slightly different conventions in different publications.

To help you evaluate this abstract, and others like it, answer these questions:

1. What is the research question that these investigators are trying to answer?

2. What is their answer to the research question?

3. What were the methods they used to collect data?

4. Is the conclusion appropriate for the methods used to collect data?

5. To what population do the conclusions apply?

6. Have the results been replicated (that is, reproduced) in other articles? Are the results consistent with what other researchers have suggested?

The sixth question is important, although it is often difficult for a layperson to answer. Research is a difficult activity, in part because of the great variability in nature. Studies are often done with samples that do not allow generalizing to populations or are subject to bias, and sometimes researchers just make mistakes. For that reason, you should not believe in the claims of a single study unless it is consistent with currently accepted theory and supported by other research. For all these reasons, wait until there is some accumulation of knowledge before changing your lifestyle or making a major decision on the basis of scientific research.

Let's see how we would answer these questions for this abstract.

1. *What is the research question that these investigators are trying to answer?* Does the addition of boceprevir to the standard treatment lead to a higher proportion of hepatitis C patients achieving a sustained virologic response?

2. *What is their answer to the research question?* Yes, the additional drug improved patients' responses. Researchers report an increased rate of patients achieving sustained virologic response with boceprevir plus the standard therapy compared to those receiving only standard therapy.

3. *What were the methods they used to collect data?* Patients were randomly assigned to one of three treatment groups. Assignments were double-blind. Patients were examined after 24 and 44 weeks to determine whether they had achieved sustained virologic response.

4. *Is the conclusion appropriate for the methods used to collect data?* The fact that subjects were randomly assigned to treatment groups means that we can make a causal conclusion and say that the difference in virologic response rates was due to the the addition of boceprevir to the standard treatment. We also know that the observed results were too large to be explained by chance. Note that p-values are given as, for example, "P < 0.001." For instance, we know that group 2 (nonblack cohort) had a greater rate of sustained virologic response than the placebo group (group 1, nonblack) with p-value < 0.001.

5. *To what population do the conclusions apply?* The sample was not randomly selected, so although we *can* conclude that the treatment was effective, we do not know if we would see the same sized effect on other samples or in the population of all hepatitis C patients. Because it is known that African Americans and non-African Americans have different responses to the standard treatment, these two groups were treated separately in the analysis.

6. *Have the results been replicated in other articles? Are the results consistent with what other researchers have suggested?* We cannot tell from the abstract.

Note that even though we don't know which statistical procedures were carried out (we could learn this from reading the article itself), we still have a good sense of the reliability of the study. We know that the reported success rates for standard treatment for hepatitis C (40% in the nonblack cohort and 23% in the black cohort) were lower than the reported success rate for the treatment that included boceprevir (67% for the nonblack cohort after 24 weeks, 68% after 44 weeks; 42% for the black cohort after 24 weeks, and 53% after 44 weeks). We know this difference cannot be plausibly assigned to chance, and we know that, because of random assignment, that boceprevir was the cause of the difference in rates.

EXAMPLE **7** Brain Games

Read the following abstract, which discusses the effects of "brain games" on elderly Americans' cognitive functioning (Rebok et al. 2014):

Objectives: To determine the effects of cognitive training on cognitive abilities and everyday function over 10 years. **Design**: Ten-year follow-up of a randomized, controlled single-blind trial (Advanced Cognitive Training for Independent and Vital Elderly (ACTIVE)) with three intervention groups and a no-contact control group. **Participants**: A volunteer sample of 2,832 persons (mean baseline age 73.6; 26% African American) living independently. **Intervention**: Ten training sessions for memory, reasoning, or speed of processing; four sessions of booster training 11 and 35 months after initial training. **Measurements**: Objectively measured cognitive abilities and self-reported and performance-based measures of everyday function. **Results**: Participants in each intervention group reported less difficulty with instrumental activities of daily living (IADLs). At a mean age of 82, approximately 60% of trained participants, versus 50% of controls ($P < 0.05$), were at or above their baseline level of self-reported IADL function at 10 years. The reasoning and speed-of-processing interventions maintained their effects on their targeted cognitive abilities at 10 years. Memory training effects were no longer maintained for memory performance. Booster training produced additional and durable improvement for the reasoning intervention for reasoning performance and the speed-of-processing intervention for speed-of-processing performance. **Conclusion**: Each Advanced Cognitive Training for Independent and Vital Elderly cognitive intervention resulted in less decline in self-reported IADL compared with the control group. Reasoning and speed, but not memory, training resulted in improved targeted cognitive abilities for 10 years.

QUESTION   Write a short paragraph describing this research. Use the six questions on the previous page as guidance.

SOLUTION   The study attempts to determine whether cognitive training (which isn't defined in the abstract) can improve cognitive functions such as reasoning, speed of processing, and memory, over a 10-year period. The researchers randomly assigned subjects to one of three different training programs or to a control group that received no training. They concluded that, in fact, the training does improve reasoning and speed of processing, but not memory. They also found that those who received cognitive training had less difficulty with daily activities (instrumental activities of daily living, or IADLs). The p-value for comparing the IADL scores of the control subjects to those of the subjects who received the cognitive training was reported as less than 0.05, so we know that at a 5% significance level, we are confident that the differences are not explained by chance. Because the researchers used random assignment, their conclusion that the differences between the intervention groups and the control groups were caused by the cognitive training is justified. The results apply to only those 2,832 people who participated, because this is not a random sample. All the participants were older than the general population (their mean age was 73.6 years at the start of the study). The abstract does not tell us whether these results have been replicated.

TRY THIS!   Exercise 12.51

Buyer Beware

Tolstoy's novel *Anna Karenina* begins "Happy families are all alike; every unhappy family is unhappy in its own way." To (very loosely) paraphrase the great Russian novelist, there are few ways that a study can be good, but there are many ways that studies can go wrong. We've given you some tips that should help you recognize a good study, but you should also be aware of some warning signs and features that indicate that a study might not be good.

Data Dredging   Hypothesis testing is designed to test claims that result from a theory. The theory makes predictions about what we should see in the data; for example, students who write about their anxieties will do better on an exam, so the mean exam score of students who wrote about their anxieties should be greater than the mean score of those who did not. We next collect the data and then do a hypothesis test to determine whether the theory was correct. **Data dredging** is the practice of stating our hypotheses after first looking at the data. Data dredging makes it more likely that we will mistakenly reject the null hypothesis. Even when the null hypothesis is true, our data sometimes show surprising outcomes just because of chance variation. If we first look at the test statistic to decide what the hypotheses should be, we are rigging the system in favor of the alternative hypothesis.

The situation is analogous to betting on a horse race after the race has begun. You are supposed to place a bet before the race starts, so that everyone is on equal footing. The odds on which horse will win (which determine the payoffs) are meant to estimate probabilities of a horse winning. But if you wait until the horses are running, you have a better chance than everyone else of correctly predicting which horse will win. You have "snooped" at the data to make your decision. Your probability of winning is not the same as what everyone else believed it to be.

Theories should be based on data, but the correct procedure is to use data to formulate a theory and then to collect additional data in an independent study to test that theory. If it is too costly to collect additional data, one common approach is to randomly split the data set into two (or even more) pieces. One piece of the data is then set aside—"locked away"—and the researchers are forbidden to look. The researchers can then examine the first piece as much as they want and use that data set to generate hypotheses. After they have generated hypotheses, they can test these hypotheses on the second, "locked away" data set. Another possibility is to use the data for generating hypotheses and then go out and collect *more* data.

Publication Bias

Most scientific and medical journals prefer to publish "positive" findings. A positive finding is one in which the null hypothesis is rejected (with the result that the researcher concludes that the tested treatment is effective). Some journals prefer this sort of finding because these are generally the results that advance science. However, suppose a pharmaceutical company produces a new drug that it claims can cure depression, and suppose this drug does not work. If many researchers are interested in studying this drug, and they do statistical tests with a 5% significance level, then 5% of them will conclude that the drug is effective, even though we know it is not. If a journal favors positive findings over negative findings, then we will read only about the studies that find the drug works, even though the vast majority of researchers came to the opposite conclusion.

Publication bias is one reason why it is important to consider several different studies of the same drug before making decisions. A new, and somewhat controversial, form of statistical analysis called meta-analysis has been developed in recent years, in part to help with problems such as these. A **meta-analysis** considers all studies done to test a particular treatment and tries to reconcile different conclusions, attempting to determine whether other factors, such as publication bias, played a role in the reported outcomes.

Psychologists conducted a meta-analysis and concluded that violent video games are not associated with violent behavior in children, despite the fact that several studies had concluded otherwise (Ferguson and Kilburn 2010). They concluded that papers that found an association between video game violence and real-life violence were more likely to be published, so researchers who found no such association were less likely to be published. This result is not likely to settle the controversy, but it points to a potential danger that arises when journals publish only positive findings.

Profit Motive

Much statistical research is now paid for by corporations or industry groups that hope to establish that their products make life better for people. Researchers are usually required to disclose to a journal whether they are themselves making money off the drug or product that they are researching, but this does not tell us who funded the project in the first place. There is no reason to reject the conclusions of a study simply because it was paid for by a corporation or business or some other organization with a vested interest. You should always evaluate the methods of the study used and decide whether those methods are sound.

However, you should be aware that sometimes the corporation funding the research can influence whether results get published and which results get published. For example, a researcher might be funded by a pharmaceutical company to test a new drug. If he finds the drug doesn't work, the drug company might decide that it doesn't want to publicize this fact. Or perhaps the drug works on only a small subset of people. The company might then publicize that the drug is effective but fail to mention that it is effective on only a small group.

For example, a 2007 study concluded that playing "active" video games (such as the Wii) was healthier than playing "passive" games and found that children burned more calories playing the active games (Neale 2007). The study also suggested that playing real sports was much more healthful than playing any video game, but still the researchers conclude, "Nevertheless, new generation computer games stimulated positive activity behaviors. Given the current prevalence of childhood overweight and obesity, such positive behaviors should be encouraged." A press release notes that the study was funded by Cake, the "marketing arm" of Nintendo UK, which manufactures the game console Wii. This does not mean that the results of the study are wrong (the children playing the active games most likely did in fact burn more calories than those who played the passive games), but it may account for the positive spin despite the finding that active video games were not nearly as effective as playing sports at promoting weight loss.

Media   The media—newspapers, magazines, television shows, and radio broadcasts—are also profit-motivated. A good journalist strives to get at the heart of the matter. Nonetheless, scientific and medical research findings are complex, and when complex ideas are condensed into easily digested sound bites (especially in a way that entertains those who pay for the papers and magazines), the truth sometimes gets distorted.

Some media outlets favor catchy headlines, and these headlines do not always capture the true spirit of a study. The most common problem is that headlines often suggest a cause-and-effect relationship, even though the wise statistics student will quickly recognize that such a conclusion is not supported by the data. Statistician Jonathan Mueller keeps a website of such headlines. A couple of examples: "Studies say lots of candy could lead to violence." "Eating brown rice to crush diabetes risk." These headlines both suggest causality: Eating candy will make you violent. A diet high in brown rice can keep you from getting diabetes. But you now know that such a conclusion can be made only for controlled experiments, and even then we must be cautious. Ideally, you can learn what you need to know by reading the news article. But not always. The information you need to judge whether the conclusions of a study are strong is often missing from news reports.

Clinical Significance vs. Statistical Significance   An outcome of an experiment or study that is large enough to have a real effect on people's health or lifestyle is said to have **clinical significance**. Sometimes, researchers discover that a treatment is statistically significant (meaning that the outcome is too large to be due to chance) but too small to be meaningful (so it is not clinically significant). Studies with a very large sample size have large statistical power and so are capable of detecting even very small differences between treatment groups. This does not mean the differences matter. For example, a drug might truly lower cholesterol levels but not enough to make a real difference in someone's health. Or playing Wii might burn more calories but maybe not enough for it to serve as a form of exercise. Treatments that cause meaningful effects are called clinically significant. Sometimes, statistically significant results are not clinically significant.

Imagine a rare disease that only 1 person in 10 million people gets. A controlled experiment finds that a new drug "significantly" reduces your risk of getting this disease; specifically, it cuts the risk in half. Here, *significantly* means that the reduction in risk is statistically significant. But given that the disease is so rare already, is it worth taking medicine to cut your risk from 1 in 10 million to 1 in 20 million? Particularly if the drug is expensive or has side effects, most people would probably decide that this treatment is not clinically significant.

CASE STUDY REVISITED

Does Stretching Improve Athletic Performance?

At the beginning of this chapter, we reported on a study that found that stretching before a race did not increase a runner's speed. Here is the abstract:

The results of previous research have shown that passive muscle stretching can diminish the peak force output of subsequent maximal isometric, concentric and stretch-shortening contractions. The aim of this study was to establish whether the deleterious effects of passive stretching seen in laboratory settings would be manifest in a performance setting. Sixteen members (11 males, 5 females) of a Division I NCAA track athletics team performed electronically timed 20 m sprints with and without prior stretching of the legs. The experiment was done as part of each athlete's Monday work-out programme. Four different stretch protocols were used, with each protocol completed on a different day. Hence, the test period lasted 4 weeks. The four stretching protocols were no-stretch of either leg (NS), both legs stretched (BS), forward leg in the starting position stretched (FS) and rear leg in the starting position stretched (RS). Three stretching exercises (hamstring stretch, quadriceps stretch, calf stretch) were used for the BS, FS, and RS protocols. Each stretching exercise was performed four times, and each time the stretch was maintained for 30 s [seconds]. The BS, FS, and RS protocols induced a significant ($P < 0.05$) increase (approximately 0.04 s) in the 20 m time. Thus, it appears that pre-event stretching might negatively impact the performance of high-power short-term exercise.

We can apply the six questions to help us determine how strong these results are:

1. What is the research question that these investigators are trying to answer?

2. What is their answer to the research question?

3. What were the methods they used to collect data?

4. Is the conclusion appropriate for the methods used to collect data?

5. To what population do the conclusions apply?

6. Have the results been replicated in other articles? Are the results consistent with what other researchers have suggested?

The answers follow:

- The research question is, "Does stretching improve sprinting speed (20 meters) for NCAA athletes?"

- They conclude that it does not and that, in fact, stretching makes sprinters run more slowly.

- To collect these data, the researchers did not take a random sample but instead used 16 college athletes. The researchers blocked on athlete, and each athlete did every treatment, including the "no stretch" treatment. Although it does not say so in the abstract, the order in which an athlete did the treatments was randomly determined. This random assignment minimizes the effects of confounding factors, including the potential confounding factor of the order in which the routines were done. The amount of stretching was carefully controlled so that every athlete did precisely the same stretches for the same amount of time.

- The fact that random assignment was used (to determine the order of the stretching) means that the researchers are justified in making a cause-and-effect conclusion: We can conclude that, for these 16 athletes, the slower times were caused by the stretching.

- However, the athletes were a small sample from one college and were not randomly selected. This means we should not generalize these results to other athletes, especially not to athletes of other ages or in other sports. Note that the increase in time to do the 20-meter sprint was only 0.04 second. Although this slight increase might be meaningful (clinically significant) to elite runners, for most of us this difference is undetectable. (But then, most of us don't run sprints.)

- We cannot tell from the abstract whether the results have been replicated elsewhere, nor do we know the extent to which this study is building on the previous research. But we should keep in mind that one paper is not enough for us to believe in the results. We should try to find out whether studies done at other colleges, or with different levels of athletes, find that stretching does not help performance.

DATAPROJECT ▸ Keep It Real

OVERVIEW

Analyzing large data sets can be overwhelming, and in this chapter we continue our discussion on developing strategies for doing so. Once you have a strong understanding of the data, as covered in Chapter 11, the next step is to check the data for errors and prepare it for analysis. In this chapter we talk about data cleaning.

GOAL

Understand the importance of reality checks when cleaning data.

STEP 2: REALITY CHECK

In this project, we continue to work with the Los Angeles crimes data set (sampleofcrime.csv). Now that you have a basic understanding of the data, the next step is to make sure the data look like what they should. For example, if there is a variable for age, are all of the ages within a reasonable range? For any variables that provide dates, what's the first and last date? Are these reasonable values? If the data cover multiple years, are there any years that are incomplete? The goal is to look for flaws in the data.

Project The crimes data set is not quite ready for analysis. To prepare it, we'll need to examine the variables to check for errors, and also consider relationships between variables. For example, *Crime Code* and *Crime Code Description* should be related. And this means that each crime code should have exactly one and only one description. Is this true? (How can you check this?)

Often, in order to carry out a reality check, you'll need to do some data cleaning. For example, the *Victim Age* variable should be numeric, but, as you discovered if you ran the Validate command, some values are non-numeric. These values are "NA," which is a symbol that is sometimes used to mean that the data are missing. StatCrunch isn't bothered by this and will simply drop missing values designated by NA. Working with data sets that have missing values is a challenge for the data analyst, and we defer this challenge to a more advanced course. But one useful guide is that if the data are missing at random, in other words, if it's as if a random number generator was used to determine which values

to remove, then we can still get unbiased estimates. But if there was some mechanism at work to remove the data, for example, if the victim's age wasn't reported if the age were less than 18, then our analysis could be biased. So it's important to make note when you have a variable with missing values.

You might also want to recode categories. For example, there are several types of battery. Perhaps you wish to lump them together. Or perhaps you want to lump all car-related crimes into a single category.

If there are date variables, you might want to create separate variables for month, day, and year. This will allow you to consider whether there are trends across the years, whether certain days of the week have more crimes than others, and whether certain seasons have more crimes than others.

Assignment Consider *the Date Reported* and the *Date Occurred* variables. Are there any years with incomplete reports? For example, are there any years that do not have all 12 months? (You will need to compute new variables for month, year and day).

While StatCrunch silently excludes the NA values from an analysis of numerical variables, if the variable is categorical, the NAs are included. For Victim Sex and Victim Descent, recode the NA variables to blanks. How does this affect the percentage of victims who are female?

Often, the value NA represents a missing value, that is, a value that was accidentally excluded. But not always. What do you think NA means for the *Weapon Used Code* variable, and why?

Write your findings in a report.

CHAPTER REVIEW

KEY TERMS

Treatment variable, *607*	Power, *609*	Strata, *616*	Publication bias, *624*
Response variable, *607*	Controlling for a variable, *609*	Cluster sampling, *617*	Meta-analysis, *624*
Confounding variables, *607*	Blocking, *610*	Clusters, *617*	Clinical significance, *625*
Controlled experiment, *607*	Matched-pairs design, *613*	Peer review, *619*	
Random assignment, *608*	Systematic sampling, *616*	Abstract, *620*	
Observational study, *608*	Stratified sampling, *616*	Data dredging, *623*	

LEARNING OBJECTIVES

After reading this chapter and doing the assigned homework problems, you should

- Know how to form blocks of experimental units, and understand why and when these blocks are useful for comparing treatment groups.

- Understand how random assignment is used to allow cause-and-effect inference, and understand how random sampling is used to allow generalizations to a larger population.

- Be able to identify different sampling strategies and understand why they are useful.

- Be prepared to apply knowledge about collecting and analyzing data to critically evaluate abstracts in the science literature.

SUMMARY

Many research questions can be divided into two categories: those that ask causal questions and those that ask about associations between variables. Causal questions can only be answered with controlled experiments, whereas observational studies can answer questions about associations. Controlled experiments are difficult to design, in part because in real life, many different factors contribute to variation in the response variable. Blocking—the practice of creating groups of subjects/objects that are similar and then randomizing assignment within the blocks—is one method that helps control for known sources of variation. Blocking can help increase the power of a study, the probability of detecting true differences among the treatment groups.

Random sampling is necessary in order to conduct a valid poll or survey, but it can be complicated to carry out in large, complex populations. Systematic sampling and stratified sampling are two approaches that can improve a random sample. Systematic sampling

is done by regularly selecting every *n*th (such as every 10th) object to appear in the sample, and this provides a representative sample if the objects are in a random/shuffled order or if the entire population is sampled this way. Stratified sampling creates strata, which are groups of subjects that are similar to each other, and then randomly samples (without replacement) within each stratum. Cluster sampling is a method used to provide access to large populations. A cluster consists of objects that are all very different from one another, and ideally, a cluster is itself a representative sample. In cluster sampling, the clusters are selected randomly, without replacement, and every object in the cluster is included in the sample.

Interpreting conclusions in scientific studies is complex, because many things can go wrong in a study. Remember that extraordinary results require extraordinary evidence, and that you should trust studies that have been replicated (repeated, resulting in the same conclusion) over ones that have not.

SOURCES

Ballard, P. J., Lindsay T. Hoyt, and Mark C. Pachucki. 2018. Impacts of adolescent and young adult civic engagement on health and socioeconomic status in adulthood. *Child Development.* Published ahead of print, January 20. https://doi.org/10.1111/cdev.12998.

Briggs, D. C. 2002. Evaluating SAT coaching: Gains, effects and self-selection. In Rebecca Zwick, *Rethinking the SAT: The future of standardized testing in university admissions.* New York: RoutledgeFarmer, 2004.

Coggins, S. B., N. C. Coops, and M. A. Wulder. 2010. Improvement of low level bark beetle damage estimates with adaptive cluster sampling. *Silva Fennica* 44(2), 289–301.

Devinsky, O., et al. 2017. Trial of cannabidoil for drug-resistant seizures in the Dravet Syndrome. *The New England Journal of Medicine* 379, 2011–2020. doi: 10.1056/NEJMoa1611618.

Eggermont, A. M. M., et al. 2018. Adjuvant pembrolizumab versus placebo in resected stage III melanoma. *The New England Journal of Medicine* 318, 1789–1801. doi:10.1056/NEJMoa1802357.

Eriksson, K. 2012. The nonsense math effect. *Judgment and Decision Making* 7(6), 746–749.

Ferguson, C. J., and J. Kilburn. 2010. Much ado about nothing: The misestimation and overinterpretation of violent video game effects in Eastern and Western nations: Comment on Anderson et al. *Psychological Bulletin* 136(2), 174–178, discussion 182–187.

Gaudino, M., et al. 2018. Radial-artery or saphenous-vein grafts in coronary-arty bypass surgery. *The New England Journal of Medicine* 378, 2069–2077. doi:10.1056/NEJMoa1716026.

Genovese, M. C., et al. 2018. Safety and efficacy of upadacitinib in patients with active rheumatoid arthritis refractory to biologic disease-modifying anti-rheumatic drugs (SELECT-BEYOND): a double-blind, randomised controlled phase 3 trial. *The Lancet* 391 (10139), 2315–2524. https://doi.org/10.1016/S0140-6736(18)31116-4.

Halpern, S. D., et al. A pragmatic trial of e-cigarettes, incentives, and drugs for smoking cessation. *The new England Journal of Medicine* 378, 2302–2310. doi:10.1059/NEJMsa1715757.

Healy, M. 2010. New vitamin D recommendations: What they mean. *Los Angeles Times*, December 6, www.latimes.com.

Ho, M. G., et al. 2009. Smoking among rural and urban young women in China. *Tob Control* 19, 13–18.

Howard, R., et al. 2018. Antipsychotic treatment of very late-onset schizophrenia-like psychosis (ATLAS): A randomised, controlled, double-blind study. *The Lancet Psychiatry* 5(7), 553–563. https://doi.org/10.1016/S2215-0366(18)30141-X.

Huang, Y., et al. 2017. Drinking tea improves the performance of divergent creativity. *Food Quality and Preference* 66, 29–35. https://doi.org/10.1016.j.foodqual.2017.12.014.

Hurst, Y. and F. Haruhisa. 2018. Effects of changes in eating speed on obesity in patients with diabetes: a secondary analysis of longitudinal health check-up data. *BMJ Open* 8(1). http://dx.doi.org/10.1136/bmjopen-2017-019589.

Jacka, F. N., et al. 2017. A randomised controlled trial of dietary improvement for adults with major depression (the "SMILES" trial. *BMC Medicine* 15(23). https://doi.org/10.1186/s12916-017-0791-y.

Jones, P. M., et al. 2018. Association between handover of anesthesia care and adverse postoperative outcomes among patients undergoing major surgery. *JAMA* 319, 143–153. doi:10.1001/jama.2017.20040.

Kolata, G. 2018. That huge Mediterranean diet study was flawed. But was it wrong? *The New York Times*, June 13.

Mubanga, M., et al. 2017. Dog ownership and the risk of cardiovascular disease and death—a nationwide cohort study. *Scientific Reports* 7, article number 15821. doi:10.1038/s41598-017-16118-6.

Neale, T. 2007. Even active video games not good enough for kids' fitness. MedPage Today, http://www.medpagetoday.com.

Nelson, A., et al. 2005. Acute effects of passive muscle stretching on spring performance. *Journal of Sports Sciences* 23(50), 449–454.

O'Brien, P. E., et al. 2010. Laparoscopic adjustable gastric banding in severely obese adolescents: A randomized trial. *Journal of the American Medical Association* 303(6), 519–526.

Poordad, F., et al., 2011. Boceprevir for untreated chronic HCV genotype 1 infection. *The New England Journal of Medicine* 364(13), 1195–1206, http://www.nejm.org/doi/pdf/10.1056/NEJMoa1010494.

Ramirez, G., and S. Beilock. 2011. Writing about testing worries boosts exam performance in the classroom. *Science* 331, 211.

Rebok, G. W., et al. 2014. Ten-year effects of the advanced cognitive training for independent and vital elderly cognitive training trial on cognition and everyday functioning in older adults. *Journal of the American Geriatrics Society* 62, 16–24.

Ritter, S. M., and S. Ferguson. 2017. Happy creativity: listenin to happy music facilitates divergent thinking. *PLOS One*, September 6. https://doi.org/10.1371/journal.pone.0182210.

Semler, M. W., et al. 2018. Balanced crystalloids versus saline in critically ill adults. *The new England Journal of Medicine* 378, 829–839. doi:10.1056/NEJMoa1711584.

Sheehan, W. J., et al. 2016. Acetaminophen versus ibuprofen in young children with mild persistent asthma. *The New England Journal of Medicine* 375, 619–630. doi:10.156/NEMJoa1515990.

Shkedy, Z., Aerts, M., Callaert, H., 2006. The Weight of Euro Coins. *Journal of Statistics Education,* 14(2).

Sophocleous, A., et al. 2017. Heavy Cannabis Use Is Associated With Low Bone Mineral Density and an Increased Risk of Fractures. *The American Journal of Medicine* 130(2), 214–221. https://doi.org/10.1016/j.amjmed.2016.07.034.

Taylor, J. H., et al. 2018. Ketamine for social anxiety disorder: A randomized, placebo-controlled crossover study. *Neuropsychopharmacology* 43, 325–33.

Tedeschi, S. K., et al. 2017. Relationship between fish consumption and disease activity in rheumatoid arthritis. *Arthritis Care & Research* 70(3), 327–332. https://doi.org/10.1002/acr.23295.

Thomalla, G., et al. 2018. MRI-guided thrombolysis for stroke with unknown time of onset. *The new England Journal of Medicine* 379, 611–622.

Tohlahunase, M., et al. 2017. Impact of yoga and meditation on cellular aging in apparently healthy individuals: open single-arm exploratory study. *Oxidative Medicine and Cellular Longevity* 2017, article ID 7928981. doi:10.1155/2017/7928981.

Wakefield, A. J., et al. 1998. Ileal–lymphoid-nodular hyperplasia, non-specific colitis, and pervasive developmental disorder in children. *Lancet* 351, 637–641.

Wang, X., et al. 2017. Partner phubbin gand depression among married Chinese adults: The roles of relationship satisfaction and relationship length. *Personality and Individual Differences* 110, 12–17. doi:10.1016/j.paid.2017.01.014. doi:10.1056/NEMJoa1804355.

White, R. E., and S. M. Carlson. 2015. What would Batman do? Self-distancing improves executive function in young children. *Developmental Science* 19(3), 419–426. doi:10.1111/desc.12314.

Writing Group for the TRIGR Study Group. 2018. Effect of hydrolyzed infant formal vs conventional formula on risk of type 1 diabetes: the TRIGR randomized clinical trial. *JAMA* 319(1), 38–48. doi:10.1001/jama.2017.19826.

SECTION EXERCISES

SECTION 12.1

12.1 Dairy Products and Muscle The following two headlines concern the same topic. Which one has language that suggests a cause-and-effect relationship, and which does not?

> Headline A: "Dairy Builds Muscle"
>
> Headline B: "People Who Consume More Dairy Products Tend to Have More Muscle"

12.2 Coffee and Depression The following two headlines concern the same topic. Which one has language that suggests a cause-and-effect relationship, and which does not?

> Headline A: "Women Who Drink Coffee Are Less Prone to Depression"
>
> Headline B: "Coffee Prevents Depression"

12.3 Marijuana Use and Bone Density In a 2017 study reported in The American Journal of Medicine, Sophocleous et al. studied 170 adults who smoked marijuana regularly and 114 adults who had never used the drug and found that people who regularly smoke large amounts of marijuana may be more susceptible to bone fractures than people who don't use the drug. Was this a controlled experiment or an observational study? Explain.

12.4 Civic Engagement and Health In a 2018 study reported in *Child Development*, Ballard et al. examined links between civic engagement (voting, volunteering, and activism) during late adolescence and early adulthood, and socioeconomic status and mental and physical health in adulthood. The researchers studied how civic engagement was associated with outcomes among 9471 adolescents and young adults. They found that all forms of civic engagement are positively associated with subsequent income and education level. Was this a controlled experiment or an observational study? Explain.

12.5 Marijuana Use and Bone Density Using the information from exercise 12.3, write two headlines announcing the results of the study. Make one of the headlines imply causality and make the other one not imply causality. Clearly label each headline. Which headline is appropriate for these data? Explain.

12.6 Civic Engagement and Health Using the information from exercise 12.4, write two headlines announcing the results of the study. Make one of the headlines imply causality and make the other one not imply causality. Clearly label each headline. Which headline is appropriate for these data? Explain.

12.7 Stroke Therapy In a 2018 study reported in *The New England Journal of Medicine*, Johnston et al. studied the effect of a combination of the drug clopidogrel and aspirin on reducing the rate of recurrent stroke among stroke patients. Stroke patients in the study were randomly assigned to receive clopidogrel and aspirin ($n = 2432$) or a placebo and aspirin ($n = 2449$). Of those receiving clopidogrel and aspirin, 121 had another stroke. Of those receiving the placebo and aspirin, 159 had another stroke. Researchers concluded that patients with minor ischemic stroke or high-risk TIA who received a combination of clopidogrel and aspirin had a lower risk of having another stroke.

a. Compare the percentage in each group who had another stroke. Based on these percentages, does it seem like clopidogrel might be effective in reducing the risk of recurrent stroke?

b. Was this a controlled experiment or an observational study?

c. Identify the treatment and response variables.

d. State the conclusion in terms of cause and effect or explain why cause-and-effect conclusions cannot be drawn from this study.

12.8 Smoking Cessation In a 2018 study reported in *The New England Journal of Medicine*, Halpern et al. randomly assigned smokers to one of five groups, including four smoking cessation interventions and usual care. Usual care consisted of access to information regarding the benefits of smoking cessation and to a motivational text-messaging service. The four interventions consisted of usual care plus one of the following: free cessation aids such as nicotine-replacement therapy or pharmacotherapy, free e-cigarettes, free cessation aids plus $600 in rewards for sustained abstinence, or free cessation aids plus $600 in redeemable funds deposited in an account for each participant, with money removed from the account if cessation milestones were not met. Researchers measured the percentage in each group who sustained smoking abstinence for six months. Results indicate that financial incentives added to free cessation aids resulted in a higher rate of sustained smoking abstinence than free cessation aids alone. Is this study an observational study or a controlled experiment? Explain.

a. Is this study an observational study or a controlled experiment? Explain.

b. Identify the treatment and response variables.

c. Can a cause-and-effect conclusion be drawn from this study? Why or why not?

12.9 Rheumatoid Arthritis Treatment (Example 1) In a 2018 study reported in *The Lancet*, a randomized, double-blind controlled experiment was conducted to determine the effect of the drug upadacitinib on patients with active rheumatoid arthritis. Patients were randomly assigned to receive the drug or a placebo. After 12 weeks,

patients receiving the drug had significant improvement compared to those receiving the placebo.

a. Identify the treatment and response variables.

b. Restate the conclusion of the study in terms of a cause-and-effect conclusion. Why can a cause-and-effect conclusion be made from this study?

12.10 Late-Onset Schizophrenia Treatment Very late onset schizophrenia affects people who are at least 60 years old. In a 2018 study reported in *The Lancet Psychiatry*, researchers conducted a double-blind controlled experiment to study the effect of the drug amisulpride on these patients (Howard et al., 2018). The experiment was divided into two stages, and subjects were randomly assigned to one of three groups: Group 1 received the drug for both stages, Group 2 received the drug for stage 1 and the placebo for stage 2, and Group 3 received the placebo for state 1 and the drug for stage 2. Researchers found that those subjects receiving the drug showed reduced psychosis symptoms compared with those receiving the placebo.

a. Identify the treatment and response variables.

b. Restate the conclusion of the study in terms of a cause-and-effect conclusion. Why can a cause-and-effect conclusion be made from this study?

12.11 Racket Sports and Health Harvard Women's Health Watch reported on a 2016 study on the association between various forms of exercise and health. In this study, researchers used data from large British and Scottish health studies to see if some forms of activity had greater health benefits than others. They examined the association between six different types of exercise with the overall risk of death and death from cardiovascular disease in particular. Researchers found that racket sports were associated with the greatest reduction in risk of death from any cause, including cardiovascular disease, followed by swimming, aerobics, and cycling. Whatever activity participants chose, risk of death dropped the more often they exercised.

Was this more likely to have been a controlled experiment or an observational study? How do you know? (Source: https://www.health.harvard.edu/exercise-and-fitness/large-study-indicates-racket-sports-offer-best-protection-against-cardiac-death)

12.12 Sugary Drinks and Brain Health An April 2017 headline from the nytimes.com said "Sugary Drinks Tied to Accelerated Brain Aging." Is this headline more likely to refer to a controlled experiment or an observational study? Explain.

12.13 Expression of Feelings A 2017 Pew poll asked a random sample "Are men and women basically different in how they express feelings?" The results by gender (in percentages) are shown in the table below.

	Yes	No
Males	84%	16%
Females	89%	11%

a. Assume the sample size for each gender was 100. Test the hypothesis that the proportions of men and women who agree with this statement are different. Report the test statistic, the p-value, and the conclusion.

b. The actual samples sizes for the survey were 2291 men and 2282 women. Repeat the hypothesis test using the actual sample sizes, reporting the test statistic, the p-value, and the conclusion.

c. Explain the difference in the results between parts a and b.

12.14 Epilepsy Treatment Dravet syndrome is a complex childhood epilepsy disorder. Researchers Devinsky et al., conducted a double-blind placebo-controlled trial to determine the efficacy of the drub cannabiliol on reducing seizures in children with Dravet syndrome. One hundred twenty children with this syndrome were randomly assigned to receive either the drug or a placebo. Researchers then recorded the percentage in each group who saw at least a 50% reduction in seizure frequency. Of the 60 children assigned to the drug group, 43% saw this reduction in seizure frequency; of the 60 children assigned to the placebo group, 27% saw this reduction.

a. Using a significance level of 0.05, do we have evidence the drug is effective in reducing seizures? You may use a chi-square test of homogeneity or a two-proportion z-test.

b. Suppose the sample sizes were doubled for each group and that the rates of seizure reduction remained the same. Repeat the test, reporting the new p-value and conclusion.

c. Explain the difference in the results between parts a and b.

12.15 Exercise and Power Imagine two studies of an exercise program that designers claim will make people lose weight. The first study is based on a random sample of 100 men and women who follow the exercise program for 6 months. A hypothesis test is carried out to determine whether their mean weight change from the start of the program to 6 months following the program is negative. The second study was based on a random sample of 100 men (no women) who followed the exercise program for 6 months. The same hypothesis test is carried out to determine whether their mean weight change is negative.

a. Which study will have more variability in the populations from which the samples are drawn?

b. Assuming the exercise program is more effective for men than women, which study will have more power? Explain.

*12.16 SAT Prep and Power Suppose an SAT tutoring company really can improve SAT scores by 10 points, on average. A competing company, however, uses a more intense tutoring approach and really can improve SAT scores by 15 points, on average. Suppose you've been hired by both companies to test their claims that their tutoring improves SAT scores. For both companies, you will collect a random sample of high school students to undergo tutoring. With both resulting samples, you will test the hypothesis that the mean improvement is more than 0. Suppose it is important to keep the power of both studies at 80%. Will you use the same sample size for both studies? If so, explain why you can. If not, which study would require the larger sample size, and why? Assume that both samples of students will be drawn from the same population.

TRY **12.17 Brain Games (Example 2)** Researchers are interested in testing whether a video game that is designed to increase brain activity actually works. To test this, they plan to randomly assign subjects to either the treatment group (spend 15 minutes per day playing the game) or a control group (spend 15 minutes per day surfing the Web). At the end of the study, the researchers will administer a test of "brain teasers" to see which group has the greater mental agility. Because they suspect that age might affect the outcome, the researchers will create blocks of ages: 18–25, 26–35, 36–45, and 46–55. To randomly assign subjects to treatment or control, they will place four tickets in a box. The tickets are labeled with the age groups. When an age group is selected, everyone in that age group will be assigned to the treatment group. They will select two age groups to go to treatment, and two to go to control. Is this an appropriate use

of blocking? If so, explain why. If not, describe a better blocking plan.

12.18 A Smile a Day Smiling is a sign of a good mood, but can smiling improve a bad mood? Researchers plan to assign subjects to two groups. Subjects in both groups will rate their mood at the beginning of the study. Then subjects in the treatment group will be told to smile while they are asked to recount a pleasant memory. Subjects in the control group will also be asked to recount a pleasant memory, but they will not be told to smile. Both groups will again rate their moods, and researchers will determine whether the reported moods differ between the two groups. Because the initial, baseline mood rating might affect the outcome, after the first mood rating the subjects will be broken into two groups: one group with low ratings ("bad mood") and one with higher ratings ("good mood"). Patients in each group will then be randomly assigned to either the treatment group or the control group. Is this an appropriate use of blocking? If so, explain why. If not, describe a better blocking plan.

TRY **12.19 Speed Skating Suits (Example 3)** Speed skating is a sport in which it is important to have a suit that minimizes wind drag as much as possible, as the difference between winning and losing a race can be as small as a thousandth of a second. In the 2014 Winter Olympics, U.S. speed skaters used a suit called the Mach 39, and none medaled despite high expectation before the games. For the 2018 Winter Olympics, a new suit design called the H1 was developed. Suppose the designers wanted to test if skaters would be faster in the H1 or the Mach 39. They have 10 Olympic speed skaters and 10 recreational speed skaters on whom to test the suits.

a. Identify the treatment variable and the response variable.

b. Describe a simple randomized design (not blocked) to test whether the H1 suit decreases race times. Explain in detail how you will assign skaters to treatment groups.

c. Describe a blocked design using the types of skaters that could be used to test whether the H1 suit decreases race times. What advantage does the blocked design have?

d. Describe a design that uses the skaters as their own controls to reduce variation.

*12.20 Flu Vaccines and Age Suppose you want to compare the effectiveness of the flu vaccine in preventing the flu using one of two different forms: nasal spray versus injection. Suppose you have 60 subjects available of different ages, and you suspect that age might have an effect on the outcome. Assume there are 20 children aged 2 to 15, 20 people aged 16 to 30, and 20 people aged 31 to 49.

a. Identify the treatment variable and the response variable.

b. Describe a simple randomized design (no blocking) to test the whether the injection or the nasal spray is more effective. Explain in detail how to assign people to treatment groups.

c. Describe a blocked design (blocking by age) to test whether the injection or the nasal spray is more effective. Explain in detail how you will assign people to treatment groups.

d. What advantage does the blocked design have?

*12.21 Preventing Heart Attacks with Aspirin Suppose that you want to determine whether the use of one aspirin per day for people age 50 and older reduces the chance of heart attack. You have 200 people available for the study: 100 men and 100 women. You suspect that aspirin might affect men and women differently.

To ensure an appropriate comparison group, those who do not get aspirin should get a placebo.

a. Identify the treatment and response variables.

b. Describe in detail a simple randomized design (not blocked) to test whether aspirin lowers the risk of heart attack.

c. Describe a blocked design to test whether aspirin lowers the risk of heart attack; the blocking variable is gender.

d. Explain why researchers might prefer a blocked design.

*12.22 Tomato Plants and Fertilizer Suppose you grow tomato plants in a greenhouse and sell the tomatoes by weight, so the amount of money you make depends on plants producing a large total weight of tomatoes.

You want to determine which of two fertilizers will produce a heavier harvest of tomatoes, fertilizer A or fertilizer B. There are two distinct regions in the greenhouse: one on the southern side that gets more light and one on the northern side that gets less light. There is room for 20 tomato plants on the southern side and 20 on the northern side. Assume that all the plants are beefsteak tomato plants.

a. Identify the treatment and response variables.

b. Describe a simple randomized design to test whether fertilizer A is better than fertilizer B.

c. Describe a blocked design to test which fertilizer produces a greater weight of tomatoes, blocking by southern side and northern side of the greenhouse. Explain why creating blocks based on whether plants are on the southern or northern side makes sense.

d. Explain why researchers might prefer a blocked design.

12.23 Dietary Improvement and Depression In a 2017 study, researchers investigated the effect of dietary improvement on adults with moderate to severe depression (Jacka et al. 2017). Subjects were randomly assigned to a treatment group consisting of seven individual nutritional consulting sessions with a clinical dietician or a control condition consisting of a social support protocol with the same visit schedule and length as the treatment group. There were 33 subjects in the treatment group and 34 subjects in the control group. Remission from depression symptoms was achieved by 10 subjects in the treatment group and 2 subjects in the control group.

a. Was this an observational study or a controlled experiment? Explain.

b. Find the percentage in each group that achieved remission from depression symptoms.

c. Researchers performed a test to determine if there was significant difference in outcomes between the treatment and control groups. The p-value for the test is 0.028. Based on a 0.05 significance level, choose the correct conclusion:

i. Researchers have shown that dietary improvement may be an effective treatment strategy for patients with moderate to severe depression.

ii. Researchers have not shown that dietary improvement may be an effective treatment strategy for patients with moderate to severe depression.

12.24 Stroke Treatment In a 2018 study, researchers investigated the effect of the drug alteplase in the treatment of stroke patients (Thomalla et al. 2018). Patients were randomly assigned to receive intravenous alteplase or a placebo. The patients' neurological function was assessed 90 days after treatment. Of the 246 patients who received alteplase, 131 had a positive neurological outcome. Of the 244 patients who received a placebo, 102 had a positive neurological outcome.

a. Was this an observational study or a controlled experiment? Explain.

b. Find the percentage in each group that had a positive neurological outcome 90 days after treatment.

c. Researchers performed a test to determine if there was a significant difference in the proportion of positive neurological outcomes between the treatment and control groups. The p-value for the test is 0.003. Based on a 0.05 significance level, choose the correct conclusion:

i. Researchers have shown that alteplase may be an effective treatment for stroke patients.

ii. Researchers have not shown that alteplase may be an effective treatment for stroke patients.

TRY 12.25 Reading Colored Paper (Example 4) Some people believe that it is easier to read words printed on colored paper than words printed on white paper. To test this theory, statistics student Paula Smith collected data. Subjects were timed as they read a passage printed in black ink on a sheet of salmon-colored paper and as they read the same passage printed with black ink on a sheet of white paper. Each person was randomly assigned to read material printed on either the white or the salmon paper first. The times in seconds are given in the table. We are assuming that smaller times (faster reading) imply that the reading is easier. Assume that these are a random sample of times and that the distribution of times is sufficiently close to Normal for *t*-tests.

Salmon	White	Salmon	White
79	72	140	160
49	45	64	57
47	45	67	61
112	120	73	72
65	63	64	48
66	62	126	122
67	60	67	73
59	54		

a. Compare the sample means. Does your result fit the theory given?

b. Use a two-sample *t*-test; report the *t*-statistic and p-value, and report whether you can reject the null hypothesis of no difference in times at the 0.05 level.

c. Use a paired *t*-test; report the *t*-statistic and p-value, and report whether you can reject the null hypothesis of no difference in times at the 0.05 level.

d. Which is the appropriate test for this data set?

e. Why is it essential that the researchers randomly assign the order so that some of the people read material on the salmon-colored paper first and some read material on the white paper first?

12.26 The Stroop Effect Suppose you had to identify the color of ink for a series of printed words that spelled out a color that did not match the ink color. For example, what color ink is used in the word RED? This might take longer than identifying the color when the ink and printed word matched (RED). This difference results from *interference* and is called the Stroop Effect, in recognition of a famous study conducted by research psychologist J. R. Stroop (1897–1973).

Same	Different
32	66
12	31
17	40
16	25
25	36
18	15
18	39
24	35
20	32
24	30

The data in the table were collected to see whether interference was significant. Each of the 10 people studied identified colors of ink in two situations and the time (in seconds) was recorded. There were 36 words in each trial. In column 1 (Same), the color of the ink was consistent with the meaning of the spelled-out word. In column 2 (Different) the color of the ink was different from the meaning of the spelled-out word. Each person was randomly assigned to read the "same color" or the "different color" words first. Assume that all the conditions required for t-tests are satisfied.

a. Compare the sample means. Does the result fit the theory given?

b. Use a two-sample t-test; report the t-statistic and p-value, and report whether you can reject the null hypothesis of no difference in times at the 0.05 level.

c. Use a paired t-test; report the t-statistic and p-value, and report whether you can reject the null hypothesis of no difference in times at the 0.05 level.

d. Which is the appropriate test for this data set?

e. Why is it appropriate to randomly assign the order so that some of the people read the "same color" word first and some read the "different color" word first?

SECTION 12.2

TRY **12.27 Student Records (Example 5)** Suppose a person with access to student records at your college has an alphabetical list of currently enrolled students. The person looks at the records of every 10th person (starting with a randomly selected person among the first 10) to see whether they have paid their latest tuition bill. What kind of sampling does this illustrate?

12.28 Student Records Suppose a person with access to student records at your college has a list of currently enrolled students. The person sorts the data to create two new lists. One contains all the male names, the other all the female names. The person then uses a random number generator to select 50 men and 50 women. What kind of sampling does this illustrate?

12.29 Electric Car Charging Station Suppose a college wishes to select the location of an electric car charging station on campus based on student preference. They have 3 possible locations and are asking a random sample of students to rank the locations with 1 being the most desirable location and 3 being the least desirable location. Explain why the campus might want to stratify the sampling into two groups: those campus members who own an electric car and those who do not.

12.30 Library Audio Books Suppose a college is deciding whether or not to allocate more resources to the purchase of audio books for the college library. Explain why the college might want to use a stratified sample rather than sampling the entire college before making a decision.

12.31 No Grass Suppose a homeowner is considering replacing the grass in the front yard with drought-resistant plants such as cactus. She wants to find out whether the neighbors approve of this or not, so she inquires about this at every fifth house in the subdivision. What kind of sampling is this?

12.32 Music Lovers A large concert promoter that operates several hundred concert locations around the country wants to survey the managers at these locations to ask their opinions about how to improve attendance at concerts. Because the survey is rather lengthy, the promoter does not want to ask all the managers and decides to ask a random sample of managers instead. The promoter organizes the concert locations into 20 different geographic zones, randomly selects 5 zones, and surveys all of the managers in those 5 zones. Is this an example of stratified random sampling, systematic sampling, or cluster sampling?

12.33 Professors A college administrator wants to determine whether the professors at the college are doing a good job. Each professor teaches multiple classes, and so for each professor, one of his or her classes is randomly chosen, and all the students are surveyed to find out their opinion of the teacher. What kind of sampling is this?

*12.34 Faculty-to-Student Ratio A researcher wants to determine whether the faculty-to-student ratio tends to be different in private colleges from that in public colleges. She has an almanac that lists this information for all accredited colleges. She creates two subgroups: one for private and one for public colleges. Then she selects every 20th private college and every 20th public college for her analysis. What two types of sampling are combined here?

SECTION 12.3

TRY **12.35 Fast Eating and Obesity (Example 6)** In a 2018 study by Hurst and Fukuda published in *BMJ Open*, researchers in Japan surveyed 59,717 participants in Japan who had Type 2 diabetes. Participants were asked to rate their eating speed as Slow, Normal, or Fast. Researchers found that those who rated their eating speeds as Slow or Normal were less likely to be obese than those who rated their eating speed as Fast.

a. Can we conclude that fast eating causes obesity from this study? Why or why not?

b. Can this association be generalized to the entire population of people with Type 2 diabetes? Why or why not?

12.36 Nicotine Patch Suppose that a new nicotine patch to help people quit smoking was developed and tested. Smokers voluntarily entered the study and were randomly assigned either the nicotine patch or a placebo patch. Suppose that a larger percentage of those using the nicotine patch were able to stop smoking.

a. Can we generalize widely to a large group? Why or why not?

b. Can we infer causality? Why or why not?

12.37 Hospital Rooms When patients are admitted to hospitals, they are sometimes assigned to a single room with one bed and sometimes assigned to a double room, with a roommate. (Some insurance companies will pay only for the less expensive, double rooms.) A researcher was interested in the effect of the type of room on the length of stay in the hospital. Assume that we are not dealing with health issues that require single rooms.

Suppose that upon admission to the hospital, the names of patients who would have been assigned a double room were put onto a list and a systematic random sample was taken; every tenth patient who would have been assigned to a double room was part of the experiment. For each participant, a coin was flipped: If it landed heads up, she or he got a double room, and if it landed tails up, a single room. Then the experimenters observed how many days the patients stayed in the hospital and compared the two groups. The experiment ran for two months. Suppose those who stayed in single rooms stayed (on average) one less day, and suppose the difference was significant.

a. Can you generalize to others from this experiment? If so, to whom can you generalize, and why can you do it?

b. Can you infer causality from this study? Why or why not?

12.38 College Tours A random sample of 50 college first-year students (out of a total of 1000 first-years) was obtained from college records using systematic sampling. Half of those students had a campus tour with a sophomore student, and half had a tour with an instructor. The tour guide was determined randomly by coin flip for each student. Suppose that those with the student guide rated their experience higher than those with the instructor guide.

a. Can you generalize to other first-year students at this college? Explain.

b. Can you infer causality from this study? Explain.

12.39 Intravenous Fluids Critically ill patients are often given intravenous fluids in hospital, either in the form of balanced crystalloids or saline solutions. In a 2018 study published in *The New England Journal of Medicine*, researchers investigated which of these approaches resulted in better clinical outcomes. Read this excerpt from the abstract that accompanies this study and answer the following questions (Semmler et al. 2018).

Methods: In a pragmatic, cluster-randomized, multiple-crossover trial conducted in five intensive care units at an academic center, we assigned 15,802 adults to receive saline or balanced crystalloids. The primary outcome was a major adverse kidney event within 30 days — a composite of death from any cause, new renal-replacement therapy, or persistent renal dysfunction.

Results: Among the 7942 patients in the balanced-crystalloids group, 1139 (14.3%) had a major adverse kidney event, as compared with 1211 of 7860 patients (15.4%) in the saline group ($P = 0.04$).

a. Identify the treatment variable.

b. The response variable in this study is major adverse kidney event within 30 days. Was there a significant difference in occurrence of major adverse kidney events between the two groups? Explain. Assume a significance level of 0.05.

c. Based on this study, do you think one type of intravenous fluid may be preferable over the other? Explain.

12.40 Acetaminophen and Asthma Does frequent use of acetaminophen lead to asthma-related complications among children? Excerpts from the abstract of a study published in The New England Journal of Medicine about this are given (Sheehan et al. 2016). Read them and then answer the questions that follow.

Methods: In a randomized, double-blind, parallel-group trial, we enrolled 300 children (age range, 12–59 months) with mild persistent asthmas and assigned them to receive either acetaminophen or ibuprofen when needed for the alleviation of fever or pain over the course of 48 weeks. The primary outcome was the number of asthma exacerbations that let to treatment with systemic glucocorticoids.

Results: The number of asthmas exacerbations did not differ significantly between the two groups, with a mean of 0.81 per participant with acetaminophen and 0.87 per participant in the ibuprofen group ($p = 0.67$).

a. Identify the treatment variable and the response variable.

b. Was this a controlled experiment or an observational study?

c. How does the p-value support the conclusion of the study?

d. Did this study use random sampling, random assignment, or both?

12.41 Phubbing and Relationship Satisfaction Phubbing is the practice of ignoring one's companion or companions in order to pay attention to one's phone or other mobile device. In the conclusion of a 2017 study published in *Personality and Individual Differences*, researchers (Wang et al. 2017) concluded "The results indicated that partner phubbing had a negative effect on relationship satisfaction, and relationship satisfaction had a negative effect on depression." Is this conclusion likely to be the result of an observational study or a controlled experiment? Can we conclude phubbing causes decreased relationship satisfaction from this study? Explain.

12.42 Dog Ownership and Cardiovascular Disease In a 2017 study published in *Scientific Reports*, researchers (Mubanga et al. 2017) concluded that "dog ownership was associated with a lower risk of incident cardiovascular disease in single-person households and with lower cardiovascular and all-cause mortality in the general population." Is this conclusion likely to be the result of an observational study or a controlled experiment? Can we conclude that owning a dog causes a decrease in cardiovascular disease from this study? Explain.

12.43 Yoga and Cellular Aging A 2017 study explored the impact of Yoga and Meditation based Lifestyle Intervention (YMLI) on cellular aging in healthy individuals (Tohlahunase et al. 2017). Ninety-six healthy individuals were enrolled in the 12-week YMLI course, which consisted of yoga postures, breathing exercises, and meditation. Participants attended YMLI for as a group for two weeks, 5 days per week. After the initial two-week period, participants did the program individually at home. Participation was monitored through the maintenance of a diary and telephone contact. After 12 weeks, researchers found significant improvement in biomarkers of cellular aging and longevity among participants. Does this study show that YMLI causes improvement in biomarkers of cellular aging and leads to increased longevity? Explain.

12.44 Fish Consumption and Arthritis A 2017 study reported in the Harvard Health Blog investigated the association between fish consumption and disease activity in 176 rheumatoid arthritis patients (Tedeschi et al. 2017). Frequency of fish consumption was assessed through a questionnaire. Researchers found that participants who consumed fish at least two times per week showed significantly lower disease activity (in other words, less inflammation associated with the disease) than participants who consumed never or less than one time per month. Does this study show that fish consumption causes lower disease activity in rheumatoid arthritis patients? Explain.

12.45 Yoga Study Design Refer to exercise 12.43. How could you investigate whether participation in a Yoga and Meditation based Lifestyle Intervention (YMLI) caused the improved cellular biomarkers associated in this study? Describe the design of a study assuming you have 200 healthy individuals participating in the study.

12.46 Fish Consumption Study Design Refer to exercise 12.44. How could you investigate whether consuming fish at least two times per week causes decreased disease activity in rheumatoid arthritis patients? Describe the design of a study assuming you have 100 rheumatoid arthritis patients in the study.

12.47 Music and Divergent Thinking In a 2017 study published at PLOS.org, researchers investigated the effect of music on creativity (Ritter and Ferguson 2017). Subjects were recruited for the study using an online research participation system at a university. Four pieces of music were selected with different emotional tones: calm, happy, sad, and anxious. Subjects were randomly assigned to listen to one of these four pieces or to a group that listened to no music (silence). After 15 seconds of music (or silence) subjects were given a task that assessed their creativity and divergent thinking. Read the excerpts from the study abstract and answer the following questions.

> *Results:* Our main hypothesis was that listening to happy music, as compared to a silence control condition, facilitates divergent thinking. An independent-samples *t*-test was conducted to compare the happy music condition with the silence control condition on overall divergent thinking (ODT). There was a significant difference in ODT between the happy music (M = 93.87, SD = 32.02) and silence (M = 76.10, SD = 32.62) conditions, t (57) = 2.110, p = .039. The results suggest that listening to happy music increases performance on overall divergent thinking.

a. Identify the treatment variable and the response variable.

b. Was this a controlled experiment or an observational study? Explain.

c. Can you conclude from that listening happy music enhances divergent thinking? Why or why not?

12.48 Tea and Divergent Creativity In a 2017 study published in the journal *Food Quality and Preference*, researchers investigated the effect of drinking tea on divergent creativity (Huang et al. 2017). Subjects were recruited from a campus Bulletin Board System and were paid a small stipend for their participation. Subjects were randomly assigned to be served either tea or water during the "greeting period" of the experiment. During the greeting period subjects filled out a background questionnaire so they were unaware that beverage was a key component in the study. Subjects were then told to build the most "attractive" building possible in a limited amount of time using a set of blocks. Independent observers then gave each building a creativity score. Read excerpts from the study results and answer the following questions.

> *Results:* A general linear model analysis showed that the creativity scores of the block buildings for the tea group (mean = 6.54, SD = 0.92) were significantly higher than those for the water group (mean = 6.03, SD = 0.94) after controlling for gender and volume consumed (*p* = 0.023).

a. Identify the treatment variable and the response variable.

b. Was this a controlled experiment or an observational study? Explain.

c. Can you conclude from that drinking tea leads to improved creativity? Why or why not?

*12.49 Alumni Donations The alumni office wishes to determine whether students who attend a reception with alumni just before graduation are more likely to donate money within the next two years.

a. Describe a study based on a sample of students that would allow the alumni office to conclude that attending the reception *causes* future donations but that it is *not* possible to generalize this result to all students.

b. Describe a study based on a sample of students that does *not* allow fundraisers to conclude that attending receptions causes future donations but does allow them to generalize to all students.

c. Describe a study based on a sample of students that allows fundraisers to conclude that attending the reception causes future donations and also allows them to generalize to all students.

*12.50 Recidivism Rates The 3-year recidivism rate in the United States is about 68%, which means that 68% of released U.S. prisoners return to prison within 3 years of release. There have been many attempts to reduce the recidivism rate. Suppose you want to determine whether electronic monitoring bracelets that track the location of the released prisoner reduce recidivism. Suppose that offenders released from prison are observed for 3 years to see whether they go back to prison and that the ones who wear electronic monitoring bracelets wear them for the first year only.

a. Describe a study based on a sample of released offenders that would allow the legal system to conclude that monitoring causes a reduction in recidivism but would not allow it to generalize this result to all released prisoners.

b. Describe a study based on a sample of released offenders that does *not* allow the legal system to conclude that monitoring causes a reduction in recidivism but does allow it to generalize to all released offenders.

c. Describe a study based on a sample of released prisoners that allows the legal system to conclude that monitoring causes a reduction in recidivism and also allows it to generalize to all released offenders.

For exercises 12.51–12.53, evaluate the study based on the extracts from the study abstracts by answering the following questions:

a. What is the research question that the investigators are trying the answer?

b. What is their answer to the research question?

c. What were the methods they used to collect data?

d. Is the conclusion appropriate for the methods used to collect data?

e. To what population do the conclusions apply?

f. Have the results been replicated (reproduced) in other articles?

g **12.51 Ketamine and Social Anxiety Disorder (Example 7)**
TRY According to the National Institute of Mental Health, Social Anxiety Disorder (SAD) is a mental health disorder that affects up to 7% of the population of the United States. Because many SAD patients experience inadequate symptom relief with available treatments, researchers in this study investigated the use of ketamine to treat SAD patients (Taylor et al. 2018). Read the following excerpts from the study abstract and evaluate the study using the given questions.

> *Methods:* We conducted a double-blind, randomized, placebo-controlled crossover trial in 18 adults with Social Anxiety Disorder and compared the effects between intravenous ketamine and placebo on social phobia symptoms. Ketamine and placebo infusions were administered in a random order with a 28-day washout period between infusions. Ratings of anxiety were assessed 3-hours post-infusion and followed for 14 days. Outcomes were blinded ratings on the Liebowitz Social Anxiety Scale (LSAS) and self-reported anxiety on a visual analog scale (VAS-Anxiety).

> *Results:* We found ketamine resulted in a significantly greater reduction in anxiety relative to placebo on the LSAS (*p* = 0.01) but not the VAS-Anxiety (*p* = 0.95). Participants were significantly more likely to exhibit a treatment response after ketamine infusion relative to placebo

in the first 2 weeks following infusion measured on the LSAS (33.33% response ketamine *vs* 0% response placebo, $p = 0.025$) and VAS (88.89% response ketamine *vs* 52.94% response placebo, $p = 0.034$).

Conclusion: This trial provides initial evidence that ketamine may be effective in reducing anxiety.

12.52 Dog Ownership and Risk of Cardiovascular Disease

Some researchers believe that dogs may be beneficial in reducing cardiovascular risk in their owners by providing social support and motivation for physical activity (Mubanga et al. 2017). The purpose of this study was to investigate the association of dog ownership with incident of cardiovascular disease in the population of Sweden. Read the following excerpts from the study abstract and evaluate the study using the given questions.

Methods: All Swedish residents aged 40 to 80 years on January 1, 2001 ($n = 3,987,937$) were eligible for this study. The age range was chosen to exclude younger individuals at low risk of CVD and the elderly at low odds of owning a dog. All Swedish residents are covered by the public health care system, and all hospital visits are registered in the National Patient Register. We obtained death data from the Cause of Death Register and incident disease data from the National Patient Register. The main diagnosis in inpatient and outpatient care and underlying cause of death were used to define four incident disease outcomes: (1) acute myocardial infarction, (2) heart failure, (3) ischemic stroke, and (4) hemorrhagic stroke. Any occurrence of these diagnoses was additionally considered as a composite cardiovascular disease (CVD) outcome . . . Dog ownership was defined as periods registered or having a partner registered as a dog owner in either of the two dog registers (required for all dogs in Sweden.)

Results: Dog ownership was inversely associated with risk of acute myocardial infarction, ischemic stroke, heart failure, and composite CVD. Dog ownership was inversely associated with cardiovascular mortality and all-cause mortality.

Conclusions: Dog ownership was associated with a lower risk of incident cardiovascular disease in single-person households and with lower cardiovascular and all-cause mortality in the general population. Our observational study cannot provide evidence for a causal effect of dog ownership on cardiovascular disease or mortality. Although careful attention was paid to adjusting for potential confounders in a set of sensitivity analyses, it is still possible that personal characteristics that we did not have information about affect the choice of not only acquiring a dog, but also the breed and the risk of CVD.

12.53 What Would Batman Do?

Researchers have found that psychological distance from our current situation facilitates self-control and allows individuals to transcend urgencies of a situation by taking a more distanced perspective. Executive function refers to higher-order regulatory processes such as inhibition and working memory. In this study published in the journal *Developmental Science*, researchers investigated the relationship between psychological distance and executive function in pre-school children (White and Carlson 2015). Read the following excerpts from the study abstract and evaluate the study using the given questions.

Method: Three-year old ($n = 48$) and 5-year old ($n = 48$) children were randomly assigned to one of four manipulations of distance from self and asked to perform several tasks that assessed executive function (EF). The four groups were: 1) self-immersed, in which children were told to focus on what they are thinking and how they feel when the task got hard, 2) Third person, in which children were told to talk to themselves using their own name when the task got hard, 3) Exemplar, in which children were told to pretend they were someone else who would be really good at the task, like Batman, Dora the Explorer, Bob the Builder,

or Rapunzel, and the children put on costume props before completing the task, and 4) a control group, in which children were given no instruction regarding distance from self before performing the task.

Results: Five-year-olds benefited from taking a self-distanced perspective on an executive function task through third person self-talk as well as taking the perspective of an exemplar other, such as Batman. Three-year-olds did not show increased EF performance as a function of greater distance from self.

Conclusion: The current study revealed the power of self-distancing to facilitate reflective, goal-directed action in the context of a cool EF task for young children. Children's ability to improve EF by mentally transcending their context underscores the critical role that representational capacities play in the development of self-control.

12.54 Autism and MMR Vaccine

An article in the British medical journal *The Lancet* claimed that autism was caused by the measles, mumps, and rubella (MMR) vaccine (Wakefield et al. 1998). This vaccine is typically given to children twice, at about the age of 1 and again at about 4 years of age. The article reports a study of 12 children with autism who had all received the vaccines shortly before developing autism. The article was later retracted by *The Lancet* because the conclusions were not justified by the design of the study.

Explain why *The Lancet* might have felt that the conclusions were not justified by listing potential flaws in the study, as described earlier.

12.55 Anesthesia Care and Adverse Postoperative Outcomes

Handing over the care of a patient from one anesthesiologist to another occurs during some surgeries. A study was conducted to determine if this transfer of care might increase the risk of adverse outcomes (Jones et al. 2018). Read the excerpt from the study abstract published in *JAMA* below and answer the questions that follow.

Methods: A retrospective population-based cohort study was conducted of adult patients undergoing major surgeries expected to last at least two hours and requiring a hospital stay of at least one night. The primary outcome measured was a composite of all-cause death, hospital readmission or major postoperative complications all within 30 postoperative days.

Results: A total of 5941 patients underwent surgery with complete handover of anesthesia care. The primary outcome (death, readmission, or major postoperative complications) occurred in 2614 of these patients. A total of 307,125 patients underwent surgery without complete handover of anesthesia care. Of these, the primary outcome occurred in 89,066 patients. The complete handovers were statistically significantly associated with an increased risk of the primary outcome ([95% CI, 4.5% to 9.1%]; $P < 0.001$), all-cause death ([95% CI, 0.5% to 2%]; $P = 0.002$), and major complications ([95% CI, 3.6% to 7.9%]; $P < 0.001$), but not with hospital readmission within 30 days of surgery ([95% CI, -0.3% to 2.7%]; $P = 0.11$).

a. Compare the percentage of each group who experienced the primary outcome (death, readmission, or major postoperative complications). Based on the abstract, can you reject the null hypothesis that there is no difference in the rates of primary outcome?

b. If you were a hospital administrator, would you recommend that complete handover of anesthesia care during operations be limited? Why or why not?

c. A difference between the two groups was found for all of the primary care outcomes except hospital readmission within 30 days of surgery. How do the confidence interval and p-values provided support this conclusion?

12.56 Coronary Artery Bypass Grafting A study reported in the *New England Journal of Medicine* was conducted to compare outcomes for radial arterial grafts and saphenous-vein grafts in coronary artery bypass surgeries (Gaudino et al. 2018). Read this excerpt from the study abstract and answer the questions that follow.

Methods: We performed a patient-level combined analysis of randomized, controlled trials to compare radial-artery grafts and saphenous-vein grafts for coronary artery bypass grafting (CABG). Six trials were identified. The primary outcome was a composite of death, myocardial infarction, or repeat revascularization.

Results: A total of 1036 patients were included in the analysis (534 patients with radial-artery grafts and 502 patients with saphenous-vein grafts). After a mean (±SD) follow-up time of 60 ± 30 months, the incidence of adverse cardiac events was significantly lower in association with radial-artery grafts than with saphenous-vein grafts (95% confidence interval [CI], 0.49 to 0.90; $P = 0.01$). As compared with the use of saphenous-vein grafts, the use of radial-artery grafts was associated with a nominally lower incidence of myocardial infarction (95% CI, 0.53 to 0.99; $P = 0.04$) and a lower incidence of repeat revascularization (95% CI, 0.40 to 0.63; $P < 0.001$) but not a lower incidence of death from any cause (95% CI, 0.59 to 1.41; $P = 0.68$).

a. Which graft method had more positive outcomes? Explain.

b. There was an outcome for which one method did not have significantly better outcomes than the other. What outcome was this and how does the p-value support this conclusion?

12.57 Infant Formula and Diabetes Risk Early exposure to complex dietary proteins may increase risk of Type 1 diabetes in children with genetic susceptibility to this disease.

In a double-blind randomized clinical trial reported in *JAMA* infants identified to be genetically at risk for developing Type 1 diabetes, were randomly assigned to a conventional formula group or a hydrolyzed formula group after weaning (Writing Group for the TRIGR Study Group 2018). Conventional formula contains proteins while hydrolyzed formula contains no intact proteins. Of the 1079 infants in the conventional formula group, 82 developed the disease by age 11.5 years. Of the 1983 infants assigned to the hydrolyzed formula group, 91 developed the disease by that age.

a. Identify the treatment and response variables.

b. Test the hypothesis that type of formula and development of Type 1 diabetes are independent using a significance level of 0.05.

c. Based on this study, do you think dietary recommendations for infants at risk for Type 1 diabetes should be revised to recommend hydrolyzed formula over conventional formula? Explain.

12.58 Melanoma In a study published in *The New England Journal of Medicine*, researchers investigated the effectiveness of the drug pembrolizamab on increasing survival rates in patients with advanced melanoma (Eggermont et al. 2018). In this randomized double-blind study, 514 patients received the drug and 505 patients received a placebo. Recurrence-free survival rates for both groups were measured after 15 months. After 15 months, 388 patients in the drug group and 308 patients in the placebo group experienced recurrence-free survival.

a. Find and compare the percentages that experienced recurrence-free survival.

b. Test the hypothesis that a greater proportion of patients taking the drug experienced recurrence-free survival than those taking the placebo. Use a significance level of 0.05.

c. Based on this study, do you think the drug pembrolizamab may be effective in treating patients with advanced melanoma? Explain.

UIDED PRACTICE

12.51 Ketamine and Social Anxiety Disorder (Example 7) According to the National Institute of Mental Health, Social Anxiety Disorder (SAD) is a mental health disorder that affects up to 7% of the population of the United States. Because many SAD patients experience inadequate symptom relief with available treatments, researchers in this study investigated the use of ketamine to treat SAD patients (Taylor et al. 2018). Read the following excerpts from the study abstract and answer the questions.

Methods: We conducted a double-blind, randomized, placebo-controlled crossover trial in 18 adults with Social Anxiety Disorder and compared the effects between intravenous ketamine and placebo on social phobia symptoms. Ketamine and placebo infusions were administered in a random order with a 28-day washout period between infusions. Ratings of anxiety were assessed 3-hours post-infusion and followed for 14 days. Outcomes were blinded ratings on the Liebowitz Social Anxiety Scale (LSAS) and self-reported anxiety on a visual analog scale (VAS-Anxiety).

Results: We found ketamine resulted in a significantly greater reduction in anxiety relative to placebo on the LSAS ($p = 0.01$) but not the VAS-Anxiety ($p = 0.95$). Participants were significantly more likely to exhibit a treatment response after ketamine infusion relative to placebo in the first 2 weeks following infusion measured on the LSAS (33.33% response ketamine *vs* 0% response placebo, $p = 0.025$) and VAS (88.89% response ketamine *vs* 52.94% response placebo, $p = 0.034$).

Conclusion: This trial provides initial evidence that ketamine may be effective in reducing anxiety.

Step 1 ▶ What is the research question that these investigators are trying to answer?

Step 2 ▶ What is their answer to the research question?

Step 3 ▶ What were the methods they used to collect data? (controlled experiment or observational study)

Step 4 ▶ Is the conclusion appropriate for the methods used to collect data?

Step 5 ▶ To what population do the conclusions apply?

Step 6 ▶ Have the results been replicated in other articles? Are the results consistent with what other researchers have suggested?

Inference without Normality

THEME

Many of the techniques discussed so far have required either that the population be Normally distributed or that we work with the sample mean and with such a large sample size that the Central Limit Theorem allows us to assume the sampling distribution is Normal. Nonparametric methods provide techniques that can be used when the population distribution is not Normal or is unknown.

Many of the techniques you've learned for inference so far have relied on a single important mathematical idea: the Normal distribution. In many cases, when the population we're studying follows a Normal distribution, we can compute accurate confidence levels and compute p-values for our hypothesis tests. Confidence levels and p-values are important because they measure our level of uncertainty. Inference is an uncertain business and might seem completely useless if it were not for the fact that we can measure how far from the truth our estimates are likely to be.

However, not every distribution is Normal. When the population is not Normal, the Central Limit Theorem (CLT) tells us that if the sample size is large enough, the distribution of the sample mean or sample proportion is still approximately Normal. But what if we are not estimating a mean or proportion? The CLT doesn't help us if we want to estimate the population median, for example. In some contexts, such as skewed distributions, the median is a more natural measure of center than is the mean. At other times, we might have too small a sample size to rely on the CLT. For example, the Case Study shows that even a large number of people in a study may not be enough if the expected proportion of responses in one category is small.

Several approaches to inference will work regardless of whether the population follows the Normal distribution. The formal name for these approaches is **nonparametric inference**. This term covers a variety of techniques and procedures, and this chapter focuses on hypothesis tests that do not depend on the Normal distribution. Certain conditions need to be satisfied in order for these tests to provide valid inference, but generally these conditions are less strict than those that must be satisfied for the tests presented in earlier chapters.

CASE STUDY

Contagious Yawns

Are yawns contagious? Many people believe that they are. After all, if you see someone else yawn, don't you usually fight the urge to yawn yourself?

The television show *Myth Busters* sought to determine whether this myth was true. They led unsuspecting participants into a small, featureless room and instructed them to sit on a chair and wait for further instructions. Some of these participants were led to the room by an assistant who yawned as she seated them. For others, the assistant did not yawn. A hidden camera recorded whether the participants yawned soon after the assistant left.

Of the 34 people who received the yawn stimulus, 10, or 29%, yawned, and 25% of the 16 people in the control group yawned. Is the difference in percentage of yawners big enough to support the conclusion that the myth is true? You might think of analyzing these data using a chi-square test of independence. However, that approach is intended for two-sided hypotheses. We don't simply want to know whether a yawn stimulus is associated with yawning; we want to know whether people who receive the stimulus are more likely to yawn than those who don't. For this, we need a test to perform with a one-sided hypothesis. A z-test for two proportions is another possibility, but the sample sizes are too small for the p-value approximation to be accurate, because we don't have ten expected success and ten expected failures in both the yawn-stimulus group and the control group.

In this chapter you'll learn how a test called a randomization test can yield a very good approximation to the p-value to help us determine whether or not yawns are contagious. A randomization test is actually a useful way of carrying out Fisher's Exact Test, which you saw in Chapter 10.

Transforming Data

One approach to dealing with data from a non-Normal population is to change the data so that they look more like they *did* come from a Normal population. Strictly speaking, this is not a nonparametric technique, because we still rely on an identifiable population distribution (the Normal, in this case) to extend our conclusions to the population. Still, this commonly used technique can be quite effective.

Suppose we want to estimate the mean income of residents in the United States on the basis of a random sample. The distribution of income is famous for being right-skewed, as you can see in Figure 13.1, which displays a histogram for a random sample of 72 U.S. residents' annual income. The fact that the distribution of the sample is skewed makes us strongly suspect that the population distribution is also skewed.

We can use technology to find a 95% confidence interval for the mean income (in dollars per year) of all U.S. citizens, based on the data in this sample. The 95% confidence interval is

$$\$26,140 \text{ to } \$74,504$$

However, one assumption behind this calculation is that either income is Normally distributed (and we know it's not) or that the sample size is "large enough." One rule of thumb for "large enough" is that the sample size should be 25 or higher. However, if the population distribution is extremely skewed, then we may need a very large sample size. So how can we be sure this is indeed large enough?

First, we examine a tool that helps us better judge whether a distribution is Normal or non-Normal. Then we will show you how to transform the data so that the distribution is closer to a Normal distribution. We can then analyze these new, transformed data and get more precise results than we would get if we relied on the original, untransformed data.

> **Details**
>
> **Nonparametric**
> The term *nonparametric* is used to describe analyses that do not make assumptions about the population distribution.

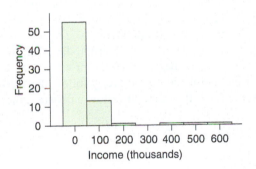

◄ **FIGURE 13.1** The distribution of a random sample of incomes is skewed right, which is a typical feature of distributions of income.

Interpreting QQ Plots

A new tool that is useful here is the **QQ plot**. A QQ plot (the Qs stand for "quantile") is a graphical tool that is helpful for deciding whether a sample was drawn from a distribution that is approximately Normal. A QQ plot graphs the actual values of a variable on the *y*-axis against the values on the *x*-axis that the variable would have if, in fact, it followed a standard Normal distribution. If the distribution of the variable is a Normal distribution, then this plot is a straight line.

Several variations of the QQ plot are common, and different software packages display different variations. For example, some software uses the *y*-axis to display the values of the variable converted to standard units, rather than plotting the actual values. Some software packages display QQ plots using the *x*-axis to plot the standard units and the *y*-axis to plot the observed values. Others show a variation of the QQ plot called

> **Looking Back**
>
> **Central Limit Theorem (CLT)**
> The CLT says that the sampling distribution of the sample mean, $\bar{x}$, is approximately $N\left(\mu, \dfrac{\sigma}{\sqrt{n}}\right)$, where μ is the mean of the population and σ is the standard deviation of the population. The larger the sample size, n, the better the approximation. If the population is Normal to begin with, then the sampling distribution is exactly a Normal distribution.

a probability plot, in which the probability of seeing a value less than or equal to the displayed value, if the distribution is Normal, is plotted. No matter which variation your technology produces, the interpretation of the plot is the same. If the data are Normal, then the points on the plot fall along a straight line. If they are non-Normal, they follow a curve.

Figure 13.2 shows the QQ plot for the incomes displayed in Figure 13.1. The black line shows where the points would fall if the distribution were Normal. As we read from left to right, the points are fairly flat from −2 to −1 on the *x*-axis, and then they start to rise. This feature is commonly seen in right-skewed distributions.

▶ **FIGURE 13.2** The QQ plot for the sample of incomes in Figure 13.1. If the data were Normally distributed, they would follow the straight line.

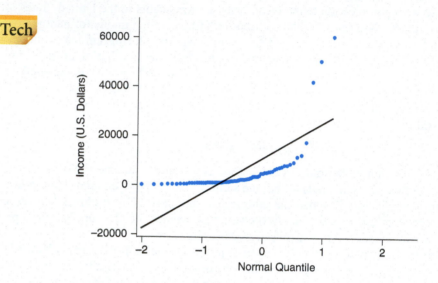

Reading a QQ plot takes a little practice. Even samples from true Normal distributions do not produce QQ plots that exactly follow the straight line. Indeed, samples from true Normal distributions typically have points that meander above and below the line in a QQ plot, but overall, the points follow the line.

Figure 13.3 shows a QQ plot from a true Normal distribution created using simulated data. (Because we simulated them, we know for a fact that the data come from a Normal distribution.) The points do not perfectly follow the straight line. However, unlike the points in Figure 13.2, they tend to follow a constant increasing trend.

▶ **FIGURE 13.3** The QQ plot for data simulated from a true Normal distribution (with mean 0, standard deviation 1). Note that the points "stray" a bit but for the most part follow the straight line.

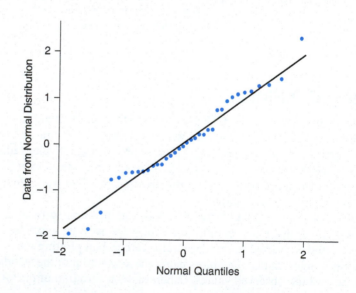

The Log Transform

The **log transform** is an approach to analyzing data that can sometimes turn right-skewed distributions into something more closely resembling a Normal distribution. The log transform can be applied only to data with all positive (greater than 0) values. This approach changes—transforms—the data to new values. If all goes well, the distribution of the new values follows the Normal distribution, so the Normal condition required of so many of the techniques presented in earlier chapters will be satisfied. Then, if we want to, we can "undo" the transform once we've learned what we need from the data.

The log transform is particularly useful when the values of the data range across orders of magnitude. In other words, if you have some values in the 10s, and others in the 100s (or 1000s, or 10,000s), then using a log transform is worth a try. As you can see in the histogram of Figure 13.1, this is certainly the case for the income data: Some of the values are in the $10s, some are in the $100s, many are in the $10,000s, and some are in the $100,000s.

To carry out the log transform, simply take the logarithm of each value. It doesn't matter whether you use log base 10 or the natural logarithm, just as long as you know which one you used and are consistent with your choice. We will use log base 10 for simplicity. Table 13.1 shows the first Three rows of the income data beside the log-transformed (log base 10) incomes.

Total_Personal_Income	Log_{10}(Total_Personal_Income)
3,900	3.5910646
175,000	5.243038
68,000	4.8325089

Tech

◀ **TABLE 13.1** Log (base 10) transforms of the first three rows of a data set containing incomes of U.S. residents.

As you can see from Figure 13.4, the transformed data come closer than the original data to following a Normal distribution. Compare these figures to the original distributions in Figures 13.1 and 13.4. Admittedly, it can be difficult to tell from the histogram whether the distribution of the log-transformed data is Normal, but this is exactly why the QQ plot is useful.

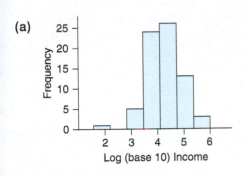

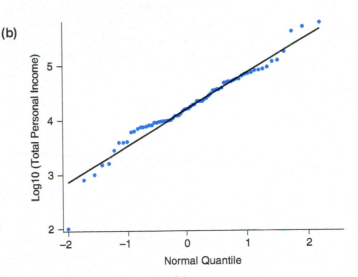

▲ **FIGURE 13.4 (a)** The histogram of the transformed data is more symmetric than that of the untransformed data. Note that with such a small sample size, it is difficult to tell from the histogram whether the distribution is truly Normal. **(b)** The QQ plot for the log-transformed data is much closer to the straight line than the QQ plot for the untransformed data shown in Figure 13.2. This plot shows us that the transformed data are fairly close to being Normally distributed.

EXAMPLE 1 Examining Normality

The histogram in Figure 13.5 shows the distribution of sizes (in square feet) of houses for sale in Beverly Hills, California. Beverly Hills is well known for extravagant lifestyles, and this is quite apparent in the size of some of the homes represented. To put these data in perspective, the average detached home (in other words, not an apartment, townhouse, or condominium) in the United States contains about 2300 square feet. The largest house in this data set measures 28,000 square feet.

▶ **FIGURE 13.5** Distribution of the sizes (in thousands of square feet) of 232 homes for sale in Beverly Hills.

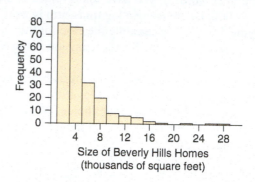

QUESTION   Describe the distribution of house sizes. Why might a log transform make the distribution of house sizes more like a Normal distribution? Does it? (Refer to Figure 13.6.)

SOLUTION   The distribution is right-skewed, with the median home size somewhere between 2500 and 5000 square feet. (In fact, by looking at the actual data set, we found that the median is 3573 square feet.) The first quartile is around 2500 square feet, and the third is just about 5000 square feet (reading off the histogram), so the interquartile range is about 2500 square feet. Figure 13.6a shows the QQ plot of the data. The very clear curve in the QQ plot tells us we have a skewed, non-Normal distribution.

Because the house sizes cover several orders of magnitude (1000s and 10,000s), a log transform might make the data more Normally distributed. Figure 13.6b shows the QQ plot of the log-transformed data. The points in this QQ plot stay closer to the line here than in Figure 13.6a, which tells us that the distribution of the transformed data is closer to a Normal distribution. Figure 13.6c shows a histogram of the transformed data with a Normal distribution superimposed.

▶ **FIGURE 13.6a** A QQ plot of the original data (sizes of houses) shows that the data are highly skewed and very non-Normal.

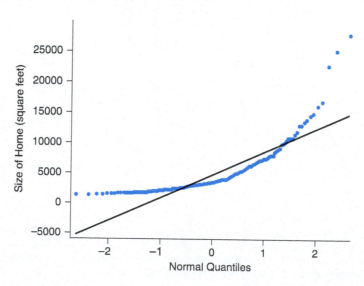

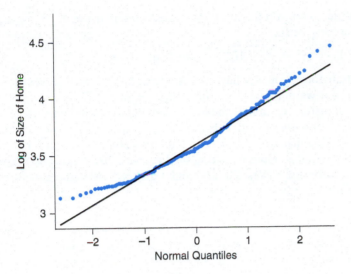

◀ **FIGURE 13.6b** QQ plot for the log transform of the sizes of the houses: $\log_{10}$(size). The data come close to following the straight line, which suggests that they are approximately Normal but not perfectly Normal.

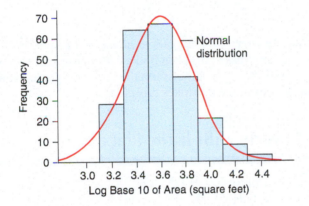

◀ **FIGURE 13.6c** A histogram of the transformed data shows that the distribution is now only slightly right-skewed.

The transformed distribution is not perfectly symmetric, but it's closer to a Normal distribution than is the distribution of the original data.

 TRY THIS! Exercise 13.5

Analyzing Log-Transformed Data

If the distribution of the log-transformed data is approximately Normal, then it makes good sense to base your analysis on the mean of the transformed data. The sample mean of the transformed data provides a useful estimate of the population mean of the log-transformed data.

One difficulty with the sample mean of the log-transformed data is that the units might be strange or unfamiliar. For example, we measured the size of Beverly Hills homes using square feet, a unit of measurement that many people (particularly those hunting for apartments or homes) are quite familiar with. But when we take the log transform, we now have measurements in log-square-feet. The sample mean of the transformed data is about 3.6 log-square-feet.

To interpret this value, it is best to **back transform**. In other words, we "undo" the transform. To undo a base 10 log transform, raise 10 to the power of the transformed value. That is, if y is a number in log units, we back transform to a number x in regular units as follows:

$$x = 10^y$$

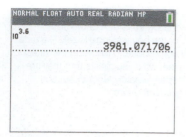

▲ **FIGURE 13.7** TI-84 output. The button used is 10^x, which is the second function of the Log button.

📌 **Details**

Geometric Mean Formula
Those of you with a mathematical inclination might remember a different definition of the geometric mean: $(x_1 x_2 \cdots x_n)^{(1/n)}$, where each x represents an observation. This definition is equivalent to ours, and if you *truly* have a mathematical inclination, you might enjoy figuring out why.

❗ **Caution**

Round After, Not Before
Do not round off before performing back transformations. With no rounding, the sample mean of the transformed home-size data is 3.5970366. Using all of the digits gives a more precise value of $10^{3.5970366} = 3954$ square feet.

For example, to convert the log-transformed sample mean of 3.6 log-square-feet back to square feet, we find

$$10^{3.6} = 3981 \text{ square feet}$$

To back transform, you will have to use a calculator or the calculator function of your statistical software. Figure 13.7 shows the output of a TI-84.

The back-transformed sample mean is called the **geometric mean**. The geometric mean is another measure of center that, like the median, is useful for summarizing the center of skewed distributions. To compute the geometric mean for a sample of data, first convert all the values by taking the log (base 10). Next, find the sample mean of these log-transformed values. Finally, back transform the sample mean of the log-transformed values.

KEY POINT   The geometric mean is a measure of center that, like the median, is sometimes useful for skewed distributions. The geometric mean is particularly useful when the log transform of the data results in a Normal (or nearly Normal) distribution.

To find a confidence interval for the geometric mean, first convert the original data by taking the log transform (base 10). Then find a confidence interval for the mean, as you have learned to do. Finally, back transform the left and right endpoints of the confidence interval. The next example illustrates this process.

EXAMPLE **2** Beverly Hills Is the Place to Be

Figure 13.6a on page 644 illustrates that the sample mean is not a good measure of the typical home size in Beverly Hills. The right-skew to the distribution "pulls" the mean up, so the mean home seems much larger than what we might think of as "typical." (Only about 1/3 of the homes are bigger than the mean size of 4858 square feet.) From Figures 13.6b and 13.6c, we know that the distribution of the log transform of these data is approximately Normal. This means that we can use the geometric mean of our sample to estimate the geometric mean of the population.

We transformed the Beverly Hills home sizes by taking the log base 10 of every value. We then used StatCrunch to calculate a 95% confidence interval for the mean of the log-transformed data. On the next page, the first few rows of the data set, including the transformed data, are shown in Figure 13.8a. The results are shown in Figure 13.8b.

QUESTION   Use the StatCrunch output to find a 95% confidence interval for the population geometric mean. Report the geometric mean. Interpret the confidence interval.

SOLUTION   To find the 95% confidence interval for the population geometric mean, we simply back transform the lower and upper limits of the 95% confidence interval for the mean of the transformed data.

$$\text{Lower limit} = 10^{3.5630028} = 3655.971$$

$$\text{Upper limit} = 10^{3.6310706} = 4276.324$$

After rounding these results to the nearest square foot, we find that a 95% confidence interval for the geometric mean of all for-sale home sizes in Beverly Hills is 3656 square feet to 4276 square feet.

To find the geometric mean, we back transform the sample mean of the transformed data:

$$\text{Geometric mean} = 10^{3.5970366} = 3953.999, \text{ or about 3954 square feet.}$$

beveryhills.txt

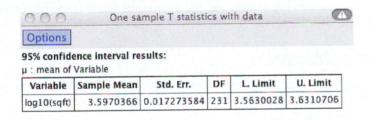

Row	city	type	sqft	log10(sqft)	bed	bath
1	Beverly Hills	Condo/Twh	1500	3.1760912	2	2.5
2	Beverly Hills	Condo/Twh	1617	3.20871	2	2.5
3	Beverly Hills	Condo/Twh	1910	3.2810333	2	2.5
4	Beverly Hills	Condo/Twh	1961	3.2924776	2	2.5
5	Beverly Hills	Condo/Twh	2512	3.4000196	2	2.5
6	Beverly Hills	Condo/Twh	2526	3.4024334	2	2.5
7	Beverly Hills	Condo/Twh	2662	3.425208	2	2.5
8	Beverly Hills	Condo/Twh	2759	3.4407518	2	2.5
9	Beverly Hills	Condo/Twh	1856	3.268578	2	3
10	Beverly Hills	Condo/Twh	2210	3.3443923	2	3
11	Beverly Hills	Condo/Twh	2283	3.358506	2	3
12	Beverly Hills	Condo/Twh	1857	3.268812	2	3
13	Beverly Hills	Condo/Twh	2265	3.3550682	3	2.5
14	Beverly Hills	Condo/Twh	2265	3.3550682	3	2.5
15	Beverly Hills	Condo/Twh	2285	3.3588862	3	2.5
16	Beverly Hills	Condo/Twh	2659	3.4247184	3	2.5

◄ **FIGURE 13.8a** The first 16 rows of data concerning homes for sale in Beverly Hills. The log-transformed data are in the fourth column, headed log10(sqft). For example, log (1500) = 3.1760912.

One sample T statistics with data

Options

95% confidence interval results:

μ : mean of Variable

Variable	Sample Mean	Std. Err.	DF	L. Limit	U. Limit
log10(sqft)	3.5970366	0.017273584	231	3.5630028	3.6310706

◄ **FIGURE 13.8b** StatCrunch output for a 95% confidence interval for the population mean of the log-transformed house sizes. The units are log-square-feet.

We interpret the confidence interval to mean that we are 95% confident that the true population geometric mean lies between 3656 square feet and 4276 square feet.

TRY THIS! Exercise 13.11

SNAPSHOT ► The Geometric Mean

WHAT IS IT? ► A numerical summary of a distribution.

WHAT DOES IT DO? ► Provides a measure of center for some right-skewed distributions.

HOW DOES IT DO IT? ► If the log transform of the data follows a Normal distribution, then inference based on the mean log-transformed data can be successfully back-transformed and interpreted in terms of the geometric mean.

HOW IS IT USED? ► To estimate a measure of center for skewed data when (a) the data have all positive values and (b) the distribution of the log-transformed data is approximately Normal.

In general, how does the 95% confidence interval for the geometric mean compare to the 95% confidence interval for the mean (of the untransformed distribution)? Sometimes, a confidence interval for the geometric mean will be narrower—and thus will provide a more precise estimate—than will an interval for the population mean. A 95% confidence interval for the mean Beverly Hills house size is given in Figure 13.9. Note that the width of the confidence interval for the geometric mean found in Example 2 is approximately 620 square feet: $4276 - 3656 = 620$. The confidence interval for the mean shown in Figure 13.9 is wider and therefore less precise: $5353 - 4364 = 989$ square feet.

► **FIGURE 13.9** StatCrunch output showing an approximate 95% confidence interval for the population mean house size. Because the distribution of data is very right-skewed, the true confidence level could differ substantially from 95%, even though the sample size is large.

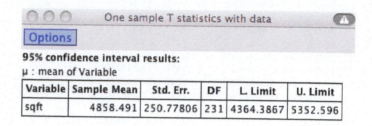

One sample T statistics with data

Options

95% confidence interval results:

μ : mean of Variable

Variable	Sample Mean	Std. Err.	DF	L. Limit	U. Limit
sqft	4858.491	250.77806	231	4364.3867	5352.596

Because the distribution of house sizes is not Normal, the confidence level for the mean could in fact be lower than 95%. For both intervals, the 95% confidence level is only an approximation, but because the nontransformed data are right-skewed and the transformed data are roughly symmetric, the approximation is much better for the geometric mean confidence interval.

Comparing Means

The confidence interval for the geometric mean (3656 to 4276) does not overlap the confidence interval for the mean (4364 to 5353). This is because the intervals are not estimating the same parameter. The geometric mean is a different creature from the "regular" mean. In fact, the geometric mean is always less than the mean. (The only exception to this occurs when every item in the data set is the same value. In that case, the geometric mean and the mean are equal.)

KEY POINT If the distribution of the log-transformed data is approximately Normal, then confidence intervals for the geometric mean can be more precise (smaller margin of error) than confidence intervals for the mean. The confidence level can also be more accurate for confidence intervals of the geometric mean.

Median or Geometric Mean?

Looking Back

Sample Median
In Chapter 3 you learned that the sample median is better than the sample mean as a measure of what is "typical" for a data set if the distribution of the data is highly skewed. The sample median is the value that has (about) half of the sorted data above and half below.

The median is another useful measure of the center of a skewed distribution. If both the geometric mean and the median are measures of the center for skewed distributions, when should you use the median and when the geometric mean? There are no hard-and-fast rules here. The geometric mean is useful only when the log transform results in a Normal (or nearly Normal) distribution. Also, finding confidence intervals for the median can be complicated and may require very particular conditions; these are left for a more advanced statistics book. On the other hand, if the distribution of the log-transformed data is *not* Normal, then the confidence intervals for the geometric mean will not be reliable, and you should rely on the median. Advanced statistics courses discuss these approaches, so if you find yourself in this situation and do not want to take additional statistics courses, it is best to consult a statistician.

Figure 13.10 shows all three measures of center on the same histogram of Beverly Hills home sizes, marked with vertical lines. The smallest measure is the median: 3573

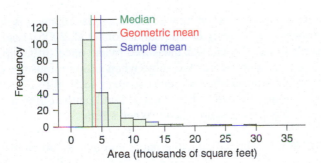

square feet. Next is the geometric mean, which, at 3954 square feet, is only slightly larger than the median. The mean is to the right of both and is substantially larger than both: 4858 square feet. (Note that, in general, the geometric mean will never be bigger than the arithmetic mean.)

SECTION 13.2

The Sign Test for Paired Data

Recall that sometimes, when we are comparing two groups, the observations in the groups are paired. Paired groups occur, for example, when we are comparing subjects' responses before and after getting a drug and thus have measured each subject twice. Also, in studies of twins where one twin is in each group, the groups are dependent. In Chapter 9, we discussed using the paired *t*-test to compare the means of dependent groups.

The **sign test** is a nonparametric test that can be used in place of the paired *t*-test. The sign test is particularly useful when the Normal condition of the paired *t*-test is not satisfied. Unlike the *t*-test, the sign test is based on the median of the population, not the mean. The sign test is similar to the paired *t*-test in that both are based on examining the differences between pairs of observations. Therefore, the first step in both tests is to subtract one observation from the other in each pair.

The sign test is useful for comparing two groups when these conditions are met:

1. The two groups are paired (sometimes called matched) so that every observation in one group is coupled with one particular observation in the other.

2. The pairs are independent of other pairs.

The sign test is a nonparametric test because it makes no assumptions about the distributions of the populations. Because it works whether or not the data come from a Normal distribution, the sign test is helpful if you do not know whether your data are Normally distributed.

Interestingly, the sign test doesn't rely on the values of the differences in the pairs. As the name suggests, it relies only on the *signs* (negative or positive).

We'll begin with a brief overview and then show how to use the sign test with real data.

Overview of the Sign Test

The paired *t*-test and the sign test share the same first step: Find the differences within each pair. The differences between the paired observations in group 1 and group 2 are found by subtracting one value from the other in each pair. Figure 13.11 shows the first

> **↻ Looking Back**
>
> **The Paired *t*-Test**
> In Section 9.5 we explained how to compare two means when the observations in each group were paired.

few rows of data from a study examining whether the sense of smell is different when we are sitting than when we're lying down (Lundstrom et al. 2006). Each row is a person, and the person's sense of smell was measured twice: once sitting up and once lying down. The *Difference* variable records the value when sitting minus the value when lying down. For the first subject, the difference score is $13.5 - 13.25 = 0.25$. Bigger values indicate a better ability to detect odors. The fact that the difference here was a positive value means this person scored better when sitting.

Stating Hypotheses The null and alternative hypotheses are based on the median value of these differences:

$$H_0: \text{The median difference} = 0.$$
$$H_a: \text{The median difference} \neq 0.$$

Because the median is the value with half of the observations above it and half below it, if we select an observation at random, then there is a 50% chance that it will be bigger than the median and a 50% chance that it will be smaller. Therefore, the null hypothesis is saying that in the population, half of the differences are positive and half are negative.

We have shown a two-sided alternative hypothesis, but the test can be modified easily enough to perform it with a one-sided alternative.

▶ **FIGURE 13.11** Measures of the sense of smell for the first 16 people in the data table. The Difference variable measures the difference "Sitting Up minus Lying Down."

StatCrunch	Edit	Data	Stat	Graphics	Help
Row	**Sitting Up**	**Lying Down**	**Sex**		**Difference**
1	13.5	13.25	woman		0.25
2	13.5	13	woman		0.5
3	12.75	11.5	woman		1.25
4	12.5	12.5	man		0
5	9.25	10	man		−0.75
6	12.5	13	woman		−0.5
7	14	11	woman		3
8	5	2.75	man		2.25
9	12.75	11.5	woman		1.25
10	13.75	13.5	woman		0.25
11	8.75	8.75	man		0
12	12.75	13.75	man		−1
13	9	10.75	man		−1.75
14	10	3.75	woman		6.25
15	7.5	9	man		−1.5
16	12.25	9.75	woman		2.5

Calculating the Test Statistic The test statistic is quite simple: Count the number of positive signs in the difference scores. Ignore any differences of 0. We will call this test statistic S (for "sign"). For example, Figure 13.11 shows nine positive signs, five negative signs, and two values of 0 for the variable *Difference*. So we find that $S = 9$. You also need to know the number of nonzero observations; we call this number n. Here $n = 14$.

If the null hypothesis is correct and the median is 0, then half of the values should be above 0 and half below. In other words, we should see about half of the pairs, $n/2$, with positive signs. If we see too many positive signs, or too few, then we suspect that the null hypothesis is not true.

Finding the p-Value After finding S, we need to know whether the observed value of S is unusually large or unusually small. The p-value is then the probability of getting a value of S as far or farther from $n/2$ as the observed value is, assuming the true median is 0.

Details

A Positive Spin
There is nothing special about counting positive signs. You could also base the test on the number of negative signs.

We can find an approximation of this probability through a simulation. If the null hypothesis is correct that the true median is 0, then about half of the observations will be bigger than 0, and the probability that a randomly selected observation is positive is 0.50. Thus we can simulate probabilities by flipping a coin. If the coin lands "heads," we record a positive sign. We flip the coin once for each of the n observations and count the number of heads. If we do this many times, say, 1000, then we can get a good sense of the probability of getting a value as extreme as or more extreme than our actual outcome when the null hypothesis is true.

As you learned in Section 6.3, we don't need to do a simulation. The probability distribution for the number of heads in n flips of a coin is modeled by the binomial model. We can use the binomial model to find the exact p-value.

Applying the Sign Test

Let's look at some examples to illustrate how to use the sign test.

EXAMPLE 3 Testing Sense of Smell

Does our sense of smell vary on the basis of whether we are sitting or lying down? Thirty-six subjects had their sense of smell measured in both positions (which position was measured first was determined randomly). Three of the subjects provided differences of 0, so these observations will not be used. The StatCrunch output for a sign test is provided in Figure 13.12.

Looking Back

The Binomial Model
The binomial model (Section 6.3) provides probabilities for random experiments in which you are counting the number of successes and these four characteristics hold:
1. There is a fixed number of trials, n (the number of pairs for which the difference is not 0).
2. The only two possible outcomes are success and failure (positive sign and negative sign).
3. The probability of success p at each trial is the same for all trials ($p = 0.50$ for the sign test).
4. The trials are independent (which means that the pairs are independent of each other).

Tech

◄ **FIGURE 13.12** StatCrunch output for a sign test of the data on sense of smell.

StatCrunch	Edit	Data	Stat	Graphics	Help			
Row	Sitting Up	Lying Down	Sex	Difference	var5	var6	var7	
1	13.5	13.25	woman	0.25				
2	13.5	13	woman	0.5				
3	12.75	11.5	woman	1.25				
4	12.5	12.5	man	0				
5	9.25	10	man	-0.75				
6	12.5	13	woman	-0.5				
7	14							
8	5							
9	12.75							
10	13.75							
11	8.75							
12	12.75							
13	9							
14	10							
15	7.5							
16	12.25							

Sign Test

Options

Hypothesis test results:
Parameter : median of Variable
H_0 : Parameter = 0
H_A : Parameter ≠ 0

Variable	n	n for test	Sample Median	Below	Equal	Above	P-value
Difference	36	33		10	3	23	0.0351

QUESTION   Is one's sense of smell different when one is lying down than when one is sitting? Use the provided StatCrunch output to perform a sign test. Follow the four-step procedure for hypothesis tests.

SOLUTION   Most of the heavy lifting was done by the software. However, it is your responsibility to make sure the test was appropriate for these data and to interpret the output in a meaningful context.

Step 1: Hypothesize (Ask questions)

H_0: The median difference in smelling ability is 0.
H_a: The median difference in smelling ability is not 0.

Step 2: Prepare (Consider data)

The data clearly are paired, because each subject was measured twice. We assume that each pair is independent of every other.

Step 3: Compute to compare (Analyze data)

We'll test these hypotheses using a 5% significance level.

The value of the test statistic is $S = 23$. This is found in StatCrunch under the Above column. "Above" means the number of observations that were above, or greater than, the null hypothesis value of 0. The sample size n is 33. (There were 36 observations, but three had differences of 0 and so were discarded.)

If the null hypothesis is true, then about half, or 16 to 17, of the observations should be positive. We instead saw 23. Is this unusual?

The reported p-value is 0.0351. This is a two-tailed p-value, as we can see in the output in Figure 13.12, where the alternative hypothesis is given.

Figure 13.13 shows the sampling distribution for S. This is the binomial distribution with $n = 33$ and $p = 0.5$. The probability that S will be "as extreme as or more extreme than" 23 means it will be as far as or farther above the mean value of $33/2 = 16.5$ than 23 is, or as far as or farther below 16.5 than 10 is. (Why 10? Because $16.5 - 10 = 6.5$ and $23 - 16.5 = 6.5$. The value of 10 is as far away from the mean as is the observed value of 23.) These values are indicated in Figure 13.13a (the probability of getting 23 or more) and Figure 13.13b (the probability of getting 10 or fewer.) The p-value is the sum of these two probabilities: $0.01754 + 0.01754 = 0.03508$.

(a) 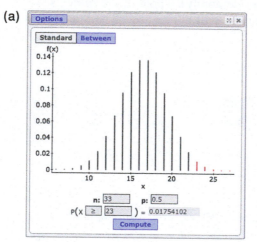 **(b)**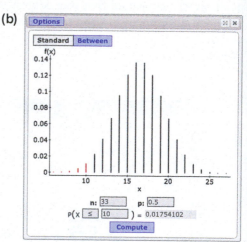

▲ **FIGURE 13.13** The sampling distribution of *S*. **(a)** The probability of getting 23 or more positive signs when the null hypothesis is true and **(b)** the probability of getting 10 or fewer positive signs. The p-value is the sum of these two probabilities. (StatCrunch)

Step 4: Interpret

Using a significance level of 5%, we reject the null hypothesis because the p-value of 0.035 is less than 0.05. We conclude that the sense of smell is in fact different when one is lying down than when one is sitting up.

Our conclusion here is based on comparing the median values. A paired *t*-test would allow us to compare the mean values. For these data, a paired *t*-test also rejects the null hypothesis and concludes that mean smelling ability is greater when sitting than when lying down.

TRY THIS! Exercise 13.15

The sign test is very useful and can even be used for one-sample tests of the median exercises. A disadvantage of the sign test is that it makes a certain type of mistake more often than the *t*-test. If the alternative hypothesis actually *is* true, and if the conditions of the *t*-test are satisfied, then the sign test is less likely to reject the null hypothesis, as it is supposed to do. If you know that the population distribution is Normal, or if your sample size is large, then the paired *t*-test is the best choice because you'll have a higher probability of correctly rejecting the null hypothesis. However, if you are not sure whether the sample size is large enough, or you know that it is not large enough, then the sign test is a more reliable test than the paired *t*-test.

SNAPSHOT ▶ The Sign Test

WHAT IS IT? ▶	A hypothesis test of whether the median difference among paired data is 0.
WHAT DOES IT DO? ▶	The sign test is based on the number of pairs with positive differences.
HOW DOES IT DO IT? ▶	If the null hypothesis is true, then about half of the pairs should be positive. If many more than half, or many fewer, are positive, it suggests that the null hypothesis is wrong.
HOW IS IT USED? ▶	It can be used whenever the paired *t*-test is used, but in particular when the sample size is too small to apply the paired *t*-test or when the distribution is known to be non-Normal.

SECTION 13.3

Mann-Whitney Test for Two Independent Groups

The **Mann-Whitney test** can be used wherever you can use a *t*-test for two independent groups (see Section 9.5). It can also be used in many situations in which you can't use the *t*-test—for example, when the population distributions are not Normal or when the sample sizes are too small for the Central Limit Theorem to provide a useful approximate result.

In order for us to use the Mann-Whitney test, the following conditions must be satisfied.

1. There are two independent groups.

2. The response variable is numerical and continuous.

3. Each group is a random sample from some population.

4. The observations are independent of one another.

5. The population distributions of the groups have the same shape.

In practice, the Mann-Whitney test also works for noncontinuous (that is, discrete) response variables, as long as the values can be ordered. Thus, the Mann-Whitney test is often applied to ranks—such as first place, second place, and so on—in competitions.

Even though the Mann-Whitney test requires that the two population distributions have the same shape, we don't care what this shape is. In other words, we don't care whether both groups come from a Normal distribution, a uniform distribution, or *any* particular distribution. We only care that they both come from a distribution with the same shape.

> **⚲ Details**
>
> **Mann-Whitney Test**
> The Mann-Whitney test is also called the Mann-Whitney-Wilcoxon test and the Wilcoxon Rank Sum test.

Overview of the Mann-Whitney Test

The Mann-Whitney test requires some pre-processing before you begin. Rather than using the actual data values, the Mann-Whitney test is based on the **ranks** of the values. This means that the smallest value in the data set gets the rank of 1, the next smallest is ranked 2, and so on.

Stating Hypotheses The Mann-Whitney test is another approach for asking the question, Do two groups differ? For example, do critics tend to rate black-and-white movies differently than they do movies in color? In previous chapters we answered questions such as these by comparing means, for example with a two-sample *t*-test to compare the mean critical rating of black-and-white movies with color movies. The Mann-Whitney test instead compares medians. This is particularly compelling when comparing skewed distributions; in this context, the median is a more natural measure of center than the mean is.

H_0: The median of population 1 is equal to the median of population 2.
H_a: The medians are not equal.

We can also have one-sided alternatives.

> **KEY POINT**   The Mann-Whitney test is based on the ranks of the observations, not on their actual values.

If the medians of the populations are equal, then each group should have a mix of low and high ranks. If the null hypothesis is not true, then you'd expect most of the low-rank values in one group and the high-rank values in the other.

Finding the Test Statistic

The first step is to compute the ranks for each observation. When doing this, *you ignore which group the observations belong to*. To illustrate, let's compare the critics' ratings for movies in black-and-white with movies in color, using the very small data set shown in Table 13.2. (In fact, the Mann-Whitney test does not work well with such very small data sets, but the small size makes it easier to understand how the test works.) Why would we ask this question? Many film lovers believe that black-and-white movies have a special beauty about them that color movies just can't capture. And perhaps film critics have a special love for this older, classic look. And perhaps movies made in the days of black-and-white were just better (or maybe they were worse).

The full data set consists of all movies since 1960, and was compiled, by statistician James Molyneux, from several different Internet sites that record information about movies. The ratings are on a scale of 0 to 100, and the numerical value represents a summary of all available ratings by professional movie critics. The higher the score, the more favorable the reviews. Table 13.2 shows a very small random sample from these data. The values in this data set are not continuous, but it turns out that they still provide for a good illustration of how to use the Mann-Whitney test. So let's examine the question: Do the critics' ratings for black-and-white movies differ from their ratings of color movies?

▶ **TABLE 13.2** Six movies with their critical rating score.

Title	Critics Rating	Color	Year
Undisputed	48	B&W	2002
Judgement at Nuremberg	90	B&W	1961
Dazed and Confused	94	Color	1993
The Sitter	21	Color	2011
The Adjustment Bureau	72	Color	2011
Happy Gilmore	60	Color	1996

The easiest way to rank the observations is to sort them from smallest to largest, ignoring whether the rating belongs to a black-and-white or color movie, and then assign the first the rank of 1, the second the rank of 2, and so on. This means that low-ranked movies correspond to movies with low critics' ratings. Table 13.3 shows the results of this ranking of the original data.

Title	Critics Rating	Color	Rank
The Sitter	21	Color	1
Undisputed	48	B&W	2
Happy Gilmore	60	Color	3
The Adjustment Bureau	72	Color	4
Judgment at Nuremberg	90	B&W	5
Dazed and Confused	94	Color	6

◀ **TABLE 13.3** Ranks based on critics' rating. The smallest rating is ranked 1.

The test statistic, represented by the letter W, is simply the sum of the ranks of one of the groups. Technically, it doesn't matter which group you choose, because if you know the sum of the ranks of one group, it is possible to determine the sum of the others (because the ranks must always sum to the same value). Most packages assume you used the group that produced the smallest sum, but other packages will give you results for both groups.

For example, when you add up the ranks of the movies in color, we get $W = 1 + 3 + 4 + 6 = 14$.

The intuition behind the Mann-Whitney W-statistic is this: If the best movies were those in color, then they would get the top four ranks: 3, 4, 5, and 6. If that were the case, then $W = 3 + 4 + 5 + 6 = 18$. At the other extreme, if the movies in color were the worst movies, they would get the lowest ranks: 1, 2, 3, and 4. Then $W = 1 + 2 + 3 + 4 = 10$.

However, if the null hypothesis is true, and both groups are really the same, then it will be as though the ranks were randomly assigned to groups. So we would expect each group to have a mix of low and high ranks. This means that if the null hypothesis is true, then W should be somewhere close to the midpoint between 10 and 18: about 14.

In our data, we observed $W = 14$, which is the value the null hypothesis might lead us to expect. We now ask how likely a value as or more extreme than 14 is, if the ranks were really just distributed by chance.

Finding the p-Value The distribution of W does not have a simple formula that allows us to compute probabilities, so we rely heavily on statistical software to compute p-values. For large sample sizes, and particularly when there are values that are tied (in other words, several values are the same and get the same rank), approximate probabilities are calculated by statistical software.

Figure 13.14 shows output from StatCrunch that gives the p-value as 1, indicating that the observed value of the test statistic is consistent with the null hypothesis (and so we should not reject the null). Because the sample size was not terribly large and there were no ties, an exact p-value was computed. (StatCrunch, like some other statistical packages, automatically decides whether to compute an exact p-value or an approximation.)

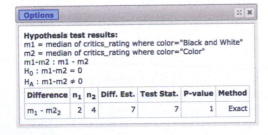

◀ **FIGURE 13.14** Output from the Mann-Whitney test on StatCrunch. $W = 7$ and the p-value $= 1$ for a two-sided alternative hypothesis.

Applying the Mann-Whitney Test

Many statisticians prefer to use the Mann-Whitney test, rather than the *t*-test, for many situations. This is because the Mann-Whitney is more robust (it doesn't need to know whether the populations are Normally distributed or whether the sample size is "large enough").

To see how to apply the Mann-Whitney test, let's consider a larger sample of movies.

EXAMPLE 4 Rating Movies

Figure 13.15 shows StatCrunch output from a Mann-Whitney test to determine whether movies in black-and-white are rated differently by critics than are movies in color. A preliminary check of the histograms (not shown) of the critics' ratings for the two types of movies indicated that both distributions were left-skewed. The data themselves consist of 150 movies randomly sampled from a large database of all movies released since 1916. Due to missing values, the effective sample size is 134.

▶ **FIGURE 13.15** StatCrunch output of a Mann-Whitney test to compare the typical critical rating of movies in black-and-white with movies in color.

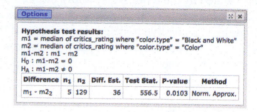

Options

Hypothesis test results:
m1 = median of critics_rating where "color.type" = "Black and White"
m2 = median of critics_rating where "color.type" = "Color"
m1-m2 : m1 - m2
H_0 : m1-m2 = 0
H_A : m1-m2 ≠ 0

Difference	n_1	n_2	Diff. Est.	Test Stat.	P-value	Method
$m_1 - m2_2$	5	129	36	556.5	0.0103	Norm. Approx.

QUESTION Do movies in black-and-white tend to get a different rating from critics than do movies in color? Perform a Mann-Whitney test, using the computer output provided. Use a 5% significance level.

SOLUTION

Step 1: Hypothesize (Ask questions)

H_0: the median critics' rating for black-and-white movies equals the median rating for color movies

H_a: the median critics' rating for black-and-white movies is not equal to the median rating for color movies.

Step 2: Prepare (Consider data)

The Mann-Whitney test is appropriate because

1. The samples are independent.

2. The variable measured (critics' rating) is numerical.

3. The movies are a random sample from a population.

4. We are told the sample distributions have the same shape and so assume the population distributions do as well.

Step 3: Compute to compare (Analyze data)

We read from the output that the test statistic is $W = 556.5$ and the p-value is 0.0103 (for a two-sided alternative).

Step 4: Interpret

Because the p-value is smaller than 0.05, we reject the null hypothesis and conclude that the median critics' rating is different for black-and-white movies and color movies.

TRY THIS! Exercise 13.23

Sample Size and the Mann-Whitney Test   The Mann-Whitney test has relatively low power. This means that even if the alternative hypothesis is true (the groups really are different), the Mann-Whitney test can be less likely than other tests to reject the null hypothesis. For example, suppose that the three highest-rated films in our initial sample of six films were all black-and-white. This would mean the black-and-white movies had ranks 4, 5 and 6 and the color movies 1, 2 and 3. In this extreme case you might think we would reject the null hypothesis, but the p-value for the two-sided alternative is a fairly high 0.10. Be aware that when the sample size is small (under ten total observations) then it may not be possible for the Mann-Whitney test to reject the null hypothesis.

For larger sample sizes, the *W*-statistic follows, approximately, a Normal distribution. For this reason, some software packages automatically produce an approximate p-value using the Normal distribution, rather than an exact p-value, because this makes the computation faster. The StatCrunch display in Figure 13.15 tells us that the method used was "Norm. Approx"; that is, StatCrunch made the decision to use the Normal approximation for this sample size.

What Can Go Wrong?   Although it is true that you don't have to worry about whether the data follow the Normal distribution, you still need to check the conditions carefully. In Section 13.2 we noted that if you do not have independent groups, then you need to use the sign test. You might also have a few other things to worry about.

Ties in the Data. In theory, when the response variable is continuous it is impossible to have two values that are exactly the same. In practice, however, data are recorded only to a certain level of precision. The movie critics' ratings, for instance, were actually averages of many critics for each movie, and so were continuous valued. But the website that reported them rounded to the nearest integer. And so in real life, ties can occur: two or more movies with the same rating.

When ties occur, it is not possible to give each observation its own rank. Instead, several observations share the same rank. One approach is to give tied observations their average rank. For example, if we saw the data

$$15.5, 16.0, 16.0, 16.0, 18.2$$

we would assign them the ranks

$$1, 3, 3, 3, 5$$

because the three values of 16.0 were assigned the ranks 2, 3, 4, and the average of these ranks is 3.

Doing this averaging leads to "conservative" p-values: p-values that tend to be too high. Most statistical software packages apply an adjustment that gives a more accurate p-value. Some software packages display a warning that says something like "adjusted for ties." If your software gives you a choice of different p-values, use the one that has been adjusted for ties.

Different-Shaped Distributions. The Mann-Whitney test requires that the shapes of the two population distributions be the same because if they are different (such as one being symmetric and the other left-skewed), then the Mann-Whitney test might reject the null hypothesis even when the medians are the same. In practice, any difference in shape, center, or spread between the two distributions can lead to rejecting the null hypothesis. Most statistical software packages treat the Mann-Whitney test as a test based on the median values, and this is how we presented it. However, this interpretation is valid only when the shapes and spreads of the distributions are the same.

SNAPSHOT ▸ The Mann-Whitney Test

WHAT IS IT? ▸	A hypothesis test to compare the centers of two groups of numerical variables.
WHAT DOES IT DO? ▸	It tests the hypothesis that the medians of two populations are different.
HOW DOES IT DO IT? ▸	The original values are pooled and ranked from smallest to largest and then sorted back into their original groups. If the null hypothesis is true, then the sums of the ranks in the two groups should be roughly equal. If the sum of one group is much larger than the other, then it suggests that the null hypothesis might not be true.
HOW IS IT USED? ▸	It can be used whenever the *t*-test for two independent samples is used, but it can also be used when the Normal condition of the *t*-test is not met.

t-Test or Mann-Whitney Test?

How do you decide whether to use the two-sample *t*-test for independent samples or the Mann-Whitney test? To some extent, it is a matter of preference. There are a few things to consider, though.

Because the Mann-Whitney test can be used to compare medians, it is preferred in situations where the distributions are skewed or outliers are present.

If you don't know whether the distributions of the populations are Normal, and are not willing to assume that they are, then the Mann-Whitney is useful. If your sample sizes are large, then the Mann-Whitney and two-sample *t*-tests should lead you to the same conclusion. But when the sample sizes are too small for the Central Limit Theorem to apply (less than 25 in most cases, larger if the distributions are severely skewed), then the Mann-Whitney test will produce more reliable results.

Even though one condition of the tests is that the data be continuous, the Mann-Whitney test also works when the data are ordinal (can be ordered) but not necessarily continuous. The most common example is when the data are ranks to begin with. This is often the situation when data from contests are examined. For example, a panel of judges might rank 12 wines using the numbers 1 (best) to 12 (worst), and we might want to compare two judges to see whether they tend to give the same ranks. The *t*-test would not be good for these data, which are very non-Normal. The Mann-Whitney would be a more suitable test.

Many statisticians feel that for most situations, the Mann-Whitney test can be used instead of the two-sample *t*-test.

SECTION 13.4

Randomization Tests

 Looking Back

Fisher's Exact Test
Fisher's Exact Test is an example of a randomization test for categorical variables. In this chapter, we focus on numerical variables.

Randomization tests include a wide variety of different types of tests in different types of situations. What these tests have in common is a general strategy for answering the question, "Can this be due to chance?" The strategy is to use a computer (or some other simulation-based procedure) to shuffle observations together in order to simulate a world in which chance is, in fact, the only reason for differences between groups. Fisher's Exact Test, which was introduced in Chapter 10, is a randomization test used to test for associations between two categorical variables. In this section, we examine randomization tests for comparing two groups of independent, numerical observations.

Randomization tests can be used whenever the two-sample *t*-test or Mann-Whitney test can be used. The advantages of randomization tests over *t*-tests and the Mann-Whitney test are that they are more versatile and that they can be used to compare statistics other than the mean and median. For example, randomization tests are also useful for comparing proportions. Randomization tests have three requirements:

1. The two groups are independent of each other, and the observations within each group are independent of other observations within the group.

2. Either the data are a random sample from some population, or the observations were assigned to groups randomly, as in a randomized controlled experiment.

3. The variability of both groups is approximately the same.

> **Details**
>
> **Permutations**
> Another term for the randomization tests we are presenting is *permutation tests*.

Overview of Randomization Tests

One way to understand randomization tests is to think of them as comparing two worlds, or two alternative realities, if you prefer. One world is the real world. This is the world that produces the data we see. The other world we'll call Chance World. We create Chance World by simulation, and we do it so that every outcome depends only on chance.

Consider a well-known study on the effects of cloud seeding (Simpson et al. 1975). Cloud seeding is the practice of using airplanes to drop silver nitrate into clouds, with the intent of producing rain. In this study, 26 clouds were randomly chosen to be seeded with silver nitrate. Another 26 clouds were selected to receive a "placebo." The pilot flying the plane did not know which was which; he or she simply acted as always, not knowing what was released into the cloud. The amount of rainfall in each area was recorded in acre-feet. The boxplots in Figure 13.16 show the results.

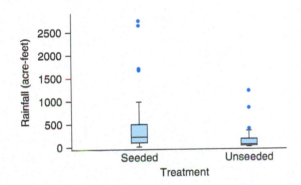

◄ **FIGURE 13.16** Rainfall amounts (in acre-feet) produced by 26 clouds seeded with silver nitrate (left) and 26 clouds seeded with a placebo (right).

Figure 13.16 shows the *real-world* results, and we summarize these real-world results in Table 13.4. We can see that the median rainfall amounts in the two groups are different. But can this difference be due to chance? Clouds were chosen at random, and perhaps if a different subset of the clouds had been chosen for the treatment, rainfalls would have turned out differently.

Group	n	Mean	SD	Median	IQR
Seeded	26	533.2	779.9	221.6	396.7
Unseeded	26	164.6	278.4	44.2	138.6

◄ **TABLE 13.4** Summary statistics for rainfall from clouds, comparing seeded and unseeded clouds.

To create Chance World, imagine a collection of slips of paper. Each slip represents a cloud. On each slip we write a rainfall amount from our actual data. This represents the rainfall from that cloud. Now imagine two stacks. One stack represents the seeded group of clouds, and the other represents the unseeded group. Shuffle all the slips of paper (from both stacks) together very well, and randomly deal them into the two new stacks so that each stack has 26 slips of paper. For each stack, find the median rainfall, and then calculate the difference between the medians. This is the Chance World difference.

In the real world, the differences between the two groups might have been caused by the seeding. But in Chance World, the only differences are due to chance, caused by our shuffling of the slips of paper. If we repeat this shuffling and dealing, we get another, different outcome. If we do this many times, we obtain an understanding of what the difference in medians looks like in a world in which everything is due to chance.

We now compare our real-world result to the many simulated results from Chance World. If we can't tell the difference between the real world and Chance World—in other words, if the real-world outcome is also a common outcome in Chance World—then we conclude that there is no evidence of a real difference; there is no evidence that cloud seeding results in a change in rainfall.

If, on the other hand, the real-world result is unusual in Chance World, then perhaps the outcome was not due to chance. We can use the simulated Chance World outcomes to compute the probability of getting, in our Chance World, a result as extreme as or more extreme than the real-world result. If the probability is small, less than our chosen significance level, we reject the null hypothesis.

Figure 13.17 illustrates this situation. The distribution shows the Chance World outcomes under 1000 shuffles for the cloud-seeding data. After each shuffle, the difference in median rainfall is computed, so the histogram shows us many possible differences in Chance World. If the real-world outcome were near where the green line indicates, then we would conclude that the real world looks a lot like Chance World, and we would not reject the null hypothesis. If, on the other hand, the real-world outcome fell where the red line indicates, then we would have to admit that our real-world outcome looks nothing at all like the sort of outcome seen in Chance World. We would reject the null hypothesis.

▶ **FIGURE 13.17** Histogram of a Chance World simulation. The green line represents one possible real-world outcome for which we would *not* reject the null hypothesis. The red line represents one possible real-world outcome for which we *would* reject the null hypothesis.

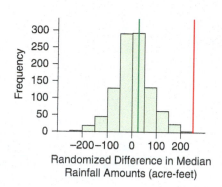

Of course, not all situations are as clear-cut as the red-line situation. Soon we will discuss how to find the p-value for a particular situation so that we can decide whether to reject the null hypothesis.

KEY POINT   Chance World is a simulated world in which every outcome is determined solely by chance. In Chance World, the null hypothesis is true. The real world may or may not be the same as Chance World. It is your job to decide whether the data in the real world look as though they could have been generated in Chance World.

The most unusual feature of the randomization test is that you get to choose which statistics you want to use to compare the two groups. You can choose either the means or the medians. Or you can choose the third quartiles. Anything you want, really. The trick is to choose a statistic that will help you answer the question that has been asked. We'll call this "Step 0" in our list of hypothesis testing steps. This is a step that exists only for these randomization tests.

Because we are comparing the "typical" rainfall, we should probably choose to look at either the difference in mean rainfall amounts or the difference in median rainfall amounts. The boxplots shown in Figure 13.16 on page 659 show us that the distributions of rainfalls are skewed, so the median might be a better choice. For this reason, our test statistic will be the median rainfall of the seeded clouds minus the median rainfall of the unseeded clouds.

Stating Hypotheses Because we've chosen the difference in medians to compare the groups, our hypotheses look like this:

H_0: The median rainfall of the seeded clouds is equal to the median rainfall of the
    unseeded clouds.
H_a: The median rainfall of the seeded clouds is greater than the median rainfall of
    the unseeded clouds.

If you choose different statistics, you will need to reword the hypotheses accordingly.

Finding the p-Value To find the p-value, we need to find the sampling distribution of the median difference when the null hypothesis is true. We obtain this through simulation. We don't have time to write down the rainfall amounts on slips of paper, shuffle, deal into two stacks, compute, and repeat this thousands of times; we can have a computer do this. (This is easier with some software packages than with others. For guidance, see Example E on page 688.)

Each time the computer shuffles the values and then deals them into two groups, it computes the difference in median values. At the end of many repetitions, we can create a histogram of these median differences. This histogram will help us to see whether the real-world difference is unusual. We can use this histogram to find approximate p-values.

Applying Randomization Tests

Let's carry out a randomization test to see if the cloud seeding increased rainfall. The null hypothesis is that the median rainfall is the same in both seeded and unseeded clouds. The alternative hypothesis is that the rainfall is greater in the seeded clouds.

In the real world, as Table 13.5 shows us, we saw a difference in medians of

$$221.6 - 44.2 = 177.4 \text{ acre-feet}$$

Is this unusually large? Or is this the sort of value that happens just through chance alone?

Table 13.5 shows a few outcomes from our Chance World simulation. Each row of this table represents shuffling and dealing out the slips of paper (but using the computer to do it). We actually did 1000 simulations, and we show the first four. This table shows us that large differences in typical rainfall can sometimes occur simply by chance.

Median Rainfall "Seeded"	Median Rainfall "Unseeded"	Difference
57.90	243.40	−185.50
93.70	155.40	−61.70
133.40	71.15	62.25
101.14	118.65	−17.51

◀ **TABLE 13.5** The first four outcomes from a Chance World simulation of rainfall amounts.

Figure 13.18 shows the histogram of all 1000 simulations. This is the approximate sampling distribution of the difference between the medians, under the assumption that the null hypothesis is true (and therefore all differences are due to chance).

▶ **FIGURE 13.18** The distribution of the difference of medians when rainfall amounts are distributed to two groups by chance. The histogram represents 1000 randomizations. These are the sorts of rainfall differences we see in Chance World, when chance is the only mechanism that determines how much rainfall occurs.

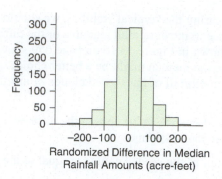

▶ **FIGURE 13.18** The distribution of the difference of medians when rainfall amounts are distributed to two groups by chance. The histogram represents 1000 randomizations. These are the sorts of rainfall differences we see in Chance World, when chance is the only mechanism that determines how much rainfall occurs.

> **! Caution**
>
> **Random p-Values**
>
> When using simulations to find p-values, the p-value will be slightly different if you repeat the test. That's why it's important to do a large number of simulations (we did 1000) so that the p-value does not change too much if you do the test again.

> **⟳ Looking Back**
>
> **Random Sampling and Randomized Assignment**
>
> Table 12.2 compared random sampling with randomized assignment. Random sampling is necessary in order for us to generalize to a larger population. Random assignment is required in order for us to make cause-and-effect conclusions.

One quick look at the sampling distribution in Figure 13.18 shows us that in Chance World, a difference of 177.4 is relatively large. It's difficult to get exact values from the histogram, but it looks as though a difference like this occurred less often than about 25 out of 1000 times. If so, then a difference as extreme as or more extreme than 177.4 occurs less than 0.025, or 2.5%, of the time. This is a rather rare outcome.

In this case, we don't need to estimate the p-value by reading off the histogram. We can ask the software to actually sum up the number of times the Chance World median difference was 177.4 or bigger. From this we learn that in 13 out of 1000 simulations, the median difference was 177.4 or greater, so the approximate p-value is 0.013.

This means that with a 5% significance level, we should conclude that this is a rare event and reject the hypothesis that the median rainfalls were the same. The observed real-world median difference is too large for us to believe it was due to chance.

Our sample of clouds was not random, so we could not extend our conclusions to all clouds. But because the clouds were randomly assigned to be seeded or unseeded, we can conclude causality; that is, we can infer that the cloud seeding caused the additional rainfall.

EXAMPLE 5 Slightly under Weight

The Youth Behavior Risk Study (Department of Health and Human Services, Centers for Disease Control and Prevention 2011) is a nationwide study of behaviors of people 12 to 18 years old in the United States. The purpose of the study is to assess their beliefs, attitudes, and practices concerning their health. In this example, we compare the weights of 17-year-old women who believe that they are "slightly under" the proper weight with the weights of those who believe they are "about right" in weight. How do beliefs match reality? Do those who believe they are "slightly under" actually weigh less, typically, than those who feel they are "about right"?

Figure 13.19 shows the approximate sampling distribution of the difference in mean weights. This distribution was achieved by shuffling the weights (in kilograms) and dealing them into two piles, recording the difference in means, and repeating 1000 times. Table 13.6 shows summary statistics for these two groups (55 kilograms is about 121 pounds).

Tech

▶ **FIGURE 13.19** The approximate sampling distribution, under the condition that only chance determines any differences between groups, of difference in mean weight between 17-year-olds who reported that they are "about right" in weight and those who reported being "slightly under" weight.

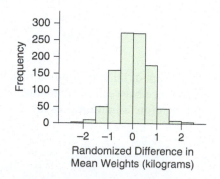

Group	n	Mean	Median	Standard Deviation	IQR
About right	985	57.65	56.70	7.90	9.08
Slightly under	166	51.79	49.90	8.07	7.71

◄ **TABLE 13.6** Summary statistics for weights (in kilograms) of 17-year-old girls who consider themselves "about right" in weight and those who consider themselves "slightly under" the proper weight.

QUESTION   Using the outputs in Figure 13.19 and Table 13.6, carry out a randomization test to determine whether differences in *mean* weight are due to chance. What is your conclusion? (Note that the p-value may have to be roughly estimated from the histogram.)

SOLUTION

Step 0: Choose a test statistic
We will use the difference in mean weight: the mean for the women responding "about right" minus the mean for the women responding "slightly under."

Step 1: Hypothesize
The hypotheses are the same as before:

H_0: The typical weight (measured by the mean) of the "about right" women is the same as that of the "slightly under" women.

H_a: The typical weight of the "about right" women is higher than that of the "slightly under" women.

Step 2: Prepare
The conditions are satisfied for the same reasons as given before. We should check, though, that the standard deviations and interquartile ranges (IQRs) are approximately the same for the two groups.

Step 3: Compute to compare
From Table 13.6, the observed value of the test statistic is 57.65–51.79 = 5.86 kilograms.

We compare this to the approximate sampling distribution summarized by the histogram in Figure 13.19. If chance is the only cause for differences, how often do we get values as large as, or larger than, 5.86? We see that we never had a value of 5.86 or larger. The largest difference in the simulations is just over 2 kg. The p-value is approximately 0.

Step 4: Interpret
We reject the null hypothesis and conclude that the typical weight of the "about right" women is higher than that of the "slightly under" women, which suggests that their perceptions match reality, at least on average. The real-world outcome difference of 5.86 does not appear to be consistent with the Chance World outcome.

TRY THIS!   Exercise 13.31

SNAPSHOT ▸ Randomization Tests

WHAT ARE THEY? ▸	A strategy for comparing two independent groups. A chance model is created by randomly assigning the observed values to two groups, computing the test statistic, and repeating this procedure many times.
WHAT DO THEY DO? ▸	Randomization tests can be used to compare means, medians, or other statistics between two groups. Their use does not require any particular shape for the population distributions, so they can be used in a variety of situations.
HOW DO THEY DO IT? ▸	By repeating the randomization simulation many times, we can estimate the p-value, because we can see the distribution of the test statistic when the null hypothesis is true and there is no difference between the groups.
HOW ARE THEY USED? ▸	Randomization tests can be used in place of *t*-tests for two independent samples, but they are much more versatile, because they can also be used with statistics other than means and in situations where population distributions are not known.

Randomization tests can be particularly useful for randomized experiments, as in the cloud-seeding example. You probably suspect that these clouds were not a random sample of all clouds in the world, and you are correct. Note that condition 2 does not require that the data be a random sample from the population, and freedom from this restriction is useful for many real-life randomized studies. Most medical studies, for example, in which subjects are randomly assigned to treatment or control groups, do not use a random sample of subjects. It is too difficult to get a random sample of patients from a large population to visit a strange doctor to partake in a study. Instead, medical researchers recruit subjects from among those available. As long as the subjects are randomly assigned to treatment groups, researchers can still make inferences about whether the treatment under investigation caused any observed changes.

CASE STUDY REVISITED

Contagious Yawns

Are yawns contagious? The folks on the television show *Myth Busters* conducted an experiment in which some people were given a "yawn stimulus" and others were not. The table shows the results.

	Stimulus	Control	Total
Yawn	10	4	14
No Yawn	24	12	36
Total	34	16	50

For a statistic, let's focus on the difference between the proportion who yawned in the Stimulus group, 0.29 (or 10/34), and the proportion who yawned in the Control group, 0.25 (or 4/16). This difference is $0.29 - 0.25 = 0.04$. This is the observed, real-world value of our test statistic.

The null hypothesis is that the stimulus has no effect. If this is true, then if we did an experiment like this many times, we would get about the same proportion of yawns in the Stimulus group as in the Control group, so the difference in proportions would be about 0. We wouldn't get a difference of exactly 0 every time, but these nonzero outcomes would be due to chance, not because yawns are contagious. This is "Chance World," where differences are due only to chance.

Under the alternative hypothesis, we will see a greater proportion of yawns in the Stimulus group, so the difference in proportions should be positive.

The randomization tests in this chapter focused on numerical outcomes, but we can also apply them to categorical outcomes such as this. We'll use a significance level of 5%. To carry out this test, we imagine that we have a deck of 50 blank cards—one for each participant. On 14 of the cards, we write "Yawn." On the rest, we write "No Yawn." We then shuffle the cards very thoroughly.

Next, we ask our 50 participants to form two lines. Obediently, 34 of them go in the line that we will call "Stimulus," and the rest, 16, go in the line we will call "Control." As each participant steps forward, she or he draws one of our cards (without looking). Thus we have created a Chance World situation in which the proportion of people who yawn in each group is due only to chance. When all have chosen their cards, we find the proportion of those in the "Stimulus" group who have a yawn card, and from this we subtract the proportion of those in the "Control" group who have a yawn card.

If we repeat this many times, we will see whether a value such as 0.04 is, or is not, unusual as a chance outcome. Of course, real people will not let us do this 1000 times or more, so we will rely on a computer. Figure 13.20 shows a histogram of the difference in proportions for 1000 such trials where only chance determined whether or not "a person" "yawned."

From the histogram, we see that an outcome of 0.04 is not very unusual under the null hypothesis. In fact, it is close to the typical outcome of 0.

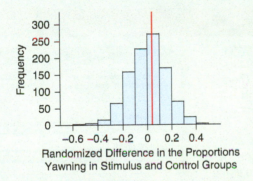

◄ FIGURE 13.20 Randomization distribution of the difference in the proportion of people yawning in a simulated "Stimulus" group and in a simulated "Control" group. The actual observed difference is 0.04. We can see that this is not an unusual outcome in Chance World. Test statistics as extreme as, or more extreme than, 0.04 happen about half the time.

We can approximate the p-value by the proportion of chance outcomes that are equal to or greater than 0.04. This turns out to be 0.53, which is larger than our significance level of 0.05. We therefore do not reject the null hypothesis, and we conclude that there's not enough evidence to say whether yawns are contagious.

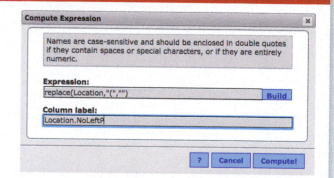

DATAPROJECT ▶ Making Maps and Slicing Strings

1 OVERVIEW

We take a short break from discussing strategies for large data sets so that we can discuss one of the interesting variables in the crime data set: *Location*. (See *sampleofcrime.csv*.) This is an interesting variable because, in some sense, it is categorical (because it represents a place), but in another sense it is numerical, because it consists of a latitude and longitude. In this Data Project we'll discuss some of StatCrunch's mapping tools and how to use the Compute Expressions menu to clean the *Location* variable.

2 GOALS

Learn to apply the slice function to a character string, learn to slice columns, and map locations identified by latitude and longitude.

3 YOU ARE HERE

Some variables have values that consist of "character strings," so-called because they consist of characters laid one after the other — a string of characters. There are several examples in this data set, and the one we wish to consider is the *Location* variable in the *sampleofcrime* data set.

We'd like to make a map of the locations at which crimes occurred. But to do this, StatCrunch needs the latitude and longitude in separate columns. There's another complication: the *Location* variable includes left and right parentheses, and we don't need these at all. Once we get the latitude and longitude in separate columns, we'll be able to map the crimes!

A quick refresher on geography. Latitude and longitude allow us to identify any point on the earth. Latitude is like the *x*-axis of a scatterplot and longitude is like the *y*-axis. They are not exactly the same thing because the earth is curved and a scatterplot is flat.

Project Our strategy for creating latitude and longitude variables from the *Location* variable has two steps. Step 1: delete the parentheses using the replace function. Step 2: Slice the Location variable into two columns.

To delete the parentheses, choose Data > Compute > Expression and type the expression exactly as it appears in the following figure.

This command might look nonsensical. Here's how to make sense out of it. The *replace* function wants three pieces of input: (1) the variable to act on, (2) the character to be replaced, and (3) the character to replace it with.

The variable to act on is *Location*. The character we wish to replace is a left-parenthesis, indicated here within quotation marks: "(". We wish to replace it with nothing, and so we place to quotation marks with no spaces or characters between them: ""

The results are saved in a new variable named *Location.NoLeftP*. (Feel free to choose another name, if you wish.)

Now, apply the *replace* function on *Location .NoLeftP* to create a new variable named *Location.NoP* that has no parentheses at all.

The second step is quite straightforward. Select Data > Arrange > Slice. You need to tell StatCrunch which column to slice (*Location.NoP*), where to do the slicing (at the comma), and where to put the results (into new columns named *Latitude* and *Longitude.*). The following figure shows how.

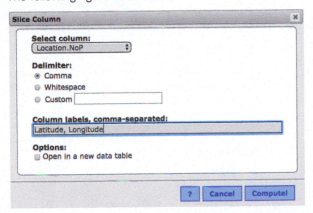

The data are nearly ready. Before making a map, however, you should do a reality check for these new variables. Are there outliers? Any values that don't make sense? (Hint: Yes, there are. You should exclude them from the map because they are crimes for which no location information was available.)

Assignment Use the Graph > Map > U.S. Location option to create a map of where the crimes occurred. A very useful option is to put in values for upper/lower latitude and longitude in the "zoom" fields. You can compute summary statistics to determine which values to use.

Note that if you wish to use Google maps instead of the StatCrunch mapping tool, you'll need to create a new location variable. Start with the original *Location* variable, and remove the left and right parentheses as before. Then, replace the "," with "|" and use this new variable to indicate the location. Then select Graph > Map > Google Map.

CHAPTER REVIEW

KEY TERMS

Nonparametric inference, *640*
QQ plot, *641*
Log transform, *643*

Back transform, *645*
Geometric mean, *646*

Sign test, *649*
Mann-Whitney test, *653*

Rank, *653*
Randomization tests, *658*

LEARNING OBJECTIVES

After reading this chapter and doing the assigned homework problems, you should

- Understand why and when a log transform is appropriate for analysis.

- Know how to apply the sign test to compare populations on the basis of dependent samples of data, and understand how the sign test compares to the paired *t*-test.

- Be able to apply the Mann-Whitney test to compare two populations on the basis of two independent groups of data, and understand how the Mann-Whitney test compares to the two-sample *t*-test.

- Know when and how to perform a randomization test to compare medians or means of two populations, and understand how this procedure compares to the two-sample *t*-test.

SUMMARY

The tests you have seen in this chapter—including tests on transformed data, the sign test, the Mann-Whitney test, and randomization tests—are often used when assumptions about the distribution of the population are not met or when you are unwilling to make assumptions about the shape of the sampling distribution. For example, the distributions of your sample may be strongly skewed (with a small sample size) so that *t*-tests are not appropriate, or the data may consist of ranks.

If all the values of the variable are greater than 0, if the distribution of your data is right-skewed, and if a log-transformation results in a distribution that is approximately Normal, then you can base your analysis on the geometric mean. Simply find a confidence interval for the mean of the log-transformed data, and then back transform.

When comparing two groups, if you're not willing to make assumptions about the shape of the distributions, then you can use the sign test (for paired data), the Mann-Whitney test (for two independent groups), or a randomization test.

In many situations, statisticians prefer the Mann-Whitney test over the two-sample *t*-test. A randomization test can also be used, particularly if you want to use a statistic other than the mean or median to compare the groups.

Parametric Test	Nonparametric Equivalent
Paired *t*-test	Sign test
Two-sample *t*-test	Mann-Whitney test, randomization test

SOURCES

Buchanan, L., et al. 2017. Exposure to digital marketing enhances young adults' interest in energy drinks: An exploratory investigation. *PLoS One.* https://doi.org/10.1371/journal.pone.0171226.

Department of Health and Human Services, Centers for Disease Control. 2011. Youth Risk Behavior Study. www.cdc.gov/Healthy Youth/yrbs/index.htm (viewed March 24, 2014).

Lundstrom, J., J. Boyle, and M. Jones-Gotman. 2006. Sit up and smell the roses better: Olfactory sensitivity to phenyl ethyl alcohol is dependent on body position. *Chemical Senses* 31, 249–252.

Simpson, J., Olsen, A., Eden, J.C. 1975. A Bayesian analysis of a multiplicative treatment effect in weather modification. *Technometrics* 17, 161–166. (Used in J. M. Chambers et al. 1983. *Graphical methods for*

data analysis, lib.stat.cmu.edu/DASL/Datafiles/Clouds.html [viewed April 25, 2010]).

Schwitzgebel, R. 1964. *Streetcorner research: An experimental approach to the juvenile delinquent.* Cambridge, MA: Harvard University Press.

Scwitzgebel, E., and J. Rust. 2009. The Self-reported moral behavior of ethics professors. Unpublished manuscript. http://schwitzsplinters.blogspot.com.

Toptaş, B. 2016. Effect of daily work on student's memorization ability in piano education. *Educational Research and Review* 11(7), 371–376.

Trumbo, B. 2001. *Learning statistics with real data.* North Scituate, MA: Duxbury Press.

SECTION EXERCISES

SECTION 13.1

13.1 What is the fundamental condition required for inference from a sample to a population?

13.2 In what situations are nonparametric statistics useful?

13.3 In addition to random samples, what other conditions are required for using the two-sample *t*-test?

13.4 What summary statistics are best used to report the "typical" value of a data set when the distribution is strongly skewed?

13.5 QQ Plot Matching (Example 1) Refer to the following two histograms and two QQ plots of the same data.

(A)
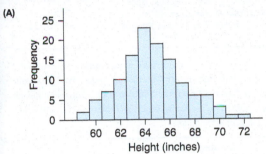

(B)
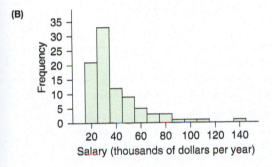

(C)
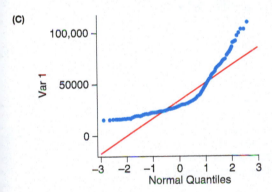

(D)
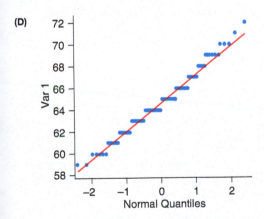

a. Match each of the histograms with the corresponding QQ plot.

Histogram A goes with QQ plot _____.

Histogram B goes with QQ plot _____.

b. Describe the shape of the histograms.

c. For which sample might a log transform be useful? (There are no zeros or negative values in either data set.)

13.6 QQ Plot Matching Refer to the following two histograms and two QQ plots of the same data.

(A)
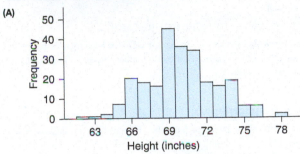

(B)
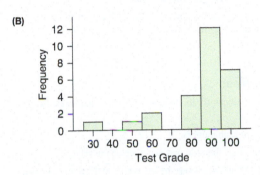

(C)
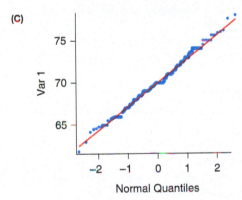

(D)
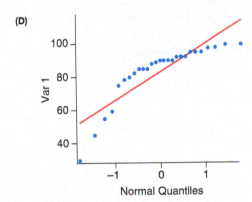

a. Match each of the histograms with the corresponding QQ plot.

Histogram A goes with QQ plot _____.

Histogram B goes with QQ plot _____.

b. Describe the shape of the histograms.

c. Is a log transform likely to be useful for either sample?

13.7 Log Transforms

a. Find the log (base 10) of each number. Round off to one decimal place as needed.

10, 100, 1000, 6500

b. The following numbers are in log units. Do the back transformation by finding the antilog (base 10) of these numbers. Round off to one decimal place as needed.

3, 5, 2.4, 3.2

13.8 Log Transforms

a. Find the log (base 10) of each number. Round off to one decimal place as needed.

10, 10000, 1500, 5

b. The following numbers are in log units. Do the back transformation by finding the antilog (base 10) of these numbers. Round off to one decimal place as needed.

2, 3, 1.5, 2.4

13.9 Geometric Mean

a. Find the geometric mean for the numbers 10, 1000, and 10000 by using the following steps:

   i. Find the log of each number.

   ii. Average the 3 logs found in part a and report the value.

   iii. Find the antilog of the average by raising 10 to the power obtained in part ii. The result is the geometric mean. Round it to the one decimal place as needed.

b. Find the mean and the median of the original 3 numbers. Then write the values for the geometric mean, the mean, and the median from smallest to largest.

13.10 Geometric Mean

a. Find the geometric mean for the numbers 125, 260, 1000, and 15000 by using the following steps:

   i. Report the log of each number. Round to 1 decimal place as needed.

   ii. Average the 3 logs found in part a.

   iii. Find the antilog of the average by raising 10 to the power obtained in part ii. The result is the geometric mean.

b. Find the mean and median of the original 3 numbers. Then write the values for the geometric mean, the mean, and the median from smallest to largest.

TRY **13.11 Credit Card Balances (Example 2)** A consumer wanted to estimate her average monthly credit card debt. She took a random sample of monthly credit card statements and recorded the total monthly balance. A histogram of the balances is shown:

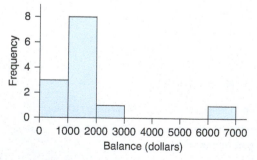

a. A 95% confidence interval for the mean credit card balance is shown. Interpret this confidence interval.

```
One sample T confidence interval:
μ : Mean of variable

95% confidence interval results:
       Variable  Sample Mean   Std.Err  DF   L.Limit    U.Limit
Balance (dollar)   1666.7585  401.54564  12  791.86568  2541.6512
```

b. A log transformation of the data shows a more Normally distributed data set. A 95% confidence interval for the log (base 10) of the credit card balances is shown. Convert the limits back into dollars by raising 10 to the power given as the limit. Report and interpret the confidence interval for the geometric mean.

```
One Sample T confidence interval:
μ : Mean of variable

95% confidence interval results:
Variable       Sample Mean    Std.Err.   DF   L.Limit    U.Limit
Log Balance      3.1323077  0.073355071  12  2.9724807  3.2921347
```

c. Which interval is narrower?

d. Which interval would you use if you wanted to report on the mean consumer credit card balance for all months? Explain.

13.12 Morning Routine
A statistics student conducted a survey to determine how much time students at her school spent getting ready to leave the house after they got up in the morning. Figure A shows a histogram of the times for men. Assume that we have a random sample of 20 college men.

(A)

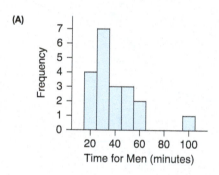

a. Figure B shows a 95% confidence interval for the mean time. Interpret the confidence interval, and explain why it might not be accurate.

(B)

```
One-Sample T: TimeMen

Variable   N   Mean   StDev   SE Mean      95% CI
Men        20  37.25  19.77     4.42   (28.00, 46.50)
```

b. After a log transform of the times is taken, a histogram of the log of the data suggests that the distribution of the transformed data is Normal. Figure C shows a 95% confidence interval for the log (base 10) of the times. Convert the boundaries back into minutes by raising 10 to the powers given as the end points of the confidence interval. Interpret the confidence interval for the geometric mean.

(C)

```
One-Sample T: LogMen

Variable    N   Mean    StDev   SE Mean      95% CI
LogMen      20  1.5223  0.2073   0.0464  (1.4253, 1.6193)
```

c. Which interval is narrower?

d. Which interval would you report if the goal was to understand the typical amount of time spent getting ready in the morning for men at this college? Explain.

*13.13 Exercise Hours A statistics student was interested in the amount of time that community college students exercise each week. He gathered data from a random sample of students at his community college and excluded those who did not exercise (those who reported 0 hours per week); this left 45 in the sample. All values were rounded to the nearest hour.

The table shows the data. Figure A shows a histogram of the data, and Figure B shows a histogram of the log transform (base 10) of the data.

1	2	3	5	8
1	2	3	6	9
1	2	4	6	9
2	2	4	6	9
2	3	4	6	10
2	3	4	7	11
2	3	5	7	11
2	3	5	8	12
2	3	5	8	20

(A)

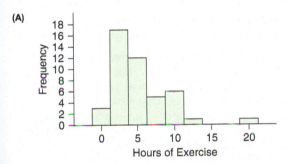

(B)

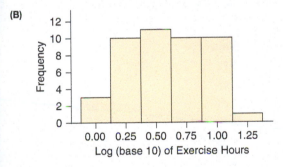

a. Describe the distribution of the untransformed sample.

b. Find a 95% confidence interval for the mean of the number of hours of exercise per week for all students at this college.

c. Describe the distribution of the transformed data, and compare it with the distribution of the original data in part a.

d. Perform a log transform on the observations. Find the boundaries for a 95% confidence interval for the mean of the log-transformed times.

e. Convert the log interval boundaries back to units of hours. Interpret the resulting interval.

f. Which interval would you report: the interval for the population mean or the interval for the population geometric mean? Explain.

*13.14 Television Viewing A Nielsen poll asked people the number of hours of television they watched in the last week. Assume that Nielsen obtained a random sample. We are analyzing the data for the 39 college students in the sample. The figure shows a dotplot of the distribution.

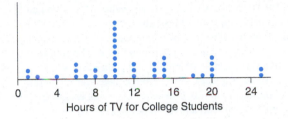

a. Find a 95% confidence interval for the mean number of hours spent watching television per week.

b. Perform a log transform on the observations, and find a 95% confidence interval for the mean of the log-transformed data.

c. Back transform the boundaries of the confidence interval into units of hours. Interpret the resulting interval.

d. Which interval is better to report and why? Consider the shape and the sample size ($n = 39$).

SECTION 13.2

TRY **13.15 Lead Exposure (Example 3)** Excessive lead levels can negatively affect brain functions; lead poisoning is particularly dangerous to children. A study was conducted to find out whether children of battery factory workers had higher levels of lead in their blood than a matched group of children. Each child in the experimental group was matched with a child in the control group of the same age, living in the same neighborhood. Although these children were not a random sample, we can test the hypothesis that the difference is too large to occur by chance if the child from the control group was randomly chosen. The figure shows a histogram of the differences in lead level.

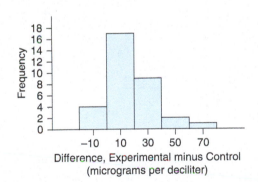

The differences were found by calculating "factory child's lead level minus matched control child's lead level." Lead levels were concentrations in the blood, measured in micrograms per deciliter. A positive difference means the child of a factory worker has a higher lead concentration than the child in the control group. The data consisted of 1 tie (the same value for the child and the matching child), 28 pairs in which the factory worker's child had a higher level, and 4 pairs in which the factory worker's child had a lower level of lead.

Carry out a sign test to determine whether children whose parents are exposed to lead at work have a higher lead level than children whose parents are not exposed to lead at work. Use a significance level of 0.05 to see whether the experimental group had higher levels of lead.

(These data appear in Trumbo [2001].)

13.16 Juvenile Delinquents Dr. Kirkland R. Gable (in Schwitzgebel 1964) studied 20 male juvenile delinquents who had each spent 6 months or more in a Massachusetts juvenile detention center. He wondered whether simply asking the juvenile delinquents to talk would help them stay out of jail in the future. The subjects were paid to talk into a tape recorder about anything they wanted for one hour, 3 to 5 days a week for 6 months; there was no therapist present. A control group was formed by matching each subject in the experimental group with a juvenile delinquent who was the same age, had the same ethnic background, grew up in the same town, had committed the same types of offenses, and had spent the same amount of time incarcerated. The control group received no treatment. The experimental and control groups were followed for three years.

The data are available at the website for the number of months of incarceration in the 3-year period following the 6-month-long experiment. The histogram shows the differences for the entire data set: experimental minus control. A negative difference means a subject in the experimental group spent less time in jail than did his control (which is the outcome the researcher is hoping for). Although these subjects were not a random sample, we can test to see whether the difference is too large to attribute to chance if we assume the matched subjects in the control group were chosen at random.

a. Summarize the months of incarceration for both groups in one or two sentences. Include appropriate numerical summaries.

b. Perform a sign test to determine whether the typical amount of jail time after the experiment was less for the treatment group than for the control group. Use a significance level of 0.05.

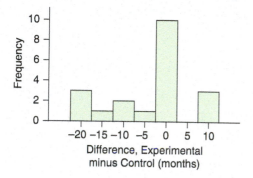

Difference, Experimental minus Control (months)

13.17 The Stroop Effect Suppose you had to identify the color of ink for a series of printed words, but the printed word appeared in a color of ink that did not match the name of the color. For example, if you were shown "RED" then you should say "Blue," but you might incorrectly say "RED" because that is what the word spells. It might take longer to correctly identify a color when it was used to print a word whose meaning did not match that color than to identify the color when the ink color and printed word matched. This difference in times is an example of something psychologists call *interference* and is called the Stroop effect, after research psychologist J. R. Stroop (1897–1973).

Same	Diff
32	66
12	31
17	40
16	25
25	36
18	15
18	39
24	35
20	32
24	30

The data in the table were collected by a student conducting research on the Stroop effect. Each of 10 subjects identified colors of ink in two different situations, and the time (in seconds) was recorded. There were 36 words in each trial. In column 1, the ink was the same color as the word. In column 2, the ink and the words were different colors. Whether the subject started with the color matching the word or with the color that was different was randomly determined. Treat the data as though they came from a random sample.

a. Write a sentence comparing the median time to identify the color for the two groups. Did the task tend to take longer when the colors were different?

b. Do a sign test to test whether those who see the ink in the "wrong" color tend to take longer to identify the color. Use a significance level of 0.05.

13.18 Reading Material on Colored Paper In the past, some people believed it was easier to read words printed on colored paper than words printed on white paper, while other people believed it was easier to read words printed on white paper. To test these theories, researchers asked a sample of 15 subjects to read a passage printed in black ink twice: once on salmon-colored paper and once on white paper. The time it took to read each passage, in seconds, is given in the table. Whether the person read the salmon-colored paper first or the white paper first was determined randomly. A histogram of the differences is also shown.

Salmon	White
79	72
49	45
47	45
112	120
65	63
66	62
67	60
59	54
140	160
64	57
67	61
73	72
64	48
126	122
67	73

a. Compare the typical values for the two groups.

b. Refer to the histogram of differences. Why is the *t*-test potentially not appropriate for these data?

c. Carry out an appropriate hypothesis test that the typical reading time is not the same for words printed on salmon-colored paper (at least for these subjects and this passage). Use a significance level of 0.05 (with a two-sided alternative), and interpret your results.

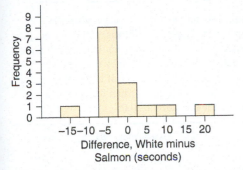

Difference, White minus Salmon (seconds)

13.19 Males' Pulse Rates Students in a statistics class were asked to measure their resting pulse rates. After that, the instructor unexpectedly screamed and ran from one side of the class to the other. Students again measured their pulse rates. The pulse rates (in beats per minute) were recorded before and after the scream for the male students in the class. Perform a sign test to see whether the pulse rate went up significantly, using a significance level of 0.05 and treating the sample as random.

Men	
Pulse Before	Pulse After
50	64
84	72
96	88
80	72
80	88
64	68
88	100
84	80
76	80

13.20 Females' Pulse Rates Refer to exercise 13.19. This time, the data (beats per minute) came from female students before and after the scream. Perform a sign test to see whether the pulse rate went up significantly, using a significance level of 0.05 and treating the sample as random.

Women		Women	
Pulse Before	Pulse After	Pulse Before	Pulse After
64	68	84	88
100	112	80	80
80	84	68	92
60	68	60	76
92	104	68	72
80	92	68	80
68	72		

*** 13.21 Ages of Brides and Grooms** A random sample of the ages of 14 brides and their grooms showed that in 10 of the pairs the grooms were older, in 1 pair they were the same age, and in 3 pairs the bride was older. Perform a sign test with a significance level of 0.05 to test the hypothesis that grooms tend to be older than their brides.

*** 13.22 Textbook Prices** A student was interested in comparing textbook prices at two universities. She matched the textbooks by subject and compared prices from the University of California at Santa Barbara (UCSB), which is on the quarter system (10 weeks) and California State University at Northridge (CSUN), which is on the semester system (16 weeks).

a. Test the hypothesis that the books for UCSB tend to cost less than the books for CSUN, using a significance level of 0.05. For 17 of the pairs the prices were higher at CSUN, and for 7 of the pairs they were higher at UCSB. There were no ties.

b. Why is the sign test probably a good choice for these data?

SECTION 13.3

TRY **13.23 Meat-Eating Behavior (Example 4)** A researcher was interested in the ethics of eating meat, so he studied and compared ethicists (philosophy professors who taught ethics) with professors who taught other subjects to find out whether ethicists eat less meat (Schwitzgebel and Rust 2009). The subjects were asked how many meals they eat per week that include meat. The output provided are from a random sample from the full study. Vegetarians (who eat no meat) were excluded. Assume the shapes and spreads of the distributions are the same.

a. Refer to the output given. Compare the sample medians. What do they tell us about the research question?

b. Refer to the output to perform a Mann-Whitney test using a significance level of 0.05. State all four steps of a significance test.

```
Mann-Whitney Test and CI
              N    Median
Ethicists     16    4.500
Control       12    6.500

Point estimate for ETA1-ETA2 is -2.000
95.2 Percent CI for ETA1-ETA2 is (-3.999, 0.001)
W = 198.0
Test of ETA1 = ETA2 vs ETA1 < ETA2 is significant at 0.0599
The test is significant at 0.0582 (adjusted for ties)
```

13.24 Credit Card Debt A statistics student who was interested in credit card debt asked a random sample of students for the total amount of their credit card debt. We eliminated the two women and the one man who had a debt of 0, which left 18 women and 19 men.

Women's Debt
(thousands of dollars)

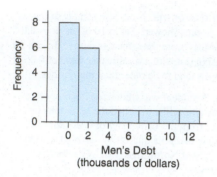

a. By looking at the histograms of the data, determine whether it would be appropriate to do a two-sample *t*-test using this data set. Explain.

b. Refer to the Minitab output. Which group had a higher median debt in this sample?

c. Refer to the Minitab output provided to test whether the typical credit card debt (as measured by the median) is different for men and women, using a significance level of 0.05.

```
Mann-Whitney Test and CI: Women, Men
          N   Median
Women    18    725.0
Men      19   1250.0


Point estimate for ETA1-ETA2 is -212.0
95.3 Percent CI for ETA1-ETA2 is (-1100.0, 700.1)
W = 322.5
The test is significant at 0.5635 level
```

13.25 Run Scored Is there a difference in the runs scored by winning teams in the American League and the National League in professional baseball? The winning scores for all games played on two randomly selected days were recorded and the league of the winning team was noted. The histogram below shows the winning scores for teams in the National League.

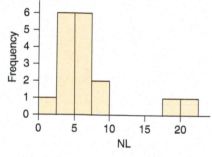

a. Describe the shape of this histogram. Are there any outliers? Would it be more appropriate to compare the means or the medians if we wanted to compare the typical number of runs scored by winning teams in each league? Explain.

b. The results of a Mann-Whitney test to determine whether the median number of runs scored by winning teams differ between the two leagues is shown. Test the hypothesis that there is a difference in the population median runs scored using a significance level of 0.05.

```
Hypothesis test results:
m1 = median of AL
m2 = median of NL
m1-m2 : m1 - m2
H₀ : m1-m2 = 0
Hₐ : m1-m2 ≠ 0

Difference  n₁  n₂  Diff.Est. Test Stat. P-value    Method
 m₁-m2₂     13  17      0        216     0.5524   Norm.Approx.
```

13.26 Sleep Typically, do men and women sleep different amounts? At a small private college in California, a random sample of students were asked how many hours of sleep they got last night. The figure shows the output for a Mann-Whitney test. Test the alternative hypothesis that the median number of hours of sleep for all men at this college is not equal to the median for all women at the same college, using a significance level of 0.05.

```
Mann-Whitney Test and CI: SleepMen, Sleep Women

                N    Median
SleepMen       24     7.250
Sleep Women    94     6.750


Point estimate for ETA1-ETA2 is 1.000
95.0 Percent CI for ETA1-ETA2 is (-0.000, 1.000)
W = 1657.5
Test of ETA1 = ETA2 vs ETA1 not = ETA2 is significant at 0.1258
The test is significant at 0.1207 (adjusted for ties)
```
Minitab output

13.27 Cell Phone Bills Cell phone bills (rounded to the nearest dollar) for the most recent month for random samples of college men (M) and college women (F) were studied. Histograms for the numbers of dollars for men and women (A) and output from a Mann-Whitney test (B) are given.

a. Why would it be inappropriate to compare means with a *t*-test?

b. A student feels that women talk more on their phones than men and therefore have higher bills. Test this hypothesis at a 5% significance level.

(A)

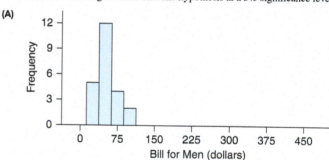

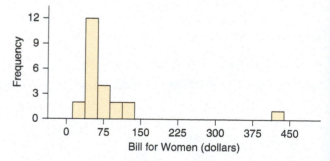

(B)

```
Mann-Whitney Test and CI: M, F
      N  Median
M    23   48.00
F    24   57.50

Point estimate for ETA1-ETA2 is -5.50
95.1 Percent CI for ETA1-ETA2 is (-18.00, 5.00)
W = 501.0
Test of ETA1 = ETA2 vs ETA1 < ETA2 is significant at 0.1413
The test is significant at 0.1411 (adjusted for ties)
```

13.28 Weights of Athletes Data were collected on the weights of 25 male baseball players and 25 male soccer players. Assume that these are random samples from all college baseball and soccer players. Refer to the Mann-Whitney output. Assume the shapes and spreads of the two distributions are the same.

a. Which group tends to weigh more in this sample? Compare the median weights by referring to the output.

b. Are these data paired or independent? Explain.

c. Test the hypothesis that the median weights are different. Use a 0.05 significance level, and use the Mann-Whitney test.

```
Mann-Whitney Test and CI: Baseball, Soccer
           N  Median
Baseball  25  192.00
Soccer    25  174.00

Point estimate for ETA1-ETA2 is 17.00
95.2 Percent CI for ETA1-ETA2 is (9.99, 25.00)
W = 857.5
Test of ETA1 = ETA2 vs ETA1 not = ETA2 is significant at 0.0000
```

13.29 Happiness A StatCrunch survey of happiness measured the happiness level for males and females. Each person selected a number from 1 (lowest) to 100 (highest) to measure her or his level of happiness. Figure A shows the output for a Mann-Whitney test to compare the median happiness level for males and females.

a. Figure B is a histogram of happiness level for the males. Describe the shape of the distribution of the sample, and comment on whether it would be better to compare means or medians and explain.

b. Compare the sample medians descriptively.

c. Assuming we have a random sample of male and female users of Stat-Crunch, test the hypothesis that the typical happiness level (as measured by the median) is different for males and females. Use a significance level of 0.05.

(A)

```
Mann-Whitney Test and CI: Happy_Female, Happy_Male
                N  Median
Happy_Female  297  80.000
Happy_Male    380  75.000

Point estimate for ETA1-ETA2 is 3.000
95.0 Percent CI for ETA1-ETA2 is (0.000, 5.000)
W = 106108.0
Test of ETA1 = ETA2 vs ETA1 not = ETA2 is significant at 0.0317
The test is significant at 0.0315 (adjusted for ties)
```

(B)

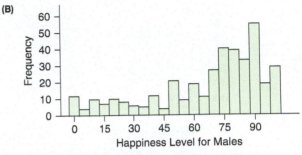

(Source: StatCrunch: Responses to Happiness survey. Owner: Webster West)

13.30 Soda A StatCrunch survey was done asking what percentage of liquid intake was in the form of soda. Figure A shows the output given.

a. Figure B is a histogram of the percentage for the females. Describe the shape of the distribution of the sample, and comment on whether it would be better to compare means or medians and explain why.

b. Descriptively compare the percentage of liquid intake that was soda for men and the percentage for women. Who consumes the larger percentage of soda?

c. Assuming we have a random sample of StatCrunch users, test the hypothesis that the medians are not equal, using a significance level of 0.05.

(A)

```
Mann-Whitney Test and CI: Soda_Female, Soda_Male
               N  Médian
Soda_Female  169   3.00
Soda_Male    163   5.00

Point estimate for ETA1-ETA2 is 0.00
95.0 Percent CI for ETA1-ETA2 is (-0.00, 0.00)
W = 28046.0
Test of ETA1 = ETA2 vs ETA1 not = ETA2 is significant at 0.9162
The test is significant at 0.9145 (adjusted for ties)
```

(B)

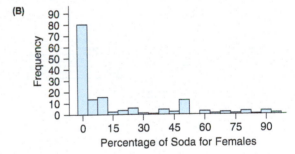

(Source: StatCrunch Responses to Soda survey. Owner: scsurvey)

SECTION 13.4

TRY *13.31 **Sports and Extraversion (Example 5)** Are students who participate in sports more extraverted than those who do not? A random sample of students at a small university were asked to indicate whether they participated in sports (yes or no) and to rate their level of extraversion. Extraverts are outgoing, are talkative, and don't mind being the center of attention. Students were asked whether they agreed with the statement that they were extraverts, using a scale of 1 to 5 with 1 meaning "strongly disagree" and 5 meaning "strongly agree." There were 51 students who participated in sports and 64 who did not.

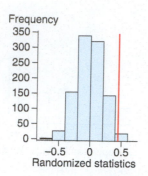

The mean extraversion score for those who participated in sports was 3.618, and the mean for those who did not participate was 3.172, a difference of 0.446 point. To determine whether the mean difference was significant, we performed a randomization test to test whether the mean extraversion level was greater for athletic students.

a. The histogram shows the results of 1000 randomizations of the data. In each randomization, we found the mean difference between two groups that were randomly selected from the combined group of data: the combined data for the sporty and nonsporty. Note that, just as you would expect under the null hypothesis, the distribution is centered at about 0. The red line shows the *observed* sample mean difference in extraversion for the sporty minus the nonsporty. From the graph, does it look like the observed mean difference is unusual for this data set? Explain.

b. The software output gives us the probability of having an observed difference of 0.446 *or more*. (See the column labeled "Proportion => Observed"). In other words, it gives us the right tail area, which is the p-value for a one-sided alternative that the mean extraversion score is higher for athletes than for nonathletes. State the p-value.

c. Using a significance level of 0.05, can we reject the null hypothesis that the means are equal?

d. If you did not have the computer output, explain how you would use the histogram to get an approximate p-value.

Statistic: mean(subset(Extraverted,Sports = 1)) = mean(subset(Extraverted,Sports = 0))								
Observed	n	Mean	Std. Dev.	2.5th	5th Per.	50th Per.	95th Per.	97.5th Per.
0.44577205	1000	-0.010929994	0.2030901	-0.4084712	-0.35761335	-0.0062806373	0.32643995	0.38028494

The table below includes the observed permutation with resamples.

Observed	Proportion <= Observed	Proportion => Observed
0.44577205	0.989011	0.011988012

13.32 Happiness Are women happier than men? A StatCrunch survey asked respondents to select a number from 1 (lowest) to 100 (highest) to measure their level of happiness. The sample mean for the 297 females was 71.15, and the sample mean for the 380 males was 67.08. To determine whether the population mean for women was higher than the population mean for men, we used a randomization test.

a. The histogram shows the results of 1000 randomizations of the data. In each randomization, 297 observations from the combined data were randomly marked "female," and the rest were randomly marked "male." We calculated the mean difference in happiness between these two randomly determined groups. Note that the distribution is centered at about 0, just as it should be, since we carried out the randomization in such a way that the null hypothesis is true. The red line shows the *observed* sample mean difference in happiness for the women minus the happiness for the men. From the graph, does it look like the observed mean difference is unusual for this data set? Explain.

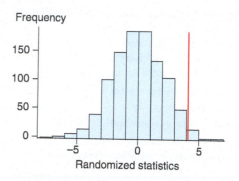

Statistic: mean(subset(Happy,Gender = Female)) - mean (subset(Happy,Gender = Male))								
Observed	n	Mean	Std. Dev.	2.5th	5th Per.	50th Per.	95th Per.	97.5th Per.
4.0718324	1000	-0.025935093	2.074339	-4.1773231	-3.4035881	-0.077020505	3.406544	3.8641454

The table below includes the observed permutation with resamples.

Observed	Proportion <= Observed	Proportion => Observed
4.0718324	0.97902098	0.021978022

b. The software output estimates the probability of having an observed difference of 4.07 *or more*. (See the output column labeled "Proportion => Observed"). Where does the value of 4.07 (or 4.0718) come from?

c. State the p-value.

d. Using a significance level of 0.05, can we reject the null hypothesis that the means are equal and so conclude that women StatCrunch users tend to be happier than men StatCrunch users? (Assume the sample was randomly selected from the population of all StatCrunch users.)

(Source: StatCrunch: Responses to Happiness survey. Owner: Webster West)

13.33 College Students and Credit Card Debt In exercise 13.24 you compared credit card debts for college men and women using the Mann-Whitney test to compare medians. We'll use the same data again, but this time you will apply a randomization test to determine whether men and women in college have different median credit card debts. There were 19 men and 18 women in this data set. As part of the randomization test, the credit card debts were randomly assigned the label "man" or "woman," and the difference of the medians (men minus women) was calculated and recorded. This was repeated 1000 times. Figure A shows a histogram of the resulting differences of medians, and Figure B shows the resampling output.

(A)

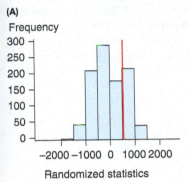

Frequency

Randomized statistics

a. Why are medians a better choice than means for comparing typical debts of men and women? Refer to the histograms given in Exercise 13.24.

b. The sample medians were $1250 for men and $725 for women, so the difference of men minus women was $525. This value is indicated with a red vertical line in the histogram. Is the observed difference unusually large? Explain.

c. The proportion of differences in the histogram that were greater than or equal to $525 is 0.245, as shown in Figure B. Our alternative hypothesis, that the medians are different, requires a two-sided test. However, with a significance level of 0.05, and knowing that the two-sided p-value must be bigger than the one-sided value of 0.245, will we reject the null hypothesis? Explain.

(B)

Statistic: median(subset(Debt,Gender = 1)) = median(subset(Debt,Gender = 2))								
Observed	n	Mean	Std. Dev.	2.5th	5th Per.	50th Per.	95th Per.	97.5th Per.
525	1000	-23.6625	635.99457	-1150	-1000	-50	956.25	1075

The table below includes the observed permutation with resamples.

Observed	Proportion <= Observed	Proportion => Observed
525	0.75624377	0.24475524

13.34 Soda Does soda constitute a larger part of the diet for women than it does for men? A StatCrunch survey asked people to report the percentage of their liquid intake that is soda. The sample mean for the 169 females was 19.51%, and the sample mean for the 163 males was 17.74% To determine whether the mean for all women StatCrunch users was more than the mean for all men, we performed a randomization test.

a. The histogram shows the results of 1000 randomizations of the data. In each randomization, 169 observations from the merged "men" and "women" values were randomly determined to be from "women" and the rest from "men." We calculated the mean difference in the percentage of sodas between these randomly determined groups. Note that the distribution is centered at about 0, because the randomization forces the null hypothesis to be true. The red line shows the *observed* sample mean percentage of soda for the women minus the mean percentage of soda for the men. From the graph, does it look like the observed mean difference is unusual for this data set? Explain.

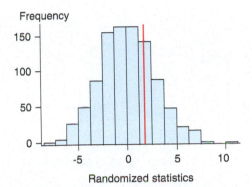

Frequency

Randomized statistics

Statistic: mean(subset(Soda,Gender = Female)) - mean (subset(Soda,Gender = Male))								
Observed	n	Mean	Std. Dev.	2.5th	5th Per.	50th Per.	95th Per.	97.5th Per.
1.7718481	1000	0.047691779	2.889647	-5.4256192	-4.679208	-0.050892857	4.8949046	6.2259471

The table below includes the observed permutation with resamples.

Observed	Proportion <= Observed	Proportion => Observed
1.7718481	0.73626374	0.26473526

b. The software output estimates the probability of having an observed difference of 1.77 *or more*. (See the column labeled "Proportion => Observed"). Where does the value of 1.77 come from?

c. Report the p-value for the one-sided alternative that the mean for the women is greater than the mean for the men.

d. Using a significance level of 0.05, can we reject the null hypothesis that the means are equal and so conclude that these women StatCrunch users tend to report a higher soda intake percentage than these men? (Assume the sample was randomly selected

e. from the population.)

(Source: StatCrunch survey results. Owner: scsurvey)

13.35 Rainfall In a well-known study on the effects of cloud seeding to produce rainfall (cited on page 659 of the text by Simpson et. al), experimenters randomly assigned airplanes to release either silver nitrate (which is believed to increase the amount of rainfall from a cloud) or a placebo. Fifty-two clouds were chosen at random; half were "seeded" with silver nitrate, and half were not. In the text, we compared the *median* rainfall amounts for the seeded (silver nitrate) and unseeded clouds. In this exercise, you will compare the *mean* rainfall amounts for the seeded and unseeded clouds.

In the study, the seeded clouds produced more rain, on average, by 368.9 acre-feet. To determine whether such differences could occur by chance, a statistician could have written the 52 rainfall amounts on separate slips of paper and randomly dealt them into two stacks. He or she would then have computed the mean of each stack and found the difference. This was actually done by a computer and repeated 1000 times. The results are shown in the histogram.

Carry out a hypothesis test to determine whether cloud seeding increased the mean rainfall. By referring to the histogram, choose from the following possible p-values (one-tailed):

$$0.50, 0.25, 0.15, 0.025, \text{less than } 0.0001$$

Use a 5% significance level for your test.

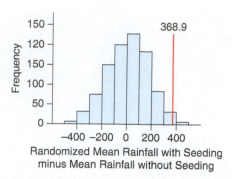

Randomized Mean Rainfall with Seeding minus Mean Rainfall without Seeding

13.36 Rainfall Refer to exercise 13.35, which discussed a study on the effects of cloud seeding to produce rainfall. Some researchers think that cloud seeding has little effect on "low rain potential" clouds. Instead, they claim, most of the action is with clouds that would produce lots of rain even without seeding. In this scenario, clouds that would produce little rain without seeding will produce little rain with seeding. However, the clouds that would produce the most rain without seeding will produce much, much more rain with cloud seeding. To test this, researchers carried out a randomization test to find out whether the third quartile of rainfall increased under cloud seeding. The table gives summary statistics.

Treatment	Minimum	First Quartile	Median	Mean	Third Quartile	Maximum
Seeded	4.10	98.13	221.60	533.20	474.30	2747.00
Unseeded	1.00	24.82	44.20	164.60	159.20	1203.00

a. Explain what it means to say that the third quartile of rainfall is 474.30 acre-feet.

b. Why is the third quartile an appropriate statistic to answer the researchers' question?

c. What is the observed difference in third-quartile rainfall between the seeded and unseeded clouds?

d. To determine whether such differences could occur by chance, a statistician could have written the 52 rainfall amounts on separate slips of paper and randomly dealt them into two stacks. He or she would then have computed the third quartile of each stack and found the

difference. A computer actually did this 1000 times, each time finding the difference between the third quartile for the seeded clouds minus the third quartile for the unseeded clouds. The results are shown in the histogram. Referring to the histogram, carry out a hypothesis test to test whether cloud seeding increased the third-quartile rainfall. (You will have to get approximate p-values by reading the histogram.) (Remember that you need only decide whether the p-value is larger or smaller than 0.05.)

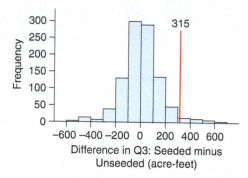

Difference in Q3: Seeded minus Unseeded (acre-feet)

13.37 Randomization Exercise 13.35 describes a simulation exercise. Which of the following is the best explanation for the process described?

a. This process allows researchers to compare the actual result to what could have happened by chance (assuming the null hypothesis is true).

b. This process allows researchers to determine the percentage of time the seeded clouds would outperform the unseeded clouds for all possible scenarios.

c. This process allows researchers to determine how many times they need to replicate the experiment for valid results.

13.38 Randomization Exercise 13.35 describes a simulation exercise. Under which of the following assumptions were the data simulated?

a. Cloud seeding typically produces more rain than not seeding the cloud.

b. Cloud seeding typically produces the same amount of rain as not seeding.

c. Cloud seeding typically produces less rain than not seeding.

CHAPTER REVIEW EXERCISES

For exercises 13.39 through 13.46, choose from the following tests, as appropriate: paired t-test, sign test, two-sample t-test, and Mann-Whitney test. There may be two acceptable choices.

13.39 Sleep You have recorded the time slept on a Tuesday and the time slept on a Sunday for a random sample of 15 students. You want to investigate whether students tend to sleep more on weekends than on weekdays. Which test(s) can you use? Answer for each circumstance.

a. Assume the distribution of sleep hours for both Tuesday and Sunday are approximately Normal.

b. Assume the distribution of sleep hours for both Tuesday and Sunday are not approximately Normal and assume that the distribution of differences in sleep hours for each student is not Normal.

c. Assume the distributions of sleep hours for both Tuesday and Sunday are not Normal but assume the distribution of the differences in sleep hours is approximately Normal.

13.40 Comparing GPAs You know the grade point averages (GPAs) of a random sample of 10 full-time college students and 10 part-time college students. You want to test the hypothesis that the typical GPAs for full-time and part-time college students are different. Which test(s) should you choose for each situation?

a. Suppose your preliminary investigation lead you to conclude that the distributions of GPAs for both groups are approximately Normal.

b. Suppose your preliminary investigation lead you to conclude that the distributions of GPAs for both groups are not Normal but have the same shape.

13.41 Starting Salaries Suppose a college career center was interested in the starting salaries of recent graduates in Communications Studies and Sociology. The center randomly samples 15 recent graduates from each of these fields and records the starting salary for the graduates. The center wants to determine whether there is a difference in the starting salaries for graduates in these majors. Which test(s) should be used in each of these situations?

a. Assume the starting salary for both majors is approximately Normally distributed.

b. Assume that one of the salary distributions is strongly right-skewed.

13.42 Pulse Rates Suppose you want to determine whether meditation can cause a decrease in pulse rate. You randomly select 15 students, teach them a meditation technique, and then measure their pulse rates before and after meditation. Which test(s) should you choose for each situation?

a. Assume that your analysis shows that the differences in pulse rates are Normally distributed.

b. Assume that the distributions of differences in pulse rates are strongly skewed.

13.43 DMV Wait Times Suppose you want to determine whether there is difference in wait times at two Department of Motor Vehicles offices. A random sample of customer wait times is obtained for each office. Which test(s) can be used for each situation below?

a. The populaton distribution of wait times is approximately Normal for both groups.

b. The population distribution of wait times for one of the offices is skewed left.

13.44 Ticket Prices Tickets for high demand events are often resold on a secondary market, such as StubHub or VividSeats. Suppose a consumer is interested in determining if there is a significant difference in ticket prices between the two resale sites. She picks an upcoming event being held at a large venue and randomly selects 15 seating areas in the venue. She then records the resale ticket prices for each area from the two sites. Which test(s) can be used for each situation below?

a. The distribution of the differences in prices suggests the population of price differences is approximately Normal.

b. The differences in price follow a right-skewed distribution.

13.45 Online Grocery Prices Suppose a random sample of online grocery prices (15 items) is obtained at Amazon and Walmart. The same items were sampled at both sites, and a student wants to determine whether the typical price differs at the two online sites. Which test(s) can be used for each situation below?

a. Suppose the distribution of prices at each site is strongly right-skewed and the distribution of differences in prices is also strongly right-skewed.

b. Suppose the distribution of prices at each site is right-skewed but the distributions of the differences in prices is approximately Normal.

c. Suppose the sample size was 150 rather than 15 and the distribution of differences is skewed.

13.46 Extraverts Suppose you give a random sample of students a questionnaire about extraversion, and some (10) are classified as extraverts and some (15) as not extraverts. You want to determine whether the typical GPA is higher for extraverts than for those who are not extraverts. Which test(s) can be used for each situation below?

a. Both distributions are strongly skewed.

b. Both distributions are nearly Normal.

c. You have 100 extraverts and 150 who are not extraverts, and both distributions are skewed. Explain your choice of test.

g **13.47 **Ice Cream Cones** McDonald's claims that its ice cream cones typically weigh 3.18 ounces (converted from grams). Here are the weights, in ounces, of cones purchased on different days from different servers:

4.2, 3.6, 3.9, 3.4, and 3.3

Carry out a sign test to determine whether the median is greater than 3.18 ounces. Use a significance level of 0.05. Why is the sign test a more appropriate choice than the one-sample t-test? *See page 681 for guidance.*

** **13.48 Average Body Temperatures** Many people believe that healthy people typically have a body temperature of 98.6°F. We took a random sample of 10 people and found the following temperatures:

98.4, 98.8, 98.7, 98.7, 98.6, 97.2, 98.4, 98.0, 98.3, and 98.0

Use the sign test to test the hypothesis that the median is not 98.6.

13.49 Contacting Mom Random samples of 30 professors of ethics and 30 professors in other disciplines (not ethics) were asked how many days it had been since they contacted their mothers; this included phone calls and face-to-face meetings. Professors whose mothers were not living were not included. The resulting data are shown.

a. Describe the shapes of the distributions of the samples.

b. Find and compare the sample means.

c. Find and compare the sample medians.

d. Perform a two-sample t-test to determine whether the population means are different at the 0.05 significance level. Assume that the sample sizes are large enough so that the approximate p-value will be good.

e. Perform a Mann-Whitney test to determine whether the medians are significantly different at the 0.05 level.

(Source: Data from Eric Schwitzgebel)

Eth	Other	Eth	Other
28	3	5	0
4	9	5	1
7	1	14	7
6	1	14	1
100	0	1	0
1	0	10	0
1	1	5	1
70	1	2	3
5	3	5	1
7	6	4	12
10	0	3	20
2	4	2	4
2	1	60	1
1	4	1	7
3	1	4	3

13.50 through 13.54 Texts Sent and Received StatCrunch did a survey of users to find out about their texting. They were asked their gender, their age, and how many texts they send in a day and how many texts they receive in a day. The data are available at this text's website. Assume the samples are random samples.

(Source: StatCrunch Survey Results. Owner: scsurvey)

13.50 Sent and Received: Paired *t*-test Determine whether the population mean number of texts sent and the population number received (for all the respondents) is different at the 0.05 level by using a paired *t*-test.

13.51 Sent and Received: Sign test Determine whether the median number of texts sent and the median number received for all StatCrunch users are different at the 0.05 level by using a sign test.

13.52 Sent and Received: Women Determine whether the number of texts sent by females and the number received by females are significantly different at the 0.05 level using a paired *t*-test.

13.53 Sent and Received: Men Determine whether the number of texts sent by males and the number received by males are significantly different at the 0.05 level using a paired *t*-test.

*13.54 Differences: Men vs. Women Find the difference in number of texts received and the number sent for females. Do the same for males. Then determine whether the differences are significantly different at the 0.05 level using the two-sample *t*-test.

13.55 Geometric Mean The dotplot shows the number of classes missed in a month for a random sample of 23 students from a private college in California. Explain why you cannot find the geometric mean for the numbers.

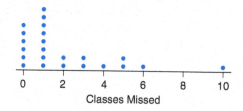

13.56 Looking at the data about contacting mom (exercise 13.49), for which group (ethicists or other) could you perform a log transform, and for which group could you not? Explain.

13.57 Resampling Moms We performed a randomization test to determine whether the mean number of days since an ethics professor contacted his or her mother is different from the mean number of days for a professor in a field other than ethics. The data consisted of a random sample of 30 ethics professors and 30 professors in fields other than ethics.

The histogram shows the results of 1000 randomizations of the data. In each randomization, 30 values were randomly determined to be in the "Ethics" group and the other 30 in the "Other" group. The mean difference was calculated and recorded. Note that the distribution of the differences of these means is centered at about 0, because the randomization forced the null hypothesis to be true.

The *observed* sample mean time since last contact for professors in other fields minus the sample mean time for ethics professors

is shown by the red vertical line. The p-value for the one-sided hypothesis that the mean time since professors in other fields contacted their mothers is less than the mean time for ethics professors is shown in the numerical output. Does this show that ethicists have less recent contact or not, using a significance level of 0.05? Comment on both the histogram and the table of output.

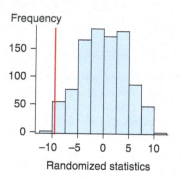

Statistic: mean(subset(Days,Profs = Other)) - mean (subset(Days,Profs = Eth))								
Observed	n	Mean	Std. Dev.	2.5th	5th Per.	50th Per.	95th Per.	97.5th Per.
-9.5333333	1000	-0.011666667	4.5617387	-8.5333333	-7.8333333	0	7.8333333	8.3333333

The table below includes the observed permutation with resamples.

Observed	Proportion <= Observed	Proportion => Observed
-9.5333333	0.008991009	0.99200799

(Source: Data are from Eric Schwitzgebel)

13.58 Resampling Runs Scored Using data from Exercise 13.25, we used a randomization test to find out whether the typical number of winning runs in the American League differed from that of the National League in professional baseball. In each randomization we found the mean difference between the two groups that were randomly selected from the combined group of data: the combined data for the American and National Leagues. Note that, just as you would expect under the null hypothesis, the distribution is centered at about 0. The observed difference in means (American League − National League) was 0.1086 and is shown by the red vertical line. Use the output provided and a significance level of 0.05 to determine if typical number of winning runs differs between the two leagues. How does this answer compare with your conclusion when using the Mann-Whitney test in exercise 13.25?

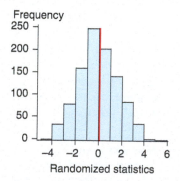

Statistic: mean(subset(Runs,League = AL)) - mean (subset(Runs,League = NL))								
Observed	n	Mean	Std. Dev.	2.5th	5th Per.	50th Per.	95th Per.	97.5th Per.
0.10859729	1000	-0.064479638	1.7254934	-3.1493213	-2.8778281	-0.16289593	2.959276	3.3665158

The table below includes the observed permutation with resamples.

Observed	Proportion <= Observed	Proportion => Observed
0.10859729	0.56443556	0.43656344

13.59 Energy Drinks Studies have shown that young adults experience faster weight gain and consume more unhealthy food than any other age group. In a 2017 study published in *PLOS One*, Buchanan et. al examined the effects of online marketing on young adults' consumption behaviors (Buchanan et al. 2017). Sixty young adults aged transformation 18 to 24 were recruited for the study and were randomly assigned to an experimental group or a control group. Participants in each group completed a pre and post-tests designed to assess their attitudes towards and purchase intention of two popular energy drinks: Red Bull and V Energy. The experimental group was exposed to the websites and social media of these two energy drinks. The control group was exposed to the websites and social media of two nut bars: Carman's and Go Natural. A portion of the data analysis using a Mann-Whitney test are shown in the table.

Between group comparisons at post-test.

	Brand/Product	Experimental	Control	*P* value
Purchase intention[a] (−2 to 2) *Median (mode)*	Red Bull	0.0 (−2.0)	−1.0 (−2.0)	.087
	V Energy	−0.5 (−2.0)	−2.0 (−2.0)	.044*
	Energy drink products (general)	−1.0 (1.0)	−1.5 (−2.0)	.108

a. Based on the results shown, using a 0.05 significance level, was there a significant difference in the purchase intention towards the products shown in the table between the experimental and control groups? If so, for which product(s)? Explain.

b. In the study analysis section of the research article the researchers state: "Based on the data distribution, independent sample *t*-tests or Mann-Whitney tests were utilized to determine the changes in attitudes towards, and purchase intention of, energy drinks." Explain how researchers would use the data distribution to determine which of these tests to use.

13.60 Piano Practice In a 2016 study by Toptaş published in *Academic Journals*, a music educator conducted research to determine if students could successfully memorize a piano piece using only short periods of daily practice (Toptaş 2016). In addition the researcher was interested in learning if short periods of daily practice would be more effective in memorizing the piece than one long period of practice. The study group consisted of 12 university students who had studied piano for 2 years. Students were shown a piece of music that they had not seen before and were given one hour to memorize the piece. Students were then randomly assigned to one of two groups. The experimental group practiced the piece for 15 minutes on four consecutive days. The control group practiced the piece for 60 minutes on the fifth day after their first exposure to the piece. Students were given a pre- and posttest performance assessment. A higher ranking indicates greater proficiency. Tables showing some of the data analysis from the study are provided.

The Mann-Whitney U-Test result of the post-test scores by groups.

Groups	N	Rank Average	Rank Total	U	Z	P
Experiment	6	9.50	57.00	0.000	−2.887	0.004
Control	6	3.50	21.00			
Total	12					

The Wilcoxin Signed-Rank test result of the pre-test and post-test scores by the experiment group

Pretest/ Posttest	N	Rank Average	Rank Total	Z	p
Negative Ranking	0	0.00	0.00	−2.226	0.026
Positive Ranking	6	3.50	21.00		
Equal	0				
Total	6				

a. What statistical test was used to determine if there a difference in performance between the experimental and control groups? Explain why this was an appropriate test to use. Based on the results provided, what conclusion would the researcher draw from this analysis?

b. What statistical test was sued to determine if short periods of daily practice was an effective method to learn this piece of music? Explain why this was an appropriate test to use. Based on the results provided, what conclusion would the researcher draw from this analysis?

gUIDED EXERCISES

g **13.47 Ice Cream Cones** McDonald's claims that its ice cream cones typically weigh 3.18 ounces (converted from grams). Here are the weights, in ounces, of cones purchased on different days from different servers):

4.2, 3.6, 3.9, 3.4, and 3.3

Carry out a sign test to determine whether the median is greater than 3.18 ounces. Use a significance level of 0.05. Why is the sign test a more appropriate choice than the one-sample *t*-test?

Step 0 ▶ Fill in the missing numbers to find the difference between the observed value and the value we would expect if the claim that McDonald's makes is correct.

Obs	Null Value	Difference
4.2	3.18	1.02
3.6	3.18	
3.9		0.72
3.4		
3.3	3.18	0.12

Step 1 ▶ H_0: _____

H_a: The median weight is more than 3.18 ounces (or the median difference is more than 0.00).

Step 2 ▶ Choose the sign test as instructed. What assumptions should be checked?

Step 3 ▶ Find the p-value. You may use a computer to do a sign test to find out whether the median difference is more than 0. Or you may use binomial probabilities, because the p-value is equivalent to the probability of getting five out of five heads in five tosses of a fair coin: b (5, 5, 0.5).

Step 4 ▶ Can you reject the null hypothesis that the median is 3.18 ounces?

Why Use the Sign Test? Why is the sign test more appropriate than the one-sample t-test for this data set?

TechTips

General Instructions for All Technology

The TI-84 is not programmed to do nonparametric tests. A randomization test using StatCrunch is shown at the end of the steps.

EXAMPLE A: QQ PLOTS (OR NORMALITY PLOTS) ▶ Make a QQ plot, or normality plot, of the following ages:

18, 22, 76, 21, 25

EXAMPLE B: TRANSFORMATIONS ▶ Transform all of the following ages to logarithm base 10, find the mean of the logs, and back transform that mean to age to find the geometric mean.

18, 22, 76

EXAMPLE C: SIGN TEST ▶ Suppose that eight people go on a diet, and the following weights show what happened. Determine whether you can reject the hypothesis of no weight change.

Person	Weight Before	Weight After	Change in Weight
1	188	182	6
2	225	225	0
3	203	200	3
4	154	150	4
5	140	142	−2
6	235	230	5
7	372	365	7
8	280	275	5

EXAMPLE D: MANN-WHITNEY TEST ▶ Suppose you have a random sample of GPAs for three men and three women, and you want to use a nonparametric test (the Mann-Whitney test) to determine whether you can reject the hypothesis of no difference.

Men	Women
2.30	3.30
3.45	3.10
2.96	3.02

TI-84

Example A: Normality Plot (QQ Plot)

1. Enter the five ages into L1.
2. Press **2ND STATPLOT** and **1**.
3. Turn on **Plot 1** by pressing **ENTER** when **On** is flashing. (The other plots should be **Off**.)
4. See Figure 13a: Use the arrows on the keypad to get to the last option for **Type:**. Press **ENTER** to select this option. Be sure the **Data List** is correct (**L1**). Change the **Data Axis** to **Y** so that the data are on the y-axis as described in the chapter.

▲ FIGURE 13a TI-84 Input for Normality Plot

5. Press **GRAPH, ZOOM** and **9** to see the Normality plot. Press **TRACE** to see the numbers. Figure 13b shows the Normality plot.

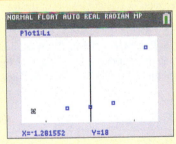

▲ FIGURE 13b TI-84 Normality Plot

Example B: Transformation and Geometric Mean

1. Enter the three ages into **L1**.
2. See Figure 13c: Move the cursor up to the label for **L2**. Press the **LOG** button, and press **2ND 1** (for **L1**) and **ENTER**.

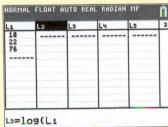

▲ FIGURE 13c TI-84 Taking Logs of a List

L2 will now contain the list of logs.

3. Find the mean of the logs by using **STAT, CALC, 1-Var Stats** and pressing **2ND 2** (for **L2**), **ENTER, ENTER, ENTER**.

The mean of the logs should be **1.4928**.

4. Back Transforming (finding the antilog)

Press **2ND 10^x** (with the **LOG** button), and put in **1.4928** and press **ENTER**.

You should get 31.1 for the geometric mean.

MINITAB

Example A: Normality Plot

1. Enter the five ages into, **C1**. Label **AGES**.

2. **Stat > Basic Statistics > Normality Test**

3. Double click **C1** to put it in the **Variable:** box. Click **OK**.

Figure 13d shows the Normality plot. Note that the data are on the *x*-axis, which is different from the plots in the chapter.

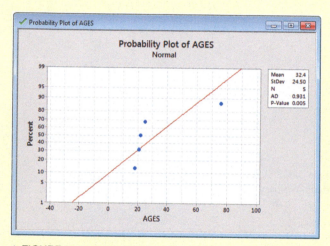

▲ **FIGURE 13d** Minitab Normality Plot

Example B: Transformation and Geometric Mean

1. Enter the three ages in **C1**. Put a label at the top, **AGES**.

2. Click **Calc** and **Calculator**.

3. See Figure 13e: Enter **C2** in the small top box. Delete any existing entry in the **Expression** box. Select **Log base 10** from the list of **Functions**. Double-click **AGES**. Click **OK**.

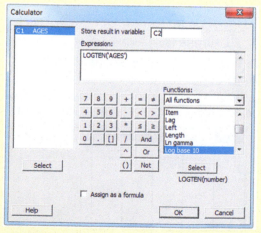

▲ **FIGURE 13e** Minitab Calculator

The logs will be in column **C2**.

4. Find the mean by using **Stat, Basic Statistics, Display Descriptive Statistics**, choosing **C2**, and clicking **OK**.

The mean of the logs will be 1.493.

Back Transforming (finding the antilog)

5. Enter **1.493** in column **C3**.

6. Click **Calc** and **Calculator**. Enter **C4** in the small top box. Choose **antilog** from the list of functions, and double-click **C3**. Click **OK**.

You will see the answer in column **C4: 31.1172**.

Example B: Geometric Mean (Directly)

Complete steps 1 and 2 above. Then select **Geometric mean** from the list of **Functions** and double-click **AGES**. Click **OK**.

Example C: Sign Test

Minitab uses the column of differences for the sign test. Start with either step 1 or step 2.

1. Enter the differences into column **C3** and go to step 3.

2. If you have entered two columns of data (labeled **Before** and **After**), click **Calc** and **Calculator**. Figure 13f shows what to do. Enter **Store result in variable: C3**. Enter **Expression: 'Before'-'After'**. Click **OK**.

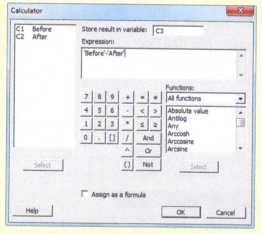

▲ **FIGURE 13f** Minitab Input for Calculating Differences

3. **Stat > Nonparametrics > 1-Sample Sign**

4. See Figure 13g: Enter **C3** in the **Variables** box. Choose **Test median** and leave the default **0.0**. If you wanted to use a one-sided alternative expecting weight loss, you would change the **Alternative** to **greater than**. Click **OK**.

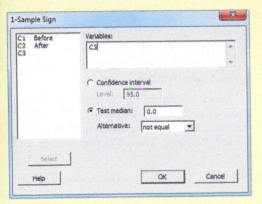

▲ FIGURE 13g Minitab One-Sample Sign Input

The output for a two-sided alternative is shown in Figure 13h.

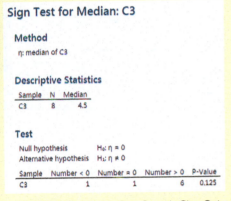

▲ FIGURE 13h Minitab One-Sample Sign Output

Example D: Mann-Whitney Test

1. Enter the six GPAs unstacked (as given previously) into columns C1 and C2. Add labels (**Men** and **Women**) above the columns.

2. **Stat > Nonparametrics > Mann-Whitney**

3. Refer to Figure 13i. Double-click one of the columns for the **First Sample,** and double-click the other column for the **Second Sample.** You may change the **Alternative** if you want it to be one-sided.

▲ FIGURE 13i Minitab Mann-Whitney Input

4. Click **OK**.

The p-value comes out to be **0.663**.

Example A: Normality Plot (QQ Plot)

1. Enter "Ages" and then the five numbers in column A.

2. **XLSTAT > Describing data > Normality tests**

3. See Figure 13j. Click in the **Data:** box to make it active; then select column A by clicking on the **A**. Click on the **Charts** tab and select **Normal Q-Q plots**.

4. Click **OK, Continue,** and **OK**.

5. Scroll down to see the plot. Note that the data are on the *x*-axis, which is different from the QQ plots in the chapter.

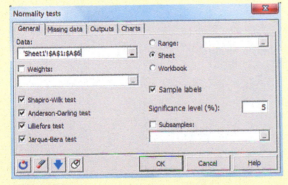

▲ FIGURE 13j XLSTAT Normality Test Input

Example B: Transformation and Geometric Mean

1. Enter the three ages in column A.

2. Click in cell **B1**. Click f_x.

3. Scroll down to **LOG10** and select it.

4. Double-click on cell A1, which contains the 18. The log will show up in cell B1 when you click **OK**.

5. To get the remaining logs, put your cursor precisely in the lower right corner of the cell (B1) containing the first log until it becomes a dark cross (+), and drag the cross down. You will see the remaining logs.

6. Click in cell B4. Click f_x and select **Average** Select the column of logs (B1:B3). Click **OK**. You will get an average of the logs of **1.492836** in B4.

7. To back transform (the antilog), click in an empty cell, click f_x, and select **Power**. For **Number** put in **10**, for the **Power** click on cell B4. Click **OK**. You should get **31.10542**, which is the geometric mean.

Example B: Geometric Mean (Directly)

1. Enter "Ages" and the three numbers in column A.

2. **XLSTAT > Describing data > Descriptive statistics**

3. See Figure 13k. Click in the **Quantitative data** box to make it active, then click on column **A**. Check **Sample labels**.

4. Click **Options** and check **Descriptive statistics**.

5. Click **Output,** click **None,** then scroll down and select **Geometric mean.**

6. Click **OK** and then **Continue**.

You should get **31.105** for the geometric mean.

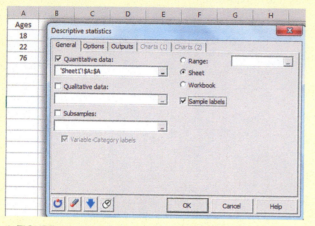

▲ **FIGURE 13k** XLSTAT Descriptive statistics Input

Example C: Sign Test

1. Enter "**Before**" and the list of before weights in column A. Enter "**After**" and the list of after weights in column B.

2. **XLSTAT > Nonparametric tests > Comparison of two samples (Wilcoxon, Mann-Whitney, . . .)**

3. See Figure 13l. Click in the **Sample 1:** box to make it active, and then select column **A**. Click in the **Sample 2:** box to make it active, and then select column **B**. For **Data Format:** select **Paired samples**. Select **Sign test**. Click **OK**.

4. Click **Continue** and then **OK**, if needed. The two-tailed p-value is shown in the output to be 0.125.

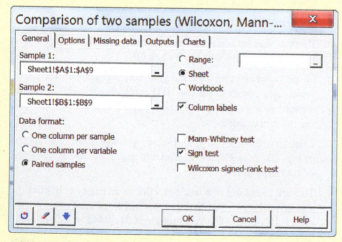

▲ **FIGURE 13l** XLSTAT Sign Test Input

Example D: Mann-Whitney Test

1. Enter the GPAs into columns A and B, including the headings "Men" and "Women" at the top.

2. **XLSTAT > Nonparametric tests > Comparison of two samples (Wilcoxon, Mann-Whitney, . . .)**

3. See Figure 13m. Click in the **Sample 1:** box to make it active, and then select column **A**. Click in the **Sample 2:** box to make it active, and then select column **B**. For **Data Format**, select **One column per sample**. Select **Mann-Whitney test**. Click **OK**.

4. Click **Continue** and then **OK** if needed. The two-tailed p-value is shown to be **0.700**.

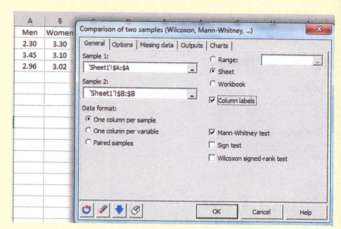

▲ **FIGURE 13m** XLSTAT Mann-Whitney Test Input

Example A: QQ Plot

1. Enter the five ages into the list for **var1**.
2. **Graph > QQ Plot.**
3. Click **var1** and **Compute!**

The output is shown in Figure 13n.

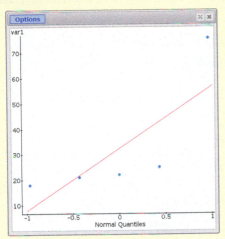

▲ **FIGURE 13n** StatCrunch QQ Plot

Example B: Transformation and Geometric Mean

1. Enter the ages (18, 22, 76) into the first list, **var1**.
2. **Data > Compute > Expression**
3. See Figure 13o. For **Expression,** enter **log10(var1)**.

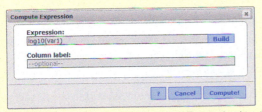

▲ **FIGURE 13o** StatCrunch Compute Expression Input

4. Click **Compute!** and a list of logs will be added to the table.
5. **Stat > Summary Stats > Columns**
6. Select the column with the logs in it.
7. Click **Compute!** You will get a table showing the mean to be **1.4928362**.

Back Transforming (finding the antilog)

8. **Data > Compute > Expression**
9. In the **Expression** box enter **10******1.4928362**; then click **Compute!** and you will get the answer **31.105429** in a new column.

Example C: Sign Test

You may start with step 1 or step 2.

1. Enter the differences in one column, and then go to step 5.
2. Enter in the two columns of data given in Example C (weight before and weight after).

To calculate the difference:

3. **Data > Compute Expression**

4. See Figure 13p. For **Expression:** enter **var1-var2**. Click **Compute!** There will be a column of differences added to your spreadsheet.

▲ **FIGURE 13p** StatCrunch Input for Finding Differences

5. **Stat > Nonparametrics > Sign Test**
6. Click the column of differences.
7. Click **Compute!**

Figure 13q shows the output for a two-sided alternative.

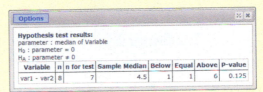

▲ **FIGURE 13q** StatCrunch Sign Test Output

Example D: Mann-Whitney Test (unstacked data)

1. Enter the GPAs unstacked (as shown previously) into **var1** and **var2**. You can put labels above the columns if you want.
2. **Stat > Nonparametrics > Mann-Whitney**
3. For **Sample 1: Values in:** select one of the columns, and for **Sample 2: Values in:** select the other column.
4. Select the appropriate alternative hypothesis. Click **Compute!**

Figure 13r shows the output.

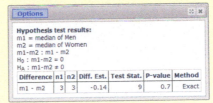

▲ **FIGURE 13r** StatCrunch Mann-Whitney Output

Example D: Mann-Whitney Test (stacked data)

1. Enter the GPAs stacked (as shown in Figure 13s), including the headings (here GPA and Gender).

Row	GPA	Gender	var3
1	2.3	male	
2	3.45	male	
3	2.96	male	
4	3.3	female	
5	3.1	female	
6	3.02	female	
7			
8			

▲ **FIGURE 13s** StatCrunch Stacked GPA Data

2. **Stat > Nonparametrics > Mann-Whitney**

3. See Figure 13t. For **Sample 1: Values in:** select **GPA**, and in the box for **Where:** enter **Gender = "male"**.

4. For **Sample 2: Values in:** select **GPA**, and in the box for **Where:** enter **Gender = "female"**.

5. Select the alternative hypothesis. Click **Compute!**

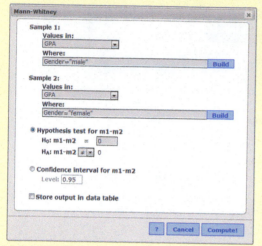

▲ **FIGURE 13t** StatCrunch Input for Mann-Whitney for Stacked Data

You will get a p-value of 0.7.

Example E: Randomization Test

Suppose we have data on the income of 12 men (m) and 8 women (f), shown in the table.

Income	Gender
$ 57000.00	m
$ 40200.00	m
$ 21450.00	f
$ 21900.00	f
$ 45000.00	m
$ 32100.00	m
$ 36000.00	m
$ 21900.00	f
$ 27900.00	f
$ 24000.00	f
$ 30300.00	f
$ 28350.00	m
$ 27750.00	m
$ 35100.00	f
$ 27300.00	m
$ 40800.00	m
$ 46000.00	m
$103,750.00	m
$ 42300.00	m
$ 26250.00	f

Use resampling to determine whether the mean difference in incomes between the men and women is significant.

1. Enter your list of numbers into one column (**var1**). Add the label **Income** in the label region at the top.

2. Enter the corresponding category code letters (m or f) into a second column, **var2**. Add the label **Gender** at the top.

3. **Stat > Resample > Statistic**

4. See Figure 13u. Select **Income** for **Columns to resample**. (Do not select Gender).

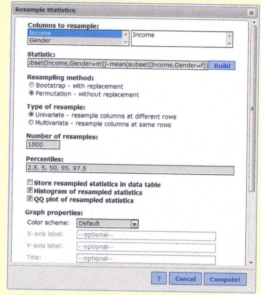

▲ **FIGURE 13u** StatCrunch Resampling Input

5. Select **Permutation-with replacement**.

6. The most difficult part is what you enter into the **Statistic:** line. It has to be perfect. If it isn't, you will get an error message. The line below shows the details.

$$\text{mean(subset(Income,Gender} = m)) - \text{mean(subset(Income,Gender} = f))$$

7. Click **Compute!**

Figure 13v shows the numerical output; yours may vary from this.

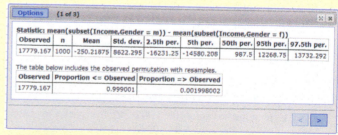

▲ **FIGURE 13v** StatCrunch Resampling Output

Be aware that two people using the same data and the same steps may end up with slightly different outcomes because randomization tests are chance processes.

8. Click the **>** in the lower right corner (see Figure 13v) to see a histogram somewhat like the one shown in Figure 13w. Note the red vertical line at the observed mean difference in incomes.

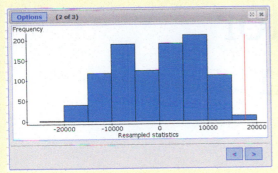

▲ **FIGURE 13w** StatCrunch Resampling Histogram

9. If you had set up this example as a two-sided test, you would double the right tail area (0.001998 above) to get your p-value.

Resampling with Medians. If you decided to use the medians instead of the means, your line of text in step 6 would be

median(subset(Income,Gender = m)) − median(subset(Income,Gender = f))

Vizualizing Resampling with an Applet. If you want to see how the resampling happens with two means:

1. Enter your data, stacked, in two columns labeled **Income** and **Gender**.
2. Click **Statcrunch > Applets > Resampling > Randomization test for two means**.
3. **Sample 1 in: Income, Where: Gender = m**. **Sample 2 in: Income, Where: Gender = f**.
4. **Compute!**
5. Click **1000 times**.
6. Click **Reset**. Click **1000 times**.

14 Inference for Regression

THEME

Using the linear model to make inferences about populations requires checking conditions carefully, as well as paying attention to how the data were collected. If the required conditions hold, then inferences can be made about the intercept and slope, as well as about predicted values of the regression line.

This chapter builds on concepts introduced in Chapter 4. There, we explained that linear regression can be used to understand associations between numerical variables, and we discussed how to use these associations to make predictions about future observations. In the meantime, you've also learned how to take knowledge about a sample and extend it to the population. In this chapter, we bring these two concepts together and show how we can use regression to make inferences about the population.

We examine two types of inference. The first is inference regarding the slope of the linear model. This is important because if the slope is 0, then there is no (linear) association between the variables. For example, biologists are interested in knowing about which factors contribute to high mercury levels in fish in freshwater lakes. If the slope of a regression between the mercury levels of fish caught in lakes and the acidity levels of the water in those lakes is not zero, then they can conclude that the acidity level of the water is related in some way to the mercury levels of the fish in those waters.

The other use for inference is to make predictions. You saw several examples of this in Chapter 4, such as predicting the end-of-year grade point average of a first-year college student on the basis of her SAT score. We're now ready to present techniques that will allow you to describe the uncertainty in these predictions.

CASE STUDY

Another Reason to Stand at Your Desk?

As we grow older, our brains shrink. In particular, an area called the medial temporal lobe (MTL) tends to get thinner. Is there anything we can do about it? Some research has suggested that the more time we spend sitting per day, the thinner the MTL tends to be. To reach this conclusion, researchers examined 35 older adults in good mental and physical health. The thickness of their MTL, measured through Magnetic Resonance Imaging (MRI) techniques, averaged 2.53 millimeters (mm), with a standard deviation of 0.19 mm. The distribution was approximately Normal. The participants were also interviewed about their levels of daily physical activity and, in particular, about how much time they spent sitting each day. This time ranged from 2 to 15 hours, with an average of 7.2 hours and a standard deviation of 3.32 hours.

In this chapter, you'll see how the method of regression can be used to determine whether an association between sitting and MTL thickness exists and, if so, how regression can be used to measure it. The same approach can be used to predict a given person's MTL thickness once we know how much time they have spent sitting.

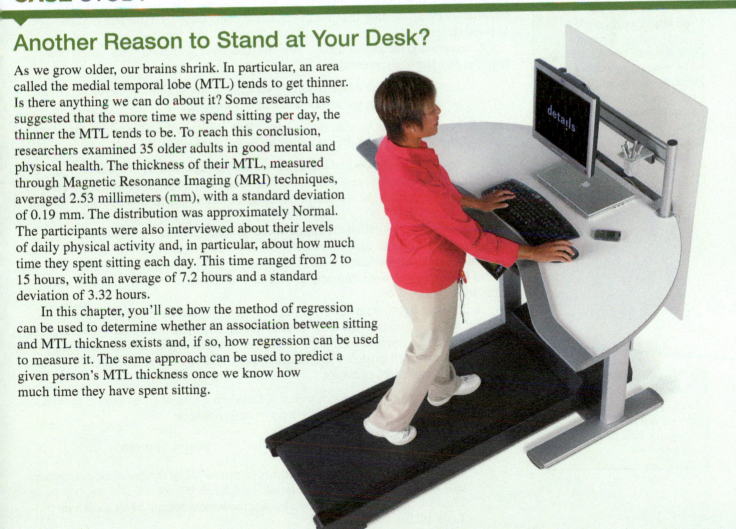

The Linear Regression Model

A statistical model is a set of features that we expect our data to have. The closer our model fits reality, the better will be our inference from the model to reality. An important statistical skill is to determine how well the data fit the model. The adage "garbage in, garbage out" applies. If we don't have good agreement between model and reality, then we won't make useful inferences about reality.

Components of the Model

Recall from Chapter 4 that the regression line is a method used to summarize the association between two numerical variables. In Chapter 4, we required only that the association be linear. This was all that was needed for us to summarize, describe, and make predictions about associations via the regression line. Requiring linearity remains the most important condition of the model in this chapter. If the trend is not linear, then our estimates and predictions will be biased, perhaps severely.

However, we now wish to do more than simply describe an association between variables. We now want to be able to measure our uncertainty in making predictions so that we can make inferences about the population. The first step is to provide a statistical model that describes this uncertainty. With this model, we can then quantify our uncertainty by calculating margins of errors for confidence intervals and p-values for hypothesis tests.

Conceptually, the model tells us how the data were generated. This generation happens in a two-step procedure:

1. For a given value of the predictor variable, a straight-line equation determines what the value of the response variable *should* be.

2. Next, some random "noise" is added that results in the *observed* value of the response variable falling a bit above or below the line.

The first step is often called the **deterministic component** of the model—deterministic because it says, "This is the real shape that determines which observations we shall see." For all of the models in this text, the deterministic component is always a straight line.

The second step is called the **random component**. The random component acknowledges that, even though the deterministic step might be the true structure of the association, randomness also appears that partially hides this structure. This randomness creates uncertainty in our inference. The random component tells us what that uncertainty looks like.

To illustrate how the model generates data, let's say we have selected four books of different lengths. We know the total number of pages in each book exactly, because we can just open the book and look at the final page number. These numbers are 203, 317, 409, and 765. We will exploit what we learned in Chapter 4 about the relation between page numbers and width of books.

1. The regression line from Chapter 4, based on a sample of 24 books, says that the width of a book (in millimeters (mm)) should be

$$\text{Predicted width} = 6.22 + 0.0366 \, (\text{number of pages})$$

This is the deterministic component of our model. The deterministic component tells us that our 203-page book *should* be $6.22 + 0.0366 \times 203 = 13.6498$ mm wide. For our four books, the widths should be as shown in Table 14.1.

Looking Back

Describing Associations
In Chapter 4 you learned that in describing associations, you should examine the trend (positive, negative, nonexistent), strength (strong trends have little vertical scatter), and shape (linear or nonlinear trend) and report them in the context of the data.

Looking Back

Finding the Regression Line
Formula 4.2 in Section 4.3 shows you how to find the regression line.

Number of Pages	Width according to Straight-line Model (mm)
203	13.6
317	17.8
409	21.2
765	34.2

◀ **TABLE 14.1** The number of pages for four books and their predicted width.

These points are shown in Figure 14.1a, along with the straight-line equation that tells us where other books *should* appear.

2. In statistics, as in real life, what should happen is rarely what does happen. To account for this, the random component of the model adds random noise to these widths. Instead of a book with width 13.6 mm, we see 13.6 mm plus-or-minus some random amount. We could simulate this observed width with a random number generator on a computer. We would instruct the computer to draw a number from a Normal distribution with mean 0 and standard deviation of, say, 5 mm. The value we actually observe, then, is the deterministic value (13.6) plus this random amount. If this random amount is positive, then the book we see is wider than the expected width. In this case, the observation will be above the regression line on a scatterplot. If the amount is negative, then the book is narrower than what we expected, and the observation will fall below the regression line.

In real life, where does this randomness come from? It comes from a variety of sources. Different books with the same number of pages have different widths because of the type of paper used, the way the book is bound, the type of cover on the book (hard or soft), and other unknown reasons. We also introduced some measurement error. Measurement error occurs because measurements vary from measure to measure, so if we measured the width of a book the next day, we might get a slightly different value. Thus, measurements of the same object vary but are centered on the true value. The results of adding this random noise are shown in Figure 14.1b.

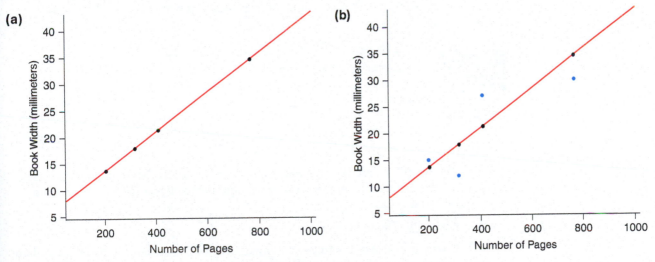

▲ **FIGURE 14.1** **(a)** The first step of the model determines the book widths so that the points fall on the straight line that they *should* follow. **(b)** The second step adds random noise, which moves each point slightly above or below the line by an amount randomly sampled from a Normal distribution.

EXAMPLE 1 Model of Mercury Level in Fresh Water

Biologists studying the relationship between the mercury levels in freshwater lakes and the mercury levels of fish living in those lakes believe that the deterministic component of the relationship is a straight line. A scatterplot (not shown) shows that even though the general trend is linear, the points do not fall exactly on a straight line.

QUESTION   What factors might account for the random component of this regression model?

SOLUTION   Variability might appear in the instruments used to measure mercury levels. The size of a fish might also affect the amount of mercury in its system. Different breeds of fish might process mercury differently, and this might affect the amount of mercury found.

TRY THIS!   Exercise 14.3

Table 14.2 lists, for each book in our sample, the width we should have seen (if the widths were completely determined by the straight line), the widths we actually did see, and the deviation of the actual widths from the straight-line widths.

▶ **TABLE 14.2** Each row represents one book, and we list the number of pages, the width the book should have if all of the points perfectly followed the linear model, the actual width, and the deviations (actual width minus straight-line width).

Pages	Predicted Width	Actual Width	Deviation
203	13.6	15	1.4
317	17.8	12	−5.8
409	21.2	27	5.8
765	34.2	30	−4.2

The deviations from the straight line have a technical name: **residuals**. *Residual* means "what's left over after something is taken away" or "remainder." The idea here is that the residual is the excess that doesn't fit on the line. Sometimes this deviation is referred to as the error, because it represents how far away the observed value is from what it should be (according to the straight line). The total deviation is measured by computing the standard deviation of all of the residuals, and is represented with the Greek symbol σ (sigma). You can think of σ as measuring the amount of random variation present.

> **KEY POINT**
>
> A residual is the difference between the actual (observed) y-value and the predicted y-value that lies on the line. In other words, it is the actual value minus predicted. By examining the collection of residuals, we can understand how the real-world data differ from our straight-line predictions.

Up to now, we've described the linear model conceptually, but to be more precise, we now list the particular conditions that must hold for the linear model to be a valid description of the data.

1. Linearity: The trend is linear.

2. Normality: The errors follow a Normal distribution with mean 0 and a standard deviation represented by σ. In shorthand notation, the distribution of the residuals must be $N(0, \sigma)$.

3. Constant standard deviation: The standard deviation σ must be the same for all values of the predictor variable. This will be explained in more detail shortly. (For example, see Figure 14.9).

4. Independence: The errors must be independent of one another. Knowing that one observation is, say, above the line, should tell us nothing about whether the next observation is above or below the line or by how much.

Checking the Conditions of the Model

A very useful tool for checking conditions 1 and 3 of the model is the **residual plot**. A residual plot is a scatterplot that has the residuals on the vertical axis and the original *x*-values on the horizontal axis. Residual plots are like magnifying glasses, in that they zoom in on potential problems in the model and make it easier to spot deviations from the linearity condition and the constant standard deviation condition.

Most statistical software packages either produce residual plots automatically as part of the regression analysis or will produce them upon request. Figure 14.2 shows the residual plot for our four books.

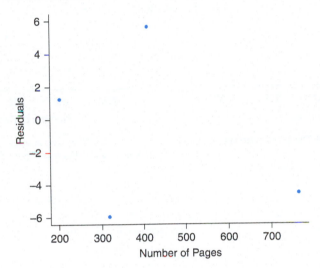

◀ **FIGURE 14.2** A residual plot for the four data points in Table 14.2. The vertical axis indicates the residuals (the true width minus the width if the point fell exactly on the straight line), and the horizontal axis marks the values of the predictor variable (the number of pages).

Figure 14.3a shows a scatterplot of the cost of a plane trip versus the distance traveled for four trips. The regression line is included on the scatterplot. The colored vertical lines represent the residuals. The value of the residual is the length of the vertical line, which is the distance between the point and the regression line. If the point is above the line, then the residual has a positive value. If the point is below the line, then the residual has a negative value.

Figure 14.3b shows these residuals plotted on the vertical axis. It illustrates that the residual plot magnifies the residuals so that they can be seen more clearly. The scale of Figure 14.3b makes the residuals easier to see.

> **! Caution**
>
> **Residuals or Data?**
> Check the label on the vertical axis. If it says "Residuals," this is a residual plot and should have no pattern. If it instead names a variable, then it is a scatterplot and should show a linear trend.

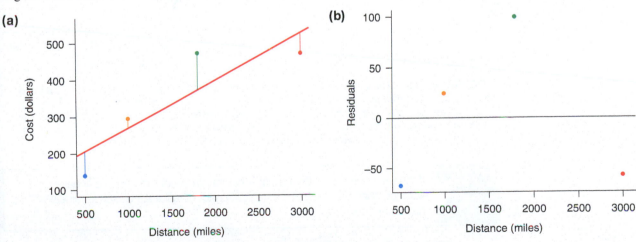

▲ **FIGURE 14.3** **(a)** The original scatterplot with the regression line. **(b)** The resulting residual plot.

Tech

Checking Linearity   If the trend in the original data is linear, then the residual plot should have *no* slope. In other words, the residual plot should have a flat trend. For example, Figure 14.4a shows a scatterplot for the larger collection of books; width is plotted against length, measured in terms of number of pages. The regression line has been added to help you see that the trend is linear. Figure 14.4b shows the residual plot. The residual plot has no trend. This is good, because it indicates that the first condition of the linear model is satisfied.

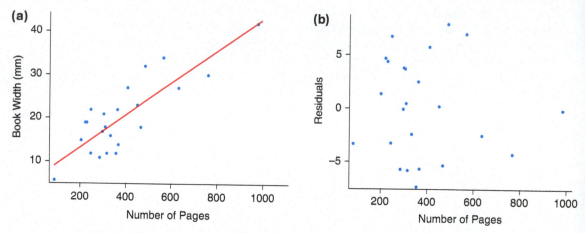

▲ **FIGURE 14.4** **(a)** The width of books versus the number of pages. The trend is linear. **(b)** The residual plot for the regression line in Figure 14.4a. The lack of a trend suggests that the first condition of the model, that *pages* and *width* have a linear trend, is satisfied.

When the trend is nonlinear, as it is for the relationship between the heights of the tallest buildings in the world and the numbers of floors in those buildings (shown in Figure 14.5a), the residual plot will display some sort of trend or pattern, as you can see in Figure 14.5b.

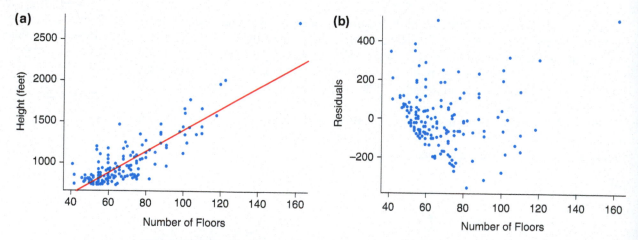

▲ **FIGURE 14.5** **(a)** The association between the height of buildings and the number of floors. The trend has a slight curve, particularly for buildings with between 40 and 60 floors, so the regression line is not a good fit. **(b)** The residual plot shows structure. For buildings with fewer floors, most of the residuals are positive, but the residuals decrease to mostly negative for the mid-sized buildings. (Source: http://architecture.about .com/od/skyscrapers/a/Worlds-Tallest-Buildings.htm, accessed September 13, 2013.)

KEY POINT   If the first condition of the model—that the shape of the trend be linear—is satisfied, then the residual plot (a scatterplot of residuals versus *x*-values) will have no slope.

You may be wondering why we need a residual plot to tell us whether a trend is linear or nonlinear. Can't you tell just by looking at the scatterplot? It is true that in many situations, you *can* tell from the scatterplot. But the residual plot serves as a sort of microscope and makes it easier to see subtle nonlinear patterns.

Figure 14.6 shows data collected by Dan Teague, an instructor at the North Carolina School of Science and Mathematics. He filled an urn with lemonade, opened the urn's spigot, and recorded the depth of the lemonade in the urn over time (seconds). At first glance, the trend looks fairly linear.

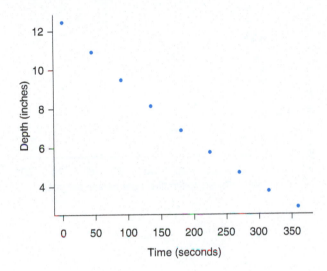

◀ **FIGURE 14.6** The depth of lemonade in an urn as the urn is drained. Does this trend look linear?

Now consider the residual plot (Figure 14.7).

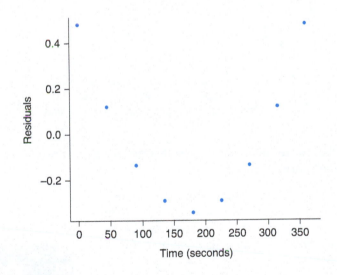

◀ **FIGURE 14.7** The residual plot for Figure 14.6. The trend in the residual plot tells us that the trend between depth of lemonade in the urn and elapsed time is nonlinear and does not satisfy the linearity condition.

The residuals show a very strong trend. The residual plot highlights the deviations from a straight line that are hard to see without very careful scrutiny. Look back at Figure 14.6, and you can see that the points do not exactly follow a straight line. (In fact, they follow a quadratic curve.) But without the residual plot, we might not have noticed that these data do not satisfy the linearity condition.

EXAMPLE 2 To Life!

In Chapter 4 we presented a scatterplot (Figure 4.29, repeated below) showing the association between mortality in a country and the wine consumption for that country in liters per person per year. Figure 14.8 shows the residual plot that results from fitting a linear regression model.

▶ **FIGURE 4.29 (repeated from Chapter 4)** Mortality (deaths per thousand) and annual wine consumption (liters per person per year) for several countries.

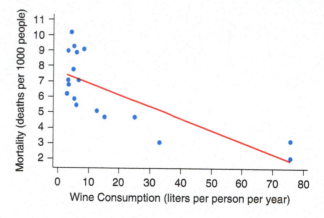

▶ **FIGURE 14.8** Residuals from a linear regression to predict mortality for a country, based on per-capita wine consumption.

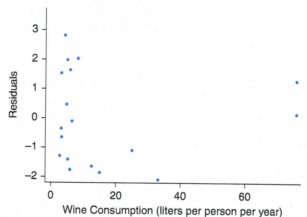

QUESTION   Does the residual plot suggest that the association between mortality and wine consumption is linear? Explain.

SOLUTION   The residual plot shows a trend: From left to right, the residuals drop sharply at first and then rise. This indicates that the association between mortality and wine consumption is nonlinear. The linearity condition of the model does not hold.

TRY THIS!   Exercise 14.7

Checking the Constant Standard Deviation (SD) Condition The

constant SD condition of the linear model says that the standard deviation in the *y*-values must be the same for all values of the predictor variable. Essentially, this means that the amount of vertical spread about the line is the same across the entire line. This is easier to explain with a picture of what happens when the condition fails.

Figure 14.9a shows an association with simulated data. We used simulated data so that we can exaggerate the problem. As you can see, the association shows greater and greater spread as you read from left to right. This association does not have the same standard deviation across all values of the predictor variable, x. For small values of x (say, between 0 and 5), the vertical spread is quite small. But for larger values of x (between 15 and 20), the y-values vary anywhere from 20 to 100.

Figure 14.9b is the residual plot created from this scatterplot. Note that it has a "fan" shape. A fan shape, whether it gets wider or narrower going from left to right, is a telltale sign that the condition of constant standard deviation has not been satisfied. We can see a fan shape even in the original scatterplot (Figure 14.9a) for these simulated data, because we exaggerated the increasing spread. But the residual plot (Figure 14.9b) magnifies this shape and makes it easier to see. In general, a residual plot highlights the existence of nonconstant SD even when it's hard to see in the original scatterplot. A good residual plot will not show any trends or any fan shapes; it will be featureless.

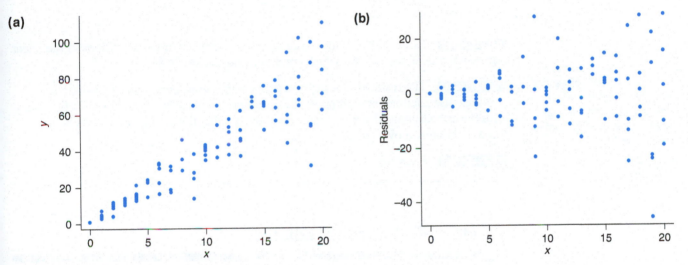

(a)

(b)

▲ **FIGURE 14.9 (a)** Simulated data that exhibit a nonconstant standard deviation. This violates one of the conditions that must hold for the linear model to be a valid description of the data. **(b)** Residual plot for Figure 14.9a. The increasing fan shape is a sure sign that the condition requiring constant standard deviation for all values of x has been violated. The residual plot clearly shows little vertical variation for low values of x, where residuals range from about -5 to $+5$. However, there is quite a bit of variation for larger values of x (where residuals range from about -40 to 25).

EXAMPLE 3 Predicting the Cost of a Vacation Rental

The website AirBnB lets homeowners rent out their homes to travelers. What factors determine the cost of these short-term rentals? One reasonable possibility is the number of bedrooms in the home. AirBnb provides data on its website, and we used data from the city of Austin, Texas, to predict the cost of a rental based on the number of bedrooms. The resulting model was

$$\text{Predicted Price} = 33.6 + 102.2 \, (\text{Number of Bedrooms})$$

Figure 14.10 provides a residual plot for this model.

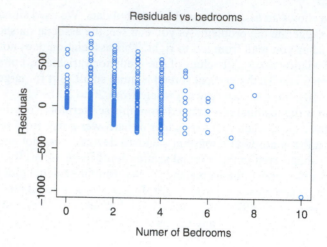

 Data Moves ► The price data from AirBnB contain a common frustration to data analysts everywhere. We want to analyze price as a numerical variable, but the prices including a dollar sign, and so the software interprets the prices as characters, not numbers. To fix this, we need to apply a data move to remove the dollar sign from the prices and to convert the values to numbers. The original data are the "listings" file from http://insideairbnb.com/get-the-data.html.

File ► airbnbaustin.csv

QUESTION   Evaluate the residual plot to determine whether the conditions of the linear regression model hold. If not, which conditions do not hold? Why?

SOLUTION   The residual plot shows a decreasing trend. This means that there is a nonlinear association between price of a rental and the number of bedrooms. For homes with fewer bedrooms, the model tends to under-predict the price, while for homes with more bedrooms, it over-predicts.

TRY THIS!   Exercise 14.8

EXAMPLE 4   A Better Predictor?

The size of a house might be a better predictor of its price on AirBnB than were the number of bedrooms. The regression model predicts

$$\text{Predicted Price} = 203.6 + 0.10 \text{ size}$$

QUESTION   The residual plot is shown in Figure 14.11. Comment on the validity of the linear model.

► **FIGURE 14.11** Residual plot for predicting the price of an AirBnB rental in Austin, TX based on the size of the house.

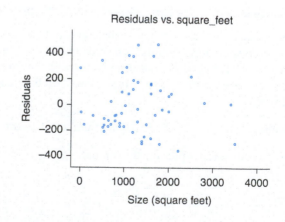

SOLUTION The residual plot shows a slight fan shape. For smaller homes (0 to 1000 square feet) there is little vertical spread. For larger homes, there is a greater amount of spread. This suggests that larger homes have more variability in price than do smaller homes.

TRY THIS! Exercise 14.9

> **KEY POINT** A residual plot with no structure is a good thing: It indicates that the linearity and constant SD conditions are satisfied. A fan shape in a residual plot indicates that the constant SD condition does not hold. A trend in the residual plot indicates that the linearity condition does not hold.

> **↻ Looking Back**
>
> **Checking Normality**
> Recall from Chapter 13 that the QQ plot is used to check whether a sample distribution is approximately Normal. If the points follow (more or less) a straight line, then the sample distribution is approximately Normal.

Checking Normality To check whether the errors follow a Normal distribution, examine the distribution of the residuals. You can look at either a histogram or a QQ plot, but the QQ plot is more useful in most situations.

Figure 14.12 shows a histogram (a) and a QQ plot (b) for the residuals from the home prices in Long Beach. A Normal curve is superimposed over the histogram, which shows us that the distribution of residuals is more "peaked" than a Normal and has a much longer right tail. The QQ plot is not a straight line, which confirms that these residuals are not Normal.

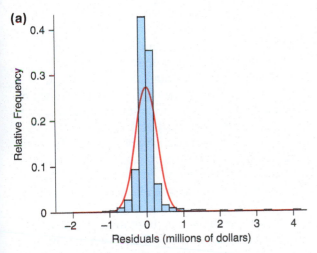
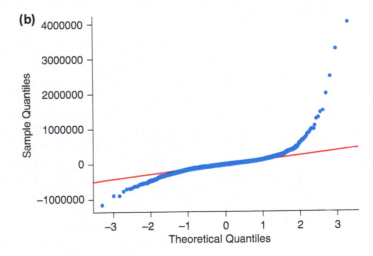

▲ **FIGURE 14.12** **(a)** The histogram of the residuals shown in Figure 14.9 (from the prices of homes, given house size) indicates that these residuals do not follow a Normal distribution. **(b)** This is confirmed by the QQ plot, which does not follow a straight line.

EXAMPLE 5 Book Widths

Figure 14.13a shows the residual plot from predicting the width of a book on the basis of the length of the book (measured in terms of number of pages). Figure 14.13b is a QQ plot of these residuals. The residual plot has no features: no trend and no fan shape. This suggests that two of the conditions of the linear model—linear trend and constant standard deviation—are satisfied.

Tech

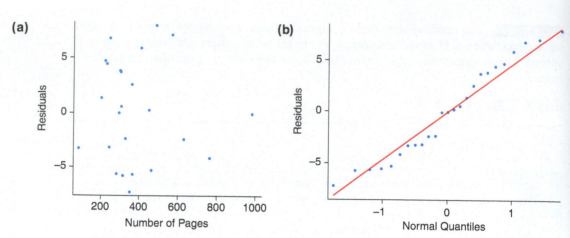

▲ FIGURE 14.13 **(a)** Residual plot for predicting the width of a book on the basis of the number of pages in the book. **(b)** The QQ plot for residuals that comes from predicting the width of a book on the basis of the number of pages in the book, using a straight line.

QUESTION   Is the Normality condition of the linear regression model satisfied?

SOLUTION   Yes. Because the points of the QQ plot generally follow a straight line, we can conclude that the residuals are approximately Normally distributed.

TRY THIS!   Exercise 14.12

Checking the Independence Condition   The independence condition usually cannot be checked by looking at the data. If you know that the data were collected in a particular order, then you might try plotting the data against the order in which they were collected. If you see a trend, it indicates that the observations might not be independent. However, you generally do not have this information. Sometimes, those who collected the data can give you enough information to determine whether this condition might be violated. For example, if you are studying the effects of a weekly music class on math scores for students in a school, you might suspect that scores of students who are in the same classroom with the same math teacher might not be independent, because if one student does very well in math, it might indicate that others in that class did very well in math, too.

In most situations, however, you will have to assume that this condition holds and be prepared to accept that your conclusions might not be sound if it is later discovered that the data are not independent.

KEY POINT   You should always check that the conditions of the linear regression model hold before interpreting the regression line.
1. Linearity: Use a residual plot. If it has no trend, the linearity condition is satisfied.
2. Constant standard deviation: Use a residual plot. If it has no fan shape, then the constant SD condition is satisfied.
3. Normality: Make a QQ plot of the residuals. If it (mostly) follows a straight line, the normality condition is satisfied.
4. Independence: Review any information you have about how the data were collected to decide whether the independence condition is satisfied.

Using the Linear Model

Marcus Vitruvius Pollio was a Roman architect and engineer who lived in the first century B.C.E. He proposed a theory of architecture concerning the qualities that make buildings beautiful. Vitruvius claimed that a sense of proportion was part of natural beauty. This belief led him to study the proportions of natural creatures, including humans. Vitruvius claimed, for example, that the typical human's arm span (the length from fingertip to fingertip with both arms extended to the side) was the same as his or her height. Over 1000 years later, Leonardo da Vinci was interested in this idea and made a famous drawing known as "The Vitruvian Man" (Figure 14.14). Variations of this picture appear on the euro coin and in many other places as well.

However, is the Vitruvian Man's proportion (arm span equals height) true? To find out, we could collect data on a class of adult students, measuring their arm spans and heights. We could then plot arm span against height. If Vitruvius was correct, what would be the intercept and slope of a regression line for these data? (We must assume that the linear model conditions hold, of course.) Given the amount of variability in such data (measurement error, differences between people), we probably won't see exactly the same values as Vitruvius would have liked, even if his theory is correct. So how can we confirm or discredit his theory?

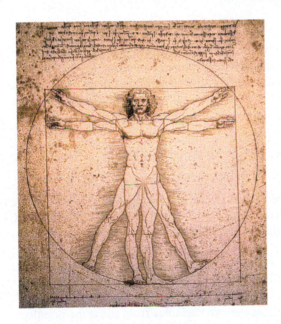

◄ **FIGURE 14.14** The length of the Vitruvian Man's arm span is the same as his height, as this drawing by Leonardo da Vinci illustrates.

Inference comes to the rescue. Our regression model will allow us to do hypothesis tests on the intercept and slope so that we can reject (or fail to reject) Vitruvius's theory.

Estimators for the Intercept and Slope

We now must be careful to distinguish between the *population* values of the slope and the intercept and *estimates* of those values. The population values are the values an intercept and a slope would have if we measured the entire population with perfect accuracy.

The population intercept and slope are parameters, and we estimate them with our sample data.

To represent the population values of intercept and slope, we follow standard statistical practice and use Greek characters for population parameters:

$$\beta_0 \text{ represents the population intercept.}$$

$$\beta_1 \text{ represents the population slope.}$$

Our method for estimating the population values uses the same estimates that we used in Chapter 4. However, now we have a different notation to represent them. The "hat" over the character tells us that the symbol now represents an estimate of the population parameter:

$$\hat{\beta}_0 \text{ represents the estimator of the population intercept.}$$

$$\hat{\beta}_1 \text{ represents the estimator of the slope.}$$

In addition,

$$\hat{y} \text{ represents the predicted value of } y.$$

The formulas for these estimators are the same as in Chapter 4, even though the notation is different:

$$\hat{\beta}_1 = r \frac{s_y}{s_x} \quad \text{and} \quad \hat{\beta}_0 = \bar{y} - \hat{\beta}_1 \bar{x}$$

We can then write the equation of the regression line:

$$\hat{y} = \hat{\beta}_0 + \hat{\beta}_1 x$$

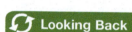

Looking Back

Regression Line Formulas
Formulas 4.2a, 4.2b, and 4.2c in Chapter 4 represent the same regression line formulas but use a different notation:

$$b = r \frac{s_y}{s_x}$$

$$a = \bar{y} - b\bar{x}$$

Predicted $y = a + bx$

As in Chapter 4, we expect that you will use these formulas only on rare occasions. Most of the time, your software will compute them for you. Your job is to interpret the output correctly and to evaluate whether the conditions of the regression model hold sufficiently well for you to learn anything from the computer output.

If the conditions for the linear model hold, then two useful facts about these estimators follow:

1. The sampling distributions of the estimators follow the Normal model.

2. The estimators are unbiased.

Even if the Normality condition is violated so that the errors are not Normal, the Central Limit Theorem (CLT) comes to the rescue. Because of the CLT, if the sample sizes are large enough, the sampling distributions of the intercept and the slope will be approximately Normal, and this approximation is better for larger sample sizes.

The fact that the estimators are unbiased means that typically our calculated values for the intercept and the slope, based on our data, will be about the same as the population values. What we mean by "typical" is that if each of a large number of researchers drew a random sample of the two variables from the same population and found the regression line, then the mean of all their intercepts and slopes would be the population values.

How close *our* estimates of the slope and intercept are likely to be to the true population values depends on how much these estimates vary from sample to sample. This variability is measured by the standard error. The estimated standard errors for the estimators will be given to you by the software package.

KEY POINT   The estimators for the intercept and slope of a regression line are unbiased. If the Normality condition holds, then the sampling distributions of these estimators are Normal. If the Normality condition does not hold, then the sampling distributions are approximately Normal (and the larger the sample size, the better).

Hypothesis Tests for Intercept and Slope

The test statistics for both the intercept and the slope have a familiar form, which was introduced in Chapter 8 for the one-proportion z-test and was also used in Chapter 9 for the one-sample t-test:

$$t = \frac{\text{estimator} - \text{null value}}{SE}$$

where SE stands for the estimated standard error (the estimated standard deviation of the estimator).

For the intercept,

$$t = \frac{\hat{\beta}_0 - \text{null}}{SE_{\hat{\beta}_0}}$$

For the slope,

$$t = \frac{\hat{\beta}_1 - \text{null}}{SE_{\hat{\beta}_1}}$$

The value for the estimated standard error is given in the software output. If the conditions for the linear regression model hold, then both of these test statistics follow a t-distribution with $n - 2$ degrees of freedom (n is the number of observations.)

Most statistical software is designed to automatically produce a test of a very specific hypothesis about the slope:

> H_0: The slope equals 0.
> H_a: The slope does not equal 0.

Or, in symbols,

> $H_0: \beta_1 = 0$
> $H_a: \beta_1 \neq 0$

This is a very important test. If the true slope is 0, then there is no linear association between the two variables—and no reason to do any further analyses. This means that this test could also be phrased as

> H_0: There is no linear association between the two variables.
> H_a: There is a linear association between the two variables.

or as

> H_0: The correlation is 0.
> H_a: The correlation is not 0.

The next example will show you how to find the information you need to do a test on the slope from the output provided.

↻ **Looking Back**

The t-Test
The test statistic for the one-sample t-test (Formula 9.2) is

$$t = \frac{\bar{x} - \mu_0}{SE_{\text{est}}}$$

where $SE_{\text{est}} = \dfrac{s}{\sqrt{n}}$

EXAMPLE **6** Mario Kart Auction

Suppose you want to buy a game on eBay, the online auction site. You make a bid, and if the next person to come along wants the item, he or she can bid more than you did. At the end of the scheduled time, the person who has made the highest bid gets to purchase the item. Suppose you want to buy a Mario Kart Wii game on eBay. (This is a popular video game that runs on the Wii system.) You might suspect that the more people there are bidding, the more expensive the item will be. David Diez, a statistician, collected data from a sample of auctions for Mario Kart on eBay.

QUESTION   Is there an association between the number of people who bid for these items and the final cost of the item? Use Figure 14.15, which shows the results from a regression model in which the total price for a Mario Kart was predicted from the number of bids for that item. State the null and alternative hypotheses, the observed value of the test statistic, and the p-value. State your conclusion using a 5% significance level. Assume that the conditions for the linear regression model are satisfied.

► **FIGURE 14.15** StatCrunch output from a regression model that predicts the price (totalPr), in dollars, for a Mario Kart game on an eBay auction for a given number of bids (nBids).

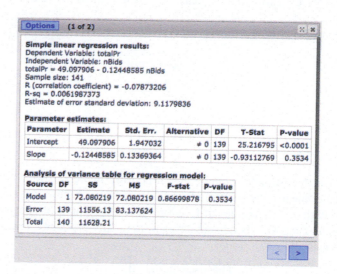

SOLUTION   To test whether there is an association between price (TotalPr) and number of bids (nBids), we test whether the slope of the regression line is 0. Usually, we would begin with a scatterplot to make sure there were no unusual observations and as a preliminary check that the association between these variables is linear. However, the problem statement tells us to assume that the conditions for the linear regression model hold, so we proceed accordingly.

H_0: $\beta_1 = 0$. There is no association between price and number of bids.
H_a: $\beta_1 \neq 0$. There is an association between price and number of bids.

The *t*-statistic is given at the intersection of the column labeled T-Stat and the row labeled Slope. The value is

$$t = -0.9311$$

The p-value is given as 0.3534. Because the p-value is larger than 0.05, we fail to reject the null hypothesis.

CONCLUSION   There is not enough evidence to conclude that the price of these items is associated with the number of bids placed on the item.

TRY THIS!   Exercise 14.15

If you wish to test a value other than 0 in the null hypothesis, you may need to do so by hand, since some statistical software packages don't support testing any value other than 0. StatCrunch *does* support this feature, but it's instructive to see how to carry out such tests by hand. The Vitruvian Man provides us with a useful example.

EXAMPLE 7 Vitruvian Students

One of the authors collected data from a statistics class. Students helped each other measure their arm spans and reported their heights (both in inches). The scatterplot is shown in Figure 14.16. Output from fitting a regression line is shown in Figure 14.17. We can use these data to test Vitruvius's theory that people's arm spans are generally equal to their height.

Tech

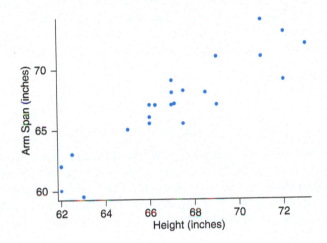

◀ **FIGURE 14.16** Data from college students showing arm span and height (both in inches).

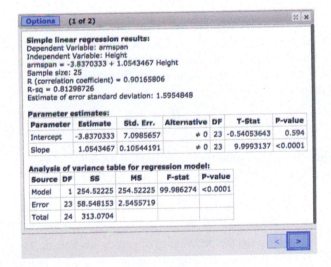

◀ **FIGURE 14.17** Output from the regression analysis, using height to predict arm span.

If Vitruvius was correct, then an equation of the line for the association between arm span and height (the deterministic component of the model) would be

$$\text{Predicted Armspan} = \text{Height}$$

or

$$\text{Predicted Armspan} = 0 + 1 \text{ Height}$$

In other words, if Vitruvius is correct, then the intercept should be 0, and the slope should be 1.

QUESTION   Was Vitruvius correct? Use the regression output provided in Figure 14.17 to test Vitruvius's theory. Be sure to carefully state the null and alternative hypotheses. State your conclusion using a significance level of 5%.

SOLUTION   From the output in Figure 14.17, we see that the equation for the regression line is

$$\text{Predicted Armspan} = -3.84 + 1.05 \text{ Height}$$

If Vitruvius was correct, then a person's arm span is equal to his or her height. This means that Vitruvius' theory says that the intercept should be 0 and the slope should be 1. We observed an intercept of -3.84 and a slope of 1.05. But are the differences between these observed values and those predicted by the Vitruvian theory simply due to chance?

First, let's examine the intercept.

$$H_0: \beta_0 = 0$$

$$H_a: \beta_0 \neq 0$$

The computer output assumes we are testing whether the intercept is equal to 0, and we can read the t-statistic directly from the output: $t = -0.54$. The p-value is large: p-value = 0.594. This is bigger than 0.05, so we do *not reject* the hypothesis that the intercept is 0. We conclude that the intercept is consistent with Vitruvius's theory. (Or, at the very least, there is not enough evidence to conclude that the data are *inconsistent* with his theory.)

As for the slope,

$$H_0: \beta_1 = 1$$

$$H_a: \beta_1 \neq 1$$

We can't simply use the output from the software, because the output is testing whether the slope is 0 or not. But we have all the ingredients we need to find the observed value of the test statistic, and we can use a calculator to find the p-value.

Recall that the test statistic is

$$t = \frac{\hat{\beta}_1 - \text{null}}{SE_{\hat{\beta}_1}}$$

Here, the null value is 1. The value for the estimated SE is found from the computer output, in the column that says "Std. Error" in the row for the slope. We find

$$t = \frac{\hat{\beta}_1 - \text{null}}{SE_{\hat{\beta}_1}} = \frac{\hat{\beta}_1 - 1}{SE_{\hat{\beta}_1}} = \frac{1.054347 - 1}{0.105442} = 0.5154$$

To find the p-value, we need the probability of getting a t-statistic as extreme as or more extreme than 0.5154. The output table tells us that we use a t-statistic with 23 degrees of freedom. We can look up this probability in a table or use a statistical calculator. Figure 14.18 shows output from StatCrunch's calculator feature. This tells us that the probability of getting a t-statistic equal to or greater than 0.5154 is about 0.306.

This is only half the story. Because our alternative hypothesis is a two-sided hypothesis, we also need the probability of getting a t-statistic equal to or smaller than -0.5154. Because the t-distribution is symmetric, the p-value will simply be twice 0.306; that is, p-value = 0.612. We again fail to reject the null hypothesis and find insufficient evidence that the slope is anything other than 1.

CONCLUSION   We lack evidence to refute Vitruvius's theory.

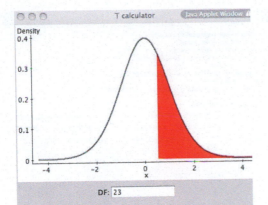

◀ **FIGURE 14.18** StatCrunch calculator showing the probability of getting a *t*-statistic greater than or equal to 0.5154 in a *t*-distribution with 23 degrees of freedom.

TRY THIS!   Exercise 14.17

Confidence Intervals for Intercept and Slope

Confidence intervals for the intercept and slope follow a familiar form:

$$\text{estimate} \pm t^* SE_{est}$$

This is the same form for all of the confidence intervals we've presented. The constant t^* is chosen from a *t*-distribution so that we get an interval with the desired confidence level. (Almost always, you'll use a 95% confidence level.) The estimate here can be either the intercept or the slope. The SE_{est} is the standard error and is obtained from the regression output.

For instance, to predict the price of a Mario Kart Wii game (totalPr) from the number of bids (nBids), we found this regression line from the output in Figure 14.15:

$$\text{Predicted totalPr} = 49.1 - 0.12 \text{ nBids}$$

The standard error for the slope is 0.134 (after rounding) and is found in the StatCrunch output column labeled *Std. Err*.

The best way to find the confidence intervals for the intercept and slope is to read them from the output. Most software provides confidence intervals for both parameters. Figure 14.19 shows StatCrunch output for the linear regression to predict the total price of the Mario Kart on the basis of the number of bids.

The lower and upper bounds of the confidence intervals are taken directly from the output. For example, we are 95% confident that the true intercept (the value of the intercept that we would get if we saw all eBay auctions and not just our small sample) is between 45.2 and 52.9. Perhaps more interestingly, we are 95% confident that the true slope is between −0.39 and 0.14. Because this interval includes 0, we cannot rule out the possibility that the true slope is 0. In other words, we cannot rule out the possibility that there is no linear association between the number of bids the Mario Kart receives at auction and the price that is finally paid.

> **↻ Looking Back**
>
> **Confidence Intervals**
> The confidence intervals for a slope and intercept follow the same structure as the confidence intervals for the population mean, Formula 9.1:
>
> $$\bar{x} \pm t^* SE_{est}$$

► **FIGURE 14.19** Regression output including confidence intervals for predicting the price of the Mario Kart on the basis of the number of bids received in an eBay auction. A 95% confidence level for the intercept is 45.2 to 52.9. A 95% confidence interval for the slope is −0.39 to 0.14.

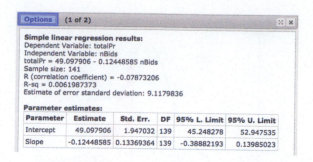

Options (1 of 2)

Simple linear regression results:
Dependent Variable: totalPr
Independent Variable: nBids
totalPr = 49.097906 - 0.12448585 nBids
Sample size: 141
R (correlation coefficient) = -0.07873206
R-sq = 0.0061987373
Estimate of error standard deviation: 9.1179836

Parameter estimates:

Parameter	Estimate	Std. Err.	DF	95% L. Limit	95% U. Limit
Intercept	49.097906	1.947032	139	45.248278	52.947535
Slope	-0.12448585	0.13369364	139	-0.38882193	0.13985023

Interpreting Confidence Intervals for Regression

The interpretation of the confidence intervals for the intercept and slope of a regression equation is exactly the same as the interpretation of the confidence intervals of other parameters. The confidence interval gives us a range of plausible population values. We can use this to test hypotheses, as the next example shows.

EXAMPLE 8 Vitruvian Confidence Intervals

The output in Figure 14.20 repeats the regression for predicting arm span from height for the class of statistics students used in Example 6. This time, the software was asked to provide confidence intervals for the intercept and slope.

► **FIGURE 14.20** Regression output to predict arm span from height for the same data as in Example 6.

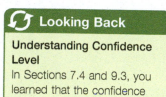

Looking Back

Understanding Confidence Level

In Sections 7.4 and 9.3, you learned that the confidence level is about the process of estimation, not about any particular interval.

Options (1 of 2)

Simple linear regression results:
Dependent Variable: armspan
Independent Variable: Height
armspan = -3.8370333 + 1.0543467 Height
Sample size: 25
R (correlation coefficient) = 0.90165806
R-sq = 0.81298726
Estimate of error standard deviation: 1.5954848

Parameter estimates:

Parameter	Estimate	Std. Err.	DF	95% L. Limit	95% U. Limit
Intercept	-3.8370333	7.0985657	23	-18.521535	10.847469
Slope	1.0543467	0.10544191	23	0.8362235	1.2724699

QUESTION Do these intervals support Vitruvius's theory that the intercept is 0 and the slope 1? State the 95% confidence intervals for the intercept and the slope, and explain whether they support or refute Vitruvius's theory.

SOLUTION The 95% confidence interval for the intercept is −18.5 to 10.8. The 95% confidence interval for the slope is 0.84 to 1.27.

Because the confidence interval for the intercept includes 0, we have no reason to reject the theory that the population intercept is 0. Similarly, the 95% confidence interval for the slope includes 1.

CONCLUSION   We cannot reject Vitruvius's theory, because the confidence interval for the intercept includes the value 0, and the confidence interval for the slope includes the value 1. This is consistent with the hypothesis tests in Example 6.

TRY THIS!   Exercise 14.21

SECTION 14.3

Predicting Values and Estimating Means

The regression line is also useful for making predictions about future observations, as you saw in Chapter 4. But there are two different types of questions that we can ask of a regression line, and it is important to be able to tell these two types of questions apart.

One type of question involves groups of individuals and is concerned with the mean value of the group. The other type of question is about an isolated individual and is concerned with an individual's value. We use the same number to answer both questions, but our *uncertainty* about this answer is different for the different questions.

Figure 14.21 shows a regression line that can be used to predict the amount of mercury found in largemouth bass (a freshwater fish) on the basis of the pH of the lake where they are caught. pH measures the acidity of the water. A value of 7 is neutral. A value of 0 represents the highest level of acidity, and a value of 14 represents the lowest level of acidity (the most alkaline reading). The data are based on samples of fish caught in 53 lakes in Florida (Lange et al. 1993).

> **⟳ Looking Back**
>
> **Predicting Observations**
> Review Example 3 in Chapter 4 to see how we predicted the width of a book on the basis of the number of pages.

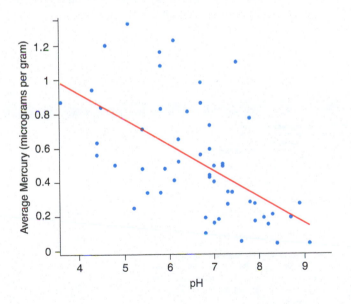

◄ **FIGURE 14.21** Mercury levels found in largemouth bass as a function of the pH of the water where they are caught. Mercury is measured in micrograms of mercury per gram of fish.

Evidence suggests that humans can get mercury poisoning from eating fish with mercury in their tissue. For this reason, it is important to understand and predict the mercury content of fish that swim in these lakes.

The equation of the regression line shown is

$$\text{Predicted Mercury Level} = 1.53 - 0.152 \, \text{pH}$$

There are two questions we might ask:

1. If the acidity of a lake measures pH 7, what is the mean level of mercury found in all fish in the lake?

2. I've just caught a fish from a lake that measures pH 7. What mercury level should we predict that this fish will have?

The answer to both questions is the same and is found with the regression line:

Predicted Mercury Level = $1.53 - 0.152 \times 7 = 0.466$ micrograms/gram

Question 2 has greater uncertainty, however, because it is asking about an individual fish. Question 1 has less uncertainty because it is asking about a collection of fish.

To express our uncertainty for questions such as question 1, we use a confidence interval. But to express uncertainty for predictions about individuals, we use something called a **prediction interval**.

For all practical purposes, prediction intervals function the same way as confidence intervals. The difference is that confidence intervals are always concerned with estimating population parameters—characteristics of a *population*. Prediction intervals are concerned with predicting values for *individuals*.

Prediction intervals are wider than confidence intervals, because there is more uncertainty in predicting an individual's value than in predicting the mean of an entire group. Individual scores have more variability than sample means. We know this because the sampling distribution for the mean has a smaller standard deviation than the population distribution. This increased uncertainty means that prediction intervals can be quite a bit wider than confidence intervals.

Figure 14.22 shows output from StatCrunch. Most statistical software either will show you both a prediction and a confidence interval or will ask you to choose which you want. StatCrunch shows you both (and allows you to input any value for *x* that you please). Note that the information you are looking for is given last, after the summary information about the regression model.

▶ **FIGURE 14.22** StatCrunch regression output for predicting mercury level in fish based on the water's pH. The last table shows a confidence interval (C.I.) for estimating the mean mercury content of fish caught in a lake with pH = 7, and the prediction interval (P.I.) for an individual fish caught in a lake with pH = 7. The prediction interval is (−0.11, 1.04) mg/g.

Options   (1 of 2)

Simple linear regression results:
Dependent Variable: Avg_Mercury
Independent Variable: pH
Avg_Mercury = 1.5309187 - 0.15230086 pH
Sample size: 53
R (correlation coefficient) = -0.57540012
R-sq = 0.3310853
Estimate of error standard deviation: 0.28164478

Parameter estimates:

Parameter	Estimate	Std. Err.	Alternative	DF	T-Stat	P-value
Intercept	1.5309187	0.20349287	≠ 0	51	7.5232055	<0.0001
Slope	-0.15230086	0.030313264	≠ 0	51	-5.0242318	<0.0001

Analysis of variance table for regression model:

Source	DF	SS	MS	F-stat	P-value
Model	1	2.0023627	2.0023627	25.242905	<0.0001
Error	51	4.0455128	0.079323781		
Total	52	6.0478755			

Predicted values:

X value	Pred. Y	s.e.(Pred. y)	95% C.I. for mean	95% P.I. for new
7	0.46481267	0.040628992	(0.38324656, 0.54637877)	(-0.10646573, 1.0360911)

Note that the confidence interval is much narrower than the prediction interval. In fact, the prediction interval is so wide that it includes negative values, even though negative mercury levels are impossible. This suggests that there might be too much variability in the data for us to make a useful prediction of how much mercury an individual bass might contain when the pH level is 7, although we can get a possibly useful estimate of the mean mercury level of all bass in the lake.

> **KEY POINT**
>
> Prediction intervals are used to express the uncertainty in predicting an individual observation. They are interpreted the same way as confidence intervals. A 95% prediction interval of (−0.11, 1.04) means that we are 95% confident that the amount of mercury found in a single fish caught from a lake with pH = 7 will be in this interval.

EXAMPLE 9   Frozen Dinners

A consumer group plots the number of calories advertised by the manufacturer for frozen dinners (single portion) against the actual caloric content based on laboratory testing. The group's regression line predicts that a frozen dinner advertised to have 1000 calories actually has, on average, 1100 calories.

QUESTIONS   Which of the two questions that follow can be answered with a confidence interval? Which can be answered with a prediction interval?

A. For all frozen dinners, what is the mean calorie content if the advertised content is 1000 calories?

B. The Frozen Man TV Dinner advertises that it contains 1000 calories. What is the actual calorie content of one dinner?

SOLUTIONS   Question A requires a confidence interval, because we are estimating the mean of a large group of frozen dinners. Question B requires a prediction interval, because we are predicting the true calorie content for an individual product.

TRY THIS!   Exercise 14.27

EXAMPLE 10   Step by Step

One of the authors has a Fitbit, a device that he keeps in his pocket that collects data about daily activity levels. You might expect the number of steps walked and the amount of calories expended in a day to be associated, and the output in Figure 14.23 provides evidence that, in fact, these are associated. (We verified, by examining the residuals, that the conditions required for the linear regression model hold.) One day the author misplaced his Fitbit, and so after a walk to the corner market, he estimated that he had taken 1000 steps.

QUESTION   About how many calories did the author expend that day? (The numbers might seem high for such a short walk, but Fitbit estimates calories based on the entire day and so includes basic metabolism expenditures.) Should he use a prediction interval or a confidence interval?

SOLUTION   We would predict that he will burn 2574.7 calories. Because we wish to predict calorie expenditure for a particular day, we use a prediction interval. We are 95% confident that the true calorie expenditure was between 2204 and 2945 calories.

▶ **FIGURE 14.23** Regression output for predicting the day's total calorie expenditure on the basis of the amount of steps walked in a day.

Options   (1 of 2)

Simple linear regression results:
Dependent Variable: Calories Burned
Independent Variable: Steps
Calories Burned = 2538.7392 + 0.035973125 Steps
Sample size: 78
R (correlation coefficient) = 0.58656135
R-sq = 0.34405422
Estimate of error standard deviation: 182.89085

Parameter estimates:

Parameter	Estimate	Std. Err.	Alternative	DF	T-Stat	P-Value
Intercept	2538.7392	38.17279	≠ 0	76	66.506513	<0.0001
Slope	0.035973125	0.0056975993	≠ 0	76	6.3137338	<0.0001

Analysis of variance table for regression model:

Source	DF	SS	MS	F-stat	P-value
Model	1	1333387.9	1333387.9	39.863234	<0.0001
Error	76	2542128.8	33449.064		
Total	77	3875516.7			

Predicted values:

X value	Pred. Y	s.e.(Pred. y)	95% C.I. for mean	95% P.I. for new
1000	2574.7123	33.529218	(2507.9331, 2641.4915)	(2204.3829, 2945.0417)

TRY THIS!   Exercise 14.31

What Can Go Wrong?

What happens if the conditions are not satisfied? Sometimes, disaster. Other times, not so much.

If the Linearity Condition Is Not Satisfied   Disaster. Estimates of slope and intercept may be biased. Predicted values and estimated means will be biased. Confidence intervals and prediction intervals are not reliable.

One thing to consider is a transformation. Sometimes, taking the log of all values for either *x* or *y*, or both, will "straighten up" a nonlinear trend. Figure 14.24a repeats the scatterplot given in Example 2. It shows a nonlinear trend between mortality in a country and the amount of wine consumed per person. Figure 14.24b shows the same data, only this time we plot the log of mortality against the log of wine consumption. Although the trend for the association of mortality and wine is not linear, the trend for the association of the log of mortality and the log of wine *is* linear.

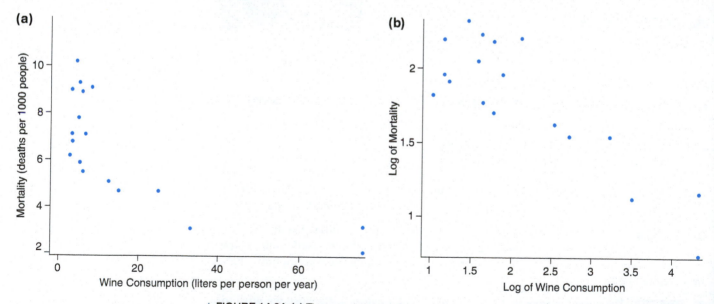

▲ **FIGURE 14.24** **(a)** The association between mortality in a country and per-capita wine consumption (in liters per person per year) is nonlinear (Leger et al. 1979). **(b)** The association between the log of mortality and the log of wine consumption is relatively linear, and a regression line can be fitted and interpreted.

Transforming the variables can make interpretation of the output tricky. We will not get into details here, but we advise you to consult a statistician if you find yourself in this situation.

If the Errors Are Not Normal If the errors are not Normal, all is not lost. Assuming that the other conditions hold, you will still get unbiased estimates of the slope and the intercept and of predicted values of y (or mean values of y) for a given value of x.

Because of the Central Limit Theorem, the p-values for hypothesis tests of the slope and intercept will be approximately correct. If the sample size is large, then this approximation can be quite good. Confidence intervals for the mean value of y at any given x-value will also be approximately correct. However, prediction intervals may not be accurate.

In some situations, the distribution of the errors is so far from Normal that the Central Limit Theorem doesn't rescue us. In many of these situations, more advanced techniques exist that can provide better p-values and confidence intervals than we can get simply by relying on the Central Limit Theorem. For example, if the response variable has only the values 0 and 1, then the linear regression model should be abandoned.

If the Standard Deviation Is Not the Same for All Values of x You will still have unbiased estimates of the slope and intercept and unbiased estimates of means and predicted values. However, the confidence intervals, prediction intervals, and p-values will not be accurate. Statisticians have other techniques for dealing with this situation. Transformations sometimes fix this problem, but sometimes they do not.

If the Errors Are Not Independent of Each Other Disaster. If the errors are not independent, then the regression model is not appropriate. One common situation in which errors are not independent is when dealing with time-series data. **Time-series data** occur when the predictor variable is time. For example, we might plot daily maximum temperature against the date, intending to predict tomorrow's temperature. Inference based on the linear regression model will fail for several reasons, but one is that the errors are not independent. If it is hotter than usual on one day (so the error is positive), there is more likely to be a hotter-than-usual temperature the next day. That is, if one day has a positive error, the next day is more likely to have a positive error as well.

Influential Points You should also be aware that regression models can be greatly influenced by outliers. Not all outliers, but outliers that appear at the extremes of the x-variable's distribution, can have a strong effect on the estimated slope. Beware of very low or very high x-values that also do not follow the linear trend. Don't delete these observations, but be aware that if they were not in the data set, or if they had different y-values, then your conclusions could change. Investigate these points to see whether you can learn anything about why they are exceptional. Perhaps they are mistaken entries; if so, you can delete them. Or perhaps they are evidence that an exception to the rule requires that you reconsider your model.

In general, it is good practice to take influential points out, find the regression line, and, by comparing your results with and without the influential point, determine whether your conclusions are greatly affected.

Interpreting r-squared Recall from Chapter 4 that the square of the correlation coefficient, r-squared or r^2, also called the coefficient of determination, is used to measure "goodness of fit." Roughly speaking, goodness of fit is a number that tells us how close the points come to our model. In order for us to interpret r-squared, the linearity condition for the linear regression model must be satisfied. A common mistake is to think that a high r-squared value means that the conditions are satisfied. In fact, you should check the conditions using the methods discussed in this chapter before you even *look* at the r-squared value.

 Looking Back

Coefficient of Determination
The correlation coefficient squared (r-squared) is called the coefficient of determination. In Section 4.4, you learned that r-squared measures how much of the variation in the response variable is explained by the explanatory variable.

The coefficient of determination ranges from 0% to 100% and represents the amount of variability in y that we have "explained" with our regression line. The higher the value, the better. How do we know whether r-squared is high enough?

Surprisingly, this is a question that shouldn't concern you. After you've checked that the conditions for the model are satisfied, the most important thing to check in the model is whether the slope is 0 or not. Look at the hypothesis test for whether the slope is 0. If you reject the null hypothesis and conclude that the slope is not 0, then congratulations! You have discovered an association between x and y. The value of r-squared is not sufficient for deciding whether the linear relationship is strong enough. We can decide whether the relationship is strong enough (and hence r-squared is high enough) by determining whether the regression line is useful for our purposes.

Recall that r-squared tells us how closely the points are clustered about the line. When you have a high value for r-squared, it means that your estimates and predictions will be more precise. But you shouldn't trust r-squared to tell you whether you should find prediction intervals or confidence intervals. Go ahead and compute them, and decide for yourself whether they are useful. If the intervals are too wide, then your r-squared is too low.

For example, suppose an admissions officer at a large college uses applicants' SAT scores to determine how successful they would be if admitted to the school. She has a large sample of data from current students that includes their SAT scores and their first-year GPAs. She carries out a regression to predict GPA on the basis of critical reading SAT score. After confirming that the conditions of the linear model are satisfied, she finds a regression line:

$$\text{Predicted GPA} = 1.108 + 0.00256\,(\text{Critical Reading SAT Score})$$

with r-squared $= 0.104$ (or 10.4%). From the output, she sees that the p-value for the hypothesis test for the slope is 0.02.

Now, this r-squared value seems low, but she can still learn quite a bit from the data. The output tells her to reject the null hypothesis that the slope is 0. This means she concludes that there really is a positive association between critical reading SAT score and first-year GPA. Now she's interested in what the mean GPA will be for all the students with a 650 critical reading score.

The 95% confidence interval for the mean GPA of all students with a 650 critical reading score is (2.7, 2.9). This is useful information for her, because it tells her that these students will, on average, have a GPA of about C+ at the end of their first year. Even though the r-squared value was low, it still yielded useful information: (1) There is a real association between critical reading SAT score and GPA, and (2) we can find useful confidence intervals for mean GPAs for groups of students.

On the other hand, suppose a worried parent calls the college. The parent's son got a score of 650 on his critical reading SAT. What will his first-year GPA be? The same regression line provides a 95% prediction interval of 1.7 to 3.9. In other words, we are confident that this student will end his first year with a GPA somewhere between a D+ and an A. Well, we didn't really need a regression analysis to tell us that. There is too much variability in the data to make predictions for individual students.

**CASE STUDY
REVISITED**

Another Reason to Stand at Your Desk?

If sitting for long periods affects the thickness of our brain's Medial Temporal Lobe (MTL), then we should see that people who report that they spend more time sitting also tend to have thinner MTLs. Researchers, who measured MTL thickness using MRI scans and measured time spent sitting based on self-reports, found, in fact, that just such an association exists. Figure 14.25 shows the association for the 35 subjects in this study. (The graphs were relabeled for ease of interpretation. The variable in the data set that measures MTL Thickness is named *Total* and the times reported sitting are in the *Sitting* variable.) The subjects were mentally and physically healthy adults and ranged in age from 45 to 75 years old.

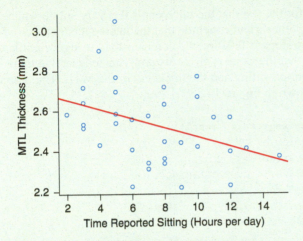

◄ **FIGURE 14.25** The scatterplot shows the association between hours spent sitting and MTL thickness for a sample of 35 healthy, older adults. The regression line is also shown and indicates a negative slope.

A residual plot (Figure 14.26) against Time Reported Sitting shows no obvious patterns. The lack of a trend in the residual plot means the assumption of linearity is satisfied, and the lack of a fan-shape means the assumption of constant standard deviation is satisfied. We see from the QQ plot shown in Figure 14.27 that the residuals are reasonably close to being Normally distributed.

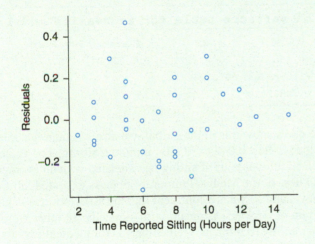

◄ **FIGURE 14.26** The residual plot shows no trend or fan-shaped, which suggests the assumption of linearity and constant standard deviation are satisfied.

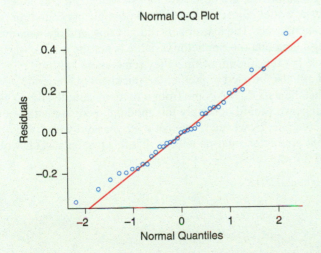

◄ **FIGURE 14.27** The QQ plot shows a fairly straight line, which suggests that the condition that the residuals follow a Normal distribution is satisfied.

Because we are satisfied that the conditions of the linear model are satisfied, we can interpret the model itself:

$$\text{Predicted MTL Thickness} = 2.700 - 0.023 \text{ (Hours Sitting)}$$

The p-value for the test that the intercept is 0 is very small, which means we can reject the null hypothesis and conclude that the intercept is non-zero. More important, the p-value for the slope is 0.0186, which means that, using a significance level of 0.05, we can conclude that the slope is also not zero. This tells us that there is an association between time reported sitting and the thickness of the MTL. The output, as produced by StatCrunch, is shown in Figure 14.28.

▶ **FIGURE 14.28** StatCrunch output from a linear regression model to predict thickness of MTL from Time Reported Sitting.

Simple Linear regression results:
Dependent Variable: TOTAL
Independent variable: Sitting
TOTAL = 2.6995057 - 0.022881901 Sitting
Sample size:35
R(correlation coefficient)= -0.39586138
R-sq = 0.15670623
Estimate of error standard deviation:0.17907882

Parameter estimates:

Parameter	Estimate	Std. Err.	Alter-native	DF	T-Stat	P-value
Intercept	2.6995057	0.073091938	≠0	33	36.933016	<0.0001
Slope	-0.022881901	0.0092402015	≠0	33	-2.4763422	0.0186

Analysis of variance table for regression model:

Source	DF	SS	MS	F-stat	P-value
Model	1	0.19665717	0.19665717	6.1322708	0.0186
Error	33	1.0582844	0.032069225		
Total	34	1.2549416			

Although we don't show this in the output, we also calculated a 95% confidence interval for the slope to be −0.042 to −0.004 mm/hour sitting. The value of the slope suggests that each additional hour of sitting is associated with an average decrease of 0.023 mm of thickness of the Medial Temporal Lobe. As expected, all the values in the interval are negative, and so we are confident that the association is negative. Whether a change of 0.023 mm of thickness is meaningful clinically (in other words, whether it has a real affect on our ability to think and remember) is for medical experts to determine. It's worth noting that the data are observational, so we cannot conclude cause and effect. A stronger analysis would adjust for known and possible confounding factors, such as age and education levels. The researchers did just this, and the negative association remained. We also note that the predictor variable is not the actual time people spent sitting each week, but the amount of time they remembered sitting.

This model can also be used to predict the thickness of someone's MTL, as long as we remember that the data do not come from a random sample and so might not apply to others. Certainly, we should not apply this model to people much younger or older than those in the sample. However, if we met a 50-year-old who spent about 7 hours per day sitting, a 95% prediction interval tells us that this person's MTL would likely be between 2.2 and 2.9 mm thick.

Source: Siddarth et al. (2018).

1 OVERVIEW

In this last Data Project, we offer a final bit of strategic advice for analyzing large data sets: think small.

2 GOAL

Learn to approach the analysis of big data sets by asking questions about subsets of the data.

3 ZOOMING IN

When confronted with large, complex data sets, the temptation is to feel that you have to do everything and use every variable. But this can be disastrous! It's much better to carve your analysis into small, solvable chunks and attack them separately.

For example, suppose your initial statistical question, when considering the very large *sampleofcrime* data set, is something along the lines of "Has Crime Increased in Los Angeles since 2010?" One approach is to just count the crimes each year and make a plot of this count against the year. But this produces a rather simplistic picture. It seems plausible that some crimes change in different ways than others. For example, perhaps when burglaries are down, car thefts go up. Or perhaps some regions of the city change in different ways. Maybe wealthier regions have seen a decrease in crime and poorer areas an increase. An overly simplistic analysis will miss this.

Project: Once you've understood the structure of your data, performed a reality check, and cleaned the data, the next step is to narrow your focus. Do you need to consider all observations, or is it better to focus on a subset of interest? For example, do you need to consider all years? All districts? All crimes? Write a statistical question that is narrow and specific to a particular area of interest and decide which features you can do without.

Assignment: Using the sample of crime data set, write one or two statistical questions using a narrow scope and answer them, providing the necessary numerical and graphical summaries in support.

Note that in narrowing your focus, you might find it helpful to recode the data or to create subsets. For example, if you want to examine burglaries, then there are several categories of burglaries that you might merge. Or perhaps you wish to focus only on crimes in winter. In that case, you should create a subset of the data that includes only winter months.

CHAPTER REVIEW

KEY TERMS

Deterministic component, *692*
Random component, *692*

Residuals, *694*
Residual plot, *695*

Prediction interval, *712*
Time series data, *715*

LEARNING OBJECTIVES

After reading this chapter and doing the assigned homework problems, you should

- Be able to check that the conditions for the linear model hold, and understand the consequences if the conditions do not hold.

- Know how to use the output from a regression analysis to perform and interpret hypothesis tests of the slope and intercept,

and know how to interpret confidence intervals for the slope and intercept.

- Be able to use the regression line to make predictions, and understand when a situation calls for a confidence interval and when it calls for a prediction interval.

SUMMARY

Making inferences from the regression model requires that the linearity condition, the Normality condition, the constant standard deviation condition, and the independence condition be satisfied. A residual plot, which plots the residuals against the *x*-values, can be used to evaluate the linearity and constant standard deviation conditions. If the residual plot has no trend, the linearity condition is satisfied. If the residual plot does not have a fan shape, the constant standard deviation condition is satisfied. A QQ plot of the residuals can be used to evaluate the Normality condition. The independence condition is harder to verify, but you should be wary of time series (regressions using time as the *x*-axis), because these are often not independent.

If the conditions are satisfied, then the standard computer output for a regression analysis can be used to determine whether there is a linear association between the variables (test whether the slope is 0) or whether the intercept is 0. Confidence intervals for the intercept and slope can be found if the situation requires an estimate of the value of these parameters. A confidence interval can be used to estimate the mean of a group for any given value of *x*, and a prediction interval can be used to predict the value for an individual who has any given value of *x*. Prediction intervals are wider than confidence intervals because there is more uncertainty in predicting an individual outcome than in predicting a mean.

SOURCES

Lange T. L., H. E. Royals, and L. L. Connor. 1993. Influence of water chemistry on mercury concentration in largemouth bass from Florida lakes. *Transactions of the American Fisheries Society* 122, 74–84, http://wiki.stat.ucla.edu/socr/index.php/NISER_081107_ID_Data (viewed April 19, 2010).

Leger, A., A. Cochrane, and F. Moore. 1979. Factors associated with cardiac mortality in developed countries with particular reference to the consumption of wine. *The Lancet* 313(8124), 1017–1020.

Siddarth, P., et al. 2018. "Sedentary behavior associated with reduced medial temporal lobe thickness in middle-aged and older adults. *PLoS ONE* 13(4): e0195549. https://doi.org/10.1371/journal.pone.0195549.

SECTION EXERCISES

SECTION 14.1

14.1 Predicting Test Scores A professor tells his class that he knows their second exam score without their having to take the test. He tells them that the second exam score can be predicted from the first with this equation:

Predicted second exam score $= 5 + 0.75$ (first exam score)

This tells us that the deterministic part of the regression model that predicts second exam score on the basis of first exam score is a straight line. What factors might contribute to the random

component? In other words, why might a student's score not fall exactly on this line?

14.2 Used-Car Values A student wishes to buy a used car. He finds a consumer website that says the price of a used car is determined by its age according to the following formula:

Predicted price in thousands of dollars $= 17 - 0.8$ (age in years)

This is the deterministic component of a regression model for predicting price on the basis of the age of the car. What factors might contribute to the random component? In other words, why might the price of the car he buys not fall exactly on this line?

14.3 Height and Age (Example 1) A doctor says he can predict the height (in inches) of a child between 2 and 9 years old from the child's age (in years) by using the equation

$$\text{Predicted Height} = 31.78 + 2.45\,\text{Age}$$

This tells us the deterministic part of the regression model. What factors might contribute to the random component? In other words, why might a child's height not fall exactly on this line?

14.4 Units and Semesters The registrar at a small college says she can predict the number of units that a full-time student has accumulated on the basis of the number of semesters the student has attended the college by using the equation

$$\text{Predicted Units} = 0 + 14\,\text{Semesters}$$

This tells us the deterministic part of the regression model. What factors might contribute to the random component? In other words, why might a student's number of units not fall exactly on this line?

14.5 Tallest Buildings in Dubai The table shows the number of floors and the height (in feet) of the five tallest buildings in Dubai. The regression model for predicting the height of a building from the number of floors is

$$\text{height} = -558.45931 + 19.934126\,\text{Floors}$$

Complete the table by finding the predicted heights and the residuals. (Remember that if the predicted value is greater than the actual value the residual will have a negative sign.) (Source: worldatlas.com)

Floors	Height (in feet)	Predicted Height	Residual
163	2717		
101	1398		
101	1358		
88	1289		
89	1250		

14.6 Tallest Buildings in China The table shows the number of floors and the height (in feet) of five of the tallest buildings in China. The regression model for predicting the height of a building from the number of floors is

$$\text{height} = 20.888067 + 16.124191\,\text{Floors}$$

Complete the table by finding the predicted heights and the residuals. (Remember that if the predicted value is greater than the actual value the residual will have a negative sign.) (Source: skyscrapercenter.com)

Floors	Height (in feet)	Predicted Height	Residual
128	2073		
115	1969		
111	1740		
101	1614		
89	1480		

TRY **14.7 Used Acuras (Example 2)** Figure A shows a scatterplot of the price and age of a random sample of used Acura MDX cars and includes the regression line. Figure B shows a residual plot based on the regression line. (Source: cars.com)

a. Is the regression model appropriate for these data? Explain.

b. How old is the car that is farthest from the regression line?

(A)

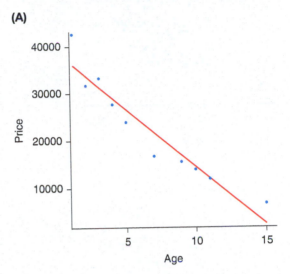

(B)

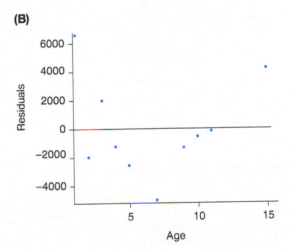

TRY **14.8 Law School Graduates (Example 3)** Figure A shows a scatterplot of the percent passing the bar exam within two years of graduation and the percent employed at graduation for U.S. law school graduates in 2015. Figure B shows the residual plot of the data.

Evaluate the residual plot to determine whether the conditions of the linear regression model hold. If not, which conditions do not hold? Explain.

(A)

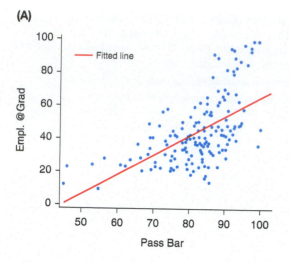

(B)

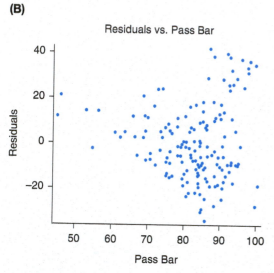

TRY **14.9 Semesters and Credits (Example 4)** Figure A shows a scatterplot for the number of semesters that students have attended a community college and the number of credits they have accumulated. Figure B shows a residual plot of the same data. Is the linear regression model appropriate for these data? Why or why not? Assume the observations are independently measured.

(A)

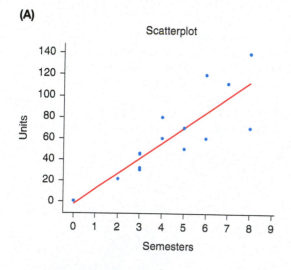

(B)

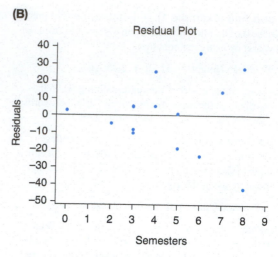

14.10 U.S. Population Figure A shows a scatterplot for the U.S. population (in millions) from 1950 to 1990. Figure B shows a residual plot for the same data. The linear model based on these data for predicting the U.S. population given the year is also given. Should the linear model be used to predict the U.S. population in 2020? If so, predict the 2020 U.S. population. If not, explain why use of this linear model would be inappropriate. (Source: demographia.com)

$$\text{pop} = -4518.89 + 2.397 \text{ year}$$

(A)

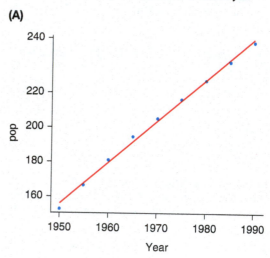

(B)

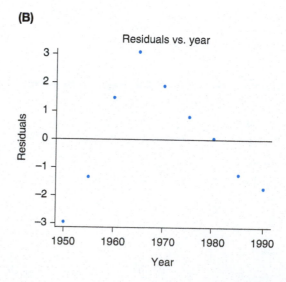

14.11 Salaries Figure A shows a scatterplot of the current salary (in thousands of dollars per year) and the beginning salary for many employees at one company. Figure B shows a residual plot of the same data. Is linear regression appropriate for these data? Why or why not?

(A)

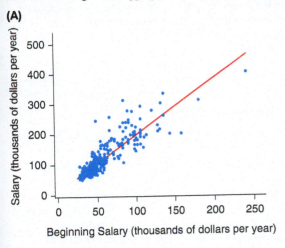

(B)

Scatterplot of Residuals vs. Beginning Salary

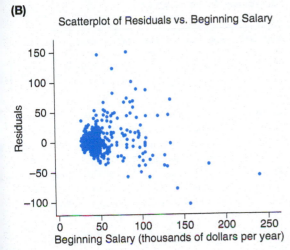

RY 14.12 Library Holdings (Example 5) Figure A shows a scatterplot of the number of print and electronic holdings for a sample of large U.S. libraries. Figure B shows a QQ plot of the residuals. Is the Normality condition of the linear regression model satisfied? Explain.

(A)

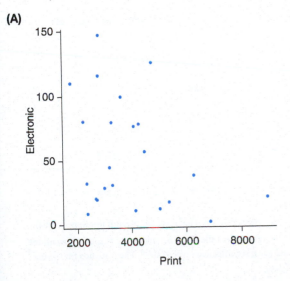

(B)

QQ Plot of Residuals

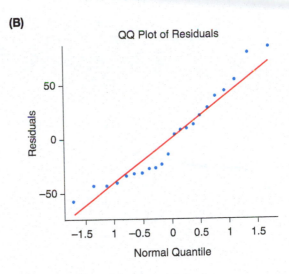

14.13 Wages of Twins Figure A shows a scatterplot of wages of twins for a group of 183 pairs of twins. Figure B shows a residual plot of the same data. Figure C shows a QQ plot of these residuals. Is the linear regression model appropriate for these data? Why or why not? Assume the observations are independently measured.

(A)

Scatterplot

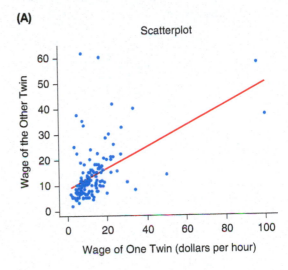

(B)

Residual Plot

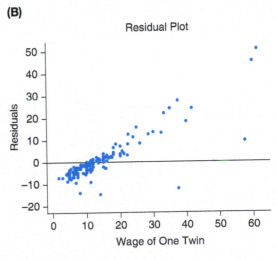

(C)

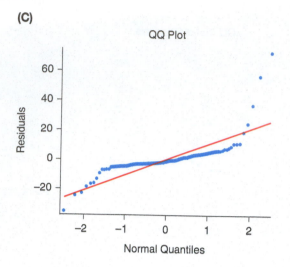

14.14 Simulated Data Figure A shows a scatterplot of some simulated data, and Figure B shows a residual plot of the same data. Is the linear regression model appropriate for these data? Why or why not? Assume the observations are independently measured.

(A)

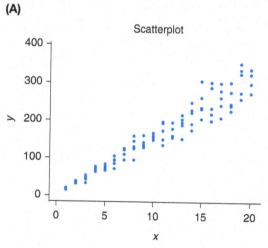

(B)

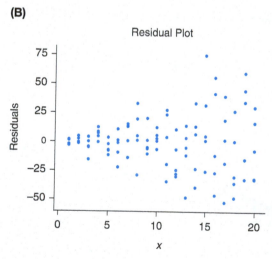

SECTION 14.2

TRY **14.15 Education Spending (Example 6)** Is there a correlation between state and federal spending on education? The following output shows the regression analysis on state and federal spending on education from a sample of states. Assume the conditions for the linear regression model are satisfied. Show all 4 steps of the hypothesis test.

Parameter estimates:

Parameter	Estimate	Std. Err.	Alter-native	DF	T-Stat	P-value
Intercept	-334.92627	162.7315	≠0	11	-2.058156	0.0641
Slope	0.29531596	0.021777028	≠0	11	13.560894	<0.0001

14.16 Movie Budgets Do movies with bigger budgets tend to make more money? The following output shows the regression analysis on movie budgets and box office gross (how much money the movie made) for a sample of movies released in 2016. Use the output to test if the slope is equal to 0. Assume the conditions for the linear regression model are satisfied. Show all for steps of the hypothesis test. (Source: the-numbers.com)

Parameter	Estimate	Std. Err.	Alter-native	DF	T-Stat	P-value
Intercept	240.75345	64.811772	≠0	12	3.7146562	0.003
Slope	0.2551418	0.42194617	≠0	12	0.60467856	0.5567

TRY **14.17 Education of Parents (Example 7)** Each of 29 randomly sampled community college students reported the number of years of formal education for his or her mother and father. For example, a value of 12 means that the parent completed high school but had no further education. (The numbers ranged from 0 to 18.) The regression equation shown in the output is for predicting years of education for the father on the basis of years of education for the mother. Assume that all the conditions required for regression are fulfilled.

a. Identify the intercept. Explain what it would mean if the intercept were zero.

b. Use the output given to test the hypothesis that the intercept is 0. Show all the steps, and use a significance level of 0.05. The statistics for the test of the hypothesis that the intercept is 0 can be found in the row labeled "Constant."

```
Regression Analysis: FatherEd versus MotherEd

The regression equation is
FatherEd = 2.78 + 0.637 MotherEd

Predictor   Coef    SE Coef   T     P
Constant    2.785   1.396     2.00  0.056
MotherEd    0.6370  0.1483    4.30  0.000


S = 3.63005  R-Sq = 40.6%  R-Sq(adj) = 38.4
```

14.18 Education of Parents Refer to exercise 14.17 and the output given.

a. Identify the slope. Explain what it would mean if the slope were 0.

b. Test the hypothesis that the slope is 0 using the output. Show all the steps, and use a significance level of 0.05. The statistics for the test of the hypothesis that the slope is 0 can be found in the row labeled "MotherEd."

c. What would it mean if the slope were 1 and the intercept 0?

* d. Test the hypothesis that the slope is 1 using a significance level of 0.05. (Hint: The df is 27.)

Use the table here to decide whether the p-value is less than 0.01, between 0.01 and 0.05, or more than 0.05 for part d.

alpha	0.05	0.01
tails	2	2
df		
26	2.056	2.779
27	2.052	2.771
28	2.048	2.763

*14.19 Student and Parent Heights A random sample of 29 community college students were asked their height in inches and the height of their biological parent of the same gender. The output of a regression analysis for predicting student height from parent height is shown. Assume that all the conditions required are fulfilled, and use a significance level of 0.05 for any tests.

a. What is the slope of the regression line? Interpret the value of the slope in context.

b. Test the hypothesis that the slope is 0 (significance level is 0.05), and explain what your answer means.

c. If the slope were 1, what would that say about the association?

*d. Test the hypothesis that the slope is 1 using a significance level of 0.05. Use the table for Exercise 14.18 to decide whether the p-value is less than 0.05. The degrees of freedom (df) is 27.

```
Regression Analysis: Student versus Parent
The regression equation is
Student = 5.63 + 0.936 Parent

Predictor   Coef   SE Coef    T     P
Constant    5.631  7.366    0.76  0.451
Parent      0.9355 0.1116   8.38  0.000


S = 1.97692  R-Sq = 72.2%  R-Sq(adj) = 71.2%
```

14.20 Trash The weight of trash (in pounds) produced by a household and the number of people living in the household were obtained for 13 houses. Refer to the Minitab regression output. Assume that all the conditions necessary for regression analysis are met.

```
Regression Analysis: Trash versus People
The regression equation is
Trash = 2.34 + 11.3 People

Predictor   Coef  SE Coef    T     P
Constant    2.340 6.869    0.34  0.740
People      11.300 1.867   6.05  0.000


S = 11.8519  R-Sq = 76.9%  R-Sq(adj) = 74.8%
```

a. What is the observed intercept, according to the equation?

b. What would it mean if the intercept were 0?

c. Test the hypothesis that the intercept is 0 using a significance level of 0.05.

TRY 14.21 Chicago Libraries (Example 8) Chicago public libraries collect data on the number of visitors and the number of library computer wifi sessions. A regression analysis was conducted using yearly totals for 2013. Assume the conditions for the linear model regression model are satisfied. (Source: data.cityofchicago.org)

a. Interpret the slope in context.

b. State the 95% confidence interval for the slope.

c. Based on the confidence interval for the slope, is there a relationship between the number of visitors and the number of library wifi sessions? Explain.

```
Simple Linear regression results:
Dependent Variable: ≠ wifi results:
Independent variable: ≠ Visitors
≠ Wifi sessions = -440092.705 + 0.21811867 ≠ visitors
Sample size:12
R(correlation coefficient)= 0.62936487
R-sq = 0.39610013
Estimate of error standard deviation:40080.425
```

Parameter estimates:

Parameter	Estimate	Std. Err.	DF	95%L.Limit	95%U.Limit
Intercept	-44092.705	67254.465	10	-193944.99	105759.58
Slope	0.21811867	0.085167288	10	0.02835417	0.40788321

14.22 Success and Retention A regression analysis was conducted using data on success rates and retention rates in transfer courses at a sample of California community colleges. Assume the conditions for the linear model regression model are satisfied.

a. Interpret the slope in context.

b. State the 95% confidence interval for the slope.

c. Based on the confidence interval for the slope, is there a relationship between success rate and retention rate in these courses? Explain.

```
Simple Linear regression results:
Dependent Variable: ≠ Transfer Retention
Independent variable: ≠ Transfer Success
Transfer Retention = 0.42111474 + 0.61582462 Transfer Success
Sample size:115
R(correlation coefficient)= 0.70232601
R-sq = 0.49326182
Estimate of error standard deviation:0.024295177
```

Parameter estimates:

Parameter	Estimate	Std. Err.	DF	95%L.Limit	95%U.Limit
Intercept	0.42111474	0.042023324	113	0.33785896	0.50437053
Slope	0.61582462	0.5871795	113	0.49949377	0.73215547

14.23 Trash and Confidence Intervals The output provided for exercise 14.20 provides a regression line to predict the amount of trash produced by a household on the basis of the number of people living in the household. Suppose you found a 95% confidence interval for the intercept of the regression line. Would that confidence interval include 0? (Refer to the output in exercise 14.20.)

14.24 Movie Budgets In exercise 14.16 you examined the association between movie budgets and box office gross income. Would a 95% confidence interval for the slope, based on the same data, include 0? Explain.

SECTION 14.3

14.25 Predicted GPA A student who has been accepted by two colleges wants to estimate what GPA he might get at the two colleges. His data consist of SAT scores and GPAs from random samples of recent graduates at each college. He wishes to predict his GPA at each school, using separate linear regression models for the two colleges. Should he use prediction intervals or confidence intervals for the prediction of his own GPA? Explain.

14.26 Used BMWs A used-car dealer is purchasing 50 used BMWs from one dealer in order to sell them for a profit. Working with collected data, the dealer has found a regression model to predict the selling price on the basis of the car's age. He wants to predict the total amount he will get for these 50 cars in order to make sure he does not lose money. All 50 cars are three years old (they were turned in after their leases expired). Should he use a confidence interval or a prediction interval for the mean selling price of these 50 cars? Explain.

TRY **14.27 Predicted Height (Example 9)** A mother wants to predict the height of her full-grown son on the basis of his height at the age of 8 years. Should she use a prediction interval or a confidence interval? Why?

14.28 Predicted GPA A dean of students at a college wishes to estimate the typical future cumulative GPA for all first-year students who earned a 2.0 GPA during their first year. Should she use a prediction interval or a confidence interval? Why?

14.29 Loggers A logging company has the diameter of each of a large number of trees and wants to estimate the mean number of usable cubic feet of wood the company will get if it cuts the trees down. Working with a sample of trees, company planners find the regression line that predicts the volume of lumber (in cubic feet) on the basis of the diameter of the tree. Should they use a prediction interval or a confidence interval? Explain.

14.30 Height of a Child A mother is interested in predicting the adult height of her 6-year-old daughter. She has information on the heights of mothers and daughters in her community. Should she use a prediction interval or a confidence interval for her prediction? Explain.

TRY **14.31 House Prices (Example 10)** Figure A contains the selling price and area (in square feet) of 81 recently sold homes in a region where a buyer wants to purchase a home.

   a. Use the equation to estimate the price of a home with 2500 square feet.

   b. The buyer wants to know the uncertainty in the prices he might pay for a home with 2500 square feet in this region. Should he use a prediction interval or a confidence interval? Explain.

(A)

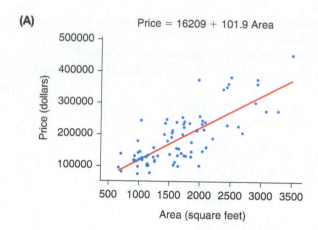

Price = 16209 + 101.9 Area

(B)

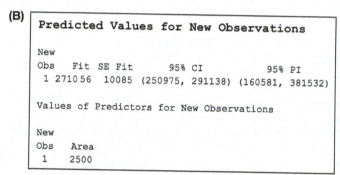

```
Predicted Values for New Observations

New
Obs   Fit  SE Fit      95% CI            95% PI
 1 271056   10085  (250975, 291138)  (160581, 381532)

Values of Predictors for New Observations

New
Obs    Area
 1     2500
```

   c. Report the correct 95% interval, confidence or prediction as determined in part b, for the predicted price. See Figure B.

   d. The buyer has prequalified for a loan up to $400,000. Is he likely to be able to afford a house in this area with 2500 square feet?

14.32 Math SAT Score and GPA Figure A shows information about a random sample of students' math SAT scores and GPAs at an unidentified four-year college.

   a. Use the formula on the graph to predict the GPA for a person with a math SAT score of 600.

   b. Figure B shows both a prediction interval and a confidence interval for a new SAT of 600; report both.

   c. One student wants to estimate the GPA he will achieve if he attends that school. Should he use the prediction interval or the confidence interval? Explain.

   d. Report the interval obtained for part c. Is it very useful? Explain.

(A)

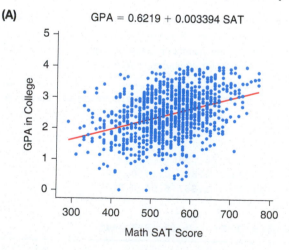

GPA = 0.6219 + 0.003394 SAT

(B)

```
Predicted Values for New Observations

New
Obs  Fit    SE Fit   95% CI            95% PI
 1 2.6582  0.0259  (2.6073, 2.7091)  (1.316, 4.000)



Values of Predictors for New Observations

New
Obs SATM10
 1   600
```

14.33 Height and Weight A scatterplot of the heights and weights of 500 people was shown in Chapter 4. The accompanying scatterplot uses the same data but displays 95% prediction intervals. From the graph, estimate the upper and lower values for the prediction interval used for predicting the weight of someone who is 70 inches tall.

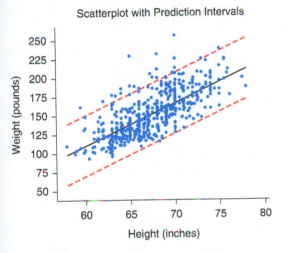

14.34 Waist Size and Weight A scatterplot of the waist sizes and weights of the same 500 people mentioned in exercise 14.33 is shown. The accompanying scatterplot uses the same people and displays 95% prediction intervals. From the graph, estimate the upper and lower values for the prediction interval for predicting the weight of someone who has a 30-inch waist.

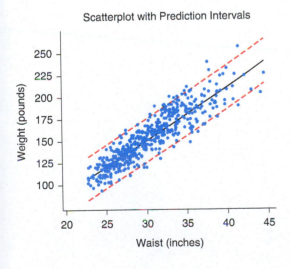

14.35 GPA and SAT The figure shows 95% prediction intervals for predicting GPA at a certain college from math SAT score, based on data from 196 students. From the graph, give approximate prediction interval boundaries for predicting GPA from a math SAT score of 750. State whether this prediction seems useful.

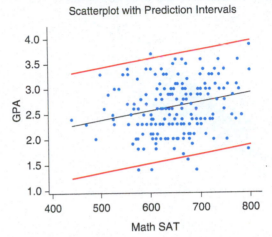

14.36 Shoes and Heights The scatterplot shows shoe size and height for 83 people. From the graph, state approximate values for the prediction interval for predicting the shoe size for someone who is 68 inches tall.

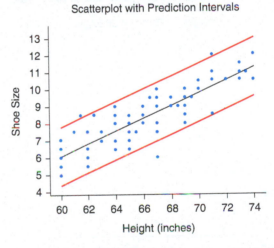

14.37 Height and Weight A scatterplot of the heights and weights of 500 people was shown in exercise 14.33. The accompanying scatterplot shows both confidence intervals and prediction intervals. Which is which? Explain what this shows.

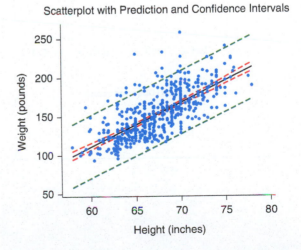

14.38 Waist Size and Weight A scatterplot of the waist sizes and weights of 500 people was shown in exercise 14.34. The accompanying scatterplot shows both confidence intervals and prediction intervals. Which is which? Explain how you can tell.

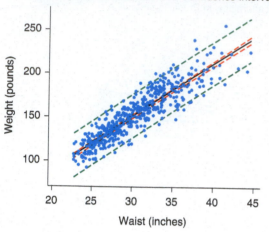

Scatterplot with Prediction and Confidence Intervals

14.39 Baseball Players Figure A shows a scatterplot with the regression line for the ages and weights of a random sample of 19 college baseball players. Figure B gives a prediction (Fit), a prediction interval, and a confidence interval for a new observation at 20 years old.

(A)

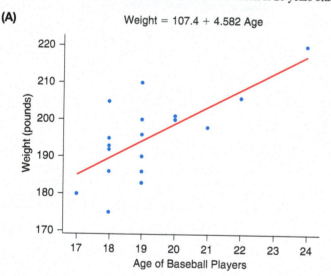

Weight = 107.4 + 4.582 Age

(B)

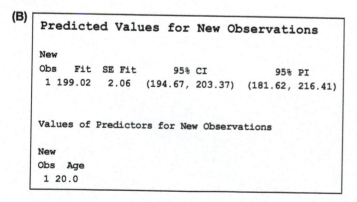

```
Predicted Values for New Observations

New
Obs   Fit   SE Fit      95% CI            95% PI
 1  199.02  2.06  (194.67, 203.37)  (181.62, 216.41)


Values of Predictors for New Observations

New
Obs  Age
 1  20.0
```

a. Identify which interval is which, and explain why one of the intervals is larger than the other.

b. Interpret the confidence interval.

c. Interpret the prediction interval.

14.40 Predicting Education Figure A shows a scatterplot with the regression line for predicting the father's education from the mother's education for a random sample of 29 students. Figure B shows the confidence interval and the prediction interval for the father for a new observation when the mother has 10 years of education.

(A)

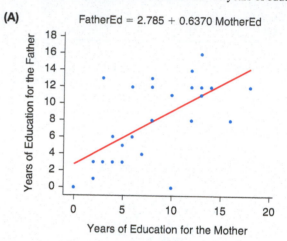

FatherEd = 2.785 + 0.6370 MotherEd

(B)

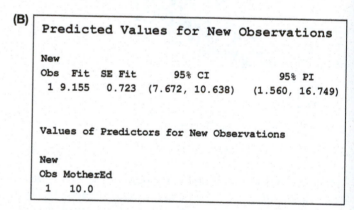

```
Predicted Values for New Observations

New
Obs   Fit   SE Fit      95% CI            95% PI
 1  9.155  0.723  (7.672, 10.638)  (1.560, 16.749)


Values of Predictors for New Observations

New
Obs MotherEd
 1   10.0
```

a. Identify which interval is which. Explain what they show and why one interval is larger than the other.

b. Interpret the confidence interval.

c. Interpret the prediction interval.

CHAPTER REVIEW EXERCISES

14.41 Life Expectancy and Gestation Periods for Animals
Data were collected showing the gestation period (in days) and the average longevity (in years) for 32 animals. Assume that all conditions of the linear regression model hold. The data are available on this text's website.

a. Make a scatterplot with gestation period as the *x*-variable and average longevity as the *y*-variable, and insert the correct regression line. Report the regression equation.

b. One animal lives much longer than we would expect, given its gestation period. Identify this animal.

c. From the data, find a confidence interval for the mean average longevity of animals with a gestation period of 266 days.

d. Humans have a gestation period of about 266 days. Does the confidence interval for the average life span for humans seem to fit what you know about humans' life spans? Explain.

14.42 School Accountability The California Department of Education assesses progress of K–12 students in meeting grade level standards in English/Language Arts and Mathematics yearly. A regression analysis was performed using 2016 assessment data from a sample of 195 California schools. The data consisted of the percentage of students at each school who met or exceeded the grade-level standards in English/Language Arts (SELA) and Mathematics (SMATH). Assume all conditions for the linear regression model hold.

a. Using the information in the regression analysis output, is there a linear association between these two variables? Explain.

b. Interpret the slope of the regression equation.

c. Using a 95% confidence interval, what is the estimated mean percent of mathematics proficient students at schools in which 50% of students are proficient in English/Language Arts?

```
Simple Linear regression results:
Dependent Variable: Sample(SMATH_Y2)
Independent variable: Sample(SELA_Y2)
Sample(SMATH_Y2)=-5.8334238 + 0.89861091 Sample(SELA_Y2)
Sample size:195
R(correlation coefficient)= 0.86834159
R-sq = 0.75401711
Estimate of error standard deviation:11.062276
```

Parameter estimates:

Parameter	Estimate	Std. Err.	Alternative	DF	T-Stat	P-value
Intercept	-5.8334238	1.8530804	≠0	193	-3.1479604	0.0019
Slope	0.89861091	0.036944933	≠0	193	24.322981	<0.0001

Analysis of variance table for regression model:

Source	DF	SS	MS	F-stat	P-value
Model	1	72397.343	72397.343	591.60742	<0.0001
Error	193	23618.174	122.37396		
Total	194	96015.518			

Predictes values:

X value	Pres.Y	s.e.(pred.y)	95% C.I. for mean	95% P.I. for new
50	39.097122	0.81064996	(37.498251,40.695992)	(17.220139,60.974104)

TechTips

General Instructions for All Technology

The steps for making a scatterplot and finding and graphing a linear regression equation are given in Chapter 4 TechTips. The basic steps for making a Normality (QQ) plot are given in Chapter 13 TechTips.

EXAMPLE ▶ The table shows the heights (in inches) and the weights (in pounds) of six women. Make a scatterplot of the data and graph the linear regression equation. Make a plot of the residuals. Create a Normality (QQ) plot of the residuals. Perform a regession test. Calculate confidence intervals for the regression.

Ht	Wt
61	104
62	110
63	141
64	125
66	170
68	160

TI-84

Residual Plot

1. Press **STAT**, choose **EDIT**, and enter heights in **L1** and weights in **L2**.
2. Press **STAT**, choose **CALC**, **8: LinReg (a + bx)**.
3. For **Xlist:** press **2ND L1**. For **Ylist:** press **2ND L2**. Scroll down to **Calculate** and press **ENTER**. (This step also calculated the residuals and stored them in the list, RESID.)
4. Turn all plots off. Press **2ND STAT PLOT**, choose **4:PlotsOff**, press **ENTER**. Press **Y=**. If **Plot1** is active (highlighted), arrow to it and press ENTER to turn it off.
5. Now, to create the residual plot, use Plot2 (assuming that you used Plot1 for the scatterplot). Choose **2ND STATPLOT**, **Plot2**, **On**, **Type:** scatterplot (first choice), **Xlist: L1**, **Ylist: RESID**. (Do this by pressing **2ND LIST**, **RESID**.)
6. To make the graph, press **GRAPH**, **ZOOM**, and **9**.
7. Press **TRACE** to see the numbers, and use the arrows on the keypad to move to other points.

Figure 14a shows a residual plot of the six points.

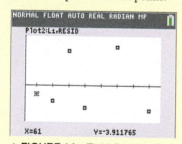

▲ **FIGURE 14a** TI-84 Residual Plot

Normality (QQ) Plot

1. Complete steps 1–3 above.
2. Turn off/clear all plots, as described in step 4 above.
3. Now use Plot3 and the list, **RESID**, for the **QQ** plot by following the steps in the Chapter 13 TechTips.

Hypothesis Test

With the data set, you can test the hypothesis that the population slope, β, is 0. This is equivalent to testing that the population correlation, ρ, is 0. (See β and ρ in Figure 14b.)

▲ **FIGURE 14b** TI-84 Input for a Regression Test

1. Press **STAT**, choose **EDIT**, and enter the heights into **L1** and the weights into **L2**.
2. Press **STAT**, choose **TESTS**, and choose **F: LinRegTTest**.
3. See Figure 14b. Be sure the predictor (**L1** in our example) is the **Xlist** and that the response (**L2** in our example) is the **Ylist**. Here we are testing the hypothesis that the slope is greater than 0. You can use $\neq$ if you want a two-sided hypothesis.
4. Scroll down to **Calculate** and press **ENTER**.

Figure 14c shows the output.

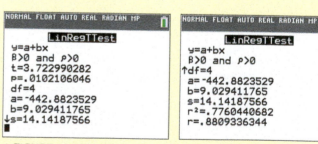

▲ **FIGURE 14c** TI-84 Output for a Regression Test

The only confidence interval available for regression is for the slope. To find it, press **STAT**, choose **TESTS**, and choose **G:LinRegTInt**. For the slope of the six heights and weights given in the example, the 95% confidence interval comes out to be (2.2957, 15.763).

Residual and Normality (QQ) Plots

1. Enter the heights and weights into columns **C1** and **C2**. Add the labels **Ht** and **Wt** in the label area above the data.
2. **Stat > Regression > Fitted Line Plot**
3. Put the dependent variable (here, **Wt**) in the **Response (Y)** box and the independent variable (here, **Ht**) in the **Predictor (X)** box.
4. Click **Graphs.**
5. See Figure 14d. Choose **Regular**, select **Normal plot of residuals**, and in the **Residuals versus the variables** box put **Ht**. Click **OK** and **OK** again.

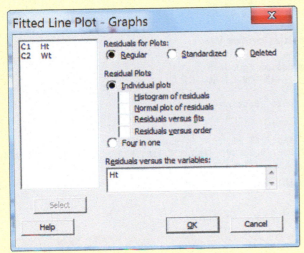

▲ **FIGURE 14d** Minitab Input for a Residual Plot

You will get a scatterplot with the regression line, a residual plot like Figure 14e, and a Normality plot.

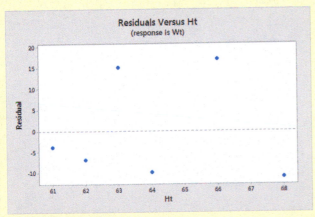

▲ **FIGURE 14e** Minitab Residual Plot

A Scatterplot with the Regression Line with Prediction and Confidence Intervals

1. The heights should be in column **C1** and the weights in column **C2**, with the labels **Ht** and **Wt** at the top.
2. **Stat > Regression > Fitted Line Plot**
3. After entering the **Response (Wt)** and the **Predictor (Ht)**, click **Options**.
4. Click **Display confidence interval** or **Display prediction interval**, or both, as shown in Figure 14f. Click **OK**, and click **OK** again on the next screen.

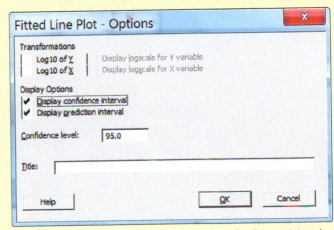

▲ **FIGURE 14f** Minitab Input for Prediction and Confidence Intervals

Figure 14g shows the result.

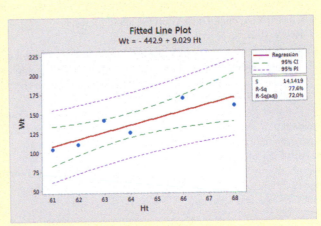

▲ **FIGURE 14g** Minitab Prediction and Confidence Intervals

Regression Numbers and Numerical Confidence and Prediction Intervals

1. Your data should be in columns **C1** and **C2**, labeled **Ht** and **Wt**.
2. **Stat** > **Regression** > **Regression** > **Fit Regression Model**
3. In the **Responses:** box put Y, here **Wt**. In the **Continuous predictors:** box put X, here **Ht**. Click **OK**.
4. **Stat** > **Regression** > **Regression** > **Predict**
5. In the empty box under **Ht**, insert **67**. Click **OK**.

The output is shown in Figure 14h and includes the prediction and confidence intervals at the bottom. These are for prediction of ranges of weights with 95% confidence from a new woman's height of 67 inches. The output in Figure 14h can also be used to test whether the null hypothesis that the slope is 0 can be rejected; because the p-value is 0.020, you can reject the hypothesis that the slope is 0. This output can also be used to test whether the intercept is 0; because the p-value is 0.046, you can reject the hypothesis that the intercept is 0.

Regression Analysis: Wt versus Ht

Analysis of Variance

Source	DF	Adj SS	Adj MS	F-Value	P-Value
Regression	1	2772.0	2772.0	13.86	0.020
Ht	1	2772.0	2772.0	13.86	0.020
Error	4	800.0	200.0		
Total	5	3572.0			

Model Summary

S	R-sq	R-sq(adj)	R-sq(pred)
14.1419	77.60%	72.01%	40.24%

Coefficients

Term	Coef	SE Coef	T-Value	P-Value	VIF
Constant	-443	155	-2.85	0.046	
Ht	9.03	2.43	3.72	0.020	1.00

Regression Equation

Wt = -443 + 9.03 Ht

Prediction for Wt

Regression Equation

Wt = -443 + 9.03 Ht

Variable	Setting
Ht	67

Fit	SE Fit	95% CI	95% PI
162.088	9.28824	(136.300, 187.877)	(115.113, 209.064)

▲ **FIGURE 14h** Minitab Output for Regression, Including Intervals

---

EXCEL

Excel can give numerical regression output and residual plots at the same time.

1. Enter the heights into column **A** and the weights into column **B**.
2. Click **DATA**, click **Data Analysis**, and double click **Regression**.
3. See Figure 14i. For **Input Y Range**, highlight the data for the weights (without the label, wt), and for **Input X Range**, highlight the data for the heights (without the label, ht); then check **Residual Plots** and click **OK**.

You will get numerical output like that shown in Figure 14j and a residual plot like that shown in Figure 14k. To see the numerical output, you may have to click **HOME**, **Format** (in the **Cells** group), and **AutoFit Column Width**. Note in Figure 14j that confidence intervals are given for the intercept (−874.14, −11.6247) and for the slope (2.295667, 15.76316). These can be used to test hypotheses about the slope and intercept.

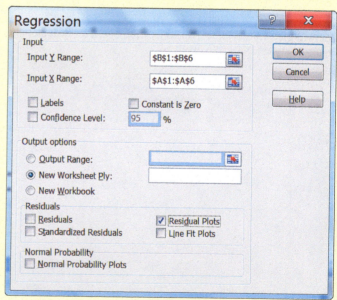

▲ **FIGURE 14i** Excel Input for Regression and Residual Plots

SUMMARY OUTPUT

Regression Statistics

Multiple R	0.880934
R Square	0.776044
Adjusted R Square	0.720055
Standard Error	14.14188
Observations	6

ANOVA

	df	SS	MS	F	Significance F
Regression	1	2772.029	2772.029	13.86066	0.020421
Residual	4	799.9706	199.9926		
Total	5	3572			

	Coefficients	Standard Error	t Stat	P-value	Lower 95%	Upper 95%	Lower 95.0%	Upper 95.0%
Intercept	−442.882	155.3273	−2.85129	0.046336	−874.14	−11.6247	−874.14	−11.6247
X Variable 1	9.029412	2.425312	3.72299	0.020421	2.295667	15.76316	2.295667	15.76316

▲ **FIGURE 14j** Excel Regression Output

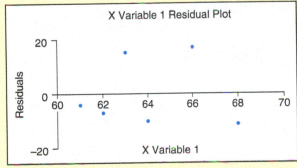

▲ **FIGURE 14k** Excel Residual Plot

STATCRUNCH

1. Enter the heights and weights into **var1** and **var2**. Include labels at the top, **Ht** and **Wt**.

2. **Stat > Regression > Simple Linear**

3. For the **X variable:** pick **Ht**, for the **Y variable:** pick **Wt**.

4. Click **Confidence intervals** and leave the default **0.95**. (This means you will get confidence intervals for the slope and intercept. If you wanted to do hypothesis tests on the slope and intercept, you would instead leave the default **Hypothesis tests** selected.)

5. For **Prediction of Y: X value(s):**, type in **67**.

6. Scroll down to **Graphs:**, see Figure 14l. Select **Fitted line plot, --- with mean interval, --- with prediction interval, QQ plot of residuals**, and **Residuals vs. X-values.** Click **Compute!**

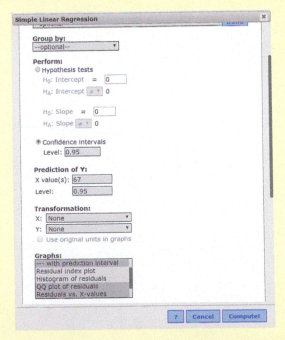

▶ **FIGURE 14l** StatCrunch Regression Input (partial view)

The numerical output is shown in Figure 14m.

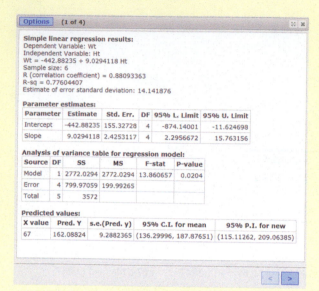

▲ **FIGURE 14m** StatCrunch Regression Output

7. If you click the **>** located at the bottom right (see Figure 14m), you will see the scatterplot with the regression line shown in Figure 14n.

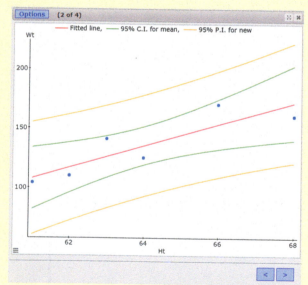

▲ **FIGURE 14n** StatCrunch Prediction and Confidence Intervals

8. If you click the **>**, you will see the QQ plot of Residuals, not shown here.

9. Click the **>** to see the residual plot shown here in Figure 14o.

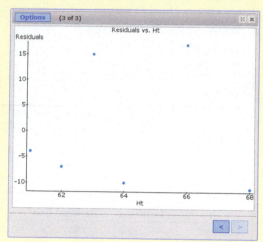

▲ **FIGURE 14o** StatCrunch Residual Plot

Appendix A: Tables

Table 1: Random Numbers

Line						
01	21033	32522	19305	90633	80873	19167
02	17516	69328	88389	19770	33197	27336
03	26427	40650	70251	84413	30896	21490
04	45506	44716	02498	15327	79149	28409
05	55185	74834	81172	89281	48134	71185
06	87964	43751	80971	50613	81441	30505
07	09106	73117	57952	04393	93402	50753
08	88797	07440	69213	33593	42134	24168
09	34685	46775	32139	22787	28783	39481
10	07104	43091	14311	69671	01536	02673
11	27583	01866	58250	38103	35825	94513
12	60801	04439	58621	09840	35119	60372
13	62708	04888	37221	49537	96024	24004
14	21169	14082	65865	29690	00280	35738
15	13893	00626	11773	14897	37119	29729
16	19872	41310	65041	61105	31028	80297
17	29331	36997	05601	09785	18100	44164
18	76846	74048	08496	22599	29379	11114
19	11848	80809	25818	38857	23811	80902
20	85757	33963	93076	39950	29658	07530
21	71141	00618	48403	46083	40368	33990
22	47371	36443	41894	62134	86876	18548
23	46633	10669	95848	69055	49044	75595
24	79118	21098	63279	26834	43443	38267
25	91874	87217	11503	47925	13289	42106
26	85337	08882	68429	61767	18930	37688
27	88513	05437	22776	17562	03820	44785
28	31498	85304	22393	21634	34560	77404
29	93074	27086	62559	86590	18420	33290
30	90549	53094	76282	53105	45531	90061
31	11373	96871	38157	98368	39536	08079
32	52022	59093	30647	33241	16027	70336
33	14709	93220	89547	95320	39134	07646
34	57584	28114	91168	16320	81609	60807
35	31867	85872	91430	45554	21567	15082
36	07033	75250	34546	75298	33893	64487
37	02779	72645	32699	86009	73729	44206
38	24512	01116	49826	50882	44086	87757
39	52463	30164	80073	55917	60995	38655
40	82588	59267	13570	56434	66413	99518
41	20999	05039	87835	63010	82980	66193
42	09084	98948	09541	80623	15915	71042

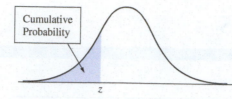

Cumulative Probability for z is The Area Under the Standard Normal Curve to The Left of z

Table 2: Standard Normal Cumulative Probabilities

z	.00
−5.0	.000000287
−4.5	.00000340
−4.0	.0000317
−3.5	.000233

z	.00	.01	.02	.03	.04	.05	.06	.07	.08	.09
−3.4	.0003	.0003	.0003	.0003	.0003	.0003	.0003	.0003	.0003	.0002
−3.3	.0005	.0005	.0005	.0004	.0004	.0004	.0004	.0004	.0004	.0003
−3.2	.0007	.0007	.0006	.0006	.0006	.0006	.0006	.0005	.0005	.0005
−3.1	.0010	.0009	.0009	.0009	.0008	.0008	.0008	.0008	.0007	.0007
−3.0	.0013	.0013	.0013	.0012	.0012	.0011	.0011	.0011	.0010	.0010
−2.9	.0019	.0018	.0018	.0017	.0016	.0016	.0015	.0015	.0014	.0014
−2.8	.0026	.0025	.0024	.0023	.0023	.0022	.0021	.0021	.0020	.0019
−2.7	.0035	.0034	.0033	.0032	.0031	.0030	.0029	.0028	.0027	.0026
−2.6	.0047	.0045	.0044	.0043	.0041	.0040	.0039	.0038	.0037	.0036
−2.5	.0062	.0060	.0059	.0057	.0055	.0054	.0052	.0051	.0049	.0048
−2.4	.0082	.0080	.0078	.0075	.0073	.0071	.0069	.0068	.0066	.0064
−2.3	.0107	.0104	.0102	.0099	.0096	.0094	.0091	.0089	.0087	.0084
−2.2	.0139	.0136	.0132	.0129	.0125	.0122	.0119	.0116	.0113	.0110
−2.1	.0179	.0174	.0170	.0166	.0162	.0158	.0154	.0150	.0146	.0143
−2.0	.0228	.0222	.0217	.0212	.0207	.0202	.0197	.0192	.0188	.0183
−1.9	.0287	.0281	.0274	.0268	.0262	.0256	.0250	.0244	.0239	.0233
−1.8	.0359	.0351	.0344	.0336	.0329	.0322	.0314	.0307	.0301	.0294
−1.7	.0446	.0436	.0427	.0418	.0409	.0401	.0392	.0384	.0375	.0367
−1.6	.0548	.0537	.0526	.0516	.0505	.0495	.0485	.0475	.0465	.0455
−1.5	.0668	.0655	.0643	.0630	.0618	.0606	.0594	.0582	.0571	.0559
−1.4	.0808	.0793	.0778	.0764	.0749	.0735	.0721	.0708	.0694	.0681
−1.3	.0968	.0951	.0934	.0918	.0901	.0885	.0869	.0853	.0838	.0823
−1.2	.1151	.1131	.1112	.1093	.1075	.1056	.1038	.1020	.1003	.0985
−1.1	.1357	.1335	.1314	.1292	.1271	.1251	.1230	.1210	.1190	.1170
−1.0	.1587	.1562	.1539	.1515	.1492	.1469	.1446	.1423	.1401	.1379
−0.9	.1841	.1814	.1788	.1762	.1736	.1711	.1685	.1660	.1635	.1611
−0.8	.2119	.2090	.2061	.2033	.2005	.1977	.1949	.1922	.1894	.1867
−0.7	.2420	.2389	.2358	.2327	.2296	.2266	.2236	.2206	.2177	.2148
−0.6	.2743	.2709	.2676	.2643	.2611	.2578	.2546	.2514	.2483	.2451
−0.5	.3085	.3050	.3015	.2981	.2946	.2912	.2877	.2843	.2810	.2776
−0.4	.3446	.3409	.3372	.3336	.3300	.3264	.3228	.3192	.3156	.3121
−0.3	.3821	.3783	.3745	.3707	.3669	.3632	.3594	.3557	.3520	.3483
−0.2	.4207	.4168	.4129	.4090	.4052	.4013	.3974	.3936	.3897	.3859
−0.1	.4602	.4562	.4522	.4483	.4443	.4404	.4364	.4325	.4286	.4247
−0.0	.5000	.4960	.4920	.4880	.4840	.4801	.4761	.4721	.4681	.4641

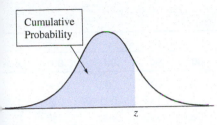

Cumulative Probability

Cumulative Probability for z is The Area Under
the Standard Normal Curve to The Left of z

Standard Normal Cumulative Probabilities (*continued*)

z	.00	.01	.02	.03	.04	.05	.06	.07	.08	.09
0.0	.5000	.5040	.5080	.5120	.5160	.5199	.5239	.5279	.5319	.5359
0.1	.5398	.5438	.5478	.5517	.5557	.5596	.5636	.5675	.5714	.5753
0.2	.5793	.5832	.5871	.5910	.5948	.5987	.6026	.6064	.6103	.6141
0.3	.6179	.6217	.6255	.6293	.6331	.6368	.6406	.6443	.6480	.6517
0.4	.6554	.6591	.6628	.6664	.6700	.6736	.6772	.6808	.6844	.6879
0.5	.6915	.6950	.6985	.7019	.7054	.7088	.7123	.7157	.7190	.7224
0.6	.7257	.7291	.7324	.7357	.7389	.7422	.7454	.7486	.7517	.7549
0.7	.7580	.7611	.7642	.7673	.7704	.7734	.7764	.7794	.7823	.7852
0.8	.7881	.7910	.7939	.7967	.7995	.8023	.8051	.8078	.8106	.8133
0.9	.8159	.8186	.8212	.8238	.8264	.8289	.8315	.8340	.8365	.8389
1.0	.8413	.8438	.8461	.8485	.8508	.8531	.8554	.8577	.8599	.8621
1.1	.8643	.8665	.8686	.8708	.8729	.8749	.8770	.8790	.8810	.8830
1.2	.8849	.8869	.8888	.8907	.8925	.8944	.8962	.8980	.8997	.9015
1.3	.9032	.9049	.9066	.9082	.9099	.9115	.9131	.9147	.9162	.9177
1.4	.9192	.9207	.9222	.9236	.9251	.9265	.9279	.9292	.9306	.9319
1.5	.9332	.9345	.9357	.9370	.9382	.9394	.9406	.9418	.9429	.9441
1.6	.9452	.9463	.9474	.9484	.9495	.9505	.9515	.9525	.9535	.9545
1.7	.9554	.9564	.9573	.9582	.9591	.9599	.9608	.9616	.9625	.9633
1.8	.9641	.9649	.9656	.9664	.9671	.9678	.9686	.9693	.9699	.9706
1.9	.9713	.9719	.9726	.9732	.9738	.9744	.9750	.9756	.9761	.9767
2.0	.9772	.9778	.9783	.9788	.9793	.9798	.9803	.9808	.9812	.9817
2.1	.9821	.9826	.9830	.9834	.9838	.9842	.9846	.9850	.9854	.9857
2.2	.9861	.9864	.9868	.9871	.9875	.9878	.9881	.9884	.9887	.9890
2.3	.9893	.9896	.9898	.9901	.9904	.9906	.9909	.9911	.9913	.9916
2.4	.9918	.9920	.9922	.9925	.9927	.9929	.9931	.9932	.9934	.9936
2.5	.9938	.9940	.9941	.9943	.9945	.9946	.9948	.9949	.9951	.9952
2.6	.9953	.9955	.9956	.9957	.9959	.9960	.9961	.9962	.9963	.9964
2.7	.9965	.9966	.9967	.9968	.9969	.9970	.9971	.9972	.9973	.9974
2.8	.9974	.9975	.9976	.9977	.9977	.9978	.9979	.9979	.9980	.9981
2.9	.9981	.9982	.9982	.9983	.9984	.9984	.9985	.9985	.9986	.9986
3.0	.9987	.9987	.9987	.9988	.9988	.9989	.9989	.9989	.9990	.9990
3.1	.9990	.9991	.9991	.9991	.9992	.9992	.9992	.9992	.9993	.9993
3.2	.9993	.9993	.9994	.9994	.9994	.9994	.9994	.9995	.9995	.9995
3.3	.9995	.9995	.9995	.9996	.9996	.9996	.9996	.9996	.9996	.9997
3.4	.9997	.9997	.9997	.9997	.9997	.9997	.9997	.9997	.9997	.9998

z	.00
3.5	.999767
4.0	.9999683
4.5	.9999966
5.0	.999999713

Table 3: Binomial Probabilities

n	x	.01	.05	.10	.20	.30	.40	.50	.60	.70	.80	.90	.95	.99	x
								p							
2	0	.980	.902	.810	.640	.490	.360	.250	.160	.090	.040	.010	.002	0+	0
	1	.020	.095	.180	.320	.420	.480	.500	.480	.420	.320	.180	.095	.020	1.
	2	0+	.002	.010	.040	.090	.160	.250	.360	.490	.640	.810	.902	.980	2
3	0	.970	.857	.729	.512	.343	.216	.125	.064	.027	.008	.001	0+	0+	0
	1	.029	.135	.243	.384	.441	.432	.375	.288	.189	.096	.027	.007	0+	1
	2	0+	.007	.027	.096	.189	.288	.375	.432	.441	.384	.243	.135	.029	2
	3	0+	0+	.001	.008	.027	.064	.125	.216	.343	.512	.729	.857	.970	3
4	0	.961	.815	.656	.410	.240	.130	.062	.026	.008	.002	0+	0+	0+	0
	1	.039	.171	.292	.410	.412	.346	.250	.154	.076	.026	.004	0+	0+	1
	2	.001	.014	.049	.154	.265	.346	.375	.346	.265	.154	.049	.014	.001	2
	3	0+	0+	.004	.026	.076	.154	.250	.346	.412	.410	.292	.171	.039	3
	4	0+	0+	0+	.002	.008	.026	.062	.130	.240	.410	.656	.815	.961	4
5	0	.951	.774	.590	.328	.168	.078	.031	.010	.002	0+	0+	0+	0+	0
	1	.048	.204	.328	.410	.360	.259	.156	.077	.028	.006	0+	0+	0+	1
	2	.001	.021	.073	.205	.309	.346	.312	.230	.132	.051	.008	.001	0+	2
	3	0+	.001	.008	.051	.132	.230	.312	.346	.309	.205	.073	.021	.001	3
	4	0+	0+	0+	.006	.028	.077	.156	.259	.360	.410	.328	.204	.048	4
	5	0+	0+	0+	0+	.002	.010	.031	.078	.168	.328	.590	.774	.951	5
6	0	.941	.735	.531	.262	.118	.047	.016	.004	.001	0+	0+	0+	0+	0
	1	.057	.232	.354	.393	.303	.187	.094	.037	.010	.002	0+	0+	0+	1
	2	.001	.031	.098	.246	.324	.311	.234	.138	.060	.015	.001	0+	0+	2
	3	0+	.002	.015	.082	.185	.276	.312	.276	.185	.082	.015	.002	0+	3
	4	0+	0+	.001	.015	.060	.138	.234	.311	.324	.246	.098	.031	.001	4
	5	0+	0+	0+	.002	.010	.037	.094	.187	.303	.393	.354	.232	.057	5
	6	0+	0+	0+	0+	.001	.004	.016	.047	.118	.262	.531	.735	.941	6
7	0	.932	.698	.478	.210	.082	.028	.008	.002	0+	0+	0+	0+	0+	0
	1	.066	.257	.372	.367	.247	.131	.055	.017	.004	0+	0+	0+	0+	1
	2	.002	.041	.124	.275	.318	.261	.164	.077	.025	.004	0+	0+	0+	2
	3	0+	.004	.023	.115	.227	.290	.273	.194	.097	.029	.003	0+	0+	3
	4	0+	0+	.003	.029	.097	.194	.273	.290	.227	.115	.023	.004	0+	4
	5	0+	0+	0+	.004	.025	.077	.164	.261	.318	.275	.124	.041	.002	5
	6	0+	0+	0+	0+	.004	.017	.055	.131	.247	.367	.372	.257	.066	6
	7	0+	0+	0+	0+	0+	.002	.008	.028	.082	.210	.478	.698	.932	7
8	0	.923	.663	.430	.168	.058	.017	.004	.001	0+	0+	0+	0+	0+	0
	1	.075	.279	.383	.336	.198	.090	.031	.008	.001	0+	0+	0+	0+	1
	2	.003	.051	.149	.294	.296	.209	.109	.041	.010	.001	0+	0+	0+	2
	3	0+	.005	.033	.147	.254	.279	.219	.124	.047	.009	0+	0+	0+	3
	4	0+	0+	.005	.046	.136	.232	.273	.232	.136	.046	.005	0+	0+	4
	5	0+	0+	0+	.009	.047	.124	.219	.279	.254	.147	.033	.005	0+	5
	6	0+	0+	0+	.001	.010	.041	.109	.209	.296	.294	.149	.051	.003	6
	7	0+	0+	0+	0+	.001	.008	.031	.090	.198	.336	.383	.279	.075	7
	8	0+	0+	0+	0+	0+	.001	.004	.017	.058	.168	.430	.663	.923	8

NOTE: 0+ represents a positive probability less than 0.0005.

(continued)

Binomial Probabilities (continued)

n	x	.01	.05	.10	.20	.30	.40	.50	.60	.70	.80	.90	.95	.99	x
9	0	.914	.630	.387	.134	.040	.010	.002	0+	0+	0+	0+	0+	0+	0
	1	.083	.299	.387	.302	.156	.060	.018	.004	0+	0+	0+	0+	0+	1
	2	.003	.063	.172	.302	.267	.161	.070	.021	.004	0+	0+	0+	0+	2
	3	0+	.008	.045	.176	.267	.251	.164	.074	.021	.003	0+	0+	0+	3
	4	0+	.001	.007	.066	.172	.251	.246	.167	.074	.017	.001	0+	0+	4
	5	0+	0+	.001	.017	.074	.167	.246	.251	.172	.066	.007	.001	0+	5
	6	0+	0+	0+	.003	.021	.074	.164	.251	.267	.176	.045	.008	0+	6
	7	0+	0+	0+	0+	.004	.021	.070	.161	.267	.302	.172	.063	.003	7
	8	0+	0+	0+	0+	0+	.004	.018	.060	.156	.302	.387	.299	.083	8
	9	0+	0+	0+	0+	0+	0+	.002	.010	.040	.134	.387	.630	.914	9
10	0	.904	.599	.349	.107	.028	.006	.001	0+	0+	0+	0+	0+	0+	0
	1	.091	.315	.387	.268	.121	.040	.010	.002	0+	0+	0+	0+	0+	1
	2	.004	.075	.194	.302	.233	.121	.044	.011	.001	0+	0+	0+	0+	2
	3	0+	.010	.057	.201	.267	.215	.117	.042	.009	.001	0+	0+	0+	3
	4	0+	.001	.011	.088	.200	.251	.205	.111	.037	.006	0+	0+	0+	4
	5	0+	0+	.001	.026	.103	.201	.246	.201	.103	.026	.001	0+	0+	5
	6	0+	0+	0+	.006	.037	.111	.205	.251	.200	.088	.011	.001	0+	6
	7	0+	0+	0+	.001	.009	.042	.117	.215	.267	.201	.057	.010	0+	7
	8	0+	0+	0+	0+	.001	.011	.044	.121	.233	.302	.194	.075	.004	8
	9	0+	0+	0+	0+	0+	.002	.010	.040	.121	.268	.387	.315	.091	9
	10	0+	0+	0+	0+	0+	0+	.001	.006	.028	.107	.349	.599	.904	10
11	0	.895	.569	.314	.086	.020	.004	0+	0+	0+	0+	0+	0+	0+	0
	1	.099	.329	.384	.236	.093	.027	.005	.001	0+	0+	0+	0+	0+	1
	2	.005	.087	.213	.295	.200	.089	.027	.005	.001	0+	0+	0+	0+	2
	3	0+	.014	.071	.221	.257	.177	.081	.023	.004	0+	0+	0+	0+	3
	4	0+	.001	.016	.111	.220	.236	.161	.070	.017	.002	0+	0+	0+	4
	5	0+	0+	.002	.039	.132	.221	.226	.147	.057	.010	0+	0+	0+	5
	6	0+	0+	0+	.010	.057	.147	.226	.221	.132	.039	.002	0+	0+	6
	7	0+	0+	0+	.002	.017	.070	.161	.236	.220	.111	.016	.001	0+	7
	8	0+	0+	0+	0+	.004	.023	.081	.177	.257	.221	.071	.014	0+	8
	9	0+	0+	0+	0+	.001	.005	.027	.089	.200	.295	.213	.087	.005	9
	10	0+	0+	0+	0+	0+	.001	.005	.027	.093	.236	.384	.329	.099	10
	11	0+	0+	0+	0+	0+	0+	0+	.004	.020	.086	.314	.569	.895	11
12	0	.886	.540	.282	.069	.014	.002	0+	0+	0+	0+	0+	0+	0+	0
	1	.107	.341	.377	.206	.071	.017	.003	0+	0+	0+	0+	0+	0+	1
	2	.006	.099	.230	.283	.168	.064	.016	.002	0+	0+	0+	0+	0+	2
	3	0+	.017	.085	.236	.240	.142	.054	.012	.001	0+	0+	0+	0+	3
	4	0+	.002	.021	.133	.231	.213	.121	.042	.008	.001	0+	0+	0+	4
	5	0+	0+	.004	.053	.158	.227	.193	.101	.029	.003	0+	0+	0+	5
	6	0+	0+	0+	.016	.079	.177	.226	.177	.079	.016	0+	0+	0+	6
	7	0+	0+	0+	.003	.029	.101	.193	.227	.158	.053	.004	0+	0+	7
	8	0+	0+	0+	.001	.008	.042	.121	.213	.231	.133	.021	.002	0+	8
	9	0+	0+	0+	0+	.001	.012	.054	.142	.240	.236	.085	.017	0+	9
	10	0+	0+	0+	0+	0+	.002	.016	.064	.168	.283	.230	.099	.006	10
	11	0+	0+	0+	0+	0+	0+	.003	.017	.071	.206	.377	.341	.107	11
	12	0+	0+	0+	0+	0+	0+	0+	.002	.014	.069	.282	.540	.886	12

NOTE: 0+ represents a positive probability less than 0.0005.

(continued)

Binomial Probabilities (*continued*)

n	x	.01	.05	.10	.20	.30	.40	.50	.60	.70	.80	.90	.95	.99	x
13	0	.878	.513	.254	.055	.010	.001	0+	0+	0+	0+	0+	0+	0+	0
	1	.115	.351	.367	.179	.054	.011	.002	0+	0+	0+	0+	0+	0+	1
	2	.007	.111	.245	.268	.139	.045	.010	.001	0+	0+	0+	0+	0+	2
	3	0+	.021	.100	.246	.218	.111	.035	.006	.001	0+	0+	0+	0+	3
	4	0+	.003	.028	.154	.234	.184	.087	.024	.003	0+	0+	0+	0+	4
	5	0+	0+	.006	.069	.180	.221	.157	.066	.014	.001	0+	0+	0+	5
	6	0+	0+	.001	.023	.103	.197	.209	.131	.044	.006	0+	0+	0+	6
	7	0+	0+	0+	.006	.044	.131	.209	.197	.103	.023	.001	0+	0+	7
	8	0+	0+	0+	.001	.014	.066	.157	.221	.180	.069	.006	0+	0+	8
	9	0+	0+	0+	0+	.003	.024	.087	.184	.234	.154	.028	.003	0+	9
	10	0+	0+	0+	0+	.001	.006	.035	.111	.218	.246	.100	.021	0+	10
	11	0+	0+	0+	0+	0+	.001	.010	.045	.139	.268	.245	.111	.007	11
	12	0+	0+	0+	0+	0+	0+	.002	.011	.054	.179	.367	.351	.115	12
	13	0+	0+	0+	0+	0+	0+	.001	.010	.055	.254	.513	.878	13	
14	0	.869	.488	.229	.044	.007	.001	0+	0+	0+	0+	0+	0+	0+	0
	1	.123	.359	.356	.154	.041	.007	.001	0+	0+	0+	0+	0+	0+	1
	2	.008	.123	.257	.250	.113	.032	.006	.001	0+	0+	0+	0+	0+	2
	3	0+	.026	.114	.250	.194	.085	.022	.003	0+	0+	0+	0+	0+	3
	4	0+	.004	.035	.172	.229	.155	.061	.014	.001	0+	0+	0+	0+	4
	5	0+	0+	.008	.086	.196	.207	.122	.041	.007	0+	0+	0+	0+	5
	6	0+	0+	.001	.032	.126	.207	.183	.092	.023	.002	0+	0+	0+	6
	7	0+	0+	0+	.009	.062	.157	.209	.157	.062	.009	0+	0+	0+	7
	8	0+	0+	0+	.002	.023	.092	.183	.207	.126	.032	.001	0+	0+	8
	9	0+	0+	0+	0+	.007	.041	.122	.207	.196	.086	.008	0+	0+	9
	10	0+	0+	0+	0+	.001	.014	.061	.155	.229	.172	.035	.004	0+	10
	11	0+	0+	0+	0+	0+	.003	.022	.085	.194	.250	.114	.026	0+	11
	12	0+	0+	0+	0+	0+	.001	.006	.032	.113	.250	.257	.123	.008	12
	13	0+	0+	0+	0+	0+	0+	.001	.007	.041	.154	.356	.359	.123	13
	14	0+	0+	0+	0+	0+	0+	0+	.001	.007	.044	.229	.488	.869	14
15	0	.860	.463	.206	.035	.005	0+	0+	0+	0+	0+	0+	0+	0+	0
	1	.130	.366	.343	.132	.031	.005	0+	0+	0+	0+	0+	0+	0+	1
	2	.009	.135	.267	.231	.092	.022	.003	0+	0+	0+	0+	0+	0+	2
	3	0+	.031	.129	.250	.170	.063	.014	.002	0+	0+	0+	0+	0+	3
	4	0+	.005	.043	.188	.219	.127	.042	.007	.001	0+	0+	0+	0+	4
	5	0+	.001	.010	.103	.206	.186	.092	.024	.003	0+	0+	0+	0+	5
	6	0+	0+	.002	.043	.147	.207	.153	.061	.012	.001	0+	0+	0+	6
	7	0+	0+	0+	.014	.081	.177	.196	.118	.035	.003	0+	0+	0+	7
	8	0+	0+	0+	.003	.035	.118	.196	.177	.081	.014	0+	0+	0+	8
	9	0+	0+	0+	.001	.012	.061	.153	.207	.147	.043	.002	0+	0+	9
	10	0+	0+	0+	0+	.003	.024	.092	.186	.206	.103	.010	.001	0+	10
	11	0+	0+	0+	0+	.001	.007	.042	.127	.219	.188	.043	.005	0+	11
	12	0+	0+	0+	0+	0+	.002	.014	.063	.170	.250	.129	.031	0+	12
	13	0+	0+	0+	0+	0+	0+	.003	.022	.092	.231	.267	.135	.009	13
	14	0+	0+	0+	0+	0+	0+	0+	.005	.031	.132	.343	.366	.130	14
	15	0+	0+	0+	0+	0+	0+	0+	0+	.005	.035	.206	.463	.860	15

NOTE: 0+ represents a positive probability less than 0.0005.

From Frederick C. Mosteller, Robert E. K. Rourke, and George B. Thomas, Jr., *Probability with Statistical Applications*, 2nd ed., copyright © 1970 Pearson Education. Reprinted with permission of the publisher.

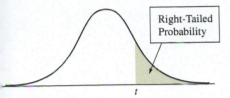

Right-Tailed Probability

Table 4: *t*-Distribution Critical Values

	Confidence Level					
	80%	90%	95%	98%	99%	99.8%
	Right-Tailed Probability					
df	$t_{.100}$	$t_{.050}$	$t_{.025}$	$t_{.010}$	$t_{.005}$	$t_{.001}$
1	3.078	6.314	12.706	31.821	63.656	318.289
2	1.886	2.920	4.303	6.965	9.925	22.328
3	1.638	2.353	3.182	4.541	5.841	10.214
4	1.533	2.132	2.776	3.747	4.604	7.173
5	1.476	2.015	2.571	3.365	4.032	5.894
6	1.440	1.943	2.447	3.143	3.707	5.208
7	1.415	1.895	2.365	2.998	3.499	4.785
8	1.397	1.860	2.306	2.896	3.355	4.501
9	1.383	1.833	2.262	2.821	3.250	4.297
10	1.372	1.812	2.228	2.764	3.169	4.144
11	1.363	1.796	2.201	2.718	3.106	4.025
12	1.356	1.782	2.179	2.681	3.055	3.930
13	1.350	1.771	2.160	2.650	3.012	3.852
14	1.345	1.761	2.145	2.624	2.977	3.787
15	1.341	1.753	2.131	2.602	2.947	3.733
16	1.337	1.746	2.120	2.583	2.921	3.686
17	1.333	1.740	2.110	2.567	2.898	3.646
18	1.330	1.734	2.101	2.552	2.878	3.611
19	1.328	1.729	2.093	2.539	2.861	3.579
20	1.325	1.725	2.086	2.528	2.845	3.552
21	1.323	1.721	2.080	2.518	2.831	3.527
22	1.321	1.717	2.074	2.508	2.819	3.505
23	1.319	1.714	2.069	2.500	2.807	3.485
24	1.318	1.711	2.064	2.492	2.797	3.467
25	1.316	1.708	2.060	2.485	2.787	3.450
26	1.315	1.706	2.056	2.479	2.779	3.435
27	1.314	1.703	2.052	2.473	2.771	3.421
28	1.313	1.701	2.048	2.467	2.763	3.408
29	1.311	1.699	2.045	2.462	2.756	3.396
30	1.310	1.697	2.042	2.457	2.750	3.385
40	1.303	1.684	2.021	2.423	2.704	3.307
50	1.299	1.676	2.009	2.403	2.678	3.261
60	1.296	1.671	2.000	2.390	2.660	3.232
80	1.292	1.664	1.990	2.374	2.639	3.195
100	1.290	1.660	1.984	2.364	2.626	3.174
∞	1.282	1.645	1.960	2.326	2.576	3.091

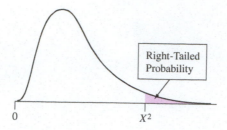

Table 5: Chi-Square Distribution for Values of Various Right-Tailed Probabilities

df	Right-Tailed Probability						
	0.250	0.100	0.050	0.025	0.010	0.005	0.001
1	1.32	2.71	3.84	5.02	6.63	7.88	10.83
2	2.77	4.61	5.99	7.38	9.21	10.60	13.82
3	4.11	6.25	7.81	9.35	11.34	12.84	16.27
4	5.39	7.78	9.49	11.14	13.28	14.86	18.47
5	6.63	9.24	11.07	12.83	15.09	16.75	20.52
6	7.84	10.64	12.59	14.45	16.81	18.55	22.46
7	9.04	12.02	14.07	16.01	18.48	20.28	24.32
8	10.22	13.36	15.51	17.53	20.09	21.96	26.12
9	11.39	14.68	16.92	19.02	21.67	23.59	27.88
10	12.55	15.99	18.31	20.48	23.21	25.19	29.59
11	13.70	17.28	19.68	21.92	24.72	26.76	31.26
12	14.85	18.55	21.03	23.34	26.22	28.30	32.91
13	15.98	19.81	22.36	24.74	27.69	29.82	34.53
14	17.12	21.06	23.68	26.12	29.14	31.32	36.12
15	18.25	22.31	25.00	27.49	30.58	32.80	37.70
16	19.37	23.54	26.30	28.85	32.00	34.27	39.25
17	20.49	24.77	27.59	30.19	33.41	35.72	40.79
18	21.60	25.99	28.87	31.53	34.81	37.16	42.31
19	22.72	27.20	30.14	32.85	36.19	38.58	43.82
20	23.83	28.41	31.41	34.17	37.57	40.00	45.32
25	29.34	34.38	37.65	40.65	44.31	46.93	52.62
30	34.80	40.26	43.77	46.98	50.89	53.67	59.70
40	45.62	51.80	55.76	59.34	63.69	66.77	73.40
50	56.33	63.17	67.50	71.42	76.15	79.49	86.66
60	66.98	74.40	79.08	83.30	88.38	91.95	99.61
70	77.58	85.53	90.53	95.02	100.43	104.21	112.32
80	88.13	96.58	101.88	106.63	112.33	116.32	124.84
90	98.65	107.57	113.15	118.14	124.12	128.30	137.21
100	109.14	118.50	124.34	129.56	135.81	140.17	149.45

Appendix B
Answers to Odd-Numbered Exercises

CHAPTER 1

Section 1.2

1.1 Eight (8)

1.3 a. Categorical **b.** Numerical **c.** Numerical

1.5 Answers will vary but could include such things as number of friends on Facebook or foot length. *Don't copy these answers.*

1.7 0 = male, 1 = female. The sum represents the total number of females in the data set.

1.9 Female is categorical with two categories. The 1s represent females, and the 0s represent males. If you added the numbers, you would get the number of females, so it makes sense here.

1.11 a. Stacked

b. 1 = male, 0 = female

c.

Male	Female
1916	9802
183	153
836	1221
95	
512	

1.13 a. Stacked and coded

Calories	Sweet
90	1
310	1
500	1
500	1
600	1
90	1
150	0
600	0
500	0
550	0

Alternatively, you could label the second column above "Salty," and then the 1s would become 0s and the 0s would become 1s.

b. Unstacked

Sweet	Salty
90	150
310	600
500	500
500	550
600	
90	

Section 1.3

1.15 Yes. Use *College Units Acquired* and *Living Situation*

1.17 No. Data on number of hours of study per week are not included in the table.

1.19 a. Yes. Use *Date*. **b.** No. Data on temperature are not included in the table. **c.** Yes. Use *Fatal* and *Species of Shark*. **d.** Yes. Use *Location*.

Section 1.4

1.21 a. $33/40 = 82.5\%$ **b.** $32/45 = 71.1\%$ **c.** $33/65 = 50.8\%$
d. 82.5% of 250 = 206 men

1.23 $15/38 = 39.5\%$ of the class were male **b.** $0.641(234) = 149.994$, so 150 men in the class

c. $0.40(x) = 20$

$20/0.4 = 50$ people in the class

1.25 The frequency of women is 6, the proportion is $6/11$, and the percentage is 54.5%.

1.27 a and b

	Male	Female	Total
Dorm	3	4	7
Commuter	2	2	4
Total	5	6	11

c. $4/6 = 66.7\%$

d. $4/7 = 57.1\%$

e. $7/11 = 63.6\%$

f. 66.7% of 70 = 47

1.29 127,244

1.31

State	Prison	Rank Prison	Population	Population (thousands)	Prison per thousand	Rank Rate
California	136,088	1	39,144,818	39145	3.48	4
New York	52518	2	19,795,791	19796	2.65	5
Illinois	48278	3	12,859,995	12860	3.75	3
Louisiana	30030	4	4,670,724	4671	6.43	1
Mississippi	18793	5	2,992,333	2992	6.28	2

California has the highest prison population. Louisiana has the highest rate of imprisonment. The two answers are different because the state populations are different.

1.33

Year	% Uncovered
1990	13.9%
2000	13.1%
2015	9.4%

The percentage of uninsured people has been declining since 1990.

1.35

Year	% Older Population
2020	16.4%
2030	19.6%
2040	21.4%
2050	22.1%

The percentage of older population is projected to increase.

1.37 We don't know the percentage of female students in the two classes. The larger number of women at 8 a.m. may just result from a larger number of students at 8 a.m., which may be because the class can accommodate more students because perhaps it is in a large lecture hall.

Section 1.5

1.39 Observational study

1.41 Controlled experiment

1.43 Controlled experiment

1.45 Anecdotal evidence are stories about individual cases. No cause-and-effect conclusions can be drawn from anecdotal evidence.

1.47 This was an observational study, and from it you cannot conclude that the tutoring raises the grades. Possible confounders (answers may vary): 1. It may be the more highly motivated who attend the tutoring, and this motivation is what causes the grades to go up. 2. It could be that those with more time attend the tutoring, and it is the increased time studying that causes the grades to go up.

1.49 a. The sample size of this study is not large (40). The study was a controlled experiment and used random assignment. It was not double-blind since researchers new what group each participant was in. **b.** The sample size of the study was small, so we should not conclude that physical activity while learning caused higher performance.

1.51 a. Controlled experiment. Researchers used random assignment of subjects to treatment or control groups. **b.** Yes. The experiment had a large sample size,; was controlled, randomized, and double-blind; and used a placebo.

1.53 No, this was not a controlled experiment. There was no random assignment to treatment/control groups and no use of a placebo.

1.55 a. Intervention remission %: $11/33 = 33.3\%$; Control remission %: $3/34 = 8.8\%$ **b.** Controlled experiment. There was random assignment to treatment/control groups. **c.** While this study did use random assignment to treatment/control groups, the sample size was fairly small (67 total) and there was no blinding in the experimental design. The difference in remission may indicate that the diet approach is promising and further research in this area is needed.

1.57 No. This is an observational study.

Chapter Review Exercises

1.59 a. $61/98 = 62.2\%$ **b.** $37/82 = 45.1\%$ **c.** Yes, this was a controlled experiment with random assignment. The difference in percentage of homes adopting smoking restrictions indicates the intervention may have been effective.

1.61 a. Gender (categorical) and whether students had received a speeding ticket (categorical)

b.

	Male	Female
Yes	6	5
No	4	10

c. Men % yes: 60%; Women % yes: $5/15 = 33.3\%$; a greater percentage of men reported receiving a speeding ticket.

1.63 Answers will vary. *Students should not copy the words they see in these answers.* Randomly divide the group in half, using a coin flip for each woman: Heads she gets the vitamin D, and tails she gets the placebo (or vice versa). Make sure that neither the women themselves nor any of the people who come in contact with them know whether they got the treatment or the placebo ("double-blind"). Over a given length of time (such as three years), note which women had broken bones and which did not. Compare the percentage of women with broken bones in the vitamin D group with the percentage of women with broken bones in the placebo group.

1.65 a. The treatment variable is mindful yoga participation. The response variable is alcohol use. **b.** Controlled experiment (random assignment to treatment/control groups) **c.** No, since the sample size was fairly small; however, the difference in outcomes for treatment/control groups may indicate that further research into the use of mindful yoga may be warranted.

1.67 No. There was no control group and no random assignment to treatment or control groups.

1.69 a. LD: 8% tumors; LL: 28% tumors A greater percentage of the 24 hours of light developed tumors. **b.** A controlled experiment. You can tell by the random assignment. **c.** Yes, we can conclude cause and effect because it was a controlled experiment, and random assignment will balance out potential confounding variables.

CHAPTER 2

Sections 2.1 and 2.2

2.1 a. 4 people; **b.** $4/125 = 3.2\%$

2.3 The vertical scale would read 0, 0.04, 0.08, 0.12, 0.16, and 0.20.

2.5 Yes, since only about 7% of the pulse rates were higher than 90 bpm. Conclusion might vary, but students must mention that 7% of pulse rates were higher than 90 bpm.

2.7 a. Both cereals have similar center values (about 110 calories). The spread of the dotplots differ. **b.** Cereal from manufacturer K tend to have more variation.

2.9 Roughly bell shaped. The lower bound is 0, the mean will be a number probably below 9, but a few students might have slept quite a bit (up to 12 hours?) which creates a right-skew.

2.11 It would be bimodal because the men and women tend to have different heights and therefore different arm spans.

2.13 About 75 beats per minute.

2.15 The BMI for both groups is right skewed. For the men it is maybe bimodal (hard to tell). The typical values for the men and women are similar although the value for the men appears just a little bit larger than the typical value for the women. The women's values are more spread out.

2.17 a. Multimodal with modes at each education level; 12 years (high school), 14 years (junior college), 16 years (bachelor's degree), and 18 years (possible master's degree). It is also left-skewed with numbers as low as 0. **b.** Estimate: $300 + 50 + 100 + 40 + 50$, or about 500 to 600, had 16 or more years. **c.** This is between 25% (from $500/2018$) and 30% (from $600/2018$) have a bachelor's degree or higher. This is very similar to the 27% given.

2.19 Ford typically has higher monthly costs (the center is near 250 dollars compared with 225 for BMW) and more variation in monthly costs.

2.21 1. is B, **2.** is A, **3.** is C

2.23 1. B **2.** A **3.** C

2.25 Students should display a pair of dotplots or histograms. One graph for Hockey and one for Soccer. The hockey team tends to be heavier than the soccer team (the typical hockey player weighs about 202 pounds while the typical soccer player weighs about 170 pounds). The soccer team has more variation in weights than the hockey team because there is more horizontal spread in the data. Statistical Question (answers may vary): Are hockey players heavier than soccer players? Which type of athlete has the most variability in weight?

2.27 How much do textbooks tend to cost at these colleges? See histogram.

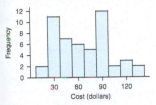

The histogram is bimodal with modes at about $30 and about $90.

2.29 See histogram. The histogram is right-skewed. The typical value is around 12 (between 10 and 15) years, and there are three outliers: Asian elephant (40), African elephant (35), and hippo (41). Humans (75 years) would be way off to the right; they live much longer than other mammals.

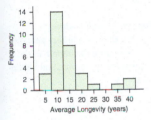

2.31 Both graphs are multimodal and right-skewed. The Democrats have a higher typical value, as shown by the fact that the center is roughly around 35% or 40%, while the center value for the Republicans is closer to 20% to 30%. Also note the much larger proportion of Democrats who think the rate should be 50% or higher. The distribution for the Democrats appears more spread out because the Democrats have a greater proportion of people responding with both lower and higher percentages.

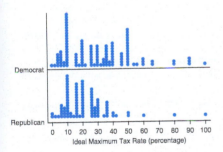

2.33 (Answers vary) Typically how much should one expect to spend per year at these law schools? The distribution appears left-skewed because of the low-end outlier at about $20,000 (Brigham Young University). The center is about $45000. Values ranges from $20K to $55K

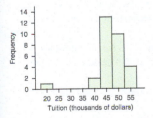

2.35 Typically, how many calories in 12 ounces of these beers? With this grouping the distribution appears bimodal with modes at about 110 and 150 calories. (With fewer—that is, wider—bins, it may not appear bimodal.)

There is a low-end outlier at about 70 calories. Values range from below 80 to more than 200 calories.

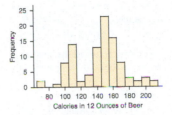

Sections 2.3 and 2.4

2.37 No, the largest category is Wrong to Right, which suggests that changes tend to make the answers more likely to be right.

2.39 A bar graph with the least variability would be one in which most children favored one particular flavor (like chocolate). A bar graph with most variability would be one in which children were roughly equally split in their preferences, with 1/3 choosing vanilla, 1/3 chocolate, 1/3 strawberry.

2.41 a. "All of the Time" was the most frequent response. **b.** The difference is about 10%. It is easier to see in the bar chart.

2.43 a. 40–59 year olds **b.** The obesity rates for women are slightly higher in the 20–39 and 60+ groups. The obesity rate for men is higher in the 40–59 age group.

2.45 Bar graph or pie chart. Chrome controls the highest market share.

2.47 This is a histogram, which we can see because the bars touch. The software treated the values of the variable *Garage* as numbers. However, we wish them to be seen as categories. A bar graph or pie chart would be better for displaying the distribution.

2.49 Hours of sleep is a numerical variable. A histogram or dotplot would better enable us to see the distribution of values. Because there are so many possible numerical values, this pie chart has so many "slices" that it is difficult to tell which is which.

2.51 Those who still play tended to have practiced more as teenagers, which we can see because the center of the distribution for those who still play is about 2 or 2.5 hours, compared to only about 1 or 1.5 hours for those who do not. The distribution could be displayed as a pair of histograms or a pair of dotplots.

Chapter Review Exercises

2.53 Since the data are numerical, a pair of histograms or dotplots could be used, one for the males and one for the females. A statistical question is Who slept more on average, men or women?"

2.55 a. The diseases with higher rates for HRT were heart disease, stroke, pulmonary embolism, and breast cancer. The diseases with lower rates for HRT were endometrial cancer, colorectal cancer, and hip fracture. **b.** Comparing the rates makes more sense than comparing just the numbers, in case there were more women in one group than in the other.

2.57 The vertical axis does not start at zero and exaggerates the differences. Make a graph for which the vertical axis starts at zero.

2.59 The shapes are roughly bell-shaped and symmetric; the later period is typically warmer but the spread is similar. This is consistent with theories on global warming. The difference is $57.9 - 56.7 = 1.2$, so the difference is only a bit more than 1 degree Fahrenheit.

2.61 a. A smaller percentage favor nuclear energy in 2016. **b.** The Republicans show the most change. **c.** A side-by-side bar graph (Republican 2010 adjacent to Republican 2016) would make the comparison easier.

2.63 The created 10-point dotplots will vary. The dotplot for Exercise 2.63 should have skew, and the dotplot for Exercise 2.64 should not.

2.65 Graphs will vary. Histograms and dotplots are both appropriate. For the group without a camera the distribution is roughly symmetrical, and for the group with a camera it is right-skewed. Both are unimodal. The number of cars going through a yellow light tends to be less at intersections with cameras. Also, there is more variation in the intersections without cameras.

2.67 Both distributions are right-skewed. The typical speed for the men (a little above 100 mph) is a bit higher than the typical speed for the women (which appears to be closer to 90 mph). The spread for the men is larger primarily because of the outlier of 200 mph for the men.

2.69 It should be right-skewed.

2.71 a. The tallest bar is Wrong to Right, which suggests that the instruction was correct. **b.** The largest group is Wrong to Right, so it appears that changes made tend to raise the grades of the students.

2.73 a. Facebook (only about 5% used it less often than weekly) **b.** LinkedIn (only about 20% used it daily) **c.** Facebook (around 75% were in one category—daily)

2.75 a. Histogram or dotplot **b.** Side-by-side barplots showing gender frequencies separately for each zip code.

CHAPTER 3

IQRs vary with different technology.

Section 3.1

3.1 c

3.3 The mean is between about 4 and 6 hours.

3.5 a. The typical number of floors is 118.6. **b.** The standard deviation of the number of floors is 26.0. **c.** Dubai is farthest from the mean.

3.7 a. The typical river is 2230.8 miles long. **b.** The standard deviation is 957.4 miles. The Mississippi-Missouri-Red Rock River contributes most to the standard deviation because it is farthest from the mean. **c.** The mean would decrease, and the standard deviation will increase. New mean: 1992.3 miles, new standard deviation: 1036.5 miles.

3.9 a. For the early 1900s the mean is 22.02 seconds; the standard deviation is 0.40 seconds. **b.** For recent Olympics, the mean is 19.66 seconds; the standard deviation is 0.35 seconds. **c.** Recent winners are faster and have less variation in their winning times.

3.11 a. A total of 185 people were surveyed. **b.** Comparing the means, men thought more should be spent on a wedding. Comparing the standard deviations, men had more variation in their responses.

3.13 a. The mean for longboards is 12.4 days, which is more than the mean for shortboards, which is 9.8 days. So longboarders tend to get more surfing days. **b.** The standard deviation of 5.2 days for the longboarders was larger than the standard deviation of 4.2 days for the shortboarders. So the longboarders have more variation in days.

3.15 San Jose tends to have a higher typical temperature; Denver has more variation in temperature.

3.17 a. 20.3 to 40.1 pounds. **b.** Less than one standard deviation from the mean

3.19 a. The mean is $3.66 and represents the typical price of a container (59 to 64 ounces) of orange juice at this site. **b.** The standard deviation is 0.51. Most orange juice of this size is within 51 cents of $3.66.

3.21 The standard deviation for the 100-meter event would be less. All the runners come to the finish line within a few seconds of each other. In the marathon, the runners can be quite widely spread out after running that long distance.

3.23 South Carolina: Mean is $198.1 thousand; standard deviation is $68.0 thousand. Tennessee: Mean is $215.8 thousand; standard deviation is $75.1 thousand. Houses in Tennessee are typically more expensive and have more variation in price than houses in South Carolina.

3.25 a. The mean for the men was 10.5, and for the women it was 4.7, showing that the male drinkers typically drank more (on average, about six drinks more) than the female drinkers. **c.** The standard deviation for the men was 11.8, and the standard deviation for the women was 4.8, showing much more variation in the number of drinks for the men. **c.** The mean for the men is now 8.6, and the mean for the women is still 4.7. Thus, the mean for the men is still larger than the mean for the women but not by as much. **d.** The standard deviation would be smaller without the two outliers.

This is because the contribution to the standard deviation from these two outliers is large because they are farthest from the mean, and that contribution would be removed.

3.27 A standard deviation can be zero if all the values are the same (no variation between the data values).

Section 3.2

3.29 a. 68% **b.** $(16 + 34 + 27)/120 = 64\%$. The estimate is very close to 68%. **c.** Between 555 and 819 runs

3.31 a. Approximately 95% **b.** Approximately 68% **c.** No, because it is not more than 2 standard deviations from the mean

3.33 a. -2 **b.** 67 inches (or 5 feet 7 inches)

3.35 An IQ below 80 is more unusual because 80 is 1.33 standard deviations from the mean while 110 is only 0.67 standard deviations from the mean.

3.37 a. $z = \dfrac{2500 - 3462}{500} = \dfrac{-962}{500} = -1.92$

 b. $z = \dfrac{2500 - 2622}{500} = \dfrac{-122}{500} = -0.24$

c. A birth rate of 2500 grams is more common (the z-score is closer to 0) for babies born one month early. In other words, there is a higher percentages of babies with low birth weight among those born one month early. This makes sense because babies gain weight during gestation, and babies born one month early have had less time to gain weight.

3.39 a. 75 inches **b.** 64.5 inches

Section 3.3

3.41 Two measures of the center of data are the mean and the median. The median is preferred for data that are strongly skewed or have outliers. If the data are relatively symmetric, the mean is preferred, but the median is also okay.

3.43

363	384	389	408	423		434	471	520	602	677
		Q1			Med			Q3		

a. Median: 428.5 million; about half the top 10 Marvel movies made more than $428.5 million. **b.** Q1 = 389 million, Q3 = 520 million; IQR = $520 - 389 = 131$ million. This is the range of the middle 50% of the sorted incomes in the top 10 Marvel movies. **c.** Range = $677 - 363 = 314$ million. The IQR is preferred over the range because the range depends on only two observations and because it very sensitive to any extreme values in the data.

3.45 Median = 471 million. About 50% of the top 7 Marvel movies made more than $471 million.

3.47 a. 25% **b.** 75% **c.** 50% **d.** IQR = 152.3. The range of the middle 50% of the sorted data is 152.3 million BTUs.

Section 3.4

3.49 a. Outliers are observed values that are far from the main group of data. In a histogram they are separated from the others by space. If they are mistakes, they should be removed. If they are not mistakes, do the analysis twice: once with and once without outliers. **b.** The median is more resistant, which implies that it changes less than the mean (when the data with and without outliers are compared).

3.51 The corrected value will give a different mean but not a different median. Medians are not as affected by the size of extreme scores, but the mean is affected.

3.53 a. The distribution is right-skewed. **b.** The median and the IQR should be used to describe the distribution.

3.55 a. The distributions are right-skewed. **b.** The medians **c.** The interquartile ranges **d.** The typical Democratic senator has been in office 9 years, while the typical Republican senator has been in office 6 years. There is more variability in the experience of Democratic senators, with an IQR of 12 years compared to an IQR of 10 years for Republican senators.

3.57 a. The median is 48. 50% of the southern states have more than 48 capital prisoners. **b.** Q1 = 32, Q3 = 152, IQR = 120 **c.** The mean is 90.8. **d.** The mean is pulled up by the really large numbers, such as Texas (243) and Florida (374). **e.** The median is unaffected by outliers.

3.59 Since the data for the West is left-skewed, use the median and IQR to compare the groups. The CPI for the West is higher than that of the Midwest (West median 244.6; Midwest median 222.8). There is more variability in the West CPI (West IQR 18.6; Midwest IQR 7.3). The West has one potentially low outlier.

3.61 a. The balancing point of the histogram, the mean is approximately 80 millimeters. **b.** 83.0 millimeters **c.** It is an approximation because we used the left-hand side of the bin to estimate the data values contained in the bin.

Section 3.5

3.63 a. South, West, Northeast, Midwest. **b.** Northeast, South, Midwest, West. **c.** The South and the Northeast each have one state with an high-priced low poverty rate. **d.** The Northeast has the least amount of variability (smallest IQR). **e.** The IQR is better because is it not influenced by unusually high or low values and because the data is not symmetric.

3.65 a. The NFL has the highest ticket prices and the most variability in ticket prices (highest median and greatest IQR). The MLB has the lowest ticket prices. **b.** Hockey tickets tend to be more expensive than basketball tickets (higher median). Both sports have some unusually high-priced tickets, and hockey has more variability in ticket prices (greater IQR).

3.67 a. Histogram 1 is left-skewed, histogram 2 is roughly bell-shaped and symmetric (not very skewed), and histogram 3 is right-skewed.

b. Histogram 1 goes with boxplot C.
Histogram 2 goes with boxplot B.

c. Histogram 3 goes with boxplot A.

A long left tail on a histogram goes with observations going down on a boxplot, because smaller numbers are to the left in a histogram and on the bottom in boxplots like these.

3.69 The maximum value (756) is greater than the upper fence (237 + 1.5(237 − 63) = 498) and is considered a potential outlier. The minimum value (1) is not less than the lower fence.

3.71 The whiskers are drawn to the upper and lower fences or to the maximum and minimum values if there are no potential outliers in the data set. There are no values less than the lower fence, so the left whisker is drawn to the minimum. Since the maximum value (128.016) is beyond the upper fence (33.223 + 1.5(33.223 − 8.526) = 70.3) the right whisker would be drawn to the upper fence (70.3).

3.73 The IQR is 90 − 78 = 12. 1.5 × 15 = 18, so any score below 78 − 18 = 60 is a potential outlier. We can see that there is at least one potential outlier (the minimum score of 40), but we don't know how many other potential outliers there are between 40 and 60. Therefore, we don't know which point to draw the left-side whisker to.

Chapter Review Exercises

3.75 a. The median is 41.05 cents/gallon. 50% of the southern states have gas taxes greater than 41.05 cents/gallon. **b.** The middle 50% of the southern states have gas taxes in a range of 11.35 cents/gallon. **c.** The mean is 43.6 cents/gallon. **d.** The data may be right-skewed.

3.77 Summary statistics are shown in the following. The 5 p.m. class did better, typically; both the mean and the median are higher. Also, the spread (as reflected in both the standard deviation and the IQR) is larger for the 11 a.m. class, so the 5 p.m. class has less variation.

The visual comparison is shown by the boxplots. Both distributions are slightly left-skewed. Therefore, you can compare the means and the standard deviations *or* the medians and the IQRs.

Minitab Statistics								
Variable	N	Mean	Median	StDev	Min	Max	Q1	Q3
11am	15	70.73	72.5	19.84	39	100	53	86
5pm	19	84.78	86.5	11.95	64.5	104.5	73	94

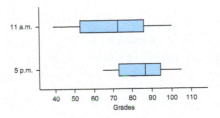

3.79 The graph is bimodal, with modes around 65 inches (5 feet 5 inches) and around 69 inches (5 feet 9 inches). There are two modes because men tend to be taller than women.

3.81 Students should provide histograms for males and females; because of lack of symmetry, they should compare medians and IQRs. Both groups typically watch about the same amount and have similar variability.

3.83 a. The mean is approximately 1000 calories. **b.** An estimate of the standard deviation is (2200 − 100)/6 = 350 calories.

3.85 Answers will vary.

3.87 Answers will vary.

3.89 Students should make a histogram for western states and southern states. Because the distributions are not symmetric, students should compare the medians and the IQRs. The western states tend to have a higher percentage of the population with a bachelor's degree. The southern states tend to have more variability.

3.91 a. Since the distributions are slightly right-skewed, compare the medians. **b.** Football ticket prices tend to be higher than hockey tickets. Football tickets also show more variation in price. (IQR: Football $84, Hockey $67).

3.93 a. 88 **b.** 74

3.95 The z-score for the SAT of 750 is 2.5, and the z-score for the ACT of 28 is 1.4. The score of 750 is more unusual because its z-score is farther from 0.

3.97 a. The distribution is right-skewed. **b.** Because the data are right-skewed, the mean would be greater than the median. **c.** The majority of ticket prices would be less than the mean price.

3.99 Answers will vary. "Who ran faster, grade 11 or grade 12 boys?" "Which group had the most consistent times, grade 11 or grade 12 boys?"

3.101 Answers vary. Possible answers include "Which buildings are typically taller: those made of concrete or those made of steel?" "Is there more variability in the heights of building constructed in the 2000s or before 2000?"

CHAPTER 4

Section 4.1

4.1 The critical reading score is a somewhat better predictor because the vertical spread is less, suggesting a more accurate prediction of GPA.

4.3 The trend appears roughly linear and positive up to about age 24, and then it starts to curve.

4.5 There is very little trend. It appears that number of credits acquired is not associated with GPA.

4.7 The trend is positive. Students with more sisters tended to have more brothers. This trend makes sense, because large families are likely to have a large number of sons and a large number of daughters.

4.9 There is a slightly negative trend. The negative trend suggests that the more hours of work a student has, the fewer hours of TV the student tends to watch. The person who works 70 hours appears to be an outlier because that point is separated from the other points by a large amount.

4.11 There is a slight negative trend that suggests that older adults tend to sleep a bit less than younger adults. Some may say there is no trend.

Section 4.2

4.13 a. You should not find the correlation because the trend is not linear. **b.** You may find the correlation because the trend is linear.

4.15 The correlation coefficient is positive since the graph shows an upward trend.

4.17 0.767 A
      0.299 B
     −0.980 C

4.19 0.87 A
     −0.47 B
      0.67 C

4.21 a. $r = 0.69$; **b.** $r = 0.69$; the correlation coefficient stays the same. **c.** $r = 0.69$. Adding a constant to all y-values does not change the value of r. **d.** $r = 0.69$; the correlation coefficient stays the same.

4.23 The value of r would stay the same.

4.25 The correlation is 0.904. The professors that have high overall quality scores tend to also have high easiness scores.

4.27 Higher gym usage is associated with higher GPA.

Section 4.3

4.29 a. The independent variable is median starting salary, and the dependent variable is median mid-career salary. **b.** Salary distributions are usually skewed. Medians are therefore a more meaningful measure of center. **c.** Between $110,000 and $120,000 **d.** $111,641 **e.** Answers will vary. The number of hours worked per week, the amount of additional education required, gender, and the type of career are all factors that might influence mid-career salary.

4.31 a. The median pay for women is about $690 when pay for men is $850. **b.** predicted women $= −62.69 + 0.887(850) = 691.26$. **c.** The slope is 0.887. Each additional dollar in mens' pay is associated with an average increase of $0.887 in the median womens' pay. **d.** The y-intercept is −62.69. It is not appropriate to interpret it in this context because median income for men (x) cannot be zero.

4.33 a. Predicted Arm span $= 16.8 + 2.25$ Height
b. $b = 0.948(8.10/3.41) = 2.25$ **c.** $a = 159.86 − 2.25(63.59) = 16.8$
d. Arm span $= 16.8 + 2.25(64) = 160.8$, or about 161 centimeters

4.35 a. Predicted Arm span $= 6.24 + 2.515$ Height (Rounding may vary.)
b. Minitab: slope $= 2.51$, intercept $= 6.2$
    StatCrunch: slope $= 2.514674$, intercept $= 6.2408333$
    Excel: slope $= 2.514674$, intercept $= 6.240833$
    TI-84: slope $= 2.514673913$, intercept $= 6.240833333$

4.37 The association for the women is stronger because the correlation coefficient is farther from zero.

4.39 a. Based on the scatterplot there is not a strong association between these two variables. **b.** The numerical value of the correlation would be close to zero because there is not an association between these variables. **c.** Since there is not an association between these variables we cannot use singles percentage to predict doubles percentage.

4.41 Explanations will vary.

	x	y
a.	odometer reading	price
b.	household size	monthly water bill
c.	time spent in gym	weight

4.43 a. The higher the percentage of smoke-free homes in a state, the lower the percentage of high school students who smoke tends to be.
b. $56.32 − 0.464(70) = 23.84$, or about 24%

4.45 a. As driver age increases, insurance prices decrease but then begin to increase again at around 65 years of age. Younger drivers and older drivers tend to have more accidents so they are charged more for insurance.

b. It would not be appropriate to do a linear regression analysis on these data because the data do not follow a linear trend.

4.47 The answers follow the guided steps.

1: Graph.

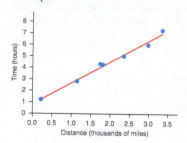

2: The linear model is appropriate. The points suggest a straight line.
3: In the formula, the time is in hours and the distance in thousands of miles.

$$\text{Predicted Time} = 0.8394 + 1.838 \text{ Distance}$$

4: Each additional thousand miles takes, on average, about 1.84 more hours to arrive.
5: The additional time shown by the intercept might be due to the time it takes for the plane to taxi to the runway, delays, the slower initial speed, and similar delays in the landing as well. The time for this appears to be about 0.84 hours (or 50 minutes).
6: Predicted Time $= 0.8394 + 1.838(3) = 0.8394 + 5.514 = 6.35$ hours. The predicted time from Boston to Seattle is 6.35 hours.

4.49 a. Positive: the greater the population, the more millionaires there tend to be. **b.** See scatterplot. **c.** $r = 0.992$ **d.** For each additional hundred thousand in the population, there is an additional 1.9 thousand millionaires.

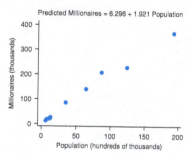

e. It does not make sense to look for millionaires in states with no people.
4.51 a.

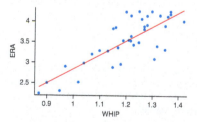

b. ERA $= −0.578 + 3.436$ WHIP
c. Each additional point in WHIP rating is associated with an average increase of 3.436 in ERA.
d. It would be inappropriate to interpret the y-intercept because there are no pitchers with a 0 WHIP rating.

Section 4.4

4.53 a. Influential points are outliers in the data that can have large effects on the regression line. When influential points are present in the data, do the regression and correlation with and without these points and comment on the difference. **b.** The coefficient of determination is the square of the correlation coefficient; it measures the percentage of variation in the y-variable that is explained by the regression line. **c.** Extrapolation means using the regression equation to make predictions beyond the range of the data. Extrapolation should not be used.

4.55 Older children have larger shoes and have studied math longer. Large shoes do not cause more years of studying. Both are affected by age.

4.57 Square 0.67 and you get 0.4489, so the coefficient of determination is about 45%. Therefore, 45% of the variation in weight can be explained by the regression line.

4.59 Part of the poor historical performance could be due to chance, and if so, regression toward the mean predicts that stocks turning in a lower-than-average performance should tend to perform closer to the mean in the future. In other words, they might tend to increase.

4.61 a. The salary is $2.099 thousand less for each year later that the person was hired, or $2.099 thousand more for each year earlier. **b.** The intercept ($4,255,000) would be the salary for a person who started in the year 0, which is ridiculous.

4.63 a. See graph.

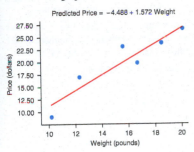

Predicted Price = −4.488 + 1.572 Weight

b. $r = 0.933$

A positive correlation suggests that larger turkeys tend to have a higher prices. **c.** Predicted Price = −4.49 + 1.57 Weight **d.** The slope: for each additional pound, the price goes up by $1.57. The interpretation of the intercept is inappropriate because it is not possible to have a turkey that weighs 0 pounds. **e.** The 30-pound free turkey changes the correlation to −0.375 and changes the equation to

$$\text{Predicted Cost} = 26.87 - 0.553 \text{ Weight}$$

This implies that the bigger the turkey, the less it costs! The 30-pound free turkey was an influential point, which really changed the results. **f.** $r^2 = 0.87$. 87% of the variation in turkey price is explained by weight.

4.65 a. Positive **b.** Slope = 0.327; each additional $1 in teacher pay is associated with an increase in per pupil spending of 0.327. **c.** The y-intercept is −5922; it would not be appropriate to interpret the y-intercept because there is no state an average teacher salary of $0. **d.** $13,698

4.67 a. Correlation is negative because the trend is downhill, or because the slope's sign is negative. **b.** For each additional hour of work, the score tended to go down by 0.48 points. **c.** A student who did not work would expect to get about 87 on average.

4.69 There is a stronger association between home runs and strikeouts ($r = 0.64$ compared to $r = -0.09$).

4.71 a. $r = 0.85$; Math = 33.03 + 0.70 Reading; 33.03 + 0.70(70) = 82.03. Predicted math score is 82. **b.** $r = 0.85$; Reading = −15.33 + 1.03 Math; −15.33 + 1.03(70) = 57. Predicted reading score is 57. **c.** Changing the choice of the dependent and independent variables does not change r but does change the regression equation.

4.73 The answer follows the guided steps.

1: **a.** Slope: $b = r\dfrac{s_y}{s_x} = 0.7\dfrac{10}{10} = 0.7$

b. Intercept: $a = \bar{y} - b\bar{x}$
$$a = 75 - 0.7(75) = 22.5$$

c. Equation
$$\text{Predicted Final} = 22.5 + 0.7 \text{ Midterm}$$

2: Predicted Final = 22.5 + 0.7 Midterm
$$= 22.5 + 0.7(95)$$
$$= 89$$

3: The score of 89 is lower than 95 because of regression toward the mean.

Chapter Review Exercises

4.75 a. $r = 0.941$
Predicted Weight = −245 + 5.80 Height

b.

Height	Weight
60(2.54) = 152.4	105/2.205 = 47.6190 kilograms
66(2.54) = 167.64	140/2.205 = 63.4921 kilograms
72(2.54) = 182.88	185/2.205 = 83.9002 kilograms
70(2.54) = 177.8	145/2.205 = 65.7596 kilograms
63(2.54) = 160.02	120/2.205 = 54.4218 kilograms

c. The correlation between height and weight is 0.941. It does not matter whether you use inches and pounds or centimeters and kilograms. A change of units does not affect the correlation because it has no units. **d.** The equations are different.
Predicted Weight Pounds = −245 + 5.80 Height (in inches)
Predicted Weight Kilograms = −111 + 1.03 Height (in centimeters)

4.77 a. $r = 0.86$; calories = 3.26 + 10.81 Carbs; Slope = 10.81; Each additional gram of carbohydrates is associated with an increase of 10.81 calories; 3.26 + 10.81(55) = 597.81 calories **b.** $r = 0.79$; calories = 198.74 + 32.75(sugar); 198.74 + 32.75(10) = 526.24 calories **c.** While both are fairly good, the number of carbs is a better predictor ($r = 0.86$ compared to $r = 0.79$).

4.79 a. There were no women taller than 69 inches, so the line should stop at 69 inches to avoid extrapolating. **b.** Men who are the same height as women wear shoes that are, on average, larger sizes. **c.** The mean increase in shoe size based on height is the same for men and women.

4.81 Among those who exercise, the effect of age on weight is less. An additional year of age does not lead to as great an increase in the average weight for exercisers as it does for non-exercisers.

4.83 a.

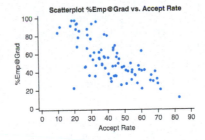

Scatterplot %Emp@Grad vs. Accept Rate

There seems to be a linear trend. The less-selective law schools also tend to have a lower employment rate at graduation.

b. I. % Employed = 97.59 − 1.03 Acceptance Rate II. Each additional percentage point in acceptance rate is associated with an average decrease of 1.03 percentage points in the rate of employment at graduation. III. It would be inappropriate to interpret the y-intercept because there is no school with an acceptance rate of 0%. IV. $r^2 = 52\%$. 52% of the variation in employment rate at graduation can be explained by the acceptance rate. V. The regression equation predicts an employment rate of about 32.6% for a school with an acceptance rate of 50%.

4.85 A comparison of scatterplots shows a stronger association between calories and fat compared with calories and carbohydrates. The correlation coefficient for fat and calories is $r = 0.82$ while the correlation coefficient for carbohydrates and calories is 0.39. Fat is a better predictor of the number of calories in these snack foods.

4.87

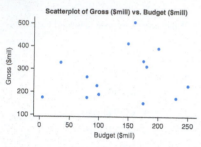

It is not appropriate to do a regression equation because the scatterplot does not show a linear trend in the data.

4.89 a. The positive trend shows that the more stories there are, the taller the building tends to be. **b.** $115.4 + 12.85(100) = 115.4 + 1285 = 1400.4$ feet, or about 1400 feet **c.** Buildings that have one additional story tend to have an average height of 12.85 additional feet. **d.** Because there are no building with 0 stories, the interpretation of the intercept is not appropriate. **e.** About 71% of the variation in height can be explained by the regression, and about 29% is not explained.

4.91 through 4.93 Answers will vary.

4.95 The trend is positive. In general, if one twin has a higher-than-average level of education, so does the other twin. The point that shows one twin with 1 year of education and the other twin with 12 years is an outlier. (Another point showing one twin with 15 years and the other with 8 years is a bit unusual, as well.)

4.97 There appears to be a positive trend. It appears that the number of hours of homework tends to increase slightly with enrollment in more units.

4.99 Linear regression is not appropriate because the trend is not linear; it is curved.

4.101 The cholesterol going down might be partly caused by regression toward the mean.

CHAPTER 5

Section 5.1

5.1 a. 55185   74834 **b.** TTHTT   THTHH   The longest streak was 2 heads. **c.** $4/10 = 40\%$ were heads.

5.3 Theoretical probability, because it is not based on an experiment.

5.5 Empirical probability because it is based on an experiment.

Section 5.2

5.7 a. The seven equally likely outcomes are Town, Wu, Hein, Lee, Marland, Penner, and Holmes. **b.** $4/7 = 57.1\%$ **c.** $3/7 = 42.9\%$ **d.** Yes, they are complements because the event "male doctor" is equivalent to the event "not a female doctor."

5.9 a. 0.26 can be the probability of an event. **b.** -0.26 cannot be a probability because it is negative. **c.** 2.6 cannot be the probability of an event since 2.6 is greater than 1; **d.** 2.6% can be the probability of an event; **e.** 26 cannot be the probability of an event since 26 is greater than 1.

5.11 a. A heart: $13/52$ or $1/4$ **b.** A red card: $26/52$ or $1/2$ **c.** An ace: $4/52$ or $1/13$ **d.** A face card: $12/52$ or $3/13$ **e.** A three: $4/52$ or $1/13$. Answers may also be in decimal or percentage form.

5.13 a. P(guessing correctly) $= 1/2$ **b.** P(guessing incorrectly) $= 1/2$

5.15 a. 16 outcomes **b. i.** $1/16 = 6.25\%$ **ii.** $4/16 = 25\%$ **iii.** $5/16 = 31.25\%$

5.17 The probability of being born on a Friday OR Saturday OR Sunday is $3/7$, or 42.86%.

5.19 a. i. $250/500 = 50\%$ **ii.** $335/500 = 67\%$ **iii.** $200/500 = 40\%$ **b.** $200/250 = 80\%$ of the college graduates took a vacation, and $135/250 = 54\%$ of non–college graduates took a vacation. College graduates were more likely to have taken a vacation.

5.21 a. $278/1028 = 27\%$ **b.** $308/1028 = 30\%$ **c.** $586/1028 = 57\%$ **d.** "Republican" and "Democrat" are mutually exclusive because a person cannot be both a Democrat and a Republican at the same time. **e.** "Republican" and "Democrat" are not complementary events because "not Republican" includes Democrats and Independents.

5.23 a. Probability the person is a man or said "about the same" or both $= (280 + 259 - 174)/430 = 84.9\%$ **b.** Probability that the person is a man and said "about the same" $= 174/430 = 40.5\%$

5.25 a. i. Not mutually exclusive; some people have travelled to both Mexico and Canada. **ii.** Mutually exclusive; a person cannot be single and married at the same time. **b.** Answers vary.

5.27 Answers vary. Being a college graduate or not being a college graduate are mutually exclusive. Saying Yes and No are mutually exclusive. Naming any two rows (or two columns) gives you mutually exclusive events.

5.29 a. The numbers on a die that are even or greater than 4 are 2, 4, 5, and 6, so the probability is $4/6 = 66.7\%$. **b.** The numbers on a die that are odd or less than 3 are 1, 2, 3, and 5, so the probability is $4/6 = 66.7\%$.

5.31 a. A OR B: $0.18 + 0.25 = 0.43$ **b.** A OR B OR C: $0.18 + 0.25 + 0.37 = 0.80$ **c.** Lower than a C: $1 - 0.80 = 0.20$

5.33 Since the total of all responses must equal 100%, $100\% - (42\% + 23\%) = 35\%$.

5.35 a. $17\% + 15\% = 32\%$ **b.** $20\% + 29\% = 49\%$; $49\% \times 1200 = 588$ people.

5.37 a. $0.6(0.6) = 0.36$ **b.** $0.6(0.4) + 0.4(0.6) = 0.24 + 0.24 = 0.48$ **c.** $0.36 + 0.48 = 0.84$ **d.** $1 - 0.36 = 0.64$

5.39 a. More than 8 mistakes: $1 - 0.48 - 0.30 = 0.22$ **b.** 3 or more mistakes: $0.30 + 0.22 = 0.52$ **c.** At most 8 mistakes: $0.48 + 0.30 = 0.78$ **d.** The events in parts a and c are complementary because "at most 8 mistakes" means from 0 mistakes up to 8 mistakes. "More than 8 mistakes" means 9, 10, up to 12 mistakes. Together, these mutually exclusive events include the entire sample space.

Section 5.3

5.41 a. i. P(has been about right | female) **b.** $34/100 = 34\%$

5.43 a. $319/550 = 58\%$ **b.** $195/500 = 39\%$ **c.** $(514 + 550 - 319)/1050 = 745/1050 = 71\%$

5.45 They are associated. Tall people are much more likely to play professional basketball than are short people. To say it another way, basketball players are much more likely to be tall than are those who don't play basketball.

5.47 Eye color and gender are independent because eye color does not depend on gender.

5.49 They are not independent because the probability of saying "Hasn't Gone Far Enough" if the person is female is $57/100$ or 57% whereas the overall probability of saying "Hasn't Gone Far Enough" is $99/200$ or 49.5% and these are not the same.

5.51 The answers follow the guided steps.

	M	W	
Right	18	42	60
Left	12	28	40
	30	70	100

1: See table.

2: 60/100, or 60%

3: 18/30 = 60

4: The variables are independent because the probability of having the right thumb on top given that a person is a man is equal to the probability that a person has the right thumb on top (for the whole data set).

5.53 a.

	Local TV	Network TV	Cable TV	Total
Men	66	48	58	172
Women	82	54	56	192
Total	148	102	114	364

b. P(cable | male) = 58/172 = 33.7%; P(cable) = 114/364 = 31.3%. Since these probabilities are not equal, the events are not independent.

5.55 a. 1/8 **b.** 1/8

5.57 They are the same. Both probabilities are $\left(\frac{1}{6}\right)^5$ or $\frac{1}{7776}$.

5.59 a. (0.62)(0.62) = 38.4% **b.** (0.38)(0.38) = 14.4% **c.** 100% − 14.4% = 85.6%

5.61 P(have C AND test positive) = P(have C) P(test positive | have C)
= 0.00008(0.84)
= 0.000067

Section 5.4

5.63 a. Trials 2, 3, and 4 had at least one 6. **b.** The empirical probability is 3/5 or 0.6.

5.65 a. and **b.** are both valid methods because the probability of a correct choice is 1/3. **c.** is invalid because the probability of a correct choice is 3/10.

5.67 a. The random numbers represent the following outcomes: HHHTH HTHHT TTHTH HTTTH. **b.** The simulated probability of getting heads is 11/20 = 55%, which is close to the theoretical probability 50%. **c.** The theoretical probability is 50%, so in this case the simulated probability is likely to be close to the theoretical probability.

5.69 Histogram B was for 10,000 rolls because it has nearly a flat top. In theory, there should be the same number of each outcome, and the Law of Large Numbers says that the one with the largest sample should be closest to the theory.

5.71 The proportion should get closer to 0.5 as the number of flips increases.

5.73 Betty and Jane are betting more times (100 times), so they are more likely to end up with each having about half of the wins, compared to Tom and Bill. The Law of Large Numbers says that the more times you repeat a random experiment, the closer the experimental probability comes to the true probability (50%). The graph shows that the proportion of wins settles down to about 50% by 100 trials. But at 10 trials, the percentage of wins has not settled down and will vary quite a bit.

5.75 You are equally likely to get heads or tails (assuming the coin is fair) because the coin's results are independent of each other—that is, the coin does not "keep track" of its past.

5.77 The probability of selecting a digit from 0 to 5 is 6/10, or 60%, so it does not represent the probability we wish to simulate.

5.79 a. You could use the numbers 1, 2, 3, and 4 to represent the outcomes and ignore 0 and 5 through 9, but answers to this will vary. **b.** The empirical probabilities will vary. The theoretical probability of getting a 1 is 1/4; remember that the die is four-sided.

Chapter Review Exercises

5.81 627/1012 = 62.0%

5.83 a. Gender and shoe size are associated because men tend to wear larger shoe sizes than women. **b.** Win/loss record is independent of the number of cheerleaders. The coin does not "know" how many cheerleaders there are or have an effect on the number of cheerleaders.

5.85 a. (0.55)(0.43) = 23.7% **b.** (0.45)(0.57) = 25.7% **c.** This corresponds with the man supports and woman doesn't or woman supports and the man doesn't. (0.55)(0.57) + (0.43)(0.45) = 50.7%

d. P(at least one supports the death penalty) = 100% − P(neither support the death penalty) = 100% − 25.7% = 74.3%

5.87 a. (0.27)(0.27) = 7.3% **b.** Because the adults are Facebook friends, they may have similar attitudes regarding online dating.

5.89 a. Both born Monday: $\frac{1}{7}\left(\frac{1}{7}\right) = \frac{1}{49}$

b. Alicia OR David was born on Monday:
P(A OR B) = P(A) + P(B) − P(A AND B)

$$= \frac{1}{7} + \frac{1}{7} - \frac{1}{7}\left(\frac{1}{7}\right)$$
$$= \frac{7}{49} + \frac{7}{49} - \frac{1}{49}$$
$$= \frac{13}{49}$$

5.91 a. 61% × 2500 = 1525 **b.** 31% × 2500 = 775 **c.** 36% × 2500 = 900; **d.** The responses are mutually exclusive because those surveyed can fall into only one of these categories.

5.93 a. 44% **b.** 56% × 400 = 224 people

5.95 Democrats: 72% × 1500 = 1080. Republicans: 36% × 1500 = 540

5.97 a. 0 heads: 1/4, or 0.25. **b.** Exactly 1 head: 1/2, or 0.50 **c.** Exactly 2 heads: 1/4, or 0.25. **d.** At least one head: 3/4, or 0.75. **e.** Not more than 2 heads (which means 2 or fewer heads): 1

5.99 a.

	Democrats	Republicans	Independents	Total
Yes	300	153	134	587
No	100	147	66	313
Total	400	300	200	900

b. 300/900 = 33.3% **c.** 147/900 = 16.3% **d.** 147/300 = 49.0% **e.** 147/313 = 47% **f.** 700/900 = 77.8%

5.101. a. (0.70)(0.70) = 49% **b.** (0.70)(0.30) + (0.30)(0.70) = 42% **c.** (0.30)(0.30) = 9%

5.103 Recidivism and gender are not independent. If they were independent, the recidivism rates for men and women would be the same.

5.105 Answers vary.

5.107 The smaller hospital will have more than 60% girls born more often because, according to the Law of Large Numbers, there's more variability in proportions for small sample sizes. For the larger sample size ($n = 45$), the proportion will be more "settled" and will vary less from day to day. Over half the subjects in Tversky and Kahneman's study said that "both hospitals will be the same." But *you* didn't, did you?

5.109 a.

	Conservative Republican	Moderate/ Liberal Republican	Moderate/ Conservative Democrat	Liberal Democrat	Total
Yes	65	77	257	332	731
No	368	149	151	88	756
Total	433	226	408	420	1487

b. 433/1487 = 29.1% **c.** 731/1487 = 49.2%

5.111 65/1487 = 4.4%

5.113 a. (408 + 420)/1487 = 55.7% **b.** These events are mutually exclusive because a person cannot be in both categories.

5.115 a. (420 + 756 − 88)/1487 = 1088/1487 = 73.2% **b.** These events are not mutually exclusive because a person can belong to both categories.

5.117 a. 332/420 = 79.0% **b.** 65/433 = 15.0% **c.** 65/731 = 8.9%

5.119 a. Not mutually exclusive because a person can belong to both categories. **b.** Mutually exclusive because a person cannot belong to both categories.

5.121 a. Trials 1 and 3 had at least 3 of the dice land on the same number. **b.** The empirical probability is 2/5, or 0.4.

5.123 a. The probability that the student will correctly guess an answer is 0.20 (1 out of 5). The probability of randomly selecting a 0 or a 1 is $2/10 = 1/5 = 0.20$.

b.

1	1	3	7	3		9	6	8	7	1
R	R	W	W	W		W	W	W	W	R

c. Yes. There were 3 correct.

d. WWRWW   WWRWW. No. The student scored only 2 correct.

e. RWWRW   WWWWR. 3 correct. Yes.
WRWWW   WWWWW 1 correct. No.

f. There were four trials, and two had a successful outcome. Thus, the empirical probability is 2/4, or 0.50.

5.125 a. The action is to arrive at a light, which is either green or not. The probability of a success is 60%. **b.** Answers will vary. Our method: Let the digits from 0 to 5 represent a green light, and let the digits from 6 to 9 represent yellow or red. (Any assignment that gives six digits to green and four digits to non-green will work.) **c.** The event of interest is "get three out of three green." **d.** A single trial consists of reading off three digits. **e.** Outcomes (non-green are labeled R). Three greens in a row are underlined.

275	830	186	658	250	381	033	582	594	513
GRG	RGG	GRR	RGR	<u>GGG</u>	GRG	<u>GGG</u>	GRG	GRG	<u>GGG</u>

608	010	443	958	621	098	403	511	960	372
RGR	<u>GGG</u>	<u>GGG</u>	RGR	RGG	GRR	<u>GGG</u>	<u>GGG</u>	RRG	GRG

Number of greens (with successful events in red)

2 2 1 1 3 2 3 2 2 3

1 3 3 1 2 1 3 3 1 2

f. P(all three green) is estimated as 7/20.

CHAPTER 6

Answers may vary slightly due to rounding or type of technology used.

Section 6.1

6.1 a. Continuous **b.** Discrete

6.3 a. Continuous **b.** Continuous

6.5

Number of Spots	1	2	3	4	5	6
Probability	0.1	0.2	0.2	0.2	0.2	0.1

The table may have a different orientation.

6.7

UU	UD	DU	DD
0.6(0.6)	0.6(0.4)	0.4(0.6)	0.4(0.4)
0.36	0.24	0.24	0.16

6.9

0	1	2
0.16	0.48	0.36

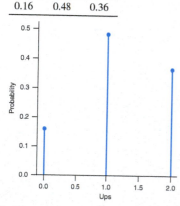

6.11 $(6 - 3)(0.2) = 0.6$, or 60%, and the area between 3 and 6 should shaded.

Section 6.2

6.13 a. ii., 95% **b.** i., Almost all **c.** iii., 68% **d.** iv., 50% **e.** ii., 13.5%

6.15 a. iii., 50% **b.** iii., 68% **c.** v., About 0% **d.** v., About 0% **e.** ii., 95% **f.** v., 2.5%

6.17 Output B is correct. 21.2%

6.19 a. 0.8708 **b.** $1 - 0.8708 = 0.1292$

6.21 a. 0.9788 or about 98% **b.** 0.9599 or about 96% **c.** 0.8136 or about 81%

6.23 a, b, and **c** are all 0.000. **d.** The proportion to the right of 4.00 would be the largest of the three, and the proportion to the right of 50.00 would be the smallest. **e.** Below -10.00

6.25 a. $z = (170 - 150)/2 = 2$. **b.** Because the Empirical Rule states that the area between -2 and 2 is 95%, the area outside -2 and 2 must be $100\% - 95\% = 5\%$. This means half of the 5% must be below -2 and half must be above 2. Therefore, 2.5% must be above a standard score of 2. This corresponds with the probability an adult St. Bernard weighs more than 170 pounds. **c.** Technology confirms this is correct. **d.** Almost all the values will lie between 3 standard deviations below the mean and 3 standard deviations above the mean, so almost all adult St. Bernards will have weights between $150 \pm 3(10)$ or between 120 and 180 pounds.

6.27 About 1.2%

6.29 About 69% of boys in this age group will have feet that are 24.6 to 28.8 cm long. The percentage that will not have feet in this range is $100\% - 69\% = 31\%$.

6.31 a. About 13.4% **b.** About 62.7% **c.** About 5.3%

6.33 a. About 72.8 % **b.** About 4.3% **c.** About 4.3%

6.35 a. About 33% **b.** About 15.6% **c.** A score of 720

6.37 a. About 86.7% **b.** $z = (89 - 71.4)/3.3 = 5.33$. About 0% of males have an arm span at least as long as 89 inches.

6.39 About 80% of the days in February have minimums of 32 °F or lower.

6.41 a. Probability; about 4.8% **b.** Measurement (inverse); 75 inches

6.43 $z = 0.43$

6.45 a. $z = 0.56$ **b.** $z = -1.00$

6.47 a. About 7.7% **b.** 281 days **c.** Almost 0%

6.49 a. 94th percentile **b.** A score of 517

6.51 62.9 inches (5 feet 2.9 inches)

6.53 a. 567 **b.** 433 **c.** $567 - 433 = 134$ **d.** The IQR is 134, and the standard deviation is 100, so the interquartile range is larger.

6.55 a. 77th percentile **b.** 66 inches (5 feet 6 inches)

6.57 a. 3.5 ounces **b.** 2.5 ounces

Section 6.3

6.59 Conditions

Two complementary outcomes: boy or girl

Fixed number of trials: 3 children

Same probability of success on each trial: 1/2 probability of a boy

All trials independent: Because there are no twins, the gender of each child is independent of the gender of the others.

6.61 Each trial has six possible outcomes (1, 2, 3, 4, 5, 6). In a binomial experiment, each trial only has two possible outcomes.

6.63 There is not a fixed number of trials because he is going to take as many free throws as he can in one minute.

6.65 a. $b(20, 0.53, 12)$ **b.** $b(20, 0.47, 10)$

6.67 a. $b(10, 0.36, 4) = 0.242$ or about 24% **b.** $b(10, 0.36, 4 \text{ or fewer}) = 0.729$ or about 73%

6.69 a. $b(50, 0.42, \text{fewer than } 20) = 0.336$ or about 34% **b.** $b(50, 0.42, \text{at most } 24) = 0.842$ or about 84% **c.** $b(50, 0.42, \text{at least } 25) = 0.158$ or about 16%

6.71 a. 0.387 **b.** 0.264 **c.** 0.736

6.73 a. $b(40, 0.52, 20) = 0.121$ or about 12% **b.** $b(40, 0.52$, fewer than 20) $= 0.340$ or about 34% **c.** $b(40, 0.52$, at most 20) $= 0.461$ or about 46% **d.** $b(40, 0.52$, between 20 and 23) $= 0.463$ or about 46%

6.75 a. $b(50, 0.59$, at least 25) $= 0.924$ or about 92% **b.** $b(50, 0.59$, more than 30) $= 0.390$ or about 39% **c.** $b(50, 0.59$, between 30 and 35) $= 0.463$ or about 46% **d.** $b(50, 0.41$, more than 30) $= 0.002$ or about 0.2%

6.77 a. TT, TN, NT, NN **b.** P(TT) $= (0.29)(0.29) = 0.0841$, P(TN) $= (0.29)(0.71) = 0.2059$, P(NT) $= (0.71)(0.29) = 0.2059$, P(NN) $= (0.71)(0.71) = 0.5041$ **c.** P(NN) $= 0.5041$ or about 50% **d.** P(TN) $+ P$(NT) $= 0.2059 + 0.2059 = 0.4118$ or about 41%

6.79 a. 200(0.53)106; $\sqrt{200 \times 0.53 \times (1 - 0.53)} = 7.1$ **b.** Since 190 is more than two standard deviations from the mean, it would be surprising.

6.81 a. 55 **b.** $\sqrt{100(0.55)(0.45)} = 5$ **c.** 45, 65 **d.** Yes, 85 would be surprising because it is so far from the 95% interval given in part c.

Chapter Review Exercises

6.83 a. discrete **b.** continuous **c.** continuous **d.** discrete

6.85 95.2%, or about 95%

6.87 a. 0.7625, or about 76% **b.** 98.6 °F

6.89 a. 0.683 or about 68% **b.** 0.159 or about 16% **c.** 0.023 or about 2%

6.91 a. 0.106 or about 11% **b.** Systolic blood pressures between 113 and 127 **c.** 0.394 or about 39% **d.** Systolic blood pressure above 128

6.93 a. $b(200, 0.44$, fewer than 80) $= 0.113$ or about 11% **b.** $b(200, 0.44$, at least 90) $= 0.414$ or about 41% **c.** $b(200, 0.44$, between 80 and 100) $= 0.849$ or about 85% **d.** $b(200, 0.44$, at most 75) $= 0.037$ or about 4%

6.95 a. $b(50, 0.46$, more than 25) $= 0.239$ or about 24% **b.** $b(50, 0.46$, at most 20) $= 0.240$ or about 24% **c.** $50(0.46) = 23$ **d.** $\sqrt{50 \times 0.46 \times (1 - 0.46)} = 3.52$; it would be surprising because 10 is more than 3 standard deviations from the mean.

6.97 a. $b(500, 0.67$, more than 500) $=$ about 100% **b.** $500(0.67) = 335$ **c.** The standard deviation is $\sqrt{500 \times 0.67 \times (1 - 0.67)} = 10.5$; 450 is more than 3 standard deviations from the mean so it would be unusual to find that more than 450 out of 500 randomly selected Americans held this belief.

6.99 a. 3.45% **b.** 0.0345(0.0345) $= 0.0012$ **c.** $b(200, 0.0345$, 7 or fewer) $= 0.614$ **d.** $200(0.0345) =$ about 7 **e.** $\sqrt{200(0.0345)(0.9655)} = 2.58$ **f.** $7 + 2(2.58) =$ about 12, and $7 - 2(2.58) =$ about 2; about 2 to 12 **f.** Yes, 45 would be surprising because it is so far from the interval.

6.101 a. 19.7 inches **b.** 20.5 inches **c.** They are the same. The distribution is symmetric and so the mean is right in the middle at the 50th percentile.

CHAPTER 7

Answers may vary slightly due to rounding or type of technology used.

Section 7.1

7.1 a. The population is adults in the United States. The sample consists of the 1021 adults surveyed. **b.** The parameter of interest is the percentage of all adults who support a ban on smoking in public places. The statistic is the 57% of the sample who supported such a ban.

7.3 a. The number 30,000 is a parameter because it describes all of Ross' paintings. The population is the collection of all of Ross' paintings. **b.** The number 10 percent is a statistic. It is based on a survey of PBS viewers who watched *The Joy of Painting*. The population is all PBS viewers who watched *The Joy of Painting*.

7.5 a. $\bar{x}$ is a statistic, and μ is a parameter. **b.** $\bar{x}$

7.7 μ is 218.8 pounds, the mean of the population. $\bar{x}$ is the 217.6 pounds, the mean of the sample.

7.9 a. 32470 is a sample mean, so the correct notation is $\bar{x}$. **b.** 28% is a sample proportion, so the correct notation is $\hat{p}$.

7.11 The sample is the 1207 survey participants. The population of interest is all Americans. The value 16% is a measure of the survey participants who are a sample of the population of interest. For this reason it is a statistic. We would use the symbol for a sample proportion, $\hat{p}$.

7.13 You want to test a sample of batteries. If you tested them all until they burned out, no usable batteries would be left.

7.15 First, all 10 cards are put in a bowl. Then one is drawn out and noted.
 "With replacement": The card that is selected is replaced in the bowl, and a second draw is done. It is possible that the same student could be picked twice.
 "Without replacement": After the first card is drawn out, it is not replaced, and the second draw must be a different card.

7.17 Chosen: 7, 3, 5, 2

7.19 Assign each student a pair of digits 00–29 (or 01–30). Read off pairs of digits from the random number table. The students whose digits are called are in the sample. Skip repeats. Stop after the first 10 are selected.

7.21 One reason the district should be cautious because of the low survey response rate.

7.23

	With Persuasion	Without Persuasion
Support Cap	6 + 2	13 + 5
Oppose	9 + 8	2 + 5

a. With persuasion: $8/25 = 32\%$
b. Without persuasion: $18/25 = 72\%$
c. Yes, she spoke against it, and fewer who heard her statements against it (32%) supported capital punishment, compared with those who did not hear her persuasion (72%).

Section 7.2

7.25 a. The sketch should show bullet holes consistently to the left of the target and close to each other. If the bullets go consistently to the left, then there is bias, not lack of precision. **b.** The sketch should show bullet holes that are all near the center of the target.

7.27 No, you cannot see bias from one numerical result. Bias is in the procedure, not the result. It is possible that the high mean happened by chance. "Statisticians evaluate the method used for a survey, not the outcome of a single survey."

7.29 a. 16/30, or about 53.3%, odd digits **b.** 16/30, is $\hat{p}$ (p-hat), the sample proportion. **c.** The error is 16/30, minus 14/30, (or 2/30), or about 6.7%.

7.31 a. 72%, the same as the population proportion.
b. $\sqrt{\dfrac{p(1 - p)}{n}} = \sqrt{\dfrac{0.72 \times 0.28}{100}} = 0.45$ or 4.5% **c.** 72%, give or take 4.5% **d.** The standard error would decrease:
$$\sqrt{\dfrac{p(1 - p)}{n}} = \sqrt{\dfrac{0.72 \times 0.28}{500}} = 0.02 \text{ or } 2\%.$$

7.33 The largest sample is the narrowest (the graph on the bottom), and the smallest sample is the widest (the graph on the top). Increasing the sample size makes the graph narrower.

7.35 The top dotplot has the largest standard error because it is widest, and the bottom dotplot has the smallest standard error because it is narrowest.

7.37 Graph A is for the fair coin, because it is centered at 0.50.

7.39 a. No. We expect the sample proportion will be close to the population proportion (20%), but the sample proportions will vary from sample to sample with a standard error of
$$\sqrt{\dfrac{p(1 - p)}{n}} = \sqrt{\dfrac{0.20 \times 0.80}{50}} = 0.057 \text{ or } 5.7\%;$$
b. The sample proportion for a sample of size 500 will be closer to the population proportion because the standard error for samples of size 500 will be smaller than that of samples of size 50 $\left(\sqrt{\dfrac{0.20 \times 0.80}{500}} = 0.018 \text{ or } 1.8\% \text{ compared with } 5.7\%\right)$.

Section 7.3

7.41 Conditions for the CLT: The sample is random. The population of high school seniors is greater than $10(100) = 1000$. $100(0.72)$ and $100(0.28)$ are both greater than 10. The standard error is

$\sqrt{\dfrac{p(1-p)}{n}} = \sqrt{\dfrac{0.72 \times 0.28}{100}} = 0.045$. The sampling distribution: N(0.72, 0.045). Use technology to find $P(x \geq 0.75)$ or find the z-score for 0.75:

$z = \dfrac{0.75 - 0.72}{0.045} = 0.67$ and find $P(z \geq 0.67)$. Answer: 0.252 or 25.2%

7.43 a. 80% **b.** Sample is random; the population of Americans is greater than $10(1000) = 10,000$; $1000(0.80)$ and $1000(0.20)$

are both greater than 10. **c.** standard error $= \sqrt{\dfrac{p(1-p)}{n}} =$

$\sqrt{\dfrac{0.80(1-0.80)}{1000}} = 0.013$ **d.** $0.80 \pm 2(0.013)$; between 0.774 and 0.826

7.45 a. 60% **b.** sample is random; the population of Americans aged 18–29 is greater than $10(200) = 2000$; $200(0.60)$ and $200(0.40)$ are both greater than 10. **c.** To determine if this is surprising find the

z-score: standard error $= \sqrt{\dfrac{p(1-p)}{n}} = \sqrt{\dfrac{0.60(1-0.60)}{200}} = 0.035$;

$125/200 = 0.625; z = \dfrac{0.625 - 0.60}{0.035} = 0.71$; no, this is not surprising.

d. $z = \dfrac{0.74 - 0.60}{0.035} = 4$; yes, this is surprising.

7.47 a. We expect the percentage of Americans who feel this way to be about 0.59. Because 0.55 is less than this and because the sampling distribution is approximately Normal, there would be a greater than 50% chance of seeing a sample proportion greater than 55%. **b.** The sampling distribution is Normal, the mean is 0.59, and the standard error is

$\sqrt{\dfrac{p(1-p)}{n}} = \sqrt{\dfrac{0.59(1-0.59)}{200}} = 0.035; z = \dfrac{0.55 - 0.59}{0.35} = -1.14$.

The area to the right of a z-score of -1.14 is 0.873.

7.49 Sample is random; the population of Americans is greater than $10(120) = 1200$; $120(0.45)$ and $120(0.55)$ are both greater than 10 so the conditions for the Central Limit Theorem are met. Sampling distribution

is approximately Normal, mean $= 0.45$, standard error $= \sqrt{\dfrac{p(1-p)}{n}} =$

$\sqrt{\dfrac{0.45(1-0.45)}{120}} = 0.045; z = \dfrac{0.50 - 0.45}{0.045} = 1.11$. The area to the right of a z-score of 1.11 is 0.133.

7.51 This probability cannot be calculated because the sample is too small to satisfy the conditions of the Central Limit Theorem. $np = 100(0.005) = 0.5 < 10$.

Section 7.4

7.53 a. 64.0% and 70.0% **b.** Yes, because the interval only includes values that are greater than 50%.

7.55 a. $\hat{p} = 617/1028 = 0.60$; **b.** (1) Random and independent (2) Large sample: $n\hat{p} = 1028(0.60) = 617$ and $n(1 - \hat{p}) = 1028(1 - 0.60) = 411$. Both of these values are greater than 10. (3) Big population: The population of the United States is more than $10(1028) = 10,280$. **c.** I am 95% confident the population proportion of adults living in the United States who feel the laws covering the sale of firearms should be more strict is between 0.570 and 0.630; **d.** Yes, because the confidence interval only includes values greater than 50%.

7.57 a. I am 95% confident that the population percentage of voters supporting Candidate X is between 53% and 57%. **b.** There is no evidence that he could lose, because the interval is entirely above 50%. **c.** A sample from New York City would not be representative of the entire country and would be worthless in this context.

7.59 a. standard error $= \sqrt{\dfrac{0.26(1-0.26)}{1207}} = .0013$ **b.** (0.235, 0.285) **c.** 0.025 **d.** No because 0.247 is contained in the confidence interval. It is possible the proportion is still the same.

7.61 a. (0.602, 0.679) **b.** We are 99% confident that the proportion of all U.S. adults who believe marijuana should be legalized is between 60.2% and 66.79%. (0.611, 0.669) **c.** We are 95% confident that the proportion of all U.S. adults who believe marijuana should be legalized is between 61.1% and 66.9%. **d.** A 90% confidence interval will be narrower than a 95% or 99% confidence interval; The 90% confidence interval is (0.615; 0.665). This interval is narrower than the 95% and 99% intervals.

7.63 a. $0.74(1000) = 740$ people **b.** $1000(0.74) = 740$ and $1000(1 - 0.74) = 260$. Since both of these values are greater than 10 the Central Limit Theorem can be applied. **c.** (0.713, 0.767) **d.** The width of the 95% confidence interval is 0.054 or about 5.4%. **e.** (0.726, 0.754); the width of the interval is 0.028. **f.** The width of the interval decreased when the sample size was multiplied by 4.

7.65 a. (0.37, 0.43); we are 95% confident that the percentage of American adults who were very likely to watch some of the Winter Olympic coverage on television is between 37% and 43%. **b.** We would expect at least 95 of them to include the true population proportion. **c.** This interpretation is incorrect because a confidence interval is about a population not a sample.

7.67 a. 34226731/68531917, or 49.9%, of the voters voted for Kennedy. **b.** No, you should not find a confidence interval. We have the population proportion. We need a confidence interval only when we have a sample statistic such as a sample proportion and want to generalize about the population from which the sample was drawn.

7.69 a. $438/1019 = 0.43$ **b.** (0.40, 0.46) **c.** A 90% confidence interval would be narrower than the 95% interval because a 90% confidence interval includes less of the sampling distribution. **d.** (0.404, 0.455). The 90% interval is narrower than the 95% interval.

7.71 a. $1/(0.06)^2 = 278$ Americans **b.** $1/(0.04)^2 = 625$ Americans **c.** As the margin of error is decreased, the required sample size increases.

Section 7.5

7.73 The confidence interval contains 0. This tells us that the proportion of people who reported being happy in 2016 and 2017 were not significantly different.

7.75 a. No, because these are sample proportions, not population proportions. **b.** (1) Random and independent samples: Gallup takes random samples and

the samples are independent; (2) Large samples: $\hat{p} = \dfrac{582 + 828}{1200 + 1200} = 0.565$;

$1200(0.565) = 678; 1200(1 - 0.565) = 522$; both these values are greater than 10; (3) Big populations: there are more than $10(1200) = 12,000$ Democratic voters. **c.** $(-0.288, -0.212)$ **d.** The interval does not contain 0. Since both numbers are negative, the $p_2 > p_1$. A greater percentage of Democratic voters thought the FBI did a good or excellent job in 2017 than in 2003. We are 95% confident the difference in the population proportions is between 0.212 and 0.288.

7.77

1. *Random and Independent:* Although we do not have random samples, we have random assignment to groups.

2. *Large Samples:* $n_1\hat{p}_1 = 57(0.65) = 37, n_1(1 - \hat{p}_1) = 57(0.35) = 20,$ $n_2\hat{p}_2 = 64(0.45) = 29, n_2(1 - \hat{p}_2) = 64(0.55) = 35$; all four numbers are more than 10.

3. *Big Populations:* There are far more than $10(57)$ African-American boys in the United States who go to preschool and more than $10(64)$ African-American boys who do not go to preschool.

4. *Independent Samples:* The boys in the Preschool group are not related to the boys in the No Preschool control group. The confidence interval is (0.022, 0.370).

7.79 a. Fish oil: 58/346 = 0.168 or 16.8%; Placebo: 83/349 = 0.238 or 23.8%; **b.** (1) Random and independent: Although we do not have random samples, we have random assignment to groups. (2) Large samples: $\hat{p} = \dfrac{58 + 83}{346 + 349} = \dfrac{141}{695} = 0.203$ or 20.3%; 346(0.203) = 70; 346(1 − 0.203) = 339; 349(0.203) = 71; 349(1 − 0.203) = 278; all of these values are greater than 10. (3) Big population: The population of children in Denmark is larger than 10(695) = 6950. **c.** (−0.130, −0.011); since the interval does not contain 0, there is a difference in the population proportions. Since both numbers are negative, $p_2 > p_1$. A smaller proportion of children whose mothers took fish oil during pregnancy developed asthma. We are 95% confident the difference in the population proportions is between 0.013 and 0.011.

7.81 a. 62.5% of men and 74.5% of women used turn signals. **b.** (−0.166, −0.073); I am 95% confident that the population percentage of men using turn signals minus the population percentage of women using turn signals is between −16.6% and −7.3%. The interval does not capture 0, so we are confident that the percentages are different. This shows that women are more likely than men to use turn signals. **c.** 62.8% of the men and 73.8% of the women used turn signals. CI: (−0.2575, 0.03739). I am 95% confident that the population percentage of men using turn signals minus the population percentage of women using turn signals is between −25.8% and 3.7%. The interval captures 0, showing it is plausible that the percentages are the same in the population. **d.** With more data (part b), we had a narrower margin of error and thus a more precise estimate of the true difference in proportions. We could be confident that the percentages were different in the population. In part d, the very wide interval did not allow us to make this call.

7.83 a. No, the rate of miscarriages was higher for the unexposed women, which is the opposite of what was feared. **b.** There was not a random sample, and there was not random assignment.

Chapter Review Exercises

7.85 a. 0.45 **b.** standard error = $\sqrt{\dfrac{0.45(1 − 0.45)}{500}} = 0.022$

c. 45%, give or take 2.2% **d.** $z = \dfrac{0.55 − 0.45}{0.022} = 4.55$; Yes, this would be surprising because 0.55 is more than 4 standard errors from the mean. **e.** Decreasing the sample size would increase the standard error.

7.87 a. 0.52 ± 0.04 = (0.48, 0.56) **b.** If the sample size were larger, the standard error would be smaller. The resulting interval would be narrower than the one in part a. **c.** If the confidence level were 90% rather than 95%, the interval would be narrower than the one in part a. **d.** The population size does not have any effect on the width of the interval.

7.89 The sample proportion must be 44% because the interval is symmetric around the sample proportion, which is in the middle.

7.91 The margin of error must be 4%. From the sample proportion to find the upper boundary you go up one margin of error and to find the lower boundary you go down one margin of error. Therefore, the boundaries are separated by two margins of error, and half of 8% is 4%.

7.93 1: *Random and Independent*: given. 2: *Large Sample*: np = 200(0.29) = 58 and n(1 − p) = 200(0.71) = 142, and both of these are more than 10. 3: *Big Population*: The population of dreamers is more than 10(200).

$$SE = \sqrt{\dfrac{p(1 − p)}{n}} = \sqrt{\dfrac{0.29(0.71)}{200}} = \sqrt{\dfrac{0.2059}{200}} = \sqrt{0.00103} = 0.0320858$$

$$z = \dfrac{0.50 − 0.29}{0.0320858} = \dfrac{0.21}{0.0320858} = 6.54$$

The area of the normal curve to the right of a z-value of 6.54 is less than 0.001. Therefore, the probability that a sample of 200 will contain 50% or more dreaming in color is less than 0.001.

7.95 The confidence interval for Clinton would be 0.45 ± 0.052 = (0.398, 0.502). Since the interval contains values that are less than 50%, we cannot predict with confidence that she would win in an election with Sanders.

7.97 a. 2008: 623/1022 = 0.610; 2017: 460/1022 = 0.450; **b.** (0.117, 0.202); Since the interval does not contain 0, there is a difference in the population proportions. Since both numbers are positive, $p_1 > p_2$. A greater proportion of people trusted the executive branch in 2008 than in 2017. We are 95% confident the difference in the population proportions is between 0.117 and 0.202.

7.99 $n = \dfrac{1}{m^2} = \dfrac{1}{(0.03)^2} = \dfrac{1}{0.0009} = 1111.11$

Thus we would need a sample size of 1111 or 1112 to get a margin of error of no more than 3 percentage points.

7.101 Marco took a convenience sample. The students may not be representative of the voting population, so the proposition may not pass.

7.103 No, the people you met would not be a random sample but a convenience sample.

7.105 The small mean might have occurred by chance.

7.107 $m = 2\sqrt{\dfrac{\hat{p}(1 − \hat{p})}{n}}$, so $m = 2\sqrt{\dfrac{0.50(1 − 0.50)}{n}}$.

Simplifying, we have $m = 2\sqrt{\dfrac{0.25}{n}}$.

CHAPTER 8

Answers may vary slightly due to type of technology or rounding.

Section 8.1

8.1 Population parameter

8.3 H_0: The proportion of American adults who are vegetarian is 0.033 ($p = 0.033$); H_a: The proportion of American adults who are vegetarian is greater than 0.033 ($p > 0.033$).

8.5 a. i **b.** ii **c.** i

8.7 ii

8.9 The null hypothesis is written correctly; the alternative hypothesis should be $p > 0.15$, since the promotion is successful if the proportion of customers ordering a soda with their meal has increased.

8.11 The probability of saying the current flu vaccine is less than 73% effective against the flu virus when, in fact, it isn't is 0.01.

8.13 a. H_0: $p = 0.37$; H_a: $p \neq 0.37$ **b.** $z = 0.6807$

8.15 a. $\hat{p} = \dfrac{11}{150} = 0.073$ **b.** $p_0 = 0.033$ **c.** $\dfrac{\hat{p} − p_0}{\sqrt{\dfrac{p_0(1 − p_0)}{n}}} =$

$\dfrac{0.073 − 0.033}{\sqrt{\dfrac{0.033(1 − 0.033)}{150}}} = 2.74$. The value of the test statistic tells us that

the observed proportion was 2.74 standard errors above the null hypothesis value of 0.033.

8.17 a. You would expect about 10 right out of 20 if the person is guessing. **b.** The smaller p-value will come from person B. The p-value measures how unusual an event is, assuming the null hypothesis is true. Getting 18 right out of 20 is more unusual than getting 13 right out of 20 when you are expecting 10 right under the null hypothesis. The larger difference in proportions ($\hat{p} − p$) results in a smaller p-value.

8.19 The probability of finding 11 or more vegetarians in a random sample of 150 American adults assuming the population proportion is 0.033. Since the probability 0.0028 is very small, the nutritionist should not believe the null hypothesis is true.

8.21 a. $\hat{p} = \dfrac{11}{70} = 0.157$ **b.** H_0: $p = 0.17$; H_a: $p < 0.17$

c. $\dfrac{\hat{p} − p_0}{\sqrt{\dfrac{p_0(1 − p_0)}{n}}} = \dfrac{0.157 − 0.17}{\sqrt{\dfrac{0.17(1 − 0.17)}{70}}} = −0.29$. The value of the test statistic tells us that the observed proportion was 0.29 standard errors below the null hypothesis value of 0.17. **d.** The probability of 11 or fewer pneumonia patients out of a sample of 70 patients being readmitted to the hospital is 0.39 assuming the population proportion is 0.17. Since the probability is not close to zero, we do not doubt the null hypothesis is true.

Section 8.2

8.23 Random sample was mentioned. Independence is assumed.
Large sample: $np_0 = 113(0.29) = 32.77 > 10$, and
$n(1 - p_0) = 113(0.71) = 80.23 > 10$.
Large population: There are more than 1130 people in the population of dreamers.
So the conditions are met.

8.25 Figure A correctly matches the alternative hypothesis $p > 0.30$ because it is one-tailed. The p-value is 0.08. If the population proportion of young Americans who would be comfortable riding in a self-driving car is 30%, there is about an 8% chance of getting 152 or more out of 461 randomly selected adults from this age group who feel this way.

8.27 Figure B is correct because the alternative hypothesis should be one-sided because the person should get better than half right if she or he can tell the difference.

8.29 *Step 1:* $H_0: p = 0.73$, $H_a: p > 0.73$. *Step 2:* One-proportion z-test: $0.73(200) = 146 > 10$; $0.27(200) = 54 > 10$, sample random and assumed independent. Entries: $p_0 = 0.73$; $x = 160$, $n = 200$. One sided, $>$.

8.31 *Step 3:* Significance level $= 0.05$; $z = 2.23$, p-value $= 0.01$. *Step 4:* Reject H_0. The proportion of Americans who report working out one or more times each week has increased.

8.33 In Figure (A), the shaded area could be a p-value because it includes tail areas only; it would be for a two-sided alternative because both tails are shaded. In Figure (B) the shaded area would not be a p-value because it is the area between two z-values.

8.35 a. *Step 1:* $H_0: p = 0.47$, $H_a: p > 0.47$. *Step 2:* 1-proportion z-test; $0.47(3635) = 1708 > 10$; $0.53(3635) = 1927 > 10$; sample random and assumed independent; large population: # Facebook users $> 10(3635)$ *Step 3:* Significance level $= 0.05$. $z = 25.37$, p-value approximately 0. *Step 4:* Reject H_0. The proportion of Facebook users who get their world news on Facebook has increased since 2013. **b.** (0.667, 0.693). The interval supports the hypothesis test conclusion because it only includes values that are greater than 0.47, supporting a conclusion that the population proportion has increased.

8.37 a. $\frac{692}{1018} = 0.68$. **b.** *Step 1:* $H_0: p = 0.57$, $H_a: p \neq 0.57$.
Step 2: 1-proportion z-test. $0.68(1018) = 692 > 10$; $0.32(1018) = 326 > 10$; sample random and assumed independent; large population: # Americans $> 10(1018)$. *Step 3:* Significance level $= 0.01$; $z = 7.07$, p-value approximately 0. *Step 4:* Reject H_0. **c.** Choice ii is correct: In 2017, the percentage of Americans who believe global warming is caused by human activities has changed from the historical level of 0.57.

8.39 *Step 1:* $H_0: p = 0.20$, $H_a: p \neq 0.20$, where p is the population proportion of dangerous fish. *Step 2:* One-proportion z-test, $0.2(250) = 50 > 10$ and $0.8(250) = 200 > 10$, population large, assume a random and independent sample. *Step 3:* $\alpha = 0.05$. $z = 1.58$, p-value $= 0.114$. *Step 4:* Do not reject H_0. We are not saying the percentage is 20%. We are only saying that we cannot reject 20%. (We might have been able to reject the value of 20% if we had had a larger sample.)

8.41 *Step 1:* $H_0: p = 0.09$, $H_a: p \neq 0.09$, where p is the population proportion of t's in the English language. *Step 2:* One-proportion z-test, the sample is independent, random, and $0.09(600) = 54 > 10$ and $0.91(600) = 546 > 10$, $\alpha = 0.10$. *Step 3:* $z = -0.86$, p-value $= 0.392$. *Step 4:* Do not reject H_0. We cannot reject 9% as the current proportion of t's because 0.392 is more than 0.10.

Section 8.3

8.43 iv. $z = 3.00$. It is farthest from 0 and therefore has the smallest tail area.

8.45 The first kind of error is concluding more than 50% of eligible voters from this age group voted in the 2016 election when, in fact, the percentage who voted was not more than 50%. The second type of error is concluding that the percentage from this age group who voted isn't more than 50% when, in fact, it is.

8.47 The first type of error is having the innocent person suffer (convicting an innocent person). The second type of error is "ten guilty persons escape" (letting guilty persons go free).

8.49 With a significance level of 0.01 and a p-value of 0.02, we would not reject H_0; however, we have not proved the null hypothesis is true. The conclusion for this hypothesis test is that, using a 1% significance level, there is not enough evidence to conclude the proportion of Americans who would pick invisibility as their superpower has increased.

8.51 Choose hypothesis testing and the one-proportion z-test, because he only wants to know whether or not it will pass; he is not interested in knowing the proportion who will vote for it. Suppose p is the population proportion supporting the proposition.
$H_0: p = 0.50$
$H_a: p > 0.50$
$z = 5.06$
p-value < 0.001
Reject H_0. The proposition is likely to pass.

8.53 Interpretation iii.

8.55 No; we don't use *prove* because we cannot be 100% sure of conclusions based on chance processes.

8.57 It is a null hypothesis.

8.59 Interpretations b and d are valid. Interpretations a and c are both "accepting" the null hypothesis claim, which is an incorrect way of expressing the outcome.

Section 8.4

8.61 Far apart. Assuming the standard errors are the same, the farther apart the two proportions are, the larger the absolute value of the numerator of z, and therefore the larger the absolute value of z and the smaller the p-value.

8.63 a. ritonavir-boosted darunavir: $\frac{306}{382} = 0.801$, dorovirine: $\frac{321}{382} = 0.840$ **b.** *Step 1:* $H_0: p_{rbd} = p_{dorovine}$. *Step 2:* Two-proportion z-test. Because the two sample sizes are equal ($n_1 = n_2$) the numbers below are the same for both samples. $\hat{p} = \frac{306 + 321}{382 + 382} = 0.82$; $n\hat{p} = 382(0.82) = 313 > 10$, $n(1 - \hat{p}) = 382(0.18) = 69 > 10$. Random assignment and independence within and between samples. *Step 3:* Significance level $= 0.01$. $z = -1.41$, p-value $= 0.16$. *Step 4:* Do not reject H_0. There is no evidence that there is a difference in the proportion of patients who achieve a positive outcome between the two treatments. Based on this study we have no evidence that dorovirine might be a more effective treatment option for HIV-1 than ritonavir-boosted darunavir.

8.65 a. 2015: $\frac{1201}{1906} = 0.63$; 2018: $\frac{1341}{2002} = 0.67$ **b.** $\hat{p} = \frac{1201 + 1341}{1906 + 2002} = 0.65$ **c.** $SE = \sqrt{0.65(1 - 0.65)\left(\frac{1}{1906} + \frac{1}{2002}\right)} = 0.015$; $z_{observed} = \frac{0.63 - 0.67}{0.015} = -2.67$.

8.67 a. *Step 1:* $H_0: p_{2016} = p_{2017}$, $H_a: p_{2016} \neq p_{2017}$. *Step 2:* Two-proportion z-test. $\hat{p} = \frac{2489 + 1808}{3072 + 2014} = 0.845$; $n_1\hat{p} = 3072(0.845)$, $n_1(1 - \hat{p}) = 3072(0.155)$, $n_2\hat{p} = 2014(0.845)$, $n_2(1 - \hat{p}) = 2014(0.155)$; all expected counts are greater than 10. Samples are random and assumed independent. *Step 3:* Significance level: 0.05. $z = -8.43$; p is approximately 0.

Step 4: Reject H_0. The proportion of college students who believe that freedom of the press is secure or very secure in the country changed from 2016. **b.** The 95% confidence interval for $p_{2016} - p_{2017} = (-0.107, -0.068)$. Since the interval does not include 0, the population proportions are significantly different. The proportion of college students who believe that freedom of the press is secure declined from 2016 and the difference is between 6.8% and 10.7%.

8.69 *Step 1:* H_0: $p_{2016} = p_{2018}$, H_a: $p_{2016} > p_{2018}$. *Step 2:* Two-proportion z-test. There were 983 sampled in 2016 and 993 sampled in 2018.
$\hat{p} = \dfrac{543 + 461}{983 + 993} = 0.508$, $n_1\hat{p} = 983(0.508)$, $n_1(1 - \hat{p}) = 983(0.492)$, $n_2\hat{p} = 993(0.508)$, $n_2(1 - \hat{p}) = 998(0.492)$; all expected counts are greater than 10. Samples are random and assumed independent. *Step 3:* Significance level: 0.05. $z = 3.91$, p-value is approximately 0. *Step 4:* Reject H_0. The proportion of Americans who are satisfied with the quality of the environment has declined.

Chapter Review Exercises

8.71 a. One-proportion z-test. Population is voters in California; **b.** Two-population z-test. Populations are residents of coastal states and residents of non-coastal states.

8.73 a. $p =$ the population proportion of correct answers. H_0: $p = 0.50$ (he is just guessing), H_a: $p > 0.50$ (he is not just guessing). One-proportion z-test. **b.** p_a is the population proportion of athletes who can balance for at least 10 seconds. p_n is the population proportion of nonathletes who can balance for at least 10 seconds.
H_0: $p_a = p_n$
H_a: $p_a \neq p_n$
Two-proportion z-test.

8.75 *Step 1:* H_0: $p = 0.50$, H_a: $p > 0.50$. *Step 2:* One-proportion z-test. $20(0.50) = 10$. Samples are random and independent; large population: # possible samples $> 10(20)$. *Step 3:* Significance level: 0.05. $z = 1.34$, p-value $= 0.09$. *Step 4:* Do not reject H_0. We have no evidence the student can tell tap water from bottled water.

8.77 0.05 (because $1 - 0.95 = 0.05$)

8.79 5% of 300, or 15

8.81 No because these are measures of *all* students at the college, so inference is not needed or appropriate.

8.83 a. *Step 1:* H_0: $p_{2012} = p_{2018}$, H_a: $p_{2012} \neq p_{2018}$. *Step 2:* Two-proportion z-test; $\hat{p} = \dfrac{66 + 76}{100 + 100} = 0.71$; $n\hat{p} = 100(0.71)$, $n(1 - \hat{p}) = 100(0.29)$, all expected counts greater than 10. Assume samples are random and independent. *Step 3:* Significance level $= 0.01$; $z = -1.56$, p-value $= 0.12$. *Step 4:* Do not reject H_0. We cannot conclude the proportion of people who reported using Facebook was different in 2012 and 2018. **b.** *Step 1:* H_0: $p_{2012} = p_{2018}$, H_a: $p_{2012} \neq p_{2018}$. *Step 2:* Two-proportion z-test; $x_1 = 0.66(1500) = 990$; $x_2 = 0.76(1500) = 1140$; $\hat{p} = \dfrac{990 + 1140}{1500 + 1500} = 0.71$; $n\hat{p} = 100(0.71)$, $n(1 - \hat{p}) = 100(0.29)$, all expected counts greater than 10. Assume samples are random and independent. *Step 3:* Significance level $= 0.01$; $z = -6.04$, p-value is approximately 0. *Step 4:* Reject H_0. The proportions of Facebook users were different in 2012 and 2018. **c.** With the larger sample size (more evidence) we got a smaller p-value and were able to reject H_0.

8.85 It would not be appropriate to do such a test because the data were for the entire population of people who voted. Since the data are for a population, no inference is needed.

8.87 a. *Step 1:* H_0: $p = 0.101$, H_a: $p \neq .101$. *Step 2:* One-proportion z-test; $np = 500(0.101)$, $n(1 - p) = 500(.899)$, all expected counts greater than 10; sample random and assumed independent; large population: # workers assumed $> 10(500)$. *Step 3:* Significance level: 0.05; $z = 1.71$, p-value $= 0.09$. *Step 4:* Do not reject H_0. We cannot conclude the proportion of self-employed workers in his area is different from 10.1%. **b.** (0.095, 0.153). This interval supports the hypothesis test conclusion because it contains 0.101.

8.89 a. Quinnipiac: $\dfrac{824}{1249} = 0.66$; NPR: $\dfrac{754}{1005} = 0.75$. **b.** *Step 1:*
H_0: $p_{\text{Quin}} = p_{\text{NPR}}$, H_a: $p_{\text{Quin}} \neq p_{\text{NPR}}$. *Step 2:* Two-proportion z-test;
$\hat{p} = \dfrac{824 + 754}{1249 + 1005} = 0.70$, $n_1\hat{p} = 1249(0.70)$, $n_1(1 - \hat{p}) = 1249(0.30)$, $n_2\hat{p} = 1005(0.70)$, $n_2(1 - \hat{p}) = 1005(0.30)$, all expected counts greater than 10. Assume samples are random and independent. *Step 3:* Significance level: 0.05, $z = -4.66$, p-value approximately 0. *Step 4:* Reject H_0. The population proportions are not equal. **c.** $(-0.128, -0.053)$. The interval does not contain 0, supporting a conclusion that the population proportions are not equal. The difference between the population proportions is between 5.3% and 12.8%.

8.91 a. The misconduct rate was higher for those in the sample who did *not* have three strikes (30.6%) than for those in the sample who had three strikes (22.2%). This was not what was expected. **b.** *Step 1:* H_0: $p_{\text{three-strikers}} = p_{\text{others}}$, H_a: $p_{\text{three-strikers}} > p_{\text{others}}$. *Step 2:* Two-proportion z-test, expected counts (213, 924, 521, 2264) are all larger than 10, assume random samples and assume independence *Step 3:* $\alpha = 0.05$. $z = -4.49$ (or 4.49), p-value > 0.999. *Step 4:* Do not reject H_0. The three-strikers do not have a greater rate of misconduct than the other prisoners. (If a two-sided test had been done, the *p*-value would have been <0.001, and we would have rejected the null hypothesis because the three-strikers had *less* misconduct.)

8.93 *Step 1:* H_0: $p = 0.52$, H_a: $p < 0.52$. *Step 2:* One-proportion z-test. $np = 1024(0.52)$, $n(1 - p) = 1024(0.48)$, all expected counts greater than 10; sample random and assumed independent; large population: #Americans $> 10(1024)$. *Step 3:* Significance level: 0.05, $z = -4.47$, p-value is approximately 0. *Step 4:* Reject H_0. Satisfaction with the quality of the environment among Americans has decreased.

8.95 a. i **b.** iii

8.97 $z = 0.89$

8.99 The p-value tells us that if the true proportion of those who text while driving is 0.25, then there is only a 0.034 probability that one would get a sample proportion of 0.125 or smaller with a sample size of 40.

8.101 He has not demonstrated ESP; 10 right out of 20 is only 50% right, which you should expect from guessing.

8.103 H_0: The death rate after starting hand washing is still 9.9%, or $p = 0.099$ (p is the proportion of all deaths at the clinic.)

H_a: The death rate after starting hand washing is less than 9.9%, or $p < 0.099$.

8.105 *Step 3:* $z = 2.83$, p-value $= 0.002$. *Step 4:* Reject H_0. The probability of doing this well by chance alone is so small that we must conclude that the student is not guessing.

CHAPTER 9

Answers may vary due to rounding or type of technology used.

Section 9.1

9.1 a. They are parameters, because they are for all the students, not a sample. **b.** $\mu = 20.7$, $\sigma = 2.5$

9.3 a. See the accompanying figure.

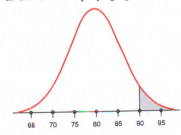

b. 2.5% (By the Empirical Rule, 95% of the observed values are between 70 and 90 which leave 5% outside of these boundaries. We only want the right half, so the answer is half of 5%, 2.5%.)

9.5 a. Answers will vary. We might expect it to be right-skewed if we think most people take relatively short showers and a few take very long showers. Some might expect it to be Normal because they think very few people take short showers and very few people take long showers. The summary statistics do not contradict a Normal distribution since it is possible to have 99% of the observations within three SDs of the mean. It is unlikely the distribution is left-skewed because the majority of people probably do not take extremely long showers. **b.** By the Central Limit Theorem for sample means, the distribution of sample means will be approximately Normal because the sample size is greater than 25. **c.** The mean will be the same as the population mean (8.2 minutes) and the standard deviation of the sample means will be $\dfrac{\sigma}{\sqrt{n}} = \dfrac{2}{\sqrt{100}} = 0.2$.

9.7 The distribution of *a* sample (*one* sample).

9.9 a. 22,306 miles **b.** standard error $= \dfrac{\sigma}{\sqrt{n}} = \dfrac{5500}{\sqrt{200}} = 388.9$

Section 9.2

9.11 a. 68% from the Empirical Rule. 6.4 is one standard deviation below the mean, and 7.6 is one standard deviation above the mean. **b.** $\dfrac{0.6}{\sqrt{4}} = 0.3$, so the standard error for the mean is 0.3. Now 7.6 is two standard errors above the mean, and 6.4 is two standard errors below the mean. The answer is 95% from the Empirical Rule. **c.** The distribution of means is taller and narrower than the original distribution and will have more data in the central area.

9.13 a. Yes, the sample size (100) is greater than 25; **b.** the mean is 75,847; the standard error $= \dfrac{\sigma}{\sqrt{n}} = \dfrac{32000}{\sqrt{100}} = 3200$; **c.** 32% from the Empirical Rule (if 68% of the data is within one standard deviation (error) of the mean, then 32% lie beyond one standard deviation (error) of the mean.

9.15 Figure B is the original distribution; it is the least Normal and widest. Figure A is from samples of 5. Figure C is from samples of 25; it is the narrowest and the least skewed. The larger the sample size, the narrower and more Normal is the sampling distribution.

9.17 a. $\mu = 3.1$ (population mean), $\sigma = 2.7$ (population standard deviation), $\bar{x} = 2.7$ (sample mean), and $s = 2.1$ (sample standard deviation); **b.** $\mu = 3.1$ (population mean) and $\sigma = 2.7$ (population standard deviation) are parameters; $\bar{x} = 2.7$ (sample mean) and $s = 2.1$ (sample standard deviation) are statistics; **c.** Yes, sample is random and sample size (35) is greater than 25; the shape will be Normal.

Section 9.3

9.19 a. (ii) We are 95% confident that the population mean is between 3.71 and 3.79. (i) is incorrect because a confidence interval is for a population parameter, not a sample statistic; (iii) is incorrect because the confidence level is not the probability of the interval containing the population parameter. It is a measure of our confidence in the method used to obtain the interval. Note: Answer is given using technology. If using a t-distribution table, confidence interval limits are 3.61 and 3.89.

9.21 a. i. is correct: (10.125, 10.525). Both ii. and iii. are incorrect. **b.** No, it does not capture 10. Reject the claim of 10 pounds, because 10 is not in the interval.

9.23 b. is the correct interpretation. **a.** is incorrect because the confidence level is not the probability of the interval containing the population parameter. It is a measure of our confidence in the method used to obtain the interval.

9.25 Use $t^* = 2.056$.

9.27 a. sample size = 30, sample mean = 170.7, sample standard deviation = 11.5; **b.** I am 95% confident that the population mean height of 12th grade students is between 166 and 175 cm.

9.29 a. $m = t^* \dfrac{s}{\sqrt{n}} = 2.064 \dfrac{13}{\sqrt{25}} = 5.37$

$72 + 5.27 = 77.37$

$72 - 5.37 = 66.63$

I am 95% confident that the mean is between 67 and 77 beats per minute.

b. $m = t^* \dfrac{s}{\sqrt{n}} = 2.797 \dfrac{13}{\sqrt{25}} = 7.27$

c. I am 99% confident that the mean is between 65 and 79 beats per minute. The 99% interval is wider because a greater level of confidence requires a bigger value for t^*.

9.31 a. (22.0, 42.8) is the 90% interval because it is narrower than (19.9, 44.0); **b.** If a larger sample size was used, the intervals would be smaller because the standard error of the distribution would be smaller.

9.33 a. Wider **b.** Narrower **c.** Wider

9.35 a. I am 95% confident that the population mean is between 20.4 and 21.7 pounds. **b.** The interval does not capture 20 pounds. There is enough evidence to reject 20 pounds as the population mean.

Section 9.4

9.37 Answers follow the guided steps given. *Step 1:* H$_0$: $\mu = 98.6$, H$_a$: $\mu \neq 98.6$. *Step 2:* One-sample *t*-test: random and independent sample, not strongly skewed, $\alpha = 0.05$. *Step 3:* $t = -1.65$, p-value = 0.133. *Step 4:* Do not reject H$_0$. Choose i.

9.39 a. You should be able to reject 20 pounds because the confidence interval (20.4 to 21.7) did not capture 20 pounds. **b.** *Step 1:* H$_0$: $\mu = 20$, H$_a$: $\mu \neq 20$. *Step 2:* One-sample *t*-test: Normal, random, and independent, $\alpha = 0.05$. *Step 3:* $t = 5.00$, p-value = 0.015. *Step 4:* Yes, reject H$_0$. Choose ii.

9.41 *Step 1:* H$_0$: $\mu = 200$, H$_a$: $\mu > 200$. *Step 2:* One-sample *t*-test: conditions are met, $\alpha = 0.05$. *Step 3:* $t = 1.44$, p-value = 0.079. *Step 4:* Do not reject H$_0$. We have *not* shown that the mean is significantly more than 200.

9.43 a. *Step 1:* H$_0$: $\mu = 38$, H$_a$: $\mu \neq 38$. *Step 2:* One-sample *t*-test: Normal and random, $\alpha = 0.05$. *Step 3:* $t = -1.03$, p-value = 0.319. *Step 4:* Do not reject H$_0$. The mean for non-U.S. boys is not significantly different from 38. **b.** *Step 3:* $t = -1.46$, p-value = 0.155. *Step 4:* Do not reject H$_0$. The mean for non-U.S. boys is not significantly different from 38. **c.** Larger n, smaller standard error (narrower sampling distribution) with less area in the tails, as shown by the smaller p-value.

9.45 a. *Step 1:* H$_0$: $\mu = 12.5$, H$_a$: $\mu < 12.5$. *Step 2:* One-sample *t*-test. Assume we have a random sample and the distribution is Normal. *Step 3:* Significance level: 0.05; $t = -11.46$, p-value < 0.001. *Step 4:* Reject H$_0$. The pressure is less than 12.5 psi. This shows the balls are deflated. **b.** (10.84, 11.38); the interval only contains values less than 12.5.

9.47 a. *Step 1:* H$_0$: $\mu = 8.97$, H$_a$: $\mu \neq 8.97$; *Step 2:* One-sample *t*-test; random, $n \geq 25$; *Step 3:* Significance level: 0.05; $t = 4.91$, p-value = approximately 0. Reject H$_0$. *Step 4:* The mean price of a movie ticket in the San Francisco Bay Area is different from the national average. **b.** (10.88, 13.66); The interval does not include 8.97, confirming the mean price is different from the national average.

9.49 *Step 1:* H$_0$: $\mu = 0$, H$_a$: $\mu > 0$. *Step 2:* One-sample *t*-test: Normal, not random (don't generalize), $\alpha = 0.05$. *Step 3:* $t = 3.60$, p-value = 0.003. Reject H$_0$. There was a significant weight loss, but don't generalize.

9.51 Expect $0.95(200) = 190$ to capture and 10 to miss.

Section 9.5

9.53 a. Paired **b.** Independent

9.55 a. The samples are random, independent, and large, so the conditions are met. **b.** I am 95% confident that the mean difference (OC minus MC) is between -0.40 and 1.14 TVs. **c.** The interval for the difference captures 0, which implies that it is plausible that the means are the same.

9.57 Answers follow the guided steps given. *Step 1:* H$_0$: $\mu_{\text{oc}} = \mu_{\text{mc}}$, H$_a$: $\mu_{\text{oc}} \neq \mu_{\text{mc}}$, where μ is the population mean number of TVs. *Step 2:*

Two-sample t-test: samples large ($n = 30$), independent, and random, $\alpha = 0.05$. *Step 3:* $t = 0.95$, p-value $= 0.345$. *Step 4:* Do not reject H_0. Choose i. Confidence interval: $(-0.404, 1.138)$. Because the interval for the difference captures 0, we cannot reject the hypothesis that the mean difference in number of TVs is 0.

9.59 a. The men's sample mean triglyceride level of 139.5 was higher than the women's sample mean of 84.4. **b.** *Step 1:* H_0: $\mu_{men} = \mu_{women}$, H_a: $\mu_{men} > \mu_{women}$, where μ is the population mean triglyceride level. *Step 2:* Two-sample t-test: assume the conditions are met, $\alpha = 0.05$. *Step 3:* $t = 4.02$ or -4.02, p-value < 0.001. *Step 4:* Reject H_0. The mean triglyceride level is significantly higher for men than for women. Choose output B: Difference $= \mu_{female} - \mu_{male}$, which tests whether this difference is less than 0, and that is the one-sided hypothesis that we want.

9.61 $(-82.5, -27.7)$; because the difference of 0 is not captured, it shows there is a significant difference. Also, the difference $\mu_{female} - \mu_{male}$ is negative, which shows that the men's mean (triglyceride level) is significantly higher than the women's mean.

9.63 *Step 1:* H_0 $\mu_{American} = \mu_{National}$, H_a: $\mu_{American} \neq \mu_{National}$. *Step 2:* Two-sample t-test; random, independent, Normal; *Step 3:* Significance level: 0.05, $t = -0.25$, p-value $= 0.803$. Do not reject H_0. *Step 4:* We cannot conclude there is a difference in mean salaries for the two leagues.

9.65 a. Yes, it would contain 0 since there we cannot conclude the means are different. It is possible they are the same. **b.** $(-1423.40, 1110.88)$; the confidence interval does contain 0 which supports the hypothesis test conclusion.

9.67 a. $\bar{x}_{UCSB} = \$61.01$ and $\bar{x}_{CSUN} = \$75.55$, so the sample mean at CSUN was larger. **b.** *Step 1:* H_a: $\mu_{UCSB} \neq \mu_{CSUN}$ where μ is the population mean book price. *Step 2:* Paired t-test, matched pairs, assume random and Normal (given). *Step 3:* $t = -3.21$ or 3.21, p-value $= 0.004$. *Step 4:* You can reject H_0. The means are significantly different.

9.69 The answers follow the guided steps. *Step 1:* H_0: $\mu_{before} = \mu_{after}$, H_a: $\mu_{before} < \mu_{after}$, where μ is the population mean pulse rate. *Step 2:* Paired t-test: each woman is measured twice (repeated measures), so a measurement in the first column is coupled with a measurement of the same person in the second column, assume random and Normal, $\alpha = 0.05$. *Step 3:* $t = 4.90$ or -4.90, p-value < 0.001. *Step 4:* Reject H_0. The sample mean before was 74.8, and the sample mean after was 83.7. The pulse rates of women go up significantly after they hear a scream.

9.71 Choose Figure B. The items are paired because the same items were priced at each store. *Step 1:* H_0: $\mu_{target} = \mu_{wholefoods}$, H_a: $\mu_{target} \neq \mu_{wholefoods}$. *Step 2:* Paired t-test: assume random, sample size large, $\alpha = 0.05$. *Step 3:* $t = -1.26$ or 1.26, p-value $= 0.217$. *Step 4:* Do not reject H_0. The means are not significantly different.

9.73 a. *Step 1:* H_0: $\mu_{ales} = \mu_{IPAs}$, H_a: $\mu_{ales} \neq \mu_{IPAs}$. *Step 2:* 2-sample t-test; assume random, Normal; *Step 3:* Significance level: 0.05, $t = -2.57$, p-value $= 0.025$. Reject H_0. *Step 4:* There is a difference between the mean calorie content of ales and IPAs. **b.** *Step 1:* H_0: $\mu_{ales} = \mu_{IPAs}$, H_a: $\mu_{ales} < \mu_{IPAs}$. *Step 2:* 2-sample t-test; assume random, Normal; *Step 3:* Significance level: 0.05; $t = -2.57$, p-value $= 0.013$. Reject H_0. *Step 4:* The mean caloric content of IPAs is significantly larger than that of Ales. The p-value is about half the p-value of the two-tailed test.

9.75 a. 95% CI $(-1.44, 0.25)$ captures 0, so the hypothesis that the means are equal cannot be rejected. **b.** *Step 1:* H_0: $\mu_{measured} = \mu_{reported}$, H_a: $\mu_{measured} \neq \mu_{reported}$, where μ is population mean height of men. *Step 2:* Paired t-test: each person is the source of two numbers, assume conditions for t-tests hold, $\alpha = 0.05$. *Step 3:* $t = 1.50$ or -1.50, p-value $= 0.155$. *Step 4:* Do not reject H_0. The mean measured and reported heights are not significantly different for men or there is not enough evidence to support the claim that the typical self-reported height differs from the typical measured height for men.

9.77 a. $(-1.67, 3.24)$. Since the interval contains 0, there is not a significant difference in the heart rates of females and males. **b.** *Step 1:* H_0: $\mu_{males} = \mu_{females}$, H_a: $\mu_{males} \neq \mu_{females}$; *Step 2:* 2-sample t-test; assume random, Normal; *Step 3:* Significance level: 0.05; $t = 0.63$,

p-value $= 0.53$. Do not Reject H_0; *Step 4:* We cannot conclude there is a significant different in the mean heart rates of males and females.

Chapter Review Exercises

9.79. a. The distribution is Normal, so this question can be answered. $N(65, 2.5)$, $p(x < 63) = 0.212$; **b.** Since the population distribution is Normal, we can apply the Central Limit Theorem. For samples of size 5, standard error $= \dfrac{2.5}{\sqrt{5}} = 1.12$; $N(65, 1.12)$, $p(x < 63) = 0.037$;

c. Since the population distribution is Normal, we can apply the Central Limit Theorem. For samples of size 30, standard error $= \dfrac{2.5}{\sqrt{30}} = 0.456$; $N(65, 0.456)$, $p(x < 63) =$ approximately 0.

9.81 a. One sample t-test **b.** Two-sample t-test **c.** No t-test (two categorical variables)

9.83 a. H_a: $\mu \neq 3.18$, $t = 3.66$, p-value $= 0.035$, reject H_0. The mean is significantly different from 3.18. **b.** H_a: $\mu < 3.18$, $t = 3.66$, p-value $= 0.982$, do not reject H_0. The mean is not significantly less than 3.18 ounces. **c.** H_a: $\mu > 3.18$, $t = 3.66$, p-value $= 0.018$, reject H_0. The mean is significantly more than 3.18 ounces.

9.85 *Step 1:* H_0: $\mu_{men} = \mu_{women}$, H_a: $\mu_{men} > \mu_{women}$. *Step 2:* Two-sample t-test: near Normal, random, $\alpha = 0.05$. *Step 3:* $t = 5.27$ (or -5.27), p-value < 0.001. *Step 4:* Reject H_0. The mean for brain size for men is significantly more than the mean for women.

9.87 *Step 1:* H_0: $\mu_{before} = \mu_{after}$, H_a: $\mu_{before} < \mu_{after}$, where μ is the population mean pulse rate (before and after coffee). *Step 2:* Paired t-test (repeated measures): assume conditions hold, $\alpha = 0.05$. *Step 3:* $t = 2.96$ or -2.96, p-value $= 0.005$. *Step 4:* Reject H_0. Heart rates increase significantly after coffee. (The average rate before coffee was 82.4, and the average rate after coffee was 87.5.)

9.89 The typical number of hours was a little higher for the boys, and the variation was almost the same. $\bar{x}_{girls} = 9.8$, $\bar{x}_{boys} = 10.3$, $s_{girls} = 5.4$ and $s_{boys} = 5.5$. See the histograms.

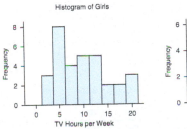 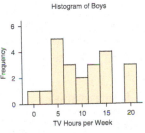

Step 1: H_0: $\mu_{girls} = \mu_{boys}$, H_a: $\mu_{girls} \neq \mu_{boys}$, where μ is the population mean number of TV viewing hours. *Step 2:* Two-sample t-test: random samples, assume the sample sizes of 32 girls and 22 boys is large enough that slight non-Normality is not a problem, $\alpha = 0.05$. *Step 3:* $t = -0.38$ or 0.38, p-value $= 0.706$. *Step 4:* You cannot reject the null hypothesis. There is not enough evidence to conclude that boys and girls differ in the typical hours of TV watched.

9.91 a. A two-sample t-test since the data are not paired; **b.** Random, independent samples; population Normal; 95% confidence interval for $\mu_{ales} - \mu_{lagers}$: $(11.88, 47.76)$. Since the interval does not contain 0, there is a significant different in the mean number of calories contained in ales and lagers. Since both numbers in the confidence intervals are positive, we can conclude that ales contain more calories than lagers.

9.93 a. Paired t-test because the data are paired (2 prices associated with one food item). **b.** *Step 1:* H_0: $\mu_{difference} = 0$, where $\mu_{difference} = \mu_{Amazon} - \mu_{Walmart}$, H_a: $\mu_{difference} \neq 0$; *Step 2:* Paired t-test, random, Normal populations; *Step 3:* Significance level: 0.05, $t = -1.50$, p-value $= 0.162$. Do not reject H_0. We cannot conclude there is a difference in the mean prices for the two delivery services.

9.95 The table shows the results. The average of s^2 in the table is 2.8889 (or about 2.89), and if you take the square root, you get about 1.6997 (or about 1.70), which is the value for sigma (σ) given in the TI-84 output shown in the exercise. This demonstrates that s^2 is an unbiased estimator of σ^2, sigma squared.

Sample	s	s^2
1, 1	0	0
1, 2	0.7071	0.5
1, 5	2.8284	8.0
2, 1	0.7071	0.5
2, 2	0	0
2, 5	2.1213	4.5
5, 1	2.8284	8.0
5, 2	2.1213	4.5
5, 5	0	0
		Sum 26.0

$$26/9 = 2.8889$$

9.97 Answers will vary.

CHAPTER 10

Answers may vary slightly due to type of technology or rounding.

Section 10.1

10.1 a. Proportions are used for categorical data. **b.** Chi-square tests are used for categorical data.

10.3

	Boy	**Girl**
Violent	10	11
Nonviolent	19	4

The table may have a different orientation.

10.5 *Mean GPA*: numerical and continuous. *Field of Study*: categorical.

10.7 a. 72 **b.** $72/120 = 0.6$ **c.** $0.53(120) = 63.6$

10.9 a.

Eat breakfast at least 3 × weekly	Females	Males	Total
Yes	206	94	300
No	92	49	141
Total	298	143	441

b. $300/441 = 68.0\%$

c. Females: $0.68(298) = 202.64$, Male: $0.68(143) = 97.24$

d. Females: $0.32(298) = 95.36$, Male: $0.32(143) = 45.76$; expected counts shown in the table:

Eat breakfast at least 3 × weekly	Females	Males	Total
Yes	202.64	97.24	299.88
No	95.36	45.76	141.12
Total	298	143	441

e. $X^2 = \dfrac{(206 - 202.64)^2}{202.64} + \dfrac{(94 - 97.24)^2}{97.24} + \dfrac{(92 - 95.36)^2}{95.36}$
$+ \dfrac{(49 - 45.76)^2}{45.76} = 0.51$

10.11 a. 40% of 16 is 6.4, so 6.4 should have had heart disease and 9.6 should not.

b. $X^2 = \dfrac{(9 - 6.4)^2}{6.4} + \dfrac{(7 - 9.6)^2}{9.6} = \dfrac{6.76}{6.4} + \dfrac{6.76}{9.6}$
$= 1.056 + 0.704 = 1.76$

10.13 a. One-fifth of 152 is 30.4 (expected correct), and four-fifths of 152 is 121.6 (expected incorrect).

b. $X^2 = \dfrac{(39 - 30.4)^2}{30.4} + \dfrac{(113 - 121.6)^2}{121.6} = \dfrac{73.96}{30.4} + \dfrac{73.96}{121.6}$
$= 2.433 + 0.608 = 3.04$

Section 10.2

10.15 One categorical variable

10.17 The answers follow the guided steps.

1: H_a: Humans are not like random number generators and are not equally likely to pick all the integers.

2: If the integers are equally likely to be picked and there are 5 integers, then $38/5 = 7.6$, so the expected counts for all the cells are 7.6, which is larger than 5, $\alpha = 0.05$.

3: $X^2 = \dfrac{(3 - 7.6)^2}{7.6} + \dfrac{(5 - 7.6)^2}{7.6} + \dfrac{(14 - 7.6)^2}{7.6} + \dfrac{(11 - 7.6)^2}{7.6}$
$+ \dfrac{(5 - 7.6)^2}{7.6} = 11.47$

11.47 is the same in the output. p-value $= 0.022$.

4: Reject the null hypothesis and pick option ii.

10.19 *Step 1:* H_0: $p(\text{heads}) = 0.50$, H_a: $p(\text{heads}) \neq 0.50$, or the coin is biased. *Step 2:* Chi-square GOF: both expected counts are $25 > 5$, $\alpha = 0.05$. *Step 3:* $X^2 = 8.00$, p-value $= 0.005$, $\alpha = 0.05$. *Step 4:* Reject H_a. The coin is biased.

10.21 *Step 1:* H_0: All four outcomes are equally likely. H_a: The four outcomes are not equally likely. *Step 2:* Chi-square for goodness of fit; all expected values are $10 > 5$. *Step 3:* Significance level: 0.05, Chi-square $= 40.40$, p-value $=$ approximately 0. *Step 4:* We can reject the null hypothesis. We have shown a significant difference. The wooden dreidel is biased.

10.23 *Step 1:* H_0: The die is fair and produces a proportion of 1/6 in each possible outcome, H_a: The die is *not* fair and does not produce proportions of 1/6 for each possible outcome. *Step 2:* Chi-square goodness-of-fit test: expected counts all 1/6 of 120, or $20 > 5$, assume random sample, $\alpha = 0.05$. *Step 3:* $X^2 = 9.20$, p-value $= 0.101$. *Step 4:* We cannot reject H_0. The die has not been shown to be unfair.

10.25 a. *Step 1:* H_0: $p = 0.20$, H_a: $p \neq 0.20$, where p is the population proportion of people who could correctly identify a Stradivarius violin. *Step 2:* Chi-square goodness-of-fit test: not a random sample, test to see whether the results could easily occur by chance, the smallest expected count is $30.4 > 5$, $\alpha = 0.05$. *Step 3:* Chi-square $= 3.04$, p-value $= 0.081$. *Step 4:* Because the p-value is 0.081, which is more than 0.05, we conclude that the results are not significantly different from guessing, which would produce about 20% correct identifications. **b.** *Step 1:* H_0: $p = 0.20$, H_a: $p > 0.20$, where p is the population proportion of people who could correctly identify a Stradivarius violin. *Step 2:* One-proportion z-test: the smallest expected count is $30.4 > 10$, not a random sample, test whether the result could occur by chance, $\alpha = 0.05$. *Step 3:* $z = 1.74$, p-value $= 0.041$. *Step 4:* Because the p-value is 0.041, we can reject H_0 and conclude that there was a significantly higher proportion of correct identification than 20%. **c.** The one-sided hypothesis has a p-value that is half that of the two-sided hypothesis, and you can therefore reject H_0 with the one-sided hypothesis.

Section 10.3

10.27 Independence: one sample

10.29 Independence, one sample

10.31 The data are for the entire population, not a sample, and so there is no need for inference. The data are given as rates (percentages), not frequencies (counts), and there is not enough information given to convert these percentages into counts.

10.33 *Step 1:* H_0: The variables fitness app use and gender are independent (not associated), H_a: The variables fitness app use and gender are not independent (are associated); *Step 2:* Chi-square test of independence, all expected counts are greater than 5; *Step 3:* Significance level: 0.05, $X^2 = 3.36$, p-value $= 0.067$; *Step 4:* Do not reject H_0. Fitness app use has not been shown to be associated with gender.

10.35 *Step 1:* H_0 Vaccination rates and race are independent (not associated), H_a: Vaccination rates and race are not independent (are associated); *Step 2:* Chi-square test of independence, all expected counts are greater than 5; *Step 3:* Significance level: 0.05, $X^2 = 58.96$, p-value $=$ approximately 0; *Step 4:* Reject H_0. Vaccination rates and race are associated.

10.37 a. Independence: one sample with two variables. **b.** *Step 1:* H_0: Gender and happiness of marriage are independent, H_a: Gender and happiness of marriage are associated (not independent). *Step 2:* Chi-square test of independence: one sample, random sample, the smallest expected count is $11.88 > 5$, $\alpha = 0.05$. *Step 3:* Chi-square $= 10.17$, p-value $= 0.006$. *Step 4:* You can reject H_0. Gender and happiness have been shown to be associated. **c.** The rate of happiness of marriage has been found to be significantly different for men and women.

10.39 a. HS Grad rate for no preschool: 29/64, or 45.3%. The preschool kids had a higher graduation rate. **b.** *Step 1:* H_0: Graduation and preschool are independent, H_a: Graduation and preschool are not independent (they are associated). *Step 2:* Chi-square test of homogeneity: random assignment, not a random sample, the smallest expected count is $25.91 > 5$, $\alpha = 0.05$. *Step 3:* $X^2 = 4.67$, p-value $= 0.031$. *Step 4:* Reject H_0. Graduation and preschool are associated.

10.41 a. For preschool, 50% graduated, and for no preschool, $21/39 = 53.8\%$ graduated. It is surprising to see that the boys who did not go to preschool had a bit higher graduation rate. **b.** *Step 1:* For the boys, graduation and preschool are independent, H_a: For the boys, graduation and preschool are associated. *Step 2:* Chi-square test for homogeneity: random assignment, not a random sample, the smallest expected count is $15.32 > 5$, $\alpha = 0.05$. *Step 3:* $X^2 = 0.10$, p-value $= 0.747$. *Step 4:* Do not reject H_0. For the boys, there is no evidence that attending preschool is associated with graduating from high school. **c.** The results do not generalize to other groups of boys and girls, but what evidence we have suggests that although preschool might be effective for girls, it may not be for boys, at least with regard to graduation from high school.

10.43 a.

	Supports Marijuana Legalization	
Generation	Yes	No
Millennial	140	60
GenX	132	68

b. *Step 1:* H_0: Support of marijuana legalization and generation are independent (not associated), H_a: Support of marijuana legalization and generation are not independent (are associated); *Step 2:* Chi-square test of independence, all expected counts greater than 5; *Step 3:* Significance level: 0.05, $X^2 = 0.74$, p-value $= 0.391$, *Step 4:* Do not reject H_0. Support for marijuana legalization and generation have not been shown to be associated. **c.** Support of marijuana legalization is not significantly different among these two generations.

10.45 a. drug: $383/1161 = 33.0\%$, placebo: $419/1164 = 36.0\%$

b.

Adverse Outcome	Drug	Placebo	Total
Yes	383	419	802
No	778	745	1523
Total	1161	1164	2325

c. *Step 1:* H_0: Treatment and adverse outcome are independent (not associated), H_a: Treatment and adverse outcome are not independent (are associated); *Step 2:* Chi-square test of independence, all expected counts greater than 5; *Step 3:* Significance level: 0.05, $X^2 = 2.33$, p-value $= 0.127$; *Step 4:* Do not reject H_0. Treatment and adverse outcome have not been shown to be associated

10.47 *Step 1:* H_0: Political party affiliation and education are independent (not associated), H_a: Political party affiliation and education are not independent (are associated); *Step 2:* Chi-square test of independence, all expected counts greater than 5; *Step 3:* Significance level: 0.05, $X^2 = 14.70$, p-value $= 0.002$; *Step 4:* Reject H_0: Political party affiliation and education are associated.

Section 10.4

10.49

	Alcohol Use		
Age	None	1–9 days	10+ days
18–20	182	100	31
21–24	142	109	39
25–29	49	41	7
30+	76	32	10

Step 1: H_0: Age group and alcohol use are independent (not associated), H_a: Age group and alcohol use are not independent (are associated); *Step 2:* Chi-square test for independence, all expected counts greater than 5; *Step 3:* Significance level: 0.05, $X^2 = 13.60$, p-value $= 0.034$; *Step 4:* Reject H_0; There is an association between age group and alcohol use.

10.51 a. The smallest expected count is 24.22 from male and other party, so the original table could have been used.

b.

	Male	Female
Dem	278	421
Rep	198	244
Other	403	416

c. Women: $421/1081 = 38.9\%$ Democrats. Men: $278/879 = 31.6\%$ Democrats. Thus, these women are more likely to be Democrats than these men. **d.** *Step 1:* H_0: Gender and political party are independent, H_a: Gender and political party are not independent. *Step 2:* Chi-square test of independence: assume random sample, the smallest expected count is $198.22 > 5$, $\alpha = 0.05$. *Step 3:* Chi-square $= 13.57$, p-value $= 0.001$. *Step 4:* Reject H_0. The difference is significant. Gender and political party are not independent.

10.53 a. You would get a smaller p-value, because it is more extreme in the direction of the antivenom working. **b.** You would get a larger p-value, because the results are less extreme. **c.** The p-value for the test in part a is 0.0002 (both one-tailed and two-tailed); yes, it is smaller. The p-value for the test in part b is 0.1002 (one-tailed) and 0.1319 (two-tailed); yes, it is larger.

10.55 a. Consume peanuts: $5/272 = 1.8\%$; Avoid peanuts: $37/270 = 13.7\%$; **b.** *Step 1:* H_0: Treatment group and peanut allergy are independent (not associated); H_a: Treatment group and peanut allergy are not independent (are associated); *Step 2:* Chi-square test for independence, all expected counts greater than 5; *Step 3:* Significance level: 0.05, $X^2 = 26.69$, p-value $=$ approximately 0; *Step 4:* Reject H_0: There is an association between treatment group and peanut allergy. **c.** Fisher's Exact Test p-value < 0.0001 **d.** Chi-square p-value $=$ approximately 0. Fisher's Exact Test p-value < 0.0001. Fisher's Exact test is correct. The chi-square p-value is a large-sample size approximation and this sample is fairly small.

Chapter Review Exercises

10.57 Chi-square goodness-of-fit test

10.59 Chi-square test of independence (one sample, two variables)

10.61 Chi-square test of independence (one sample, two variables)

10.63 No test because the data are a population, not a sample

10.65 a. 31/65, or 47.7%, of those in the control group were arrested, and 8/58, or 13.8%, of those who attended preschool were arrested. Thus, there was a lower rate of arrest for those who went to preschool.

b.

	Preschool	No Preschool
Arrest	8	31
No Arrest	50	34

Step 1: H_0: The treatment and arrest rate are independent, H_a: The treatment and arrest rate are associated. *Step 2:* Chi-square for homogeneity: random assignment, not a random sample, the smallest expected count is 18.39 > 5, $\alpha = 0.05$. *Step 3:* Chi-square = 16.27, p-value = 0.000055 (or < 0.001). *Step 4:* We can reject the hypothesis of no association at the 0.05 level. Don't generalize. We conclude that preschool attendance affects the arrest rate. **c.** Two-proportion z-test. *Step 1:* H_0: $p_{pre} = p_{nopre}$, H_a: $p_{pre} < p_{nopre}$ (p is the rate of arrest). *Step 2:* Two-proportion z-test: the smallest expected count is 18.39 > 10, $\alpha = 0.05$. *Step 3:* $z = -4.03$, p-value = 0.000028 (or < 0.001). *Step 4:* Reject the null hypothesis. Preschool lowers the rate of arrest, but we cannot generalize. **d.** The z-test enables us to test the alternative hypothesis that preschool attendance *lowers* the risk of later arrest. The chi-square test allows for testing for some sort of association, but we can't specify whether it is a positive or a negative association. Note that the p-value for the one-sided hypothesis with the z-test is half the p-value for the two-sided hypothesis with the chi-square test.

10.67 It would not be appropriate because the data are percentages (not counts), and we cannot convert them to counts given the information in the problem.

10.69 a. Men: 120/500 = 24.0%, Women: 182/519 = 35.1%; **b.** *Step 1:* H_0: Gender and experience of sexual harassment in the workplace are independent; H_a: Gender and experience of sexual harassment in the workplace are not independent; *Step 2:* Chi-square test of independence; all expected counts greater than 5; *Step 3:* Significance level: 0.05, $X^2 = 14.96$, p-value = 0.0001; *Step 4:* Reject H_0: Gender and experience of sexual harassment in the workplace are not independent.

10.71 a. *Step 1:* H_0: The presence or absence of robots is independent of grouping, H_a: The presence or absence of robots is not independent of grouping. *Step 2:* Chi-square test for homogeneity: random, two expected counts are 5, which is on the low side for this approximation, $\alpha = 0.05$. *Step 3:* Chi-square = 4.32, p-value = 0.038. *Step 4:* Reject H_0. The proportions are not the same. The robots have a significant effect. **b.** With Fisher's Exact Test, using a two-sided alternative, the p-value is 0.080. You cannot reject H_0. The robots do not have a significant effect. **c.** For chi-square, the p-value was 0.038, and with Fisher's Exact Test, the p-value was 0.080. Fisher's Exact Test is accurate, and the chi-square test is a large-sample approximation. The approximation is not very good with two expected counts of 5, and that is why the two p-values are so different. The p-value of 0.080 is the accurate value.

10.73 a. 19/67 = 28.4% of the minority defendants were convicted. **b.** 38/118 = 32.2% of the white defendants were convicted. **c.** *Step 1:* H_0: Race and conviction are independent, H_a: Race and conviction are not independent. *Step 2:* Chi-square test of independence: random sample, smallest expected count is 20.6 > 5, $\alpha = 0.05$. *Step 3:* Chi-square = 0.30, p-value = 0.586. *Step 4:* Do not reject H_0. Race and conviction have not been shown to be associated.

CHAPTER 11

Correct answers may vary slightly due to rounding and type of technology used. All *t*-statistics are reported as positive values, although the sign of your

t-statistics (positive or negative) must be consistent with which group you chose for group 1.

Section 11.1

All calculations for this section were done without assuming equal variances.

11.1 a. ANOVA **b.** Two-sample *t*-test

11.3 a. 5(5 − 1)/2 = 20/2 = 10 AB, AC, AD, AE, BC, BD, BE, CD, CE, DE **b.** 0.05/10 = 0.0050

11.5

Comparison	*t*-statistic	p-value	Conclusion
Seattle − SF	$t = -5.17$	0.0003	Significantly different
Seattle − Santa Monica	$t = -3.36$	0.006	Significantly different
SF − Santa Monica	$t = 2.27$	0.043	Not significantly different

11.7 a. 0.05/3 = 0.0167 **b.** Means: LA − $3.46, Chicago − $3.13, Boston − $2.94. The closest means are Chicago and Boston. **c.**

Comparison	*t*-statistic	p-value	Conclusion
LA − Chicago	$t = 20.05$	< 0.0001	Significantly different
LA − Boston	$t = 25.68$	< 0.0001	Significantly different
Chicago − Boston	$t = 7.78$	< 0.0001	Significantly different

11.9 a. 3 **b.** 0.05/3 = 0.0167 **c.** 1 − 0.0167 = 98.33%

11.11

LA − Chicago	(0.282, 0.378)	Does not capture 0. Reject equality.
LA − Boston	(0.463, 0.585)	Does not capture 0. Reject equality.
Chicago − Boston	(0.126, 0.262)	Does not capture 0. Reject equality.

These conclusions are the same as those for exercise 11.7.

All of the means are significantly different from each other. These are the same conclusions reached in exercise 11.7.

11.13 a. Means: Shortstop − 50.6, Left Field − 51.3, First Base − 52.3 **b.** 3 comparisons, Bonferroni-corrected significance level: 0.05/3 = 0.0167 **c.**

Comparison	*t*-statistic	p-value	Conclusion
SS − LF	$t = -0.16$	0.875	Not significantly different
SS − FB	$t = -0.47$	0.648	Not significantly different
LF − FB	$t = -0.23$	0.820	Not significantly different

Section 11.2

11.15 The *F*-value of 9.38 goes with A, B, and C. And the *F*-value of 150.00 goes with L, M, and N. The reason for the difference is that the variation between groups (the separation between means) is larger for L, M, and N, relative to the variation within groups (which is the same in all groups).

$$F = \frac{\text{Variation between Groups}}{\text{Variation within Groups}}$$

11.17 a. H_0: The population means are all equal, or marital status and cholesterol levels are not associated. H_a: At least one mean is different from another, or marital status and cholesterol levels are associated. **b.** $F = 9.14$ **c.** Largest mean cholesterol: divorced. Smallest mean cholesterol: never married. **d.** This was an observational study, from which you cannot conclude causality. One possible confounder is age. For example, the "never married" may tend to be young, and youth may cause the low cholesterol.

11.19 a. SS Error $= 7790.5 - 893.5 = 6897$ **b.** $6897/106 = 65.066$, which rounded is 65.1. **c.** $297.8/65.066 = 4.5769$, which rounded is 4.58. **d.** When MS factor is more than MS Error, the F-value will be more than 1.

11.21 a. Highest number of hours was for the freshmen, and the lowest was for the seniors. **b.** μ is the population mean number of hours of schoolwork per week.

$H_0: \mu_1 = \mu_2 = \mu_3 = \mu_4$

H_a: At least one mean is different from another, or class has an effect on schoolwork.

c. $F = 4.58$ **d.** No. There was no random assignment. There could be confounding factors, such as age, hours of work for money, or living situation.

Section 11.3

11.23

Price	Code
2.75	1
2.74	1
2.72	1
2.73	1
2.75	1
2.70	2
2.68	2
2.66	2
2.71	2
2.74	2
2.85	3
2.83	3
2.84	3
3.01	3
2.97	3

11.25 p-value $= 0.005$. Reject H_0. The mean amount of schoolwork does vary by class.

11.27 The pulse rates are not in three independent groups, so the condition of independent groups fails.

11.29 The ratio of the largest to the smallest is $34.02/14.36$, or 2.37, which is larger than 2. Do not use ANOVA, because the standard deviations are too different.

11.31 *Step 1:* $H_0: \mu_{Seattle} = \mu_{SF} = \mu_{SantaMonica}$; H_a: At least one population mean is different from another; *Step 2:* One-way ANOVA. Distributions satisfy ANOVA conditions; *Step 3:* Significance level: 0.05, $F = 14.16$, p-value $= 0.0001$; *Step 4:* Reject H_0. The mean rents for one-bedroom apartments in these cities are not equal.

11.33 *Step 1:* $H_0: \mu_{IL} = \mu_{LA} = \mu_P = \mu_S$, H_a: At least one mean is different from another. *Step 2:* ANOVA: Assume samples are random and Normal and independent. SD ratio $4369/3219 = 1.36 < 2$, $\alpha = 0.05$. *Step 3:* $F = 39.02$, p-value < 0.001. *Step 4:* Reject H_0. Type of school does affect median starting salary.

11.35 a. The medians and interquartile ranges are all similar, although the median for the athletes was a little larger (slower!) than the others, and the interquartile range was a bit larger for the moderate group. Also, the shapes are not strongly skewed. There are no potential outliers. **b.** $F = 0.10$, p-value $= 0.903$. The sample means were not significantly different, and we conclude that we cannot reject the hypothesis that the population means are the same.

11.37 *Step 1:* $H_0: \mu_{front} = \mu_{middle} = \mu_{back}$, H_a: At least one group mean differs from another. *Step 2:* ANOVA: Assume random and Normal and independent. SD ratio $0.4637/0.2574 = 1.80 < 2$, $\alpha = 0.05$. *Step 3:* $F = 7.50$, p-value $= 0.003$. *Step 4:* Reject H_0. The mean GPA is not the same for all rows.

11.39 *Step 1:* H_0: All four population means are equal (suggesting health status and hours of sleep are independent), H_a: At least one mean is different from another. *Step 2:* ANOVA: Assume Normality, random sample, independent observations. SD ratio $1.786/1.373 = 1.30 < 2$, $\alpha = 0.05$. *Step 3:* $F = 4.44$, p-value $= 0.0053$. *Step 4:* Reject H_0. The means are not all equal. Health status and hours of sleep are not independent at this company.

Section 11.4

All calculations for post hoc tests were done with pooled variances.

11.41 The $UT - CA$ interval does not contain 0. Since both limits are negative, CA has longer travel times than UT. The NY-UT interval also does not contain 0. Since both limits are positive, NY has longer travel times than UT. So UT has the shortest travel times of these 3 states.

11.43 $F = 7.50$, p-value $= 0.003$. Conclusion: Reject H_0. GPA and row level appear to be associated.

Post Hoc:

Comparison	CI Tukey	TI-84	Reject H_0?
Middle – front	(–0.98, –0.10)	(–1.07, –0.01)	Yes
Back – front	(–1.08, –0.20)	(–1.06, –0.22)	Yes
Back – middle	(–0.54, 0.34)	(–0.54, 0.34)	No

<u>Back</u>   <u>Middle</u>   Front

Those who sit in the front row tend to have higher GPAs, on average, than do those who sit either in the back or in the middle. There is no distinguishable difference in mean GPAs between those in the back and those in the middle.

11.45 The p-value for ANOVA was 0.903. We cannot reject the null hypothesis of no differences in population mean reaction distances, so you should not do post hoc tests. In other words, the sample means are not significantly different.

11.47 Since the confidence interval limits of $SF - Seattle$ are both positive, SF is more expensive than Seattle. Since the confidence interval limits of Santa Monica $-$ Seattle are both positive, Santa Monica is also more expensive than Seattle. Since the confidence interval limits of Santa Monica $-$ SF are both negative, SF is more expensive than Santa Monica. The cities (in order of least expensive to most expensive): Seattle, Santa Monica, San Francisco.

11.49 *Step 1:* H_0: The mean level of concern over nuclear power is the same for these political parties. H_a: At least one mean is different from another. *Step 2:* One-way ANOVA; assume ANOVA conditions are satisfied. *Step 3:* Significance level: 0.05, $F = 5.98$, p-value: p $= 0.0005$. *Step 4:* Reject H_0. The mean level of concern over nuclear power is not the same for these political parties. At least one is different. Post hoc analysis: Republican is significantly less concerned than all other parties. There are no other significant differences.

Chapter Review Exercises

11.51 No, since multiple tests were performed. There are a total of 7 groups and therefore 21 possible comparisons. The Bonferroni-corrected significance level would be $0.05/21$, which is 0.0024. The p-value of 0.012 is not less than 0.0024.

11.53 Because all the intervals capture 0 [$(-5.21, 17.81)$, $(-1.28, 20.35)$, and $(-1.91, 8.37)$], we have not found any significant differences in the mean length of time since contacting mother for the three groups of professors.

11.55 $F = 7.33$, p-value $= 0.001$. Reject H_0.

Phil   <u>Other</u>   <u>Ethicist</u>

Philosophy professors report giving a significantly lower percentage to charity than ethicists and professors in other fields. There is no significant difference between ethicists and other professors.

11.57 The p-values are the same. The F-statistic is the square of the t-statistic. You get the same conclusion either way: Do not reject the null hypothesis of equal population mean triglyceride levels for men and women.

CHAPTER 12

Answers are given using a chi-square approach for hypothesis tests as appropriate unless the problem specifically says to use a 2-prop-z-test approach.

Section 12.1

12.1 a. Cause and effect **b.** No cause and effect

12.3 This was an observational study. Researchers could not assign subjects to smoke marijuana regularly.

12.5 Regular marijuana use leads to increased likelihood of bone fractures (causality). Regular marijuana use associated with increased bone fracture susceptibility (no causality). The second is correct because we cannot infer cause and effect from an observational study.

12.7 a. clopidogrel and aspirin: $121/2432 = 5.0\%$, placebo and aspirin: $159/2449 = 6.5\%$ **b.** controlled experiment (random assignment) **c.** The treatment variable is the drug (clopidogrel); the response variable is the percentage of subjects who had another stroke. **d.** A combination of clopidogrel and aspirin lowers the risk of recurrent stroke in stroke patients compared with treatment with aspirin alone.

12.9 a. The treatment variable is the drug (upadacitinib); the response variable is improvement in rheumatoid arthritis symptoms. **b.** The drug upadacitinib causes improvement in symptoms among rheumatoid arthritis patients.

12.11 This was an observational study because subjects were not assigned to exercise groups.

12.13 a. Two-sample proportion test; $z = -1.03$, p-value $= 0.301$. Do not reject H_0. We cannot conclude the proportion of men and women who agree this statement differ. **b.** Men: $0.84(2291) = 1924$, Women: $0.89(2282) = 2031$; two-sample proportion test, $z = -4.96$, p-value $=$ approximately 0. Reject H_0. The proportion of men and women who agree with this statement are different. **c.** The larger sample size gave us more power, leading us to reject the null hypothesis.

12.15 a. The first will have more variability because it draws from both men and women. **b.** The second study is drawing a sample from a population with less variability and so it will have more power.

12.17 This is not an appropriate use of blocking. In this design, the researchers randomized the blocks, not the subjects. Randomization should happen within blocks. For this study, each patient in a block should be assigned a number, and these numbers should be put into a bowl, mixed up. There will be four bowls, one for each age group. From each bowl (each block), half of the subjects' numbers are chosen, and these people will receive the treatment. The others receive a placebo.

12.19 a. The treatment variable is the speed skating suit; the response variable is the race speed. **b.** Randomly assign each skater to race either wearing a Mach 39 suit or an H1 suit. To do this, in a bag put 10 tickets that say "Mach 39" and 10 tickets that say "H1." Each skater chooses a ticket and will use that type of suit. Record the race times. **c.** Block on whether the skater is an Olympic skater or not. Randomly assign half the Olympic skaters to the Mach 39 suits by putting in a bag 5 tickets that say "Mach 39" and five tickets that say "H1." Each Olympic skater chooses a ticket and will use that type of suit. Do the same process with the recreational speed skaters, randomly assigning half to use the Mach 39 suit and half to use the H1 suit. The blocked design will prevent having uneven groups with more Olympic skaters in one group than the other. **d.** Have each skater race with a Mach 39 suit and also with an H1 suit. Use a paired t-test for the comparison. Randomly assign half of all skaters to wear the Mach 39 first and the other half to wear the H1 first. (Otherwise some skaters might skate more slowly in their second race because they were tired from their first race)

12.21 a. The treatment variable records whether subjects get aspirin or placebo. The response variable is whether the person has a heart attack. **b.** You could put 100 slips of paper marked A and 100 slips of paper marked B in a bag. Each person would draw out a slip of paper. Those who got A would get the aspirin, and those who got B would get the placebo. Then observe the subjects for a given time interval to see whether they have a heart attack, and compare the percentages with heart attacks for the two

groups. **c.** Randomly assign half of the men and half of the women to use the aspirin, and assign the rest to use a placebo. You could use two separate bags, one for the men and one for the women. Each bag would have 100 slips of paper: 50 marked A and 50 marked B. Each woman draws randomly from the women's bag, and each man draws randomly from the men's bag. Then observe the subjects for a given time interval to see whether they have a heart attack, and determine the percentage of aspirin-taking men who had a heart attack, the percentage of placebo-taking men who had a heart attack, the percentage of aspirin-taking women who had a heart attack, and the percentage of placebo-taking women who had a heart attack. **d.** The blocked design improves statistical power, in part by preventing an uneven distribution of men and women in the two groups. With the blocked design, we have a higher probability of determining whether aspirin reduces the risk of heart attack, if it actually does so.

12.23 a. This is a controlled experiment (used random assignment). **b.** Treatment group: $10/33 = 30.3\%$, Control group: $2/34 = 5.9\%$. **c.** (i) Since the p-value is less than 0.05 the null hypothesis is rejected. Researchers have shown that dietary improvement may be an effective treatment strategy for patients with moderate to severe depression.

12.25 a. The mean for white paper was smaller (the means were 74.3 and 76.3). This contradicts the idea that reading material written on colored paper is easier to read. **b.** Two sample t-test: t-value $= 0.19$, p-value $= 0.573$ (for a one sided hypothesis). Do not reject H_0. **c.** Paired t-test: t-value $= 0.97$, p-value $= 0.825$ (for a one-sided hypothesis). Do not reject H_0. **d.** The paired t-test is appropriate because each person is tested twice, so the numbers are coupled. **e.** People might get faster if they had read the passage previously, and you don't want the order of reading to affect the answer.

Section 12.2

12.27 Systematic sampling

12.29 Those who own an electric car may feel the location of the charging station is more important than students who do not have an electric car and who would be unlikely to use this resource.

12.31 It is systematic sampling.

12.33 It is cluster sampling.

Section 12.3

12.35 a. No, because this is an observational study, not a controlled experiment with random assignment. **b.** This association can only be generalized to the entire population of people with Type 2 diabetes if the participants represented a random sample of people with Type 2 diabetes. The sample in this study were all from Japan, so the association cannot be generalized to all people with Type 2 diabetes.

12.37 a. You can generalize to other people admitted to this hospital who would have been assigned a double room because of the random sampling from that group. **b.** Yes, you can infer causality because of the random assignment.

12.39 a. The treatment variable is type of intravenous fluid (balanced crystalloids or saline). **b.** Yes. 14.3% of the balanced crystalloids group and 15.4% of the saline group had a major adverse kidney event. This difference is significant (p-value $= 0.04$). **c.** Yes. This controlled experiment used random assignment so a causal conclusion can be made. Use of balanced crystalloids for an intravenous fluid results in a lower rate of major adverse kidney events.

12.41 This is likely the result of an observational study because researchers would not randomly assign subjects to "phub" or "not phub" if the possible outcomes were negative effects on relationship satisfaction and depression. A causal conclusion cannot be made from an observational study.

12.43 This study did not use a control group and random assignment. A causal conclusion cannot be made.

12.45 Randomly assign 100 subjects to participate in YMLI and 100 subjects to do some other type of physical activity for the same amount of time (5 days per week). All study subjects do the activity for 12 weeks. At the end of 12 weeks, measure the biomarkers of cellular aging and longevity in both groups.

12.47 a. The treatment variable was type of music; the response variable was overall divergent thinking (ODT). **b.** This was a controlled experiment. Subjects were randomly assigned to one of the treatment conditions. **c.** The study used random assignment but not random selection. A causal conclusion can be made but we cannot conclude that everyone in general would respond similarly.

12.49 a. Take a nonrandom sample of students and randomly assign some to the reception and some to attend a "control group" meeting where they do something else (such as learn the history of the college). **b.** Take a random sample of students and offer them the choice of attending the reception or attending a "control group" meeting where they do something else (such as learn the history of the college). **c.** Take a random sample of students. Then randomly assign some of the students in this sample to the reception and some to the "control group" meeting.

12.51 a. Is ketamine an effective treatment for Social Anxiety Disorder? **b.** Yes, ketamine may be effective in reducing anxiety. **c.** Patients were given ketamine and placebo infusions in a random order with a 28-day period between infusions. Patients' anxiety was rated 3 hours after the infusion and for a period of 14 days using blinded ratings as well as a self-reported scale. **d.** Yes, the conclusion is appropriate for the study. This was a controlled experiment that used random assignment, so a causal conclusion is appropriate. **e.** Because there was no random sampling from the population of all people with Social Anxiety Disorder, we cannot generalize widely, and the results apply only to these patients. **f.** There was no mention of other articles.

12.53 a. Does self-distancing improve executive function in 3-year old and 5-year old children? **b.** Self-distancing improved executive function in 5-year olds but not in 3-year olds. **c.** This was a controlled experiment. Children were randomly assigned to one of three treatment groups or to a control group. Children in each group did a task and were assessed on their executive function in completing the task by researchers. **d.** The study used random assignment but not random selection. A causal conclusion can be made, but we cannot conclude that all 3-year olds and 5-year olds would respond similarly. **e.** Because there was no random sampling from the population, we cannot generalize widely, and these results apply only to these children. **f.** There was no mention of other articles.

12.55 a. Handover of anesthesia care: $2614/5941 = 44.0\%$; no handover of anesthesia care: $89,066/307,125 = 29.0\%$; The difference in the two groups is statistically significant (p-value < 0.001). **b.** Yes, because the group without complete handover had a lower percentage of bad outcomes. **c.** The confidence interval $(-0.3\%$ to $2.7\%)$ contains 0 and the p-value is 0.11, leading to a conclusion that there was no significant different in this outcome between the two groups.

12.57 a. The treatment variable was formula type and the response variable was development of Type 1 diabetes by age 11.5 years.

b.

Developed Disease	Conventional Formula	Hydrolyzed Formula	Total
Yes	82	91	173
No	997	1892	2889
Total	1079	1983	3062

Step 1: H_0: Formula type and disease development are independent, H_a: Formula type and disease development are not independent. *Step 2:* All expected counts greater than 5. *Step 3:* Significance level: 0.05; $X^2 = 11.88$; p-value < 0.001. *Step 4:* Reject H_0. Formula type and development of Type 1 diabetes are associated. **c.** Yes, since infants in this randomized clinical trial who were given hydrolyzed formula were significantly less likely to develop Type 1 diabetes than infants who were given a conventional formula. However, because there was no random sampling from the population, we cannot generalize widely, and these results apply only to these infants.

CHAPTER 13

Section 13.1

13.1 You should have a random sample from the population.

13.3 The data should be drawn from Normal distributions, or the sample sizes should be large (typically at least 25 from each population). Observations must be independent of each other. The groups must be independent of each other.

13.5 a. Histogram A goes with D, and Histogram B goes with C.

b. A is roughly bell-shaped (Normal), and B is right-skewed.

c. Use a log transform on the data that are shown in histogram B.

13.7 a. 1, 2, 3, 3.8; **b.** 1000, 100,000, 251.2, 1584.9

13.9 a. i. 1, 3, 4; ii. 2.67; iii. 467.7; **b.** mean = 3670, median = 1000; From smallest to largest: 467.7, 1000, 3670 (geometric mean, median, mean)

13.11 a. We are 95% confident that the population mean credit card balance is between $791.87 and $2541.65; **b.** $(10^{2.97}, 10^{3.29}) = (933.25, 1949.84)$. We are 95% confident that the population mean credit card balance is between $933.25 and $1949.84. **c.** $2541.65 - 791.87 = 1749.78$; $1949.84 - 933.25 = 1016.59$. The confidence interval for the geometric mean (from the logs) is narrower. **d.** The confidence interval for the geometric mean is more appropriate because the distribution of the log-transformed data is more symmetric and so the confidence interval for the geometric mean is more accurate. Also we can get a more precise (smaller margin of error) estimate for the geometric mean than for the mean based on this sample.

13.13 a. Right-skewed **b.** (4.04, 6.31) **c.** It is closer to Normal than the distribution of untransformed data. **d.** (0.515, 0.7025) **e.** (3.27, 5.04); We are 95% confident that the population geometric mean number of hours of exercise per week for all students at this college is between 3.27 hours and 5.04 hours. **f.** Answers may vary. The interval for the geometric mean is more precise (smaller margin of error), which suggests that the geometric mean might be a better measure of center. On the other hand, the sample size is large, so both confidence intervals are valid and the population mean could also be used. (If the sample size were under 25, the reported confidence level for the resulting confidence interval of the population mean would not be accurate.)

Section 13.2

13.15 *Step 1:* H_0: The median levels of lead in the blood are the same, H_a: The median level of lead in the blood is larger for the experimental group (one-sided). *Step 2:* Sign test: paired, $\alpha = 0.05$. *Step 3:* $S = 28$ or $S = 4$, p-value < 0.001. *Step 4:* Reject H_0. The median level of lead in the blood was larger for the children in the experimental group. For a p-value from a TI-84: $b(32, 0.5, 4$ or fewer) $= 9.7 \times 10^{-6}$, or about 0.00001.

13.17 a. The medians were 19 (for same) and 33.50 (for different). Yes, it tended to take longer for the different colors. **b.** *Step 1:* H_0: Median time for same equals median time for different, H_a: Median time for same $<$ median time for different. *Step 2:* Sign test: paired, $\alpha = 0.05$. *Step 3:* $S = 1$ or 9, p-value $= 0.0107$. *Step 4:* Reject H_0. The difference is significant. It took longer to identify the "wrong" color.

13.19 *Step 1:* H_0: The median pulse rate does not change, H_a: The median pulse rate goes up (one-sided). *Step 2:* Sign test: paired, $\alpha = 0.05$. *Step 3:* $S = 4$ or 5, p-value $= 0.500$. *Step 4:* Do not reject H_0. The median pulse rate did not go up significantly for men.

13.21 *Step 1:* H_0: Median age for grooms $=$ median age for brides, H_a: Median age for grooms $>$ median age for brides. *Step 2:* Sign test: paired, $\alpha = 0.05$. *Step 3:* $S = 10$ or 3, p-value $= 0.046$ from $b(13, 0.5, 3$ or fewer). *Step 4:* Reject H_0. The median age for the grooms is significantly greater than the median age for the brides.

Section 13.3

13.23 a. The median for these ethicists was 4.5 meals per week, which is lower than the median for this control group (6.5). In the sample, ethicists reported eating fewer meals with meat per week than nonethicists reported.

b. *Step 1:* H_0: Median meals of meat per week of control $=$ median of ethicists, H_a: Median meals of meat per week of control $>$ median of ethicists. *Step 2:* Mann-Whitney test: independent, $\alpha = 0.05$. *Step 3:* $W = 198$, p-value $= 0.058$. *Step 4:* Do not reject H_0. We have not found a significant difference in behavior.

13.25 a. This shape is right-skewed. It appears there may be two outliers at about 17 runs. Because of the skew and the outliers, it would be more appropriate to compare the medians. **b.** *Step 1:* H_0: The median for winning runs is the same for both the American and National leagues. H_a: The median for winning runs is different for the two leagues. *Step 2:* Mann-Whitney test. *Step 3:* Significance level: 0.05, $W = 216$, p-value $= 0.552$; *Step 4:* Do not reject H_0. There is not sufficient evidence to conclude the medians are significantly different.

13.27 a. We don't compare means because there is an outlier of around $400 for the women, and the samples are not large. **b.** *Step 1:* H_0: Median for all men $=$ median for all women. H_a: Median for all men $<$ median for all women. *Step 2:* Mann-Whitney test: independent, $\alpha = 0.05$. *Step 3:* $W = 501.0$, p-value $= 0.143$. *Step 4:* Do not reject H_0. The medians for men and women with regard to cell phone bills are not significantly different.

13.29 a. The histogram is strongly left-skewed, so it would be more appropriate to compare medians. **b.** The sample median of 80 for the women is higher than the sample median of 75 for the men. **c.** *Step 1:* H_0: The median for all men is equal to the median for all women. H_a: The median for all men is not equal to the median for all women. *Step 2:* Mann-Whitney: not paired, independent observations, $\alpha = 0.05$. *Step 3:* $W = 106108.0$, p-value $= 0.032$. *Step 4:* Reject H_0. The medians are significantly different.

Section 13.4

13.31 a. The red line looks like it is pretty far out in the tail of the data and suggests visually that there is a real difference in extraversion between the sporty and the nonsporty students. **b.** The p-value is 0.011988012, or 0.012. **c.** We reject the null hypothesis and conclude that "sporty" students typically have a higher level of extraversion than "nonsporty" students. **d.** We could get an approximate p-value using the histogram, by finding the (approximate) proportion of observations to the right of the red vertical line.

13.33 a. The distributions are strongly right-skewed, and the median is often a better choice for skewed data than the mean. **b.** 525 is not far out in the tail, so it is not unusually large. **c.** The one-tailed p-value is larger than 0.05, so the two-tailed p-value will be even larger. Do not reject H_0. The population median for the men has not been shown to be different from the population median for the women.

13.35 Choose 0.025 by looking at the graph, knowing that the observed number is 368.9 acre-feet. Reject the null hypothesis. The mean rainfall is greater for the seeded clouds.

13.37 Explanation a

Chapter Review Exercises

13.39 a. Paired t-test or sign test; **b.** Sign test; **c.** Paired t-test or sign test.

13.41 a. Two-sample t-test or Mann-Whitney test; **b.** Mann-Whitney test

13.43 a. Two-sample t-test or Mann-Whitney test; **b.** Mann-Whitney test

13.45 a. Sign test; **b.** Paired t-test or the sign test; **c.** The paired t-test or the sign test (sample size is large enough to use the t-test).

13.47 *Step 0:*

Obs	Null Value	Difference
4.2	3.18	1.02
3.6	3.18	0.42
3.9	3.18	0.72
3.4	3.18	0.22
3.3	3.18	0.12

1: H_0: The median weight of the cones is 3.18 ounces, as advertised (or the median difference is 0). H_a: The median weight is more than 3.18 ounces (or the median difference is more than 0).

2: Use the one-sample sign test. The cones are independent because they are purchased from different servers, and although we don't have a truly random sample, we hope we have a representative sample.

3: p-value $= 0.031$

4: Reject H_0 and conclude that the population median is more than 3.18 ounces (or the population median difference is more than 0.00).

Why? We have a small sample and do not know whether the distribution is Normal or not.

13.49 a. They are both right-skewed. **b.** The mean for the ethicists is 12.7 days, and the mean for other professors is 3.2 days. The non-ethics professors tend to have been in more recent contact with their mothers. **c.** The median for the ethicists is 5 days, and the median for other professors is 1 day, so other professors tend to have been in more recent contact with their mothers. **d.** *Step 1:* H_0: $\mu_{eth} = \mu_{other}$, H_a: $\mu_{eth} \neq \mu_{other}$. *Step 2:* Two-sample t-test: the samples are random and large ($n = 30$ for each), $\alpha = 0.05$. *Step 3:* $t = 2.23$, p-value $= 0.033$. *Step 4:* Reject H_0. There is a significant difference in means. **e.** *Step 1:* H_0: $\text{Med}_{eth} = \text{Med}_{other}$, H_a: $\text{Med}_{eth} \neq \text{Med}_{other}$. *Step 2:* Mann-Whitney test: not paired, independent observations, $\alpha = 0.05$. *Step 3:* $W = 1126.5$, p-value $= 0.002$. *Step 4:* Reject H_0. There is a significant difference in medians. The typical time since contact with mothers is different for ethics professors than for other professors.

13.51 *Step 1:* H_0: $\text{Med}_{sent} = \text{Med}_{received}$, H_a: $\text{Med}_{sent} \neq \text{Med}_{received}$. *Step 2:* Sign test: samples random, $\alpha = 0.05$. *Step 3:* $S = 40$ or 11, p-value < 0.001. *Step 4:* Reject H_0. The median number of texts sent and the number received are significantly different.

13.53 *Step 1:* H_0: $\mu_{sent} = \mu_{received}$ for males, H_a: $\mu_{sent} \neq \mu_{received}$ for males. *Step 2:* Paired t-test: samples are large and random. *Step 3:* $t = 1.20$ or -1.20, p-value $= 0.234$. *Step 4:* Do not reject H_0. We do not have enough evidence to conclude that the means are significantly different.

13.55 You cannot find the geometric mean because you cannot find a logarithm of 0. And there are several students who missed no classes.

13.57 The red line is far out in the left tail, which suggests a significant difference. The p-value for the one-sided hypothesis is 0.00899. The conclusion is that ethicists tend to have less recent contact with their mothers than other professors. (The mean number of days since contact is greater for the ethicists than for the non-ethicists.)

13.59 a. There was a significant difference in purchase intention toward V Energy drink. The p-value was less than 0.05 indicating there was a difference between the control group and experimental group for this item. **b.** To use the independent sample t-test, the population distributions must be Normal or the sample sizes must be large enough for the Central Limit Theorem to be applied. The Mann-Whitney test only requires that the two population distributions have the same shape, not necessarily a Normal distribution.

CHAPTER 14

Section 14.1

14.1 Answers will vary. Some possibilities: the amount of time the student could study, the amount of sleep the student got the night before, the particular choice of questions on the exam, the noise level in the room, the health of the student.

14.3 Answers will vary. Random factors: Food, genetics, drugs

14.5

Floors	Height (in feet)	Predicted Height	Residual
163	2717	2690.8	26.2
101	1398	1454.9	−56.9
101	1358	1454.9	−96.9
88	1289	1195.8	93.3
89	1250	1215.7	34.3

14.7 a. The residual plot shows curvature, indicating that the linear condition fails. The linear model is not appropriate. **b.** There are two points that look farthest from the regression line: a car that is about 7 years old and a car that is about 15 years old.

14.9 The residual plot is fan-shaped, showing more variation in the number of units for students who have attended for many semesters, and less variation for those who have attended for few semesters. This shows that the constant standard deviation condition does not hold, so inference would not be appropriate.

14.11 Linear regression is not appropriate, because the constant-SD condition does not hold; there is more variation with the larger beginning salaries than with the smaller beginning salaries.

14.13 The residual plot shows an increasing trend, and the QQ plot does not follow a straight line. Linear regression is inappropriate for this data set, because the linearity condition and the Normality condition are not met.

Section 14.2

14.15 *Step 1:* H_0: Slope $= 0$; H_a: Slope $\neq 0$. *Step 2:* t-test for slope. *Step 3:* Significance level: 0.05, $t = 13.6$, p-value < 0.0001. *Step 4:* Reject H_0. There is a linear association between state and federal spending on education.

14.17 a. The intercept is 2.78. If the intercept were 0, it would mean that if a mother had 0 years of education, then the father would be predicted also to have 0 years of education. **b.** *Step 1:* H_0: Intercept $= 0$, H_a: Intercept $\neq 0$. *Step 2:* t-test for intercept. The conditions are assumed met, $\alpha = 0.05$. *Step 3:* $t = 2.00$, p-value $= 0.056$. *Step 4:* We cannot reject H_0. We don't have enough evidence to reject an intercept of 0.

14.19 a. The slope is 0.936. On the average, for each inch taller a parent is, the child is about 0.94 inch taller, *in the sample.* **b.** *Step 1:* H_0: Slope $= 0$, H_a: Slope $\neq 0$. *Step 2:* t-test for slope, $\alpha = 0.05$. *Step 3:* $t = 8.38$, p-value < 0.001. *Step 4:* Reject H_0. There is a significant linear relationship between parent height and student height. **c.** If the slope were 1, it would mean that on the average, for every inch taller the parent was, the student would be 1 inch taller also. **d.** *Step 1:* H_0: Slope $= 1$, H_a: Slope $\neq 1$. *Step 2:* t-test for slope, $\alpha = 0.05$. *Step 3:* $t = -0.58$, see calculation below. p-value $=$ larger than 0.05 (because 0.578 is less than 2.052). *Step 4:* Do not reject H_0.

$$t = \frac{\hat{\beta}_1 - \text{null}}{SE_{\hat{\beta}_1}} = \frac{0.9355 - 1}{0.1116} = -0.58$$

14.21 a. Slope $= 0.22$. The slope tells us that for each additional library visitor, the number of wifi sessions increase on average by about 0.22. In other words, for 10 additional library visitors, the number of wifi session increase on average by about 2. **b.** (0.03, 0.41); **c.** The interval does not contain 0 so there is a relationship between number of visitors and number of wifi sessions.

14.23 Yes. The output shows that we cannot reject the null hypothesis that the intercept is 0 (because the p-value is 0.740, which is larger than 0.05). This means that the confidence interval will include 0 pounds of trash.

Section 14.3

14.25 Use a prediction interval. This concerns prediction of a single value, not a mean.

14.27 She should use a prediction interval because she is predicting one value, not a mean.

14.29 Use a confidence interval. This concerns prediction of a mean, not a single value.

14.31 a. 270,959 **b.** Prediction interval, because we are predicting the value for one house, not the mean of a group of houses. **c.** ($160,581, $381,532) **d.** Yes, he is likely to be able to afford a house because he has access to enough money to pay the price at the top of the 95% interval.

14.33 (About 125, about 200)

14.35 (about 1.8, about 3.8) This is a very wide interval, and it is not very useful to find that a student with a 750 on the math SAT will have a GPA between a D and an A.

14.37 The green lines are prediction intervals (for individuals), and the red lines are confidence intervals (for means). Means tend to be more stable than individual measurements and give more precise results, which is why their intervals are narrower.

14.39 a. PI is (182, 216); CI is (195, 203). The confidence interval is narrower because it is estimating a population mean, and there is less uncertainty in this than in the prediction interval, which is predicting the weight of an individual. **b.** We are 95% confident that the mean weight of all baseball players who are 20 years old is between 195 pounds and 203 pounds. **c.** We are 95% confident that one 20-year-old baseball player will weigh between 182 and 216 pounds.

Chapter Review Exercises

14.41 a. See graph. The equation is

Longevity $= 7.84 + 0.0327$ Gestation

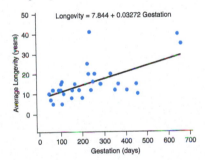

b. The hippopotamus is predicted to live about 10 years but lives over 40 years, on average.

c. CI is 13.92 to 19.18 years

```
Predicted Values for New Observations

New
Obs   Fit  SE Fit      95% CI         95% PI
  1  16.55  1.29  (13.92, 19.18)  (1.75, 31.35)
```

d. Humans do not fit into this pattern and are not included in the data set. The average human life expectancy is much more than the top of the confidence interval, which is 19.18 years.

Appendix C: Credits

PHOTO CREDITS

Cover: Paul Scott/EyeEm/Getty Images
Throughout: John Lund/Blend Images/Getty Images
Page iv: (top) Pearson Education, Inc.; (middle) Rebecca Wong; (bottom) Courtesy of Colleen Ryan

CHAPTER 1

Page 1: Teoh Chin Leong/123RF; **Page 2:** Leungchopan/Shutterstock; **Page 6** (top): Chad Bontrager/Shutterstock; (bottom): NASA; **Page 7:** S. Kuelcue/Shutterstock; **Page 9:** Tiler84/Fotolia; **Page 13:** James Steidl/123RF; **Page 15:** Laborant/Shutterstock; **Page 17:** Art_zzz/Fotolia; **Page 21:** Netsuthep/Fotolia; **Page 24:** Shutswis/123RF; **Page 26** (top): Yulia Davidovich/Shutterstock; (bottom): Lukas Gojda/Shutterstock; **Page 28:** LSqrd42/Shutterstock.

CHAPTER 2

Page 40: Galina Peshkova/123RF; **Page 41:** Playfair 1786; **Page 46:** Junial Enterprises/Shutterstock; **Page 50:** Pashin Georgiy/Shutterstock; **Page 51:** PCN Chrome/PCN Photography/Alamy Stock Photo; **Page 55** (top): George Doyle & Ciaran Griffin/Stockbyte/Getty Images; (bottom): Iofoto/Shutterstock; **Page 57:** Marben/Shutterstock; **Page 61:** Stephen VanHorn/Shutterstock; **Page 63:** Cla78/Shutterstock; **Page 67:** Rvlsoft/123RF.

CHAPTER 3

Page 90: Rawpixel.com/Shutterstock; **Page 91:** Andy Dean/Fotolia; **Page 94:** Saurabh13/Shutterstock; **Page 95:** Rafa Irusta/Shutterstock; **Page 99:** El Greco/Shutterstock; **Page 104:** Shutterstock; **Page 105:** OLJ Studio/Shutterstock; **Page 110:** BW Folsom/Shutterstock; **Page 112:** Jon Delorey; **Page 115:** JRP Studio/Shutterstock; **Page 116:** Ciaran Griffin/Stockbyte/Getty Images; **Page 118:** Arka38/Shutterstock; **Page 121:** Ilona Ignatova/Shutterstock.

CHAPTER 4

Page 149: Scanrail/123RF; **Page 150:** Rena Schild/Shutterstock; **Page 155:** Birute Vijeikiene/Shutterstock; **Page 167:** Joingate/Shutterstock; **Page 172:** SpeedKingz/Shutterstock; **Page 175:** Gino Santa Maria/Shutterstock; **Page 176:** ImagePixel/iStock/Getty Images; **Page 186:** Wavebreakmedia/Shutterstock.

CHAPTER 5

Page 213: Cylonphoto/123RF; **Page 216:** Bilder/Shutterstock; **Page 221:** Pearson Education, Inc.; **Page 222:** Pearson Education, Inc.; **Page 224:** Pearson Education, Inc.; **Page 226** (left): Koncz/Shutterstock; **Page 226** (right): HomeStudio/Shutterstock; **Page 230:** Bedrin/Shutterstock; **Page 236:** Radlovsk Yaroslav/Shutterstock; **Page 238:** Michael Rosskothen/Shutterstock; **Page 240:** Joyce Vincent/Shutterstock; **Page 244:** Gjermund/Shutterstock; **Page 248:** Kenishirotie/123RF.

CHAPTER 6

Page 266: Lajo_2/123RF; **Page 267:** Jupiterimages/Photos.com/Getty Images; **Page 268:** Studiotouch/Shutterstock; **Page 273:** Claudio Divizia/Shutterstock; **Page 285:** Alexander Kharchenko/123RF; **Page 286:** Michaeljung/Shutterstock; **Page 289:** Levent Konuk/Shutterstock; **Page 301** (bottom): PaulPaladin/Shutterstock; **Page 301** (top): Alekss/Fotolia; **Page 303:** Keeton Gale/Shutterstock.

CHAPTER 7

Page 323: Bloomicon/Shutterstock; **Page 324:** Elena Yakusheva/Shutterstock; **Page 326:** Syda Productions/Shutterstock; **Page 327:** Joy Brown/Shutterstock; **Page 327:** Shutterstock; **Page 328:** Stockbroker/MBI/Alamy Stock Photo; **Page 332:** Orla/Shutterstock; **Page 333:** Pearson Education, Inc.; **Page 338:** Pearson Education, Inc.; **Page 340:** Pashin Georgiy/Shutterstock; **Page 346:** Paulo Williams/Shutterstock; **Page 353:** Wavebreakmedia/Shutterstock; **Page 354:** Feverpitched/123RF; **Page 356:** O.Bellini/Shutterstock; **Page 359:** Eurobanks/Shutterstock; **Page 362:** Anaken2012/Shutterstock; **Page 364:** Fotofreaks/Shutterstock.

CHAPTER 8

Page 382: Koi88/Fotolia; **Page 383:** Robert Daly/OJO Images/Getty Images; **Page 384:** Yellowj/Fotolia; **Page 386:** Mipan/Fotolia; **Page 387:** Helder Almeida/Shutterstock; **Page 390:** Lindasj22/Shutterstock; **Page 391:** Mircea Maties/Shutterstock; **Page 400:** Julian Rovagnati/Shutterstock; **Page 404:** Shotgun/Shutterstock; **Page 407:** Ajt/Shutterstock; **Page 411:** Popaukropa/123RF; **Page 414:** Gabriele Maltinti/Fotolia; **Page 416:** Ted Foxx/Alamy Stock Photo.

CHAPTER 9

Page 435: Susazoom/123RF; **Page 436:** Primagefactory/123RF; **Page 440:** Mmaxer/Shutterstock; **Page 442:** Christoffer Hansen Vika/Shutterstock; **Page 445:** Sashkin/Shutterstock; **Page 450:** Elnur/Shutterstock; **Page 454:** John Baran/Alamy Stock Photo; **Page 455:** Africa Studio/Shutterstock; **Page 457:** Martin Allinger/Shutterstock; **Page 468:** Africa Studio/Fotolia; **Page 474:** Brian A Jackson/Shutterstock; **Page 476:** Bajinda/Fotolia; **Page 481:** Valentyn Volkov/Shutterstock; **Page 484:** Manfredxy/Shutterstock.

CHAPTER 10

Page 507: Sergey Rasulov/123RF; **Page 508:** Chris Curtis/Shutterstock; **Page 513:** Koncz/Shutterstock; **Page 515:** Akugasahagy/Shutterstock; **Page 522:** Dimedrol68/Shutterstock; **Page 524:** Cathy Yeulet/123RF; **Page 526:** Andriy Popov/123RF; **Page 527:** DPS/Shutterstock; **Page 533:** Picsfive/Shutterstock; **Page 538:** Maxpayne222/123RF; **Page 543:** Serezniy/123RF.

CHAPTER 11

Page 558: Nikola Bilic/Shutterstock; **Page 559:** Julia photo/Shutterstock; **Page 563:** Eric Isselee/Shutterstock; **Page 564:** Bhathaway/Shutterstock; **Page 570:** Nito/Fotolia; **Page 572:** Take Photo/Shutterstock; **Page 576:** Elena Galach'yants/Shutterstock; **Page 577:** Agorohov/Shutterstock; **Page 581:** Digital Storm/Shutterstock; **Page 583:** Paul Paladin/123RF; **Page 586:** Kevin L Chesson/Shutterstock.

CHAPTER 12

Page 604: Poznukhov Yuriy/Shutterstock; **Page 605:** Wavebreakmedia/Shutterstock; **Page 609:** Michal Bednarek/123RF; **Page 611:** Mihalec/Shutterstock; **Page 612:** Lalito/Shutterstock; **Page 614:** Naruedom Yaempongsa/123RF; **Page 620:** Jim Barber/Shutterstock; **Page 623:** Sunabesyou/Shutterstock; **Page 628:** Image Source/Getty Images.

CHAPTER 13

Page 639: Vega Gonzalez/Shutterstock; **Page 640:** Milosl]ubicic/Fotolia; **Page 645:** Kuzmaphoto/Shutterstock; **Page 652:** Schankz/Shutterstock; **Page 656:** Toho Company/Ronald Grant Archive/Alamy Stock Photo; **Page 663:** Luis Louro/Shutterstock; **Page 666:** Rainer Lesniewski/123RF.

CHAPTER 14

Page 690: RossHelen/Shutterstock; **Page 691:** Shutterstock; **Page 694:** Potapov Alexander/Shutterstock; **Page 698:** Indigolotos/Fotolia; **Page 702:** Studiovin/Shutterstock; **Page 703:** Reeed/Shutterstock; **Page 713:** Mikeledray/Shutterstock; **Page 719:** Noppawan09/Shutterstock.

TEXT CREDIT

CHAPTER 8

Page 385: By permission. From Merriam-Webster.com © 2018 by Merriam-Webster, Inc. https://www.merriam-webster.com/dictionary/hypothesis.

Subject Index